P9-CJT-109

Zeolite Chemistry and Catalysis

JULE A. RABO, *Editor*

Union Carbide Corp.
Tarrytown Technical Center
Tarrytown, N. Y. 10591

ACS Monograph **171**

AMERICAN CHEMICAL SOCIETY

WASHINGTON, D. C. 1976

University of Texas
At San Antonio

Library of Congress CIP Data

Zeolite chemistry and catalysis.
(ACS monograph; 171)

Includes bibliographical references and index.
1. Zeolites. 2. Catalysis.
I. Rabo, Jule A., 1924- II. Series: American Chemical Society. ACS monograph; 171.

TP245.S5Z38 546'.683'595 76-17864
ISBN 0-8412-0276-1 ACMOAG 171 1-796 (1976)

PRINTED IN THE UNITED STATES OF AMERICA

GENERAL INTRODUCTION

American Chemical Society's Series of Chemical Monographs

By arrangement with the interallied Conference of Pure and Applied Chemistry, which met in London and Brussels in July 1919, the American Chemical Society undertook the production and publication of Scientific and Technologic Monographs on chemical subjects. At the same time it was agreed that the National Research Council, in cooperation with the American Chemical Society and the American Physical Society, should undertake the production and publication of Critical Tables of Chemical and Physical Constants. The American Chemical Society and the National Research Council mutually agreed to care for these two fields of chemical progress.

The Council of the American Chemical Society, acting through its Committee on National Policy, appointed editors and associates (the present list of whom appears at the close of this sketch) to select authors of competent authority in their respective fields and to consider critically the manuscripts submitted. Since 1944 the Scientific and Technologic Monographs have been combined in the Series. The first Monograph appeared in 1921, and up to 1972, 168 treatises have enriched the Series.

These Monographs are intended to serve two principal purposes: first to make available to chemists a thorough treatment of a selected area in form usable by persons working in more or less unrelated fields to the end that they may correlate their own work with a larger area of physical science; secondly, to stimulate further research in the specific field treated. To implement this purpose the authors of Monographs give extended references to the literature.

American Chemical Society

F. Marshall Beringer
Editor of Monographs

Contents

CATALYSIS: TECHNOLOGY

Preface

AFTER THE DISCOVERY of zeolite catalysts in the late 1950s it appeared for a while that hydrocarbon catalysis over zeolites could be interpreted, although not rigorously, by the characterization of structural details, particularly the cations, in the zeolite crystal. Given the advantage of the structural analysis of a crystal matrix over that of a crystal surface, the difference between zeolites and the surface of many other catalysts, the chance to decipher catalysis by characterizing the zeolite structure looked excellent. Now, after two decades of intensive work, as reflected in the many chapters and over 1000 references in this book, several of the basic questions of zeolite catalysis remain unanswered. As usual, the discovery of new phenomena runs far ahead of lasting interpretation.

While a fully satisfying interpretation of the catalytic phenomena over zeolites remains to be spelled out in the future, the past work on zeolite catalysis and zeolite chemistry greatly enriched the science of catalysis and surface chemistry. The study of the chemical environment within the pores and cavities of the zeolite crystal showed that the zeolitic surface is different from either the ordinary crystal surface or the ordinary crystal lattice; it has properties of both. Accordingly, a substantial part of zeolite chemistry resembles phenomena found in solid state. In addition to new and rich chemistry, zeolite catalysis introduced several new catalytic processes into petroleum refining with major economic impact.

This book has been conceived to give a rounded view on all important aspects of zeolite catalysis including structure, chemistry, and technology. The chapters on technology also give insight into several major processes in petroleum refining for the first time. We have left out only those areas of zeolite technology which have been already adequately reviewed in recent publications (zeolite synthesis, cation exchange).

Each chapter critically reviews its particular area, providing a full review of significant contributions, giving ample space for interpretation and discussion. The discussions reflect the personal view of the authors, and therefore, interpretation of the same phenomena may differ somewhat from chapter to chapter. This will help the reader to consider both pro and con and to help him to form his own opinion. The information in the text and in the many tables and figures is consistent, and we have adopted a uniform nomenclature throughout the book for zeolite types and products, structural sites, and for other phenomena.

In recent years, as far as I can remember, almost every major symposium or congress on catalysis has ended on a note promising a great leap forward in the understanding of catalytic phenomena in the very near future. Zeolite catalysis and its interpretation had several leaps and bounds forward in the past. The remaining black or dark areas of knowledge were relatively easy to tolerate since progress, choosing its path sometimes in a mercurial way, has been steady and exciting all the way.

Union Carbide Corp. JULE A. RABO
Tarrytown, N.Y.
October 1974

Structure and Chemistry

Chapter

1

Origin and Structure of Zeolites

J. V. Smith
Department of the Geophysical Sciences, University of Chicago, Chicago, Ill. 60637

MACROSCOPIC DESCRIPTIONS OF ZEOLITE chemistry and catalytic phenomena are adequate for many purposes, but a scientific understanding is strongly augmented by detailed knowledge of the underlying crystal structure. This chapter reviews the key structural factors ranging from topology of the aluminosilicate framework to the problems of cation location and surveys various topics involving the crystallization and thermodynamics of zeolites. Emphasis is placed on crystal structural data obtained from x-ray analysis, but information from other techniques such as ESR, NMR, and IR is utilized where desirable. I am indebted to the comprehensive treatise by D. W. Breck (*1*) on the structure, chemistry, and use of zeolite molecular sieves and, whenever possible, I refer to the Proceedings of the International Conferences on Molecular Sieves (*2, 3, 4, 5*) in preference to scattered articles in journals.

Because of the complexity of the physical properties of zeolites, Smith (*6*) proposed a loose definition: "a zeolite is an aluminosilicate with a framework structure enclosing cavities occupied by large ions and water molecules, both of which have considerable freedom of movement, permitting ion-exchange and reversible dehydration." The framework structure consists of corner-linked tetrahedra in which small atoms (collectively denoted T atoms) lie at the centers of tetrahedra and oxygen atoms lie at the corners. The T sites of all natural zeolites are dominated by Al and Si atoms, but chemically related atoms such as Ga, Ge, and P can be incorporated into synthetic zeolites. The large ions in the cavities of natural zeolites are mono- or divalent, and the principal species (Na, Ca, K, Mg, and Ba) reflect the geochemical abundance and the competition with other minerals during geochemical differentiation. In the laboratory a wide range of other ions can be exchanged into zeolites, or incorporated by direct synthesis. The ideal formula of a zeolite is $M_pD_q[Al_{p+2q}Si_rO_{2p+4q+2r}] \cdot sH_2O$: the infinite corner-sharing of tetrahedra requires that there are twice as many framework oxygens as T atoms, while charge balance requires that the number of trivalent Al ions equals the sum of p (monovalent) ions and twice q (divalent) ions.

Table I. Properties of Selected Zeolites and Feldspathoids

Name	*Crystallographic Data*	*Selected Chemical Composition*
A	isometric: *a* 12.3A; Pm3m(pseudo)	$Na_{12}Al_{12}Si_{12}O_{48} \cdot 27H_2O$
cancrinite	hexagonal: *a* 12.7 *c* 5.1A; $P6_3$	$Na_6Al_6Si_6O_{24} \cdot CaCO_3 \cdot 2H_2O$
chabazite	rhombohedral: *a* 9.4A *α* 94.5°; $R\bar{3}m$	$(Ca,Na_2)_{\sim 2}Al_4Si_8O_{24} \cdot 13H_2O$
erionite	hexagonal: *a* 13.3 *c* 15.1A; $P6_3/mmc$	$(Ca,K_2,Na_2)_{\sim 4}Al_8Si_{28}O_{72} \cdot 27H_2O$
faujasite	isometric: *a* 24.7A; Fd3m	$\sim Na_{13}Ca_{11}Mg_9K_2Al_{55}Si_{137}O_{384} \cdot 235H_2O$
X	isometric: *a* 25.0	$Na_{86}Al_{86}Si_{106}O_{384} \cdot 264H_2O$
Y	isometric: *a* 24.7	$Na_{56}Al_{56}Si_{136}O_{384} \cdot 250H_2O$
gmelinite	hexagonal: *a* 13.7 *c* 10.0A; $P6_3/mmc$	$(Na,etc.)_{\sim 8}Al_8Si_{16}O_{48} \cdot 24H_2O$
L	hexagonal: *a* 18.4 *c* 7.5A; P6/mmm	$K_9Al_9Si_{27}O_{72} \cdot 22H_2O$
mazzite	hexagonal: *a* 18.4 *c* 7.6A; $P6_3/mmc$	$K_{2.5}Mg_{2.1}Ca_{1.4}Na_{0.3}Al_{10}Si_{26}O_{72} \cdot 28H_2O$
Ω	do. but *a* 18.2	$(Na,etc.)_8Al_8Si_{28}O_{72} \cdot 21H_2O$
mordenite	orthorhombic: *a* 18.1 *b* 20.5 *c* 7.5A Cmcm	$Na_8Al_8Si_{40}O_{96} \cdot 24H_2O$
offretite	hexagonal: *a* 13.3 *c* 7.6A; $P\bar{6}m2$	$KCaMgAl_5Si_{13}O_{36} \cdot 15H_2O$
sodalite	isometric: *a* 8.9; $P\bar{4}3n$	$Na_6Al_6Si_6O_{24} \cdot 2NaCl$
ZK5	isometric: *a* 18.7A; Im3m	$Na_{30}Al_{30}Si_{66}O_{192} \cdot 98H_2O$

For brevity I concentrate on the zeolites listed in Table I. Those with proper names occur as natural minerals. The synthetic zeolites X and Y have the same framework topology as faujasite while Ω is probably related to mazzite. No natural analogs of A, L, and ZK5 have been discovered. Two feldspathoids—cancrinite and sodalite—are included because their frameworks are topologically related to the zeolite frameworks.

All the chosen zeolites have large pores into which molecules can be introduced after dehydration. The exchangable cations (and the aluminosilicate frameworks) can be modified by chemical treatment, thereby permitting control of the chemical forces on the sorbed molecules. Favorable choice of the host zeolite as a catalyst results in valuable chemical transformations of the sorbed molecule.

Geochemistry and Mineralogy of Natural Zeolites

Zeolites of Volcanic Rocks. The first zeolite minerals were obtained from vesicles and fractures in basalts, and museum displays of zeolites are dominated by this type. The vesicles resulted from bubbles arising during emplacement of the basaltic liquid, and the zeolites formed by later precipitation from fluids which permeated the basalts. The chemical and physical factors which govern the precipitation of these zeolites are complex and poorly understood. In some occurrences, more than one episode of zeolite crystallization occurred. Thus, at Sasbach, Kaiserstuhl, Germany, faujasite crystals lined the cavities prior to growth of tufts of phillipsite (*7*). The phillipsite and faujasite each vary in chemical composition from grain to grain, and later ionic exchange may have modified the primary compositions. Despite these chemical variations which occur over distances of centimeters or less, some broad generalizations apply over regions of tens or hundreds of kilometers, as in thick piles of basalt flows in Northern Ireland and Iceland.

In Northern Ireland, the zeolite zones systematically transgress the basalt layers except in local variations attributed to rock faulting (*8*). The zeolites apparently crystallized late, and the zones may result from variation of temperature or pore-water chemistry. However, here and elsewhere, the bulk composition of the host rock correlates with that of the zeolites (*9*). Thus, mordenite and other Si-rich zeolites occur in rocks supersaturated in silica while faujasite, chabazite, gmelinite, and other Si-poor zeolites occur preferentially in rocks deficient in silica.

Zeolites of Sediments. The great bulk of zeolites occur in certain sediments and low-grade metamorphic rocks with a grain size near the limit of optical microscopy. X-ray diffraction and electron microscopy have allowed accurate characterization and demonstrated the existence of huge deposits of zeolites.

Six reviews (*1, 9, 10, 11, 12, 13*) of zeolites in sediments constitute the basis of most of the following statements:

(a) Zeolites on the sea floor grow from volcanic debris, much occurring as highly reactive glass. Phillipsite, first reported in 1891 from the sea bed by the Challenger expedition (*14*), occurs preferentially in the upper cores from the Pacific Ocean where it locally makes up half of the sediment while clinoptilolite occurs deeper (*15*). Clinoptilolite dominates all the Atlantic Ocean cores whereas clinoptilolite and phillipsite are comparable in Indian Ocean cores. Phillipsite tends to occur in younger deposits than clinoptilolite. Clinoptilolite, consistent with its Si-rich composition, tends to occur in Si-rich sediments, where chert is a common companion. Phillipsite is more alkalic (*16*) and tends to occur in slowly deposited clays and in sediments rich in volcanic ash. In general the ratio of zeolite to unaltered glass increases with age and burial depth (*10*).

(b) Zeolites commonly occur in deposits from saline, alkaline lakes generally by alteration of volcanic debris, especially unstable glass. They also occur in alkaline soils in arid climates. These environments are characterized by high pH (commonly near 9.4) and very high salinities (generally associated with sodium carbonates and borates). Zeolites are crystallizing now in suitable environments such as the saline lakes of Western United States and Eastern Africa (*e.g.*, Searles and China Lakes, Calif.; Lake Natron, Tanzania; Lake Magadi, Kenya). Most deposits, however, are the products of deposition over millions of years producing beds up to hundreds of meters thick with quite variable proportions of minerals in the different beds as a result of intermittent volcanic activity, geographic location, and chemistry of ephemeral lakes. Such accumulation of thick beds results in compaction, increasing temperature, and greater time for thermodynamic equilibration. Recent near-surface sediments contain up to about 33% zeolites. Analcime is the commonest zeolite in these environments, but phillipsite, erionite, chabazite, and clinoptilolite also occur.

(c) Zeolite-bearing tuff beds (*i.e.* originally composed of volcanic ash) reach thicknesses of several kilometers in certain continental areas away from the sea. Such non-marine tuffs are abundant in the Basin and Range Province of the Western United States: *e.g.*, the John Day Formation of Oregon contains zeolites, principally clinoptilolite, 1 km thick over an area of more than 3000 km^2. The conversion of ash to zeolite increases with depth and time. The type of zeolite and its chemical composition tend to correlate with the chemistry of the volcanic debris and the groundwater chemistry. Common zeolites are analcime, chabazite, clinoptilolite, erionite, mordenite, and phillipsite.

In certain beds the sediment may be composed almost entirely of zeolite, and in some beds there is only one zeolite. Such deposits are particularly useful for commercial exploitation. They can be located only by systematic mineralogical exploration of likely sedimentary sequences.

Zeolites of Metamorphosed and Metasomatised Rocks. Contrasting with the near-surface deposits of zeolites are the deep-seated deposits which have undergone prolonged low-grade metamorphism and metasomatism. Crucial to the analysis of these deposits is the degree of approach to true or highly metastable thermodynamic equilibrium. Metamorphic petrologists naturally describe an ideal situation in which a mineral assemblage equilibrated at some particular temperature, pressure, and chemical environment, but actually in the real world a lack of equilibrium results from sluggishness of reactions. The whole subject is further complicated by the effects of deformation of the earth's crust from differential erosion, volcanic activity, and relative movements of the earth's crust and mantle.

The simplest situation concerns the isochemical reactions of thick beds of sedimentary rocks described in the preceding section. In general, the volcanic glass reacts before the anhydrous silicates. With increasing depth the glass is supplanted by water-rich zeolites over a scale of about 1 km. At greater depth these water-rich zeolites themselves tend to be supplanted by water-poor zeolites or by anhydrous feldspars, but the situation is even more complex. The percentage of water-rich zeolites tends to decrease with increasing age of the sediment; thus analcime tends to replace the water-rich zeolites (half-life of about 20 million years.)

A more complex situation involves the introduction of hydrothermal fluids in active thermal areas such as Wairakei, New Zealand and the Geysers, Calif. (*9*).

Finally, the most complex situation involves the major geosynclines at the active margins of continents. These sediment beds receive many kinds of debris. Strong tectonic forces cause major deformation and strong metamorphism of the beds. However, all the zeolite occurrences can be qualitatively interpreted in terms of the principles outlined in the next section. Heulandite and analcime are common in the first 6 km of some geosynclinal sediments whereas laumontite is favored by greater depth.

Coombs (*17*) concluded that analcime, wairakite, mordenite, heulandite, and laumontite occurred so repeatedly in association with other minerals in the system $CaO–Al_2O_3–(Mg,Fe)O–SiO_2–H_2O$ that each zeolite has a range of thermodynamic stability (or, at least, strong metastability). Most synthetic zeolites are thermodynamically unstable, but fortunately for chemical industry the breakdown to the more stable zeolites is often very sluggish as demonstrated by the prolonged occurrence of natural zeolites. Roughly speaking, the natural zeolites tend to be less hydrous than synthetic ones and breakdown reactions tend to move to less hydrous varieties.

Commercial Exploitation of Natural Zeolites. Although over 30 zeolites occur naturally, only the eight zeolites (analcime, chabazite, clinoptilolite, erionite, ferrierite, laumontite, mordenite, and phillipsite) which dominate the sedimentary deposits permit commercial exploitation. Exploration is somewhat tedious since visual identification in the field is imprecise even after x-ray diffraction and electron microscopic analyses of selected specimens in the laboratory. Deffeyes (*11*) estimated in 1968 that 10^8 tons of clinoptilolite, 10^7 tons of erionite, 5×10^6 tons of phillipsite and mordenite, and 10^5 tons of chabazite had been found in the Basin and Range Province of the western United States. For the entire world, the potential deposits of exploitable zeolites must be many orders of magnitude greater when account is taken of unlocated deposits: a guide to potential locations is given by Munson and Sheppard (*13*). Unquestionably the major problems are economic rather than scientific. Particularly unfortunate is the total absence of type A zeolite in nature, and the extreme scarcity of faujasite, for which there are only two fully verified localities (Sasbach, Germany and Oahu, Hawaii), each containing less than 1 kg. Synthesis will continue to furnish these zeolites for commercial use. Although the ingenuity of chemists allows synthesis of many species which do not occur naturally, the natural materials seem most suited for applications where purity is not important or where large tonnages are necessary.

Commercial utilization of natural zeolites was reviewed by Mumpton (*18*). Zeolite-rich rocks have been used as building stone for thousands of years and in cement and concrete at least since Roman times. Zeolite expanded by heating may become used as a light-weight aggregate. About 10^5 tons per year are used as a filler in paper. Natural zeolite is used to extract radioactive Sr and Cs from nuclear wastes, and experimental work on extraction of NH_4^+ from sewage and agricultural wastes may lead to enormous quantities being used in commercial treatment plants. Natural mordenite is used to produce oxygen- and nitrogen-rich fractions from air by a pressure-swing process. In agriculture, zeolites are being tested as a food additive and as an adjunct to antibiotics. Purification of gases (including methane produced from organic waste) may become another major use. Natural zeolites can also be used as catalysts, but so far the extra cost of synthetic zeolites has been outweighed by the greater efficiency and quality control. Probably 10^5–10^6 tons of natural zeolites were consumed in 1974, and these values may increase by one or two orders of magnitude over the next few decades.

Crystallization Kinetics and Thermodynamic Stability of Zeolites

The simplest way to crystallize a zeolite in the laboratory is to produce a highly supersaturated aqueous solution of appropriate composition at a relatively low temperature in the range 0°–300°C. Under such conditions the actual product is determined by kinetic factors, and the truly stable situation may be completely irrelevant. Chapter 4 by Breck (*1*) is a thorough account of synthesis conditions of important systems such as Na_2O–Al_2O_3–SiO_2–H_2O while innumerable patents and papers provide additional details. Because of the lack of thermodynamic equilibrium, there is infinite scope for modifying the reactants and physical conditions to produce new zeolites or to modify the chemical composition (*e.g.*, SiO_2/Al_2O_2 ratio) and physical properties (*e.g.*, grain size). Instead of equilibrium phase diagrams with geometry topologically compatible with the phase rule, reaction diagrams are used to record the products of synthesis.

Typical syntheses utilize highly unstable reactants, such as young coprecipitated gels, in aqueous solutions of high pH containing an alkali hydroxide. High supersaturation at low temperature results in abundant nucleation. Under such conditions, the zeolite should inherit structural units (*e.g.*, four-rings of linked TO_4 tetrahedra with associated cations and water molecules). Numerous studies have provided evidence that coprecipitated gels undergo an aging in which the bulk physical nature (and consequently the intimate atomic linkages) changes. After this aging process, which presumably produces the appropriate structural units (or building blocks), nucleation and growth of the zeolite can be accomplished. Commonly the aging process is carried out at a lower temperature (~ 25°C) than the crystallization (~ 50°–200°C). Reactions with smaller changes of entropy favor zeolites with high disorder, which tend to crystallize initially. Those zeolites with wide pores and consequent disorder among the water molecules and exchangeable cations are closer in structural properties and entropy to the highly disordered gels, and tend to form initially in preference to the compact zeolites. With increasing temperature of synthesis, the more compact zeolites dominate because of the greater reaction rate towards true equilibrium. The yield of zeolite depends on the type of reactant: switching the source of SiO_2 from sodium silicate to colloidal silica produces marked changes in the products even for the same bulk composition and temperature.

Unfortunately it is not possible to observe the detailed atomic movements during gel formation and crystallization, and all theories are speculative. Most zeolites will not grow from a system with the same bulk composition: X and Y grow only from a system with lower Al_2O_3 than their own composition (Ref. *1*, Figure 4.5). The zeolites which grow in the Na_2O–Al_2O_3–SiO_2–H_2O system (gismondine, chabazite, gmelinite, A, X, Y: Ref. *1*,

p. 270) can be built entirely from four- and six-rings of tetrahedra, which may be basic building blocks inherited from the gel. However, the A, X, and Y zeolites contain the sodalite unit, another possible building block, especially since hydroxysodalite is a common product. Large cations (including organic ones) have been used in the hope that they would act as a template around which the aluminosilicate might polymerize to produce large pores—*e.g.*, synthesis of Ω and ZK5 (Ref. *1*, pp. 309–310). Hydrated cations occur in many zeolites, and the growth of natural mazzite may be affected by the presence of the hydrated magnesium complex. The composite crystals of L, offretite, and erionite may result from different ways of joining hexagonal prisms and cancrinite units which had already developed around cations and hydrated cations (*19*). Phosphorus was incorporated into zeolite frameworks by simultaneous copolymerization and coprecipitation of all components into the intermediate gel (*20*). All these data support theories of structural inheritance during zeolite syntheses. Nevertheless, there are many complications, and readers are referred to Flanigen (*21*) for a review with new perspectives in crystallization of zeolites.

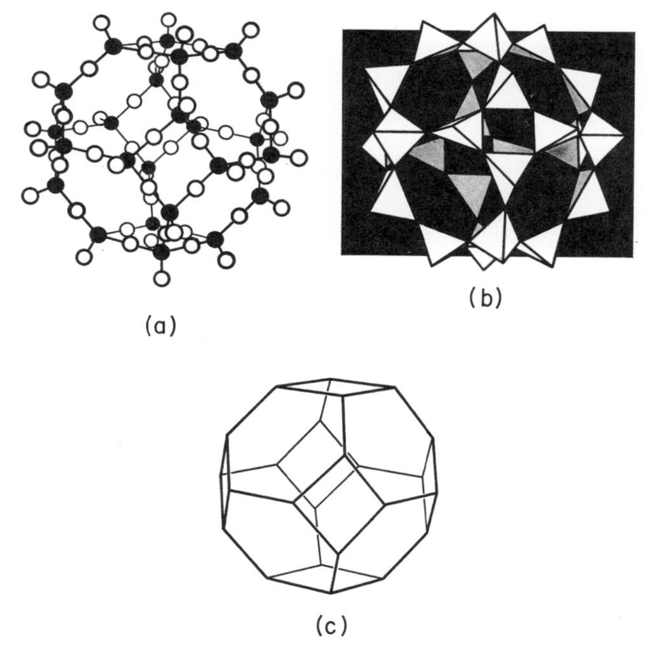

Figure 1. Three ways of depicting the truncated octahedron (or sodalite unit) in aluminosilicate frameworks

Data on the growth mechanisms of natural zeolites are sparse, but scanning electron micrographs of sedimentary zeolites show textural relations consistent with several mechanisms. Direct solution of glass with subsequent precipitation of zeolite appears dominant (*22*).

Topology of Zeolite Frameworks

The only item of a zeolite structure which can be precisely specified is the topology of the aluminosilicate framework. All other items, such as the distribution of atoms on the crystallographically equivalent sites, are complex and uncertain. An infinite number of frameworks is possible theoretically, but some 32 or so different topologies have been identified in zeolites while some 20 or so zeolites have unidentified framework topologies.

For mathematical description of the topology, it is desirable to reduce the atomic pattern to the simplest features. Instead of using the chemical concept of four large oxygen anions (radius ~ 1.35 A) lying at the vertices of a tetrahedron occupied by a small cation (Al or Si), it is convenient to envisage merely the center position [1 A = 0.1 nm]. Linked tetrahedra can be represented by joining the centers of adjacent tetrahedra. A zeolite framework thus becomes represented by a four-connected three-dimensional network. The oxygen atoms lie near but not at the midpoints of each branch. Having reduced the silicate framework to a four-connected net, it is now possible to recognize polygons or polyhedra as subunits. Figure 1 shows three levels of abstraction for part of a framework—actually the sodalite unit. Diagram a shows the centers of the oxygen and T atoms by open and filled circles, respectively (the T atoms are about the correct size, but the O atoms have only one-third the proper ionic radius). Diagram b shows imaginary silicate

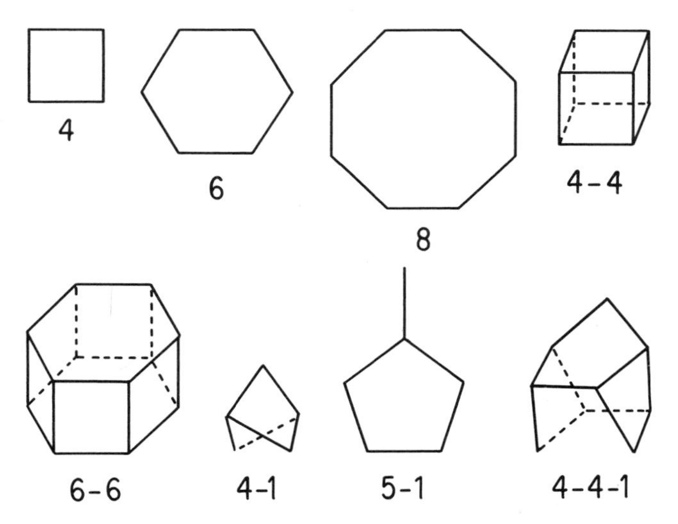

Figure 2. Secondary building blocks of zeolite frameworks as chosen by Meier. T atoms occur at the intersections of the lines. The code is the simplest description: from the viewpoint of symmetry, 4-4 would be described as a cube or six 4-rings (e.g., 4^6). Unit 6-6 contains two 6-rings and six 4-rings, 4-1 contains two 4-rings, and 4-4-1 contains two 5-rings as well as two 4-rings (23).

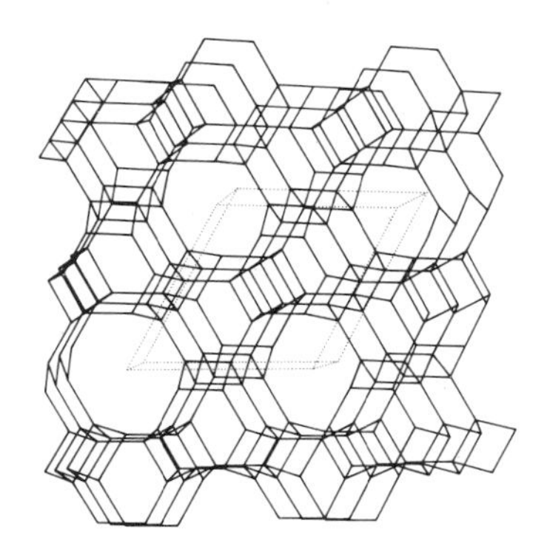
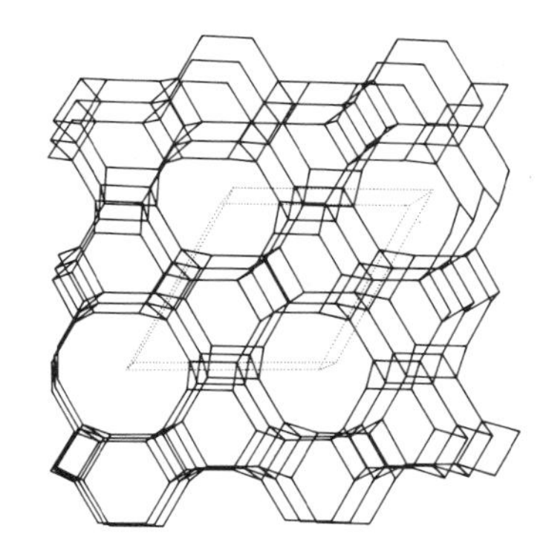

Advances in Chemistry Series

Figure 3. Stereodiagram of framework topology of cancrinite. The dotted lines show one unit cell (214).

tetrahedra with shared corners, while c shows merely the T-T linkages. At this final stage of abstraction, the T-T linkages outline a truncated octahedron which is one of the Archimedean semiregular solids.

Smith (*6*) classified zeolites on the basis of common structural units such as parallel six-rings or Archimedean polyhedra. Meier (*23*) presented a modified classification in which seven groups were recognized, and he went on to dissect the frameworks into secondary building units (Figure 2) which might have been assembled during crystallization. These units (single 4-ring, single 6-ring, single 8-ring, cube, hexagonal prism, 4 + 1 combination, 5 + 1 combination, 4 + 4 + 1 combination) are sufficient to construct zeolite frameworks but could be polymerized into larger building blocks.

The next series of figures are mostly stereodiagrams of T-T linkages of the zeolites listed in Table I. Many readers may wish to construct models from tetrahedral stars and plastic spaghetti which can be purchased from several supply houses. Each model will take several hours to construct, but the increased understanding of the principles well repays the time.

Six of the materials in Table I have structures related by the crosslinking of parallel 6-rings (cancrinite, sodalite, offretite, gmelinite, chabazite, and erionite). Cancrinite (Figures 3 and 4) can be constructed from 4- and 6-rings. Looking perpendicular to Figure 3 (*i.e.* down the hexagonal axis of the unit cell), the 6-rings are linked by tilted 4-rings. The 6-rings lie at different heights and can be labeled as in Figure 4 (upper left) in which the increments of height are 2.5 A. In projection, one set of hexagonal rings superimposes at heights 0, 2, 4, etc. and the other set at 1, 3, 5, etc. Another way of describing the structure utilizes the unit shown in Figure 4 (upper right) —the "cancrinite cage." This consists of two planar 6-rings and three pairs of adjacent 4-rings. In addition, the cancrinite cage has three boat-shaped 6-rings which complete the 11-sided irregular polyhedron. Returning to Figure 3, the cancrinite framework can be built by stacking together the cages

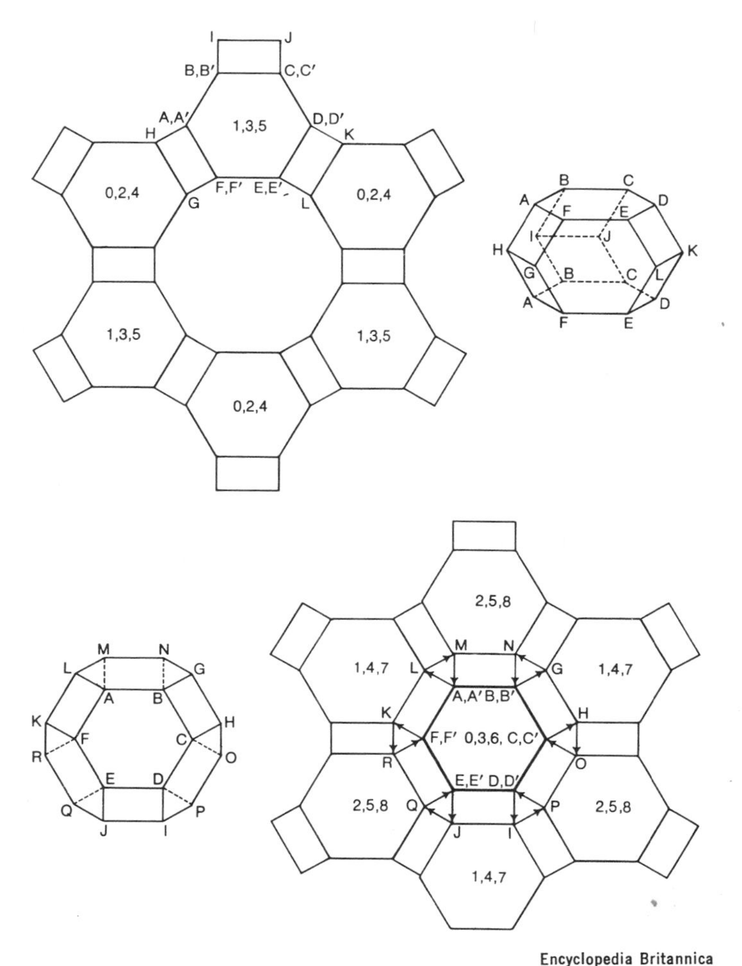

Encyclopedia Britannica

Figure 4. Two-dimensional projections of three-dimensional frameworks of cancrinite and sodalite. In cancrinite (upper left) hexagonal rings superimpose in projection at heights 0, 2, 4, etc. or at 1, 3, 5, etc. The edges of the rings form tilted squares which project as rectangles. A cancrinite cage formed from the labeled intersections of A is shown at upper right. In sodalite (lower right), the hexagonal rings superimposed at 0, 3, 6, etc., 1, 4, 7, etc., and 2, 5, 8, etc., form truncated octahedra as shown at lower left. The upper hexagonal face of a truncated octahedron is emphasized at lower right and descending lines are shown by arrows (215).

such that all 4-rings and planar 6-rings are shared. Finally, the stacks of cancrinite cages do not utilize all the available volume; left over are channels centered on the vertical edges of the unit cell and bounded by crown-shaped, non-planar, 12-rings.

Imagine the positions of the oxygen atoms which lie near the centers of the edges. Remembering that the T-T distances for an aluminosilicate are near 3.1 A and that the radius for an oxygen anion is about 1.35 A, one can begin to appreciate the amount of space available for the exchangeable cations and water molecules. The cancrinite cages each have enough room for several guests while the long channels provide much more space. Diffusion will be easiest through the crown-shaped 12-rings and relatively difficult through the regular and boat-shaped 6-rings. If a long channel is blocked by occluded species, the intervening spaces could be reached only by the boat-shaped 6-rings. If exchangeable cations lie in the 6-rings, access would be possible only *via* the crown-shaped 12-rings. These matters are discussed again later and are mentioned here as a guide to topologic and geometric features that should be borne in mind.

After looking at the upper-left diagram of Figure 4, one might ask if the bits and pieces could be reassembled in another pattern. The lower-right diagram is one possibility; instead of leaving open the hexagonal channels, connect the 4- and 6-rings so that they stagger in a three-fold zig-zag such that the 6-rings fall into three sets at heights 0, 3, 6, etc., 1, 4, 7, etc., and 2, 5, 8, etc. (again at spacings of 2.5 A). Instead of cancrinite cages, the structure now is composed of "sodalite cages" as shown in the lower-left diagram. Each sodalite cage consists of eight 6-rings and six 4-rings. The two-dimensional projections of Figure 4 do not bring out the full elegance of the sodalite structure, but the stereodiagram of Figure 5 should do so. Here one can see that the space is entirely occupied by sodalite units which nest together so that every 4- and 6-ring is shared by adjacent sodalite units. Note that the stereodiagram does not have the same orientation as the projection in Figure 4. The reason for the new orientation is that the true symmetry is isometric and that the unit cell is a cube with sodalite cages at the corners and the body center. Indeed the projection in Figure 4 can be obtained by looking along any of the body diagonals of Figure 5. The cubic and hexago-

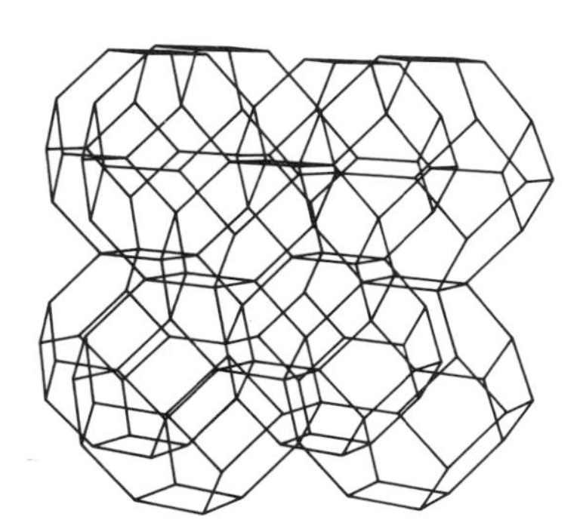
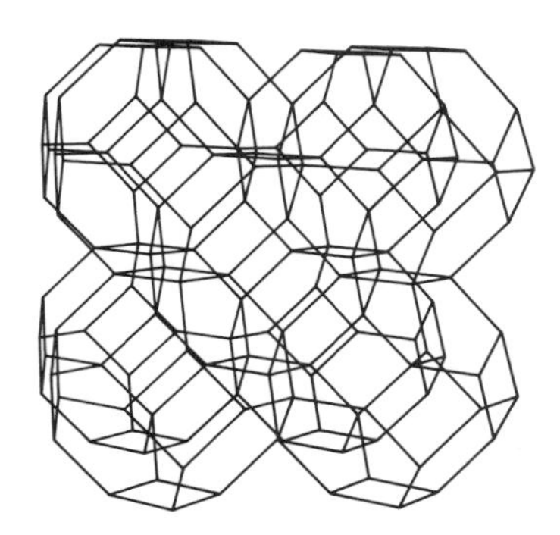

Advances in Chemistry Series

Figure 5. Stereodiagram of framework topology of sodalite (214)

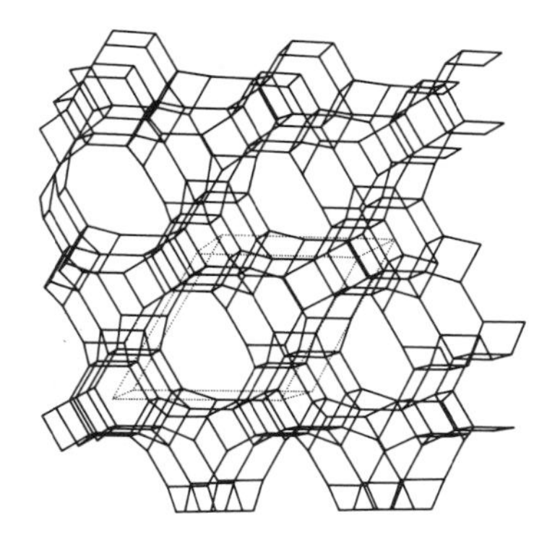
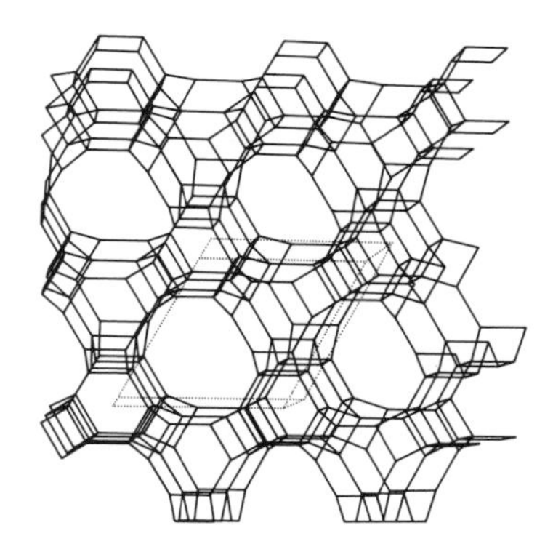

Advances in Chemistry Series

Figure 6. Stereodiagram of framework topology of offretite (214)

nal close packing of spheres is analogous with the linkage of the 6-rings of sodalite and cancrinite. The sodalite cage is actually one of the semiregular (Archimedean) polyhedra and is often called the truncated octahedron; it is also one of the Fedorov space-filling polyhedra.

In cancrinite, the 6-rings of each column are separated by two units of height (*e.g.* 1, 3, 5, etc.) whereas in sodalite the 6-rings of each column of parallel rings are separated by three units (*e.g.* 1, 4, 7, etc.). In offretite (Figure 6) the pattern of the 6-rings is more complicated. Around the vertical sides of the hexagonal unit cell, the 6-rings lie at heights 1, 2, 4, 5, 7, 8, etc., forming columns of alternate hexagonal prisms (the 6-6 secondary building unit in Figure 2) and cancrinite cages. These columns are joined by another type of column composed of "offretite cages" which share parallel 6-rings at the top and bottom. [Note that offretite cages have also been denoted "gmelinite" cages]. Each offretite cage also has three triplets of adjacent 4-rings plus three boat-shaped 8-rings. It can be constructed from the cancrinite cage

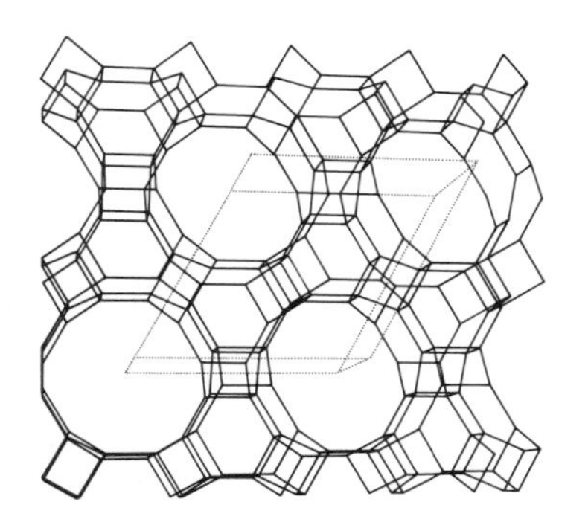
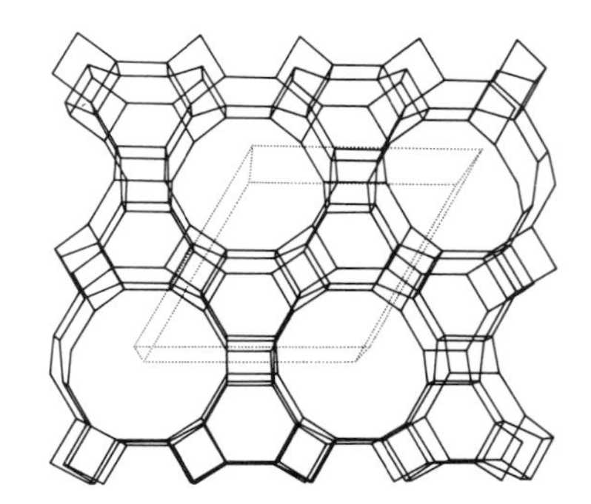

Advances in Chemistry Series

Figure 7. Stereodiagram of framework topology of gmelinite (214)

in Figure 4 (upper right) by replacing the edges GH, IJ, and KL by 4-rings. Just as in cancrinite, there are long channels bounded by crown-shaped 12-rings, but the channels are now interconnected by 8-rings instead of 6-rings.

Gmelinite (Figure 7) is composed of columns of hexagonal prisms alternating with offretite cages. Note that each hexagonal prism shares its 6-rings with offretite cages instead of the cancrinite cages in the offretite structure. In both gmelinite and offretite, the hexagonal prisms share 4-rings with offretite cages. Again there are infinite channels bounded by crown-shaped 12-rings and interconnected *via* boat-shaped 8-rings. The gmelinite structure has 6-rings at height 1, 2, 5, 6, 9, 10, etc. and 3, 4, 7, 8, etc. in adjacent columns. Returning now to Figure 4, one notes that sodalite was obtained from cancrinite by adding a third column of 6-rings in the vacant position. Chabazite is derivable from gmelinite by the same trick: its 6-rings (Figure 8) lie in

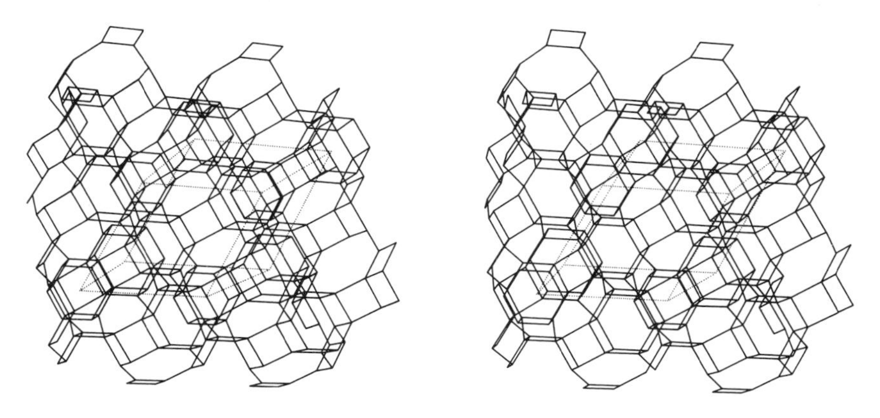

Advances in Chemistry Series

Figure 8. Stereodiagram of framework topology of chabazite (214)

three columns at 1, 2, 7, 8, 13, 14, etc., 3, 4, 9, 10, 15, 16, etc. and 5, 6, 11, 12, 17, 18, etc. The unit cell is now a rhombohedron instead of the hexagonal prism of gmelinite, and the structure can be obtained merely by placing a hexagonal prism at each corner of the unit cell and linking them by 4-rings. [The dotted lines in Figure 8 outline the triply primitive hexagonal unit cell. To obtain the primitive rhombohedral cell, connect the centers of adjacent hexagonal prisms.] After subtracting the space occupied by the hexagonal prisms, the remainder consists of identical, near-spherical chabazite cages, each composed of two 6-rings at top and bottom, six 8-rings in rhombohedral positions and six pairs of adjacent 4-rings. This structure does not have one-dimensional channels as in cancrinite, offretite, and gmelinite but consists of large pores interconnected to six adjacent pores by the near-planar, chair-shaped 8-rings.

Finally for this group of zeolites, Figure 9 shows the framework topology of erionite. Centered on the vertical edges of the hexagonal unit cell are

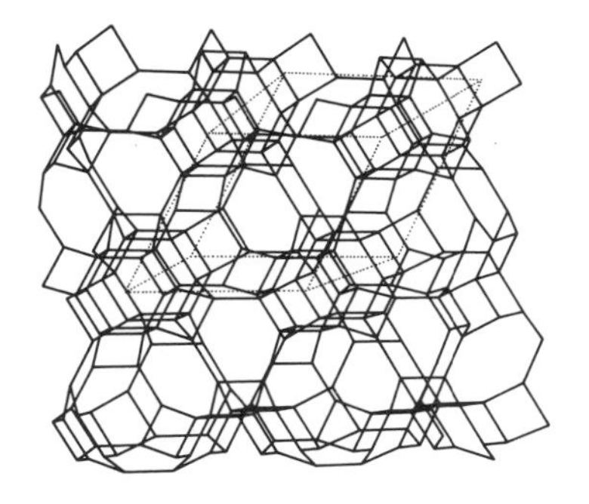
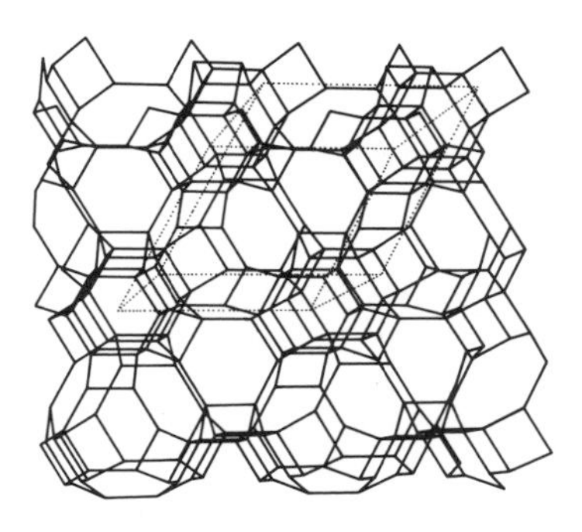

Advances in Chemistry Series

Figure 9: Stereodiagram of framework topology of erionite (214)

alternating hexagonal prisms and cancrinite units which are crosslinked by 4-rings and single 6-rings to form a complex pore system interconnected by 8-rings.

By now the reader will have suspected that there is an infinite family of frameworks based on parallel 6-rings. Indeed there is, as can be seen algebraically (*6*). Returning to Figure 4 (lower right), note that the parallel 6-rings of sodalite lie at three sets of positions in projection. Label the three sets A, B, C. Then the vertical sequence of parallel 6-rings is 1 = A, 2 = B, 3 = C, 4 = A and so on, or briefly . . . ABC . . . . For cancrinite (upper left), the vertical sequence is 1 = A, 2 = B, 3 = A, 4 = B, or briefly . . . AB . . . . The possible stacking arrangements can be systematized as follows:

AB	cancrinite
ABC	sodalite
AAB	offretite
AABB	gmelinite
AABC	
ABAC	LOSOD, a synthetic zeolite (*1*)
ABABC	
AABBC	
AABAC	
AABBCC	chabazite
AABAAC	erionite
etc.	

Obviously the list continues indefinitely (note that AABCCABBC is the probable code for levyne). Stacking errors are also possible (*see* later).

The next group has frameworks whose T atoms are related to Archimedean polyhedra. Sodalite (Figures 4 and 5) has already been described. The A zeolite (Figure 10) is obtained by replacing each 4-ring of the sodalite cage at the corner of the unit cell by a cube (the 4-4 secondary building unit, Figure 2). Each sodalite cage at the body center of the unit cell transforms into a truncated cuboctahedron with twelve 4-rings, eight 6-rings, and six 8-rings. The truncated cuboctahedra form a system of pores linked in three directions by windows of 8-rings. The sodalite units are separated from each other, and access to them is *via* 6-rings.

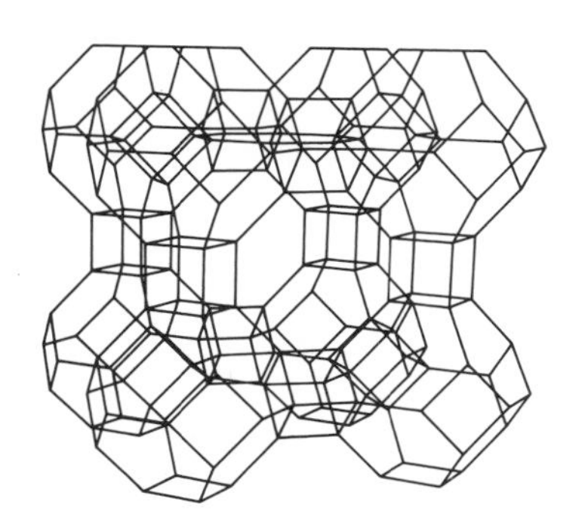

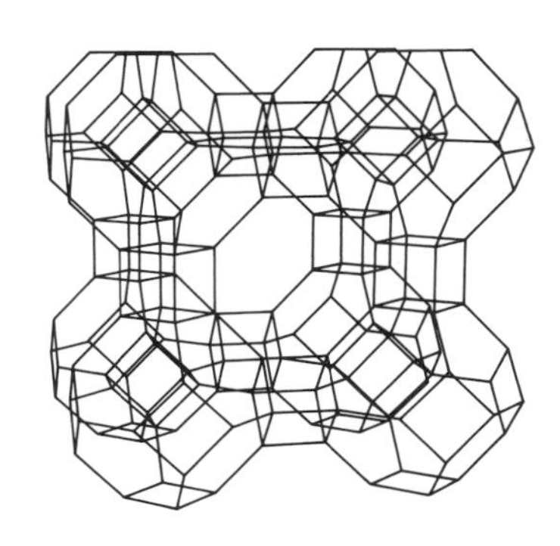

Advances in Chemistry Series

Figure 10. Stereodiagram of framework topology of type A (214)

The topology of faujasite (and of the related X and Y zeolites) is obtained by linking sodalite units with hexagonal prisms (the 6-6 secondary building unit, Figure 2). Each sodalite unit (Figure 11) is linked to four sodalite units in a tetrahedral configuration by hexagonal prisms which are attached to four of the eight hexagonal faces; the other four hexagonal faces are unshared, as are the six 4-rings. Actually the maximum number of sodalite neighbors in Figure 11 is three, but this results merely from termination of the drawing. There are two ways of linking sodalite units by a hexagonal prism: in faujasite, the 4-rings bordering on the hexagonal prism alternate

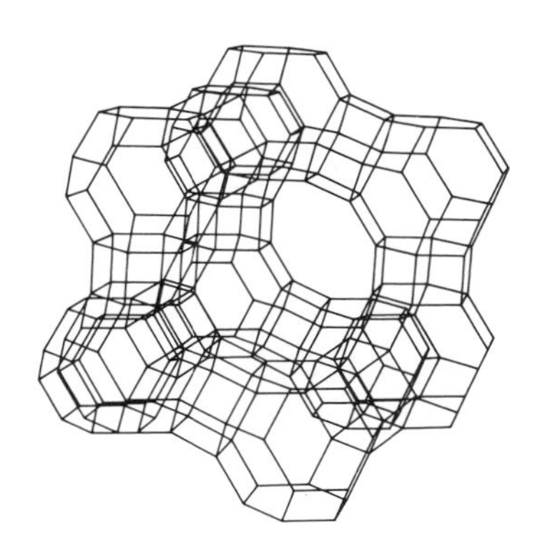

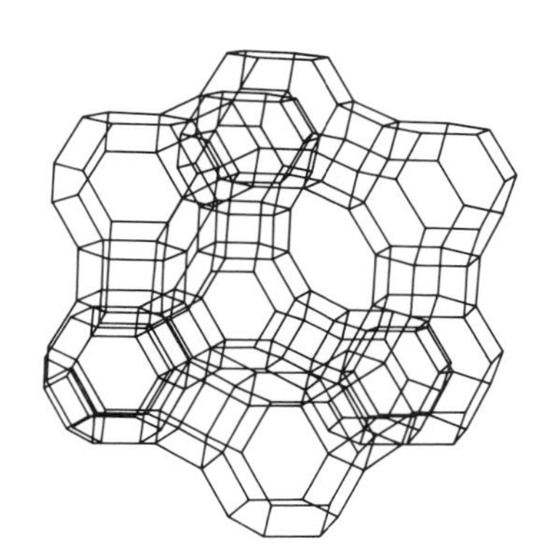

Advances in Chemistry Series

Figure 11. Stereodiagram of framework topology of faujasite (214)

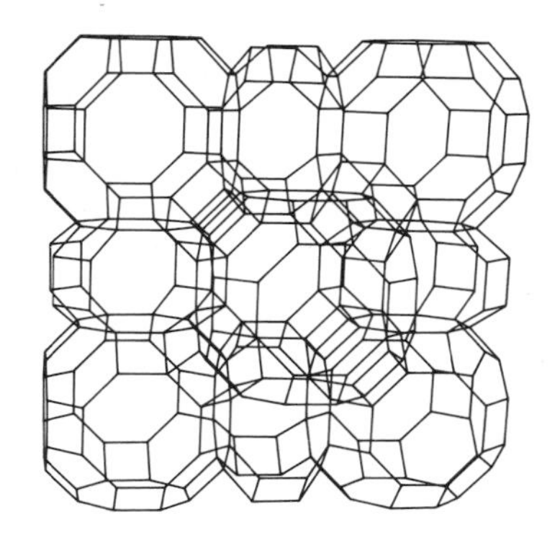

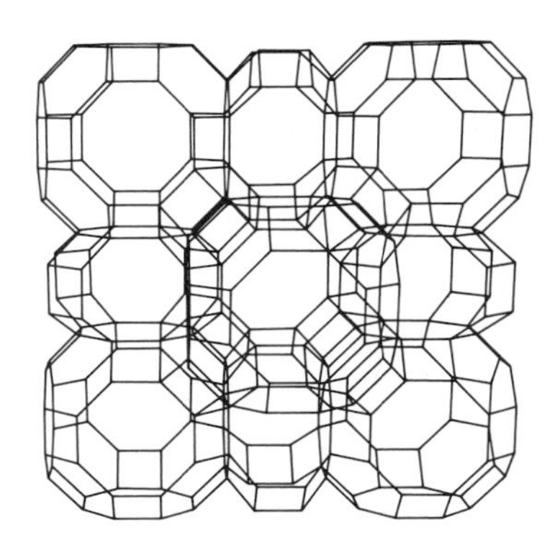

Advances in Chemistry Series

Figure 12. Stereodiagram of framework topology of ZK-5 (214)

it give an axis of inverse three-fold symmetry. Other structures can be built in which some, but not all, of the 4-rings face each other. The sodalite units and the hexagonal prisms are topologically equivalent to the Zn and S atoms in the sphalerite variety of ZnS. The framework of faujasite is very open and encloses a system of large cages linked by four windows of 12-rings to adjacent pores. Each of the faujasite cages is bounded by eighteen 4-rings, four 6-rings, and four 12-rings.

Returning to the A structure (Figure 10), it can be envisaged as the linkage of truncated cuboctahedra by cubes. The ZK5 structure (Figure 12) consists of the linkage of truncated cuboctahedra by hexagonal prisms. The remaining volume is occupied by a polyhedron consisting of twelve 4-rings, four boat-shaped 8-rings, and two planar 8-rings. The pore system is interconnected in three dimensions with access controlled by the 8-rings.

This group of zeolite is completed by Rho which consists of truncated cuboctahedra linked by octagonal prisms (*24*).

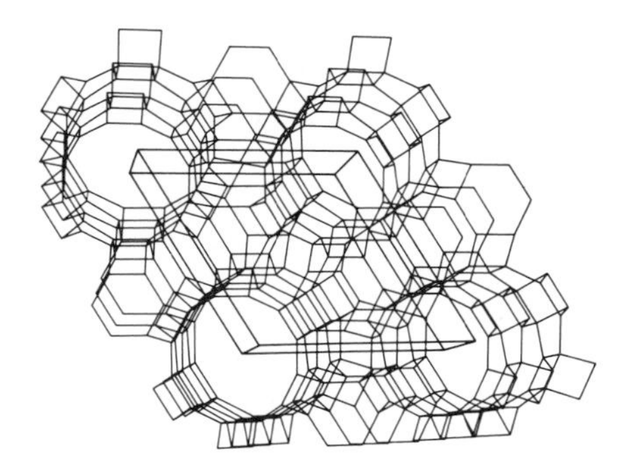

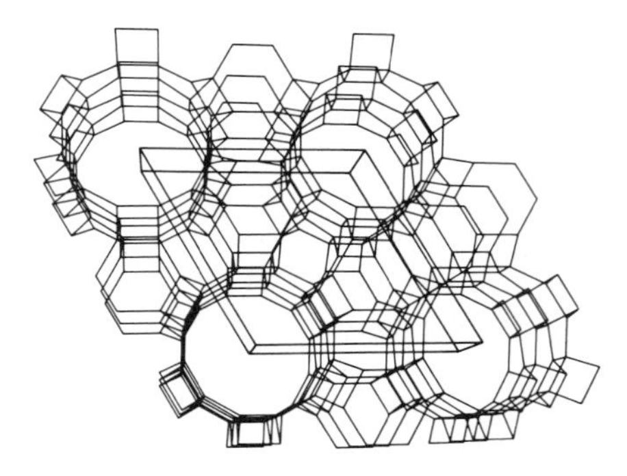

Figure 13. Stereodiagram of framework topology of mazzite (prepared by R. Rinaldi and J. J. Pluth)

The framework of mazzite (*25, 26*), on the basis of x-ray powder patterns, probably applies also to the synthetic zeolite Ω. This framework (Figure 13) is related to that of offretite (Figure 6) in that both contain infinite chains of offretite cages sharing hexagonal faces. However, these chains are crosslinked by hexagonal prisms in offretite and by strips of 5-rings in mazzite. Each 5-ring is composed of an edge from three offretite cages plus two horizontal joins shared with adjacent 5-rings. Six of the horizontal joins plus six horizontal edges from offretite cages define a horizontal 12-ring. Large cylindrical channels are bounded by the 12-rings and walled by a continuous linkage of 4- and 5-rings. Between the offretite cages there are irregular flattened spaces permitting access by non-planar 8-rings. The pore system is therefore extremely complex with one-dimensional tubes bounded by 12-rings separated by the walls of 4- and 5-rings from the irregular three-dimensional system of linked offretite cages.

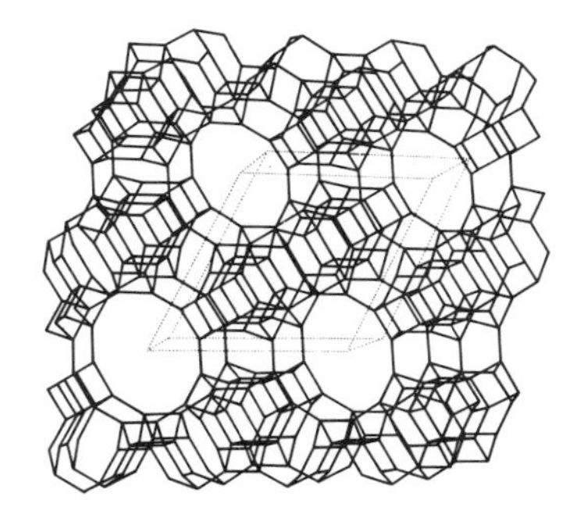
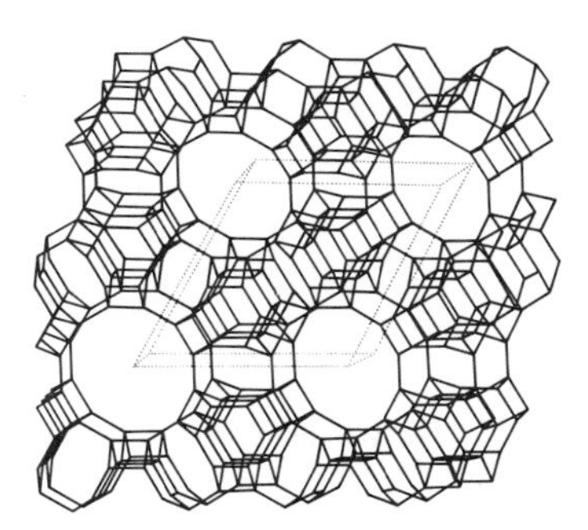

Advances in Chemistry Series

Figure 14. Stereodiagram of framework topology of type L (214)

The framework of L (Figure 14) consists of columns of alternating cancrinite cages and hexagonal prisms as in erionite (Figure 9); however, the crosslinkages are single horizontal joins instead of hexagons. Six of the horizontal joins plus six horizontal edges from cancrinite cages define a horizontal 12-ring. Large cylindrical channels are bounded by 12-rings and walled by boat-shaped 8-rings and triple 4-rings, the former providing access into the spaces between the cancrinite cages.

Mazzite and L have parallel 6-rings and could be classified with the cancrinite–erionite group given earlier. However, they do not fit into the ABC classification.

Mordenite (Figure 15) is a complex structure composed of horizontal 4-rings interspersed by pairs of tilted 5-rings sharing an edge. These combine together to form twisted 12-rings which span vertical, near-cylindrical channels. The walls of these channels contain 5-, 6-, and 8-membered rings, the latter providing access through zig-zag passages between adjacent cylinders. The pairs of tilted 5-rings are connected up and down to other pairs

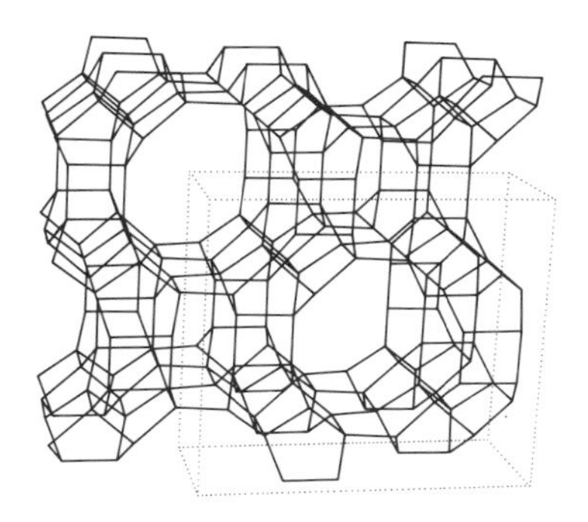
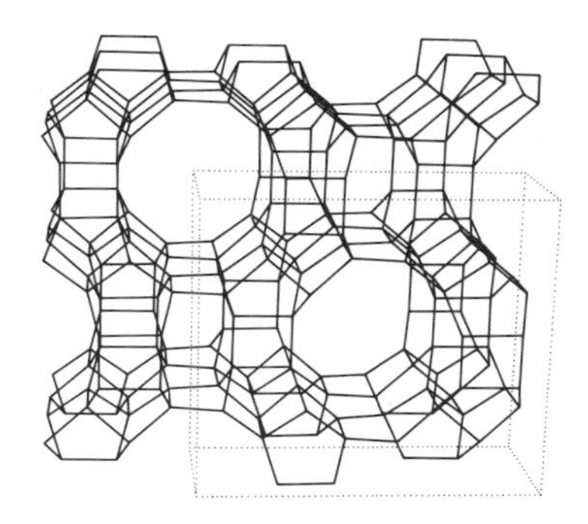

Advances in Chemistry Series

Figure 15. Stereodiagram of framework topology of mordenite (214)

of 5-rings to form vertical strips of high rigidity. These strips can be cross-linked in different ways to form other members of the mordenite group of zeolites (*23*).

Technical Problems of Crystal Structure Determination by X-ray and Electron Diffraction

All crystal structure determinations are based in principle on a trial-and-error fitting of observed and calculated diffraction intensities because of the loss of phase in the diffraction event. When single crystals wider than 20–50 μm are available, a full set of x-ray intensities can be obtained, but when a powder is used, many x-ray intensities are lost in the background or confused by overlapping diffraction lines. It is especially difficult to determine the space-group symmetry by x-rays. Electron diffraction is very useful for fine-grained zeolites because of the ease of detecting subsidiary diffractions, but multiple diffraction poses problems. Studies on mordenite, erionite, and L zeolites have proved the value of the technique, but many fine-grained zeolites remain unstudied by electron diffraction. Fischer (*27*) reviewed the problems of refinement of crystal structures from x-ray data. Each diffraction wave integrates the vector amplitude from all unit cells of each coherent piece of crystal. Hence random occupancy at slightly different positions by two chemical species of one crystallographic site results in only one electron density peak. Thus if Ca and Na lie near the same type of 6-ring of oxygens, one peak will result and the cation–oxygen distance will be the weighted average of Ca–O and Na–O distances. Random occupancy of T sites by Al and Si will cause displacements of adjacent O atoms, but only the average position will be measured. In a simple crystal structure such as NaCl, the electron density peaks broaden with temperature in response to thermal vibration. In a zeolite the broadening is composed of thermal vibration and static displacements of the centers of vibration. For framework oxygens the displacements are small ($\sim$ 0.15 A), but for molecules in large cages the displacements appear to be large (perhaps 0.2–1 A). It is very difficult to distinguish experimentally between reduction of an electron density peak by a lower occupancy factor or by a combination of thermal and static positional displacements. Crystallographers utilize a Gaussian displacement model in which the experimental parameter B $\sim 8\pi^2\overline{u}^2$, where $\overline{u}$ is the root-mean square displacement in Angstroms. When B is less than 3 (*i.e.* $\overline{u} < 0.2$ A), refinement is reasonably straightforward, but when B is over 12 (*i.e.*, $\overline{u} > 0.4$ A), refinement is uncertain.

Complex Problems Involving Zeolite Frameworks

The first problem involves the correctness of the topology. Even when the calculated and observed intensities fit well, it is still possible that another structure with rearrangement of the building blocks may be the correct one. Of course, the proposed structure must be capable of explaining the chemical properties. There is no reason to doubt seriously the validity of any of the frameworks given in this review. However, it should be noted that:

(a) Offretite and erionite were confused for many years, and even now, there are severe problems because they occur intergrown. Careful single-crystal diffraction studies (especially electron-optical) provide accurate characterization (*28, 29, 30, 31*).

(b) The x-ray powder data for Ω fit better with those for the mineral mazzite (*32*) than for those calculated earlier for a hypothetical structure (*33*) based on a different space group, probably derived incorrectly from the x-ray powder pattern. It also contained offretite columns but linked in a different way.

(c) Whereas the topological structure of mordenite (Figure 15) should be correct for the single crystal of Na-exchanged ptilolite (a mineral variety of mordenite) used by Meier (*34*), there are three related theoretical structures with similar diffraction properties (*35*).

Figure 16, a projection down the *c* axis, can be related easily to the stereodiagram of Figure 15. In mordenite, the 4-rings labeled A, B, C, and D

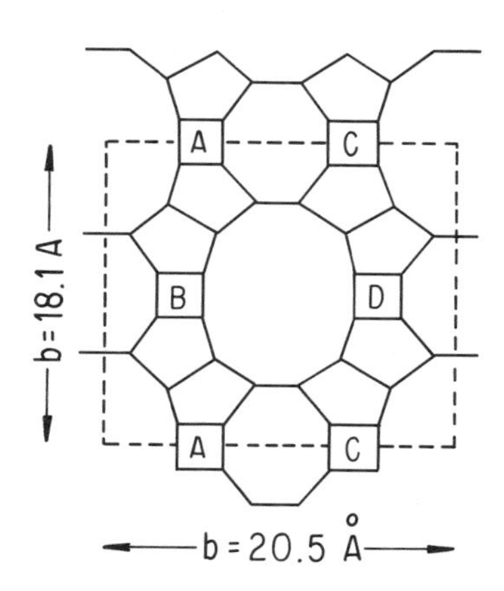

Advances in Chemistry Series

Figure 16. Mordenite framework projected down c *axis. A, B, C, and D refer to the heights of the 4-rings (*see *text)* (35).

lie at heights $0,\frac{1}{2},\frac{1}{2},0$. Sherman and Bennett noticed that three other structures could be obtained: $0,0,0,0$; $0,\frac{1}{2},0,\frac{1}{2}$; and $0,0,\frac{1}{2},\frac{1}{2}$. All should have similar cell dimensions, though the space-group symmetries are different: Cmcm (mordenite), Cmmm, Immm, and Imcm. The theoretical x-ray powder patterns (*see* discussion in Ref. *5*, pp. 5–14) have many features in common, though they differ in detail. Whereas the Cmcm and Imcm structures have a one-dimensional system of large pores, the others have a two-dimensional system with a second set of smaller channels. Sherman and Bennett found that nearly all synthetic and most mineral specimens have the Cmcm structure of Figure 15 but that several mineral specimens were mixtures with the other types.

The second problem involves stacking faults. In mordenite, any one set of 4-rings can be moved independently of other sets by $c/2$ and reassembled into the framework by rebuilding the 5-rings. In the cancrinite–erionite group, stacking faults can occur for layers of types A, B, or C: for example, layers of chabazite type . . . AABBCC . . . could be interspersed between layers of gmelinite type, and vice-versa. Actually there is no evidence for stacking faults in chabazite, but in gmelinite (*36*) the strong streaks on single-crystal x-ray patterns require the presence of stacking disorder. Furthermore the sorption of gmelinite appears to be limited by 8-rings. This is readily explained by blocking of the one-dimensional channels by layers of chabazite type. Intergrowths of erionite and offretite can be similarly explained by stacking faults in the sequence of A, B, and C (*19, 28*). The well-known simple twins of faujasite (spinel twin law on {111}: Ref. *1*) can be explained by a stacking fault at one layer of hexagonal prisms.

The third problem involves the location of Si and Al atoms on the tetrahedral nodes. For several zeolites (*e.g.* natrolite) there is unequivocal x-ray evidence that the Si and Al atoms are ordered, but for most zeolites the x-ray and other evidence are consistent with completely random order. Unfortunately, long-range order may be overlooked if the x-ray crystallographer assumes a space group with too high a symmetry. Thus in chabazite the assumption that the symmetry is $R\bar{3}m$ automatically yields a disordered Al,Si configuration. Optical evidence suggests that at least some chabazite crystals consist of six-fold twins with triclinic symmetry $P\bar{1}$. The latter space group would allow Al,Si order.

The most detailed evidence on Si,Al ordering in aluminosilicates is for feldspar minerals (summarized by Smith (*37*)). Naive crystal-chemical theory suggests that in aluminosilicate tetrahedral frameworks no oxygen should be bonded to 2Al when Si is available. However, there is evidence of deviation from this expectation, especially for material grown rapidly at high supersaturation. In general, ordering should be more likely as the Al,Si ratio approaches one. Long-range order is apparently lacking however in recently crystallized Na,Ca feldspars for Al/Si up to 0.9. Only for Al/Si over 0.95 does long-range order occur upon crystallization. Probably, submicroscopic coherent domains with short-range order occur in feldspars lacking long-range order. The tetrahedral mean distances Si–O and Al–O are near 1.61 and 1.75 A, respectively. When ordering occurs, the tetrahedral nodes tend to lie at positions which would obey the symmetry of a disordered arrangement. The oxygen atoms tend to be displaced from average positions by about 0.1 A. If anti-phase domains exist, it is very difficult to detect Si,Al ordering. Twinning may also cause problems.

Of the zeolites listed in Table I, only A and X have been shown conclusively by x-ray methods to have long-range Si,Al order. Gramlich and Meier (*38*) measured weak diffractions in 70 μm crystals of hydrated A which required a unit cell with $a = 24.61$ A and space group Fm$\bar{3}$c instead of the pseudo cell with $a = 12.3$ and Pm3m. Refinement yielded a framework with alternating mean T–O distances 1.608 (*2*) and 1.728 (*2*) A, which correspond to Si- and Al-rich occupancy. [*Note:* crystallographers *conventionally* express the standard deviation in brackets: 1.608(2) means 1.608 $\pm$ 0.002 A.] However Seff and co-workers (*39*) were unable to find extra diffractions in dehydrated A crystals, and refined all their structures on the small cell thereby automatically obtaining Si,Al disorder (*see* later). Both preparations were made by the method of Charnell (*40*), but apparently some difference, perhaps involving the Si/Al ratio, changed the Si,Al order. Barrer and Meier (*41*) also demonstrated order in a Ge-substituted A-type zeolite. Seff (*42*) recently observed the 531 diffraction for a somewhat larger crystal (0.08 mm) of a new batch of hydrated Na-A.

Olson (*43*) found that hydrated X crystals had symmetry Fd3 instead of Fd3m for the ideal faujasite framework. The cell content of 88 Al and 104 Si (from bulk chemical analysis) prohibits complete Al,Si order in Fd3 but the mean T–O distances 1.619(4) and 1.729(4) A indicate strong segregation between Si and Al. All other x-ray structural data for faujasite-type zeolites are consistent with long-range disorder of Al,Si in Fd3m. Plots of cell dimensions against Si/Al ratio show discontinuities (*44, 45, 46*) which may indicate changes of ordering patterns. Such changes may occur in the short-range order and do not require long-range order.

Mordenite (*34, 47*), when refined in Cmcm, showed mean T–O distances of 1.622(3), 1.607(2), 1.639(4), and 1.625(4) which indicate a small concentration of Al into the third tetrahedron and perhaps into the first and fourth one as well. W. M. Meier (*48*) recently concluded that C2cm is more probable and that Al is ordered into the third and fourth sites.

Chabazite (*49*) shows optical anomalies which indicate triclinic symmetry rather than R$\bar{3}$m used for structure refinement. Further study is in progress by J. J. Pluth and J. V. Smith to test the possibility of Al,Si ordering.

Positions of Cations and Molecules in Zeolites

This section reviews the more accurate of the experimental data (x-ray, IR, Mössbauer, NMR, etc.) which provide evidence on the location of cations and molecules.

Zeolite A. When fully dehydrated, the A zeolite is rather unsatisfactory for ionic bonding to cations. The 4 Ca and 4 Na of the unit cell of dehydrated 5A variety (x-ray powder data) lie near the centers of the eight six-

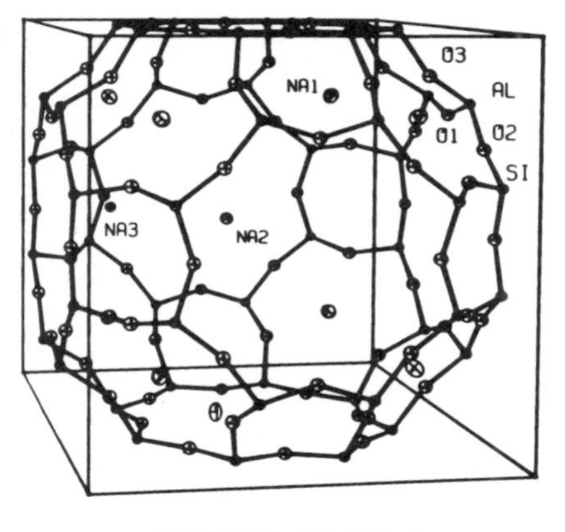

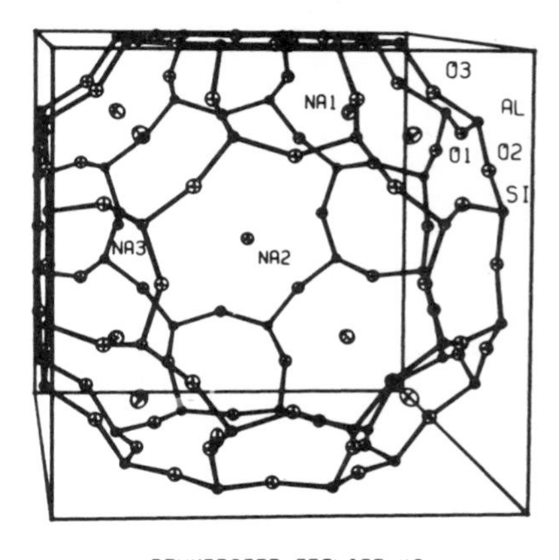

(a)

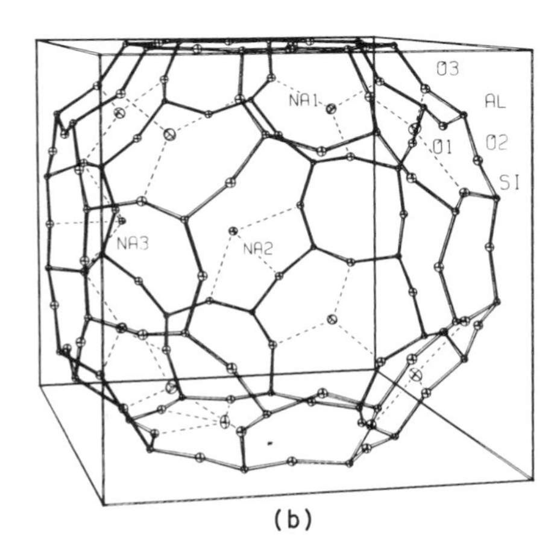

(b)

Journal of Physical Chemistry

*Figure 17. Dehydrated Na-A zeolite. (a) Stereodiagram. Each atom lies at the center of a displacement ellipsoid showing the 20% level of probability. The framework is shown by T-O linkages (*not *T-T linkages as in Figure 10). Although separate positions are shown for Al and Si, the structure refinement was based on complete Al,Si disorder. All eight 6-rings are occupied by Na(1). Only two of the eight-rings are shown occupied by Na(2) whereas all of them should contain one Na(2) in one of the four possible sites displaced from the center as shown for the atom labeled Na2. One of the 12 Na(3) sites is shown occupied. (b) Left-hand diagram of (a) modified to show coordination of Na* (39).

rings (Figure 10). Three of the oxygens (type O(3)) are pulled towards the center of the 6-ring to yield a Ca,Na–O distance of 2.32 A while the other three (O(2)) move away to 2.84 A (Seff and Shoemaker (*50*)). (The bracketed number is merely an arbitrary label to distinguish crystallographically-distinct positions.) This trigonal distortion of a 6-ring to accommodate small cations like Ca and Na is found in many zeolites. Apparently, such a trigonal coordination is preferred to hexagonal coordination with a longer distance ($\sim$ 2.6 A needed for an ideal 6-ring, which is longer than the usual value of $\sim$ 2.4 A for octahedral coordination of Na and Ca). The separate locations of Ca and Na could not be distinguished accurately, but Seff and Shoemaker (*50*) suggested that Ca lay near the plane of the 6-ring while Na was displaced about 0.4 A into the sodalite cage. However, the locations of the Na are probably rather uncertain.

The 12 Na of the dehydrated 4A variety (x-ray single crystal data) face severe problems. In the latest refinement, Yanagida, Amaro, and Seff (*51*) placed eight Na(1) near the centers of the 6-rings, three Na(2) offset from the centers of the 8-rings, and one Na(3) statistically over a 12-fold equipoint (Figure 17). Each Na(1) is bonded to three O(3) at 2.32(1) A, and is displaced 0.2 A *out* of the sodalite unit, perhaps because of electrostatic repulsion from other Na(1) ions. Each Na(2) randomly occupies four positions offset from the center of an eight-ring, and is bonded only on one side to one O(2) at 2.40(6) A and two O(1) at 2.64(3) A. The Na(3) is also bonded on one side and has two O(3) at 2.47(7) A and two O(1) at 2.51(7) A. This coordination is highly unfavorable because of close approach to Na(1) ions.

These atomic distributions explain the sharp increase of N_2 adsorption when four Na per unit cell are exchanged by two Ca. The remaining eight Na plus the two Ca will occupy the eight 6-rings and only two out of the three 8-rings, thereby allowing free passage in three dimensions without the need to wait for Na diffusion from the 8-rings (*52*).

Dehydrated Tl-exchanged A provides a useful comparison with Na-A because of the larger size of the Tl ion. The crystal structure (Riley, Seff, and Shoemaker (*53*)) is shown in Figure 18. Instead of 12 Tl expected for ion-exchange of a single crystal from the batch used for determination of the Na-A structure, only 11 were found. Each 6-ring has one Tl associated with it, but the larger Tl ions are forced to project almost 2 A from the plane of the O(3) atoms (compared with 0.2 A for Na). Seven of these Tl (type 1) project into the larger cage, and one (type 2) projects into the smaller (sodalite) cage. The remaining three Tl (type 3) lie statistically in positions offset from the centers of the 8-rings. Apparent coordinations are: Tl(1)-three O(3) 2.64(2); Tl(2)-three O(3) 2.82(3); Tl(3)-one O(2) 2.60(3) and two O(1) 3.11(5) A. Note that the true bond distances may be different because of the averaging in the electron density maps.

Dehydrated partially Mn-exchanged A (*54*) revealed 4.5 Mn(II) and three Na near the centers of 6-rings. The former were displaced 0.1 A into the sodalite unit from the plane of three O(3) at 2.11(1) A while the latter were apparently displaced 0.5 A into the larger cage. The apparent Na–O(3) distance of 2.16(5) A is implausibly short, probably because of errors caused by averaging of two types of O(3) bonded to either Mn or Na. Dehydrated partially Co-exchanged A (*55*) showed a similar structure with four Co at 2.08(2) A to three O(3), and four Na apparently at 2.12(2) A. However, the former were displaced 0.3 A *into* the larger cage, and the latter 0.6 A into the sodalite unit. These x-ray data agree with the IR data for dehydrated

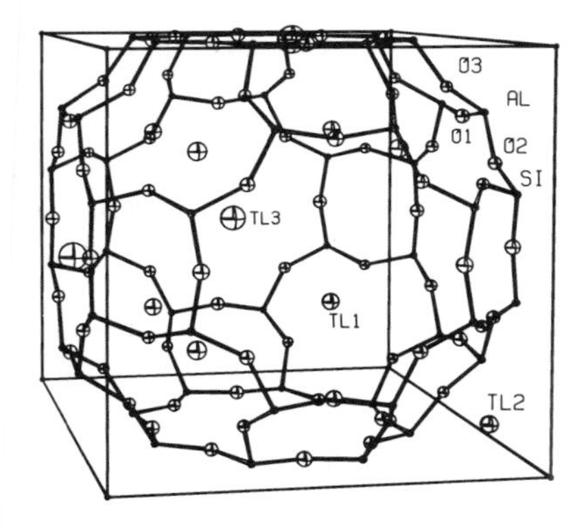

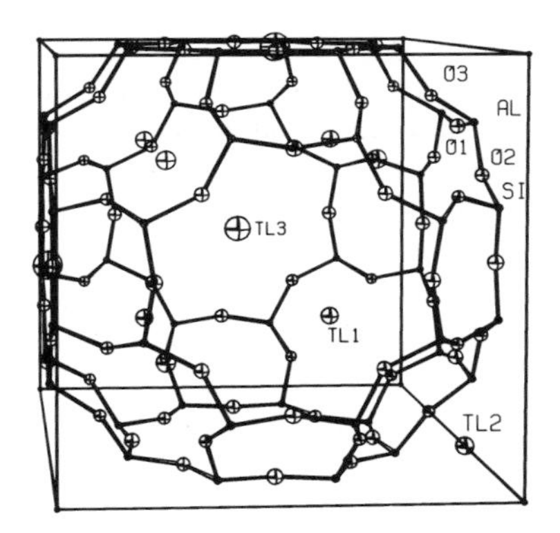

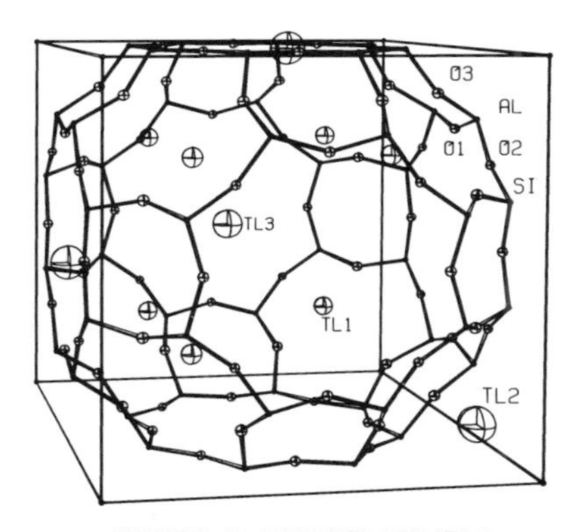

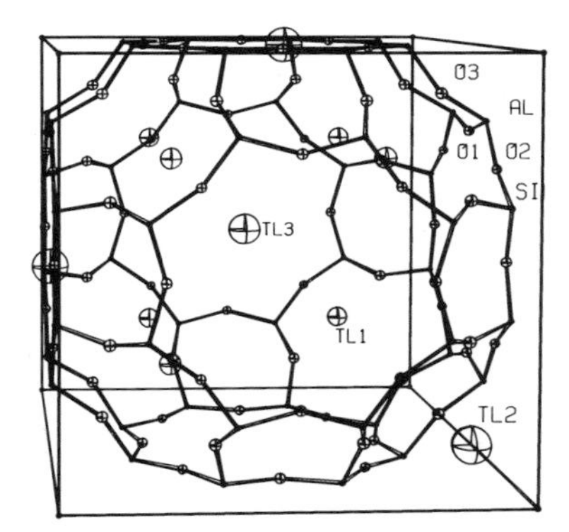

Journal of Physical Chemistry

Figure 18. Stereodiagrams of dehydrated and hydrated Tl-exchanged Na-A zeolite. Displacement ellipsoids show 50% probability. Note that each 6-ring has one Tl associated with it. Tl(3) are shown in three of the 8-rings; the other three also have a Tl(3) when the lattice symmetry is applied (53).

Co-exchanged A: Klier (*56*) had deduced that Co lay in almost planar trigonal coordination. Klier and Ralek (*57*) found similar IR data for Ni-exchanged A.

Comparison of these structures shows that the cations prefer to lie at the six-ring, and that the O(3) atoms are pulled towards the triad axis especially by the small cations. The resulting strain causes geometrical displacements of the whole aluminosilicate framework, principally by change of T–O–T angles; the details are too complex for this review.

Amaro *et al.* (*58*) briefly reported data for dehydrated Co_4Na_4-A, $Ni_{2.5}Na_7$-A and Zn_5Na_2-A. Upon dehydration, all cations in Co_4Na_4-A adopted near-planar, trigonal coordination to three O(3) of a 6-ring. Dehydration of $Ni_{2.5}Na_7$-A at 350°C yielded a black color and metallic luster, indicative of Ni metal. Water may have been disproportioned with loss of H_2 and formation of complexes between O^{2-} or OH^- and Ni^{3+} or Ni^{4+}. A similar result was indicated from Mössbauer studies of Fe^{2+}-Y (*59*).

Even for the dehydrated structures, there are some serious problems in interpreting the crystallographic data for Si,Al order, the exact number of cations, and the averaging of atomic coordinates. The situation for the hydrated structures is even more complex.

A very detailed x-ray study by Gramlich and Meier (*38*) of a single crystal of hydrated Na-A revealed 90 weak diffractions consistent with space group $Fm\bar{3}c$ and *a* 24.61(1) A. Refinement confirmed the expected alternation of Si and Al (*see* earlier). Sixty-four of the expected 96 Na lay 0.6 A from the centers of the 6-rings (site 1) with three O(3) at 2.36(4) A as nearest neighbors [note that the true cell is eight times as large as the pseudo cell]. Each sodalite unit appeared to contain four H_2O in a distorted tetrahedron with bonding to O(3) at 2.8 A and Na(1) at 2.9 A. Each large cage contained two sets of water molecules which could correspond to twenty H_2O lying at the

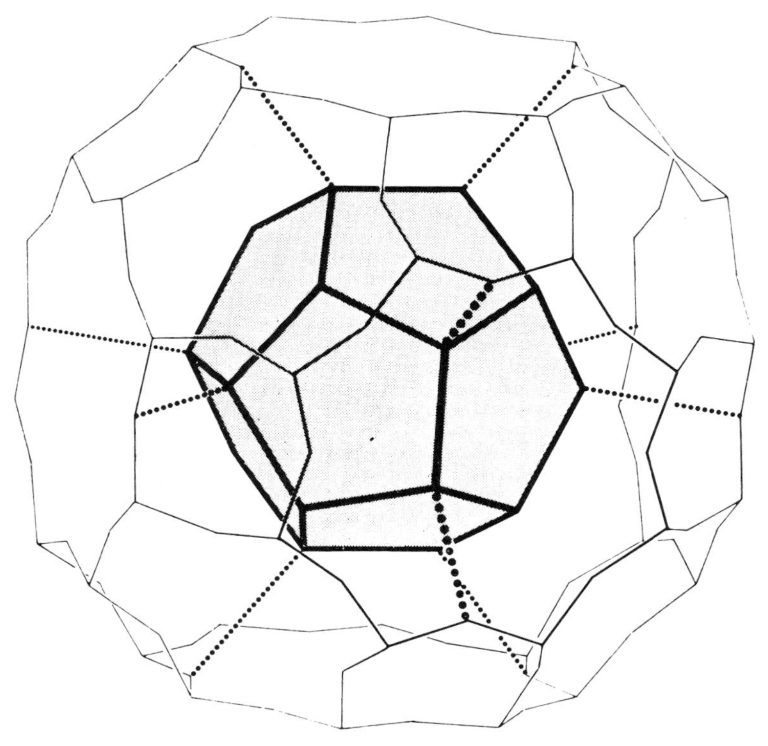

Zeitschrift fuer Kristallographie

Figure 19. Cluster of water molecules in hydrated Na-A. The water molecules lie at the corners of the pentagonal dodecahedron, and the framework T and O atoms of the large cage lie at the intersections of the thin lines. Dotted lines show shortest H_2O–O distances (38).

corners of a pentagonal dodecahedron (Figure 19). [This arrangement was first reported by Seff (*60*) for hydrated Ca_4Na_4-A.] Distances indicated bonding of H_2O to Na(1) and O(3). Such an arrangement of water molecules is well known in clathrate hydrates. Each 8-ring may contain one Na and one H_2O. The root-mean-square displacements of the cage contents ranged from 0.25 to 0.7 A, indicating strong deviations from the ideal structure either on a space or time average or both. At least some of the displacement is thermal, but another part must arise from the asymmetric crystal fields emanating from oxygen atoms bonded to the ordered pattern of Si and Al atoms.

The positions of the framework and Tl atoms of hydrated Tl-exchanged Na-A (*53*) are shown in Figure 18. The framework geometry is more regular than for the dehydrated form. The Tl atoms occupy similar positions, but they are slightly nearer the 6-rings. These become closer to regular hexagonal geometry as the Tl(1)–O(3) distance increases from 2.64(2) to 2.75(2) A during hydration while Tl(2)–O(3) increases from 2.60(3) to 2.81(2) A. No water molecules were located.

The structure of hydrated $Mn_{4.5}Na_3$-A was determined by Yanagida, Vance, and Seff (*54*). Each of the 4.5 Mn(II) atoms lies in a 6-ring where it is bonded to three O(3) at 2.28(1) A and two H_2O which complete a trigonal bipyramid (Figure 20): $H_2O(1)$ projects into the sodalite unit at 2.03(6) and $H_2O(2)$ into the larger cage at 2.06(7). The bipyramid is slightly distorted, with Mn displaced 0.2 A into the larger cage from the plane of O(3). This displacement is much less than the values of 0.5, 1.5, 1.6, and 0.4 A, respectively, for hydrated Na, Ni(II), Co(II), and Zn(II) varieties. Interpretation of the remainder of the contents of the larger cage is equivocal, and

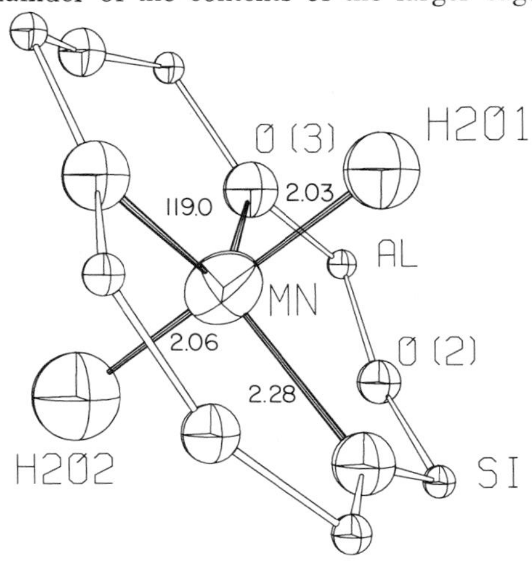

Inorganic Chemistry

Figure 20. Coordination of Mn in hydrated $Mn_{4.5}Na_3$-A zeolite. Displacement ellipsoids show 50% probability (54).

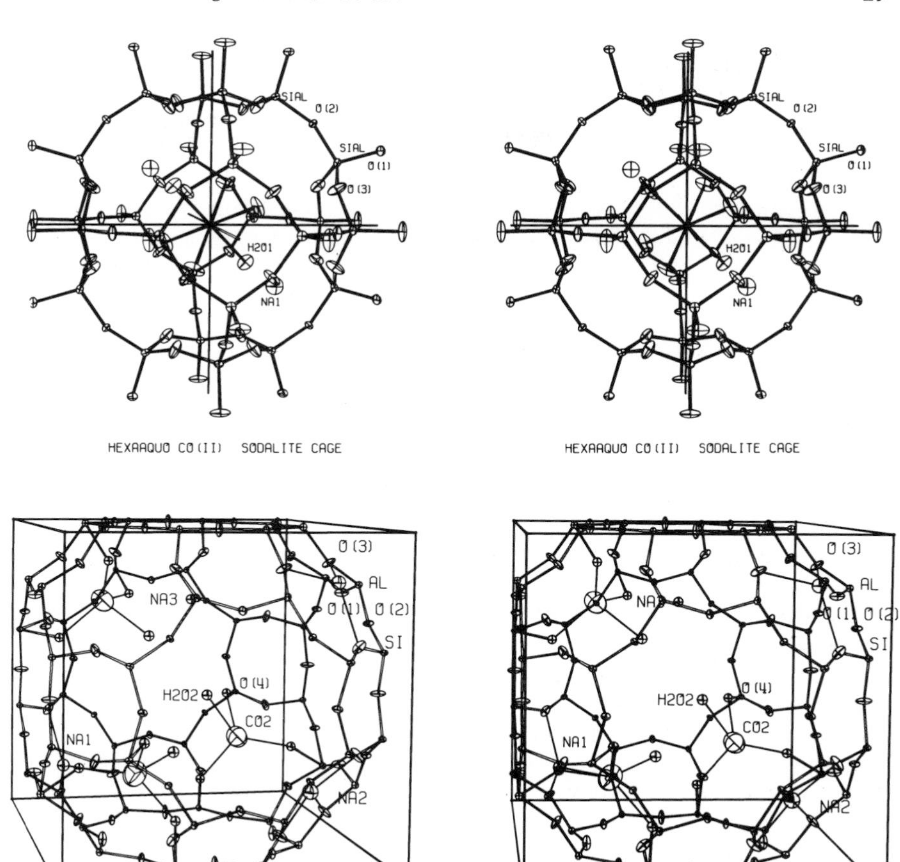

Figure 21. Stereo plots of hydrated Co_4Na_4-A. The upper diagram shows a Co ion (unlabeled) at the center of a sodalite cage where it is surrounded by six $H_2O(1)$. The three perpendicular lines are the crystallographic axes. The lower diagram shows three Co ions (type 2) each bonded to one $H_2O(2)$ and three oxygen species (type 4). Each of the latter is bonded to a T atom (probably Al) changing its coordination from 4 to 5 (42).

the original paper should be consulted. In contrast to the hydrated Tl-A, here the framework is somewhat distorted because of the short Mn–O(3) distance.

Amaro *et al.* (*58*) reported brief data for hydrated Co_4Na_4-A, $Ni_{2.5}Na_7$-A, and Zn_5Na_2-A. In the Co_4Na_4 variety (Figure 21), one Co^{2+} ion lies at the center of the sodalite cage where it is surrounded by six water molecules each hydrogen-bonded to two framework oxygens. Three Co(2) ions lie on or near a triad axis displaced 1.6 A into the large cage from the plane of a 6-ring. The distance to the nearest framework oxygens is 2.7 A, much too long for bonding. Electron density at 2.36 A from Co(2) and 1.74 A from the T site was ascribed to an oxygen species. Each Co(2) is tetrahedrally coordinated to one $H_2O(2)$ and three oxygens. This indicates extensive hydrolysis with transformation of a T atom (probably Al) from four- to five-coordination. Such a coordination occurs for half of the Al atoms in the mineral andalusite. (Full details have not been published so far, and I am especially indebted to K. Seff for providing Figure 21 from a preprint.)

Fully hydrated $Ni_{2.5}Na_7$-A (*58*) has Ni occupying positions similar to those in Co_4Na_4-A, with 1Ni at the origin and 1.5 Ni in the large cage. Hydrated $Ni_{0.6}Na_{10.8}$-A probably has Ni at the center of the large cage.

A series of studies on Zn_5Na_2-A (*58*) showed how the Zn ions adjust to progressive loss of water. The fully hydrated variety has one Zn at the center of the sodalite unit where it is coordinated tetrahedrally to four H_2O. The other four Zn project into the supercage where they coordinate to three framework oxygens of a 6-ring and one H_2O. Dehydration at 350°C yielded three Zn in the above 6-ring site, while two Zn were bridged by one H_2O in the sodalite unit. Dehydration at 380°C resulted in loss of H_2O for the first three Zn, which adopted near-planar coordination to the three framework oxygens; the other two Zn retained the bridging species. This series of experiments shows that dehydration is a complex process involving breakdown of hydration complexes, migration of cations, and sometimes hydrolysis.

The transition metals are suitable for study by resonance and spectroscopic techniques. As an illustration, Heilbron and Vickerman (*61*) showed by EPR and reflectance spectroscopy of Co-exchanged A, X, and Y zeolites that with decreasing dehydration Co changed its coordination from involvement with H_2O and framework oxygens to involvement with framework oxygens and either a residual H_2O or a hydroxyl or oxide ion. It is obvious that a wide range of intermediate structures can occur during dehydration.

Turning to sorption complexes of A, the data become even more complex and uncertain. The first two x-ray determinations were made on powders of $Na_{12}\text{-A} \cdot \sim 6\,Br_2$ (*62*) and $Ca_4Na_4\text{-A} \cdot \sim 6\,I_2$ (*50*). In the first structure, eight Na were located near the 6-rings, but the 12 Br atoms were represented by a 48-fold general position in the larger cage. It is chemically inconceivable that the Br_2 molecules would break up into quarter-atoms at distances ranging down to 1 A or less. The only possible solution is that there are several types of distributions occurring randomly throughout the structure. In any one cage, the Br_2 molecules occupy 12 of the 48 indicated positions. Selection of the sets of 12 is purely speculative, and Meier and Shoemaker (*62*) invented plausible positions with Br atoms forming sets of six dumbbells lying near the corners of a cuboctahedron. Seff and Shoemaker (*50*) faced a similar problem though the 48 positions were different from those in the Br_2 variety. The only plausible solution yielded reasonable distances, but the large displacement factors ($\sim$ 0.5 A) preclude detailed interpretations. Nevertheless these data, as well as similar data obtained by Fang and Smith (*63*) for a chlorine–chabazite complex, show that molecules occupy definite positions in the cages and do not occur as a free fluid.

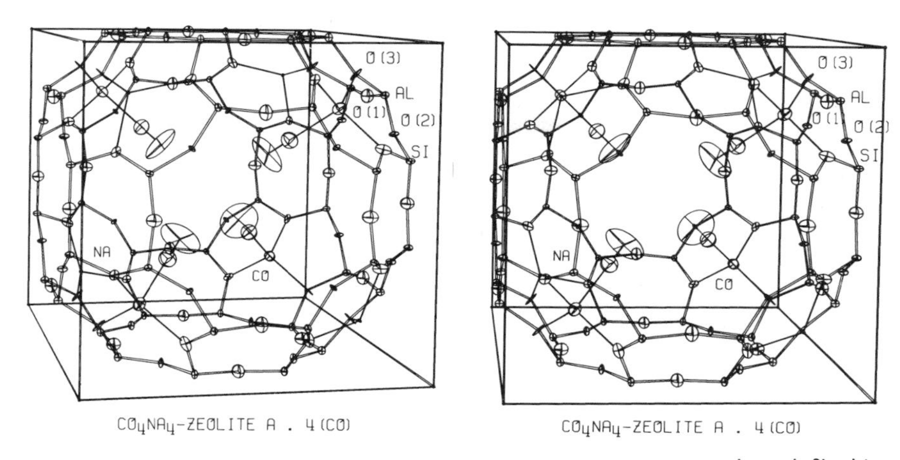

Inorganic Chemistry

Figure 22. Stereodiagram of $Co_4Na_4A \cdot 4$ CO complex. Displacement ellipsoids show 20% probability (55).

Recent x-ray studies of complexes with zeolite A used single crystals. Riley and Seff (*55*) followed their study of dehydrated Co_4Na_4-A (*see* earlier) with a similar one on the CO adduct produced at 716 torr. The four CO molecules (Figure 22) are each bonded to one Co atom which moves 0.2 A from its position in the dehydrated form. Concomitantly, the Co–O(3) distance probably lengthens slightly from 2.08(2) to 2.11(2) A, and the O(3)–Co–O(3) angle decreases slightly from 117(1)° to 114(1)°, as the coordination moves from near-planar trigonal to near tetrahedral. Whereas the C atom is fairly well defined at 2.29(16) A from Co, the O atom is poorly defined. Strong thermal motions are expected because no other atom is within van der Waals bonding distance (4.5 A to the three other oxygens of the CO molecules). The Na atoms do not bond to CO. Infrared data by Angell and Schaffer (*64*) for CO sorbed on X and Y zeolites exchanged with transition metals showed C–O stretching frequencies higher than for CO gas. Electrostatic interaction of the metal ion with the CO dipole ultimately results in charge transfer in the CO molecule with strengthening of the bonds.

The acetylene molecule approaches broadside-on to the exchangeable cation in A zeolite, indicating interaction between the cation charge and the laterally polarizable π electron system of the C_2H_2. Crystal structures have been determined for Na_{12}-A $\cdot \sim 6\,C_2H_2$ (*65*), $Mn_{4.5}Na_3$-A $\cdot$ 4.5 C_2H_2 (*66, 67*) and Co_4Na_4-A $\cdot$ 4 C_2H_2 (*67*). When transition metals are present, the acetylene molecules attach themselves to these highly polarizing entities rather than to the Na ions. Figure 23 shows the spatial relations for the Mn-exchanged variety. Each Mn is bonded to three O(3) at 2.18(1) A with O(3)–Mn–O(3) angles of 114.9(7)°. A C_2H_2 molecule completes a tetrahedral coordination around each Mn cation, with an Mn–C distance of 2.63(17) A. For the Co variety, the spatial relations are quite similar, with distances of 2.19(1) and 2.54(7) A and an angle of 113.3(4)°. The Na

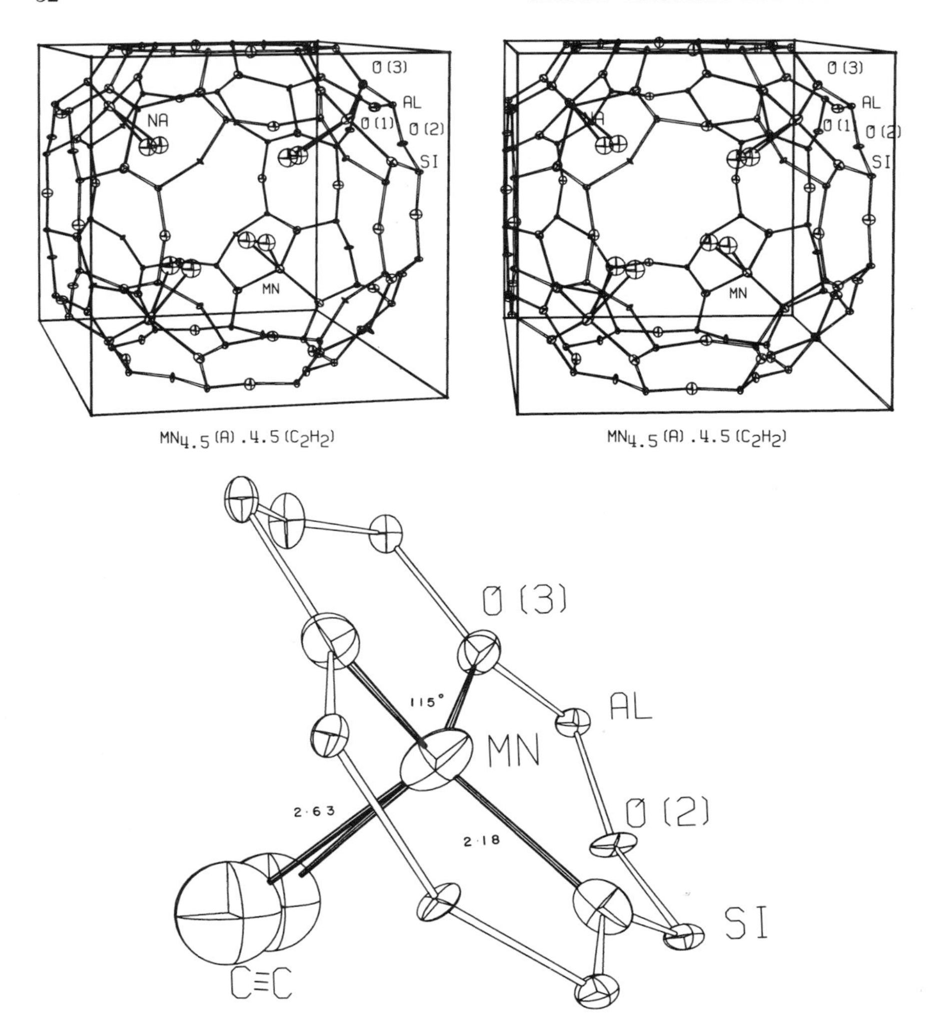

Inorganic Chemistry

Figure 23. Structure of $Mn_{4.5}Na_3$-A·4.5 C_2H_2 complex. The upper diagram is a stereo plot using ellipsoids of 20% probability. The lower diagram shows the coordination of Mn with atoms represented by ellipsoids of 50% probability. See *text for description of disorder of C_2H_2 molecule* (67).

ions are slightly recessed into the sodalite units in both structures and are too far (4.4 A) from the C_2H_2 molecules to give significant interaction. Sorption of C_2H_2 into dehydrated $Mn_{4.5}Na_3$-A causes the Mn ions to move 0.6 A from a position slightly inside the sodalite unit to one slightly inside the larger cage. Simultaneously the Mn–O(3) distance lengthens from 2.11(1) to 2.18(1) A in response to interaction with the C_2H_2 molecule. Although Figure 23 shows that the C_2H_2 molecules occupy distinct positions, this is merely for clarity. Actually the electron density consists of three broad peaks lying symmetrically about the three-fold axis, and the molecules must be disordered, probably because of hindered rotation. The data for Na_{12}-A · ~ 6 C_2H_2 are rather complex, probably because of the weak interaction between Na ions and C_2H_2 molecules. Nevertheless they indicate a strong tendency for the molecules to lie broadside-on to the cations.

The sulfur complex of Na_{12}-A produced by exposure to dry sulfur at 270°C was interpreted by Seff (*68*) to contain two crown-shaped S_8 molecules per large cage (Figure 24) but the displacements of the S atoms are

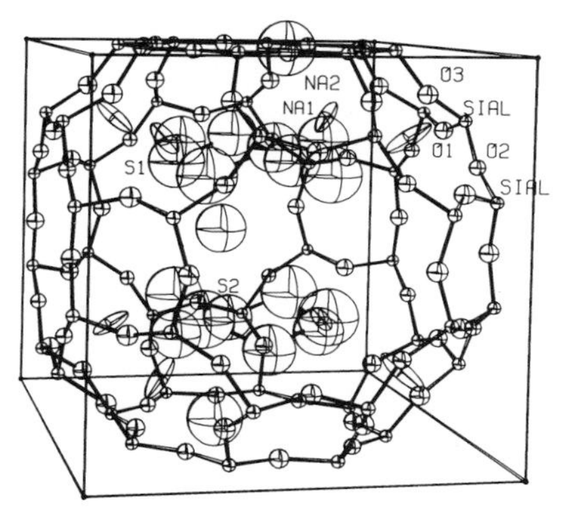

ZEOLITE 4A . $2S_8$ ZEOLITE 4A . $2S_8$

Journal of Physical Chemistry

Figure 24. Stereodiagram of $Na_{12}A{\cdot}2\ S_8$ complex. Displacement ellipsoids show 50% probability. This is one of three arrangements needed to obey the cubic symmetry (68).

very large (*ca.* 0.5–0.6 A). Whereas the framework and Na(1) atoms are well-defined and have similar positions to those in hydrated Na and Tl varieties, the Na(2) position is poorly defined (displacements *ca.* 0.5 A).

Finally, Yanagida and Seff (*69, 70*) reported structures for Na_{12}-A loaded first with 32 NH_3 at 604 torr and then with 8 NH_3 at 12 torr. No electron density peaks for NH_3 were located in the partly loaded zeolite, but the framework coordinates showed shifts to the relaxed state typical of A complexes. Figure 25 shows the complicated arrangement of NH_3 molecules assigned to the strongly loaded zeolite. Of the 12 molecules in the sodalite unit, eight were coordinated to Na and four were hydrogen-bonded to framework oxygens and the other eight NH_3. The 20 NH_3 form a distorted pentagonal dodecahedron in the larger cavity and interact with Na and framework oxygens. High displacement factors (~ 0.3–0.6 A) for N and Na(2) indicate caution in making detailed interpretation.

Chabazite. Although the crystallographic data are technically obsolete, they provide interesting information on the positions of cations when hexagonal prisms are available. Data on hydrated Ca-exchanged chabazite ($Ca_{1.95}$-

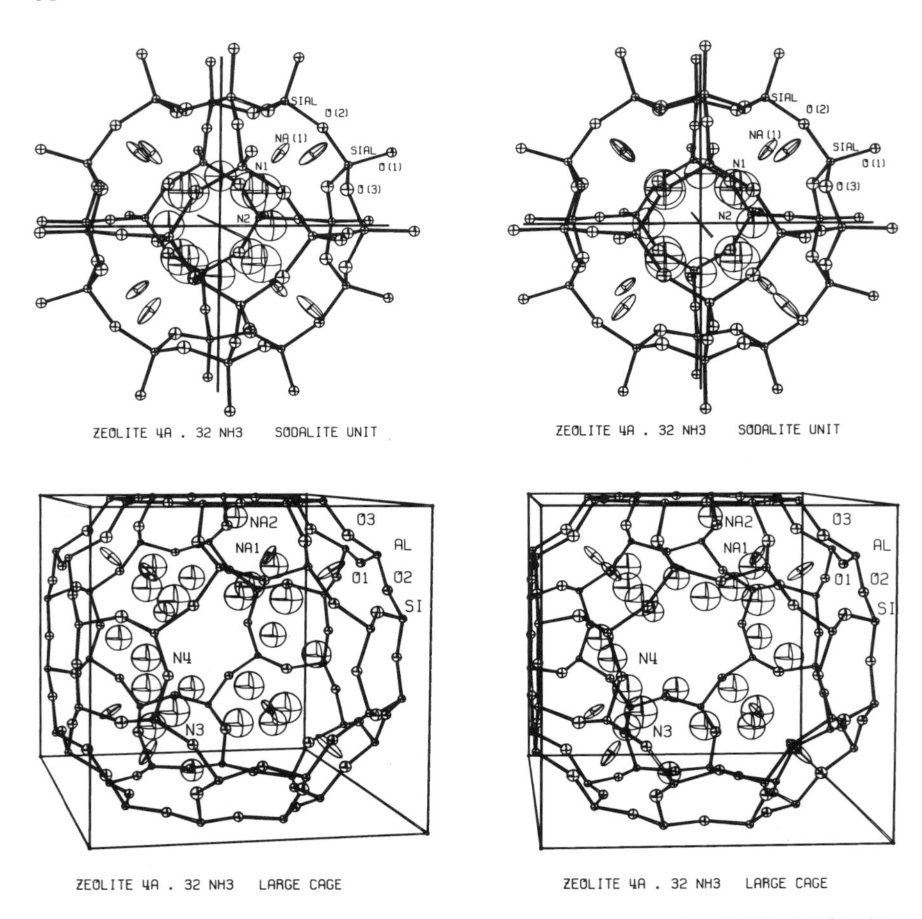

Journal of Physical Chemistry

Figure 25. Stereodiagrams of $Na_{11}A \cdot 32\ NH_3$ complex. The lower diagram shows the unit cell frame whereas the upper diagram shows the axes at the origin. Al and Si were assumed to be disordered. H not shown. Probability level 50% (70).

$Al_{3.9}Si_{8.1}O_{24} \cdot \sim 13\,H_2O$) at both room temperature and −150°C (*71, 72*), dehydrated Ca-chabazite (*73*), and the chlorine complex at room temperature and −150°C (*74*) were reviewed by Smith (*75*).

Natural chabazites range in Si/Al from 3 to 1.6 while synthetic ones vary at least from 2.3 to 1.1. Many natural specimens have Si/Al of *ca.* 2 and have optical anomalies indicating triclinic symmetry of the framework (Figure 8). These crystals may have Si.Al order with two Al in each 6-ring. On the average natural specimens contain 1.5 Ca, 0.5 Na, and 0.2 K atoms per cell.

Calcium atoms in silicates usually are bonded to six or seven oxygens at about 2.4 A in an octahedral or distorted-octahedral coordination. In dehydrated chabazite, such an octahedral coordination can be obtained by placing Ca at the center of the hexagonal prism (site 1) where it can pull in three oxygens from each 6-ring. There is only one such site per cell. The next best site (labeled 2), close to the single 6-rings, is twofold, but the coordination is near-planar rather than octahedral. All other sites (type 3) on the wall of the large cavity provide poorer coordination. Simultaneous occupancy of sites 1 and 2 is unlikely because of strong electrostatic repulsion between the cations: hence p(1) + 0.5 p(2) should be less than or equal to 1, where p(1) and p(2) are the populations of the two sites. The x-ray analysis of dehydrated Ca-chabazite indicated 0.6 Ca(1), 0.7 Ca(2), and perhaps 0.7 Ca(3); this can be interpreted as two populations in which approximately two-thirds of the hexagonal prisms were occupied by Ca(1) and the other one-third by two Ca(2). Note that the above avoidance prediction is obeyed. Perhaps the Al,Si pattern controls the location of the Ca atoms.

In the hydrated variety a complex pattern of small peaks in the large cavity could be explained by triclinic symmetry and formation of a pentahedron of five H_2O around each of the two Ca atoms: the details, however, are uncertain. Almost exactly the same atomic coordinates were found at room temperature and −150°C, ruling out possible complications from thermal transitions.

Studies of the chlorine complex revealed six diffuse peaks at the corners of a distorted octahedron centered in the large cavity. The structures were essentially the same at room temperature and −150°C. No Ca occurred at the center of the hexagonal prism, and the occupied sites near the two 6-rings were displaced into the large cavity.

Figure 26 shows the large changes in the shape of the 8-ring which occur upon dehydration and sorption. The shifts of up to 0.5 A in the framework atoms can be explained by trigonal distortions of the hexagonal prism in response to movement of the Ca atom, together with rotation about the single four-rings, which can act as hinges between the hexagonal prisms.

No data are available for varieties of chabazite with univalent ions, but occupancy of the 8-rings by large ions such as Rb^+ and Cs^+ is likely. Such sentinels, to use the excellent term of R. M. Barrer, could block sorption of those molecules which are insufficiently polar to cause the sentinels to move.

Space prohibits description of the extensive physicochemical data on chabazite obtained by R. M. Barrer and colleagues, but the reviews by Smith (*75, 76*) provide access to the literature and structural interpretations.

Faujasite, X, and Y. The mineral faujasite has a range of compositions and combines features found separately in A and chabazite. It tends to have about 55 Al atoms per unit cell together with a mixture of Ca, Na, Mg, and K exchangeable cations. The synthetic analogs, X and Y, were synthesized in Na-bearing systems and have compositions $Na_pAl_pSi_{192-p}O_{384} \cdot q\,H_2O$, where p ranges from 96 to 74 for X and 74 to 48 for Y, while q drops from about 270 to about 250 as Al decreases.

There are so many papers by so many techniques on the structural properties of chemically treated members of the faujasite family that this review

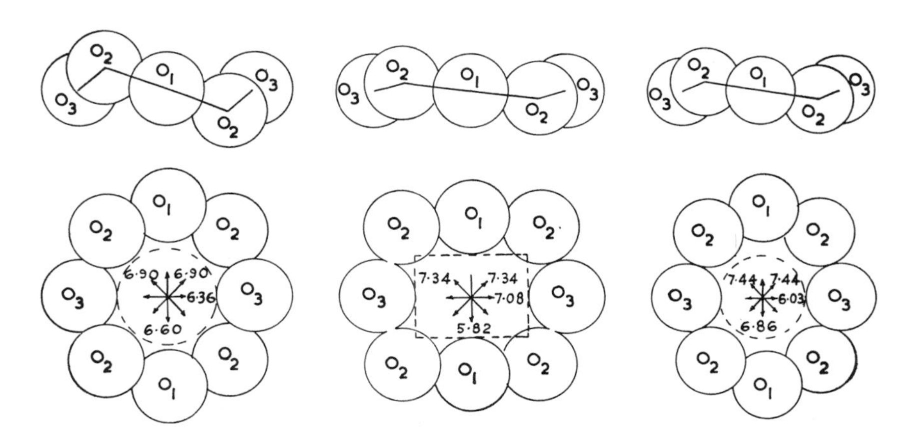

Journal of the Chemical Society, London

Figure 26. Elevation and plan of the 8-ring in chabazite for the hydrated variety (left), dehydrated (center), and chlorine complex (right). The distances between framework oxygens in the directions of the arrows are in A; these oxygens were assigned a radius of 1.35A. The dashed circles and rectangle are merely guides for viewing (75).

cannot do justice. For data up to January 1970 the review by Smith (77) covers the properties of the aluminosilicate framework, the positions of cations and molecules, and the nomenclature. Breck (Ref. *1*, pp. 92-107) gave detailed tables of cation distribution for structures determined till then. The present review lists all papers with significant x-ray data on the crystal structures, and Tables II, III, and IV show selected features. Again, there are serious problems in interpreting x-ray diffraction data, especially those obtained for powders. In particular, protons cannot be detected directly by x-ray methods, though their presence as OH may be inferred from interatomic distances. Resonance and absorption methods are much more sensitive for detection of OH and for the environment of certain elements. No evidence is available however on the x,y,z coordinates. Consequently it is necessary to combine the data from several techniques to elucidate many of the problems.

With the experience gained from A and chabazite and a general knowledge of ionic principles, it is possible to predict fairly well the positions of cations in the various members of this family. Referring to Figure 11, the centers of the 16 hexagonal prisms (site I) are the obvious first choice for cations that prefer an octahedral bonding. Figure 27 (top) shows an idealized perspective drawing of the sodalite unit with the adjacent hexagonal prisms while Figure 27 (bottom) shows a section. Occupation of site I causes six O(3) to be pulled inwards to yield a near-octahedral coordination while the six O(2) move away. By analogy with chabazite, site I′ provides one-sided coordination to three O(3) of a 6-ring of the hexagonal prism. Simultaneous occupancy of two I′ sites of the same prism is electrostatically reasonable, but simultaneous occupancy of adjacent I and I′ sites is very unlikely even for monovalent cations. Hence $p(I) + p(I') \leq 1$ where $p(1)$ and $p(I')$ are

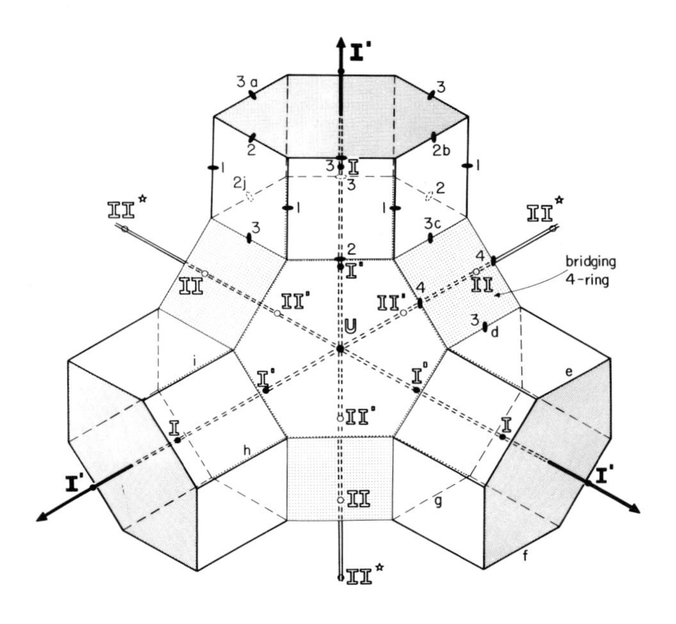

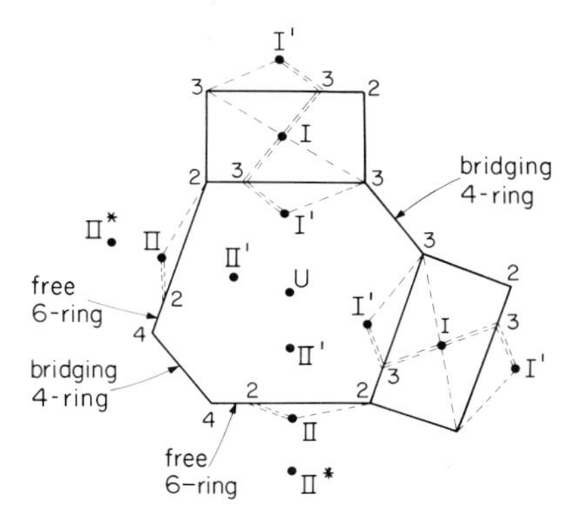

Advances in Chemistry Series

Figure 27. Idealized projection and section through a sodalite unit of faujasite. Top: T sites lie at the intersections of the polyhedral edges. Oxygens, labeled 1 through 4, lie near the mid-points of the edges but displaced to give a tetrahedron around each T site. Four axes of inverse three-fold symmetry pass through the hexagonal prisms (one hidden) to the center (site U) of the truncated octahedron (sodalite unit), but one is hidden because it passes vertically downwards. Bottom: this is a section through abcdefghij in top diagram showing the labeling of oxygens (1-4) and of cation sites (Roman symbols). The dashed lines show principal cation-oxygen bonds when I or I′ sites are occupied; note that some of these are out of the plane of the section (46).

Table II. Assignments of Cations and Water

Specimen[b]	*X-Ray*[c] *Technique*	*Site Population*			
		I	*I′*	*II′*	*II*
Natural faujasite	sc		17?	32 Ox	
Natural faujasite	sc		11.5 Ca	41 Ox	
Na-faujasite	sc	1.9 Na	16.7 Na	9.8 Ox	28.8 Na
Ca-faujasite	sc		9.7 Ca		11.5 Ca
La-faujasite	sc		3.3 La	28 Ox	
Ce, etc.-faujasite	sc		18 Na	32 Ox	
K-Y(48)	p		13.6 K		17.8 K
K-Y(55)	p	1.3 K	13.3 K		20.0 K
$Ni_{14}Na_{23}H_5$-Y(56)	p	3.5 Ni	6 Ni	1.9 Ni	24 Na
Na-X(88)	sc	9 Na	8 Na	26 Ox	24 Na
			12 Ox		8 Ox
Na-X(81)[b]	sc	9 Na	11 Na	32 Ox	22 Na
			12 Ox		
K-X(70)[b]	p	7 K	12 K		24 K
K-X(86)[b]	p	9 K	7 K		23 K
Ca-X(80)[b]	sc		20 Ca	40 Ox	
$Sr_{42}Na_1$-X(80)[b]	p	2.1 Sr	11.1 Sr	32 Ox	15.0 Sr
$Sr_{30}Na_{24}$-X(80)[b]	p	12 Na	7.3 Sr	26 Ox	11.5 Sr
La_{29}-X(87)[b]	p		12 La	32 Ox	17 La
La_{29}-X(87)[b]	p		30 La	32 Ox	
$La_{26}Na_5$-X(82)[b]	p	32 Ox	14 La	24 Ox	13 La
Na_1Ce_{26}-X(88)[b]	p	11 Ox	9 Ce	>32 Ox	21 Ce

[a] X-ray diffraction yields electron density, and the above assignments to cations and water oxygens may be incorrect (*see* text).

the *fractional* populations of the 16-fold site I and 32-fold site I′. Each sodalite unit contains four more 6-rings, each giving access to a supercage. Here there are 32-fold sites labelled II′ or II according to whether displacement occurs into the sodalite unit or into the supercage. For this 6-ring, three O(2) are pulled inwards to provide bonding to cations. In hydrated faujasite, electron density occurs at II*, too far for simple bonding to O(2). Anions in certain partly dehydrated zeolites may lie in sites II′ or I′ or at site U at the center of the sodalite unit.

The supercage (Figure 11) is much bigger than the cages in chabazite and A. The 16 12-rings are too large for unhydrated cations, but site V, at or near the center, is occupied by certain hydration complexes. One-sided coordination is possible on the walls of the supercage, for example to four oxygens of a bridging 4-ring. However, actual x-ray data are complex and confused, and the term site III is used here in a broad sense for all positions near the supercage wall (*see* Hseu (*78*); Mikovsky *et al.* (*79*); Mortier (*80*) and Mortier, Bosmans, and Uytterhoeven (*81*) for details of site III′). Finally site IV is used in a very broad sense to cover sites too far inside the supercage to permit simple bonding to framework oxygens (*78*).

Molecules in Hydrated Faujasite-type Zeolites[a]

Site Population				
II*	III	IV	V	*Reference*
11 Ox				Baur (*82*)
25 Ox	8.5 Na	37 Ox		Hseu (*78*)
				Hseu (*78*)
23 Ox		29 Ox	2.2 Ca	Bennett and Smith (*83*)
14 Ox			10.3 La	Bennett and Smith (*84*)
26 Ox			5.8 Ce	Olson *et al.* (*85*)
				Mortier and Bosmans (*86*)
				Mortier and Bosmans (*86*)
				Gallezot *et al.* (*87*)
	several sites			Olson (*43*)
6 Ox	20 Na	22 Ox		Hseu (*78*)
				Mortier and Bosmans (*86*)
				Mortier and Bosmans (*86*)
26 Ox		67 Ox		Pluth and Smith (*88*)
				Olson and Sherry (*89*)
				Olson and Sherry (*89*)
			4 La	Olson *et al.* (*85*)
	(calcined then rehydrated)			Olson and Sherry (*89*)
			3 La	Bennett *et al.* (*90*)
				Hunter and Scherzer (*91*)

[b] Bracketed number is Al atoms per cell.
[c] Legend: sc, single crystal; p, powder.

The x-ray data on hydrated varieties are open to various interpretations. The site populations in Table II are merely the mathematical approximations used to fit the observed electron densities. Obviously, the greater the electron density of a cation, the greater is the chance of distinguishing it from a water molecule. Unfortunately Na is only slightly denser than water. All the x-ray crystallographers had great difficulty in finding enough electron density in the supercage to explain the expected number of cations and water molecules. Baur (*82*) concluded that "most of the water molecules and exchangeable cations in faujasite float freely through the aluminosilicate framework." Further investigations show that a better model involves preferred positions which are occupied on a temporary basis. Five structure determinations in Table II revealed electron density at V, which can be assigned certainly or almost certainly to a cation which cannot float in space and must be at the center of a hydration complex. Site I is apparently occupied by cations in several varieties (the oxygen listed in the last two structures is almost certainly a mathematical fiction for cations). There is too much electron density at I′, II′, and II in some of the structures to allow for the maximum possible concentration of 32 water molecules per site. Hence cations must occur in

Table III. Assignment of Cations (and Residual

Specimen[b]	*X-Ray*[c] *Technique*	*Site Population*				
		I	*I′*	*U*	*II′*	*II*
Na-faujasite	sc	10 Na	9 Na			32 Na
Na-faujasite	sc	9.3 Na	16.7 Na			31 Na
K-faujasite	sc	8.6 K	12.9 K			32 K
Ca-faujasite	sc	14.2 Ca	2.6 Ca			11.4 Ca
$Ni_{27}Na_4$-faujasite	sc	10.6 Ni	3.2 Ni		1.9 Ni	6.4 Ni
			5.8 Ox		1.9 Ox	
Ba-faujasite	sc	7.3 Ba	5.0 Ba			11.3 Ba
La-faujasite	sc	11.8 La	2.5 La			1.5 La
La-faujasite	sc	11.7 La	2.5 La			1.4 La
Ce, etc.-faujasite	sc	3.4 La	11.5 Ce		16 Ox	10.7 Na
Natural faujasite	sc	12.8 Ca				10.4 Ca
Na-Y(57)	p	7.8 Na	20.2 Na			31.2 Na
$Na_{27}H_{29}$-Y	p	7 Na	7 Na			13 Na
$Na_{14}H_{42}$-Y	p	4 Na	3 Na			10 Na
Na_2H_{54}-Y	p	2 Na	3 Na			3 Na
K-Y(48)	p	6.4 K	14 K			26 K
K-Y(55)	p	5.4 K	18 K			27 K
K-Y(57)	p	12.0 K	14 K			30 K
Ag-Y(57)	p	16 Ag	11 Ag			28 Ag
$La_{16}Na_{13}$-Y	p	13 Ox	16 La	3 Ox	29 Ox	
$La_{16}Na_{13}$-Y	p	5.2 La	8.9 La			5.5 La
$Ce_8Na_8H_{23}$-Y	p	5.8 Ce	1.9 Ce			10 Na
$Ce_8Na_8H_{23}$-Y	p	1.5 Ce	8.3 Ce			7 Na
$Ni_{19}Na_{15}$-Y	p	11.3 Ni	1.9 Ni			20 Na
$Ni_{14}Na_{23}$-Y	p	11.7 Ni	1.1 Ni			21 Na
$Ni_{10}Na_{31}$-Y	p	8.8 Ni	1.7 Ni			27 Na
$Cu_{16}Na_{24}$-Y	p	3.2 Cu	11.1 Cu			20 Na
$Cu_{12}Na_5H_{27}$-Y	p	1.7 Cu	9.9 Cu			8 Na
Ca-X	p	7.5 Ca	17 Ca		9.0 Ca	17 Ca
					10 Ox	
Sr-X	p	11.2 Sr	7.0 Sr		4.2 Sr	19 Sr
					5 Ox	
Sr-X	p	6.1 Sr	12.0 Sr		6.4 Sr	20 Sr
					8 Ox	
Na-X(81)	sc	3.1 Na	32 Na	3 Ox		32 Na
Na_1Ce_{26}-X(88)	p	4.2 Ce	23 Ce	1 Ox		
Na_1Ce_{26}-X(88)	p	3.2 Ce	24 Ce	6 Ox		

[a] Data for Co_{14}-Y and Co_{19}-Y dehydrated at 200° and 600°C given by Gallezot and Imelik (*106*).

these sites but not necessarily at full occupancy. Ion-exchange data (*see* later) also indicate occupancy of sites inside the sodalite unit. Pluth and Smith (*88*) presented arguments for mixed occupancy in hydrated Ca-faujasite with 13 Ca and 17 H_2O in I′, six Ca and 26 H_2O in II′, 27 H_2O in II*, and mixed occupancy in supercage sites. Perhaps the best data on the contents of the supercage were obtained by Hseu (*78*) for hydrated faujasite. Sets of peaks

Oxygen Species) in Dehydrated Faujasite-type Zeolites[a]

Heat Treatment, °C; Comments	*Reference*
350; *vacuo*	Dodge (*94*)
420; *vacuo*	Hseu (*78*)
400; *vacuo*	Pluth (*92*)
475; *vacuo*	Bennett and Smith (*95*)
400; *vacuo*	Olson (*96*)
500; *vacuo*	Pluth (*92*)
475; *vacuo* (data at RT)	Bennett and Smith (*93*)
475; *vacuo* (data at 420°C)	Bennett and Smith (*97*)
350; *vacuo*	Olson *et al.* (*85*)
420; *vacuo*	Hseu (*78*)
350; *vacuo*	Eulenberger *et al.* (*98*)
400; *vacuo*	Gallezot and Imelik (*99*)
400; *vacuo*	Gallezot and Imelik (*99*)
	Gallezot and Imelik (*99*)
300; *vacuo*	Mortier *et al.* (*81*)
300; *vacuo*	Mortier *et al.* (*81*)
350; *vacuo*	Eulenberger *et al.* (*98*)
350; *vacuo*	Eulenberger *et al.* (*98*)
350; *vacuo*	Smith *et al.* (*100*)
725; *vacuo* (data at 725°C)	Smith *et al.* (*100*)
380; *vacuo*	Gallezot and Imelik (*101*)
380; ambient	Gallezot and Imelik (*101*)
600 (also data for 200°C)	Gallezot and Imelik (*102*)
600 (also data for 140-300°C)	Gallezot and Imelik (*102*)
600 (also data for 200°C)	Gallezot and Imelik (*102*)
500; *vacuo*	Gallezot *et al.* (*103*)
300; *vacuo*	Gallezot *et al.* (*103*)
400; *vacuo*	Olson (*104*)
400; *vacuo*	Dempsey and Olson (*105*)
680; *vacuo*	Dempsey and Olson (*105*)
420; *vacuo*: 8 Na in III_1	Hseu (*78*)
540; N_2 (also density at V)	Hunter and Scherzer (*91*)
540; air	Hunter and Scherzer (*91*)

[b] Bracketed number is Al atoms per cell.
[c] Legend: sc, single crystal; p, powder.

were ascribed to water molecules at the corners of a hexakaidecahedron. Olson (*43*) reported various sites in the supercage of hydrated NaX. However, all these data are complex and open to alternative interpretations.

The data for dehydrated varieties (Table III) can be interpreted more reliably, but complications arise from the presence of residual oxygen species. Infrared data clearly show that highly polarizing cations split residual water molecules into hydroxyl anions (which bond to the cations) and protons which condense with framework oxygens. The positions of monovalent ions can be explained readily without invoking residual molecules. The excellent single-crystal data of Pluth (*92*) and Hseu (*78*) for dehydrated K-faujasite and Na-faujasite yield cations in the obvious positions I, I′, and II. Probably the site occupancies correspond to minimum electrostatic potential when account is taken both of intercation repulsions and cation–anion attraction (*see* later). Actually, the populations of sites I and I′ in Na-faujasite slightly exceed the rule given earlier, implying adjacent occupancy of sites I and I′. This might result from some residual heavy cation or water or from experimental error.

The data on multivalent cations are complex. In Ca-faujasite, dehydrated at 475°C, most of the cations occupy sites I and II, which positions yield minimum cation–cation repulsion. A few occupy I′ (2.6 *vs.* 14.2(I) and 11.4(II)), because of bonding to a residual oxygen species or because of bonding to a 6-ring with a high content of Al (*see* later). In La-faujasite, dehydrated at 475°C, most of the cations occupy site I in a distorted octahedral coordination (*see* Bennett and Smith (*93*)), but a few lie in I′ and II. The distribution was the same whether the data were collected at 475°C or at room temperature (*97*). Several other sets of data in Table III for rare earth ions in Y and faujasite varieties show strong occupancy of I′ (*e.g.*, Ce,Na-faujasite studied by Olson *et al.* (*85*)). The incomplete exchange causes some uncertainty in the interpretation, but at least some sodalite units must contain two rare earth ions in I′. This is highly unfavorable electrostatically unless there are bridging anions. Such anions could occur either in II′ or U, thereby providing strong stabilization. The direct x-ray evidence for such anions is rather poor, but the indirect evidence plus infrared and other data are compelling (*see* later).

Gallezot and Imelik (*102, 103, 106*) presented x-ray powder data which show that Ni and Co prefer I whereas Cu prefers I′ in dehydrated Y zeolites.

In X, the higher Al and cation content lead to further complications. For Na–X, Hseu (*78*) filled all the I′ and II sites (64 Na in all) and placed a further eight near the wall of the supercage and three at I. For divalent ions in X, all four sites I, I′, II′, and II are apparently occupied to some level (*104, 105*), and it is not certain what combinations of sites are actually occupied in the individual sodalite units. The high occupancy of I′ by Ce in X calcined either in N_2 or air at 540°C is best explained by formation of a complex with residual oxygen species, either at U or elsewhere in the sodalite unit (*91*).

Sorption with various molecules results in shifts of cations (Table IV). By 1974 no x-ray data had given any direct indication of the position of sorbed molecules. Simpson and Steinfink (*107, 108*) assumed that *m*-dichlorobenzene and 1-chlorobutane existed as homogeneous liquids in the supercage while other workers ignored the scattering from the molecules. However, all the structure determinations revealed insufficient electron density in the sodalite unit to account for the number of cations, and it is assumed that some cations moved into the supercage to interact with sorbed molecules; the one exception is the complex of NH_3 with $Cu_{16}Na_{24}$-Y, in which it appears that NH_3 is small enough to enter the sodalite unit and interact with Cu ions in I′ (*103, 109*).

Table IV. Assignments of Cations in Complexes with Faujasite-Type Zeolites

(a) $Ni_{28.5}Al_{57}Si_{133}O_{364}$·24-32 *m*-dichlorobenzene, single crystal, 12.5 Ni(I′), 24 H_2O(II′) 2.6 Ni(II), liquid scattering functions used for 32 molecules in supercage and seven, five, and two unlocated Ni, respectively, in supercage, sodalite unit and hexagonal prism (Simpson and Steinfink (*107*)).
(b) $Mn_{28.5}Al_{57}Si_{133}O_{364}$·24 1-chlorobutane, single crystal, 12.2 Mn(I′), 27 H_2O(II′), 5.1 Mn(II), liquid scattering functions used for 24 molecules and 11.5 Mn in supercage (Simpson and Steinfink (*108*)).
(c) $Cu_{16}Na_{24}$-Y· *x* pyridine, powder, 1.9 Cu(I), 2.3 Cu(I′), 26 Na(II); therefore 10 Cu gone into supercage: with *x* naphthalene, 2.3 Cu(I), 3.5 Cu(I′), 22 Na(II); therefore 8.5 Cu to supercage: with *x* NH_3, 2.1 Cu(I), 12.1 Cu(I′), 9 NH_3(II′), 18 Na(II); therefore NH_3 goes into sodalite unit, together with slight movement of Cu from I to I′ (Gallezot *et al.* (*103, 109*)).
(d) $Cu_{12}Na_5H_{27}$-Y·*x* naphthalene, powder, 2.1 Cu(I), 3.2 Cu(I′), 7 Na(II); therefore 6.3 Cu to supercage: with *x* *n*-butene, 1.9 Cu(I), 2.7 Cu(I′), 6 Na(II); therefore 7 Cu to supercage (Gallezot *et al.* (*103, 109*)).
(e) Te/Na-X, single crystal, 32 Na(I′), 9 Na(II′), 23 Na(II), 21 Na assumed to lie in III, 1.3 Te(U), 3·7 Te in large cavity suspended between two Na (Olson *et al.* (*110*)).
(f) Na-Y hydrocarbon, powder, cell dimension changes with increasing multiple bond character of hydrocarbon (Gallezot and Imelik (*111*)).
(g) Hydrated K, alkylammonium-Y, powder, complex data (*see* Mortier *et al.* (*112*)).
(h) Pd-exchanged Y zeolite, powder, before and after reduction with H_2, (*see* text and Gallezot and Imelik (*174*)).
(i) $Ni_{14}Na_{23}H_5$-Y, powder, first dehydrated at (a)140, (b)200, (c)300 and (d)600°C yielding 3.7, 4.2, 10.0 and 11.9 Ni, respectively, in site I, then subjected to sorption at room temperature at 100 torr with the following molecules for the specified time in hours: (a) with NO, 12 hours, 1.7 Ni in site I; (b) with pyridine, 24, 1.9; (c) with pyridine, 72, 9.0; (c) with NO, 12, 6.5; (c) with NO, 96, 3.4; (c) with NH_3,24,2.7; (d) with pyridine, 72, 11.1 (Gallezot, Ben Taarit and Imelik (*87*)).

The data for $Ni_{14}Na_{23}H_5$-Y reported in Table IV(i) are complicated by sluggish diffusion but indicate cation migration upon sorption of molecules.

The Te complex with Na-X, produced by heating Te with Na-X at 540°C followed by heating in flowing H_2 at 475°C, is a remarkably effective catalyst for aromatization reactions (*79*). The crystal structure determination (*110*) indicates that 1.3(2) Te occupy site U where they are suspended between

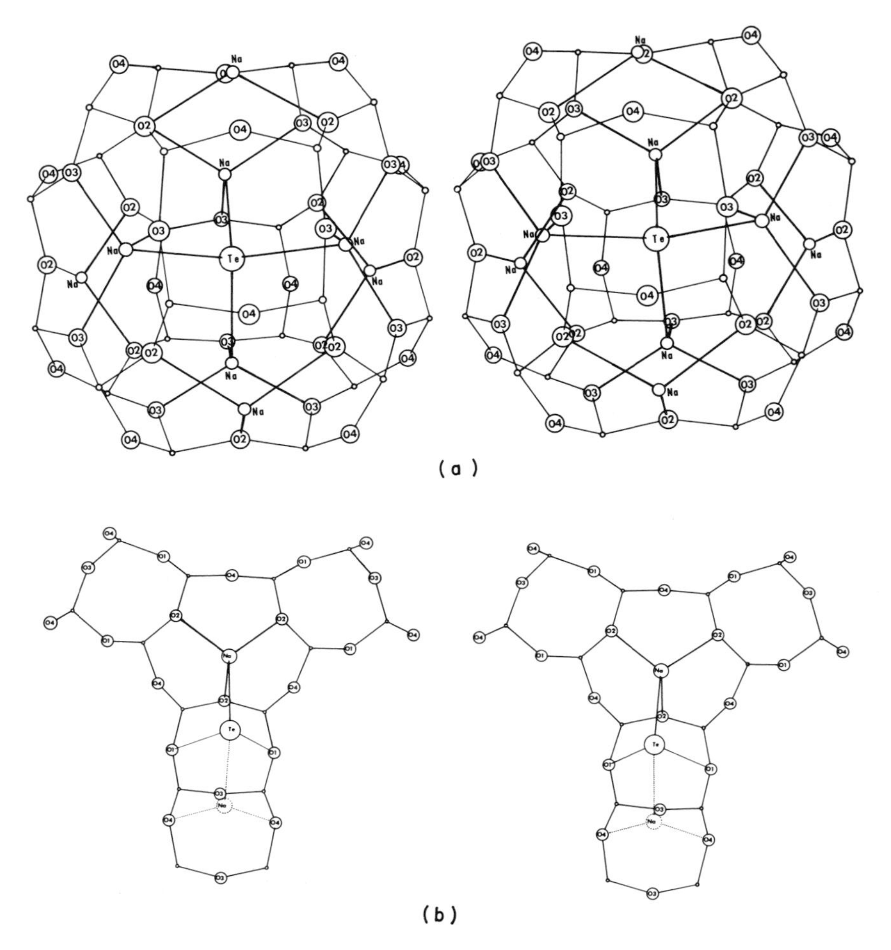

Journal of Catalysis

Figure 28. Stereodiagrams of coordination of telluride ions in Te/NaX. In (a), a Te anion in U is suspended between 4 Na in I′: each Na cation is tetrahedrally bonded to three O(3) at 2.41 and one Te at 2.59 A. In (b) a Te anion near the wall of the supercage lies at 3.2 A to Na in II and a similar distance to a hypothetical Na in III: unfortunately, there is no x-ray evidence for Na in site III, perhaps because of its low electron density (110).

four Na in I (Figure 28a) while 3.7(6) Te lie between one Na at II and a hypothetical Na in III (Figure 28b). These coordinations are consistent with Te occurring as an anion. Indeed, diffuse infrared reflectance spectra showed a strong band at 3650 cm^{-1}, which indicates occurrence of a framework hydroxyl to balance the reduction of the Te.

Space precludes detailed review of bond lengths and framework geometry. Let it suffice that the framework geometry and bond lengths in hydrated faujasite zeolites are consistent with a relaxed state in which there is little local deviation from electrostatic charge balance. Dehydration causes distortions which correlate with the local deviations from electrostatic balance and with steric considerations. In particular, distortion of 6-rings to permit bonding to cations results in twisting of the sodalite unit. A detailed mathematical analysis is given by Hseu (*78*). (*See* later for data on H-faujasite, ultrastable faujasite, and complexes in the sodalite unit.)

Erionite, Gmelinite, L, Mazzite, and Offretite. The hydrated offretite structure ($K_{1.1}Ca_{1.1}Mg_{0.7}Si_{12.8}Al_{5.2}O_{36} \cdot 15\ H_2O$) contains a K ion in each cancrinite cage (*113*) (*see* Figure 29). Most gmelinite cages contain Mg surrounded by five H_2O. The wide channels are occupied by a mixture of

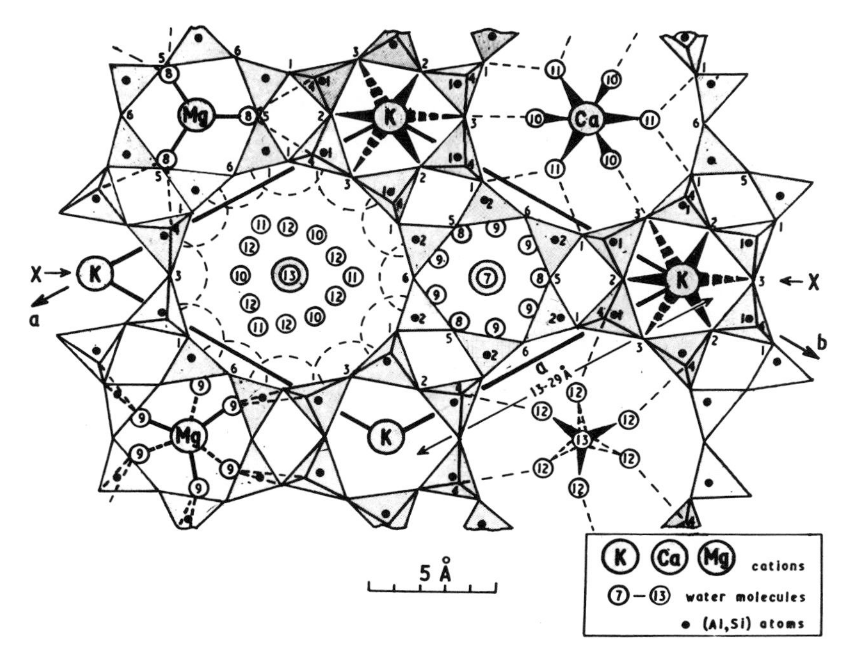

Figure 29. Projection of offretite structure down the c*-axis. Heavy lines show the* a *and* b *edges of the unit cell. One K cation occupies the center of each cancrinite cage, for which dots labeled 1 (upper center) show the T atoms at the centers of the 6-rings. Each K is bonded at 2.96 A to six O(2) at the corners of a trigonal prism (solid spokes), and less strongly at 3.3 A to six O(3) at the corners of another trigonal prism (broken spokes). Most gmelinite cages contain one Mg ion with one $H_2O(7)$ vertically above, one $H_2O(7)$ vertically below at 2.0 A, and three H_2O in the same plane at 2.1 A, either three $H_2O(8)$ as at the top left or three $H_2O(9)$ in one of two orientations (bottom left). In a few gmelinite cages, only water molecules occur (one of the combinations of water molecules shown to the right of center). The large channels contain compatible combinations of water molecules of type 10, 11, 12, and 13, some of which form a hydration complex around Ca(1) as at top right, while others form water clusters as at bottom right and left center. The few Ca(2) in hexagonal prisms are not shown* (113).

hydrated Ca ions and water complexes, but five out of the 15 H_2O per cell were unlocated. A few Ca ions occupy double 6-rings. The hydrated erionite structure ($K_{2.0}Na_{1.9}Ca_{1.3}Mg_{0.6}Al_{9.3}Si_{26.2}Fe_{0.5}O_{72} \cdot > 10\ H_2O$) also contains a K ion in each cancrinite cage (*114*). The positions of the other ions were uncertain, but bonding to water molecules and framework oxygens was indicated. The hydrated L structure ($K_6Na_3Al_9Si_{27}O_{72} \cdot 21\ H_2O$) was determined only from powder data, and the cation and water sites are uncertain. Again each cancrinite cage is apparently occupied by a K ion (*115*). Hydrated (K,Ba)-G,L ($K_{2.7}Ba_{7.6}Al_{18}Si_{18}O_{72} \cdot 23\ H_2O$) has a similar but slightly expanded framework while Ba(1) lies in the center of a cancrinite cage, Ba(2) lies between two adjacent cancrinite cages, and K may lie near the boat-shaped 8-ring; the latter ion would be the only one with easy access for ion exchange (*116*). Probably the Si and Al atoms alternate, but no evidence was revealed from the x-ray powder data. Hydrated P-L ($K_{23}Al_{33}Si_{26}P_{13}O_{144} \cdot 42\ H_2O$) has a doubled *c*-axis (single-crystal electron-diffraction data), perhaps because of ordering of T atoms (*117*). Hydrated mazzite ($K_{2.5}Mg_{2.1}Ca_{1.4}Na_{0.3}Al_{9.8}Si_{26.5}O_{72} \cdot 28\ H_2O$) contains a hydrated Mg ion in each gmelinite cage while K,Ca ions lie in the narrow irregular channels between the columns of gmelinite cages. Some additional Ca occurs in hydration complexes at the centers of the wide channels (*118*). In mazzite dehydrated at 600°C, each Mg ion has moved to either the upper or lower 6-ring of each gmelinite cage (Figure 30) where it is tetrahedrally bonded to three framework oxygens and one residual water or hydroxyl molecule (*119*). A residual water molecule lies in all three 8-rings of the gmelinite cage. Calcium (probably with K and H_2O) is coordinated to framework oxygens and

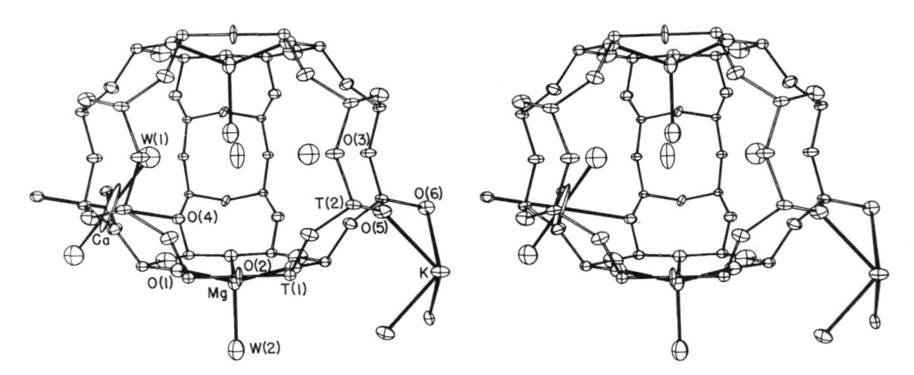

Acta Crystallographica

*Figure 30. Coordination of cations with the gmelinite cage of partly dehydrated mazzite. One Mg per cage is bonded to three O(1) from a distorted 6-ring at ~ 2.2 A and one $H_2O(2)$ at 2.4 A: each 6-ring is blocked by one Mg. About 1.5 K per unit cell occur at a 12-fold site inside the large channel, where they are bonded to four framework oxygens, O(5) and O(6), two each from two gmelinite cages at 2.6 and 2.8 A. About 1.4 Ca, ~ 1.0 K and ~ 2 H_2O lie in the channels between the gmelinite cages where they are surrounded by four O(4) at ~ 2.8 A and two $H_2O(1)$ at ~ 2.6 A. Each 8-ring is occupied by $H_2O(1)$ (*see *original paper for complications)* (119).

residual water molecules in the narrow irregular channels, while 1.5 K atoms have one-sided coordination to four framework oxygens on the wall of the large channel. The presence of four exchangeable cations caused uncertainty in the above site assignment, but crystal-chemical considerations favor the assignment shown in Figure 30. No data are available for dehydrated erionite, gmelinite, L, or offretite.

All the above data fall into a single pattern in which a cancrinite cage contains a large cation (K or Ba) while the larger gmelinite cage contains hydrated Mg. This can be explained readily by these ions fitting snugly into the available space. Indeed these ions may act as templates around which the zeolites crystallized. Thus K^+, Ba^{2+}, or tetramethylammonium ions (the latter being similar in size to the hydrated Mg^{2+} ion) are essential ingredients of gels from which synthetic varieties (and omega, the probable analog of mazzite) grow (*see* Breck (*1*) for references). Kerr *et al.* (*120*) provided detailed data and interpretation of the crystallization and structural relations between L, offretite, and erionite, together with electron diffraction and morphological data. (*See* also Wu *et al.* (*121*) for physicochemical aspects of the crystallization of offretite from systems containing tetramethylammonium.) The K^+ ion is rather firmly locked inside the cancrinite cage and cannot be removed from erionite and offretite by ion-exchange with NH_4^+ at 80°C.

Many more structure determinations of ion-exchanged and dehydrated varieties of these zeolites are needed to clarify the ion-exchange and sorption properties fully, but preliminary interpretation is satisfactory (*e.g.*, Gard and Tait (*113*)). In detail, there are uncertainties caused by stacking faults and by lack of knowledge of cation positions. The structure of hydrated gmelinite (*122*) was not solved completely, but cations lie near the 6-rings.

Mordenite. Structural interpretation of the properties of mordenite-type zeolites is incomplete. Ideally, mordenite should have a one-dimensional channel system bounded by 12-rings, with crosslinks bounded by 8-rings. Some mordenites (the large-pore type) have sorption properties consistent only with 8-rings. Of course, the obvious interpretation is that the wide channels are blocked in the small port variety, but there are no convincing structural data on the nature of the blockages. The only published crystal structure determinations are for a natural mordenite ($Na_{4.0}K_{0.2}Ca_{1.9}Al_8Si_{40}O_{96} \cdot 24\ H_2O$) ion-exchanged to give eight Na and 24 H_2O. The structure determination by Meier (*34*) was refined by Gramlich (*47*). Unfortunately Na and H_2O have similar scattering powers, and atomic assignments are equivocal. Two well-defined sites probably correspond to Na bonded to H_2O while five ill-defined sites remain unassigned. Data for hydrated Tl-mordenite are unpublished (*123*). Upon dehydration, the cations must bond to framework oxygens, but there are no data on their positions. Electron and x-ray diffraction studies (*see* Sherman and Bennett (*35*)) revealed the occurrence of intergrowths and stacking faults in mordenite, but conversion of small-pore to large-pore varieties by simple chemical treatment, such as acid leaching, suggests that channel blocking by occluded species is also contributing (*e.g.*, Sand (*124*)). Perhaps cations on channel walls hinder or restrict diffusion;

this could explain why H-mordenite has wider channels than ones with Na, Ca, or Ba (*e.g.*, Barrer and Oei (*125*)). The situation is highly confusing, and definitive structural interpretations await further study.

ZK-5. The positions of the cations and water molecules in hydrated ZK-5 are not known, though 16 Na probably lie near 6-rings (*126*). X-ray powder data of the ZK-5 analogs, species P and O, loaded with $CaCl_2$ or $BaBr_2$, were interpreted by Barrer and Robinson (*127*). The non-framework species Ba^{2+}, Br^-, Cl^-, and H_2O were placed in both types of cages, but the details are complex and uncertain.

Electrostatic Bonding Models for Zeolites

Electrostatic bonding models provide a guide for interpreting the flood of data on zeolite structures and yield estimates of the electrostatic field gradient applied to sorbed molecules. Probably the models are fairly realistic for many exchangeable cations but must be inadequate for cations with strong directional bonding (*e.g.*, the first group of transition metals). The simplest models interpret the experimental data on the basis of local electrostatic effects while more complex models sum together the electrostatic contributions from the entire crystal. For clarity, the following discussion is based on faujasite-type structures.

Consider first how closely the framework oxygens with a formal charge of -2 are balanced electrostatically by the contributions from nearest neighbors. For simplicity, assume that each T atom distributes its charge equally to the four oxygen neighbors O(1), O(2), O(3), and O(4).

Al and Si may be distributed over T sites in several ways. In an X-type with $Al/Si = 1$, regular alternation of Al and Si would maximize the local charge balance, but experimental data do not rule out occurrence of Al atoms sharing the same oxygen. Instead of all oxygens with incoming charges of $3/4 + 4/4 = 1.75$, there would be a mixture of oxygens with 1.5, 1.75, and 2.0 incoming charges. In faujasite-types with $Al/Si < 1$ there is no simple theoretical answer. At one extreme the Al and Si can be assumed to be fully disordered whereas at another extreme with local order and no oxygen shared by 2 Al, incoming charges are either 1.75 or 2.0. (*See* Smith (*46*) for review of various considerations concerning the state of Si,Al order–disorder.) More than one type of local order can occur for a chosen Si/Al ratio (except for unit ratio). In Y zeolites, 6-rings might have either two or three Al atoms separated by Si atoms.

Next consider the problem of distribution of charges from the large cations. In site I, a cation has six O(3) as near neighbors while the six O(2) are more distant. The simplest approximation distributes the cation charge only to O(3). Similarly for sites I′, II′, and II, the charge is distributed to three O(3), three O(2), and three O(2), respectively. The situation for site III is unclear, but in dehydrated NaX, two O(1) and two O(4) and perhaps a fifth oxygen are near neighbors (*78*).

In dehydrated Ca-faujasite (*95*), the Ca distribution ($I = 14.2$, $I' = 2.6$, $II = 11.4$ atoms/cell) and Al/Si ratio 0.292/0.708 produces the following statistical average charges at the oxygens:

	Incoming Charges	*T–O Distances, A*
O(1)	$2(0.708 \times 4/4 + 0.292 \times 3/4) = 1.854$	1.633(5)
O(2)	$1.854 + 11.4/32 \times 2/3 = 2.091$	1.651(5)
O(3)	$1.854 + 14.2/16 \times 2/6 + 2.6/32 \times 2/3 = 2.204$	1.671(5)
O(4)	1.854	1.620(5)

The charge imbalance is adjusted by lengthening of T–O(3) and T–O(2) and shortening of T–O(1) and T–O(4). For comparison note that at room temperature, the thermal displacements of T and O atoms in silicate frameworks are about 0.1 A.

Local effects also influence the bonding. Instead of being bonded to the statistical fiction of 0.292 Al and 0.708 Si, each oxygen is bonded to either two Si, one Al + one Si, or two Al with incoming charges of 2.0, 1.75, or 1.50 from the adjacent T atoms. Therefore the above variation of charges from 1.854 to 2.204 can undergo further local perturbations of +0.15 to −0.35, though any short-range or long-range order will reduce this range. In addition each Ca exists as a single ion rather than a statistical fiction. It is very unlikely that adjacent I and I′ sites are occupied. Instead of receiving the statistical average of 0.35 charge from Ca in I and I′, oxygens O(3) will receive either 0, 1/3, or 2/3 charge from Ca. From the viewpoint of local charge balance, Ca should choose sites near to positive fluctuations of Al; in addition, Ca atoms probably are displaced from the symmetry element towards undersaturated oxygen atoms [this may explain why La atoms in dehydrated La-faujasite are displaced from the center of symmetry (*93*)]. No matter how one shuffles the ions, it is impossible in the Si-rich varieties of faujasite-type zeolites to avoid local charge unbalances ranging from 0 to about 0.5. This is the simplest explanation why zeolites act as solid-state electrolytes.

The next simplest electrostatic model adds in the repulsion between the exchangeable cations. For dehydrated Ca-faujasite with 28 Ca ions, minimum repulsions occur when only sites I and II are occupied. Actually sites I and II accommodate most of the Ca^{2+}, but a few go into I′. The simplest explanation of occupancy of I′ is that the adjacent 6-ring has an unusual amount of Al (probably three atoms, but perhaps even higher). Another possibility is formation of a complex with residual water or hydroxyl molecules (*see* later).

Using these simple ideas as a guide, it is now possible to appreciate the computer calculations of Dempsey (*128, 129*). Faujasite-type zeolites with different Al/Si ratios were constructed from identical pairs of sodalite units with ordered Al and Si atoms resulting in zero net charge and dipole moment. Framework coordinates were taken from existing structure determinations for Na- and Ca-faujasites, and imaginary cations were placed in I, I′, II, and III at distances from the framework oxygens determined by the sum of the ionic radii. The Madelung potential was calculated as a function of cation radius for monovalent, divalent, and trivalent cations in frameworks with Si/Al ratios of 1 and 2 which could be modeled by ordered patterns. Site I was filled to 100% occupancy, then site II was loaded with the remaining cations, and finally site III took the excess of monovalent cations. A second series of calculations used partial occupany of I, I′, and II for Si/Al ratios of 1.18 and 2.43. I and I′ were not occupied simultaneously, and the ordered Si,Al pattern compensated for occupancy of three I, one I′, and either three or one II, respectively.

For divalent ions, the Madelung potential is lower for I than II, especially as the cation radius increases (Figure 31a,b). This results from the increasing displacement of type II cations into the supercage away from the oxygens. Note that the two curves in (b) result merely from the lower symmetry caused by Si,Al ordering on a 2:1 ratio. Figure 31c,d shows that mixed occupany of I, I′, and II is unstable for divalent cations smaller than *ca.* 1.5 A and that readjustment to occupancy of only I and II should occur for the ordered arrangements selected by Dempsey. These arrangements do not include the possibility of unusually high Al contents in some of the 6-rings.

For small monovalent ions, site III has the highest Madelung potential, site II has the lowest, and sites I and I′ have intermediate values. The experimental data showing full occupancy of II, mixed occupancy of I and I′, and overflow occupancy of III are explainable by these calculations, especially when the effects of local excess of Al are considered. For monovalent ions larger than *ca.* 1.2 A, site I is favored over II.

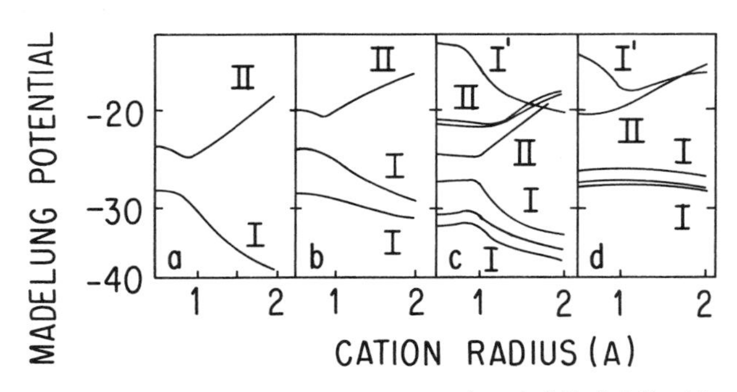

Journal of Physical Chemistry

Figure 31. Variation of Madelung potential (normalized units) with cation radius for divalent cations.

(a) X with Al/Si = 1: regular alternation of Al and Si: full occupancy of I and II.

(b) Y with Al/Si = 2: ordered pattern of Al and Si results in degradation of symmetry so that there are two distinct I and II sites: full occupancy of both sub-sites I and only one sub-site II.

(c) X with Al/Si = 1.18: complex ordered pattern of Al and Si with triple degradation of symmetry: I, 75% occupancy; II, 75%; I′, 25%.

(d) Y with Al/Si = 2.43: complex ordered pattern of Al and Si with triple degradation of symmetry: I, 75%; II, 25%; I′, 25% (129).

All trivalent ions are small, and the calculations show that site I should be fully occupied (16 atoms), followed by overflow occupancy into II (maximum occupancy of 16 out of the 32 possible sites). Site I′ should not be occupied in Dempsey's models, and observed occupancies must be ascribed to complexes with residual molecules and/or to local fluctuations of Al content.

Lechert (*130*) calculated the effect of induced dipoles and quadrupoles on the electrostatic field gradient in order to interpret the NMR spectra of ^{23}Na in dehydrated X and Y. The multipoles arise almost entirely from the oxygens and seriously affect the electrostatic field gradient. Unfortunately data are available only for site I. Similar studies on feldspars have shown quite good agreement between calculated and observed electrostatic field gradient tensors (reviewed by Smith (*37*)).

Th above models neglect covalent bonding, but reduction of the formal charge by about one-half should allow the ionic model to give semiquantitative answers. Flexing of the aluminosilicate framework around the T–O–T angles would change the overlap functions in molecular orbital calculations, but no quantitative calculations have been made. Some transition metals prefer trigonal and tetrahedral coordination to octahedral coordination, thereby favoring single 6-rings over the octahedral site I. Residual molecules cause major effects especially for highly polarizing cations. Finally, the infrared data on stretching and bending motions of zeolite frameworks show that there are subtle differences which apparently depend on the type and interlinkage of building blocks (*131*). In addition, infrared bands correlate with the type of cation in X and Y zeolites (*see* Chapter 3). It is quite obvious that electrostatic models provide a useful first approximation, but many more calculations are needed on multipole effects. Resonance techniques will be hampered until single crystals rather than powders have been evaluated.

Hydroxyl Anions in the Framework and in Cation Complexes

Most of the information on hydroxyls in zeolites was obtained from infrared absorption spectra, and the literature is extremely confused because of numerous speculations on the location of the hydroxyls. The present account deliberately selects only those proposals which are plausible from a crystal-chemical viewpont. [For a detailed account *see* the reviews by Ward (*132*), Rabo and Poutsma (*133*), and Breck (*1*).] Chapter 3 reviews the present status of infrared data.

Hydroxyl stretching bands occur at 3000–3800 cm^{-1} while water-bending bands occur at 1600–1700 cm^{-1}. The latter vary in a complex way from one zeolite to another and from one cation variety to another, but the details have not been interpreted in structural terms. Of course, the water bands fade away as dehydration proceeds.

All zeolite spectra show a weak, sharp band at 3740–3750 cm^{-1}. This is reasonably ascribed to a hydroxyl at the external surface. Mathematical calculations indicate about 10^{20} OH per cm^3 for crystals 1 μm across (*134, 135*), many orders of magnitude too low to be detected by x-ray methods. Alternatively the band may arise from occluded silica residue (*136*).

Ammonium-ion exchange of faujasite-type zeolites produces one-to-one replacement of Na^+, and heating in an open system at 300°–450°C results in escape of NH_3, leaving H^+ in the zeolite. The infrared pattern shows two strong peaks near 3650 and 3550 cm^{-1}, the former being perturbed by many sorbed molecules whereas the latter is perturbed only by some sorbed molecules at elevated temperature where protons should be mobile. Furthermore, the 3650 cm^{-1} band occurs even when there are residual cations, but the 3550 cm^{-1} band is suppressed by the presence of small cations which can enter the sodalite cage. Consequently the 3650 and 3550 cm^{-1} bands are reasonably ascribed to framework hydroxyls projecting into the supercage and the sodalite–prism complex, respectively. Apparently the hydroxyl stretching frequency (and hence the bond energy) increases as the proton projects into a larger volume. Detailed speculation on the locations of the framework hydroxyls was clarified by the single-crystal x-ray study of H-faujasite (*137*). This was prepared by exchange with 1*M* ammonium acetate solution at 100°C for 38 days, followed by calcination up to 376°C for 17 hr at 1×10^{-6} torr. The T–O distances definitely show that hydroxyls occupy the O(1) and O(3) sites (0(1) 1.653(2), O(2) 1.634(1), O(3) 1.663(2), O(4) 1.623(2)A), and the lengthening of the T–O(1) and T–O(3) distances is semiquantitatively consistent with expectations for *ca.* 59 OH per cell. Olson and Dempsey's model (Figure 32) places H(1) projecting into the supercage from O(1) of the wall of a hexagonal prism and H(2) projecting from O(3) in the plane of the shared 6-ring. Approximately one-sixth of the framework oxygens must be hydroxyls, which corresponds to one-third of O(1) and O(3) on the Olson-Dempsey model. Naively, the longer T–O(3) distance suggests that the H(2) population is greater than the H(1) population, but the evidence from feldspars (*37*) indicates that T–O bond lengths can vary by *ca.* 0.01 A merely from linkage-related effects. The infrared peaks have comparable intensities. but the 3550 cm^{-1} peak has a greater area (*e.g.*, Ward (*132*), Figure 2), also suggesting that the H(2) population is greater. Olson and Dempsey proposed from electrostatic considerations that (a) each of the 32 6-rings contains one H(2) lying in the plane of the ring, while 27 H(1) project into the cavity (b) each H condenses with an oxygen bonded to one Al and one Si (*i.e.*, half of the available oxygens of this type) (c) if Al and Si are ordered according to the theoretical meta-para combination for Si,Al = 2, there would be only one type of electrostatic field for H(1) and H(2), corresponding to the observation of only two infrared bands. Gallezot and Imelik (*99*) found the following T–O distances for Na_2H_{54}-Y powder heated finally to 400°C at 10^{-5} torr for 12 hr: O(1) 1.622(4), O(2) 1.645(4), O(3) 1.643(5), O(4) 1.618(5). Probably these errors are underestimated by a factor of two or three because of the overlap of spectra in cubic powder patterns, but at face value, these distances indicate occupancy of O(2) and O(3). Both of these atoms belong to the same type of 6-ring, and it is not possible to devise such an elegant model as that of Olson and Dempsey.

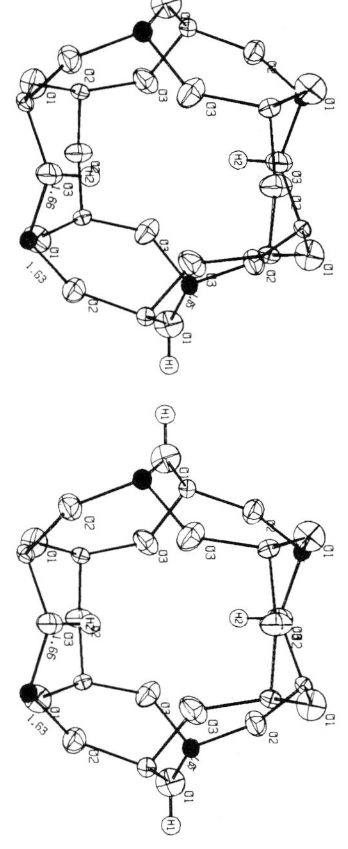

Journal of Catalysis

Figure 32. Stereo plot of proposed model for hexagonal prism of H variety of faujasite with Si/Al = 2. Small black circles are Al. Ellipsoids show atomic displacements from x-ray analysis (137).

Jacobs and Uytterhoeven (*138*), from an infrared study of Na,H varieties of X and Y zeolites assigned the narrow 3650 cm^{-1} band to a proton attached to O(1), as in the Olson-Dempsey model, but assigned the broad 3550 cm^{-1} band to protons attached to O(2), O(3), and O(4). Obviously there is scope for further work, including single-crystal infrared and neutron-diffraction studies. Probably the number of residual cations and the degree of Al,Si order–disorder affect the OH positions.

Highly polarizing cations form hydration complexes in hydrated zeolites, but upon dehydration the last water molecules are held tenaciously. For a given temperature and framework, the amount of retained water correlates with the ionization potential of the cation. Infrared absorption bands indicate formation of cation–hydroxyl complexes plus framework hydroxyls. Alkali metals and Ba are unable to split water molecules, but weak hydroxyl bands can occur from partial exchange with H^+ under acid conditions. Small divalent cations such as Ca^{2+} and Mg^{2+} produce hydroxyl groups in proportion to their polarizing power (*139, 140*). The spectra are very complex, showing several poorly resolved bands from 3200–3700 cm^{-1}. Bands near 3640 and 3540 cm^{-1} were assigned to framework hydroxyls, perhaps comparable with those in H-faujasite. A band at 3600–3560 cm^{-1} was assigned to hydroxyl bonded to a divalent cation while a band near 3690 cm^{-1} was ascribed to water. Rehydration below 200°C augmented the 3690 band, but upon heating above 200°C the 3690 band was replaced by the 3640 and 3600–3560

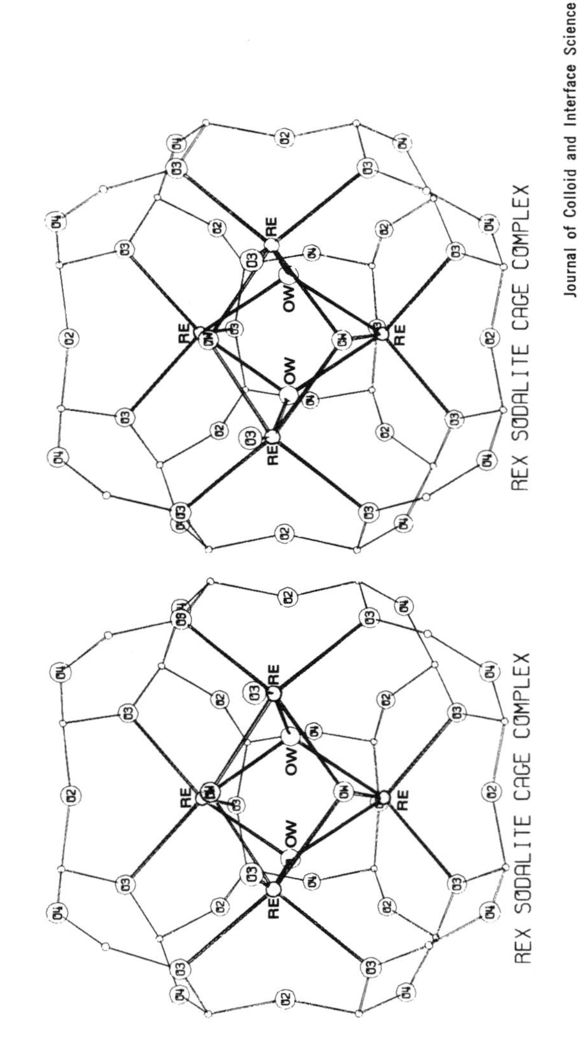

Journal of Colloid and Interface Science

Figure 33. Stereo plot of complex in sodalite cage of REX, showing four RE in I′ and four H_2O in II′. For clarity, two T and four O were omitted from the top of the cage (85).

bands, presumably because the higher temperature allowed cations to polarize mobile water molecules. The frequency of the 3600–3560 cm^{-1} band increases with the polarizing power of the cation. Details are given in Chapter 3.

The nature of hydroxyls in rare earth zeolites goes to the heart of models for catalysis (*see* later). Partly dehydrated RE-Y yields infrared bands at 3740 (weak), 3640 (strong), and 3520 cm^{-1} (strong). Interpretation of most data is complicated by use of mixed rare earths together with incomplete exchange of the Na.

The 3520 cm^{-1} band is unique to RE-zeolites (*141*), and the frequency increases linearly with ionic radius (*142*). Ammonia and benzene do not show hydrogen bonding with the 3520 cm^{-1} hydroxyl. X-ray diffraction data (Table III) show occupancy of I′ by rare earth ions when dehydration is incomplete. All these data indicate that the 3520 cm^{-1} band results from hydroxyl bonded to RE ions at I′ in the sodalite unit. The 3640 band has the same frequency as a band in H-Y, and shows strong hydrogen bonding with water, ammonia, and benzene. It can be ascribed to hydroxyl projecting into the supercage (*141*) not necessarily at O(1).

Comparison of spectra taken at room and elevated temperatures shows that the 3500 cm^{-1} band shifts to lower frequency and that the 3640 cm^{-1} band is weaker at high temperature with respect to the 3520 cm^{-1} band. This may indicate weakening of bonding together with mobility of the 3640 cm^{-1} proton at high temperature.

Heating above 680°C results in loss of all hydroxyl bands except the 3740 cm^{-1} one. The x-ray data indicate that RE ions move from I′ to I, which is the preferred site when there is no interaction with residual molecules (*e.g.*, Bennett and Smith (*93, 97*)). Unfortunately the data in Table III are not fully convincing, probably because of the difficulty of controlling the atmosphere in the x-ray equipment, but the trend is quite clear. The simplest model is that a RE–OH complex in the sodalite unit loses hydroxyl which strips away the proton from a framework hydroxyl to yield escaping water, while the bare RE ion goes to site I. It is more plausible (*1*) than the model in which a framework hydroxyl combines with a proton from the RE hydroxyl leaving an oxide ion bonded to RE ions and an oxygen vacancy in the framework (*141*). The latter proposal conflicts with the x-ray data since the proposed RE–O^{2-}–RE complex must lie in the sodalite unit thereby preventing RE ions moving to site I. Evidence of Lewis acidity can be explained by interaction with the few RE ions in II and does not require defects in the framework. Hydroxyl groups could not be regenerated by adding water to RE–Y at room temperature but could be regenerated by subsequent heating above 200°C (*141*). This can be explained by penetration of the water molecules into the sodalite unit where they react with RE ions from site I.

The nature of the RE-hydroxyl complex is not fully known, but a plausible model can be based on the observation by Bolton (*143*) that one water molecule is released per RE ion upon heating from 300° to 700°C a series of Na,RE-Y zeolites (actually the data show 1.03–1.13 H_2O per RE). Olson *et al.* (*85*) found electron density consistent with 30 La in I′ and 32 H_2O in

II′ in an x-ray study of $La_{29.1}Na_{0.4}$-X powder, calcined at unspecified temperature and rehydrated at room temp. Figure 33 shows that this atomic assignment produces a plausible complex with each La at 2.5 A to three O(3) and three H_2O. Returning to Y zeolite, pairs of RE ions in adjacent I′ sites could bond to a pair of OH molecules in the two adjacent II′ sites. Each RE ion would be bonded to three O(3) and two OH while each OH would be suspended between two RE and three O(2). This $RE_2(OH)_2$ complex would account for 16 RE per cell; of course, complexes with three or four RE might also occur. For less severe dehydration, complexes such as $RE_2(OH)_2(H_2O)_2$ could occur. There are insufficient x-ray data to test this model in detail, but no data in Table III conflict when account is taken of experimental error and the uncertainty caused by incomplete exchange of Na.

Chemical Modification of Frameworks

Many studies provide evidence that Al can be removed from zeolite frameworks without complete loss of crystallinity, but the mechanisms of the structural changes are purely speculative. Chapter 4 contains a review on zeolite stability and ultrastable zeolites, which involves the Al content of the zeolite framework.

Aluminum can be removed from clinoptilolite and mordenite merely by treatment with cold HCl. Barrer and Coughlan (*144*) reported complete removal of Al from clinoptilolite using 2*N* HCl, and many workers, including Kranich *et al.* (*145*), gave data on mordenite. The most obvious structural mechanism was replacement of an AlO_4 tetrahedron by an $(OH)_4$ cluster, but Thakar and Weller (*146*) emphasized that Al-poor mordenite gave the same x-ray powder pattern after heating to 500°C as after heating to 110°C and that an $(OH)_4$ cluster could not survive heating to 500°C. The only reasonable alternative seems to be removal of Al from the framework combined with healing of the defects to give a continuous, Si-rich framework. There are no detailed x-ray structure analyses.

The data for faujasite-type zeolites are very complex. McDaniel and Maher (*147*) reported on an ultrastable variety of Y formed by heat treatment after NH_4^+ exchange. The literature is confusing, but reviews are given by Breck (Ref. *1*, pp. 507–519) and Kerr (*148*). Mathematically, removal of NH_3 and H_2O from NH_4-exchanged Y results in loss of oxygen from the framework:

$$Na_x\text{-}Y + xNH_4^+ \rightarrow (NH_4)_x\text{-}Y + xNa^+ \rightarrow$$

$$xNH_3\uparrow + H_x\text{-}Y \rightarrow \frac{x}{2}\,H_2O\uparrow + \left(Y \text{ less } \frac{x}{2}\,O\right)$$

Early speculation was based on simple removal of oxygen, leaving three-coordinated Al and Si (*149, 150*). Oxygens bonded either to two Al or one Al + one Si should be less tightly held than ones bonded to two Si, and the simple model predicts that both three-coordinated Al and Si should be formed because oxygens bonded to two Al should be rare. Unfortunately there is a severe stability problem since two highly-charged cations would now be unshielded. The experimental evidence (*see* Chapter 4) can be interpreted in terms of readjustment of the silicate framework. Calcination of NH_4-Y in an environment in which H_2O remains in contact with the zeolite results in a thermally stable product whereas calcination with rapid extraction of H_2O results in an unstable product. The simplest explanation is that protons catalyze healing of the framework to a defect-free, Si-rich variety, and that the greater thermal stability results merely from the higher proportion of strong Si–O bonds. However, the fate of Al must also be considered since it could form a stabilizing complex inside the sodalite cage. Mathematically, each vacant oxygen defect can be healed by removal of one Al_2O_3. If possible effects from such a complex are ignored, the Si/Al ratio of the new framework can be estimated from the mean T–O distance and cell dimension, and indeed measured values (*151, 152*) are consistent with higher Si/Al ratios. Infrared bands from framework atoms move to higher frequencies, again consistent with a higher Si content (*153*). Dempsey (*154*) suggested that removal of Al should be relatively easy when more than one Al occur in a ring of tetrahedra but difficult when only one remains. The product should then become acid resistant. Kerr (*155*) removed Al from Na-Y by extraction with the chelating agent H_4EDTA, and many other agents have been discovered.

In spite of the plausibility of the above suggestions, the data from x-ray and other structural studies pose problems. Maher *et al.* (*151*), made analyses on four powders produced by various heat treatments of NH_4-exchanged Y. All were studied at room temperature after rehydration. The details are too complex to give here, but the most puzzling feature is the apparent absence of 15 ± 7 O(3) and 13 ± 8 O(4) atoms in the ultrastable variety. Bennett and Smith (*46*) made powder x-ray studies of a material kindly supplied by C. V. McDaniel. Measurements were made directly at 25°, 400°, 700°, and 900°C to study the effect of H_2O removal. Whereas the data at 25°C indicated absence of 27 ± 8 O(1) and 17 ± 9 O(4), the data at 700° and 900°C indicated a full complement of atoms in the framework. Perhaps recrystallization occurred while the x-ray data were obtained, but another possibility is that the x-ray data are too inaccurate to measure the population of framework sites correctly, especially when H_2O is present. Line overlap in powder patterns probably results in errors two or three times greater than those obtained by computer. A drastic movement of nearby atoms as the exposed T atoms repelled each other poses problems in attempting to locate vacant oxygens. I believe that the x-ray powder data are not inconsistent with a model in which unstable intermediates with defect structures ultimately transform into a topologically continuous tetrahedral framework richer in Si.

The fate of the Al is unknown, but the wavelength of the AlK_β fluorescence line shifted in the direction expected for six-coordination and away from the direction expected for three-coordination. The SiK_β line remained at the wavelength for four-coordination whatever the type of treatment (*156*). Thus the simplest explanation is that the stoichiometry of the reaction increment is:

$$H_xAl_xSi_{192-x}O_{384} \rightarrow (x/2)\ H_2O + (x/2)\ Al_2O_3 + (192\text{-}x)SiO_2$$

Here the Al adopts six-fold coordination as in corundum. However, there are structural problems unless the Al goes into microcrystallites undected by x-ray diffraction. A common suggestion is that the Al forms a complex inside the sodalite unit, but there are severe difficulties in finding a suitable model. Maher *et al.* and Bennett and Smith found evidence for electron density at I′ in all eight structure analyses. Maher *et al.* ascribed the electron density in I′ to Al bonded to three O(3) at 2.7 A and three hydroxyls in II′ at 2.0 A. The 2.7 A distance is much too long for Al–O, unless the O(3) atoms are pulled inwards when Al occupies I′. The density at I′ might result from H_2O in the hydrated varieties. Unfortunately the situation is quite unclear, and it is uncertain whether an aluminum–oxide complex occurs as a separate entity inside the sodalite unit or whether the Al actually bonds to framework oxygens. The latter would be unlikely from a bonding viewpoint for framework oxygens already satisfied electrostatically by two Si. The cell dimensions and T–O distances should increase if the Al atoms caused compensating weakening of Si–O bonds. Jacobs and Uytterhoeven (*138*) interpreted infrared spectral bands at $\sim$ 3680 and $\sim$ 3600 cm^{-1} in terms of surface OH ions near Al-defect locations, but detailed structural interpretation was not given.

In conclusion, there seems little doubt that recrystallization to a Si-rich framework does occur in ultrastable varieties of faujasite-type zeolites and that this enhances the thermal stability. The nature of defects and Al-O complexes is nevertheless highly speculative.

Inorganic Complexes

Zeolite-Salt Adducts. Under suitable conditions, the entire volume inside an aluminosilicate framework can be filled with salt. Well-known mineral examples are the sodalite, cancrinite, and scapolite families whose compact aluminosilicate frameworks contain small interstices completely filled by cations and anions. Some of the cations balance the framework Al while the others balance the anions. Formally, the chemical composition of sodalite can be expressed as $Na_6Si_6Al_6O_{24} \cdot 2\,NaCl$, but all the Na^+ cations are crystallographically identical. From a structural viewpoint a salt guest and the host feldspathoid or zeolite are indistinguishable, and the entire assemblage can be viewed as a chemical entity in which the aluminosilicate framework acts as a relatively rigid electrostatic matrix enclosing the combined set of cations and anions. There are no detailed single-crystal x-ray data on zeolite–salt adducts, though powder data are available. Consequently the single-crystal data on the sodalite and cancrinite groups are reviewed.

Salt can be occluded in zeolites either directly during synthesis or by subsequent exposure to aqueous salt solution or molten anhydrous salt. Breck lists data on synthesis (Ref. *1*, pp. 331-3) and on imbibition (pp. 585-8). A detailed review of salt occlusion in zeolites is given in Chapter 5. This section emphasizes the structural data for zeolites and considers comparable data for feldspathoids.

Natural sodalites show complex chemical substitutions (*e.g.*, Taylor (*157*)) while synthetic ones can contain many different salts. Of particular interest are the photochromic ones (*see*, for example, the ESR study of Hodgson, Brinen, and Williams (*158*) which suggests the presence of a trapped electron associated with an S_2^{2-} ion). Single-crystal x-ray data for sodalite $Na_8Si_6Al_6O_{24}Cl_2$ (*159*) revealed alternating Si and Al atoms. Each sodalite unit contains one Cl at the center tetrahedrally coordinated at 2.73 A to four Na displaced inwards from four of the eight 6-rings. Each Na is bonded tetrahedrally to one Cl and to three framework oxygens at 2.35 A. Each 6-ring has only one Na neighbor and is strongly distorted such that the other three framework oxygens are more than 3 A from Na. This distortion results from the Na being pulled 1 A into the sodalite unit by the Cl from the mean plane of the 6-ring.

Nosean, $Na_8Al_6Si_6O_{24}(SO_4)$, was assigned the space group $P\bar{4}3m$. This requires Al,Si disorder in spite of the 1:1 ratio (*160*). Diffuse streaks were weakened irreversibly by heating to 750°C. Single-crystal x-ray refinement showed that half of the sodalite units contain S at the center, but well-defined peaks were not found for the expected four oxygens in tetrahedral coordination (*161*). Two types of sodium positions lay either near the plane of the 6-ring or displaced about 1 A towards the center of the sodalite unit.

Complexities were also found for hauyne $\sim (Na_5K_1Ca_2)Al_6Si_6O_{24}$-$(SO_4)_{1.5}$ by Löhn and Schulz (*162*). A volcanic specimen showed strong diffractions obeying $P\bar{4}3n$ and faint diffuse streaks and superstructure diffractions. The SO_4 tetrahedron statistically occupies two positions with some displacement of oxygens from the triad axes. The cations occupy two positions.

The structures and composition ranges of synthetic sodalites are poorly known. Chemical data indicate about 14 ± 1 extra-framework species per sodalite cage (*e.g.*, sodalite hydrate $Na_6Al_6Si_6O_{24} \cdot \sim 7.5\, H_2O$ and basic sodalite $Na_8Al_6Si_6O_{24}(OH)_2 \cdot 4\, H_2O$). Each sodalite cage shares eight 6-rings with neighboring cages. The simplest model involves eight positions displaced inwards from the 6-rings with about six other positions forming a cluster at the center of the cage. Either disorder or a complex ordering scheme must occur.

Bukin and Makarov (*163*) concluded from neutron diffraction data of hydroxysodalite, $Na_8(Al_6Si_{5.5}H_2O_{24})$ 0.4 NaCl · 0.7 NaOH, that 1/12th of the SiO_4 tetrahedra were substituted by tetrahedral H_4O_4 groups with definite hydrogen bonds and that the OH and Cl anions occupied different sites. Galitskii, Shcherbakov, and Gabuda (*164*) concluded from proton magnetic resonance that a tetrahedral cluster of $OH \cdot 3\, H_2O$ occupied the Cl position, and that circular diffusion and reorientation of H bonds occurred. Using nuclear magnetic resonance of Na from −140 to 350°C, Galitskii, Shcherbakov, and Gabuda (*165*) concluded that the symmetry was trigonal polar at −120°C and that motion increased with temperature until at 275°C the Na was diffusing with an estimated barrier energy 75 kJ/mole.

The structure of a natural cancrinite, approximately $Na_8Al_6Si_6O_{24} \cdot CaCO_3 \cdot 2\, H_2O$, consists of a framework with alternating Al and Si in which the Na,Ca cations lie in two positions (*166*). One type lies at the center of a horizontal 6-ring (Figure 3) between three framework oxygens at 2.46 A and two water molecules directly above and below at 2.58 A. The 6-ring is distorted by the pulling in of three oxygens. The second type lies in the periphery of the large channels surrounded by an octahedron of three framework oxygens and three oxygens from CO_3^{2-} groups which lie at the center of the channels. Jarchow, Reese, and Saalfeld (*167*) observed superstructure diffractions in synthetic carbonate–cancrinites, and Brown and Cesbron (*168*) found superstructures with $c = 5$, 8, 11, 16, and 27C in natural cancrinites from slowly-cooled rocks. The diffractions became diffuse and tended to disappear upon heating to 300°–450°C. A Cl-cancrinite from a volcanic rock did not show superstructure.

Barrer and Cole (*169*) synthesized many varieties of sodalite and cancrinite, including ones with NaCl, NaBr, $NaClO_3$, $NaClO_4$, and Na_2CO_3. Using a Guinier-Lenné camera they showed that Na_2SO_4– sodalite and –cancrinite were stable above 850°C whereas the other varieties broke down at lower temperatures. Barrer, Cole, and Villiger (*170*) attempted to determine the structural details of NaOH–, $NaNO_3$–, Na_2CrO_4–, and Na_2MoO_4–cancrinites from powder patterns, but the conclusions are rather uncertain.

These complexities in the sodalite and cancrinite minerals can probably be explained by two or more chemical species occupying similar crystallographic sites, by packing difficulties, and/or by thermal motion (*e.g.*, hindered rotation).

When synthesized from an Al-rich system, type A zeolite has $Si/Al < 1$ (Ref. *1*, p. 87), and occlusion of $NaAlO_2$ may have occurred (*171*). In zeolites from the system $Na_2O–Al_2O_3–B_2O_3–H_2O$, Barrer and Freund (*172*) found no replacement of framework Al by B. Analcime did not incorporate B, but sodalite hydrate incorporated up to one B per sodalite cage, perhaps as $NaBO_2$ or $NaB(OH)_4$. The H_2O dropped from 14 to 10 wt %. In type A, the maximum uptake was 1.7 B per unit cell, but washing removed all but 0.16 B. In type X, washing left three B per unit cell. Presumably the residual B occupied the sodalite unit and the leachable B the larger cage. Barrer and Marcilly (*173*) and Barrer and Robinson (*127*) synthesized salt complexes of several zeolites.

Barrer and Meier (*171*) prepared salt complexes of A and X by treatment with either concentrated brine or molten salt. Treatment of Ag-A with molten $AgNO_3$ gave the adduct $Ag_{12}Al_{12}Si_{12}O_{48} \cdot AgAlO_2 \cdot 9\,AgNO_3$, of which 65% of the $AgNO_3$ was leached at room temperature by washing for 10 days while all was removed by hot water wash. X-ray powder study of the adduct indicated a tetragonal superstructure with $a = 2A$, $c = C$. Structure refinement was not attempted, but Barrer and Meier proposed that the nine $AgNO_3$ formed a complex in the supercage with the NaCl type of packing. The complexes would rotate 90° between adjacent cages so that Ag^+ and NO_3^- ions would face each other through the 8-rings. Liquornik and Marcus (*174*) found the following amounts of salt in type A after treatment at the listed temperature: 11.7 $LiNO_3$, 300°C; 10 $NaNO_3$, 330°C; 10 $AgNO_3$, 285°C; ~0

KNO_3, 375°C. They proposed that K ions act as sentinels which block the 8-rings. In addition, they proposed that the Barrer-Meier model for the $AgNO_3$ adduct was too simple and invented a more complex one to accommodate 10 $AgNO_3$. Unfortunately this model is unreasonable since it places pairs of Ag directly above and below the center of an NO_3^- group thereby giving a very short Ag–N distance. In the structure of $AgNO_3$, the Ag ions are suspended between the oxygens of six NO_3^- triangles thereby giving a long Ag–N distance (Ref. *175*, Vol. 2, p. 362). Actually the $AgNO_3$ structure is merely a distorted version of the well-known NaCl structure type, and the original Barrer-Meier model is more plausible than the Liquornik-Marcus model. At high temperature, the NO_3^- groups of $NaNO_3$ are disordered, probably because of hindered rotation. Similar disorder may occur in zeolite adducts.

Barrer and Cole (*169*) suggested that salt occlusion should stabilize zeolites, and Rabo, Poutsma, and Skeels (*176*) proved this for halide and nitrate salts deliberately inserted into the sodalite unit of Y zeolite. The salt was introduced into the supercages by adding concentrated salt solution to anhydrous zeolite. Heating caused loss of water and provided activation energy for halide ions to pass through free 6-rings into the sodalite unit. Washing did not remove the migrated anions. A maximum content of about one halide per sodalite unit was achieved for NaCl, NaBr, and $LaCl_3$ in Na-Y. The migration rate was insignificant below 300°C but increased with temperature and decreased for larger ions. Thus a full complement of NaCl was achieved in a few hours at 550°C whereas NaBr required a day at 700°C. Migration of NaI was very slow. Some non-halide salts led to structural collapse, but 1.3 $NaNO_3$ and 0.9 $NaClO_3$ were introduced into the sodalite cages of Na-Y at 650°C. X-ray powder study of Ca-Y · NaCl showed $\sim$0.5 Cl near the center of the sodalite unit and cation occupancy of I, I′, and II (only brief details were given). Infrared and Raman spectra of the $NaNO_3$ adduct showed bands at 1385 and 1045 cm^{-1} consistent with NO_3 groups. Rabo *et al.* (*176*) suggested that the NO_3^- groups were too large to penetrate the 6-rings intact, and that they dissociated into NO_2 and oxygen which recombined into an NO_x species after entering the sodalite unit. During heat treatment, H-halide was evolved, and infrared spectra showed loss of bands attributed to framework hydroxyls. Concomitantly the adduct lost the ability to catalyze various hydrocarbon transformations usually attributed to framework hydroxyls. All the data were consistent with the exchange of a framework proton by an alkali ion. Infrared bands remaining after heat treatment of the adducts are consistent with hydroxyls bonded to cations (*e.g.*, strong band at 3520 cm^{-1} in La-Y · NaCl fired at 550°C) and occurring at the surface (3740 cm^{-1} band).

Detailed x-ray structure analyses are needed of zeolite–salt adducts, but reasonable interpretations of the structural phenomena can be made by using general structural knowledge of zeolites and feldspathoids.

Reduced Species. Breck (Ref. *1*, pp. 519-523) reviewed the available data. Chapter 10 deals with catalytic properties of metal-containing zeolites.

Incorporation of H_2 at elevated temperature into appropriate varieties of Y can lead to formation of reduced metal cations such as Cu^+ from Cu^{2+} or even formation of uncharged metal atoms which can be either lost by vaporization (Hg,Cd,Zn: *e.g.*, Yates (*177*)) or assembled into small clusters enclosed in the zeolite (Pd,Pt) or expelled as a separate metal phase (transition metals, Pd,Pt). Charge balance requires formation of one framework hydroxyl for every charge unit of reduction.

Gallezot and Imelik (*178*) determined the crystal structure of $Pd_{12.5}$-$Na_{19.5}H_{11.5}Al_{56}Si_{136}$-Y after dehydration at 600°C and after adsorption with H_2 at 25°, 200°, and 300°C. Interpretation of the powder data was complex, and liquid scattering functions were used for Pd metal atoms assumed to lie in the sodalite cage:

	Dehydrated	*25°C*	*200°C*	*300°C*
I	1.3(2)Pd	1.7(2)Pd	1.7(2)Pd	1.9(2)Pd
I′	10.6(2)Pd	5.5(4)Pd[a]	4.8(4)Pd[a]	1.8(4)Pd[a]
II	19(1)Na	18(1)Na	17(1)Na	18(1)Na
sodalite cage	—	10(3)Pd	9(3)Pd	1(3)Pd

[a] Distributed over two adjacent I′ sites.

Gallezot and Imelik concluded that Pd^{2+} ions prefer I′ upon dehydration at 600°C and that the three Pd-O(3) bond distances of 2.0 A indicated partial covalency. After treatment with dried H_2 at 25°C, 1.5 Pd remained at I′, but 4 Pd moved to a new I′ site at 2.7 A to three O(3), presumably bonded to a residual water molecule. The remaining five Pd were assumed to be reduced to Pd° and occur dispersed inside the sodalite units (actually 10 ± 3 were obtained from the x-ray analysis). Hydrogen adsorption at 300°C resulted in reduction of most of the Pd with migration to the outer surface. Diffuse x-ray lines indicated crystallites of Pd metal about 18 A across. No chemisorption was found for H_2 and O_2 on the sample reduced at 25°C.

Lewis (*179*) concluded from x-ray absorption edge spectroscopy that Pt in reduced Ca-Y (0.5 wt % Pt) consisted of a mixture of tiny crystallites, small enough to enter supercages, and larger crystallites about 60 A across. Egerton and Vickerman (*180*) from a magnetic study of Ni-Y showed that octahedrally coordinated Ni^{2+} (*i.e.*, in site I) is more easily reduced by H_2 at 700°K than tetrahedrally coordinated Ni^{2+} (presumably bonded in I′ to three framework oxygens and perhaps one hydroxyl).

Rabo *et al.* (*181*) found that of several reducing agents alkali metal vapor was particularly potent. Ni^{2+}-Y was reduced by exposure to alkali metal vapor at temperatures up to 575°C producing color changes. EPR spectra (*see* also Chapter 6) were interpreted in terms of Ni^+ occupying octahedral site I and trigonal site II. The former is thermally stable at 400°C and inert to H_2, NH_3, and CO but is oxidized by O_2 yielding O_2^-. The latter reacts easily with sorbed gases and is thermally unstable.

Alkali-metal X and Y zeolites incorporate alkali metal to yield colored complexes. Na-Y incorporated Na metal at 580°C to give a red complex whose 13-peak EPR spectrum indicated a Na_4^{3+} complex. This complex was stable to over 500°C and reacted reversibly with oxygen to yield O_2^- radicals. Na-X yielded a blue complex whose 19-peak EPR spectrum indicated a Na_6^{5+} complex. The location of the Na species is unknown, but Rabo *et al.* suggested that the first complex involved an electron shared by four Na in a tetrahedral pattern in sites II while the second complex involved Na in sites III. Ben Taarit *et al.* (*182*) used EPR study of O^{17} to identify O^- produced from Na_4^{3+} complexes in γ-irradiated and Na-reduced NaY zeolite.

Kasai and Bishop (*183*) from an EPR study showed that NO adsorbed on Na-, Ba-, Zn-, and Ni-Y zeolites disproportioned into NO^+ and NO_2^- ions and that NO reduced Ni^{2+} at site II. Barrer and Whiteman (*184*) gave detailed data on Hg uptake in A, X, chabazite, and gmelinite zeolites.

General Comment. Kasai and Bishop (*183*) reviewed ionization and electron transfer reactions in Y zeolite (*see* also Chapter 6). The common feature of all the following reactions is the increase in the number of cations and anions in the cages:

$$Na \rightarrow Na^+ + e^- \qquad NaO_2 \rightarrow Na^+ + O_2^- \qquad NaCl \rightarrow Na^+ + Cl^-$$

$$Cu^{2+} \text{ (or } Ni^{2+}) + NO \rightarrow Cu^+ \text{ (or } Ni^+) + NO^+$$

$$Cu^+ \text{ (or } Ni^+) + NO_2 \rightarrow Cu^{2+} \text{ (or } Ni^{2+}) + NO_2^-$$

Many of the reactions are endothermic outside the zeolite but proceed easily inside the zeolite. They regard zeolites as solid-state electrolytes whose electrostatic imbalances favor ionization of the incoming species.

Dehydration and Thermal Stability

Further discussion of zeolite stability is given in Chapter 4. Dehydration of a zeolite is a complex process involving adjustments between all the structural components and the external environment. At one extreme the zeolite may be heated in a dynamic vacuum in which escaping water is removed rapidly. At the other extreme a constant humidity is maintained. Application of 1 atm of H_2O can retard loss of water at 200°C. Most large-pore zeolites are thermodynamically unstable over most or all of the temperature range over which measurements are made, and strictly speaking, "stability" means "metastability." Thermal breakdown is usually determined under dry conditions from minutes to hours. Prolonged exposure, as in a catalytic cracker, leads to breakdown at lower temperature, especially if a trace of water is available.

Van Reeuwijk (*185*) summarized many data on thermal dehydration of natural zeolites and obtained new data using (a) continuous x-ray powder diffraction patterns (Guinier-Lenné technique) and (b) thermogravimetric analysis (TG) and differential thermal analysis (DTA). Mordenite lost its water mainly in a single broad episode centered on ~170°C, apparently with complete dehydration by ~400°C (Figure 34); the x-ray pattern (Figure 35) showed no apparent change until structural collapse at 900°C. Chabazite

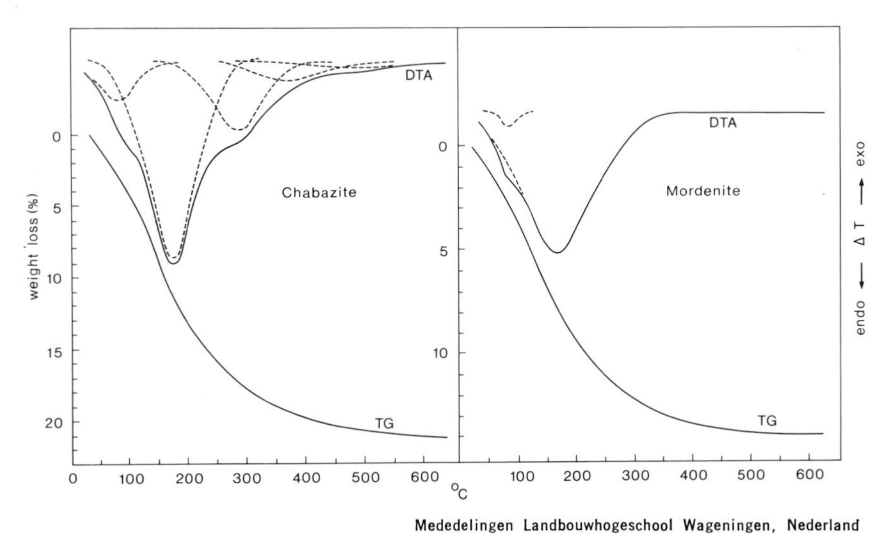

Mededelingen Landbouwhogeschool Wageningen, Nederland

Figure 34. DTA and TG curves of chabazite and mordenite at $P_{H_2O} \sim 0.03$ *atm. The dashed lines show peaks interpreted by a curve resolver. The heating rate was 10°/mm* (185).

lost its water in a complex manner, perhaps with four or five resolvable components (Figure 34), and the x-ray pattern showed complex changes of position and intensity up to 650°C, after which the pattern remained steady up to 870°C (Figure 35). The crossover of lines results from decrease of *a* from 13.78–13.37 A and increase of *c* from 14.99–15.75 A. Gmelinite

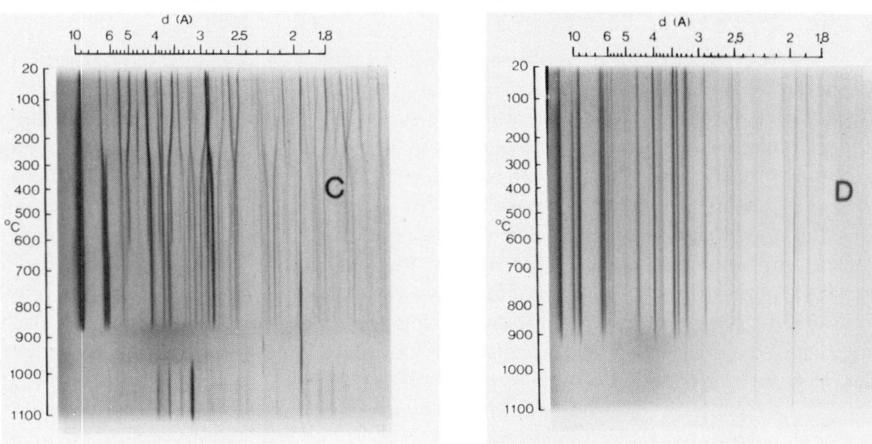

Mededelingen Landbouwhogeschool Wageningen, Nederland

Figure 35. Continuous-heating Guinier-Lenné x-ray powder photographs of (C) chabazite and (D) mordenite. Heating rate 0.3°/mm. Initial relative humidity over 51%. Continuous lines from top to bottom are from the platinum holder (185).

showed even greater changes of intensity in the x-ray pattern. Aiello and Barrer (*186*) and Aiello *et al.* (*187*) found complex phenomena in DTA, TG, and continuous x-ray powder data of sodalite, omega, erionite, and offretite zeolites crystallized from systems containing tetramethylammonium ion. Breakdown of the complex ion caused major changes in the region of 300°–400°C.

Breck (Ref. *1*, pp. 447–459) summarized the extensive DTA and TG data, most of which lack parallel x-ray data. Highlights for large-pore zeolites follow:

(1) Na- and Ca-varieties of Type A zeolite are structurally stable in dry air to 700°C but at 800°C transform to a cristobalite-type structure, probably carnegieite.

(2) Chabazites with divalent Ca, Sr, and Ba lose water continuously up to 600°C while those with large monovalent ions lose most water by 400°C (*188*). The Li variety retains some water to about 650°C. Probably the highly polarizing cations bond to the water molecules, perhaps causing hydrolysis. The breakdown temperature tended to increase as the Si/Al ratio increased, but the data are complex. Most varieties were stable to at least 800°C, but Rb and Cs varieties were stable to over 1000°C, perhaps because Rb^+ and Cs^+ occupied the 8-rings thereby giving a good charge compensation (*75*). All existing data are for natural chabazites from volcanic rocks or for synthetic varieties. Natural chabazites from sedimentary rocks are more siliceous and may tend to be more stable than volcanic ones.

(3) Natural erionite breaks down between 825° and 1000°C (*189*) while synthetic offretite, L, and omega are stable to various temperatures between 900° and 1000°C (*1*). However, there are no data on the effects of cation substitution and Si/Al ratio.

(4) For the faujasite family (*1*), most water is lost by about 400°C, but strongly polarizing cations hold the last water tenaciously. Thermal breakdown for Na-X is near 760°C, for Na-Y between 760° and 800°C, and for ultrastable varieties over 1000°C. In general, greater stability accompanies increase of the Si/Al ratio. Rare-earth cations augment the stability of X and Y, probably because of formation of a complex inside the sodalite cage. Ca and Sr reduce the stability.

(5) Natural mordenite is stable to about 900°C (Figure 35), but there are no data on the effect of cations or of Si/Al ratio.

The above macroscopic observations give little knowledge of the chemical processes during dehydration. Piecing together the scattered submicroscopic evidence leads to the following picture of dehydration:

(1) At low temperature under high humidity, the aluminosilicate framework adopts a relaxed configuration as the cations and water molecules assume positions consistent with chemical equilibrium. The unbalanced charge from the framework is compensated either directly by bonding of cations to framework oxygens or indirectly by water molecules forming hydration complexes around the cations. In large pores, water molecules near the periphery are more tightly bonded to framework oxygens than ones near the center. Hydration complexes are favored by highly polarizing cations (*e.g.*, EPR spectra of Co-A)-X, and -Y (*61*); EPR spectra of Cu-Y (*90*); Mössbauer spectra of Fe-Y (*191*); EPR spectra of Mn-bearing varieties of X

(*192*). Water molecules jump frequently from one site to another, but the aluminosilicate framework gives an overall crystallographic control which distinguishes zeolitic water from amorphous liquid water.

(2) With increasing temperature and decreasing humidity, water molecules escape. For weakly polarizing cations, there appear to be no major complications. As dehydration proceeds, the cations move towards framework oxygens and the final configuration is strongly governed by minimization of electrostatic energy as indicated earlier. The water molecules no longer smooth out the electrostatic fields, and the aluminosilicate framework goes into a tensed configuration. Both topological and chemical factors govern the tensed configuration. In chabazite, the framework changes its shape greatly (Figure 26) as the hexagonal prisms distort in response to cation bonding and rotate about the linking 4-rings (Figure 8) which act as hinges. In the zeolites containing sodalite units, there are no hinges, and the tensed configuration involves local twists which still amount to atomic movements up to 0.5 A. In mazzite, the rigid columns of linked gmelinite cages inhibit major changes during dehydration. The same probably occurs for the columns of 5-rings in mordenite. Cations affect the shape of the tensed framework by adopting different sites and by pulling framework oxygens to the required bond distances. The lumpy nature of thermogravimetric curves probably results from the formation of a series of intermediate structures. No x-ray structure data are available to test this.

(3) For strongly-polarizing cations, major complications occur. Polarization of a water molecule yields a hydroxyl bonded to the cation plus a proton which attaches itself to a framework oxygen. Transition metal cations bond to three oxygens of a 6-ring and a hydroxyl (*e.g.*, x-ray data given earlier: EPR spectra of Co-A, -X, and -Y (*61*)); Mössbauer spectra of Fe-Y (*193*). Rare earth ions form a complex in the sodalite unit of faujasite-type zeolites. Ultimately as the temperature increases, the hydroxyls are destroyed with expulsion of water.

(4) No simple rules can be given for the thermal stability of a zeolite. Theoretically, one can envisage silicate frameworks containing only Si. Under dry conditions, tridymite is stable up to 1550°C. Large-pore frameworks should tend to be less stable because of the weaker long-range interactions, but the effects of the Si–O–Si angle on the contribution from covalent bonding have not been investigated. Such large-pore frameworks might be highly persistent because of the high activation energy for breaking Si–O bonds. The actual reaction velocity should depend greatly on the presence of defects. Entrance of Al causes further complications since the chemical nature of the compensating cation is another variable. The experimental data show a tendency for the thermal stability to decrease as the Al/Si ratio increases but the idiosyncratic effect of cations confuses the pattern. Breakage of an Al–O bond should require less activation energy than of an Si–O bond. Aluminous zeolites tend to break down into feldspathoids which are stuffed derivatives of the silica minerals. There is no scientific understanding of the topologic effect on stability, except for crude speculations in which frameworks are envisaged in engineering terms. Building framework models from tetrahedral stars and plastic spaghetti gives instinctive understanding of instability when T–O–T angles become too small.

(5) Finally, a definite distinction must be made between simple thermal stability and stability in a complex chemical system such as is found in a

catalytic cracker. There appear to be no published data on the scientific factors governing the latter situation. Traces of water probably act as catalysts for hydrothermal degradation. Numerous data on silicates testify to enormous increases of reaction velocities in the presence of water.

Diffusion Phenomena, Ion Exchange, Molecular Sorption

The rest of this chapter deals with time-dependent phenomena which are difficult to characterize even with resonance techniques. Because Chapter 7 deals thoroughly with diffusion, the following treatment is deliberately brief and rather qualitative. The main aim is to provide a general guide tied as closely as possible to the crystallographic properties of zeolites. Details are given in Breck (*1*). Many aspects of intracrystalline diffusion were reviewed by Barrer (*194*) and of cation exchange by Sherry (*195*).

Diffusion of Al and Si is insignificant up to about 500°C because of the high activation energy but should occur in oxygen-depleted frameworks at high temperature. Cation diffusion in large-pore zeolites is very easy in comparison with anhydrous silicates, such as feldspar; comparable rates are attained at temperatures lower by 400° to 700°C. Diffusion of water is very rapid, as shown by complete rehydration in a few seconds of a dehydrated zeolite upon exposure to air. However, even in large-pore zeolites, it is one or two orders of magnitude slower than for pure H_2O or brine. Resing and Thompson (*196*) measured relaxation times from NMR studies of protons in hydrated X almost free of Fe^{3+} impurities. Data at 200°–500°K could be explained by an intercrystalline fluid some thirtyfold more viscous than water at 300°K and with proton mobility some 1000-fold higher than for ice at 273°K. Further data are given in a review by Resing and Murday (*197*). Cations are less mobile in a fully dehydrated zeolite than in a hydrated one at the same temperature. For a constant chemical composition, diffusion rates increase with temperature according to an Arrhenius-type equation. An exponential increase of the number of the diffusing species with enough energy to overcome barriers occurs as temperature increases. Interactions between the framework, the cations, and the sorbed molecules complicate all the diffusion phenomena, and complex explanations are needed. Unfortunately all experimental data have been obtained on powders and thereby lack three-dimensional information.

The simplest idea involves steric hindrance for diffusion between sites separated by a window. Naively, the size of a window can be obtained from models such as in Figures 3-16 with the assumption that framework oxygens have 1.35 A radius. This yields free diameters of *ca.* 2.7 A for 6-rings, *ca.* 4.3_5 A for 8-rings, and $\sim 7.7_3$ A for 12-rings. Substitution of Al for Si increases the theoretical window size because Al–O is *ca.* 1.75 A and Si–O *ca.* 1.61 A. However, the windows are usually distorted in zeolites either because of topological requirements (*e.g.*, 8-ring in gmelinite cage) or because of chemical factors (*e.g.*, bonding to cations) or both. Windows change shape in response to ion-exchange and degree of molecular sorption (*e.g.*, Figure 26). A molecule faces an energy barrier at a window because of interaction (mostly electrostatic) with framework oxygens. The potential energy can be expressed by a Lennard-Jones formula involving an attractive dipole–dipole component at long distances which is overwhelmed at short disstances by a repulsive component. Non-spherical molecules may need to adopt a special orientation in order to pass through a window. Long molecules may need to twist into a special shape so that the head can pass through one window while the tail is passing through another. Thus long-chain hydrocarbons diffuse more slowly than short-chain ones even though their diameter is the same.

Actually atoms and molecules are not composed of rigid spheres, and the effective pore size is greater by a few tenths of an Angstrom for molecular diffusion than the rigid-sphere window. With increasing temperature, the effective size of a window increases about 0.3 A in response to greater kinetic energy of the molecule and probably to greater vibration of framework oxygens. Breck (Ref. *1*, Table 8.14 and Figure 8.15) gives the size of various molecules and a chart of effective pore sizes for several zeolites and describes the channel system and pore sizes of zeolites (Ref. *1*, Tables 2.18-2.56). Occupancy of a window by a firmly bound cation stops molecular diffusion until the cation jumps to a new site. Diffusion can also be hindered by cations which lie on channel walls. This blocking can be eliminated or mitigated by ion exchange with smaller cations or ones with higher valence. Particularly useful is exchange with protons (perhaps *via* NH_4-exchange followed by NH_3 expulsion) which can result in complete absence of blocking cations. Channels which interconnect in three dimensions are almost immune to blocking by accidental impurities whereas isolated one-dimensional channels can be blocked completely by just a few impurities. Diffusion blocks can result from salt occlusion either during synthesis, by later addition, or by deliberate partial destruction of part of the zeolite framework. Of course, such blocks can occur haphazardly in catalytic crackers. Stacking faults in the framework may result in blockage or narrowing of channels. In general, zeolites provide almost unlimited opportunity for adjusting the selectivity of molecular sorption by choice of framework topology and exchangable cation.

There are many other subtle effects on molecular adsorption. Highly polar molecules, especially water, interact so strongly with cations that they form tightly bound complexes. Some cations adopt new positions altering the number of windows blocked by cation sentinels. Water molecules may split up into hydroxyl and proton which condenses with a framework oxygen. The pre-adsorption of a small amount of a polar molecule may strongly affect subsequent sorption of a non-polar molecule. At equilibrium between a zeolite and a gas mixture, the zeolite will adsorb a greater amount of the molecule which has the greater heat of sorption. In general for Al-bearing zeolites, the heat of sorption is greater for the more polar and more polarizable molecules as a result of electrostatic interactions. Thus water is favored over unsaturated organic molecules which in turn are favored over saturated ones.

The isosteric heat of sorption of various molecules in chabazite, A, X, and Y is very high initially, probably because the first molecules attach themselves to cations and unsaturated framework oxygens. A steep drop to a minimum value after loading of about one-quarter of the molecules is followed by a small rise probably resulting from molecule–molecule interaction. Finally there is a small drop to the latent heat of condensation as molecules condense on the outer surface. Probably the molecules adjust their positions as sorption proceeds, causing yet another complication.

Cation diffusion is also governed by many complex factors. Steric hindrance is complicated by the strong interaction with framework oxygens. Naively one can imagine a large cation such as Cs^{+} being blocked geometrically by a 6-ring. However, the passage of a small cation such as Ca^{2+} or Ni^{2+} through a 6-ring cannot be governed merely by size. Almost certainly the cation becomes bonded temporarily to the framework oxygens, and the diffusion process can be modeled as a two-step process governed by jump frequencies to other sites and a residence time. Theoretically, the residence time must be much greater for cations surrounded by framework oxygens (such as ones in site I of faujasite, and the cages of the erionite–offretite–mazzite group) than for cations with waisted coordination, especially for those lying on walls of supercages. Data on electrical conductivity of anhydrous zeolites can be explained largely in terms of movement of those cations with the weakest bonding. Of course, the diffusion rate increases with temperature. Some ions which will move at high temperature cannot be moved in any reasonable length of time at low temperature. Rare-earth ions which enter a sodalite unit of a faujasite-type zeolite at high temperature stay there indefinitely at room temperature even when the zeolite is hydrated. In general, the diffusion and ion-exchange properties of dehydrated zeolites correlate fairly well with the ionic potential (*i.e.*, formal charge/ionic radius) but the details are complex. Ion exchange with fused salts may lead to simultaneous, irreversible salt occlusion.

Addition of water to a zeolite causes major complications. Highly polarizing cations tend to surround themselves with water molecules or even to disproportion a water molecule into a hydroxyl which bonds to the cation and a proton which condenses with a framework oxygen. Hydration energies tend to correlate with the ionic potential and the total charge of the cations (*i.e.* charge on the framework) (*see* data on heats of immersion of many zeolites by Barrer and Cram (*198*)). Cation diffusion now involves simultaneous movement of water molecules. The latter are in a continual state of readjustment, which might formally be regarded as being autocatalyzed by movement of protons. In large channels, cations probably move mainly by jumping from one temporary coalition to another rather than by migration of a complete hydration complex. Simultaneously, water molecules are engaged in similar movements. Weakly polarizing cations tend to be bonded to framework oxygens either entirely or in part, but diffusion and ion-exchange also involve the water molecules. Migration of a cation through a small window can be envisaged formally as stripping away of a hydration complex followed by growth of another one after passage through the window.

However, the rate-determining step may be the long residence time when the cation is bonded to framework oxygens.

Some recent papers are: diffusion and sorption of light hydrocarbons and other non-polar molecules in A (*199, 200*), self-diffusion and mobility of Zn in X and Y (*201, 202*), ion-exchange of Mn, Co, Zn, and Ni in X and Y zeolites (*203, 204*), electrical properties of X and Y exchanged with monovalent cations (*205, 206*), NMR studies of ^{23}Na in dehydrated X and Y zeolites modified by small amounts of H_2O, NH_3, and H_2S (*131, 207*), and NMR studies of molecules adsorbed on A, X, and Y (*208*).

Proton mobility is extremely important in relation to catalytic activity (*see* later). Early infrared studies showed an apparent weakening of OH infrared bands at elevated temperature, but interpretation in terms of mobile protons is equivocal (*see* Mestdagh, Stone, and Fripat (*209*)). Interpretation of NMR data in terms of second moment, relaxation times, and jump frequency is complex and model dependent. Earlier conclusions by Mestdagh *et al.* for H-Y at −180° to 400°C were criticized by Freude *et al.* (*210*). Taken at face value, they yielded a frequency of $3.3 \times 10^{10} \exp(-10^4/RT)$ sec^{-1} for protons jumping between framework oxygens. This corresponds to 3.2×10^6 sec^{-1} at 300°C and 2×10^7 sec^{-1} at 450°C. Freude *et al.* used both stationary and spin-echo techniques to study H,Na-X, and -Y zeolites at 20° to 400°C. They concluded that (1) activation energy for proton jumping between framework oxygens is 20–40 kJ/mole compared with about 250 kJ/mole for dehydroxylation, (2) proton jump frequency is 5×10^4 sec^{-1} at 200°C in H varieties compared with similar values at 300°C for dehydrated varieties with divalent and trivalent ions, (3) proton jump frequency increased 60-fold at 200°C when two pyridine molecules per supercage were added but only twofold for toluene. Thus, in spite of the problems over detailed interpretation, it seems reasonable to invoke considerable proton mobility, especially in the presence of molecules which can accept protons reversibly.

Chemical Environment of Sorbed Molecules

In spite of many statements in the literature that sorbed molecules form an intercrystalline *fluid*, the sum total of the experimental data definitely indicates that the molecules occupy crystallographically controlled sites and that they form part of a crystalline *solid*. The original statements are not seriously misleading, however, because of the high thermal motion and large number of defects. The greater the interaction between the molecules and the framework plus cations, the better defined will be the positions of the molecules. X-ray diffraction data show distinct sites for molecules (*see* earlier), though the positional displacements are large. Probably the clearest evidence of distinct positions is for the CO molecule attached end-on by the C atom to the transition metal ion Co in Type A zeolite (Figure 22). Whereas the C atom shows little positional displacement, the O atom swings in a wide arc inside the supercage. Almost certainly only part of the apparent positional displacement results from thermal motion. The remainder should derive from the atom (or molecule) adopting a different center-of-movement in response to different electrostatic fields in different cages. Such field variations must occur when Si and Al atoms are disordered and when cation sites are not fully occupied. Obviously the stabilizing features are reduced for molecules at the centers of large cages. Thus the crystallographic model for molecules consists of a statistical assemblage of similar but not identical sites occupied at least most of the time under the control of chemical forces which are relatively weak.

Resonance and absorption techniques provide information on lifetime and bonding effects while calorimetric data provide macroscopic data. The latter clearly show that the first molecules find unusually energetic sites while the last molecules must be satisfied by molecule–molecule interactions when the zeolite has wide pores. Barrer and Wasilewski (*211*) concluded that the energy heterogeneity increases with the polarizability of the molecule, its dipole and quadrupole moments, and with the polarizing power of the cation. All these factors are consistent with interaction between cations and molecules.

Indeed the x-ray structural analyses definitely show that cations move from inaccessible positions in dehydrated zeolites in order to bond to sorbed polar molecules. Innumerable IR absorption data testify to cation–molecule interaction (*see* later chapter). One example is the observation by Rabo *et al.* (*181*) of cation-dependent absorption band for CO sorbed on bivalent-cation X and Y zeolites.

Particularly important are experiments using nuclear interactions. Interpretation is usually model dependent, and open to subjective argument. Neutron scattering spectroscopy by Egelstaff, Downes, and White (*212*) of K-A zeolite loaded with water, heavy water, methyl alcohol, ammonia, or methyl cyanide revealed that the molecules occupy definite positions for about 10^{-11} sec at room temperature during which they vibrate in resonance with the zeolite framework. This type of quasi-elastic scattering experiment actually examines the vibration of the proton (or deuterium in exchanged specimens), and it is necessary to ask whether the entire molecule moves or just the proton(s). Since it is chemically unreasonable to propose significant dissociation of protons from CH_3CN at room temp, and since CH_3OH and CH_3CN have similar residence times, it is concluded that the neutron scattering spectroscopy measures the jump time for molecular translation. The calculated diffusion constant for a random-walk model is about 10^{-5} cm^2 sec^{-1} assuming a jump of 4 A and assuming that passage through a window is equally probable to jumping to a new site in the same cage. [The relation between macroscopic diffusion coefficient D, residence time τ, and jump distance a is $D = a^2/6\tau$]. No macroscopic diffusion rates have been measured for the above molecules in K-A zeolite, but Barrer (discussion in Egelstaff *et al.* (*212*), pointed out that self-diffusion coefficients of zeolitic water tend to be about 10^{-7} to 10^{-8} cm^2 sec^{-1} at 45°C. The difference of three orders of magnitude might be resolved if the residence time between jumps through a window is much greater than for jumps between sites in the same cage. Interpretation of the vibrational spectra showed that the molecular vibrations were anisotropic with components ranging from below 0.3 to about 1 A. These displacements correspond to B factors for x-ray diffraction varying from below 7 to about 80, a range which brackets the experimental data. Actually when B increases beyond about 20, the x-ray data become too uncertain to give accurate information on atomic positions, and the electron density becomes so smeared out that it corresponds nearly to a liquid.

Resing and Murday (*197*) reviewed NMR relaxation data on molecular motion in zeolites. Interpretation is model dependent, and interaction occurs with paramagnetic species in the zeolite framework including Fe^{3+} impurity. Again there is a serious problem in interpreting the macroscopic diffusion coefficient in terms of the jump frequency. Furthermore the relation between the various relaxation processes and molecular rotation and translation is not clear. For non-spherical molecules, reorientation may occur mainly with translation while for spherical molecules independent rotation may occur. The available data indicate that cyclohexane and sulfur hexafluoride in Na-X begin to rotate at least by 80°K whereas benzene begins to rotate at 223°C. Apparent diffusion coefficients for various molecules in X and Y zeolites range from 10^{-10} to 10^{-5} cm^2/sec. For example, an apparent diffusion rate of 10^{-7} cm^2/sec was found at 125°K for SF_6 in Na-X and $\sim$ 300°K for benzene. This indicates the much higher mobility at a chosen temperature of an inert molecule. Activation energies ranged from about 9 to 25 kJ/mole. Diffusion was apparently slower for high loading than for low loading, perhaps because of fewer empty sites to which jumping could occur.

Hydrocarbon molecules with double bonds are strongly sorbed, and the π electrons interact with a cation as the molecule adopts a broadside position (*e.g.*, Figure 23). Pfeifer *et al.* (*208*) summarized proton spin relaxation of benzene, cyclohexadiene, cyclohexene, and cyclohexane absorbed in NaY which demonstrate decreasing mobility with increasing number of π electrons.

In spite of uncertainties in the above and other data, sorbed molecules can be envisaged jumping rapidly from site to site in the cages with an occasional escape through a window into the next cage. The greater the chemical attraction between a molecule and the cations or framework species, the greater will be the residence time. The more asymmetric a molecule, the lower will be the frequency of rotation. The larger a cage, the greater will be the delocalization of the molecules.

Structural Aspects of Catalytic Processes

Finally, a few comments are desirable on structural aspects of proposed catalytic processes. Because a zeolite is a crystal with chemical and physical interaction between *all* components, no item can be arbitrarily isolated except in a crude theoretical model. Catalytic activity has been ascribed to (a) a framework hydroxyl acting as a Brönsted acid by reversible transfer of a proton to a sorbed molecule, (b) electrostatic field gradient acting as a Lewis acid by polarization of a bond in a sorbed molecule, (c) interaction with metal clusters, and (d) vacant oxygen in the aluminosilicate framework which acts as a Lewis acid because of the residual positive charge. None of these processes can act in total isolation. Whatever the actual mechanism, the sorbed molecule is under the influence of a crystal field wherever it lies. Furthermore the interaction with the framework and cations, even for inert molecules, increases the residence time, thereby increasing the opportunity for catalysis. Zeolites used in catalysis have substitutional disorder of Al and Si and of the exchangeable cations. Consequently the catalytic sites cannot be homogeneous. Of course, the perturbation caused by structural disorder may be trivial for some catalytic processes, but in others it may result in major changes of efficiency between sites.

Currently the best structural model for hydroxyl in faujasite-type zeolites is that of Dempsey and Olson, and it is reasonable to suppose that the hydroxyl projecting into the supercage from O(1) is responsible for Brönsted activity. However, the experimental data do not preclude some protons projecting into the supercage from other framework oxygens.

An electrostatic field gradient must occur in all zeolites, and it must reach particularly large values near multivalent cations. Dempsey (*128*) calculated that the electrostatic field could reach 1 to 3 V/A at 2.5–3 A from a surface cation. This value is probably too high because of various processes reducing the effectiveness of the formal charge. Nevertheless such a field should result in significant polarization of an adjacent molecule, perhaps causing a displacement of up to 0.1 electronic charge in a C–H bond of a hydrocarbon. Polarization of residual water molecules by multivalent cations with formation of OH groups tends to reduce the effectiveness of cations, especially as the cation–hydroxyl complexes tend to become hidden. Although cations in contact with molecules would have the greatest effect, all cations contribute to the overall elecrostatic field of the crystal.

Particularly difficult to evaluate are speculations involving missing oxygens of the framework. In the early days of zeolite catalysis there was a strong tendency to carry over ideas from amorphous catalysts. It was natural to ascribe observations of Lewis acid activity to framework defects, especially the unproven concept of three-coordinated aluminum. Later, it was realized that Lewis activity could arise merely from interaction with any cation. There is no need now to invoke three-coordinated aluminum except for the unstable materials produced by dehydrating H-faujasite. Mathematically, the removal of two OH must result in loss of a framework oxygen, thereby removing the electrostatic shield between two T atoms. All structural ideas are speculative without any reliable experimental control. Because the Al–O tetrahedral bond is weaker than the Si–O bond, the lost oxygen should have been bonded to either two Al or one Al + one Si. The former is unlikely in Si-rich zeolites but may occur in Al-rich ones. Repulsion between the two T atoms should cause drastic movements, presumably with formation of three-coordination if more drastic events do not occur. Crystal chemically, there should be a tendency for recrystallization into an Si-rich tetrahedral framework with expulsion of Al into a six-coordinated species of composition Al_2O_3.

Metal clusters can be treated by various theoretical approximations to Schrödinger's equation. Slater and Johnson (*213*) reviewed recent approaches using the self-consistent field x-alpha scattered-wave method.

From the viewpoint of a structural crystallographer, catalysis by zeolite molecular sieves can be regarded as the combined effect of electrostatic fields, reversible proton and electron transfer, and both spatial and temporal factors. All of these contribute to the combined processes of chemisorption and bond weakening which constitute the heart of heterogeneous catalysis.

Addendum. The following structure determinations, all by W. J. Mortier, J. J. Pluth, and J. V. Smith, fill in several gaps mentioned in the text: dehydrated offretite, *Nature* (1975) **256,** 718 and *Z. Krist.*, in press; dehydrated Ca-exchanged mordenite, *Mat. Res. Bull.* (1975) **10,** 1037; dehydrated H-exchanged mordenite, *Mat. Res. Bull.* (1975) **10,** 1319; hydrated Ca-mordenite, *Mat. Res. Bull.*, in press; CO-complex of offretite, *Z. Krist.*, in prep.; dehydrated Ca- and Na-chabazite, *Acta Cryst.*, in prep.; CO-complex of chabazite, *Acta Cryst.*, in prep. The structure of dehydrated mazzite is published: R. Rinaldi, J. J. Pluth, and J. V. Smith, *Acta Cryst.* (1975) **B31,** 1603.

Acknowledgments

I thank the following: J. J. Pluth, J. A. Rabo, and K. Seff for constructive comments; J. A. Gard, W. M. Meier, D. H. Olson, J. J. Pluth, R. Rinaldi, K. Seff, and L. P. van Reeuwijk for originals of the figures; W. M. Meier, D. H. Olson, and K. Seff for unpublished data; D. W. Breck and E. Flanigen for their long-term advice; J. M. Bennett, L. S. Dent Glasser, J. J. Pluth, F. Rinaldi, and R. Rinaldi for their collaboration in research; Petroleum Research Fund administered by the American Chemical Society, Union Carbide Corporation and Materials Research Laboratory of the University of Chicago (funded by National Science Foundation) for financial assistance; I. Baltuska for secretarial help. V. Schomaker and D. W. Breck provided constructive criticism of the manuscript.

Literature Cited

1. Breck, D. W., "Zeolite Molecular Sieves," Wiley, New York, 1974.
2. "Molecular Sieves," Society of the Chemical Industry, London, 1968.
3. "Molecular Sieve Zeolites, I and II," *Advan. Chem. Ser.* (1971) **101** and **102.**
4. "Molecular Sieves," *Advan. Chem. Ser.* (1973) **121.**
5. "Molecular Sieves," J. B. Uytterhoeven, Ed., Leuven University Press, Belgium, 1973.
6. Smith, J. V., *Amer. Mineral Soc. Spec. Paper* (1963) **1,** 281.
7. Rinaldi, R., Smith, J. V., *Int. Mineral. Soc. Mtg.* (abstr.), Berlin (1974).
8. Walker, G. P. L., *Mineral Mag.* (1960) **32,** 503.
9. Coombs, D. S., Ellis, A. J., Fyfe, W. S., Taylor, A. M., *Geochim. Cosmochim. Acta* (1959) **17,** 53.
10. Hay, R. L., "Zeolites and Zeolitic Reactions in Sedimentary Rocks," *Geolog. Soc. Amer. Spec. Paper No.* **85** (1966).
11. Deffeyes, K. S., "Molecular Sieves," Soc. Chem. Ind., London, 1968, p. 7.
12. Sheppard, R. A., *Advan. Chem. Ser.* (1971) **101,** 279.

13. Munson, R. A., Sheppard, R. A., *Minerals Sci. Engng.* (1974) **6,** 19.
14. Murray, J., Renard, A. F., "Report of the Scientific Results of the Voyage of H. M. S. Challenger during the Years 1873-76," Neill and Co., Edinburgh, 1891.
15. Stonecipher, S. A., *Geolog. Soc. Amer., Abstr. Progr.* (1974) **6,** 262.
16. Sheppard, R. A., Gude, A. J. III, Griffin, J. J., *Amer. Mineral.* (1970) **55,** 2053.
17. Coombs, D. S., *Advan. Chem. Ser.* (1971) **101,** 317.
18. Mumpton, F. A., "Industrial Minerals and Rocks," 4th ed., Society of Mining Engineers, in press.
19. Kerr, I. S., Gard, J. A., Barrer, R. M., Galabova, I. M., *Amer. Mineral.* (1970) **55,** 441.
20. Flanigen, E. M., Grose, R. W., *Advan. Chem. Ser.* (1971) **101,** 76.
21. Flanigen, E. M., *Advan. Chem. Ser.* (1973) **121,** 120.
22. Mumpton, F., "Molecular Sieves," p. 156, Leuven University Press, Belgium, 1973.
23. Meier, W. M., "Molecular Sieves," p. 10, Society of the Chemical Industry, London, 1968.
24. Robson, H. E., Shoemaker, D. P., Ogilvie, R. A., Manor, P. C., *Advan. Chem. Ser.* (1973) **121,** 106.
25. Galli, E., *Cryst. Chem. Comm.* (1974) **3,** 339.
26. Rinaldi, R., Pluth, J. J., Smith, J. V., *Amer. Cryst. Ass. Summer Mtg.* (1974) 272.
27. Fischer, K. F., *Advan. Chem. Ser.* (1973) **121,** 31.
28. Bennett, J. M., Gard, J. A., *Nature* (1967) **214,** 1005.
29. Gard, J. A., Tait, J. M., *Advan. Chem. Ser.* (1971) **101,** 230.
30. Gard, J. A., Tait, J. M., *Acta Crystallogr.* (1972) **B 28,** 825.
31. Gard, J. A., Tait, J. M., "Molecular Sieves," p. 94, Leuven Univ. Press, Belgium, 1973.
32. Galli, E., Passaglia, E., Pongiluppi, D., Rinaldi, R., *Contr. Mineral. Petr.* (1974) **45,** 99.
33. Barrer, R. M., Villiger, H., *Chem. Comm.* (1969) 659.
34. Meier, W. M., *Z. Kristallogr.* (1961) **115,** 439.
35. Sherman, J. D., Bennett, J. M., *Advan. Chem. Ser.* (1973) **121,** 52.
36. Fischer, K., *Neues Jahrb. Mineral. Monatsch.* (1966) 1.
37. Smith, J. V., "Feldspar Minerals," Springer-Verlag, New York, 1974.
38. Gramlich, V., Meier, W. M., *Z. Kristallogr.* (1971) **133,** 134.
39. Yanagida, R. Y., Amaro, A. A., Seff, K., *J. Phys. Chem.* (1973) **77,** 805.
40. Charnell, J. F., *J. Cryst. Growth* (1971) **8,** 291.
41. Barrer, R. M., Meier, W. M., *Trans. Faraday Soc.* (1958) **54,** 1074.
42. Seff, K., personal communication; a preprint submitted to *J. Phys. Chem.*
43. Olson, D. H., *J. Phys. Chem.* (1970) **74,** 2758.
44. Kühl, G. H., *J. Inorg. Nucl. Chem.* (1971) **33,** 3261.
45. Dempsey, E., Kühl, G. H., Olson, D. H., *J. Phys. Chem.* (1969) **73,** 387.
46. Smith, J. V., *Advan. Chem. Ser.* (1971) **101,** 171.
47. Gramlich, V., Diss. no. 4633 (1971), Eidg. Techn. Hochschule, Zürich, Switzerland.
48. Meier, W. M., personal communication.
49. Smith, J. V., "Molecular Sieves," p. 28, Soc. Chem. Ind., London, 1968.
50. Seff, K., Shoemaker, D. P., *Acta Cryst.* (1967) **22,** 162.
51. Yanagida, R. Y., Amaro, A. A., Seff, K., *J. Phys. Chem.* (1973) **77,** 805.
52. Reed, T. B., Breck, D. W., *J. Amer. Chem. Soc.* (1956) **78,** 5972.
53. Riley, P. E., Seff, K., Shoemaker, D. P., *J. Phys. Chem.* (1972) **76,** 2593.
54. Yanagida, R. Y., Vance, T. B. Jr., Seff, K., *Inorg. Chem.* (1974) **13,** 723.
55. Riley, P. E., Seff, K., *Inorg. Chem.* (1974) **13,** 1355.
56. Klier, K., *J. Amer. Chem. Soc.* (1969) **91,** 5392.
57. Klier, K., Ralek, M., *J. Phys. Chem. Solids* (1968) **29,** 951.

58. Amaro, A. A., Kovaciny, C. L., Kunz, K. B., Riley, P. E., Vance, T. B. Jr., Yanagida, R. Y., Seff, K., "Molecular Sieves," p. 113, Leuven Univ. Press, Belgium, 1973.
59. Garten, R. L., Delgass, W. N., Boudart, M., *J. Catal.* (1970) **18,** 90.
60. Seff, K., Ph.D. thesis, Mass. Inst. of Techn., 1964.
61. Heilbron, M. A., Vickerman, J. C., *J. Catal.* (1974) **33,** 434.
62. Meier, W. M., Shoemaker, D. P., *Z. Krist.* (1966) **123,** 357.
63. Fang, J. H., Smith, J. V., *J. Chem. Soc.* (1964) 3749.
64. Angell, C. L., Schaffer, P. C., *J. Phys. Chem.* (1965) **69,** 3463.
65. Amaro, A. A., Seff, K., *J. Phys. Chem.* (1973) **77,** 906.
66. Riley, P. E., Seff, K., *J. Amer. Chem. Soc.* (1973) **95,** 8180.
67. Riley, P. E., Seff, K., *Inorg. Chem.*, in press.
68. Seff, K., *J. Phys. Chem.* (1972) **76,** 2601.
69. Yanagida, R. Y., Seff, K., *J. Phys. Chem.* (1973) **77,** 138.
70. Yanagida, R. Y., Seff, K., *J. Phys. Chem.* (1972) **76,** 2597.
71. Smith, J. V., Rinaldi, F., Glasser, L. S. Dent, *Acta Cryst.* (1963) **16,** 45.
72. Smith, J. V., Knowles, C. R., Rinaldi, F., *Acta Cryst.* (1964) **17,** 374.
73. Smith, J. V., *Acta Cryst.* (1962) **15,** 835.
74. Fang, J. H., Smith, J. V., *J. Chem. Soc.* (1964) 3749.
75. Smith, J. V., *J. Chem. Soc.* (1964) 3759.
76. Smith, J. V., "Molecular Sieves," p. 28, Soc. Chem. Ind., 1968.
77. Smith, J. V., *Advan. Chem. Ser.* (1971) **101,** 171.
78. Hseu, K., Ph.D. thesis, Univ. of Wash., 1972.
79. Mikovsky, R. J., Silvestri, A. J., Dempsey, E., Olson, D. H., *J. Catal.* (1971) **22,** 371.
80. Mortier, W. J., Ph.D. thesis, Univ. of Leuven, Belgium, 1972.
81. Mortier, W. J., Bosmans, H. J., Uytterhoeven, J. B., *J. Phys. Chem.* (1972) **76,** 650.
82. Baur, W. H., *Am. Mineral.* (1964) **49,** 697.
83. Bennett, J. M., Smith, J. V., *Mater. Res. Bull.* (1968) **3,** 933.
84. Bennett, J. M., Smith, J. V., *Mater. Res. Bull.* (1969) **4,** 343.
85. Olson, D. H., Kokotailo, G. T., Charnell, J. F., *J. Colloid Interface Sci.* (1968) **28,** 305.
86. Mortier, W. J., Bosmans, H. J., *J. Phys. Chem.* (1971) **75,** 3327.
87. Gallezot, P., Taarit, Y. Ben, Imelik, B., *J. Catal.* (1972) **26,** 481.
88. Pluth, J. J., Smith, J. V., *Mater. Res. Bull.* (1973) **8,** 459.
89. Olson, D. H., Sherry, H. S., *J. Phys. Chem.* (1968) **72,** 4095.
90. Bennett, J. M., Smith, J. V., Angell, C. L., *Mater. Res. Bull.* (1969) **4,** 77.
91. Hunter, F. D., Scherzer, J., *J. Catal.* (1971) **20,** 246.
92. Pluth, J. J., Ph.D. thesis, Univ. of Wash., 1971.
93. Bennett, J. M., Smith, J. V., *Mater. Res. Bull.* (1968) **3,** 865.
94. Dodge, R. P., unpublished data.
95. Bennett, J. M., Smith, J. V., *Mater. Res. Bull.* (1968) **3,** 633.
96. Olson, D. H., *J. Phys. Chem.* (1968) **72,** 4366.
97. Bennett, J. M., Smith, J. V., *Mater. Res. Bull.* (1969) **4,** 7.
98. Eulenberger, G. R., Shoemaker, D. P., Keil, J. G., *J. Phys. Chem.* (1967) **71,** 1812.
99. Gallezot, P., Imelik, B., *J. Chim. Phys. Physicochim. Biol.* (1971) **68,** 816.
100. Smith, J. V., Bennett, J. M., Flanigen, E. M., *Nature* (1967) **215,** 241.
101. Gallezot, P., Imelik, B., *J. Chim. Phys. Physicochim. Biol.* (1971) **68,** 34.
102. Gallezot, P., Imelik, B., *J. Phys. Chem.* (1973) **77,** 652.
103. Gallezot, Y., Taarit, Y. Ben, Imelik, B., *J. Catal.* (1972) **26,** 295.
104. Olson, D. H., *J. Phys. Chem.* (1968) **72,** 1400.
105. Dempsey, E., Olson, D. H., *J. Phys. Chem.* (1970) **74,** 305.
106. Gallezot, P., Imelik, B., *J. Chim. Phys. Physicochim. Biol.* (1974) **71,** 155.
107. Simpson, H. D., Steinfink, H., *J. Amer. Chem. Soc.* (1969) **91,** 6225.
108. Simpson, H. D., Steinfink, H., *J. Amer. Chem. Soc.* (1969) **91,** 6229.

109. Gallezot, P., Taarit, Y. Ben, Imelik, B., *C.R. Acad. Sci.* (1971) **C, 272,** 261.
110. Olson, D. H., Mikovsky, R. J., Shipman, G. F., Dempsey, E., *J. Catal.* (1972) **24,** 161.
111. Gallezot, P., Imelik, B., *J. Phys. Chem.* (1973) **77,** 2364.
112. Mortier, W. J., Costenoble, M. L., Uytterhoeven, J. B., *J. Phys. Chem.* (1973) **77,** 2880.
113. Gard, J. A., Tait, J. M., *Acta Cryst.* (1972) **B28,** 825.
114. Gard, J. A., Tait, J. M., "Molecular Sieves," p. 94, Leuven Univ. Press, 1973.
115. Barrer, R. M., Villiger, H., *Z. Kristallogr.* (1969) **128,** 352.
116. Baerlocher, C., Barrer, R. M., *Z. Kristallogr.* (1972) **136,** 245.
117. Flanigen, E., Grose, R. W., *Advan. Chem. Ser.* (1971) **101,** 76.
118. Galli, E., *Cryst. Struct. Comm.* (1974) **3,** 339.
119. Rinaldi, R., Pluth, J. J., Smith, J. V., *Amer. Cryst. Assoc. Summer Mtg.* (1974) **2,** 272; submitted *Acta Crystallogr.*
120. Kerr, I. S., Gard, J. A., Barrer, R. M., Galabova, I. M., *Amer. Mineral.* (1970) **55,** 441.
121. Wu, E. L., Whyte, T. E. Jr., Rubin, M. K., Venuto, P. B., *J. Catal.* (1974) **33,** 414.
122. Fischer, K., *Neues Jahrb. Mineral. Monat.* (1966) 1.
123. Meier, W. M., personal communication.
124. Sand, L. B., "Molecular Sieves," p. 71, Soc. Chem. Ind., London, 1968.
125. Barrer, R. M., Oei, A. T. T., *J. Catal.* (1973) **30,** 460.
126. Meier, W. M., Kokotailo, G. T., *Z. Kristallogr.* (1965) **121,** 211.
127. Barrer, R. M., Robinson, D. J., *Z. Kristallogr.* (1972) **155,** 374.
128. Dempsey, E., "Molecular Sieves," p. 293, Soc. Chem. Ind., London, 1968.
129. Dempsey, E., *J. Phys. Chem.* (1969) **73,** 360.
130. Lechert, H., *Advan. Chem. Ser.* (1973) **121,** 74.
131. Flanigen, E. M., Khatami, H., Szymanski, H. A., *Advan. Chem. Ser.* (1971) **101,** 201.
132. Ward, J., *Advan. Chem. Ser.* (1971) **101,** 380.
133. Rabo, J. A., Poutsma, M. L., *Advan. Chem. Ser.* (1971) **102,** 284.
134. Uytterhoeven, J. B., Christner, L. G., Hall, W. K., *J. Phys. Chem.* (1965) **69,** 2117.
135. Kerr, G. T., Dempsey, E., Mikovsky, R. J., *J. Phys. Chem.* (1965) **69,** 4050.
136. Angell, C. L., Schaffer, P. C., *J. Phys. Chem.* (1965) **69,** 3463.
137. Olson, D. H., Dempsey, E., *J. Catal.* (1969) **13,** 221.
138. Jacobs, P. A., Uytterhoeven, J. B., *J. Chem. Soc. Faraday Trans.* (1973) **69,** 359 and 373.
139. Ward, J., *J. Catal.* (1968) **10,** 34.
140. Ward, J., *J. Phys. Chem.* (1968) **72,** 4211.
141. Rabo, J. A., Angell, C. L., Schomaker, V., *Actes Congr. Intern. Catalysis, 4th, Moscow,* 1968.
142. Eberly, P. E. Jr., Kimberlin, C. N. Jr., *Advan. Chem. Ser.* (1971) **102,** 374.
143. Bolton, A. P., *J. Catal.* (1971) **22,** 9.
144. Barrer, R. M., Coughlan, B., "Molecular Sieves," p. 141, Soc. Chem. Ind., London, 1968.
145. Kranich, W. L., Ma, Y. H., Sand, L. B., Weiss, A. H., Zwiebel, I., *Advan. Chem. Ser.* (1971) **101,** 502.
146. Thaker, D. K., Weller, S. W., *Advan. Chem. Ser.* (1973) **121,** 598.
147. McDaniel, C. V., Maher, P. K., "Molecular Sieves," p. 186, Soc. Chem. Ind., London, 1968.
148. Kerr, G. T., *Advan. Chem. Ser.* (1973) **121,** 219.
149. Szymanski, H. A., Stamires, D. N., Lynch, G. R., *J. Opt. Soc. Amer.* (1960) **50,** 1323.
150. Rabo, J. A., Pickert, P. E., Stamires, D. N., Boyle, J. E., *Int. Congr. Catal.,* Paris, 1961.
151. Maher, P. K., Hunter, F. D., Scherzer, J., *Advan. Chem. Ser.* (1971) **101,** 266.

152. Kerr, G. T., *J. Phys. Chem.* (1969) **73,** 2780.
153. Scherzer, J., Bass, J. L., *J. Catal.* (1973) **28,** 101.
154. Dempsey, E., *J. Catal.* (1974) **33,** 497.
155. Kerr, G. T., *J. Phys. Chem.* (1968) **72,** 2594.
156. Kühl, G. H., "Molecular Sieves," p. 227, Leuven Univ. Press, Belgium, 1973.
157. Taylor, D., *Contr. Min. Petr.* (1967) **16,** 172.
158. Hodgson, W. G., Brinen, J. S., Williams, E. F., *J. Chem. Phys.* (1967) **47,** 3719.
159. Löns, J., Schulz, H., *Acta Cryst.* (1967) **23,** 434.
160. Saalfeld, H., *Neues Jahrb. Mineral. Monat.* (1958) 38.
161. Schulz, H., Saalfeld, H., *Tschermak's Mineral. Petrog. Mitt.* (1965) **10,** 225.
162. Löhn, J., Schulz, H., *Neues Jahrb. Mineral. Abt.* (1968) **109,** 201.
163. Bukin, V. I., Makarov, Ye. S., *Geochem. Intern.* (1967) **4,** 19.
164. Galitskii, V. Yu., Shcherbakov, V. N., Gabuda, S. P., *Sov. Phys. Cryst.* (1973) **17,** 691.
165. Galitskii, V. Yu., Shcherbakov, V. N., Gabuda, S. P., *Sov. Phys. Cryst.* (1974) **18,** 620.
166. Jarchow, O., *Z. Kristallogr.* (1965) **122,** 407.
167. Jarchow, O., Reese, H. H., Saalfeld, H., *Neues Jahrb. Mineral. Monat.* (1966) 289.
168. Brown, W. L., Cesbron, F., *C.R. Acad. Sci. Paris, Ser. D* (1973) **276,** 1.
169. Barrer, R. M., Cole, J. F., *J. Chem. Soc. A* (1970) 1516.
170. Barrer, R. M., Cole, J. F., Villiger, H., *J. Chem. Soc. A* (1970) 1523.
171. Barrer, R. M., Meier, W. M., *J. Chem. Soc.* (1958) 299.
172. Barrer, R. M., Freund, E. F., *J. Chem. Soc. Dalton Trans.* (1974) 1049.
173. Barrer, R. M., Marcilly, C., *J. Chem. Soc. A* (1970) 2735.
174. Liquornik, M., Marcus, Y., *Israel J. Chem.* (1968) **6,** 115.
175. Wyckoff, R. W. G., "Crystal Structures," 2nd ed., vol. 2, p. 362, John Wiley, New York, 1964.
176. Rabo, J. A., Poutsma, M. L., Skeels, G. W., *Catal. Proc. Int. Congr., 5th* (1972) **2,** 1353.
177. Yates, D. C. Y., *J. Phys. Chem.* (1965) **69,** 1676.
178. Gallezot, P., Imelik, B., *Advan. Chem. Ser.* (1973) **121,** 66.
179. Lewis, P. H., *J. Catal.* (1968) **11,** 162.
180. Egerton, T. A., Vickerman, J. C., *J. Chem. Soc. Faraday Trans.* (1973) **69,** 39.
181. Rabo, J. A., Angell, C. L., Kasai, P. H., Schomaker, V., *Disc. Faraday Soc.* (1966) 328.
182. Taarit, Y. Ben, Naccache, C., Che, M., Tench, A., *J. Chem. Phys. Lett.* (1974) **24,** 41.
183. Kasai, P. H., Bishop, R. J. Jr., *J. Amer. Chem. Soc.* (1972) **94,** 5560.
184. Barrer, R. M., Whitman, J. L., *J. Chem. Soc. A* (1967) 19.
185. Van Reeuwijk, L. P., *Mededelingen Landbouwhogeschool Wageningen, Nederland* (1974) **74,** 9.
186. Aiello, R., Barrer, R. M., *J. Chem. Soc. A* (1970) 1470.
187. Aiello, R., Barrer, R. M., Davies, J. A., Kerr, I. S., *Trans. Faraday Soc.* (1970) **66,** 1610.
188. Barrer, R. M., Langley, D. M., *J. Chem. Soc.* (1958) 3804, 3811, 3817.
189. Staples, L. W., Gard, J. A., *Mineral. Mag.* (1959) **32,** 261.
190. Naccache, C., Taarit, Y. Ben, *Chem. Phys. Lett.* (1971) **11,** 11.
191. Delgass, W. N., Garten, R. L., Boudart, M., *J. Chem. Phys.* (1969) **50,** 4603.
192. Barry, T. I., Lay, L. A., *J. Phys. Chem. Solids* (1966) **27,** 1821.
193. Delgass, W. N., Garten, R. L., Boudart, M., *J. Phys. Chem.* (1969) **73,** 2970.
194. Barrer, R. M., *Advan. Chem. Ser.* (1971) **102,** 1.
195. Sherry, H. S., *Advan. Chem. Ser.* (1971) **102,** 350.
196. Resing, H. A., Thompson, J. K., *Advan. Chem. Ser.* (1971) **101,** 473.
197. Resing, H. A., Murday, J. S., *Advan. Chem. Ser.* (1973) **121,** 414.

198. Barrer, R. M., Cram, P. J., *Advan. Chem. Ser.* (1971) **102,** 105.
199. Barrer, R. M., Clarke, D. J., *J. Chem. Soc., Faraday Trans. I* (1974) **70,** 535.
200. Ruthven, D. M., Loughlin, K. F., Derrah, R. I., *Advan. Chem. Ser.* (1973) **121,** 330.
201. Dyer, A., Salford, E., *J. Inorg. Nucl. Chem.* (1973) **35,** 3001.
202. Dyer, A., Townsend, R. P., *J. Inorg. Nucl. Chem.* (1973) **35,** 2993.
203. Maes, A., Cremers, A., *Advan. Chem. Ser.* (1973) **121,** 230.
204. Gallei, E., Eisenbach, D., Ahmed, A., *J. Catal.* (1974) **33,** 62.
205. Jansen, F. J., Schoonheydt, R. A., *J. Chem. Soc., Faraday Trans., I* (1973) **69,** 1338.
206. Jansen, F. J., Schoonheydt, R. A., *Advan. Chem. Ser.* (1973) **121,** 96.
207. Lechert, H., Habilitationsschrift, Universität Hamburg, 1973.
208. Pfeifer, H., Schirmer, W., Winkler, H., *Advan. Chem. Ser.* (1973) **121,** 430.
209. Mestdagh, M. M., Stone, W. E., Fripiat, J. J., *J. Phys. Chem.* (1972) **76,** 1220.
210. Freude, D., Oehme, W., Schmiedel, H., Staudte, B., *J. Catal.* (1974) **32,** 137.
211. Barrer, R. M., Wasilewski, J., *Trans. Faraday Soc.* (1971) **57,** 1140, 1153.
212. Egelstaff, P. A., Downes, J. S., White, J. W., "Molecular Sieves," p. 306, Society of the Chemical Industry, London, 1968.
213. Slater, J. C., Johnson, K. H., *Physics Today* (1974) **34.**
214. Meier, W. M., Olson, D. H., *Advan. Chem. Ser.* (1971) **101,** 155.
215. Encyclopaedia Britannica, 15th ed., 1974.

Chapter

2

Structural Analysis by Infrared Spectroscopy

Edith M. Flanigen, Union Carbide Corp., Linde Division, Tarrytown Technical Center, Tarrytown, N. Y. 10591

AMONG THE MOST widely utilized techniques for structural analysis of crystalline solids are x-ray and electron diffraction. X-ray powder diffraction and single crystal x-ray analysis have been applied extensively to zeolites and, more recently, electron diffraction has been used more frequently. During the last decade there has been increasing recognition in structural investigations of zeolites that infrared spectroscopy can yield information not only on short range bond order and characteristics but also on long range order in crystalline solids caused by lattice coupling, electrostatic, and other effects and can serve as a very rapid and useful structural technique.

An extensive literature exists on the mid-infrared spectra of silica, silicates, and aluminosilicates (*1*, *2*). Much of this has been devoted to the assignment of fundamental vibrations as in the early work of Lippincott *et al.* (*3*). However, many literature reports have applied structure–infrared relationships for silica and aluminosilicates to the solution of a variety of structural problems. These include identification of crystalline species, coordination number, degree of ordering, polymorphic transformation, isomorphic substitutions, study of solid state reactions, and, in general, structural transformations [*see*, for example, White and Roy (*4*), where other work is cited, and especially the reviews by Dutz (*1*) and Lazarev (*2*)].

Zhdanov *et al.* (*5, 6, 7*), Wolf *et al.* (*8, 9, 10*), Wright *et al.* (*11*), and Avdeeva and Vorsina (*12*) reported on the mid-infrared spectra of various zeolites, including zeolites A, X, and Y, and developed structural interpretations from the infrared spectra. A systematic study of the infrared–structure relationship of a large number of synthetic zeolites was carried out by Flanigen, Khatami, and Szymanski (*13*) based on an empirical correlation of the mid-infrared spectral features and the known crystal structures as determined by x-ray diffraction techniques. The treatment developed by these authors will serve as the main basis of discussion in this chapter. Emphasis will be on the application of their interpretation and correlation rather than any reassessment or revision of spectral assignments. The zeolites discussed will be limited in general to synthetic zeolites of particular relevance to catalysis.

Mid-Infrared Spectra of Zeolites

Techniques for Determination of IR Spectra of Zeolites. Several techniques for determining the mid-infrared transmission spectra for fine powders have been applied to zeolites, including the pressed pellet matrix or KBr pellet technique (*14*), mineral oil (Nujol mull) or other liquid mull (*14*), or the self-supported wafer technique (*15*). Synthetic zeolites which typically crystallize as micron-sized crystals or particles require no sample treatment such as grinding to avoid large particle type effects (*14*). Since only milligram quantities of sample are required, any of the techniques offers much advantage in terms of sample size, but care is required that such a small quantity is representative of the bulk zeolite sample. Homogeneity of sample should be established by duplicate or triplicate runs on different portions of sample. Mixing to homogenize is also advantageous, but high intensity mixing such as achieved in Wig-L-Bug type analytical homogenizers should be avoided since it can result in crystallographic degradation in the case of many zeolites (*16*). Indeed, extended hand grinding in an agate mortar and pestle can cause similar crystal degradation. The simplest, and most rapid and widely used experimental technique is that of the pressed pellet matrix where about 1 mg of zeolite powder is mixed in with ~300 mg of KBr or CsI powder and pressed under vacuum at 15,000 to 20,000 psig. Pressures greater than 20,000 psig applied to many synthetic zeolites can also cause crystallographic degradation (*17*). CsI offers some advantage over KBr in that the resolution in the 200 to 400 cm^{-1} region is improved. Matrix and ion exchange effects from diluent salt are known to occur (*14*). Cation exchange reactions with the salt matrix are especially facilitated with zeolites. Therefore, spectra should be checked in at least two matrix compositions. Spectra have been determined with several types of high resolution double beam grating spectrometers that cover the spectral region from 200 to 1300 cm^{-1}, *e.g.*, Perkin Elmer models 225, 521, or 621, and Beckman model IR-12. Resolution should be of the order of ± 5 cm^{-1} using the same technique with experimental care on the same instrument but can be as high as ± 10 cm^{-1} with other measurement variations. For the most precise quantitative work, an internal standard should be run with each spectrum, and exactly weighed quantities of zeolite and pellet matrix should be used. Two concentration levels should be chosen to give good resolution of both the strongest absorption bands (without bottoming out) and the medium-to-weak bands. Typical concentration ranges are 0.25 to 1 mg of zeolite in 300 mg of KBr. When spectral shifts are to be measured, the last several comments are especially important. Indeed, additional accuracy can be gained in the latter case by using slower scanning speeds and other well-known methods (*14*).

Attenuated total reflectance (ATR) techniques have been applied to zeolites (*18, 19*) but tend to give spectra of poorer quality than the transmission techniques described above. The technique of choice in terms of optimum analysis time and quality of spectra is the KBr pellet. (The possible advantage of CsI as halide matrix in the lower frequency spectral region was noted above.) Besides the procedural precautions cited above (*14*), additional problems may be encountered with zeolites. For example, the state of hydration of the zeolite should be known and kept constant, since spectral changes can occur on dehydration (*13*). Pressed pellets can be prepared by mulling in solvent with subsequent heating to remove solvent or by pellet pressing carried out in a vacuum press. Either technique can cause dehydration of zeolites. Heating in the IR beam during spectral measurement can also effect zeolite dehydration. Spectral measurements with zeolite/KBr pellets are limited to room temperature or lower temperatures since there is a high probability that cation exchange reactions will occur between the zeolite and halide matrix at higher temperatures, especially in the presence of adsorbed H_2O. In addition, salt occlusion phenomena can be encountered in the temperature range of 400°–500°C (*20*). Any desired thermal treatment should be carried out prior to blending and pelletizing.

The use of self-supported wafers (*15*) is limited to extremely thin wafers (1 to 5 mg) if sufficient transmission is to be obtained in the mid-infrared region. Even then, only the medium to weak bands can usually be resolved.

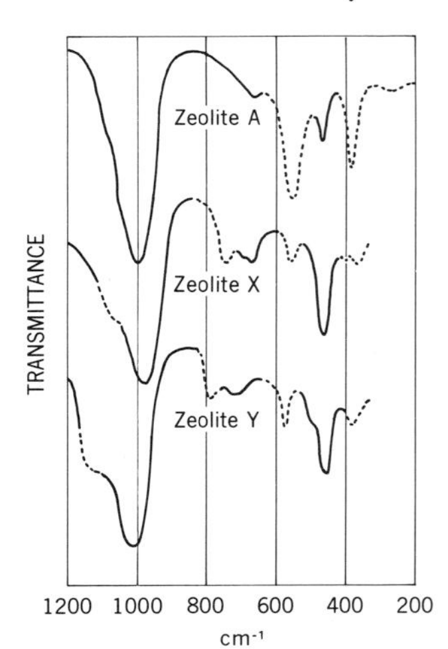

Advances in Chemistry Series

Figure 1. Infrared spectra of zeolites A, X, and Y. Si/Al in X is 1.2 and in Y, 2.5 (13).

Spectral Results. The typical mid-infrared spectra determined for a large number of hydrated synthetic zeolites as determined by Flanigen, Khatami, and Szymanski (*13*) are shown in Figures 1, 2, and 3. Table I contains the spectral frequencies, and Table II lists typical compositions of some synthetic zeolites (*13*). The spectra were determined using the pressed pellet matrix technique discussed previously with either KBr or CsI as the pellet matrix. Several investigators have determined the mid-infrared spectra of many of the mineral zeolites (*21, 22, 23*). Because the mineral zeolites are not of general interest in catalysis, they are not included in this discussion.

Spectral/Structure Correlations

Early Interpretations. The mid-infrared region of 200–1300 cm^{-1} contains the fundamental vibrations of the Si,AlO_4 or TO_4 units in all zeolite frameworks and, therefore, may be expected to contain useful information on the structural characteristics of the zeolite frameworks. Chapter 1 of this volume contains a comprehensive review of zeolite structures and details of structural nomenclature.

Zhdanov *et al.* (*5, 6, 7*) applied mid-infrared spectroscopy to a study of the framework of a series of X and Y zeolites as a function of framework Si/Al ratio, cation type, and state of hydration. In the region of framework vibration investigated, 400–800 cm^{-1}, they identify two types of vibrations, the first related to individual bonds within the Al,SiO_4 tetrahedra, and the second to the tetrahedron as a whole. The zeolite spectra are interpreted by

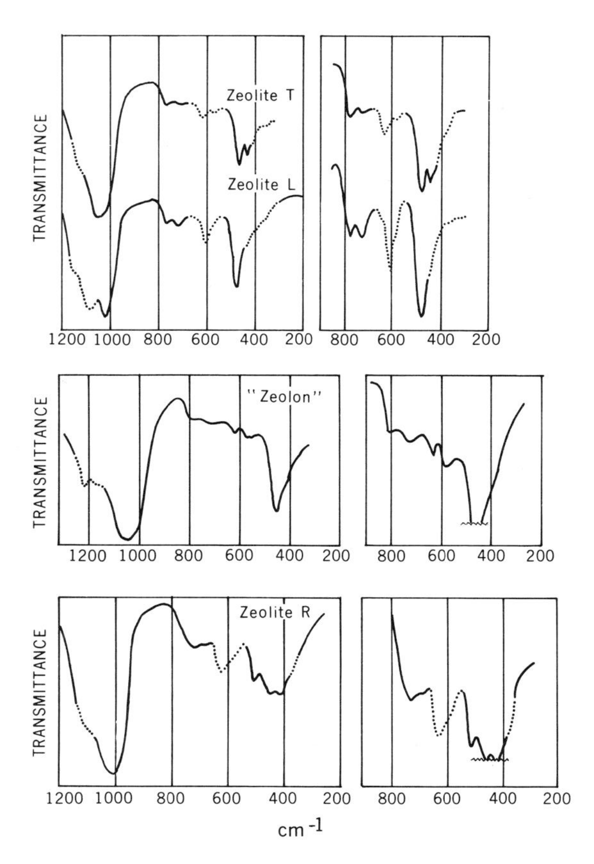

Advances in Chemistry Series

Figure 2. Infrared spectra of zeolites T, L, R and Zeolon (13). *Right portion represents a higher zeolite concentration in the pellet than in left portion.*

Table I. Infrared Spectral Data for

Zeolite	$\frac{SiO_2}{Al_2O_3}$	*Asym. Stretch*			*Sym. Stretch*			
A	1.88	1090vwsh	1050vwsh	995s			660wv	
X	2.40		1060msh	971s	746m	690wsh	668m	
Y	3.42	1135msh		985s	760m		686m	
Y	4.87	1130msh		1005s	784m		714m	635vw
Y	5.63	1130msh		1017s	789m		718m	645vw
B(P)	2.8	1105mwsh		995-1000s	772mwsh	738mw	670mw	
Ω	7.7	1130wsh		1024s	805mw	772mw		
R	3.25	1136mwsh		1007s	738w	678w		
G	5.44	1138mwsh		1027s	720w	696wsh		
S	2.5	1140wsh		1020s	770vwsh	722mw	690vwsh	
T	7.0	1156wsh	1059s	1010s	771w	718w		
L	6.0	1160wsh	1080s	1015s	767mw	721mw		642wsh
Zeolon	9.95	1216w	1180vwsh	1046s	795, 772 } wb	715, 690 } wb		

[a] s = strong; ms = medium strong; m = medium; mw = medium weak; w = weak; vw = very weak; sh = shoulder; b = broad.

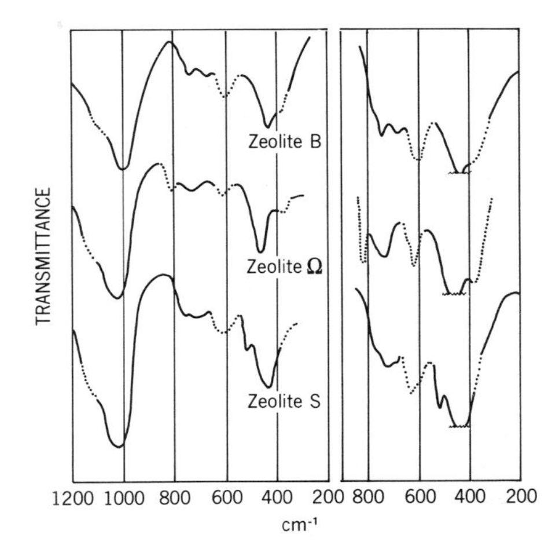

Advances in Chemistry Series

Figure 3. Infrared spectra of zeolites B (P1), Ω, and S (13). *Right portion represents a higher zeolite concentration in the pellet than in left portion.*

comparison with the spectra of quartz and aluminosilicates. In NaX zeolite, bands at 405, 465, 683, and above 920 cm^{-1} are related to similar absorption bands in quartz; the band near 760 cm^{-1} is assigned to an Al—O vibration commonly found in other aluminosilicates. Bands near 568, 610, and 758 cm^{-1} were found to be sensitive to changes in framework Si/Al composition. With increasing Si content, the intensity of the band near 568 cm^{-1} decreased and shifted to higher frequency. The framework vibrations were sensitive to cation type and charge, especially the band near 760 cm^{-1} assigned to an

Synthetic Zeolites (cm^{-1} [a]) (*13*)

Dbl. Rings			*T–O Bend*			*Pore Opening?*		
	550ms			464m			378ms	260vwb?
	560m			458ms		406w	365m	250vwb?
	564m		508vwsh	460ms			372m	
	572m		500wsh	455ms			380m	260vwb?
	575m		504mwsh	456ms			383m	315vwsh
600m					435ms		380mwsh	
610mw				451ms			372m	
625m		508mw		452m	426m		370vwsh	
632m		515m		460m	408m		378vwsh	
623, 595sh } mb		518mb		448m	424ms		370vwsh	
623mw	575w			467ms	433ms	410vwsh	366wsh	
606m	580wsh			474ms	435wsh		375vwsh	
621w	571, 555 } w			448ms			370vwsh	

Advances in Chemistry Series

Al—O vibration. All of the spectra reported by Zhdanov *et al.* (*5, 6, 7*) were determined on self-supported wafers after dehydration in vacuum at 400°C.

In a systematic study of the infrared spectra of zeolite A and several other synthetic and mineral zeolites in the mid-infrared region, Wolf and Fuertig (*8*) assigned a structure specific band in the region 545 to 630 cm^{-1} to a chain frequency of alternating SiO_4 and AlO_4 tetrahedra present in the crystal lattice. The chain frequency band was characteristic of all of the synthetic and natural zeolites studied, and its frequency depended upon Si/Al ratio and crystal structure of the zeolite. Decrease in intensity of this band accompanied crystallographic decomposition. [The chain frequency band in

Table II. Compositions of Synthetic Zeolites (*13*)

Zeolite	*Cation* [a]	*SiO_2/Al_2O_3*
A	Na	2
X	Na	2.0–3.0
Y	Na	$>$ 3–6
B (Pl)	Na	2–5
Omega (Ω)	Na, TMA [b]	5–12
S	Na	4.6–5.9
R	Na	3.5–3.7
G	K	2–6
T	K, Na	6.4–7.4
L	K; K, Na	5.2–6.9
Zeolon	Na	10–11

[a] Cation composition as synthesized.
[b] TMA = tetramethylammonium ion $(CH_3)_4N^+$.

Advances in Chemistry Series

general corresponds to the double ring assignment in the treatment of FKS, *vide infra,* Table IV.] In a later study of mordenite-type zeolites by Wolf *et al.* (*10*), comparison with the spectra of zeolites A, X, and Y showed an increase in the chain frequency with increasing Si/Al (565 cm^{-1} in zeolite A to 635 cm^{-1} in mordenite).

Wright *et al.* (*11*) determined the framework infrared spectra in the region of 300–1200 cm^{-1} for a series of type X and Y zeolites with varying Si/Al ratios. They assigned bands near 1140 cm^{-1} to a symmetric Si—O—Si stretching mode, 1075 cm^{-1} to a symmetric Si—O—Al stretching mode, the strongest band near 1000 cm^{-1} to both Si—O—Si and Si—O—Al asymmetric stretching modes, which because of strong coupling of adjacent groups results in an averaging effect rather than discrete bands. The shoulder near 505 cm^{-1} found in NaY zeolite, but absent in NaX, is assigned to an Si—O—Si out-of-plane bending mode. The remainer of the bands in the region of 350–800 cm^{-1} are attributed to various more or less strongly coupled bending modes. In addition, they showed a linear correlation of frequency with mole fraction of aluminum in the framework for three absorption bands near 1000, 770, and 570 cm^{-1}.

Avdeeva and Vorsina (*12*) reported differences in the IR spectra in the Si,Al—O framework region of 400–1200 cm^{-1} among NaA zeolite, sodalite, and cancrinite-type phases which can be used to identify those structural types in synthesis products.

Later Shikunov *et al.* (*24*) in a study of the framework vibrations of synthetic zeolites A, X, Y, erionite, and mordenite assign the band near 770–800 cm^{-1} to an Si—O bond, and bands in the region 680 to 780 cm^{-1} to an Al—O vibration. The bands in the 540 to 640 cm^{-1} region are as-

Table III. Structural Characteristics

Zeolite	*Structure Type/Group*	*Ideal Unit Cell Symmetry*
A	A/faujasite	Cubic, Pm3m (pseudo)
X, Y	Faujasite/faujasite	Cubic, Fd3 / Fd3m
B(Pl)	B(Pl)/phillipsite	Tetragonal, $\bar{I4}$, near cubic
Ω	Ω/chabazite	Hex, P6/mm
S	Gmelinite/chabazite	Hex, $P6_3$/mmc
R, G	Chabazite/chabazite	Trigonal, $R\bar{3}m$
T	Offretite-erionite/chabazite	Offr: Hex., $P\bar{6}m2$; Erion: Hex., $P6_3$/mmc
L	L/chabazite	Hex., P6/mmm
Zeolon	Mordenite/mordenite	Ortho., Cmcm

[a] See Table IV for definition. Ideal size and symmetry of polyhedral units: D-4, hedra, D_{3h}; T.O., 24 tetrahedra, T_d; Gmel., 24 tetrahedra, D_{3h}. S = single, D = S-4 = single ring of 4 tetrahedra, D-6 = double 6 ring, S-8 = single 8 ring.

signed to chainlets of tetrahedra in the Si,Al—O skeleton following the chain frequency assignment of Wolf and Fuertig (*8*). The chainlet vibration was specific to the type of skeleton and the Si/Al ratio. Vibrations near 440 to 460 cm^{-1} are assigned to deformation vibrations of Si—O bonds in the SiO_4 tetrahedron.

Correlation of FKS. Flanigen, Khatami, and Szymanski (*13*), abbreviated FKS, further extended the structure/IR spectra interpretations for a large number of synthetic zeolites. Their interpretation of the mid-infrared spectra was based on an empirical correlation of the known framework structures of synthetic zeolites as determined by x-ray diffraction techniques and the IR spectral features. Each zeolite had a typical IR spectrum, and spectral specificity was observed for zeolites with the same structure type and structure group and containing the same types of structural subunits such as double rings, polyhedral groupings of tetrahedra, and large pore openings. According to the FKS treatment, the mid-infrared vibrations are classified into two types of vibrations, those related to the internal vibrations of the TO_4 tetrahedra or the primary building unit in zeolite frameworks, which tend to be insensitive to variations in framework structure, and vibrations primarily related to external linkages between tetrahedra, which are sensitive to the framework topology and to the presence of symmetrical clusters of tetrahedra in the form of larger polyhedra. No vibrations were specifically assigned to SiO_4 groups or AlO_4 groups, but rather to the vibrations of TO_4 groups and T—O bonds where the vibrational frequencies represent the average Si,Al composition and bond characteristics of the central T cation.

The structural characteristics and classification of zeolite frameworks have been reviewed comprehensively by Smith in Chapter 1 of this volume.

of Synthetic Zeolites (*13*)

Building Units[a]			
S-R	*D-R*	*Pore Opening*	*Polyhedra*
4,6,8	D-4	S-8, planar	T.O.
4,6	D-6	S-12, nonplanar	T.O.
4,8	—	S-8, nonplanar	—
4,6,8	D-12	D-12, planar	Gmel.
4,6,8	D-6	S-12, nonplanar	Gmel.
4,6,8	D-6	S-8, nonplanar	—
4,6,8	D-6	S-12, nonplanar	Cancr., Gmel.
4,6,8	D-6	S-8, nonplanar	Cancr.
4,6,8	D-6	S-12, planar	Cancr.
4,5,8	—	S-12, nonplanar	—

Advances in Chemistry Series

8 tetrahedra, cube, T_d; D-6, 12 tetrahedra, hexagonal prism, D_{6h}; Cancr., 18 tetra-
double, R = ring (of tetrahedra); 4, 6, etc., = number of tetrahedra in ring; *e.g.*,

Table IV. Building Units in Zeolite Structures[a]

Primary Building Unit—Tetrahedron (TO_4)
Tetrahedron of 4 oxygen ions with a central tetrahedral ion (T) of Si^{4+} or Al^{3+}
All oxygens shared between 2 tetrahedra, $(TO_2)_n$
Secondary Building Units, SBU
Rings, S-4, 6, 8,
Double rings, D-4, 6, 8
Larger Symmetrical Polyhedra[b]
Truncated octahedron (T.O.) or sodalite unit
11-Hedron or Cancrinite (Cancr.) unit
14-Hedron II or gmelinite (gmel.) unit
Zeolite Structure
Packing of SBU's and polyhedra in space

[a] *See* Chapter 1 of this volume and Ref. *13* for additional description.
[b] Larger polyhedral units (> 24 tetrahedra) are not considered here because they are not believed to be important in determining mid-infrared spectral characteristics.

Advances in Chemistry Series

In addition, a summary of the structural characteristics used here is given in Table III for a selected number of synthetic zeolites and framework building units are described in Table IV.

Zeolite structures in this interpretation are classed according to their primary building unit which is the tetrahedral configuration of four oxygens surrounding a central cation of Si or Al, secondary building units (SBU) consisting of rings of tetrahedra and double rings of tetrahedra, and larger symmetric polyhedra containing up to 24 tetrahedra such as the truncated octahedron (T.O.) or sodalite unit. The zeolite framework structure results from the packing of the SBU's in space. In terms of framework volume and space filling, the structure can be more properly represented as a packing of oxygen ions (ionic radius r, for $O^{2-} = 1.35$ A, r for $Si^{4+} = 0.26$ A, r for $Al^{3+} = 0.39$ A).

A summary of the infrared assignments are contained in Table V and illustrated for NaY zeolite in Figure 4. In the spectra shown (*13*), the internal tetrahedral linkages are drawn in full line, and the structure-sensitive external vibrations with broken lines.

Table V. Zeolite Infrared Assignments, Cm^{-1} (*13*)

Internal Tetrahedra	
Asym. stretch	1250–950
Sym. stretch	720–650
T–O bend	420–500
External Linkages	
Double ring	650–500
Pore opening	300–420
Sym. stretch	750–820
Asym. stretch	1050–1150 sh

Advances in Chemistry Series

The assignment of class of vibrations was chosen following the assignments of Lippincott *et al.* (*3*). Since that assignment, de Kanter *et al.* (*25*) have pointed out, from comparison of the IR with the Raman spectra of zeolites, that the assignment of the medium intensity band in the region of 480–500 cm^{-1} common to all zeolites to a TO_4 bending vibration, ν_4 (T_2), and the assignment of the symmetric stretching modes, ν_1 (A_1), in the region of 650–720 cm^{-1}, may not be correct. Raman spectra of zeolites published (*25, 26, 27*) subsequent to the original IR interpretation of FKS show that in all zeolites the intensity of bands in the ν_1 symmetric stretching region are consistently orders of magnitude lower than those in the ν_4 bending region. deKanter *et al.* (*25*) point out that one would normally expect more relative intensity from symmetric modes in Raman spectra. Since those authors did not suggest alternate assignments, and since there has been no rigorous spectroscopic treatment of band assignment in zeolites, the tentative assignments presented in Table V will be used for this discussion. Several authors prefer to assign the band at about 730 cm^{-1} to a pure isolated AlO_4 tetrahedron rather than a T—O symmetric stretch as in Table V. Pichat, Beaumont, and Barthomeuf (*28*), in a study of dealuminized forms of Y zeolite, show that

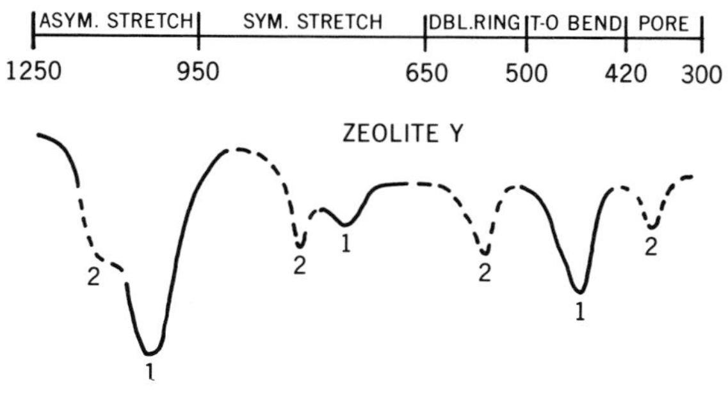

Advances in Chemistry Series

Figure 4. Infrared assignments illustrated with the spectrum of zeolite Y, Si/Al of 2.5 (13)

the integrated intensity of that band decreases linearly with decreasing Al content in this series of modified Y zeolites. However, in reviewing the intensity of this band among all of the zeolite spectra published and their Al content, a generalization of that assignment for all zeolites does not seem warranted until further data on other zeolite structures confirm a reassignment to an Al specific frequency. Indeed, the intensity of this band in the spectra for synthesized sodium forms of zeolites X and Y with varying aluminum contents shown in Figure 5 also does not tend to support the AlO_4 assignment.

The presence of a medium to medium-strong band near 500 to 650 cm^{-1} was assigned to the presence of double rings of tetrahedra. This is especially evident in the spectra of zeolites A, faujasite-, chabazite-, and gmelinite-type, and offretite/erionite-type frameworks. Zeolites devoid of double rings show bands in this region of medium-weak to weak intensity. The assignment is inconsistent with the presently accepted structures for zeolite B(P) and zeolite Ω. Those structures do not contain double rings, yet their infrared spectra contain medium bands in the 600 to 620 cm^{-1} region.

The tentative assignment of bands near 300–420 cm^{-1} to the larger pore openings (rings of 8 or more tetrahedra) is less certain. These bands are clearly evident in the case of zeolites A, X, Y, chabazite, and Ω, but are much less prominent in erionite/offretite and in the large pore zeolites L, and the large pore mordenite Zeolon.

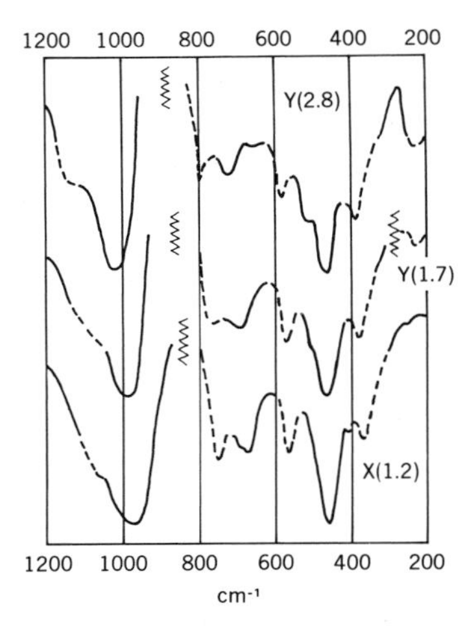

Advances in Chemistry Series

Figure 5. Infrared spectra for zeolites X and Y with different Si/Al contents; numbers in parenthesis refer to Si/Al in the zeolite (13). *Right portion represent a higher zeolite concentration in the pellet than in left portion.*

A weak band near 250–300 cm^{-1} was considered to be related to the presence of the larger symmetrical polyhedra building units in some zeolite frameworks such as the truncated octahedron, cancrinite, and gmelinite units. Vibrations of such large groups (18 to 24 tetrahedra, ~8 A in cross section) might be expected to appear well below 300 cm^{-1} and perhaps be detected in the far infrared region (*13*). The assignment has not been substantiated. Although bands in the region of 300 cm^{-1} appear in the Raman spectra of zeolites (*25, 26, 27*) which were not given any structural interpretation, there is no obvious correlation of these bands with the polyhedral units.

In interpreting zeolite framework structures, it is important that the overall pattern in the spectrum be considered as well as the building units discussed above. Slight structural differences can often be detected by changes in band symmetry and band shifts. Common types and classes of zeolites have similar spectral characteristics. In general, the spectra tend to be simple, with relatively few main vibrations resembling the IR spectra of silicate glasses. This indicates a high degree of lattice coupling of the tetrahedral vibrations in the framework. The uniqueness and structure specificity of the infrared spectrum of zeolites increases with increasing openness of the framework and increasing unit cell symmetry. deKanter *et al.* (*25*) suggest that the broad bands or the observed absorption profiles for zeolite framework vibrations in both the IR and the Raman are an envelope of unresolved vibrations from SiO_4 and AlO_4 tetrahedra. Application of a duPont model 310 Curve Resolver to the IR spectrum of NaY (*19*) and other synthetic zeolites shows clearly that the relatively simple broad bands are indeed envelopes of bands, and that instead of the apparent 9 to 10 bands present, for example, in the IR of NaY between 200–1200 cm^{-1}, the spectrum contains at least 18 to 20 absorption maxima resolved with good fit (but with some degree of arbitrariness) with the curve resolution technique used. However, the envelope may not necessarily be related to differentiation of AlO_4 and SiO_4 tetrahedra, but may alternately be caused by additional vibrational modes.

The structure–infrared relationships drawn where specific frequency regions were correlated with structural groups in the framework is not intended to represent an assignment of vibrational mode or species of the structural group. The problems associated with the application of factor group and other types of spectral analysis techniques to zeolites are formidable (*13*). In addition, because of the large number of atoms in zeolite unit cells, the number of calculated vibrations will always exceed the number of observed vibrations by manyfold, and coincidence of an observed infrared vibration with a calculated vibrational mode cannot serve as a basis of assignment of spectral species. The publication of the Raman spectra of zeolites has added further information to serve as the basis of spectral assignments. However, the structure specific IR vibrational modes are, in general, absent in the Raman. Indeed, the general weakness or absence of Raman bands in the 500–1200 cm^{-1}, and the absence of any significant specificity to framework structure type suggests that Raman spectroscopy may be a less sensitive tool in solving zeolite framework structural problems.

The classification of bands into structure sensitive external linkages and structure insensitive internal vibrational modes of single tetrahedra appears to be valid (*13*) and has been substantiated by other investigators. This classification is analogous to that utilized in molecular crystals or complex ionic crystals where vibrations are classified as intramolecular and intermolecular (*29, 30*). Zeolite spectra represent an interesting and perhaps unique combination of many of the spectral features of molecular crystals but contain more ionic bonding and complex ions or clusters as subgroups in the crystal. In typical molecular crystals, there is usually a clear distinction between the intramolecular and lattice vibrations because the forces connecting atoms within a molecule are usually stronger than forces acting between molecules. However, in the infinitely extended crystalline anionic framework of a zeolite, both intramolecular vibrations and intermolecular or lattice-type vibrations assume prominence in the spectral features. In molecular crystal treatment, not only translational vibrations but also librational vibrations are considered and assume prominence.

Lazarev gives an excellent and comprehensive review and discussion of the vibrational spectra of silicates (*2*). He notes that interpretation of the vibrational spectrum of crystals containing complex ions is usually semiempirical in character. In treating the vibrational spectrum of such crystals, Lazarev classified vibrations into two groups, external and internal, analogous to the classification of FKS. The external vibrations correspond to rotational and translational motions of complex ions in the crystal lattice. When the interaction between groups of complex ions in the lattice is weak, such vibrations are low in frequency; the frequencies of the internal vibrations are usually much higher. However, when there are strong interactions among complex ions in the crystal lattice, as is the case with zeolites, the external or lattice vibrations may occur in the same region as the internal vibrations. Lazarev further suggests that combinations of internal and external lattice vibrations, *e.g.*, translational, vibrational and acoustic modes, may assume prominence in the vibrational spectrum of crystals containing complex ions, which are normally inactive in the form of fundamental vibrations.

The structure specific bands observed in the infrared spectra of zeolites from 250 to 1300 cm^{-1} are probably related to movements of groups of oxygens, or TO_4 tetrahedra, which are librational or translational modes, lattice vibrations normally found at frequencies less than 300 cm^{-1} occurring at higher frequencies, or combination vibrations as suggested by Lazarev (*2*). The effect of the large void volumes present in the zeolites on the spectral characteristics has been qualitatively correlated (*13*) but not considered theoretically. Electrostatic interactions of the framework and the cation-framework, as pointed out by de Kanter *et al.* (*25*) and Maxwell and Baks (*27*), also play an important role. Until a detailed rigorous spectroscopic analysis of zeolites is attempted, the original FKS spectral assignments (*13*) appear to be adequate to apply mid-infrared spectroscopy to zeolite structural problems. Some of these interesting applications are discussed in the rest of the chapter.

Other Applications of Mid-Infrared to Zeolite Structural Problems

Framework Composition. The frequency shift of infrared stretch bands with variation of Si/Al ratio in the tetrahedral frameworks of a wide variety of tectosilicates has been reported (*21, 31, 32, 33*). In all cases, a nearly linear decrease was shown for the position of the main asymmetric stretch band near 980 to 1100 cm^{-1} with increasing atom fraction of Al in the tetrahedral sites. Similar shifts for other T—O stretch bands were also observed. Since the mass of Al and Si are nearly the same, the decrease in frequency with increasing Al concentration appears to be related to variation in bond length and bond order. The longer bond length of Al—O and decreased electronegativity of Al result in a decrease in force constant.

Similar shifts in several bands in the mid-infrared have been reported for zeolites (*5, 6, 11, 13*). A correlation of shift in the main asymmetric stretch band for all synthetic zeolites is shown in Figure 6 (*13*). Although a linear relationship was found (solid line) with a slope of 18.1 cm^{-1}/0.1 atom fraction of Al, there is considerable scatter, suggesting some structure-sensitive characteristics in the main asymmetric stretch band in synthetic zeolites. The dashed line in Figure 6 is the best line through the points when

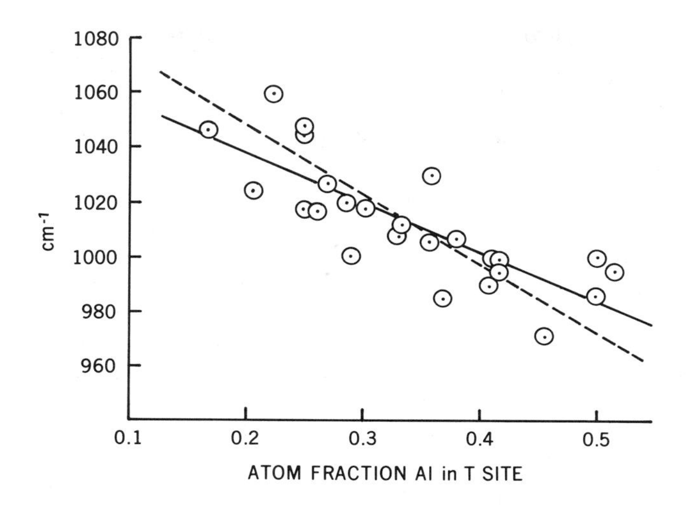

Advances in Chemistry Series

Figure 6. Frequency of the main asymmetric stretch band vs. *the atom fraction of Al in the framework for all synthetic zeolites* (13)

the y intercept is 1100 cm^{-1}, the average main stretch frequency observed for pure silica frameworks *(3)*.

Frequency shifts with Si/Al ratio for other classes of vibrations have been found for synthetic zeolites with faujasite-type frameworks *(5, 6, 11, 13)* and A-type frameworks *(5, 6, 13)*. The spectra of zeolites X and Y with varying Si/Al ratio are shown in Figure 5 and plots of frequency *vs.* fraction of Al in the framework for several classes of vibrations are shown in Figure 7 *(13)*. A linear decrease in frequency with increase in fraction of Al is shown

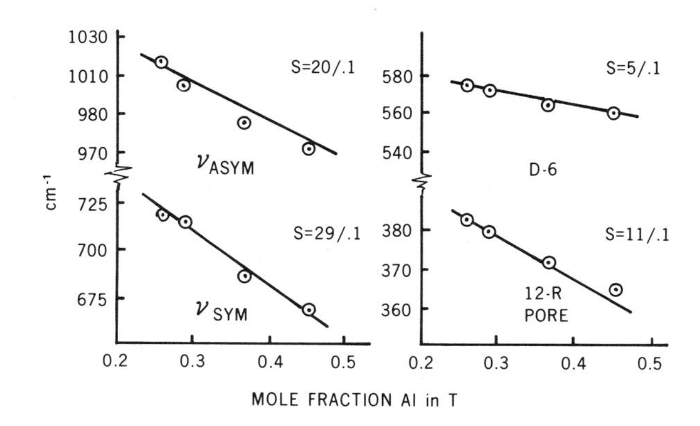

Advances in Chemistry Series

Figure 7. Frequency vs. *atom fraction of Al in the framework for zeolites X and Y for several infrared bands.* $S = cm^{-1}/0.1$ *mole fraction Al* (13).

for the main asymmetric stretch band (970 to 1020 cm^{-1}), a symmetric stretch band (670–725 cm^{-1}), the D-6 ring band (565–580 cm^{-1}), and the 12-R pore opening band (360–385 cm^{-1}). The T—O bending frequencies near 450 to 460 cm^{-1} did not vary. Since bend frequencies tend to be less sensitive to T—O distance, the latter observation further suggests that the IR band near 450–500 cm^{-1} is a fundamental bending mode. The data are in very good agreement with those of Wright *et al.* (*11*). (Note an apparent plotting error in the number legend on the *y* axis in their Figure 3; band B increments should vary from 750 to 790 cm^{-1} rather than the 770 to 810 cm^{-1} shown based on comparison of their Figure 3 with their Figure 2.) However, the absolute frequencies given by Wright *et al.* are displaced a constant ~5 to 8 cm^{-1} higher than the data in Figure 7. Such differences are well within the precision of measurement and may be related to difference in measurement technique

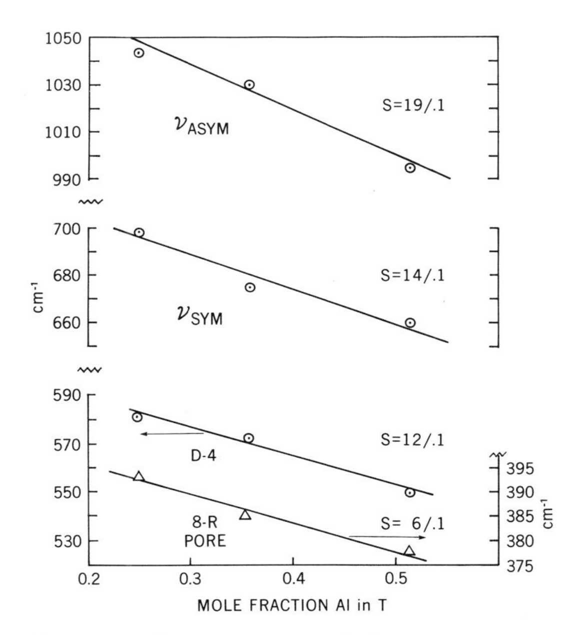

Figure 8. Frequency vs. *mole fraction Al in the framework of zeolites A and N–A for several infrared bands; S = cm⁻¹/0.1 mole fraction Al* (19)

(Wright *et al.*, Nujol Mull; Flanigen *et al.*, KBr pellet). Zhdanov *et al.* (*5, 6*) noted shifts in frequency of the framework infrared with varying Si/Al for a series of X and Y zeolites. Their data were, however, determined on self-supported wafers which were dehydrated in vacuum before determination of the spectra. Since dehydration affects shifts in bands (*13*), their results cannot be compared directly to those cited above.

The quantitative relationship between IR frequency and Si/Al ratio shown for the faujasite-type zeolites X and Y, will probably differ for other zeolite structure types. Shifts in frequencies of several bands for zeolites with A type structure and varying Si/Al ratio are shown in Figure 8 (*19*). The slopes for all but the main asymmetric stretch band differ significantly from those of X and Y zeolites.

Shikunov *et al.* (*24*) also reported spectral shifts as a function of zeolite Si/Al ratio. From the IR spectra of the synthetic zeolites NaA, NaX, NaY, Na,K-erionite, and Na-mordenite, the authors drew a quantitative relationship between the position of the main asymmetric stretch band near 1000 cm^{-1} and zeolite Si/Al ratio. Their reported frequencies are consistently lower than those cited previously since they are based on the inflection point of the band rather than the absorption maxima.

Several authors have extended the quantitative correlation of infrared spectral shift to determine the Al content of zeolite frameworks modified by thermal or chemical treatment to remove Al, including stabilized and aluminum-deficient forms of zeolite A and zeolite Y. Lahodny-Sarc and White (*34*) studied the removal of Al from NaA, NaX, and NaY zeolite by treatment with H_2CO_3 or ethylenediaminetetraacetic acid (EDTA) and, from the shifts in the main asymmetric band near 1000 cm^{-1}, assigned an Al content to the aluminum-deficient frameworks. Pichat, Beaumont, and Barthomeuf (*35, 36*) showed a similar linear variation of IR frequency shifts for aluminum-deficient forms of Y zeolite prepared by EDTA extraction of aluminum from ammonium exchanged forms. These authors found good agreement in the slopes of spectral shift with Al content for the aluminum-deficient Y samples compared to the synthesized NaX and NaY shown in Figure 7. This would suggest that the linear increase in IR frequency with decreasing framework aluminum content in X- and Y-type zeolites is the same for samples obtained by direct synthesis (Na forms) and for chemically dealuminated samples. Similar applications of IR spectral shifts to interpret framework Al content in stabilized forms of zeolites have been reported by Eberly *et al.* (*37*), Peri (*38*), Scherzer and Bass (*18*), Kubasov *et al.* (*39*), and Pichat *et al.* (*36*). These will be discussed in detail below.

Some caution should be exercised in quantitatively applying spectral shifts to Al content in aluminum-deficient and stabilized zeolite framework modifications. Spectral changes from decationization, tetrahedral substitution by $[H_4O_4]$ groups, the presence of structural defects and cationic aluminum species, all well-recognized as present in this class of zeolites (*40*), must be carefully weighed in any interpretation. A more straightforward interpretation should result by limiting the spectral shift interpretation of zeolite framework composition to synthesized zeolites with the same cation species and in the fully hydrated state.

Flanigen and Grose have used mid-infrared spectroscopy to characterize the framework composition of phosphate zeolites and establish proof of phosphorus substitution (*41*). Because of the shorter tetrahedral P—O bond distance of 1.54 A, compared to a T—O distance of 1.61 A for Si—O and 1.75 A for Al—O, a shift to higher frequencies for the main asymmetric stretch band was observed as a function of phosphorus substitution in zeolite frameworks. In principle, IR frequency shifts can generally be applied to proofs of frame-

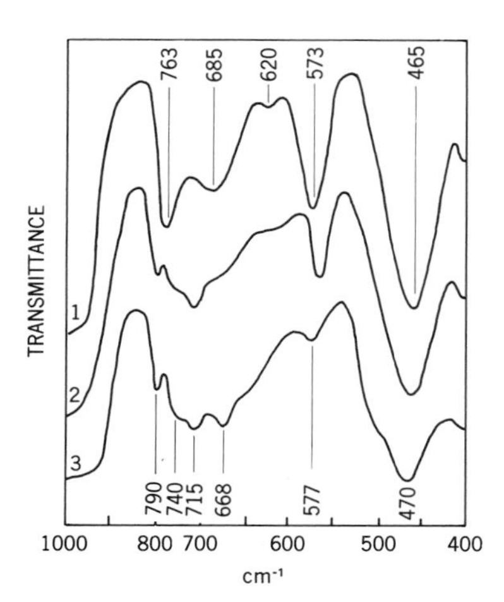

Russian Journal of Physical Chemistry

Figure 9. Framework infrared spectra for zeolite X with different exchangeable cations after vacuum treatment at 400°C for four hours. NaX–curve 1, SrX–curve 2, CaX–curve 3 (5).

work substitution with tetrahedral cations other than Si, Al, and P, as long as sufficient information exists on tetrahedral bond distance and bond order for the cation substituent and the concentration in the framework is high enough to be detected in the infrared (probably $\geq$ 5 wt % as oxide).

Cation Type and Cation Sites. The nature and crystallographic sites of cations in zeolites have been reported to be reflected in the mid-infrared, far-infrared, and Raman spectra of zeolites. Zhdanov *et al.* (5) showed the strong sensitivity of framework IR vibrations with cation-type and charge

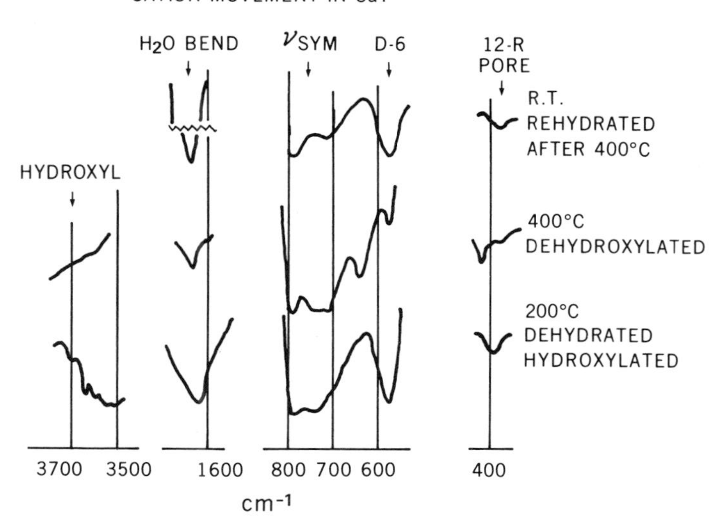

Advances in Chemistry Series

Figure 10. Infrared spectra for Ca-exchanged Y zeolite (Si/Al of 2.5) after dehydration, dehydroxylation, and rehydration (13)

for a series of Na, Sr, Ca-ion exchanged forms of zeolite X in the *dehydrated* state (self-supported wafers, evacuated at 400°C). Their data are shown in Figure 9. Flanigen *et al.* (*13*) noted that for the limited number of cation exchanged forms of A, X, and Y zeolite studied, where the spectra were determined on fully hydrated zeolites in KBr pellets, there was no significant change in spectral characteristic upon cation exchange. The exchanged forms included Ca, La, and NH_4. Dehydration of the synthesized forms of zeolites A, X, Y, L, and Ω, all containing principally alkali metal cations, showed only minor spectral changes. However, the same authors reported significant changes in the spectra of dehydrated zeolites in polyvalent cation form, such as CaY, as shown in Figure 10. The spectral changes were interpreted as cation movement or resiting of cations as a result of dehydration, dehydroxylation, and rehydration reactions. Using the assignments shown in Table V, those authors interpreted the changes in the D-6 band near 570 cm^{-1}, the symmetric stretch band near 710–750 cm^{-1}, and the pore opening band at 390 cm^{-1} upon dehydration and dehydroxylation, in terms of migration of Ca^{2+} cations from inside of the sodalite unit into a position near the center of the D-6 ring (Site I). They reported analogous spectral changes for La-exchanged Y zeolite.

Kubasov *et al.* (*42*) studied the spectral characteristics of the framework for hydrogen, lanthanum, and yttrium cation forms of zeolite Y after thermal treatment in air at 550°C and related the spectral changes to catalytic activity in cumene cracking. Prolonged calcination of the HY form caused loss in catalytic activity and changes in the IR spectra indicative of crystalline decomposition. They noted that the 580 cm^{-1} band (D-6 ring) is most sensitive to the degree of conversion of zeolite into the amorphous form. Exchanging La^{3+} into HY causes a shift in the D-6 ring vibration from 593 cm^{-1} to 581 cm^{-1}. Direct exchange of yttrium and lanthanum with NaY shows little spectral differences between the yttrium exchanged and initial NaY after heating in vacuum at 400°C, but high levels of lanthanum exchange under the same conditions of activation shift the 580 and 790 cm^{-1} bands toward lower frequency. The changes are interpreted in terms of lanthanum ions being located in sites near or within the hexagonal prism (D-6 ring) in Y, whereas yttrium ions are not normally located at these sites. The increased crystal stability of the lanthanum decationated forms as evidenced from an intense IR band at 580 cm^{-1} correlated with high catalytic activity. They conclude that the IR spectra of zeolite Y catalysts in the regions of 580 cm^{-1} and 790 cm^{-1} serve as a criterion for retention of crystallinity and as an indicator of positions of cations in the structure.

Dyrkheev *et al.* (*43*) in studies of the dehydration of CaX by thermal evacuation, similarly interpreted IR spectral changes in the framework region as related to dehydration phenomena at low temperatures. Higher temperature changes were attributed to the migration of Ca^{2+} ions to new sites.

The effect of cation exchange on the mid-infrared spectra of NaA zeolite for a large number of cationic species was studied by Wolf and Fuertig (*8*). In almost all cases, the cation forms were activated at 450°C before the

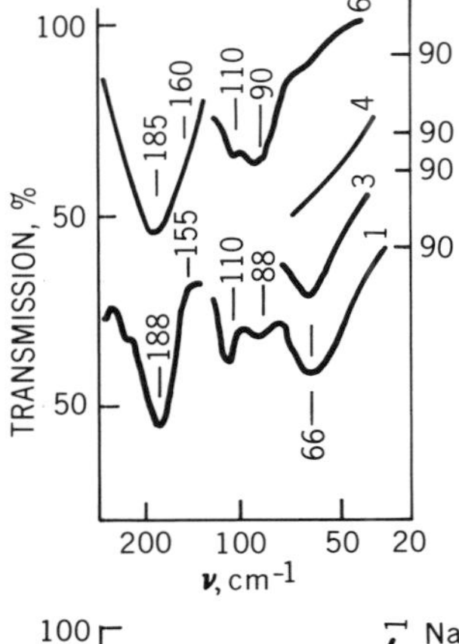

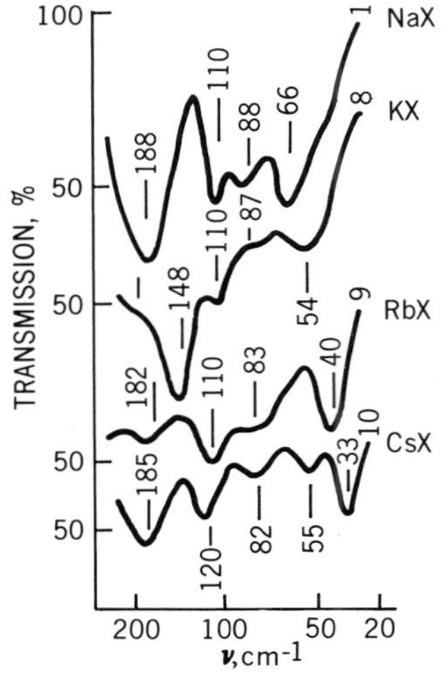

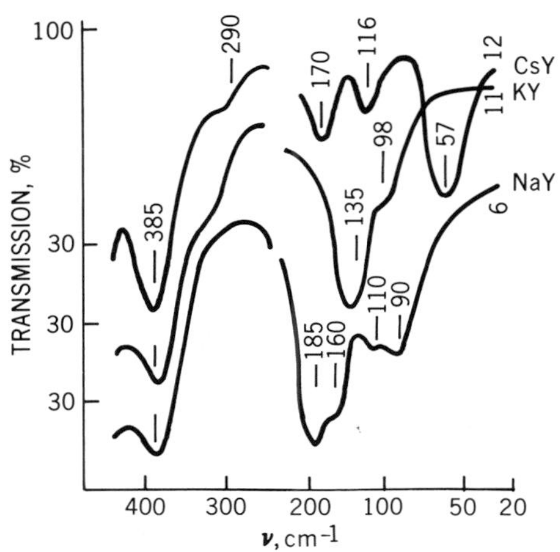

Soviet Physics, Solid State

Figure 11. (Top) The spectra of NaX and NaY zeolites with different values of the ratio Si/Al; 1) 1.14: 3) 1.38: 4) 1.66: 6) 2.56. The thickness of the samples was 2–5 mg/cm² (46).

Soviet Physics, Solid State

(Middle) Spectra of type X zeolites. The thickness of all the samples was 5–6 mg/cm² (46).

Soviet Physics, Solid State

(Bottom) Spectra of type Y zeolites with the Si/Al ratio equal to 2.56. The thickness of all the samples in the 470–220 cm^{-1} range was 2 mg/cm². In the region 220–20 cm^{-1}, it was 5 mg/cm² (46).

spectral determination. Therefore the spectra are for thermally treated and dehydrated forms. The position of bands assigned to the SiO_4 valence bands near 470, 1000, and 1100 cm^{-1} by those authors depended upon the nature of the cation and its field strength and radius, with the higher field strength cations (Mg^{2+} and Li^+) giving the most significant shifts and changes in the valence bands compared to the starting Na form. The chain frequency at 565 cm^{-1} was found to be relatively insensitive to cation form. Based on a decrease in intensity (and broadening) of the chain frequency near 565 cm^{-1} at very high exchange levels and after activation at 450°C, the Co^{2+}, Ni^{2+}, Sr^{2+}, Zn^{2+}, and Ba^{2+} forms were found to be crystallographically degraded, whereas the Li^+, K^+, Mg^{2+}, Mn^{2+}, Ag^+, and Na^+ forms maintained a high level of crystallinity.

Wolf *et al.* (*10*) in an infrared study of a mordenite-type zeolite and its ion exchanged forms showed no shift in the chain frequency at 635 cm^{-1} with ion exchange, except with Cs^+.

Vishnevskaya *et al.* (*44*) state that for the Na-forms of zeolites X and Y with varying SiO_2/Al_2O_3 ratios, the position of the 580 cm^{-1} band in general is independent of heat treatment in vacuum from 25°C to 550°C (self-supported wafers). However, dehydration of specimens having high SiO_2/Al_2O_3 ratios had a significant effect on the displacement of the 580 cm^{-1} band to higher frequencies.

Brodskii, Zhdanov, and Stanevich (*45, 46*) studied the effect of dehydration on a series of alkali metal cation forms of zeolites X and Y with varying SiO_2/Al_2O_3 ratios in the far infrared. Their spectral data include the region of 20 to 500 cm^{-1} (Figure 11) and clearly show that spectral changes in that region are sensitive to the nature of the substituting cation and the cation site, as well as the adsorbed water content. In type X zeolites, bands assigned to interionic vibrations between the cation and framework are observed in the Na form at 66 cm^{-1} for Site II cations, and 188 cm^{-1} for Site I′ or Site II. Shifts in the 66 cm^{-1} band for K, Rb, and Cs-exchanged zeolites to 54, 40, and 33 cm^{-1}, respectively, show that the frequency is proportional to $m^{-1/2}R^{-3/2}$, where m is cation mass and R is cation radius. The shift in the band at 188 cm^{-1} appears to be related to interionic vibrations of cations in Site I′ or Site II, but in this case the correlation with cation mass is not as clear. The spectrum of NaY shows a doublet band at 185–169 cm^{-1} which shifts to 135–98 cm^{-1} in KY and to a single band at 57 cm^{-1} in CsY. Again the shift correlates with a $\nu \propto m^{-1/2}R^{-3/2}$ relationship. Absorption bands at 110, 80–90 cm^{-1}, and 385 cm^{-1} are attributed to crystal lattice vibrations.

deKanter *et al.* (*25*) found a linear relationship between the frequency shifts of some lattice vibrations and the reciprocal of the sum of the cation and framework oxygen ionic radii for a series of hydrated monovalent forms of zeolite A in both the infrared and Raman spectra. Their data are shown in Figure 12. No experimental detail describing the method of sample preparation or spectral measurement technique is given. The frequency shifts of the lattice vibrations of the ν_1 (A_1) symmetric stretching vibration and the ν_4 (T_2) bending vibrations of the $Si(Al)O_4$ tetrahedra indicated that the perturbation

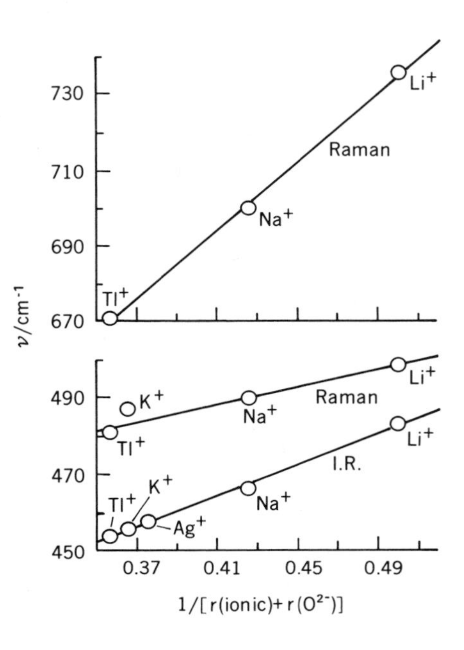

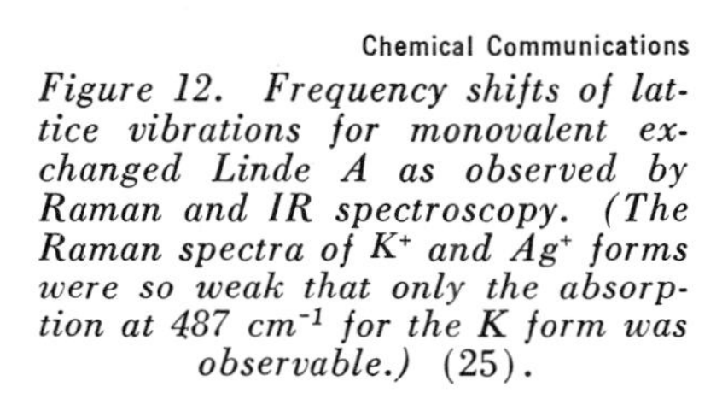

Chemical Communications

Figure 12. Frequency shifts of lattice vibrations for monovalent exchanged Linde A as observed by Raman and IR spectroscopy. (The Raman spectra of K^+ and Ag^+ forms were so weak that only the absorption at 487 cm^{-1} for the K form was observable.) (25).

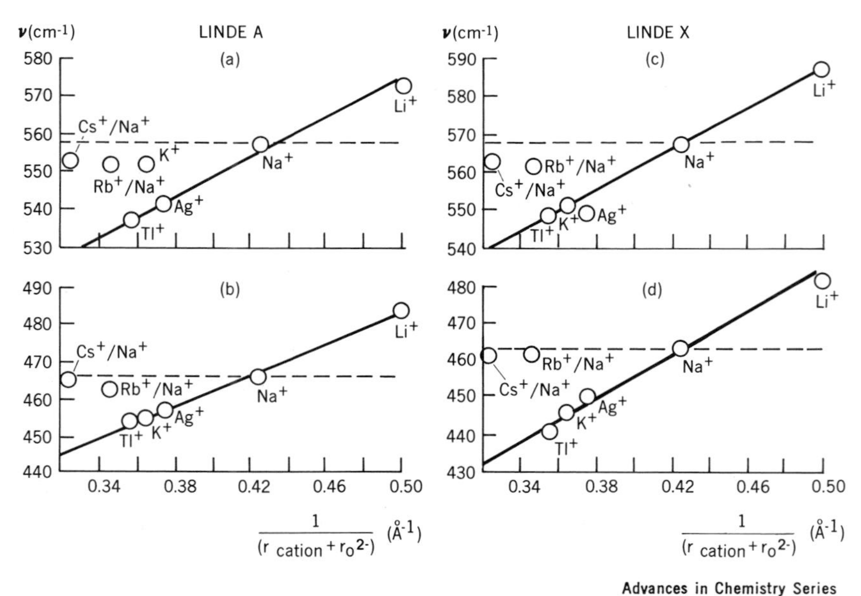

Advances in Chemistry Series

Figure 13. Plots of infrared frequency, v, against the reciprocal of the sum of the cation and oxygen ionic radii, $1/(r_{cation} + r_{O^{2-}})$, for various monovalent cation exchanged forms of Linde A and X: (a) 530–580-cm^{-1} infrared band, (b) 450–490-cm^{-1} infrared band (c) 540–590-cm^{-1} infrared band, (d) 440–485-cm^{-1} infrared band (27).

is caused by electrostatic interaction of the cation with the framework. The frequencies were shifted to higher wavenumbers with decreasing ionic radius.

Maxwell and Baks (*27*) in an extension of the work of deKanter *et al.* (*25*) show a similar relationshift in the shift of infrared framework vibrations with the reciprocal of the cation radius for a series of monovalent forms of hydrated A and X zeolites. The spectral shifts in framework vibrations are attributed to cations which interact strongly with the zeolite framework and are interpreted as similar cation siting in the zeolite framework. The strongly interacting cations are Li, Na, Ag, K, and Tl. Rb and Cs deviate from the linear relationship because only partial exchange is achieved because weak coulombic interactions exist between these cations and the zeolite framework. The linear relationship of infrared frequency with the reciprocal of the sum of the cation and oxygen ionic radii are shown in Figure 13 for the band near 530–580 cm^{-1} assigned to a double ring vibration in Table V and the bending mode at 450–490 cm^{-1}. The frequency increases with the

Table VI. Proposed Cation Siting for Rb^+ and Cs^+ Exchanged Forms of Hydrated A and X Zeolites (*27*)

A. Hydrated Zeolite A:

Formula per Unit Cell	*Large Cavity Cations*	*Six-Membered Ring Cations*
$Na_{12}(AlO_2)_{12}(SiO_2)_{12}$	4 Na^+	8 Na^+
$Rb_8Na_4(AlO_2)_{12}(SiO_2)_{12}$	8 Rb^+	4 Rb^+, 4 Na^+
$Cs_6Na_6(AlO_2)_{12}(SiO_2)_{12}$	4 Cs^+	2 Cs^+, 6 Na^+

B. Hydrated Zeolite X:

	Site I	*Site I′*	*Site II*	*Site III*
$Rb_{62}Na_{26}(AlO_2)_{88}(SiO_2)_{104}$ [a]	9 Na^+	8 Na^+	9 Na^+, 15 Rb^+	47 Rb^+
$Cs_{49}Na_{39}(AlO_2)_{88}(SiO_2)_{104}$ [a]	9 Na^+	8 Na^+	22 Na^+, 2 Cs^+	47 Cs^+

[a] Based on chemical analyses.

Advances in Chemistry Series

reciprocal of the cation radius, consistent with a simple electrostatic model of the cation-framework oxygen interaction. There is also considerable change in the position and character of the bands in the symmetric stretch region of 650–760 cm^{-1} and the pore ring opening near 370 cm^{-1} as a function of monovalent cation species, but no quantitative correlation was drawn by the authors. The cation siting based on the infrared interpretation is shown to be consistent with published x-ray data on similar cation forms of A and X zeolites. Cation sites for Rb and Cs in hydrated zeolites A and X are proposed based on the infrared interpretation and are shown in Table VI.

The spectral data of Maxwell and Baks (*27*) were measured on pressed pellets of CsI (3 mg of zeolite in 300 mg of CsI) with instrument calibration checked. The zeolites were dried under vacuum and hydrated over a saturated ammonium chloride solution. Kubasov (*47*), in the discussions of the paper by Maxwell and Baks, stated that, in measurements of the infrared spectra of different polyvalent forms of zeolites in the hydrated state, they found no significant effect of cation on the IR spectra in the region of 400–800 cm^{-1}, but that definite changes occurred on dehydration. Maxwell, in response, commented that they had observed interesting effects of divalent cations on the IR spectra of zeolites A and X in the hydrated states (*48*).

Comparison of the infrared spectra published on similar zeolites show differences in band positions and spectral features that probably reflect variation in technique of sample preparation, IR measurement, and sample thermal history and state of hydration. Even in the cases where the data are reported for hydrated zeolites, some dehydration may have occurred during preparation of a KBr or CsI pellet by vacuum pressing techniques or during spectral measurement (beam heating). Therefore, quantitative comparisons among different authors should be approached with caution, although quantitative differences cited in any one reference usually appear to be valid.

Despite the experimental variations discussed, the mid- and far-infrared spectra and the Raman spectra of zeolites have been shown to yield abundant information on cation type and crystallographic position that should serve as a supplemental structural tool to x-ray diffraction studies. The far-infrared region and Raman spectra appear to offer the most information on cation specific bands.

Framework Modification, Decomposition, and Ordering. Extensive application of mid-infrared spectroscopy has been made to the complex structural and chemical problems in zeolite framework modification, where stabilization and framework dealumination reactions are generally assumed to be involved. Kerr (*40*) and McDaniel and Maher (in Chapter 4 of this volume) have recently reviewed the chemistry and structure of the stabilization reactions of ammonium forms of zeolites involving deammoniation, decationization, the formation of hydrogen or acid forms, and framework dealumination and stabilization reactions. The formation of aluminum-deficient forms of zeolites by chemical removal of aluminum is also reviewed.

Zhdanov *et al.* (*7*) studied the mid-infrared spectra of thermally decationized forms of NH_4X zeolite. Changes in the skeletal vibrations in the region of 400–900 cm^{-1} after heating in vacuum to 200°–300°C are attributed to a slight stabilization of the structure from the formation of a maximum concentration of structural hydroxyl groups. After evacuation at 400°C, where the concentration of hydroxyls drops sharply, there is an abrupt decrease in the intensity of the skeletal vibrations indicating a total loss of crystallinity. They conclude that infrared spectroscopy gives valuable information on the effect of decationization on the structure of the zeolite.

McDaniel and Maher (*49*) prepared an ultrastable form of zeolite Y by controlled thermal treatment of the ammonium exchanged form. The ultrastable Y had a substantially enhanced thermal stability and a contracted unit cell. Kerr (*50*) prepared a stabilized zeolite Y by controlled calcination of the hydrogen form accompanied by removal of aluminum from the tetrahedral framework sites. The resulting cationic aluminum could be removed by ion exchange with NaOH solution. Kerr (*51, 52*) also prepared aluminum-deficient forms of type Y zeolite from the Na- and NH_4-exchanged forms by chemical extraction of framework aluminum with ethylenediaminetetraacetic acid solutions (EDTA), followed by calcination. Kerr showed an increase in thermal stability of the aluminum-deficient forms and suggested that the increased stability might be attributed to the formation of new Si—O—Si bonds (*51, 52*). In addition, aluminum removal was accompanied by a contraction of the unit cell. Kerr (*52*) states that the ultrastable Y of McDaniel and Maher is an aluminum-deficient framework similar to the aluminum-deficient Y forms prepared by EDTA extraction. Neither McDaniel and

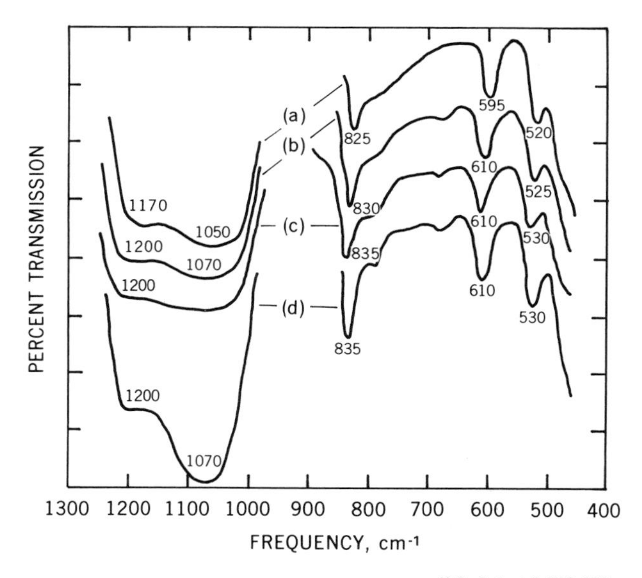

U.S. Patent 3,506,400

Figure 14. Infrared spectra of acid forms of Y after steaming. Starting NaY has $Si/Al_2 = 4.70$; NH_4-exchanged forms wet air calcined at 538°C, followed by steaming at 649°C. Cation composition before calcination and steaming: (a) $[(NH_4)_{0.56}K_{0.38}Na_{0.06}]Y$; (b) $[(NH_4)_{0.84}K_{0.11}Na_{0.05}]Y$; (c) $[(NH_4)_{0.91}K_{0.04}Na_{0.05}]Y$; and (d) $[(NH_4)_{0.92}Na_{0.08}]Y$. Corresponding absorption bands for sample (d) before calcination and steaming found at 1150, 1010, 780, 720, 570, and 500 cm^{-1}. All spectra determined in KBr pellets (37).

Maher (*49*) nor Kerr (*50, 51, 52*) reported any mid-infrared spectral data on the samples.

Eberly *et al.* (*37*) showed shifts in the mid-infrared bands to higher frequencies in siliceous zeolites derived from NH_4-exchanged forms of zeolite Y, erionite, and mordenite by high temperature steam calcination and subsequent removal of aluminum by mineral acid or an organic chelating agent. Typical infrared spectra for steamed NH_4Y products are shown in Figure 14.

These siliceous zeolites appear to represent stabilized, aluminum-deficient forms of the respective zeolites with SiO_2/Al_2O_3 ratios greater than 20 (from wet chemical analysis).

Pichat *et al.* (*35*) also prepared a series of aluminum-deficient forms of Y zeolite by the method of Kerr (*52*) using EDTA extraction of the NH_4-exchanged forms and investigated their mid-infrared spectra in the region of 300–1300 cm^{-1}. They found a linear increase in frequency for five infrared absorption bands (1150, 1050, 800, 585, and 385 cm^{-1}) with decreasing aluminum fraction in the framework, as long as a high degree of crystallinity was maintained (Figure 15). The slope of the line in Δ cm^{-1}/0.1 atom fraction of Al was compared to the same values published by various authors (*5, 11, 13*) for synthesized sodium forms of types X and Y zeolite with varying SiO_2/Al_2O_3 ratios (Figure 15). The authors note the close resemblance of the spectral shifts in the aluminum-deficient Y zeolites compared to the synthesized sodium forms. In general, the frequencies reported by Pichat *et al.* (*35*) are somewhat higher than those reported for the synthesized Na forms.

Lahodny-Šarc and White (*34*) studied the removal of Al from NaA, NaX, and NaY zeolite by treatment with H_2CO_3 or ethylenediaminetetraacetic acid

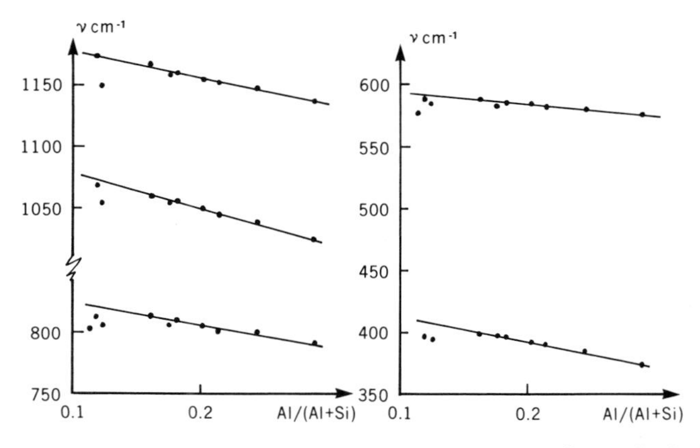

Comptes Rendu, C

Figure 15. Relationship of absorption frequency with aluminum concentration for aluminum deficient Y-zeolites from EDTA extraction of NH_4Y (35). Comparison of slopes of lines with as-synthesized Na forms of zeolites X and Y from various authors is shown below.

Slopes of lines, in cm^{-1}/0.1 atom fraction of Al:

		Frequencies (cm^{-1})				
	Al/(Al + Si)	*1150.*	*1050.*	*800.*	*585.*	*395.*
Present work	*0.292–0.16*	*21*	*23.5*	*17*	*9.5*	*10.5*
Ref 5	*0.475–0.282*	—	—	*15.5*	*7.5*	—
Ref. 11	*0.455–0.292*	—	*23.5*	*22*	*9.5*	—
Ref. 13	*0.455–0.261*	—	*20*	*23*	*9*	*11*

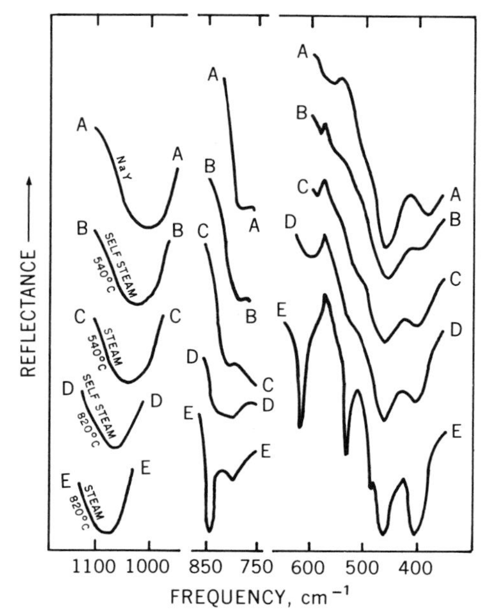

Journal of Catalysis

Figure 16. Framework bands of Y zeolites: (A) NaY, $Si/Al_2 = 4.2$; NH_4^+-exchanged, and (B) self-steamed, 540°C, (C) steamed, 540°C, (D) self-steamed, 820°C, ultrastable Y, Type I, (E) steamed, 820°C, high silica ultrastable zeolite (18).

(EDTA). From the shifts in the main asymmetric band near 1000 cm^{-1} and the slopes shown in Figures 6 and 7, they assigned an Al content to the aluminum-deficient frameworks. All of their spectra show broadening and/or decrease in intensity of the structure-sensitive bands, strongly suggesting significant crystallographic decomposition accompanying the acid treatment. Application of spectral shift to framework Al content, therefore, seems inappropriate. The importance of using infrared in conjunction with x-ray analysis and adsorption to fully characterize the crystallinity of zeolites cannot be overemphasized.

Peri (*38*), in a study of ultrastable Y, reported that nearly all infrared bands from 400 to 1200 cm^{-1} in ultrastable Y were 15 to 30 cm^{-1} higher in frequency than the starting NH_4Y. No spectra were shown. Scherzer and Bass (*18*), in a more detailed study of the infrared spectra of ultrastable forms of zeolite Y, show similar shifts to higher frequencies of the framework bands betwen 300–1200 cm^{-1} which are a function of the nature of the steaming treatment (Figure 16 and Table VII). The shifts are accompanied by a decrease in the unit cell. The IR shifts and unit cell decrease are both interpreted in terms of removal of tetrahedral aluminum from the framework. In addition, they observe a sharpening of the framework vibrations and the appearance of new bands (Figure 17). They attribute the sharper absorption bands to a higher degree of crystallinity or a higher degree of framework ordering in the ultrastable and high silica ultrastable zeolites. Peri, and Scherzer and Bass conclude that during formation of ultrastable Y, aluminum migrates from framework tetrahedral sites to cation positions and that silicon replaces the removed aluminum by recrystallization of the framework. It should be noted that Peri and Scherzer and Bass, determined the spectra by an ATR technique and the spectral quality and resolution is somewhat lower

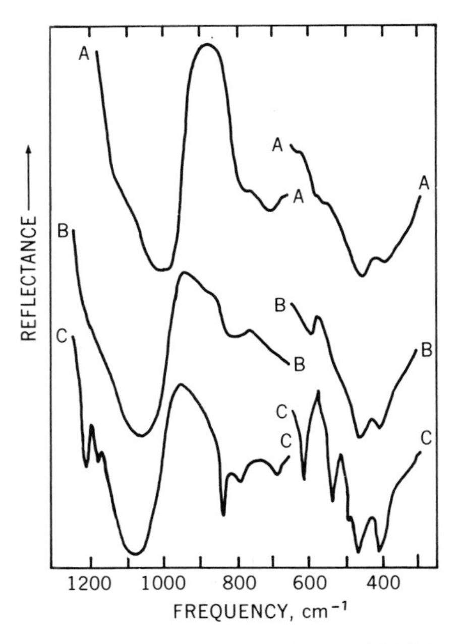

Journal of Catalysis

Figure 17. Framework bands of (A) NaY, Si/Al_2 = 4.2, (B) ultrastable Y zeolite, Type 1, and (C) high silica ultrastable Y zeolite (18).

than that for spectra using transmission techniques such as KBr or CsI pressed pellets.

Vishnevskaya *et al.* (*44*) show that the position of the infrared absorption band near 580 cm^{-1} varies linearly with SiO_2/Al_2O_3 for both dealuminized and synthesized forms of zeolites X and Y. Kubasov *et al.* (*39*) studied the decomposition of NH_4Y zeolite in water vapor at high temperatures using infrared spectroscopy and report similar shifts of the framework absorption bands in the region of 400–850 cm^{-1} to higher frequencies and relate the shift to partial loss of aluminum from the crystal structure. Kubasov *et al.* (*42*), in a study of the framework infrared and catalytic activity of a series of HY zeolites and lanthanum exchanged HY forms, report similar spectral

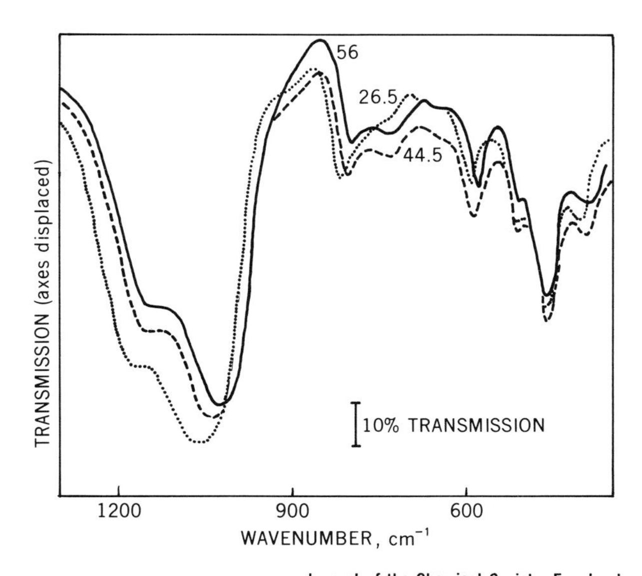

Journal of the Chemical Society, Faraday I

Figure 18. Infrared spectra of some aluminum-deficient Y zeolites containing the indicated numbers of Al atoms per unit cell. Prepared by EDTA extraction of NH_4-exchanged Y ($Si/Al_2 = 4.9$), followed by treatment in dry air flow at 380°C and 550°C (36).

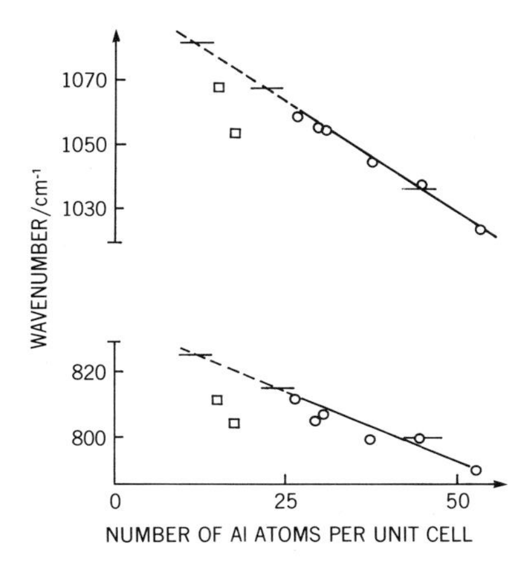

Journal of the Chemical Society, Faraday I

Figure 19. Variation of band positions as a function of the number of Al atoms per unit cell. ○, crystalline aluminum-deficient zeolites; □, partially amorphous aluminum-deficient zeolites; ——, ultrastable zeolites (36).

shifts in the lanthanum containing forms compared to decationated forms. They relate these shifts to increased crystallographic stability and higher catalytic activity of the lanthanum form. In a comparison of the infrared spectra of zeolite catalysts, both the positions and intensities of the bands in the region of 580 to 790 cm^{-1} can serve as a criterion of retention of crystal structure and positions of cations. The 580 cm^{-1} is most sensitive to both loss in crystallinity and lanthanum position.

Pichat *et al.* (*36*), in an infrared structural study of aluminum-deficient Y zeolites, include an excellent review of previously published data on the framework infrared of aluminum-deficient Y zeolites prepared by chemical extraction of aluminum, and of ultrastable zeolites (Figure 18). They compare the band positions for the two classes of zeolites for the asymmetric stretch frequency near 1050 cm^{-1} and the symmetric stretch frequency near 800 cm^{-1} and show the same linear relationship for both (Figure 19). Their results tend to confirm the similarities between both types of modified zeolites. The quantitative linear relationship yields the number of framework aluminum atoms for the two classes. (Caution is suggested in assigning IR spectral shifts to framework Si/Al content in modified zeolites when other recognized structural and chemical changes are occurring simultaneously.) Pichat *et al.* studied the integrated intensities of all framework bands and found that the intensity of the 730 cm^{-1} in the aluminum-deficient zeolites decreases linearly

Table VII. Framework Vibrational Frequencies (*18*)

Zeolite Type[a]	*T–O Asym. Stretch (cm^{-1})*	*T–O Sym. Stretch (cm^{-1})*	*Double 6 Ring (cm^{-1})*	*T–O Bend (cm^{-1})*	*12-Ring Pore Opening (cm^{-1})*	*Unit Cell (A)*
NaY ($Si/Al_2 = 4.2$)	995	775	580	463	387	24.70
H, NaY (540°C vacuum)	1005	780	583	456	388	24.64
H, NaY (540°C, moist air)	1008	780	583	457	393	24.63
H, NaY (540°C, self-steamed)	1023	780	583	456	393	24.61
H, NaY (540°C, 100% steam)	1035	805	590	460	398	24.57
US-YI (self-steamed)	1067	810	600	460	403	24.37
HS US-YI (100% steam)	1082	837	614	465	408	24.24
H, NaY (760°C, moist air)	1024	795	584	455	402	24.61
H, NaY (760°C, self-steamed)	1037	800	586	460	404	24.56

[a] The formulas used for the different samples do not reflect the true composition and are only used to distinguish the compounds according to their preparation. All forms were derived by various thermal treatments of NH_4^+-exchanged forms. US = ultrastable; HS US = high silica ultrastable, as defined by the authors.

Journal of Catalysis

with aluminum content and therefore could be attributed to isolated AlO_4 tetrahedra. The intensities of the other bands are not affected by aluminum removal. They concluded that during aluminum removal using EDTA at 100°C, silicon atoms replace aluminum atoms in the framework in the same manner as proposed for ultrastable zeolites.

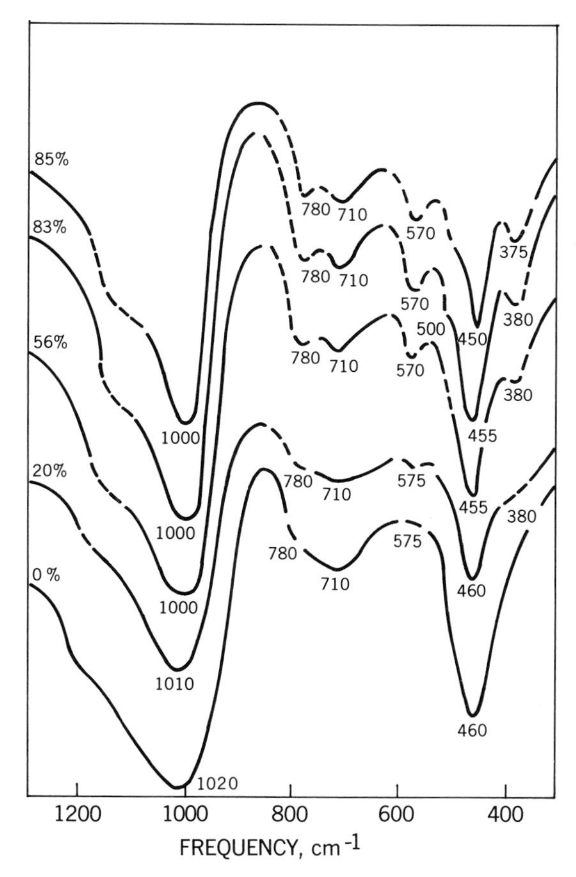

Advances in Chemistry Series

Figure 20. Infrared spectra of thermally decomposed Y zeolite (Si/Al of 2.5). The percentages are % residual x-ray crystallinity of the sample (13).

In an infrared structural study of framework modifications of type L zeolite, Pichat *et al.* (*53*), using both EDTA chemical extraction of aluminum and thermal stabilization techniques, found little shift in the framework vibrations and a decrease in thermal stability after certain treatments. Apparently type L zeolite, unlike zeolites Y, erionite, and mordenite, does not form stabilized aluminum-deficient ultrastable modifications, at least under the treatment conditions studied. The IR spectra were more sensitive to structural changes than x-ray diffraction patterns. The intensity of structure-sensitive bands gives useful information on structure degradation.

The mid-infrared region containing the framework vibrations can in general be used to determine the extent of crystallographic degradation of zeolites. By careful spectral measurements and the use of integrated intensities, a quantitative measure of crystallinity can be obtained. Figure 20 shows the decrease in intensity of the structure-sensitive or external linkage vibrations (Table V) for NaY zeolite during thermal decomposition. Other useful information can be derived from the position of the internal tetrahedral vibrations. The absence of significant shifts in the latter vibrations show that there is little change in the average Si/Al composition in the decomposed amorphous network; the absence of octahedral aluminum with a characteristic band near 550 cm^{-1} is also shown. Recrystallization products are evident by the appearance of their structure-sensitive bands.

Wolf and Fuertig (*8*) showed crystallographic degradation of NaA zeolite at 750°C and formation of nepheline at 900°C, based on the decreased intensity (and broadening) of the chain frequency for zeolite A near 565 cm^{-1} (D-4 ring) and the appearance of the typical absorption spectrum of nepheline. Wolf *et al.* (*9*) used the same chain frequency IR band to determine the crystallographic stability of various ion exchanged forms of type A and type X zeolite, where decrease in intensity of the band correlated with the loss in crystallinity as determined independently by x-ray and adsorption measurements.

Application of mid-infrared spectroscopy appears to be a most fruitful and sensitive structural technique in yielding information on framework structural changes occurring during stabilization and dealumination reactions, ordering, and decomposition. Retention of intensity of structure-sensitive bands reflects maintenance of high levels of crystallinity. Shifts in the framework vibrations can yield quantitative information on the Si/Al composition of the framework. Broadening and reduction of intensity of the structure-sensitive bands indicates framework disruption, decomposition, and a high concentration of defects. Sharpening of the framework bands shows recrystallization, an increase in the degree of crystal perfection, and a reduction in the number of defects.

Zeolite Synthesis Mechanism. Mid-infrared spectroscopy can also be applied to investigation of the mechanism of crystallization of zeolites from hydrous aluminosilicate gels or other aluminosilicate systems, such as conversion of clay to zeolite (*54*). An example of the kind of results achievable is shown in Figure 21 for the crystallization sequence for sodium X zeolite from a sodium aluminosilicate gel (*55*). Spectra were obtained on the solid phase separated from the gel and show the appearance of the structure sensitive bands near 360, 555, 665, 745, and 1060 cm^{-1} as NaX formation proceeds, as evidenced in the x-ray powder pattern of the solids. The absorption band near 850 cm^{-1} is assigned to an Si—OH bending vibration which decreases as crystallization of NaX proceeds. Although the presence of a broad band

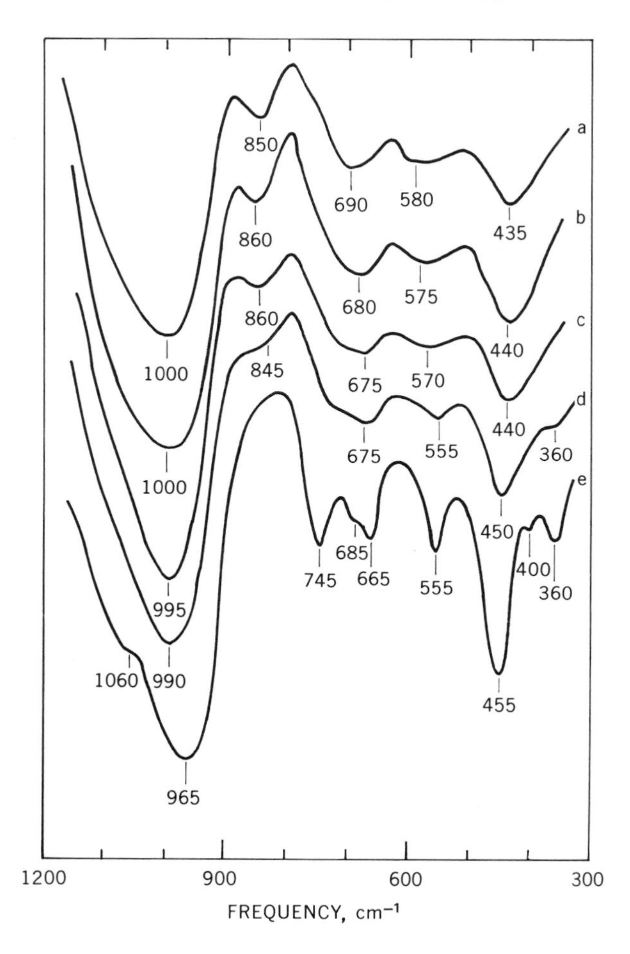

Figure 21. Mid-infrared spectra of the crystallization stages of NaX zeolite from a sodium aluminosilicate gel. Spectra of solids separated from the gel at: a) initial gelation; b) 1 hr, 100°C, no x-ray crystallinity; c) 2 hr, 100°C, 3% NaX by x-ray; d) 4 hr, 100°C, 20% NaX by x-ray, and e) 5 hr, 100°C, 96% NaX by x-ray. Spectra determined as KBr wafers (55).

near 575 cm^{-1} may suggest the presence of structural subunits such as double six rings in the initial gel, the strong absorption of H_2O in that spectral region makes such an assignment tentative. In the same study, spectra were obtained on gels and the separated solid and liquid phases, during the stages of crystallization for NaA, NaX, NaY, KL, and Ω zeolites. Useful information on the Si/Al ratio in the gel network, species present in the liquid phase, and the onset of crystallization are readily obtained from the mid-infrared spectra. However, such investigations require extremely careful techniques in handling the metastable gel and in making the infrared measurements, including careful establishment of the reproducibility of the spectra.

McNicol *et al.* (*56, 57*) applied Raman and phosphorescence spectroscopy to a study of the mechanism of crystallization of zeolites A and X from aluminosilicate gels. Tetramethylammonium (TMA) ion was added as a probe molecule in laser Raman studies of zeolite A crystallization (Figure 22). A shift in the Raman band of the TMA ion from 754 cm^{-1} in the amorphous gel to 768 cm^{-1} at the onset of NaA crystallization was observed and related to its incorporation into the zeolite A framework. Similarly, the Raman band at 485 cm^{-1} typical of the NaA framework vibration appeared in the Raman spectrum at the onset of crystallization. Raman spectra of the

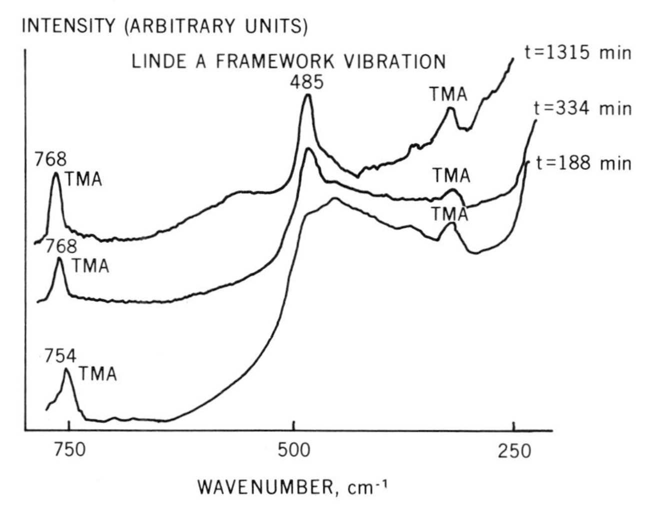

Journal of Physical Chemistry

Figure 22. Raman spectra of crystallizing gels: $(CH_3)_4N^+$/ *Linde A system* (56).

separated liquid phase from the A gel showed the presence of $Al(OH)_4^-$, $SiO_2(OH)_2^{2-}$, Na^+, and OH^- ions whose concentration did not change significantly during the induction and crystallization period, and no evidence of the presence of any soluble aluminosilicate anions. As pointed out by Zhdanov (*57*), the concentration of such soluble aluminosilicate species may be below the limit of detection of the experimental technique (cited as $\sim$0.1% by McNicol). The same authors (*56, 57*) doped the aluminosilicate gels with trace amounts of Fe^{3+} substituting for tetrahedral Al^{3+}, and followed the phosphorescence spectra of the Fe^{3+} during crystallization to zeolites. A characteristic phosphorescence spectrum of Fe^{3+} in the crystalline zeolite allowed them to follow the course of zeolite crystallization by this technique. McNicol *et al.* conclude from the Raman and phosphorescence studies that crystallization of zeolites from gels occurs within the solid phase of the gel. This consists of a disordered amorphous aluminosilicate network.

Beard (*59*) studied the mid-infrared spectra of silicate solutions used in zeolite synthesis to characterize the molecular weight of the silicate species. Increasing molecular weights with decreasing Na_2O/SiO_2 ratios were observed in sodium silicate solutions from shifts of the main Si—O asymmetric stretch frequency near 950 cm^{-1} for monomeric and dimeric species to 1120 cm^{-1} for colloidal silica sols with molecular weights as high as a million. Tetramethylammonium silicate solutions showed an unusual equilibrium mixture of at least two species, one of low molecular weight (1025 cm^{-1}) and the other of high molecular weight (1120 cm^{-1}).

Borowiak (*60*) studied the Raman spectra of aqueous sodium aluminate solutions to establish the effect of aluminate species on the kinetics of zeolite Y crystallization. The relative concentrations of a monomeric tetrahedral $Al(OH)_4^-$ species (623 cm^{-1}, Raman) and a dimeric $Al_2O(OH)_6^{2-}$ species (543 cm^{-1}, Raman) present in solution were determined from their Raman intensities. The concentration of the 543 cm^{-1} species increased with increasing water content. The species and concentration of aluminate ion present in solution prior to the gel precipitation step affected the kinetics of crystallization of NaY zeolite.

Raman spectroscopy offers distinct advantages over infrared in determining the vibrational spectra of aluminate, silicate, and aluminosilicate species in aqueous solution because of the very weak Raman scattering of water. The intensities of the Raman spectra of aluminate and silicate anions in aqueous solution are strong compared to the weak Raman scattering observed for such anions when present in solid zeolite frameworks. Experimental problems are encountered in determining the Raman spectra of solutions because of high background from fluorescent impurities and Tyndall scattering from the presence of colloidal sized particles. However, good quality Raman spectra can be obtained by combining low impurity content and beam burning techniques (*see* below) with multiple filtration of the solution through submicron (Millipore-type) filters before spectral measurements (*19*).

Raman Spectra of Zeolite Frameworks

Angell (*26*) determined the Raman spectra of a series of mineral and synthetic zeolites (Figures 23a and b). In general, the Raman scattering of zeolites is very weak, and the Raman spectra do not show the degree of structural specificity for zeolite framework type exhibited in the mid-infrared spectra. However, some differences in the position of the strongest band near 400 to 500 cm^{-1} appear to be structure related. Difficulty in obtaining the Raman spectra of zeolites because of their weak Raman scattering and high excessive background (the latter related to the presence of fluorescent impurities) was overcome by using high purity samples and prolonged exposure to the laser beam (burnout) to decrease the background. Because of the weak Raman spectra of zeolites, the Raman spectra of adsorbed species on zeolites can readily be obtained (*26*). However, it appears that Raman spectroscopy is a less sensitive tool than mid-infrared in determining framework structural features.

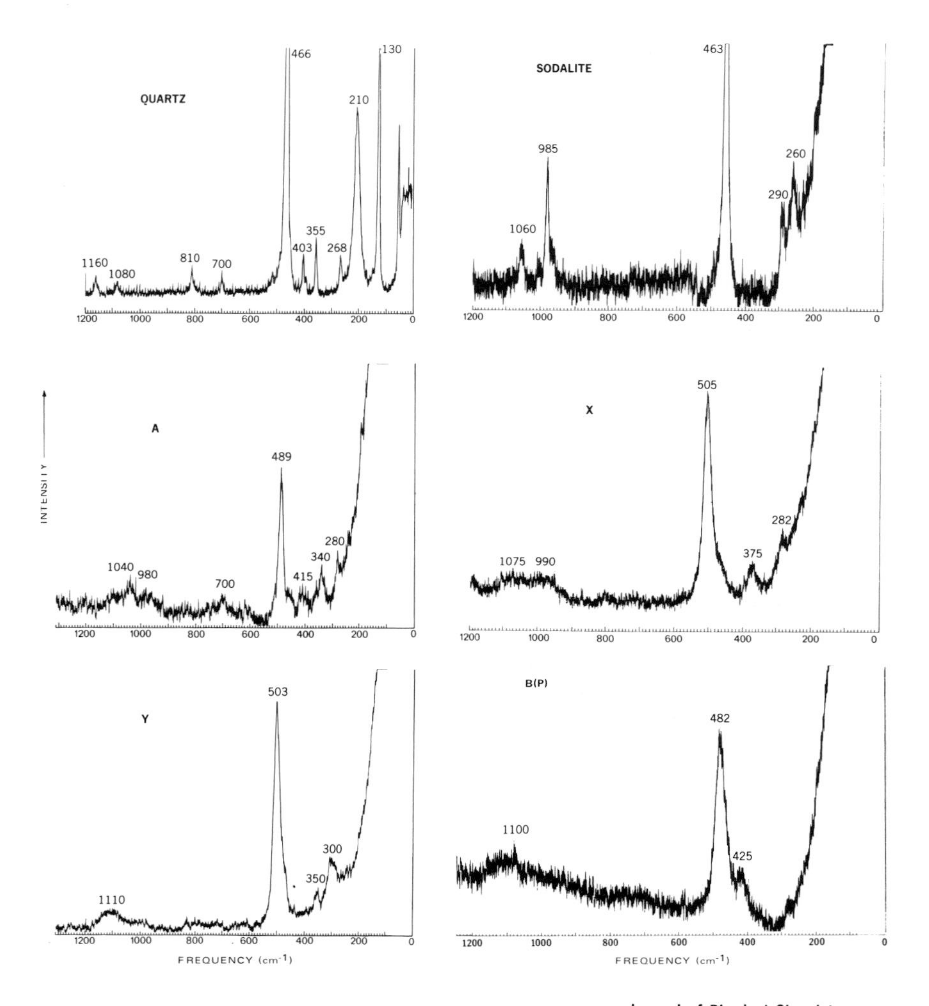

Journal of Physical Chemistry

Figure 23a. Raman spectra of quartz, sodalite, and some zeolites obtained with 4880 A excitation (26)

The application of Raman spectroscopy to determination of cation-framework oxygen interactions and cation position has been discussed previously. Raman appears to be more sensitive to this aspect of zeolite structure than to framework topology. Similar information on cation position and interactions can also be obtained from far infrared spectra.

Summary

The application of mid-infrared spectroscopy to zeolite structural problems has been shown to yield abundant information on many aspects of zeolite

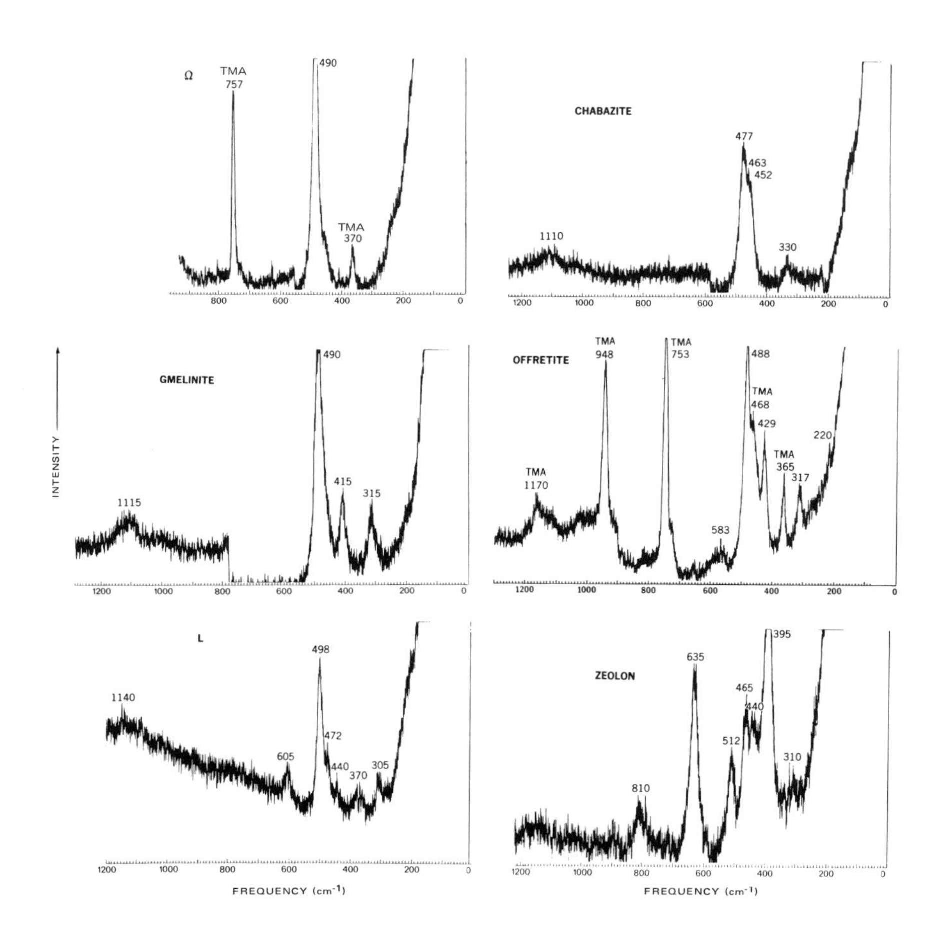

Journal of Physical Chemistry

Figure 23b. Raman spectra of some natural and synthetic zeolites obtained with 4880 A excitation (26)

chemistry and structure. The application of vibrational spectroscopy to the study of inorganic crystals has been developed largely during the last fifteen years (*2*). The high degree of structural specificity in the vibrational spectra of zeolites, particularly in the mid-infrared region, results from an almost unique combination of their chemistry and structure and is largely related to the presence of complex anions in the form of highly symmetric polyhedral cages within the zeolite framework. Two areas of importance for future investigation are apparent. The first is a theoretical treatment of vibrational assignments, particularly the structure specific frequencies tentatively correlated to structural subunits in zeolite frameworks. This is required before many of the interpretations of structural changes discussed in this chapter can be confirmed. Second, additional determination of the far infrared spectra of zeolites can result in new information on cation-framework interactions and long-range structural order in zeolite frameworks. Further Raman investigations in the area of cation interactions also appear to be fruitful.

Finally, the use of infrared and Raman spectroscopy for structural studies should be complementary to other structural techniques, such as x-ray and electron diffraction. X-ray powder diffraction, chemical analysis, and adsorption measurements should always accompany spectroscopic studies to completely characterize and interpret the crystal chemical properties of zeolites.

Literature Cited

1. Dutz, Von H., *Ber. Deut. Keram. Ges.* (1969) **46,** 75.
2. Lazarev, A. N., "Vibrational Spectra and Structure of Silicates," Consultants Bureau, New York, 1972.
3. Lippincott, E. R., Van Valkenberg, A., Weir, C. E., Bunting, E. N., *J. Res. Natl. Bur. Std.* (1958) **A 61,** 61.
4. White, W. B., Roy, R., *Amer. Mineralogist* (1964) **49,** 1670.
5. Zhdanov, S. P., Kiselev, A. V., Lygin, V. I., Totova, T. I., *Russ. J. Phys. Chem.* (1964) **38,** 1299.
6. Kiselev, A. V., Lygin, V. I., "Infrared Spectra of Adsorbed Species," L. H. Little, Ed., pp. 361-367, Academic, London, 1967.
7. Zhdanov, S. P., Lygin, V. I., Titova, T. I., "Zeolites, Their Synthesis, Properties, and Applications," M. M. Dubinin, Ed., p. I-37, U.S.S.R. Academy of Sciences, 1964.
8. Wolf, F., Fuertig, H., *Tonind, Ztg. Keram. Rundschau* (1966) **90,** 310.
9. Wolf, F., Fuertig, H., Haedicke, V., *Chem. Tech. (Berlin)* (1966) **18,** 524.
10. Wolf, F., Fuertig, H., Knoll, H., *Chem. Tech. (Leipzig)* (1971) **23,** 368.
11. Wright, A. C., Rupert, J. P., Granquist, W. T., *Amer. Mineralogist* (1968) **53,** 1293.
12. Avdeeva, T. I., Vorsina, I. A., *Acad. Sci. U.S.S.R., Bull. Div. Chem. Sci.* (1969) **10,** 2051.
13. Flanigen, E. M., Khatami, H., Szymanski, H. A., *Adv. Chem. Series* (1971) **101,** 201.
14. Alpert, N. L., Keiser, W. E., Szymanski, H. A., "IR, Theory and Practice of Infrared Spectroscopy," 2nd ed., especially Chap. 2, pp. 9-64 and Chap. 7, pp. 329-342, Plenum Press, New York, 1970.
15. Angell, C. L., Schaffer, P. C., *J. Phys. Chem.* (1965) **69,** 3464.
16. Flanigen, E. M., Mumpton, F. A., Union Carbide Corp., unpublished results.
17. Freeman, D. C., Jr., Stamires, D. N., *J. Chem. Phys.* (1961) **35,** 799.
18. Scherzer, J., Bass, J. L., *J. Catalysis* (1973) **28,** 101.
19. Flanigen, E. M., Union Carbide Corp., unpublished results.
20. Rabo, J. A., Poutsma, M. L., Skeels, G. W., *Proc. Vth Intl. Congr. Catalysis* (1973) **II,** 1353.
21. Milkey, R. G., *Amer. Mineralogist* (1960) **45,** 990.
22. Oinuma, K., Hayashi, J., *J. Tokyo Univ.* (1967) **8,** 1.
23. Belytskii, I. A., Golubova, G. A., *Mater. Genet. Mineral.* (1972) **7,** 310.
24. Shikunov, B. I., Lafer, L. I., Yakerson, V. I., Mishin, I. V., Rubinshtein, A. M., *Acad. Sci. U.S.S.R., Bull. Div. Chem. Sci.* (1972) **21,** 201.
25. deKanter, J. J. P. M., Maxwell, I. E., Trotter, R. J., *J. Chem. Soc. Chem. Commun.* (1972) 733.
26. Angell, C. L., *J. Phys. Chem.* (1973) **77,** 222.
27. Maxwell, I. E., Baks, A., *Adv. Chem. Series* (1973) **121,** 87.

28. Pichat, P., Beaumont, R., Barthomeuf, D., *J. Chem. Soc. Faraday Trans. I.* (1974) **70,** 1402.
29. Brasch, J. W., Mikawa, Y., Jakobsen, R. J., *Appl. Spectroscopy Rev.* (1968) **1,** 187.
30. *See* for example, review by Nakagawa, I., *Appl. Spectroscopy Rev.* (1974) **8,** 229.
31. Stubican, V., Roy, R., *Amer. Mineralogist* (1961) **46,** 32.
32. Stubican, V., Roy, R., *J. Amer. Ceram. Soc.* (1961) **44,** 625.
33. Stubican, V., Roy, R., *Z. Krist.* (1961) **115,** 200.
34. Lahodny-Sarc, O., White, J. L., *J. Phys. Chem.* (1971) **75,** 2408.
35. Pichat, P., Beaumont, R., Barthomeuf, D., *Compt. Rend. C.* (1971) **272,** 612.
36. Pichat, P., Beaumont, R., Barthomeuf, D., *J. Chem. Soc. Faraday Trans. I.* (1974) **70,** 1402.
37. Eberly, P. E., Laurent, S. M., Robson, H. E., U.S. Patent **3,506,400** (1970).
38. Peri, J. B., *Proc. Vth Intl. Congr. Catal.* (1973) **1,** 18-329.
39. Kubasov, A. A., Topchieva, K. V., Burenkova, L. N., Mitichenko, M. G., *Russ. J. Phys. Chem.* (1973) **47,** 1363.
40. Kerr, G. T., *Adv. Chem. Ser.* (1973) **121,** 219.
41. Flanigen, E. M., Grose, R. W., *Adv. Chem. Ser.* (1971) **101,** 76.
42. Kubasov, A. A., Topchieva, K. V., Ratov, A. N., *Russ. J. Phys. Chem.* (1973) **47,** 1023.
43. Dyrkheev, V. V., Kiselev, A. V., Lygin, V. I., *Russ. J. Phys. Chem.* (1974) **48,** 418.
44. Vishnevskaya, L. M., Kubasov, A. A., Tkhoang, K. S., Topchieva, K. V., *Russ. J. Phys. Chem.* (1973) **47,** 873.
45. Brodskii, I. A., Zhdanov, S. P., Stanevich, A. E., *Optics Spectr.* (1971) **30,** 30.
46. Brodskii, I. A., Zhdanov, S. P., Stanevich, A. E., *Sov. Phys. Solid State* (1974) **15,** 1771.
47. Kubasov, A., "Proc. Third Intern. Conf. Molec. Sieves," p. 17, Leuven University Press, Leuven, 1973.
48. Maxwell, I. E., "Proc. Third Intern. Conf. Molec. Sieves," p. 17, Leuven University Press, Leuven, 1973.
49. McDaniel, C. V., Maher, P. K., "Molecular Sieves," p. 186, Society of Chemical Industry, London, 1968.
50. Kerr, G. T., *J. Phys. Chem.* (1967) **71,** 4155.
51. Kerr, G. T., *J. Phys. Chem.* (1968) **72,** 2594.
52. Kerr, G. T., *J. Phys. Chem.* (1969) **73,** 2780.
53. Pichat, P., Franco-Parra, C., Barthomeuf, D., *J. Chem. Soc., Faraday Trans. I.* (1975) **71,** 991.
54. Breck, D. W., "Zeolite Molecular Sieves: Structure, Chemistry and Use," Chap. 4, p. 245, John Wiley and Sons, New York, 1974.
55. Khatami, H., Flanigen, E. M., Szymanski, H. A., Union Carbide Corp., unpublished results.
56. McNicol, B. D., Pott, G. T., Loos, K. R., *J. Phys. Chem.* (1972) **76,** 3388.
57. McNicol, B. D., Pott, G. T., Loos, K. R., Mulder, N., *Adv. Chem. Series* (1973) **121,** 152.
58. Zhdanov, S. P., "Proc. Third Intern. Conf. Molec. Sieves," p. 24, Leuven University Press, Leuven, 1973.
59. Beard, W. C., *Adv. Chem. Series* (1973) **121,** 162.
60. Borowiak, M. A., Berak, J. M., *Rocz. Chem. Ann. Soc. Chim. Polo.* (1973) **47,** 1565.

Chapter

3

Infrared Studies of Zeolite Surfaces and Surface Reactions

John W. Ward, Union Oil Company of California, Union Research Center, Brea, Calif. 92621

NUMEROUS TECHNIQUES have been used to examine the surface, gas-solid interactions, adsorption and catalysis on molecular sieve zeolites. Many of these involve the measurement of bulk changes, such as volumes of adsorbed gases, heats of adsorption, conductivity, reactants, and products, etc. The interactions and events occurring at the surface are then deduced from these observations.

Spectroscopic techniques can ideally furnish information directly about the nature of surfaces and species adsorbed on surfaces. They also provide information on a molecular level rather than on a total system level. For example, whereas a kinetic study is limited to measurements of reactants and products, a spectroscopic method can characterize the nature of the surface species, how these change with reaction conditions and which surface structural groups interact and which do not interact.

In most applications of zeolites, surface properties and reactivity are of major importance. Infrared spectroscopy can give useful and often definitive information on the structure and surface properties of zeolites and how these are modified by reaction and treatment. Changes in the spectra of the zeolite itself and of molecules adsorbed on the surface can yield direct information about the surface, how molecules adsorb, where they adsorb and how they interact.

The application of infrared spectroscopy to surface chemistry has been extensively covered by Little (*1*) and Hair (*2*). The subject has also been covered by several reviews (*3, 4, 5, 6, 7, 8, 9, 10, 11, 12, 13, 14*). Discussions of the applications of infrared spectroscopy to zeolites have appeared recently (*1, 2, 9, 16, 17, 18*).

Although many molecular sieve zeolites are known, most of the spectroscopic studies have been of the commercially important large pore X and

Y zeolites. Other more limited studies of zeolites such as A, L, Ω, mordenite, offretite, erionite, and some naturally occurring zeolites have been made. The application of mid-infrared spectroscopy to zeolite structural analysis has been reviewed in Chapter 2. This review concentrates mainly on the characterization of surface structural sites, the interaction sites of adsorbates with the surface, the nature of the adsorbed species, the types of events occurring on the surface under reaction conditions and the modification of the surface by chemical treatment.

Information from Infrared Studies, Experimental Techniques, and Interpretation

Information from Infrared Spectra. Infrared measurements can potentially give information on the vibrations of any species through which the beam passes. Hence, it is possible to obtain in principle information on the zeolite structure and any material in contact with it. Types of information which can be obtained are as follows.

(1) The structure of the zeolite. Information can be derived from the structural vibrations concerning:

(a) the nature of zeolite lattice framework such as the framework type, the SiO_2/Al_2O_3 ratio, cation types and locations and structural changes due to thermal treatment; and

(b) the nature of surface structural groups which are often centers of adsorption and catalysis.

(2) Zeolite–adsorbate interactions. Information can be obtained concerning:

(a) the identity of the physically or chemically adsorbed species on the zeolite;

(b) the type and nature of the adsorbate–zeolite interactions occurring;

(c) the identification of the surface sites at which adsorption is occurring;

(d) changes in the nature of adsorbed molecules and their interaction with the surface as a function of time, temperature, pressure, and interaction with an additional species, etc.; and

(e) catalytic reactions occurring on the surface.

Unfortunately, these are ideal goals, and their attainment is often limited by other considerations. Observations are often confined to limited spectral regions, primarily because of the strong absorption of radiation by the solid adsorbent. Thus detailed absorption bands of the solid support are not always observed, and not all of the absorption bands of the adsorbed molecules can be obtained. Sometimes the absorption bands of prime interest are outside the range scanned by the spectrophotometer, are beyond the transmission range of the cell windows or are obscured by strong vibrations of the zeolite lattice. For example, it is often difficult to observe the direct bonding of molecules to metal atoms.

In certain systems, other spectral techniques such as Raman, visible or ultraviolet spectra can provide additional data. However, little use has yet been made of these techniques.

Care must be exercised in interpreting the spectra of surfaces. Generally, species detected will be stable forms and not initial or transient adsorbed species which might be of greatest interest in catalysis. However, very detailed and specific information can be obtained concerning the stable chemisorbed species, physically adsorbed molecules, and how these species interact with the surface functional groups. It is also possible to observe different stable structures at different temperatures.

Experimental Techniques. The experimental methods used in studying surfaces and surface reactions have been reviewed recently (*1, 2, 9, 10*). Hence, only brief discussion will be given. Nearly all techniques used have employed absorption spectroscopy.

Infrared Spectrophotometers. Studies of zeolite samples place high demands on spectrophotometric equipment. Because most samples used in surface chemistry and catalytic applications have very poor transmission, relatively little energy passes through the sample and reaches the detector. Furthermore, in most studies of adsorbate–adsorbent interactions, the absorption bands, with a few exceptions, are weak.

In general, to obtain good quality spectra, good spectrophotometers are required which have high energy and good signal to noise ratios. Higher than normal energy can often be obtained with a given instrument by increasing the source radiation and by increasing slit widths. Grating spectrophotometers are superior to prism types for recording spectra of zeolites. Improved spectra can also be obtained by using spectrophotometers which operate on ratio recording rather than optical null systems. Originally it was desirable to build a spectrophotometer or to modify a commercial spectrophotometer. Nowadays, there are several commercial instruments available which can be used without modification. For samples which have very low transmission, a standard technique to increase the recorder deflection is to use wire mesh screens in the reference beam of a double beam instrument to compensate for scattering losses in the sample beam.

An additional desirable feature, especially for catalysis studies is the ability to observe the spectra at non-ambient temperatures. To make such measurements, a spectrophotometer is required which does not chop the radiation after it has passed through the sample. Attempts to observe spectra at elevated temperatures with chopping of the radiation after the sample result in many problems because of emission of radiation from the hot sample and cell components. When it is desired to control accurately the sample temperature, sample heating in the infrared beam must be minimized by suitable optical filters.

Sample Preparation. One of the more critical areas in the study of zeolites is sample preparation. The most serious problem is the minimization of radiation loss from scattering. Generally scattering problems are not too severe below 2000 cm^{-1}. However, at shorter wavelengths, serious problems can be encountered, particularly with zeolites which have been formed in relatively large crystallites. Samples must also be made thin enough so that adequate transmission is obtained (at least about 2 %) and yet remain thick enough so that reasonably strong absorption bands are observed. In many cases compromises are necessary, or two separate samples are used to optimize conditions for the desired study.

Many techniques have been used to prepare samples. These have been reviewed in detail (*1*, *11*). Since infrared studies of zeolites are relatively recent, most work has involved the use of the newer and more refined techniques.

NUJOL MULL AND ALKALI HALIDE TECHNIQUES. The nujol mull (also fluorocarbon mull) and alkali halide pellet methods have been used in many studies. In the nujol mull method, a small quantity of zeolite is dispersed in nujol. A lake of liquid is then spread uniformly between two infrared

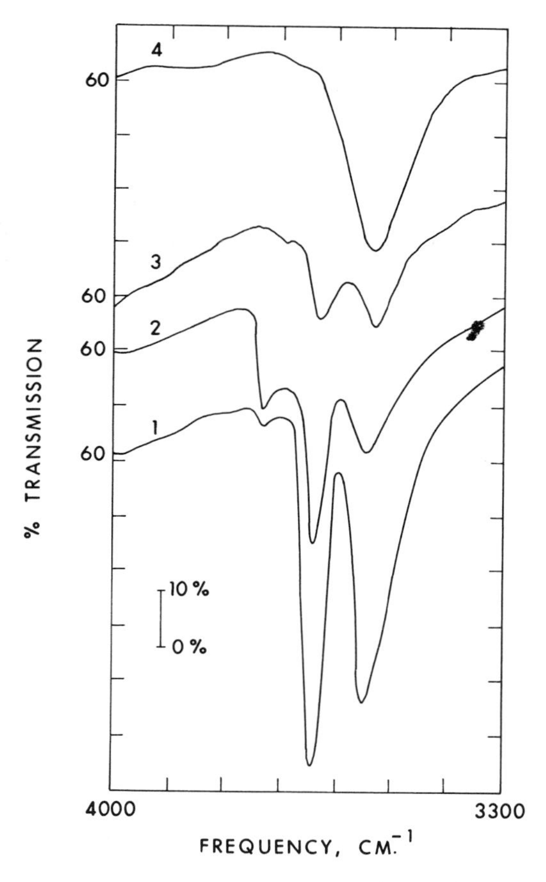

Journal of Catalysis

Figure 1. (1) Decationized-Y calcined in vacuum at 350°C; (2) As (1) but hydrated in air and recalcined at 350°C; (3) Calcined as powder and then mulled with Convalex; (4) Calcined as powder and then mulled with Nujol (20)

transparent windows and the spectra are recorded. To cover the complete spectral range, both mulling fluids are used.

Halide pellets are made by mixing the zeolite with about 500–1000 times its weight of alkali halide and then compacting into a clear, transparent wafer in a metallurgical die at pressures up to 20,000 lbs/in^2. The actual halide used depends on the spectral region to be investigated. Potassium bromide and cesium iodide wafers allow the whole spectral range of interest to be covered.

These methods have various advantages and disadvantages. Disadvantages include: (1) The samples cannot be treated after their incorporation into the halide or nujol. Thus, unless precautions are taken, the types of measurement are restricted, *e.g.*, by physically adsorbed water. Samples must be in the desired state before incorporation into the matrix. Unless pretreated, adsorbate–adsorbent interactions cannot be studied; and (2) Interactions can often occur with the halide matrix. Either intermolecular interactions or ion-exchange can occur unless suitable precautions are taken.

Advantages include: (1) the samples are easily prepared; and (2) very low or high concentrations of zeolite in the matrix can be prepared. Hence it is much easier to obtain absorption bands of the most desirable intensity. In particular, suitable samples can readily be prepared for investigation of the strong lattice bands around 1300–800 cm^{-1} (*see* Chapter 2).

The type of problem sometimes encountered when mulling techniques are used is illustrated in a recent study (*20*). Figure 1 shows spectra of decationized Y (1) after calcination, (2) as a wafer after rehydration and recalcination, (3) after suspension in convalex, and (4) after suspension in nujol. Only a single band is observed at 3550 cm^{-1} for the nujol sample compared with the usual three bands for the convalex sample. However for the convalex samples, the band intensities are somewhat different to those of the pressed wafers, indicating that mild interactions are occurring.

SELF SUPPORTING WAFERS. Most studies of gas–solid or liquid–solid adsorption have used self supporting wafers of zeolite formed by compacting the zeolite in the same type of die, although possibly larger in diameter, as used for halide wafer preparation. The thickness of the wafer is controlled by the weight of material placed in the die. Pressures used in compacting the sample range from about 1000 to 100,000 psi. In general, the lowest possible pressure should be used so that interactions and loss of crystal structure can be avoided.

Zeolites are moderately difficult to form into wafers. The ease of pressing varies markedly from batch to batch of zeolite. Zeolites are generally easier to compact than alumina but more difficult than silica. Scattering can be a serious problem, particularly with X zeolite and mordenite. Use of zeolites with small particle size can help overcome this problem.

Another method to overcome detection problems, particularly when studying hydroxyl groups, is to convert the groups to deuteroxyl groups by deuterium exchange and then to study these groups.

Advantages of self supporting wafers include: (1) scattering losses are only a fraction of those found with powders; (2) they can be easily used in vacuum and purge systems; and (3) since there is no matrix, meaningful adsorption, desorption, and surface reaction measurements can be made.

Disadvantages include: (1) sometimes they are difficult to form; and (2) it can be difficult or impossible to obtain the best sample thickness. In the lattice vibration region, the absorption bands are usually much too strong for good measurements to be made.

SAMPLE CELLS. The type of cell used depends on the investigation being conducted. Little and Hair have recently reviewed a variety of cells used for studying spectra of surfaces and surface reactions (*1*, *2*). Studies using nujol mulls and halide discs can be made in conventional infrared equipment.

Desirable features of cells used to study supported wafers are: (1) the sample temperature can be controlled over a wide range both above and below room temperature; (2) the sample treatment and gas adsorption are possible while the sample is in the infrared beam; (3) the samples can be calcined at temperatures up to about 800°C; (4) short path length is possible within the cell; (5) the gas phase can be examined independent of the sample; (6) the amount of adsorbate taken up can be measured independently; (7) cell windows do not limit frequency range; (8) cells can be easily dismantled for sample change and cleaning; and (9) cells can be used in a wide range above and below atmospheric pressure.

No cell description published has fulfilled all of these design criteria. Usually several cells are used which have various limitations but have the features necessary for the particular experiments. There is a trend now for measurements to be made at non-ambient temperatures and pressures. This trend is particularly apparent in catalytic reactions for which more meaningful results are obtained from studies near reaction temperatures.

Most cells used for zeolite studies have two or more of the following features; they can (1) be heated to 600°C for sample pretreatment; (2) be evacuated; (3) have monitored volumes of adsorbates added; (4) be used with a continuous gas flow; and (5) be used to record spectra at non-ambient temperatures.

OTHER FORMS OF INFRARED SPECTROSCOPY. Infrared reflection spectra and emission spectra have potential use in zeolite studies. So far, these techniques have been used only to a limited extent. The applications of the techniques to other surface systems have been discussed (*1*, *2*, *19*).

Interpretation of Infrared Spectra. The initial interpretation of spectra is made by comparison with characteristic group frequencies of typical chemicals. For instance, an absorption band in the 3750–3500 cm^{-1} region would be attributed to hydroxyl groups or in the 3200 cm^{-1} region to acetylenic C—H groups.

Interpretations are usually made by comparison with similar molecules in the liquid or solution phase. For example, spectra of adsorbed hydrocarbons are usually compared with spectra of liquid hydrocarbons or solutions of hydrocarbons. Thus spectra derived from adsorbing propane can be compared with hexane (dipropyl). In other cases, environments or bonding approximations are made to simulate the surface interaction. For example, the interaction of pyridine with aluminum chloride can simulate pyridine interaction with Lewis acid sites on a surface.

Because of the nature of the comparisons involved in making these deductions, discretion is often needed in making interpretations. Also, often no comparable frequencies have been observed in model systems so that chemical intuition is needed in attempts to interpret absorption bands. Further complications can arise from shifts in the frequencies of bands by the large fields present on the zeolite surfaces.

Structural Hydroxyl Groups

Cation-Containing Zeolites. ALKALI- AND ALKALINE EARTH-CATION ZEOLITES. The earliest spectroscopic study of zeolite surfaces is probably due to Frohnsdorff and Kingston (*21*) who investigated Na-, Ca-, and Li-A zeolites. The dehydrated zeolites contained no hydroxyl groups detectable under the conditions used. Szymanski *et al.* (*22*) observed hydroxyl bands at 3500 and 3400 cm^{-1} on Na-A. However, the H_2O band near 1650 cm^{-1} was detected, so it is not clear whether these bands are from structural OH groups or adsorbed water.

The initial work on the faujasite-type zeolites (X and Y) was done by Szymanski *et al.* (*22*). They observed Na-X after dehydration at a series of temperatures between 25 and 500°C. After treatment at 500°C, two hydroxyl absorption bands were observed at 3500 and 3400 cm^{-1}. Since these bands were accompanied by the water-bending vibration near 1660 cm^{-1}, as in the case of Na-A, the OH bands may be caused by structural hydroxyl groups or by physically adsorbed water. In several studies, Zhdanov *et al.* (*23, 24, 25*) found only a weak band near 3690 cm^{-1} on Na-X after treatment under vacuum at 300°C. Since the band was absent at 400°C, they considered that this band was related to the Na cation rather than to structural hydroxyl groups. More recent studies have suggested that the band is probably caused by small amounts of molecular water interacting with the cation directly. A more detailed study of Li-, Na-, and K-X zeolite was made by Bertsch and Habgood (*26*). After pretreatment at 500°C, no hydroxyl groups were detected for any of the samples. In contrast, Carter *et al.* (*27*) found hydroxyl groups on a series of X zeolites (Li, Na, K, Ag, Ba, Ca, Sr, Cd). The zeolites were dehydrated at a series of increasing temperatures from 150° to 600°C under vacuum. Typical frequencies are listed in Table I for samples dehydrated at 450°C. They also showed that the OH groups could be exchanged by deuterium. The results for the monovalent cations are in direct conflict with the observations of Bertsch and Habgood. The number and frequencies of the hydroxyl groups vary with the cation exchanged. The 3750 cm^{-1} band was attributed to silica-like hydroxyl groups, the 3715-3685 cm^{-1} band to AlOH groups, and the lowest frequency band ($\sim$3600 cm^{-1}) to OH groups affected by lattice cations since this frequency varied with cations. A band was also observed in most cases near 3650 cm^{-1},

Table I. Frequencies of Structural OH and OD Groups on X Zeolites (*27*)

Cation	*Band*	*Band Position (cm^{-1})*			
Li	OH	3740		3660	
Na	OH	3740	3695	3665	
	OD	2765	2725	2700	
K	OH	3750	3715	3650	
	OD	2760	2735	2690	
Ag	OH	3750	3685	3630	
Ba	OH	3750	3695	3620	
Ca	OH	3750	3695	3590	
Sr	OH	3750	3700	3660	3605
Cd	OH	3750	3690	3600	
	OD	2760	2715	2640	

Journal of Physical Chemistry

similar in frequency to the strong band in decationized Y, although this was not assigned by Carter *et al.* In a subsequent study as part of an attempt to clear up the discrepancy, Habgood (*28*), by using thicker samples, observed bands at 3750 cm^{-1}, 3695 cm^{-1}, and 3655 cm^{-1} in Na-X. The same bands, however, could be produced by extensive washing or treatment of thin samples with dilute acid. It was, therefore, concluded that traces of structural hydroxyl groups could be present as cation deficiencies or impurities in the zeolite. Whether or not hydroxyl bands are detected appears to depend on the extent of cation deficiency and the spectral sensitivity. Hattori and Shiba (*29*) observed only a single band at 3750 cm^{-1} on Ca-X after treatment at 450°C. Bands at 3650 and 3570 cm^{-1} were observed at 300°C.

A series of Group IA and IIA Y, Na- and Ca-X zeolites were studied by Angell and Schaffer (*30*) after pretreatment at about 500°C. They observed hydroxyl groups in all cases which could be exchanged with D_2O.

Table II. Frequencies of Structural OH Groups on Zeolites (*30*)

Zeolite	*OH Band cm^{-1}*				*OD Band cm^{-1}*		
Decat. Y	3744		3636	3544	2758	2686	2617
Na-Y	3748		3652				
Li-Y	3744						
Mg-Y	3745	3688	3643	3540	2762	2686	2616
Ca-Y	3746		3645		2762	2690	
Sr-Y	3746	3691					
Ba-Y	3744		3647				
Mn-Y	3748		3644	3545		2685	2616
Co-Y	3748		3646	3540			
Ni-Y	3746	3682	3643	3544			
Zn-Y	3744	3673	3642	3542			
Ag-Y	3745		3634	3550	2762		2610
Na-X	3744						
Ca-X	3744						

Journal of Physical Chemistry

Table II summarizes the data. The band at 3745 cm^{-1} was observed on all samples and was attributed to silica-type hydroxyl groups present either as lattice terminators or siliceous impurities. On Na- and Ca-X and Li-Y, this was the only band detected. A second band at 3652 cm^{-1} was observed on Na-Y. These observations seem to be intermediate between those of Bertsch *et al.* and Carter *et al.*, suggesting that sample differences may be the reason for the variation in results. No distinct pattern was observed in the Group II zeolites studied. Bands were observed at or near 3690, 3640 and 3540 cm^{-1}. The bands at 3640 and 3540 cm^{-1} were attributed, by analogy with decationized Y, to cation deficiencies with the consequential introduction of hydrogen into the structure. A band near 3690 cm^{-1} in Mg- and Sr-Y was attributed to AlOH groups. However, in view of more recent studies (*26, 31, 32, 33, 34, 35, 36*) this band may be caused by small amounts of water interacting with the cations. Contrary to the above studies, Eberly (*34*) reported no hydroxyl groups on Na- and Ca-Y after dehydration at 427°C. Similarly, Ward (*32, 35*) and Hall *et al.* (*31, 36*) also reported no structural hydroxyl groups on group IA zeolites. However, they did observe

Table III. Frequencies (cm^{-1}) of

Temp., °C	*Li*	*Na*
100	3740, 3718, 3670, 3560, 3420, 3280	3735, 3690, 3600, 3380, 3280
225	3740, 3718, 3660, 3560, 3400, 3280	3735, 3690, 3635, 3544
360	3740, 3715, 3580, 3450, 3270	3738, 3695, 3588
480	3740, 3714, 3600-3300	—
Ambient	3740, 3714, 3628, 3550, 3420, 3230	3735, 3694, 3460, 3360, 3242
250	3740, 3714, 3600-3300	—
	Mg	*Ca*
110	3740, 3690, 3640, 3548, 3460-3300	3800-3100
230	3740, 3688, 3642, 3550	3740, 3690, 3640, 3590, 3560, 3520
350	3740, 3688, 3642, 3540	3740, 3690, 3640, 3590, 3520
480	3740, 3688, 3642	3738, 3690, 3640, 3585, 3540
Ambient	3740, 3700, 3690, 3550	3740, 3683, 3640, 3550
105	3740, 3688, 3642, 3545	3740, 3700, 3640, 3560, 3400
225	3740, 3688, 3642, 3540	3740, 3690, 3640, 3585, 3520
350	3740, 3688, 3642, 3540	3740, 3690, 3640, 3585

hydroxyl bands on group II A-cation-exchanged zeolites. For Mg- and Ba-Y, Hall *et al.* (*36*) observed frequencies analogous to those of Angell and Schaffer (*30*). By correlating with other data, they found that the frequency of the 3650 and 3545 cm^{-1} bands decreased with increasing electron affinity of the cation. In a subsequent study (*31*), they attributed a band near 3690 cm^{-1} to water bound to sodium ions in the structure in agreement with others (*26, 32*) and a band near 3605 cm^{-1} to $CaOH^{+}$ groups. This second study involved the K^{+}, Mg^{2+}, Ca^{2+}, and Ba^{2+} forms of both X and Y zeolites. In most cases, bands were observed at 3740, 3690–3695, 3650 and 3550 cm^{-1}. A systematic study of Group IA and IIA Y zeolites and Na-X has been made (*32*). No hydroxyl groups were detected on Group IA, Ba-Y and Na-X zeolites apart from Li-Y after evacuation at 360°C. For lithium, bands were observed at 3740, 3714, and 3600–3300 cm^{-1}. The two lower frequency bands are probably caused by water remaining adsorbed on the zeolite interacting with the zeolite cations and lattice.

For the other Group IIA zeolites, hydroxyl bands were detected, in agreement with others (*29, 30, 31, 34, 35, 36*). The frequencies after calci-

Various Cation Y Zeolites (*32*)

K	*Rb*	*Cs*
3740, 3610, 3540, 3440, 3290	3740, 3620, 3584, 3520, 3440, 3280	3740, 3610, 3520, 3500-3200
3740, 3590, 3520, 3400, 3290	3740, 3640-3400	3740, 3656-3200
3740, 3700-3400	3740, 3650-3400	3740-3200
—	—	3740
3668, 3490, 3420, 3255	3654, 3580, 3420, 3260	3640, 3450, 3270
—	—	—

Sr	*Ba*	*RE*
3740, 3690, 3638, 3610, 3540, 3460	3740, 3700-3100	3740, 3620, 3560
3740, 3690, 3639, 3570, 3500-3200	3740, 3700-3100	3740, 3630, 3618, 3555
3738, 3700, 3639, 3570	3740, 3690-3300	3740, 3636, 3540
3740, 3690, 3640	3740	3740, 3640, 3522
3735, 3679, 3639, 3590-3200	3740, 3682, 3630, 3430	3740, 3636, 3610, 3580, 3520
3735, 3680, 3639, 3572, 3500-3200	3740, 3682, 3500, 3260	3740, 3636, 3610, 3580, 3520
3735, 3690, 3639, 3570	3740,3682, 3500, 3200	3740, 3636, 3610, 3580, 3520
—	—	3740, 3640, 3528

Journal of Physical Chemistry

nation at several temperatures are listed in Table III. In general, the observed frequencies are similar to those observed in other studies but are probably more comprehensive and internally consistent. Absorption bands were observed at 3740, 3688–3690, 3640, and in the case of Ca, at 3585 and 3540 cm^{-1}. The exact frequencies of the absorption bands varied systematically with the properties of the exchanged cation, such as electrostatic field, electrostatic and ionization potential. Typical spectra for the Group IA and IIA-cation forms are shown in Figures 2 and 3.

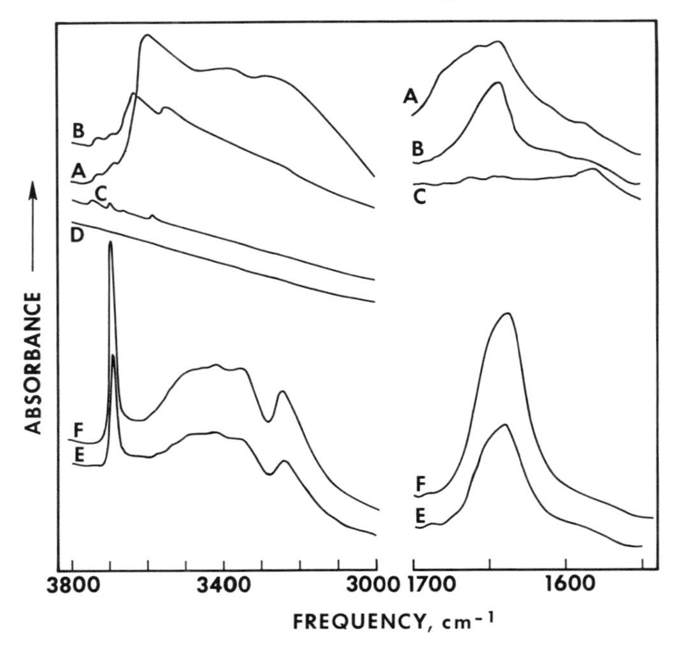

Journal of Physical Chemistry

Figure 2. Na-Y zeolite (A) evacuated overnight at 100°C; (B) 2 hr at 225°C; (C) 2 hr at 380°C; (D) 2 hr at 480°C; (E) 6μ mol of H_2O added; (F) 12μ mol of H_2O added (32)

If interpretation of the more recent data is emphasized, which is justifiable because of the superior spectroscopic techniques used, the properties of cation-exchanged zeolites can be rationalized to give a uniform description of the surface groups. The frequencies observed in various reports are summarized in Tables I-IV.

Although hydroxyl groups were detected in a few instances on Group IA zeolites, it appears that non-cation deficient monovalent zeolites contain no structural hydroxyl groups. Hydroxyl groups on some Group IA zeolites are probably due to cation deficiency caused by partial hydrolysis, multivalent cation impurities and slight amounts of siliceous impurities. The only hydroxyl groups expected from structural considerations would be those terminating the giant lattice with a frequency near 3740 cm^{-1}. These conclusions are

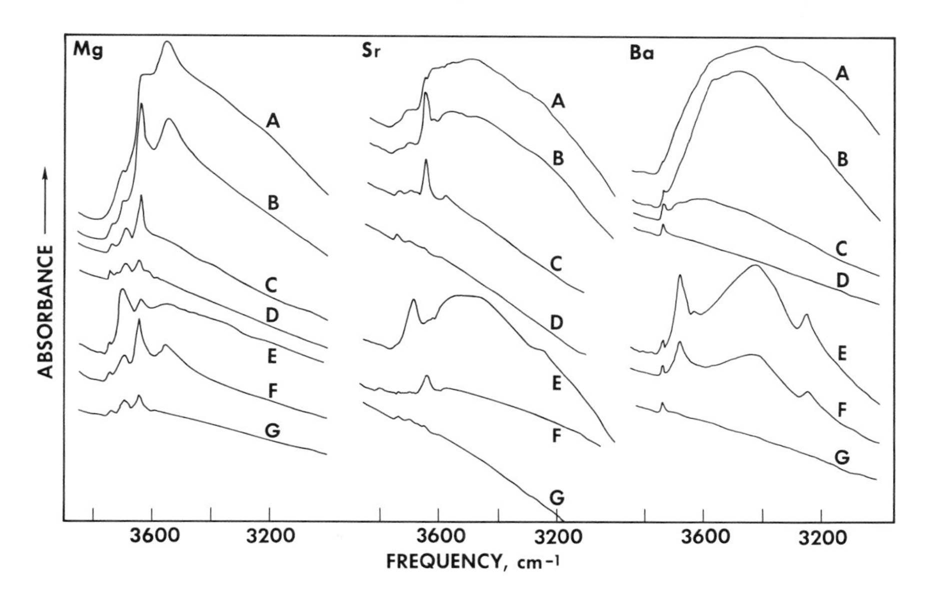

Journal of Physical Chemistry

Figure 3. Mg-, Sr-, and Ba-Y zeolites (A) evacuated overnight at 110°C; (B) 2 hr at 215°C; (C) 2 hr at 350°C; (D) 2 hr at 480°C; (E) 12μ mol of H_2O added; (F) evacuated at 150°C; (G) evacuated at 330°C (32)

confirmed by the data of Habgood (*28*) who showed that treatment of Na-zeolite with water or dilute acid could introduce structural hydroxyl groups. Barium-exchanged zeolite behaves like Group IA zeolites in that no significant hydroxyl groups are detected. Some structural hydroxyl groups must be present since Bronsted acidity is detected, whereas no acidity is detected on the Group IA-exchanged zeolites.

For the remaining zeolites (Mg, Ca, Sr), the absorption bands at 3650 and 3540 cm^{-1} are similar in frequency to those observed in decationized

Table IV. Frequencies of Structural OH Groups on X and Y Zeolites

Zeolite	*Band Position, cm^{-1}*					*Ref.*
Ba-Y			3645			*36*
Mg-Y	3745	3685	3642	3550		*36*
Zn-Y	3745	3675	3645	3550		*36*
Ca-Y	3745	—	3640	3550		*36*
$Ca_{0.64}$-X	3740	3695	3650	3550	3605	*31*
$Ca_{0.42}$-X	3740	3695	3650	—	3610	*31*
Ca-Y	3740	3695	3650	3550		*31*
Mg-X	3740	—	3650	3550		*31*
Mg-Y	3740	3690	3650	3550	3610	*31*
Ba-X	—	—	3650	3550	—	*31*
Ba-Y	—	—	3650	3550	—	*31*

Y zeolites (below) and are probably the same type of hydroxyl groups. Confirming this are the observations that pyridine and ammonia interact with these groups in the same manner as those on decationized Y zeolite. The concentration of the hydroxyl groups is much less than that on decationized Y but is much greater than observed for cation-deficient group IA zeolites. The concentration also varies with the cation present in the zeolite. Therefore the presence of the hydroxyl groups may not be due to cation deficiency. A possible explanation is that the simple divalent cation cannot satisfy the charge distribution requirements of the zeolite lattice in the absence of much water. During dehydration, the divalent cation becomes localized, and its associated electrostatic field may induce dissociation of coordinated water molecules to produce MOH^+ and H^+ species. The proton would react with the lattice oxygen at a second exchange site to produce the hydroxyl present in the decationized zeolites with absorption frequencies near 3650 and 3540 cm^{-1} (*30, 31, 32, 33, 34, 35, 36*).

```
     M2+(OH2)                          MOH+                   H
                                                              |
   O     O     O                  O     O     O     O     O
    \   / \   /                    \   / \   / \   / \   /
2    Si    Al-      ---------->     Si    Al-   Si    Al
    /   \ /   \                    /   \ /   \ /   \ /   \
   O     OO    O                  O     OO    OO    OO    O
```

Other bands in the spectra of the divalent cation zeolites are not found in the spectra of the alkali cation and decationized zeolites. They are probably specific to the presence of divalent cations. The band around 3690 cm^{-1} was originally assigned to AlOH groups. Although this cannot be ruled out, the assignment to traces of physically adsorbed water may probably be more correct (*31, 32*). As discussed below, a band is usually found near this frequency when water is adsorbed on the zeolites and is generally attributed to the interaction of molecular water with the cations. In these cases, the water-bending band near 1650 cm^{-1} is detected. However, depending upon the pretreatment conditions of the zeolite, the band could be caused by structural hydroxyl groups similar to those found in stabilized zeolite (*40, 41, 42, 44*). The second band occurs between 3600 and 3570 cm^{-1} depending on the cation. The frequency increases with decreasing cation-size indicating stronger interaction with the stronger fields of the small cations. This type of hydroxyl group has been attributed to the MOH^+ groups formed in the hydrolytic scheme above (*31, 32*). It is very sensitive to the level of hydration and disappears on mild dehydration possibly with the formation of MO or M^+–O–M^+ groups (*38, 39*). Again, these hydroxyl groups were non-acidic to pyridine. In a study of hydrated Ca-Y and Mg-Y (*32, 43*), the disappearance of the band near 3690 cm^{-1} was observed with increasing temperature and the formation of bands near 3590 and 3640 cm^{-1} was observed. The assignment of the band in the 3600–3570 cm^{-1} region to MOH^+ groups has recently been questioned since a band near 3600 cm^{-1} is observed in stabilized Y zeolites (*40, 41, 42*). Whether the band is augmented by stabilization products is not clear, but it appears clear that since the band can be formed by treatment of the zeolite at 250°C and eliminated by higher temperature treatment, some other type of group, presumably MOH^+ groups, must be involved.

Alkali Cation L Zeolites. In contrast to X and Y zeolites, other zeolites have received little attention. Two investigations of L zeolite have been reported. Group IA cation zeolites were reported to contain no structural hydroxyl groups apart from the usual band near 3740 cm^{-1} (*18*). Tsitsishvili (*45*) has reported the spectra of potassium L zeolite as a function of evacuation at different temperatures. Typical spectra are shown in Figure 4. After treatment at 40°C, absorption bands from physically adsorbed water

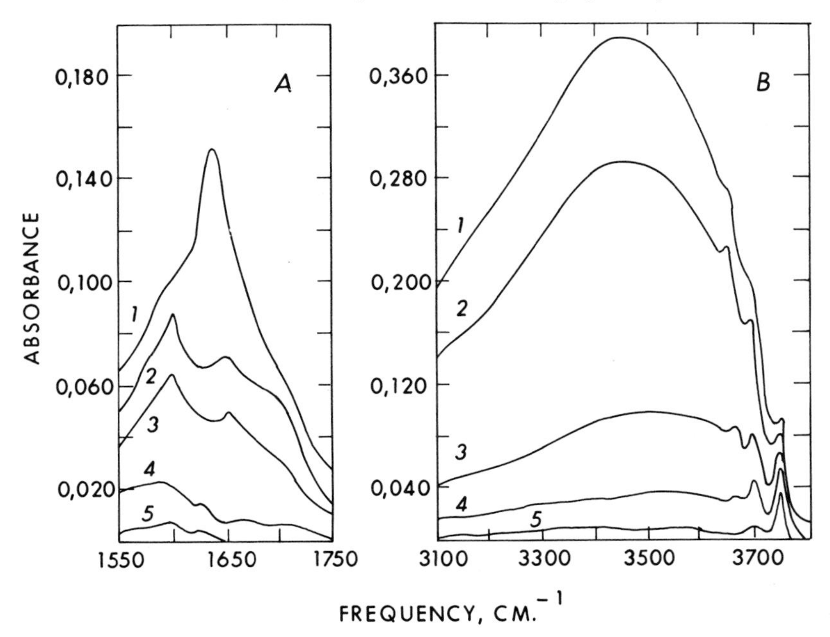

Advances in Chemistry Series

Figure 4. K-L zeolite after evacuation at (1) 40°C; (2) 100°C; (3) 320°C; (4) 410°C; (5) 500°C (45)

near 1650 cm^{-1} and 3600–3200 cm^{-1} are observed. Treatment between 100 and 500°C reveals three absorption bands near 1602, 1630, and 1650–1660 cm^{-1}, suggesting different locations for adsorbed water. Similarly bands from hydrogen-bonded water are seen near 3665, 3685, and 3700 cm^{-1}. As the temperature is raised, the water band disappears until just the band near 3740 cm^{-1} remains. This latter band may be caused by trace amounts of residual water bound directly to the potassium cations or by hydroxyl groups attached to amorphous impurities in the zeolite.

A detailed study of several cation-exchanged L zeolites has been made (*46*). Up to 120°C, the water bending vibration near 1640 cm^{-1} is detected. Above this temperature, no physically adsorbed water is observed. For K-L,

Table V. Frequencies (cm^{-1}) of

Temp., °C	*Mn*	*Co*
≈100	3740, 3680, 3635, 3530	3740, 3630, 3615, 3540
≈220	3740, 3680, 3636, 3530	3740, 3638, 3540
350	3740, 3680, 3636, 3530	3740, 3638, 3540
450	3740, 3680, 3636	3740, 3705, 3638
Ambient	3740, 3680, 3520	3740, 3680, 3638, 3615, 3540
100	3740, 3680, 3630, 3525	3740, 3680, 3638, 3540
250	3740, 3675, 3630, 3540	3740, 3638, 3540
350	3740, 3680, 3630	3740, 3638, 3540
	Cu	*Zn*
110	3740, 3635, 3600-3200	3740, 3610, 3570, 3510
≈220	3740, 3625, 3550	3740, 3670, 3625, 3580, 3530
≈310	3740, 3625, 3550	3740, 3625, 3530
450	3740, 3625, 3550	3738, 3670, 3640, 3560
Ambient	3700-3000	3680, 3625, 3530
≈120	3740, 3625, 3550	3740, 3630, 3540
250	3740, 3625, 3550	3740, 3630, 3540
350	—	3740, 3670, 3630, 3560

bands are observed at 3740, 3682 and a broad band near 3600–3200 cm^{-1}. Only the bands at 3740 and 3682 cm^{-1} are observed after treatment at 330°C. These results are similar to those reported by Tsitsishvili. Similar results were obtained for Na-L and Cs-L, except the band near 3682 cm^{-1} for K-L was displaced to 3689 cm^{-1} in Na-L and 3677 cm^{-1} in Cs-L. This behavior is similar to that of the X and Y zeolites.

Transition Metal Cation Zeolites. Several studies of transition metal cation zeolites have been made. The initial studies of Carter *et al.* (*27*) of Ag- and Cd-X zeolites found hydroxyl bands at 3750, 3695 (3690), and 3630 (3600) cm^{-1}. Hattori and Shiba (*29*) extended the study to Zn-X finding absorption bands at 3750, 3650 and 3570 cm^{-1}. Angell and Schaffer studied Mn-, Co-, Ni-, Zn- and Ag-Y (*30*). These data are summarized in Table II for the zeolites after calcination at 500°C. Later Zn-Y was studied (*36*) and bands were found at 3745, 3675, 3645, and 3550 cm^{-1}, these frequencies being almost the same as those reported by Angell *et al.* Mn-, Co-, Ni-, Cu-, Zn-, Ag-, and Cd-Y zeolites have recently been investigated (*47*) as a function of dehydration temperatures between 100 and 450°C. The data obtained are summarized in Table V. Bands are observed in all cases at 3740 and near 3640 cm^{-1}. For most forms a band is observed near 3530 cm^{-1} and near 3680 cm^{-1} for Mn, Co, Ni, Zn, and Cd. In the case of Cd, an additional band is observed near 3460 cm^{-1}. Absorption bands were detected on Ni-X at 3650, 3560, 3610 and 3590 cm^{-1} after dehydration at 250°C by Guilleux *et al.* (*48*). Only a single band at 3650 cm^{-1} was detected at 450°C. Somewhat different results were obtained for Cu-Y by Ben Taarit *et al.* (*49*). For the initial zeolite, bands were observed at 3745 and 3680 cm^{-1} after treatment at 500°C. After treatment at 200°C, bands

Various Cation Y Zeolites (*47*)

Ni
3740, 3640, 3535
3740, 3680, 3635, 3540
3740, 3680, 3635, 3540
3740, 3680, 3635, 3540
3740, 3675, 3635, 3600, 3540
3740, 3690, 3635, 3600, 3540
3740, 3690, 3635, 3540
3740, 3680, 3635, 3540

Ag	*Cd*
3740, 3680, 3638, 3585	3740, 3636, 3610, 3530
3740, 3680, 3638	3740, 3636, 3530
3638	3740, 3630, 3530, 3460
3638	3740, 3680
3685, 3550, 3480	3700, 3650-3300
3638	3740, 3636, 3530
3638	3740, 3636, 3530
—	3740, 3636

Transactions of the Faraday Society

are observed at 3740, 3615, and 3550 cm^{-1}. At 300°C, the 3615 cm^{-1} band has disappeared. At higher temperatures, the 3640 and 3550 cm^{-1} bands were no longer observed. The band at 3615 cm^{-1} is probably caused by $CuOH^+$ groups. This band and those near 3640 and 3550 cm^{-1} are probably generated by the reaction $Cu(OH_2)^{2+} \rightarrow CuOH^+ + H^+$. The high stability of the band near 3680 cm^{-1} can be attributed to hydroxyl groups in a stabilized structure. Typical spectra are shown in Figure 5 for transition metal zeolites.

In common with the Group IIA cation-containing zeolites, the bands near 3650 and 3540 cm^{-1} are observed in all cases. The probable origin of these bands is the same as that discussed above—namely, the hydrolytic fission of adsorbed water: $zeolite^{2-} + M^{2+} + H_2O \rightarrow MOH^+ + zeolite—OH^-$. The ability of the zeolite to dissociate adsorbed water varies with the cation. Chemisorption of bases like pyridine and ammonia indicates that these two hydroxyl groups are acidic. The hydroxyl groups are similar to those present on decationized Y and are probably associated with the O_1 and O_3 oxygens of the structure. The band observed near 3680 cm^{-1} in all cases except Cu is probably caused by MOH^+ groups. It varies in frequency by about 20 cm^{-1} with the cation. Because of this variation with cation, because it is observed after dehydration at low temperatures (100°C) and because the intensity decreases with the treatment temperature, the assignment to the MOH^+ group is more reasonable than to structures similar to those formed in stabilized zeolites.

Ni-A zeolite was studied by Guilleux and Tempere (*50*). Absorption bands were observed near 3490 cm^{-1} after evacuation at room temperature and at 3610 cm^{-1} after treatment at 150°–200°C. Above 200°C, no bands

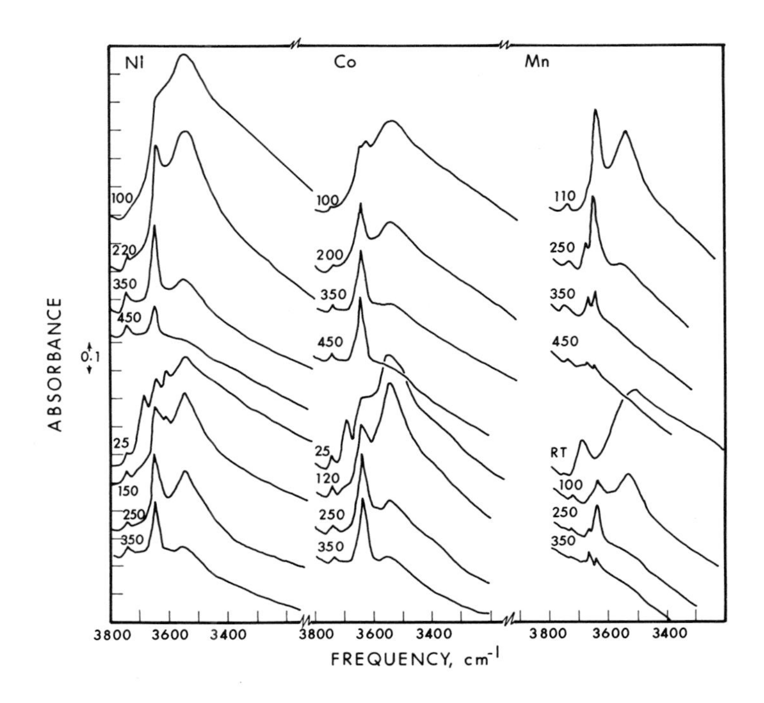

Transactions of the Faraday Society

Figure 5. Hydroxyl groups on Ni-, Co-, and Mn-Y zeolites after calcination and rehydration (47)

were observed, suggesting that the bands observed at lower temperatures are from physically adsorbed water.

Trivalent Cation Zeolite. The only trivalent ions reported so far are the rare earth-exchanged zeolites and iron exchanged zeolites. Typical spectra for a mixed rare earth-exchanged Y zeolite are shown in Figure 6. In general, after removal of physically adsorbed water, hydroxyl bands are observed near 3740, 3640, and 3530 cm^{-1} (*29, 32, 36, 51, 52, 53, 54, 55*) after evacuation of about 450°C. At lower temperatures, bands are observed at 3550 and 3618 cm^{-1}. Eberly and Kimberlin (*55*) studied most rare earth forms individually. They found that the frequency of the absorption band near 3500 cm^{-1} decreased with the cation radius from 3522 cm^{-1} for La-Y to 3470 cm^{-1} for Yb-Y. Although several studies reported that addition of water to the zeolite calcined at a temperature sufficient to bring about dehydroxylation resulted in reconstitution of the hydroxyl group, Rabo *et al.* (*52*) showed that this did not occur at room temperature but that the zeolite required heating to about 200°C.

Most of the above investigations have resulted in two proposed possible mechanisms for the formation of the hydroxyl groups. One scheme initially proposed by Venuto, Hamilton and Landis (*56*) visualizes the following hydrolysis, $RE^{3+}(H_2O)\ \bar{O}Z \rightleftarrows RE^{2+}OH + H^{+}\bar{O}Z$ giving the structure

```
RE OH2+                                   H
                                          |
O     O     O                       O     O     O
 \   / \   /                         \   /     /
  Si     Al̄          +                Si     Al
 /   \ /   \                         /   \ /   \
O     OO     O                      O     OO     O
```

Thus the SiOH group would be responsible for the 3640 cm^{-1} band and the $RE^{2+}OH$ for the 3520 cm^{-1} band.

Rabo *et al.* (*52*) proposed a somewhat different scheme based on observations that only one water molecule is present in each sodalite cage and that the hydroxyl group associated with the water molecule is shared by two rare earth cations.

```
                                                                      H
                                                                      |
                O     O     O                                   O     O     O
                 \   / \   /                                     \   /     /
2 RE + H2O +      Si     Āl   ——→ RE—OH—RE +                      Si     Al
                 /   \ /   \                                     /   \ /   \
                O     OO     O     (sodalite cage)              O     OO     O

                                                                O           O
                                                                 \         /
                            ——→ Re—O—Re +                         Si     Al
                                                                 /   \ /   \
                                (sodalite cage)                 O     OO     O
```

From a consideration of acidity and the suggested structures for divalent cation systems, Ward (*57*) suggested that the additional reaction occurred $RE(OH_2)OH^{2+} + H^+O^-—Z + O^-—Z \rightarrow RE(OH)_2^+ + 2\ H^+O^-—Z$. Dehydroxylation at higher temperature would eliminate the hydroxyl group.

Because of the similar frequencies, the 3640 cm^{-1} band has been assigned to hydroxyl groups in the supercages similar to those of decationized Y. The unique frequencies and the variation in frequency with cation suggest that the band near 3520 cm^{-1} is from hydroxyl groups associated directly with the cations. Because these groups do not interact with pyridine or piperidine they are probably located in the hexagonal prisms or sodalite cages. Since the 3520 cm^{-1} hydroxyl groups do not interact with piperidine, they must be more tightly bonded than those present in divalent cation zeolites. The x-ray diffraction data suggest that the cations are located in the sodalite cages near the S'_I sites. On severe dehydroxylation, x-ray diffraction suggests that the cations may move to the S_I positions inside the hexagonal prisms (*58*). These studies have been extended (*53*). By observing the changes in relative intensities of the two hydroxyl bands as a function of exchange level, the 3520 cm^{-1} band increased in intensity relative to the 3640 cm^{-1} band as the extent of exchange was increased. As shown thermogravimetrically, one molecule of water is associated with each rare

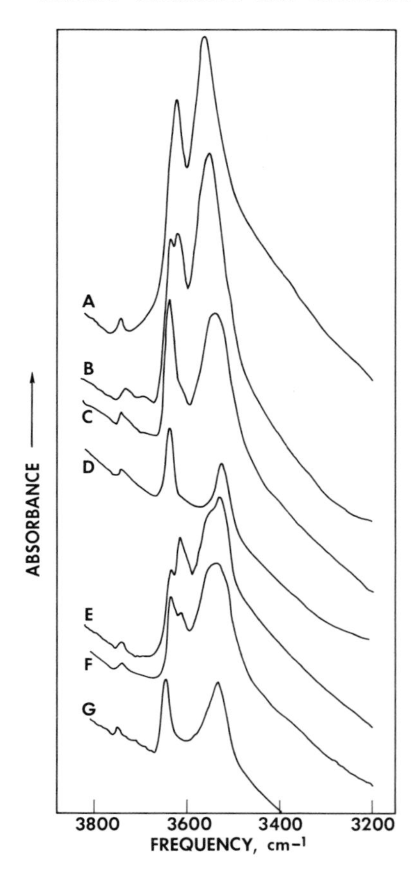

Journal of Physical Chemistry

Figure 6. Rare earth Y zeolite (A) evacuated overnight at 100°C; (B) 2 hr at 225°C; (C) 2 hr at 330°C; (D) 2 hr at 450°C; (E) 12μ mol of H_2O added; (F) evacuated at 150°C; (G) evacuated at 450°C (32)

earth cation up to 970°C for samples of various exchange levels. Because the ratio of the hydroxyl group bands changes with exchange level, a substantial contribution to the 3640 cm^{-1} band may result from exchange directly with protons from the acidic rare earth solutions, particularly at low levels of exchange.

Structural hydroxyl groups have been observed on iron zeolites by Ermolenko *et al.* (*59*). Zeolites (X and Y) prepared by exchange and heated in vacuum at 350°C contain structural hydroxyl groups with absorption bands near 3740 and 3640 cm^{-1}. These bands are typical of those found in polyvalent cation-exchanged zeolites. If the exchanged zeolite is reduced and then oxidized, an additional band is observed near 3685 cm^{-1}. This could be caused by structural changes similar to those observed for ultra-stable zeolites or by interaction of water with the iron species.

Tetravalent Cation Zeolites. A brief study of a tetravalent cation (thorium) exchanged into Y zeolite has been made (*55*). Examination of the surface area and x-ray diffraction pattern indicated that some loss in crystallinity had occurred. Hydroxyl bands near 3630 cm^{-1} were detected, but hydroxyl groups with frequencies between 3470 and 3520 cm^{-1} were not detected. The 3740 cm^{-1} band was about three times as intense as the 3640 cm^{-1} band. The intensity of this band supports the conclusion that substantial structure had been lost. No bands were detected in the 3470–3550 cm^{-1} region.

Hydrogen Zeolites and Their Derivatives. Zeolites which are most active catalytically and most closely related to amorphous silica-alumina are those based on hydrogen or decationized zeolite.

Hydrogen has been introduced into zeolites in several ways:

1. By exchange with ammonium ions followed by thermal decomposition

$$MZ \xrightarrow{NH_4^+} NH_4Z \xrightarrow{heat} HZ$$

2. By treatment with dilute acid

$$MZ \xrightarrow{HA} HMZ$$

3. By washing with water

$$MZ \xrightarrow{H_2O} HMZ$$

4. By treatment with organic ammonium salts—*e.g.*, TMA—followed by thermal decomposition

$$MZ \xrightarrow{TMA} TMAMZ \xrightarrow{heat} HMZ$$

5. By exchange with multivalent cations

$$MZ \xrightarrow{M_1^{2+}} (M_1OH)^{+}HZ\text{: } MZ \xrightarrow{M_1^{3+}} (M_1OH)^{2+}HZ \rightarrow [M_1(OH)_2]^{+}H_2Z$$

Method 5 has already been discussed, and 4 is discussed later in this section.

As discussed in Chapter 1, zeolites produced by processes 1 and 4 will be termed decationized zeolites, while those produced by processes 2 and 3 will be known as hydrogen zeolites. Ammonium X and Y zeolites have been extensively studied, probably because of their ease of preparation and important relationship to commercial catalysts. Ammonium zeolites are readily prepared by exchange of the metal cation form with ammonium salts. The extent of sodium removal from the parent zeolite depends greatly on the type of zeolite and on the severity of the exchange treatments, *viz.*, strength of solution, temperature and number of exchanges. It is relatively easy to obtain an exchange level of about 85% but exchange to lower levels of sodium requires on the order of ten or more repeated exchanges at elevated temperatures for X and Y zeolites. On the other hand, mordenite can be readily exchanged to a low sodium level.

Most studies of X and Y zeolites have used samples containing 1-2% sodium, *i.e.*, they have been exchanged to 80 to 90% by ammonium ions. Probably the earliest reported investigation is that of Szymanski, Stamires and Lynch (*22*) who studied ammonium X zeolite. In contrast to most other infrared studies of zeolites, nujol mull techniques as well as the usual compacted powder techniques were used. They also used an interesting technique of hydrating the samples with D_2O. This procedure significantly reduced the scattering problems encountered with thin discs of zeolite. Ammonium-X was studied as a function of pretreatment temperature between 25° and 600°C and observed at room temperature in nujol mulls. The disappearance of ammonium ion bands and the formation of a hydroxyl band near 3550 cm^{-1} was observed. This latter band was attributed to structural hydroxyl groups but since the band is always accompanied by a band near 1670-1700 cm^{-1} probably from the water bending vibrations, the assignment may be erroneous, and the band could be caused by physically adsorbed water.

They suggested a scheme for the formation of hydroxyl groups in which a proton, formed by decomposition of an ammonium ion, attacks the lattice to give at least two possible products.

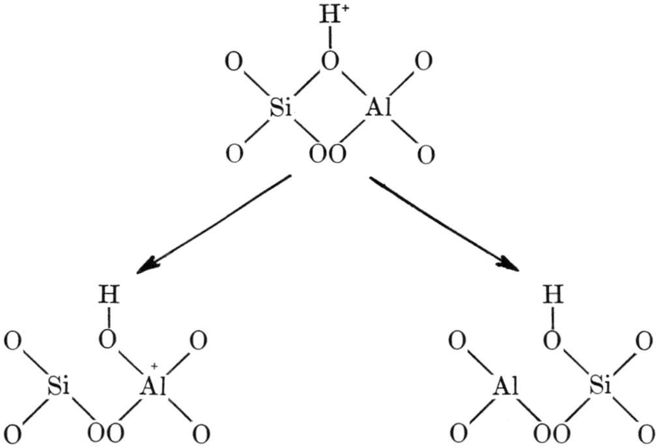

H^+
O O O
Si Al
O OO O

H
O O O
Si $\overset{+}{Al}$
O OO O

H
O O O
Al Si
O OO O

At higher temperatures, dehydroxylation occurs, resulting in a structure different from the initial starting zeolite.

Subsequently, Zhdanov *et al.* (*23, 24, 25*) using high vacuum techniques observed the disappearance of ammonium ions, as shown by the elimination of the band near 1450 cm^{-1} and the formation of hydroxyl groups with a band near 3655 cm^{-1} as NH_4-X was progressively heated under vacuum up to 300°C. Above 400°C, dehydroxylation occurred. The OH group could be exchanged with D_2O to give a band at 2695 cm^{-1}. Similarly for NH_4-Y, a hydroxyl absorption band was detected at 3645 cm^{-1} after decomposition of the ammonium ions at temperatures above 200°C.

The first reported systematic study of NH_4-X and -Y samples is that of Uytterhoeven, Christner, and Hall (*60*) who studied the deammoniation and dehydroxylation process in considerable detail. An extensive infrared study was made in conjunction with careful measurements of the zeolite hydroxyl contents using D_2 exchange methods. The Y zeolite was about 80% exchanged, and the exchange level of the X zeolite varied from about 50 to 90%. Observations were made in the 2800–3800 cm^{-1} and 1300–1700 cm^{-1} regions. The

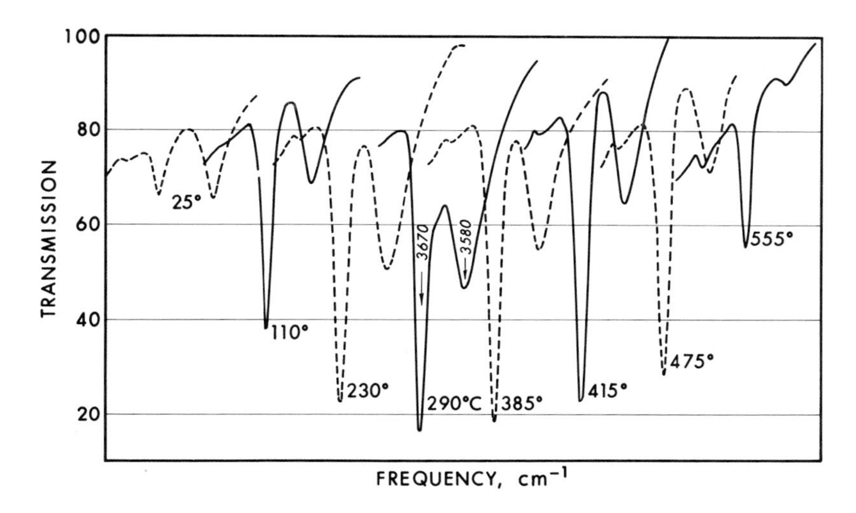

Journal of Physical Chemistry

Figure 7. Development of OH bands on NH_4-Y zeolite as a function of calcination temperature (60)

appearance of hydroxyl bands and the disappearance of NH bands as a function of the calcination temperature was studied under vacuum conditions. Observed spectra changes for NH_4-Y are shown in Figures 7 and 8. On progressive heating from room temperature to 300°C, they observed the disappearance of ammonium ion bands between 3450 and 300 cm^{-1} and near 1450 and 1670 cm^{-1} and the formation of structural hydroxyl groups with frequencies near 3740, 3660 and 3570 cm^{-1}. On heating to higher temperatures, the band intensities decreased progressively. All of the NH_4^+ bands were eliminated by treatment at 290°C. Readdition of ammonia to a decationized Y heated at 415°C resulted in the elimination of the hydroxyl bands and the

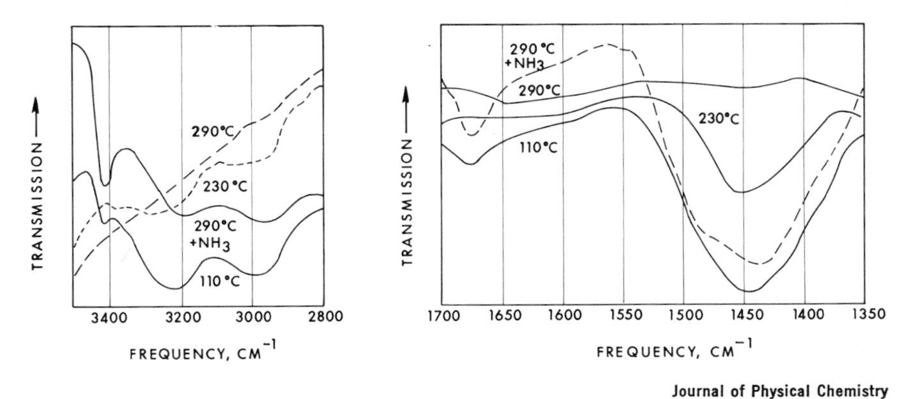

Journal of Physical Chemistry

Figure 8. Elimination of NH bands on NH_4-Y zeolite as a function of calcination temperature (60)

reformation of bands from ammonium ion vibrations. Lewis-bonded ammonia was detected on samples calcined above 300°C but not on those calcined below as expected from the initial calcination studies. However, addition of water resulted in an increase in the ammonium ion bands and a decrease in the Lewis-bound ammonia bands.

The observations of X zeolite were similar in most respects. The band near 3570 cm^{-1} was weaker, and a band near 3605 cm^{-1} was observed for samples exchanged less than 10%. Studies of X zeolite as a function of exchange level showed that the zeolite dehydroxylated more readily as the exchange level increased and that the X zeolite dehydroxylated more easily than the Y zeolite.

On the basis of the experimental data, the following scheme, which is similar to that of Szymanski *et al.* (*22*) and Rabo (*62*) was suggested.

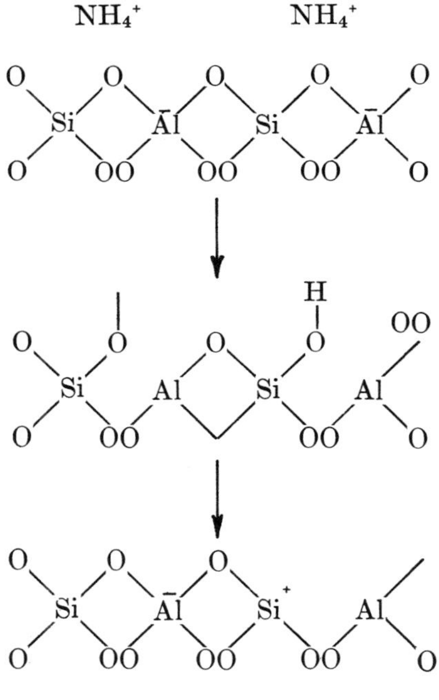

NH_4^+ NH_4^+

O O O O O
Si $\bar{Al}$ Si $\bar{Al}$
O OO OO OO O

↓

H
O O O O OO
Si Al Si Al
O OO OO O

↓

O O O
Si $\bar{Al}$ Si^+ Al
O OO OO OO O

This scheme explains the generation and elimination of hydroxyl groups, the relationship between hydroxyl groups and ammonium ions and the generation of Bronsted and Lewis acidity.

Angell and Schaffer (*30*) observed similar spectra for NH_4-Y after treatment at 500°C. The zeolite was about 90% exchanged. Three different activation techniques were used: (1) vacuum activation in which the sample was evacuated overnight at 5×10^{-5} torr, heated to 500°C over 4 hr and maintained at 500°C for 4 hr; (2) air activation in which the sample was heated to 500°C for 2 hr, maintained at 500°C for 3 hr, and then evacuated at 5×10^{-6} torr; and (3) flash activation in which the sample was

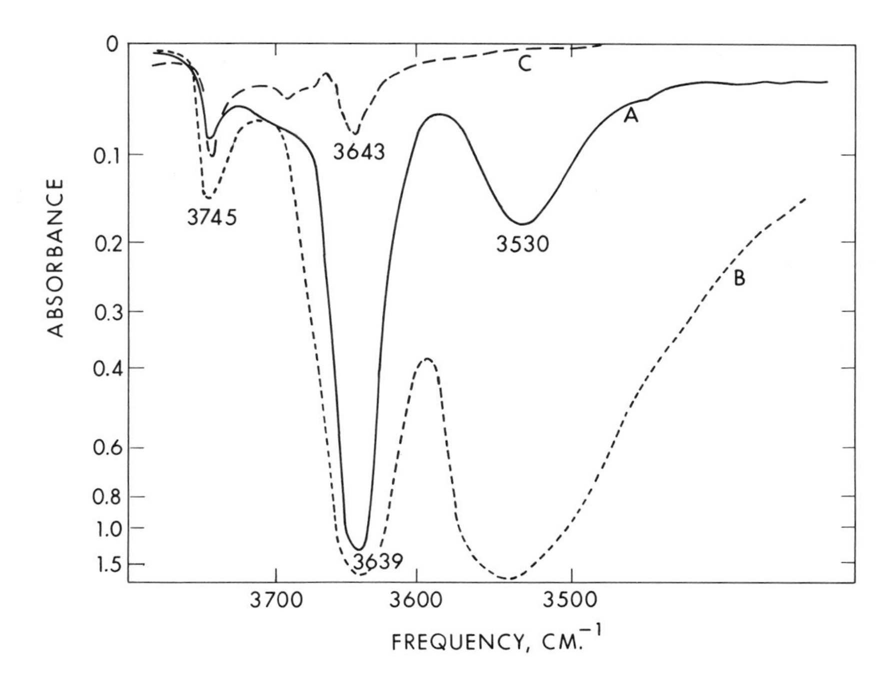

Journal of Catalysis

Figure 9. NH_4-Y zeolite calcined at 500°C (A) 7% exchanged; (B) 32% exchanged; (C) 84% exchanged (113)

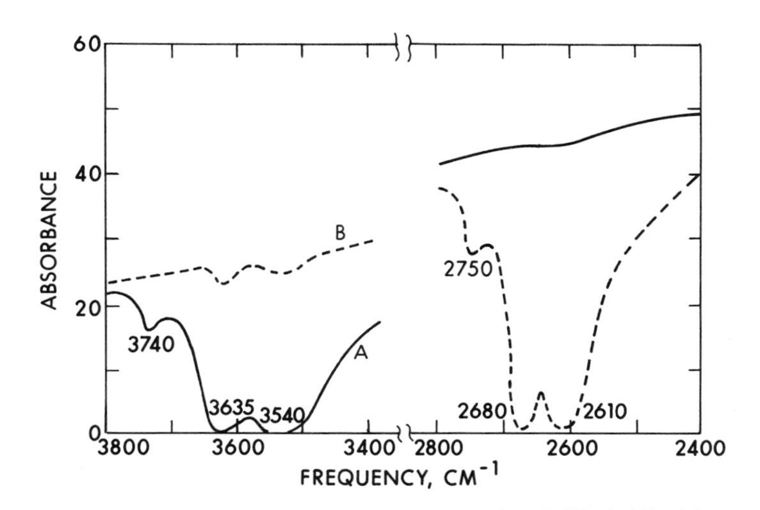

Journal of Physical Chemistry

Figure 10. NH_4-Y zeolite calcined at 427°C (A) decationized Y; (B) after treatment with D_2 at 427°C (64)

briefly evacuated at room temperature, heated to 500°C in less than 10 min, and evacuated at 500°C for 3 hr. The bands reported by Uytterhoeven *et al.* were observed. They found that the absorption bands, particularly the 3540 cm^{-1} band varied in intensity with the degree of exchange as shown in Figure 9. Flash heating of the zeolite produced weaker absorption bands than slow heating. This was attributed to self-steaming by physically adsorbed water of the rapidly heated zeolite. However, in the light of present knowledge, an ultrastable type zeolite was probably formed by the flash activation. Uytterhoeven *et al.* (*63*) compared flash activation with slow activation. They found that the flash activated samples were more stable and the 3550 cm^{-1} band had increased in intensity relative to 3650 cm^{-1} band. Eberly observed the spectrum of decationized Y at 427°C (*64*). The spectrum was similar to those reported at room temperature. The three hydroxyl groups exchanged with deuterium to give bands near 2750, 2680, 2610 cm^{-1} as shown in Figure 10. This conflicts with the observations of Angell *et al.* (*30*) who found that the 3740 cm^{-1} band did not interact. Liengme and Hall (*65*), White *et al.* (*66*) and Kermarec *et al.* (*67*) obtained similar results to Eberly (*64*) using room temperature measurements. Hattori and Shiba (*29*) observed bands at 3660 and 3570 cm^{-1} for ammonium X treated at 300°C.

Hughes and White (*68*) and Ward (*69*) carried out detailed studies of the deammination process and the variation in the hydroxyl group intensities as a function of temperature. Although their samples were more highly exchanged (90%), their results were similar to those of Uytterhoeven *et al.* (*60*), in that the hydroxyl group concentration reached a maximum after heating to about 300°C and then declined until few hydroxyl groups were left at 700°C. Typical data are shown in Figure 11. Above 550°C, the band near 3740 cm^{-1} increased markedly in intensity as the 3640 and 3540 cm^{-1} bands decreased rapidly. This probably reflects the loss of crystal structure of the zeolite. The hydroxyl groups could be readily exchanged with deuterium oxide to give OD bands with frequencies near 2754, 2685, and 2617 cm^{-1}.

Bolton and Lanewala (*70*) obtained similar results. By using a combination of techniques, they showed that there was considerable overlap of temperatures over which deammination and dehydroxylation occurred. They also found that the temperature for dehydroxylation increased with increasing silica to alumina ratio and decreased with increasing extent of exchange. They also showed that the ammonium Y zeolite could be partially reconstituted by treatment of the dehydroxylated zeolite with water prior to reammination.

The calcination of ammonium Y can be summarized as follows. Deammination appears to start at about 100°C. The temperature at which ammonia evolution ceases depends markedly on the duration of calcination at a particular temperature. In most cases all but traces of ammonia are removed at 400°C. The resulting decationized Y zeolite has at least three major types of hydroxyl groups with frequencies near 3740, 3640, and 3540 cm^{-1}, formed by decomposition of ammonium ions and attack of the lattice structure by the liberated protons. As ammonium ions are decomposed, hydroxyl groups are formed. The population of hydroxyl groups remains relatively constant from 400°C to 550°C. At this point dehydroxylation starts, resulting finally in structural collapse. The increase in the 3740 cm^{-1} band intensity probably indicates the formation of amorphous silica or silica–alumina. The loss of structure can be confirmed by x-ray diffraction and adsorption capacity. A pronounced weight loss is observed by the thermogravimetric analysis and an endothermic heat change at 550°–650°C is also observed. Readdition of water restores the hydroxyl group intensity if the samples have only been heated to around 300°–500°C, but at higher temperatures the hydroxyl groups are not restored (*60, 68*).

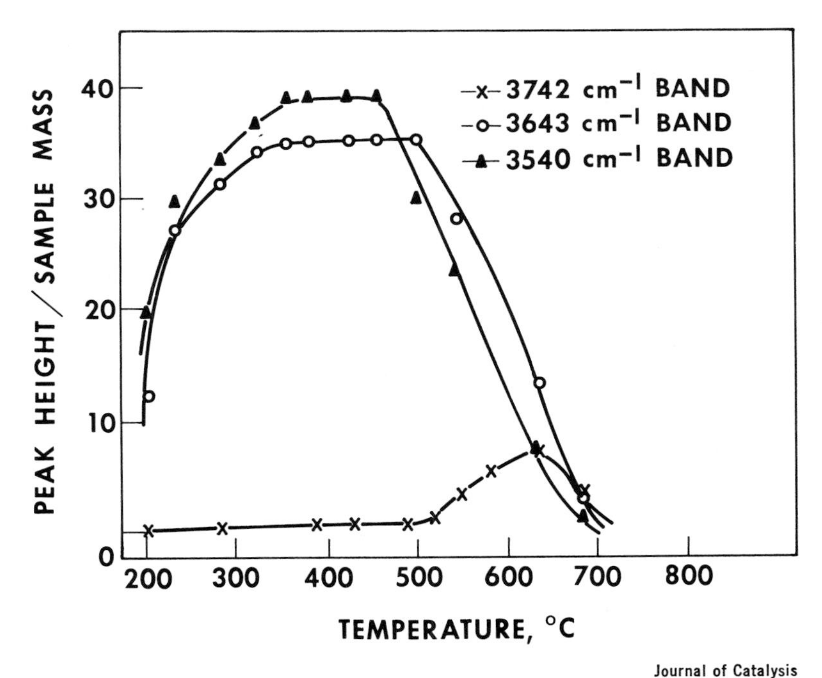

Journal of Catalysis

Figure 11. Intensity of hydroxyl bands on decationized and dehydroxylated Y zeolite as a function of calcination temperature (69)

The location and origin of the three major hydroxyl bands has been the subject of much discussion. The band at 3740 cm^{-1} has been attributed to hydroxyl groups terminating the crystal structure and to occluded siliceous impurity in the zeolite. The frequency suggests that they are similar to those on silica or silica–alumina. Possibly both origins are correct since in some cases these groups are inert while in other cases they interact, at least with deuterium.

The 3650 and 3550 cm^{-1} bands have no reported analogue in other oxide systems. Since they can be deuterated and the frequencies of the O–D absorption bands are displaced from the OH band frequencies by the expected isotope shift, they probably do not represent coupled vibrations (*66*). Because the 3650 cm^{-1} band interacts with most adsorbents, most investigations have located these hydroxyl groups in the accessible supercage parts of the structure.

There has been less agreement as to the location of the 3550 cm^{-1} band. This band interacts with some molecules at all pressures and other molecules depending upon the adsorbate molecule pressure. An early suggestion (*30*) was that the band could be attributed to interaction between neighboring hydroxyl groups, possibly through hydrogen bonding. Alternatively, hydrogen bonding to an oxygen of the lattice could occur (*68*). In contrast, Hall and coworkers (*60, 65*) suggested that the 3550 cm^{-1} band represented groups in a different location to those corresponding to the 3650 cm^{-1} band possibly in the sodalite cages of the structure. White *et al.* (*66*) concluded that the 3650 cm^{-1} band represented hydroxyl groups in the supercages and the 3550 cm^{-1} band represented groups in the sodalite cages from a study of the behavior of physically adsorbed molecules.

Eberly initially concluded that the 3650 cm^{-1} band represented hydroxyl groups in the supercages near S'_{II} positions since these groups interact readily with adsorbate molecules and that the 3550 cm^{-1} band represented hydroxyl groups in the hexagonal prisms or near S_I sites since these groups did not readily interact with moderately large molecules (*64*). From a study of pyridine and piperidine adsorption, Hughes and White (*68*) concluded that under certain conditions, both types of hydroxyl groups are accessible to large molecules but that the 3650 cm^{-1} band type are more accessible. The O_1 and O_4 oxygen atoms were believed to be the hydroxyl sites. Others reached similar but less specific conclusions (*31, 39, 69*). The hydroxyl group locations have probably been clarified by single crystal x-ray diffraction data for decationized faujasite by Olson and Dempsey (*71*). By careful measurements, they deduced a systematic lengthening of the T—O bond length for the O_1 and O_3 oxygen atoms. Hence, the O_1H-group represented the 3650 cm^{-1} band and the O_3H group represented the low frequency band residing in the supercage and hexagonal prisms respectively from consideration of the location of the O atoms and the properties of the hydroxyl groups. Thus the original conclusions of Eberly (*64*) appear to have been correct. By studying the changes in the hydroxyl band intensities as a function of the degree of exchange and utilizing the ion-exchange data of Sherry (*72*), further evidence was obtained to support the conclusions of Olson and Dempsey (*71*). On initial exchange of Na-Y zeolite, only the 3650 cm^{-1} band is formed. This band increases in intensity, as shown in Figure 12, with increasing exchange and is essentially the only band detected until about 55–60% exchange has been achieved (*77*). Beyond this point, the 3540 cm^{-1} band appears. These data correlate well with the exchange data of Sherry (*72*) who showed that the last 16 cations are more difficult to exchange than the initial cations, the former probably being located in the hexagonal prisms and the latter in the supercages and possibly the sodalite cages. Additional confirmation is obtained from exchange of cesium ions into NH_4-Y. If the 3540 cm^{-1} band represents OH groups in the hexagonal

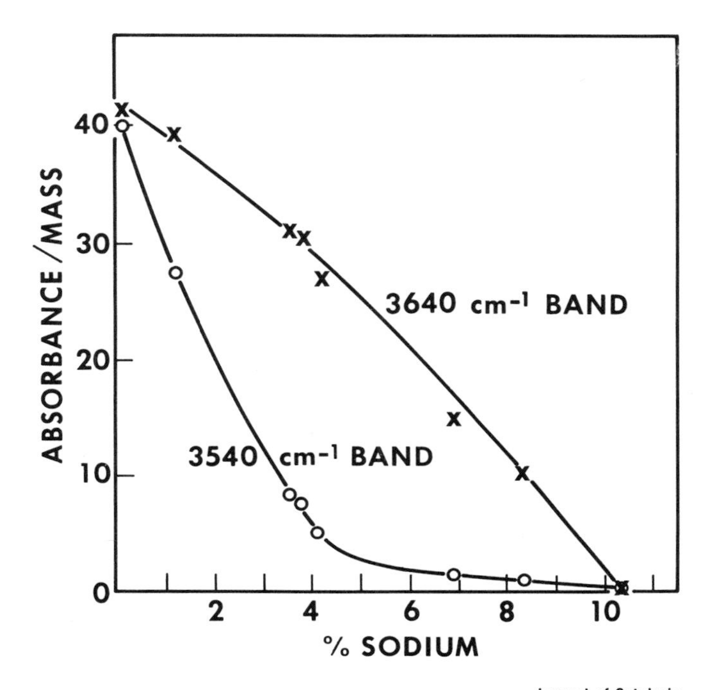

Journal of Catalysis

Figure 12. Absorbance of the 3640 and 3540 cm⁻¹ hydroxyl bands of decationized Y zeolite as a function of zeolite sodium content (77)

prisms, cesium should not influence those groups since it is too large to enter the hexagonal prisms. On the other hand, cesium should be able to replace groups in the supercages. The data shown in Figure 13 confirm the location of the hydroxyl groups, since the 3550 cm^{-1} band is unaffected while the 3650 cm^{-1} band is largely eliminated.

The proposed OH group assignments pose one difficulty: namely, if the 3550 cm^{-1} band represents groups in the hexagonal prisms, the group should be inaccessible to molecules which cannot enter the hexagonal prisms. In several instances this applies, but, to the contrary, molecules like piperidine, pyridine and cumene can interact with the 3550 cm^{-1} hydroxyl groups. This apparent contradiction has been rationalized in terms of proton mobility (*71, 73*) which will be discussed later. Because piperidine can interact with the hydroxyl groups at room temperature, strong bases can induce proton movement without the introduction of thermal effects.

The question of the number of types and the assignment of hydroxyl groups on X and Y zeolites has been studied in more detail by Jacobs and Uytterhoeven recently (*74*). They studied a series of zeolites with different Si/Al ratios and different degrees of exchange. The samples were outgassed at 320°–375°C, and the spectra were recorded at room temperature. The overlapping bands were resolved using a curve analyser assuming six com-

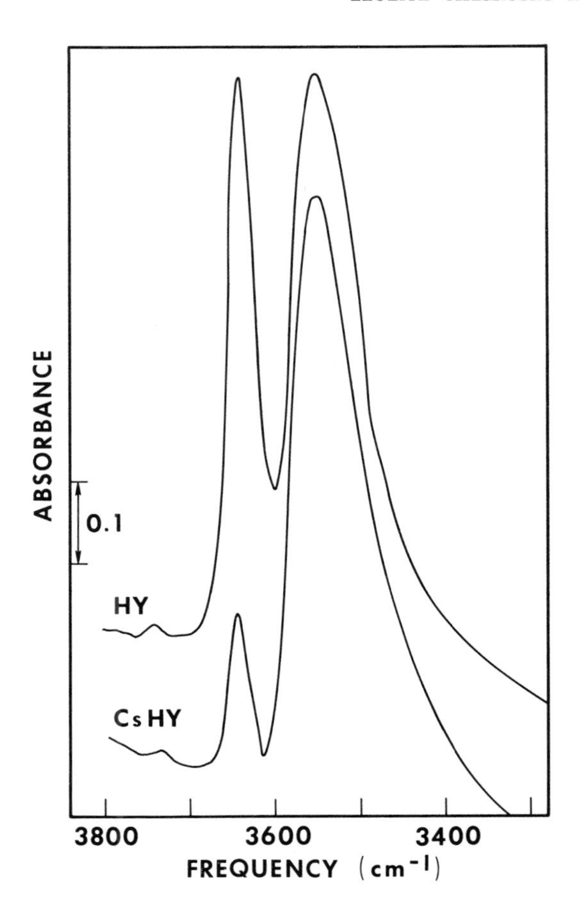

Journal of Physical Chemistry

Figure 13. Hydroxyl groups on decationized and cesium decationized Y zeolites (73)

ponents and the bands were of 65 ± 2.5% Lorentzian shape. Typical resolved frequencies based on the presence of six bands are given in Table VI.

The data show that bands II and VI are not influenced by the Al or Na content of the samples. Bands IV and V are at a lower frequency in Type Y than in Type X but are independent of the Na content. At elevated temperature, bands I, II and V move to lower frequency and decrease in intensity, bands III and VI decrease in intensity. Band IV moves to higher frequency and increases slightly in intensity.

On deuterium exchange, the six resolved hydroxyl bands collapsed into five deuteroxyl bands. These resolved components were assigned as follows: Band I was attributed to OH groups inside the supercages. Band II is attributed to unspecified non-lattice OH groups and Band VI may be an artifact. Band III is assigned to O_3-H groups. Bands IV and V are attributed to coupled

Table VI. Frequencies (cm^{-1}) of the Various Band Components (*74*)

$(SiO)_2/(AlO_2)$ % NH_4 Exchange	$(SiO_2)_{137}/(AlO_2)_{55}$ 58	$(SiO_2)_{122}/(AlO_2)_{71}$ 52 and 76	$(SiO_2)_{137}/(AlO_2)_{55}$ 51 and 58	$(SiO_2)_{144}/(AlO_2)_{49}$ 41 and 70
I	3664.5 3668.3	3659.2 3661.5	3652.0 ± 1.5	3651.0 ± 1.0
II	3626.9 ± 3.2	3626.9 ± 3.2	3626.9 ± 3.2	3629.9 ± 3.2
III	3599.0 3595.2	3586.3 3582.5	3572.0 3568.0	3567.0 3562.5
IV	3575.2 ± 4.3	3578.0 ± 3.9	3546.6 ± 2.6	3551.0 ± 3.2
V	3543.0 ± 4.5	3537.5 ± 4.7	3515.9 ± 5.1	3516.5 ± 4.8
VI	3442.0 ±11	3455.0 ± 11	3455.0 ± 11	3455.0 ± 11

Journal of the Chemical Society, Faraday I

vibrations of O_2-H and O_4-H groups. The presence of O_2-H groups is supported by x-ray data of Gallezot and Imelik (*75*). Although this is an interesting attempt to unravel the spectra of decationized Y zeolite, more certainty should be introduced into the curve analysis procedure particularly as to whether a band VI is real or an artifact. In several spectra band VI is more intense than band II.

A Zeolite. Ammonium A zeolite was studied by Zhdanov *et al.* (*25*) at several different levels of exchange varying from 3 to 75%. After evacuation at 400°C, hydroxyl bands were observed at 3450 and 3290 cm^{-1}. The spectra always contained a band in the water bending region, so it is difficult to distinguish whether the observed bands are due to physically adsorbed water or to structural hydroxyl groups. The sample, 3 % exchanged, was studied as a function of treatment temperature between 25 and 400°C. The broad bands between 3600 and 3200 cm^{-1} and the band near 1660 cm^{-1} were progressively removed.

Erionite. Ammonium erionite has been studied as a function of calcination temperature at 300, 400, 500, 600 and 700°C (*76*). All water is removed at 400°C, as indicated by the absence of the water bending vibration band at 1630 cm^{-1}. Typical spectra are shown in Figure 14. The presence of bands at 3380 and near 3200 cm^{-1} up to 500°C indicates the presence of ammonium ions up to these temperatures along with hydroxyl bands at 3612 and 3565 cm^{-1}. At 600°C, the ammonium ion bands are gone and sharp hydroxyl bands are observed at 3745, 3612, and 3565 with a shoulder at 3660 cm^{-1}. At 700°C, most of the hydroxyl bands have been eliminated. Compared with Y zeolite, the erionite seems to be more stable to deammination and dehydroxylation. Because the 3612 cm^{-1} hydroxyl band interacts with propylene whereas the 3565 cm^{-1} band does not, the former groups point towards the large cavities whereas the latter groups are inside the hexagonal prism.

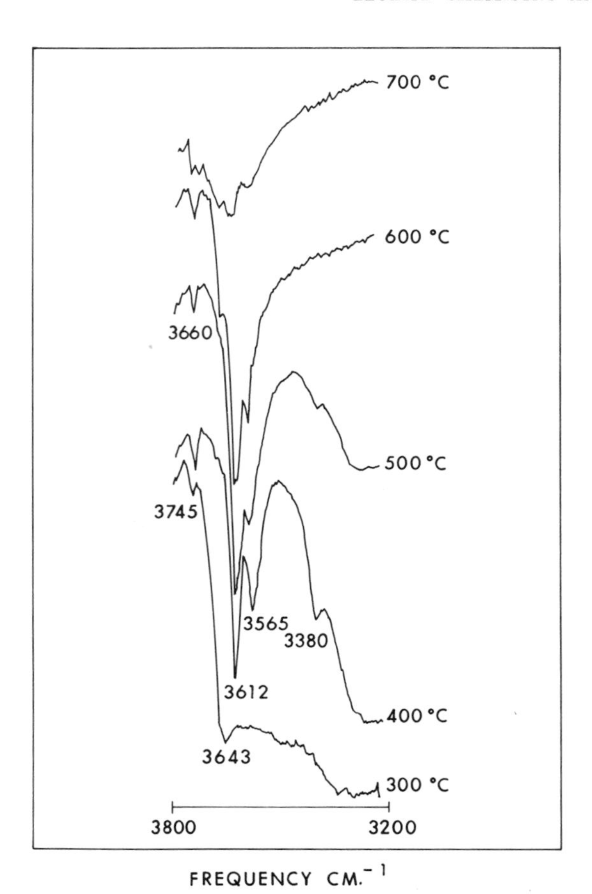

Journal of Physical Chemistry

Figure 14. NH_4-erionite after calcination in air for 1 hr at each temperature (76)

Decationized L Zeolite. Spectra of decationized L (40% exchange) are shown in Figure 15. On evacuation at 260°C, sharp bands are observed at 3740, 3630, and 3400 cm^{-1} and a broad band at 3300–2700 cm^{-1}. On heating to 350°C, the sharp band at 3400 cm^{-1} disappears, and partially resolved bands are observed at 3400, 3300, 3150, and 2930 cm^{-1}. The band at 3630 cm^{-1} remains unchanged.

Evacuation at higher temperature simply decreases the band intensities without improving the resolution until at 540°C, only a weak band at 3740 cm^{-1} is observed. On readdition of water, a broad hydrogen-bonded hydroxyl absorption is again observed. Since no discrete hydroxyl bands are observed, extensive hydrogen bonding of surface hydroxyl groups and physically adsorbed water must be occurring. Because the broad absorption bands are observed even at relatively low degrees of exchange, the hydroxyl groups

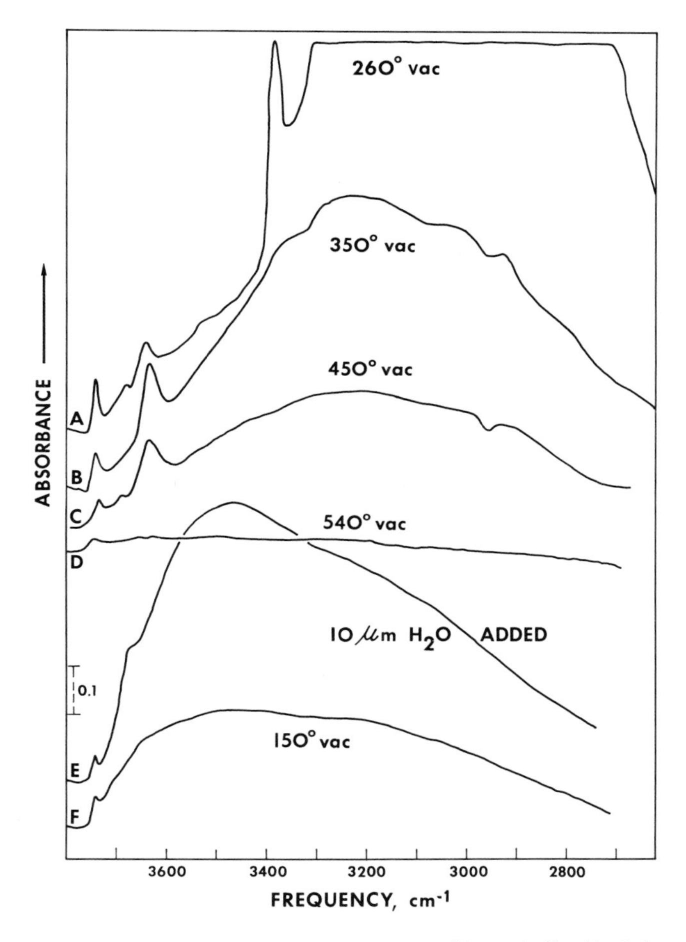

Advances in Chemistry Series

Figure 15. NH_4-L zeolite after (A) calcination at 260°C; (B) 350°C; (C) 450°C; (D) 540°C; (E) 10 mol of H_2O added; (F) evacuation at 150°C (18)

in the large pore structure must be involved. Spectra of more highly exchanged L zeolite (>80%) are similar but more intense. The sharp band near 3400 cm^{-1} is probably caused by some type of adsorbed water.

Cation-Decationized Zeolites. Several studies of mixed cation-decationized zeolites have been made to investigate the surface structure as the ratio of cations to hydrogen is altered and the population of the various crystallographic sites is changed. Both mono- and divalent systems have been studied.

Initial studies of Na-H system in this area are probably those of Uytterhoeven, Christner, and Hall (*60*) who showed from a study of three different exchange levels of X zeolite that the intensity of the hydroxyl band increased with extent of exchange and that the 3540 cm^{-1} band became more intense relative to the 3650 cm^{-1} band. Angell and Schaffer obtained similar results for Y zeolite (*30*). Ward and Hansford (*77*) studied sodium-decationized Y in detail. Sodium ions were removed from the zeolite by exchange with ammonium nitrate. Samples with sodium contents of 0.2 to 10.3% were studied representing a range of exchange from 99.8 to 0%. These samples were pretreated at 480°C. Typical spectra are shown in Figure 16, and the variation in hydroxyl group band intensities are shown in Figure 12. Starting from the sodium form, the 3640 cm^{-1} band progressively increases in intensity almost linearly with increasing extent of exchange.

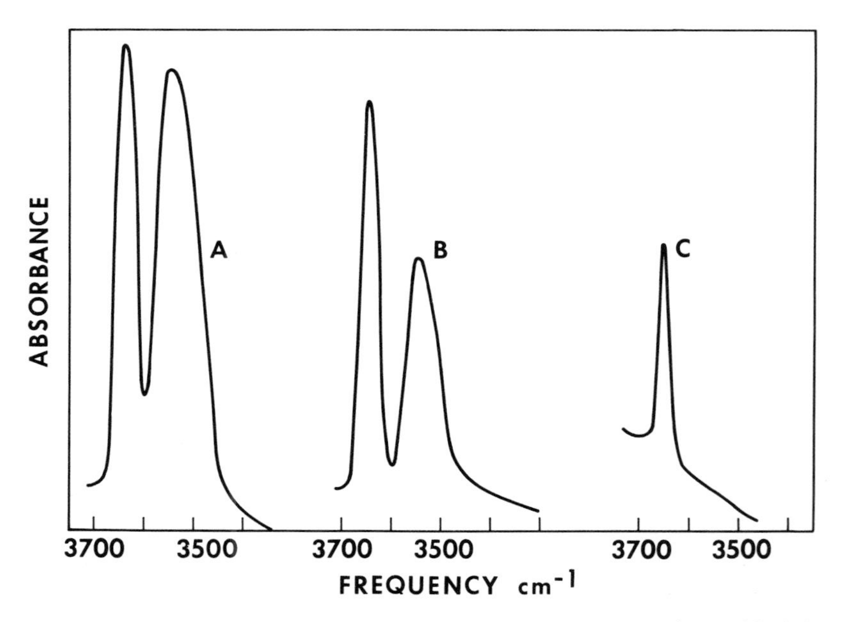

Figure 16. Na decationized Y zeolite after calcination at 480°C (A) 0.2% Na, (B) 3.6% Na, (C) 6.9% Na (77)

On the other hand, the 3540 cm^{-1} band increases very little until the sodium content has reached 5 wt %. Up to this level of exchange, the zeolite can be prepared with essentially one type of hydroxyl group. As the level of exchange is increased further, the 3540 cm^{-1} band intensity increases rapidly, as shown in Figure 12, until at low sodium content, the two bands are comparable in intensity. This data shows that the 3640 cm^{-1} band is formed at least initially by replacement of the easily removed sodium ions, and the hydroxyl groups are situated on the O_1 oxygen atoms as discussed above. Because the 3540 cm^{-1} band only forms at higher degrees of exchange, these groups are probably formed by introduction of hydroxyl groups into more inaccessible parts of the structure and correspond to the 16 or so cations found to be more difficult to replace by ion-exchange. These cations are

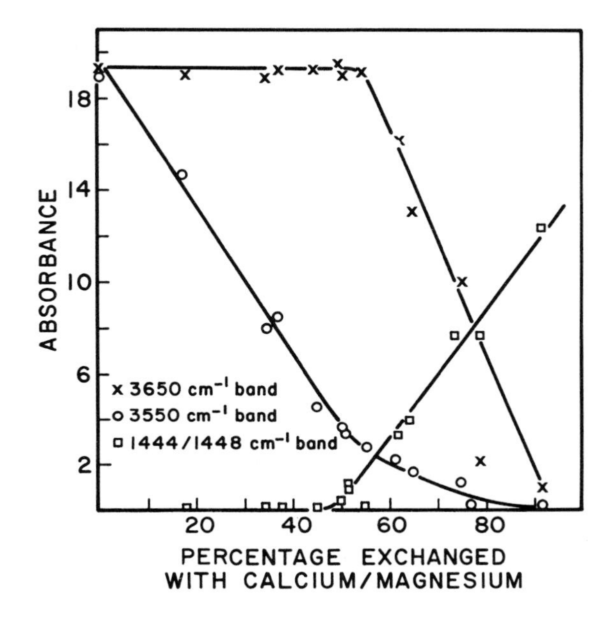

Journal of Physical Chemistry

Figure 17. Absorbance of 3650 (OH), 3550 (OH), and 1444(1448) cm^{-1} (cation–pyridine) bands of cation-decationized Y zeolite as a function of divalent cation content (78)

believed to be located in hexagonal prism portions of the structure near S_I sites.

The Mg and Ca systems have also been studied (*78, 79*). In these instances, a highly exchanged ammonium form was prepared (0.1% Na) and then reexchanged with the various cations to different extents. The samples were calcined at 450°C. Analogous results were obtained in both cases. The intensity of the 3540 cm^{-1} band decreased linearly with increasing divalent cation content to about 50% exchange, then the band intensity decreased more slowly to essentially zero. Meanwhile the 3650 cm^{-1} band remained essentially constant to about 55% exchange and then decreased rapidly. The observations are shown in Figure 17. The data show dramatically the selective replacement of ammonium ions by divalent cations. At about 55% exchange, x-ray diffraction data suggest that cations have filled the hexagonal prisms and then spill over into the supercages. Hence the first cations are replacing ions in the hexagonal prism and then after 50% exchange are replacing cations in the supercages.

This conclusion is confirmed by the observation that above 50% exchange, cations are available in positions in the structure which enable them to interact with pyridine molecules. Nickel cations have been similarly studied. The changes in the OH group spectra as a function of the nickel content are shown in Figure 18. Here the 3540 cm^{-1} band decreases linearly

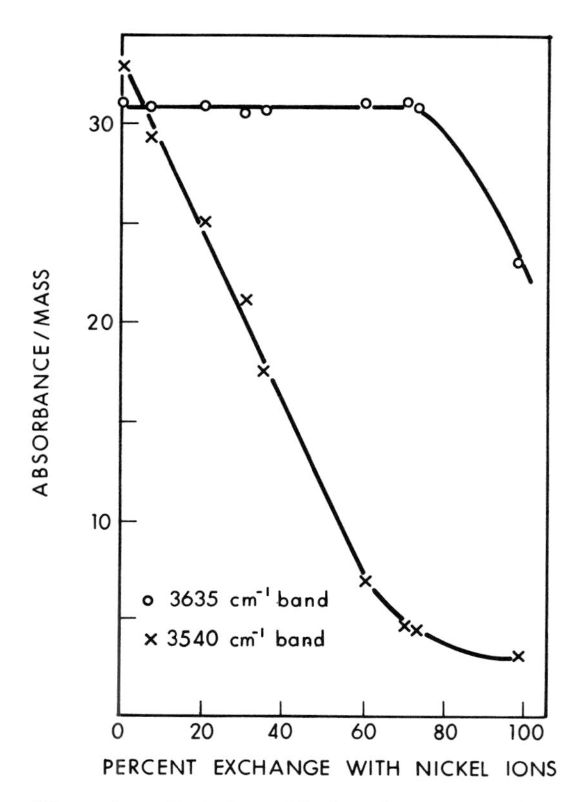

Figure 18. Variation of hydroxyl group band intensity of Ni decationized Y as a function of Ni content (81)

to about 60% exchange and the 3650 cm^{-1} band is unaffected to about 65% exchange. These results are reasonably similar to those observed for Mg- and Ca-Y zeolites. In marked contrast, as shown in Figure 19, nickel ions appear to interact with pyridine at all levels of exchange. This means that nickel ions must be in the supercages at all levels or can be drawn from other positions into the supercages by the pyridine molecules. Because of the observations with the hydroxyl groups, the latter is probably more likely.

The deammination of a magnesium–ammonium Y zeolite was studied in considerable detail (*57*). The sample was stable up to at least 800°C. The variation of the hydroxyl group bands with temperature is shown in Figure 20. The behavior of the 3660 cm^{-1} band is similar to the corresponding band on decationized Y except that the maximum intensity is reached after calcination at 425°C and starts to decline at 575°C; both temperatures are somewhat higher than those observed for decationized Y. The 3560 cm^{-1} band behaves completely differently in that its intensity declines constantly with increasing calcination temperature above 150°C.

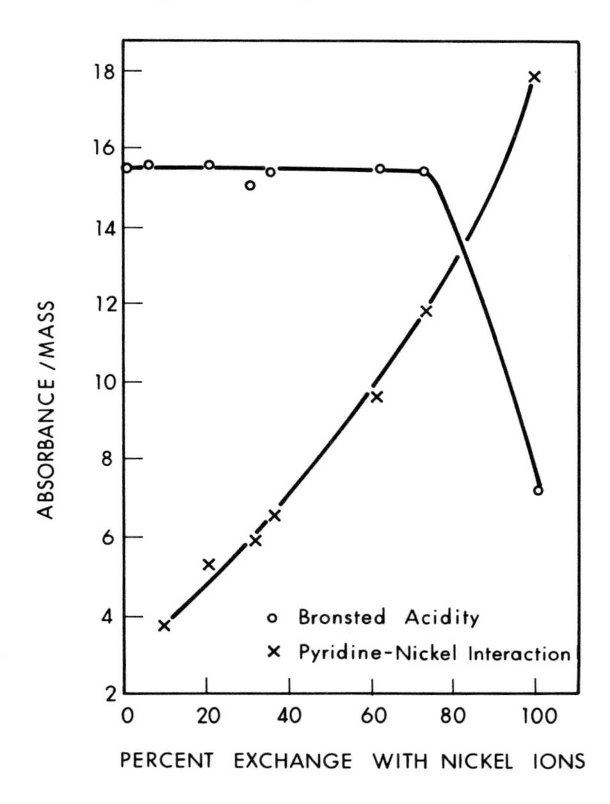

Figure 19. Variation of Bronsted acidity and pyridine–Ni interaction of Ni decationized Y as a function of Ni content (81)

The 3740 cm^{-1} band also behaves differently from that observed in decationized Y. Over the temperature range studied (250°–750°C), its intensity remains constant. The latter observation is probably related to the greater resistance to structural collapse of the zeolite. Similarly, the shift in temperature for maximum hydroxyl band intensity probably relates to a greater stability of the hydroxyl groups. There does not appear to be any explanation for the 3560 cm^{-1} band. The data suggest that even at 250°C, deammination must have occurred to produce these hydroxyl groups.

The various hydroxyl groups on several decationized sodium zeolites with different sodium and silica to alumina ratios, La-X, La-Y and deep bed calcined Y zeolites were found to exchange at the same rate with deuterium over the 200°–400°C temperature range. The 3750 cm^{-1} band was an exception and exchanged more slowly. The activation energy for the exchange process increased with decreasing silica to alumina ratio and decreasing sodium content. It was also dependent on the nature of the exchanged cation (*81*).

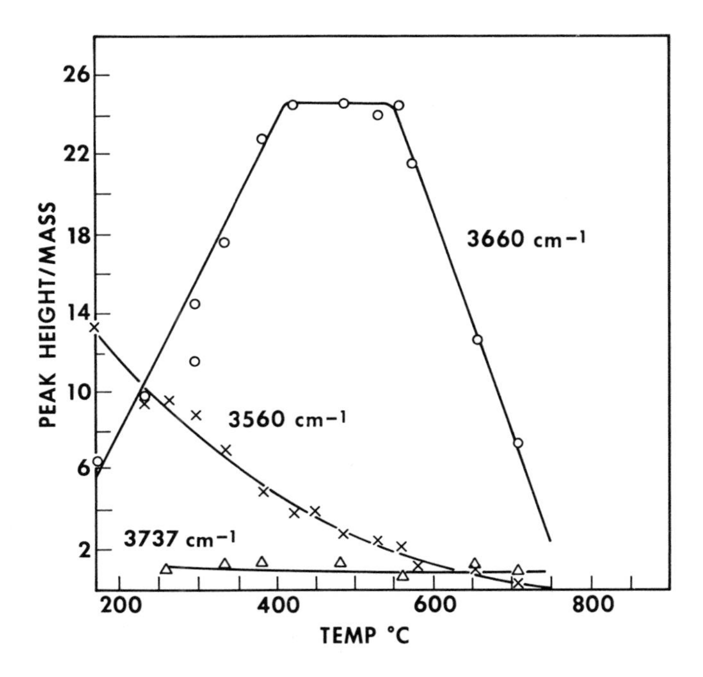

Journal of Catalysis

Figure 20. Intensity of hydroxyl bands on Mg decationized Y zeolite as a function of calcination temperature (57)

A brief study of the palladium–ammonium Y zeolite has been made. The two hydroxyl bands, at 23 and 49% exchange, decreased in intensity in a parallel manner suggesting the palladium ions, when introduced as the tetra-amino complex, exchange randomly into the zeolite (Figure 21).

Schoonheydt and Uytterhoeven (*82*) studied Li-, Na-, K-, and Mg-Y partially exchanged with ammonium ions. They found that there was little influence if any on the frequency of the 3650 cm^{-1} band from the differences in electrostatic field produced by different cations. However, they found that, at least for $NaNH_4$-Y, the extent of cation exchange could vary the frequency by several wavenumbers.

Aluminum–hydrogen Y, prepared by exchanging Na-Y with $Al(NO_3)_3$ to different extents has been studied (*83*). No aluminum atoms are removed from the lattice during the exchange treatment. These materials are hydrothermally unstable. Spectra of two AlH-Y zeolites containing 15.6 and 3.6 Al ions per unit cell and ammonium Y are shown in Figure 22. Bands are observed for the 3.6 Al ion sample near 3740, 3640, and 3540 cm^{-1}. The strong band near 3740 cm^{-1} probably indicates some structural destruction during the exchange. Only very weak bands were observed in the 15.6 Al ions per unit cell sample. Hence, the surface hydroxyl groups are those typical of decationized Y.

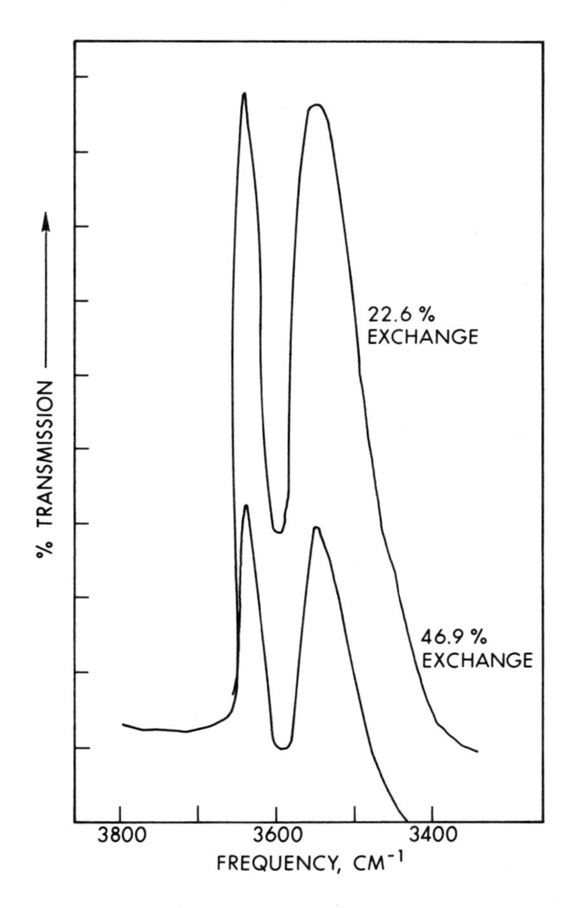

Figure 21. Palladium decationized Y zeolite calcined at 460°C (81)

Stabilized Hydrogen Zeolites. Ammonium Y zeolite can be decomposed into different products depending upon the calcination conditions. The first reports resulting in zeolites with improved stability are those of McDaniel and Maher (*84, 85*) and Hansford (*86*) in the patent literature. Conditions which allow easy escape of water and ammonia during calcination or predrying of the samples before calcination usually tend to produce the expected decationized Y zeolite, whereas calcination in NH_3 or H_2O vapor atmospheres results in the production of an enhanced stability product, the so-called ultrastable zeolite. The ultrastable zeolite can be prepared by calcination such that water and/or ammonia do not escape from the system readily, *i.e.*, in a thick bed or closed vessel or by calcination in an added atmosphere of steam or ammonia. The essential experimental step is the calcination under hydrothermal conditions. Stabilized products can also be made by removal of alumina from the zeolite lattice by treatment with a

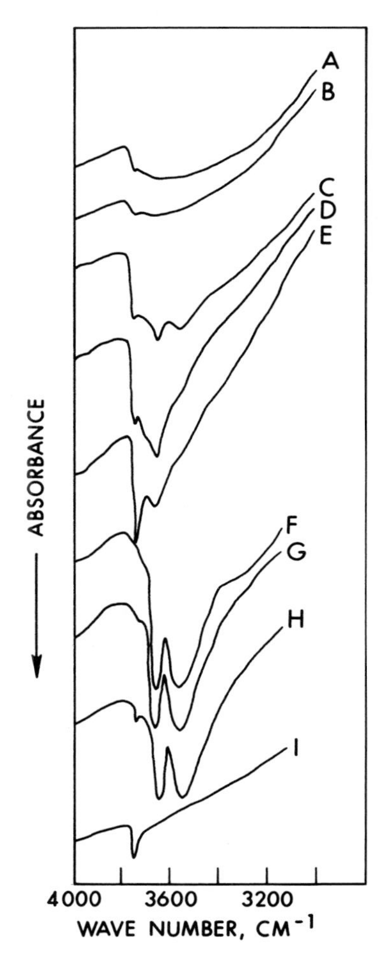

Journal of Catalysis

Figure 22. AlH-Y (15.6 Al ions/unit cell) (A) evacuated at 300°C; (B) evacuated at 400°C. AlH-Y (3.6 ions/unit cell) (C) evacuated at 300°C; (D) evacuated at 400°C; (E) evacuated at 500°C. NH_4-Y (F) evacuated at 300°C; (G) evacuated at 400°C; (H) evacuated at 500°C; (I) evacuated at 600°C (83).

chelating agent such as ethylenediaminetetraacetic acid (EDTA) (*87*). Since the stabilization of zeolites is treated in detail in Chapter 4, this discussion will be limited to spectroscopic studies.

Decationized Y and stabilized Y zeolites can be distinguished by thermograms, differential thermal analyses, x-ray diffraction patterns, ion exchange capacities, and thermal stabilities. They can also be distinguished by their infrared spectra. Not only does infrared spectra distinguish between the two gross forms but also shows that the ultrastable Y zeolites are not very well defined materials. From their spectra, they constitute a family of modified zeolites with improved stability. Figures 23 and 24 compare spectra of decationized Y with seven samples of stabilized Y produced under different conditions. All of these samples met the criteria generally associated with ultrastable Y (*44*). The differences in the absorption band frequencies and intensities are readily seen between the decationized Y and the ultrastable Y samples. In Figure 23, the ultrastable Y samples (b) and (c) were prepared by the procedure of Hansford (*86*) using steaming temperatures of 500° and 600°C respectively and (d), (e), and (f) by the procedure of McDaniel and Maher (*84*) with second calcination temperature of 600°, 700° and 800°C respectively. In Figure 24, the zeolites were prepared (b) by the procedure of McDaniel and Maher and (c) by the procedure of Kerr (*88*).

Comparison of the spectra of ultrastable and decationized Y shows that there are many fewer hydroxyl groups on the ultrastable Y zeolites, particularly of the 3650 cm^{-1} and 3540 cm^{-1} types. Instead dominant bands are observed near 3740, 3675, and 3600 cm^{-1}. The 3740 cm^{-1} band is much stronger than in decationized Y and appears to increase in intensity with increasing calcination temperature, suggesting that amorphous silica or silica–alumina type hydroxyl groups are formed during the stabilization process. The other two major bands near 3685 and 3600 cm^{-1} are not usually found in other Y zeolites (except some cation Y zeolites when partially hydrated). These bands are peculiar to the stabilization process and may be attributed to SiOH or AlOH groups. The band at 3600 cm^{-1} is influenced only to a small extent by pyridine adsorption indicating that these groups are non-acidic or very inaccessible.

A detailed study of the so-called deep bed (*41*) calcined ammonium Y zeolites was made by Jacobs and Uytterhoeven. As shown above, these materials are probably related to the ultrastable zeolite described by McDaniel and Maher. Four principal samples were prepared: (1) YDB: by exchange of Na-Y to about 70% with NH_4^+ ions followed by calcination in deep bed conditions; (2) $YDBNH_4$: by NH_4^+ exchange of (1) to about 99%; (3) YAlD: by treatment of Na-Y with EDTA; and, (4) $YAlDNH_4^+$ by NH_4^+ exchange of (3). The YDB sample was also treated with NaOH and then with NH_4^+ ions.

The YAlD sample was exceptional in that its spectra resembled that of Na-Y. Weak bands were observed at 3700 and 3620 cm^{-1}. Typical spectra for YDB are shown in Figure 25, after various outgassings. Characteristic bands are observed near 3675, 3660, 3605, and 3560 cm^{-1}. On exchange with NH_4^+ ions, ($YDBNH_4$), and calcination in vacuum, the 3660 and 3560 cm^{-1} bands became much more intense indicating the formation of decationized Y.

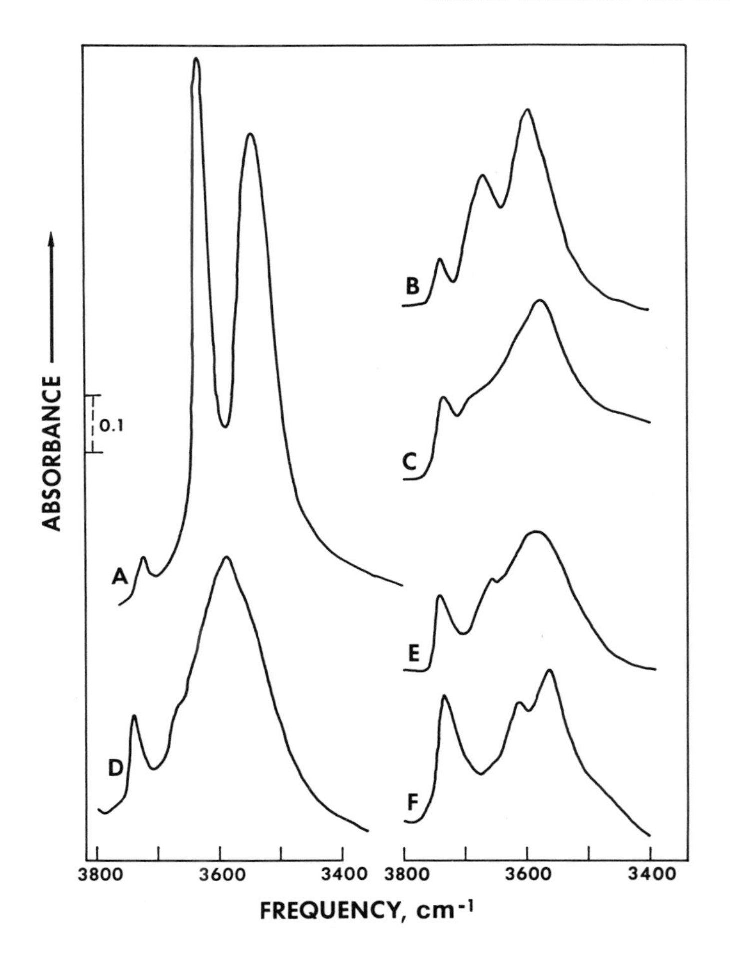

Figure 23. Decationized Y after evacuation at 500°C (A) simple evacuation; (B) steamed at 500°C (86); *(C) steamed at 650°C* (86); *(D) calcined by procedure of* (84), *final calcination at 600°C; (E) as (D) but final calcination at 700°C; (F) as (D) but final calcination at 815°C* (121)

Similarly the $YAlDNH_4^+$ has strong bands near 3660 and 3560 cm^{-1}. These bands and the same bands in YDB are almost certainly from hydroxyl groups of the type found in decationized Y. The YDB sample seems to contain more decationized Y type material than the samples of stabilized Y reported above. Maybe this is because in using the deep bed type of calcination, the calcination probably produces material of variable composition, the top of the sample being more like decationized Y and the bottom more like ultrastable Y. The absorption bands near 3620–3595 and 3675–3705 cm^{-1} appear to be characteristic of stabilized zeolites. In contrast to the 3660 and 3560 cm^{-1}, these bands are stable to hydrothermal conditions

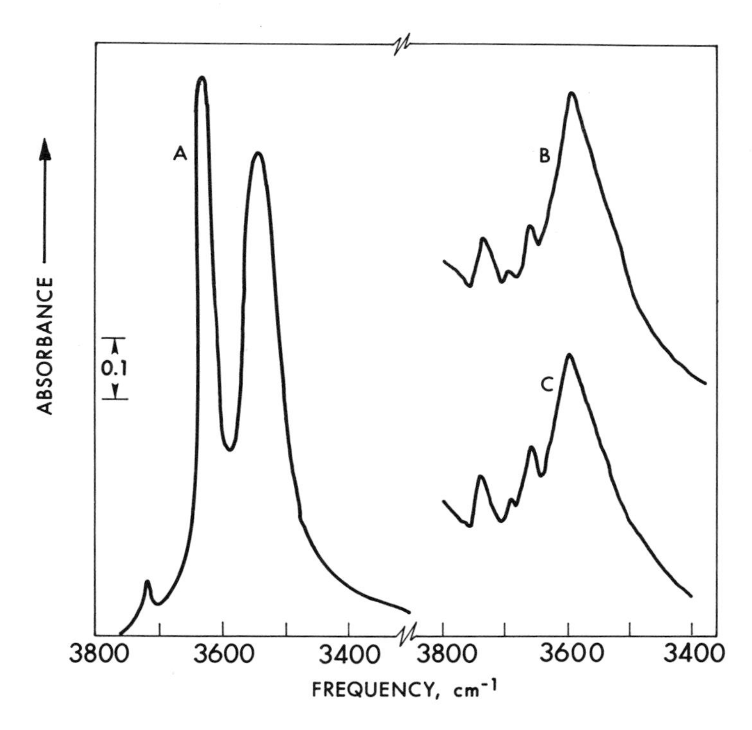

Preprints, Division of Petroleum Chemistry, 161st American Chemical Society Meeting

Figure 24. (A) Decationized Y; (B) Stabilized Y (84); *(C) Stabilized Y* (88) *after dehydration at 460°C* (181)

and do not interact with pyridine or ammonia. The band near 3700 cm^{-1} persists even after severe evacuation indicating that it is not from water interacting with cations and hence probably of a different origin from the band observed in hydrated cation Y zeolites. The band near 3600 cm^{-1} shows similar properties. Since the bands are not influenced by extraction of alumina by NaOH, they cannot be attributed to alumina which has been removed from the lattice during the stabilization process.

Spectra of the stabilized zeolites have been subjected to graphical resolution *(89)*. The same components as used for decationized Y plus an additional band at 3690 cm^{-1} (I′) were employed to resolve the spectra. A typical spectrum with its resolved components is shown in Figure 26. Comparable frequencies for ammonium Y, deep bed Y, and ammonium-exchanged deep bed Y are given in Table VII. Comparing the resolved spectra, the major changes are an increase in intensity of bands near 3685 and 3600 cm^{-1}, a decrease in intensity of the band near 3650 cm^{-1} and the occurrence of a band near 3530 cm^{-1} instead of two bands near 3550 and 3516 cm^{-1}. The interpretation and graphical resolution of the spectra are subject to considerable uncertainty.

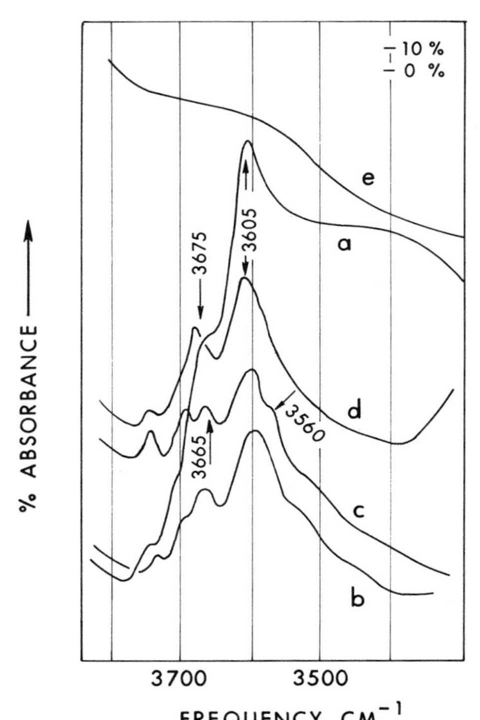

Figure 25. Y zeolite deep bed calcined (a) evacuated at room temperature; (b) 150°C; (c) 450°C; (d) NH_3 added; (e) sample after deuteration (41)

Stabilized zeolites prepared by extraction of aluminum with EDTA from NH_4-Y zeolites have been studied by Beaumont *et al.* (*90*). The samples were then treated in dry air at 380°C and 550°C. Samples studied had 56, 37.5, 26.5, and 17.4 Al atoms per unit cell. For spectral studies, the zeolites were heated at 450°C under vacuum. Absorption bands were observed at 3742, 3670–3695, 3630, and 3555 cm^{-1} for the zeolite containing 37.5 Al atoms per unit cell and at 3742, 3670–3695, and 3605 cm^{-1} for the zeolite containing 26.5 Al atoms per unit cell. As observed for the stabilized Y zeolites produced by hydrothermal treatment, the bands near 3600 and 3670 cm^{-1} do not interact with pyridine. The appearance of these spectra are similar in most respects to those reported for hydrothermally treated Y zeolites. The presence of the bands near 3600 and 3670 cm^{-1} in these samples in which the aluminum has been removed from the structure supports the conclusions of Jacobs *et al.* (*41*) that the bands do not represent hydroxy-aluminum ions removed from the lattice and deposited in the structure.

Table VII. Frequency (cm^{-1}) of the Different OH Bands in X and Y Zeolites, Deammoniated in Different Ways (74)

Sample	Calcination Conditions	I	I′	II	III	IV	V	IV	VI
F 49/70[a]									
	vacuum[b]	—	3650	3626	3563	3553	3520	—	3465
	AD[c]	—	3546	3620	3560	3550	3510	—	3461
F 85/73[d]									
	vacuum[b]	—	3666	3626	3594	3575	3543	—	3472
	AD[c]	3688	3664	3623	3584	3577	3558	—	3470
	WD[e]	3687	3662	3622	3585	3573	3563	—	3471
F 55/70[f]									
	vacuum[b]	—	3652	3627	3565	3547	3516	—	3455
	YDB	3684	3649	3602	3561	—	—	3530	—
	$YDBNH_4$	3684	3649	3602	3561	—	—	3530	—

[a] 49 Al atoms per unit cell with 70 % NH_4^+ exchange.
[b] Vacuum indicates calcined in vacuum.
[c] AD indicates calcined in air.
[d] 85 Al atoms per unit cell with 73% NH_4^+ exchange.
[e] WD indicates calcined in water vapor.
[f] 55 Al atoms per unit cell with 70% NH_4^+ exchange.

Journal of the Chemical Society, Faraday I

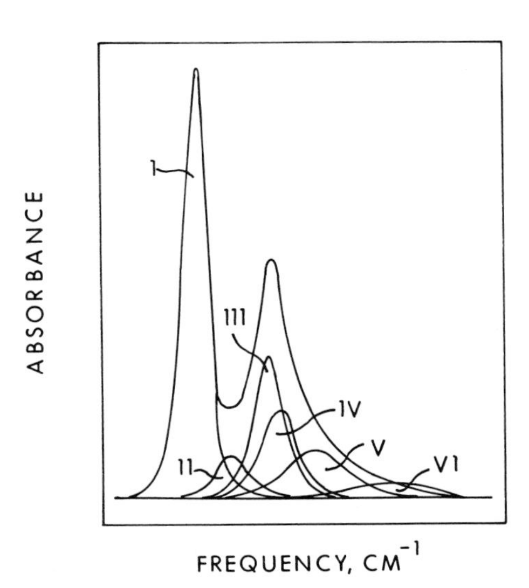

Transactions of the Faraday Society

Figure 26. Graphical resolution of decationized Y spectrum (89)

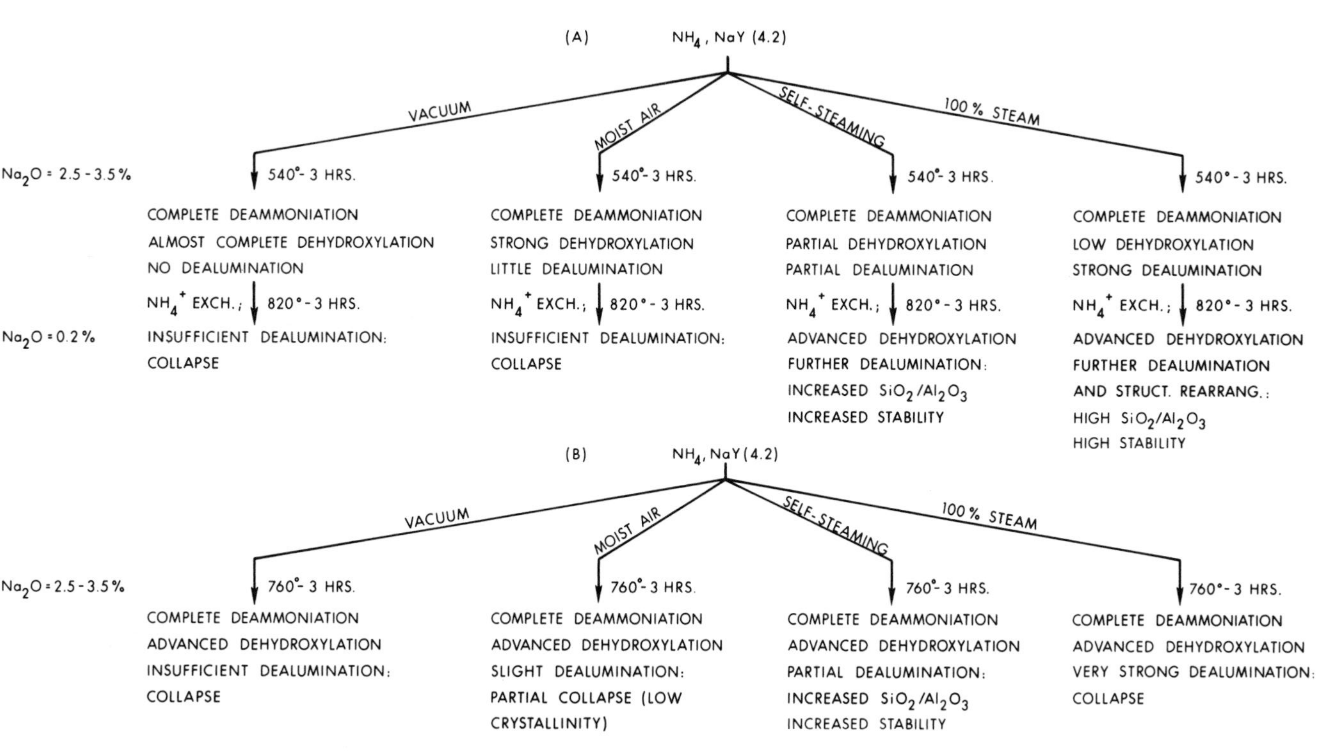

Journal of Catalysis

Figure 27. Preparation steps and schematic representation of the processes taking place in NH_4-Y during calcination under different condition: (A) ultrastable type I; (B) ultrastable type II (40)

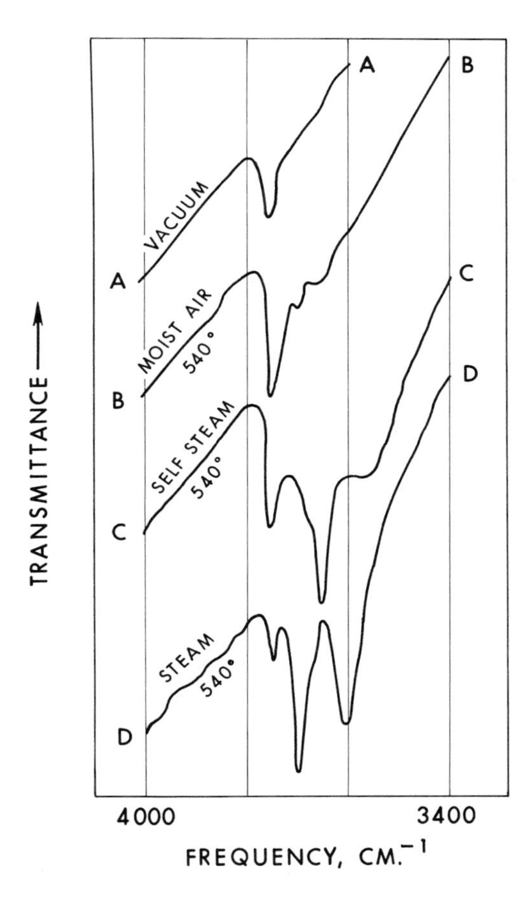

Journal of Catalysis

Figure 28. Calcination of $NH_{40.67}$-exchanged Na-Y zeolite in (A) vacuum; (B) moist air; (C) self-steam; (D) steam (40)

Table VIII. Hydroxyl Stretching Frequencies[a] (*40*)

Zeolite Type	*cm^{-1}*	*cm^{-1}*	*cm^{-1}*
H, Na-Y (540°C, vacuum)	3650 (sh)		
H, Na-Y (540°C, moist air)	3650 (w)	3690 (w)	
H, Na-Y (540°C, self-steamed)	3575 (m)	3650 (s)	3680 (sh)
H, Na-Y (540°C, 100% steam)	3615 (s)	3670 (sh)	3705 (s)
US-YI (self-steamed)	3600 (sh)	3670 (w)	
H, Na-Y (760°C, moist air)	3620 (sh)	3685 (sh)	3705 (m)
H, Na-Y (760°C, self-steamed)	3600 (m)	3680 (sh)	3707 (s)

[a] w = weak; m = medium; s = strong; sh = shoulder.

Journal of Catalysis

Scherzer and Bass (*40*) have made a detailed study of ultrastable Y zeolite and its precursors. They prepared ultrastable Y type I and type II by the methods of McDaniel and Maher (*84*). Type I zeolite was prepared by exchanging Y-zeolite with ammonium ions, calcining at 540°C, exchanging with ammonium ions, and calcining at 820°C. Type II was prepared by exchanging with ammonium ions, calcining at 760°C, and exchanging with ammonium ions. An additional eight materials were prepared as shown in Figure 27. Two sets of four samples were prepared using conditions of (a) vacuum, (b) moist air, (c) self steaming, and (d) introduced steam. Spectra obtained under these conditions at 540°C are shown in Figure 28 for samples containing 3.5% Na_2O. Spectra of stable zeolites produced at 760°C are shown in Figure 29. The frequencies for the six samples and for that of an ultrastable zeolite type I are given in Table VIII. The spectra of Figure 28 for samples calcined at 540°C show the dramatic effect of the atmosphere around the zeolite during the calcination. The sample calcined in vacuum shows only a weak band near 3650 cm^{-1}. Increasing the amount of water vapor by calcining in moist air or self steaming results in more hydroxyl groups being detected. When the calcination is carried out in 100% steam, the 3650 cm^{-1} band is absent and strong bands are detected near 3615, 3670 and 3705 cm^{-1}. The moist air and self-steam samples have a strong band near 3740 cm^{-1}, suggesting substantial structural collapse. Changes are also observed in the lattice vibration.

The bands observed in the spectra of the self-steamed and steamed materials near 3600 and 3700 cm^{-1} can probably be attributed to OH groups resulting from the dealumination process as suggested above. The increase in intensity of the bands with the severity of steaming supports this assumption. The absence of the water bending band near 1640 cm^{-1} indicates that these bands are from structural hydroxyl groups and not from physically adsorbed water. As discussed above, Scherzer and Bass find that these bands are unaffected by pyridine adsorption and only the 3600 cm^{-1} band is slightly affected by treatment of the zeolite with sodium hydroxide, which removes alumina from the structure. These data show that the OH groups are non-acidic and are probably not directly associated with the alumina removed from the structure. Increasing the extent of removal of alumina from the structure shifts the absorption bands to higher frequencies, further indicating that they are connected with groups remaining in the structure after alumina has been removed.

Ammonium ion exchange and calcination at 820°C of the samples calcined initially under vacuum and in moist air results in their collapse. The samples calcined under self-steaming conditions and 100% steam are stable. These later two samples are of the ultrastable Type I zeolite type. The spectra of these zeolites show only weak hydroxyl bands despite the treatment in excess steam. No spectra are shown and no absorption bands reported. These spectra should be similar to those shown in Figure 29, and the reasons for the differences are not obvious.

Type II ultrastable zeolites were prepared by calcining ammonium Y sieve containing 2.5–3.5% Na_2O at 760°C. Compared with the samples treated at 540°C, the higher temperature results in more intense bands near 3600 and 3700 cm^{-1} (Figure 29).

Thus the increase in temperature increases the dealumination process. Confirmation is found in the reduced unit cell obtained at higher calcination temperature. Comparison of spectra obtained at 540°C under 100% steam

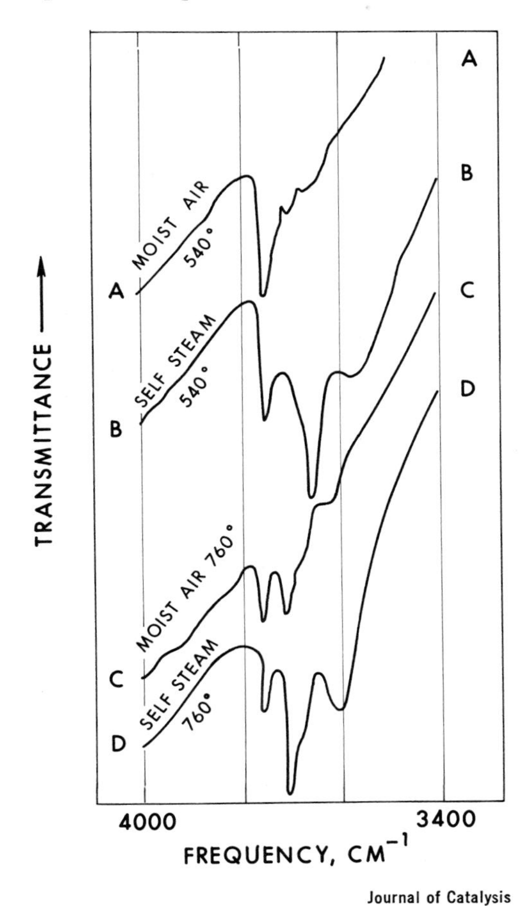

Journal of Catalysis

Figure 29. Calcination of $NH_{40.67}$-exchanged Na-Y zeolite in (A) moist air at 540°C; (B) self-steam at 540°C; (C) moist air at 760°C; (D) self-steam at 760°C (40)

and at 760°C under self-steaming conditions suggests that a similar amount of dealumination can be achieved at low temperature and high moisture content or higher temperature and lower moisture content.

At 760°C, calcination under vacuum or in 100% steam results in structural collapse. Hence stabilization requires some, but not too extensive, dealuminization. The optimum stabilization appears to be under self-steaming conditions resulting in a material of fairly high crystallinity. Under the hydrothermal conditions, silica may migrate in the lattice to positions from which aluminum ions have been removed. Hence the hydroxyl groups peculiar to stabilized zeolites may result from SiOH groups in these new positions in the lattice.

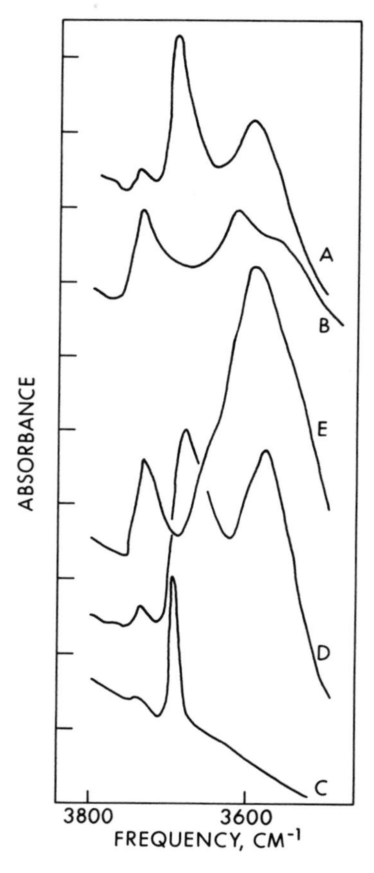

"Catalysis"

Figure 30. (A) Ultrastable Y precursor (3.4% Na). (B) Ultrastable Y (0.3% Na). (C) Hydrothermally treated Y (8.2% Na). (D) Hydrothermally treated Y (5.8% Na). (E) Hydrothermally treated Y (0.1% Na). (91).

Further studies have been reported by Peri (*42*). Absorption bands characteristic of ultrastable Y at 3750, 3700, and 3625 cm^{-1} were obtained by heating ammonium Y in static air at 700°C. Calcination at 500°C produced a zeolite containing bands both characteristic of decationized Y

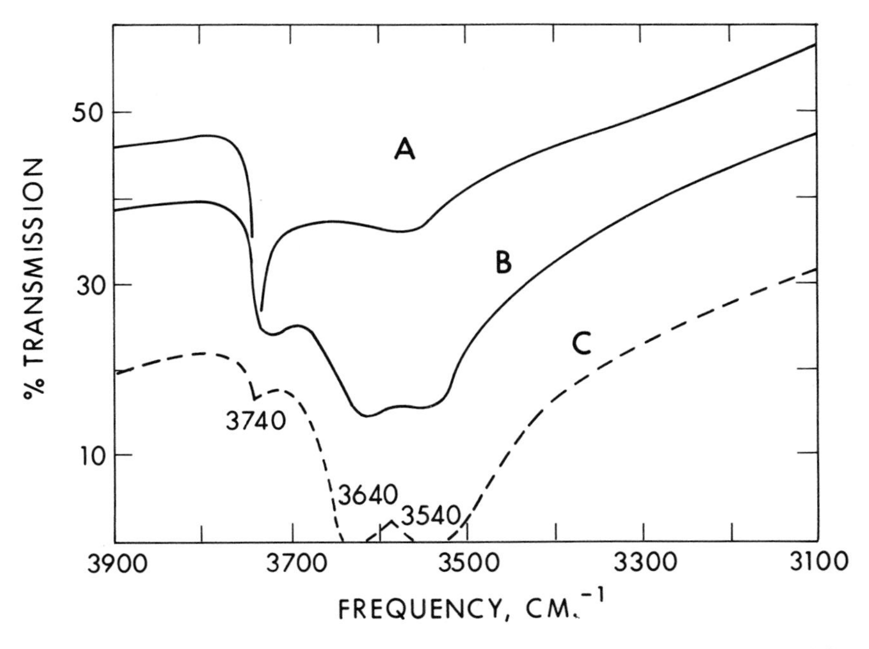

Figure 31. $NH_{4\,0.92}$-exchanged Na-Y after calcination (A) evacuated at 427°C; (B) wet air at 538°C; (C) wet air at 538°C + steam at 649°C (92)

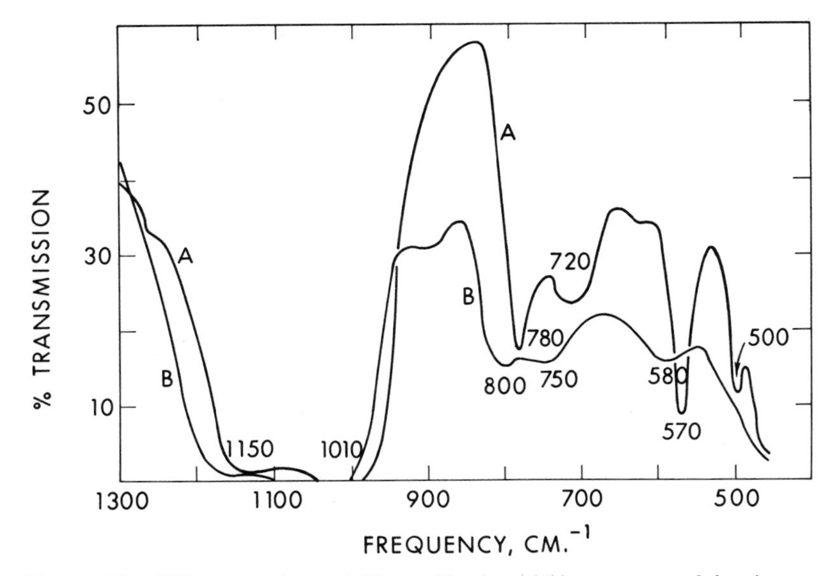

Figure 32. $NH_{4\,0.92}$-exchanged $Na_{0.08}$-Y after (A) vacuum calcination at 427°C; (B) wet air calcination at 538°C (92)

and ultrastable Y. Treatment in flowing air at 600°C had the same effect. These results are similar to those reported by Jacobs *et al.* at 540°C (*41*). In contrast to other concepts, the bands near 3700 and 3625 cm^{-1} could be attributed to AlOH groups. The band near 3700 cm^{-1} was removed by extraction with acetylacetone, and both the 3625 and 3700 cm^{-1} bands were reduced by about 50% on treatment with a NH_4Cl solution.

The hydrothermal treatment of ammonium Y containing from 0.1 to 8.2 % Na (*i.e.*, 99+ to 20% exchanged) has been studied (*91*). The samples were heated in flowing steam at 700°C. Results are shown in Figure 30. Samples containing more than 7 % Na (*e.g.*, C) exhibit only one band near 3700 cm^{-1}. Samples extensively exchanged to 0.1 % Na (*e.g.*, E) have a very strong band near 3590 cm^{-1} and no absorption band near 3700 cm^{-1}. At intermediate degrees of exchange, both bands are observed at slightly lower frequencies. Samples of ultrastable Y and its precursor show similar bands (a and b). The observation of only the 3700 cm^{-1} for the 7% Na sample is of interest since decationized Y containing the same amount of Na shows only one band near 3650 cm^{-1} (Figure 11). Hence these two hydroxyl groups may be formed from the decomposition of ammonium ions at the same site. The 3700 cm^{-1} band may represent hydroxyl groups in the supercages and may even involve the O_1 framework oxygens. Similarly the 3590 cm^{-1} band grows progressively more intense in the same manner as the 3550 cm^{-1} band in decationized Y as the sodium level is decreased. Hence these two types of hydroxyl groups may result from the same type of site, possibly in the hexagonal prisms and associated with the O_3 atoms. These different locations may be an explanation of Peri's results in which the 3700 cm^{-1} band was eliminated by acetylacetone extraction. Both bands are shifted upwards in frequency by 40–50 cm^{-1} from the position in decationized Y.

To summarize, the stabilized zeolites made by several different procedures contain hydroxyl bands near 3700 and 3600 cm^{-1}. The precise frequencies, relative intensities and intensities relative to other absorption bands depend on the preparation conditions and properties of the starting material.

Eberly *et al.* (*92*) have studied the influence of hydrothermal treatments on several different zeolites. Figures 31 and 32 compare spectra of 92% exchanged ammonium Y after calcination in vacuum at 427°C, moist air at 538°C and moist air at 538°C plus steam at 649°C. Distinct changes in the hydroxyl bands are seen and also in the lattice vibration. In particular, the band near 3740 cm^{-1} has increased substantially in intensity and the lattice vibration frequencies have shifted upwards. Similar spectra are shown for 72% NH_4-Y in Figures 33 and 34. Although there is a dramatic loss of hydroxyl groups, the materials still retained high crystallinity. The lattice unit cell constant also decreases.

Changes in structure on hydrothermal treatment for erionite are shown in Figures 35 and 36. In Figure 35, because of poor transmission in the OH region, the OD bands are shown. Mordenite also exhibits marked changes in its spectra on hydrothermal treatment as shown in Figures 37 and 38.

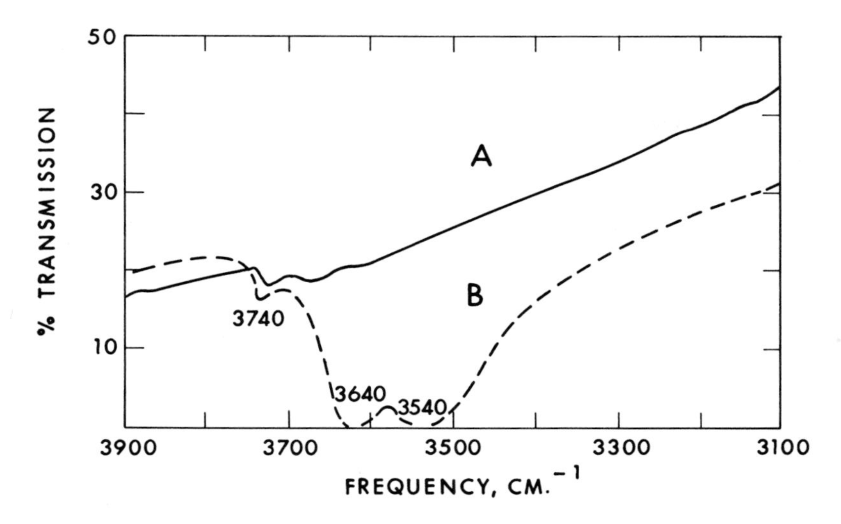

Figure 33. (A) $NH_{4\,0.72}$-exchanged $Na_{0.28}$-Y after steaming at 649°C; (B) $NH_{4\,0.92}$-exchanged $Na_{0.08}$-Y after evacuation at 427°C (92)

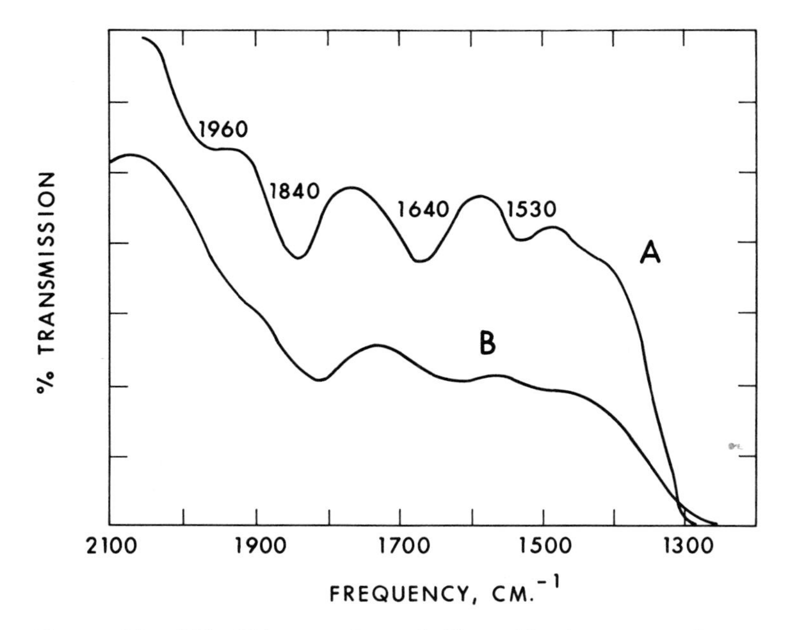

Figure 34. (A) $NH_{4\,0.92}$-exchanged $Na_{0.08}$-Y after evacuation at 427°C; (B) $NH_{4\,0.72}$-exchanged $Na_{0.28}$-Y after steaming at 649°C (92)

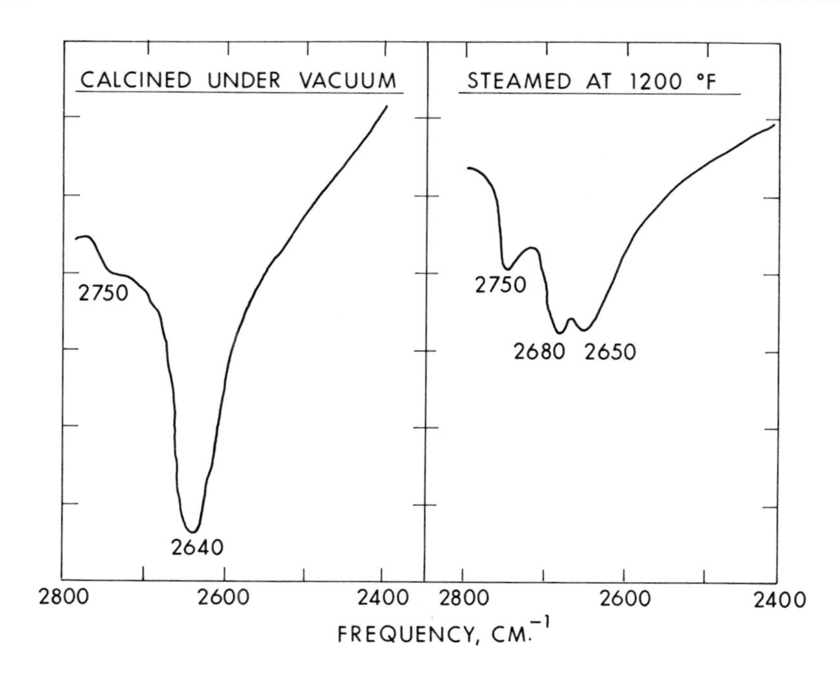

Figure 35. Acid erionite (92)

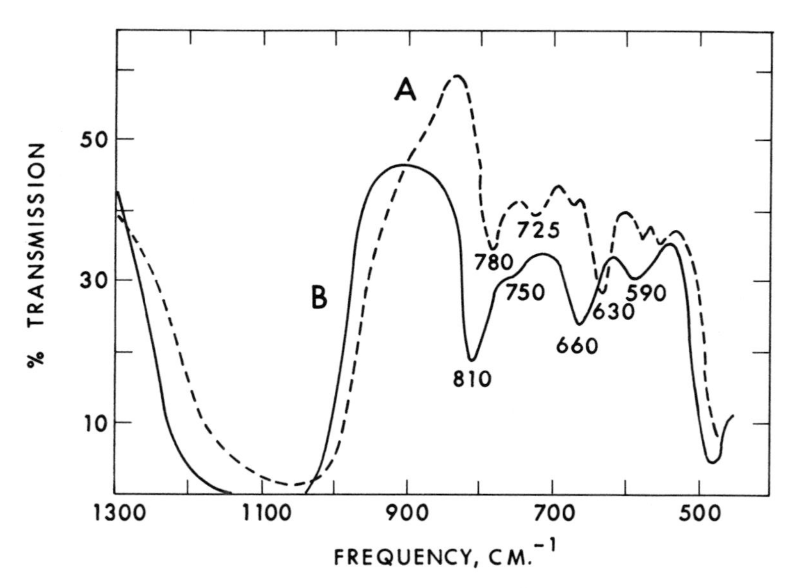

Figure 36. NH_4*-erionite (A) before steaming; (B) after steaming* (92)

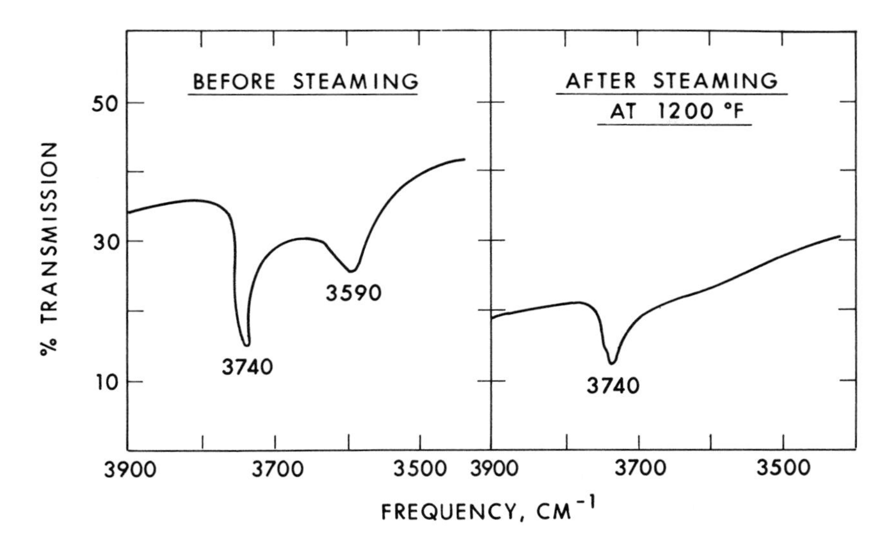

Figure 37. Acid mordenite (92)

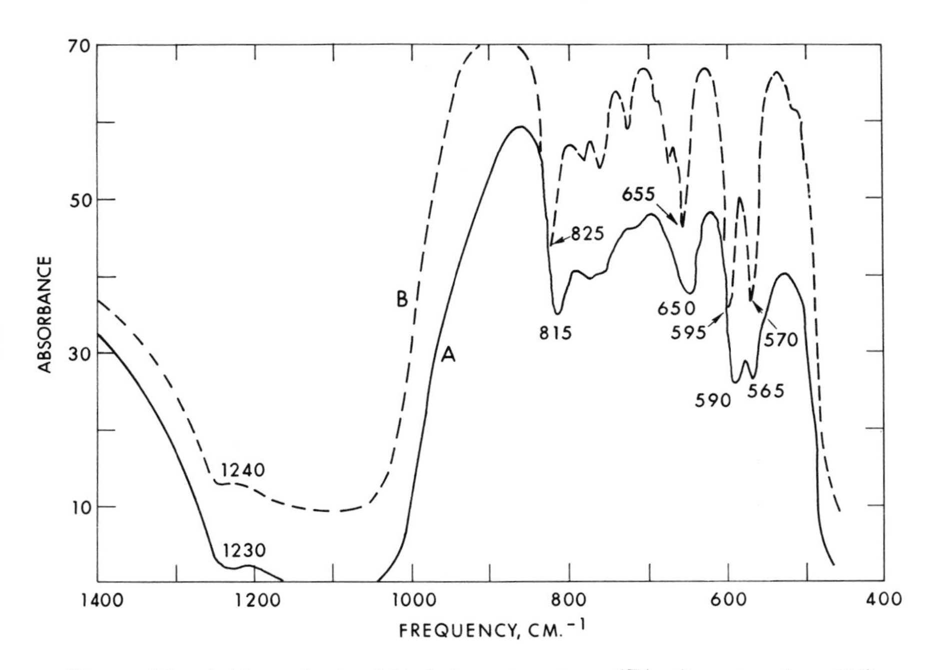

Figure 38. Acid mordenite (A) before steaming; (B) after steaming (92)

The lattices of both zeolites shrink in a manner similar to that observed for faujasite.

The influence of steaming on rare earth Y has been investigated and compared with similarly treated ammonium-Y (*93*). The samples were steamed at one atmosphere for 1 hr. The RE-Y, after pretreatment has bands at 3640 and 3520 cm^{-1}. At high levels of exchange, a shoulder is observed at 3550 cm^{-1}, as shown in Figure 39. On steaming at 600°C, spectra shown

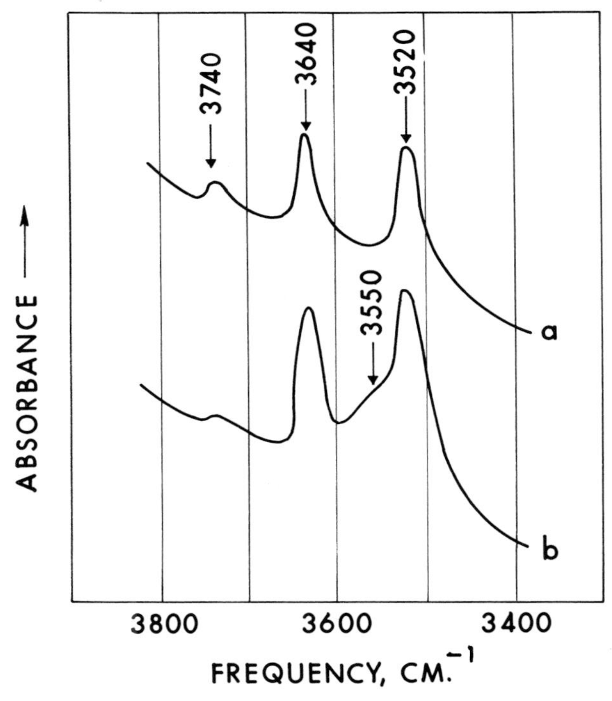

"Proceedings of the Third International Conference on Molecular Sieves"

Figure 39. Zeolite spectra after evacuation at 400°C (a) $RE_{0.70}$-exchanged-Na-Y; (b) $RE_{0.97}$-exchanged-Na-Y (93)

in Figure 40 were obtained. New bands are found at 3685 and 3615 cm^{-1}. The band near 3640 cm^{-1} had decreased substantially in intensity. Essentially no change was observed on resteaming at 675°C. Observations for RE-Y (97% exchanged) are shown in Figure 41. Weak bands are observed near 3685 and 3640 cm^{-1}. The band near 3520 cm^{-1} was unaffected by the steaming treatment. Spectra of steamed ammonium-Y are shown in Figure 42. Typical bands are observed near 3685 and 3615 cm^{-1}. From the intensities of the 3685 and 3615 cm^{-1} bands, it is seen that the presence of rare earth ions reduces the tendency for aluminum deficiency introduced by steaming. La- and Ca-Y zeolites have been subjected to hydrothermal conditions by

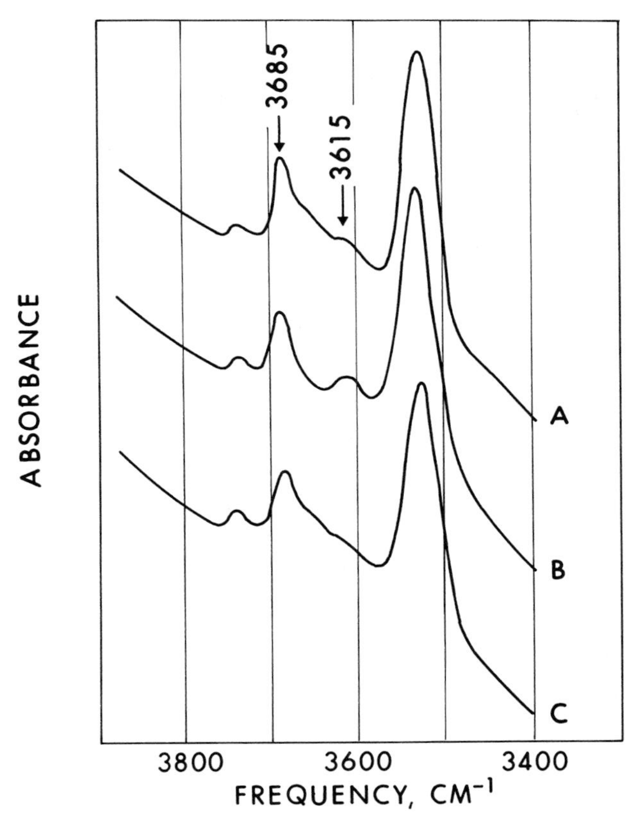

"Proceedings of the Third International Conference on Molecular Sieves"

Figure 40. $RE_{0.70}$-exchanged-Na-Y (A) steamed at 600°C, evacuated at 400°C; (B) A) + pyridine; (C) A) + steam at 675°C (93)

Jacobs and Uytterhoeven (*89*). Spectra are shown in Figure 43 comparing La-Y calcined in a deep bed at several conditions. Absorption bands are observed at 3740, 3690, 3650, and 3520 cm^{-1}. For Ca-Y, bands were observed at 3740, 3690, 3650, and 3575 cm^{-1}. Steaming results in the formation of additional bands at 3680 and 3605 cm^{-1}. The bands in the region of 3690–3670 and 3600–3530 cm^{-1} are non-acidic. These represent hydroxyl groups located at defect points in the structure from which aluminum has been extracted.

Stabilization by Treatment with Ammonia. Y zeolite can also be stabilized by treatment with ammonia (*94, 95*). Passage of NH_3 gas over decationized zeolites results in the replacement of chemical water by ammonia without any final weight change. The ammonia is much more strongly held to the zeolite than are the usual ammonium ions and is expelled over the 600–800°C temperature range, the same range where chemical water is lost. The hydroxyl absorption bands of the original decationized zeolite at 3630

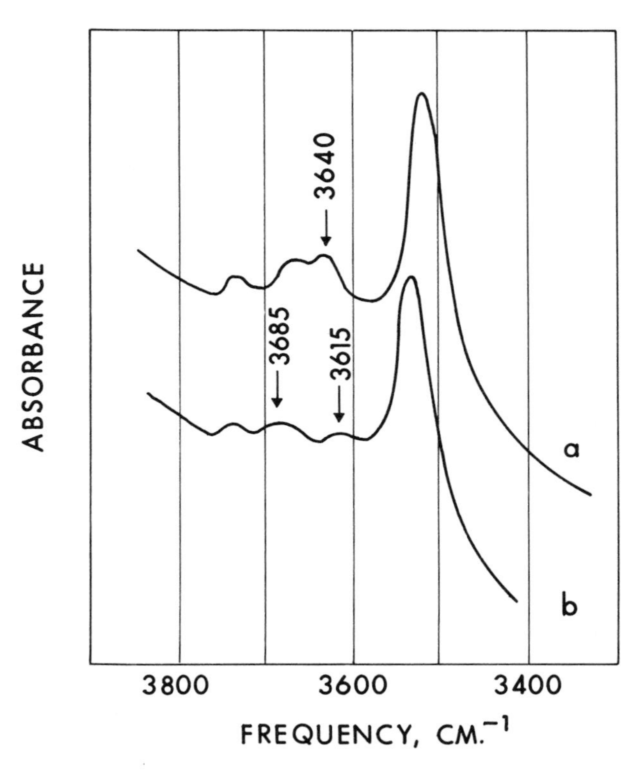

"Proceedings of the Third International Conference on Molecular Sieves"

Figure 41. $RE_{0.97}$-exchanged-Na-Y (a) steamed at 600°C, evacuated at 400°C; (b) (a) + pyridine (93)

and 3540 cm^{-1} are removed and new bands appear between 3350 and 2940 cm^{-1}. These bands are poorly resolved except for a shoulder at 3311 and bands at 3184 and 3067 cm^{-1}. Little evidence for ammonium ions or for physically adsorbed ammonia was found. The observed bands are at frequencies similar to those observed for 1:10 decanediamine and the proposed structure is of the form:

```
     O      H  O
     |      |  |
   O—Al     N—Si—O
     |      |  |
     O      H  O
```

Other Zeolites. CLINOPTILOLITE. The thermal decomposition of NH_4-clinoptilolite has been investigated over the temperature range of 25° to 550°C (*96*). Typical spectra are shown in Figure 44. In contrast to the studies of Breger *et al.* (*97*) on natural clinoptilolite, the NH_4 form exhibited structural hydroxyl groups. The ammonium ions are essentially completely decomposed (as indicated by the 1440 cm^{-1} band) by 400°C while physically adsorbed water is removed by 300°C. Broad bands near 3620, 3200, 3050, and 2850 cm^{-1} are observed for pretreatment temperatures below 210°C. The first band can probably be attributed to hydroxyl groups and the latter three to ammonium ions. For temperatures above 300°C, a sharp strong band develops near 3620 cm^{-1} and a weaker band near 3740 cm^{-1}. A shoulder is also observed near 3560 cm^{-1}. The 3620 cm^{-1} band goes through a maximum near 400°C and then declines steadily to zero at about 650°C. Readdition of water to the clinoptilolite after pretreatment at 550°C had no influence on the 3620 cm^{-1} band.

These data suggest that the deammination and dehydroxylation of NH_4-clinoptilolite proceeds like NH_4-Y zeolite, first deammination occurring with

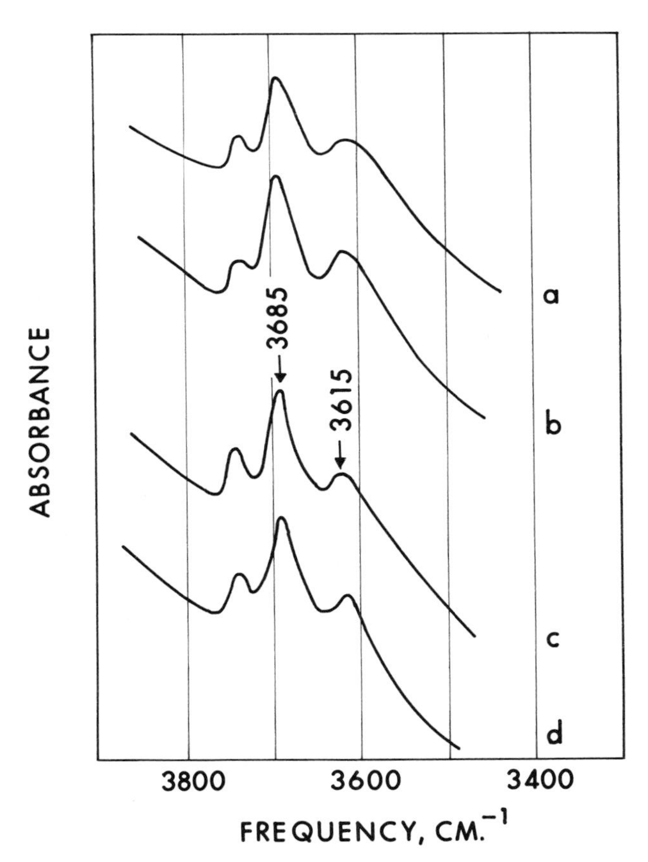

"Proceedings of the Third International Conference on Molecular Sieves"

Figure 42. Steamed $NH_{40.85}$-exchanged Na-Y (a) steamed at 600°C, evacuated at 400°C; (b) (a) + pyridine, evacuated at 210°C; (c) (a) + steamed at 675°C; (d) (c) + pyridine (93)

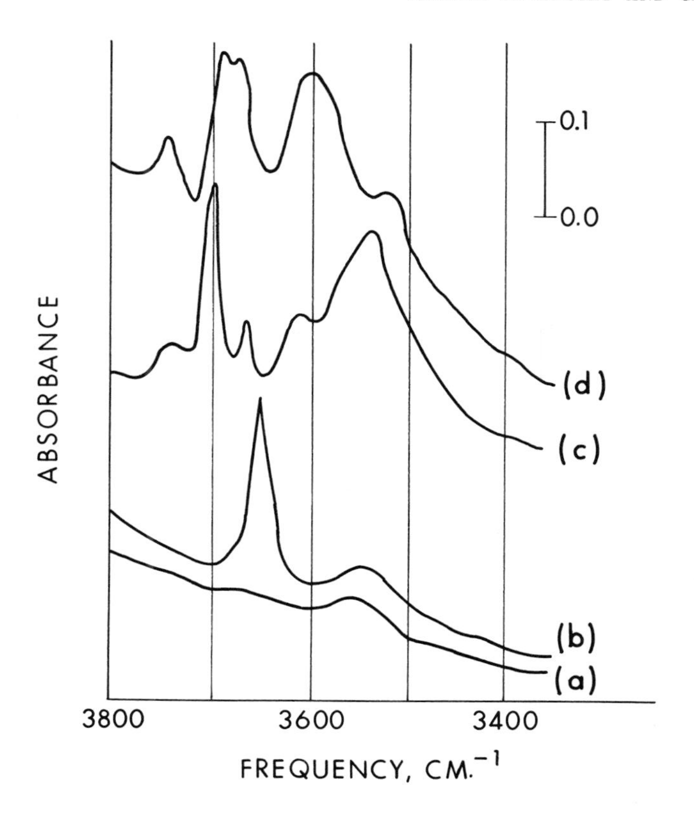

Transactions of the Faraday Society

Figure 43. La-Y zeolite (a) outgassed at 420°C; (b) (a) + NH_3; (c) (a) + steaming; (d) Y-deep bed La outgassed at 450°C (89)

the formation of structural hydroxyl groups and then dehydroxylation. One hydroxyl group is formed for each ammonium ion decomposed. Above 400°C, the sum of the hydroxyl group concentration and twice the Lewis acid site concentration remains constant. At about 600°C, lattice destruction starts to occur. In contrast to decationized Y, readdition of water does not result in structural collapse. This increased hydrothermal stability may be expected from the higher silica–alumina ratio. Because the 3620 cm^{-1} band type hydroxyl groups interact with both ammonia and pyridine, they are probably located on the external surfaces of the zeolite. The 3560 cm^{-1} band hydroxyl groups interact only with ammonia and are therefore probably located inside the zeolite channels.

NATROLITE AND STILBITE. The surface structure of several natural occurring zeolites was studied by Yukhnevich *et al.* (*98*). Spectra of natrolite, thomsonite, laumonite, chabazite, stilbite, and analcime were studied as mulls at room temperature. According to their spectra, they can be divided into three groups: (1) stilbite and chabazite, whose spectra have broad overlapping bands from adsorbed water; (2) natrolite, thomsonite, and laumonite, whose spectra have sharp bands; and (3) analcime, whose spectrum has a single very broad band. All of the minerals have bands at 1620 cm^{-1} in the spectra. Thomsonite and laumonite also have bands at 1675 and 1410 cm^{-1}. Thermal analysis studies show that water is gradually removed from the stilbite group and suddenly near 400°C from the natrolite group. These observations suggest that on stilbite and chabazite, there are a large number of molecules in a great variety of force fields whereas in the natrolite group, there is less water influenced by a strictly defined field. Spectra of natrolite were determined under vacuum as a function of evacuation temperature. At room temperature, a series of bands were observed near 3615, 3533, 3460, 3318, 3230, 3175, and 2140 cm^{-1}. At 160°C, all these bands are removed except those at 3533 and 3318 cm^{-1}; these bands are much weaker. All of the bands are attributable to adsorbed water.

Spectra of thomsonite show that between 60° and 140°C, water characterized by bands near 3500 and 1600 cm^{-1} is removed from the zeolite. Between 170° and 200°C, a second type of water characterized by absorption bands near 3370, 3270, and 1675 cm^{-1} is removed. However, after heating to 220°C, the spectra show that water is still present on the zeolite.

MORDENITE. Spectra of mordenite dealuminized to different extents have been investigated. Eberly *et al.* (*99*) investigated dealuminized hydrogen mor-

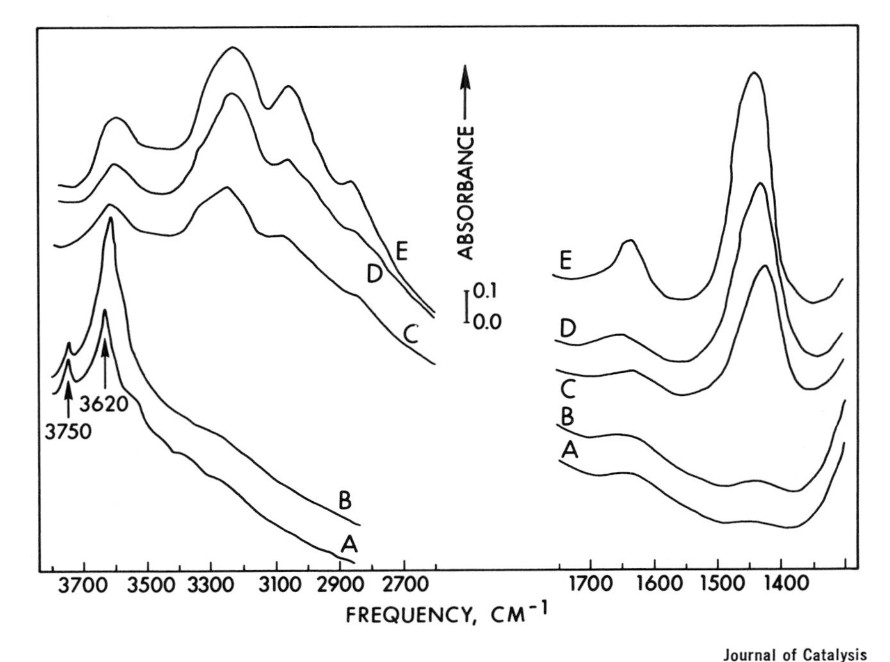

Journal of Catalysis

Figure 44. NH_4-clinoptilolite outgassed at (C) 25°C; (D) 125°C; (C) 210°C; (B) 400°C; and (A) 550°C (96)

denite prepared by extraction of hydrogen mordenite with 6*N* HCl solution. A series of materials with SiO_2/Al_2O_3 ratios of 12 to 97 were prepared. The acid leaching of mordenite changes the lattice vibrations significantly (*see* Chapter 2). Typical spectra obtained in the hydroxyl stretching region are shown in Figure 37. For the starting material (SiO_2/Al_2O_3 = 12) hydroxyl bands are observed at 3740 and 3590 cm^{-1} while for a leached material (SiO_2/Al_2O_3 = 66), only a single band at 3740 cm^{-1} is observed. More detailed spectra were obtained by Shikunov *et al.* (*100*). Hydrogen mordenite with SiO_2/Al_2O_3 ratios of 10, 17, 30, and 73 were used. To improve the spectral operation, the samples were deuterium exchanged. Evacuation at room temperature resulted in bands near 2550, 2700, and 2720 cm^{-1}. Samples with SiO_2/Al_2O_3 of 30 and 73 also had a band near 2760 cm^{-1}. Heating to high temperature (*e.g.*, 100°C) shifted the band from 2550 to 2610 cm^{-1}. At 300–450°C, the 2610 cm^{-1} band disappeared leaving a spectrum with bands near 3670, 2700, and 2760 cm^{-1} corresponding to hydroxyl groups at 3600, 3650 and 3740 cm^{-1}. As the SiO_2/Al_2O_3 ratio is increased, the 2670 and 2700 cm^{-1} bands decrease in intensity and the 2760 cm^{-1} band increases. The bands near 2670 (3600) and 2700 (3650) cm^{-1} appear to be associated with the presence of alumina tetrahedra in the structure and the 3740 cm^{-1} is associated with silanol groups.

The deammination of ammonium mordenite follows a somewhat similar pattern to that of ammonium Y zeolite (*101, 102*). The ammonium ion decomposition appears to be complete at about 380°C *vs.* 260–300°C for ammonium-Y zeolite (*69*). After treatment at 350°C, two hydroxyl group frequencies are detected at 3740 and 3610 cm^{-1}. Analogous to Y zeolites, the intensity of the hydroxyl group bands is a function of the spectral measurement temperature: a 20% decrease occurs between room temperature and 350°C.

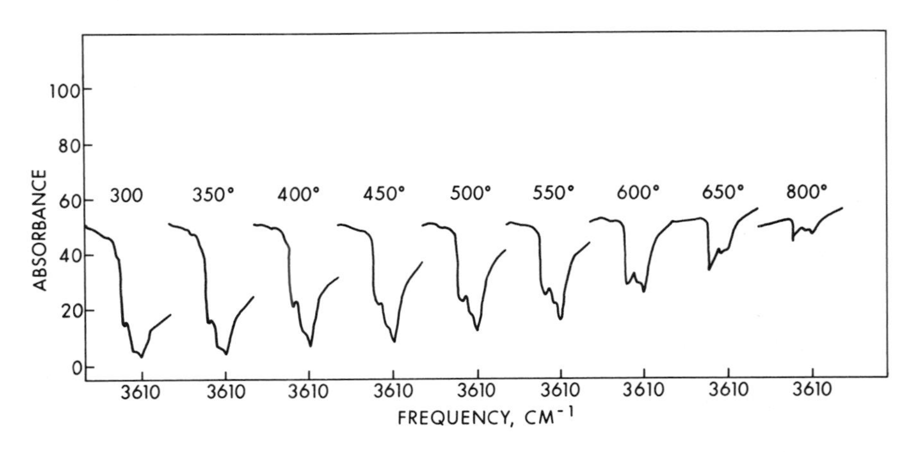

Zeitschrift für Physikalische Chemie, Neue Folge

Figure 45. Hydrogen-mordenite calcined at various temperatures between 300° and 800°C (103)

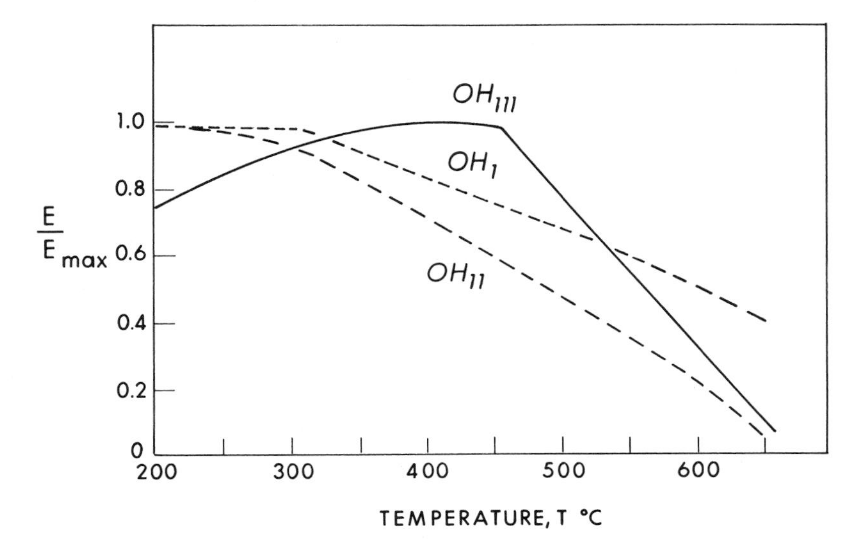

Zeitschrift für Physikalische Chemie, Neue Folge

Figure 46. Temperature dependence of hydroxyl bands on hydrogen-mordenite (103)

Hydrogen mordenite calcination studies have also been investigated (*103*). Figure 45 shows the spectra over a 500°C calcination temperature range. In contrast to decationized mordenite obtained from ammonium mordenite decomposition, three hydroxyl bands are observed in the spectrum at 3740, 3650 and 3610 cm^{-1}. These observations agree with Shikunov *et al.* (*102*) but differ from that of Eberly *et al.* (*99*). The temperature dependence of the bands is shown in Figure 46. The 3737 cm^{-1} band corresponds to the OH_1 group, the 3650 cm^{-1} band to the OH_2 group, and the 3610 cm^{-1} band to the OH_3 group. The 3740 and 3610 cm^{-1} bands interact with pyridine molecules but the 3650 cm^{-1} band does not. The 3610 cm^{-1} band represents hydroxyl groups in the main channels and the 3650 cm^{-1} groups represent hydroxyl groups in the side channels.

GERMANIC ZEOLITE. The dehydration of germanic near-faujasite zeolites was studied by Fripiat *et al.* (*104*). In these zeolites, germanium is substituted for silica in the lattice. Typical spectra are shown for the sodium form as a function of temperature in Figure 47. After treatment at 100°C, strong bands are observed at 3690, 3340, 3280, 1664, 1480, and 1432 cm^{-1}. Washing of the zeolite before calcination at 100°C introduces strong bands near 3650 and 3590 cm^{-1}. On heating, the two bands at 3340 and 3280 cm^{-1} disappear at about 160°C along with the band near 1664 cm^{-1} and the bands near 1480 and 1432 cm^{-1} become weaker. Three new bands near 1615, 1572, and 1375 cm^{-1} are observed. At 350°C, the 1664 and 1375 cm^{-1} bands are almost completely eliminated. The bands at 3690, 3340, 3280, and 1664 cm^{-1} are probably from physically adsorbed water, the band at 3690 cm^{-1} corresponding to interaction of the water with the sodium cations. The

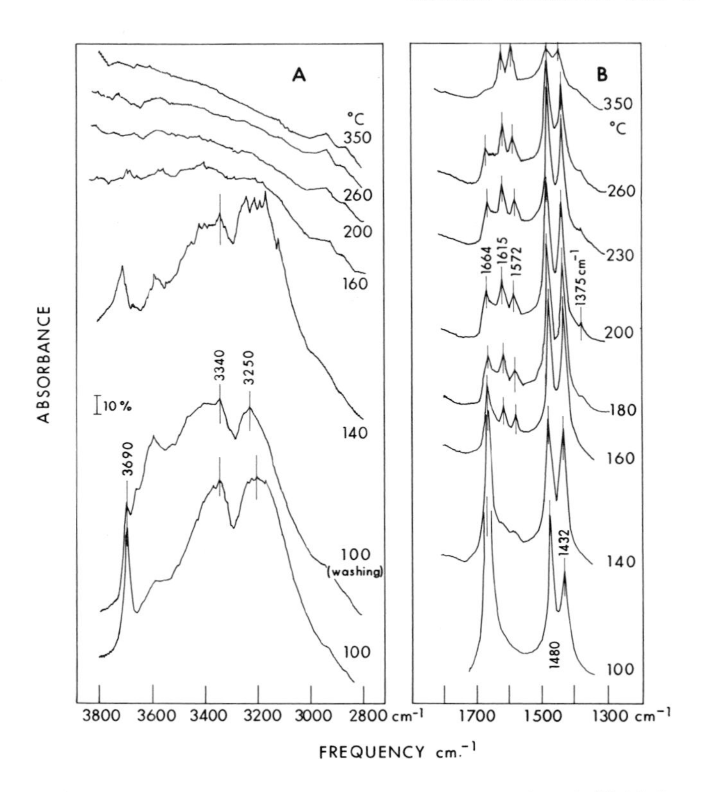

Journal of Catalysis

Figure 47. Na-germanic zeolite calcined for 2 hr at progressively increasing temperatures (104)

other bands observed in the spectra are probably from chemisorbed carbon dioxide.

The ammonium exchanged form was also studied. Because of high instability, the species studied was only 3% exchanged with ammonium ions. After evacuation at 100°C, absorption bands were observed at 3690, 3658, 3605, 3357, and 3240 cm^{-1}. The bands at 3690, 3357, and 3240 cm^{-1} are from absorbed water although the latter two bands also contain probably some contribution from the ammonium ions. The 3658 and 3605 cm^{-1} bands are from structural hydroxyl groups. At higher calcination temperatures, only bands near 3658 and 3605 cm^{-1} are observed. The spectra are interpreted in terms of the schematic structure:

NH_4^+ H

O O O O O O

Al Ge ⟶ Al Ge $+ NH_3$

O OO O O OO O

Two main bands near 3655 and 3580 cm^{-1} are observed in the spectra of the calcium form. By analogy with X and Y zeolites, the 3580 cm^{-1} band is assigned to $Ca(OH)^+$ groups and the 3655 cm^{-1} band to GeOH groups. These groups could be formed by the scheme:

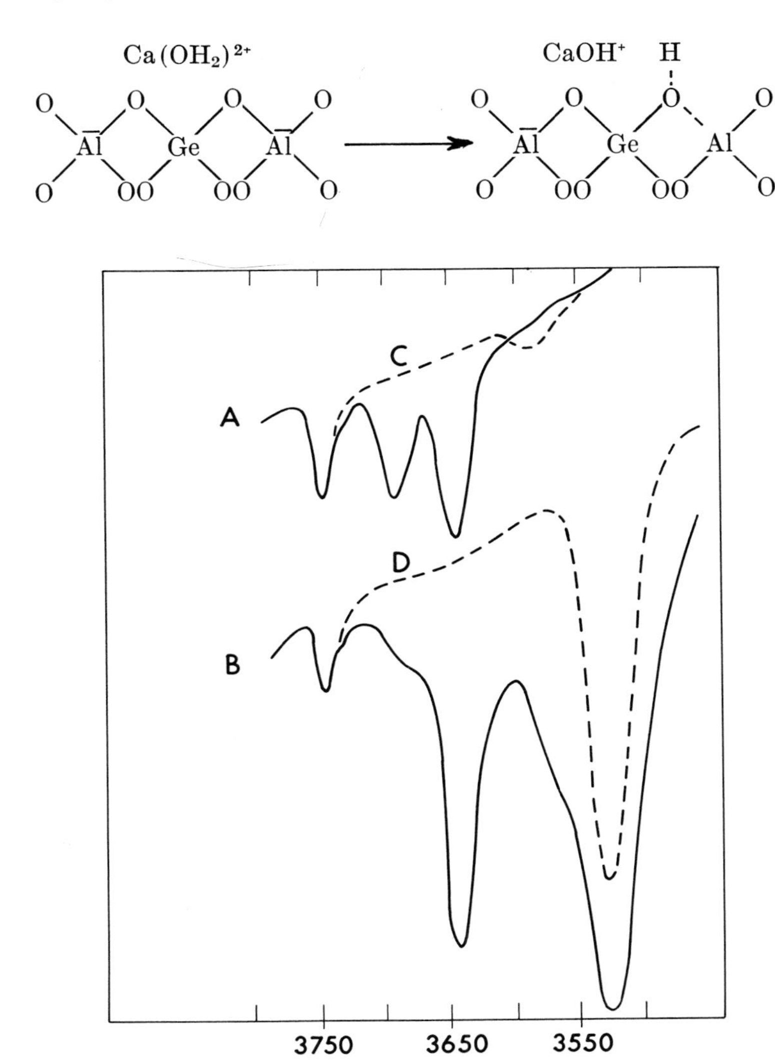

"Catalysis"

Figure 48. Salt-loaded zeolites (A) Ca-Y calcined at 550°C; (B) La-Y calcined in air at 550°C; (C) Ca-Y–7% NaCl calcined in air at 550°C; and (D) La-Y–7% NaCl calcined in air at 550°C (105)

ZEOLITE-SALT ADDUCTS. Rabo *et al.* (*105*) reported spectra of Ca-Y and La-Y which had been loaded with 7% NaCl and calcined at 550°C. The spectra, Figure 48, show that the hydroxyl groups which are probably present in the supercages are eliminated while those probably located in the sodalite cages are unaffected, *e.g.*, the 3520 cm^{-1} band in La-Y.

Decomposition of Organic Cation Containing Zeolites. Additional information is now being obtained on zeolites and their mechanisms of decomposition from studies of organic cation-exchanged zeolites. In particular, ion-exchanged zeolites containing mono-, di-, tri., and tetramethylammonium, mono-, di-, and triethylammonium cations have been studied. Naturally for zeolites which are synthesized containing these ions, it is logical to study the parent zeolite, *e.g.*, in the cases of synthetic offretite and Ω zeolite.

Wu *et al.* (*106*) studied the decomposition of tetramethylammonium Y (TTMA) zeolite by several techniques. Infrared spectra, taken *in situ*, were recorded from 150° to 350°C. At about 200°C, absorption bands typical of the amine near 3028, 2965, 2930, and 1480 cm^{-1} started to decrease and absorption bands near 3735, 3637, and 3550 cm^{-1} typical of decationized Y appeared. After treatment at 350°C, the spectrum was similar to that of decationized Y formed by decomposition of ammonium Y zeolite. The hydroxyl band at 3550 cm^{-1} was detected before significant decomposition of the tetramethyl ammonium ion had occurred, suggesting that the hydroxyl groups arose from hydronium ions introduced into the sodalite cages during the initial ion exchange. The 3637 cm^{-1} hydroxyl band is derived from TTMA cations located in the supercages. Its similarity to the hydroxyl band in decationized Y zeolite leads to the conclusion that the protons formed interact with the O_1 oxygens. Chemisorption of ammonia showed that both types of hydroxyls are acidic and similar to those of decationized Y. The surface groups behave like those of decationized Y since as the surface is dehydroxylated by calcination at increasingly higher temperatures, the acidity changes from Bronsted to Lewis.

The combination of infrared, thermal analyses, and product analysis show that the organic cations decomposed in two distinct steps: at 150°–275°C and at 275°–475°C. Detailed study shows that the mechanisms by which the decationized Y zeolite is reached vary with the decomposition temperature. Less detailed studies of mono-, di-, and trimethylammonium Y zeolites led to similar conclusions.

Jacobs and Uytterhoeven (*107*) made a similar study of mono-, di-, and trimethylammonium, mono-, di- and triethylammonium, propyl- isopropyl- and butylammonium, piperidinium and pyridinium ions exchanged into Y zeolite. Spectra were observed after evacuation at room temperature, 125°, 225°, 350°, and 450°C. In general, similar results were obtained. With the primary alkylammonium ions, one hydroxyl group was formed for each alkylammonium ion decomposed and, unless heated further, only Bronsted acid sites were present. Decomposition of the secondary and tertiary alkylammonium ion-containing zeolites results in a simultaneous partial dehydroxylation even at low temperatures. The decomposition temperature increased with increasing chain length of the exchanged cation. During the decomposition, the amines tended to interact preferentially with the 3650 cm^{-1} hydroxyl groups as judged by the changes in the ratio of the hydroxyl group band intensities with temperature. The spectra of the pyridine-exchanged zeolites exhibit three bands near 1540 cm^{-1}, which can be attributed to three different locations for pyridinium ions. However since the bands are equally spaced, it is possible that a mechanical resonance between two pyridine molecules adsorbed on the same proton is being observed. Two bands are also observed at 1465 and 1455 cm^{-1} from pyridine adsorbed on Lewis acid sites. The first band is at a considerably higher frequency than usually observed for decationized Y zeolites and suggests the presence of stronger than usual Lewis acid sites.

Many reaction products were detected by Wu *et al.* (*106*) and Jacobs *et al.* (*107*). Fripiat and Lambert-Helsen (*108*) have also investigated the nature of the products. Wu *et al.* (*106*) suggest decomposition of the cations to give trimethylamine, methanol, and decationized zeolite followed by secondary reactions including the formation of a methoxy zeolite at low temperatures or decomposition *via* ylide mechanisms at higher temperatures. Jacobs and Uytterhoeven also suggest that some cations may decompose via a transalkylation mechanism. Fripiat and Lambert-Helsen confirmed this mechanism for ethyl-, diethyl-, and triethylammonium Y and drew the

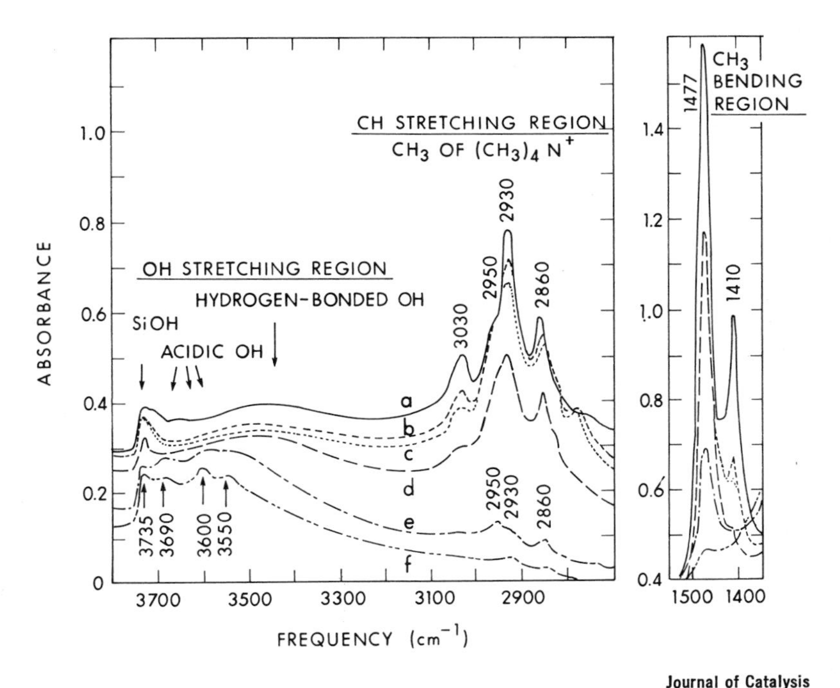

Journal of Catalysis

Figure 49. Offretite calcined at (a) 100°C; (b) 200°C; (c) 300°C; (d) 350°C; (e) 400°C; and (f) 425°C (110)

analogies with mechanisms previously proposed for alkylammonium montmorillonites (*109*). Hence the mechanisms of decomposition depends greatly on the nature of the alkyl substituent.

Wu *et al.* (*110*) have also studied the decomposition of synthetic offretite. The tetramethylammonium (TTMA) cations were decomposed by heating the catalyst under vacuum. The decomposition rate was particularly rapid above 300°C. Typical spectra are shown in Figure 49. As the bands attributed to TTMA near 3030, 2950, 2930, and 2860 cm^{-1} declined in intensity, hydroxyl bands near 3735, 3690, 3600, and 3550 cm^{-1} (3690, 3615, and 3550 cm^{-1} at 100°C) grew. The relative intensity variations for the 2950 and 2860 cm^{-1} bands compared with the 2930 cm^{-1} band indicated that the decomposition of the zeolite probably went *via* a methoxy species, Me–O–Zeol. These species were formed at about 400°C. Wu *et al.* (*110*) concluded that the hydroxyl groups were similar in nature to those of decationized Y but that the observed frequencies differ because of the different lattice environments. Jacobs and Uytterhoeven (*107*) in contrast have suggested that because of more severe pore opening restrictions, the offretite becomes partially hydrolyzed by the higher concentration of decomposition products within the zeolite structure. The evidence for this hypothesis is that under certain calcination conditions, decationized Y zeolite exhibits absorption bands near the same frequencies. The surface hydroxyl groups reacted with ammonia showing that they were proton acids. On dehydroxylation, Lewis acid sites were formed. Analysis showed that the organic cations decomposed into 21 gaseous products.

Influence of Temperature on Hydroxyl Bands—Proton Delocalization and Mobility. Hydrogen is believed to be mobile on the surface of many catalysts under reaction conditions. Because the 3550 cm^{-1} hydroxyl groups, located in the inaccessible parts of the structure, can interact with large molecules like cumene at high temperatures and large polar molecules like piperidine at ambient temperature, mobility of hydrogen in zeolites seems probable.

Infrared band intensity changes can be used to observe proton movements, but it is probably inferior to nuclear magnetic resonance. The major problems are that the observations could be caused by factors other than proton mobility and that the residence time required to observe an infrared vibration is very short.

Several studies have been made to investigate the question of whether hydrogen on the molecular sieve surface is mobile or localized (*111, 112, 113*). The change in intensity of the hydroxyl absorption bands is observed as a function of temperature. A decrease in intensity of the absorption bands indicates a decrease in the population of hydroxyl groups and hence a decrease in the number of hydrogen atoms attached to the surface. However, thermal vibrational changes can be responsible for the intensity changes.

Studies made have involved the deammination of ammonium Y zeolite at 450° to 500°C followed by the study of the intensities and frequency of the hydroxyl groups as a function of sample temperature. All of the detailed studies show a decrease in the integrated intensity of the two hydroxyl bands as the temperature was increased. The magnitude of the intensity change varies among the studies. The 3650 cm^{-1} band is more sensitive to temperature than the 3550 cm^{-1} band. Typical spectra are shown in Figure 50 and the change in intensity is shown in Figure 51. The frequency of

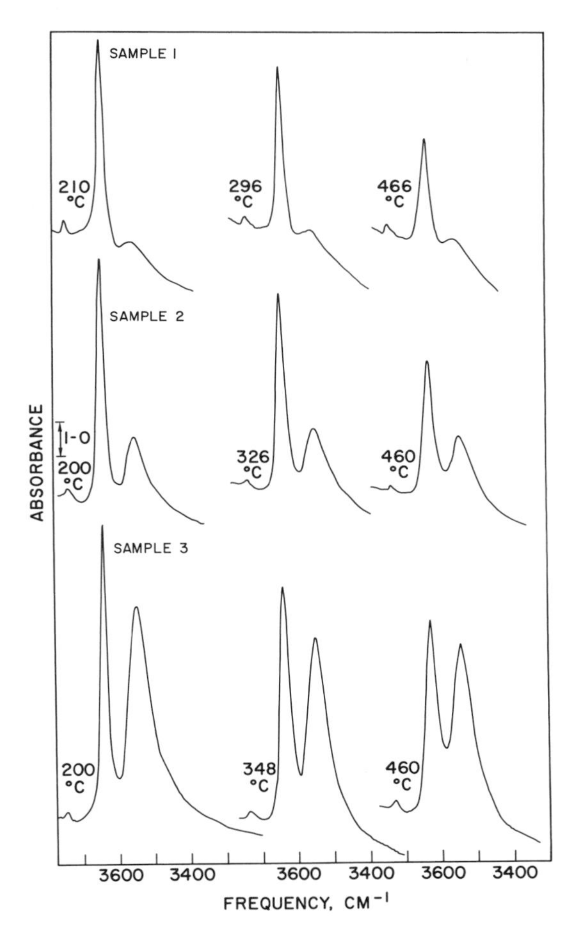

Journal of Catalysis

Figure 50. Hydroxyl bands of decationized-Y-zeolite calcined at 470°C as a function of sample temperature, (1) $NH_{4\,0.32}$-exchanged Na-Y; (2) $NH_{4\,0.63}$-exchanged Na-Y; (3) $NH_{4\,0.87}$-exchanged Na-Y (113)

the hydroxyl group vibrations decreases with increasing temperature. Again the 3650 cm^{-1} band is more sensitive to temperature changes than the 3550 cm^{-1} band. These observations have been interpreted in terms of proton mobility: the decrease in frequency can be associated with greater interaction

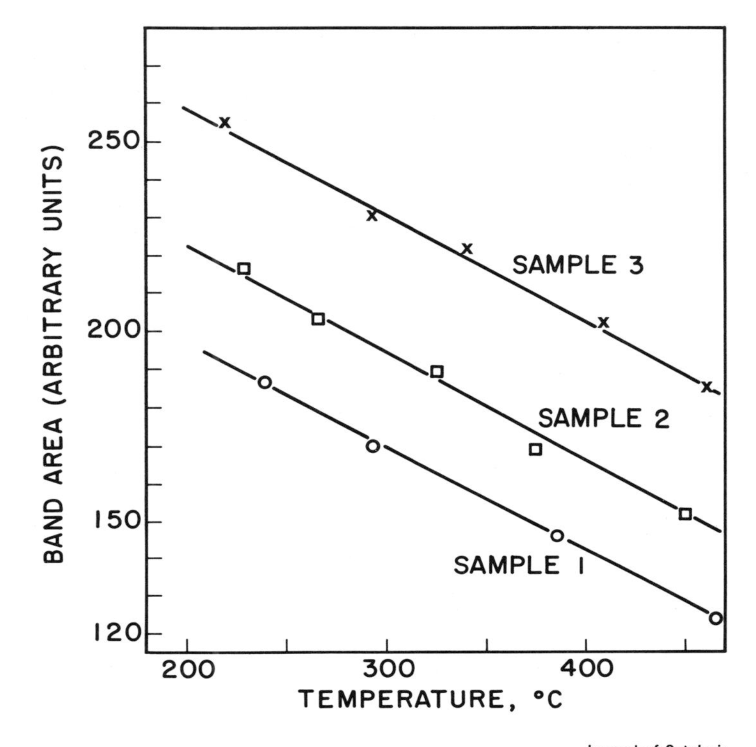

Journal of Catalysis

Figure 51. Band area of 3640 cm^{-1} band on decationized Y zeolite as a function of sample temperature (113)

of the hydroxyl group with the adjacent aluminum atom as typified by the following scheme.

```
           H                                  H
           |                                  ¦
  O        O       O                O         O       O
    \            /                    \     ,'  \    /
     Al        Si       <------->      Al         Si
    /   \     /   \                   /   \      /   \
  O      OO        O                O      OO         O
```

The next step can readily be envisaged as the breaking of the H–O bond. This type of mechanism is supported by the known details of the dehydration of decationized Y zeolite.

If the proton moves from oxygen to oxygen atom, the lifetime of a given OH bond must be very short ($<10^{14}$ sec). Because of this, Cant and Hall consider that their observations did not represent delocalization but rather the intensity changes from increased thermal energy of the lattice (*111*). Fripiat *et al.* (*116*) have pointed out that temperature dependency of band intensities is a necessary but not sufficient criterion for proton delocalization. That is, band intensities can change as a function of temperature without proton delocalization occurring, but if proton delocalization does occur, there must be a change in band intensity. Magnetic resonance measurements are probably more suitable than infrared for observing proton movements because the time scale is more suitable. Using various techniques, mobility has been shown. Mestdagh and Fripiat (*117*) used wide-line and pulse methods between —200° and 400°C. Above 200°C, the proton jump frequency (ν) was found to obey the relationship $\nu = 10^{10} \exp(-10^4/RT)$ sec^{-1} and the jump distance is 4 A.

More recently, Freude, Oehme, Schmiedel, and Staudte (*118*), using a pulse method, found jump frequencies of 2–10 $\times$ 10^4 sec^{-1} at 200°C and activation energies of 5–10 kcal/mole.

Schoonheydt and Uytterhoeven have also reported changes in intensity of the hydroxyl bands as a function of temperature of comparable magnitude with those of other studies (*82*) and attributed the intensity changes to thermal motions in the solids. However, they extended their data further than others and found that the ratio of the 3640 cm^{-1} band to the 3550 cm^{-1} band changed with temperature and concluded that the protons were redistributing themselves between the various crystallographic sites as a function of temperature.

Although, as suggested by Fripiat (*116*), infrared spectra are probably not sufficient evidence to establish proton movement or delocalization, the data, in conjunction with the more definitive magnetic resonance and conductivity data, strongly indicate that there is some movement at elevated temperatures. Other chemical evidence also supports proton mobility. Thus, if the locations of hydroxyl groups in the structure discussed above are correct, the ability of both types of hydroxyl groups to interact with molecules such as piperidine, pyridine (*68, 69*) and cumene (*119*) necessitates proton mobility. Furthermore, the interaction of cumene with the hydroxyl groups is a marked function of the zeolite temperature, suggesting enhanced mobility of the zeolite hydrogen at higher temperatures.

The occurrence of mobile protons on solid oxides possibly has important applications in catalysis. Thus the number of delocalized protons on the surface of decationized Y is the same order of magnitude as the number of active sites estimated on these materials in acid-catalysed reactions. These protons could function as strong acid centers and the nature of the zeolite surface may form a unique substrate on which the protons can move. In contrast to the decationized zeolite, silica was found to have no proton delocalization by infrared studies. It is well known that silica is a poor catalyst for acid-catalyzed reactions. The relationships between zeolite surface chemistry and their catalytic properties are discussed below.

Interaction with Water

Cation-Containing Zeolites. Zeolites have a strong affinity for water. Since in many catalytic processes feedstocks contain water in traces to substantial amounts, it is important to know how, if at all, the zeolite interacts with water. Furthermore, x-ray and magnetic studies have shown that the locations of the cations are influenced by water. Many studies of water adsorption on zeolites have been made spectroscopically. Early studies have been reviewed by Kiselev and Lygin (*120*). One of the earliest systematic studies was that by Bertsch and Habgood of small amounts of water adsorbed on Group IA X zeolites (*26*). The zeolites were pretreated in vacuum at 500°C. They observed three major bands near 3720–3650, 3400, 3250, and 1650 cm^{-1}. The frequencies of the band depended on the cation exchange. The band in the 3700 cm^{-1} region increased and the band near 1650 cm^{-1} decreased in frequency with decreasing cation radius. Previously, Szymanski *et al.* (*22*) had observed bands near 3500 and 3400 cm^{-1} for Na-X and Na-A and Frohnsdorff and Kingston (*21*) observed bands near 3450 cm^{-1} for Na-A. The band near 3700 cm^{-1} observed by Bertsch and Habgood was sharp whereas the other bands were broad and poorly resolved. The studies suggest that the water molecule interacts directly with the cation through the oxygen atom, *e.g.*

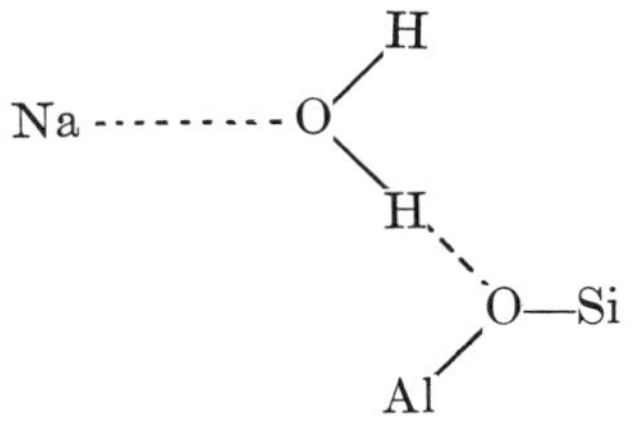

```
                   H
                  /
Na --------- O
                  \
                   H
                     \
                      O—Si
                     /
                   Al
```

The free hydrogen attached to the oxygen gives rise to the sharp cation dependent band near 3700 cm^{-1}, while the hydrogen attached to the oxygen and also interacting with the silica-alumina framework gives rise to the broad bands. The presence of more than one broad band probably indicates that different locations in the structure are involved and that different interacting forces prevail.

Zhdanov *et al.* (*24*) showed that the hydroxyl frequencies decreased with increasing cation radius, indicating the polarizing influence of the cations on the adsorbed water. The existence of a linear relationship between the frequency shift of the OH vibration and the ionic radius of the cation was shown by Kiselev and Lygin. Recently, Kiselev *et al.* made similar studies to Bertsch and Habgood on Li., Na-, K-, and Cs-X. Spectra were obtained as a function of readsorbed water content (*120*). For all the zeolites except Cs-X, a sharp band near 3700 cm^{-1} and broad bands between 3650 and 3000 cm^{-1} were observed analogous to the results of Bertsch *et al.* The water bending band near 1650 cm^{-1} was sensitive to the cation, varying in frequency from 1660 to 1643 cm^{-1} when the water content is about two molecules per cavity. However, at greater coverage, the band in all cases is near 1640–1645 cm^{-1}. In contrast to the Li, Na and K-X zeolites for which the water seems to interact strongly with the cations, Cs-X spectra exhibit no sharp bands—only broad bands. The interaction with the cation is weaker, and both hydrogens of the water seem to be interacting with the oxygen atoms of the lattice.

Y zeolites have been extensively studied. Early work (*30*) showed that for divalent cation Y zeolites, a broad band near 3540 cm^{-1} appeared, with little effect on the 3640 cm^{-1} band. Eberly (*64*) reported formation of bands near 3650 and 3550 cm^{-1} when water was added to well-dehydrated Ca-Y zeolite. The readsorption of water onto Group IA, IIA, and rare earth Y zeolites has been studied in considerable detail (*32*). For Group IA zeolites, the results of readsorption of water are similar to those reported for Na-X (*26*): that is, a sharp cation dependent band near 3700 cm^{-1} and several broad bands between 3650 and 3000 cm^{-1}. The frequency of the sharp band varied between 3720 and 3640 cm^{-1} as the cation was changed from Li to Cs-Y in a systematic manner with the electrostatic field and potential of the zeolite and the ionization potential of the cation. In contrast to X zeolite, a reasonably sharp band was observed for Cs-Y. Figures 52 and 53 show the relationship between the electrostatic properties of the zeolite and the frequency. The water bending vibration near 1640 cm^{-1} decreased in frequency as the electrostatic and ionization potentials increased. However, the trend was small and somewhat random. During the water adsorption phase, the intensity of all the bands increased uniformly thus indicating no preferential adsorption sites. The adsorbed water could be removed simply by evacuation at elevated temperature. Water, as detected by the 1640 cm^{-1} band was removed from Li by 550°C, Na by 370°C, K by 360°C, Rb by 220°C and Cs by 210°C. The water appears to be adsorbed *via* the cations in a manner analogous to that proposed by Bertsch and Habgood for X zeolites (*26*). The three broad bands observed between 3650 and 3200 cm^{-1} can be attributed to hydrogen of the water molecule bound to crystallographically different oxygen atoms or to the adsorption of water on cations located in different positions. As shown by Figure 2, the spectra observed on initial dehydration are different from those observed during rehydration and subsequent dehydration suggesting possible changes in position or coordination of the cation or the occurrence of extensive hydrogen bonding. The greatest changes are probably in the intensity of the sharp band near 3700 cm^{-1} and the broad band near 3600 cm^{-1}. The former band increases while the latter band decreases markedly in intensity. However, no new structural hydroxyl groups are detected. Because the 3700 cm^{-1} band represents direct interaction with the cations, some cation migration is occurring during the dehydration. A rough approximation would suggest that about 10 times as many cations are interacting with the water in the rehydrated sample *vs.* the initial sample. The reason for the decline in the intensity of the band near 3000 cm^{-1} is not clear.

Spectra of water adsorbed on Group IIA Y zeolites are more complicated in behavior than on Group IA zeolites with the exception of Cs-Y. Ba-Y resembles the Group IA zeolite in behavior and Cs-Y in particular. On readsorption of water, a moderately sharp band is observed near 3680 cm^{-1} and broad bands near 3500 and 3260 cm^{-1}. The bands are removed by evacuation at 300°C and no new structural hydroxyl groups are detected. In contrast to the other alkaline earth zeolites discussed below, barium zeolite is unable to dissociate adsorbed water to any appreciable extent.

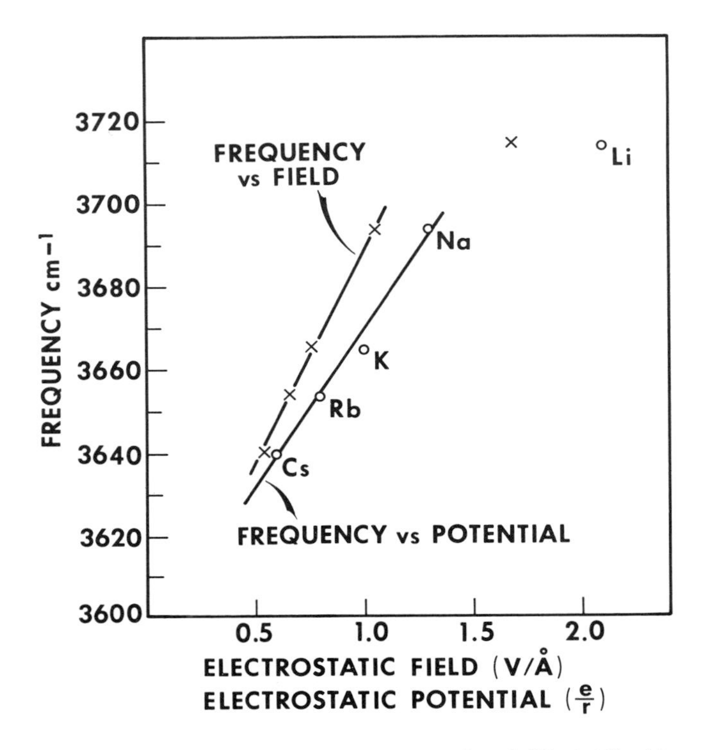

Journal of Physical Chemistry

Figure 52. Frequency of the 3715–3640 cm^{-1} band vs. *electrostatic field and potential of the cation* (32)

The readsorption of water onto Mg, Ca, and Sr-Y zeolites is more complex. Readsorption below 100°C on Ca-Y produces bands near 3690, 3640, ~3580, and a broad band near 3550 to 3400 cm^{-1}. On heating to 200°C, the 3690 cm^{-1} band decreases markedly in intensity while the 3640 cm^{-1} band grows. After evacuation at 470°C, the spectra resemble that of the initial species. Similar results were obtained with Mg-Y and Sr-Y. Compared with Mg and Ca-Y, the Sr-Y had a much broader and intense band near 3570 cm^{-1} and was somewhat similar in contour to Ba-Y. On heating to about 250°C, the band at 3680 cm^{-1} and the broad bands at lower frequencies were removed and the bands near 3640 cm^{-1} and 3540 cm^{-1} increased in intensity. On careful dehydration, a band near 3585 cm^{-1} is also observed. The band near 3690 cm^{-1} formed on addition of water probably arises from interaction of the water with the divalent cations. Its precise frequency depends on the nature of the cation. The other broad bands observed indicate interaction of adsorbed water with the zeolite lattice similar to that occurring in Group IA zeolites. On heating, the adsorbed water dissociates under the influence of the zeolite crystal field. Thus the bands from molecular water are eliminated. Probably the water interacts with the lattice by the scheme:

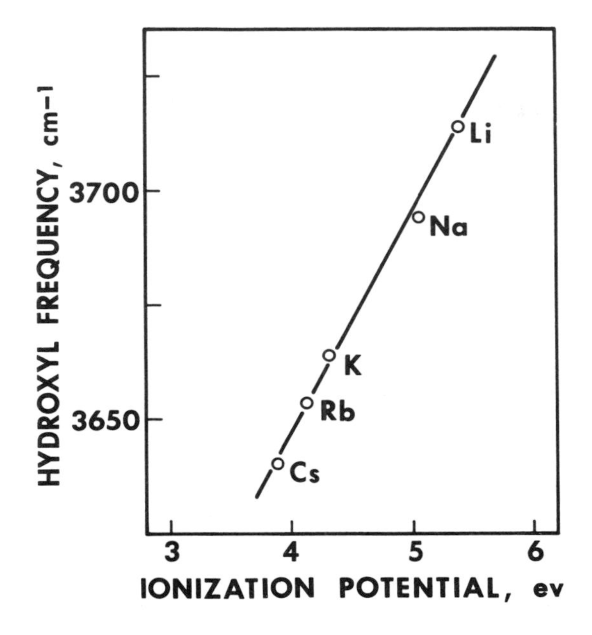

Journal of Physical Chemistry

Figure 53. Frequency of the 3715–3640 cm^{-1} band vs. *ionization potential of the cation* (32)

$$SiO + M(OH_2)^{2+} \rightleftarrows 2MOH^{+} + SiOH^{+}$$

Thus the band due to silanol groups near 3640 cm^{-1} grows. It would be expected that a band due to MOH^+ groups might be detected. A cation-dependent band has been observed at 3595 cm^{-1} for Mg-Y, 3585 cm^{-1} for Ca-Y and 3570 cm^{-1} for Sr-Y. No comparable band was observed on Ba-Y. This type of hydroxyl group was non-acidic to pyridine. The intensities of these bands are very sensitive to the degree of hydration and temperature and upon mild dehydration, the bands disappear, suggesting the formation of MO or M^+—O—M^+ groups (*44, 60, 73*). Similar observations have been made for Ca-X zeolites (*31*).

Several transition metal exchanged Y zeolites have been studied (*47*). Typical spectra are shown in Figure 5. Mn-, Co-, Ni-, Cu-, Zn-, Ag-, and Cd-Y zeolites were studied. For Cu- and Ag-Y, no discrete absorption bands were observed on readsorption of water but broad bands between 3700 and 3000 cm^{-1} were formed as shown in Figure 54. The characteristic bending band near 1635 cm^{-1} was also detected. Reevacuation removed the adsorbed water and restored the spectrum to that observed after the initial evacuation at 450°C. Thus Cu-Y behaves very much like Ba-Y zeolite in that the zeolite appears unable to dissociate physically adsorbed water. This is somewhat unexpected since the ionizing power of the divalent Cu ion is one of the largest of the cations studied. However, the Cu ions may exist in the zeolite as monovalent ions. As such, the polarizing power would be much lower. Silver might have been expected to behave similar to Group IA zeolites. Addition of water results in interaction with the 3633 cm^{-1} band on the zeolite surface, indicating that some hydrogen bonding occurs with the structural hydroxyl groups. The rest of the spectrum is similar to water adsorbed on a Group IA zeolite, namely, a sharp band near 3685 cm^{-1} and a broad band near 3250 cm^{-1}. These bands arise from interactions similar to those discussed for the Group IA zeolites and can be represented by a structure such as:

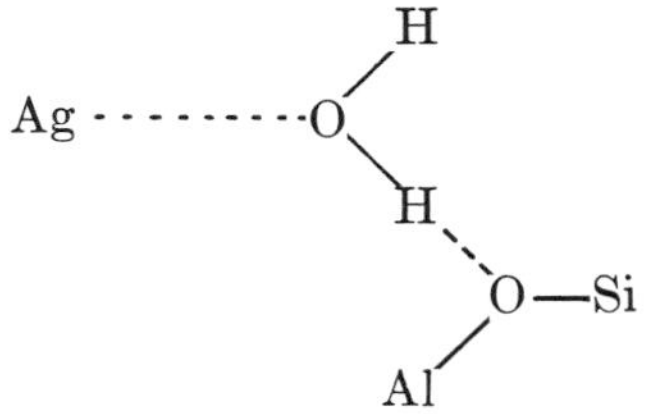

```
              H
             /
Ag - - - - -O
             \
              H
               `.
                 O—Si
                /
              Al
```

The 3685 cm^{-1} band probably represents the vibrations of the free hydroxyl group, and the other band represents the interaction of the hydroxyl group with the lattice. No band representing hydroxyl groups associated with the cation was detected. All the adsorbed water could be removed at 120°C, leaving only the band at 3633 cm^{-1}. The band is only about 50% as intense as in the initially dehydrated zeolite.

The other transition metal cation zeolites studied behave similarly to the Group IIA zeolites. Readdition of water to Ni-Y zeolites followed by evacuation at room temperature resulted in the bands at 3635 and 3540 cm^{-1} becoming stronger and new bands being observed at 3675 and 3600 cm^{-1}. By heating at 110°C, the 3675 cm^{-1} band is eliminated, at 250°C, the 3600 cm^{-1} band is eliminated. Treatment at 350°C results in the spectrum returning to that of the originally calcined zeolite. Adsorption on Mn-Y results in broad bands near 3680 and 3520 cm^{-1}. However, on evacuation at 250°C, discrete bands are observed at 3675, 3630, and 3540 cm^{-1}. The spectrum is similar to, but weaker than, that of the initial zeolite calcined at 250°C. Co-Y spectra are similar except a strong band is observed near 3615 cm^{-1} after readsorption of water. This band is readily removed by evacuation at 100°C. A strong band near 3680 cm^{-1} and broad bands between 3600 and 3400 cm^{-1} are observed on readsorption of water on Zn-Y. The bands are readily removed by evacuation at 100°C, resulting in spectra similar to those of the initially dehydrated material. Cd-Y behaved similarly.

On these four transition metal ion zeolites, the hydroxyl groups, with frequencies near 3640 and 3540 cm^{-1}, which are lost during dehydration, are reformed when water is readsorbed. The readsorbed water is first associated with the cations giving rise to the bands near 3600–3300 cm^{-1}. On heating, this adsorbed water dissociates and interacts with the lattice to give the structural hydroxyl group with bands near 3640 and 3540 cm^{-1}. This transformation of physically adsorbed water into structural hydroxyl groups is similar to that observed on the Group IIA zeolites.

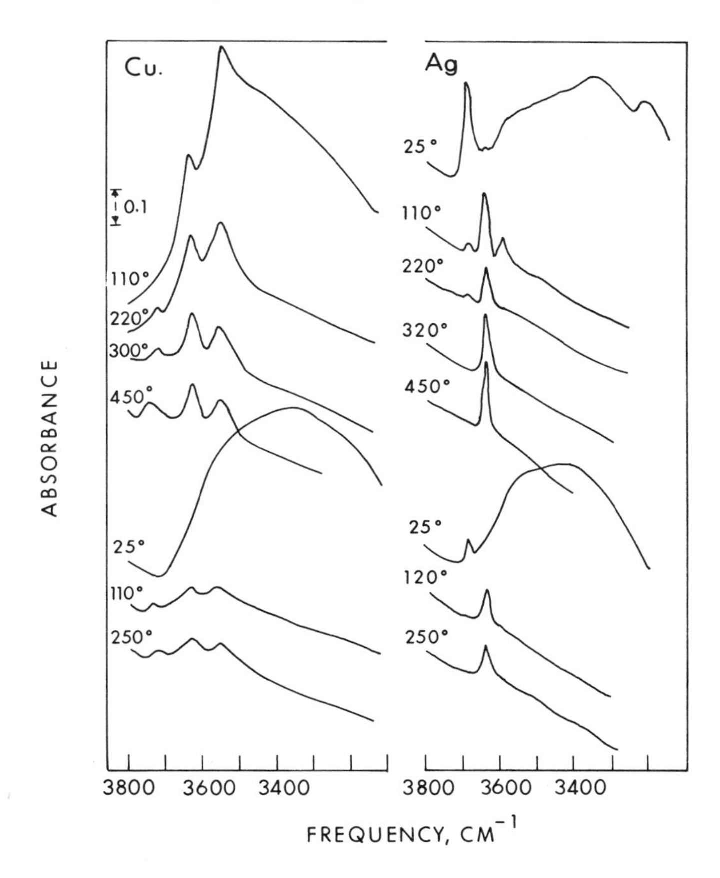

Transactions of the Faraday Society

Figure 54. Hydroxyl bands and adsorbed water on Cu-Y and Ag-Y after evacuation at 110°, 220°, about 320°, and 450°C, rehydrated at 25°C, and reevacuated at 25°, about 110°, and 250°C (47)

Water readsorption on rare earth Y zeolites gives rise to absorption bands near 3640, 3610, and 3580–3510 cm^{-1} as shown in Figure 6. No band near 3700 cm^{-1} was detected. Subsequent evacuation at 350°C removes the 3610 cm^{-1} band and shifts the 3580–3510 cm^{-1} band to 3528 cm^{-1}. The spectrum at this stage resembles that of the initial spectrum of the dehydrated zeolite. It appears that the adsorbed water is simply physically adsorbed. No new hydroxyl bands were detected. The nature of the groups responsible for the 3610 cm^{-1} band is not clear. If the zeolite is dehydrated at 700°C instead of 450°C, no structural hydroxyl group remains. Readdition of water at room temperature does not result in the formation of new hydroxyl groups, but subsequent heating at 200°C restores the 3640 and 3520 cm^{-1} bands. A completely dehydroxylated rare earth zeolite is apparently capable of dissociating adsorbed water at elevated temperature, but readsorbed water is simply physically adsorbed on less severely dehydroxylated zeolites.

The degree of hydration of the zeolite markedly influences the type and quantity of acidity present on the zeolite surface. X-ray diffraction measurements, gas adsorption studies, and magnetic susceptibility studies show that the locations of the cations in the zeolite lattices are influenced by the water content. These aspects are considered below.

Decationized and Dehydroxylated Zeolites. Several studies have been made of water readsorption on decationized zeolites. Most of the studies have been concerned with the influence of water on zeolite acidity and are discussed in the appropriate sections.

Concern with the influence of water on the hydroxyl groups has been relatively limited. Liengme and Hall (*65*) report that the 3640 and 3540 cm^{-1} bands can be restored if the zeolite has not been heated above 300°C. Above this temperature, the water is irreversibly removed. Hughes and White (*68*) found that readdition of water to NH_4-Y previously calcined at 600°C did not increase the intensity of the 3650 cm^{-1} band although the concentration of Bronsted acid sites was increased by interconversion of some Lewis acid sites. If the zeolite is preheated at 800°C, the readdition of water results only in a broad absorption band near 3700–2800 cm^{-1}. If water is added back to a zeolite calcined at 520°C, broad hydrogen-bonded OH bands are detected. On removal of the excess water, the hydroxyl bands near 3650 and 3550 cm^{-1} reappear, but they are much weaker than in the original zeolite. NH_4-X (45% exchanged) zeolite was rehydrated after pretreatment at 490°C. A sharp band at 3695 cm^{-1} was formed. If the sample was highly exchanged, no band was detected (*31*).

Thus, water is only partially reversibly adsorbed and desorbed from decationized zeolites. The ratio of reversible to irreversible effects depends on the calcination temperature.

Interaction with Inorganic Molecules

Carbon Monoxide. Carbon monoxide adsorption on zeolites is a powerful method of investigating the surface. The stretching vibration of the carbon monoxide molecule gives rise to a strong infrared absorption band: thus a detailed examination of the surface can be made. Furthermore, carbon monoxide molecules are unable to enter the hexagonal prism and sodalite cage portions of the X and Y zeolite structure. Although this prevents generation of information about these portions of the structure, it allows specific study of the supercage surface.

Angell and Schaffer (*121*) have carried out a detailed study of carbon monoxide adsorption on a large number of cation exchanged X and Y zeolites: sodium, lithium, magnesium, calcium, strontium, barium, manganese, iron, cobalt, nickel, zinc, and cadmium. None of the zeolites chemisorbed carbon monoxide. However, bands from physically adsorbed carbon monoxide were observed, with a few exceptions, near 2170 and 2120 cm^{-1}. Sodium X (2164 and 2121 cm^{-1}) barium Y (2105 cm^{-1}) and strontium Y (2098 cm^{-1}) were exceptions. The divalent cation containing zeolites also showed a third band at higher frequency (about 2200 cm^{-1}). All three bands represent weakly bonded carbon monoxide. The intensity of the 2200 cm^{-1} band was independent of pressure. Furthermore the intensity did not increase when the zeolites were cooled from room temperature in the presence of gas whereas the other bands were influenced by these parameters. Typical spectra are shown in Figure 55. The 2200 cm^{-1} band varied in frequency with the

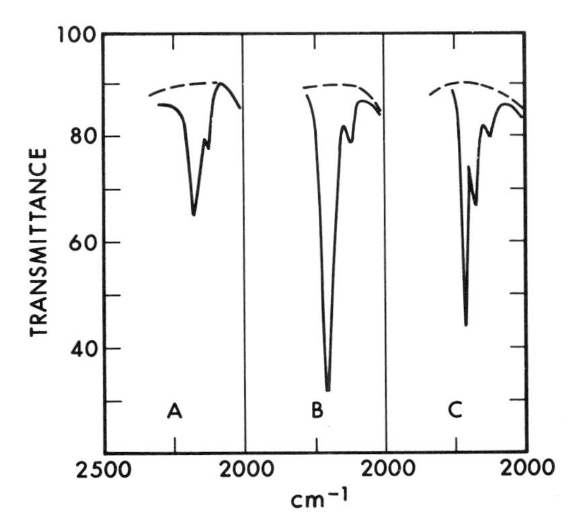

Journal of Physical Chemistry

Figure 55. Spectra of carbon monoxide on various cation zeolites (A) Na-Y; (B) Ca-Y; (C) Zn-Y (121)

cation type. Thus the 2200 cm^{-1} band can be attributed to carbon monoxide interacting directly with the cation. Supporting the assignment of the specific band to interaction with the cations is the observation that the band is not formed in the presence of a small amount of water, probably because water is more strongly bound to the cation. The stronger the electrostatic field of the cation, the greater is the frequency shift of the specific absorption band. Figure 56 shows the good relationship between the field strength and the band frequency. Angell and Schaffer showed that the adsorption of carbon monoxide could be used as a tool for locating the positions of divalent cations in zeolites by using the cation specific band (*121*). Two important observations permit the diagnostic use of carbon monoxide. First, carbon monoxide, because of steric reasons, can only interact with sites in the supercages. Secondly, only multivalent cations, and not monovalent cations, interact to give the specific band. Thus, for sodium Y 35% exchanged with calcium ions, no carbon monoxide–cation band was observed, indicating that all of the calcium ions are located in inaccessible parts of the structure and not in the supercages. At the higher levels of exchange, the cation specific band is observed. The data correlates with x-ray and

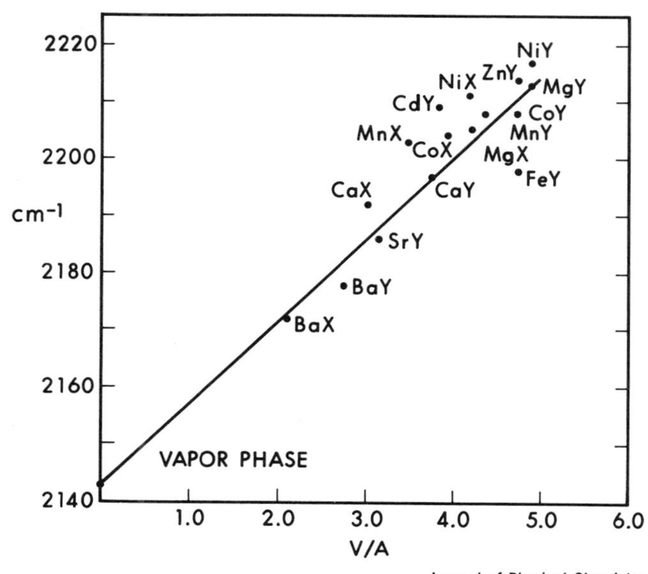

Journal of Physical Chemistry

Figure 56. CO frequencies vs. *cation electrostatic field* (121)

gravimetric adsorption studies of carbon monoxide using a series of different exchange levels which confirm that the first sixteen calcium ions go exclusively into the hidden hexagonal prism portions of the structure (*122, 123*). Cobalt and nickel ions were less selectively located since these cation zeolites gave rise to specific bands when only about 20% exchanged. These latter observations have since been confirmed by other techniques (*24, 125, 126, 127*). Of course, it is possible that interaction with carbon monoxide may result in the migration of cations into the supercages. The two non-cation specific bands were attributed to the interaction of the carbon monoxide with the zeolite lattice.

The interaction of carbon monoxide with sodium- and calcium-A and -X zeolites at pressures of about 10 torr has recently been studied by Fenelon and Rubalcava (*128*). Using isotopic carbon monoxide and by analyzing the absorption band contours of the adsorbed species, and of gaseous and liquid CO, they concluded that the carbon monoxide molecules freely rotate in the sodium zeolites until they collide with the cage walls. For the calcium zeolites, the adsorbed carbon monoxide indicated strongly hindered rotation. The is plausible since carbon monoxide adsorbs preferentially on the multivalent ions and is held more strongly than on univalent ions.

The influence of carbon monoxide on copper Y zeolite in the absence and presence of nitrogen bases has been investigated (*129*). Copper(I) zeolites were prepared by reducing copper(II) zeolite with carbon monoxide at 400°C. Cu^{+}–CO complexes are formed on the addition of carbon monoxide at room temperature. In these complexes, the copper ions are believed to be situated in the sodalite cages and hexagonal prisms of the structure and the carbon monoxide molecules are located in the supercages. When ammonia, ethylenediamine, or pyridine is added to the zeolite before carbon monoxide, the band frequency for the Cu–CO complex is significantly lower. Apparently, the addition of the base results in movement of the coppor ions out into the supercages resulting in a change in the interactions with pyridine or ethylenediamine. For ammonia, a band is observed on addition of CO. The band is readily removed by evacuation, whereas readdition of carbon monoxide gives rise to an absorption band close to that observed when no ammonia has been added to the sample. For pyridine and ethylenediamine, no absorption bands from CO are observed until some of the base has been removed by evacuation. Another possible explanation of the results is that carbon monoxide and bases are competing for coordination sites around the copper ions and causing different Cu–CO frequencies. Carbon monoxide was not adsorbed on ultrastable Y zeolite and decationized Y zeolite at 40 torr pressure (*42*).

Carbon monoxide adsorption has been investigated on silver X and Y zeolites (*130*). The zeolites, greater than 90% exchanged, were evacuated at 350–400°C and then heated in oxygen. An absorption band was detected at 2195 cm^{-1} on both zeolites, in contrast to 2160 cm^{-1} for Cu(I)-Y. No carbon monoxide bands were detected in the presence of preadsorbed ammonia. However, on partial removal of ammonia by evacuation at 25°C, a band was observed at 2170 cm^{-1}; after evacuation at 110°C, a band was observed at 2180 cm^{-1} and at 2200 cm^{-1} after treatment at 380°C. The carbon monoxide is probably interacting with the zeolite cation by π-bonding. In contrast to Cu-Y for which a shift of 80 cm^{-1} in the carbon monoxide frequency was observed in the presence of ammonia, a shift of only 25 cm^{-1} was observed in the case of Ag-Y. Probably the ammonia adsorption did not change the location of the silver ions in the zeolite in the manner proposed for Cu-Y.

Carbon monoxide adsorption has also been studied on Co^{2+}-, Cu^{2+}-, Fe^{2+}-, and Mn^{2+}-Y zeolites (*131*). The specific cation–carbon monoxide band was observed near 2200 cm^{-1}. Treatment of the zeolite with oxygen resulted in the formation of physically adsorbed carbon dioxide and surface carbonates. Ni^{2+}-Y zeolite can be reduced to Ni^{+}-Y by treatment with sodium (*51*). Carbon monoxide is strongly adsorbed on Ni^{+}-Y and is removed only by evacuation at 200°C. The C—O stretching frequency is 29 cm^{-1} lower on Ni^{+} than on Ni^{2+}, suggesting that the C—O bond is weaker on Ni^{+}-Y.

Carbon Dioxide. Carbon dioxide adsorption has been extensively studied. The addition of carbon dioxide to X zeolites significantly enhances the activity for alcohol dehydration (*132*). Thus, an understanding of the mode of interaction with the zeolite surface might contribute significantly to knowledge of the catalytic sites. The adsorption on highly exchanged alkali and alkaline earth X sieves was complex—several different types of adsorbed species were detected (*26*). At pressures of about 100 torr, physically adsorbed CO_2 was detected. A chemisorbed carbonate type species and CO_2 adsorbed by ion-dipole interaction were also detected. An absorption band near 2350 cm^{-1} was observed in all cases and attributed to physically adsorbed species. This species was readily removed by evacuation. A second species gave rise to a single sharp band in the 2349 to 2374 cm^{-1} region which varied in frequency in a regular manner with changes in the cation. The frequency increased with increasing cation electrostatic field and with decreasing cation size. It was attributed to a direct interaction with the cation by an ion-dipolar interaction such as:

$$M^{n+} \ldots O^{\delta-}{=}C{=}O^{\delta+}$$

Angell (*133*) showed that the band frequency for both X and Y zeolites was a linear function of the calculated electrostatic field, confirming the polarization effect. The frequency field plots do not fall on the same line, indicating that the field calculations do not completely consider the differences between X and Y zeolites. These data are shown in Figure 57.

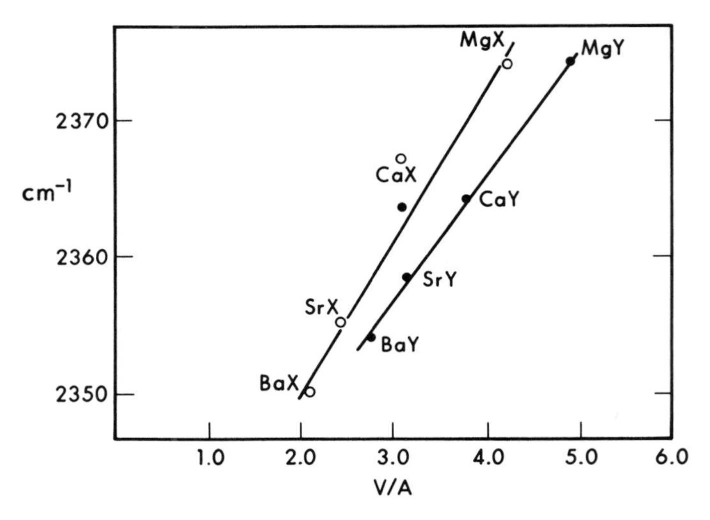

Journal of Physical Chemistry

Figure 57. *CO_2 frequencies* vs. *cation electrostatic field* (133)

Carbonate-type species from the interactions of CO_2 with the lattice oxygen were observed on the alkali metal forms and magnesium (*134*). Time dependent studies showed that the population of the carbonate type species grew as that of the ion-dipole species declined. The transformation was speeded by the addition of water. The precise frequencies varied with the cation, indicating the interaction of the adsorbed species with the cations may be with the oxygen atom. On Na-X zeolites, two types of carbonate type species were detected. Their relative abundances depended on the CO_2 pressure. No carbonate type species were observed on Ca, Sr, and Ba zeolites. The contrasting results of a single carbonate species found on Li, K, Rb, and Mg zeolites, two species on Na zeolite, and none on Ca, Sr, and Ba zeolites indicate a strong influence of the exchanged cation on the formation of carbonate species. The reasons for the lack of carbonate formation on these three zeolites are not clear. A possible explanation is that at some stages in the adsorption process, a cation and a lattice site with specific relationships to each other are needed and that the nature of the cation can strongly influence this relationship. The original explanation (*137*) that few cations would be accessible does not seem to be applicable, since at the level of exchange used, there are probably 12–14 cations per supercage. Furthermore since the absorption band in the region 2374–2349 cm^{-1} is observed, caused by the ion–dipole interaction, there must be cations accessible to the carbon dioxide.

At higher pressures, up to four additional bands are observed in the region of 2320 to 2425 cm^{-1}. Possible interpretations of these absorption bands are adsorption of carbon dioxide on cations in different environments, adsorption of more than one molecule per cation site, or simple spectral combination bands.

Somewhat different results are obtained by reacting zeolites at 500°C in air or CO_2, compared with pretreatment in vacuum followed by carbon dioxide addition (*135*). Carbonate type species were formed on Mg-, Ca-, Ba-, Mn-, Co-, and Zn-Y zeolites and Mg-, Co-, and Zn-X zeolites. The species were stable up to at least 700°C. Under these conditions, Ni-, Na-, and Ag-Y did not produce the carbonate type bands. However, the results of Bertsch and Habgood (*26*) were confirmed at room temperature with Na-X. The species appear to be carbonate type species, similar to those observed at room temperature, formed by interaction with the lattice oxygen and the surface cations. Addition and removal of water and ammonia eliminated and restored the species whereas benzene had no effect.

The early studies have recently been extended and developed as a method for indicating the presence of cations in the supercages (*136*). By studying Ca- and Mg-Y zeolites exchanged to different levels, it was possible to determine when cations appeared in the supercages by observing the appearance of the ion–dipole interaction band near 2360 cm^{-1} since CO_2 does not enter into the small cavities. The bands at 1385 and 1275 cm^{-1} can also be used. Thus, for a sample exchanged to 29% with Ca^{2+} ions, no bands were observed near 2367 and 1385 cm^{-1}, indicating that no Ca^{2+} cations are present in the supercages. Adsorption was detected on Na^{+} ions in the supercages by the observation of a band at 2357 cm^{-1}. With samples exchanged to 46% and higher, the absorption band at 2367 cm^{-1} is always observed. At 41% exchange, the observation of the 2367 cm^{-1} band depended on the pretreatment conditions. For short pretreatment times, the band is observed but at longer pretreatments it is not. These results indicate the migration of cations from the supercages to the small cage system, probably as a result of more extensive dehydroxylation. The migration of cations as a function of outgassing treatment has been observed for several systems previously (*138, 139*). These infrared results indicate the appearance of calcium in the supercages at a lower level of exchange (about 46%) than shown by the gas adsorption (*123*) and x-ray diffraction results (*122*) (about 50-55%). A previous infrared study using pyridine as the indicator had indicated about 50% exchange for the calcium–ammonium system

(*79*). Carbon dioxide adsorption on magnesium Y gave similar results, except that some cations were detected in the supercages after partial evacuation at 24% exchange. Furthermore, because the absorption bands for the various cations occur at different frequencies, the relative amounts of the types of cations present in the supercages can be determined.

By studying the adsorption of pyridine, Hall *et al.* have shown that the addition of carbon dioxide to Y zeolites increases the proton acidity (*65*). This increase in acidity may be the reason for the enhanced catalytic activity for isopropyl alcohol dehydration shown by addition of CO_2 to X zeolites (*132*). Carbon dioxide adsorbed on ultrastable Y zeolite gave rise to a band near 2360 cm^{-1} (*42*).

Adsorption of carbon dioxide has been investigated on the Na-, NH_4-, and Ca-germanic near-faujasite (*104*). The absorption bands observed on the Na form at various temperatures are listed in Table IX. Except for the

Table IX. CO_2 Absorption Band Frequencies at Different Temperatures on Na-Germanic-Faujasite (*104*)

100°C [a]	*200°C* [a]	*350°C* [a]
1664 [b]	1664 (w)	—
—	1615 (w)	1615 (w)
—	1572 (w)	1572 (w)
1480 (s)	1480 (s)	1480 (w)
1432 (s)	1432 (s)	1432 (w)
—	1375 (sh)	—

[a] w = weak; s = strong; sh = shoulder.
[b] Overlapped with the H_2O deformation band.

Journal of Catalysis

1664 cm^{-1} band, all of the bands can be attributed to chemisorbed carbon dioxide. The absence of a band near 1260 cm^{-1} rules out a bidentate carbonate species. As observed by others, the 1480 and 1432 cm^{-1} bands probably belong to the same carbonate species. The 1615 and 1574 cm^{-1} bands probably belong to a single carbonate or bicarbonate species in a different environment. The 1664 and 1375 cm^{-1} bands are attributed to a less stable carbonate species since they are readily removed on heating.

No absorption bands are observed when carbon dioxide is adsorbed on the ammonium germanic near-faujasite zeolite. Carbon dioxide adsorption is inhibited, possibly by OH groups blocking that adsorption site.

Carbon dioxide adsorption on the calcium form produces two strong bands near 1475 and 1430 cm^{-1}. These bands are attributed to an almost symmetrical carbonate species.

Nitrogen Oxides. The interaction with nitrogen oxides is important since zeolites offer possibilities for removal of NO_x pollutants either by adsorption or catalytic decomposition. Studies of the interaction have been made by both infrared and electron spin resonance techniques involving sodium-, calcium-, decationized-, dehydroxylated-, and chromium-Y zeolites and sodium-A zeolites. The decomposition of nitric oxide into nitrous oxide and nitrogen dioxide on A zeolite was observed several years ago.

Early studies of nitric oxide on sodium- and calcium-A- and -X-sieves were made by Alekseev *et al.* (*140*). Natural zeolites, natrolite and desmine were also studied. No adsorbed NO was detected on the two natural zeolites and on a specially synthesized Na-X. On all the other zeolites, absorption bands from NO and N_2O were detected. Absorption bands from nitrate formation and NO_2^+ from NO_2 molecules were detected when NO was adsorbed on Ca-A. A more recent study by Chao and Lunsford extended this work by investigating in detail Na-Y, Ca-Y, decationized-Y and hydrogen-Y (*141*). For Na-Y, on addition of NO, no adsorbed species were detected at room temperature, but on cooling to —78°C, addition species were detected. Some of the new species remained even after outgassing the sample for 1 hr at 400°C. Similar absorption bands were detected on Ca-Y after treatment with NO, suggesting that the same species are formed but that Ca-Y is a better catalyst.

Very little of the adsorbed species were detected on decationized Y even at —78°C indicating that it is a very poor catalyst. Ben Taarit *et al.* (*142*) also found no adsorption on decationized Y. Dehydroxylated-Y was about as active a catalyst as Na-Y for the decomposition of NO. However the detailed species formed were somewhat different. Table X summarizes the absorption bands observed, and Table XI outlines the assignment of the frequencies. Depending on the zeolite, several of the various possible species are formed. The study was later extended to a series of temperatures between 73° and 190°C (*143*). In most cases, six or more absorption bands were observed. The absorption bands observed are listed in Table XII. Three types of species are observed, adsorbed NO, adsorbed N_2O_2, and disproportionation reaction products of NO. The N_2O_2 was found to exist in both cis and trans forms. The types of species observed at the various temperatures are given in Table XIII along with the assigned bands. No disproportionation of NO occurred on dehydroxylated Y and decationized Y at any temperature studied. For Na-Y and Ca-Y, no reaction occurred at —190°C, but disproportionation was observed at —150°C on Ca-Y and —130°C on Na-Y. Products detected were N_2O, N_2O_3, NO_2^+, and NO^-. The disproportionation is believed to proceed by interaction of NO with adsorbed N_2O_2.

Nitric oxide and nitrogen dioxide adsorption have also been studied on Y zeolite and mordenite exchanged with copper, chromium, and nickel ions (*141*). Nitric oxide adsorption on nickel-Y zeolite dehydrated at room temperature or at elevated temperatures gave rise to a single absorption band near 1892 cm^{-1} attributable to a Ni^+–NO^+ complex. Confirming this assignment, ESR spectra typical of an ion with a $3d^9$ electron configuration were observed. Detailed discussion of ESR data is given in Chapter 6.

Combining the observations with x-ray data suggests that the complex is located in the sodalite cages of the zeolite structure near to the S'_I position. Different species are observed on chromium-Y and mordenite depending on whether the chromium has been oxidized to the Cr^{5+} species or reduced to the Cr^{2+} species. Cr^+–NO^+ complexes were formed with the reduced species.

Table X. Absorption Bands (cm^{-1}) Induced

	23°	*Cooled to −78°*
Na-Y	None	1260, 1305, 1555, 1920, 2000, 2100, 2240
Ca-Y	1305 (unresolved), 1565, 1935, 2040, 2200, 2250	1305, 1565, 1935, 2040, 2200, 2250
Dehydroxylated Y	None	1300, 1570, 1910, 1980, 2120, 2240
Decationized Y	None	2200

Table XI. Assignment of Bands (cm^{-1}) Produced by Reactions of NO on Zeolites (*141*)

	Zeolite			
Adsorbed Species	*Na-Y*	*Ca-Y*	*Dehydroxylated Y*	*Decationized Y*
]—(N_2O)	2240, 1260	2250	2240	2200
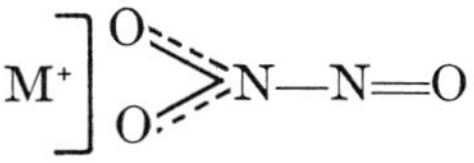 M^+] (O,O)⋛N—N=O	3105, 1555, 1920	1305 (unresolved), 1565, 1935	1305, 1570, 1910	
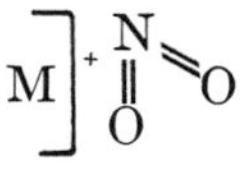 M]$^+$ N(=O)(‖O)	2000, 2100	2040, 2200	1980, 2120	
Si—O—N=O			1585, 1650	
]—ONO		1470		
NO_3^-	1400	1400		
NO_2^-	1260			

Journal of the American Chemical Society

by Absorbing NO on Zeolites (*141*)

Degassed at 23°	*Degassed at 200°*	*Degassed at 300°*	*Degassed at 400°*
1260, 1400, 3650	1400	1400	none
2040, 2200, 3640	1400, 1470	1400, 1470	none
1535, 1650, 2000, 2120	1585, 1650	none	none
none	none	none	none

Journal of the American Chemical Society

Table XII. Absorption Bands Induced by Adsorbing NO on Zeolites Near −130° and −150°C (*143*)

At −130°C

Substrate	*Pressure of NO*	
	1 torr	*10-20 torr*
	Line Positions, cm⁻¹	
Decationized Y	1750, 1780, 1910	1750, 1780, 1895, 1910
Na-Y	1900	1750, 1780, 1790, 1825, 1870, 1880, 1900, 1930
Ca-Y	1845, 1917, 1970, 2000, 2130	1770, 1790, 1840, 1885, 1935, 1970, 2000, 2140

At −150°C

Substrate	*Line Positions, cm⁻¹*
Decationized Y	1790 (with br sh down to 1740), 1825, 1900 (with br sh down to 1870)
Dehydroxylated Y	1770, 1790, 1890, 1900
Na-Y	1730, 1795, 1840, 1900
Ca-Y	1555, 1750, 1770, 1850, 1890, 1917, 1970, 2000, 2150

Journal of the American Chemical Society

The diamagnetic complex:

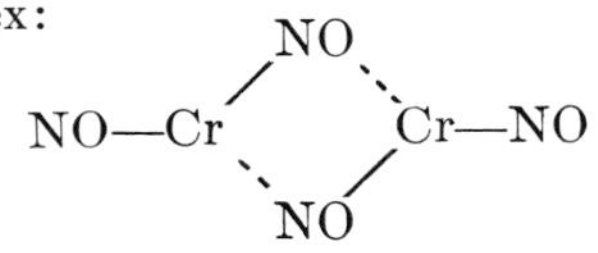

```
          NO
         /  `.
NO—Cr         Cr—NO
      `.    /
          NO
```

was also believed to be formed. Surprisingly, the Cr^{+}–NO^{+} complexes were found to be unreactive to oxygen up to 200°C, suggesting that they are located in the sodalite cages of Y zeolite and the narrow channels of mordenite. No chemisorbed species were detected on the oxidized chromium

Table XIII. Absorption Bands of Adsorbed NO on Zeolites (*143*)

Adsorbed Species	*Temp. of Adsorption °C*	*Substrate*			
		Decation-ized Y	*Dehydroxyl-ated Y*	*Na-Y*	*Ca-Y*
NO^+	−130	1910		1900	1917
	−150	1900	1900	1900	1917
	−190	1900		1900	1917
NO^-	−130			1825	1845
	−150		1840	1825	1850
	−190	1820		1825	1845
trans-N_2O_2	−130	1750		1750	
	−150		1730	1750	1750
	−190			1750	1750
cis-N_2O_2 (Species A)	−130				1770
					1885
	−150	1770		1770	1780
		1890		1880	1890
	−190	1765		1775	1770
		1870		1880	1884
cis-N_2O_2 (Species B)	−73				1788
	−130	1780		1780	1790
		1895		1880	1902
	−150	1790	1795	1790	1790
		1890	1900	1890	1902
	−190	1790		1795	1790
		1890		1890	1902
NO_2^+	−130				1970
					2000
					2140
	−150				1980
					2000
					2150

Journal of the American Chemical Society

zeolite. When NO_2 was adsorbed on the reduced chromium form, nitrates and nitro complexes were detected. Similar NO^+ species were detected on Cu-Y zeolite and Cu- and hydrogen-mordenite. The hydrogen-mordenite had to be heated above 450°C for a signal to be detected.

NO on Ag-Y. Two absorption bands are observed when nitric oxide (pressure <56 torr) is adsorbed on Ag-Y pretreated in vacuum at 500°C (*145*). The bands are observed at 1884 and 1844 cm^{-1} compared to 1876 cm^{-1} for the gas phase. These bands could be removed by evacuation at room temperature. At higher pressure, the initial spectrum is the same, but on standing a band appears at 1876 cm^{-1} in place of the 1884 cm^{-1} band.

Removal of the new band required evacuation for greater than 30 min at 200°C. The addition of oxygen caused no change. The 1884 cm^{-1} band was attributed to $[Ag(I)NO]^{+}$ and the 1876 cm^{-1} band to $[Ag(I)_2NO]^{2+}$. The 1844 cm^{-1} is from adsorbed NO at some unknown site.

Adsorption of nitric oxide on ferrous ion-Y zeolite initially gives rise to a spectrum having bands near 1890 cm^{-1} (*146*). At higher pressures, bands at 1822 and 1930 cm^{-1} are detected. Evacuation at room temperature decreased the intensity of the 1930 and 1822 cm^{-1} bands but a new band

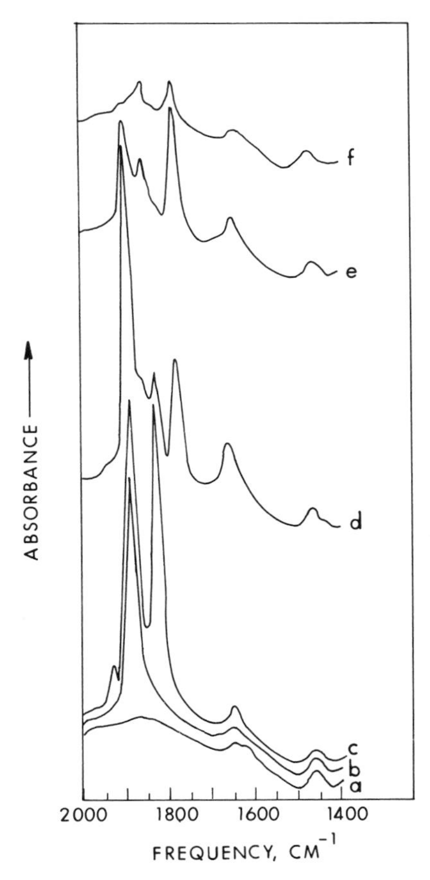

Journal of Physical Chemistry

Figure 58. Fe(II)Y zeolite: (a) after calcination at 400°C; (b) after addition of 1 torr of NO; (c) after addition of 10 torr NO, evacuated 15 sec; (d) sample under vacuum 4 hr at 25°C; (e) under vacuum 1 hr at 50°C; (f) under vacuum 1 hr at 100°C (146)

appeared at 1778 cm^{-1}. Evacuation at 50°C decreased the 1890 cm^{-1} band but the 1778 cm^{-1} band increased further. Typical spectra are shown in Figure 58. By analogy with other compounds, NO stretching frequencies above 1700 cm^{-1} can be attributed to NO^+ species coordinated to the surface. The band at 1778 cm^{-1}, when considered in conjunction with ESR evidence, is probably due to $[Fe(I)NO]^{2+}$ groups. Very little nitric oxide was adsorbed by ultrastable Y or by decationized Y (*42*).

The spectra of adsorbed N_2O on sodium-A zeolite depend markedly on temperature (*147*). The band near 2260 cm^{-1} from fundamental vibrations of the N_2O molecule is broad at 40°C but becomes much narrower at —50°C. The other fundamental vibration near 1300 cm^{-1} appears as a doublet at 40°C with frequencies of 1303 and 1264 cm^{-1}. On cooling, the bands are more distinct with frequencies near 1305 and 1255 cm^{-1}. The spectra suggest that at elevated temperature, the N_2O molecule is moving about the cavities whereas at —50°C, the molecule is closely associated with the cations.

Hydrogen Chloride. The addition of hydrogen chloride to fully dehydrated magnesium-Y zeolite results in the formation of hydroxyl bands with frequencies near 3643 and 3533 cm^{-1}. These bands are the same as those found with decationized Y zeolite, indicating the reaction of the hydrogen chloride with the lattice to give hydroxyl groups. When deuterium chloride was used, deuteroxyl bands appeared near 2684 and 2605 cm^{-1} (*30*).

Interaction with Sulfur-Containing Molecules

Zeolites have found an important use in the removal of sulfur from sour gases. They have also been suggested as catalysts for the Claus reaction and for hydrodesulfurization (*148*).

As part of a study of the mechanism of and intermediates in the Claus reaction, Deo, Dalla Lana, and Habgood (*149*) studied the interaction of hydrogen sulfide and sulfur dioxide with sodium- and decationized Y zeolites. Addition of 98 torr of hydrogen sulfide at room temperature to sodium-Y zeolite dehydrated at 600°C resulted in an absorption band near 2575 cm^{-1} from physical adsorption. Appearance of bands near 3690 and 1650 cm^{-1} also indicated the formation of a small amount of water, possibly by the oxidation of the hydrogen sulfide by chemisorbed oxygen. Heating of the sample in contact with hydrogen sulfide resulted in more water formation. Evacuation at room temperature removed all of the hydrogen sulfide. However, an unknown species with absorption bands near 1720 cm^{-1} remained on the surface. It could be removed by evacuation at 400°C.

In this study, ammonium-Y was pretreated at 600°C. Thus the decationized Y was partially dehydroxylated. Hydrogen sulfide hydrogen-bonded to the surface hydroxyl groups at 3650 and 3550 cm^{-1}. Similar results were observed for a sample outgassed initially at 400°C. In contrast to sodium-Y, no oxidation of hydrogen sulfide to water was detected, suggesting a lesser degree of chemisorption of oxygen on decationized Y. Although both hydroxyl groups hydrogen-bond with hydrogen sulfide, addition of hydrogen sulfide to deuterated Y zeolite resulted only in the deuterium exchange of the 2689 cm^{-1} OD (3650 cm^{-1} OH) band. The 3650 cm^{-1} band represents the most acidic and accessible hydroxyl groups and, hence, the groups most strongly interacting with the hydrogen sulfide. Other studies have shown that this hydroxyl group most readily protonates basic molecules such as pyridine, propylene, etc. No evidence for the chemisorption of hydrogen sulfide was obtained.

Adsorption of sulfur dioxide on sodium-Y zeolite gave rise to absorption bands near 1330 and 2470 cm^{-1} attributed to physically adsorbed gas. No change occurred on heating to 400°C. Adsorption of sulfur dioxide onto sodium-Y which had first been exposed to hydrogen sulfide and then evacuated gave similar results. However, addition of hydrogen sulfide to the sodium-Y zeolite containing sulfur dioxide resulted in the production of water (detected by the absorption band near 1650 cm^{-1}).

Addition of sulfur dioxide to decationized-Y zeolite resulted in hydrogen bonding to the 3650 cm^{-1} band of the zeolite. Bands were observed near 2470 and 1330 cm^{-1} as on sodium-Y indicating physical adsorption. No evidence for chemisorption was found. Addition of a 2:1 mixture of hydrogen sulfide and sulfur dioxide resulted in the rapid formation of sulfur and the appearance of absorption bands from water. No chemisorbed species were detected. Hence, the major role of the zeolite is simply to bring the reactants together in a high concentration.

Zeolites are possible supports for desulfurization catalysts (*148*). Lygin *et al.* (*150*) have studied the adsorption of thiophene on calcium- and decationized Y zeolites which had been heated in air at 550°C and then in vacuum at 400°C. At room temperature thiophene is both physically and chemically adsorbed. On evacuation at 20°C the bands from physically adsorbed thiophene disappear. Absorption bands are observed up to 400°C near 2975, 2935, 2870, 1450, and 1355 cm^{-1} which can be assigned to CH_3, CH_2, and CH groups formed by the decomposition of the thiophene on catalytic sites. An absorption band is also observed near 1595 cm^{-1}, up to 300°C, which has been attributed to thiophene-ring vibrations (*151*). This suggests that thiophene is chemisorbed on the surface without destruction of the ring. At higher temperatures, a band attributed to C=C groups is also detected.

Comparison of the spectra on the two zeolites with those of alumina shows marked similarities suggesting that the adsorption sites are similar in nature. It is well known that alumina contains Lewis acid sites and that decationized zeolites contain a substantial concentration of Lewis acid sites after calcination at 550°C. Although not studied in detail, changes in the spectra of thiophene adsorbed on the calcium-Y zeolite suggest that specific interactions with the cations may be occurring.

Interaction with Organic Molecules

Olefins. Several studies of olefin adsorption on zeolites have been made under widely different conditions. Olefins are interesting in that they are more basic than saturated hydrocarbons but less basic than the ammonia derivatives discussed later. Initial studies by Carter *et al.* (*152*) involved the adsorption of about 10 torr of ethylene on Li-, Na-, K-, Ag-, Ca-, Ba-, and Cd-X zeolites at room temperatures after pretreatment at 450°C. After equilibrating and recording of spectra, the samples were studied as the excess ethylene was removed. Heats of adsorption were also determined calorimetrically on the same sample. No evidence for dissociation or polymerization of the ethylene was detected. Thus the ethylene was adsorbed molecularly. In all cases, the C-H stretching bands near 3100 cm^{-1}, the C═C stretching bands near 1600 cm^{-1} and the deformation bands near 1500 cm^{-1} were detected. Since these latter two bands are usually infrared forbidden, their observance as strong bands is indicative of an interaction occurring between the ethylene and the zeolite structure. With the exception of silver, cadmium, and calcium, the adsorbed ethylene can be removed by evacuation at room temperature and hence was held physically or by weak chemisorption forces. A temperature of 100°C was required to remove the ethylene from the cadmium form and higher than 200°C from the silver form. A typical spectrum for Ag-X is shown in Figure 59.

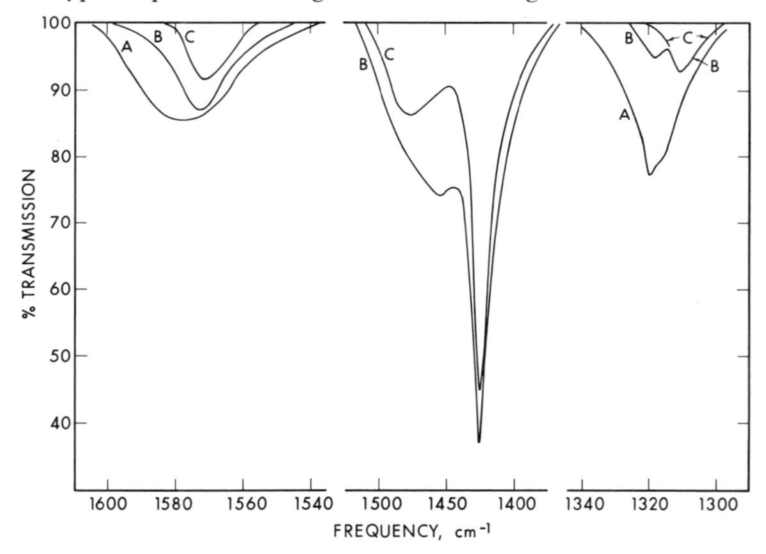

Figure 59. Ethylene adsorbed on Ag-X (A) 10.9 cm C_2H_4; (B) evacuated 5 min at room temperature; (C) evacuated 15 min at 100°C (152)

The C═C stretching frequency varies with the nature of the exchanged cation. A relationship exists, as shown by Figure 60, between the shift in cm^{-1} from the gas phase frequency and the heat of adsorption: the greater the heat of adsorption, the greater the frequency shift, and the higher the temperature required for desorption. The order of increasing frequency shift is the same as for carbon monoxide interaction (*121*). These observa-

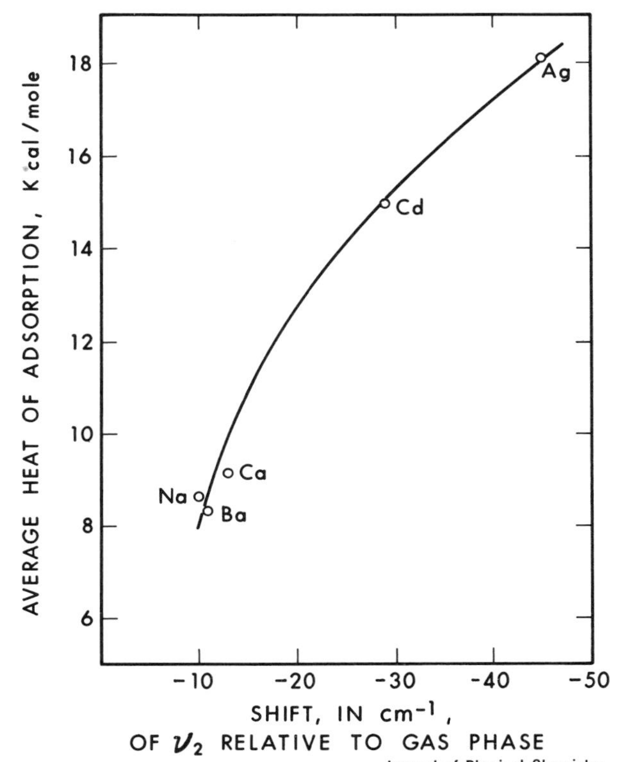

Figure 60. Heat of ethylene adsorption and the double bond C=C frequency shift as a function of cation (152)

tions strongly indicate that the ethylene is specifically interacting with the cations in the supercages and agree with other observations that ethylene interacts strongly with, for example, silver salts. The ethylene probably is adsorbed by π-bonding to the cations. The absorption bands at 1600 and 1340 cm^{-1} are much wider in silver zeolites than for the other zeolites studied. From an analysis of the band widths, it was determined that the ethylene was freely rotating in all cases except silver. For silver, compared to *e.g.*, cadmium, the bonding involves overlap of the filled π-orbital of the olefin with a vacant 5 *sp*-orbital of the cation as shown in Figure 61. Examination of the data in Figure 59 especially the bands in the 1300 cm^{-1} region indicates the presence of two different strongly adsorbed species on Ag-X. The second species is attributed to additional bonding to the surface by means of back donation from the filled 4*d* orbital of the cation to the vacant π^* orbital of the olefin. The strongly chemisorbed ethylene is held by interaction with cations in the supercages while the moderately strong adsorption represents interaction with cations in the hexagonal windows of the sodalite cages

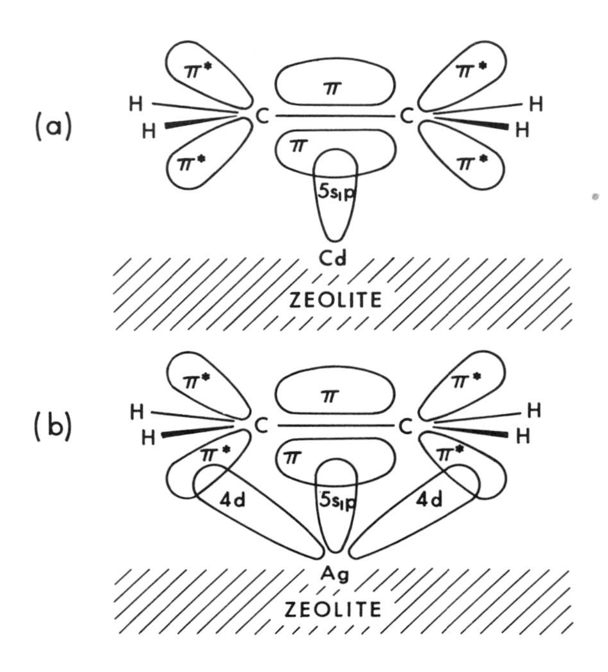

Journal of Physical Chemistry

Figure 61. Schematic diagram showing the spatial arrangement of the orbitals involved in the bonding between ethylene and (a) Cd-X and (b) Ag-X (152)

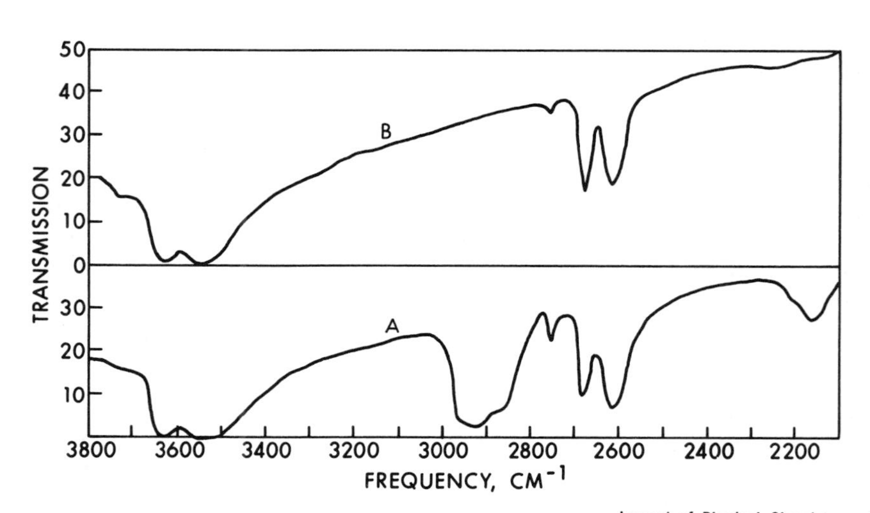

Journal of Physical Chemistry

Figure 62. Spectra of D-Y after exposure to 1-hexene at (A) 93°C and (B) after heating to 427°C (64)

(*153*). The π-bonding interaction between the olefin and the zeolite cations is supported by magnetic resonance studies (*154*). An interesting but unexplained observation is the absence of C-H stretching bands in the cases of Cd- and Ca-X. Under the same conditions, the other zeolites gave rise to strong bands.

An entirely different study was carried out by Eberly (*64*). He studied the adsorption of several olefins on Group IA-, IIA-, Cd-, Ag-, and decationized Y zeolites at elevated temperatures. The samples were initially treated at 427°C. On adsorption of ethylene at 93°C, no bands attributable to adsorbed species were detected.

A detailed study of 1-hexene was made. At 93°C and 2 torr pressure, hexene had lost its double bond character in 15 min as evidenced by the absence of an absorption band near 3090 cm^{-1}. When adsorbed on deuterated Y zeolite, rapid exchange of hydrogen for deuterium occurred. C—D and OH groups were detected by their absorption frequencies. The OD band at 2680 cm^{-1} (equivalent to the OH band at 3650 cm^{-1}) decreases in intensity on adsorption of the olefin and is restored by removal of the olefin from the surface indicating interaction of this particular group with the adsorbing hexene. The adsorption of 1-hexene was studied on decationized Y as a function of temperature. Detailed spectra are shown in Figure 62. Raising the temperature from 93°C results in desorption of hydrocarbon from the zeolite as shown by the decrease in the intensity of the CH bands. No absorption bands from olefinic species were detected and, as in the case of the deuterated zeolite, interaction appears to be preferentially with the 3635 cm^{-1} hydroxyl groups. On heating to 150°C, polymerization and dehydrogenation start to occur. A band at 1600 cm^{-1} indicative of olefinic groups in conjugated polyenes is detected. A decrease in the saturated C—H group bending frequencies is observed simultaneously. At higher temperatures (200°–260°C), the 1600 cm^{-1} band becomes more intense and shifts to 1580 cm^{-1} indicative of the formation of aromatic ring structures. This species cannot be removed from the surface by evacuation at 427°C. Similar results were obtained with 1-pentene, 1-butene, isobutylene and propylene.

Hexene was also studied on Na-, Li-, K-, Ca-, Mg-, Cd-, and Ag-Y (*64*). As found for ethylene by Carter *et al.* (*152*), the degree of interaction depends on the exchanged cation. The interaction with Group IA cations appears to be very weak. The main observation is that the hexene loses some or all of its double bond character. Table XIV shows the ratio of the intensity of the C═C band at 1630 cm^{-1} to the CH band at 1460 cm^{-1}. The ratio is much lower for divalent cation and zero for silver and decationized Y zeolites. These are probably the more strongly adsorbing systems. Adsorption of hexene on Ag and decationized zeolites results in the formation of aromatic structures.

Liengme and Hall (*65*) studied ethylene and propylene on decationized Y. The zeolite contained about 55% of the original sodium and was pretreated at 480°C. Ethylene interacted reversibly with the 3650 cm^{-1} band. The band intensity decreased, and a new band near 3300 cm^{-1} was detected. This band is characteristic of hydrogen-bonded species. The 3550

Table XIV. 1-Hexene Adsorption at 93°C and 2 torr on Y Zeolites. Amount of Double Bond Character in Adsorbed Phase (*64*)

Solid	*Percent Exchange*	A_{1630}/A_{1450}
Na-Y	100	1.53
Li-Y	64	1.45
K-Y	95	1.34
Ca-Y	75	0.84
Mg-Y	67	0.35
Cd-Y	73	0.20
Ag-Y	100	0
Decationized Y	92	0

Journal of Physical Chemistry

cm^{-1} band did not interact at low pressures, but at pressures greater than 200 torr, some interaction may have occurred, but the data is not definitive. No exchange occurred between C_2D_4 and the zeolite below 150°C and no polymerization was detected at or below 240°C. About 30% more ethylene

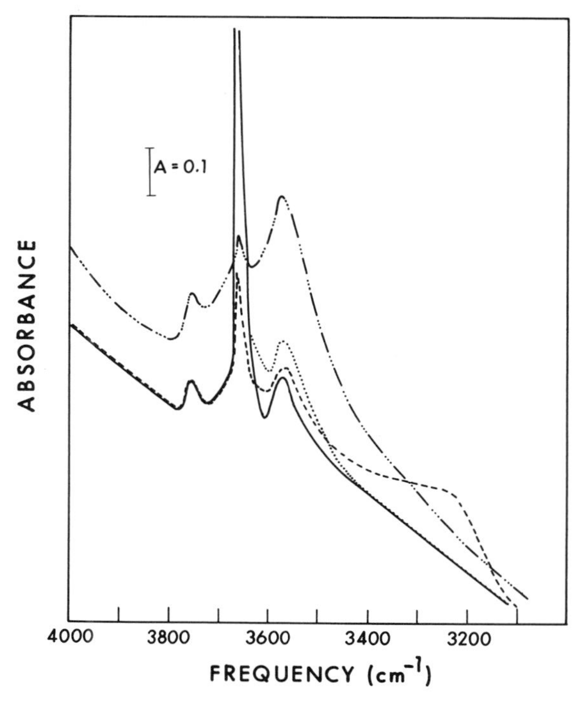

Transactions of the Faraday Society

Figure 63. Propylene adsorbed on decationized Y zeolite: —— evacuated at 300°C; – – – 5 torr propylene added; · · · · · evacuated; —· · —· · on standing at 5 torr for 44 hr (65)

is adsorbed than can be accounted for by the acidic hydroxyl groups. The ethylene interacts both with the hydroxyl groups and also with sodium cations in the supercages. At higher pressures, physical adsorption of ethylene may also be occurring. When the zeolite was dehydroxylated by thermal treatment at 600°C, a large fraction of the ethylene was irreversibly adsorbed and polymerization products were detected. Propylene adsorption was qualitatively similar to ethylene, except that the interactions were stronger and not completely reversible. It was found that 5 torr of propylene was as effective as 200 torr of ethylene in attenuating the 3650 cm^{-1} band and that the hydrogen-bonding shift was about 100 cm^{-1} greater than in the ethylene study. Both these observations indicate stronger interactions with the zeolite.

Over a period of several days, the propylene pressure decreased, the hydrogen-bonding band at 3200 cm^{-1} disappeared, the 3650 cm^{-1} band was not restored, and a new band was formed at 3560 cm^{-1} as shown in Figure 63. Since similar results could be obtained by adsorbing hexane, it was concluded that the propylene had polymerized at room temperature. When the surface species was desorbed by evacuation at 200°C, isobutane was the major product, suggesting that a polymerization–isomerization–cracking process must have occurred. In contrast to ethylene, propylene also underwent deuterium exchange at room temperature with both types of hydroxyl groups.

The results suggest that the major mode of adsorption is a hydrogen-bonding interaction between the olefin and the acidic hydroxyl group, which can be represented in terms of charge transfer theory by:

```
R CH═CH2                      RĊH—CH2   (+ on C)
                                    |
      D                             D
      |
O     O      O               O     O      O
 \     \    /                 \ − /  \   /
  Al    Si                     Al     Si
 /  \  /  \                   /  \   /  \
O    OO    O                 O    OO     O
```

This type of model makes an explanation of the observed differences between ethylene and propylene possible.

The ethylene observations were extended by Cant and Hall (*155*) who studied various pressures of ethylene between 25° and 150°C. They were able to calculate the heat of formation of the hydrogen-bonded ethylene to be about 9 kcal/mole for interaction with the 3650 cm^{-1} band.

In contrast to the earlier measurements at room temperature, deutero-ethylene exchanged readily with both hydroxyl groups at temperatures between 200° and 350°C. No significant differences in the rate of exchange were observed for the two major hydroxyl groups. Activation energy was calculated for the exchange reaction. It varied from 16 kcal/mole for a sample pretreated at 290°C to 19 kcal/mole for samples pretreated at 475°C and rehydrated at 350°C.

Best *et al.* (*76*) have studied the interaction of propylene on decationized erionite after pretreatment at 600°C. The propylene interacted with the 3612 cm^{-1} band but not with the 3565 cm^{-1} band. From this observation, the 3612 cm^{-1} band must represent hydroxyl groups in the large channels and the 3565 cm^{-1} band must represent hydroxyl groups inside a six-membered ring which would be inaccessible to propylene.

Acetylene. Supporting the concepts for adsorption of olefins and aromatics on zeolites is the study of acetylene on Na-, Ca-, Mn-, Ni-, Co-, and decationized A zeolites and Na-, Ni-, and Co-X zeolites (*156*). On all forms, the C—H stretching band is observed in the region 3210 to 3228 cm^{-1}. Although the band frequency varied with the cation form, there seemed to be no systematic variation. It is possible that the shifts are within the experimental accuracy. Only one C—H stretching vibration is observed indicating that the acetylene molecule is remaining in a very symmetrical environment. The acetylene is much more strongly held on the zeolite surface than on alumina. Possibly the molecules are interacting *via* the π-electrons of the triple bond with the cations or the hydroxyl groups on the surface by a charge transfer mechanism.

The cyclotrimerization of acetylene on Ni-Y zeolites has been investigated (*157*). Zeolites containing 10, 14, and 19 cations per unit cell were investigated after treatment in oxygen at 200° and 600°C; the acetylene was added at room temperature. The zeolites contained 0, 1.2, and 5.8 Ni^{2+} ions in the supercages after calcination at 200°C as determined by x-ray diffraction. Adsorption of acetylene on all six samples of nickel zeolite gave spectra similar to those observed on Na-Y zeolite. Absorption bands were observed near 3240, 3215, and 1950–1955 cm^{-1}. These frequencies are similar to those of gaseous acetylene. Because the relative intensities of the 3240 and 3215 cm^{-1} bands in the spectrum are different from those of the gaseous phase alone, acetylene must be adsorbing to give rise to an enhancement in intensity of the 3240 cm^{-1} band.

Spectra of acetylene adsorbed on the 19 cation per unit cell Ni sample after calcination at 600°C are shown in Figure 64. It is seen that three absorption bands in the 3100–3000 cm^{-1} region and four bands in the 2000–1400 cm^{-1} region are readily formed. These bands are characteristic of benzene molecules. By detailed analysis of the spectrum and by analogy with previous studies of benzene adsorption (*158, 159*), the benzene is adsorbed by π-bonding to the surface. The benzene can be readily removed by evacuation at 100°C. The trimerization of acetylene is much slower on the zeolites containing lesser amounts of nickel. In fact, only traces of benzene are formed on Ni-10 after 24 hr of contact. The reaction is substantially slower on the 200°C calcined samples. On Ni-19, the reaction is slow at 5 torr pressure, and on Ni-14, only traces of benzene were produced. At higher pressures (100 torr), a faster reaction is observed on Ni-14 and Ni-19 after a 20 min induction period. The difference in activity was attributed to inhibition of the reaction by adsorbed water remaining on the catalyst after the lower calcination treatment. Nitric oxide and pyridine were also found to inhibit the reaction probably by interacting

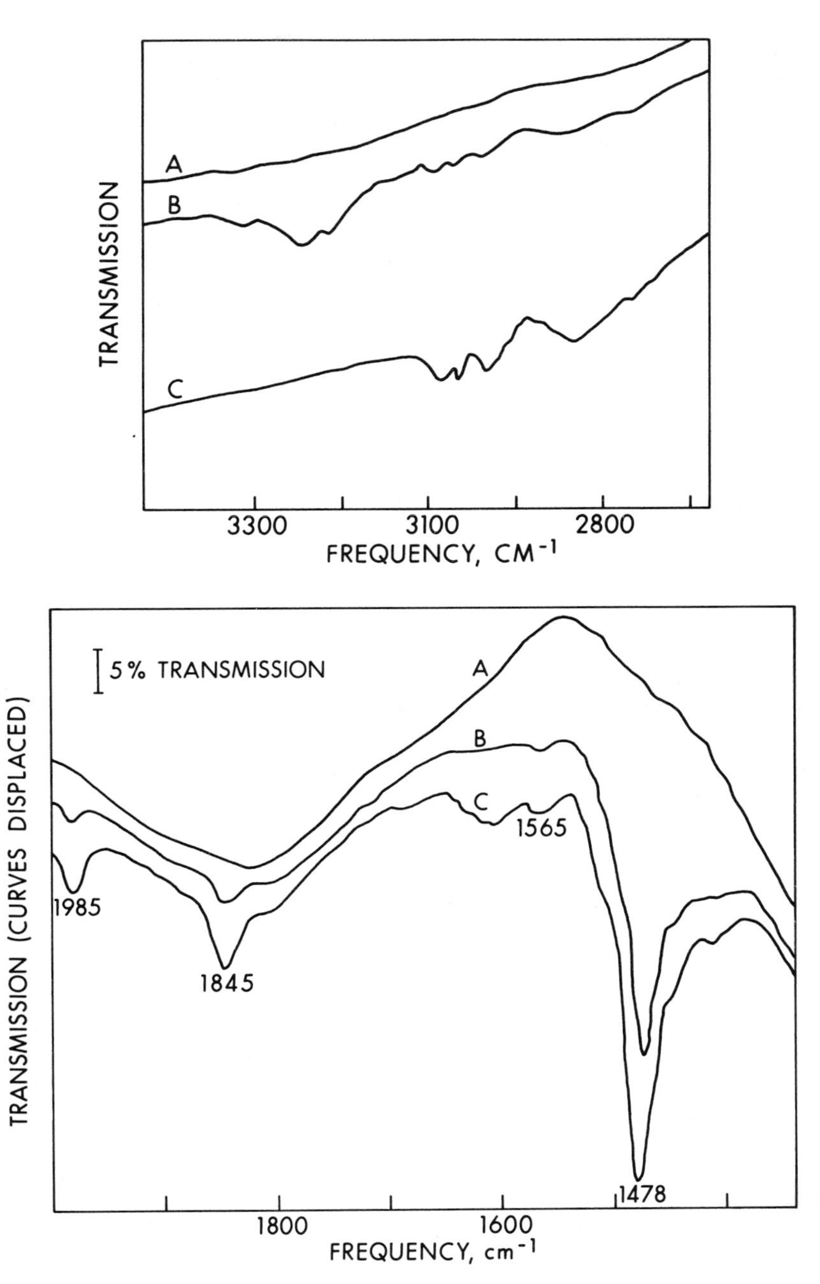

Journal of Catalysis

Figure 64. $Ni_{0.68}$-exchanged Na-Y (A) after evacuation at 600°C; (B) 20 min after introduction of about 5 torr C_2H_2; (C) 3 hr after introduction of C_2H_2 (157)

with the nickel ions and thus preventing the acetylene from adsorbing on the ions. Excess ammonia also inhibited the reaction. However, when outgassed at 100°C, the catalysts were active. Absorption bands are also observed near 2430 and 2865 cm^{-1} which were assigned to CH_2 groups. By comparison with studies of nickel supported on silica (*158*), these groups probably represent hydrocarbon species formed by linear polymerization of the acetylene on traces of nickel metal. Ni° and Ni^{+} were detected by ESR studies. The samples were also observed to darken, which is also suggestive of Ni°.

Propyne was also observed to trimerize on Ni-19. Of the possible isomers, the 1, 2, 4-isomer appeared to be most abundant. With both acetylene and propyne, no evidence for interaction of the hydrocarbons with Ni^{2+} ions was detected. Possibly the concentration is too low or the species too unstable. The C═C frequency of propyne was the same as that observed for Na-Y and only 300 cm^{-1} shifted from the gas phase value.

In a microreactor, the main products obtained by interaction of methylacetylene with sodium germanic sieve were allene and acetylene (*104*). Only traces of propylene were found.

Adsorption at room temperature immediately yields an OH band with a frequency near 3638 cm^{-1} probably from Ge-OH groups. Strong broad bands are observed in the region 3200–3400 cm^{-1} and sharp bands are observed near 2112 and 1643 cm^{-1}. No band is observed near 1260 cm^{-1}. These spectral results can be explained by the following scheme:

$$\mathrm{Al{-}O{-}Ge} \longrightarrow \begin{array}{c} \mathrm{CH_2{-}C{\equiv}CH} \\ | \\ \mathrm{Al} \\ \updownarrow \\ \mathrm{Al} \\ | \\ \mathrm{CH{=}C{=}CH_2} \end{array} + \mathrm{GeOH}$$

Because the 1260 cm^{-1} band is not detected, the adsorbed species is probably Al—CH═C═CH_2. On heating above 150°C, the OH band grows, the 2112 cm^{-1} band decreases in intensity and new bands appear near 2972, 1944 and 1643 cm^{-1}. These bands are probably caused by adsorbed allene type species. Allene could be formed by the reaction:

$$\begin{array}{c} \mathrm{CH{=}C{=}CH_2} \\ | \\ \mathrm{Al} \end{array} \xrightarrow{\mathrm{H_2}} \begin{array}{c} \mathrm{H} \\ | \\ \mathrm{Al} \end{array} + \mathrm{CH_2{=}C{=}CH_2}$$

Above 235°C bands appear near 1238, 1092, and 1706 cm^{-1}, indicating the formation of acetone. Also a band near 1643 cm^{-1} characteristic of propylene is observed. The observed species can be generated by the following scheme:

$$\text{Al}\overset{\text{O}}{\frown}\text{Ge} + CH_3C{\equiv}CH \longrightarrow \underset{\text{Al}}{\underset{|}{CH{=}C{=}CH_2}} + \text{GeOH}$$

$$\underset{\text{Al}}{\underset{|}{CH_2{=}C{=}CH_2}} + \text{GeOH} \longrightarrow \text{Al}\overset{\text{O}}{\frown}\text{Ge} + CH_2{=}C{=}CH_2$$

$$\underset{\text{Al}}{\underset{|}{CH{=}C{=}CH_2}} + H_2 \longrightarrow \underset{\text{Al}}{\underset{|}{CH_2 - HC{=}CH_2}} + \text{GeOH} \longrightarrow$$

$$\text{Al}\overset{\text{O}}{\frown}\text{Ge} + CH_3{-}CH{=}CH_2$$

$$\underset{\text{Al}}{\underset{|}{CH{=}C{=}CH_2}} + \text{GeOH} + H_2O \longrightarrow \text{Al}\overset{\text{O}}{\frown}\text{Ge} + CH_3{-}CO{-}CH_3$$

$$\underset{\text{Al}}{\underset{|}{CH{=}C{=}CH_2}} \longrightarrow CH{\equiv}CH + H_2O$$

Adsorption of methylacetylene on Ca-germanic sieve produces absorption bands near 3280, 2112, and 3638 cm^{-1} (*104*). The 3638 cm^{-1} band is probably caused by GeOH groups. On heating to 100°–200°C, new bands appear near 2183 and 1946 cm^{-1} attributed to allene or acetylene. The band observed near 2183 cm^{-1} could be caused by GeH groups. Above 200°C, acetone is formed in small amounts and probably some water.

Cyclic Hydrocarbons. The interaction of benzene with zeolites was first observed by Abramov, Kiselev, and Lygin (*158*). They studied sodium- and calcium-X after pretreatment at 450°C in vacuum. Compared with liquid benzene, they observed that the C—C stretch absorption near 1486 cm^{-1} was much more intense than the C—H stretch bands near 3000 cm^{-1}. Other bands in the spectrum changed in both intensity and frequency relative to the liquid phase, suggesting that the zeolite is interacting with the benzene ring structure rather than with the C—H groups. Similar results were obtained with silica (*160, 161*). The spectra can be reasonably interpreted by considering that the benzene molecules are adsorbed on the large cavity walls. The changes in band intensity and position were attributed to changes in the electron distribution by the zeolite electrostatic field. Ultraviolet studies confirmed that the adsorption involved the π-electron system of the adsorbed benzene (*161*). That the benzene ring remains intact was confirmed by the detection of $C_6H_6H^+$ species after irradiation of benzene adsorbed on decationized Y zeolite.

Angell *et al.* studied benzene and toluene adsorbed on a series of cation exchanged Y zeolites (*30, 159*). Mg-, Ni-, Co-, Zn-, Ag-, La-, and Ce-Y

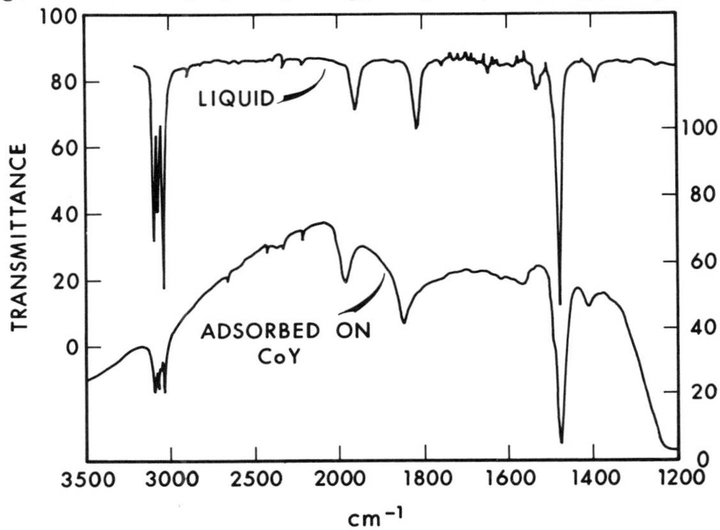

Journal of Colloid and Interface Science

Figure 65. Benzene adsorption on Co-Y (a) liquid; (b) adsorbed on zeolite (159)

zeolites were studied. Typical spectra observed for liquid benzene and benzene adsorbed on Co-Y are shown in Figure 65. The CH stretching frequencies were the same as those of liquid benzene. Also, bands from fundamental and combination vibrations that are in the plane of the ring did not change in frequency whereas those from combinations of out of plane vibrations moved to higher frequencies. Examination of the hydroxyl region of the spectrum showed that the zeolite hydroxyl groups hydrogen-bond to the benzene. The spectra were interpreted in terms of interaction between the structural hydroxyl groups, cations, and surface oxygens and the π-electrons of the aromatic ring, assuming that the molecule lies parallel to the surface. Because no strong band frequency dependence is seen with changes in cations, the interactions with the zeolite cations must be relatively weak. Deuterobenzene and toluene studies confirmed the benzene results. An interesting comparison of the adsorption of benzene, cyclohexane, cyclohexene and cumene on decationized Y has been made (*162*). The mode of adsorption is confirmed by the observation that cyclohexane could be desorbed at room temperature, whereas 200°C was necessary to desorb benzene. Furthermore, no H—D exchange was observed when either molecule was adsorbed on D-Y. In contrast, cyclohexene underwent exchange with all three surface O—D groups on D-Y. The spectra of adsorbed cyclohexene showed =CH or

C═C bands and resembled that of liquid cyclohexane. Cumene interacted by both the aromatic ring and the isopropyl group. Strong interaction with the 3650 cm^{-1} OH group occurred. At 150°C or higher, H—D exchange of the isopropyl group occurs. Calculations based on the band intensities indicate that the isopropyl group interacts more strongly than the aromatic ring.

Phenol, Aniline, and Nitrobenzene. Phenol, aniline and nitrobenzene adsorption on Na-X evacuated at 400°C has been reported (*163*). The major changes, compared to the solution phase, for phenol and aniline are that the OH and NH frequency shifts are greater than for adsorption on silica indicating stronger interaction with the zeolite groups. The cation electrostatic field probably interacts with the lone pair electrons of the oxygen and nitrogen atoms. Since only slight changes in the frequency of the aromatic nucleus are observed, the interactions of the nucleus must be relatively weak. In contrast, the major interaction was with the aromatic ring and not the nitro group, when nitrobenzene was adsorbed.

These results show how the electron availability distribution in the adsorbate dramatically affects the mode of adsorption, and are confirmed by ultraviolet spectral studies (*161*). Similar conclusions were reached from a study of the adsorption of aniline and diphenylaniline from carbon tetrachloride and dodecane solutions onto Na- and Ca-X (*166*). Studies of the electronic spectra of adsorbed *N,N*-dimethylaniline also showed that the primary adsorption was *via* the nitrogen atom (*165*).

Acetone and Acetaldehyde. Acetone and acetaldehyde interact strongly with the exchangeable cation *via* the C═O group. In their study of group IA and IIA zeolites, Geodakyan, Kiselev, and Lygin (*166, 167*) found that the C═O frequency decreased with increasing cation radius from Li to Cs and from Ca to Sr. A second band, observed in the case of acetone and removed by evacuation at 200°C is believed to indicate interaction with the zeolite lattice. Both molecules hydrogen-bond to the 3650 cm^{-1} band of decationized Y. Adsorbed acetone also has bands at 1715 cm^{-1} (from hydrogen-bonded C═O) and a pair of bands at 1640 and 1575 cm^{-1} possibly from adsorption on dehydroxylated sites. After evacuation of excess acetone, bands remain at 1455 and 1410 cm^{-1} attributed to formation of a carbonate structure. Apart from bands from physically adsorbed acetaldehyde, an absorption band attributed to C═O vibrations is observed caused by interaction of the CO group with dehydroxylated sites on the surface. A residual species remains on the zeolite surface which has an infrared spectrum typical of a carboxylate species.

Organic Acids. The investigation of organic acids (formic, acetic, and benzoic) on decationized Y zeolites has shown that the zeolites can have basic properties (*168*). $NH_{4_{0.30}}$-exchanged Na-Y was studied after pretreatment at 420°C. At room temperature adsorbed acetic acid was detected by absorption bands near 1415, 1720, and 1750 cm^{-1} and acetate ions by bands near 1480 and 1600 cm^{-1}. Water was also detected. Initially, the 3660 cm^{-1} OH band decreased in intensity and the 3560 cm^{-1} band was unaffected. After larger additions of acid, both hydroxyl bands decreased in intensity.

On increasing the temperature to 100°C, more acetate and less adsorbed acid were detected. At 350°C, the acid bands decreased further. Furthermore, no hydroxyl groups were detected, while at 400°C, only acetate bands were present. Under similar conditions, no acetate species were detected on Na-Y zeolite. Similarly, acetate groups were observed on silica–alumina but not on silica.

Formic acid behaved similarly. Formate ions and formic acid were observed on decationized Y after treatment at room temperature. On heating to 200°C, both species were removed from the surface. However the hydroxyl groups were not restored. Benzoic acid gave rise to benzoate ions along with benzoic acid. As the temperature was increased, more benzoate was formed.

The adsorption of organic acids and the subsequent formation of anions apparently proceeds by abstraction of hydroxyl groups from the lattice with the formation of water:

$$\text{Zeol—OH} + \text{HOOC—R} \rightarrow \text{Zeol}^{+} + \text{RCOO}^{-} + H_2O$$

Confirmation of the type of mechanism comes from observations of pyridine chemisorption after treatment with acid. Little Bronsted acidity is detected and substantial Lewis acidity is observed. These properties are similar to those of zeolites which have been dehydroxylated at high temepratures. Acid treatment appears to be a simple method of converting a Bronsted acid into a Lewis acid, since an analogous process is found to occur with silica–alumina.

Methanol and Other Alcohols. Studies of methanol adsorption (*166, 169*) on Li-, Na-, K-, Rb-, Cs-, and Sr- and decationized X also show interaction of the oxygen atom lone pair electrons with the exchanged cations and of the hydrogen atom hydrogen bonding to the lattice oxygen in a manner similar to that proposed for water. Observations of the shifts in band frequencies showed a greater shift in the OD frequency with increasing cation field. At elevated temperature, the structural hydroxyl groups can be converted into methoxyl groups (*170*). At higher concentrations of adsorbed methanol, deuteroxyl frequencies from interactions between the methanol groups themselves are observed.

Methanol has been adsorbed on hydrogen mordenite calcined at 400°C. The methanol was adsorbed at 150°C and the excess removed by evacuation. The hydroxyl bands at 3740 and 3650 cm^{-1} are eliminated and bands typical of methoxy groups are formed. The reaction $\text{—SiOH} + CH_3OH \rightarrow SiOCH_3 + H_2O$ occurs (*103*).

The interaction of 2-propanol with a sodium germanic sieve pretreated at 200°C has been investigated over the temperature range of 20° to 300°C (*104*). Microreactor studies show that propylene is by far the major product. Smaller amounts of saturated C_4 hydrocarbons, acetone and diisopropylether are also formed.

Below 148°C, absorption bands near 1160, 1131, and 1114 cm^{-1} are observed. These are attributed to the skeletal vibrations of the isopropyl group. A band observed near 1378 cm^{-1} is from CH_3 groups and a band near 1312 cm^{-1} is from COH groups. On raising the temperature above 100°C, these bands decrease in intensity while a new band is formed at 1238 cm^{-1}. At 300°C, a weak band appears near 1092 cm^{-1} and a broad band at 1600–1700 cm^{-1} with a shoulder at 1707 cm^{-1}. Also, an OH band appears near 3690 cm^{-1}. These spectral changes can be interpreted in terms of the disappearance of 2-propanol and the formation of acetone (1707 and 1238 cm^{-1}), water, and propylene. In contrast to the microreactor experiments, propylene is not detected on the surface. Hence the propylene formed must have a low affinity for the surface and readily desorb. In contrast, acetone is strongly adsorbed on the surface. Bands from acetone can be detected after outgassing at 300°C.

Propyl Chloride. The decomposition of organic halides on zeolites has been studied by Angell and Howell (*171*). They investigated the interaction of propyl chloride and bromide with 11 different cation forms of Y zeolite and with sodium-X at room temperature. The zeolites were initially activated at 500°C in vacuum. As shown in Figure 66, exposure of Ca-Y

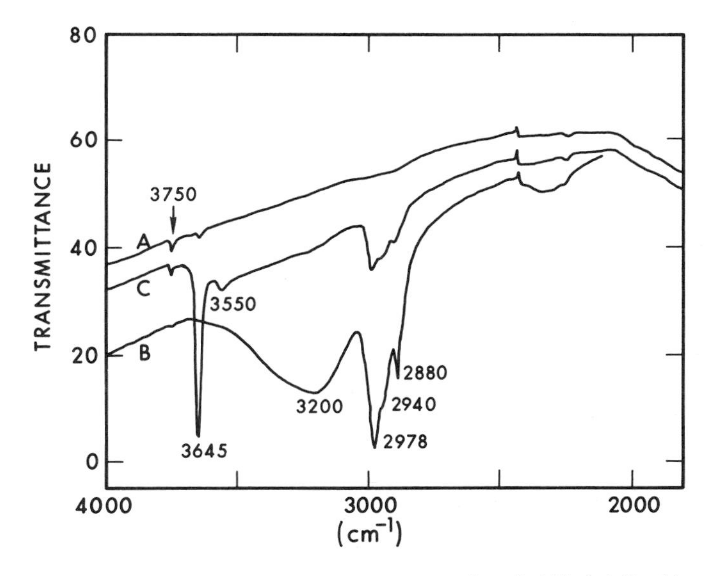

Figure 66. n-*Propyl chloride adsorbed on Ca-Y. (A) zeolite calcined at 500°C; (B) 100 torr* n-*propyl chloride added; (C) evacuated for 60 min* (171)

zeolite to propyl chloride resulted in a broad absorption band near 3200 cm^{-1}, indicating hydrogen-bonded hydroxyl groups and bands between 2980 and 2850 cm^{-1} from the CH stretching vibrations of the propyl group. On evacuation at room temperature, most of the CH bands and the 3200 cm^{-1} band were removed, and hydroxyl bands appeared near 3640 and 3540 cm^{-1}. However, no free water was detected. Magnesium-, strontium-, barium-, cobalt-, nickel-, and zinc-Y zeolites behaved similarly except that the ratio of the two hydroxyl bands varied with the cation. In the case of sodium-Y, the bands at 3200 cm^{-1} and the 3640 cm^{-1} were very much weaker. Silver-Y-zeolite behaved similar to the divalent cation forms. No new hydroxyl bands were observed over lanthanum-Y zeolite.

In this study, the zeolites catalyze the decomposition of the organic halide into HCl and propylene. The HCl then reacts with the zeolite surface to produce a hydroxyl group. Since the sodium zeolite does not interact with the halides to give hydroxyl groups whereas the divalent forms do, the electrostatic field associated with the cations probably polarizes the molecules, resulting in the splitting out of HCl. The inability of lanthanum to generate hydroxyl groups is in agreement with earlier data which indicated that there are no cations in the zeolite supercages at room temperature. Carbon monoxide could not be adsorbed on the various zeolites after addition of propyl chloride to give a cation specific band. This suggests that the chloride is associated closely with the zeolitic cations. If the zeolites are heated sufficiently to remove all or part of the hydroxyl groups, the band representing carbon monoxide–cation interaction can be detected. The simultaneous removal of hydroxyl groups and chloride suggests that HCl and not water is actually removed from the surface. The hydroxyl groups of these samples can be removed by heating to 450°C whereas hydroxyl groups on decationized Y formed by decomposition of ammonium-Y are only partially removed by treatment at 550°C and temperatures near 700°C are needed for complete removal. The removal of HCl rather than H_2O probably reflects these differences in stability. Further evidence that the chloride is associated with the cations is that the populations for the two hydroxyl groups vary with the nature of the cation.

The spectroscopic study demonstrates the possibility of introducing hydroxyl groups into zeolites while retaining the desired cation population. Thus, the properties of the hydroxyl groups and their population can be controlled by the exchanged cations.

Nitriles. The interaction of nitriles, (hydrogen cyanide, acrylonitrile and acetonitrile) on sodium-Y and decationized Y has been studied and compared to the interaction with silica (*172*). The interaction with zeolites is much stronger. The nitriles interact with the cations in Na-X and with the hydroxyl groups and dehydroxylated sites on decationized Y.

A more extensive study of acetonitrile has been made by Angell and Howell (*173*). Acetonitrile-d_3 and benzonitrile were also studied. Sodium-, magnesium-, calcium-, strontium-, barium-, manganese-, cobalt-, nickel-, silver-, lanthanum-, copper-, and decationized Y were studied. The adsorption of nitriles on zeolites is interesting since the interaction of molecules containing triple bonds with the zeolite surface can be investigated. The polarity of the C≡N bond, in contrast to the acetylene bond, results in a strong absorption band readily detectable by infrared. These strong bands are near 2254 cm^{-1} (C≡N stretching vibration) and 2293 cm^{-1} (a Fermi resonance band) in the liquid phase. On adsorption on zeolite, these two bands move to higher frequencies. The frequency shifts can be related to the calculated electrostatic field of the cations: the higher the field, the greater the shift. For magnesium and the transition metal cations, a third band is observed whose frequency is also cation dependent. The two higher frequency bands are attributed to adsorption directly on the cations because of their frequency dependence on the cation. The behavior is analogous to that observed for carbon monoxide (*121*), pyridine (*34*), and ethylene (*152*). The low frequency band is attributed to nonspecific adsorption within the zeolite cavity because of its easier removal by evacuation (*173*). However, because the frequency moves significantly with the cation, it may correspond to a weaker interaction with the cations. The frequency shifts for acetonitrile adsorbed in lanthanum-Y and decationized Y are relatively small. Acetonitrile interacts with the 3650 cm^{-1} but not the 3550 cm^{-1} band of decationized Y and with the 3650 cm^{-1} but not the 3520 cm^{-1} band of lanthanum-Y. The nitrile is interacting with the hydroxyl groups of the zeolites. The lower frequency shift indicates that the hydrogen-bonding interaction is weaker than the cation interaction. Since lanthnum cations are not in accessible surface positions, no interaction of acetonitrile with the 3520 cm^{-1} band is expected. Acetonitrile-d_3 and benzonitrile showed similar phenomena, confirming the interaction of the nitrile group with the surface. Although not investigated, less extensively exchanged zeolites (*e.g.*, $<40\%$) should not show any specific interaction of the cation with the nitrile group caused by the absence of multivalent cations in accessible portions. Similar results were obtained by Ratov *et al.* (*174*) for adsorption on Na-, La-, Y-, and H-Y zeolite.

Triphenyl Compounds. Triphenyl compounds have been investigated on Na-, Ca-, Ce-, and decationized Y after pretreatment at 400°C (*175*). Decationized Y was also pretreated at 600°C so that the dehydroxylated species was formed. The triphenyl compounds were adsorbed at 100°C and the spectra recorded at room temperature. Triphenylchloromethane and triphenylcarbinol produced triphenylcarbonium ions on all the zeolites studied. The species were detected by their characteristic absorption band near 1357 cm^{-1}. No bands from COH or C-Cl were detected. When triphenylchloromethane was adsorbed on decationized Y, no new hydroxyl bands were observed near 3635 cm^{-1}. On Ca-Y, Ce-Y, and Na-Y, the weak hydroxyl band became more intense. About 33% fewer carbonium ions are generated on Na-Y than on Ce-Y. These carbonium ions are desorbed more readily.

Adsorption of triphenylmethane on Ca-Y produced no carbonium ions. The spectrum observed was that of the hydrocarbon. No studies of the adsorption of triphenylmethane on other zeolites has been reported.

The formation of carbonium ions was found to be inhibited by the adsorption of acetonitrile. Preadsorption of the nitrile inhibits the formation of carbonium ions while postadsorption displaces the carbonium ions. In this study, the nitrile band of the acetonitrile was found at a constant frequency of 2270 cm^{-1} displaced from the liquid phase frequency. Hence the acetonitrile and the triphenyl carbonium ions are interacting with the same sites. Since the nitrile frequency was invarient with the cation, the sites involved were Lewis acid sites. No Lewis acid sites were detected by pyridine adsorption. Hence, the sites must be in locations other than the supercages. However, the invariance of the nitrile frequency with cation is contrary to the observations of Angell and Howell (*173*), who found that the frequency was a linear function of the cation electrostatic field. If the latter results are correct, the cations may be the adsorption sites.

Miscellaneous Adsorbed Molecules. 5-Cyanopent-1-ene adsorption on sodium- and decationized Y zeolite has been investigated. On sodium-Y, the interaction appears to be simple physical adsorption. Adsorption on decationized Y results in the elimination of the olefinic band. The shift in the frequency of the C≡N band and the appearance of broad bands in the 2900–2400 cm^{-1} region suggest hydrogen bonding between the nitrile groups and the surface hydroxyl groups with frequencies at 3640 cm^{-1} (*176*).

Beckmann rearrangements of oximes have been observed on decationized zeolite surfaces (*177*). Cyclopentanone oxime adsorption results in interaction with the 3640 cm^{-1} hydroxyl band, loss of the C≡N band at 1690 cm^{-1} of the oxime, and a shift of the band near 1428 cm^{-1} to 1410 cm^{-1}. These results indicate that the oxime is protonated on the zeolite surface. On heating to 120°C, carbonyl species are formed and the spectrum resembles that of 2-piperidine. At 250°C, no oxime is detected, suggesting that the reaction has gone to completion. Similar effects are observed with cyclohexanone oxime. The product spectrum resembles 2-piperidone rather than caprolactam. The spectra suggest that ring opening occurs and an intermediate is formed.

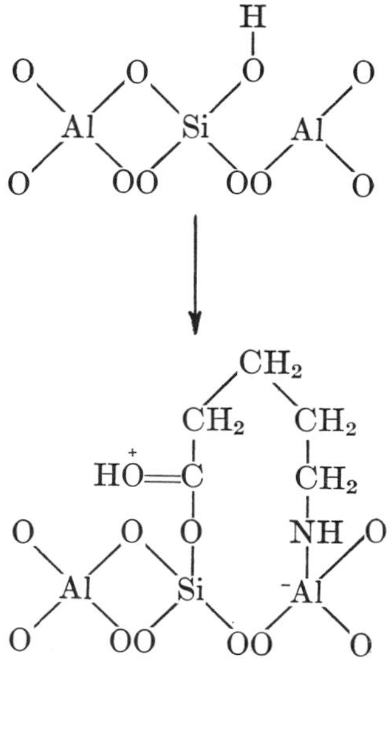

Interaction with Basic Molecules—Surface Acidity

Zeolites catalyze a variety of reactions (*178*). Since many of these reactions are of the carbonium ion type (which are usually acid-catalyzed), much effort has been spent determining if zeolites are acidic, how acidity is introduced, and what the nature of the acidity is. Although many reactions catalyzed by zeolites are the same as those catalyzed by amorphous silica–alumina, early theories of their action (*179*) suggested that the electrostatic field of the zeolites was responsible for their action. There is evidence to support this concept since the activity of zeolites in several reactions increases with the calculated electrostatic field (*179, 180, 181*) and in many cases, *e.g.*, with group IA cation zeolites which have weak fields, the reactions appear to proceed by radical rather than carbonium ion mechanisms. However, in instances where the reaction proceeds *via* carbonium ion mechanisms, studies have suggested that if electrostatic fields are involved, their role may be to promote the formation of acid sites.

Most zeolite studies have been made using basic molecules such as ammonia, pyridine, and piperidine as surface probes. These molecules interact with Bronsted acid sites, Lewis acid sites, and cations. These and their hydrogen-bonding interactions give rise to different species detectable by infrared spectroscopy. Thus, adsorption on Bronsted acid sites gives rise to ammonium, pyridinium, and piperidium ions with characteristic absorption frequencies of 1475, 1545, and 1610 cm^{-1} respectively. Adsorption on Lewis acid sites, tricoordinated aluminum ions, occurs *via* a coordinate bond $>$N-Al$<$ and gives rise to bands near 1630, 1450, and 1450 cm^{-1} respectively. The exchangable cations can also act as Lewis acid centers and give rise to similar bands. Although in some instances, interactions with cations have been mistaken for interactions with true Lewis acid sites, fortunately, the frequencies for and strengths of adsorption on aluminum centers differ from those attributed to adsorption on cations. Furthermore, the adsorption frequency varies with cation whereas the frequency for adsorption on true Lewis acid sites remains constant. Similarly, species adsorbed on the surface by hydrogen bonding can be distinguished spectroscopically by means of frequencies and ease of removal of the adsorbed species.

Other bands in the spectrum of adsorbed pyridine can also be used as diagnostic indicators. From the study of model systems (*1, 2, 9*), absorption bands near 3266, 3188, 3150, 1640, 1550, and 1490 cm^{-1} are also indicative of Bronsted acidity while bands near 3150, 3120, 1620, 1580, 1490, and 1450 cm^{-1} can be attributed to coordinately bound pyridine and are thus indicative of Lewis acid sites. Assignments of these bands are given in Table XV. There is considerable overlap in the presence of excess pyridine between the coordinately bonded and hydrogen-bonded species. However these can be differentiated readily since the hydrogen-bonded species can be removed by evacuation at relatively low temperature leaving the coordinately bonded species relatively intact.

Spectra of adsorbed piperidine have a strong band characteristic of piperidinium ions at 1610 cm^{-1} and coordinately-bonded piperidine at 1460 cm^{-1}. Another band from piperidinium ions is observed at 1450 cm^{-1}.

Table XV. Assignment of Pyridine Absorption Bands

Vibrational Assignment		*PY*[a] *(cm⁻¹)*	*LPY*[b] *(cm⁻¹)*	*BPY*[c] *(cm⁻¹)*	*HPY*[d] *(cm⁻¹)*
7a	νNH	—	—	3260	—
20b	νCH	3083	(3147)	(3147)	—
16	νCH	3054	(3114)	(3114)	3065
7b	νCH	3054	(3087)	(3087)	3043
8a	νCC (N)	1580	1620	1638	1614
8b	νCC (N)	1572	1577	1620	1593
19a	νCC (N)	1482	1490	1490	1490
19b	νCC (N)	1439	1450	1545	1438

[a] PY—physically adsorbed.
[b] LPY—adsorbed on Lewis acid sites.
[c] BPY—adsorbed on Bronsted acid sites.
[d] HPY—adsorbed on hydrogen bonding.

Ammonia adsorption gives rise to species with bands at 3423, 2990, 3200, 1435, 1485, and 1680 cm^{-1} from ammonium ions and at 3320 and 1630 cm^{-1} attributed to coordinately-bound ammonia.

Surface Acidity of Cation Exchanged Zeolites. GROUP IA ZEOLITES. The acidity of cation containing zeolites has been extensively studied. Most of the studies have concerned X and Y zeolites but more limited studies have been made of L, A and mordenite. From a theoretical viewpoint, the group IA cation-containing zeolites should be non-acidic. However the various studies of sodium zeolites have led to varying conclusions of no acidity to Bronsted acidity to Lewis acidity. The discrepancy appears to be caused by differences in the purity and pretreatment of the zeolites. Sodium-X and -Y zeolites of high purity contain no detectable acidity. When the zeolite has become partially cation-deficient by hydrolysis or contains a substantial impurity level of multivalent cations such as calcium, a small amount of Bronsted or Lewis acidity is detected, depending on the pretreatment temperature. Thus, in general, if the zeolite is of high purity the sodium and other group IA cation-exchanged zeolites derived from it are non-acidic (*33, 34, 35, 36*). Moscou confirmed these results by chemical techniques (*189*). Pyridine is chemisorbed on these zeolites but the interaction appears to be solely with the exchangable cations. An absorption band is detected in the frequency range of 1435 to 1450 cm^{-1}. This species can be removed by evacuation at 200°C. The observed frequency is always less than that observed for pyridine adsorption on Lewis acid sites on alumina, silica–alumina, and dehydroxylated molecular sieves (*33, 34, 35, 36*), and thus probably represents interaction with cations. Since the cations are electron acceptors, thus being able to function as Lewis acids in the general sense, it is not surprising that absorption bands are detected in this region. The frequency of the band is a function of many properties which influence the field strength in the neighborhood of the cation, such as cation radius, electronegativity, electrostatic potential, and electrostatic field of zeolite (*35*). The variation of the frequency with ionic radius and electro-

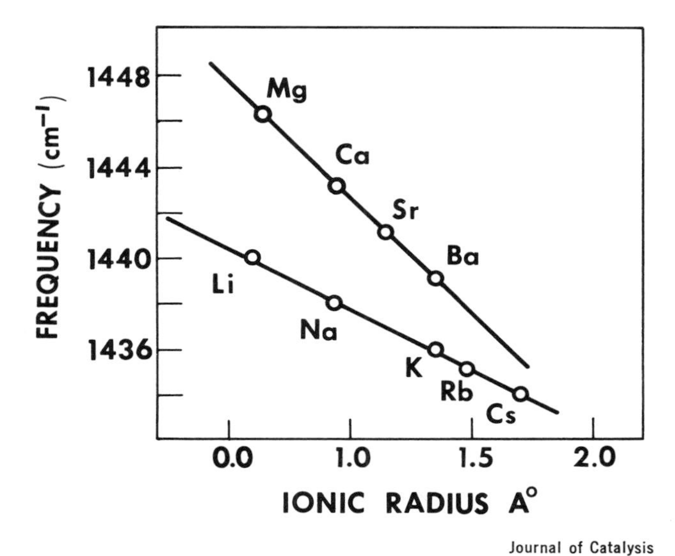

Journal of Catalysis

Figure 67. Frequency of absorption band from pyridine–cation interaction as a function of cation radius (35)

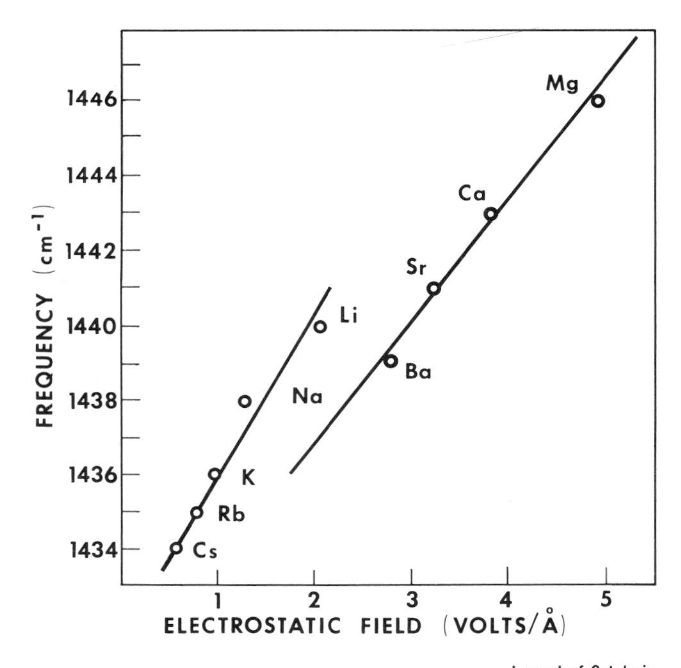

Journal of Catalysis

Figure 68. Frequency of absorption band from pyridine–cation interaction as a function of electrostatic field (35)

static field is shown in Figures 67 and 68. The greater the field, the higher the frequency, thus reflecting a stronger interaction between the adsorbed pyridine and the cations.

The strength of interaction is also reflected by the removability of the adsorbed molecules from the zeolites surface. Higher temperatures are needed to remove the pyridine from the zeolite as the cation radius is decreased (*35*). Confirming the interpretations that the basic molecules adsorb by interaction with the cations are observations which show that in mixed cation zeolites, several bands are observed at frequencies corresponding to those for single cation containing zeolites. Thus, for example, a magnesium-sodium zeolite gives rise to two bands on chemisorption of pyridine corresponding to independent adsorption on the magnesium and sodium cations. A further confirmation is that when small amounts of water are added to the system, no generation of Bronsted acid sites is observed whereas for silica–alumina, Lewis acid sites are converted into Bronsted acid sites (*35*). No absorption bands are observed corresponding to the formation of pyridinium ions indicating the absence of Bronsted acid sites. Of the Group IA zeolites, sodium-X and -Y have been most thoroughly studied, particularly in relation to whether they are acidic or not.

Nishizawa *et al.* (*183*) reported Lewis acidity on Na-X. However, because of its relatively low frequency, its ease of removal, and its lack of Bronsted acidity upon addition of water, the species is more likely caused by a cation–pyridine interaction rather than by true Lewis acidity. Studies by Ignat'eva (*184*) gave the same results. Similarly, Eberly (*34*) reports no Bronsted acidity on Na-Y pretreated at 427°C. Only a Lewis bonded species was detected. Because this species was readily removed by evacuation at 260°C it also probably represented a cation–pyridine species. Similar results were reported for Li- and K-Y zeolites. Similarly, Ward (*35*) detected only pyridine interacting with the lithium, sodium and potassium ions and no absorption bands due to Bronsted or Lewis acidity. Hattori and Shiba (*29*) also found no acidity on Na- and K-X zeolites and Ward obtained similar results for Li-, Na-, and K-X (*114*). Watanabe and Habgood (*182*) obtained the same results on thoroughly dehydrated Group IA zeolites and partially dehydrated Na-X zeolites. However Na-X zeolite, after treatment with any of several aqueous treatments, and several samples of Y zeolites, after partial hydration, gave rise to pyridinium ions on treatment with pyridine. The Bronsted acidity may result from cation deficiency introduced by hydrolysis during washing, splitting of water by divalent cation impurities to give acidic hydroxyl groups, or by formation of some metastable structure. Hydroxyl groups were detected in the acidic samples.

The addition of carbon dioxide to Li-Y zeolite introduces Bronsted acidity (*65*). The nature of the acidic species is not clear at this time.

Ammonia has been adsorbed on the Group IA-X zeolites (Li, Na, Rb, Cs). The spectra are similar to that of ammonia adsorbed on silica. The adsorbed species are readily removed by evacuation at room temperature, the ease of removal increasing with cation radius and indicating interaction with the cation analogous to the pyridine results (*185, 186*).

GROUP IIA ZEOLITES. The surface acidity of multivalent cation-exchanged zeolites is considerably more complicated and much more interesting, particularly from a catalytic viewpoint. As discussed above, introduction of divalent cations into zeolites introduces simultaneously structural hydroxyl groups. Analogous to amorphous oxides, some of these hydroxyl groups should be acidic and acidic sites should also be generated by their thermal expulsion. Studies of alkaline earth zeolites have shown that there are Bronsted and/or Lewis sites on the surface depending on the calcination conditions (*29, 34, 35, 36*). Eberly (*34*) used the spectra of pyridine adsorbed on Mg- and Ca-Y zeolites at 150°C after initial pretreatment of the samples at 427°C. Bronsted acidity was detected on the two zeolites. The quantity was about one-tenth to one-twentieth of that found on decationized Y. Addition of water increased the Bronsted acidity concentration. At the same time, an OH band appeared near 3580 cm^{-1}. This band could represent $CaOH^{+}$ groups. The concentration of acid sites on magnesium was greater than on calcium. It was not clear whether the band near 1440–1450 cm^{-1} represented adsorption on Lewis acid sites or interactions with cations. The latter is probable, at least in part, since the band is much stronger than on decationized Y zeolites calcined at the same temperature and since it is decreased in intensity by addition of water.

A systematic study of the alkaline earth zeolites has been made (*35*). Using pyridine chemisorption on Y zeolite samples calcined at 500°C, no Lewis acidity was observed. However a band, whose frequency depended on the cation, appeared near 1440–1450 cm^{-1}. The variation in frequency with properties of the cation is shown in Figures 67 and 68. The frequency and difficulty of removal increased with decreasing ionic radius and increasing electrostatic field and potential. The frequency change is similar to that reported for carbon monoxide (*121*) and carbon dioxide (*133*), thus supporting the interpretation of interaction with the cation. All the zeolites contain Bronsted acid sites. The concentration of the Bronsted sites increases with increasing electrostatic field. Typical spectra of pyridine chemisorbed on sodium-, magnesium-, and barium-Y zeolites are shown in Figure 69.

Of the hydroxyl groups present on the zeolite surface, the ones responsible for the band near 3650 cm^{-1} interacted strongly with the chemisorbed pyridine while the other bands appeared to be inert. In contrast to the alkali cation forms, the alkaline earth cation forms all exhibit absorption bands near 1545, 3190, and 3260 cm^{-1} which are indicative of the formation of pyridium ions and the presence of Bronsted acid sites. The concentration of acid sites, as indicated by the intensity of the 1545 cm^{-1} absorption band, increased with decreasing cation radius and increasing electrostatic properties. Typical values are given in Table XVI and Figure 70 shows the dependence on electrostatic field and ionic radius. Two different samples of Mg-Y are shown in Figure 70. Mg(I) was 80% exchanged with magnesium ions and Mg(II) was 50% exchanged. As might be expected, increasing the extent of exchange increased the acidity. Observations on the desorption of pyridine as a function of temperature indicate that the acid sites are of very similar strength. However, the Bronsted sites seem to be weaker than

Table XVI. Bronsted Acidity of Alkaline Earth Y Zeolites (*35*)

Cation	*Acidity (arbitrary unit)*
Mg	6.9
Ca	4.3
Sr	2.5
Ba	1.5

Journal of Catalysis

the Lewis acid sites for Mg and Ca but of equal strength for Sr and Ba. Because the 3650 cm^{-1} band interacts strongly with pyridine, this type of hydroxyl group is apparently the source of the Bronsted acidity. If this is so, calcination of the zeolite at higher temperature should result in dehydroxylation of Bronsted acid sites. Calcination of the cation Y zeolites at 650°C results in a decrease in the intensity of the hydroxyl group absorption

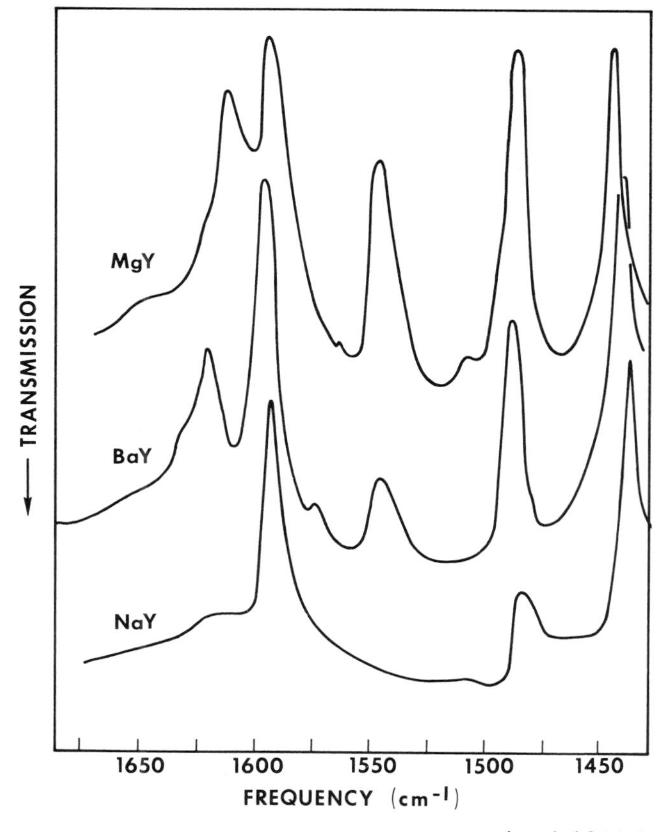

Journal of Catalysis

Figure 69. Pyridine chemisorbed on Na-, Mg-, and Ba-Y zeolites after evacuation at 250°C (35)

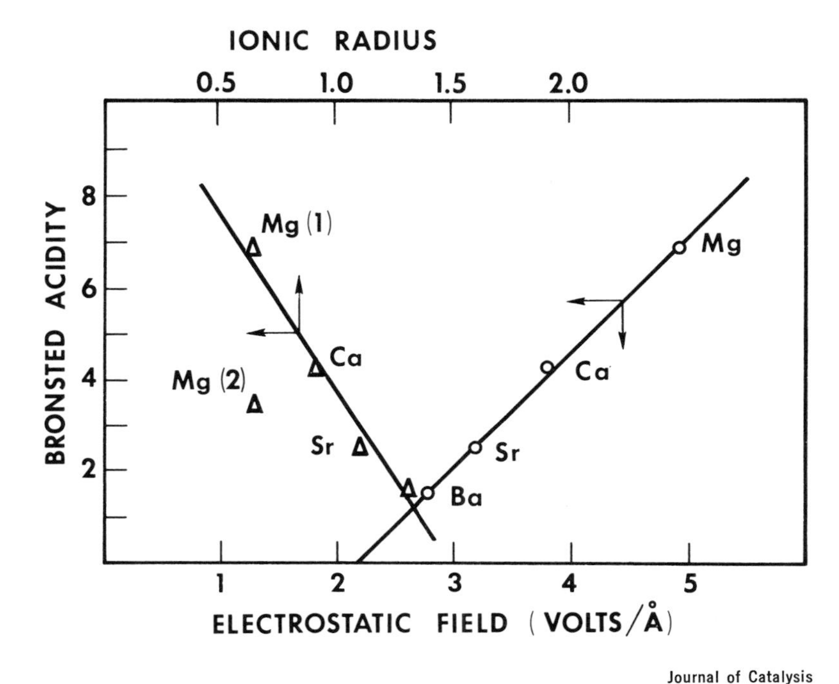

Journal of Catalysis

Figure 70. Variation of Bronsted acidity with cation radius and electrostatic field for alkaline earth cations (35)

bands and the 1545 cm^{-1} band indicative of Bronsted acidity. Simultaneously, a band at 1451 cm^{-1} characteristic of Lewis acidity is observed. Similar treatment of alkali metal zeolites had no effect. Changes in the spectra are shown in Figure 71 for Ca-Y.

Rehydration of the zeolites results in elimination of the Lewis acid sites and an increase in the concentration of Bronsted acid sites. New hydroxyl bands are also formed as discussed below. The frequency is near 3595 and 3588 cm^{-1} for Mg- and Ca-Y respectively. These bands do not interact with pyridine, indicating that they are non- or weakly acidic. Christner, Liengme, and Hall (*36*) have also studied magnesium- and barium-Y zeolite. They observed Bronsted acidity and pyridine–cation interactions. The data also seem to indicate that Lewis acidity was also detected (band near 1450 cm^{-1}) on samples pretreated at 480°C. Redistribution of pyridine was observed on the zeolite as a function of adsorption conditions. Typical data for Mg-Y is shown in Figure 72. Pyridine migrates from weaker sites (*e.g.*, Na^{+}) to stronger sites (*e.g.*, Mg^{2+} and Lewis sites). Heating to about 150°C also resulted in the formation of additional Bronsted acid sites. Formation of pyridinium ions upon activation was also observed on Ce-Y. At room temperature, little pyridinium ion was formed but on heating to 70°C, the 3650 cm^{-1} was eliminated and a strong pyridinium ion band formed.

Addition of carbon dioxide to the hydrated zeolites resulted in the production of Bronsted acid sites. It has been found independently that the addition of carbon dioxide to Na-Y increased the catalytic acivity for alcohol dehydration (*132*).

Ammonia and pyridine adsorption was studied by Yashima and Hara (*187*) on Be- and Ca-Y after pretreatment at 500°C. As expected, both zeolites contained Bronsted acid sites and Lewis acid sites.

Most of the acidity phenomena can be understood in terms of the scheme outlined previously for the generation of hydroxyl groups.

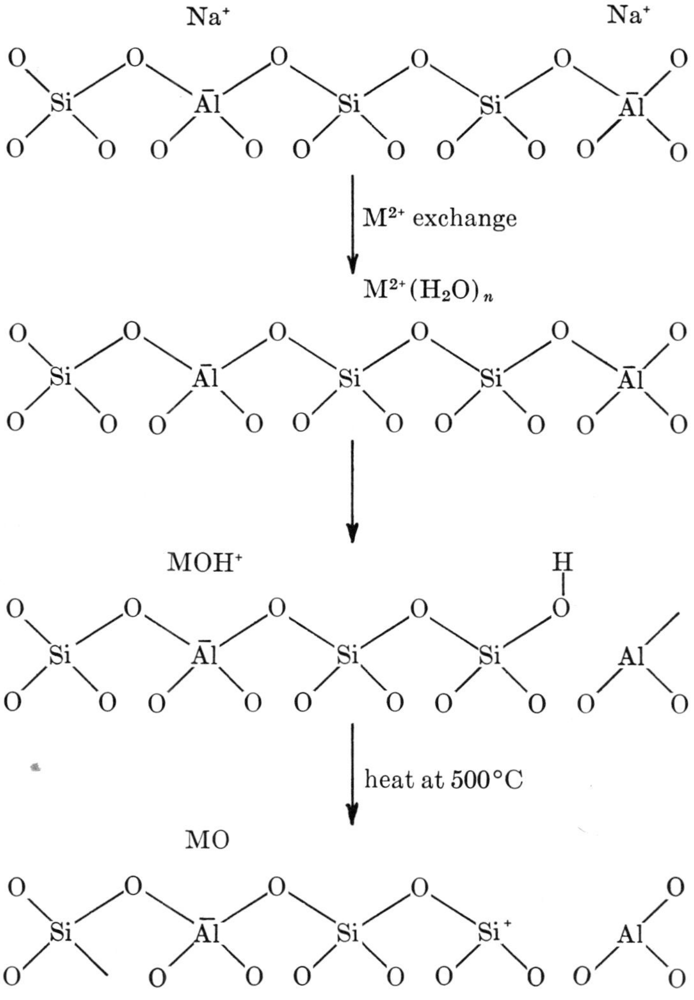

Na^+ Na^+

O O O O O O
Si $\bar{Al}$ Si Si $\bar{Al}$
O O O O O O O O O O

↓ M^{2+} exchange

$M^{2+}(H_2O)_n$

O O O O O O
Si $\bar{Al}$ Si Si $\bar{Al}$
O O O O O O O O O O

↓

MOH^+ H

O O O O O
Si $\bar{Al}$ Si Si Al
O O O O O O O O O O

↓ heat at 500°C

MO

O O O O O
Si $\bar{Al}$ Si Si^+ Al
O O O O O O O O O

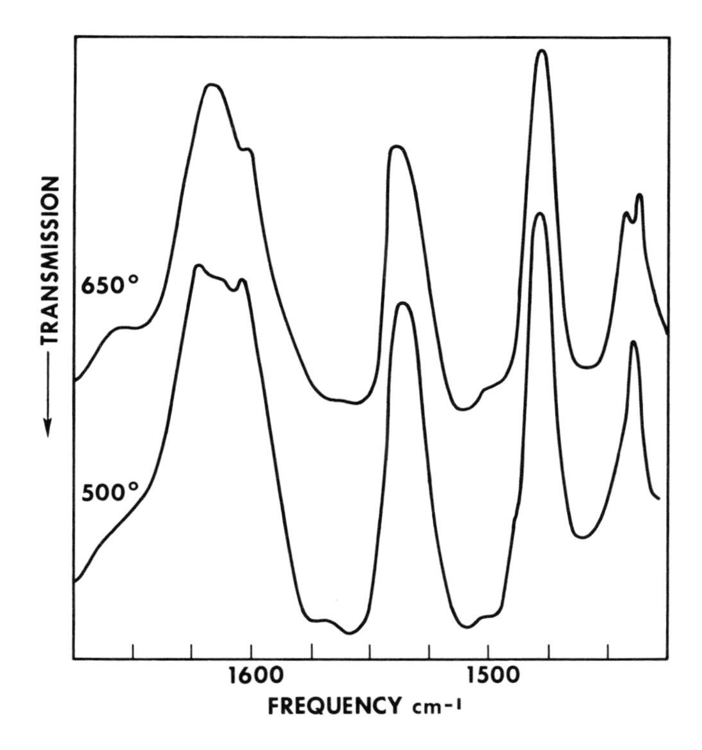

Journal of Catalysis

Figure 71. Spectra of pyridine adsorbed on Ca-Y zeolite after evacuation at various temperatures (35)

Thus, one structural hydroxyl group is initially formed for every two exchange sites. These hydroxyl groups have the same frequency as those of decationized Y (3660 and 3540 cm^{-1}) and hence would be expected to behave similarly. The variation in hydroxyl content and Bronsted acidity of the various alkaline earth zeolites may mean that an equilibrium exists:

$$M^{2+}(OH_2) \rightleftarrows MOH + H^+$$

The smaller cations, with their associated high electrostatic field and polarizing power, would result in the equilibrium moving to the right, while the larger cations would be expected to produce less dissociation. Support for the scheme is found in observation of absorption bands which can be attributed to MOH^+ groups; the frequency of these bands varies with the nature of the cation M.

Heating of the zeolite above 500°C results in dehydroxylation and hence conversion of Bronsted sites into Lewis acid sites as shown by the scheme. The final form of the multivalent cation is still uncertain. After high temperature treatment, *e.g.*, 600°C, a pyridine absorption band is

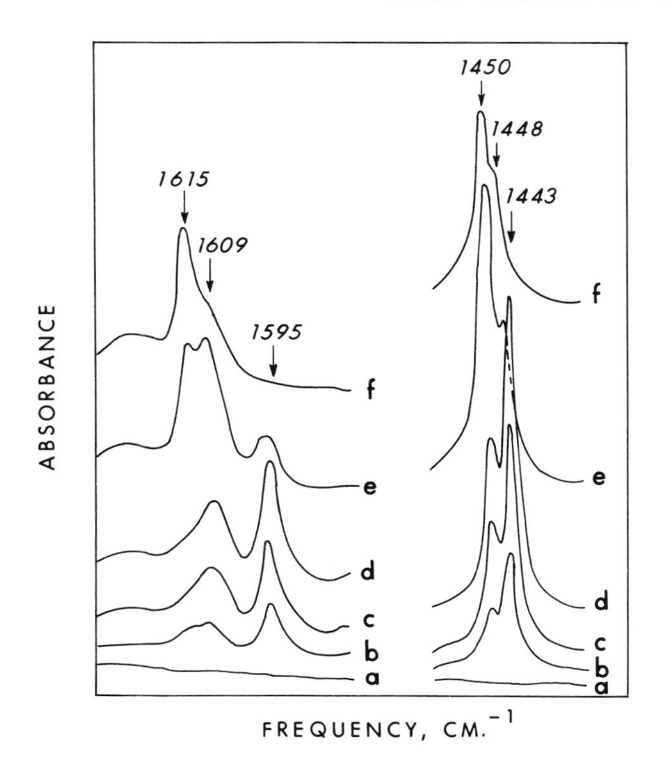

Transactions of the Faraday Society

Figure 72. Redistribution of pyridine on Mg-Y (a) evacuated at 480°C; (b),(c),(d) three successive doses of pyridine; (e) heated at 100°C for 2 hr; (f) heated at 160°C for 16 hr (65)

observed at 1451 cm^{-1}, indicating the interaction of pyridine with alumina and hence the presence of Lewis acid sites.

Readdition of water to a 600°C calcined zeolite followed by reevacuation at 400°C, results in an increase in the Bronsted acidity and also a reduction in the pyridine–cation band. Simultaneously new hydroxyl groups are formed near 3585 cm^{-1} for calcium and 3595 cm^{-1} for magnesium. Hence, water is displacing pyridine from the cation sites and then dissociating to give MOH groups and additional Bronsted acid sites.

Similar studies of X zeolites have been made. Hattori and Shiba (*29*) studied acidity by means of pyridine adsorption on magnesium-, calcium-, and strontium zeolites after pretreatment at 450°C. Only Lewis acidity was detected on Ca- and Sr-X initially, but on hydration, Bronsted acidity could be detected. Mg-X was found to have a small amount of Bronsted acidity. No detectable increase in Bronsted acidity was observed on hydration. The four alkaline earth zeolites were studied by Ward (*114*) after pretreatment at 480°C. Acidity measurements were made on both the dehydrated and partially hydrated samples. In general, the results were similar to those obtained with Y zeolites. Interaction with the cation was observed, the frequency of the band at 1440–1450 cm^{-1} increasing with decreasing cation radius or increasing electrostatic field. No band attributed to Lewis acidity was detected. All the samples exhibited Bronsted acidity, the concentration of which decreased with increasing cation radius and increased with hydration. The data are summarized in Table XVII. The Bronsted acid site concentration is much less than for the Y zeolite series (*e.g.*, Ca-X, 0.71 units, and Ca-Y, 4.3 units). In contrast to the results of Hattori and Shiba (*29*), no Lewis acidity was found and Bronsted acidity was observed on all samples. Apparently they attributed bands due to pyridine–cation interaction to Lewis acidity.

Table XVII. Bronsted Acidity of X Zeolites (*114*)

	Acidity (arbitrary unit)	
Cation	*Hydrated*	*Anhydrous*
Na	0	0
K	0	0
Mg	2.9	8.5
Ca	0.7	5.1
Sr	0.2	3.0
Ba	0.1	0.6
Mn	1.2	3.1
Co	1.1	2.1
Cu	<0.1	<0.1
Zn	0.5	3.2
Ag (1)	6.9	7.4
Ag (2)	4.1	4.7

Journal of Catalysis

Partial hydration resulted in an increase in Bronsted acidity. Because no Lewis acidity was observed in the initial samples, the increase in acidity cannot be attributed to the conversion of Lewis acid sites to Bronsted acid sites. The increase may be caused by relocation of cation with simultaneous generation of Bronsted acid sites.

Ammonia adsorption on hydrated Ca-X zeolites showed that the 3650 cm^{-1} band interacted with ammonia to give ammonium ions (1485 cm^{-1}), but the band near 3695 was not affected (*31*).

The available data indicate that the types of acid sites and their mode of formation on X zeolites are similar to those of Y zeolites. The number of sites is lower probably because the electrostatic field in the zeolite is lower and the spacing between the aluminum atoms is less. Hence the reaction

$$M(OH_2)^{2+} \rightarrow MOH^+ + H^+$$

is less complete.

Thus, Y zeolites are more acidic than the comparable X zeolites. A study of the Bronsted acidity of magnesium and calcium zeolites as a function of the zeolite alumina content shows that the acidity decreases linearly with increasing alumina content of the zeolite (*115*). The values for magnesium remain consistently higher than those for calcium as shown by Figure 73. Since the zeolites were exchanged to the same extent, the

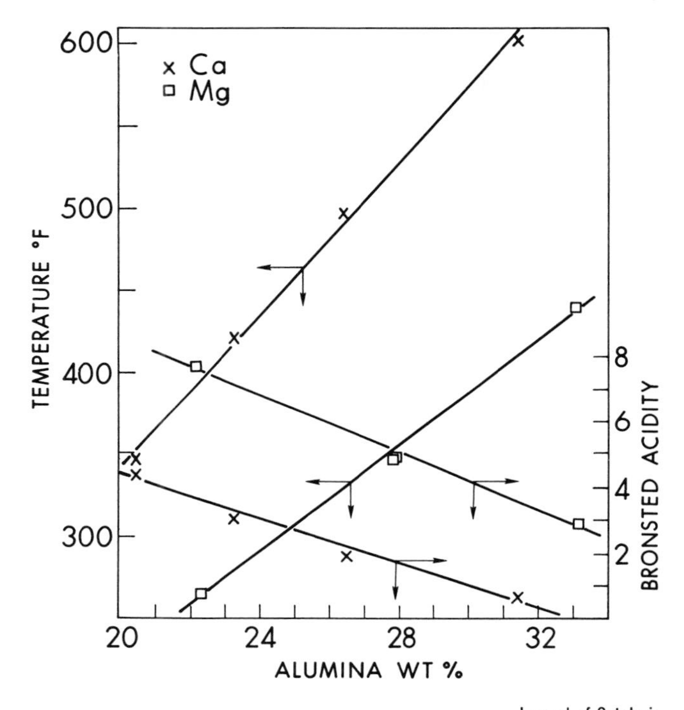

Journal of Catalysis

Figure 73. Bronsted acidity and catalytic activity of Mg- and Ca-Y zeolites as a function of zeolite alumina content (115)

zeolites with higher aluminum contents would be expected to have more acidic sites since they contain more multivalent cations. However, a compensating effect is that the spacing between the aluminum sites is less and therefore the electrostatic field is less and hence the reaction $M(OH_2)^{2+} \rightarrow MOH^+ + H^+$ is probably less complete. A greater number of the potential acid sites are apparently obtained in samples of higher silica to alumina ratio.

TRANSITION METAL CATIONS. The influence of many of the transition metal ions particularly from the third row of the periodic table on zeolite acidity has been studied mostly using pyridine adsorption as an indicator. Nishizawa, Hattori, and Shiba (*183*) studied Mn-, Zn-, Co-, Ni-, and Tl-exchanged X zeolites after pretreatment at 450°C. All samples were greater than 80% exchanged. They found no Bronsted acidity on Ni-, Co-, and Tl-X before or after rehydration. Mn- and Zn-X had small amounts of Bronsted acidity. The quantity was increased by rehydration. In all cases except Tl, a band in the 1450 cm^{-1} region was observed. However this band is probably caused by interaction of pyridine with the cations rather than with Lewis acid sites.

Acidity of $Ni_{0.15}$-exchanged Na-X has recently been studied by ammonia and pyridine adsorption (*48*), after pretreatment at 200 and 350°C. On addition of ammonia to the 200°C calcined sample, the 3650 and 3610 cm^{-1} hydroxyl bands interact and give rise to ammonium ions indicating that these hydroxyl groups are proton acids. Addition of ammonia to the 350°C calcined sample results in formation of ammonium ions by interaction with the 3650 and 3560 cm^{-1} bands. The interaction of ammonia with the nickel cations was also observed. Two other bands were attributed to interaction with the nickel ions. However, because of the low degree of exchange, these may represent interaction with the sodium ions. The presence of proton acidity on the 350°C calcined sample was confirmed by the interaction with pyridine. Other absorption bands were observed at 1455, 1451, 1448, and 1440 cm^{-1}. The first two are probably caused by adsorption on two types of Lewis acid sites while the latter two are probably caused by interaction with nickel and sodium ions or hydrogen bonding.

Mn-, Co-, Cu-, Zn-, Ag-, and Cd-X have been examined at high levels of exchange before and after rehydration (*114*). Proton acidity was found on all forms except copper. The acidity increased on hydration. Acidity may not have been detected on the copper form because of collapse of the structure. The data are summarized in Table XVII. No bands attributable to Lewis acidity were observed. All forms had an absorption band in the spectrum in the 1440–1450 cm^{-1} region from the interaction of pyridine with the cations, the specific frequency depending on the cation and generally increasing with the field. These data are in conflict with those of Nishazawa *et al.* (*183*). Differences may be attributed to differences in pretreatment and detection. Unlike the alkaline earth zeolites, there seems to be no simple relationship between acidity and cation properties or with the quantity of structural water. Since no bands are observed which are attributable to Lewis acid sites, the increase in Bronsted acidity on hydration is probably not caused by conversion of Lewis sites to Bronsted sites as on silica–alumina and dehydroxylated zeolites. However, since a decrease in the absorption bands from cation–pyridine interaction is observed, the change in acidity is probably associated with the cations. Possible explanations are the formation of acidic $M^{2+}OH^{-}$ groups or the movement of cations from sites accessible to pyridine to sites not accessible. Such movement would allow the formation of additional Bronsted acid sites by making accessible sites available. Nickel ions are known to move locations with degree of hydration (*127*) and similar movements have been suggested in other systems. Silver was found to be the most acidic of the X zeolites (apart from rare earth). The acidity was not influenced by water.

Transition metal ion-containing Y zeolites have also been studied in considerable detail. Probably the results on the Y zeolites are more reliable since the zeolite is more resistant to acid attack which leads to structural instability in X zeolites. Eberly (*34*) showed that cadmium-Y

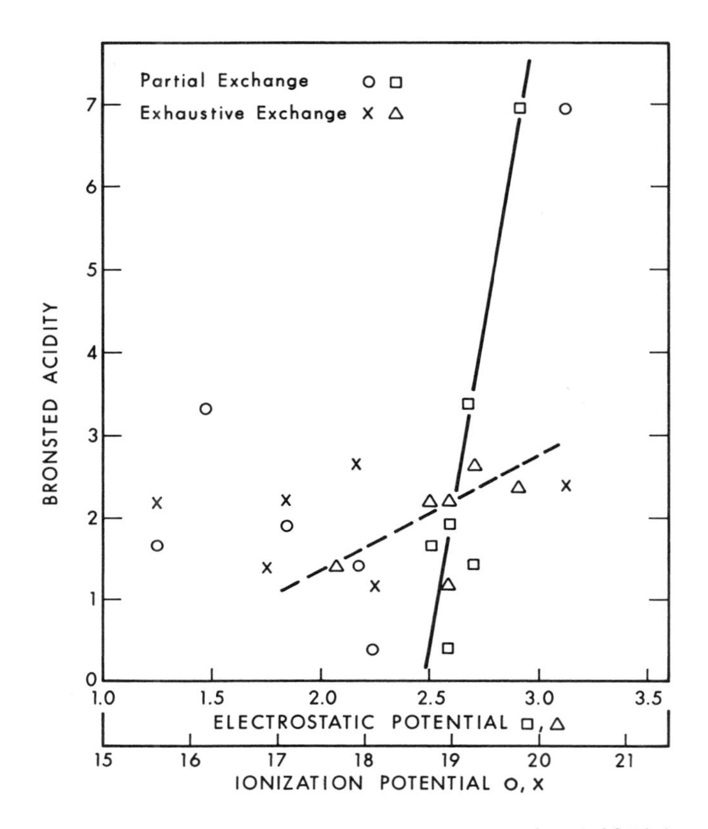

Journal of Catalysis

Figure 74. Variation of Bronsted acidity concentration with ionization potential and electrostatic potential (188)

zeolite was a proton acid and zinc-Y pretreated at 460°C contained a small amount of proton acidity. Addition of water increased the concentration of proton acidity substantially. Mn-, Co-, Ni-, Cu-, Zn-, Ag-, and Cd-Y zeolites 90–99% exchanged and Cr-, Mn-, Fe-, Co-, Ni-, Cu-, and Zn-Y zeolites 74–80% exchanged have been examined by pyridine adsorption. Bronsted acidity was observed on all samples before and after hydration (*188*). The data are summarized in Table XVIII. The acidity data are related to the cation properties in Figures 74 and 75. A reasonable relationship exists between the acidity and the electrostatic field: the stronger the field, the greater the acidity. However, the relationship is much poorer than that observed for alkaline earth-Y zeolites, suggesting that the simple consideration of cation charge and radius is too approximate. An unusual phenomenon is the observation that some of the partially exchanged zeolites have more protonic acidity than the highly exchanged forms. With the exception of copper and chromium, all of the transition cation forms had fewer acid sites than magnesium- and calcium-Y zeolite. However, all of

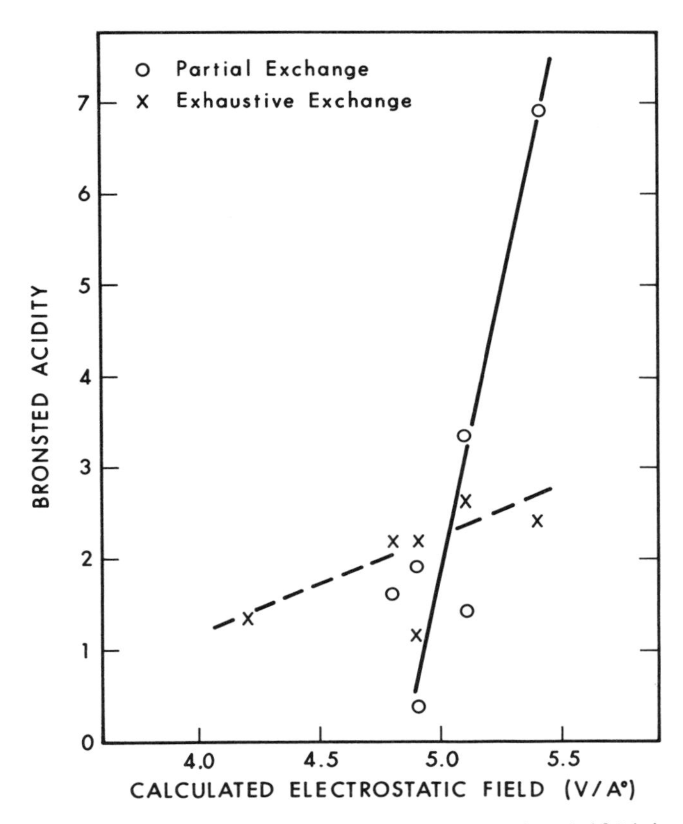

Journal of Catalysis

Figure 75. Variation of Bronsted acidity concentration with electrostatic field (188)

the forms had more sites than the corresponding X zeolites which parallels the observations on the alkaline earth forms.

None of the samples had any detectable Lewis acid sites at the conditions of examination. However, an absorption band in the spectrum of chemisorbed pyridine from cation–pyridine interaction was observed between 1440–1450 cm^{-1}, varying in frequency with the cation. There was no simple relationship between the frequency and the cation properties. The frequencies were all comparable to those of the transition metal cation-X zeolites and higher than those observed for alkaline earth zeolites indicating stronger interaction of the pyridine with the cations, possibly by involving *d*-electrons for the transition metal ion zeolites.

As in the case of the X zeolites, the increase in Bronsted acidity on hydration is probably caused by movement of cations or formation of $M^{2+}OH^{-}$ sites. The latter would seem less likely, since the addition of nickel ions to ammonium-Y zeolite decreases the Bronsted acidity after 70% exchange when Ni ions are located in the supercages (*190*). Ben Taarit *et al.* (*49*)

Table XVIII. Bronsted Acidity of Transition Metal Y-Zeolites (*188*)

Cation		*Bronsted Acidity (arbitrary units)* Anhydrous	Hydrated
Mn	exhaustive exchange	2.2	6.3
Co		2.1	8.9
Ni		1.1	4.5
Cu		2.3	4.7
Zn		2.6	5.8
Cd		1.3	5.7
Cr	partial exchange	5.0	5.6
Mn		1.6	7.4
Fe		3.3	7.0
Co		1.9	8.5
Ni		0.3	9.5
Cu		6.9	6.5
Zr		1.4	5.2

Journal of Catalysis

recently studied Cu-Y zeolite (70% exchanged) after activation at 200, 300, and 500°C. They detected Bronsted acidity by chemisorption of pyridine. A band at 1451 cm^{-1} was determined to be not caused by Lewis acidity but was attributed to pyridine–copper ion interaction. The Bronsted acidity was of two types, one corresponding to hydroxyl groups yielding protons readily and a second only yielding protons at higher temperatures.

The acidic properties of iron-containing X and Y zeolites have been studied (*59*). Two types of materials were studied. Fe-substituted zeolites were prepared by ion exchange and then pretreated in vacuum at 350°C. Fe_2O_3-containing zeolites were prepared by ion exchange, reduction in hydrogen, followed by oxidation. Pyridine adsorption studies were made by adsorbing pyridine and then removing excess by evacuation at 180°C. A strong band indicative of Bronsted acidity was observed at 1540 cm^{-1}. The intensity of the band, and therefore the numbers of acid sites, increased with the degree of exchange. Both types of zeolite had about the same number of acid sites but only about two-thirds the number contained by a similarly treated decationized Y zeolite. The hydroxyl groups formed near 3640 cm^{-1} were acidic. A band observed in the spectrum near 1443 cm^{-1} to 1445 cm^{-1} was attributed to pyridine interacting with the iron. The intensity of this band was greater for the X zeolites than for the Y zeolites and decreased with increasing extent of exchange.

CERIUM, LANTHANUM AND MIXED RARE EARTH CATION ZEOLITES. Rare earth containing X and Y zeolites have been studied extensively. Of the two characteristic hydroxyl bands, only the 3640 cm^{-1} band reacts with pyridine and piperidine. The 3520 cm^{-1} hydroxyl band, characteristic of the rare earth exchanged zeolites, did not interact. Figure 76 shows the hydroxyl

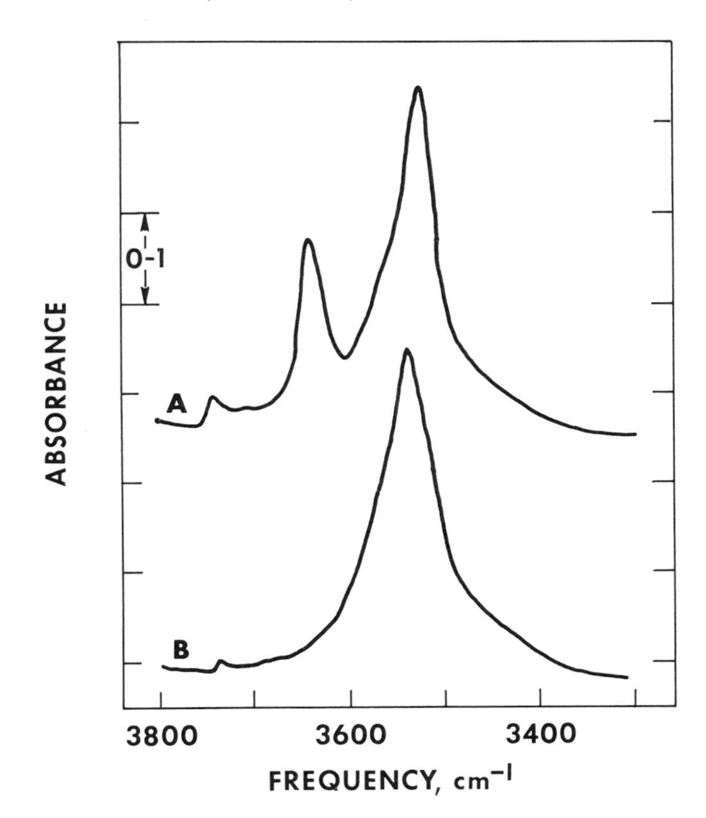

Journal of Catalysis

Figure 76. Hydroxyl groups on rare earth Y zeolite (A) after calcination at 480°C; (B) after pyridine chemisorption and evacuation for 2 hr at 250°C (39)

groups before and after interaction with pyridine—Bronsted acidity was detected in all cases. Christner *et al.* (*36*) detected both Bronsted and Lewis acidity on Ce-Y pretreated at 460°C when the pyridine adsorption was monitored after evacuation at 250°C, but only Lewis acidity could be seen after evacuation at 460°C. Similar results were obtained for the mixed rare earth Y zeolite by Ward (*39*) after pretreatment at 480°C, but at higher pretreatment temperatures, (*e.g.*, 680°C) the concentration of Lewis acid sites increased at the expense of the Bronsted acid sites. Readdition of water restores the Bronsted acid sites. Ben Taarit *et al.* (*191*) obtained analogous results for Ce- and La-Y zeolites. Bronsted acidity on cerium-exchanged Y zeolite has also been confirmed by ammonia and pyridine adsorption (*187*). In all cases, an absorption band is observed near 1440–1443 cm^{-1} from interaction of pyridine with the rare earth cations.

Lanthanum-, cerium-, and rare earth-X zeolites all show Bronsted and Lewis acidity after pretreatment at 450°C (*29, 114, 174*). All of the rare earth containing zeolites give rise to an absorption band near 1443 cm^{-1} on adsorption of pyridine. This band is almost certainly caused by interaction of the pyridine with the cation. Yashima *et al.* (*187*) have also shown that cerium mordenite contains both Bronsted and Lewis acid sites.

The rare earth-containing X and Y zeolites are much more acidic than the alkaline earth zeolites. Comparative arbitrary values are summarized in Table XIX for X and Y zeolites.

Table XIX. Acidity of Cation Exchanged Zeolites

	Bronsted Acidity (Arbitrary Units)	
Cation	*X*	*Y*
Na	0	0
Mg	2.9	6.9
RE	10.0	11.3
H	0.4	15.8

These results can readily be understood in terms of the models discussed for the generation of structural hydroxyl groups. The acidity of the decationized X samples is low because extensive structural collapse occurred.

The formation of acidity on rare earth zeolites has been supported by the different approach of Moscou *et al.* (*189*). No acidity was detected when the zeolite moisture content was greater than 14%. On calcination at temperature in the range of 200°–300°C, Bronsted acidity develops corresponding to one acid site per rare earth atom in agreement with the reaction (*56, 192*).

$$RE^{3+}(H_2O) \rightarrow [RE(OH)]^{2+} + H^+$$

On treatment at 300°–400°C, more hydroxyl groups are formed with corresponding more acid sites. Because of simultaneous dehydroxylation, the resultant increase in acidity and hydroxyl groups corresponds to about 1–½ hydroxyl groups per rare earth atom. A possible mechanism is discussed above. There is also the possibility that hydroxyl groups are introduced during exchange, as suggested by Bolton (*53*).

The interaction of pyridine with the 3640 cm^{-1} but not with the 3520 cm^{-1} band indicates that only the structural silanol groups and not the RE $(OH)_x$ groups are contributing to the acidity. The RE $(OH)_x$ groups must either be non-acidic or inaccessible to pyridine. Piperidine also did not interact with the 3520 cm^{-1} band. This tends to support the inaccessibility of the groups. Eberly *et al.* (*55*) have studied 11 rare earth cation zeolites individually which had been precalcined at 427°C. The acidity was measured by adsorption of pyridine at 2 mm pressure and 260°C and also after evacuation for 30 min. These two measurements give rise to reversible and irreversible acidity. The results are summarized in Figure 77. About 30–40% of the acidity is reversible and hence must represent weaker sites. If the hydroxyl groups are inaccessible, a study of their interaction with

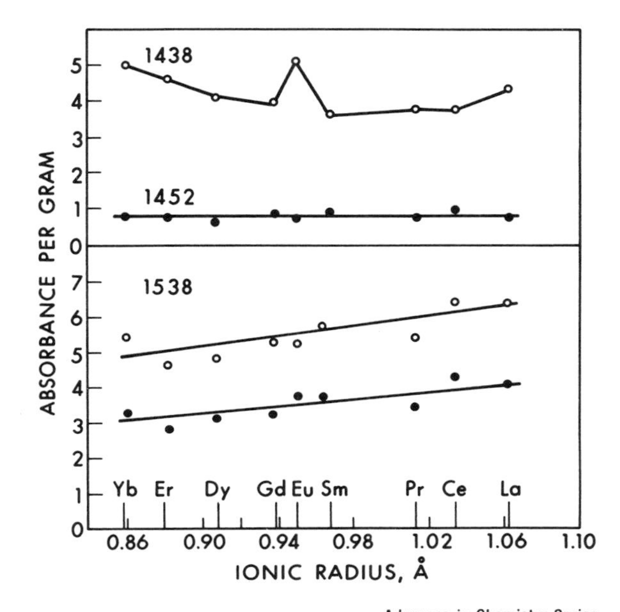

Advances in Chemistry Series

Figure 77. Absorbance of pyridine bands on rare earth forms of Y zeolite at 260°C: ○ *2 torr pyridine;* ● *evacuated 30 min after pyridine adsorption* (55)

ammonia would show whether the hydroxyl groups are acidic or not. This study apparently has not been made. The Bronsted acidity increases as the cation radius increases, in contrast to observations with the alkaline earth-zeolites. A constant amount of Lewis acidity was found on all of the samples. As would be expected, the Bronsted acidity decreased, and the Lewis acidity increased, with increasing calcination temperatures.

HYDROGEN, DECATIONIZED AND DEHYDROXYLATED ZEOLITES. Because the amorphous silica–alumina analogues of zeolites contain hydrogen and not cations, the hydrogen and decationized forms of zeolites and their various modifications have been extensively studied. Decationized Y in particular has been the subject of many studies probably because of its relative stability and ease of preparation. Early studies of Uytterhoeven *et al.* (*60*) showed that readdition of ammonia to decationized X and Y zeolites, deamminated by heating to less than 300°C, eliminated the 3660 and 3550 cm^{-1} hydroxyl bands and reformed NH_4^+ ions, indicating that both types of hydroxyl groups behave as Bronsted acids and are accessible to molecules of the size and basic strength of ammonia. No absorption bands from NH_3 adsorbed on Lewis acid sites were detected, suggesting the absence of Lewis sites. Heating of the zeolites to higher temperature, *e.g.*, 600°C, resulted in a decrease in the concentration of Bronsted acid sites and the detection of ammonia adsorbed on Lewis acid sites. Readdition of water converted some of the Lewis acid sites into Bronsted acid sites. In contrast, Geodakyan *et al.*

(*185*) detected both types of acid sites on decationized zeolite after pretreatment at 400°C by ammonia adsorption. The discrepancy between the results may be caused by the temperature chosen for pretreatment being close to the temperature at which dehydroxylation commences, differences in locations of samples relative to the measurement point, and differences in extent of ion-exchange. A detailed study of ammonium-Y was carried out by Hughes and White using pyridine and piperidine as basic probes for acid centers (*68*). They studied the acidity as a function of calcination temperature over the range of 300°–700°C. Under the conditions used (150°C and 2×10^{-6} torr) the 3650 cm^{-1} and the 3550 cm^{-1} hydroxyl groups both interacted with piperidine but only the 3650 cm^{-1} type of hydroxyl groups interacted with the weaker base, pyridine. The intensity of the 3650 cm^{-1} band decreased linearly with the amount of pyridine added, while the intensity of the 1540 cm^{-1} band from pyridinium ion increased linearly. The intensity of the 3550 cm^{-1} band remained unchanged. Piperidinium and pyridinium ions were detected over the total temperature range. These results indicate that both hydroxyl groups are acidic to the strong base piperidine but not to pyridine. The 3650 cm^{-1} band appears to represent stronger acid sites. Differences in accessibility are probably not the major factor since piperidine and pyridine are approximately the same size and react with both hydroxyl groups when excess reagent is present (*69*). With increasing calcination temperature, the concentration of Bronsted acid sites decreased and the concentration of Lewis acid sites increased. As indicated by the absorption band at 1451 cm^{-1}, maximum Lewis acid site concentration was reached at 600°C. Above this temperature, both site concentrations decreased, probably because of loss of zeolite structure. Readdition of water to a zeolite calcined at 650°C reconverted Lewis acid sites back into Bronsted acid sites, but did not restore the 3650 cm^{-1} hydroxyl band. This would suggest that the acid sites formed by rehydration may be different from those present originally. Between 300° and 600°C, the sum (Bronsted acid sites) + two (Lewis acid sites) remained constant as would be expected from the structural considerations. Two Bronsted acid sites are eliminated for each Lewis acid site formed. Above 600°C calcination temperature, the total acid site population decreased rapidly probably because of collapse of the zeolite structure.

A similar study to that of Hughes and White was carried out by Ward (*69*). Whereas Hughes and White used the same sample for the various temperatures studied, Ward used a separate sample for each temperature. The zeolite was about 90% exchanged. The temperature dependence of the hydroxyl bands was also studied (*see* above). The procedure used was to calcine the zeolite at the desired temperature, record the hydroxyl spectra, chemisorb pyridine, and observe the spectra of the adsorbed pyridine. By this technique, the behavior of the hydroxyl groups and the acidity could be related. Under the conditions used, in the presence of 2 torr pyridine, both the 3650 and the 3550 cm^{-1} hydroxyl bands interacted with pyridine, but after evacuation at 250°C, only the 3650 cm^{-1} band interacted. This behavior is shown in Figure 78. These results suggest that the hydroxyl groups of the 3550 cm^{-1} band can be accessible to pyridine but are not sufficiently acidic to protonate the molecule under the conditions used. Thus, the 3550 cm^{-1} hydroxyl groups are probably sufficiently acidic to protonate pideridine but not sufficiently acidic to protonate pyridine.

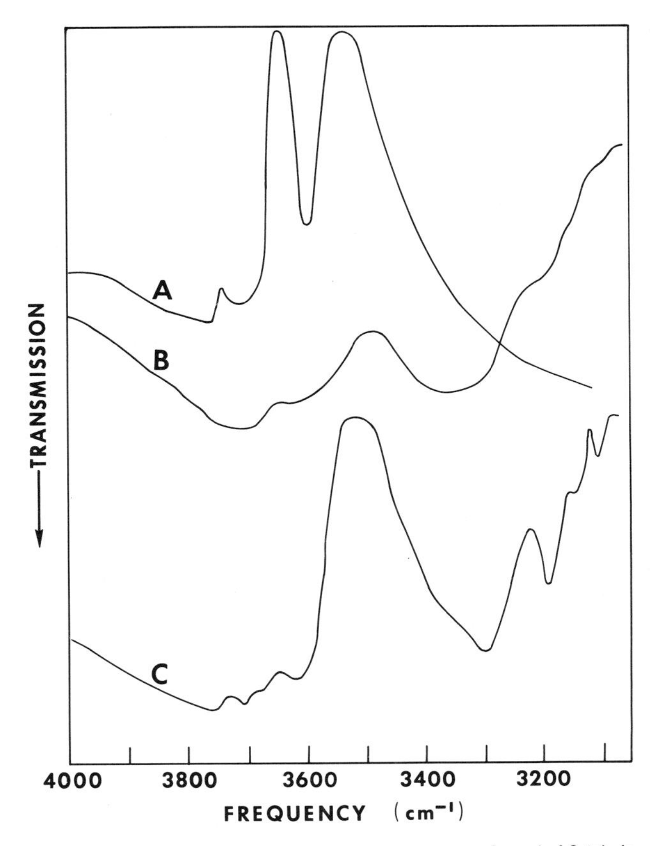

Figure 78. Pyridine adsorbed on decationized Y-zeolite (A) after calcination at 450°C; (B) excess pyridine added at 2 torr; (C) evacuated at 200°C (69)

Removal of chemisorbed piperidine by evacuation at elevated temperatures showed further the difference in acid strengths of the two types of hydroxyl groups. Piperidine interacted very strongly with the 3640 cm^{-1} band and less strongly with the 3540 cm^{-1} band. However, in contrast to pyridine, evacuation at 280°C was required to restore the 3540 cm^{-1} band intensity. At this temperature, the 3650 cm^{-1} band had not been affected again showing the stronger acidity of these hydroxyl groups (*65*). Considering the pK_a values of the bases (NH_3, 9.3; piperidine, 11.2; and pyridine, 5.2), limits can be set on the acidity of the hydroxyl groups.

The concentrations of Bronsted acid sites as detected by pyridine chemisorption (Figure 79) increase up to 350°C, remain constant to 500°C and then decrease rapidly. They agree well with those reported by Hughes and White and follow the thermal dependence of the concentration of 3650 cm^{-1} hydroxyl groups (Figure 11). The rapid decay in hydroxyl group concentration is accompanied by a similar decline in Bronsted acidity. Starting at about 450°C, the Lewis acid site concentration increases slowly to 550°C and then rapidly builds up until at 700°C, there were equal concentrations of both types of acid sites. At 800°C, the zeolite is essentially a pure Lewis acid. Analogous to Hughes and White, the sum of Bronsted acid sites plus 2 × Lewis sites remained constant.

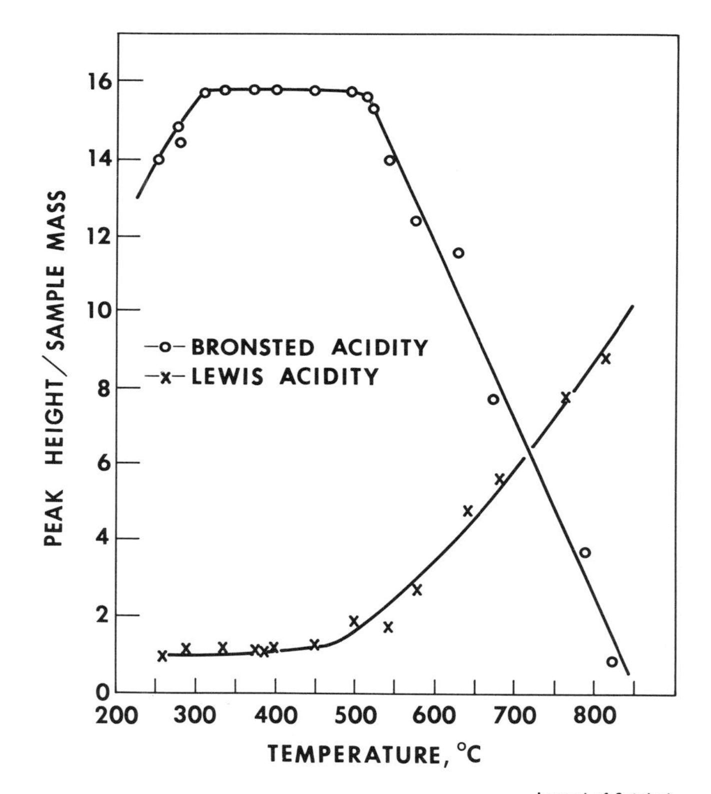

Journal of Catalysis

Figure 79. Acid site populations on decationized Y as a function of calcination temperature (69)

The adsorption of small amounts of pyridine or by desorbing pyridine as a function of increasing temperature indicates that pyridinium ions are formed as the 3650 cm^{-1} hydroxyl groups are eliminated (*68*). It is therefore reasonable to attribute the Bronsted acidity to this particular type of hydroxyl group. The scheme for the transformation of Bronsted acid sites into Lewis acid sites, is, then the same as proposed for zeolite dehydroxylation.

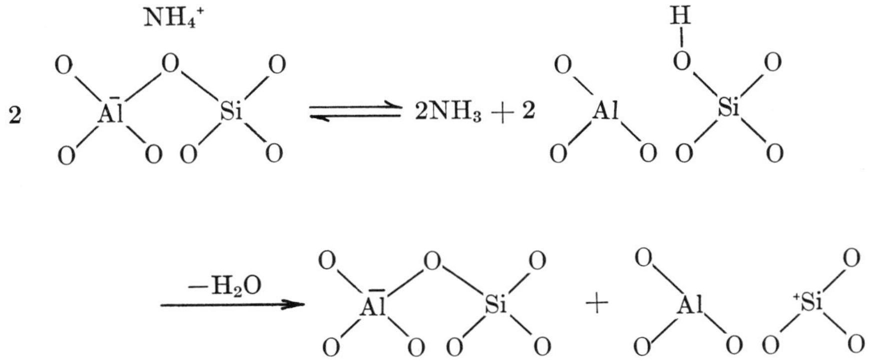

$$2\ \left[\begin{matrix} & NH_4^+ & \\ O & O & O \\ \bar{Al} & & Si \\ O & O\ O & O \end{matrix}\right] \rightleftharpoons 2NH_3 + 2\ \left[\begin{matrix} & H & \\ & | & \\ O & O & O \\ Al & & Si \\ O & O\ O & O \end{matrix}\right]$$

$$\xrightarrow{-H_2O} \left[\begin{matrix} O & O & O \\ \bar{Al} & & Si \\ O & O\ O & O \end{matrix}\right] + \left[\begin{matrix} O & & O \\ Al & & {}^+Si \\ O & O\ O & O \end{matrix}\right]$$

In this scheme, acidic hydroxyl groups are removed by the high temperature thermal treatment and tricoordinated aluminum and silicon sites are exposed. Although the frequency of the Lewis acid band (1451 cm^{-1}) is the same as found on silica–alumina and the Lewis acidity is converted into Bronsted acidity on readdition of water, the mechanism, for interconversion of acid sites must be different from that suggested by Trambouze (*173*) for silica–alumina. In that mechanism, one Bronsted site is converted into one Lewis site. The study of Hughes and White (*68*) showed that although some Bronsted acidity can be restored by rehydration, the scheme shown above is not totally reversible.

The concentration of acid sites observed on the decationized Y-zeolite is much greater than that observed on the most acidic cation-containing zeolite reported so far. More of the exchange positions in the supercages are therefore occupied by hydrogen than in cation-containing zeolites. Liengme and Hall (*65*) reported similar results for decationized Y (about 50% exchanged) calcined at 480° and 600°C. They, however, extended their measurements to include interactions with the weaker bases, ethylene and propylene. These interactions will be covered in more detail later. The olefins interacted weakly with the hydroxyl groups of the zeolite. Ethylene was reversibly adsorbed. However, with the more basic propylene, the interaction was sufficiently strong that it was not completely reversible and proton transfer occurred.

Instructive experiments were carried out by Eberly (*34*) in which ammonium-Y was calcined at 427°C, pyridine was adsorbed, and spectral measurements were made at 150°–260°C. With 0.1 torr pyridine pressure under these conditions, the 3650 cm^{-1} band, but not the 3550 cm^{-1} band, was eliminated, and most of the detected acidity was of the Bronsted type. On heating from 150°C to 260°C, the intensity of the 1545 cm^{-1} band decreased, indicating desorption of pyridine and a partial restoration of the 3650 cm^{-1} hydroxyl band occurred. These data suggest that the 3650 cm^{-1} band represents different types and strengths of hydroxyl groups. Carefully controlled desorption of pyridine or other bases from zeolites in conjunction with infrared spectroscopy should yield detailed information on the acid site strength distribution. Studies of this type on ammonium-Y during initial deammination have not been very illuminating possibly because the ammonia is too strong a base. However, comparison of the rate of removal of the 1545 cm^{-1} band (adsorption on Bronsted acid sites) and the 1451 cm^{-1} band (adsorption on Lewis acid sites) shows the pyridine is removed more readily from the former. At 500°C evacuation temperature, the 1545 cm^{-1} band has almost completely disappeared, whereas the 1451 cm^{-1} band still shows an appreciable absorption. Hence, the Lewis acid sites are the stronger of the two types of sites. As the temperature is raised, causing a decrease of the 1545 cm^{-1} band, the 3640 cm^{-1} band is gradually restored, showing a direct relationship between the Bronsted acidity and the 3640 cm^{-1} hydroxyl groups. Eberly (*64*) also studied olefin adsorption at high temperature. These observations are discussed later.

Pyridine adsorption on decationized Y has also been investigated by Kiselev *et al.* (*186*), Zdhanov *et al.* (*62*), and Ignat'eva *et al.* (*184*). After treatment at 500°C Bronsted and Lewis acid sites were detected. Pyridine adsorption has also been used to study the surface of ammonium-X after deammination. Because of instability of the zeolite, the results are probably not really representative of the X zeolite but represent a mixed zeolite–amorphous substance. Hattori and Shiba (*29*) studied NH_4 (82%)-exchanged Na-X after pretreatment at 270°–300°C. Both Bronsted and Lewis acidity were detected. Evacuation at 450°C collapsed the structure. Ward (*114*) studied NH_4 (97%)-exchanged Na-X after calcination at 480°C. Extensive structural collapse had occurred. Bronsted and Lewis acid sites were detected. The concentration of acid sites was low compared to most zeolites and similar in magnitude to those found on silica–alumina. Similar results were obtained by Zdhanov *et al.* (*62*).

Emission spectra have been reported for pyridine adsorbed on decationized Y zeolite at 400°C and 100 torr pressure (*19*). In contrast to the results of Eberly obtained at 260°C and about 10^{-5} torr (*64*), all three absorption bands interact with the pyridine. However, it has been shown that all the bands interact with pyridine at room temperature and 15 torr pressure (*69*). Hence it is unclear whether the hydroxyl bands interact because of the pressure of the pyridine or because of some new phenomena occurring at high temperatures.

Quinoline has also been used to determine catalyst acidity (*194*). Absorption bands are observed at 1640 and 1515 cm^{-1} for quinoline adsorbed on ammonium-Y deamminated at 400°C and on alumina, suggesting that these bands are characteristic of quinoline on Lewis acid sites. On calcium-Y, bands are observed at 1425 and 1480 cm^{-1}. These bands are probably caused by quinoline interacting with the cation.

The nature of the acidity on the decationized zeolites is not totally accounted for by the hydroxyl groups. A study of the adsorption of pyridine on a sieve calcined at 800°C to eliminate hydroxyl groups shows that, after heating the absorbed pyridine to 200°C, pyridinium ions are formed, indicating the presence of Bronsted acidity (*69*). It is not clear whether these sites are formed by rearrangements in the zeolite structure, migration of trace amounts of water in the system on heating, or by decomposition of adsorbed pyridine. Ammonium-L zeolite, after pretreatment at 450°C possesses both Bronsted and Lewis acid sites (*46*). Treatment of both X and L zeolites with water converted Lewis acid sites into Bronsted acid sites.

ULTRASTABLE Y ZEOLITE. The acidity of ultrastable type zeolites has been investigated by pyridine adsorption. Figure 80 shows spectra of the

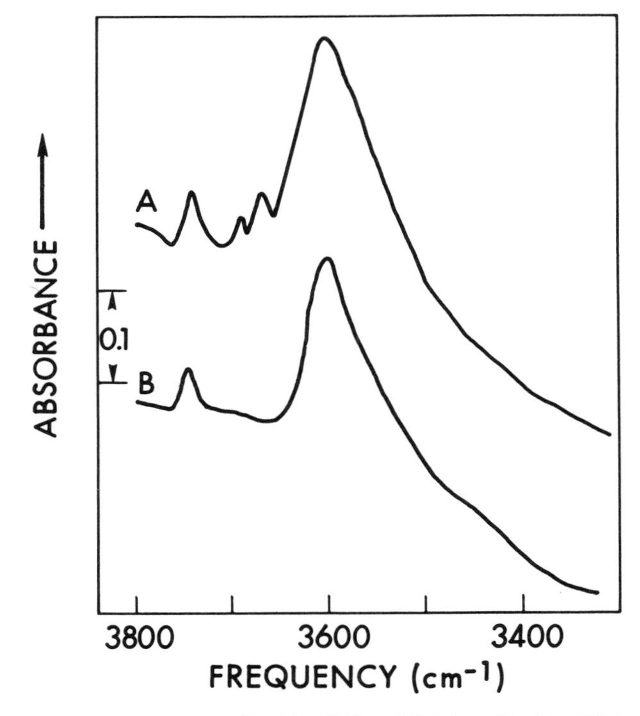

Preprints, Division of Petroleum Chemistry, 161st Meeting of the American Chemical Society

Figure 80. Stabilized Y zeolite (A) after dehydration at 460°C; (B) adsorption of pyridine at 2 torr followed by evacuation at 260°C (181)

zeolites in the hydroxyl region before and after addition of pyridine (*181*). The zeolite was prepared by the procedure A of McDaniel and Maher (*84*) and was evacuated at 460°C before adsorption of pyridine. The major changes in the spectrum are the elimination of the absorption bands near 3650 and 3680 cm^{-1}, suggesting that these are the acidic hydroxyl groups. Bronsted acidity was detected along with Lewis acidity. Compared to decationized-Y of the same sodium content (0.1% Na), the stabilized zeolite contained about half as many Bronsted acid sites. In comparison to the population of hydroxyl groups with frequencies near 3650 cm^{-1}, the acid site concentration is substantially greater than expected. Hence, some other type of surface species must contribute to the acidity. Scherzer and Bass (*40*) similarly detected Bronsted and Lewis acidity on ultrastable zeolites and also showed that the acidity is not significantly redistributed by readdition of water. In a study of various stabilized zeolites, Jacobs and Uytterhoeven (*89*) investigated the acidity by adsorption of ammonia and pyridine. For a sample containing ~3 % Na calcined in a thick bed at 550°C, the bands at 3665 and 3560 cm^{-1} interacted with ammonia, but those at 3750, 3675, and 3600 cm^{-1} did not. Pyridine did not interact with the bands at 3675 and 3600 cm^{-1}. Proton acidity was detected on this sample and its analog after ammonium ion-exchange. An aluminum deficient sample also had Bronsted acid sites. Although the presence of Lewis acid sites would be predicted, none are mentioned. In contrast to this study, Peri found that the band near 3700 cm^{-1} interacted with ammonia, but the bands at 3750 and 3620 cm^{-1} did not (*42*). Bronsted and Lewis acidity were detected. Beaumont *et al.* (*90*) investigated the acidity of Y zeolites which had been stabilized by treatment with EDTA. Samples with three different levels of aluminum extraction were studied. Lewis and Bronsted acid sites were detected after the sample had been pretreated at 450°C. The ratio of the numbers of sites was a function of the temperature of pyridine removal—the higher the temperature, the greater the proportion of Lewis acid sites.

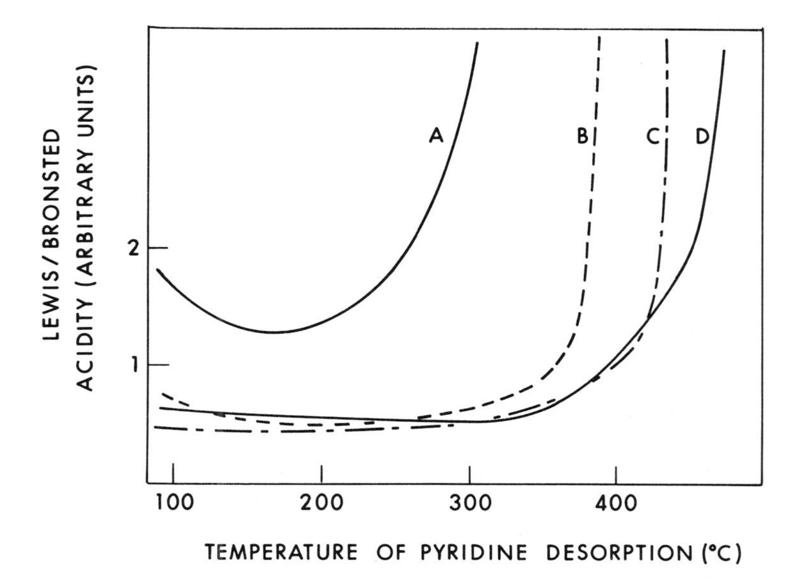

"Catalysis"

Figure 81. Ratio of Lewis to Bronsted acidity as a function of the acid strength represented by the temperature of pyridine desorption for dealuminized H-Y (A) 56 Al/unit cell; (B) 37.5 Al/unit cell; (C) 26.5 Al/unit cell; (D) 17.4 Al/unit cell (90)

At pyridine desorption temperatures below 300°, the extracted samples have a much decreased concentration of Lewis acid sites. As shown by Figure 81, the Bronsted acid site strength of the extracted samples is more constant than that of the original material. Consequently, it can be seen from Figure 81 that the number of strong acid sites has increased substantially. The number of the proton acidic sites is little influenced by the removal of aluminum but their strength is increased. However, a study of pyridine desorption as a function of temperature shows that the strength of the Bronsted acid sites increases. It is apparent that the strength of both types of acid sites increases with increasing dealumination.

The acidity of ammonium-Y extracted to different silica to alumina ratios with EDTA was investigated after calcination at 450°C (*195*). Three materials were used with SiO_2/Al_2O_3 ratios of 4.80, 7.38 and 10.29. Pyridine remaining chemisorbed on the zeolites was used to measure the acidity. The results in Table XX show that the Bronsted acidity went through a maximum despite the removal of alumina.

Table XX. Bronsted and Lewis Acid Site Concentrations (*195*)

Sample	*Bronsted Acidity*[a] *(mmole/g)*	*B/L*
$H\text{-}Y_{4.80}$	0.44	5.8
$H\text{-}Y_{7.38}$	0.66	0.9
$H\text{-}Y_{10.29}$	0.18	10.2

[a] Apparent integrated molar adsorption intensity taken as 2.5 cm/μ mole.

Bulletin of the Chemical Society of Japan

The effect of alumina removal on acid strength was studied by ammonia adsorption. Although the number of Bronsted acid sites decreases, the ratio of the number of strong acid sites to total acid sites increases. Probably these strong acid sites are the effective catalytic sites.

Analogous to the results on thermally stabilized Y zeolites (*40, 41, 42, 44*) the bands near 3600 and 3680 cm^{-1} do not appear to interact with pyridine. Since desorption of pyridine is detected before the hydroxyl bands start to reappear, some proton acid sites must be present which are not associated with the detected hydroxyl groups.

CATION-DECATIONIZED ZEOLITES. Because of the uncertainties involved in the studies of decationized Y due to structural collapse at high temperatures, a study has been made of magnesium-stabilized decationized Y zeolite (*57*). Introduction of magnesium of about 40% of the capacity results in a zeolite which showed no loss of structure up to 800°C. At this level of exchange, most, if not all, of the magnesium ions should occupy sites other than those in the supercages. However, since an absorption band near 1446 cm^{-1} was always observed on adsorption of pyridine because of pyridine–cation interactions, some of the cations must have been in the supercages. The dependence of the acid-site concentrations on temperature are shown in Figure 82. The basic results are the same as those obtained

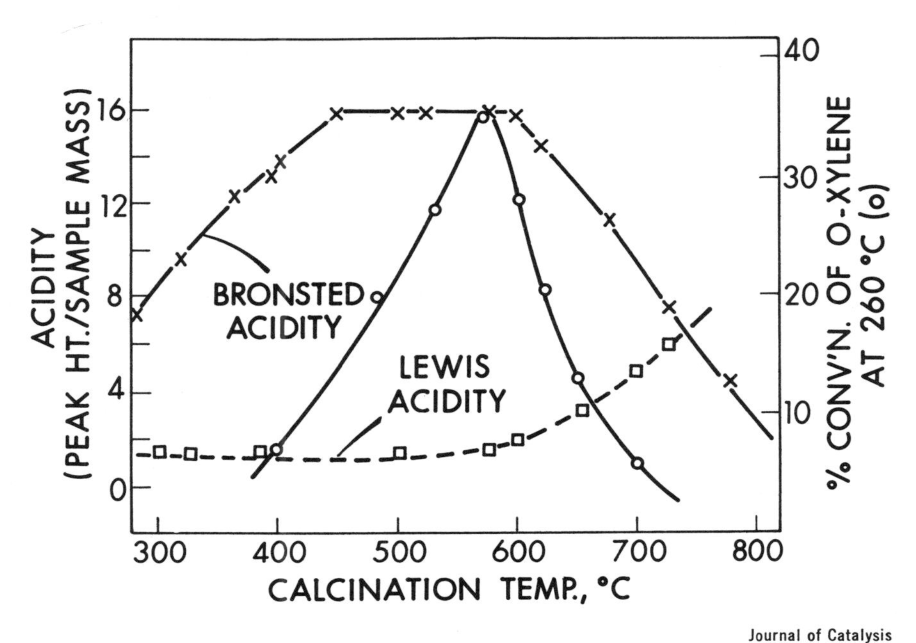

Journal of Catalysis

Figure 82. Acidity and catalytic activity for o-*xylene isomerization of magnesium decationized Y as a function of calcination temperature* (57)

for decationized Y. The concentration of Bronsted acid sites increases as the temperature is increased from 300° to 425°C, remains constant, and then declines sharply as the temperature is raised to 800°C. These temperatures are displaced upwards compared to those observed for decationized Y zeolites analogous to the results found for the hydroxyl groups. The band from Lewis acid sites remains weak until the calcination temperature reaches 550°C and then increases rapidly as the Bronsted acid site concentration decreases. Analogous to decationized Y zeolite, only the 3640 cm^{-1} band interacts with pyridine to yield pyridinium ions. In contrast to deactionized Y, the concentration of Bronsted sites plus 2 $\times$ the concentration of Lewis sites remained constant between 400° and 800°C. This verifies the proposed stoicheometry for the generation of Bronsted and Lewis acid sites when complications from structural disintegration are removed.

Mixed cation–decationized Y zeolites often behave as mixtures of the cation and decationized zeolite, if the cations are present in the supercages. Thus, for example, NaH-Y and MgH-Y behave as two-component systems under some conditions (*60, 65, 77*).

The acidity of sodium- and alkaline-earth decationized Y zeolites has been studied as a function of cation content (*77, 79*). For the calcination conditions used (480°C), the series of samples were all in the Bronsted acid form. In the case of the sodium system, the acidity was studied from 0 to 100% sodium. The concentration of Bronsted acid sites detected by pyridine adsorption increased linearly as the sodium content decreased until only about 16 sodium ions were left in the zeolite (Figure 83). This corresponds to the number of sodium ions which are easily removed by ammonium ion-exchange and subsequently form the 3650 cm^{-1} hydroxyl band. The remaining 16 cations are probably in the hexagonal prisms. At this degree of exchange, most of the hydroxyl groups have an absorption frequency near 3640 cm^{-1}. Thus, they are located in the supercages and interact to form pyridinium ions.

Further exchange resulted in a slight increase in Bronsted acidity, and about a 10% increase in the number of 3640 cm^{-1} hydroxyl groups, although the 3540 cm^{-1} hydroxyl groups increased rapidly in population (*see* above).

The behavior of the alkaline earth-decationized Y zeolites is somewhat similar. In this case, the study was made by first preparing ammonium-Y and then replacing the ammonium ions by magnesium, calcium, strontium, and barium. The acidity at moderate calcination temperatures is essentially

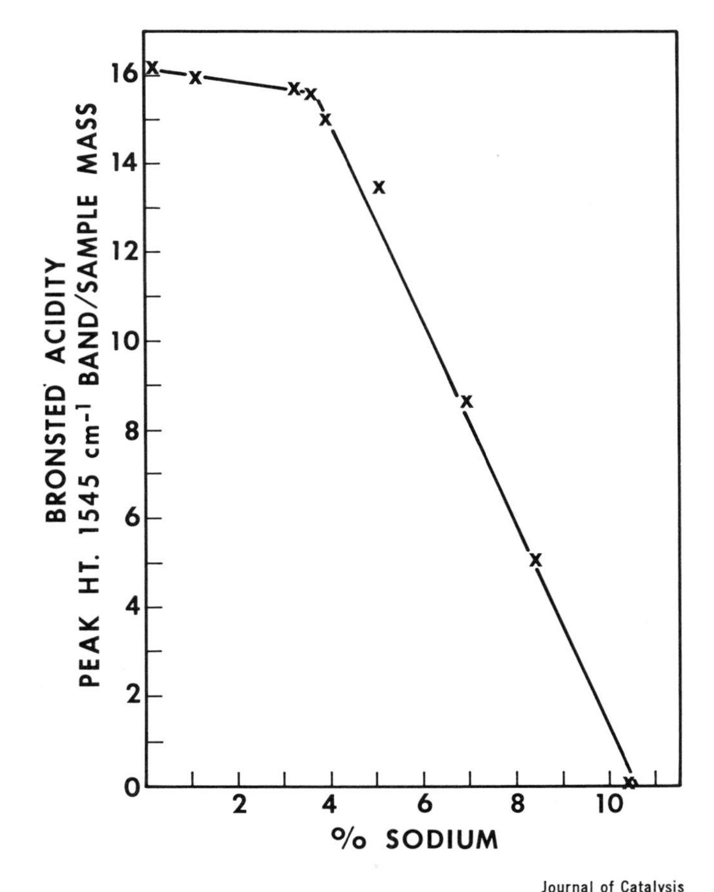

Journal of Catalysis

Figure 83. Bronsted acidity of Na-decationized Y as a function of sodium content (77)

all protonic. After calcination at 480°C, as the percent exchange with divalent cations increases from 0 to 55%, the acidity remains constant and equal to that of the initial ammonium-Y zeolite. As the degree of exchange is further increased, the acidity decreases to about 20–25% of its maximum value. The degree of exchange at which the acidity starts to decrease corresponds to the level at which cations begin to appear in the supercages. That cations are appearing in the supercages above 55% exchange is confirmed by the occurrence of the specific absorption band from cation–pyridine interaction.

Figure 84 shows that the Bronsted acidity and the 3650 cm^{-1} hydroxyl groups vary with cation in a similar manner, again confirming that the

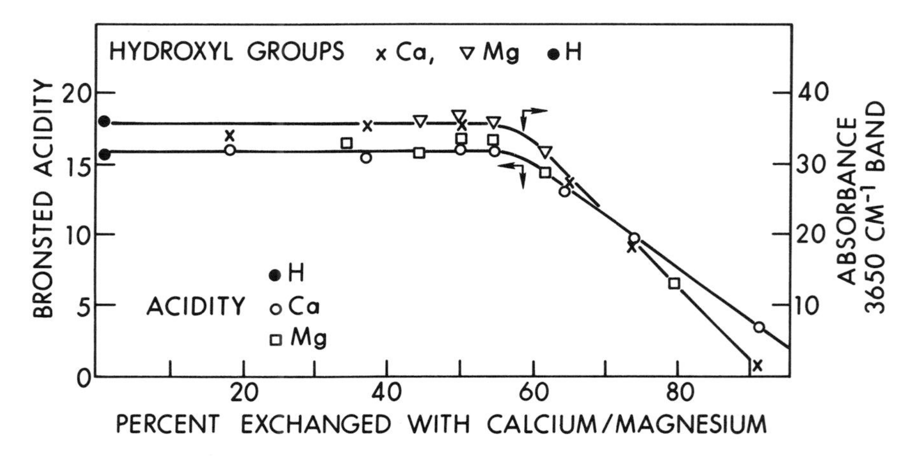

Advances in Chemistry Series

Figure 84. Dependence of Bronsted acidity and the 3650 cm^{-1} hydroxyl band on the cation content of Mg (Ca)-decationized Y zeolite (79)

acidity can be attributed to these particular groups. These data allow interesting conclusions concerning the nature of groups in the zeolite. It is generally accepted that divalent ions reside in the zeolite as a result of the reaction $M^{2+}H_2O \rightarrow M^+OH + H^+$. The question is: are the M^+OH groups acidic? If they are, they should be capable of donating a proton to pyridine. According to the equation, $2\ NH_4Y + M^{2+} + H_2O \rightarrow MOHY + HY + 2\ NH_4^+$, there should be no change in total Bronsted acidity. The results however show that as the number of divalent ions in the supercages increases, the total acidity decreases indicating that the MOH^+ groups are nonacidic or very weakly acidic.

Similar results have been obtained with rare earth cation substituted decationized Y. In this case, however, the acidity remains constant until about 90% exchange which corresponds to the point at which cations appear in the supercages (*196*).

The behavior of nickel decationized Y as a function of the nickel content is somewhat similar (*46*). The proton acidity remains constant to about

70% exchange and then decreases rapidly. In contrast to the alkaline earth cations, the band for nickel interacting with pyridine is observed from about 10% exchange with nickel. The band intensity increases steadily with increasing nickel exchange. These results are in agreement with the studies of the nickel–sodium system, in which the nickel ions are distributed non-specifically in the structure. No study of acidity was made in the recent investigation of AlH-Y (*83*).

OTHER ZEOLITES. The acidity of mordenite has been studied using the spectra of pyridine as a probe (*197*). The mordenite was pretreated between 100 and 600°C after treatment with dilute sulfuric acid to decrease the sodium level to 0.94%. Above 300°C calcination temperature, two bands are observed at 1462 and 1455 cm^{-1} which are probably caused by pyridine adsorbed on two different types of Lewis acid sites. Below 300°C, only the 1455 cm^{-1} band was observed. Cannings (*197*) suggested the following structures:

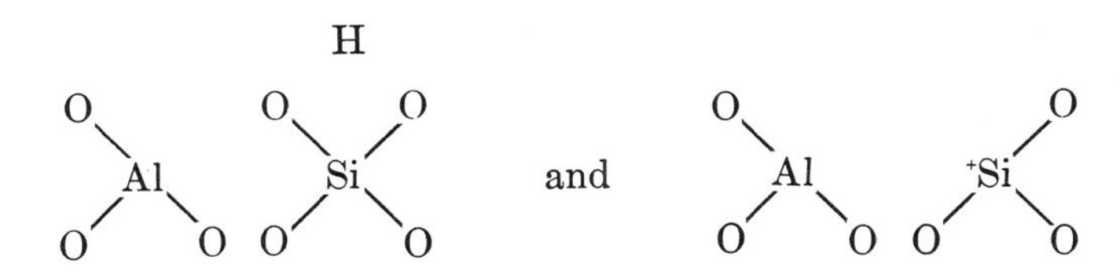

Similar structures would be expected in X and Y zeolites, but generally only one absorption band is observed. Analogous to X and Y zeolites, the Bronsted acidity decreases and the Lewis acidity increases with increasing calcination temperatures until at 600°C, no Bronsted acidity is detected. Readdition of water restores the Bronsted acidity. On calcination above 600°C, the total acidity decreased markedly, suggesting collapse or major modification of the structures.

The acidity of hydrogen mordenite has also been studied as a function of calcination temperature using pyridine adsorption (*103*). Both Lewis and Bronsted acid sites are detected. Figure 85 shows the changes in acidity with calcination temperature from 300° to 700°C. The Bronsted acid concentration declines to zero at about 660°C, while the Lewis acid concentration declines to about 630°C and then remains constant to 750°C. The 3610 cm^{-1} band hydroxyl groups are probably the source of Bronsted acidity. As found by Cannings, the total acidity declines markedly for calcination temperatures above 500°–600°C.

The acidity of H-, Li-, Na-, K-, Cs-, Mg-, Ca-, Ba-, and Zn-mordenite samples have been investigated by pyridine adsorption at 200°C (*198*). The influence of water on the acidity was also studied at 200°C. After pretreatment at 400°C, Lewis and Bronsted acidity were detected in agreement with Cannings (*197*). However, no band was detected near 1462 cm^{-1}. Addition of water converted the Lewis acidity into Bronsted acidity. If the sequence (a) pyridine addition, (b) D_2O addition, (c) evacuation, (d) H_2O addition and evacuation is followed, a band is detected near 1462 cm^{-1}. The origin of this band is uncertain. Progressive increase of calcination temperature resulted in the conversion of Bronsted sites into Lewis sites.

The cation containing mordenites have as a main feature an absorption band near 1445 cm^{-1} and a shoulder near 1455 cm^{-1}. Addition of water eliminates the 1455 cm^{-1}. The various cation forms all have bands at the same frequency but the intensity decreases with increasing cation radius. This is in marked contrast to Y zeolites, for which cation specific absorption bands were observed. Examination of sodium mordenite as a function of calcination temperature showed that the 1445 cm^{-1} band grew weaker and a shoulder appeared at 1455 cm^{-1} as the calcination temperature increased.

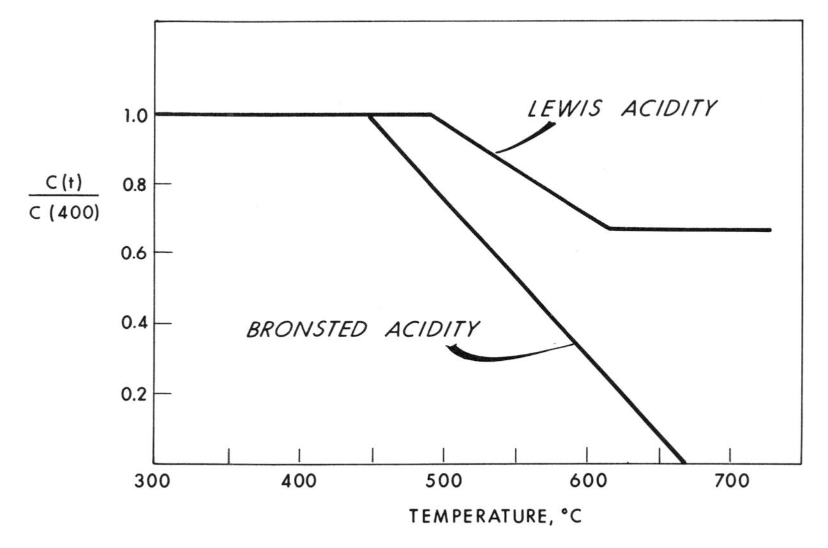

Zeitschrift für Physikalische Chemie, Neue Folge

Figure 85. Acidity of hydrogen mordenite as a function of calcination temperature (103)

Acidity studies have been made on decationized mordenite (*102*). After initial calcination at 400°C, the mordenite has both Lewis and Bronsted acid sites. From studies of the ease of removal of pyridine, the Lewis sites are much stronger than the Bronsted sites and are probably of more uniform strength. Pyridine starts to desorb from Bronsted sites at 200°C and is essentially completely desorbed by 450°C. Little pyridine is desorbed from Lewis acid sites below 450°C. The pyridine then rapidly desorbs.

ERIONITE. Natural erionite, after ammonium ion-exchange and calcination at 600°C, was found to be a Bronsted acid (*76*). The hydroxyl bands at 3612 and 3565 cm^{-1} interacted with ammonia to give ammonium ions, indicative of Bronsted acid sites. Although Lewis acid sites might be expected as a result of the calcination conditions, none were reported.

CLINOPTILOLITE. The acidic properties of ammonium clinoptilolite were examined by Detrekoy *et al.* (*96*) using ammonia and pyridine chemisorption. Up to 400°C, only Bronsted acid sites were detected. The hydroxyls represented by the 3620 cm^{-1} band interact with ammonia. However with pyridine, only about 10% of the hydroxyls interact. Pretreatment at higher temperatures results in dehydroxylation and the formation of Lewis acid sites. Up to 550°C, the sum of the hydroxyl group concentration plus twice the Lewis acid site concentration remains constant. At about 600°C, the Lewis acid concentration starts to decline and lattice destruction occurs. After the pretreatment at higher temperatures, the number of acid sites titrated by pyridine is again only 10% of those titrated by ammonia. Since ammonia is a stronger base than pyridine, pyridine molecules must not be able to enter the channels of clinoptilolite. It appears that pyridine is interacting specifically with hydroxyl groups on the external surface of the zeolite.

ALKYLAMMONIUM ZEOLITES. Alkylammonium-Y zeolites possess acid sites after decomposition at about 350°C (*106, 107*). Proton acid sites were detected by the formation of ammonium ions on adsorption of ammonia. As with conventional decationized Y zeolite, the hydroxyl bands near 3640 and 3550 cm^{-1} interacted with ammonia. Part of the ammonium ions could be decomposed by evacuation at 100°C. These observations suggest different strengths of acid sites. Heat treatment at 650°C resulted in dehydroxylation of the zeolite and the formation of Lewis acid sites.

The adsorption of different amines has been investigated on X and Y zeolites (*107*). The X zeolite was 40% exchanged with ammonium ions, and thus only showed the 3650 cm^{-1} band. The Y zeolite was 70% exchanged and contained both hydroxyl bands. Ethylamine, diethylamine, triethylamine, butylamine, isopropylamine, piperidine, and pyridine were investigated. All of these amines interact by the reaction with the acidic hydroxyls, coordination of the amine group with sodium cations, and interaction with Lewis acid sites. At room temperature, the amines show little selectivity preference for either type of hydroxyl groups. However at higher temperatures, (*e.g.*, 150°C) strong preference for the 3650 cm^{-1} groups is observed. As the amines are degassed, particularly pyridine, the absorption band from amine–cation interaction decreases but the band near 1455 cm^{-1} from amine–aluminum interaction increases. This intensity increase may be caused by dehydroxylation of the zeolite by the amine followed by readsorption on dehydroxylated or Lewis acid sites. It is also possible that the growth of the 1455 cm^{-1} band could be caused by some form of activated adsorption of the base on existing sites. The migration of pyridine from sodium to magnesium cations has previously been reported after initial adsorption has occurred and the sample heated to modest temperatures.

Tetramethylammonium offretite (*110*) behaves analogous to alkylammonium-Y zeolites. After decomposition at 425°C, the three hydroxyl bands at 3690, 3615, and 3550 cm^{-1} interacted with ammonia to give ammonium ions, indicating their proton acidity. Dehydroxylation was essentially complete after treatment at 600°C, and practically all of the acidity was of the Lewis form.

Ammonia adsorption on Na-germanic near-faujasite sieve gives four bands near 1178, 1135, 3370 and 3260 cm^{-1} (*107*). The bands at 1178, 1135, and 3260 cm^{-1} probably represent ammonia interacting with the sodium cations. The band near 3370 cm^{-1} is from physical adsorption. Hence, after dehydration at 300°C, the germanic sieve appears to be non-acidic.

The many studies of the acidity of faujasite-type zeolites can probably be broadly summarized and lead to some generalizations. The Group IA zeolites are non-acidic when pure. Cation deficiency or other impurities can lead to the presence of acidic sites. These zeolites and all others examined are Bronsted and/or Lewis acids depending on the calcination temperature. Calcination below 450°C produces mainly Bronsted acid sites while calcination above 600°C produces mainly Lewis acid sites. Increasing the calcination temperature increases the relative numbers of Lewis acid sites while readdition of water decreases the Lewis acid site population. The Bronsted acid sites on cation- and decationized-zeolites are similar and are believed to be the 3650 and 3550 cm^{-1} hydroxyl groups. The two bands represent acid sites of different strengths and/or in different locations. The Lewis acid sites are also probably of the same type. They are probably tricoordinated aluminum atoms formed by dehydroxylation. The electrostatic field promotes the formation of Bronsted acid sites on the cation zeolites by the reaction $M(OH_2)^{2+} \rightleftharpoons MOH^+ + H^+$. The MOH^+ groups do not appear to be strong acid sites. Acid sites on decationized and hydrogen-zeolites are formed by direct introduction of hydrogen by exchange with acid or by decomposition of ammonium ions.

The exchanged cations in zeolites, if they are in accessible positions, can interact with bases and, hence, function as Lewis acid sites. The strength of these sites is less than that of the conventional Lewis acid sites.

Hydrothermal treatment of the zeolites causes a structural reorganization in which the acidity cannot be accounted for only by the two types of hydroxyl groups. Furthermore, although rehydration of a calcined zeolite restores a substantial amount of the Bronsted acidity, the types of hydroxyl groups are different. The relationships between the acidic properties and catalytic properties are discussed later.

Interaction with Physically Adsorbed Gases

White *et al.* (*66*) studied the interaction of decationized Y zeolite with molecules which would be expected only physically to adsorb substances such as N_2, O_2, CH_4, Ar, and Kr. The adsorption of these gases suppresses the 3650 cm^{-1} band but does not influence the 3550 cm^{-1} band. N_2 and CH_4 shift the band the most, and O_2 and Ar shift it the least. Kr produces an intermediate shift. Simultaneously with the frequency shift, the band intensity increases. The frequency shift can be related to the dielectric constant of the adsorbed molecule by a simple electrostatic model. A further extension of the investigation has resulted in the frequency shifts being related to the heats of adsorption of Ar, Kr, O_2, and CH_4 (*199*).

Use of Adsorbed Molecules to Locate Cations in Zeolite Structures

Several methods have been used to deduce the locations of cations in zeolites. The most satisfactory and definitive technique is x-ray diffraction, although ion-exchange, Mossbauer spectroscopy, and gas adsorption data have also been used. Infrared measurements of adsorbed molecules can give rapid information on cation positions. Although not as definitive as x-ray diffraction, a gross positioning of cations in positions accessible or inaccessible to molecules of various sizes can readily be determined.

The characteristic experiment is to contact the zeolite with a molecule which will interact with the type of center one wishes to locate and then to observe the presence or absence of a characteristic infrared band from adsorbate–absorbent interaction. Typical molecules used include carbon monoxide (*121*), carbon dioxide (*136*), pyridine (*68, 78*), piperidine (*68*), nitrogen oxides (*143*), and propylene (*76*).

Studies made include determination of which sites are accessible to adsorbant or reactant molecules in zeolites of a fixed composition and determination of the composition at which certain sites become accessible instead of inaccessible and which sites are acting as adsorption centers.

Probably the earliest example of the use of physical adsorbents in measuring cation location is the use of carbon monoxide by Angell *et al.* to locate multivalent cations in zeolites (*121*). Using the assumption that cations in supercages of Y-zeolite interact with carbon monoxide to give rise to a cation specific band, they showed that the highly exchanged divalent cation zeolites all have cations in the supercages—*e.g.*, Ca-Y, exchanged to 84% level gave a specific absorption band at 2197 cm^{-1} indicative of cation –CO interaction. On the other hand, Ca-Y, 35% exchanged, gave no band, indicating that no calcium ions were in the supercages. These results are in agreement with x-ray diffraction data which indicate that the calcium ions are initially specifically exchanged into S_I positions. For Co- and Ni-Y, the specific band appeared at low degrees of exchange indicating that these ions did not specifically exchange at any site or move to accessible positions on adsorption of CO. Pyridine adsorption has been used in mixed cation or cation–decationized zeolites to show that under the appropriate conditions, two adsorption sites are present, *e.g.*, in Mg-Na-Y adsorption occurs both on sodium and magnesium cation; in MgH-Y, after calcination above 500°C, adsorption occurs on both magnesium and aluminum (Lewis acid sites (*57, 65*).

The changes in the population of sites can be followed by observing the band intensities as a function of thermal treatment or of cation exchange. Detailed studies of several zeolites have been made to determine the degree of exchange at which cations appear in the supercages. Thus, using pyridine as an indicator, in the magnesium- and calcium-decationized Y systems, the divalent cations appear in the supercages at about 50–55% exchange, *i.e.*, after introduction of 15–17 cations. These results agree well with x-ray diffraction data for the Ca-Na-Y systems in which the first 16–18 calcium ions enter the inaccessible positions. Typical data are shown in Figure 17. In this case, the behavior of the hydroxyl group band intensities confirmed the locations of the cations. Thus, the intensity of the band caused by hydroxyl groups in the supercages started to decrease at about 55% exchange, corresponding to replacement of ammonium ions (hydroxyl group precursors) by divalent cations. Similar results have been obtained by Jacobs *et al.* (*136*) using carbon dioxide for cation-position locating. The spectra of adsorbed CO_2 showed that a band at 2367 cm^{-1} from cation–CO_2 interaction appeared in the spectrum after 37% exchange, in agreement with the gas adsorption and x-ray data for the composition at which divalent cations appeared in the supercages.

Another technique used has been to observe the influence of the degree of exchange of the cations of the intensity of hydroxyl groups in cation–decationized Y zeolites. Since the gross location of the hydroxyl groups is known, the locations of the cations can be deduced. Thus, for cesium-decationized Y, the band attributed to hydroxyl groups in the supercages declines in intensity while the band from hydroxyl groups in the rest of the structure is not affected. Therefore, the cesium ions exchanged selectively into the supercages (Figure 13). On the contrary, for calcium-decationized Y, the band from hydroxyl groups in the rest of the structure declines continuously up to about 50–55% exchange when the latter type of hydroxyl groups are depleted, and the hydroxyl groups in the supercages are eliminated as more cations spill over into the supercages (Figure 17). The sodium–hydrogen system is intermediate between these two extremes.

Zeolites Containing Reducible Metals

Until recently, most studies of zeolites had been concerned with the nonreducible cation and ammonium-decationized forms. Because of the use of zeolites as bifunctional catalysts in which they can contain reducible cations such as platinum, nickel, etc., studies of these systems are now being reported. Very little information was available concerning the change which a reducible metal cation undergoes from its initial introduction by ion-exchange until after its reduction in hydrogen before the recent studies were made.

The reduction of platinum ion-exchanged into Y zeolite has been investigated (*200*). Platinum was introduced into Na-, Mg-, Ca-, and rare earth-Y zeolites by exchange with $Pt(NH_3)_4Cl_2$ solutions to give about 5 % in the zeolite. The reductions were carried out by six different techniques. Infrared results were reported on Pt on Ca-Y samples after direct reduction with hydrogen and also after pretreatment with oxygen and reduction with hydrogen. Initially, the $Pt(NH_3)_4^{2+}$ ion exists as a bipyramidal structure as indicated by the absorption bands near 1350, 3200, 3270, and 3350 cm^{-1}. On initial heating, symmetry is lost because stronger interaction between the zeolite and the complex as water is lost. The 1350 cm^{-1} band splits into components at 1357 and 1377 cm^{-1}. On further heating in hydrogen, decomposition of the platinum complex starts, as shown by the rapid decrease in the intensity of the NH_3 bands and the rapid uptake of hydrogen above about 100°C. The ammonia liberated adsorbs on the zeolite, giving rise to the bands near 1415, 1460, 3340, and 3400 cm^{-1} characteristic of NH_4^+ ions. On heating at 350°C, these bands are eliminated, leaving a strong band near 3640 cm^{-1}. Measurement of the dispersion of this material by the hydrogen titration technique showed it to be about 6–9%. When heated in oxygen, the platinum complex did not decompose until above 200°C. The ammonia released above 250°C is not observed adsorbed on the zeolite. Probably in the presence of oxygen, decomposition occurs by burning of the ammonia. When this product is reduced, good dispersion of the platinum is obtained (80–100%). However, treatment of the oxidized species with water before reduction can cause some agglomeration. Deductions from the number of OH-groups exchanged with deuterium suggested that platinum exists in clusters no larger than six atoms. The initial hydrogen reduction is believed to follow the stoichiometry:

$$Pt(NH_3)_4{}^{2+} + 2\,H_2 \rightarrow Pt(NH_3)_2H_2 + 2\,NH_3 + 2\,H^+$$

$$Pt(NH_3)_2H_2 \rightarrow Pt + 2\,NH_3 + H_2$$

The protons released in the first stage attack the lattice to form structural hydroxyl groups giving rise to the 3640 cm^{-1} band.

The reduction of palladium-Y zeolites by hydrogen has been studied at various temperatures (*201*). The zeolites were prepared by exchange of Na-Y with palladium chloride in ammonia solution and contained 12.5 Pd^{2+} ions, 19.5 Na^+ ions, and 11.5 $NH_4{}^+$ ions.

Spectra of the zeolite are shown after various treatments in Figure 86. Treatment in oxygen at 500°C gives the spectrum (a) with weak hydroxyl bands in the 3700–3500 cm^{-1} region. No hydroxyl bands which could be attributed to $Pd(OH)^+$ were detected. Hydrogen treatment at 25°C results in the formation of characteristic hydroxyl bands at 3640 and 3540 cm^{-1}. These bands became more intense on treatment for 16 hr at 200°C. Carbon monoxide adsorption on the oxidized sample has no effect on the hydroxyl groups absorption bands. Bands are seen near 2135 and 2110 cm^{-1} after excess carbon monoxide has been removed. These bands are probably caused by carbon monoxide adsorbed strongly on Pd^{2+} ions. Apparently, two CO molecules are adsorbed on the same ion.

Carbon monoxide can be used to monitor changes in the oxidation state of the palladium by observing changes in frequency. After treatment with hydrogen, absorption bands are detected near 2100, 1935, and 1895 cm^{-1} in contrast to the 2135 and 2110 cm^{-1} bands present before reduction. These bands are probably caused by interaction with Pd° species. The band at 2100 cm^{-1} is somewhat higher in frequency than those observed for palladium films and palladium on silica. The increase in frequency is ascribed to the influence of Lewis acid sites on the electron density of the palladium atoms. The intensity of the 2100 cm^{-1} band decreases with increasing hydrogen reduction, possibly because the palladium atoms migrate to form agglomerates. If the reduction temperature is above 250°C, the 2100 cm^{-1} band becomes weak and a new band at 2070 cm^{-1} appears. This band is from carbon monoxide adsorbed on well-organized crystallites on the zeolite surface. The frequency is similar to that reported for palladium on non-acidic carriers. Thus, by use of the spectra of chemisorbed carbon monoxide, the progressive reduction of palladium ions in zeolites to crystallite agglomerates can be followed.

The reduction by hydrogen at 200°C is partially reversible by treatment with oxygen. Carbon monoxide adsorption gives rise to spectra typical of those for adsorption on Pd^{2+} ions. Treatment of adsorbed carbon monoxide on reduced Pd-Y zeolite with oxygen results in the production of carbon dioxide. The interaction of nitric oxide with the Pd-Y zeolite used in the above study has been investigated before and after reduction (*202*). After treatment with oxygen at 500°C, the ESR study indicates the presence of Pd^{3+}. On addition of nitric oxide at 23°C, absorption bands are observed at 1775–1875 cm^{-1} and 2025–2175 cm^{-1}. Evacuation at 23°C removes most of the lower frequency band and all of the high frequency bands. The 2025–2175 cm^{-1} bands are caused by interaction of nitric oxide with Pd^{3+} to produce Pd^{2+}.

$$Pd^{3+} + NO \rightleftarrows Pd^{2+} + NO^{+}$$

The change in valence of the Pd on addition or removal of nitric oxide is shown by changes in the ESR signals. The observed infrared bands are caused by interaction of NO^{+} with Pd^{2+} ions. The lower frequency bands are probably caused by nitrosyl complexes formed by interaction of NO with Pd^{2+} ions. The two bands may be from NO molecules interacting with one Pd ion and others interacting with two Pd ions. No evidence for the dissociation of nitric oxide could be observed by ESR or indirectly by observation of NO_2 and N_2O molecules.

Treatment with hydrogen at 23°C reduced the palladium about 80% to Pd atoms as shown by volumetric measurements. The ESR study indicates that the Pd^{3+} present is transformed to Pd^{+}. The bands observed at 1875–1780 cm^{-1} suggest that Pd atoms are oxidized to Pd^{2+} which interact with nitric oxide. The presence of Pd^{2+} ions was confirmed by the adsorption of carbon monoxide which gave rise to a characteristic band near 2140 cm^{-1} caused by CO on oxidized Pd. The reoxidation is also shown by the disappearance of the Pd^{+} ESR signal on addition of nitric oxide. The nitric oxide probably decomposes into N_2O and O.

Reduction at 150°C mainly converts the cations to palladium metal. On adsorption of nitric oxide, it decomposes into nitrous oxide as shown by the band near 2200 cm^{-1}. The free oxygen then converts the palladium metal to Pd^{2+} ions which adsorb nitric oxide and gives rise to the bands near 1800–1865 cm^{-1}. A band observed near 1680 cm^{-1} could be caused by nitrate formed by the reactions.

$$Pd + O \rightarrow PdO$$

$$PdO + NO \rightarrow PdNO_2$$

The nitrite then interacts with the lattice to produce the nitrate.

The addition of a mixture of NO and H_2 to reduced Pd-Y yields mixtures of nitrous oxide, water and ammonia. The bands observed in the spectrum

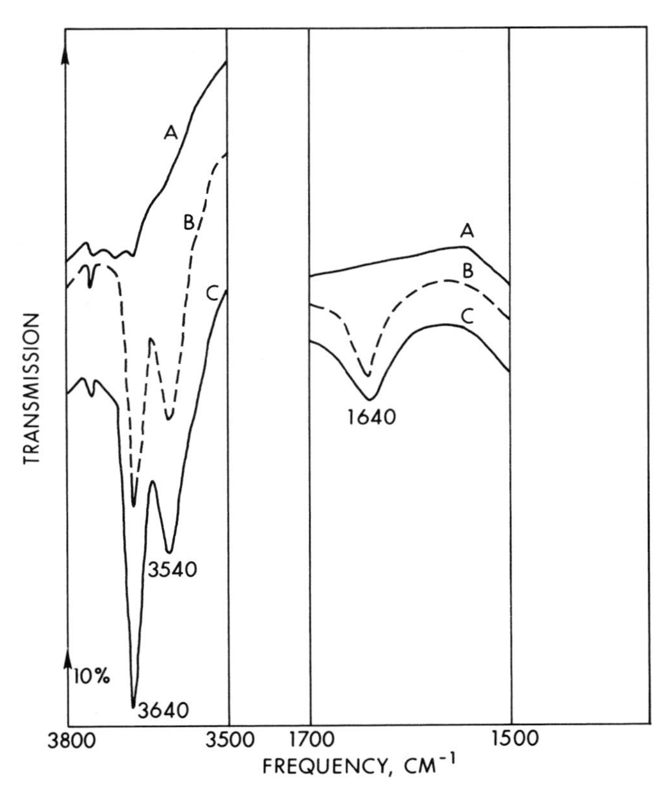

Advances in Chemistry Series

Figure 86. Reduction of Pd-zeolite by H_2 (A) initial sample; (B) 200 torr H_2 for 6 hr at 25°C; (C) 200 torr H_2 for 16 hr at 200°C, then evacuated at room temperature (201)

Table XXI. Positions of Infrared Bands of Adsorbed Carbon Monoxide (*203*)

Support	*Wt % Pd*	*Wavenumber (cm^{-1}) of Bands of Adsorbed CO*
Na-X	2	2035-1830-1815
MgO	2	2065-1965
Al_2O_3	1	2075-1970
Na-Y	1.7	2075
K_{25}[a]	1	2090-1965-1935
K_{13}	1.2	2100-1965
Decationized Y	1.2	2105-1950
Mg-Y	1.84	2100
Ca-Y	1.9	2115-1950-1900
La-Y	1.86	2105-1940-1880

[a] K_{25} and K_{13} are 25- and 13% silica–alumina.

Advances in Chemistry Series

near 1445, 1640, and 2220 cm^{-1} and the broad band near 3000–3400 cm^{-1} are attributed to ammonia, water, and nitrous oxide.

Carbon monoxide interaction with palladium supported on a number of Y zeolites and amorphous oxides containing about 1 wt % palladium has been studied. The samples were treated in oxygen at 500°C and then reduced in hydrogen at 250°C (*203*). Carbon monoxide was adsorbed at 25°C and 50 torr pressure. The absorption bands observed are listed in Table XXI. A band was observed above 2000 cm^{-1} and one or two bands below 2000 cm^{-1} between 2000 and 1800 cm^{-1}. The band above 2000 cm^{-1} is removed on evacuation, while those below 2000 cm^{-1} decrease in intensity and move to lower frequency. The band above 2000 cm^{-1} is probably caused by interaction of carbon monoxide with Pd° in the zeolite. The frequency is influenced by nearby Lewis acid sites or other changes on the support which change the electron density of the metal. The bands below 2000 cm^{-1} are still subject to interpretation. The frequency of the band above 2000 cm^{-1} was related to the catalytic activity.

Carbon monoxide adsorption has been studied on rhodium-Y zeolite. Both after evacuation at 300°C and after reduction in hydrogen at 300°C, absorption bands are observed near 2045 and 2110 cm^{-1} similar to those observed for the interaction of carbon monoxide and rhodium metal supported on alumina. Thus, rhodium ions in zeolites can be readily reduced.

The properties and reduction of copper-Y zeolite have been investigated after pretreatment in oxygen at 500°C (*204*). The observed spectrum had absorption bands near 3745 and 3680, 3640, and 3540 cm^{-1}. Adsorption of pyridine yielded a band at 1540 cm^{-1} characteristic of Bronsted acidity and at 1451 cm^{-1} caused by Lewis acidity or pyridine–cation interaction. As discussed above, the observation can be represented by the scheme

$$2\ \left[\begin{array}{c}Cu^{2+}(OH_2)\\ (O)_2Si(O)_2\bar{Al}(O)_2\end{array}\right] \longrightarrow \left[\begin{array}{c}H\\ (O)_2Si(O)\ \ Al(O)_2\end{array}\right] + \left[\begin{array}{c}CuOH^{+}\\ (O)_2Si(O)_2\bar{Al}(O)_2\end{array}\right]$$

The bands at 3745, 3640, and 3540 cm^{-1} are caused by hydroxyl groups similar to those of decationized Y. The 3640 and 3540 cm^{-1} hydroxyl groups interact with pyridine to give pyridinium ions. The band at 3680 cm^{-1} could be caused by adsorbed water or a stabilized structure.

Reduction of Cu^{2+}-Y with hydrogen produced Cu°-Y. Typical spectra of various copper-Y zeolites are illustrated in Figure 87. Hydroxyl groups are formed at 3745, 3680, 3640, and 3550 cm^{-1} which are much stronger in intensity than for other copper zeolites. The temperature dependence of the hydroxyl groups is shown in Figure 88 in the range of 500° to 700°C. The 3640 and 3550 cm^{-1} bands are much more stable than the comparable bands of decationized Y. At 700°C, the 3640 ad 3550 cm^{-1} bands have been eliminated while the bands at 3745 and 3680 cm^{-1} remain. A band at about

3600 cm^{-1} is also observed. Addition of pyridine to the reduced samples yields no band near 1451 cm^{-1}, but the 1540 cm^{-1} band is more intense than for the unreduced samples. Readdition of water further increased the intensity of the 1540 cm^{-1} band.

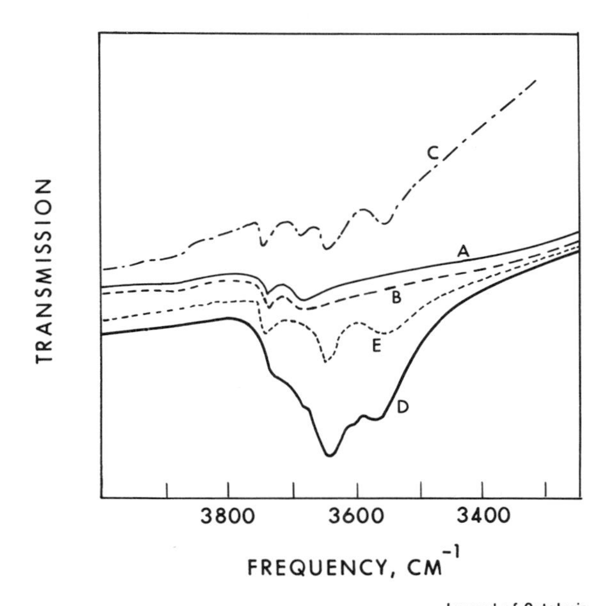

Journal of Catalysis

Figure 87. Cu^{2+}-Y (A) dehydrated at 500°C; (B) reduced by CO at 500°C; (C) reduced by CO at 500°C in the presence of a small amount of water; (D) reduced by H_2 at 500°C; and (E) dehydroxylated NH_4-Y at 495°C (204)

The bands at 3680 and 3600 cm^{-1} are probably caused by hydroxyl groups attached to cations or stabilized Y zeolite structures. The production of water during the reaction would favor the latter. The elimination of the 1451 cm^{-1} band on reduction probably indicates the removal of Cu^{2+} ions from the structure and the formation of Cu°. The increased intensity of the 1540 cm^{-1} band suggests that hydrogen ions formed in the reduction interact with the lattice. Thermal treatment of the reduced samples is similar to that of typical Y zeolite. As the temperature is increased, the hydroxyl band intensity and the Bronsted acidity decrease while the Lewis acidity increased. Readdition of water reconverts the Lewis acidity back to Bronsted acidity analogous to the silica–alumina system.

In contrast to hydrogen, carbon monoxide reduces Cu^{2+}-Y to Cu^{+}-Y. Because the hydroxyl region spectra does not change significantly on treatment with carbon monoxide, the reduction must proceed by a different route. Furthermore, no Bronsted acidity is detected, but a band near 1451 cm^{-1} attributed to Lewis acidity or pyridine–cation interaction is detected. Addi-

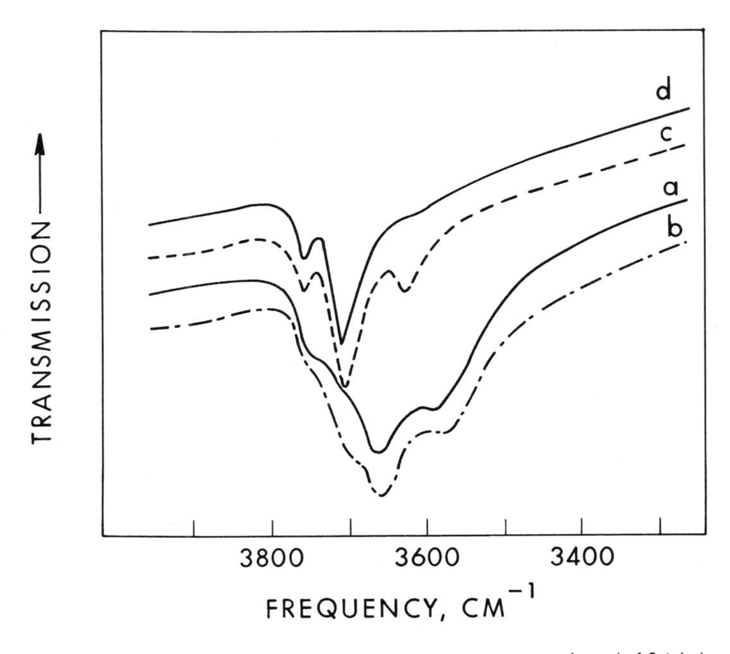

Journal of Catalysis

Figure 88. Hydrogen reduced Cu°-Y after calcination at (a) 500°C; (b) 600°C; (c) 650°C; and (d) 700°C (204)

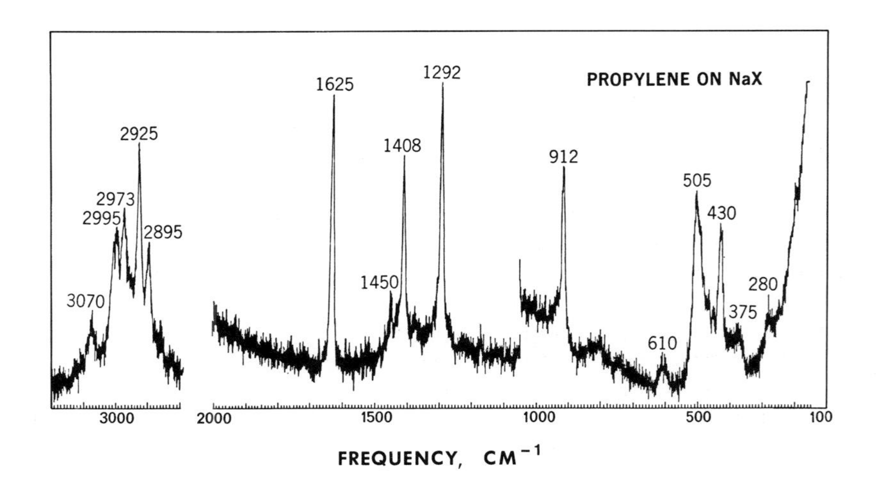

Journal of Physical Chemistry

Figure 89. Raman spectrum of propylene adsorbed on Na-X (206)

tion of water was found to generate Lewis acid sites. The reduction possibly goes *via* the following route:

$$[2\ Cu^{2+}O^{2-}]_2 + CO \rightarrow 2\ Cu_2^{+} + CO_2$$

Thus, the nature of the zeolite formed by reduction depends strongly on the reducing agent used.

Raman Spectra

Raman spectra of pyridine adsorbed on Na-Y have been reported (*205*). After removing hydrocarbons by treatment with oxygen for 16 hr at 500°C, a residual fluorescence remained near 1300 cm^{-1} (*i.e.*, displaced 2000 cm^{-1} from the exciting line). This fluorescence has been attributed to iron impurities. Raman spectra were also observed for adsorbed pyridine. Bands were observed near 3070, 1596, 1218, 1006, 1002, 654, and 613 cm^{-1}. These bands were attributed to pyridine interacting with the sodium cations.

Spectra of acetonitrile, propylene, carbon dioxide, and acrolein have been observed on the sodium forms of A, X, and Y zeolite after pretreatment in vacuum at 250°C (*206*). The spectra could be recorded over the complete range of interest, in contrast to the infrared data which is only available above 1200 cm^{-1}. A typical spectrum for propylene adsorbed on Na-X is shown in Figure 89. The spectra of the adsorbed species are similar to those of the liquid phase, suggesting that only physical adsorption is occurring. Since the spectra are unaffected by treatment under vacuum, the interactions must be relatively strong. The differences in frequencies between the adsorbed phase and the liquid phase may also indicate that some interaction of the multiple bonds with the cations is occurring.

Applications to Catalytic Systems

The application of infrared spectroscopy to the characterization of zeolite surfaces, to the sites at which adsorbates interact with the surface and to changes which occur in adsorbates on interaction with the surface allow the following generalizations to be made.

Several types of sites on zeolite surfaces are now known and fairly well characterized. These are various structural hydroxyl groups, the dehydroxylated sites, and the exchanged cations. The changes in these sites as a function of the exchanged cations and the thermal history have been extensively studied and characterized.

The sites on the surface with which adsorbate molecules interact have also been identified for many different molecules under mainly static conditions and some of the changes in the properties of the adsorbate molecules identified. This information can give increased understanding of the catalytic processes taking place on zeolite catalyst surfaces.

The application of infrared spectroscopy to the catalytic aspects of zeolites has been in two major areas:

(1) The elucidation of the structural groups on zeolites and their properties as they pertain to catalytic centers, and

(2) The observation of catalytic reactions on zeolites while the reactions are actually occurring on the zeolite surface.

Relationship between Structural Features on Zeolites and Catalytic Activity. Since the discovery of catalysis over zeolites (*62, 207, 208, 209, 210*), the nature of the surface sites and the mechanisms pertaining to zeolite catalysis has been extensively studied. The early studies showed that catalytic reaction could occur *via* either an ionic or radical type mechanism. Infrared spectroscopy is not able at this time to shed much light on the radical type reactions but has been able to contribute significantly to ionic type reactions. The main reason for this latter contribution is that most reactions proceeding *via* carbonium ion intermediates are acid catalyzed Other techniques such as magnetic resonance are probably more applicable to radical-catalyzed reactions.

From infrared and catalytic observations of molecular sieves the following generalizations can be made:

Zeolite Cation	*Hydroxyl Groups*	*Acidity*	*Catalytic Activity*
Group IA	no	no	no
Group IIA	yes	yes	yes
Transition cations	yes	yes	yes
Rare earths	yes	yes	yes
Decationized	yes	yes	yes

The relationships between zeolite structure and catalytic activity are considered in detail in Chapter 8 and therefore are not considered further here.

Observations of Zeolite Catalyst Surfaces During Catalytic Reactions. Because of the uncertainties mentioned above, some studies have been made recently under actual catalytic or approximately catalytic conditions. The surface of decationized Y zeolite was studied during cumene cracking on the zeolite (*119*). A cell was used in which the catalyst sample was maintained at elevated temperatures, and a stream of cumene diluted with helium was passed over the catalyst which had been pretreated at 500°C. The catalyst was cooled to 250°C for the initial catalytic measurements. Spectra obtained for the hydroxyl region are shown in Figure 90 at 250°C for different time periods. The 3550 cm^{-1} band is not influenced by the passage of cumene, indicating that at these reaction conditions, the cumene is interacting neither physically nor chemically with this type of hydroxyl group. On the other hand, the intensity of the 3640 cm^{-1} band progressively decreased with time until after the cumene had flowed overnight, the band intensity reached a constant value. When the temperature of the catalyst was raised to 365°C, the 3550 cm^{-1} band intensity decreased 5%. A second experiment in which cumene was passed over the catalyst for 3 hr at a series of progressively increasing temperatures is shown in Figure 91. These spectra again show the interaction of the 3640 cm^{-1} band hydroxyl groups with the cumene. However, as the temperature is raised, the 3550 cm^{-1} hydroxyl groups interact progressively more extensively. The study suggests that below 325°C, no detectable adsorption or interaction with the 3550 cm^{-1} band occurs, and hence, that these sites are probably not active catalytic centers. The interactions with the 3640 cm^{-1} band indicate that this site

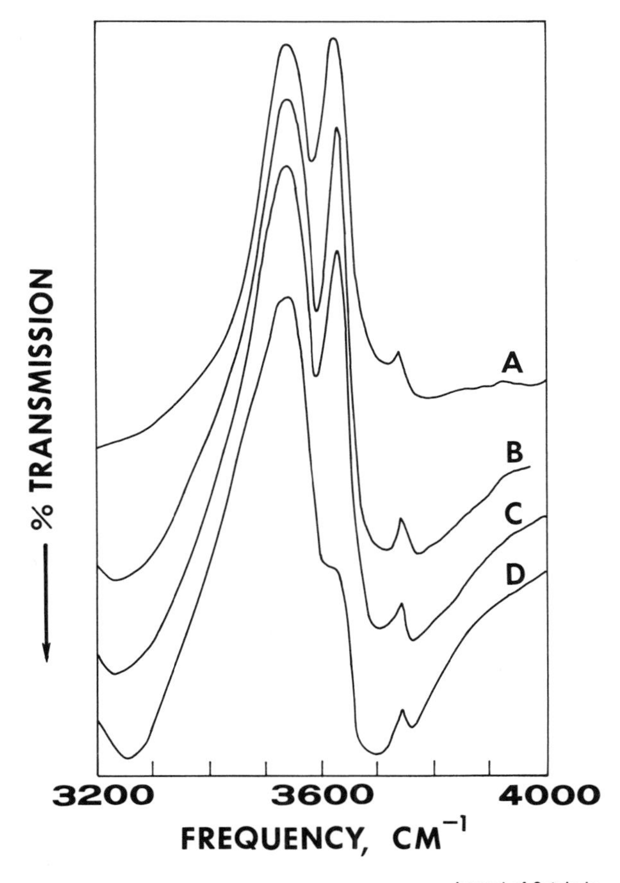

Journal of Catalysis

Figure 90. Decationized Y zeolite during cumene cracking at 250°C (A) after calcination; (B) after 1½ hr cumene flow; (C) after 3 hr cumene flow; (D) after overnight cumene flow (119)

is a possible active center. As the temperature is raised, the interaction of this hydroxyl group with the cumene progressively increases just as the catalytic activity increases. These hydroxyl groups are possibly interacting with the cumene by proton transfer. As the temperature is increased, a larger number of suitably activated hydroxyl groups become available. Similarly as the temperature is raised, the 3550 cm^{-1} band hydroxyl groups start to interact. These interactions are feasible since, as discussed above, protons may become mobile on zeolite surfaces at elevated temperatures.

An extensive study of the cracking of hexane over decationized Y zeolite deamminated at 550°C has been made (*210*). In this technique, the surface is observed in the absence of gaseous reactants and products at frequent intervals during the reaction. Thus it does not monitor the catalyst during true reaction conditions. After calcination of the sample in air at 550°C, the zeolite was treated with nitrogen at 550°C. The spectrum of the zeolite was recorded at room temperature, the sample reheated to 450°C, and hexane in nitrogen was passed through the cell for the desired time. The sample was next cooled to 200°C, and excess hydrocarbons were removed

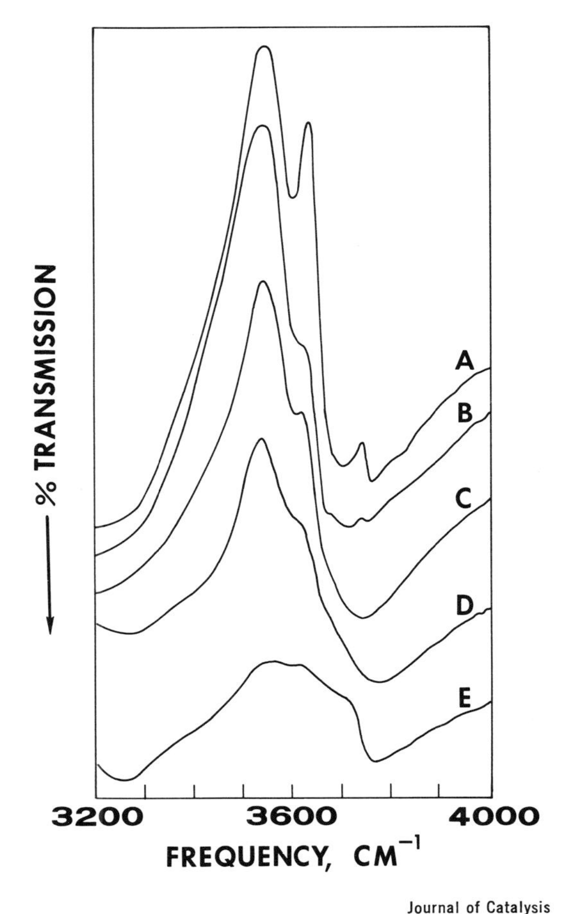

Journal of Catalysis

Figure 91. Decationized Y zeolite during cumene cracking as a function of temperature (3 hr at each temperature) (A) 325°C; (B) 365°C; (C) 400°C; (D) 450°C; (E) overnight at 450°C (119)

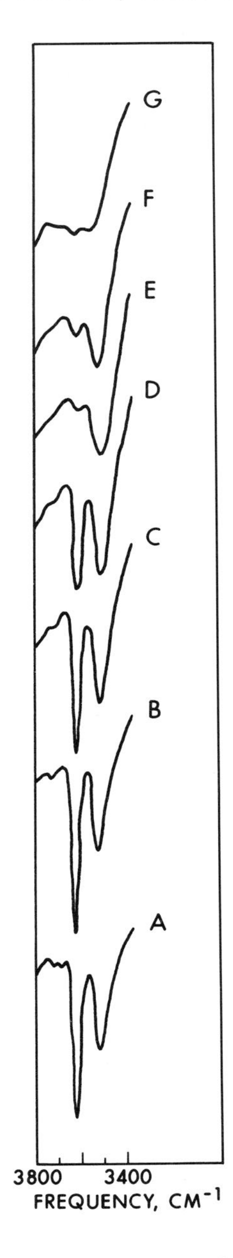

Journal of Catalysis

Figure 92. Decationized Y zeolite during hexane cracking (A) calcined at 550°C; (B) 1 hr at 450°C; (C) 3 hr at 450°C; (D) 4 hr at 450°C; (E) 6 hr at 450°C; (F) 8 hr at 450°C; (G) 14½ hr at 450°C; (219)

by evacuation. The sample was then cooled to room temperature, and the spectrum was recorded. The cell was then reheated to 450°C, and the procedure was repeated for another time interval. Typical spectra observed over 14 hr are shown in Figure 92, and the change in optical density *vs.* time is shown in Figure 93. During the reaction period the gaseous products were also analyzed. Results are shown in Figure 93. The spectral data show that the 3540 cm^{-1} band remains constant in intensity until the 3640 cm^{-1} band has been reduced to below about 20% of its original intensity. At this stage of reaction, the 3540 cm^{-1} band decreased rapidly in intensity. The product distribution changed from mainly C_3 and C_4 saturated hydrocarbons at the initial stages of reaction to C_1 and C_2 hydrocarbons when the intensity of the 3640 cm^{-1} band had been reduced by 80%. Simultaneously, the olefin to paraffin ratio changes from 2–3 to 5.

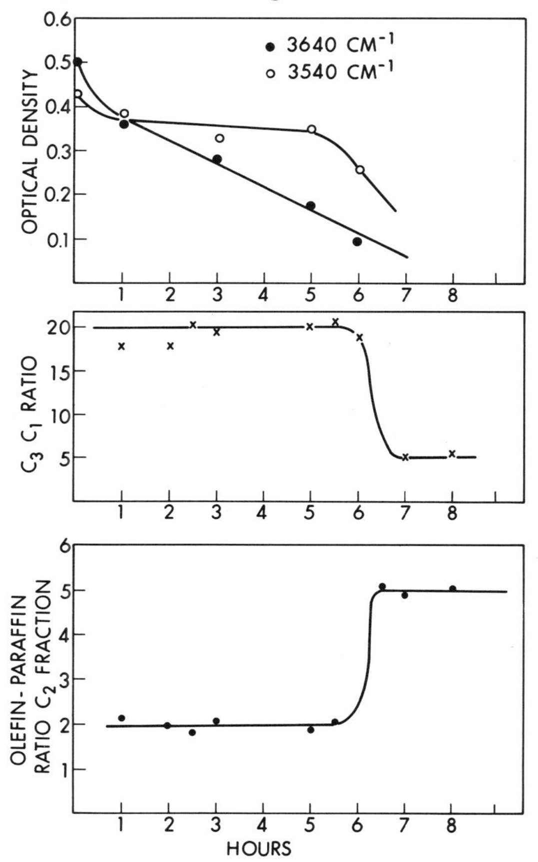

Journal of Catalysis

Figure 93. Catalytic cracking at 450°C of hexane (219)

Regeneration of the catalysts in air did not restore the catalytic selectivity or the hydroxyl groups. The addition of palladium to the zeolite and the use of hydrogen instead of nitrogen as a carrier gas resulted in the same C_3 to C_1 hydrocarbon ratio, no unsaturated hydrocarbons, and no change in the hydroxyl group absorption bands. The data show that the catalyst structure is undergoing a marked change during the reaction, and the linear decrease of the 3640 cm^{-1} band with reaction time indicates that the loss of hydroxyl groups is associated with the cracking reaction. Little carbon was deposited on the catalyst until the 3640 cm^{-1} band had been reduced by 80%. Then the sample rapidly turned black and the product distribution simultaneously changed from that of a typical ionic mechanism to that characteristic of thermal cracking. Some of the characteristics of cracking paraffins over zeolites—namely, the lower olefin production and the lower C_1 and C_2 production can probably be rationalized by the infrared data. Cracking probably occurs *via* a disproportionation mechanism:

$$\begin{aligned} 2\,C_6 \rightarrow [C_{12}] &\rightarrow 3\,C_4 \\ &\rightarrow 4\,C_3 \\ &\rightarrow C_3 + C_4 + C_5 \end{aligned}$$

Thus, small amounts of C_1 and C_2 products would be expected. The smaller amounts of olefin compared to silica–alumina can be explained if the zeolitic hydroxyl groups supply hydrogen to the olefins. When the source of hydrogen is used up, namely the hydroxyl groups, a change in mechanisms is expected. The addition of the palladium hydrogenation component to the zeolite along with the use of hydrogen carrier gas appears to be a method of producing a replacable source of hydrogen. Either the palladium directly results in hydrogenation of the olefin or it provides a means of reprotonating the zeolite.

Furthermore, the higher the extent of ammonium ion exchange of the zeolite and, hence, the greater the population of hydroxyl groups on the zeolite surface, the smaller the olefin to paraffin ratio. The C_3 to C_1 hydrocarbon ratio also increases with increasing exchange. These data seem to indicate the direct involvement of the structural hydroxyl groups in the catalytic reaction. Whether the hydroxyl groups and/or other sites are the active sites awaits clarification. Without the use of *in situ* spectral techniques, the changes occurring on the zeolite surface would be unknown.

Cumene cracking has also been studied over ammonium-Y and a steamed ammonium-Y pretreated between 400° and 800°C (*211, 212*). Pyridine adsorption at 150°C was used to determine the zeolite acidity, and cumene dealkylation was followed at 250°C. The dependence of the acidity and the amount of pyridine required to reduce the initial activity to zero as a function of calcination temperature is shown in Figure 94. The change in acidity as a function of temperature is similar to that reported by others previously (*68, 69*). Below 650°C, the number of pyridine molecules required to suppress the catalytic activity agrees well with the number of pyridine molecules adsorbed on Bronsted acid sites. After calcination at higher temperatures and for all temperatures with the hydrolyzed samples (a hydrothermally stabilized type of zeolite), the number of molecules needed to poison the reaction is greater than the number of Bronsted acid sites. It is, however, less than the total number of acid sites. The polymeric residue on the catalyst surface is greater for the high temperature calcined samples and for the hydrolyzed zeolite. Separate measurements of catalytic activity show a maximum activity near 600°–650°C for decationized Y and a gradually declining activity for hydrolyzed Y. The temperature for maximum activity is similar to that reported previously by others (*213, 214, 215, 216*).

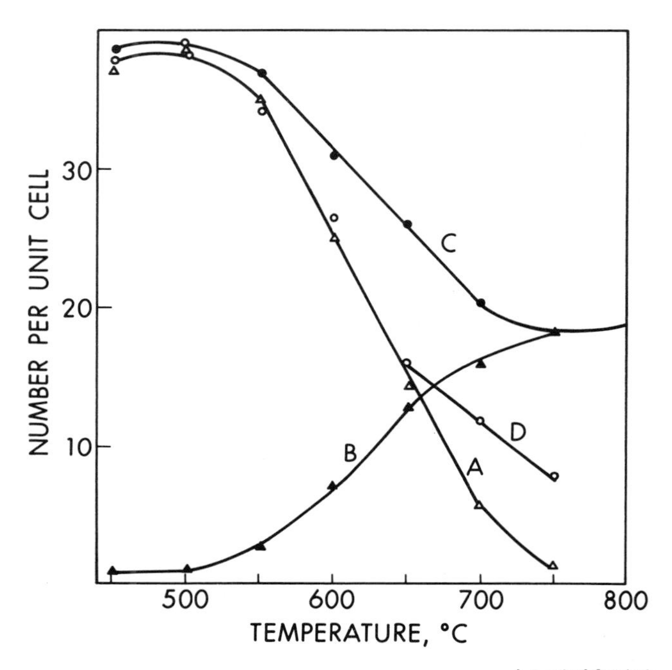

Journal of Catalysis

Figure 94. Change in amount of (A) Bronsted acid sites; (B) Lewis acid sites; (C) sum of Lewis and Bronsted sites; and (D) the amount of pyridine molecules needed to reduce the initial activity of decationized Y to zero (211)

Adsorption of cumene at room temperature on decationized Y calcined at 500°C resulted in the elimination of the 3650 cm^{-1} band and formation of a new band near 3385 cm^{-1} attributed to hydrogen bonding. Absorption bands from the benzene ring at 1603 and 1495 cm^{-1} and 1386 and 1367 cm^{-1} from isopropyl groups were observed. At 300°C, no benzene ring vibrations were observed, but new bands were detected at 1585 and 2920 cm^{-1}. The 3650 cm^{-1} band was partially restored. A sample removed from the reactor after being used to crack cumene had a similar spectrum. These results are in contrast with those discussed above which were obtained under actual catalytic conditions. Propylene was also studied on decationized Y pretreated at 400°C. Adsorption of propylene at room temperature eliminated the 3640 cm^{-1} band and produced a new band near 3250 cm^{-1}. After

2 hr, no band at 1635 cm^{-1} from the C—C was detected. However, a new band near 1460 cm^{-1} was observed, characteristic of saturated hydrocarbon. Adsorption at 300°C for 30 min resulted in no C—C, but bands at 1585 and 1540 cm^{-1} were observed. After one hr, bands at 1573, 1355, and 2940 cm^{-1} were detected. These observations are similar to those on adsorption of cumene at 300°C. Adsorption of cumene or propylene at 300°C on decationized Y or hydrolyzed Y zeolite pretreated at higher temperatures gave similar results. From these studies, the initial catalytic activity and the total hydroxyl group concentration show similar changes with pretreatment temperature. Furthermore, the amount of pyridine required to poison the initial activity and to interact with the acidic hydroxyl groups correspond. Hence, the Lewis acid sites are not the primary active sites and the variations in catalytic activity correspond to changes in acidic hydroxyl groups. Similarly, for the hydrolyzed Y samples, the initial conversion of cumene and the acidic hydroxyl group population decrease correspondingly with temperature. The acidic hydroxyl groups are the primary active sites. Since the concentration of acidic hydroxyl groups decreases much faster than the initial activity, only a small fraction of the hydroxyl groups play an active role in the dealkylation reaction. Similar results have been obtained for a number of modified zeolites (*212*).

Infrared and catalytic measurements have been made separately but on similarly treated samples for 1-butene isomerization on decationized Y zeolite after pretreatment at several temperatures between 400° and 700°C (*217*). Spectra of 1-butene after contacting decationized Y at 150°C for an hour indicate the loss of the C=C band and the formation of a diene type structure. The relative intensities of the 2940 cm^{-1} and the 2930 cm^{-1} bands also suggest that polymerization has occurred. There is also a marked reduction in intensity of the 3650 cm^{-1} band on absorption of the 1-butene. Spectra of samples of decationized Y initially calcined at 550° and 650°C and then used in a catalytic reactor for butene isomerization at 150°C show similar effects. The catalytic reaction was measured both in a flow reactor and a continuous circulation system. In the flow reactor, there is a rapid initial loss of activity followed by a slow decay. Resolution of the rate data according to the procedure of Ballivet *et al.* (*218*) yields the initial rates for two component reactions. These are shown in Figure 95 along with the steady state rate as a function of temperature. The rate v_1 decreases rapidly between 500° and 550°C and represents the activity of sites which are poisoned most rapidly at the initial stage of reaction. The hydroxyl groups on the catalyst show a similar rapid decrease in concentration in this temperature range. Therefore, the hydroxyl groups are the most active sites for butene isomerization. The rate v_2 is related to sites which are poisoned more slowly. The rate increases to a maximum near 660°C, which closely parallels the Lewis acid site concentration change. Pyridine poisoning experiments show that the amounts of pyridine required to reduce the v_1 rate to 0 and the concentration of pyridinium ions are comparable, again suggesting structural hydroxyl groups as the initial active sites. It is not clear whether the polymeric species shown to be on the catalyst surface by infrared served to furnish active sites or not in the advanced stages of the reaction.

Another study of butene adsorption has been made as part of a study of alkylation reactions over Y zeolite (*219*). The ammonium zeolites were pretreated between 300° and 700°C. In contrast to the study of Declerek *et al.* (*217*), the studies were generally made at room temperature, using 20 torr of l-butene. The excess gas was desorbed by evacuation for 0.5 hr at room temperature. Generally, 20% of the butene remained on the surface.

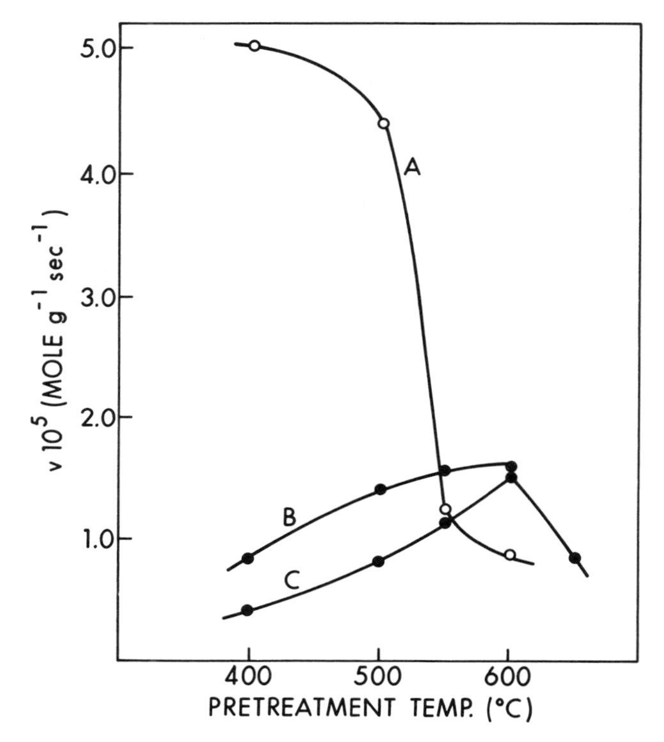

"Proceedings of the Third International Conference on Molecular Sieves"

Figure 95. Initial and steady state rate at 150°C as a function of calcination temperature (A) v_1; (B) v_2; (C) v_{ss} (217)

Immediately after addition of butene, the infrared spectra indicated that isomerization to 2-butene did occur to some extent. Absorption bands from C=C in l-butene and 2-butene were observed. On standing, the bands from C=C disappeared and absorption bands near 2960, 2930, 2870, 1465, 1380, and 1365 cm^{-1} were detected. These bands can be attributed to adsorbed saturated species which cannot be removed by evacuation at room temperature. During the formation of the saturated species, the 3640 cm^{-1}

hydroxyl bands, which had remained unaffected by the adsorption of butene, decreased in intensity. The same types of spectra are observed on Y zeolite dehydroxylated at 700°C. At 300°C, the bands in the 2900 cm^{-1} region are observed. This can be attributed to polymeric surface species.

Attempts were made to follow the alkylation reaction spectroscopically. Addition of benzene to the zeolite on which olefin was preadsorbed resulted in no change in the spectrum of the surface species at room temperature. At 100°C, a trace of *sec*-butylbenzene was formed. In contrast, addition of l-butene to zeolite on which benzene was preadsorbed yielded *sec*-butylbenzene immediately at room temperature. The butylbenzene was identified by absorption bands near 1615, 1380, and 1460 cm^{-1}. The hydroxyl groups could be restored and the butylbenzene desorbed by evacuation at 300°C. No polymer could be detected by infrared. The results were similar on both the deactionized and dehydroxylated zeolites. Since the reaction takes place equally readily over both forms of the zeolite, zeolitic hydrogen is apparently not involved, unless some form of hydrogen which is not readily removed by calcination functions as the active site. The initial reaction of the olefin is to isomerize. This reaction occurs prior to polymerization. The infrared detected species formed on the surface do not take part in the alkylation reaction. In fact instead of promoting the reaction, the species appear to inhibit alkylation. The role of the zeolite catalyst appears to be to adsorb the aromatic compound and in so doing activate the aromatic species. Infrared spectra show changes in the frequency and intensity of the benzene absorption bands, indicating interaction with the surface. This aromatic-surface complex can then be alkylated by olefins. Mass spectral examination of reaction products from the use of $^{13}CH_3CH_2{=}CH_2$ shows that isomerization of the l-butene to 2-butene must have occurred before alkylation: alkylation takes place on the 2- and 3-carbon atoms of the butene. The sites on which this isomerization takes place are not clear.

The role of zeolites in alkylation reactions was also investigated by Kirsch *et al.* (*220*), for the isoparaffin–olefin system over cerium ammonium-Y zeolite. The zeolites were pretreated in air at 450°C, and the hydrocarbons admitted at room temperature. Introduction of isobutane to the system resulted in a decrease in the intensity of the 3640 cm^{-1} band and an increase in the intensity of the 3550 cm^{-1} band. Subsequent addition of 2-butene to the system decreased the intensity of the 3640 cm^{-1} band further and did not affect the 3550 cm^{-1} band. Reactivation in dry air at 450°C of the sample restored the hydroxyl bands. The addition of olefin before paraffin was also studied. After addition of olefin, no absorption band attributed to C═CH was observed, suggesting that the olefin structure was eliminated on adsorption. Similar observations were made by Weeks *et al.* (*219*) and Declerck *et al.* (*217*). The absorption bands in the hydroxyl region were also broadened. The only observed change on addition of isobutane was the elimination of the 3740 cm^{-1} band. When smaller doses were used, 2-butene reduced the 3640 cm^{-1} band but had no effect on the 3740 and 3550 cm^{-1} bands. Again, no olefin absorption was detected. Addition of isobutane then reduced the intensity of the 3640 cm^{-1} band further and also reduced the 3740 cm^{-1}. 1-Butene behaved similarly to 2-butene but had less influence on the 3640 cm^{-1} band. Adsorption of the reaction product, isooctane, reduced the 3640 cm^{-1} band intensity. Paraffin CH stretching vibrations were observed in the 3000 to 2800 cm^{-1} region. From these data and a study of the catalytic reaction, protonic sites on the catalyst are the active centers. A possible center could comprise species such as:

$$Ce(OH)^{2+},\quad Al{-}O{-}Si;\quad Si{-}OH\quad \text{and} \;{>}Al\quad Si{<}\;\text{sites.}$$

This conclusion is somewhat different to that reached by Weeks *et al.* (*219*) who consider dehydroxylated sites and not protonic sites to be the active centers. Because the olefin and paraffin were added simultaneously in this study, it is not possible to conclude whether preadsorbed olefin inhibits the olefin–paraffin alkylation reaction as in the olefin–benzene reaction.

A similar study of butene isomerization over clinoptilolite (*96*) showed that the rate of isomerization correlated with the concentration of protons on the external parts of the zeolite. Lewis acid sites generated by high temperature calcination had no influence on the reaction.

The reaction of carbon monoxide and ammonia over calcium and iron zeolites has been studied in relation to the synthesis of amino acids and purines (*221*). After conducting the reaction for five days at 325°C, absorption bands are detected on the catalyst which can be attributed to CO and HCN. Depending upon further treatment of the catalyst, absorption bands are observed corresponding to -COOH groups, -C =O and/or N-CO amide groups, and/or C=N groups of imines, COO^-, NH_3^+ and CON-NH. Several amino acids were detected by chromatographic techniques. The results suggest that HCN is initially formed. The HCN then leads to a precursor polymer which upon hydrolysis yields amino acids. Although this study substantially described the nature of the reaction products and intermediates, little was concluded about the types of sites on the surface responsible for reaction.

The exchange of hydroxyl groups on decationized X, decationized Y and CaH-X has been investigated after the zeolites have been pretreated at 300°, 400°, and 350°C in vacuum respectively (*222*). The deuterium exchange was carried out at 260°–300°, 280°–300°, and 300°–330°C, and the spectra were measured at room temperature. Exchange periods extended up to 7 hr. For decationized X, the 3660 cm^{-1} band exchanged more rapidly than the 3560 cm^{-1} band (which is ion exchanged more rapidly than the 3745 cm^{-1} band). The relative rates of exchange of the hydroxyl groups on CaH-X was a function of pressure. Below 15 torr, the order was 3560 $>$ 3660 $>$ 3595 cm^{-1} while above 20 torr the order was 3595 $>$ 3660 $>$ 3560 cm^{-1}. The reaction rate can be expressed in terms of a Langmuir-type equation. The activation energy for exchange of the 3660 cm^{-1} OH group varies from 7.7 kcal/mole for decationized X to 41 kcal/mole for decationized Y. Addition of small amounts of water and carbon dioxide to the zeolites markedly speeds up the rate of exchange. For water, the increase in rates is two orders of magnitude. At low water additions, the reaction is accelerated by the exchange reaction between the proton of the water molecule and the deuterium molecule. At higher water contents, the water molecules shield the electrostatic field of the cations and thus deactivate the deuterium molecule.

Figueras *et al.* (*203*) have shown that the activity for benzene hydrogenation of supported palladium catalysts is markedly dependent on the support. Turnover numbers varied from 56 for 0.9 % Pd on silica to 280 for 1.77 % Pd on Ce-Y. These results were interpreted as an interaction between the support and the palladium. Using the frequency of chemisorbed carbon monoxide as a measure of the metal support interaction, a qualitative correlation was obtained between the frequency and the turnover number. No attempt was made to correlate the activity with other properties of the supports.

Although many studies have now been reported on catalytic reactions over zeolites, the types of sites responsible for reaction still remains uncertain. The infrared studies suggest that the main potential sites are hydroxyl groups which function as Bronsted acids, tricoordinated aluminum sites which can function as Lewis acid sites, and accessible cations. For carbonium ion reactions, the acidic hydroxyl groups, particularly those with an absorption band near 3650 cm^{-1} in X and Y zeolite appear to be important sites in the reaction. Whether other sites play an important role and whether acid site strength is important still requires clarification.

More generally, the influence of changes in the zeolite surface produced by ion exchange, calcination, reaction or other changes of the physical and catalytic properties have been substantially elucidated so that it is now possible to predict changes in physical and catalytic properties brought about by various treatments.

Some reactions such as H_2/D_2 exchange appear to relate more to Lewis acid site concentrations, while many reactions do not appear to relate to currently measured properties.

Literature Cited

1. Little, L. H., "Infrared Spectra of Adsorbed Species," Academic Press, London, 1967.
2. Hair, M. L., "Infrared Spectroscopy in Surface Chemistry," Marcel Dekker, Inc., New York, 1967.
3. Eischens, R. P., Pliskin, W. A., *Advan. Catal.* (1958) **X,** 1.
4. Crawford, V., *Quart. Rev.* (1960) **14,** 378.
5. Sheppard, N., *Pure Appl. Chem.* (1962) **4,** 71.
6. Kiselev, A. V., Lygin, V. I., *Russ. Chem. Rev.* (1962) **31,** 175.
7. Schurer, P., *Cesk Casopsis Fys* (1965) **15,** 449.
8. Van Hardeveld, R., *Chem. Weekblad* (1966) **62,** 261.
9. Basila, M. R., *Appl. Spectr. Rev.* (1968) **1,** 289.
10. Blyholder, G., "Experimental Methods in Catalytic Research," R. B. Anderson, Ed., p. 323, Academic Press, New York, 1968.
11. Amberg, C. H., "The Solid-Gas Interface," E. A. Flood, Ed., p. 869, Marcel Dekker, Inc., New York, 1967.

12. Sheppard, N., Avery, N. R., Morrow, B. A., Clark, M., Takenaka, T., Ward, J. W., *Institute of Petroleum, Spectroscopic Conference*, Brighton, England, 1968, p. 67.
13. Pritchard, J., *Ann. Rep.* (1969) **66,** 65.
14. Sheppard, N., Avery, N. R., Morrow, B. A., Young, R. P., "Chemisorption and Catalysis," Institute of Petroleum, London, 1970, p. 135.
15. Rochester, C. H., Scurrell, M. S., "Surface and Defect Properties of Solids," Chemical Society, London, 1973, p. 114.
16. Yates, D. J. C., *Catal. Rev.* (1968) **2,** 113.
17. Yates, D. J. C., "Molecular Sieves," Society of Chemical Industry, London, 1968, p. 334.
18. Ward, J. W., *Adv. Chem. Ser.* (1971) **101,** 380.
19. Dewing, J., "Chemisorption and Catalysis," Institute of Petroleum, London, 1970, p. 173.
20. Patzelova, V., Tvaruzkov, Z., *J. Catal.* (1974) **33,** 158.
21. Frohnsdorff, G. J. C., Kingston, G. L., *Proc. Roy. Soc. (London)* (1958) **A247,** 469.
22. Szymanski, H. A., Stamires, D. N., Lynch, G. R., *J. Opt. Soc. Amer.* (1960) **50,** 1323.
23. Zhdanov, S. P., Kiselev, A. V., Lygin, V. I., Titova, T. I., *Dokl. Acad. Nauk SSR* (1963) **150,** 442.
24. Zhandov, S. P., Kiselev, A. V., Lygin, V. I., Titova, T. I., *Russ. J. Phys. Chem.* (1964) **38,** 1299.
25. Zhandov, S. P., Kiselev, A. V., Lygin, V. I., Ovepyan, M. E., Titova, T. I., *Russ. J. Phys. Chem.* (1965) **39,** 1309.
26. Bertsch, L., Habgood, H. W., *J. Phys. Chem.* (1963) **67,** 1621.
27. Carter, J. L., Luccesi, P. J., Yates, D. J. C., *J. Phys. Chem.* (1964) **68,** 1385.
28. Habgood, H. W., *J. Phys. Chem.* (1965) **69,** 1764.
29. Hattori, H., Shiba, T., *J. Catal.* (1968) **12,** 111.
30. Angell, C. L., Schaffer, P. C., *J. Phys. Chem.* (1965) **69,** 3463.
31. Uytterhoeven, J. B., Schoonheydt, R., Liengme, B. V., Hall, W. K., *J. Catal.* (1969) **13,** 425.
32. Ward, J. W., *J. Phys. Chem.* (1968) **72,** 4211.
33. Ward, J. W., *J. Catal.* (1968) **11,** 238.
34. Eberly, P. E., Jr., *J. Phys. Chem.* (1968) **72,** 1042.
35. Ward, J. W., *J. Catal.* (1968) **10,** 34.
36. Christner, L. G., Liengme, B. V., Hall, W. K., *Trans. Faraday Soc.* (1968) **64,** 1679.
37. Hirschler, A. E., *J. Catal.* (1963) **2,** 428.
38. Olson, D. H., *J. Phys. Chem.* (1968) **72,** 4366.
39. Ward, J. W., *J. Catal.* (1969) **13,** 321.
40. Scherzer, J., Bass, J. L., *J. Catal.* (1973) **28,** 101.
41. Jacobs, P., Uytterhoeven, J. B., *J. Catal.* (1971) **22,** 193.
42. Peri, J. B., "Catalysis," J. W. Hightower, Ed., Vol. I, p. 329, North Holland, Amsterdam, 1973.
43. Ward, J. W., *J. Phys. Chem.* (1968) **72,** 2689.
44. Ward, J. W., *J. Catal.* (1970) **18,** 348.
45. Tsitsishvili, G. V., *Adv. Chem. Ser.* (1973) **121,** 291.
46. Ward, J. W., unpublished results.
47. Ward, J. W., *Trans. Faraday Soc.* (1971) **67,** 1489.
48. Guilleux, M. F., Tempere, J. F., Delafosse, D., *Proc. of 3rd Inter. Conf. Molecular Sieves*, Zurich, 1973, p. 377.
49. Ben Taarit, Y., Primet, M., Naccache, C., *Compt. Rend.* (1970) **67,** 1434.
50. Guilleux, M. F., Tempere, J. F., *Compt. Rend.* (1971) **272,** 2105.
51. Rabo, J. A., Angell, C. L., Kasai, P. H., Schomaker, V., *Discuss. Faraday Soc.* (1966) **41,** 328.

52. Rabo, J. A., Angell, C. L., Schomaker, V., *4th Intern. Congr. Catal., Moscow, Preprint 54,* 1968.
53. Bolton, A. P., *J. Catal.* (1971) **22,** 9.
54. Ben Taarit, Y., Mathieu, M. V., Naccache, C., *Adv. Chem. Ser.* (1971) **102,** 362.
55. Eberly, P. E., Jr., Kimberlin, C. N., Jr., *Adv. Chem. Ser.* (1971) **102,** 374.
56. Venuto, P. B., Hamilton, L. A., Landis, P. S., *J. Catal.* (1966) **5,** 484.
57. Ward, J. W., *J. Catal.* (1968) **11,** 251.
58. Smith, J. V., *Adv. Chem. Ser.* (1971) **101,** 171.
59. Ermolenko, N. F., Tsybulskaya, Y. A., Malashevick, L. N., *Kin. Catal.* (1973) **14,** 904.
60. Uytterhoeven, J. B., Christner, L. G., Hall, W. K., *J. Phys. Chem.* (1965) **69,** 2117.
61. Zhandov, S. P., Kiselev, A. V., Lygin, V. I., *Russ. J. Phys. Chem.* (1966) **40,** 560.
62. Rabo, J. A., Pickert, P. E., Stamires, D. N., Boyle, J. E., *Act. Congr. Intern. Catal., 2nd, Paris, 1960* (1961) 2055.
63. Uytterhoeven, J. B., Jacobs, P., Makay, K., Schoonneydt, R., *J. Phys. Chem.* (1968) **72,** 1768.
64. Eberly, P. E., Jr., *J. Phys. Chem.* (1967) **71,** 1717.
65. Liengme, B. V., Hall, W. K., *Trans. Faraday Soc.* (1966) **62,** 3229.
66. White, J. L., Jelli, A. W., Andre, J. M., Fripiat, J. J., *Trans. Faraday Soc.* (1967) **63,** 461.
67. Kermarec, J., Tempere, J. F., Imelik, B., *Bull. Soc. Chim. France* (1969) 3792.
68. Hughes, T. R., White, H. M., *J. Phys. Chem.* (1967) **71,** 2192.
69. Ward, J. W., *J. Catal.* (1967) **9,** 225.
70. Bolton, A. P., Lanewala, M. A., *J. Catal.* (1970) **18,** 154.
71. Olson, D. H., Dempsey, E., *J. Catal.* (1969) **13,** 221.
72. Sherry, H. S., *J. Phys. Chem.* (1966) **70,** 1158.
73. Ward, J. W., *J. Phys. Chem.* (1969) **73,** 2086.
74. Jacobs, P. A., Uytterhoeven, J. B., *J. Chem. Soc. Faraday I* (1973) **69,** 359.
75. Gallezot, P., Imelik, B., *Compt. Rend.* (1970) **271,** 912.
76. Best, D. F., Larson, R. W., Angell, C. L., *J. Phys. Chem.* (1973) **77,** 2183.
77. Ward, J. W., Hansford, R. C., *J. Catal.* (1969) **13,** 364.
78. Ward, J. W., *J. Phys. Chem.* (1970) **74,** 3021.
79. Hansford, R. C., Ward, J. W., *Adv. Chem. Ser.* (1971) **102,** 354.
80. Heylen, C. F., Jacobs, P. A., *Adv. Chem. Ser.* (1973) **121,** 490.
81. Ward, J. W., unpublished results.
82. Schoonheydt, R. A., Uytterhoeven, J. B., *J. Catal.* (1970) **19,** 55.
83. Wang, K. M., Lunsford, J. H., *J. Catal.* (1972) **24,** 262.
84. McDaniel, C. V., Maher, P. K., "Molecular Sieves," Society of the Chemical Industry, London, 1968, p. 186.
85. Maher, P. K., McDaniel, C. V., U.S. Patent **3,293,192** (1966).
86. Hansford, R. C., U.S. Patent **3,354,077** (1967).
87. Kerr, G. T., *J. Phys. Chem.* (1968) **72,** 2594.
88. Kerr, G. T., *J. Catal.* (1969) **15,** 200.
89. Jacobs, P. A., Uytterhoeven, J. B., *Trans. Faraday Soc.* (1973) **69,** 373.
90. Beaumont, R., Pichat, P., Barthomeuf, D., Trambouze, Y., "Catalysis," J. W. Hightower, Ed. Vol. I, p. 343, North Holland, Amsterdam, 1973.
91. Ward, J. W., *Ibid.*, p. 355.
92. Eberly, P. E., Jr., Laurent, S. M., Robson, H. E., U.S. Patent **3,506,400** (1970); **3,591,488** (1971).
93. Mone, R., Moscou, L., *Proc. 3rd Inter. Conf. Molecular Sieves,* Zurich, 1973, p. 351.
94. Kerr, G. T., *J. Catal.* (1968) **72,** 3071.
95. Young, D. A., U.S. Patent **3,644,200** (1972).

96. Detrekoy, E. J., Jacobs, P. A., Kallo, D., Uytterhoeven, J. B., *J. Catal.* (1974) **32,** 442.
97. Breger, J. A., Chandler, J. C., Zubovic, P., *Amer. Mineral* (1970) **55,** 825.
98. Yukhnevich, G. V., Karyakin, A. V., Khitarov, N. I., Senderov, E. E., *Geochem.* (1961) 937.
99. Eberly, P. E., Jr., Kimberlin, C. N., Jr., Voorhies, A., Jr., *J. Catal.* (1971) **22,** 419.
100. Shikunov, B. I., Lafer, L. I., Yakerson, V. I., Rubenshtein, A. M., *Izvest. Akad. Nauk SSSR, Ser. Khim.* (1973) 449.
101. Shikunov, B. I., Lafer, L. I., Yakerson, V. I., Mishin, I. V., *Izvest. Akad. Nauk SSSR, Ser. Khim.* (1972) 207.
102. Karge, H., Klose, K., *Zeit fur Physik. Chem. Neue Folge* (1973) **83,** 100.
103. Karge, H., *Zeit fur Physik. Chem. Neue Folge* (1971) **76,** 133.
104. Lerot, L., Poncelet, G., Debru, M. L., Fripiat, J. J., *J. Catal.* (1975) **37,** 396.
105. Rabo, J. A., Poutsma, M. L., Skeels, G. W., "Catalysis," J. W. Hightower, Ed., Vol. II, p. 1353, North Holland, Amsterdam, 1973.
106. Wu, E. L., Kuhl, G. H., Whyte, T. E., Jr., Venuto, P. B., *Adv. Chem. Ser.* (1971) **101,** 490.
107. Jacobs, P. A., Uytterhoeven, J. B., *J. Catal.* (1972) **26,** 175.
108. Fripiat, J. J., Lambert-Helsen, M. M., *Adv. Chem. Ser.* (1973) **121,** 518.
109. Durand, B., Pelet, R., Fripiat, J. J., *Clays and Clay Minerals* (1972) **20,** 21.
110. Wu, E. L., Whyte, T. E., Jr., Venuto, P. B., *J. Catal.* (1971) **21,** 384.
111. Cant, N. W., Hall, W. K., *Trans. Faraday Soc.* (1968) **64,** 1093.
112. Uytterhoeven, J. B., Schoonheidt, R., Fripiat, J. J., *Abstracts, Intern. Symp. on React. Mech. Inorg. Solids,* Aberdeen, 1966.
113. Ward, J. W., *J. Catal.* (1967) **9,** 396; (1970) **16,** 386.
114. Ward, J. W., *J. Catal.* (1969) **14,** 365.
115. Ward, J. W., *J. Catal.* (1970) **17,** 355.
116. Fripiat, J. J., *Catal. Rev.* (1971) **5,** 269.
117. Mestdagh, M. M., Fripiat, J. J., *Third N. Amer. Catal. Soc. Meet.,* 1974, Paper 32.
118. Freude, D., Oehme, W., Schmiedel, H., Staudte, B., *J. Catal.* (1974) **32,** 137.
119. Ward, J. W., *J. Catal.* (1968) **11,** 259.
120. Kiselev, A. V., Lygin, V. I., *Surface Science* (1964) **2,** 236.
121. Angell, C. L., Schaffer, P. C., *J. Phys. Chem.* (1966) **70,** 1413.
122. Bennett, J. M., Smith, J. V., *Mater. Res. Bull.* (1968) **3,** 633.
123. Egerton, T. A., Stone, F. S., *Trans. Faraday Soc.* (1970) **66,** 2364.
124. Gallezot, P., Imelik, B., *J. Phys. Chem.* (1973) **77,** 652.
125. Egerton, T. A., Stone, F. S., *J. Chem. Soc., Faraday I* (1973) **69,** 22.
126. Egerton, T. A., Hagan, A., Stone, F. S., Vickerman, J. C., *J. Chem. Soc., Faraday I* (1972) **68,** 723.
127. Egerton, T. A., Vickerman, *J. Chem. Soc., Faraday I* (1973) **69,** 39.
128. Fenelon, P. J., Rubalcava, H. E., *J. Chem. Phys.* (1969) **51,** 961.
129. Huang, Y. Y., *J. Amer. Chem. Soc.* (1973) **95,** 6636.
130. Huang, Y. Y., *J. Catal.* (1974) **32,** 482.
131. Bregadze, T. A., Seleznev, V. A., Kadushin, A. A., Krylov, O. V., *Izvest. Akad. Nauk SSSR, Ser. Khim.* (1973) 2701.
132. Frilette, V. J., Muns, G. W., *J. Catal.* (1965) **4,** 504.
133. Angell, C. L., *J. Phys. Chem.* (1966) **70,** 2420.
134. Ward, J. W., Habgood, H. W., *J. Phys. Chem.* (1966) **70,** 1178.
135. Angell, C. L., Howell, M. V., *Can. J. Chem.* (1969) **47,** 3831.
136. Jacobs, P. A., van Cauwelaert, F. H., Vansant, E. F., Uytterhoeven, J. B., *Trans. Faraday Soc.* (1973) **69,** 1056.
137. Jacobs, P. A., van Cauwelaert, F. H., Vansant, E. F., *J. Chem. Soc., Faraday I* (1973) **69.**
138. Smith, J. V., Bennett, J. M., Flanigen, E. M., *Nature* (1967) **215,** 241.

139. Olson, D. H., Kokotailo, G. T., Charnell, J. F., *Nature* (1967) **215**, 270.
140. Alekseev, A. V., Filimonov, V. N., Terenin, A. N., *Dokl. Akad. Nauk SSSR* (1962) **147**, 1392.
141. Chao, C. C., Lunsford, J. H., *J. Amer. Chem. Soc.* (1971) **93**, 71.
142. Ben Taarit, Y., Naccache, C., Imelik, B., *J. Chim. Phy.* (1973) **70**, 728.
143. Chao, C. C., Lunsford, J. H., *J. Amer. Chem. Soc.* (1971) **93**, 6794.
144. Naccache, C., Ben Taarit, Y., *J. Chem. Soc., Faraday I* (1973) **69**, 1475.
145. Chao, C. C., Lunsford, J. H., *J. Phys. Chem.* (1974) **78**, 1174.
146. Jermyn, J. W., Johnson, T. J., Vansant, E. T., Lunsford, J. H., *J. Phys. Chem.* (1973) **77**, 2964.
147. DeLara, E. C., Vincent-Geisse, J., *J. Phys. Chem.* (1972) **76**, 945.
148. Gladrow, E. M., Parker, P. T., U.S. Patent **2,967,159** (1961).
149. Deo, A. V., Dalla Lana, I. G., Habgood, H. W., *J. Catal.* (1971) **21**, 270.
150. Lygin, V. I., Romanovskii, A. V., Topchieva, K. V., Tkhoang, K. S., *Russ. J. Phys. Chem.* (1968) **42**, 156.
151. Thompson, H. W., Temple, R. B., *Trans. Faraday Soc.* (1945) **41**, 27.
152. Carter, J. L., Yates, D. J. C., Lucchesi, P. J., Elliott, J. J., Kevorkian, V., *J. Phys. Chem.* (1966) **70**, 1126.
153. Yates, D. J. C., *J. Phys. Chem.* (1966) **70**, 3693.
154. Muha, G. M., Yates, D. J. C., *J. Chem. Phys.* (1968) **49**, 5073.
155. Cant, N. W., Hall, W. K., *J. Catal.* (1972) **25**, 161.
156. Tsitsishvili, G. V., Bagratishvili, G. D., Oniashvili, N. I., *Zh. Fiz. Khim.* (1969) **43**, 950.
157. Pichat, P., Vedrine, J. C., Gallezot, P., Imelik, B., *J. Catal.* (1974) **32**, 190.
158. Abramov, V. N., Kiselev, A. V., Lygin, V. I., *Russ. J. Phys. Chem.* (1963) **37**, 613.
159. Angell, C. L., Howell, M. V., *J. Colloid Interface Sci.* (1968) **28**, 279.
160. Galkin, G. A., Kiselev, A. V., Lygin, V. I., *Russ. J. Phys. Chem.* (1962) **36**, 951.
161. Kiselev, A. V., Koupcha, L. A., Lygin, V. I., *Kin. Catal.* (1966) **7**, 621.
162. Topchieva, K. V., Kubasov, A. A., Ratov, A. N., *Dokl. Akad. Nauk SSSR* (1969) **184**, 383.
163. Abramov, V. N., Kiselev, A. V., Lygin, V. I., *Russ. J. Phys. Chem.* (1964) **38**, 575.
164. Kiselev, A. V., Lygin, V. I., Starodubcseva, R. V., *Kolloid Zh.* (1969) **31**, 68.
165. Zhdanov, S. P., Kotov, E. I., *Adv. Chem. Ser.* (1973) **121**, 240.
166. Geodokyan, K. T., Kiselev, A. V., Lygin, V. I., *Russ. J. Phys. Chem.* (1967) **41**, 227.
167. Geodokyan, K. T., Kiselev, A. V., Lygin, V. I., *Russ. J. Phys. Chem.* (1967) **41**, 476.
168. Bielanski, A., Datka, J., *J. Catal.* (1974) **32**, 183.
169. Kiselev, A. V., Kubelkova, L., Lygin, V. I., *Russ. J. Phys. Chem.* (1964) **38**, 1480.
170. Bosacek, V., Tvaruzkova, Z., *Coll. Czech. Chem. Commun.* (1971) **36**, 551.
171. Angell, C. L., Howell, M. V., *J. Phys. Chem.* (1970) **74**, 2737.
172. Geodokyan, K. T., Kiselev, A. V., Lygin, V. I., *Russ. J. Phys. Chem.* (1966) **40**, 857.
173. Angell, C. L., Howell, M. V., *J. Phys. Chem.* (1969) **73**, 2551.
174. Ratov, A. N., Kubasov, A. A., Topchieva, K. V., Rosolovskaya, E. N., Kalinin, V. P. *Kin. Catal.* (1973) **14**, 896.
175. Karge, H. G., *Surface Science* (1973) **40**, 157.
176. Butler, J. D., Poles, T. C., *J. Chem. Soc., Perkin II* (1973) 48.
177. Butler, J. D., Poles, T. C., *J. Chem. Soc., Perkin II* (1973) 41.
178. Venuto, P. B., *Adv. Chem Ser.* (1971) **102**, 260.
179. Pickert, P. E., Rabo, J. A., Dempsey, E., Schomaker, V., *Proc. Intern. Congr. Catal., Amsterdam*, 1964, p. 714.
180. Richardson, J. T., *J. Catal.* (1967) **9**, 182.

181. Ward, J. W., *Preprints, Div. Petrol. Chem., Amer. Chem. Soc.* (1971) **16,** B6.
182. Watanabe, Y., Habgood, H. W., *J. Phys. Chem.* (1968) **72,** 3066.
183. Nishizawa, T., Hattori, H., Uematsu, T., Shiba, T., *4th Intern. Congress Catal., Moscow, 1968,* paper 55.
184. Ignateva, L. A., Moskovskayan, I. F., Oppengeim, V. D., Spozhakima, A. A., Topchieva, K. V., *Kin. Catal.* (1968) **9,** 111.
185. Geodokyan, K. T., Kiselev, A. V., Lygin, V. I., *Russ. J. Phys. Chem.* (1969) **43,** 106.
186. Kiselev, A. V., Lygin, V. I., Titova, T. I., *Russ. J. Phys. Chem.* (1964) **38,** 1487.
187. Yashima, T., Hara, N., *J. Catal.* (1972) **27,** 329.
188. Ward, J. W., *J. Catal.* (1971) **22,** 237.
189. Moscou, L., Lakeman, M., *J. Catal.* (1970) **16,** 173.
190. Ward, J. W., unpublished results.
191. Ben Taarit, Y., Primet, M., Naccache, C., *J. Chim. Phys.* (1970) **67,** 37.
192. Plank, C. J., comment on paper by Pickert, P. E., Rabo, J. A., Dempsey, E., Schomaker, V., *Proc. Intern. Congr. Catal. Amsterdam* (1964) **1,** 727.
193. Trambouze, Y. J., DeMouges, L., Perrin, M., *J. Chim. Phys.* (1954) **51,** 723.
194. Boreskova, E. G., Lygin, V. I., Topchieva, K. V., *Kin. Catal.* (1964) **5,** 991.
195. Tsutsumi, K., Kajiwara, H., Takahashi, H., *Bull. Soc. Chim. Japan.* (1974) **47,** 801.
196. Ward, J. W., unpublished ıesults.
197. Cannings, F. R., *J. Phys. Chem.* (1968) **72,** 4691.
198. Lefrancois, M., Malbois, G., *J. Catal.* (1971) **20,** 350.
199. Andre, J. M., Fripiat, J. J., *Trans. Faraday Soc.* (1971) **67,** 1821.
200. Dalla Beta, R. A., Boudart, M., "Catalysis," J. W. Hightower, Ed., Vol. II, p. 1329, North Holland, Amsterdam, 1973.
201. Naccache, C., Primet, M., Mathieu, M. V., *Adv. Chem. Ser.* (1973) **121,** 266.
202. Che, M., Dutel, J. F., Primet, M., *Proc. 3rd Intern. Conf. Molecular Sieves,* Zurich, 1973, p. 394.
203. Figueras, F., Gomez, R., Primet, M., *Adv. Chem. Ser.* (1973) **121,** 480.
204. Naccache, C. M., Ben Taarit, Y., *J. Catal.* (1971) **22,** 171.
205. Egerton, T. A., Hardin, A. H., Kozirovski, Y., Sheppard, N., *J. Catal.* (1974) **32,** 343.
206. Angell, C. L., *J. Phys. Chem.* (1973) **77,** 222.
207. Mitrasch, A., Schnieder, C. L., Morawitz, H., U.S. Patent **1,215,396** (1917).
208. Weisz, P. B., Frilette, V. J., *J. Phys. Chem.* (1960) **64,** 382.
209. Seubold, F. H., U.S. Patent **2,983,670** (1961).
210. Bolton, A. P., Bujalski, R. L., *J. Catal.* (1971) **23,** 331.
211. Jacobs, P. A., Leeman, H. E., Uytterhoeven, J. B., *J. Catal.* (1974) **33,** 17.
212. *Ibid.* (1974) **33,** 31.
213. Benesi, H., *J. Catal.* (1967) **8,** 368.
214. Venuto, P. B., Hamilton, L. A., Landis, P. S., Wise, J. J., *J. Catal.* (1966) **5,** 81.
215. Hopkins, P. D., *J. Catal.* (1968) **72,** 325.
216. Hickson, D. A., Csicsery, S. M., *J. Catal.* (1968) **10,** 27.
217. Declerck, L. J., Vandamme, L. J., Jacobs, P. A., *Proc. 3rd Intern. Conf. Mol. Sieves, Zurich* (1973) 442.
218. Ballivet, D., Barthomeuf, D., Trambouze, Y., *J. Catal.* (1972) **26,** 32.
219. Weeks, T. J., Angell, C. L., Ledd, J. R., Bolton, A. P., *J. Catal.* (1974) **33,** 256.
220. Kirsch, F. W., Lauer, J. L., Potts, J. D., *Preprints, Div. Petrol. Chem., Amer. Chem. Soc., 161st Meet., Los Angeles* (1971) B24.
221. Fripiat, J. J., Poncelet, C., Van Assche, A. T., Mayaudon, J., *Clays and Clay Minerals* (1972) **20,** 331.
222. Imanaka, T., Okamoto, Y., Takahata, K., Teranishi, S., *Bull. Chem. Soc. Japan* (1972) **45,** 366.

Chapter

4

Zeolite Stability and Ultrastable Zeolites

C. V. McDaniel and P. K. Maher, Davison Chemical Division,
W. R. Grace & Co., Washington Research Center 7379 Route 32,
Columbia, Md. 21044

THIS CHAPTER provides an overall view of the nature of stability in zeolites and its role in their practical application. Until recently the complex nature of stability in zeolites had not been appreciated and its influence on zeolite behavior has often been ignored. Instability need not lead to total destruction, but it can and often does appear as an irreversible change or alteration. It is their susceptibility to such change or alteration, often without gross destruction, that makes the subject of stability in zeolites so intriguing and important. These changes sometimes drastically alter the behavior of zeolites and limit their use in specific applications. Conversely, many of the most valuable applications of these materials result from these alterations.

The Nature and Importance of Stability in Zeolites

The subject matter of zeolites encompasses a diverse range of scientific phenomena and application. The term stability takes on a somewhat different and unique connotation in each of these areas and makes definition difficult. Furthermore, the term zeolite has been historically applied to such a variety of substances as to defy precise definition. We will attempt to circumvent the problems of rigid definition by focusing attention on some of the characteristics usually associated with species generally regarded as zeolites and then by discussing stability in terms of these characteristics. Many scientists will encounter zeolites in their work and therefore would profit from an overall view of zeolite stability. To this end, we present some specific examples, generalize on the factors involved, and summarize some pertinent work reported in the literature.

The investigation of zeolites dates back many years, at least to 1858 (*1*). Until recently, however, this work was restricted to a few selected areas such as synthesis and selective adsorption. Investigations were accelerated by the advent of large-scale, commercial applications of one or two types of zeolites. Closely related to the growth in commercial application was an increased interest in the stability of these materials.

Early interest in zeolite stability was related to the limitations imposed on the particular application being considered (*2, 3*). Although considerable quantities were used, in their early use as ion exchangers in water softening, zeolites were severely limited by their lack of chemical stability. However, little was done to enhance their stability as water softeners and eventually zeolites were largely replaced by more stable organic exchangers. With their application as adsorbents, the concern shifted to their ability to withstand the conditions needed for regeneration or activation. As the application of adsorbents grew to include more corrosive adsorbates such as acidic gases, *e.g.*, H_2S, the interest in stability was expanded to include acid resistance and more acid stable varieties such as erionite and mordenite were sought.

When it was realized that zeolites have enormous potential for catalysis, the interest in stability took on an entirely new importance. However, many catalytic applications that promised the greatest rewards required, for practical operation, an unachieved degree of stability. Investigation revealed how alterations, previously unnoticed, occurred within the zeolite structure and, more importantly, how some of these alterations led to a broad array of new and interesting materials.

This new awareness and interest led to a more quantitative study of stability and of the factors influencing it. Because of the unusual promise that zeolites exhibited as catalysts for high-temperature acid-catalyzed reactions, thermal and acid stability received the attention of many investigators, and the work in this area, in particular, has led to rapid progress in the development of modified zeolites with a range of new properties.

Stability of Zeolites at Elevated Temperatures

The resistance to collapse at elevated temperatures was one of the first recognized manifestations of zeolite stability. The discovery that these materials could be heated past the point of dehydration without collapse attracted widespread attention. Literature references to this phenomenon are numerous but only recently has the thermal stability of zeolites been reported in any detail. We will consider some of the factors affecting the structural stability at elevated temperatures.

The geometry of the crystalline network is undoubtedly a major factor in stability. However, the quantitative relationship has not been determined and we are still unable to predict the stability of a given structure. Even empirical correlations between stability and structure are difficult because other factors make simultaneous contributions. Intuitively, the more dense structures might have a higher degree of stability because of their thicker walls between pores or channels. However, faujasite, which possesses one of the most open structures of any zeolite has a high degree of thermal stability. This stability is surprising and indeed fortunate since both structural openness and thermal stability are essential to many of the most important commercial uses. Any quantitative estimate of structural stability based on ring stability, bond strength, bond strain, etc., must await future advances in structural analysis.

In general, there is a relation between the silica-to-alumina ratio of a zeolite and its thermal stability: the more siliceous zeolites tend to be more stable. Many authors attribute the relatively high stability of zeolites such as mordenite, clinoptilolite, and erionite to their high silica content.

Several framework structures having different silica-to-alumina ratios can be made for a series of isostructural zeolites. Most important is the series of zeolites having the faujasite structure and varying in silica-to-alumina mole ratio from about 2.4 to about 6. Synthetic products of this group are zeolites X and Y.

Experimental data obtained by the authors and shown in Figure 1 illustrate the increase in thermal stability with silica content for X and Y zeolites

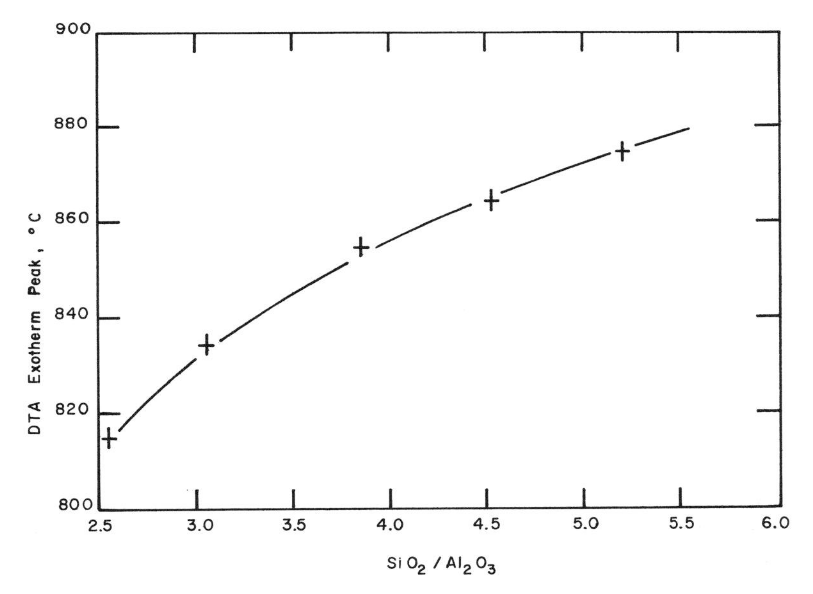

Figure 1. Thermal stability of X- and Y-type zeolites in the sodium form as a function of silica-to-alumina ratio

in the sodium form. The DTA exotherm temperature is for the peak corresponding to the collapse of the crystalline structure. These data agree qualitatively with those reported by other authors (*4*). However, as will be discussed later, quantitative agreement among authors is poor. Attempts to correlate stability and silica content for other structures over small compositional ranges such as type A zeolites are not so convincing (*5*). However, this particular series of zeolites may not have actually contained significantly different silica contents in the structure.

The nature of the experimental data in the literature establishes why it is so difficult to draw more than semiquantitative conclusions about thermal stability. Because the thermal collapse of a crystalline zeolite structure does not represent a true melting point, but rather a gradual degradation that is dependent not only on temperature but also on time, the presence of water vapor, and probably on other factors, there exists no satisfactory reproducible method to measure and accurately express the degree of thermal stability. The results of stability determinations have therefore been reported in many ways. Some authors have reported the loss in crystallinity by the decrease in x-ray diffraction peak heights of a sample heated to a predetermined temperature for a fixed time as compared to the original peak height for the unheated sample. Others have reported similar data, with the heating carried out in a vacuum or other controlled atmosphere. Still other investigators have heated samples for a fixed time at various temperatures and evaluated the products by x-ray diffraction analysis to determine the temperature where the crystallinity pattern of the zeolite disappeared.

Properties other than x-ray diffraction have been used to indicate crystalline collapse. Among these are changes in surface area, volume, electrical conductivtiy, infrared spectra and thermal behavior. The thermal collapse of a zeolite crystal is usually accompanied by a liberation of heat. The magnitude of this exothermic reaction and the temperature at which it occurs can often be determined by differential thermal analysis. Consequently, the position of this exothermic peak has often been used as an indication of thermal stability. The typical behavior of a zeolite being heated and subjected to differential thermal analysis is shown in Figure 2. The endotherm (a) that occurs near 200°C is caused by the evolution of water and possibly other volatile species, if present. The first exotherm (b) is associated with the collapse of the crystalline zeolite to an amorphous phase, and a second exotherm (c) at a still higher temperature is often observed as a result of recrystallization to a new phase. The nature of the exothermic collapse (b) is considered in the following discussion of mechanisms.

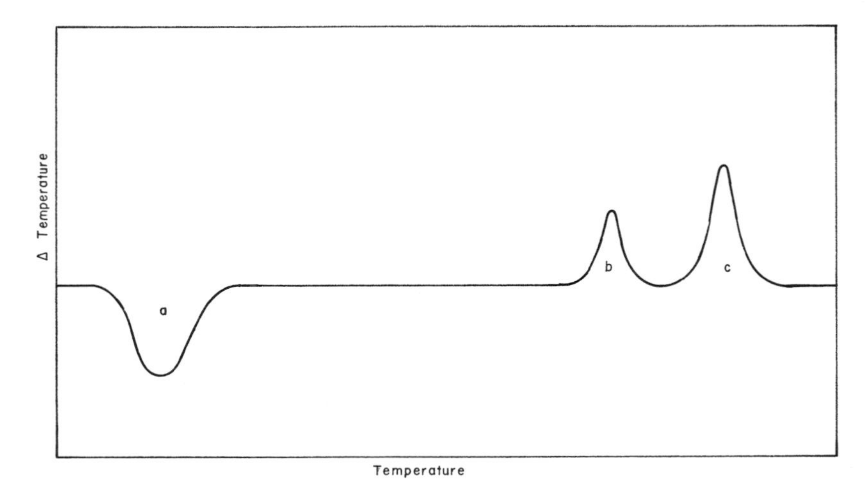

Figure 2. Differential thermal behavior of zeolites

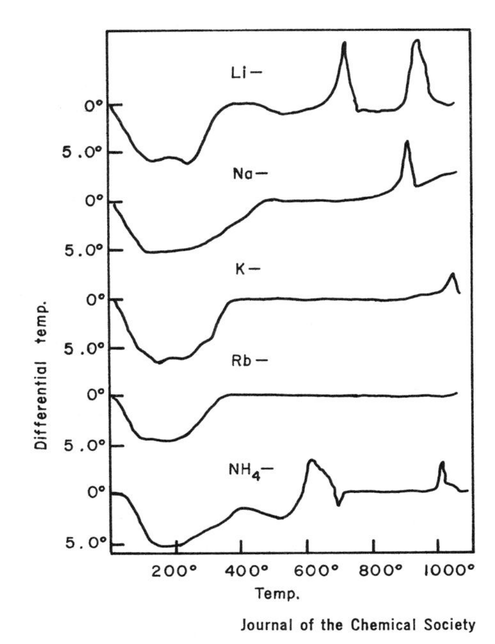

Figure 3. Differential thermograms of ion-exchanged natural chabazites (6)

Although thermal analysis gives valuable information on the changes from heating, it has not successfully provided a general, reproducible, and standard method for measuring and reporting thermal stability. Many factors influence the temperature at which the exotherm is observed— *e.g.*, heating rate, the particular instrument used, atmosphere, sample size. Consequently, although it is often quite good for direct comparisons, it is very difficult to compare results reported by different authors.

The nature and number of metal cations that are present contribute significantly to the stability exhibited by a particular zeolite structure. This fact was recognized in early zeolite work and is the subject of numerous references in the literature. The examples of the natural zeolite chabazite and the synthetic analogs of faujasite in various cationic forms typify the behavior of other zeolites.

Figure 3 shows a series of thermograms for the differential thermal analysis of alkali metal and ammonium cation-exchanged forms of chabazite presented by Barrer *et al.* (*6*). The temperature at which exothermic peaks associated with crystal collapse are observed increases through the alkali series, being at 731°C for Li, 921°C for Na, and 1064°C for K; for Rb, only the beginning of a peak is observed at the limit of the measurements at nearly 1100°C.

The same general behavior of increasing thermal stability with ion size in the alkali series has been observed for other zeolites, including analcite and type A (*5*). Some authors attribute this relation to the relative ability of the various cations to fill the voids in the crystal after dehydration (*6*). The proton and ammonium exchanged zeolites represent a class having complex and unique behavior; these materials are considered later. The thermogram for the ammonium exchanged form of chabazite included in Figure 3 illustrates the complex thermal behavior of these materials.

The synthetic analogs of faujasite provide an interesting example where both the silica-to-alumina ratio and the exchange cations can be varied for materials having the same type of framework structure. Such results from Bremer *et al.* (*4*) are shown in Figure 4, 5, and 6. Figure 4 illustrates the increase in thermal stability with an increase in silica-to-alumina ratio and the change in stability with change in cation composition. Figures 5 and 6 suggest that the relation between stability and cation composition can be varied and complex. Bremer *et al.* attribute much of the complex behavior to specific interactions between the cation and zeolite, leading in some instances to distortion and strain in the framework.

Again, to point out the hazards of any attempt to compare data obtained by different methods and different investigators, we note that the thermal stability of NaY having a silica-to-alumina ratio of between 5 and 6 was found to be about 920°C by Bremer by DTA, about 870°C by Ambs and Flank (*7*) by DTA, and about 700°C by McDaniel and Maher (*8*) by muffle heating. There are sample-to-sample differences in thermal stability, of course, but these should be rather small.

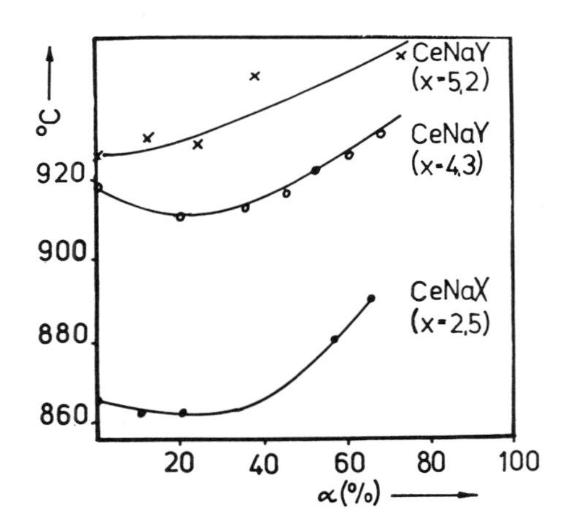

Advances in Chemistry Series

Figure 4. Thermal stability (°C) of cerium-exchanged zeolites X and Y as a function of silica-to-alumina ratio and degree of exchange (4). *Silica-to-alumina ratios indicated by X. Percent of sodium removed by exchange indicated by α (%).*

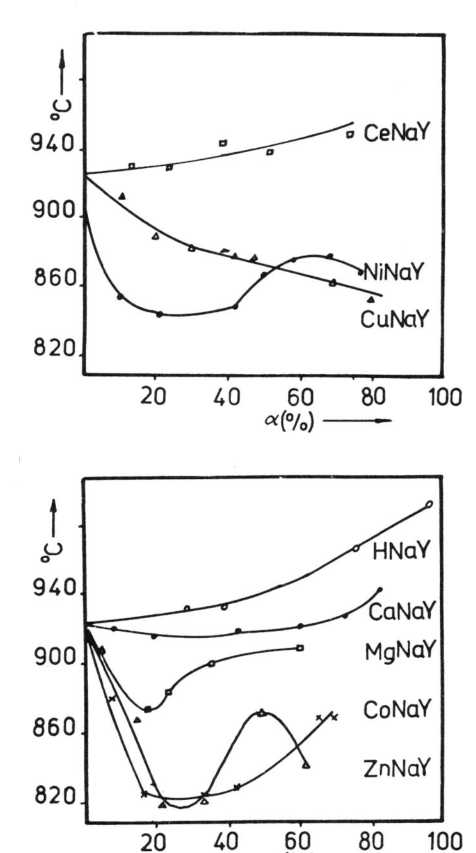

Figures 5 and 6. Thermal stability (°C) of cation-exchanged zeolite Y as a function of degree of exchange (4). *Silica-to-alumina ratios indicated by X. Percent of sodium removed by exchange indicated by α (%).*

In summary, thermal stability increases with silica content and is greatly affected by the choice of cation, with some ionic series appearing to have a fairly systematic influence while others exhibit complex behavior. Some explanations that have been advanced may be valid, but the need for additional work is obvious. Progress would be greatly facilitated by standard techniques for measuring and reporting thermal stability.

Stability of Zeolites in Synthesis Media

Zeolite stability is usually considered from the user's point of view—namely, the crystalline zeolite is a starting material. For the synthetic chemist, however, the primary interest is in the behavior of the crystalline zeolite during formation and in the medium from which it formed. These considerations determine the purity of a particular zeolite and the possible approaches to prepare new and novel types of zeolites.

Generally, zeolites are not formed from a solution but rather from a caustic aqueous mixture of slurry containing gels or other solids. The solid phase that is present just before zeolite crystallization often has the same or similar chemical composition as the crystallizing zeolite. Complete separation of the zeolite from this nonzeolite phase is often difficult or impossible. The difficulties encountered in the study of this type of system are obvious. High purity is difficult to obtain or establish, and analysis and evaluation of the synthesized zeolite are always subject to error because of possible impurity.

Synthetic zeolites, in general, are not very stable in the environment in which they are formed. Zeolite synthesis is a chemistry of metastable or non-

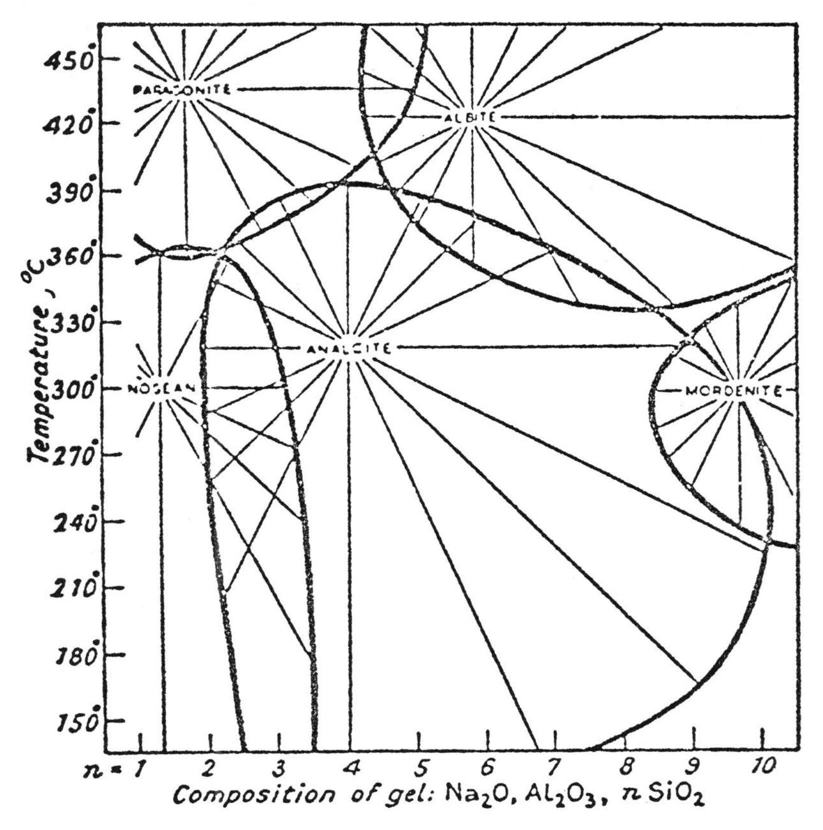

Journal of the Chemical Society

Figure 7. Diagrammatic representation of the approximate areas of formation of products of a hydrothermal synthesis (9)

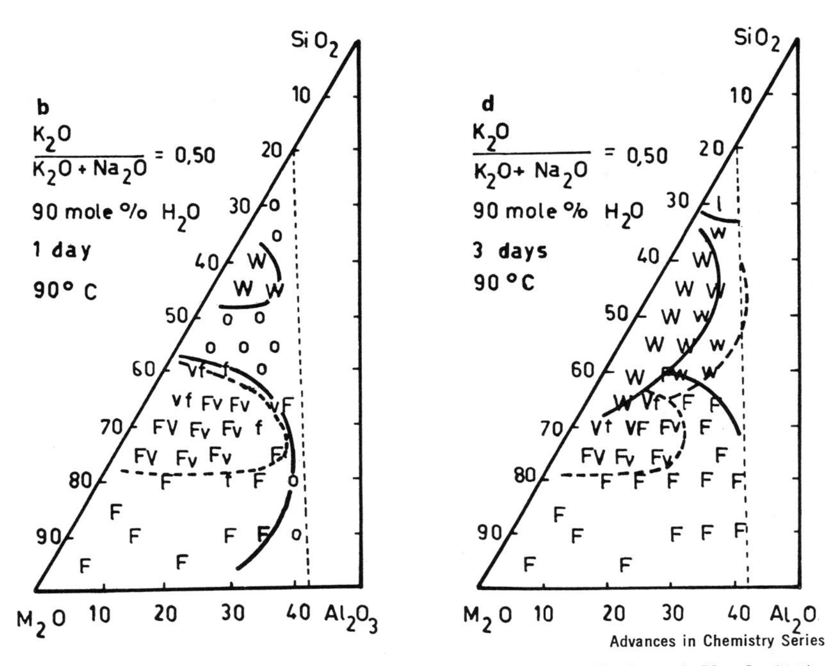

Figure 8. Zeolite syntheses in the mixed-base system K_2O and Na_2O (10). The symbols F, W, V, L, and O refer to the phases, zeolite K-F, zeolite K-M, zeolite V, zeolite L, and amorphous gel respectively. Upper case letters indicate greater than 50% yield and lower case letters indicate less than 50% yield.

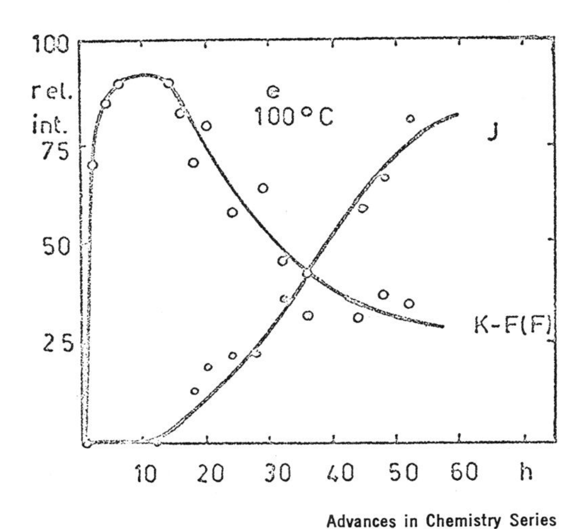

Figure 9. Kinetics of formation of zeolites F and J (10)

equilibrium phases and, in laboratory synthesis at least, the desired zeolite product is more often than not only a transitory phase.

The instability of zeolites during synthesis is seen in the nature of the compositional diagrams employed by workers in the field. Figures 7, 8, and 9 show typical examples of how different authors have graphically represented the products formed in syntheses. The curves in Figure 7 do not represent rigid phase boundaries but indicate only the general region in which the various phases are formed. Also, there are usually areas of overlap where two or more phases are formed.

Because the zeolite phase of interest is often unstable, often recrystallizing to another phase, the compositional diagrams are, of course, time dependent. The time dependence is sometimes represented by sequential diagrams as shown in Figures 8a and 8b, but often this is not done and the reader must exercise caution in interpreting such graphical data since they can be mistaken as representing phase diagrams which are independent of time.

In general, the stability of a given species varies with the exact location within the compositional region where it is formed. In some areas it may never occur in high yield, either not forming completely or degrading rapidly. Two or more competing or consecutive reactions are often possible and the product obtained will depend on the relative rates with which these reactions occur. The relative rates are influenced by reactant composition, temperature, nucleation, and many other factors that are not well understood. The kinetics of formation of two species formed in a synthesis are represented graphically in Figure 9.

When the reactant composition is not optimum for the crystallization of Y-type zeolites, the degree of crystallinity increases to a maximum before it sharply decreases as the Y-type zeolite is recrystallized to a phillipsite-type structure. A-type zeolites show similar behavior with a recrystallization to a sodalite-like phase. Barrer in his many publications on zeolite synthesis describes the recrystallization products of many zeolites.

In summary, zeolites are often unstable in the solutions present following their crystallization. Fortunately, however, we can usually segregate the zeolites from the hydrothermal crystallizing media and recover them in substantial yield. Their stability and yield can be optimized by proper selection of reactant composition and reaction conditions.

Stability of Zeolites in Acid Media

One of the greatest limitations in the application of aluminosilicate zeolites is their limited resistance to acid degradation. Although this phenomenon is often treated in the literature, many of the references do not have the necessary qualifications to make them generally useful. Only recently has an understanding developed of the nature and complexity of the subject. Consequently, caution must be exercised in the interpretation of much of the literature. The wide range of comments on the stability of zeolites to acid attack is seen in the following excerpts. The first comments on the utility of zeolites as ion exchangers:

"They (zeolites) . . . are deficient in chemical stability, decomposing in waters that are slightly acidic (below pH 6.5) or slightly basic (above pH 8.0) and in low silica waters" (*3*).

The next comments on the properties of mordenite:

"Preliminary results show that this material is a unique cation exchanger, operating over the entire pH range. Exchanges of di and tri-valent cations, such as Mg^{2+}, Ba^{2+}, Al^{3+} can be accomplished without change in the crystal structure from the parent material" (*11*).

These statements are, of course, viewed out of context, and a full relization of the author's intent makes them reasonably compatible. However, unless the reader is already thoroughly knowledgeable of all aspects of zeolite behavior, the varied statements can be confusing and misleading. The lack of stability of zeolites in acid media, though presenting severe limitations in some applications, is responsible for the creation of a wide variety of new materials having enormous commercial potential.

The aluminosilicate zeolites have limited stability in acid media because of the solubility of aluminum away from the structure. Hydrated alumina becomes appreciably soluble at a pH of about 4. In the presence of hydrated silica, the solubility of aluminum is suppressed to a slightly lower pH. The zeolites are chemically similar in nature to other aluminosilicate minerals and to the hydrated amorphous silica-aluminas. They are especially susceptible to acid attack because, in most cases, their aluminum is essentially all surface aluminum and receives little or no protection by being buried or inaccessible. Some clays, for example, resist acid leaching of the aluminum because it is largely located between silicate layers and therefore inaccessible to the aqueous phase. Heat treatment delaminates the layer structure, making the aluminum susceptible to acid extraction. Zeolites, by their nature, are extremely open structures and their aluminum is consequently subject to acid attack.

Some of the zeolites can tolerate removal of some or all of the aluminum from the framework without gross collapse, leaving the x-ray diffraction pattern substantially unchanged. This has led some authors to mistakenly conclude that the zeolite remained unchanged. Probably exposure to acid always modifies the zeolite structure somewhat. The effects of such modification are well established in some instances, suspected in others, and undoubtedly undetected in many.

There is very little quantitative data in the literature concerning the stability of zeolites to acid attack. Nevertheless, we can distinguish three categories of behavior concerning their acid stability:

1. Those that cannot be put into the acid from without collapse of the crystalline framework.

2. Those that can be put into the acid form by a conventional ion-exchange process with an acidic solution. (There is undoubtedly some removal of aluminum involved, the amount being dependent on the acid used, its concentration, etc.)

3. Those that can be conveniently put into the acid form only by an indirect method wherein an ion exchange is performed with some ion such as

ammonium, which upon subsequent heating will decompose, leaving the zeolite in the acid form.

Table I classifies a few of the zeolites into these three categories.

Table I. Acid Resistance of Some Zeolites

Group I (Unstable as HZ)	*Group II (HZ can be prepared by acid exchange)*	*Group III (HZ can be prepared indirectly)*
Zeolite A	mordenite	faujasite
Zeolite X	clinoptilolite	chabazite
Zeolite L	erionite	zeolite Y
Cancrinite	heulandite	gmelinite
Sodalite	ferrierite	

The acid environment that a particular zeolite can tolerate without destruction is often described in terms of pH. However, the ratio of hydrogen ion to other exchangeable ions present is also an important criterion. The quantitative effect of hydrogen ions on a zeolite depends on the chemical potential or activity of the hydrogen ion in the zeolite phase, which in turn is governed by the exchange equilibria involved. For the case of only two cations, hydrogen and a metal ion,

$$(M^+)_z + (H^+)_w = (M^+)_w + (H^+)_z$$

$$\left(\frac{a_{H^+}}{a_{M^+}}\right)_z = K\left(\frac{a_{H^+}}{a_{M^+}}\right)_w$$

where the subscripts z and w denote the zeolite and aqueous phases, respectively. Since the zeolite phase has a relatively fixed exchange capacity, for a material predominantly in the metal-ion form, the ratio of hydrogen-ion activity to metal-ion activity in the aqueous phase determines the hydrogen-ion activity within the zeolite.

The study of the acid treatment of mordenite by Baran *et al.* (*12*), which is discussed in more detail in a later section on altered zeolites, gives an excellent view of the nature and complexity of acid–zeolite interactions. With acid treatment, aluminum is extracted from this zeolite without collapse of the structure but with a resulting change in properties. The reaction is complex, since the degree of aluminum extraction is greatly dependent on the thermal history of the zeolite in a manner opposite that observed for clay.

In summary, the action of acids on zeolites ranges from the complete collapse of the structural framework to the formation of the acid form of zeolite, a process that is accompanied to a greater or lesser degree by some aluminum extraction or dealumination depending on the exact acid treatment used. The process of aluminum extraction obviously reduces the number of ion-exchange sites, resulting in a series of adsorbents having modified properties. The process of aluminum extraction is not well understood and depends on a variety of factors, including type of acid, treatment time, treatment temperature, and thermal history of the zeolite.

Functional Stability

The instability of zeolites, regardless of the underlying cause, is of interest insofar as it affects the various functions for which one wishes to use the zeolite. The degree of instability in many applications such as ion exchange has long been understood and discussed. Unfortunately, especially in some of the scientific studies reported (as opposed to commercial application), loss in functional stability has gone unnoticed or unreported. The lack of stability sometimes is manifested in some common applications.

Adsoprtion Applications. Although the ability to separate molecules on the basis of size was not the first property associated with zeolites, it has received such widespread attention that the terms zeolite and molecular sieve are often used synonymously although a variety of nonzeolite materials, including porous glasses and porous carbons, exhibit sieving properties. This aspect of zeolites has been studied in considerable detail by Barrer and others (*13, 14, 15, 16, 17*).

The zeolites (in particular, zeolite A developed by Union Carbide) have enjoyed widespread commercial application of their molecular sieving properties. Zeolite A is especially attractive in this application because its effective pore size can readily be varied by ion exchange. It can behave as a porous crystal having pores of about 3, 4, or 5A, depending respectively on whether it is largely in the K, Na, or Ca ion-exchanged form. Many references are concerned with the effective pore size exhibited by most of the known aluminosilicate zeolites, but few discuss how these values may be altered.

There can be many causes for an alteration in pore size characteristics, and it is often important that the user account for these changes. Effects that result in major destruction of crystalline structure, including exceeding the limit of thermal stability, acid attack leading to collapse, etc., will obviously impair the sieving ability. This degradation can be determined by x-ray analysis or by relatively simple adsorption measurements. Those effects that do not result in major destruction but rather in an alteration having little effect on the gross framework structure are more likely to go unnoticed.

The pore-size properties of most zeolites—*e.g.*, zeolite A— are affected by the nature and amount of exchange cations present. However, as may be seen in a comparison of zeolites A and X in the sodium and calcium forms, the influence is not always predictable. The calcium form of zeolite A exhibits a larger effective pore size than the sodium form while the reverse order is observed for zeolite X. The effective pore size is largely determined by the location of the ions with respect to the apertures. An excellent discussion of the internal channel structure in zeolites is given by Breck (*18*).

The sieving properties of other zeolites, such as mordenite, are greatly affected by acid exchange with its accompanying extraction of aluminum. This has affected some reported results where adsorption characteristics have been attributed to stacking faults in the crystal structure and to small-port and large-port mordenites.

The sieving properties for many zeolites in various ionic forms have been measured and reported by Barrer in his many publications on this subject (*see*, for example, Ref. *19*). Although there is a considerable amount of data on the sieving properties of various materials, essentially all of the reported measurements have been made on zeolites prepared by fairly standard procedures, and these values may be changed by other factors. In particular, heat treatment can sometimes cause a relocation of cations or an alteration of structure that can influence sieving behavior.

The high degree of catalytic activity inherent in many zeolites provides the mechanism for altering their sieving properties. For example, the catalyzed reaction may result in an occluded residue within the zeolite such as carbonaceous deposits or coke formed when zeolites are utilized in the presence of hydrocarbons. The catalyzed reaction can also have an indirect effect such as that seen in the zeolite-catalyzed decomposition of fluoromethanes that can lead in turn to a destruction of the zeolite (*20*).

The possibility of zeolite recrystallization to an entirely different structure must be considered in some applications. Barrer has observed such recrystallization in the use of zeolites for ion sieving where, in some instances, the zeolite recrystallized into other aluminosilicate species when subjected to certain ionic solutions (*21*).

In summary, the sieving properties of zeolites can be altered by the nature and amount of cation present. However, other factors can alter these properties during use. When molecular sieves are used in unusual applications where unknown changes might occur, the sieving properties of the zeolite can often be determined by fairly simple adsorption measurements using standard molecules of established size such as those described by Barrer (*22*).

Ion Exchange Applications. Zeolites found early application in the removal of undesirable ions from water, *i.e.*, for water softening. This was accomplished by ion exchange or an exchange of ions between the aqueous solution and the zeolite phase. In discussing ion-exchange materials, Kunin lists several essential features of a satisfactory exchanger. They include the following:

(1) The functional groups must be present in reasonably large numbers per unit weight or volume.

(2) The functional groups must be accessible to ions in solution.

(3) The ion exchange material must be physically and chemically durable over a wide range of conditions (*23*).

Deterioration of the ability of a zeolite to function satisfactorily as an ion exchanger is most often observed as a loss in the exchange capacity. This may arise from several causes, including fouling of the porous structure, a collapse of the porous structure, mechanical loss in attrition, or the destruction of the exchange site by an irreversible exchange or by acid attack.

The problems of acid attack and the accompanying aluminum extraction discussed earlier are probably the most common hazard encountered and must be considered whenever appreciable hydrogen-ion exchange takes place. This can occur without an intentional hydrogen exchange, especially in solutions of acidic salts. In fact, a high degree of exchange (about 20%) for sodium ions can occur by prolonged washing of NaX with distilled water (*24*). The possibility of significant hydrogen-ion exchange is often overlooked. Proof that awareness on the part of authors is increasing is evidenced by the fact that they are more frequently indicating the importance of hydrogen-ion exchange in the preparation and use of many of the so-called metal-cation forms of zeolites (*24*).

The degree of stability required is often dependent on the length of service required for a particular application. The very severe limitations of zeolites in long exposure to exchange solutions is expressed by Kunin's statement (*3*) that they are "deficient in chemical stability, decomposing in waters that are slightly acidic (below pH 6.5) or slightly basic (above pH 8.0) and in low-silica water." Subjecting zeolites to pH values outside this range for times of shorter duration is, of course, extremely commonplace.

Another fairly common occurrence that reduces the exchange capacity of zeolites is the non-reversible exchange of certain ions. The thermal history of the zeolite is an important factor influencing the degree of reversibility. Again, it is encouraging that authors are beginning to express a greater awareness (*25*).

There are many references to the destruction or recrystallization of zeolite structures by what appears to be a simple ion-exchange treatment. Barrer has shown instances of such destruction in exchange solutions containing NH_4^+, Co^{2+}, and Ba^{2+} in essentially neutral solutions. The explanation of such destruction is not clear.

In spite of limited stability in some instances, zeolites are surprisingly stable in others. Ion-exchange in salt melts has been reported by several authors. Platek and Marinski have reported fused salt exchanges with type A zeolite without measurable loss in structure as determined by x-ray (*26*). Of course, the possibility of other more subtle alterations must be considered.

Catalytic Applications. More zeolites are currently used in catalysis than in any other application. Cracking catalysts for converting petroleum into lighter products including gasoline, fuel oil, and petrochemical feeds have provided the major market for these materials. However, as our knowledge increases, other catalytic applications are rapidly expanding.

In many chemical reactions, zeolites have proved to be excellent catalysts when compared with other known catalytic materials. Their catalytic activity is estimated to be several thousandfold that of previously known catalysts. Fortunately, because of their crystalline nature, these catalysts also lend themselves to scientific investigation to a degree vastly greater than their amorphous counterparts. As a result, much of the former artistry of catalysis is being replaced by hard science.

The successful commercial application of zeolites has closely corresponded to understanding and controlling the factors that affect their stability. Many commercially important applications require a catalyst to endure extremely harsh environments. High temperatures in the presence of steam, for example, are often encountered in the regeneration of catalysts where carbonaceous materials formed as a by-product of the reaction are burned off.

The importance of structural stability of the crystalline framework to the conditions of temperature, steam, acidity, etc., needed for practical utilization of the zeolite is fairly obvious and was recognized during the earlier work on zeolite catalysts. The more subtle changes that can occur in the zeolite are only now becoming appreciated and will become increasingly important in the future as more sophisticated catalytic reactions are required.

The catalysts used in petroleum cracking are far from stable. They have a very limited lifetime, with some of the larger refinery units requiring the addition of as much as 5 tons of catalyst per day to replace the catalyst lost because of mechanical instability or withdrawn because the catalytic properties had degraded. This degradation has several causes, including changes in the zeolite component.

Zeolites, like other catalysts, can be poisoned by substances in the petroleum feedstock. Moreover, metals from the same source can build up on the catalyst and cause undesirable side reactions. Carbonaceous material (coke) formed as result of the cracking reaction rapidly deposits and blocks the pores of the catalyst, prohibiting access of the oil molecules to the active surface. This coke buildup is so extreme that only about 0.25 g of oil is cracked per gram of catalyst. The catalyst must then be regenerated by burning off the coke before it is used again. Constant exposure to the steam and high temperatures of the regeneration cycle causes a gradual destruction of the crystalline structure of the zeolite. In commercial practice, the stability of the zeolite can be affected by the nature of the matrix (*27, 28*).

Of special interest is the stability of the acid sites involved in petroleum cracking reactions. Just how unstable these sites are and how few survive the conditions of use has not been fully determined. The potential catalytic activity of some zeolites is so enormous that only a fraction of the order of 10^{-4} of the available sites are required to account for the catalytic activity actually observed in practical use (*29*). This suggests that the geometry and resulting diffusion limitations are the controlling factors in determining catalytic activity. The stability of the catalytic sites per se should therefore not be very important so long as more than one in ten thousand are maintained and the integrity of the structure is not destroyed. However, in use, the activity is drastically reduced by high-temperature treatments that do not appear to alter the channel structure. Changes in the nature of the catalytic site (or site stability) are therefore extremely important and the acid sites in some zeolites, including the zeolites X and Y, are extremely unstable. Results of cracking-activity measurements shown in Table II illustrate the loss in activity caused by a thermal-steam treatment that approximates actual conditions that the zeolite would encounter in use (*30*).

Although the acidic sites responsible for many catalytic reactions are generally regarded to be associated with the aluminum atoms, mordenite provides an interesting example where catalytic activity can be increased by acid leaching a portion of the aluminum from the structure (*31*). An activity loss may depend on close proximity of neighboring sites leading to a zipper type reaction at elevated temperatures. More isolated sites, although initially fewer in number, may be less subject to this reaction. Alternatively, the acid extrac-

Table II. Loss of Catalytic Activity with High-Temperature Steam

Zeolite Type	Cracking Activity, % Conversion[a, b] Fresh	Steamed
HY	93	50
CaY	78	40
MgY	83	40
REY	84	70

[a] Catalysts were evaluated as 10% zeolite in 90% semi-inert matrix. Values represent conversion of a light West Texas crude at 482°C, 16 WHSV, 6/1 catalyst-to-oil ratio.
[b] Catalyst pretreated 8 hr in 100% steam at 732°C.

tion of aluminum, although it reduces the number of catalytic sites, also creates a more open structure or a larger port mordenite, making the active surface more accessible.

Zeolite Properties. Different facets of stability can be observed in the various applications of zeolites. Although these factors can often be investigated and discussed separately, in practical application they are often interrelated. More sophisticated uses that require the utilization of more than one property or function of the zeolite (*i.e.*, shape-selective catalysis) are growing rapidly and the simultaneous consideration of different aspects of stability will become increasingly important (*32*).

Zeolites having 3-A pore openings are used for drying refrigerants and this application provides another example of how the dual functional nature of zeolites is used and how it influences zeolite stability. A refrigeration system having a refrigerant such as monochlorodifluoromethane (Freon 22) often includes a dehydration or desiccant cartridge that contains a zeolite desiccant to remove and hold any water that might enter the system. In this application, the zeolite desiccant properties are of prime importance. However, a zeolite can also act as a catalyst, causing decomposition of the Freon and the liberation of acids such as HCl, which in turn can attack the zeolite and destroy its desiccant properties (*20, 33*). Advantage is taken of the molecular-sieving properties of a 3-A zeolite to prohibit the entry of Freon into the desiccant, preventing the undesirable catalytic reaction and its resulting damage. Hence, in this application, to prevent the catalytic properties from destroying the desiccant properties, the sieving properties are essential.

Another excellent example of the interrelation of the properties of a zeolite is given by Lee (*34*) while discussing the utilization of 3 A molecular sieves for cracked gas drying:

> The degradation of 3A molecular sieve in cracked gas drying is not from a chemical destruction of zeolite crystals but rather from an accumulated deposit of carbon material on the zeolite. Since a regular 3A, *i.e.*, potassium exchanged type A, is not thermally stable enough to withstand an *in situ* carbon burn-off operation, it is replaced with a fresh charge of 3A molecular sieve when it has accumulated excessive carbon and other hydrocarbon derivatives. A recent patent, however, described a rare earth containing 3A mole-

cular sieve having sufficient thermal stability to withstand normal carbon burn-off conditions. This should prolong the service life of the zeolite adsorbent, and therefore, enhance the advantage of zeolite adsorbent in cracked gas drying over nonzeolitic desiccants.

The choice of a 3A molecular sieve was required to prevent the entry of olefins that would be polymerized by the catalytic action of the zeolite. The utilization of the desiccant properties of prime interest is greatly dependent on the thermal stability, molecular sieving, and catalytic properties.

Some Mechanisms for Structural Changes

We have discussed several manifestations of zeolite instability and some of the factors responsible for their destruction or alteration. We now consider some actual or proposed mechanisms that may be involved. Since the remainder of the chapter deals primarily with the materials derived from the acid form of zeolites, discussion of reaction mechanisms involved in that system is deferred.

Care must be exercised when interpreting zeolite structural stability from x-ray diffraction measurements alone. Large decreases in intensity can be

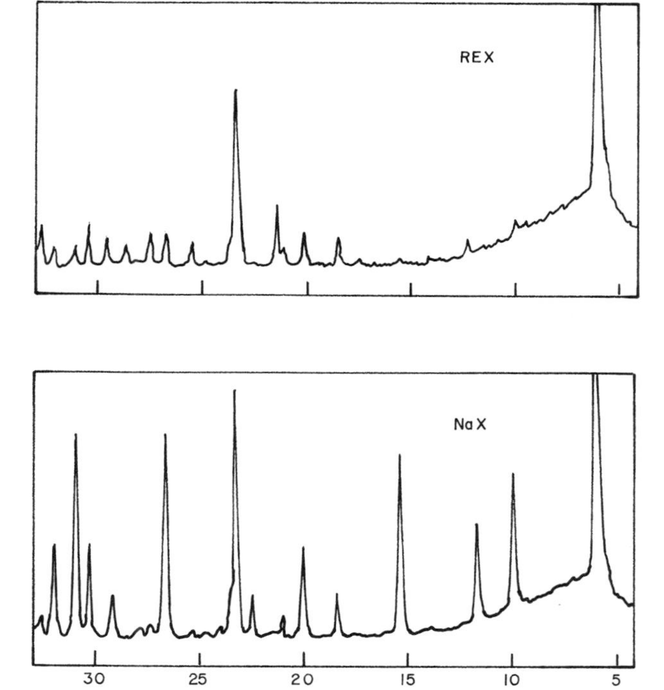

Figure 10. X-ray diffraction patterns for a Y-type zeolite in the sodium and rare-earth forms

observed that are not associated with destruction of crystallinity. Conversely, a material can be altered to exhibit quite different behavior with very little observable change in the powder x-ray pattern. Figure 10 illustrates the large decrease in x-ray intensity observed when the sodium ions of zeolite X are replaced by rare-earth ions. Other measurements indicate little or no change in the degree of crystallinity.

Since many of the mechanisms are still not understood, we attempt to point out only a few of the better understood ones, some of those proposed, and some of the questions raised.

The mechanism involved in the high-temperature collapse of zeolites is still not completely clear. This type of crystallinity loss may occur when the vibrational energy of the framework atoms exceeds the bonding energy that holds them in their ordered array. Indeed, the Si–O–Si bonds are very freely broken at high temperatures, at least in the presence of water vapor. At temperatures well below the collapse temperature, there is an extremely rapid exchange between the oxygen atoms in the zeolite framework and those in H_2O_{18} (*35*). This has also been observed in amorphous silica-aluminas. To explain the extremely rapid, almost instantaneous rate of the oxygen exchange, Oblad *et al.* suggest: "Such exchange could show a chain-like effect if the incoming oxygen caused a Walden Type inversion of the tetrahedron with effects transmitted to adjacent tetrahedra" (*36*). Thus, although the vibrational energy may be quite important, there are also other important factors.

Differential thermal analysis provides some interesting insight concerning the mechanism of thermal collapse. As described earlier, the collapse of the framework is accompanied by a liberation of heat or an exothermic peak, whereas the usual melting of a crystalline material is an endothermic reaction. This same sort of behavior is occasionally seen with some minerals such as clays. The exothermic nature of the crystal collapse has been attributed to the large amount of surface energy associated with the zeolite. At the temperature of collapse, the surface area decreases and the energy associated with the lost surface appears in the form of liberated heat (*37*). If this is correct, the same effect should be observed for the sintering or collapse of high-surface-area amorphous silica-aluminas such as cracking catalysts. This is indeed the case, as can be seen in Figures 11 and 12 where exotherms can be seen as broad but definite shoulders (a) between 900° and 1000°C preceding the carnegieite peaks above 1000°C. However, the differential thermal analysis of silica gel, which collapses at about the same temperature, failed to show such an exotherm (Figure 13). Although we have no evidence to confirm this phenomenon the collapse of the silica gels may be less sharp and consequently not observable in these particular thermograms (*38*).

As in the case of amorphous silica-aluminas, zeolites are destroyed at much lower temperatures in the presence of steam. However, the mechanism is not completely clear. Amorphous silica has two distinct modes of thermal collapse, depending on whether or not water vapor is present. Without water vapor, sintering occurs as a crystallization of silica and thus follows a moving phase boundray. In the presence of water vapor, the surface area is decreased by a movement of silica from the walls of larger capillaries to fill in the

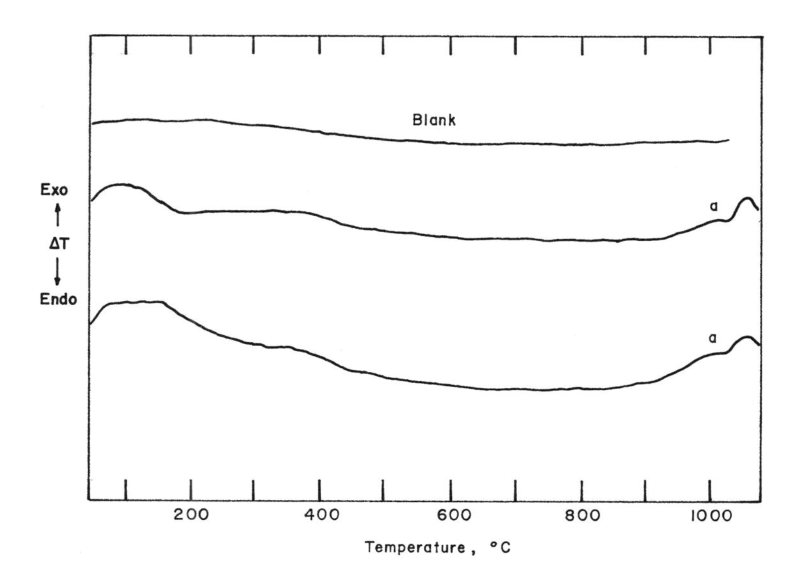

Figure 11. Differential thermograms for two 13% Al_2O_3–SiO_2 cracking catalysts

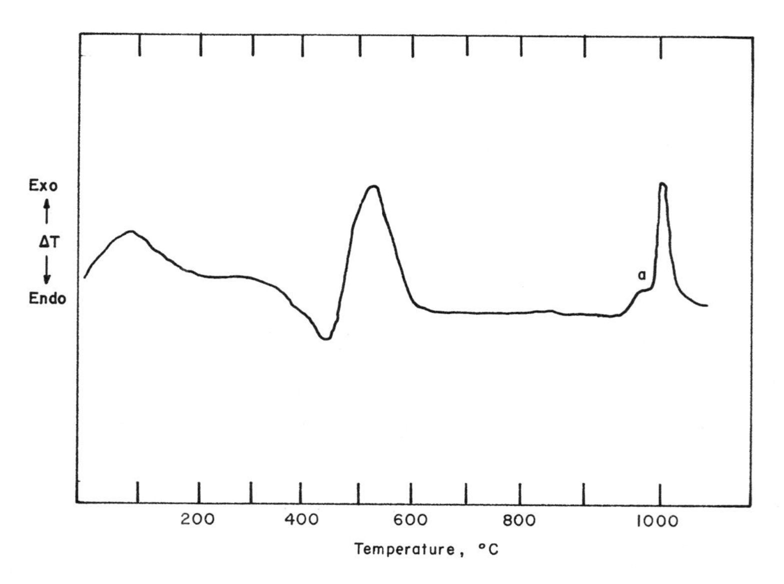

Figure 12. Differential thermograms for a 28% Al_2O_3–SiO_2 cracking catalyst

smaller capillaries. This type of collapse has also been observed in amorphous silica-alumina cracking catalysts and has been described in Iler (*39*). Since amorphous silica-magnesia catalysts are more resistant to steam sintering, the other atoms present can also influence the lability of the silica.

If the same mechanism applied to the collapse of zeolites in the presence of steam, we might expect instances where silica migration fills the zeolite pores—*i.e.*, a stuffed structure. This would result in a greater decline in adsorption capacity than the decline in crystallinity. Although zeolites with excess silica in their structure are possible and have been described (*40*), such behavior as the result of steaming has not been observed.

The behavior of Y-type zeolites during thermal collapse supports the idea of a chain or zipper type reaction wherein the entire crystal collapses rapidly

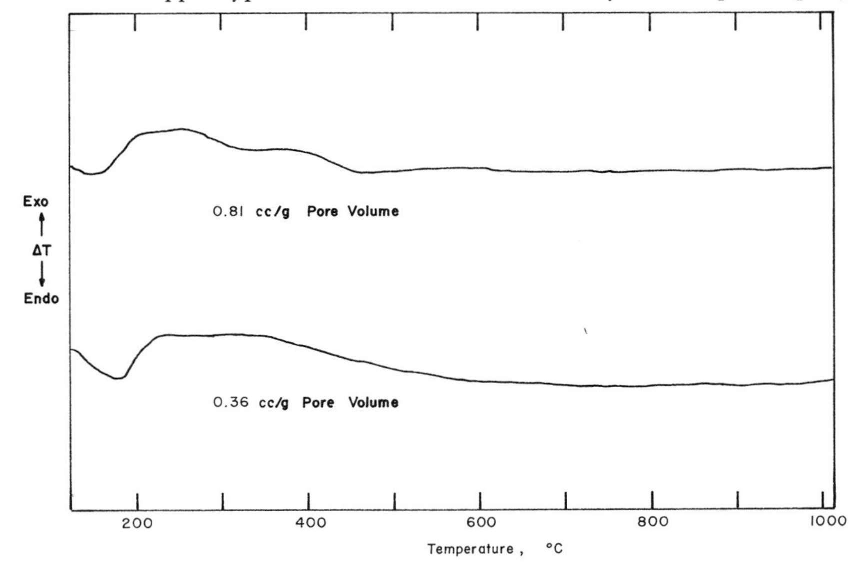

Figure 13. Differential thermograms for two silica gels

once disintegration has begun. The normal crystal size of synthetic faujasite-type zeolites (about 1 μ) is only slightly above the size where broadening of the x-ray lines occurs. If the crystal collapse within a single crystal were a slow process, a point might be observed during sintering when many very small crystal fragments would be present, resulting in substantial broadening of the x-ray lines. Although some broadening is seen, it is very small compared with that for microcrystalline zeolites of less than 0.1 μ. This suggests that once a zeolite crystal begins to degrade, the rate of collapse within that particular crystal is very rapid so that very few partially collapsed crystals are present at any given time.

The role of the cation in zeolite stability also suggests a complex mechanism. Within the alkali metal cation series, as shown in Figure 3, the thremal stability increases with the size of the unhydrated cation (from Li to Cs). An inverse relationship is seen for the melting point of the corresponding silicates and aluminates. The influence of the cation has been attributed to its size and consequently to its space-filling capacity by Barrer (*6*) and to its distorting effect on the lattice by Bremer (*4*).

Increased stability with increased silica-to-alumina ratio is well established and appears to be general whenever isostructural zeolites with different silica-to-alumina ratios occur. However, the same trend does not appear to exist, at least to any major degree, in the amorphous silica-aluminas such as cracking catalysts. In fact, the more aluminous materials may be somewhat more stable in steam atmospheres because of the lower lability of alumina. The different behavior of amorphous silica-aluminas may indicate that their structure has alumina on the surface of silica rather than a true homogenous distribution of aluminum and silicon atoms throughout the structure. The structural stability would then depend primarily on the stability of the base silica structure and would not be greatly dependent on Si–O–Al bonds. Further, amorphous silica-alumina catalysts are commonly prepared by first forming a silica gel and then causing aluminum to be adsorbed by the gelled silica.

Since the length of Al–O bonds is somewhat greater than that for Si–O bonds, the exact configuration of the various ring structures depends on aluminum content. A result of this is seen in the increase in unit cell with alumina content in the faujasites (*41*). Certain compositions might result in more strained configurations with a resulting lower stability. Evidence for the possible existence of such preferred compositional ratios in the faujasites has been presented (*42, 43*), but little or no evidence of any resulting influence on stability is available.

Two reactions that obviously contribute to thermal instability are the migration of silica caused by water vapor and the interaction between protons and the structure (discussed later in greater detail). Both factors may be indirectly related to the silica-to-alumina ratio, and the decreased stability with increasing alumina content observed in zeolites may be directly attributable to the number of exchange sites which reflects, in turn, the detrimental influence of the cation, and particularly, the proton.

The mechanism of acid attack in aqueous systems is probably simple dissolution of aluminum from the structure, depending on the activity of the attacking acid. Because of the unique geometry of the zeolites, the activity of the acid solution is influenced by somewhat different factors than dissolution from a crystal face in aqueous solutions.

Because of the negatively charged framework, the aqueous solution within the zeolite crystal behaves as though it were separated from the external aqueous phase by a semipermeable membrane. Consequently, the distribution of ions between the zeolite phase and aqueous phase is governed by a Donnan membrane type equilibrium. As a result, ionic concentrations within the zeolite are quite different from those in the aqueous phase and depend more on the ratios of competing ions rather than their absolute concentrations. This, along with the tendency to hydrolyze, explains the removal of sodium from the faujasite-type zeolites wherein a significant portion of the sodium can be removed by prolonged washing with distilled water (*24*). Another consequence of the Donnan equilibrium is that electrolyte invasion can occur. In this process, anions from the aqueous phase enter into the zeolite phase with a correspondingly equivalent number of additional cations. This effect becomes appreciable only in concentrated solutions. Many references describe the behavior of electrolytes within the sationary phase of ion exchange systems (*44, 45, 46*).

The unique geometry of zeolites allows each framework atom to be, in a sense, a surface atom, accessible to the solution without disruption of the remaining structure. This is particularly true in certain open zeolites such as faujasite and is obviously true for mordenite from which the aluminum can essentially be removed completely without crystal collapse.

Information concerning the chemical attack of zeolites by non-acid species is sparse. The exception is the considerable work reported for aluminum extraction by the use of chelating agents such as EDTA (*47, 48, 49*). The chelating agent may actually play a direct part in extracting the aluminum, a process that would require electrolyte invasion by the chelating anion. Alternately, its function may be to negate the activity of the aluminum ion. This could be determined by measuring the influence of concentration of the chelating agent. Zeolites are readily destroyed by the action of HF in essentially the same manner as any porous siliceous material.

There have been reports of zeolite destruction by what appears to be a simple ion exchange (*21*). Barrer attributes this in some instances to the removal of a suitable space filling ion which occupied the void. Although no evidence is available, an alternative explanation might be that with certain ions, which because of their size are excluded from competing in the exchange, the resulting exchange is actually a hydrogen ion exchange even in solutions of fairly high pH.

Zeolite crystallinity can often be destroyed by mechanical pressure. Some forms of zeolites can be significantly damaged by the mechanical pressure exerted by a laboratory press and even in some instances by the attempt to form pills in a laboratory pilling machine at unusually high pressures (*e.g.*, at times when the materials were difficult to pill). This mechanical destruction is caused by the high localized pressure that can occur in such operations. The crystallinity of some zeolites can also be significantly decreased by grinding operations such as ball milling. The grinding of amorphous silica gel causes a similar loss in surface area and pore volume. Based on his observation that the mode of sintering is similar to that which occurs at elevated temperatures in the absence of water vapor, Ries (*50*) has suggested that such sintering is caused by the localized frictional heat that is generated. The decomposition products of NaY given a grinding treatment have been investigated by Moraweck *et al.* (*51*) using x-ray, infrared, and adsorption techniques. In all cases the powder was completely amorphous after 10 hr of the grinding treatment.

In systems of extremely small particles, the particle size can become a significant factor in stability because of the large amount of energy associated with the surface (*52*). In the case of zeolites, the concept of surface energy is not as clear because of the existence of external and internal surfaces. In addition, the concept of radius of curvature on which the surface energy depends is difficult to define. In spite of the difficulty in making precise mathematical calculations, crystal size may exert an influence on stability in zeolites since they border on the colloidal size range. In the faujasite system microcrystalline zeolites having crystal sizes less than one-tenth that of conventional X and Y can be prepared which exhibit appreciably lower thermal stability compared to larger zeolite crystals of the same composition (*53, 54*).

Although the importance of crystal size has received some attention in the literature, the references have been concerned primarily with the influence of size on diffusion rates, etc. For example, Chutoransky and Dwyer (*55*) have described the influence of zeolite crystal size on the catalytic properties of an isomerization catalyst. An interesting similarity with zeolite behavior can be seen in the differential thermal analysis of fine-size kaolin presented by Grimshaw *et al.* (*56*) and shown in Table III. Two exothermic peaks as ob-

Table III. Effect of Particle Size on the Thermal Characteristics of Kaolin (*56*)

Average Particle Size, μ	*Endothermic Reaction*		
	T_{max}, °C	*Finishing Temp., °C*	*Exothermic T_{max}, °C*
10–44	600	670	980
0.5–1.0	605	650	980
0.25–0.5	605	630	980
0.10–0.25	600	615	980
< 0.10	600	610	945, 990[a]
< 0.10 dialyzed	605	610	955, 986[a]

[a] Two peaks.

Transactions of the British Ceramic Society

served for zeolite decomposition (Figures 2 and 3) were also observed for kaolin, but only for the smaller particle sizes. The exothermic reaction accompanying crystal collapse is probably attributable to the decrease in surface area with its associated surface energy.

The commercial use of zeolites requires the mechanical stability of the zeolite bodies or particles employed. The zeolites that have found wide commercial application occur as fine crystals about 1 μ in size. However, since this form is generally not useful, they are formed into much larger bodies (balls, extrudates, pellets, etc.) which are composite bodies of zeolite crystals held together by a binder. By their nature, the fine zeolite crystals are somewhat difficult to bind into extremely hard particles while maintaining sufficient porosity to provide access to all of the crystals. Consequently, abrasion and attrition resistance are important considerations in their manufacture and use. Several processes have been successfully applied in manufacturing zeolite bodies of acceptable strength, including clay and inorganic oxide gel binders.

Modification of Zeolites

The term modified zeolite has been used to describe a wide range of materials ranging from the zeolite products of a simple ion exchange to the decomposition products resulting from complete destruction of crystallinity. In the latter instance, authors have described such materials as amorphous type X, amorphous type A, etc. We consider here only some zeolites that have been altered in some manner without complete collapse of the crystalline framework and in an essentially irreversible manner. Such materials include products of irreversible ion exchange processes, some products obtained by modifying what would otherwise be conventional synthesis procedures, zeolites in which various species have been irreversibly occluded, and modifications involving aluminum extraction and/or interaction between protons and the zeolite framework.

Irreversible Ion Exchange. All cations in a zeolite may not be amenable to removal by a conventional ion-exchange process. Many of the natural zeolites and other related minerals having ion exchange properties often contain a portion of such ions (*57, 58*). The presence of nonexchangeable metal ions in natural minerals can arise from three sources. In some structures such as montmorillonite, the metal ion, magnesium, is a part of the rigid or framework structure isomorphously replacing aluminum or silicon atoms. In other instances, what would normally be an exchangeable ion is not accessible, as for example, when the ions are located between the sheets of layer structures. Often such layers can be exfoliated or expanded by an appropriate solvent, making such ions exchangeable. In other instances such as sodalite, metal ions can be present in the form of neutral species such as NaCl or NaOH that have been occluded or trapped within the cagelike structure. The exact nature of some of the nonexchangeable ions in some of the natural zeolites has not yet been established.

Synthetic zeolites can also contain nonexchangeable ions. In these instances we can often more easily deduce their nature since we know the zeolit's history and the circumstances under which the ion might have been incorporated. For example, we now know whether or not the particular ion was an original cation or an exchanged cation. Nonexchangeable cations can occur in synthetic zeolites in all three of the manners described.

About 15% of original sodium ions in type Y zeolite are not readily exchangeable. The exchange behavior of this material has been described by Sherry (*59, 60*), and a method for replacing these sodium ions has been described by McDaniel and Maher (*8*). In addition to nonexchangeability of original cations, ions introduced by ion exchange are sometimes irreversibly held. This suggests that some of the nonexchangeable cations found in natural zeolites might have originated from subsequent exchange of the original cations.

One mechanism by which an irreversible exchange can occur is a consequence of the structural geometry and, in particular, the cage-like units. Because of the electronic distribution in a zeolite structure, the cation favors a discrete set of preferred locations or sites of lowest energy, and a certain activation energy is required to remove a cation from such a site. If the required activation energy is high enough, the cation will not be removed by a conventional ion exchange.

For geometrical representation of this phenomenon, consider that many zeolites contain cells or cages within the structure that can be entered only through windows or rings of silicon, aluminum, and oxygen atoms. If the size of a particular cation is very nearly the size of the window, appreciable energy may be required for it to pass through. If for some reason the ion temporarily possesses the energy to enter, it may be irreversibly trapped or locked in. The sodium ions in type Y zeolite that are difficult to exchange are located in cages not easily accessible to most ions (*60*).

Maes and Cremers (*25*) have shown how transition metal cations introduced into X and Y zeolites at a higher temperature can be nonreversibly held to lower temperature exchanges. Although they present no structural analyses, the higher temperature exchange may provide sufficient energy for ions to occupy sites from which they could not be removed by a subsequent exchange at lower temperature. Maes and Cremers interpret the sites involved to be those within the sodalite cage structures present in these zeolites.

Extensive structural analysis has been directed toward the elucidation of cation location in faujasite-type zeolites. In certain instances, ions of the rare-earth metals are irreversibly held after a heat treatment. Olson *et al.* have shown that the heat treatment causes exchanged rare-earth ions to move from the large open channels into the sodalite cages (*61*). Maher and McDaniel have described a process for taking advantage of this redistribution to remove undesirable locked in ions and to irreversibly introduce desirable ions by aid of a heat treatment (*62*).

The phenomenon of irreversible exchange can depend on exchange temperature, drying, and heat treatment of the zeolite. The published information concerning such behavior for the large variety of possible ions and the large variety of zeolites is very sparse. Most exchange studies have been carried out under a single set of conditions and those who wish to use zeolites under conditions where nonreversible exchange may occur should be aware of the published exchange studies and should interpret them with appropriate caution.

Occlusion in Zeolites. Modified varieties of many zeolites can be prepared by occluding extraneous species within the zeolite crystal either during or after synthesis. Synthetic zeolites containing occluded species have their natural counterparts, such as sodalite, which may contain occluded NaCl or NaOH. Many of these minerals have been investigated by Barrer and others. Barrer suggests that the occluded species sometimes act as mineralizers by providing a form or template around which the cage-like zeolite can more readily crystallize (*63*). For a comprehensive discussion of this topic, refer to Chapter 5, "Salt Occlusion in Zeolite Crystals."

Various metal silicates and oxides have been incorporated into faujasite-type structures (*64, 65*). Unfortunately, a limited amount of structural analysis of such materials has been reported and the exact structures are not well known. X-ray structural analysis (*66*) of a faujasite-type zeolite produced in the laboratory, in which cerium was incorporated during synthesis, indicated that the cerium existed as a hydroxyl complex within the sodalite cage (*67*).

Another type of modified zeolite can result from either advertently or inadvertently decomposing adsorbed compounds, leaving occluded species within the structure. This technique has been widely used to prepare zeolites containing such occluded species as metallic Pt, Pd, etc. (*68, 69*). Additional silicon has been incorporated into a Y zeolite by McAteer and Rooney (*37*) by a process in which tetramethylsilane was adsorbed and decomposed to liberate methane. The catalytic properties of this material were determined.

A most interesting and commercially important example of this type of occluded species is that of carbon laydown or coking often observed when zeolites are used in the presence of hydrocarbons. The splitting or cracking of the hydrocarbon molecules results in a carbon-rich residue that remains occluded in the zeolite and eventually prevents the diffusion of other moleclues. Such occluded material naturally decreases the useful function of the zeolite, necessitating a periodic burn-off or regeneration. This occlusion is so severe during the catalytic cracking of petroleum that 1 lb of catalyst can effectively be used to crack only about 0.2 lb of oil, after which its activity must be renewed by regeneration.

Aluminum Extraction. A most interesting series of modified zeolites is associated with an extraction of aluminum from the framework and/or the interaction between protons and the zeolite structure. Several facets of such reactions have been observed, but only recently has the relationship of the various phenomena been recognized. The investigation of interactions between protons and zeolites was probably ignored for so long because of the widely held belief that zeolites were extremely unstable in acid media. In fact, the existence of the hydrogen form of zeolites was extensively questioned (*70*).

Aluminum can be extracted, at least in part, from certain zeolite structures without the collapse of the crystalline framework. Barrer described the essentially complete removal of aluminum from $Na_{1.0}$–$(Al_{1.0}Si_{4.65})$ clinoptilolite without loss of crystalline structure by treatment with the strong mineral acid HCl (*71*). He characterized several variously dealuminated clinoptilolites by their CO_2 and krypton adsorption. The cation originally associated with the removed aluminum was, of course, also removed, leading eventually to a crystalline silica having no ion-exchange character. Consequently, Barrer referred to this aluminum extraction process as decationization, a term that has been widely used more recently in a somewhat different context.

Mordenite, erionite and high silica zeolite L, also high silica zeolites, can be dealuminated by direct leaching with strong mineral acids such as HCl. The aluminum extraction of mordenite, in particular, has received considerable attention because of its commercial availability and application (*12, 72, 73*). Kranich *et al.* (*72*) prepared a series of aluminum-deficient mordenites ranging in silicon-to-aluminum ratio from 6 to greater than 600 by a combination of thermal acid treatments. They also investigated the catalytic and adsorption properties of these materials.

The mechanism of aluminum extraction by strong mineral acids apparently involves at least a two-step process of hydrogen-ion exchange followed by replacment of the AlO_4 tetrahedron by four OH groups as shown in

Figure 14. However, other aspects of the process are still not clearly established. For example, electrolyte invasion, wherein additional acid beyond the exchanged protons enters the zeolite structure, may play an assential role, or the acid in the external phase may merely provide the opportunity for the ionized aluminum to leave the zeolite. A third step, wherein the partially dealuminated zeolite is in the aluminum cationic form, may be involved.

We first consider the possibility that aluminum extraction occurs only at the outer boundary of structural aluminum and proceeds inward as aluminum is extracted. This would make the acid attack essentially the same as normally observed for the dissolution of aluminum at the phase boundary of a heterogeneous system; the electrolyte invasion, cation migration, etc., are therefore probably not essential. The evidence would not seem to support this type of mechanism. A direct result of aluminum extraction of this type would

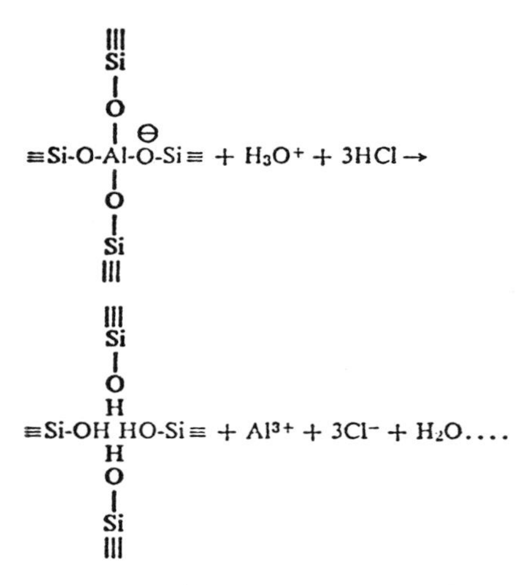

Figure 14. Aluminum extraction from a zeolite structure

be the creation of two distinct types of zeolite surface, the outer portion showing properties of a crystalline silica and the interior core showing that of the initial zeolite. Barrer's characterization of aluminum deficient mordenite, however, suggests otherwise, namely that the zeolite surface changes in a more uniform manner.

If aluminum extraction by strong acid occurs in a fairly homogeneous manner throughout the zeolite, several possible mechanisms can replace the AlO_4 by the four OH groups. In aluminum extraction from mordenite by HCl, three sources for the three additional protons may arise. They may be derived from invading HCl by the reaction:

$$Al^{3+} + 3\ HCl = AlCl_3 + 3\ H^+;$$

they may result from the liberation of three exchange protons by Al^{3+} cations;

or they may, at least, in part arise from the partial hydrolysis of Al^{3+} ions by reactions such as,

$$Al^{3+} + H_2O = [Al(OH)]^{2+} + H^+$$
$$Al^{3+} + 2H_2O = [Al(OH)_2]^+ + 2\,H^+.$$

It has not been established which mechanism or mechanisms are involved.

Electrolyte invasion may play an essential role in the acid extraction of aluminum. To obtain significant aluminum extraction from mordenite, high acid concentrations are necessary. Kranich *et al.* found that two hours boiling in 1*N* HCl was necessary to extract about 50% of the aluminum and that 6 hr boiling in 6*N* HCl was required to extract about 80% (*72*). These conditions are much more severe than would be required to extract aluminum from amorphous silica-aluminas and are expected if the aluminum extraction were dependent on electrolyte invasion.

Considering the distribution of ions between the zeolite phase and the external aqueous phase to be governed by a Donnan-type equilibrium, the activity of HCl will be equal in each phase, *i.e.*,

$$(a_{\mathrm{HCl}})_{\mathrm{z}} = (a_{\mathrm{HCl}})_{\mathrm{w}}$$

where the subscripts z and w refer to the zeolite and water phases, respectively. Expressing activities in terms of molalities and activity coefficients,

$$(m_{\mathrm{H^+}})_{\mathrm{z}}(m_{\mathrm{Cl^-}})_{\mathrm{z}}(\gamma^2_{\pm\mathrm{HCl}})_{\mathrm{z}} = (m_{\mathrm{H^+}})_{\mathrm{w}}(m_{\mathrm{Cl^-}})_{\mathrm{w}}(\gamma^2_{\pm\mathrm{HCl}})_{\mathrm{w}}$$

or

$$(m_{\mathrm{Cl^-}})_{\mathrm{z}} = \frac{(m_{\mathrm{H^+}})_{\mathrm{w}}(m_{\mathrm{Cl^-}})_{\mathrm{w}}}{(m_{\mathrm{H^+}})_{\mathrm{z}}}\;\frac{(\gamma^2_{\pm\mathrm{HCl}})_{\mathrm{w}}}{(\gamma^2_{\pm\mathrm{HCl}})_{\mathrm{z}}}$$

Ignoring the activity coefficients, the concentration of invading HCl is given roughly by

$$(m_{\mathrm{HCl}})_{\mathrm{z}} = \frac{(m^2_{\mathrm{HCl}})_{\mathrm{w}}}{(m_{\mathrm{H^+}})_{\mathrm{z}}}.$$

Since the internal molality in mordenite is roughly 6*m*.

$$(m_{\mathrm{HCl}})_{\mathrm{z}} \simeq \frac{1}{6}(m^2_{\mathrm{HCl}})_{\mathrm{w}}.$$

Electrolyte invasion will not result in appreciable HCl concentration within the zeolite until the external concentration is of the order of 1*m*. There are, as yet, no published results of experimental measurements of electrolyte invasion during aluminum extraction.

Strong evidence supporting the role of hydrogen exchange as an essential step in aluminum extraction is supplied by the work of Baran and Belen'kaya (*12*). If we assume a two-step process for which the initial step is a proton exchange, a treatment that reduces the exchangeability of the original cation should result in reduced aluminum extraction. Baran and Belen'kaya found that the heat treatment of Na-mordenite decreases the degree of exchange of Na^+ by H^+. They conclude that at least some of the Na^+ ions, as a result of the thermal treatment, have changed their positions in the lattice, moving to places whose accessibility is more difficult. This agrees with other observations that cations can be relocated by thermal treatment (*8, 61*). Baran and Belen'kaya observed a corresponding decrease in the susceptibility of the heated zeolite to dealumination.

Baran and Belen'kaya also found that heat treatment in the presence of water vapor reduces the susceptibility to dealumination to a much greater degree than a similar heat treatment in the absence of water. The role of the water vapor has not been established, but perhaps it facilitates the relocation of Na^+ ions to less accessible positions by its labilizing effect on oxygen in the structure, with the resulting decrease in ion exchangeability leading to a decreased susceptibility to dealumination.

In faujasite type zeolites which are destroyed by strong acids, dealumination has been accomplished by two other distinct methods—*i.e.*, by (1) aluminum extraction by chelating agents such as EDTA, and (2) a process involving thermal treatment, in particular, the thermal treatment of the ammonium form of the zeolite. In such instances, where aluminum extraction is not accomplished by direct attack by strong mineral acids, the process appears to be somewhat different (discussed later).

The behavior of mordenite following dealumination has not been investigated to any great extent. Thermal treatment of the hydroxylated structure resulting from dealumination may result in the elimination of water and the formation of new Si–O–Si bonds. This loss of chemical or constitutive water does, in fact, occur in zeolites (*74*) and in some instances is accompanied by or associated with the formation of a structure having an unusually high degree of stability.

Modified Y-Type Zeolites Involving Proton-Zeolite Interactions

Most of the zeolite alterations and modifications we have considered represent fairly general phenomena that occur in a wide variety of zeolites. This is probably true also of the complex set of reactions that can occur between protons and a zeolite framework, but, Y-type zeolites have been the only ones for which many of these phenomena have been investigated to any great degree. This follows from the commercial importance that Y-type zeolites have attained and to their unique combination of properties, including a relatively high degree of thermal stability, a large open pore structure, and a reasonable tolerance to acid attack. Consequently, the remainder of our discussion will be centered on modifications of Y-type zeolites.

Structures and Reactions Related to Proton-Zeolite Interactions in Y-Type Zeolites. Several terms have been used to describe the complex maze of reactions and products from these zeolites, but unfortunately the same or similar terms have been used by different authors with different meanings. The complexity of communication is compounded because the true nature of a specific material with which an investigator is working is often unknown or subject to question. When dealing with materials of uncertain character, many authors have applied many of the terms indiscriminately, and it is often not clear to which specific substance or reaction an author is referring.

A generally accepted terminology will probably have to await the firmer establishment of the validity of the presently postulated structures and reactions. Hopefully, a move toward standardization of terminology in the entire zeolite area will be made by IUPAC in the near future.

We will not, in general, attempt to characterize specific materials made by a particular process. An endless variety of modified zeolites have been reported, and the exact nature of a particular one is often open to question. Several phenomena can and do occur, and we discuss these without entering the controversial area of attempting to fit a product made by a specific process into a particular category.

Various terms have been used in discussions of modified Y-type zeolites. We briefly outline the meaning of some of these terms as used by various authors and describe their role in zeolite modification. However, a full characterization of any particular material or process described by an individual author may involve one or several of these descriptive terms. We do not attempt to establish any preferred usage since this has been and is being considered by others, but we use the set of terms as a focal point for discussing the various structures and reactions that are postulated.

HYDROGEN Y AND PROTONATED Y. These terms have been used interchangeably to denote the hydrogen-exchanged form of the zeolite. There is

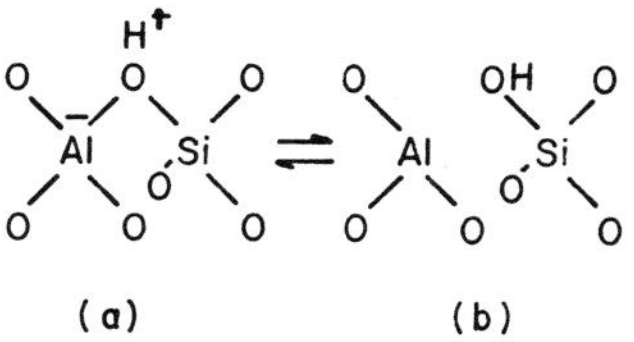

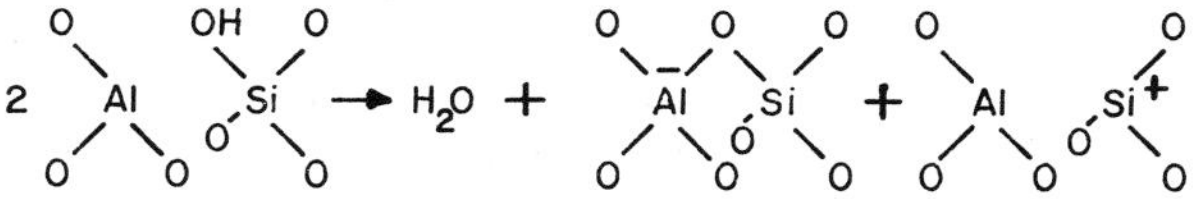

Figure 15. (1) Equilibrium between the Brønsted and hydroxyl forms of a hydrogen zeolite and (2) the reaction of dehydroxylation in zeolites

disagreement among authors as to what materials should be properly given these labels. Some feel that the term hydrogen should be reserved for only those materials made by a direct hydrogen-ion exchange. Others have applied the description to include the products of low-temperature thermal decomposition of ammonium and other nitrogenous-base-exchanged zeolites.

Regardless of nomenclature, the hydrogen form of zeolite can exist, in contradiction to earlier doubts (*70, 75*), as an equilibrium between a Bronsted acid and the hydroxyl form as represented in Equation 1 of Figure 15. For Y-type zeolites, the equilibrium at room temperature strongly favors the hydroxyl form (*74*). Hydrogen ions can be introduced into Y-type zeolites by a carefully controlled exchange or by the indirect route of the thermal decomposition of the ammonium-exchanged form. The hydrogen form is extremely active and unstable at elevated temperatures and plays an important part in subsequent reactions such as dealumination, decationation, and stabilization.

AMMONIUM Y. This term signifies simply an ammonium-exchanged form of the zeolite. Many authors have failed to specify the degree of exchange, and the term has been widely applied to zeolites containing appreciable quantities of other ions, particularly sodium. The term has also been used to describe materials that actually no longer contain nitrogen, such as calcined ammonium Y. The term is quite prevalent in the zeolite literature because of the importance of ammonium-exchanged zeolites as intermediate products in the preparation of other materials.

DEAMINATED Y. The term deamination has been used to describe the liberation of NH_3 from the ammonium-exchanged form of the zeolite by heating. The deamination reaction as described by Uytterhoeven *et al.* is shown in Figure 16. Reactions involving the liberation of NH_3 from ammonium-

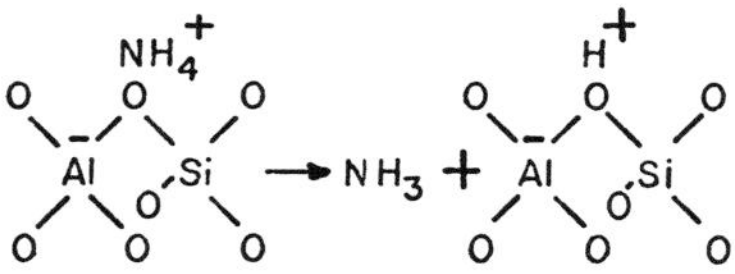

Figure 16. The reaction of deamination in zeolites

exchanged zeolite when subjected to thermal treatment have received much attention because of their role in preparing other products, including decationated, hydrogen, and stabilized zeolites (*74, 76, 77, 78, 79, 80*).

DECATIONATED Y AND DECATIONIZED Y. These terms, unfortunately, have not been used consistently by different authors. Barrer applied the term decationization in referring to the removal of aluminum from a zeolite structure, or dealumination. This is quite justified, of course, since the removal of an aluminum atom is necessarily accompanied by the removal of the associated cation. Rabo favors the term decationized to designate those zeolites which have been heat treated following an ammonium exchange. Others have used the term to denote that the equivalent number of metal cations is less than that of aluminum. This use is based on the fact that non-metal ions and, in particular, hydorgen ions are often not included in the chemical analysis of a zeolite.

The decationization reaction described by Rabo *et al.* (*81*) is the same as that described by Uytterhoven *et al.* and termed dehydroxylation (Figure 15). This reaction results in trigonally bonded aluminum and silicon atoms in the zeolite structure. Authors have generally equated the expression percent decationization to the percent of ammonium exchange, implying that the reaction in Figure 15 has proceeded to completion.

CATIONATED Y AND PARTIALLY CATIONATED Y. The term cationated has been used to refer to zeolites containing ions other than hydrogen. This usage reflects that there is generally no analysis for protons or hydorgen in zeolites and, hence, they are not included as cations. The term partially cationated has in the same manner been used to describe zeolites that may be considered to contain both protons and a metal ion such as NaHY,, KHY, etc. Thus, the term partially cationated is synonomous with the term decationated used by some authors.

DEPROTONATED Y. This term designates a reaction wherein a hydrogen zeolite has been heated, with a resulting decrease in exchange capacity. The term has been used synonomously with decationation. Although it may appear less ambiguous, it does not specify any particular mechanism or product.

ALUMINUM Y AND ALUMINUM-HYDROGEN Y. These terms have been used to signify that the zeolite is at least partially in the aluminum cationic form. Many earlier references to aluminum-exchanged zeolites are probably misleading since a high degree of aluminum exchange by direct methods does not, in general, occur. The exchanged aluminum may have been deduced from the decrease in other cations upon treatment with solutions of aluminum salts. Such decrease is largely accounted for by hydrogen-ion exchange in the acidic aluminum-salt solution. More recent work has been reported where aluminum-hydrogen Y zeolites have been prepared and characterized (*82*). Recently, the existence of Y-type zeolites containing cationic aluminum derived from a partial extraction of framework aluminum has been demonstrated by Kerr (*83*).

AMORPHOUS Y. This term has been applied to the amorphous degradation products resulting from the destruction of Y-type zeolites. It is, of course, a contradiction in terms since Y-type zeolites are, by definition, crystalline materials. Since the possible products resulting from destruction of crystallinity are indefinite and varied, there is little justification for the use of this term.

METAL-HYDROGEN Y. Such terms as sodium-hydrogen Y, rare-earth-hydrogen Y, etc., have been used to designate materials that were believed to contain significant amounts of both hydorgen and metal cations. In some instances there have been proper grounds for such belief; in most instances, however, the amount of hydrogen ion has been inferred from the difference between the aluminum and metal cation analyses. These terms have most often been applied to the calcination products of mixed ammonium-metal forms of the zeolite. Thus, the term metal-hydrogen Y is synonomous with the terms partially cationated and decationated used by some authors. Many zeolites used in commercial catalytic applications have received such treatment.

DEHYDROXYLATED Y. This term has been widely used as a result of the work of Uytterhoeven *et al.* (*74*) to signify the elimination of water from hydroxyl groups that originated from the hydrogen form of the zeolite. Their concept of the process of dehydroxylation is shown in Figure 15. Uytterhoeven *et al.* and others (*84*) present evidence for the existence of an equlibrium between the Bronsted acid form (A) of zeolite with hydroxylated acid form (B) as shown in Equation 1 by showing the extensive presence of hydroxyl groups. Equation 2 illustrates the elimination of water from two such hydroxyl groups, a process that results in trigonally coordinated aluminum and silicon atoms. The reaction in Equation 2 has been termed dehydroxylation.

ALUMINUM DEFICIENT Y AND DEALUMINATED Y. These terms have been used interchangeably to denote products resulting from a partial extraction of aluminum from the structural framework of a Y-type zeolite without collapse of the structure. An aluminum-deficient framework structure appears to be characteristic of ultrastable zeolites. Although it has not been the practice, these terms can properly be applied to those zeolites that have had aluminum extracted from the framework but not removed from the crystal. These structures are certainly aluminum deficient, even though it is not obvious from the chemical analysis.

Aluminum-deficient Y-type zeolites have been prepared by two distinct methods: (1) removal of aluminum with a chelating agent, *e.g.*, EDTA, and (2) heat treatment of the ammonium or hydrogen form of the zeolite followed by removal of the extracted aluminum by ion exchange or by complexing agents. Both of these methods appear to require a prior step of hydrogen exchange of the zeolite.

Kerr first proposed and demonstrated that aluminum extraction from the tetrahedral positions occurs in the preparation of ultrastable zeolites. He also demonstrated an alternate method of aluminum extraction from Y-type zeolites by treatment of sodium Y and ammonium Y with the acid form of EDTA. This work has been described in an excellent series of articles on zeolites related to hydrogen Y (*47, 48, 77, 83, 85, 86, 87*). The preparation and properties of aluminum deficient Y-type zeolites have also been described by others (*88, 89, 90*).

STABILIZED Y, ULTRASTABLE Y, AND Z-14-US. Y-type zeolites can undergo a transformation that results in structures with gross structural properties similar to those in the parent zeolite but with greatly enhanced stability and drastically different other properties. The chemical composition of these unusually stable structures is almost indistinguishable from that of the acid-exchanged zeolites, which are extremely unstable. Although the exact nature of these ultrastable zeolites is not yet fully understood, our understanding of their structure and the reactions involved in their formation has recently increased greatly.

Highly stable zeolites of this sort have been referred to by various authors as stabilized Y, ultrastable Y, and Z-14-US. The term stabilized Y is rather broad and has been also employed to describe zeolites whose stability is associated with a metal cation. The term ultrastable Y or ultrastable structure has been generally accepted to describe the zeolites of our discussion. Z-14-US is a specific material described in United States patent 3,293,192 (*91*).

METASTABLE Y, STABILIZABLE Y, AND UNSTABLE Y. These terms apply to the low-sodium ammonium or hydrogen zeolite which upon heating to temperatures above *ca.* 650° will either convert to the ultrastable structure with a shrinkage in unit cell size or will collapse to an amorphous phase. McDaniel and Maher applied the term metastable to the former state and unstable to the latter. However, stabilizable may be a more descriptive term than metastable.

ULTRASTABLE ZEOLITES. Since ultrastable zeolites were first described (*8*), several methods have been reported for preparing, from Y-type zeolites, materials that have ultrastable properties. A complex set of reactions either cause or accompany the stabilization of the structure and some of these reactions may vary depending on the method of preparation. However, all of the methods described involve certain common reactions and structures. Consequently, the ultrastable properties common to these materials follow from a single stabilization phenomenon.

The ultrastable structure is characterized by a significant contraction in unit cell dimensions of the order of 1% to 1.5% compared with the parent sodium zeolite. This contraction is shown graphically in Figure 17. The parent sodium zeolites have a definite relationship between silica-to-alumina ratio and unit cell dimension (*41, 8*). When converted to the ultrastable structure, the unit cell size decreases to some point within the range indicated. The exact amount of decrease depends on several factors, including sodium content, degree of aluminum extraction, and calcination conditions. The full relationship between this contraction of the structure and stabilization is not completely clear, but certain things are apparent.

The contraction of the structure is accomplished only when a significant number of the aluminum atoms have left their position in the framework. This was proposed and demonstrated by Kerr when he first showed that after stabilization by thermal treatment, cationic aluminum was present in the ultrastable structure and could be removed by ion exchange. Kerr later showed that ultrastable zeolites could be prepared by an alternate method wherein aluminum was extracted from the structure by use of a chelating agent such as EDTA (*83, 47*).

The ion exchange capacity of the ultrastable zeolite is substantially lower than that of the parent zeolite. This, of course, follows from the extraction of framework aluminum. The presence of residual cations limits the thermal stability and the sodium must essentially be removed in order to obtain maximum stability. Nevertheless, the cell shrinkage with accompanying stabilization can be achieved when an appreciable amount of sodium is present especially in the process described by Kerr (*47*).

PREPARATION OF ULTRASTABLE ZEOLITES. Although several procedures have been reported for the preparation of ultrastable zeolites, we describe only three in detail. The other methods reported are essentially variations of those described here.

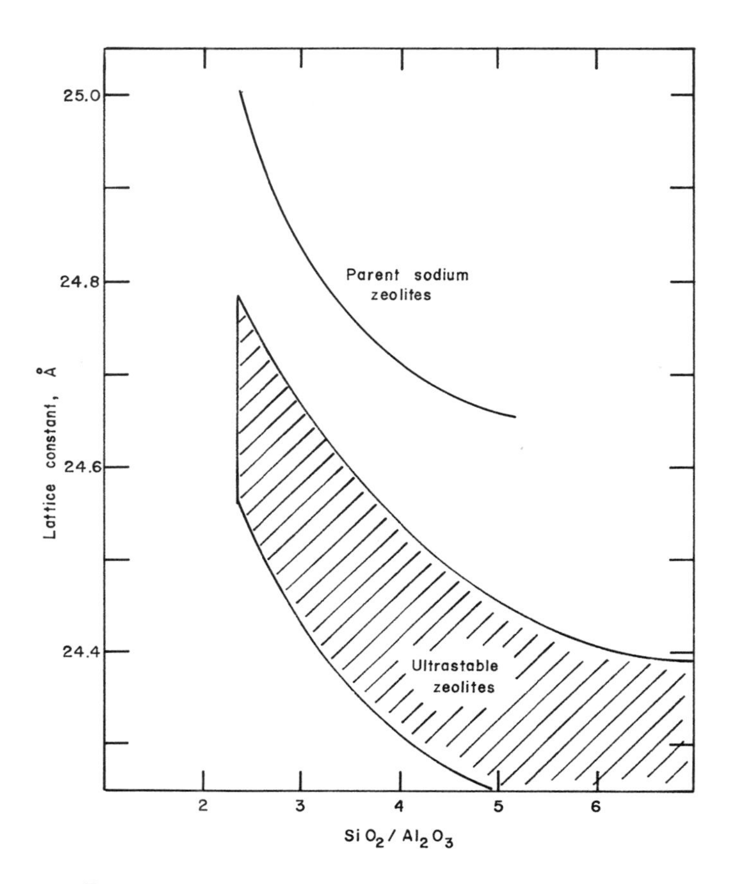

Figure 17. Unit cell contraction in ultrastable zeolites

PROCEDURE A. This procedure illustrates the stabilization of the structure following a simple ammonium exchange and while the zeolite still contains 10 to 25% of its original alkali-metal cations.

(1) A Y-type zeolite is ion-exchanged with an ammonium-salt solution to reduce the level of sodium to about 10 to 25% of its original value. The remaining sodium is increasingly difficult to remove by conventional exchange because of its location in the zeolite.

(2) The partially exchanged zeolite is washed free of excess salts and stabilized by heating in a static or steam-containing atmosphere to a temperature between 600° and 825°C, so that the unit cell shrinks about 1% or about 0.2 to 0.3 A.

(3) The thermal stability of the stabilized structure can be increased still further by removing the major portion of the remaining sodium which has now become readily exchangeable.

PROCEDURE B. This procedure illustrates the stabilization of a zeolite after essentially complete removal of the alkali-metal cation.

(1) A Y-type zeolite is ion-exchanged with an ammonium-salt solution to reduce the level of the sodium to about 10 to 25% of its original value.

(2) The partially exchanged zeolite is washed free of excess salts and heated to a temperature between 200° and 600°C so that the sodium ions have been redistributed but the structure has not undergone significant shrinkage.

(3) The major portion of the remaining sodium is removed by again ion-exchanging with an ammonium-salt solution. The zeolite at this stage is in an extremely precarious metastable state and can easily be made unstable by a variety of conditions not yet fully understood.

(4) The low-sodium, metastable zeolite is rapidly heated in a static or steam-containing atmosphere to a temperature between 600° and 800°C to minimize the time during which the hydrogen form might be present. This stabilizing calcination causes shrinkage of approximately 1 to 1.5% in the unit cell.

PROCEDURE C. This method described by Kerr (*47*, *48*) illustrates the preparation of an ultrastable structure in which framework aluminum is extracted from a Y-type zeolite with a solution of the acid form of EDTA.

(1) Approximately 0.25 to 0.50 mole of H_4EDTA per equivalent of zeolite cation is slowly added to a slurry of the zeolite in water under refluxing conditions, the complete addition requiring at least 18 hr.

(2) The aluminum-deficient zeolite is heated with a purge of inert gas to 800°C, resulting in a contraction of the lattice of about 1%.

The three procedures result in ultrastable materials with only minor differences in properties. The products of procedures A and C may have a slightly larger unit cell because the sodium present during calcination limits the lattice contraction. However, if reduced to low sodium content, the final products will undergo further lattice contraction at any future time when the material is reheated to the stabilization temperature. A procedure similar to A was described by Kerr (*86*) in which the NaY was clearly in the hydrogen-exchanged form rather than ammonium-exchanged form prior to stabilization. Procedures B can, in general, attain lower final sodium levels. For example, procedure A will result typically, in a final sodium content of about 0.1 to 0.3% by weight of sodium, expressed as Na_2O, while procedure B can easily attain sodium levels of 0.05 to 0.10%. The method of aluminum extraction given in procedure C will, of course, lead to a product containing much lower values, as indicated by chemical analysis, of aluminum. However, the amount of framework aluminum is probably similar in all three products.

It is essential that water vapor be present during the calcination step, at least in procedures A and B. This was pointed out by Kerr (*85*) in a comparison of the products formed by heating the ammonium form of zeolite Y in thick and thin beds. Ward (*92*) demonstrated the effect of calcination in beds of different geometry and different in atmospheres and showed that the presence of water vapor is critical to formation of the ultrastable structure.

PROPERTIES OF ULTRASTABLE ZEOLITES. Although properties will undoubtedly vary, depending on the method of preparation, typical properties are presented for materials produced as described above. The properties of ultrastable zeolites made by procedure A, B, and C are given in Table IV, which presents data compiled by McDaniel and Maher (*8*) and Kerr (*47*).

Table IV. Characteristics of Ultrastable Zeolites

Procedure Used	*A*	*B*	*C*
SiO_2/Al_2O_3 of starting zeolite Y	5.2[a]	5.2[a]	5.3
Chemical analysis, db			
Na_2O	0.24 %	0.10 %	6.96 %
Al_2O_3	23.18 %	21.10 %	13.30 %
SiO_2	76.82 %	78.77 %	79.75 %
NH_3	0.04 %	0.03 %	0 %
SiO_2/Al_2O_3	5.63	6.35	10.3
Unit cell of starting Y	24.65A	24.65A	—
Unit cell of ultrastable product	24.45A[b]	24.34A	—

[a] As indicated by unit cell.
[b] Decreased to 24.35 when heated to 815°C.

The data for procedures A and B give evidence of aluminum extraction (later demonstrated by Kerr) in the increased silica-to-alumina ratio of the ultra-stable product as compared with that of the parent NaY. McDaniel and Maher, however, failed to make this observation; they reported the silica-to-alumina ratio of the parent sodium zeolite on the basis of chemical analysis in obvious disagreement with the silica-to-alumina ratio indicated by unit cell (*8, 41*). Their analysis of the starting sodium zeolite was made on a sample that contained excess SiO_2 from inadequate washing. More recent work indicates that the zeolite was adequately washed before conversion to the ultrastable zeolites; consequently, the silica-to-alumina ratios determined for these products were not subject to this error.

The properties of the stabilized form are compared in Table V to those for the same zeolite that was not stabilized. Although the chemical analyses are identical, the unit cell dimensions and thermal stabilities are quite different.

Table V. Comparison of Zeolites, Stabilized and Not Stabilized

	Stabilized	*Not Stabilized*
SiO_2/Al_2O_3 of starting zeolite Y	5.2	5.2
Chemical analysis		
Na_2O	0.10 %	0.10 %
Al_2O_3	21.10 %	21.10 %
SiO_2	78.77 %	78.77 %
SiO_2/Al_2O_3	6.35	6.35
Unit cell, A	24.34	24.69
Unit cell of starting zeolite Y, A	24.65	24.65
Surface area after 2 hr @		
700°F	837 (m^2/g)	1008 (m^2/g)
1500	851	254
1550	793	132
1650	842	18
1700	743	15
1725	678	
1800	542	

Figure 18 shows thermograms for the stabilized zeolites and for a product of similar chemical composition that was not stabilized. Endotherms a correspond to the evolution of water, the exotherms b correspond to a recrystallization, and the exotherms c and d correspond to the crystal collapse of the unstabilized and the ultrastable materials, respectively. The lower temperature observed for the desorption endotherm for the ultrastable zeolite indicates a less hydrophilic surface, a finding that was also observed in equilibrium water adsorption measurements (*8*).

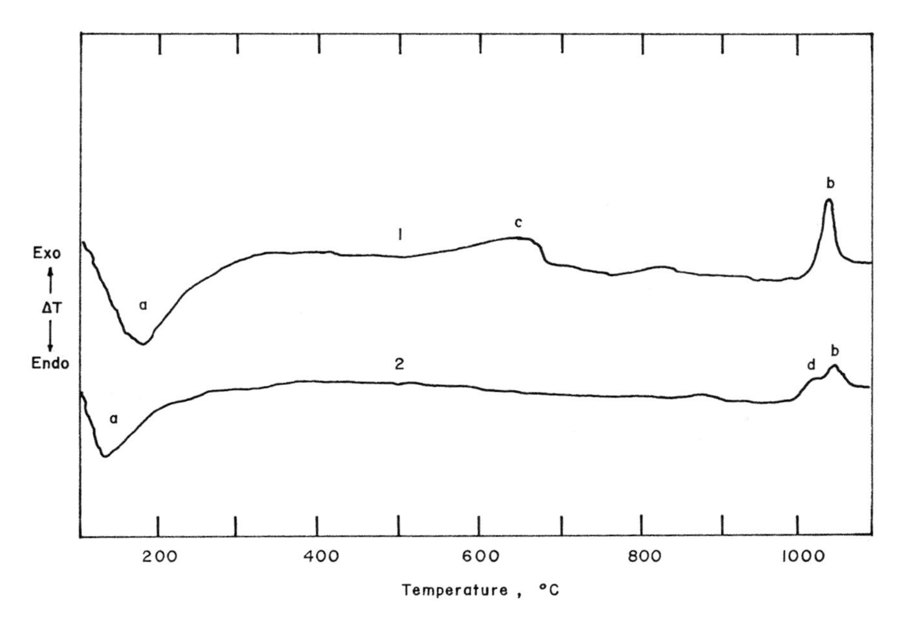

Figure 18. Differential thermograms of unstabilized and stabilized zeolites

Other than the angle shift associated with the decrease in unit cell size, the gross x-ray diffraction pattern of the ultrastable zeolite is quite similar to the unstabilized zeolite. Figure 19 shows such a pattern for a 50% physical mixture of the two materials. Each peak for the unstabilized zeolite is paired with a peak at a slightly higher angle for the ultrastable zeolite.

The cation-exchange capacities of ultrastable zeolites were shown by McDaniel and Maher to be greatly reduced when compared to the parent sodium zeolite. However, their results did not take into account the extraction of framework aluminum. Kerr subsequently indicated that when the extracted aluminum was removed from a similar material, the exchange capacity agreed fairly well with that calculated from the remaining aluminum content. More comprehensive work in this area is indicated.

Results of catalytic and acid properties of ultrastable zeolites have been reported in a few instances (*88, 93, 94, 95, 96*). Barthomeuf and Beaumont have shown that two distinct types of acidic sites are found when aluminum is

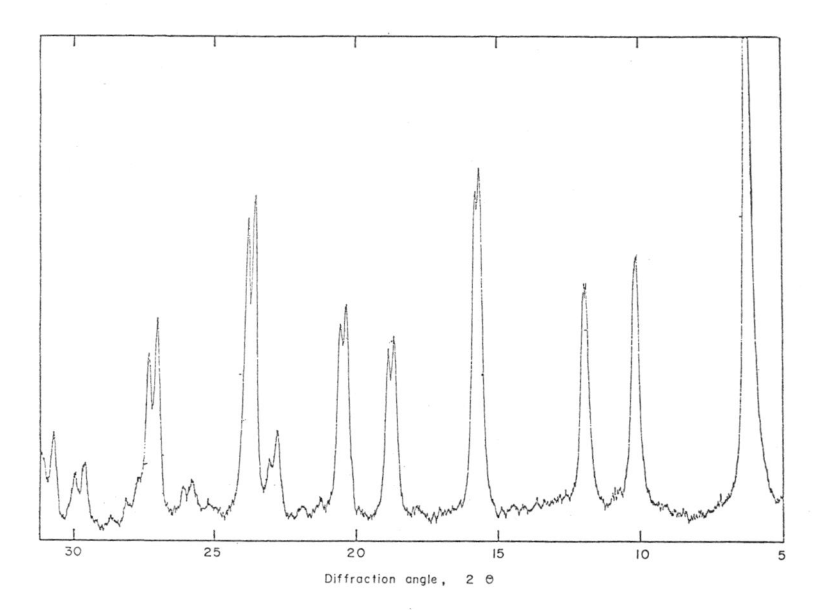

Figure 19. X-ray diffraction pattern of a mixture of an ultrastable zeolite and its parent NaY

extracted from the zeolite and that extraction proceeds by preferentially removing the aluminum atoms associated with the sites of lowest acidity. They have also demonstrated that the catalytic activity (cracking of isooctane) is almost entirely located at the strong acid sites. A decrease in activity was observed only when the amount of aluminum extracted exceeded 30%. The ultrastable structure is caused by the selective extraction of certain aluminum atoms and possibly by their replacement with silicon atoms. Dempsey (*97*) has suggested, based on the work of Beaumont and Barthomeuf, and on the geometry of the zeolite, which aluminum atoms might be most easily extracted.

A structural analysis of an ultrastable zeolite has been presented by Maher, Hunter, and Scherzer (*98*). Their analysis showed the ultrastable zeolite to contain nonframework aluminum in agreement with work by Kerr. At least in some instances, silica can migrate to fill the voids resulting from aluminum extraction. X-ray, infrared, and ESR analyses have led other investigators to conclude that silica migration does occur (*99, 100, 101, 102, 103*).

STABILIZATION PHENOMENA. Stabilization of the zeolite structure directly involves hydrogen exchange, aluminum extraction, dehydroxylation in the presence of some water vapor, and contraction of the framework structure. In addition, cationic aluminum is present in at least some and possibly all stabilized structures. The migration of silica to fill the voids left by dealumination has also been suggested as a factor in the stabilization (*98, 99, 100, 101*).

The stabilization reaction appears to consist of several steps with corresponding intermediate structures. The entire sequence is not yet clear, but we can outline some aspects that have been established and discuss some of the additional postulates that have been advanced. It has been fairly well established that the hydrogen form of zeolite is a first and necessary stage in stabilization. No reported results seem to conflict with this. The unit cell contraction that is associated with stabilization is not observed with the sodium or other monovalent metal-cation forms of the zeolite, but only with the hydrogen form or other forms that may be converted to the hydrogen form. For a subsequent essential step of aluminum extraction, the hydrogen form is required. Kerr's work showed that aluminum can be extracted with the acid form of EDTA but not with the sodium salt (*47*).

The second essential step appears to be the extraction of aluminum from the lattice framework. We have found no reported evidence that conflicts with this view. Such extraction of aluminum (*i.e.*, dealumination) creates a void in the lattice which allows a subsequent contraction of the structure to occur.

The extraction of aluminum per se does not result in unit cell contraction, nor does it ensure an ultrastable structure. The results presented in Table VI

Table VI. Aluminum Removal from Y-Type Zeolites

	Fraction Removed		*Moles H_4EDTA Used/ Formula Wt of NaY*	*Unit Cell, $a_o(A)$*
Zeolite	*Na*	*Al*		
(1) Parent NaY	0	0	0	24.65
(2) Al def. zeolite[a]	0.40	0.25	0.25	24.65
(3) Al def. zeolite[a]	0.47	0.35	0.35	24.64
(4) Al def. zeolite[a]	0.55	0.49	0.50	24.61
(5) Al def. zeolite[b]	0.95	0.35	0.35	24.62

[a] Prepared by process described by Kerr (*44*).
[b] Prepared from sample 3 by ammonium exchange.

demonstrate that zeolite Y which has been partially dealuminated by the use of EDTA, as described by Kerr, does not exhibit the contracted unit cell characteristic of the stabilized zeolite, prior to calcination. The results in Table VII lead to the same conclusion. When aluminum is extracted by a series of alternating ion exchange and mild thermal treatments, the unit cell size is not decreased. This suggests that aluminum extraction is a necessary but not sufficient condition to ensure the stable structure (the same conclusion reached by Jacobs and Uytterhoeven (*104*)).

The third essential step appears to be the contraction of the structure, which can be equated with stabilization. Although there are opposing views concerning the nature of the reactions involved in the contraction, the most likely possibility is a dehydroxylation reaction wherein water is eliminated from hydroxyl groups with the formation of new Si–O–Si bonds with silica taking positions in the vacated tetrahedral sites. The loss of hydroxyl groups

Table VII. Sodium Removal from Y-Type Zeolites

Zeolite	Na_2O, %	SiO_2/Al_2O_3 [a]	*Unit Cell*, a_o
Parent NaY	13.40	5.52	24.65
NH_4^{ex} NaY [b]	2.56	—	24.68
NH_4^{ex} NaY [b]	1.11	—	24.69
NH_4^{ex} NaY [b]	0.41	5.52	24.71
Parent NaY	13.40	5.52	24.65
NH_4^{ex} NaY [c]	2.56	—	24.68
NH_4^{ex} NaY [c]	0.61	—	24.69
NH_4^{ex} NaY [c]	0.13	6.34	24.69
Ultrastable zeolite [d]	0.10	6.35	24.34

[a] By chemical analysis.
[b] Sodium level attained by exhaustive exchange of parent NaY.
[c] Sodium level attained by series of exchanges with intermediate heat treatment at 350°C.
[d] Prepared by procedure B in text.

as shown by Uytterhoeven *et al.* (*74*) and the loss in weight shown by Kerr (*85*) tend to support this view.

The role of water vapor and NH_3 in stabilization has not been conclusively established. It is likely that water vapor aids in stabilization by increasing the lability of oxygen in the framework. The framework oxygen lability in the presence of water vapor is extremely great and this could influence or aid in the nature of the dehydroxylation, migration of silica and the formation of new Si–O–Si bonds. NH_3 may increase the lability in a manner similar to water and, in addition, may suppress formation of the hydrogen form of zeolite.

The role of the cationic aluminum in the stabilization of the structure is not yet clear. Some authors (*104*) have attributed the high degree of stability directly to the presence of cationic aluminum, which is analogous to the fairly high stability characteristics of certain other materials that contain trivalent ions, such as the rare earths. This hypothesis is extremely difficult to disprove because it may be impossible to prepare a material free of cationic aluminum or, even if the preparation is accomplished, it may be impossible to prove. If the cationic aluminum were removed from an ultrastable structure by ion exchange, an additional amount may form upon heating the zeolite to elevated temperatures. The stabilization is probably a result of those reactions that cause cell shrinkage, and the cationic aluminum may play a rather passive but not deleterious role in stability. The work of Wang and Lunsford, which shows that exchanged cationic aluminum does not result in an ultrastable structure, supports this view (*82*). Future work may show that the high degree of stability of some of the polyvalent cation forms follows from stabilization of the structure in a manner similar to that observed for ultrastable zeolites.

The presence of metal cations and particularly of sodium has a twofold effect on stability. High temperature stability is limited by the presence of sodium (as is the case with other zeolites and amorphous silica-aluminas), a fact long recognized in the catalyst art (*105*). Aside from this, the presence of more than about 25% of the original sodium prevents unit cell contraction during calcination of the partially ammonium-exchanged zeolite, probably because the presence of sodium ions prevents aluminum extraction. Kerr demonstrated that cell contraction with accompanying stabilization can be achieved, even when appreciable sodium is present, by using a chelating agent to extract aluminum from the framework (*47*).

Figure 20 illustrates the influence of sodium content on the stability of ammonium-exchanged Y-type zeolites. The stability increases as the Na_2O

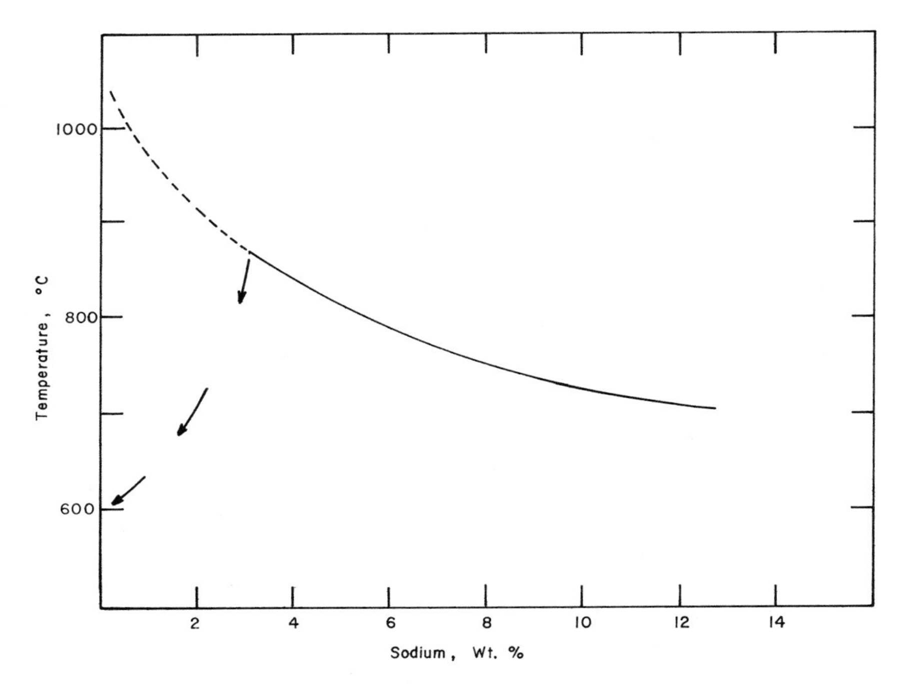

Figure 20. Thermal stability of ammonium-exchanged zeolite Y as a function of Na_2O content

level is decreased to about 3% by weight. Below this level the zeolite can either be converted to the ultrastable structure with a further increase in stability or, if not stabilized, will show sharply decreased stability.

The unit cell contraction, although influenced by sodium content, is not a direct result of sodium removal. The fact that no contraction occurs during sodium removal is illustrated by the results shown in Table VII. The sodium level was reduced to low levels in a sample of zeolite Y by two techniques: (1) an exhaustive exchange with an ammonium salt solution, and (2) a process consisting of a series of ammonium exchange steps, with intermediate heat treatment at 350°C to redistribute the residual sodium ions. Instead of decreasing, the unit cell increases slightly as the sodium ions are replaced by ammonium ions. Significant extraction of aluminum occurred with intermediate calcination but not with exhaustive exchange. Results in Figure 21 illustrate the unit cell contraction of this same zeolite with varying levels of sodium content when stabilized by calcination at 750°C.

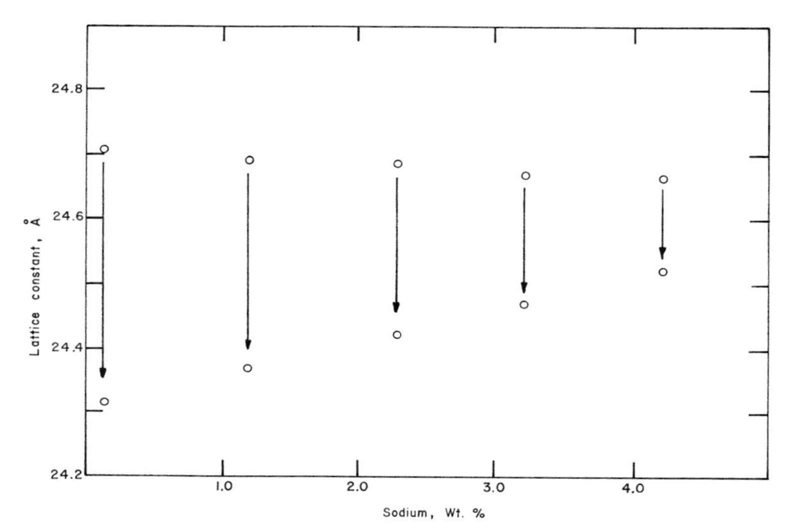

Figure 21. Unit cell contraction of ammonium-exchanged zeolite Y during stabilization as a function of sodium content

The exact role of the dehydroxylation reaction in stabilization has not yet been established. Dehydroxylation appears to occur both with and without stabilization. Kerr points out that the extreme instability of dehydroxylated zeolite Y to moisture complicates its investigation (*77*).

The nature of the stabilizable or metastable state *vs.* the unstable state presents another void in our understanding of stabilization. A Y-type zeolite that has had a substantial number of the locked in or difficultly exchangeable sodium ions removed without being stabilized can be in one of two distinct states: metastable or unstable. Upon heating to a high temperature in a static atmosphere, the metastable form will convert to the ultrastable structure while the unstable form will collapse to an amorphous phase.

Except for their different behavior during thermal treatment, no discernible difference between the forms has been established. They can be of identical chemical composition, with the metastable form transforming to the unstable form in a few hours to several weeks, depending on storage conditions.

To add to the puzzlement of the phenomenon, the unstable form can be transformed back to the metastable form by a process of annealing in which the zeolite in the unstable state is heated in water or a salt solution such as ammonium sulfate for an extended time (*106*).

Since the hydrogen form is known to be unstable, the two states may result from the distribution of cations. Unfortunately, the phenomenon of metastability has not yet attracted the attention of other investigators, and a fuller understanding will have to await such attention.

Kerr (*77*) presents an excellent review of the work reported in the area of ultrastable and similar zeolites and concepts proposed by various authors. Breck (*101*) has suggested reaction mechanisms that appear to be consistent with most of the observed behavior of zeolite Y during calcination. The process leading to the less stable decationized zeolite may involve an extraction of framework oxygen as well as aluminum, leaving a rather unstable defect structure. Conversely, the process leading to the ultrastable structure involves an intermediate state wherein the extracted aluminum atoms are replaced by four hydroxyl groups, as indicated in Figure 14. This mechanism does not require the formation of an oxygen deficient framework. He suggests the final step in stabilization to be the replacement of the H_4 group by silicon introduced as $Si(OH)_4$. The obviously important role of water in forming the hydroxyl groups and in transporting silica give these mechanisms considerable appeal. As more information becomes available, the experimental findings and postulated mechanisms begin to form a more consistent picture, but a full explanation of the nature of stabilization must await future developments.

Literature Cited

1. Eichhorn, H., *Ann. Phys. Chem.* (1858) **105,** 130.
2. Barrer, R. M., Baynham, J. W., *J. Chem. Soc.* (1956) 2892.
3. Kunin, R., "Elements of Ion Exchange," p. 25, Reinhold Publishing Corp., New York, 1960.
4. Bremer, H., Morke, W., Schodel, R., Vogt, F., *Adv. Chem. Ser.* (1973) **121,** 249.
5. Berger, A. S., Yakovlev, L. K., *Zh. Prik. Khim.* (1965) **38,** 1240.
6. Barrer, R. M., Langley, D. A., *J. Chem. Soc.* (1958) 3804.
7. Ambs, W. J., Flank, W. H., *J. Catal.* (1969) **14,** 118.
8. McDaniel, C. V., Maher, P. K., "Molecular Sieves," p. 186, Society of Chemical Industry, London, 1968.
9. Barrer, R. M., *J. Chem. Soc.* (1952) 1561.
10. Bosmans, H. J., Tambuyzer, E., Paenhuys, J., Ylen, L., Vancluysen, J., *Adv. Chem. Ser.* (1973) **121,** 179.
11. Keough, A. H., Sand, L. B., *J. Amer. Chem. Soc.* (1961) **83,** 3536.
12. Baran, B. A., Belen'kaya, I. M., Dubinin, M. M., *Izv. Akad. Nauk, SSSR, Ser. Khim.* (1973) **3,** 510-15.
13. Weigel, O., Steinhol, E., *Z. Krist.* (1925) **61,** 125.
14. McBain, J. W., *Colloid Symp. Mono.* (1926) **4,** 7.
15. Barrer, R. M., *J. Soc. Chem. Ind.* (1945) **64,** 130.
16. Barrer, R. M., *Disc. Faraday Soc.* (1949) **7,** 135.
17. Barrer, R. M., Belchetz, L., *J. Soc. Chem. Ind.* (1945) **64,** 131.
18. Breck, D. W., "Zeolite Molecular Sieves," Chap. 2, John Wiley & Sons, New York, 1974.
19. Barrer, R. M., Riley, D. W., *Trans. Faraday Soc.* (1950) **46,** 853.
20. Barrer, R. M., Brook, D. W., *Trans. Faraday Soc.* (1953) **49,** 940.
21. Barrer, R. M., *J. Chem. Soc.* (1950) 2342.
22. Barrer, R. M., *Quart. Revs.* (1949) **3,** 292.
23. Kunin, R., "Elements of Ion Exchange," p. 22, Reinhold Publishing Corp., New York, 1960.

24. Bolton, A. P., *J. Catal.* (1971) **22,** 9.
25. Maes, A., Cremers, A., *Advan. Chem. Ser.* (1973) **121,** 230.
26. Platek, W. A., Marinsky, J. A., *J. Phys. Chem.* (1962) **65,** 2118.
27. Eastwood, S. C., Plank, C. J., Weisz, P. B., *Proc. World Petrol. Congr., 8th, Moscow, 1971.*
28. Magee, J. S., Blazek, J. M., this volume, Chap. 11.
29. Richardson, J. T., *J. Catal.* (1967) **9,** 182.
30. McDaniel, C. V., unpublished data.
31. Beecher, R. G., Voorhies, A., Eberly, P. E., *Ind. Eng. Chem., Prod. Res. Develop.* (1968) **7,** 203.
32. Chen, N. Y., Garwood, W. E., *Adv. Chem. Ser.* (1973) **121,** 575.
33. Cannon, P., *J. Amer. Chem. Soc.* (1958) **80,** 1766.
34. Lee, H., *Adv. Chem. Ser.* (1973) **121,** 311.
35. Antoshin, G. V., Minachev, K. M., Sevastjanov, E. N., Kondratjev, D. A., Newy, C. Z., *Adv. Chem. Ser.* (1971) **101,** 514.
36. Oblad, A. G., Hinden, S. G., Mills, G. A., *J. Amer. Chem. Soc.* (1953) **75,** 4096.
37. Freund, F., *Ber. Dtsch. Keram. Ges.* (1961) **37,** 209.
38. Dobres, R. M., unpublished data.
39. Iler, R. K., "The Colloid Chemistry of Silica and Silicates," p. 270, Cornell Univ. Press, Ithaca, New York, 1955.
40. McAteer, J. C., Rooney, J. J., *Adv. Chem. Ser.* (1973) **121,** 258.
41. Breck, D. W., Flanigen, E. M., "Molecular Sieves," p. 47, Society of Chemical Industry, London, 1968.
42. Dempsey, E., Kuhl, G. H., Olson, D. H., *J. Phys. Chem.* (1969) **73,** 387.
43. Smith, J. V., *Adv. Chem. Ser.* (1971) **101,** 171.
44. Myers, G. E., Boyd, G. E., *J. Phys. Chem.* (1956) **60,** 521.
45. Boyd, G. E., Lindenbaum, S., Myers, G. E., *J. Phys. Chem.* (1961) **65,** 577.
46. Helfferich, F., "Ion Exchange," McGraw-Hill, New York, 1962.
47. Kerr, G. T., *J. Phys. Chem.* (1968) **72,** 2594.
48. Kerr, G. T., *J. Phys. Chem.* (1969) **73,** 2780.
49. Topchieva, K. V., T'Huoang, H. S., *Kinet. Katal.* (1970) **11,** 490.
50. Ries, H. E., "Advances in Catalysis and Related Subjects," p. 87, Vol. IV, Academic Press, New York, 1950.
51. Moraweck, B., Gallezot, P., Renouprez, A., Imelik, B., *J. Phys. Chem.* (1974) **78,** 1959.
52. Bikerman, J. J., "Surface Chemistry," p. 263, Academic Press, New York, 1958.
53. McDaniel, C. V., Duecker, H. C., U.S. Patent **3,574,538** (1971).
54. McDaniel, C. V., unpublished data.
55. Chutoransky, P., Dwyer, F. G., "*Adv. Chem. Ser.* (1973) **101,** 540.
56. Grimshaw, R. W., Heaton, E., Roberts, A. L., *Trans. Brit. Ceram. Soc.* (1945) **44,** 76.
57. Barshad, I., *Soil Sci.* (1954) **77,** 463.
58. Barrer, R. M., Rees, V. C., *Trans. Faraday Soc.* (1960) **56,** 709.
59. Sherry, H. S., *J. Phys. Chem.* (1966) **70,** 1158.
60. Sherry, H. S., in "Ion Exchange," J. A. Marinsky, Ed., p. 89, Marcel Dekker, Inc., New York, 1969.
61. Olson, D. H., Kokotailo, G. T., Charnell, J. F., *Nature* (1967) **215,** 270.
62. Maher, P. K., McDaniel, C. V., U.S. Patent **3,402,996** (1968).
63. Barrer, R. M., Meier, W. M., *J. Chem. Soc.* (1958) 299.
64. Whittam, T. V., British Patent **1,171,464** (1969).
65. Bilisoly, J. P., U.S. Patent **3,322,690** (1967).
66. Hunter, F., unpublished data.
67. Rundell, C. A., McDaniel, C. V., U.S. Patent **3,769,386** (1973).
68. Freeman, D. C., U.S. Patent **3,013,989** (1961).
69. Breck, D. W., Castor, C. R., Milton, R. M., U.S. Patent **3,013,990** (1961).

70. Hey, M. M., *Miner. Mag.* (1930) **22,** 422.
71. Barrer, R. M., Coughlan, B., "Molecular Sieves," p. 141, Society of Chemical Industry, London, 1968.
72. Kranich, W. L., Ma, Y. H., Sand, L. B., Weiss, A. H., Zwiebel, I., *Adv. Chem. Ser.* (1971) **101,** 502.
73. Zhdanov, S. P., Novikov, B. G., *Dokl. Akad. Nauk* (1966) **166,** 1107.
74. Uytterhoeven, J. B., Christner, L. B., Hall, W. K., *J. Phys. Chem.* (1965) **69,** 2117.
75. Barrer, R. M., *Nature (London)* (1949) **164,** 113.
76. Benesi, H. A., *J. Catal.* (1967) **8,** 368.
77. Kerr, G. T., *Adv. Chem. Ser.* (1973) **121,** 219.
78. Cattanach, J., Wu, E. L., Venuto, P. B., *J. Catal.* (1968) **11,** 342.
79. Stamires, D. N., Turkevich, J., *J. Amer. Chem. Soc.* (1964) **86,** 749.
80. Ward, J. W., *Adv. Chem. Ser.* (1971) **101,** 380.
81. Rabo, J. A., Pickert, P. E., Stamires, D. N., Boyle, J. E., *Act. Congr. Int. Catal.* (1960) 2.
82. Wang, K. M., Lunsford, J. H., *J. Catal.* (1972) **24,** 262.
83. Kerr, G. T., *J. Phys. Chem.* (1967) **71,** 4155.
84. Szymanski, H. A., Stamires, D. N., Lynch, G. R., *J. Opt. Soc. Amer.* (1960) **50,** 1323.
85. Kerr, G. T., *J. Catal.* (1969) **15,** 200.
86. Kerr, G. T., Shipman, G. F., *J. Phys. Chem.* (1968) **72,** 3071.
87. Kerr, G. T., Cattanach, J., Wu, E. L., *J. Catal.* (1969) **13,** 114.
88. Beaumont, R., Barthomeuf, D., *J. Catal.* (1972) **26,** 218.
89. Eberly, P. E., Laurent, S. M., Robson, H. E., U.S. Patent **3,506,400** (1970).
90. Pickert, P. E., U.S. Patent **3,761,396** (1973).
91. Maher, P. K., McDaniel, C. V., U.S. Patent **3,293,192** (1966).
92. Ward, J. W., *J. Catal.* (1972) **27,** 157.
93. Kerr, G. T., Plank, C. J., Rosinski, E. J., U.S. Patent **3,442,795** (1969).
94. Topchieva, K. V., Thuong, C. S., *Dokl. Akad. Nauk* (1971) **198,** 141.
95. Beaumont, R., Barthomeuf, D., *J. Catal.* (1972) **27,** 45.
96. Barthomeuf, D., Beaumont, R., *J. Catal.* (1973) **30,** 288.
97. Dempsey, E., *J. Catal.* (1974) **33,** 497.
98. Maher, P. K., Hunter, F. D., Scherzer, J., *Adv. Chem. Ser.* (1971) **101,** 266.
99. Gallezot, P., Beaumont, R., Barthomeuf, D., *J. Phys. Chem.* (1974) **78,** 1550.
100. Peri, J. B., *Proc. Int. Congr. Catal., 5th* (1972).
101. Breck, D. W., "Zeolite Molecular Sieves," p. 507, John Wiley & Sons, New York, 1974.
102. Scherzer, J., Bass, J. L., *J. Catal.* (1973) **28,** 101.
103. Vedrine, J. C., Abou-Kais, A., Massardier, J., Dalmai-Imelik, G., *J. Catal.* (1973) **29,** 120.
104. Jacobs, P., Uytterhoeven, J. B., *J. Catal.* (1971) **22,** 193.
105. Bond, G. R., U.S. Patent **2,617,712** (1949).
106. Maher, P. K., McDaniel, C. V., U.S. Patent **3,374,056** (1968).

Chapter

5
Salt Occlusion in Zeolite Crystals

Jule A. Rabo, Union Carbide Corp., Corporate Research Department, Tarrytown, N.Y. 10591

Zeolites as Solid Electrolytes

THE DOMINANT and probably most useful property of a zeolite is its intracrystalline pore and cavity system which readily occludes and releases small molecules into and from the entire crystal, adsorbing and desorbing on a large scale. As synthesized, the intracrystalline cavities are filled with water, but they are vacated upon dehydration. In contrast to other hydrated salts, the zeolite structure is maintained upon dehydration with minor shifts in the position of framework ions. The stability of the anhydrous zeolite crystal must be ascribed to its three-dimensional anionic framework built of SiO_4 and AlO_4 tetrahedra fully crosslinked by strong Si–O–Si and Si–O–Al bridges. The crystal framework consisting of pores and cavities is ordered throughout the crystal. These pores and cavities of the anhydrous zeolite framework, only a few angstroms in size, give rise to the well-demonstrated molecular sieve phenomena.

Upon dehydration, the zeolite cations lose their water of hydration and seek coordination with framework oxygen. In most cases the anhydrous framework does not provide enough high coordination sites, and at least a fraction of the cations have to occupy sites of low coordination and low symmetry. In faujasite and in Linde X and Y zeolites the unit cell has 16 cation sites of near octahedral symmetry in the hexagonal prism while the rest of the cation sites, in the sodalite units and in the large interconnecting cavities, are of lower symmetry and coordination. The cations in the large cavities are often called surface cations because they lie on an imagined surface between the framework and the large cavities and can coordinate directly with adsorbed gases. These surface cations are linked to oxide ions only on one side, causing a large electrostatic field adjacent to the large zeolite cavity. The large electrostatic field in the intracrystalline cavities near the surface cations and the uniformity of the pores linking the large cavities throughout the crystal are responsible for the characteristic adsorption properties of zeolites and are the main influence in zeolite chemistry.

The characteristic zeolitic adsorption can be readily interpreted on the basis of our knowledge of structure. In addition, certain zeolites show outstanding catalytic activity, mainly in carbonium ion reactions. This activity centers on strongly acid hydrogens linked to framework oxygens facing the large cavities. While these hydrogens have been identified as centers of carbonium ion activity, neither the reason for their superior efficiency nor the reaction path of the catalytic process in zeolites has been described in a satisfactory manner. There are fundamental differences of catalysis between zeolites and other solid acid catalysts.

Zeolite catalysts differ from other acid catalysts such as the amorphous silica–alumina gels by having uniform and small pores. Moreover, molecules occluded within zeolite crystal are subject to its electrostatic field. The electrostatic interaction between the zeolite and the occluded molecules must of course be mutual and it must result, if the occlusion reaction reaches equilibruim, in reaching the lowest possible free energy state for the zeolite–adsorbate system as a whole. Therefore, the catalytic function of zeolites may be better understood by examining which chemical change induced by the zeolite in the adsorbed reactant molecule can best enhance the energy of the zeolite-adsorbate system. The best possible enhancement might result if the adsorbed molecules were polarized to simulate the original water content in every respect. Such polarization will not occur with most adsorbates although some may be found to have even greater effects than water without destroying the zeolite structure. This polarizing effect will depend on the electrostatic field accessible to adsorbed molecules, the polarity and polarizibility of the adsorbate, and the steric factors controlled by the zeolite structure and the shape of adsorbed molecules. What has to be minimized clearly, is not the energy of the zeolite alone, but the energy of the whole zeolite–adsorbate system.

In any case, the intracrystalline cavities vacated by water are several angstroms in size, and they are undesirable for an ideal ionic lattice. Since the zeolite lattice has a substantial ionic character, the changes induced by the zeolite upon adsorbed molecules will tend to fill the intracrystalline voids with polar or ionic species to minimize the free energy of this system. The ability of zeolites to separate and position polar or ionic products within their crystal may contribute to the efficiency of the acidic hydrogens in the zeolite lattice. Thus, the superior catalytic activity of zeolites may rest not only on the intrinsic acidity of the structural hydrogen atoms but also on the reduction of the free energy of the whole system by further polarizing or even ionizing the occluded molecules and by arranging their position according to the preference of the zeolite crystal.

There are several examples of ionization of molecules in zeolites. First, the ionization of water through cation hydrolysis is a common phenomenon, and this reaction is one of the sources of acidic hydrogen atoms. Several other examples for the ionization of molecules are discussed in the chapter on spin phenomena. The great ability of zeolites to fix the position of charged species within the zeolite structure is shown by the well-defined orbital occupied by electrons trapped in Na_4^{3+} and Na_6^{5+} centers in Y- and X-zeolites (*see* Chapter 6 on spin phenomena).

Zeolites should have high affinity for ionic compounds such as salts, and they should be readily occluded into zeolite crystals if the salt cations and anions can be properly arranged in the intracrystalline voids without undue repulsion of ions of similar charge. The role of the zeolite crystal in salt occlusion will be more like a solid electrolyte rather than just a symmetric arrangement of rigid void spaces.

The concept of stabilization of the zeolite lattice by adsorbed molecules has been considered by Barrer (*1*). In an effort to give a quantitative account for the effect of adsorbed molecules on the chemical potential of the host zeolite, Barrer calculated the changes in the zeolite chemical potential due to adsorbed water and ammonia. Measurements based on changes in volatility of H_2O and NH_3 between the free state and the adsorbed state show substantial stabilization of the NaX zeolite in the presence of both adsorbates.

In this chapter we review several aspects of interaction between guest salts and metal aluminosilicate framework crystals in the anhydrous state with emphasis on large pore zeolites. To clarify chemistry and structural considerations a brief review is given on interactions involving small-pore or even non-porous crystals when necessary.

For reference on zeolite–salt–water systems, the reader is referred to the Proceedings of the International Conferences on Zeolites and other literature sources (*2*).

Interactions between Anhydrous Salts and Anhydrous Metal Aluminosilicate Framework Crystals

Interaction between zeolite host crystals and guest compounds in the absence of water can result in several phenomena including chemical reactions and various types of occlusions. Here, we review these interactions in the following categories: ion exchange, reversible occlusion, irreversible salt occlusion in small pore zeolites, and irreversible salt occlusion in large pore zeolites.

Ion Exchange. The exchange of cations between various ion exchangers, including zeolites, and molten, low melting salts such as nitrates has been studied extensively. Cation exchange between NH_4Cl and zeolites in an anhydrous state has been reported by Barrer (*3*). Nearly complete cation exchange resulted from extended exposure of anhydrous zeolites to vapors of NH_4Cl at 300°C. A special type of cation exchange is the exchange of a metal cation for H^+ in solids. Troup and Clearfield (*4*) observed the reaction of zirconium phosphate with chloride salts in the solid state, resulting in the elimination of HCl derived from hydrogen atoms on the solid surface and the chloride ion of the halide salt:

$$Zr(HPO_4)_2 \cdot H_2O + \frac{2}{x} MCl_x \rightleftarrows ZrM_{2/x}(PO_4)_2 + H_2O + 2\ HCl\uparrow$$

While all the reported experiments were conducted with zirconium phosphate type ion exchangers, it was suggested that the reaction between H^+ type ion exchangers and chloride salts is a general phenomenon. Similar observations were made with titanium, thorium, and cerium phosphates, with zeolites, and with Dowex-50 ion exchanger resin.

Similar observations were reported by Rabo, Poutsma, and Skeels (*5,6*) on the large pore Y zeolite. They observed that upon heating in the presence of halide salts several cation-exchanged Y zolites released HCl. They ascribed this dehydrohalogenation to cation exchange. The amount of HCl released depends on the H content of the zeolite. The elimination of zeolite H appears to be more efficient by cation exchange with anhydrous alkali halides than by the reaction: $2\ OH^-_{zeolite} \rightleftharpoons H_2O + O^{2-}_{zeolite}$. The latter requires a temperature of 700°C or higher to reach completion while the exchange with NaCl is readily completed at about 500°C (as shown by the evolved HCl and by the absence of O–H bands in the infrared spectrum of the zeolite). The efficiency of elimination is best demonstrated by catalytic experiments on the isomerization of 1-butene to 2-butene, which is a very sensitive indicator for protonic activity. This reaction is readily catalyzed even on NaY, probably because of small amounts of acidic hydrogen introduced on hydrolysis. After heat treatment of the NaCl–NaY the isomerization of 1-butene was greatly reduced (*see* Figure 1). Treatment with alkalichloride salt at and beyond 300°C is therefore a very efficient method for the elimination of both the zeolite hydrogen and the acid type activity.

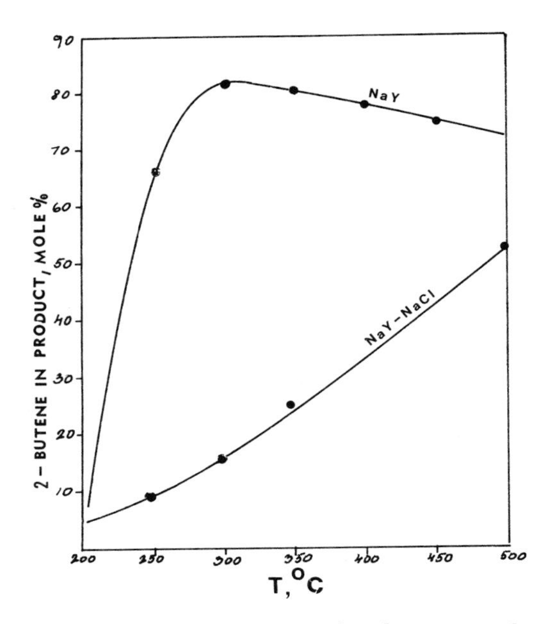

Figure 1. Isomerization of 1-butene to 2-butene on NaY and on NaCl-treated NaY

Cation exchange between HCl vapor and zeolites including NaY was studied by Barrer *et al.* (*7*). Upon treatment with HCl at room temperature, the H^+ replaced Na^+ in the zeolite. After this treatment the NaCl product was readily washed out of the zeolite, and no residual chloride content was reported for the washed product. The reaction between H halides and zeolites is temperature dependent, and low temperatures favor the protonation of the zeolite by the acid.

Cation exchange between ultramarine and salt melts of $AgNO_3$, $TlNO_3$, $LiNO_3$, KCNS, $PbBr_2$ was reported by Barrer *et al.* (*8*). Exchange between fused salts and chabazite and other ion exchange minerals was reported by Callahan *et al.* (*9,10*).

Cation exchange between molten nitrate salts and Linde A zeolites was carried out by Liquornik and Marcus (*11,12,13,14*). Based on weight measurements and chemical analyses they concluded, in agreement with Barrer and Meier (*15*), that the occluded salt molecules do not penetrate the sodalite cages. They also proposed an arrangement for the nitrate salt in the large cages of the Linde A zeolite and established the order of selectivity and the change in free energy upon cation exchange. A similar investigation was reported by Petranovic and Susic (*16*) using Linde A zeolite and molten cobalt nitrate salt. Here, diffusion through the solid was the rate determining step in the cation exchange process.

Reversible Occlusions in Zeolites. Reversible occlusion of molecules including iodine and fluorine compounds using Linde A and Linde X zeolites has been investigated by Barrer *et al.* (*17,18*). Irreversible reaction was only observed with the chemically reactive fluorine compounds which underwent reactions within the zeolite crystal. These authors also studied the thermochemical properties of the zeolite–iodine and zeolite–fluorocarbon systems and observed an enhanced high temperature stability of the zeolite lattice in the presence of the occluded molecules (*19,20,21,22*).

The occlusion of $AgNO_3$ in Linde A zeolite was first investigated by Barrer and Meier (*15*) using x-ray crystallography. They found that in a heat treatment of Ag–A zeolite with $AgNO_3$ melt at 250°-260°C for at least eight hr the nitrate salt penetrated only the large α cages, leaving the sodalite cages empty. Following this step, all $AgNO_3$ was removed by washing with hot water. According to the x-ray evidence, the silver nitrate guest salt forms an ordered cluster of $(AgNO_3)_9$ in each large cavity. The $AgNO_3$ units are probably arranged in a rock salt configuration, with a set of five $AgNO_3$ in a cluster having one orientation and a set of four others in a different orientation. There is a complementary arrangement in the next cage so that a superstructure is observed. The existence of other complexes in which lithium, sodium, or ammonium nitrate replaces silver nitrate has also been observed.

Irreversible Salt Occlusion in Small Pore Zeolites. The study of feldspathoidal minerals encourages an attempt to form irreversible zeolite-salt occlusion compounds. The crystal structure and chemical composition of sodalites, noselite, hauynite, and cancrinites show that these metal aluminosilicate framework structures favor the presence of guest molecules such as alkali chlorides, sulfates, or hydroxides in the cages. According to the early work of Pauling (*23*) the sodalite framework consists of truncated octahedral cages (sodalite cages) fully fused together, with access only through the small O_6 ring faces of the cages. In noselite and hauynite the guest anion occupies the center of the sodalite cage.

The affinity of these minerals for the occluded guest compound is best understood if we consider, as an example, the structure and cation arrangement in the sodalites. Since sodalites are found with a Si/Al ratio of near unity, 12 of the 24 framework cations belonging to a sodalite cage will be aluminum, the other 12 silicon. Since each framework cation is shared by four sodalite cages, 12/4 = 3 alkali metal cations are required within each sodalite cage to balance the charge of the AlO_4^- units. However, to have three alkali cations directly face and electrostatistically repel each other in the sodalite cage would obviously be undesirable. Therefore, the occlusion of a salt molecule such as NaCl is a desirable step, compensating for the direct cation–cation interaction adding interactions between the alkali cations and the guest anion in the center of the sodalite cage. The newly introduced cation can readily occupy an empty cation site. The new cation–cation repulsion introduced with the guest salt cation is balanced by the stabilizing effect of the guest chloride anion by providing fourfold, near tetrahedral coordination for all four cations.

Large empty spaces such as a vacant sodalite cage or an arrangement where cations directly face each other without shielding by an anion is undesirable for an ionic lattice. Therefore, the presence of guest salt molecules in these minerals is probably very essential for the stabilization of the crystal lattice. So far no mineral based on the sodalite framework has been found without a guest compound in the sodalite cage.

The structure and chemistry of salt occlusion in feldspathoid minerals has been described by Barrer and co-workers in several fundamental articles. They also successfully synthesized sodalite and cancrinite analogs by applying various guest compounds in solution during synthesis—*e.g.*, sodalites containing a variety of sodium salts (*24*). They found that in the crystallization process the occlusion of salt generally increased the size and quality of the crystals produced. Based on the efficiency of occlusion as measured by the filling of each sodalite cage the sodalite structure appears very selective toward H_2O, Cl^-, Br^-, ClO_3^-, and ClO_2^-. Each sodalite cage can accommodate four H_2O molecules, two NaOH or one anion such as Cl^- or ClO_4^-. With the preferred salt types the saturation value of one salt molecule per sodalite cage was readily obtained. They concluded that any porous crystal that is wholly or partly filled with a thermally stable salt should be more stable than its empty counterpart. The stabilization effect between the crystal and the guest salt is mutual, and encapsulation also stabilizes the occluded salt. For example, while $NaClO_4$ begins to decompose at 520°C, upon occlusion in sodalite no decomposition is observed up to 650°.

Barrer, Cole and Villiger (*25*) successfully extended the synthesis of sodalites to cancrinites and found that with cancrinite certain salts seem to catalyze the crystallization process and that salt occlusion is readily achieved with sodium nitrate, chromate, and molybdate. Substantially similar results were obtained by Barrer and Marcilly (*26*) in the synthesis of a variety of salt-bearing aluminosilicates, identified as P, Q, N, and O. They observed that occluded salts such as BaI_2 or $BaBr_2$ greatly increased the thermal stability of several host crystals. In the case of P and Q, particularly filled with barium halide, x-ray data indicated no significant change in the crystal structure even upon heating at 1000°C. Salt occlusion, applying the guest salt solution in the mother liquor during crystallization, has been recently applied to other zeolite-salt systems as well. Kuhl reported on the occlusion of phosphate in zeolite A (*27*), and Barrer and Freund reported on occlusion of borate in the sodalite cages of zeolite A, and of sodalite itself (*28*).

Based on successful synthesis of a variety of salt-doped crystals Barrer suggested that more attention should be directed to the role of the sodalite cages in zeolites A, X, and Y. If the cages of these zeolites were filled with stable salt molecules, migration of zeolite cations might be prevented and the catalytic and sorptive properties of these zeolites might be modified. In addition, crystal stability might be enhanced.

Table I. Infrared (IR) and Raman (R) Spectra of Nitrate Ions
(Frequencies in cm^{-1})

$NaNO_3$	$NaA/NaNO_3$ Complex	$NaY/NaNO_3$ Complex (5)	Na–Sodalite $NaNO_3$ Complex
1385 (IR)	1455 (IR)	1385 (IR)	1043 (R)
1064 (R)	1385 (IR)[a]	1045 (R)	
836 (IR)	1064 (R)[a]		
	1055 (R)		
	824 (IR)		
	814 (IR)		

[a] Bands which disappear upon washing the complex.

In their early work Barrer and Meier found no occlusion of silver nitrate in the sodalite units of zeolite A upon treatment of the zeolite with the salt melt at 250°–260°C. More recently, however, Barrer and Villiger (*29*) demonstrated the occlusion of nitrate salts in the sodalite cages of Linde A zeolite. They studied the occlusion of $NaNO_3$ at about 350°C through melts with several zeolites and made a detailed x-ray study of the structure of the NaA–$NaNO_3$ complex. Certain zeolites tended to decompose nitrate catalytically to $NaNO_2$ (chabazite and mordenite) and even to produce Na_2O (chabazite). The x-ray study definitely indicated more than one NO_3^- (up to two) in the sodalite cage of NaA zeolite. The x-ray study also indicated that the free volume of the sodalite cages of zeolite A with $NaNO_3$ inclusion to be $\sim$95 A^3 as compared with $\sim$68 A^3 in a typical sodalite. The NaA–$NaNO_3$ system first studied by Barrer and Villiger was also investigated by J. F. Cole and K. R. Loos (*30,31*) using infrared and Raman spectroscopy. They found that the frequencies of nitrate ions in the large cages and in sodalite cages differed appreciably, the former resembling those for solid $NaNO_3$. Upon thorough washing all nitrate was removed, leaving only those bands caused by nitrate ions trapped in the sodalite cage (*see* Table I).

The observed splitting of the nitrate bending vibration into a doublet at 824 and 814 cm^{-1} was interpreted as an indication for the presence of two ions in each sodalite cage of the NaA zeolite. Interactions between adjacent nitrate ions result in a coupling of the bending vibrations of the individual ions and account for the perturbed value found for the symmetrical nitrate stretching mode at 1055 cm^{-1} in the complex, (*cf.* 1043 cm^{-1} in sodalite and Na–Y) (*5*). In addition, a thermogravimetric analysis of the washed NaA–$NaNO_3$ complex yields two nitrate ions per sodalite cage.

Irreversible Salt Occlusion in Large Pore Zeolites. THE METHOD OF SALT OCCLUSION. The imbibing of salts in large-pore zeolites such as X, Y is a common phenomenon during cation exchange when the zeolite is in contact with a concentrated salt solution. The process is slow, particularly for anions which are repelled by the negatively charged port made of O_{12} rings, but at higher salt concentration it is readily accomplished. The salt molecules introduced from salt solution onto the zeolite crystal can be readily washed out again, and the crystal can be freed of the ions of the guest salt. Of course, there are exceptions if the guest salt anion forms insoluble salt with the zeolite cation ($NaCl + AgY \leftrightarrows AgCl + NaY$) and the insoluble salt product becomes trapped in the zeolite crystal. If the solubility problem does not apply, then the occlusion is readily reversible, suggesting that salt anions are occluded in the large interconnecting cavities but not in the sodalite cage. The cations of the applied salt are expected to exchange with the zeolite cations and to penetrate the whole crystal lattice including the sodalite cages and the hexagonal prisms. Even the large barium cation with a diameter of ~2.72 A can occupy all known cation sites in X and Y zeolites readily passing through O_6 rings of about 2.2 A free passage (*32*). Furthermore, both water and ammonia molecules can enter the sodalite cage despite their significantly larger size (H_2O = 3.4 A) than the O_6 port. However, while the passage of cations is greatly aided by the affinity of the negatively charged O_6 ring for cations, the passage of anions is difficult because of electrostatic repulsion by the O_6 ring. Anion transport through O_6 rings is further aggravated by the large size of even simple anions (diameters: F^- = 2.66, Cl^- = 3.62, Br^- = 3.92, and I^- = 4.40 A), while complex anions are even larger. The expectation that larger anions cannot readily penetrate sodalite cages follows from the Barrer and Meier (*15*) work on the occlusion of alkali nitrate melts in Linde A zeolite since at low temperatures they found no NO_3 ions in the sodalite cages. With this absence of anion occlusion in the sodalite cages one can appreciate the approach described earlier which was successfully used to synthesize feldspathoid structures (sodalites, cancrinites) by applying the guest salt solution in the mother liquor during synthesis. However, no successful application of this method has yet been reported in the synthesis of Linde X or Y zeolites.

In agreement with Barrer and Cole's work with feldspathoid crystals, Rabo, Poutsma, Skeels (*5*) and Rabo and Kasai (*6*) felt that the occlusion of salts into the sodalite cages of type X and Y zeolites will probably change several properties of these zeolites. The presence of guest salt (or NaOH) in sodalite minerals supports the simple concept that guest salts will reduce undesirable cation–cation interactions and that the filling of the empty cages by properly arranged guest ions should increase both the lattice energy and the crystal stability.

Since the thermodynamics of salt occlusion were considered favorable, the difficulty of salt occlusion into the sodalite cages of X and Y zeolites must rest on kinetics, mainly on the passage of the large guest anion through the small O_6-ring port leading into the sodalite cage. The passage of guest anion should require substantial activation energy and the activation energy should be proportional to the size of the guest anion.

The occlusion of guest salts was carried out in two steps. First, the guest salt was introduced in concentrated water solution into the large interconnecting cavities. Then the water was removed by heating at about 200°, and finally the salt loaded zeolite was heat treated at high temperatures to effect the occlusion of salt molecules into the sodalite cages. After the final heat treatment the zeolite was washed free of soluble guest salt by a large excess of hot water. All guest salt contained in the large cavities was removed in the wash step. Analysis of the product defined the amount of salt irreversibly occluded in the sodalite cages.

The amount of guest salt applied in solution to the zeolite amounted to two to three anions per sodalite unit. While it was enough to fill the sodalite cages, it was less than necessary to fill the large cavities. In choosing this salt:zeolite ratio it was hoped to supply an excess of salt relative to filling the sodalite cages throughout the whole crystal lattice and to discourage any crystal-like stacking of the guest salt in the large cavities. The water was removed by gradual heating up to 350°C in air; the sample was kept at this temperature for 8–16 hr. The sample was heated in an oven at a chosen temperature for 24 hr, cooled, washed, and analyzed to determine irreversible salt occlusion.

Following the removal of water but before heat treatment, several salt-loaded samples were x-rayed to establish by the absence of its crystal diffraction pattern that the guest salt had been loaded into the zeolite cavities. Without exception the diffraction patterns did establish the absence of the guest salt crystal phase. Considering the sensitivity of this method, which was established by calibration with physical zeolite–salt mixtures, only a very small fraction of the added salt could have been left outside the zeolite crystal upon removal of water.

OCCLUSION OF HALIDE SALTS IN ZEOLITE Y. Irreversible occlusion of NaCl, NaBr, and NaI was attempted by heat treatment at increasing temperatures. With NaCl no irreversible occlusion was observed up to 330°C. However, after heating at 440°C a significant amount of salt occlusion was observed while 550°C resulted in an apparent saturation value of one chloride ion per sodalite cage. This value was not exceeded in the following experiments with halide salts either through prolonged treatment or by increasing the temperature of the treatment. No significant occlusion of NaBr was observed below 500°C; heating at 700°C was necessary to reach one bromide per sodalite cage, and 875°C (melting temperature of NaBr) resulted in the collapse of the zeolite crystal and in a decline of salt occlusion. Occlusion of NaI was small even at 645°C, and higher temperatures resulted in partial collapse of the zeolite crystal (*see* Table II (*5*)).

Table II. Effect of Temperature on Irreversible Occlusion of Halide Salts in NaY (*5*)

Guest Salt	*Thermal Treatment* *T, °C*	*Time, hr*	*Cation Equiv. per* $Al_{zeolite}$	*Anion per Sodalite Cage*	*X-Ray Crystallinity*
NaCl	200	24	1.01	0.02	excellent
	330	24	1.04	0.04	excellent
	440	24	1.09	0.29	excellent
	550	24	1.18	1.02	excellent
NaBr	550	24		0.16	excellent
	700	24		0.98	excellent
	785	24		0.64	fair
NaI	550	24		0.06	excellent
	645	24		0.33	fair-good
	700	24		0.25	poor

Progress in Solid State Chemistry, Vol. 9

These experiments show that the rate of salt occlusion depends on the size of the guest anion and that for halides saturation is reached at one halide ion per sodalite cage. The effect of anion size on the rate of occlusion is very significant: with chloride ion (diameter = 3.4 A) saturation is readily achieved at 550°C whereas with the large iodide ion (diameter = 5.4 A) only partial filling of the sodalites was reached even at 700°C.

In addition to alkali halides various halide salts (including transition metal halides) were heated with NaY and with cation-exchanged Y zeolites. The results were generally more complex than with alkali halides, mainly because of hydrolysis of the zeolite cation or of the guest salt. Hydroxyl groups formed in the zeolite upon hydrolysis are usually stable, even following removal of the water of hydration. They are, however, unstable in the presence of occluded alkali halides, and at high temperatures they undergo cation exchange with the alkali halide liberating H halides: $OH_{zeolite} + NaCl \rightarrow ONa_{zeolite} + HCl$. In addition to hydrolysis of the zeolite cation the hydrolysis of the guest salt has to be considered too. In the case of a guest halide salt H-halide is formed: $LaCl_3 + H_2O \leftrightharpoons La(OH)Cl_2 + HCl$. A fraction of the produced HCl will react with the zeolite, forming framework hydroxyls, while the halide ion will be attached to zeolite cations. Another fraction of the HCl may vaporize before reacting with the zeolite during the removal of water. The efficiency of these various steps will effect the cation-anion ratio of the final zeolite-salt adduct. In addition, Y zeolites containing bi- or tervalent cations tend to form cation clusters linked by hydroxyl or oxide ion (La^{3+}–O–La^{3+}) in the sodalite cage (*33*). Sodalite cages already occupied by a stable cation–O–cation cluster probably do not occlude salts. Nevertheless, in spite of differences in efficiency, substantial irreversible salt occlusion was achieved with every cation-exchanged Y zeolite (*see* Table III).

Table III. Occlusion of Halide Salts in Y Zeolites

Zeolite–Salt	*Thermal Treatment* *T, °C*	*Time, hr*	*Cation Equiv. per* $Al_{zeolite}$	*Anion per Sodalite*	*X-Ray Crystal-linity*
$CaY–CaCl_2$	550	64	1.14	0.44	excellent
$ZnY–ZnCl_2$	550	24	1.28	0.50	excellent
$NaY–CuCl_2$	550	48	1.12	0.81	excellent
$NaY–NiCl_2$	550	48	1.10	0.86	excellent
$NaY–LaCl_3$	550	64	1.25	0.30	excellent
$NaY–LaCl_3$	650	24	1.24	0.98	excellent
NaY–KF	550	16	0.97	0.18	excellent
$NaY–CaF_2$	550	16	1.07	0.79	excellent
CaY–KF	550	16	1.12	0.90	excellent
KY–KF	550	16	—	—	O[a]
$KY–CaF_2$	550	16	1.09	0.81	excellent
BaY–KF	550	16	1.11	0.51	excellent

[a] Further thermal treatment developed a leucite structure.

Progress in Solid State Chemistry, Vol. 9

THE RATE OF HALIDE SALT OCCLUSION. The rate of irreversible occlusion of NaCl by NaY was studied by G. W. Skeels (*34*), who found that the initial guest salt concentration strongly affects the rate of irreversible salt occlusion. Thus, at low loadings the diffusion of the guest salt becomes a rate controlling factor (*see* Figure 2). Analysis of the rate data showed that the salt occlusion does not follow simple first-order kinetics, suggesting a more complex mechanism. This can be envisioned by assuming that anions of the size of halides are too large to pass through the O_6 rings leading to the sodalite cage without bond cleavage between framework cations (Si, Al) and oxide ions. Such bond cleavage should sufficiently enlarge the O_6 port to allow the passage of the guest anion.

Bond cleavage between framework cations and oxide ions is related to the ability of zeolite oxide ions to exchange with gaseous oxygen. Recently Antoshin *et al.* (*35*) measured this exchange with $^{18}O_2$. The rate, significant only at or above 600°C, was first order with respect to $^{18}O_2$ pressure. With NaY the reaction had an activation energy of $\Delta E = 45\text{-}50$ kcal/mole, indicating that lattice oxygen is more mobile in NaY than in γ-Al_2O_3 but less mobile than in SiO_2. More recently, Antoshin *et al,* (*36*) showed that the presence of very small amounts of Cu^{2+} has a promoting effect, and the rate of oxygen exchange increased by several orders of magnitude with increasing Cu^{2+} concentration.

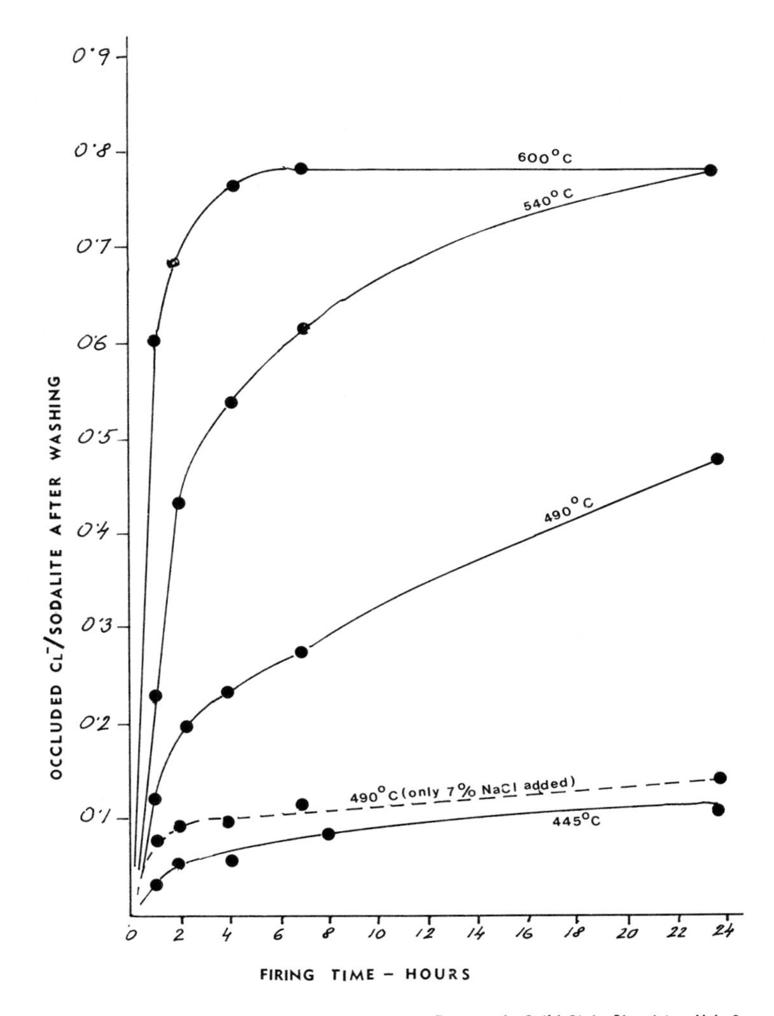

Progress in Solid State Chemistry, Vol. 9

Figure 2. Heat treatment of NaY with NaCl (27 wt %) (6)

The rate of irreversible salt occlusion in Y zeolites seem to depend on several factors. The effect of anion size and salt concentration have already been mentioned. In addition, the strength of the Si(Al)–O^{2-} bond must also play a part. The latter will be influenced by the Al/Si ratio and the details of the structure. Finally, a catalytic effect on the rate of salt occlusion may be expected in the presence of transition metal cations, either as part of the host zeolite or the guest salt, mainly because of their effect on the strength of the Si^{4+} (Al^{3+})–O^{2-} linkage.

IRREVERSIBLE OCCLUSION OF NON-HALIDE SALTS. Loading NaY with halide salts resulted in no degradation of the zeolite structure except at very high temperatures (above the melting temperature of the salt) and with the KY–KF system, which recrystallized to leucite. In experiments with non-halide salts, the zeolite structure collapsed on several occasions after the initial steps of salt loading and water removal. No loss of crystal structure was observed with salts or univalent anions, but with anions of valence of two or more, at least fractional loss of zeolite structure usually resulted. Loss of zeolite crystal structure always corresponded to high pH of the salt containing solution. Therefore, crystal degradation was ascribed to chemical reaction between the zeolite and the base at the higher temperatures used for water removal. With univalent-anion salts hydrolysis was usually small if not negligible and the crystallinity of the zeolite was always maintained.

The occlusion of non-halide salts offers several new phenomena not found with halides. Iodide ion was already found to be too large for achieving saturation of the sodalite units, but the substantially larger nitrate ion ($NaNO_3$) rather surprisingly reached anion/sodalite ratios of unity or above at 650°C (*see* Table IV).

Table IV. Occlusion of Non-Halide Salts in NaY

Zeolite–Salt	*Thermal Treatment* *T, °C*	*Time, hr*	*Cation Equiv. per $Al_{zeolite}$*	*Anion per Sodalite*	*X-Ray Crystal-linity*
NaY–$NaNO_3$	550	64	1.05	0.59	excellent
NaY–$NaNO_3$	650	24	1.18	1.32	excellent
NaY–$AgNO_3$	550	48	1.21	0.37	excellent
NaY–$NaClO_3$	650	16	—	0.9	excellent

The ease of occlusion of certain complex anions in the sodalite cage suggested that these anions may have decomposed by losing oxygen and the resulting smaller molecule may have entered the sodalite cage. Perhaps NO_3^- is reconstituted after the NO or NO_2 group has been occluded in the sodalite cage.

To characterize the NO_x group occluded in the sodalite cage Angell (*37*) examined spectroscopically the $NaNO_3$–NaY zeolite with 1.3 nitrate/sodalite. The infrared spectrum showed a broad band centered about 1385 cm^{-1}, and the Raman spectrum showed a sharp band at 1045 cm^{-1} (*see* Table I). The standard frequencies of nitrate ions in dilute water solution are 1384 cm^{-1} (infrared) and 1048 cm^{-1} (*38*). In $NaNO_3$ the Raman frequency is at 1060 cm^{-1}. Thus, NO_3^- is present in this Y structure but not as $NaNO_3$; rather, it resembles the free nitrate ion. Similar results have been reported by Cole and Loos (*30, 31*) with the NaA/$NaNO_3$ complex.

THE STRUCTURE OF ZEOLITE-SALT OCCLUSION COMPOUNDS (*39*). X-ray diffractometer traces of mono-, di- and trivalent cation forms of Y zeolites doped with halide salts show significant changes in the intensities of certain peaks characteristic of the undoped material. These changes in the x-ray diffraction pattern coincide with the presence of anions and a corresponding excess of cations within the zeolite–salt occlusion compound. These extra ions must have become part of the structure, analogous to the chloride in the mineral sodalite. Chemical analysis of a CaY, loaded with NaCl and fired at 550°C, showed eight sodium ions and eight chloride ions per unit cell. This number corresponds to one extra cation (Na^+) and one chloride ion for each sodalite cage.

The structure determination of this material from powder data determined the position within the structure of these extra sodium and chloride ions. More accurate results would be obtained by using a salt-doped single crystal, but preliminary results concerning the position occupied by the non-framework atoms can be obtained from powder data.

The NaCl-doped CaY sample was dehydrated at 350°C for 2 hr under vacuum in a MRC high-temperature attachment for a powder diffractometer. Intensities were recorded out to 65° (2θ) with copper radiation, and the integrated intensities were obtained by planimetering the recorder traces. Th space group used was Fd3m, and the structure was refined by a combination of least-square refinement and difference-map syntheses. The temperature factors of the non-framework atoms were arbitrarily fixed and not refined,

Table V. Nonframework Structural Parameters for a Calcium Exchanged Y Type Zeolite Loaded with Sodium Chloride[a]

Atom Type	*Site*	*Population*[b,c]	x[d]	y	z	$B(A^2)$[c,e]
Ca	I	6.4(2) [14.2]	0	0	0	(2.0)
Ca	I′	12.3(2) [2.6]	.063(3)	.063	.063	(3.0)
Ca	II′	1.0(3)	.170(2)	.170	.170	(3.0)
Ca	II	5.4(1) [11.4]	.231(2)	.231	.231	(3.0)
Cl	U	4.0(3)	.135(5)	.135	.135	(5.0)

[a] Cell dimension a_o = 24.53 A; Final R factor = 14.8%.
[b] [ ] show cation distribution in calcium exchanged faujasite (*40*).
[c] The values of error are shown in parentheses to the last significant level.
[d] Note that the occupancy factor depends on the assumed atomic species.
[e] Values for B in parentheses were chosen from other typical zeolite structure determination and were not refined.

Table VI. Nonframework Parameters for a Sodium Y Zeolite Loaded with Sodium Nitrate, Containing 0.8 Nitrate Ion/UC[a]

Atom Type	*Site*	*Population*[c,d]	x[c]	y	z	$B(A^2)$[c,e]
Na	I	1.0(3)	0	0	0	(2.0)
Na	I′	33.2(1.0)	.0708(7)	.0708	.0708	(2.0)
O	II′	14.9(1.3)	.2229(5)	.2229	.2229	(2.0)
N	U	8.0(3)	1/8	1/8	1/8	(2.0)

[a] Cell dimension a_o = 24.48 A; Final R factor = 14.7%.
[b] [ ] show cation distribution in calcium exchanged faujasite (*40*).
[c] The values of error are shown in parentheses to the last significant level.
[d] Note that the occupancy factor depends on the assumed atomic species.
[e] Values for B in parentheses were chosen from other typical zeolite structure determination and were not refined.

and the occupancy factors and positional parameters of these non-framework atoms were varied on alternate cycles of the least squares refinement.

The results (*see* Table V and VI) show a decrease in occupancy of both sites I and II with a corresponding increase in site I′ as compared with the data (in square brackets) for a dehydrated calcium-exchanged faujasite single crystal. The chloride ions occupied a site close to the center of the sodalite cage. (The displacement from the center–U– may not be real.) The increase in occupancy of site I′ is to be expected since the presence of an anion inside the sodalite cage must favor the occupancy of site I′ over site II. A similar movement of cations has been shown to occur for incompletely dehydrated bivalent-cation-exchanged faujasites, when the remaining few water molecules occupy site II′ (*see* Chapter 1 to identify cation sites).

The non-framework atoms were not all found: only 24 out of a total of 32 (calculated as calcium) cations and four out of the eight chloride ions were located, perhaps because the ions occupy different positions than those examined or because power diffraction is a method of very limited power.

IR studies with adsorbed carbon monoxide on the same sample showed the absence of surface–cation carbonyl complex and thus demonstrated the absence of surface cations (site II). This suggested the formation of cation–anion–cation clusters in the sodalite unit.

It was expected from the IR studies that the NaCl-loaded CaY would have full occupancy of site I′ (requiring 28 calcium atoms and four sodium atoms) with one chloride ion in the center of each of the eight sodalite cages, leaving four sodium atoms per unit cell to occupy sites in the large cavity. This was not observed; the occupancy of sites I′ and U suggest that the chloride anion may form a chloride bridge between calcium atoms in site I′. With this arrangement a reduced occupancy of sites I and II is expected since it would give the maximum charge separation. The samples loaded with sodium bromide or iodide showed similar intensity changes to those observed for CaY–NaCl, suggesting that again halide ion is located near the center of the sodalite cage.

A NaY–$NaNO_3$ sample with 0.8 nitrate ion per sodalite cage was also examined (*see* Tables V and VI) by the same method used for the CaY–NaCl sample and showed electron density in sites I′, II′, and U (at the center of the sodalite cage). The density in site I′ was elongated along the threefold axis and indicated partial occupancy of the site by two or more poorly resolved atoms separated by approximately 0.5 A. Calculating the density in site I′ as Na^+ gave 33.2 ± 1.6 atoms, indicating full occupancy of this site. The center of the sodalite cage was also fully occupied when calculated as nitrogen, and site II′ contained either 14.9 ± 1.3 oxygen atom or 10.8 ± .9 sodium atoms. The difference maps showed electron density in the large cavities which would account for the undetermined cations. This cation–anion distribution was similar to that expected for the NaCl-loaded CaY zeolite. The oxygen atoms in site II′ and the nitrogen atom at U may indicate the presence of a nitrate group in each sodalite cage although for this arrangement 24 oxygen atoms would be required in site II′. A nitrate group would have an influence on the position of one of the four sodium atoms in site I′ since the three nitrate oxygen atoms will surround it tetrahedrally; this could account for the elongation of density observed in site I′.

These two structure determinations show that under the correct conditions certain salts can be incorporated into the zeolite structure when the anion occupies a position in the center of the sodalite cage. It is not easy to see how more than one anion could occupy a sodalite cage, yet chemical evidence showed several NaY–$NaNO_3$ samples with two nitrate groups per sodalite cage. The study of these and similar zeolite–salt occlusion compounds would best be examined in the future by single crystal structure determinations.

STABILITY OF THE ZEOLITE Y-SALT OCCLUSION COMPOUNDS. To test the reversibility of salt occlusion under more strenuous conditions, samples of NaY, CaY, and LaY were first heated with NaCl at 550° to effect salt occlusion, treated with 1 atm of steam at 700°C for 1 hr, cooled, examined by powder x-ray diffraction, washed, and analyzed for halide. The diffractometer traces indicated full retention of crystallinity, and chemical analysis showed no loss of halide.

To evaluate the effect of salt occlusion on crystal stability, both the host zeolites and the zeolite–salt compounds were first heated for 16 hr at 500°C and x-rayed. This treatment was repeated, at 50° higher temperatures than in the previous step, until the peak heights of the x-ray diffraction pattern dropped *ca.* 50%, indicating substantial loss in crystallinity. In all cases studied the occlusion of NaCl increased the thermal stability of the zeolite crystal by 50° to 100°C.

The effect of salt occlusion on the stability of complex anions was studied by J. N. Francis (*41*). While imbibition of alkali nitrates and perchlorates into the supercages of zeolite Y had little effect on decomposition temperatures of these oxyanions, occlusion into the sodalite cage results in appreciable stabilization. The thermogravimetric decomposition curves of 13% sodium and lithium nitrates and perchlorates evaporated from solution onto a zeolite Y differed by less than 50° from those of the pure salts. The perchlorates, both pure and in the zeolites, decomposed cleanly to the chlorides. The completely occluded sample of 13% $LiNO_3$ on LiY, prepared by heating the components together overnight at 500°C and subsequently washing to remove excess salt, began to decompose only above 700°C with concomitant loss of zeolite structure. In comparison, the reversibly absorbed (not pre-heat treated)

Table VII. Decomposition Temperature of Occluded Salts[a]

	Pure Salt (°C)	*13 wt % on Zeolite Y (°C)*
$NaClO_4$[b]	500/542	490/535
$LiClO_4$[b]	405/445	425/470
$NaNO_3$[b]	628/740	570/675
$LiNO_3$[c]	520/	710/

[a] (10/90% decomposition).
[b] Introduced through solution but not heat treated.
[c] Introduced through solution and heat treated to effect irreversible occlusion.

$LiNO_3$/LiY began to decompose at 525°. The exclusion of the perchlorate salts by the sodalite cage can be attributed to the greater size of the perchlorate anion or to the irreversible nature of its decomposition (*see* Table VII).

CATALYTIC PROPERTIES OF ZEOLITE-SALT OCCLUSION COMPOUNDS. The effect of irreversible salt occlusion on the catalytic properties of Y zeolites was studied by Rabo, Poutsma, and Skeels (5). Using the dealkylation of cumene as measure for carbonium ion type activity, they found that in most cases the activity of zeolite–salt occlusion compounds was less than the activity of the zeolite host. The changes in activity (mainly declines) were attributed to the elimination of zeolite protons by cation exchange with the guest salt cation. While in most cases treatment with salt usually decreased protonic activity, when NaY was treated with a salt of a multivalent cation ($LaCl_3$), significant catalytic activity emerged. Here the increase in activity was attributed to an increase in proton content produced by hydrolysis of the lanthanum salt.

Literature Cited

1. Barrer, R. M., *Proc. Conf. Mol. Sieves* (1967) p. 41.
2. Barrer, R. M., Walker, A. J., *Trans. Faraday Soc.* (1964) **60,** 171.
3. Barrer, R. M., *Nature* (1949) **164,** 112.
4. Troup, T. M., Clearfield, A., *J. Phys. Chem.* (1970) **74,** 2578.
5. Rabo, J. A., Poutsma, M. L., Skeels, G. W., *Proc. Int. Congr. Catalysis 5th,* North Holland Publishing Co., pp. 1353-1363.
6. Rabo, J. A., Kasai, P. H., *Progr. Solid State Chem.* **9,** Pergamon Press, New York, in press.
7. Barrer, R. M., Kanellopoulos, A. G., *J. Chem. Soc. A* (1970) 765.
8. Barrer, R. M., Raitt, J. S., *J. Chem. Soc.* (1954) 4641.
9. Callahan, C. M., Kay, M. A., *J. Inorg. Nucl. Chem.* (1966) **28,** 233.
10. Callahan, C. M., *J. Inorg. Nucl. Chem.* (1966) **28,** 2743.
11. Liquornik, M., Marcus, Y., *J. Phys. Chem.* (1968) **72,** 2885.
12. Liquornik, M., Marcus, Y., *Isr. J. Chem.* (1968) **6,** 115.
13. Liquornik, M., Marcus, Y., *J. Phys. Chem.* (1968) **72,** 4704.
14. Liquornik, M., Irvine, J. W., *Inorg. Chem.* (1970) **9,** 1330.
15. Barrer, R. M., Meier, W. M., *J. Chem. Soc.* (1958) **58,** 299.
16. Petranovic, N. A., Susic, M. V., *J. Inorg. Nucl. Chem.* (1969) **31,** 551.
17. Barrer, R. M., Wasilewski, S., *Trans. Far. Soc.* (1961) **57,** 1140.
18. Barrer, R. M., Reucroft, P. J., *Proc. Roy. Soc. A* (1960) 431.
19. Barrer, R. M., *Trans. Far. Soc.* (1961) **57,** 1140.
20. Barrer, R. M., *Trans. Far. Soc.* (1961) **57,** 1153.
21. Barrer, R. M., Reucroft, P. J., *Proc. Roy. Soc. A* (1960) 449.
22. Barrer, R. M., *J. Phys. Chem. Solids* (1960) **16,** 84.
23. Pauling, L., *Z. Krist.* (1930) **74,** 213.
24. Barrer, R. M., Cole, J. F., *J. Chem. Soc. A* (1970) 1516.
25. Barrer, R. M., Cole, J. F., Villiger, H., *J. Chem. Soc. A* (1970) 1523.
26. Barrer, R. M., Marcilly, C., *J. Chem. Soc. A* (1970) 2735.
27. Kuhl, G. H., *Advan. Chem. Ser.* (1971) **101,** 75.
28. Barrer, R. M., Freund, E. F., *J. Chem. Soc. (Dalton)* (1974) 1049.
29. Barrer, R. M., Villiger, H., Imperial College, private communication.
30. Cole, J. F., Loos, K. R., Shell Dev. Co., private communication.
31. Loos, K. R., Cole, J. F., "Raman Spectra of Polyatomic Ions in Zeolites," *Proc. Int. Conf. Zeolites, 3rd,* pp. 230-234, Leuven University Press, 1973.

32. Pluth, J. J., doctoral dissertation, University of Washington, 1971.
33. Rabo, J. A., Angell, C. L., Schomaker, V., *Proc. Int. Congr. Catalysis, 4th, 1968,* Paper 54.
34. Skeels, G. W., Union Carbide Corp., private communication.
35. Antoshin, G. V., Minachev, Kh. M., Sevastjanov, E. N., Kondratjev, D. A., *Russ. J. P. Chem.* (1970) **44,** 1491.
36. Antoshin, G. V., Minachev, Kh. M., Sevastjanov, E. N., Kondratjev, D. A., Chan Zui *Newy, Advan. Chem. Ser.* (1971) **101,** 514.
37. Angell, C. L., Union Carbide Corp., private communication.
38. Szymanski, H., "Raman Spectroscopy," Plenum Press, New York (1967) 246.
39. Bennett, J. M., Union Carbide Corp., private communication.
40. Bennett, J. M., Smith, J. V., *Mat. Res. Bull.* (1968) **3,** 633-642.
41. Francis, J. N., Union Carbide Corp., private communication.

Chapter

6

Electron Spin Resonance Studies of Zeolites

Paul H. Kasai and R. J. Bishop, Jr., Union Carbide Corp.,
Tarrytown Technical Center, Tarrytown, N.Y. 10591

THE FUNDAMENTAL PRINCIPLES of electron spin resonance (ESR) spectroscopy are now well known (*1, 2, 3*). This technique has been extensively used to study various catalysts including zeolitic materials. The ESR studies so far carried out on zeolites can be classified into four categories: (1) transition metal cations exchanged into zeolites, (2) stable molecular free radicals adsorbed in zeolites, (3) radiation induced radicals, and (4) observation by ESR of intracrystalline ionization and redox phenomena. Of these, the ESR studies of transition metal cations in zeolites have been recently reviewed by Mikheikin *et al.* (*4*). This chapter, therefore, focuses on the remaining three subjects. Particular emphasis is given to the last subject since it presents the broadest implication for the properties and chemistry of zeolites that could occur in their interstices.

Several excellent review articles on ESR study of catalysis have recently appeared (*5, 6*). We first present a brief account of the aspects of ESR spectroscopy most frequently encountered in the study of zeolites. For more thorough accounts of the technique and the theories involved, the readers are referred to the general references and the review articles cited above.

ESR Spectroscopy

Consider a paramagnetic molecule (one unpaired electron) placed in a static magnetic field H. The magnetic moment of such species arises both from the orbital motion of the electron characterized by L and from its intrinsic spin S. The Hamiltonian representing the interaction between the magnetic moment μ of this molecule and the static field H thus takes the following form.

$$\mathcal{H} = -\mu \cdot H = (L + g_e S)\beta \cdot H \tag{1}$$

where β is the Bohr magneton ($= e\hbar/2m_e$), and g_e is the Landé factor of a free electron (equal to 2.0023).

Most of the paramagnetic species encountered in catalytic studies reside in a position of low symmetry so that its orbital ground state is non-degenerate. In this situation the orbital moment becomes nearly quenched, *i.e.*, $\langle L \rangle \cong 0$, and the Hamiltonian (Equation 1) can be simplified to

$$\begin{aligned} \mathcal{H} &\cong g_e S\beta \cdot H \\ &= g_e \beta H M_S \end{aligned} \tag{2}$$

and has the following eigenvalues.

$$E = g_e \beta H M_S$$

Here M_S is the magnetic quantum number of the spin S, and in the case of one-electron system $M_S = \pm 1/2$. In ESR spectroscopy one observes the resonance absorption $\Delta E = h\nu$ corresponding to the transition between such two states. Hence

$$\begin{aligned} h\nu &\cong g_e \beta H \\ \text{or } H &\cong \frac{h\nu}{g_e \beta} \end{aligned} \tag{3}$$

g-Tensor. In actual experiments, however, the g values are often not only larger or smaller than the free spin value g_e but also vary with the direction of the magnetic field relative to the symmetry axes of the paramagnetic sites. This is the result of the spin-orbit coupling interaction $\lambda L \cdot S$, the effect of which is to admix excited states into the non-degenerate ground state and to induce a small amount of orbital moment.

Let ψ_o denote the singlet ground state, and ψ_n the excited states of the same L and S manifold. The energy separations between ψ_o and ψ_n's are usually much larger than the spin-orbit coupling interaction. The $\lambda L \cdot S$ term thus can be treated as a perturbation. When treated by the first-order perturbation theory, it yields the ground state Ψ_o of the following form.

$$\Psi_o \cong \psi_o + \sum_{n \neq 0} \frac{\langle \psi_0 | \lambda L \cdot S | \psi_n \rangle}{E_o - E_n} \psi_n \tag{4}$$

Evaluation of the Zeeman term $(L + g_e S)\beta \cdot H$ of Equation 1 in terms of the new wavefunction Ψ_o gives a spin Hamiltonian shown below.

$$\begin{aligned} \mathcal{H}_{\text{spin}} &= \beta S \cdot \tilde{g} \cdot H \\ &= \beta(g_z S_z H_z + g_x S_x H_x + g_y S_y H_y) \end{aligned} \tag{5}$$

where $$g_i = g_e + 2\lambda \sum_{n \neq 0} \frac{\langle \psi_o | L_i | \psi_n \rangle \langle \psi_n | L_i | \psi_o \rangle}{E_o - E_n} \tag{6}$$

The g factor is now in the tensor form, and H_z, for example, represents the component of the magnetic field along the principal z axis of the g tensor. The advantage of the spin Hamiltonian is that it reduces the appearance of the problem to that of a spin-only case. The effect of the induced orbital moment is embedded in the g tensor and is responsible for its anisotropy and deviation from the free spin value g_e. The spin-orbit coupling constants λ of the relevant atoms may be assessed from the known optical spectroscopy data. Thus the symmetry of the paramagnetic site as well as the magnitude of energy level separations can be determined from the observed g tensor.

Hyperfine Coupling Tensor. A nucleus with spin I possesses a magnetic moment $\mu = g_n\beta_n I$ where g_n and β_n are the nuclear g factor and the nuclear magneton, respectively. The magnetic field $\widetilde{H}_{\text{dip}}$ caused by a magnetic dipole $\widetilde{\mu}$ at a point $\widetilde{r}$ away from the dipole is given by

$$\widetilde{H}_{\text{dip}} = -\frac{1}{r^3}\left[\widetilde{\mu} - \frac{3\,(\widetilde{r}\cdot\widetilde{\mu})\,\widetilde{r}}{r^2}\right] \tag{7}$$

Hence, when our paramagnetic molecule contains a magnetic nucleus, one needs to add to our spin Hamiltonian (5) a hyperfine coupling term which represents the interaction of the magnetic moment of the electron $g_e\beta_e S$ with the field $\widetilde{H}_{\text{dip}}$ set up by the nucleus.

$$\begin{aligned}\mathcal{H}_{\text{dip}} &= g_e\beta_e S\cdot\widetilde{H}_{\text{dip}} \\ &= g_e\beta_e g_n\beta_n\left[\frac{3\,(\widetilde{r}\cdot S)\,(r\cdot\widetilde{I})}{r^5} - \frac{\widetilde{S}\cdot\widetilde{I}}{r^3}\right]\end{aligned} \tag{8}$$

Here $\widetilde{r}$ represents the separation between the electron and the nucleus and hence can be separated by evaluating $\mathcal{H}_{\text{dip}}$ over the space coordinates of the ground state wave function ψ_o. Let us suppose that ψ_o possesses an axial symmetry about the nucleus. We then have the relation such as $\langle x^2/r^5\rangle = \langle y^2/r^5\rangle$ and $\langle xy/r^5\rangle = 0$, etc., and Equation 8 reduces to :

$$\mathcal{H}_{\text{dip}} = 2A_{\text{dip}}I_zS_z - A_{\text{dip}}\,(I_xS_x + I_yS_y) \tag{9}$$

where $$A_{\text{dip}} = g_e\beta_e g_n\beta_n\left\langle\frac{3\cos^2\alpha - 1}{2\,r^3}\right\rangle_{\psi_o}$$

The z axis is identified as the symmetry axis, and α is the angle between r and the symmetry axis.

Equation 8 or 9 is clearly inadequate when the electron is "inside" the nucleus. The magnetic interaction in this particular situation is isotropic and is known as the Fermi contact term. It is given by

$$\begin{aligned}\mathcal{H}_{\text{cont}} &= \frac{8\pi}{3}\,g_e\beta_e g_n\beta_n\,|\,\Psi(0)\,|^2\,I\cdot S \\ &= A_{\text{iso}}\,(I_zS_z + S_xI_x + S_yI_y)\end{aligned} \tag{10}$$

$|\Psi(0)|^2$ is the probability of the electron being at the nucleus and is non-zero only if the unpaired electron occupies an *s* orbital of the magnetic nucleus.

Thus the Hamiltonian for the hyperfine interaction in an axially symmetric case can be written as

$$\begin{aligned}\mathcal{H}_{\text{hfs}} &= \mathcal{H}_{\text{cont}} + \mathcal{H}_{\text{dip}} \\ &= A_{||}\, I_z S_z + A_{\perp}\, (I_x S_x + I_y S_y) \end{aligned} \tag{11}$$

where we now have

$$A_{||} = A_{\text{iso}} + 2A_{\text{dip}} \tag{12a}$$

$$A_{\perp} = A_{\text{iso}} - A_{\text{dip}} \tag{12b}$$

Combination of Equation 11 with Equation 5 gives the following spin Hamiltonian for an axially symmetric case involving one unpaired electron and one magnetic nucleus with spin of *I*.

$$\begin{aligned}\mathcal{H}_{\text{spin}} &= g_{||} S_z H_z + g_{\perp}\, (S_x H_x + S_y H_y) \\ &+ A_{||} S_z I_z + A_{\perp}\, (S_x I_x + S_y I_y) + aI \cdot H \end{aligned} \tag{13}$$

The last term in this expression represents the Zeeman interaction between the nuclear magnetic moment and the external field *H*. The magnitude of this interaction is small, and since the ESR transition does not usually involve a change in the magnetic quantum number of the nucleus, it can be omitted from the discussion. Evaluation of the remaining terms in terms of the $(2S + 1) \times (2I + 1)$ spin functions and diagonalization of the resulting matrix give the following eigenvalues for each state defined by the electronic and nuclear magnetic quantum numbers M_S and M_I.

$$W(M_S, M_I) = g\beta H M_S + A M_S M_I \tag{14}$$

where $M_S = \pm\ 1/2,\ M_I = +\ I,\ \text{----},\ -I$

$$g^2 = g_{||}^2 \cos^2\theta + g_{\perp}^2 \sin^2\theta \tag{15}$$

and

$$A^2 = \frac{A_{||}^2\, g_{||}^2}{g^2} \cos^2\theta + \frac{A_{\perp}^2\, g_{\perp}^2}{g^2} \sin^2\theta \tag{16}$$

Here θ is the angle defined by the magnetic field and the symmetry axis. Thus $(2I + 1)$ ESR transitions $(\Delta M_S = \pm 1,\ \Delta M_I = 0)$ would be observed at the following resonance fields.

$$H_{\text{res}} = \frac{h\nu}{g\beta} - \frac{A}{g\beta} M_I \tag{17}$$

In the special event where the paramagnetic species under observation possess rapid rotational motion, the anisotropies of both the g and A tensors are averaged out, and the isotropic resonance absorption would be observed at the position given by $g = (g_{||} + 2g_{\perp})/3$ and $A = A_{\text{iso}}$.

The g-Tensor and the Hyperfine Coupling Tensor of NO. As an example of the g tensor and the hyperfine coupling tensor discussed above, let us examine in some detail these tensors of nitric oxide (NO) adsorbed upon zeolite cations. NO is a stable but unique radical. In spite of its unpaired electron the free molecule exhibits no paramagnetism in its $^2\Pi_{1/2}$ ground state. This is because of the degeneracy of the π orbitals which leads to exact cancellation of the spin magnetic moment of the electron by its orbital magnetic moment. The ESR spectrum of the spin-only system, therefore, should become observable if the orbital moment of the electron is quenched—*i.e.*, if the degeneracy among the π orbitals is removed by its environment. Suppose that the electric field associated with the cations of zeolite removes such degeneracy of adsorbed NO. Figure 1 shows the valence orbitals of NO in such a state. The z axis is identified with the N–O internuclear direction, and the

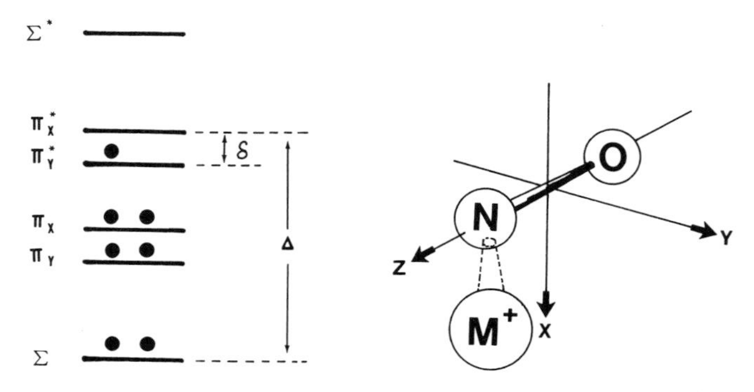

Figure 1. Valence orbitals of NO adsorbed on a cation with a bent structure

zeolitic cation is placed in the y–z plane. The unpaired electron of our concern thus resides in the antibonding $\pi_y{}^*$ orbital. In an LCAO description it can be expressed as:

$$\pi_y{}^* = C_{\text{N}}P_y\ (\text{N}) - C_{\text{O}}P_y\ (\text{O}) \tag{18}$$

where $P_y(\text{N})$, for example, represents the $2p_y$ orbital of the nitrogen atom. We see immediately $\langle L \rangle = 0$ for this orbital since $P_y = 1/\sqrt{2}(|1, +1> + |1, -1>)$. The spin orbit coupling interaction, however, mixes other states according to Equation 4. The calculation of the final g tensor using Equation 6 yields the following result.

$$g_x = g_e + 2\left(\frac{\lambda}{\Delta}\right) \tag{19a}$$

$$g_y = g_e \tag{19b}$$

$$g_z = g_e - 2\left(\frac{\lambda}{\delta}\right) \tag{19c}$$

The spin orbit coupling constant of NO can be estimated to be 0.01 eV from the interval of its $^2\Pi_{2/3}$ and $^2\Pi_{1/2}$ states. The separation Δ between the σ and π^* orbitals of this type of molecule is $3 \sim 5$ eV. The separation δ between the $\pi_x{}^*$ and $\pi_y{}^*$ orbitals would be much smaller than Δ. One thus predicts

$$g_z < g_x \cong g_y = 2.0023 \tag{20}$$

Although ^{16}O (natural abundance = 99.8%) has no magnetic moment, ^{14}N (natural abundance = 99.6%) has a nuclear spin of 1. As seen in Equations 8 and 10, the hyperfine interaction diminishes rapidly with increasing r. The essential feature of the hyperfine coupling tensor to a magnetic nucleus is, therefore, determined by the nature of the wavefunction only near the nucleus. The hyperfine coupling tensor to the ^{14}N nucleus of NO is thus determined by the $P_y(N)$ part of Equation 18 and possesses apparent axial symmetry about the y axis. We therefore expect the following relation.

$$A_y = A_{\text{iso}} + 2\,A_{\text{dip}} \tag{21a}$$

$$A_z = A_x = A_{\text{iso}} - A_{\text{dip}} \tag{21b}$$

where $$A_{\text{dip}} = C_N^2 g_e\beta_e g_n\beta_n \left\langle\frac{3\cos^2\alpha - 1}{r^3}\right\rangle_{N,\,2p}$$

For a pure p orbital $|\Psi(0)|^2 = 0$. A_{iso} in Equation 21 is non-zero, however. It is produced as the result of spin polarization of inner core orbitals by the unpaired electron in the $2p$ orbital. For an unpaired electron in a pure $2p$ orbital of nitrogen, it has been found that $A_{\text{iso}} \cong A_{\text{dip}}$. The principal elements of the hyperfine coupling tensor to ^{14}N in NO are thus expected to be:

$$A_y \cong 3\,A_{\text{dip}} \tag{22a}$$

$$A_z \cong A_x \cong 0 \tag{22b}$$

Powder Pattern Spectra. Since zeolite crystals large enough for single-crystal ESR measurements are not usually available, almost all published ESR studies on zeolites are of powder, polycrystalline samples. (The ESR spectrum of Cu^{2+} in a single crystal of chabazite has been reported by Chao and Lunsford (*7*).) The ESR powder pattern $S(H)$ expected from an ensemble of randomly oriented radicals is given by

$$S(H) = \sum_{M} \int_{0}^{2\pi} \int_{0}^{\pi} F\,[H - H_M\,(\theta,\,\psi)]\,\sin\theta\,d\theta\,d\psi \qquad (23)$$

The angles θ and ψ define the orientation of the g tensor relative to the magnetic field, and the summation over M is to be performed for all the hyperfine components. $F[H\text{-}H_M(\theta,\psi)]$ is the line shape function of the single-crystal case, and $H_M(\theta,\psi)$ defines the resonance position of the M-th hyperfine component given, for example, by Equation 17. The sin θ factor takes into account the weighting factor of the solid angle.

The dependency upon θ of H_{res} and the powder pattern $S(H)$ expected therefrom for the special case of axial symmetry ($g_{||} > g_{\perp}$) with no hyperfine interaction are shown in Figure 2a and 2b, respectively. In actual experi-

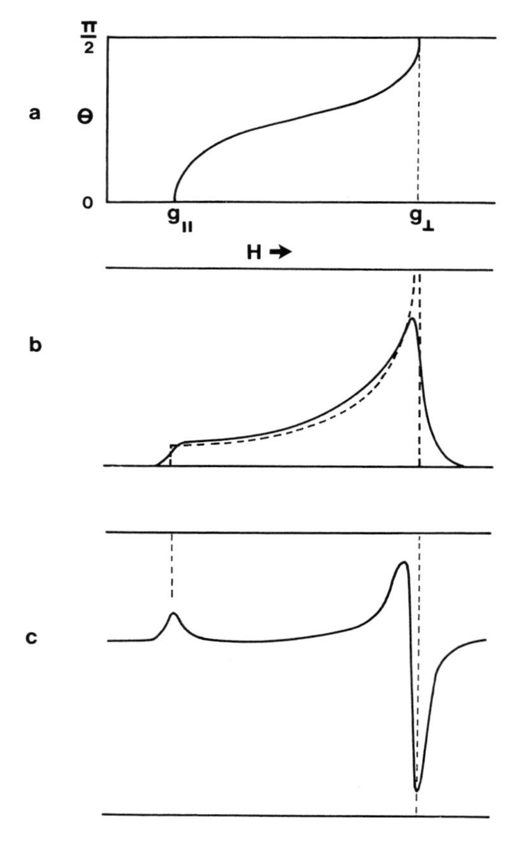

Figure 2. Dependency of the resonance field upon θ for an axially symmetric spin Hamiltonian with $\mathbf{g}_{||} > \mathbf{g}_{\perp}$ (a), the powder absorption pattern expected therefrom (b), and its derivative (c)

ments, ESR spectra are usually observed in the first derivative form of the absorption curve. Figure 2c shows the experimental derivative pattern expected from Figure 2b.

Shown in Figures 3a and 3b are the angular dependencies of the three hyperfine components of NO discussed in the preceeding section for magnetic field rotation within the x–y and the y–z planes, respectively. Figure 3c

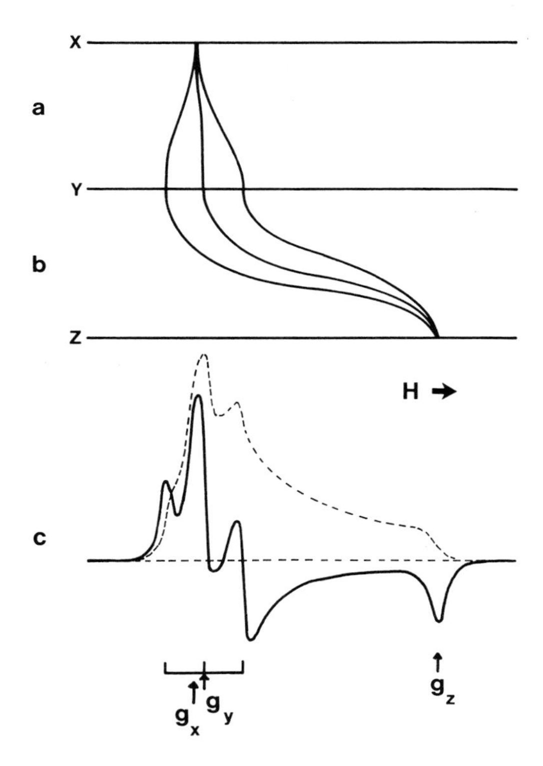

Figure 3. The orientation dependencies of the three hyperfine components of NO when the magnetic field is rotated in the x–y *and the* y–z *planes (a) and (b) (*see *Figure 1). The absorption pattern (dotted line) and its derivative curve (solid line) expected from an ensemble of randomly oriented NO molecules (c).*

shows the theoretical absorption curve and the derivative pattern expected from NO adsorbed in randomly oriented polycrystalline zeolite. All the spectra shown in this chapter are those of polycrystalline zeolites.

Stable Radicals

Soon after the molecular sieve property of large pore zeolites became known, it was realized that such zeolite might be used to isolate and store stable free radicals for ESR investigation. More recently stable radicals are being used to study the zeolitic cations and other Lewis acid sites with which the radicals interact. The observation of NO_2, NF_2, and ClO_2 adsorbed in zeolites are of the former type, and many ESR studies of NO represent the latter.

Presented below are brief discussions of the results obtained from these radicals. Unless mentioned otherwise all the zeolite samples discussed here had been vacuum activated at 300°–600°C for several hrs, and all the ESR spectra are those obtained at liquid nitrogen temperature (77°K) using an X-band (9.0–9.5 GHz) spectrometer.

NO_2. The ESR spectrum of NO_2 adsorbed in zeolite was first reported by Colburn, *et al.* (*8*). Using NaX, they observed the spectrum at room

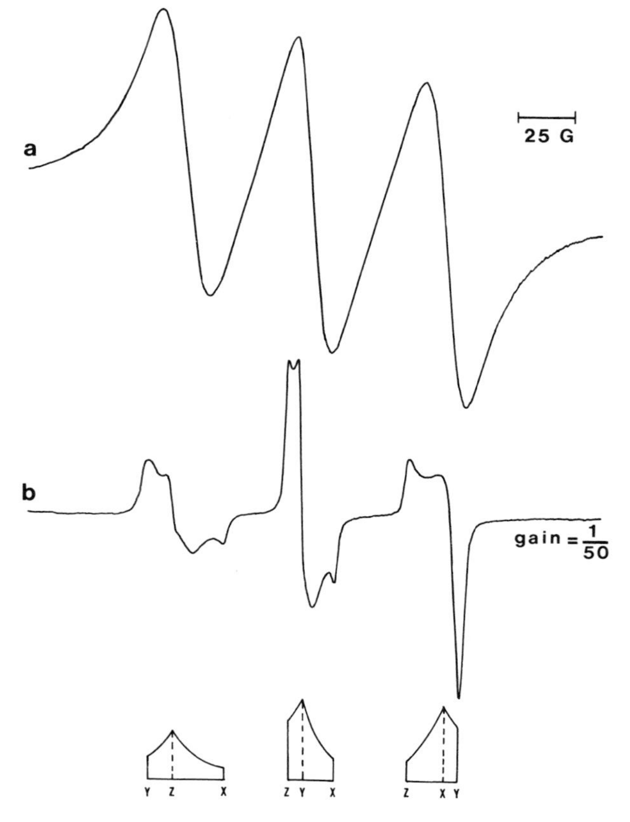

Figure 4. ESR spectra of NO_2 adsorbed in BaY observed (a) at room temperature and (b) at 77°K

temperature. A similar room temperature spectrum of NO_2 adsorbed in BaY is shown in Figure 4a (*9*). The broad triplet pattern is caused by hyperfine interaction with the ^{14}N nucleus. The anisotropy of the spectrum is averaged by the rotational motion of the radicals within the zeolite. Figure 4b is the spectrum of the same sample observed at liquid nitrogen temperature (*9*). The spectrum is extremely anisotropic, indicating freezing of the rotational motion observed at room temperature. The *g* tensor and the hyperfine coupling tensor assessed from Figure 4b are:

$$
\begin{aligned}
g_x &= 1.995\ (1.9920)\\
g_y &= 2.003\ (2.0030)\\
g_z &= 2.007\ (2.0062)\\
A_x(^{14}N) &= 49\ G\ (45.8G)\\
A_y(^{14}N) &= 69\ G\ (61.7G)\\
A_z(^{14}N) &= 53\ G\ (51.7G)
\end{aligned}
$$

The values in parentheses are determined from NO_2 isolated in a rare gas matrix (*10*). The larger coupling constant to ^{14}N in the case of BaY is attributed to the NO_2–Ba^{2+} complex (*9*). More recently Pietrzak and Wood (*11*) studied the temperature dependence of the spectrum of NO_2 adsorbed in NaX and CaX and concluded that restricted rotation of NO_2 in CaX occurs above 114°K and that the majority of NO_2 exist within the cavity as diamagnetic N_2O_4 and constitutes the rotation barrier of the monomer.

NF_2. NF_2 may be considered a stable free radical in that a substantial amount of it exists in equilibrium with its diamagnetic dimer.

$$N_2F_4 \rightleftarrows 2\ NF_2$$

Colburn *et al.* (*12*) observed the room temperature spectra of NF_2 absorbed in NaX and NaA. The well resolved triplet ($A = 56$ G) of triplets ($A = 16$ G) was readily recognized as that of NF_2 undergoing rotational motions within the zeolite cavities. The larger triplet is caused by two equivalent ^{19}F nuclei ($I = 1/2$), while the smaller triplet is caused by ^{14}N. These splitting constants are very similar to those determined for NF_2 isolated in a rare gas matrix (*13*).

Colburn *et al.* (*12*) were able to observe both the ^{19}F NMR of N_2F_4 and the ESR of NF_2 in NaX but only the ESR of NF_2 in NaA. They concluded that N_2F_4 and NF_2 are following independent adsorption isotherms in the zeolite cavities.

ClO_2. The ESR spectrum of ClO_2 adsorbed in zeolites was first studied by Coope *et al.* (*14*). The spectrum observed with H–Zeolon was very similar to that given by ClO_2 produced in crystals of $KClO_4$ by γ-irradiation (*15*) (*see* Table I). For NaX two sets of spectra were observed—one having a slightly larger hyperfine coupling interaction with the chlorine nucleus than

Table I. ESR Data of ClO_2 Adsorbed on Zeolites

	g Tensor[a]			*Hyperfine Tensor to* ^{35}*Cl (G)*[a]			
Medium	g_z	g_x	g_y	A_z	A_x	A_y	*Ref.*
$KClO_4$	2.0016	2.0167	2.0121	74.7	−10.8	−11.5	(*15*)
NaX	2.0023	2.0123	2.0115	84.5	−18.9	− 8.8	(*14*)
NaX	2.0023	2.0123	2.0115	77.5	−17.4	− 8.0	(*14*)
HZ	2.0023	2.0123	2.0115	74.9	−16.7	− 7.8	(*14*)
NaX	2.002	2.017	2.012	84.0	−11.0	−11.0	(*9*)
NaY	2.002	2.017	2.012	80.0	−11.0	−11.0	(*9*)
BaY	2.003	2.017	2.012	75.8	−11.0	−11.0	(*9*)

[a] The *z* axis is perpendicular to the molecular plane.

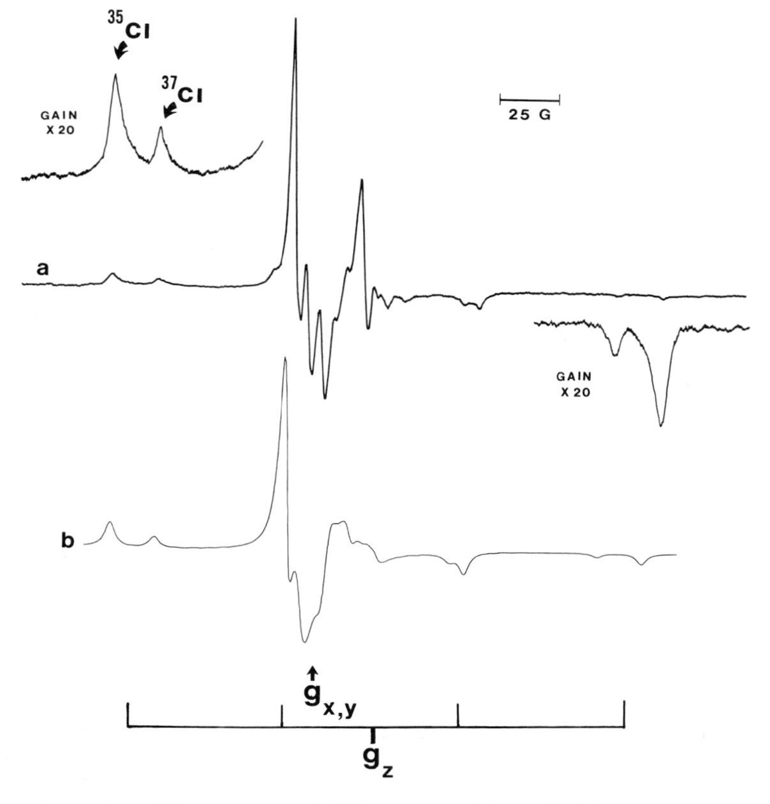

Figure 5. ESR spectrum of ClO_2 adsorbed in BaY (a). A computer-simulated spectrum of ClO_2 based upon the parameters determined for ClO_2 generated in $KClO_4$ (15) *(b). The discrepancy in the central part of the spectrum is probably caused by the forbidden transitions ($\Delta M_s = \pm 1, \Delta M_I \neq 0$).*

the other. The unpaired electron in ClO_2 is located in the antibonding π-orbital of the bent triatomic system. Assuming that the more electronegative oxygen side of the molecule lies closest to the zeolitic cation, Coope *et al.* suggested that the cation shifts the electron density of the bonding orbital toward the oxygen and hence shifts the density of the unpaired electron toward the chlorine (*14*). The ClO_2 spectrum in NaX with the larger Cl coupling constant was thus assigned to ClO_2 adsorbed by the cations at site III and the other with the smaller Cl coupling constant to those adsorbed by the cations at site II. According to the calculation by Dempsey (*16*), the electric field of a cation at site III is several times stronger than that of the same cation at site II (*see* Table II).

Table II. Electric Field (V/A) at Indicated Distance from Cation Surfaces (*16*)

	Site II Cation			*Site III Cation*		
Zeolite[a]	*1 A*	*1.75 A*	*2.5 A*	*1A*	*1.75 A*	*2.5 A*
NaX	1.8	0.43	0.10	3.1	1.2	0.43
NaY	2.4	0.94	0.43	3.2	1.3	0.64
CaX	6.1	2.7	1.3			
CaY	6.4	3.2	1.8			

[a] X and Y are assumed to have Si/Al of 1 and 2.

We have also examined the ESR spectra of ClO_2 adsorbed in NaX, NaY, and BaY zeolites synthesized in our laboratory (*9*). The *g* tensors and the hyperfine coupling tensors assessed from the spectra are given in Table I. Figure 5 compares the spectrum observed on BaY with the computer-simulated spectrum using the parameters determined for ClO_2 in $KClO_4$. The indicated quartet structure shows the hyperfine interaction with the chlorine nuclei ^{35}Cl (natural abundance = 75%, $I = 3/2$) and ^{37}Cl (natural abundance = 25%, $I = 3/2$). At variance with the earlier study, we observed only one type of adsorption site in NaX. Furthermore, the observed trend in the Cl coupling constant is in the order of NaX > NaY > BaY. It is not consistent with the analysis prescribed by Coope *et al.* One is tempted to speculate that even in the α cages of NaX the cations are located mostly at site II, and the adsorbed ClO_2 is attached to the cation through its chlorine atom. The extra adsorption site observed in the NaX of the earlier study is probably the result of multivalent impurity cations or hydroxyl sites created by cation deficiency.

NO. As discussed previously NO is a $^2\Pi$ radical, and its spin resonance signal cannot be observed unless the degeneracy among the π orbitals is removed. The resulting spectral pattern, as seen in Equation 19, is extremely sensitive to the magnitude of this separation. NO should, therefore, be an ideal probe for studying the electric field associated with the cations of zeolites. Lunsford (*17*) was the first to report on the ESR spectra of NO adsorbed on zeolites (NaY and NH_4Y). Gardner and Weinberger (*18*) also observed similar spectra of NO adsorbed on 4A, 5A, NaX, and H–Zeolon. The spectra reported by these authors were broad and ill-defined and were analyzed only in terms of an axially symmetric spin Hamiltonian.

During our independent study of NO adsorbed on zeolites (*19*) we noticed that the NO spectra displayed by Y type zeolites became much sharper if the samples sealed with NO were left standing at room temperature for several days. The rationale for this sharpening by "NO treatment" is discussed later. Figures 6a and 7a are the sharp spectra of NO observed with NO treated NaY and BaY, respectively (*19*). In the case of BaY the hyperfine structure caused by ^{14}N can be readily recognized (*cf.* Figure 3). The g tensors, the hyperfine coupling tensors, and the π orbital separations (δ in Figure 1) determined from these spectra are given in Table III together with the results obtained from other zeolites. In accordance with Equations 20 and 22 the unique axis of the g tensor and that of the hyperfine coupling tensor are indeed perpendicular to each other. BaY shows smaller Δg_z $(= g_e - g_z)$ than NaY, reflecting a larger separation of the π^* orbitals and hence the stronger electric field of the Ba^{2+} ions. The computer-simulated spectra based upon these parameters are in excellent agreement with the observed spectra (Figures 6b and 7b).

Gardner and Weinberger (*18*) proposed a linear structure for the NO–zeolitic cation complex and suggested that a non-axial component of the field gradient, $\delta^2V/\delta x^2 - \delta^2V/\delta y^2$, as the factor responsible for the separation of

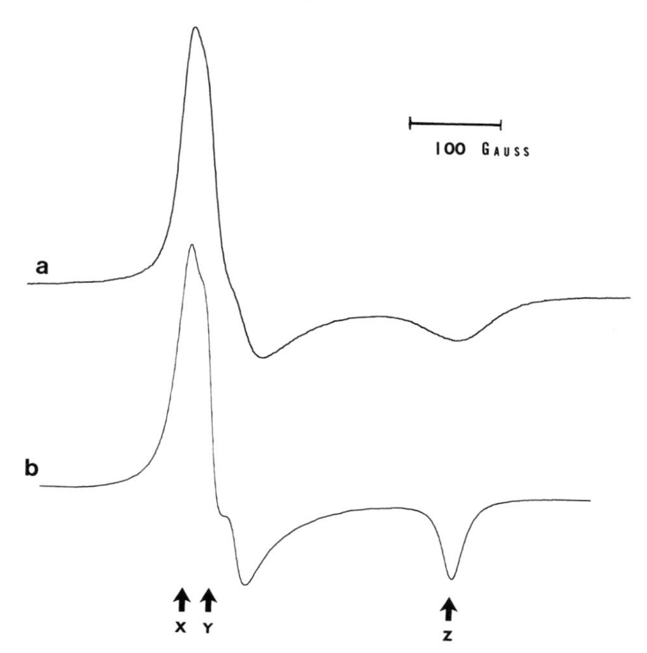

Figure 6. Observed (a) and computer-simulated (b) spectra of NO adsorbed on NO-treated NaY

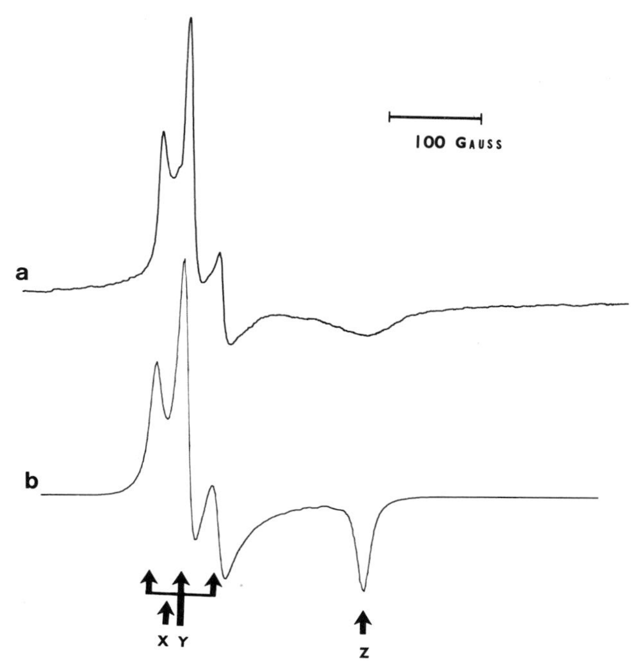

Figure 7. Observed (a) and computer-simulated (b) spectra of NO adsorbed on NO-treated BaY

the π^* orbitals. In view of the large dependency of Δg_z upon the cations, we believe that NO molecules are attached to the cations in a bent configuration.

The spectrum of NO observed by Lunsford with dehydroxylated NH_4Y was especially broad (*17*). It was assigned to NO adsorbed on trigonal aluminum at the oxygen deficient sites of the framework.

$$(O)_3\text{Si}{-}O(H)\cdots\text{Al}(O)_3 \longrightarrow (O)_3\text{Si}^+ \quad \text{Al}(O)_3$$

The extra broadening is attributed to unresolved hyperfine structure caused by ^{27}Al ($I = 5/2$, natural abundance = 100%). At 2°K Lunsford obtained a spectrum showing the structure partially resolved in the $g_\perp$ region (*17*). A similarly resolved spectrum of NO adsorbed on dehydroxylated NH_4Y was observed in our laboratory at 77°K (Figure 8) (*9*). The spectrum of NO adsorbed on activated NH_4Y is not very reproducible, however. Deep bed

Table III. ESR Data of

Medium	*g Tensor*[a]			*Hyperfine Tensor to ^{14}N (G)*[a]		
	g_z	g_x	g_y	A_z	A_x	A_y
NaX	1.79	$g_\perp$ = 1.970				
NaY	1.83	1.986	1.978	~0	~0	29
BaY	1.89	1.999	1.995	~0	~0	34
ZnY	1.93	2.000	1.998	~0	~0	30
NH_4Y	1.95	$g_\perp$ = 1.996				
Cu(I)Y	1.89	$g_\perp$ = 2.009				

[a] The z axis is parallel to the N–O bond.

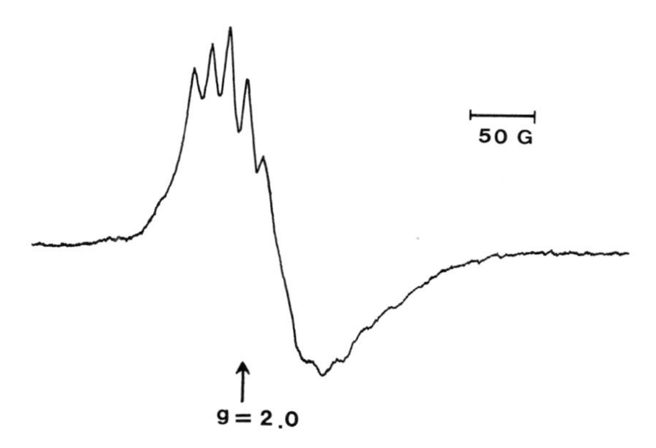

Figure 8. ESR spectrum of NO adsorbed on dehydroxylated NH_4Y

activation appears to be conducive to the revelation of the aluminum hyperfine structure. It has been shown that during activation of NH_4Y some of the aluminum atoms are removed from the framework with subsequent formation of interstitial aluminum (hydro)oxy cations (*20, 21, 22*). NO molecules adsorbed on activated NH_4Y are probably associated with such interstitial aluminum species rather than aluminum atoms at the oxygen deficient sites of the framework.

Another study by Lunsford (*23*) on NO revealed that MgY, CaY, and SrY but not BaY exhibit ESR spectra of NO similar to that seen with activated NH_4Y. This is in accord with the result of infrared study by Ward (*24, 25*), who showed that the water molecules of hydration dissociate to give hydroxy groups during activation of MgY, CaY, and SrY but not BaY. The dissociation reaction may take the form of

$$M^{2+}\ (H_2O) \rightarrow [M\ (OH)]^{+} + H^{+}$$

NO Adsorbed on Zeolites

A to Other Nucleus	δ (*eV*)	*Ref.*
	0.09	(*18*)
	0.12	(*19*)
	0.18	(*19*)
	0.28	(*19*)
$A(\mathrm{Al}) = 14$	0.38	(*17*)
$A_{\parallel}(\mathrm{Cu}) = 240$; $A_{\perp}(\mathrm{Cu}) = 190$	0.18	(*28*)

or

$$2M^{2+} + H_2O \rightarrow (M{-}O{-}M)^{2+} + 2\ H^+$$

Ward concluded that the electrostatic field associated with the barium cation is insufficient to induce the dissociation process.

The Cu^{2+} ions in CuY zeolite can be reduced to Cu^+ with CO (*26*). Naccache, Che, and Ben Tarrit (*27*) and Chao and Lunsford (*28*) observed independently the ESR spectrum of NO adsorbed on CuY reduced by CO and obtained a direct evidence for the formation of NO–cation complex. The Cu nuclei have a spin of 3/2. The observed spectrum is essentially that of NO; the $g_{\perp}$ and $g_{\parallel}$ components are split respectively into a quartet by the hyperfine interaction with the Cu nucleus. The assignment of Naccache *et al.* (*27*) is incorrect. Figure 9 shows the spectrum of the NO–Cu^+ complex in CuY obtained in our laboratory and that computer-simulated using the parameters assessed by Chao and Lunsford (*28*) (Table III). The large and essentially isotropic coupling constant to the Cu nucleus indicates a substantial delocalization of the unpaired electron into the Cu 4*s* orbital. Based upon the data available for atomic copper orbitals Chao and Lunsford (*28*) concluded that the unpaired electron spends about 20% of its time on Cu^+ distributed evenly between the $3d_{z^2}$ and 4*s* orbitals. They proposed a linear structure for the complex. No mixing can occur between the semifilled $\pi_y{}^*$ orbital of NO and the $3d_{z^2}$–4*s* orbital of Cu, nor can the degeneracy between the π^* orbitals be removed in a linear configuration. The complex must have a bent structure.

DTBN (di-*tert*-Butyl Nitroxide). Hoffman and his co-workers (*29*) showed that stable free radical di-*tert*-butyl nitroxide (DTBN), I, forms the molecular complex II with strong Lewis acid.

$$\left[\ \mathrm{(t\text{-}Bu)_2\ddot{N}{-}\dot{O}} \rightleftarrows \mathrm{(t\text{-}Bu)_2\dot{N}{-}\ddot{O}:}\ \right] \xrightarrow{+L} \mathrm{(t\text{-}Bu)_2\dot{N}{-}O:L}$$

Ia Ib II

They demonstrated that the complex can form not only in solution (*e.g.* $AlCl_3$ in CCl_4) but also on a heterogeneous system such as silica–alumina surface (*30*). The complexes are formed through the datively bonded lone pair electrons at the oxygen thus leading to stabilization of the resonance structure Ib hence an increase in the hyperfine interaction with the ^{14}N nucleus.

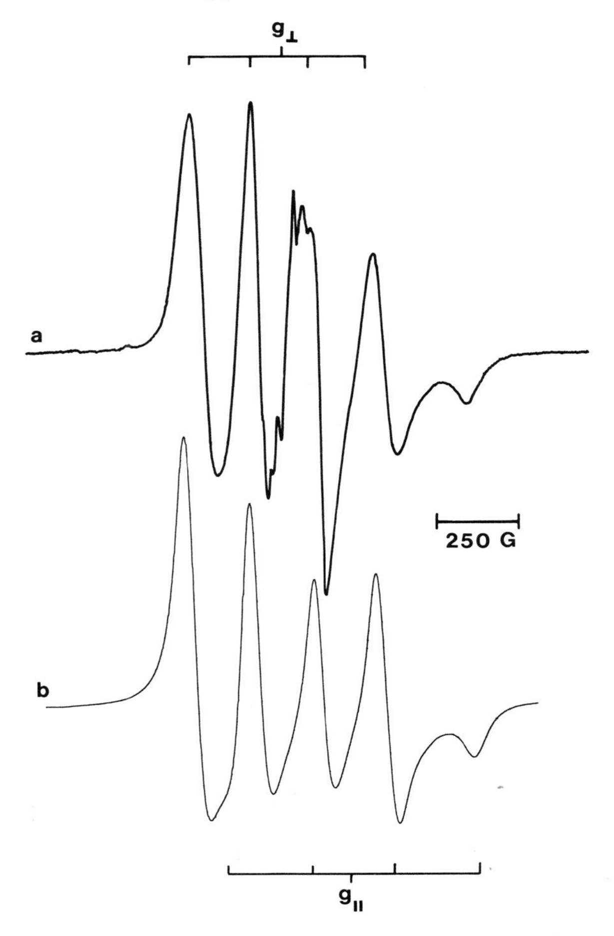

Figure 9. Observed (a) and computer-simulated (b) spectra of NO adsorbed on Cu(I)Y. The discrepancy in the central part of the $\mathbf{g}_\perp$ *region is caused by the presence of other signals (*Cu^{2+} *and NO), and probably by orthorhombicity of the* **g** *and* ***A*** *tensors.*

Figure 10 shows the ESR spectra observed at room temperature of DTBN adsorbed on NH_4Y activated at 300° and 500°C, respectively (*9*). The spectrum in Figure 10a can be readily recognized as that dominated by an uniaxially anisotropic hyperfine coupling tensor ($A_{||} \gg A_\perp \cong 0$) to the

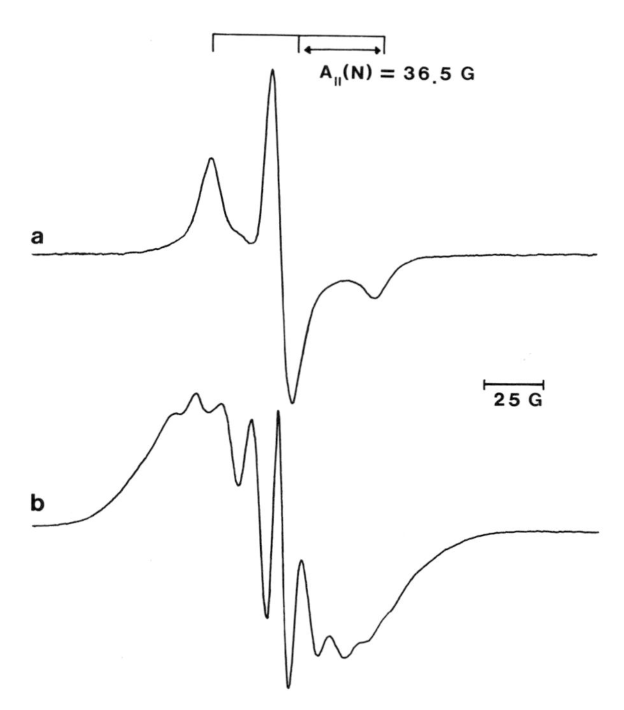

*Figure 10. ESR spectra of DTBN (di-*tert-*butyl nitroxide) adsorbed on (a) NH_4Y activated at 300°C and (b) NH_4Y activated at 500°C*

^{14}N nucleus, consistent with structure II in which the unpaired electron is in the nitrogen p_π orbital perpendicular to the molecular plane. It is attributed to DTBN hydrogen bonded to HY. The magnitude of $A_{||}$ (36.5 G) assessed from the figure is larger than that observed in frozen inert solution (34.6 G) (*30*) but is smaller than that of DTBN hydrogen bonded to the surface of silica–alumina (38.6 G) (*30*).

The complex structure in Figure 10b must be the result of the hyperfine interaction with ^{27}Al. Almost identical spectra have been reported by Hoffman for the DTBN–$AlCl_3$ complex trapped in frozen CCl_4 solution and for DTBN adsorbed on activated silica–alumina (*30*). As in the case of NO adsorbed on activated NH_4Y, we believe that interstitial aluminum species are responsible for the observed complex. Had the trigonally coordinated Al sites been formed within the framework in the manner depicted earlier, DTBN should react preferentially with the much stronger Lewis acid, the positively charged trigonally coordinated Si atoms.

Radiation-Induced Radicals

Since the early work by Stamires and Turkevich (*31*), ionizing radiation (γ- or x-ray) has been extensively used to generate paramagnetic species within zeolites. Irradiation of zeolites with adsorbed molecules often leads to formation of otherwise unstable anion radicals stabilized by the electric field of the zeolitic cations. The observed ESR spectra of the anions provided not only powerful evidence for the existence of the cationic field but also a probe to measure its strength. Irradiation of zeolites activated and sealed in vacuum without adsorbed molecules also produced paramagnetic centers of equally revealing nature, particularly with NaY and activated NH_4Y.

The most persistent difficulty encountered in the study of irradiated zeolites appears to be its reproducibility, especially among different laboratories. Stamires and Turkevich (*31*) observed two γ-radiation induced ESR signals; X_1 on NaX and NaY and X_2 on activated NH_4Y. The former was a broad singlet ($\Delta H_{peak\text{-}to\text{-}peak} = 38$ G) at $g = 2.002$, and the latter was a sextet at $g = 2.0017$ with a successive spacing of 5 G. An electron hole trapped by framework oxygen, and an excess electron trapped by trigonally coordinated aluminum of the framework were proposed to be the X_1 and X_2 centers, respectively (*31*). These assignments should be viewed with reservation; subsequent studies at other laboratories were unable to reproduce the result. The g tensor of O_2^- produced in NaY has also been disputed (*32*). We have found that the ESR spectra of irradiated zeolites are extremely sensitive to trace impurity cations and often to the temperature at which the sample was irradiated and maintained after the irradiation. Simpler and more reproducible spectra are often obtained when extra-pure laboratory synthesized zeolites are used and irradiated at low temperature (77°K). Discussed below are some of the better understood paramagnetic centers in zeolites created by γ- or x-ray irradiation.

Superoxide Ion O_2^-. An early study in our laboratory (*33*) revealed that irradiation of zeolites (NaY and BaY) in the presence of oxygen produces O_2^- ions attached to the zeolitic cations. O_2^- is also a $^2\Pi$ radical, and, as in the case of NO, the degeneracy among the π orbitals must be removed before a well-defined ESR spectrum can be observed. Figure 11 shows the valence orbitals of O_2^- attached to a cation. The unpaired electron is now in

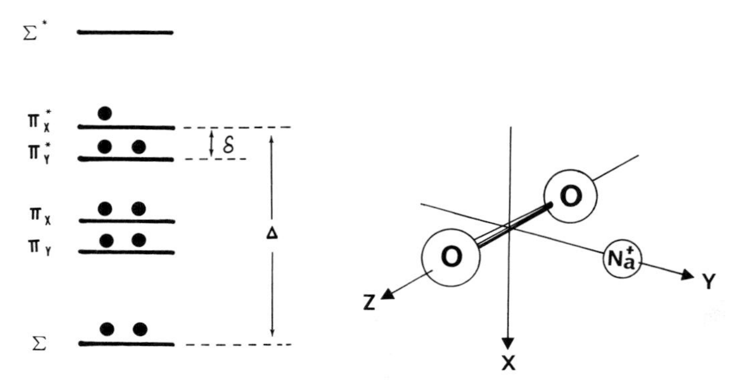

Figure 11. Valence orbitals of O_2^- associated with a cation

the π_x^* orbital. Its g tensor can be calculated according to the method prescribed for NO.

$$g_x = g_e \tag{24a}$$

$$g_y = g_e + 2\left(\frac{\lambda}{\Delta}\right) \tag{24b}$$

$$g_z = g_e + 2\left(\frac{\lambda}{\delta}\right) \tag{24c}$$

One thus expects the relation $g_z > g_y \cong g_x$.

Figure 12 shows the ESR spectrum of O_2^- observed in NaX that had been irradiated by x-rays at 77°K in the presence of oxygen (*9*). Similar spectra

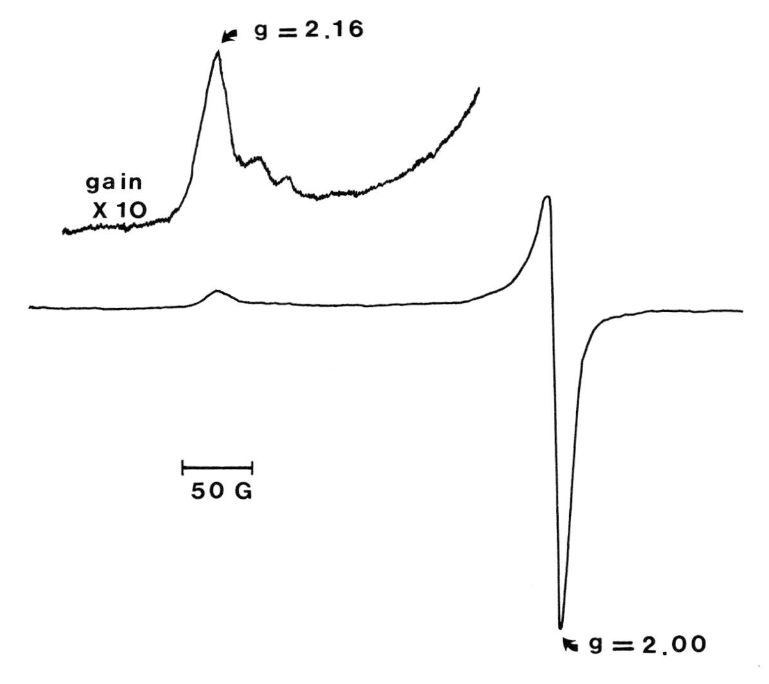

Figure 12. ESR spectrum of O_2^- generated in NaX by x-ray irradiation at 77°K in the presence of oxygen

were also obtained from NaY and BaY (*9*). When excess oxygen was introduced to these samples, the spectra broadened beyond recognition. They were restored completely, however, when the oxygen was evacuated. The O_2^-–cation complexes must be located within large cavities. Table IV lists the g tensors and the π^* orbital separations δ of the O_2^- ions assessed from these spectra. The spin orbit coupling constant of 0.014 eV (*33*) was used to calculate δ.

Our earlier paper (*33*) reported the g_z of 2.113 for O_2^- observed in NaY. The O_2^- ions were then produced by γ-rays at room temperature. When NaY containing adsorbed oxygen was irradiated by x-rays at 77°K, the O_2^- spectrum showed the most prominent z component at $g_z = 2.08$. When this sample was warmed to room temperature for several minutes, the component at $g_z = 2.08$ decayed, and several new z components appeared at $g_z = 2.10–2.15$ (*9*). We believe the spectrum characterized by $g_z = 2.08$ represents the O_2^- attached to the Na^+ ions at usual sites, and other spectra are created by diffusion of the originally formed $Na^+–O_2^-$ complexes. The amount of O_2^- produced by irradiation was $\sim 10^{18}$ ions/g, corresponding to one O_2^- ion per 100 α-cages. As discussed in a later section, when a much larger amount (10^{20} ions/g) of O_2^- was produced by a chemical method, the z component was found to be unique and stable at $g_z = 2.08$.

It is interesting that only one major adsorption site is indicated by the O_2^- spectrum of NaX. It has a much weaker field than that of Na^+ at site II in NaY. According to the calculation by Dempsey (*16*), only the cations at site II of NaX would have such a field (*see* Table II). Again as in the case of ClO_2 we are compelled to speculate if most of the cations are located at site II in the α-cages of NaX.

Table IV. ESR Data of O_2^- Generated in Zeolites by Ionizing Radiations

Medium	*g Tensor*[a]			A to Other Nucleus (G)	δ (eV)	*Ref.*
	g_z	g_x	g_y			
NaX	2.162	2.0000	2.0048		0.18	(*9*)
NaY	2.080	2.0016	2.0066		0.36	(*9*)
BaY	2.057	2.0046	2.0090		0.51	(*33*)
AlHY	2.038	2.003	2.009	5.7 (Al)	0.78	(*34, 35*)
ScY	2.030	2.002	2.009	5.1 (Sc)	1.01	(*35*)
LaY	2.044	2.005	2.009	9.7 (La)	0.67	(*35*)

[a] The z axis is parallel to the O–O bond.

Wang and Lunsford (*32*) examined the spectra of O_2^- generated in Mg, Ca, Sr, and BaY zeolites by γ- and UV irradiations. In each zeolite they noted the formation of several types of O_2^- radicals distinguished by their g_z values. Two dominant types were recognized; one type with g_z in the range of 2.062 ± 0.003 was produced by γ-rays, while the other with g_z in the range of 2.048 ± 0.002 was produced by UV irradiation. However, a larger dosage of γ-rays (10^8 rads—a typical dosage $= 10^6$ rads) produced a predominant signal with $g_z = 2.046$. Clearly the situation is quite complex, and without further definitive study little can be said about the exact origins of the various O_2^- ions. One obvious source of complexity is the dissociation of water molecules known to occur during activation of MgY, CaY, and SrY and various modes of dehydroxylation processes that could follow. Wang and Lunsford (*32*) observed the simplest O_2^- pattern with BaY consistent with the fact that no hydrolytic reaction occurs in BaY.

Wang and Lunsford (*34*) also observed O_2^- generated in dehydroxylated Y (NH_4Y activated at 600°C). The spectrum showed readily recognizable hyperfine structures arising from ^{27}Al (*see* Figure 16b). They attributed the spectrum to O_2^- associated with the trigonal aluminum at the oxygen-deficient site of the framework (*34*). It is more likely, as in the case of NO on dehydroxylated Y, that the O_2^- ions are associated with interstitial aluminum species. This view is supported by another study of Wang and Lunsford (*35*) who observed the identical spectrum of O_2^- in Al–exchanged HY activated at 400°C, although these authors suggested that the O_2^- in this zeolite was also associated with the framework aluminum. Vedrine and Naccache (*36*) obtained a completely different spectrum with no hyperfine structure when they irradiated HY (NH_4Y activated at 400°C) in the presence of oxygen (*see* the later section on Activated NH_4Y).

Using oxygen enriched with ^{17}O ($I = 5/2$) isotope, Ben Taarit and Lunsford (*37*) showed that the two oxygen nuclei of O_2^- attached to the aluminum species in dehydroxylated Y are not equivalent. It indicates a peroxy type arrangement of O_2^- at the adsorption site.

O_2^- ions were also observed in ScY and LaY (*35*). The observed spectra of O_2^- possessed well recognizable hyperfine structures caused by the respective cations ($I = 7/2$, natural abundance = 100% for both ^{45}Sc and ^{139}La) and showed the existence of only one type of adsorption site. Sc and La ions located at site II′ were proposed to be involved (*35*).

Cl_2^-. The ESR spectrum of Cl_2^- ion was first observed in a γ-irradiated KCl crystal (*38*). The spectrum showed conclusively that the unpaired electron is in the antibonding σ_u^* orbital built from the halogen p_z orbitals. Coope *et al.* (*14*) reported the formation of Cl_2^- ions in NaX and H-Zeolon that had been γ-irradiated in the presence of Cl_2. They noted, as they did in their study of ClO_2 adsorbed in NaX, the presence of two types of adsorption sites in NaX, and only one type in H-Zeolon. The central portions of their reported spectra are obscured by a strong sharp signal caused by the second unidentified species (*14*).

We also examined the spectra of Cl_2^- generated by x-rays at 77°K in NaX, NaY, and BaY (*9*). In each case we observed only one set of Cl_2^- spectra. Figure 13a shows the spectrum obtained from NaY that had been irradiated at 77°K in the presence of Cl_2, warmed to dry ice temperature to anneal the sharp singlet at $g = 2.00$, and cooled back to 77°K. The total pattern is now ascribed to Cl_2^- ions randomly oriented within the zeolite. Both the ^{35}Cl (natural abundance = 75%) and ^{37}Cl (natural abundance = 25%) nuclei have spin of 3/2. The indicated septet pattern is caused by those Cl_2^- whose internuclear axes are parallel to the magnetic field and signifies the equivalence of two nuclei at the adsorption site. The observed spectrum agrees well with the computer-simulated spectrum based upon the assessed parameters (Figure 13b).

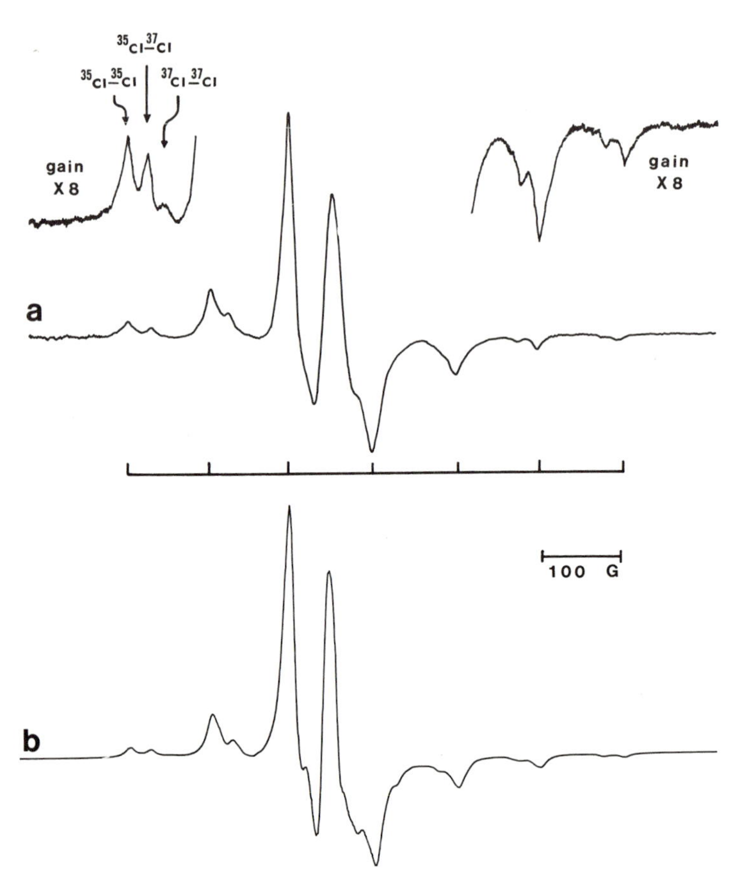

Figure 13. Observed (a) and computer-simulated (b) spectra of Cl_2^- generated in NaY

Table V. ESR Data of Cl_2^- Generated in Zeolites by Ionizing Radiations

	g Tensor			Hyperfine Tensor to ^{35}Cl (G)[a]			
Medium	g_z	g_x	g_y	A_z	A_x	A_y	*Ref.*
KCl	2.001	2.043	2.045	101	9.0	9.0	*(38)*
HZ[b]	2.002	2.035	2.035	107	9.6	9.6	*(14)*
				93	8.4	8.4	*(14)*
NaX	2.002	2.035	2.035	90	14.5	6.2	*(14)*
	2.002	2.035	2.035	101	9.0	9.0	*(14)*
NaX	2.001	2.040	2.033	105	12	12	*(9)*
NaY	2.001	2.040	2.033	103	12	12	*(9)*
BaY	2.001	2.040	2.033	99	12	12	*(9)*

[a] The z axis is parallel to the Cl–Cl bond.
[b] The Cl nuclei are found to be non-equivalent only in HZ.

The g tensors and the hyperfine coupling tensors determined for Cl_2^- generated in zeolites are compared in Table V. Coope *et al.* (*14*) showed that the effect of a strong electric field applied perpendicular to the internuclear direction of Cl_2 is to mix the ground $^2\Sigma_u$ state with the excited $^2\Pi_g$ state and thereby to reduce the hyperfine coupling constant along the z direction. The A_z values determined in our series of study are indeed in the order of NaX > NaY > BaY and are consistent with the view that the cations involved in the stabilization of Cl_2^- are located at site II. The Cl_2^- ions with A_z of 90 G observed by Coope *et al.* (*14*) in NaX are probably caused by multivalent impurity cations. In the case of H-Zeolon the spectrum was assigned to Cl_2^- having two non-equivalent chlorine nuclei (*14*).

NaY; Na_4^{3+} Centers. When irradiated in vacuum, NaY became distinctly pink and exhibited a unique ESR spectrum shown in Figure 14 (*33*).

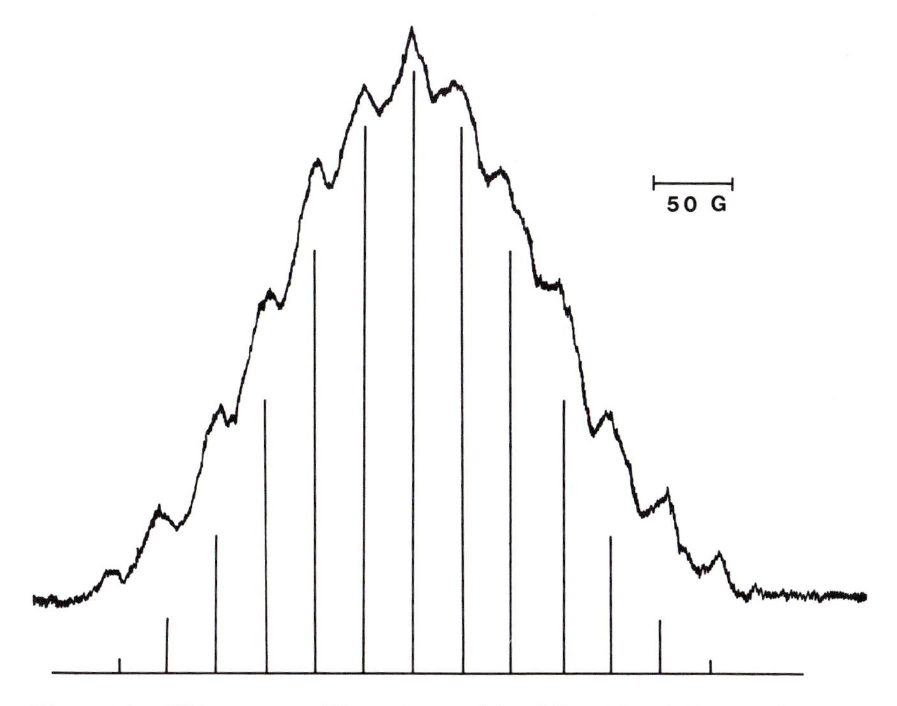

Figure 14. ESR spectrum (dispersion mode) exhibited by NaY γ–irradiated in vacuum. It is compared with the theoretical intensity pattern expected from isotropic hyperfine interactions with four equivalent nuclei of I = *3/2.*

The ESR signal was so easily saturated that it could only be seen in the dispersion mode. Both the ESR signal and the pink color were destroyed on exposure to oxygen. As illustrated in the figure, the spectrum consists of 13 hyperfine components, and its intensity pattern is in excellent agreement with that expected from the interaction with four equivalent nuclei with $I = 3/2$. The only nuclei wih $I = 3/2$ in NaY are ^{23}Na (natural abundance = 100%).

In dehydrated NaY all site II positions are occupied (*39, 40*), and there are four site II's tetrahedrally arranged within each α cage. The color center was therefore proposed to be an electron trapped within an α cage shared among four Na^{+} ions located at site II; hence the notation Na_4^{3+} center (*33*). The assessed g and hyperfine coupling constant of the Na_4^{3+} centers are:

$$g = 1.999 \pm 0.001$$

$$A = 32.3 \pm 0.2 \text{ G}$$

The amount of Na_4^{3+} centers created by irradiation was $10^{17} \sim 10^{18}$ spins/g. The centers are stable at room temperature but are destroyed when heated above 200°C.

The optical reflection spectrum of the pink Na-Y showed a broad absorption band peaking at 5000 A. Since the electron is more or less confined within a cubic space defined by four Na^{+} ions, its energy levels may be viewed as those of "a particle in a box." If one assumes that the observed optical absorption at 5000 A corresponds to the transition from the ground state to the first excited state of an electron trapped in a box, one calculates 6.7 A as the edge dimension of the cube. The edge dimension of a cube defined by the four Na^{+} ions within an α cage is $\sim$ 6 A.

Stamires and Turkevich (*31*) also noted the pink color of irradiated NaY but did not observe the ESR spectrum of Na_4^{3+} centers, probably because of the difficulty caused by saturation. Nozaki and Turkevich (*41*) observed the 13-line spectrum with little difficulty, however, by activating NaY at lower temperature (300°C) prior to the irradiation. Most interestingly Rabo *et al.* (*42*) discovered that the same Na_4^{3+} centers can be prepared chemically by treating NaY with Na vapor. Details of this chemically prepared "pink NaY" are discussed later.

Activated NH_4Y. Irradiation in vacuum of activated NH_4Y by γ- or x-rays produces several types of paramagnetic centers. The observed ESR spectra varied markedly depending upon the details of activation process used before irradiation. Figure 15 shows the spectra obtained in our laboratory from three different samples that had been irradiated in vacuum by x-rays.

The type I spectrum is observed mainly with HY (NH_4Y activated at 400°–500°C) and is characterized by an axially symmetric g tensor and a hyperfine interaction with one ^{27}Al nucleus ($A = 7.5$ G). Vedrine and Naccache (*36*) suggested that the center responsible for the spectrum is an electron hole located in a non-bonding p orbital of a lattice oxygen next to aluminum. It is created as the hydroxy bond is cleaved by the radiation. In support of this mechanism Abou-Kais *et al.* (*43*) showed the formation of H atoms in irradiated HY. On exposure to oxygen the type I signal disappeared, and a new relatively sharp signal appeared in its place (*36*). The latter signal possessed no hyperfine structure arising from ^{27}Al. An experiment using oxygen enriched with ^{17}O isotope revealed that the oxygen induced centers involve two non-equivalent oxygen nuclei (*36*). Subsequent evacuation at room temperature removed the oxygen induced signal completely and restored the original type I spectrum. Vedrine and Naccache (*36*) assigned the oxygen induced signal to the peroxy radicals formed between O_2 molecules and the type I centers. When HY was irradiated in the presence of oxygen, the peroxy signal was produced with 10 times greater intensity (*36*).

The type II and III spectra (Figures 15b and 15c) are observed when NH_4Y is activated at higher temperature (500°–600°C) in a deep-bed condition. Vedrine *et al.* (*44*) demonstrated that the paramagnetic centers responsible for these signals are associated with interstitial aluminum species created from the framework aluminum by the action of water and ammonia. The type II spectrum is characterized by a hyperfine interaction with two equivalent ^{27}Al nuclei. The type III spectrum is occasionally observed and is characterized by a coupling to one ^{27}Al. Vedrine *et al.* (*44*) assigned these spectra to an electron hole created in the non-bonding sp^2 orbital of an oxygen in a dimeric (or monomeric) form of interstitial aluminic acid.

A strong ESR signal of H atoms was indeed observed in these samples (*45*). The type II center was also found to react with oxygen to give a sharp signal possessing no hyperfine structure due to ^{27}Al. Although Vedrine *et al.* (*44*) suggest the formation of O_2^-, the formation of peroxy radicals in a manner similar to that proposed for type I centers seems more reasonable.

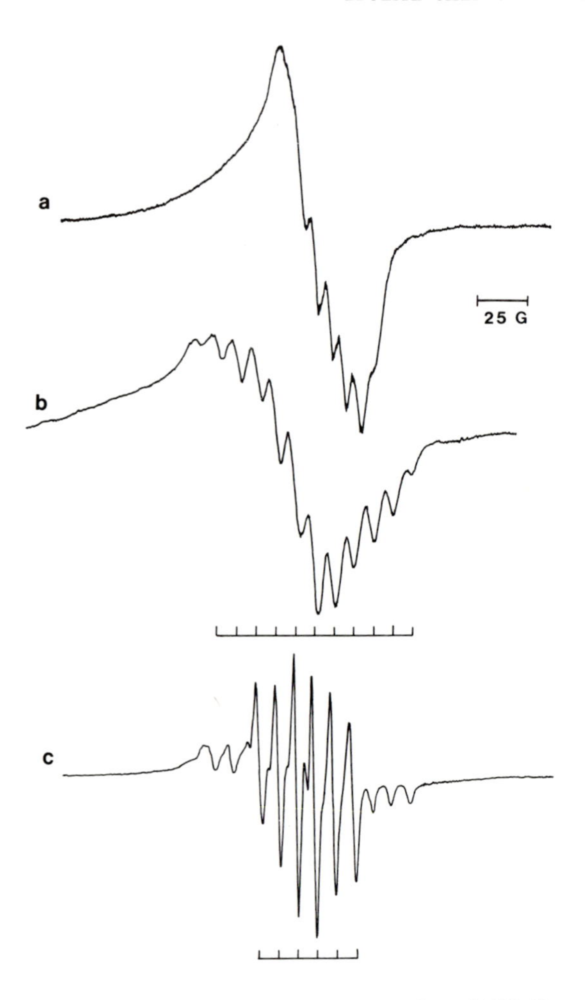

Figure 15. ESR spectra of x-ray irradiated NH_4Y. NH_4Y activated at 400°–500°C (a); NH_4Y activated at 500°–600°C (b and c).

Table VI. ESR Data of Radiation Induced

Zeolite	*Primary Radical*	g *Tensor of Primary Radical*		A *to* $^{27}Al(G)$
		$g_{\parallel}$	$g_{\perp}$	
HY	I	2.046	1.995	7.5
Dehydroxylated Y	II	2.0125	2.0030	10.0[a]
	III	2.0048	2.0048	9.5

[a] Type II radical is coupled to two equivalent Al nuclei.
[b] Oxygen adduct shows no coupling to Al.

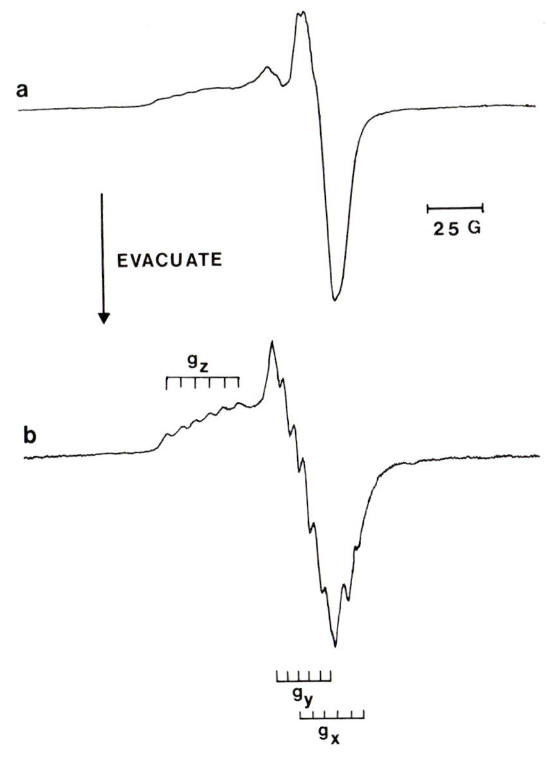

Figure 16. ESR spectra of NH_4Y (activated at 600°C) x-ray irradiated at 77°K in the presence of oxygen: (a) immediately after irradiation; (b) after subsequent evacuation at room temperature. Spectrum (b) is identical to that reported by Wang and Lunsford (34).

Radicals in NH_4Y and Their Oxygen Adducts

g Tensor of Oxygen Adduct[b]			
g_z	g_x	g_y	*Ref.*
2.032	1.9995	2.0015	*(36)*
2.022	1.9975	2.0015	
2.058	2.0026	2.0085	*(44)*
			(44)

The assessed *g* tensors and the hyperfine coupling constants of various paramagnetic centers observed in irradiated NH_4Y are listed in Table VI. Despite the abundance of experimental results many of the structures proposed for these centers should be considered tentative. Neither the axially symmetric *g* tensor of the type II center nor the isotropic *g* tensor of the type III center is consistent with the symmetry of the respective model proposed above.

When Vedrine *et al.* (*44*) irradiated their samples in the presence of oxygen, they saw no ESR spectrum possessing hyperfine structure of aluminum. On the other hand Wang and Lunsford (*34*) observed in a similar sample the O_2^- signal with a well recognizable structure caused by ^{27}Al. Figure 16a is the spectrum of NH_4Y that had been activated at 600°C and irradiated by x-rays at 77°K in the presence of oxygen (0.1 atm) (*9*). It is a superposition of the O_2^- signal observed by Wang and Lunsford (*34*) and that of the peroxy radicals reported by Vedrine *et al.* (*44*). On evacuation at room temperature the peroxy signal was removed, and the O_2^- signal was observed free from the interference (Figure 16b).

Other Radiation Induced Radicals. Other anion radicals generated and stabilized in zeolites include SO_2^- and CO_2^- ions (*9*). Vedrine and Naccache (*46*) detected CO^+ ions in HY γ–irradiated in the presence of CO.

Several neutral radicals such as $CH_3\cdot$ (*47*), $CH_3CH_2\cdot$ (*48*), and $NH_2\cdot$ (*49*) have also been observed in zeolites. They were produced by γ–irradiation of zeolites in the presence of methane, ethane, and ammonia, respectively. These radicals are thought to be stabilized by their ability to hydrogen-bond to the framework oxygen. Svejda (*50*) obtained the ESR spectrum of CF_3 radicals by UV irradiation of hexafluoroacetone adsorbed in NaX.

Ionization and Electron Transfer Reactions in Zeolites

In dehydrated zeolites many, if not all, of the cations are located near the inner surface of the crystallographically defined large channels and cavities. These cations are hence shielded only on one side, and, by reason of symmetry, many of the negative charges of the polyanionic framework $[(AlO_2)_m$-$(SiO_2)_n]^{m-}$ are also shielded unevenly. Anhydrous zeolites thus can be viewed as expanded ionic crystals having extremely irregular and spacious arrangement of the ions. The crystallinity of the material, however, assures a periodicity at least on a large scale. It follows then that the total crystalline energy (the Madelung energy) of the system can be greatly decreased (the system becomes more stable) by filling these intracrystalline void spaces with properly arranged polarized species or by additional cations and anions (*see* Figure 17). The ionizing power rendered to zeolites by this structural feature was found to be large enough not only to dissolve a strong electrolyte such as NaCl in its ionized state but also to induce an electron transfer reaction between two adsorbed molecules A and B, creating an ionic pair, A^+ and B^- (*51*).

Since the early work of Stamires and Turkevich (*52*) electron accepting capacities of dehydroxylated NH_4Y have been examined by many authors. On contact with dehydroxylated NH_4Y, cation radicals are formed from polynuclear aromatic molecules having low ionization potentials (perylene, anthracene, triphenylamine, etc.). Stamires and Turkevich suggested the tricoordinated aluminum of the framework as the electron-accepting sites (*52*). How-

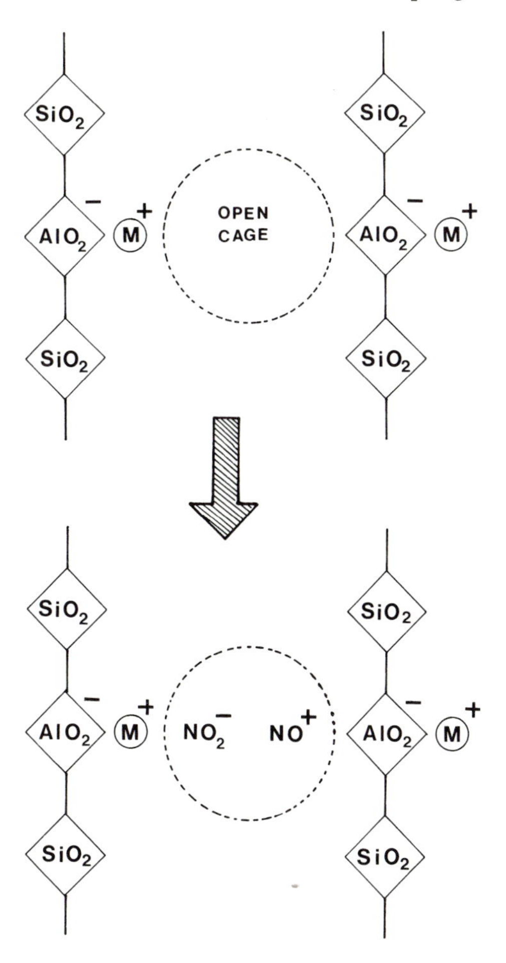

Figure 17. Schematic of irregular and extremely spacious arrangement of the polyanionic framework $[(AlO_2)_m(SiO_2)_n]^{m-}$ *and the cations in dehydrated zeolite (upper figure). A gain in the stability of the ionic lattice should be realized when the void spaces are filled with properly arranged additional cations and/or anions (lower figure).*

ever even when the adsorption occurred to the extent of one molecule per α cage, only one molecule in 20 was oxidized to the cation. Dollish and Hall (*53*) showed that the number of the cation radicals produced on dehydroxylated NH_4Y was closely related to the number of oxygen molecules present in the zeolite and suggested the formation of cation$^+$–O_2^- complexes as the mechanism involved. Using dehydroxylated NH_4Y containing various amounts of Ce^{4+} cations Neikam (*54*) showed that the amount of cation radicals produced are related stoichiometrically to the amount of Ce^{4+} ions present at the external surface of the zeolites and suggested that most of the electron accepting sites observed in dehydroxylated NH_4Y are located on the external surfaces.

The ionization or electron transfer processes expected from the electrolytic property of zeolite due to its expanded ionic lattice should not be limited to dehydroxylated NH_4Y. Furthermore these processes must be effected throughout the zeolite structure. Presented below are examples of ionization and electron transfer reactions that have been observed and attributed to this unique crystalline property of zeolites.

NO and NO_2 in Zeolites (*19*). As stated earlier the ESR spectra of NO displayed by Y–type zeolites become much sharper if the samples sealed with NO are left standing at room temperature for several days. Pumping on the aged sample at room temperature quickly (with 5 min) produced a sample which no longer gave any ESR signal. Introduction of a small amount of NO into this aged-and- pumped sample produced immediately the strong, well-resolved ESR spectrum of NO. Clearly some permanent chemical or physical modification occurs within zeolite as a result of the aging process. Let us define the NO-treatment as sealing freshly activated zeolite with NO gas, letting it stand at room temperature for one week, and finally pumping on it at room temperature until it no longer exhibits any ESR signal.

Figure 18 compares the ESR spectra of NO introduced to freshly activated BaY and that introduced to NO-treated BaY. A similarly dramatic sharpening of the ESR signal was found with NaY. The modification brought about by the NO-treatment survived exposure to air, exposure to moisture, and even vacuum activation at 300°C overnight. Vacuum activation at 400°C overnight finally returned the sample to the original state, whereupon the ESR signal of newly introduced NO appeared broad and ill-defined. When a NO-treated sample was heated to 400°C in a sealed tube, brown gas evolved. When cooled to room temperature, the gas was readsorbed, and the sample exhibited the ESR spectrum of NO_2. Thus some of the characteristics of the elements responsible for the NO treatment are: (1) they are not paramagnetic; (2) they are tenaciously held by zeolites; (3) they produce NO_2 when heated; (4) their presence results in a sharpening of the NO spectrum.

Adsorption isotherms of NO on various zeolites have been studied by Addison and Barrer (*55*). Based upon the mass balance, and the density of the gas evolved, they concluded that zeolites catalyze the disproportionation reaction $4\ NO \rightarrow N_2O + N_2O_3$ nearly to completion at temperature below 0°C, and that, while the resulting N_2O can be removed quantitatively below 150°C, N_2O_3 stays occluded in zeolites above 200°C.

It thus appears that N_2O_3 or its equivalent generated by the disproportionation reaction is responsible for the NO treatment. However N_2O_3 is known to dissociate readily into NO and NO_2, both of which would surely be

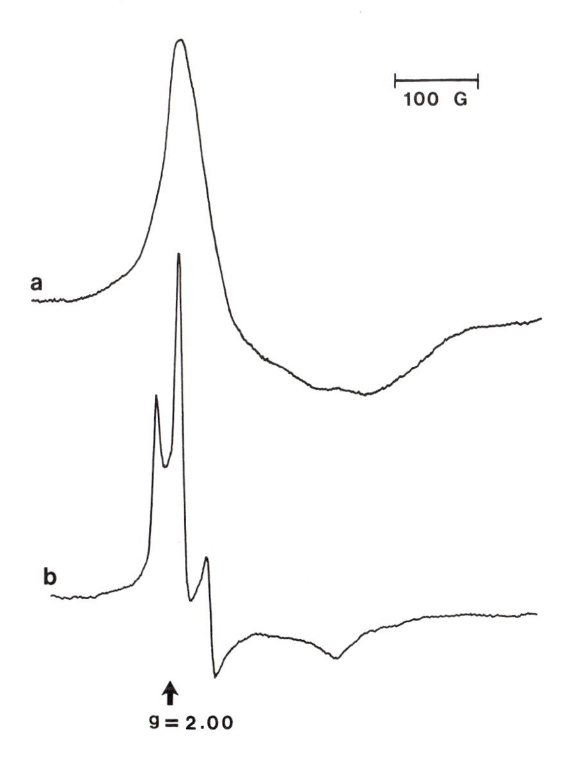

Figure 18. ESR spectra of (a) NO introduced to freshly activated BaY and (b) NO introduced to NO-treated BaY

detected by ESR. It is proposed, therefore, that the sequence of the disproportionation reaction is

$$3\ NO \rightarrow N_2O + NO_2$$

$$NO + NO_2 \rightarrow NO^+ + NO_2^-$$

and that the elements responsible for the NO–treatment are NO^+ and NO_2^- ions (*19*). Proper redistribution of these additional cations and anions must result in improved uniformity of the zeolitic crystal field and sharpening of the NO spectrum. This model also explains both the diamagnetism and the tenaceous affinity of the elements responsible for the NO treatment.

Although NO has a relatively low ionization potential (9.25 eV) and NO_2 has a large electron affinity (3.8 eV), the envisaged electron transfer reaction $NO + NO_2 \rightarrow NO^+ + NO_2^-$ is an endothermic process. We proposed that the endothermicity is offset by a gain in the Madelung energy of the crystal resulting from a proper arrangement of the additional ions produced (*see* Figure 17). According to the data of Addison and Barrer (*55*) the amount of NO^+ and NO_2^- produced in zeolite could easily amount to one cation–anion pair per α cage.

The presence of NO_2^- ions in the NO–treated zeolite is supported by the appearance of a strong ESR spectrum of NO_2 when the sample was irradiated by UV at 77°K (*19*). Unlike the spectrum obtained by direct adsorption of NO_2, the photo-induced spectrum of NO_2 decayed gradually at room temperature ($T_{1/2}$ = several hr) and disappeared completely when annealed at 50°C for 30 min. The reversible phenomenon is attributed to

$$NO_2^- + NO^+ \underset{\Delta}{\overset{h\nu}{\rightleftarrows}} NO_2 + NO$$

The capacity of NO to reduce other absorbed molecules is demonstrated clearly by the observation of the ESR spectrum of SO_2^- ions when a 1 to 1 mixture of NO and SO_2 was absorbed on BaY and heated to 300°C (*9*). That the presence of additional anions and cations leads to improved uniformity of the zeolite field is supported by immediate display of a sharp NO spectrum by zeolites containing dissolved NaCl.

Dissolution of NaCl (*19, 51*). If the electrolytic power of zeolite is indeed strong enough to offset endothermicity of ~ 5 eV, it is quite possible that an electrolyte such as NaCl might dissolve into zeolite in an ionized form. The energy required to dissociate NaCl molecule into Na^+ and Cl^- ions is ~ 6 eV. Figure 19 shows a lower angle section ($2\theta \cong 15° \sim 35°$) of the x-ray (CuK$\alpha$) powder patterns obtained from anhydrous mixture of NaA and NaCl (10 wt %). The zeolite was preactivated at 500°C and mixed thoroughly with dried NaCl. The x-ray powder pattern of the mixture was examined before heat treatment and after successive heat treatments at 350°C and 500°C for 24 hr, rspectively. In the 2θ range covered, the only prominent peak from the NaCl crystal is the (2, 0, 0) peak at 31.7° indicated by the arrows. All the other peaks belong to the powder pattern of the A zeolite. The patterns clearly demonstrate that while the crystallinity of the zeolite stays virtually unchanged throughout the heat treatments, the NaCl phase disappears gradually. The melting point of NaCl is 801°C. Thus, once the diffusion point is reached, NaCl begins to dissolve into zeolite cages in an ionized form arranging itself in a manner that would minimize the crystalline energy of the total system. A similar result was obtained from Y type zeolite. 10 wt % of NaCl corresponds to approximately 3 Na^+–Cl^- ion pairs per α cage. Most assuringly when a small amount of NO was introduced to these zeolites containing dissolved NaCl, they displayed immediately the sharp, well-resolved ESR spectra identical to those obtained from NO-treated zeolites.

Ionization of Na Atoms in NaX and NaY; $Na_m^{(m-1)+}$ Centers (*51*). The most dramatic and the simplest example of the ionization process in zeolite is probably that of Na atoms in NaY. When NaY was exposed to Na vapor at 300° ~ 500°C, its color changed from white to bright red. The colored material showed an ESR spectrum consisting of 13 hyperfine components

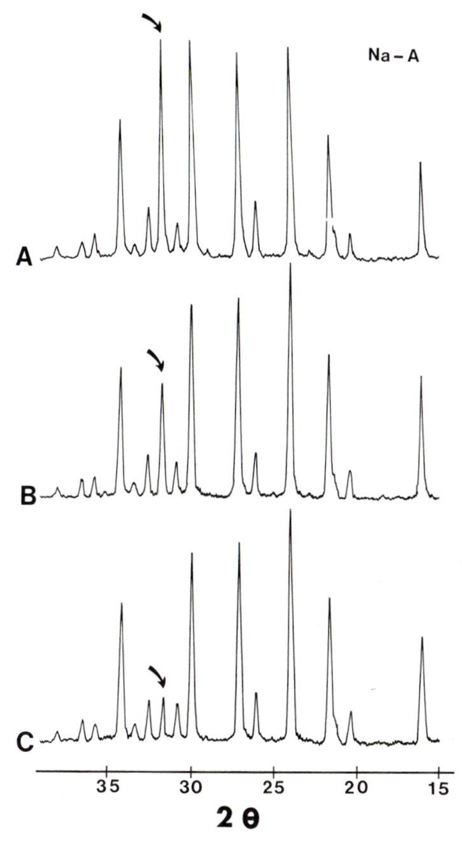

Figure 19. X-ray powder patterns of anhydrous mixture of NaA and NaCl (10 wt %) obtained from the fresh mixture (A) and after successive heat treatment at 350°C (B) and 500°C (C) for 24 hr, respectively. The arrows indicate the NaCl peak (2, 0, 0) at 31.7°.

with $A = 32.3$ G and $g = 1.999$ (Figure 20a) (*42*). The spectrum was readily recognized as that of Na_4^{3+} centers observed with γ–irradiated NaY. The amount of Na_4^{3+} centers created by irradiation was $10^{17} \sim 10^{18}$ spins/g; these centers were destroyed when the sample was heated above 200°C. The concentration of Na_4^{3+} centers produced by Na vapor amounted to $10^{20} \sim 10^{21}$ spins/g, corresponding to one electron per α cage. Spontaneous formation of the Na_4^{3+} centers by this stoichiometric amount at temperature above 200°C can be best understood in terms of the ionization of Na atoms induced by the electrolytic property of the zeolite.

$$\mathrm{Na} \rightarrow \mathrm{Na}^+ + e^-$$

The ionization potential of Na atoms is 5.14 eV.

When NaX was exposed to Na vapor, the material became deep purple and displayed an ESR spectrum consisting of 16–19 hyperfine components with $A = 25$ G (*9*). The 19-line pattern was earlier assigned to an electron shared among six Na^+ ions located at site III, octahedrally situated within each α cage (*42*). However, the exact number of hyperfine components was found to vary, depending upon the particular NaX used and the extent of the

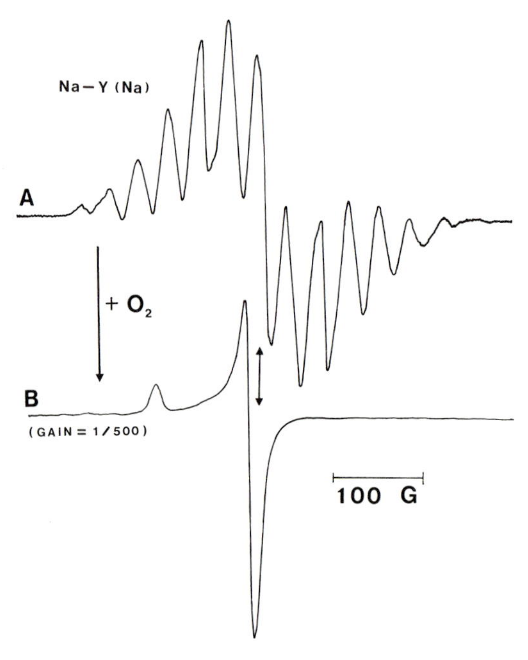

Figure 20. The ESR spectrum of Na_4^{3+} centers produced by exposing NaY to Na vapor (a). The ESR spectrum of the same sample after it had been exposed to oxygen (b). The arrow indicates the position corresponding to g = 2.00.

Na vapor treatment (*9*). Also in a recent x-ray diffraction study of KX (*56*) 26 K^+ ions were located at site II, but 38 cations were not located and were assumed to be near site III. Thus, the exact number and the locations of Na^+ ions involved in the $Na_m^{(m-1)+}$ centers of NaX must be regarded as uncertain.

Formation of O_2^- in NaX and NaY with $Na_m^{(m-1)+}$ Centers (*51*). When NaY treated with Na vapor was subsequently exposed to oxygen, the red material became instantly white, the ESR spectrum of Na_4^{3+} centers disappeared, and a new signal with a characteristic *g* tensor appeared (Figure 20b). The oxygen-induced spectrum was readily identified as that of O_2^- attached to a Na^+ ion. We should note that generation of O_2^- radicals by exposing NaY containing ionized Na atoms to oxygen is equivalent to incorporating a well known oxide, alkali superoxide, in its ionized state.

$$NaO_2 \rightarrow Na^+ + O_2^-$$

Recently using ^{17}O enriched oxygen Ben Taarit *et al.* (*57*) showed that the structure of the $Na^+–O_2^-$ complex generated in NaY containing Na_4^{3+} centers is indeed $\overset{O}{\underset{O}{||}}$ - - - Na^+ as illustrated in Figure 11. The same conclusion was reached by comparing the spectra of O_2^- produced in NaY containing ionized Na and in KY containing ionized K (*51*).

As stated earlier, the O_2^- produced in NaY by ionizing radiation showed several different g_z values depending upon its thermal history. The amount of O_2^- ions produced by irradiation is small ($\sim 10^{18}$ spins/g) corresponding to one anion per several hundred α cages. Diffusion of the electrically neutral complexes $Na^+–O_2^-$ toward more stable sites is, therefore, very probable.

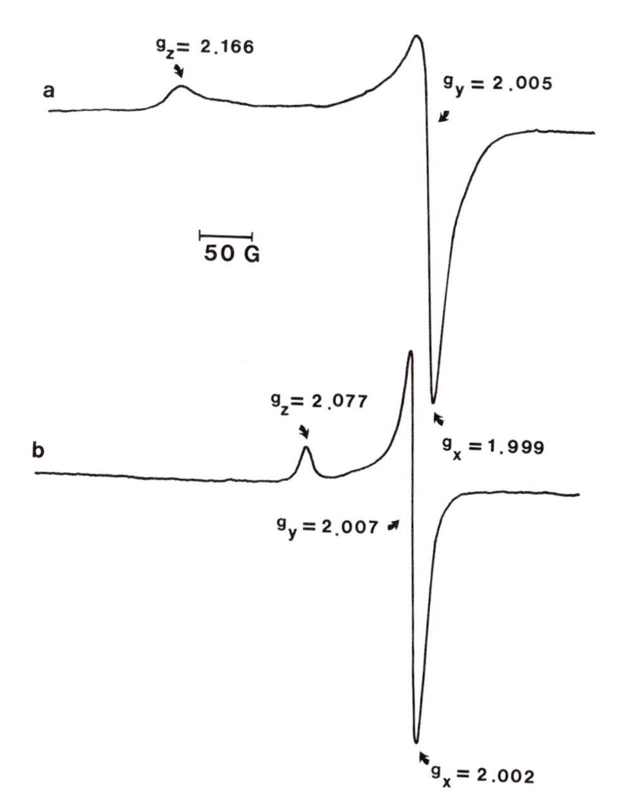

Figure 21. ESR spectra and the **g** *tensors of* O_2^- *produced by exposure to oxygen of (a) Na-vapor treated NaX, and (b) Na-vapor treated NaY*

When a much larger amount ($\sim 10^{20}$ spins/g) of O_2^- is produced in NaY containing ionized Na atoms, the resulting O_2^- spectrum is more likely to be dominated by the Na^+–O_2^- complexes located at the normal site. The spectra and the g tensors of O_2^- generated in NaX and NaY containing ionized Na atoms are shown in Figure 21. These g tensors are very similar, if not identical, to the g tensors of O_2^- generated in NaX and NaY by x-rays at 77°K (Table IV).

Reaction of CuY and NiY with NO (*19. 51*). The intracrystalline ionizing power of zeolite stems from its expanded ionic lattice. It should become even more pronounced when the usual monovalent Na^+ ions are replaced by divalent cations.

Richardson (*58*) examined the ionization of anthracene and other aromatic molecules adsorbed on Y type zeolites ion-exchanged with alkali, alkaline earth, and transition metal cations. He proposed that the zeolitic cations are the electron accepting sites and demonstrated that the number of cation radicals formed are related exponentially to the difference between the ionization potential of the aromatic molecules and the electron affinity of the zeolitic cations. The electron affinity of the cations in zeolite may be inferred from the ionization potential of the free atoms. Thus among the cations Richardson studied Cu^{2+} and Ni^{2+} ions are thought to have the highest electron affinity, the second ionization potentials of Cu and Ni atoms being 20.3 and 18.2 eV, respectively, *vs.* 11.8 eV of Ca. Indeed the largest amount of anthracene cation radicals was formed by CuY. However, its concentration was less than 1% of the exposed Cu^{2+} ions in the zeolite. This discrepancy may be caused by pore clogging caused by strong adherence of the initially formed cation radicals to the framework or more likely by the difficulty of properly orienting the large molecular cations to minimize the total ionic crystalline energy. Neikam (*54*) showed that anthracene adsorption in Y type zeolite occurs to the extent of one molecule per α–cage. As suggested by Neikam (*54*) the electron accepting sites observed by Richardson are probably those located on the external surface.

Although the ionization potential of NO (9.25 eV) is much higher than that of anthracene (7.5 eV), its small size should permit much greater freedom to achieve the optimum arrangement of the adsorbed molecules. If the gain in the stability of crystal in going from a divalent form $(Cu^{2+})Y$ to a monovalent form $(Cu^+, NO^+)Y$ is sufficiently large, adsorption of NO on CuY should result in the reduction of a substantial fraction of Cu^{2+} to Cu^+. Figure 22 shows the ESR spectra of Cu (80%) Y observed before and after exposure to NO. The observed signal is that of Cu^{2+} ($3\,d^9$). The large, spontaneous decrease of the signal on exposure to NO at room temperature is attributed to the electron transfer reaction induced by the electrolytic property of zeolite.

$$(Cu^{2+})\ Y + NO \rightarrow (Cu^+,\ NO^+)\ Y$$

According to the x-ray diffraction study of dehydrated Cu (75%) Y by Gallezot *et al.* (*59*), 75% of the Cu^{2+} ions are located at site I′ and the remaining at site I. The upper spectrum in Figure 22 is a superposition of a broad symmetric signal at $g = 2.17$ and an asymmetric spectrum characteristic of Cu^{2+} ions immobilized in a distorted field (*60, 61*). The broad symmetric signal is attributed to the Cu^{2+} ions at site I′; the signal is narrowed by an exchange process as suggested by Chao and Lunsford (*62*). The latter asymmetric signal is assigned to the Cu^{2+} ions at site I. Figure 22 shows that all the

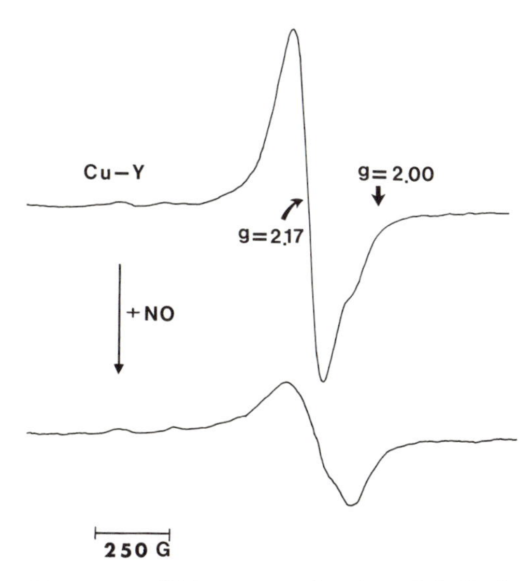

Figure 22. ESR spectra of activated CuY observed before and after exposure to NO

cations at site I′ have been reduced. The observed decrease of the Cu^{2+} signal could also be attributed to the formation of Cu^{2+}–NO complex. The result obtained with NiY (*see* below), however, strongly supports the electron transfer reaction.

No ESR signal is observed from NiY. The configuration of Ni^{2+} ion is $3d^8$. Figure 23a shows the spectrum obtained when Ni (65%) Y was exposed to NO and maintained at 100°C for 3 hr. The observed spectrum has a g tensor ($g_{||} = 2.430$ and $g_{\perp} = 2.171$) typical of $3d^9$ system and is therefore assigned to Ni^{+}. Olson (*63*) found in his x-ray diffraction study of Ni-exchanged faujasite that Ni^{2+} ions occupy site I preferentially and then site II. The number of Ni^{+} ions produced by NO in Ni (65%) Y was 1×10^{20} spins/g, in exact agreement with the number of Ni^{2+} ions expected at site II.

Reoxidation of Ni(I) and Cu(I) by NO_2 (*51*). Electron transfer reaction $NO + NO_2 \rightarrow NO^{+} + NO_2^{-}$ discussed earlier suggests that NO_2 might be able to reoxidize Ni^{+} and Cu^{+} ions produced in zeolites. Figure 23b is the spectrum obtained when (Ni^{+}, NO^{+}) Y discussed above was exposed to NO_2 and heated at 200°C for 10 min. A near total disappearance of the Ni^{+} signal and the weakness of the NO_2 spectrum are attributed to the following reaction.

$$(Ni^+, NO^+)\ Y + NO_2 \rightarrow (Ni^{2+}, NO^+, NO_2^-)\ Y$$

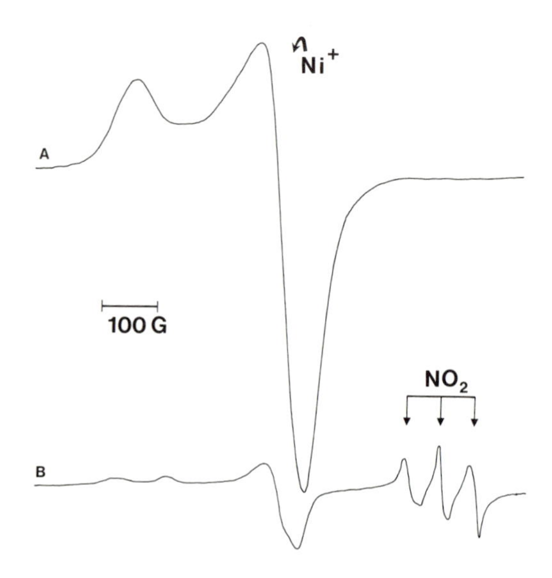

Figure 23. ESR spectrum of Ni (65%) Y reduced by NO (a). ESR spectrum of the same sample after it had been exposed to NO_2 and heated at 200°C for 10 min (b).

Naccache *et al.* (*64*) observed that Cu^+ ions produced in CuY by treatment with CO at 500°C are reoxidized to Cu^{2+} when exposed to NO_2. Figure 24 shows the spectra obtained when the same sequence of experiments was repeated. Unlike the reduction by NO at room temperature, the CO treatment at 500°C results in the reduction of all the Cu^{2+} ions. The treatment must lead to relocation of the Cu^{2+} ions at site I and possibly those at site I′ to the more exposed sites. The reduction by CO, we believe, is assisted by the presence of tenaciously held water molecules. Thus the reduction and the reoxidation steps can be written as

$$(Cu^{2+}, \frac{1}{2}\ H_2O)\ Y + \frac{1}{2}\ CO \rightarrow (Cu^+, H^+)\ Y + \frac{1}{2}\ CO_2$$

$$(Cu^+, H^+)\ Y + NO_2 \rightarrow (Cu^{2+}, H^+, NO_2^-)\ Y$$

The tenaciously held water need not exist as H_2O. It most probably is held in the ionized form—*i.e.*, the hydrolysis of the cations. The IR spectrum of activated CuY shows a strong OH band (*9*).

Catalytic properties of Y type zeolites for various organic reactions are well known and have been the subjects of many investigations. The observed activity has been attributed to the electric field associated with the exposed

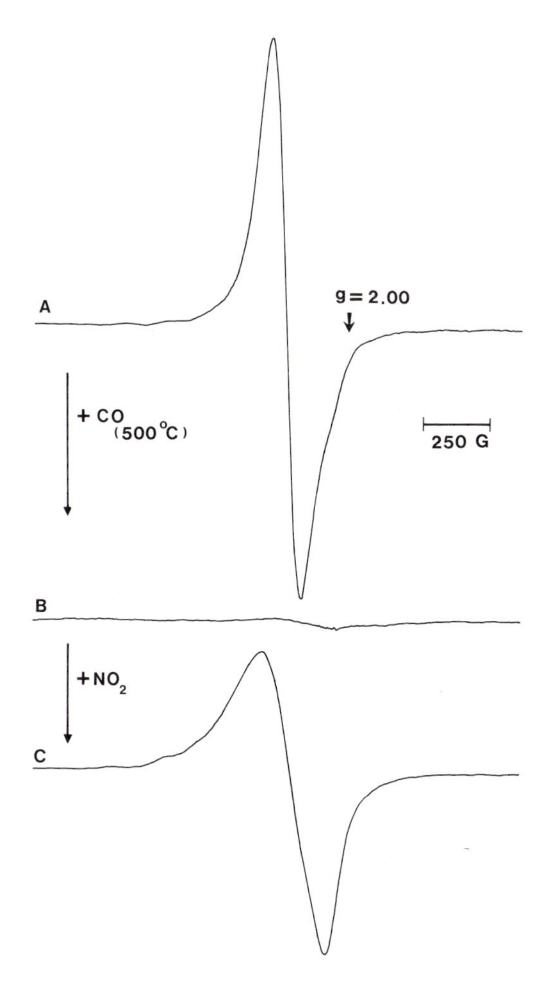

Figure 24. ESR spectra obtained (a) from freshly activated CuY, (b) after it had been reduced by CO at 500°C, and (c) after the reduced sample had been exposed to NO_2 at room temperature

cations, to the acidity of the hydroxy groups or to the combined effect of both. One of the most important properties of zeolites responsible for the observed catalytic activity must be the ionizing or polarizing power manifested throughout the structure by a macroscopic property of the crystal—the Madelung energy. The presence of such a power is amply demonstrated by the electron spin resonance studies described above.

Literature Cited

1. Ayschough, P. B., "Electron Spin Resonance in Chemistry," Methuen, London, 1967.

2. Atkins, P. W., Symons, M. C. R., "The Structure of Inorganic Radicals," Elsevier, New York, 1967.
3. Wertz, J. E., Bolton, J. R., "Electron Spin Resonance," McGraw-Hill, New York, 1972.
4. Mikheikin, I. D., Zhidomirov, G. M., Kazanskii, V. B., *Russ. Chem. Rev.* (1972) **41,** 468.
5. Kokes, R. J., "Experimental Methods in Catalytic Research," R. B. Anderson, Ed., Academic, New York, 1968.
6. Lunsford, J. H., *Adv. Catal.* (1972) **22,** 265.
7. Chao, C. C., Lunsford, J. H., *J. Chem. Phys.* (1973) **59,** 3920.
8. Colburn, C. B., Ettinger, R., Johnson, F. A., *Inorg. Chem.* (1963) **2,** 1305.
9. Kasai, P. H., Bishop, Jr., R. J., unpublished results.
10. Kasai, P. H., Weltner, Jr., W., Whipple, E. B., *J. Chem. Phys.* (1965) **42,** 1120.
11. Pietrzak, T. M., Wood, D. E., *J. Chem. Phys.* (1970) **53,** 2454.
12. Colburn, C. B., Ettinger, R., Johnson, F. A., *Inorg. Chem.* (1964) **3,** 455.
13. Kasai, P. H., Whipple, E. B., *Mol. Phys.* (1965) **9,** 497.
14. Coope, J. A. R., Gardner, C. L., McDowell, C. A., Pelman, A. I., *Mol. Phys.* (1971) **21,** 1043.
15. Byberg, J. R., Jensen, S. J. K., Muus, L. T., *J. Chem. Phys.* (1967) **46,** 131.
16. Dempsey, E., "Molecular Sieves," p. 293, Society of Chemical Industry, London, 1968.
17. Lunsford, J. H., *J. Phys. Chem.* (1968) **72,** 4163.
18. Gardner, C. L., Weinberger, M. A., *Can. J. Chem.* (1970) **48,** 1317.
19. Kasai, P. H., Bishop, R. J., *J. Amer. Chem. Soc.* (1972) **94,** 5560.
20. Kerr, G. T., *J. Phys. Chem.* (1967) **71,** 4155.
21. Kerr, G. T., *J. Catal.* (1969) **15,** 200.
22. Jacobs, P., Uytterhoeven, J. B., *J. Catal.* (1971) **22,** 193.
23. Lunsford, J. H., *J. Phys. Chem.* (1970) **74,** 1518.
24. Ward, J. W., *J. Phys. Chem.* (1968) **72,** 4211.
25. Ward, J. W., *J. Catal.* (1968) **10,** 34.
26. Naccache, C., Ben Taarit, Y., *J. Catal.* (1971) **22,** 171.
27. Naccache, C., Che, M., Ben Taarit, Y., *Chem. Phys. Letters* (1972) **13,** 109.
28. Chao, C. C., Lunsford, J. H., *J. Phys. Chem.* (1972) **76,** 1546.
29. Hoffman, B. M., Eames, T. B., *J. Amer. Chem. Soc.* (1969) **91,** 5168.
30. Lozos, G. P., Hoffman, B. M., *J. Phys. Chem.* (1974) **78,** 200.
31. Stamires, D. N., Turkevich, J., *J. Amer. Chem. Soc.* (1964) **86,** 757.
32. Wang, K. M., Lunsford, J. H., *J. Phys. Chem.* (1970) **74,** 1512.
33. Kasai, P. H., *J. Chem. Phys.* (1965) **43,** 3322.
34. Wang, K. M., Lunsford, J. H., *J. Phys. Chem.* (1969) **73,** 2069.
35. Wang, K. M., Lunsford, J. H., *J. Phys. Chem.* (1971) **75,** 1165.
36. Vedrine, J. C., Naccache, C., *J. Phys. Chem.* (1973) **77,** 1606.
37. Ben Taarit, Y., Lunsford, J. H., *J. Phys. Chem.* (1973) **77,** 780.
38. Castner, T. G., Kanzig, W., *J. Phys. Chem. Solids* (1957) **3,** 178.
39. Eulenberger, G. R., Shoemaker, D. P., Keil, J. G., *J. Phys. Chem.* (1967) **71,** 1812.
40. Dodge, R. P., unpublished result.
41. Nozaki, F., Turkevich, J., *Shokubai* (1965) **7,** 328.
42. Rabo, J. A., Angell, C. L., Kasai, P. H., Schomaker, V., *Dis. Far. Soc.* (1966) **41,** 328.
43. Abou-Kais, A., Vedrine, J. C., Massardier, J., Dalmai, G., Imelik, B., *J. Chim. Phys.* (1972) **69,** 561.
44. Vedrine, J. C., Abou-Kais, A., Massardier, J., Dalmai-Imelik, G., *J. Catal.* (1973) **29,** 120.
45. Abou-Kais, A., Vedrine, J. C., Massardier, J., Dalmai, G., Imelik, B., *C.R. Acad. Sci., Paris* (1971) **C272,** 883.

46. Vedrine, J. C., Naccache, C., *Chem. Phys. Letters* (1973) **18,** 190.
47. Noble, G. A., Serway, R. A., O'Donnell, A., Freeman, E. S., *J. Phys. Chem.* (1967) **71,** 4326.
48. Kudo, S., Hesegawa, A., Komatsu, T., Shiotani, M., Sohma, J., *Chem. Letters* (1973) 705.
49. Vansant, E. F., Lunsford, J. H., *J. Phys. Chem.* (1972) **76,** 2716.
50. Svejda, P., *J. Phys. Chem.* (1972) **76,** 2690.
51. Kasai, P. H., Bishop, Jr., R. J., *J. Phys. Chem.* (1973) **77,** 2308.
52. Stamires, D. N., Turkevich, J., *J. Amer. Chem. Soc.* (1964) **86,** 749.
53. Dollish, F. R., Hall, W. K., *J. Phys. Chem.* (1967) **71,** 1005.
54. Neikam, W. C., *J. Catal.* (1971) **21,** 102.
55. Addison, W. E., Barrer, R. M., *J. Chem. Soc.* (1955) 757.
56. Mortier, W. J., Bosmans, H. J., Uytterhoeven, J. B., *J. Phys. Chem.* (1972) **76,** 650.
57. Ben Taarit, Y., Naccache, C., Che, M., Tench, A. J., *Chem. Phys. Lett.* (1974) **24,** 41.
58. Richardson, J. T., *J. Catal.* (1967) **9,** 172.
59. Gallezot, P., Taarit, Y. B., Imelik, B., *C.R. Acad. Sci., Ser. C* (1971) **272,** 261.
60. Nicula, A., Stamires, D., Turkevich, J., *J. Chem. Phys.* (1965) **42,** 3684.
61. Richardson, J. T., *J. Catal.* (1967) **9,** 178.
62. Chao, C. C., Lunsford, J. H., *J. Chem. Phys.* (1972) **57,** 2890.
63. Olson, D. H., *J. Phys. Chem.* (1968) **72,** 4366.
64. Naccache, C., Che, M., Taarit, Y. B., *Chem. Phys. Lett.* (1972) **13,** 109.

Chapter

7

Diffusion in Zeolites

Paul E. Eberly, Jr., Exxon Research and Development Laboratories,
Exxon Co., U.S.A., Baton Rouge, La. 70821

MANY INVESTIGATIONS have been done on the sorption of gases and liquids, both in the pure state and as mixtures, in zeolitic materials. Much of this work has been concerned with equilibrium studies, the determination of sorption isotherms, and data derived therefrom (the heat, entropy and free energies of sorption). For most sorbate–zeolite systems, the shape of the isotherm is adequately described by the Langmuir equation

$$\theta = \frac{c}{c_S} = \frac{bp}{1 + bp} \tag{1}$$

where p is the pressure of the sorbate in the gas phase and c, the concentration of sorbed species in the zeolite in equilibrium with this pressure. The term, c_S, represents the limiting amount adsorbed at saturation which is attained at the higher pressure levels.

In this chapter, we will concentrate on the much less extensively studied rates and kinetics of sorption phenomena rather than equilibrium quantities. Since the theoretical basis for zeolitic diffusion—*i.e.*, diffusion in pores where the dimensions of the molecules are nearly the same as that of the pores—is complex and not too well defined, a need exists for accumulating experimental data for a theory of zeolitic diffusion. From the results already published, certain general principles have begun to emerge. However, diffusion in zeolites is very complicated and diffusivity (D) has proven to be one of the most variable of constants. A recent review of zeolitic diffusion is given by Barrer (*1*).

Diffusion was first put on a quantitative basis by Fick (*2*) who, for an isotropic substance, formulated the equation

$$J = -D \frac{\delta c}{\delta x} \tag{2}$$

where J is the rate of transfer of molecules per unit area, $\delta c/\delta x$ is the concentration gradient of the diffusing material measured in a direction perpendicular to the unit area, and D, the diffusivity (or diffusion coefficient), most generally expressed in units of cm^2/sec. It must be emphasized that this equation represents diffusion in a medium in which the structure and properties are the same in all directions. While this is true for gases and liquids, zeolites are frequently anisotropic and the diffusion coefficients differ depending on the direction of flow.

To place zeolitic diffusion in its proper context, consider the types of diffusion occurring in a fixed bed of zeolite pellets or extrudates. For example, the diffusivity applicable to the space between the pellets would approximate that for bulk gas phase. The following correlating type equation has been found to give the least error for diffusion of substance A into B (*3*).

$$D_{AB} = \frac{1.00 \times 10^{-3} T^{1.75} \left(\frac{1}{M_A} + \frac{1}{M_B} \right)^{1/2}}{p\,[(\Sigma_A v_i)^{1/3} + (\Sigma_B v_i)^{1/3}]^2} \quad (3)$$

where T is temperature in °K, M_A and M_B the molecular weights, p, pressure in atmospheres, and v_i, special diffusion volumes for each element which can be summed to give the diffusion volume for the molecule, $\Sigma_A v_i$ or $\Sigma_B v_i$. Here, the dependence of diffusivity upon temperature is$T^{1.75}$ rather than $T^{3/2}$ which had been used formerly.

At least two pore structures generally exist inside the zeolite pellet. Since zeolites rarely can be synthesized or found as large particles, reasonably sized pellets are formed by compaction of micron or submicron particle size crystallites. The macropore structure is defined as that existing in the pellet between the individual zeolite crystallites. If the dimensions of this structure are such that the mean free path of the diffusing molecule is greater than he pore diameter, Knudsen type diffusion occurs in which the diffusivity is given by

$$D_K = 9700\, r \sqrt{\frac{T}{M}} \quad (4)$$

where r is the radius of the pore in cm. For a material having a distribution of irregularly shaped pores, the average pore radius can be defined as 2 V/SA, where V and SA are the pore volume and surface area, respectively, of the macropore structure. This diffusivity, D_K, does not depend on pressure and is independent of the presence of other gases. It is only mildly dependent on the temperature ($T^{1/2}$).

The fine micropore structure in which true zeolitic diffusion occurs is found within the zeolite crystals. It is difficult to devise a generalized equation relating the zeolitic diffusion constant to the properties of the molecule, size and shape of the intracrystalline pore, pressure and temperature. Since the diffusing molecules are always in intimate contact with the pore walls, the flow depends strongly on these variables as well as the chemical nature of the pore walls themselves. These factors contribute to the general non-ideality of zeolitic diffusion. As a result, the diffusivity is generally a function of pressure or more explicitly, a function of the concentration of sorbed species. The dependence is frequently expressed as

$$D = D_o \frac{\delta \ln p}{\delta \ln c} \qquad (5)$$

As stated earlier, the relationship between pressure (p) and concentration of sorbed species (c) is most often found to be expressed by the Langmuir isotherm in which case $\delta \ln p / \delta \ln c = 1/(1 - \theta)$. Consequently,

$$D = \frac{D_o}{1 - (c/c_S)} = \frac{D_o}{1 - \theta} \qquad (6)$$

D_o, the limiting diffusivity at zero sorbate concentration, is considered to be a more fundamental constant. It is equal to D only in the case of a linear adsorption isotherm in which case $\delta \ln p / \delta \ln c = 1$. Barrer (*1*) shows the expected dependence of diffusivity upon sorbate concentration for a number of other adsorption isotherms.

One of the most striking characteristics of zeolitic diffusion is its strong dependence on temperature. It generally increases exponentially with this variable as expressed by the Arrhenius type of relationship,

$$D \text{ or } D_o = D_* e^{-E/RT} \qquad (7)$$

Activation energies in the range of 0.5 to 20 kcal/mole have been reported (*1*).

From the above considerations on the different types of diffusion occurring in a bed of zeolite, some confusion arises as to the characteristic distance or radius of particle (a) to be used for calculating the diffusivity, D. Certain authors use the radius of the particle including the macropore as well as the intracrystalline pore structure. Others define a as the characteristic radius of the zeolite crystallites themselves. Still others prefer to express the diffusivity as D/a^2 rather than specify the characteristic distance. In reading the literature, it is important to note which radius is actually used since diffusivities calculated on the basis of pellet radius can be several orders of magnitude higher than those calculated from the crystallite radius. Fundamentally, the use of crystallite radius is perhaps preferred, although it is recognized that for certain engineering calculations the use of pellet radius may have some advantages.

Since zeolite particles contain both a micropore and macropore network, either (or both) systems can control the diffusion rate under any given set of conditions. Some of the apparent discrepancies in reported diffusivity data probably arise from a failure to identify properly the controlling resistance. For an idealized system assuming spherical particles and linear isotherms,

Ruckenstein *et al.* (*4*), devised a model for transient diffusion taking into account both micropore and macropore resistances.

Since the diffusivity in zeolites varies strongly with molecular size and shape, temperature, etc., a very wide range of absolute values for this constant exists ranging from zero for molecules which are too large to enter the crystalline channels to 10^{-5} cm^2/sec or higher (using crystalline radius). Consequently, a number of techniques have been required to obtain the necessary experimental data. These are discussed in the following section.

Measurement of Diffusion

Single Component Methods. CONSTANT PRESSURE. In this technique, the zeolite particles at time = 0 are suddenly subjected to a constant pressure of a sorbate and the rate of sorption is recorded as a function of time until equilibrium is reached. An electromagnetic microbalance in which the zeolite can be directly and continuously weighed is frequently used for these studies. Here, the boundary conditions are expressed as

$$c = c_o \text{ at time } = 0$$

$$c = c_\infty \text{ at the particle surface for } t > 0$$

$$c = c_\infty \text{ throughout the particle at } t = \infty$$

where c represents the concentration of sorbed species. For this case, assuming a concentration independent diffusion coefficient, Crank (*5*) formulates the following expression for spherical particles of radius a:

$$\frac{Q_t - Q_o}{Q_\infty - Q_o} = 1 - \frac{6}{\pi^2} \sum_{n=1}^{n=\infty} \frac{1}{n^2} e^{-Dn^2\pi^2 t/a^2} \tag{8}$$

where Q represents the total amount adsorbed in the particle at the various indicated times. Modern numerical techniques (such as those used for non-linear regression) give the best fit to the entire range of experimental results permitting the evaluation of D or D/a^2.

For small values of t, however, a simpler equation can be derived

$$\frac{Q_t - Q_o}{Q_\infty - Q_o} = 6 \left(\frac{D}{\pi a^2}\right)^{1/2} \sqrt{t} \tag{9}$$

Thus, by plotting the left side against $\sqrt{t}$, the diffusivity can be directly evaluated from the slope. This manner of expressing experimental results is frequently used. Because of the approximations involved, deviations from the straight line relationship are to be expected at the larger values of time.

Barrer (*1*) notes that for larger times, Equation 8 reduces to the simpler form,

$$\ln \frac{Q_\infty - Q_t}{Q_\infty - Q_o} = \ln \frac{6}{\pi^2} - \frac{D\pi^2 t}{a^2} \tag{10}$$

Here, when the left side is plotted against t, a straight line should be obtained whose slope is equal to $-D\pi^2/a^2$. This treatment is less frequently used than Equation 9. The nature of the sorption curve is independent of particle shape at initial conditions. However, as equilibrium is approached, the dependency becomes greater, requiring a more detailed knowledge of particle shape when using Equation 10.

Strictly speaking, the above equations only apply to spherical particles. For irregularly shaped particles, the radius a is sometimes taken to be that of a sphere which is equivalent in volume to that of the irregularly shaped particle. Thus,

$$a = \frac{3\,V}{SA} \qquad (11)$$

where V is the volume of the particles and SA, their external surface area. Similar equations exist for rectangular parallelepipeds, cubes and cylinders (*5*). Ruthven and Loughlin (*6*) have treated this subject thoroughly and computed a number of sorption *vs.* time curves for spherical and cubic particles. They have also considered the changes of the respective curves caused by particle size distribution. Figure 1, taken from their article, shows the

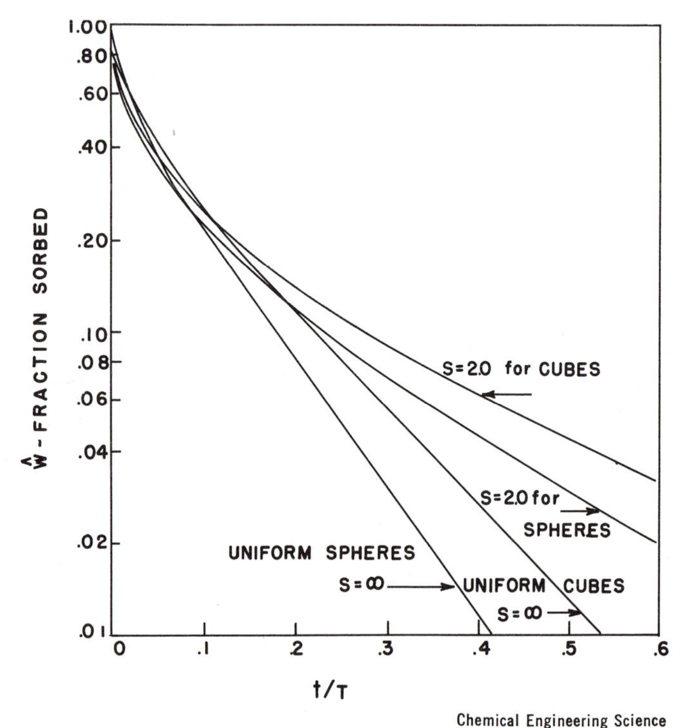

Figure 1. Comparison of sorption curves for spherical and cubic particles (6)

effect of particle size distribution for both spherical and cubic particles. Here, the ordinate W is equal to $1 - [(Q_t - Q_o)/(Q_\infty - Q_o)]$ and the abscissa is a dimensionless time variable. S is a parameter for particle size distribution. Higher values mean narrower distributions. The nature of the curve is dependent on particle shape and particle size distribution, particularly at longer time values. Some authors prefer therefore to use the region of the curve at shorter times. For best accuracy, however, it is desirable to have a detailed knowledge of the particle size distribution as obtained by electron microscopy or other techniques.

DECREASING PRESSURE–CONSTANT VOLUME. In experiments where the zeolite is not directly weighed, a widely used technique is to admit a known amount of gas to a constant volume chamber containing the zeolite. As sorption occurs, the pressure gradually decreases and the amount of material sorbed at any time t can be calculated from the pressure decrease. For this use, the solution to the diffusion equation for spherical particles (*5*) is given by

$$\frac{Q_t - Q_o}{Q_\infty - Q_o} = 1 - \sum_{n=1}^{\infty} \frac{6\alpha(\alpha + 1)\, e^{-Dq_n^2 t/a^2}}{9 + 9\alpha + q_n^2\alpha^2} \qquad (12)$$

where the q_n's are the non-zero roots of

$$\tan q_n = \frac{3\, q_n}{3 + \alpha q_n^2} \qquad (13)$$

The parameter α is defined as $p_\infty/p_o - p_\infty$ where subscripts 0 and ∞ refer to initial and equilibrium pressures. Brandt and Rudloff (*7*) used this technique for sorption on chabazite and have obtained computerized solutions for the above equations for various values of the parameter α.

OTHER METHODS. A number of other methods have been developed for measuring the diffusion of single components into the pore structure of zeolites. In the case of liquid diffusion, Satterfield and Cheng (*8*) reported on a method which appears to have general application. Figure 2 shows their apparatus. The zeolite is placed in the specially designed flask connected in series to a graduated capillary and a U-shaped tube. This entire glass vessel is heated under vacuum to liberate adsorbed impurities from the zeolite. The end of the U-tube is then sealed under vacuum and the unit placed in a constant temperature bath in which the U section of the tube is immersed in the liquid selected for sorption studies. At time = 0, the bottom of the U section is broken, and the liquid is sucked into the flask containing the zeolite. Shortly thereafter, the top of the graduated cylinder is snapped off, and readings are taken as a function of time. This records the slow movement of liquid into the zeolite crystals from the external liquid phase. Equations similar to Equations 8 and 9 are used to evaluate diffusivities.

Various nuclear magnetic resonance techniques are also being increasingly used to measure self-diffusion constants of sorbates in zeolites. These include wide line spectroscopy, spin-echo methods, pulsed field gradient technique, etc. The reader is referred to the following publications for more information: Pfeifer (*9*); Pfeifer, Shirmer and Winkler (*10*); Resing and Murday (*11*); and Resing and Thompson (*12*).

Isotopic methods have also been employed. For water, the use of HTO and D_2O have been reported for measuring intrinsic diffusion coefficients (*13, 14*). Sargent and Whitford (*15*) investigated CO_2 diffusion into zeolite 5A using $C^{14}O_2$.

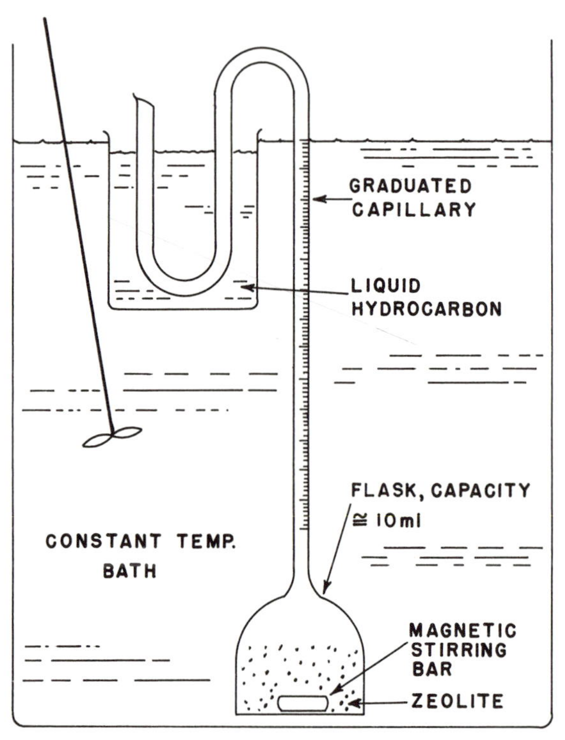

Figure 2. Apparatus for diffusion (8)

Two-Component Methods. The measurement of the diffusion of a particular substance is frequently done in the presence of another, generally inert, nonsorbable material, *e.g.*, the sorption of ethane using helium as a carrier gas. For this case the zeolite is generally packed into a fixed bed. The rate of adsorption or desorption is discerned by variations in the concentration of the sorbable gas exiting from the packed column. Gas–solid chromatography using pulse flow techniques has been developed for this purpose. Similarly, methods involving the analysis of breakthrough or elution curves are frequently used.

Gas–Solid Chromatography. The retention time and shape of the effluent pulse concentration curve can be used to evaluate both equilibrium and kinetic quantities. Frequently, the concentrations involved in these experiments are such that only the initial linear portion of the sorption isotherm is involved. Experimental results are easy to obtain, but mathematical interpreation is more difficult. In general, two approaches have been utilized: one involves application of the van Deemter, Zuiderweg, Klinkenberg (*16*) equation relating *HETP* to carrier gas velocity and the other involves a moment analysis as reported by Ma and Mancel (*17*) and Schneider and Smith (*18*).

In regard to the former approach, the *HETP* (height equivalent to a theoretical plate) is determined from the concentration profile of the effluent pulse by

$$HETP = \frac{L\, w_e^2}{8\, t_m t_e} \tag{14}$$

where L is the length of the column of zeolite; t_e and w_e are the retention time and width of the pulse at a fraction of $1/e$ (0.368) of pulse height; and t_m is the retention time of the pulse maximum. As discussed by Eberly (*19*), the *HETP* is evaluated over a wide range of carrier gas flow rates, U, and use is made of the equation developed by van Deemter *et al.* (*16*)

$$HETP = A + \frac{B}{U} + CU \tag{15}$$

For our purposes, the most important constant is C, relating to the mass transfer effects of the porous pellets themselves. This can be determined by plotting *HETP vs.* U and evaluating the limiting slope at high velocities. Such curves are shown in Figure 3 for inert gases on zeolite 3A. The effective size of the zeolite channels is too small to admit argon but large enough to accept helium atoms. The upper curve with its higher value of C reflects the diffusion of helium in the zeolite channels. The lower curve for argon pulses does not include intracrystalline diffusion and can be used as a correction factor for determining wall effects.

The constant C is related to the diffusion constants in the gas and porous solid phases, D_g and D_s, respectively, by

$$C = \frac{\dfrac{F_1^2\,(d_p)^2}{75\,(1-F_1)^2 D_g} + \dfrac{F_1 K\, d_p^2}{2\,\pi^2(1-F_1)D_s}}{\left[1 + K\left(\dfrac{F_1}{1-F_1}\right)\right]^2} \tag{16}$$

where F_1 is the fraction of voids in the packed column, d_p, the diameter of the particle and K, a distribution coefficient, defined as the ratio of concentrations in the mobile and immobile phases. Values for D_g can be computed by methods already discussed. Thus, we can evaluate the quantity, D_s/d_p^2, or if desired, the value of D_s by assuming a characteristic length (d_p) of the particle or crystallite.

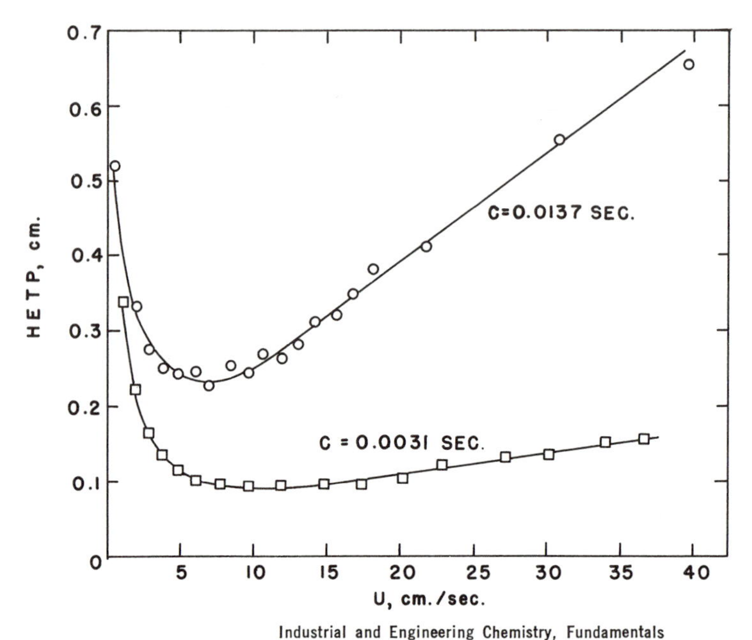

Industrial and Engineering Chemistry, Fundamentals

Figure 3. Results on 3A molecular sieve at 27°C (19)

□ *Argon pulses injected into helium carrier*
○ *Helium pulses injected into argon carrier*

For techniques using moment analysis, similar experimental procedures and data are involved. The mathematical treatment, however, is different. The first absolute moment (μ_1') of the chromatographic peak is defined as

$$\mu_1' = \frac{\int_0^\infty tc(L,t)dt}{\int_0^\infty c(L,t)dt} \tag{17}$$

As discussed by Ma and Mancel (*17*), μ_1' represents the center of gravity of the chromatographic peak and can be used to evaluate equilibrium quantities such as sorption constants. If measured at a series of temperatures, the heats, entropies and free energies of sorption can be calculated.

The second central moment (μ_2) is defined as

$$\mu_2 = \frac{\int_0^\infty (t - \mu_1')^2 c(L,t)dt}{\int_0^\infty c(L,t)dt} \tag{18}$$

Again, this term can be used to obtain various rate constants reflecting the resistances to flow in the packed column. By computerized techniques, one can evaluate diffusivities both in the macropore and intracrystalline channels.

Another method of separating diffusion resistances was reported by MacDonald and Habgood (*20*). If transport in the micropores of the zeolite crystals is assumed to be independent of carrier gas, the micropore resistance can be separated from the gas phase resistance by measurement with hydrocarbon pulses in nitrogen and hydrogen carrier gases.

BREAKTHROUGH OR ELUTION CURVES. In these experiments, the zeolite is placed in a packed column and an inert carrier gas is permitted to flow through the bed. After the proper activation procedure to remove sorbed water and other impurities, a known concentration of sorbate in the carrier gas is introduced to the zeolite column and the effluent concentration continually measured. The breakthrough curve is the concentration *vs.* time curve that appears at the column exit in which the concentration increases from zero to that of the incoming gas stream. After steady state has been reached, the introduction of sorbate is discontinued and the pure carrier gas permitted to desorb the material from the zeolite. This latter is called the elution or desorption curve.

Many analyses of such processes have been reported, each incorporating a certain set of simplifying assumptions. Anderson *et al.* (*21*) solved a set of differential equations in which diffusion was assumed to be rate controlling. This slow process was followed by rapid adsorption according to the Langmuir isotherm. Both adsorption and desorption cycles were considered. The particles were assumed to be spherical and the diffusivity was taken to be a function of temperature but not of concentration or amount adsorbed.

Later, Garg and Ruthven (*22*) solved a similar set of equations specifically applicable to molecular sieves. They assumed a Langmuir isotherm, but allowed diffusivity to vary with sorbate concentration. Curves were computed for two general cases—namely, (1) macropore diffusion control and (2) zeolitic diffusion control. Differences between these were not large. For isotherms which are linear or nearly so, the shape of the elution curve is nearly the exact inverse of the breakthrough curve. However, for highly nonlinear isotherms, the desorption is considerably slower. This is predicted mathematically and can be quantitatively accounted for by the variation of the diffusion constant with sorbate concentration (Equations 5 and 6).

OTHER METHODS. Kondis and Dranoff (*23, 24*) used a flow system in a microbalance to determine the kinetics of ethane sorption on 4A molecular sieve. The ethane is introduced to the balance by using helium as the carrier gas. The weight of the zeolite is continuously recorded during the adsorption and desorption processes. Diffusivities are calculated from Equations similar to 8 and 9. They also considered the effect of nonisothermal conditions during the sorption process (*25*). Experimentally, they observed pellet temperatures as high as 15°C above ambient during ethane sorption on 4A near room temperature. The rates, however, did not differ considerably from the isothermal case. This may be caused by the lower capacity at the higher temperature. The increased diffusion offsets the lower capacity so that the fraction sorbed remains the same. The effect of temperature changes during sorption was also considered by Eagan *et al.* (*26*).

The counterdiffusion of hydrocarbons was investigated by Satterfield *et al.* (*27*) to measure the diffusion of one sorbate in the presence of another. After proper activation the zeolite is saturated with the vapor of one hydrocarbon. This material is then immersed in the other hydrocarbon. The slurry is stirred and samples periodically removed and analyzed for concentrations of both species. Fick's law of diffusion is used to evaluate the diffusivities. Other sorption studies from binary liquid solutions are reported by Weaver (*28*).

Chabazite

Chabazite was one of the first naturally occurring zeolites to be extensively studied. Its typical composition is $(Ca, Na_2)O \cdot Al_2O_3 \cdot 4\,SiO_2 \cdot 6\,H_2O$. Its pore structure consists of a three dimensional network of sorption cavities (*29*). The structure of each cavity is shown in Figure 4. It has a length of

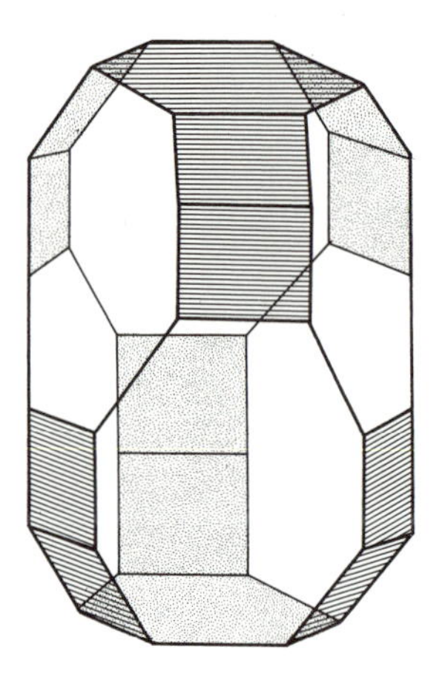

Figure 4. Structure of chabazite cavity

11 A and a diameter of 6.5 A. Entrance into the cavity is attained through six openings of somewhat elliptical shape with a major and minor diameter determined crystallographically as 4.4 and 3.1 A, respectively.

In a study of its sorption properties, Barrer and Ibbitson (*30*) classified sorbates into three different categories, namely, (1) small molecules rapidly occluded (*e.g.*, O_2, H_2, N_2, He, Ar, CH_4 and C_2H_6); (2) *n*-hydrocarbons which were occluded at moderate rates; and (3) larger molecules such as isoparaffins, cycloparaffins, and aromatics which were excluded from the sorption cavities. On the basis of these sorption properties and molecular dimensions derived from Pauling (*31*), the effective size of the openings was deduced to be between 4.89 and 5.58 A.

The rates of *n*-hydrocarbon sorption were of special interest. The fractional amount sorbed was linear with the square root of time within certain limits, so Fick's law of diffusion as expressed by Equation 9 could be utilized. The rate of sorption was influenced by the calcination conditions which, if too mild, could leave extraneous impurities in the crystal or, if too severe, could produce some collapse of the structure. Rates increased with pressure, temperature, and degree of subdivision.

The dependence on temperature could be expressed by an Arrhenius-type of relationship, as shown by Equation 7. Activation energies for gases in

Table I. Activation Energies in Chabazite (*30*)

Hydrocarbon	*E, kcal/mole*	*Temperature Range, °C*
C_3H_8	4.5	23–225
n-C_4H_{10}	8.9	23–156
n-C_5H_{12}	7.1	224–295
n-C_7H_{16}	11.1	182–300

Transactions of the Faraday Society

natural chabazite are given in Table I. The energies increase with length of hydrocarbon chain. The heavier hydrocarbons, n-C_5H_{12} and n-C_7H_{16}, are appreciably sorbed only at elevated temperatures.

The effect of pressure or concentration of sorbed species on the diffusivity was explored more thoroughly by Barrer and Brook (*32*). Results are listed

Table II. Some Average Diffusion Coefficients in Chabazite (*32*)

Sorbate	*Temp, °C*	Q_0, *cm³ (STP)/g*	Q_∞, *cm³ (STP)/g*	$\bar{D}$, *cm²/sec*
C_3H_8	150	1	11.4	17.6×10^{-13}
		1.60	13.0	8.3
		4.00	15.4	3.6
		8.60	20.0	2.0
		15.10	26.5	1.5
CH_2Cl_2	0	0	70.00	81.2×10^{-15}
		6.44	70.04	8.0
		12.89	70.00	2.6
		19.43	70.03	1.2
		26.27	69.97	0.6
$(CH_3)_2NH$	0	0	87.00	77.0×10^{-19}
		2.25	87.05	18.8
		4.47	86.97	10.6
		6.69	87.99	7.1
		8.93	87.03	5.3
		10.61	87.01	4.5

Transactions of the Faraday Society

in Table II. With these sorbates, the diffusivity decreases as the amount of material initially sorbed increases. This effect complicates the computation of true activation energies from the above data. However, apparent activation energies can be calculated and used on a relative basis to determine general trends. In this sense, the authors computed activation energies using values

Table III. Apparent Energies of Activation for Diffusion in Chabazite (*32*)

Sorbate	*E, kcal/mole*
C_3H_8	3.1
n-C_4H_{10}	7.3
CH_2Cl_2	6.4
$(CH_3)_2NH$	15.6–19.0

Transactions of the Faraday Society

of the diffusivity obtained with $Q_o = 0$. Table III lists their results. The effect of chemical nature of the sorbate is clearly revealed by these results. The molecules, C_3H_8, CH_2Cl_2 and $(CH_3)_2NH$ have approximately the same size and shape, but their polarity increases in the order given. The dipole interaction with cations in the zeolite channels causes diffusion to become hindered and the activation energy increases with polarity of the molecule. Also, comparing the results with CH_2Cl_2 and $(CH_3)_2NH$ at 0°C, we see that the latter, more polar molecule has a diffusivity nearly four orders of magnitude less than that of CH_2Cl_2.

Other diffusion studies on natural chabazite were reported by Brandt and Rudloff (*7, 33*). They studied primarily the diffusion of rapidly occluded gases, such as He, H_2, Ar, and CH_4 using a decreasing pressure–constant volume technique. CO_2 and Freon-21 were also investigated. The diffusion curves were analyzed to examine effect of grain boundaries having high permeability. Freon-21 was useful in this case since it cannot enter the intracrystalline channels and diffuses only in the grain boundaries. The need for great accuracy in determining sorption curves was stressed to clearly define the effect of grain boundaries and/or compact amorphous regions.

Diffusion into H-chabazite was investigated by Barrer and Davies (*34*). The H-form was prepared from natural Ca-rich chabazite by exchange with NH_4Cl and subsequent calcination in dry oxygen. In a detailed study of n-C_4H_{10} sorption on this material, the diffusivity increased with the amount of material sorbed. This is opposite to the effect previously observed with n-C_4H_{10} on Ca-rich chabazite. This difference might be caused by changes in the sorption isotherm since diffusivity depends strongly upon the shape of the isotherm as shown by Equation 5. With an equation of this type, they were able to evaluate a constant $\lambda = D_o/RT$ which was shown to be essentially independent of sorbate concentration. This was suggested to be a more fundamental constant of the sorption process. Activation energies for n-hydrocarbons were evaluated and are listed in Table IV.

Table IV. Activation Energies in H-Chabazite (*34*)

Hydrocarbon	*E, kcal/mole*	*Temperature, °C*
C_3H_8	4.00	50–200
n-C_4H_{10}	4.16	55–200
n-C_5H_{12}	4.92	100–200

Proceedings of the Royal Society of London

Comparing these results with those in Table I for Ca-rich chabazite, it is seen that the activation energies increase more slowly with the H-form. This might be expected if the cavities are free of the larger Na and Ca cations.

Erionite

Erionite is a naturally occurring mineral closely related to chabazite. However, in its natural form it generally contains a greater variety of alkali and alkaline earth cations as shown by the formula for its typical composition $(Ca, Mg, Na_2, K_2)O \cdot Al_2O_3 \cdot 6\,SiO_2 \cdot 6\,H_2O$. The structure of erionite is best described by reference to Figure 5 taken from an article by Whyte *et al.*

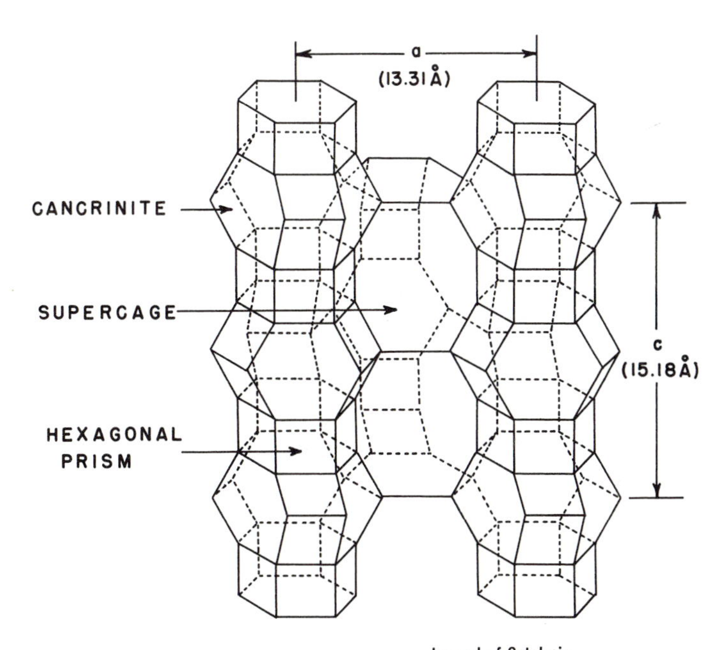

Figure 5. Line drawing of erionite (35)

(*35*). The sorption cavity (supercage) is enclosed by rows consisting of alternating units of cancrinite cages and hexagonal prisms. It should be noticed that every other cancrinite cage is rotated 60° out of position with the one preceding it. As in chabazite, the sorption cavities constitute a three-dimensional network of pores. Each cavity has a length of 15.1 A and a free cross-sectional diameter of 6.3 to 6.6 A (*29*). Sorbate molecules can enter the cavity through six elliptical openings formed by the eight-membered oxygen rings. Each opening has a minimum and maximum diameter of 3.5 and 5.2 A, respectively.

In working with a sample of erionite obtained from Rome, Oregon, Eberly (*36*) observed that the sorption properties were profoundly affected by ion exchange even though only a small amount of exchange was achieved. Data are given in Table V.

Table V. Sorption Properties of Erionite from Rome, Ore. (*36*)

	Surface Area, m^2/g	*Sorption Capacity at 95°C and 300 mm Hg, mmoles/g*		
		n-C_5	n-C_6	n-C_7
Original sample	203	0.26	0.29	0.15
KCl treated	9	0	0	0
$CaCl_2$ treated	171	0.48	0.38	0.11

American Mineralogist

The original erionite exhibits molecular sieve action in that *n*-paraffins are sorbed but branched, cycloparaffins and aromatics are essentially excluded. The original sample had appreciable capacity for *n*-C_5 and *n*-C_6. However, contrary to experience with wider pore zeolites, capacity for the heavier molecule, *n*-C_7, was only half that of the lower molecular weight molecules. Exchange with K ions prevented the sorption of *n*-hydrocarbons and sharply reduced the surface area as measured by N_2 sorption. An analogous effect of K exchange in decreasing effective pore diameter is observed with chabazite and the synthetic zeolite A. This is attributed to the larger ionic radius of the K ion (1.33 A *vs.* 0.95 A for Na). On the other hand, Ca exchange which reduces the total number of ions opened up channels and permitted more of the *n*-hydrocarbons to be sorbed. On the basis of these results, it is not sur-

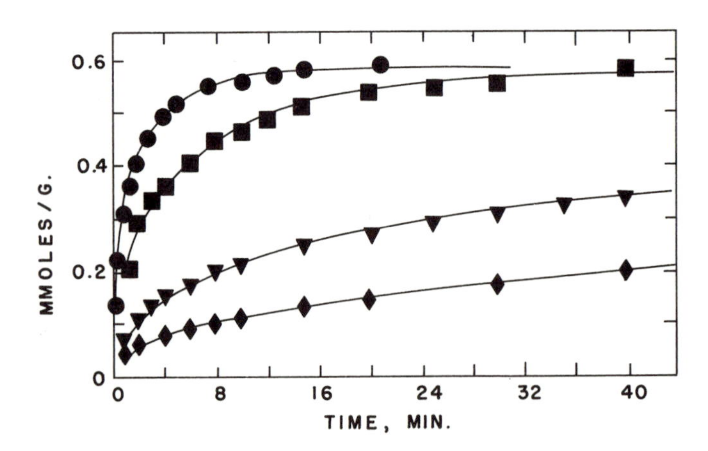

American Mineralogist

Figure 6. Normal paraffin adsorption on erionite (20–35 mesh) at 93°C (37)

● n-*Pentane at* p = *0.8 mm*
■ n-*Hexane at* p = *0.7 mm*
▼ n-*Heptane at* p = *0.6 mm*
◆ n-*Octane at* p = *0.4 mm*

prising that erionites from various natural deposits have different sorption properties depending on their composition and the location of cations in the structure. K ions inside the cancrinite cages or hexagonal prisms should have little effect on the aperture dimensions into the sorption cavity.

In a subsequent study on a better quality erionite sample (surface area $= 456 \text{ m}^2/\text{g}$) from Jersey Valley, Nevada, Eberly (*37*) determined the rates of sorption of several *n*-paraffins in an electromagnetic microbalance under constant pressure conditions. These are shown in Figure 6. The rate is seen to drop off markedly as the molecular weight is increased. Thus, the low capacities for $n\text{-}C_7$ previously discussed for the Rome, Oregon, sample may have been partially due to kinetic rather than equilibrium effects. In desorption of these materials under vacuum at 93°C, only partial removal of $n\text{-}C_5$ and $n\text{-}C_6$ is realized. $n\text{-}C_7$ and $n\text{-}C_8$ could not be removed at all under these conditions.

Using Fick's law and Equation 8, values of diffusivities were calculated and are listed in Table VI. Since the erionite particles are formed by com-

Table VI. Sorption on Jersey Valley Erionite Sample at 93°C (*37*) (20–35 Mesh Particles)

Compound	Q_∞, *mmoles/g*	D/a^2 (*sec*$^{-1}$) × *10*^{4}	
		Adsorption	*Desorption*
n-Pentane	0.59	5.9	0.48
n-Hexane	0.59	2.3	0.032
n-Heptane	0.39	0.59	—
n-Octane	0.23	0.24	—

Industrial and Engineering Chemistry, Product Research and Development

paction of the powdered zeolite and hence have an inhomogeneous pore structure, the diffusivities are expressed as D/a^2 rather than as absolute values. The desorption rates are much slower than adsorption and the difference appears to increase with molecular weight. Part of this effect can be attributed to the dependence of the diffusivity upon concentration of sorbed species as seen in Equations 5 and 6. However, a certain amount of sorption is irreversible as seen by the desorption attempt with $n\text{-}C_7$ and $n\text{-}C_8$. By raising the temperature, increased rates of sorption can be achieved as seen in Table VII.

Over this temperature range, there is a tenfold increase in diffusivity. This variation follows an exponential relationship with temperature, and the apparent activation energy is found to be between 6.9 and 9.0 kcal/mole. Because of the rate limitations, there is an apparent increase in capacity, Q_∞ as the temperature is raised from 93° to 149°C. Above this, however, the capacity begins to decrease (as expected in these constant pressure runs). Considerable desorption of the *n*-heptane is only possible at the higher temperatures. The adsorption and desorption diffusivities tend to approach one another as the temperature is raised or, as previously discussed, the molecular weight is lowered. The pertinent portions of the isotherms become more nearly

Table VII. *n*-Heptane Sorption on Jersey Valley Erionite at 0.6 mm Hg (*37*) (20–35 Mesh Particles)

Temp, °C	Q_∞, *mmoles/g*	D/a^2 (*sec*$^{-1}$) $\times$ *10*^{4}	
		Adsorption	*Desorption*
93	0.39	0.59	—
121	0.47	0.97	—
149	0.54	1.9	—
171	0.50	3.2	0.65
207	0.45	5.4	2.2

Industrial and Engineering Chemistry, Product Research and Development

linear under these conditions making the diffusion process more in accord with Fick's law. Robson *et al.* (*38*) compared the kinetics of *n*-hexane sorption on several erionite samples and other zeolites as seen in Table VIII.

Table VIII. Sorption of *n*-Hexane at 93°C and 0.7 mm Hg (*38*) (20–35 Mesh Particles)

Zeolite	Q_∞, *mmoles/g*	D/a^2 (*sec*$^{-1}$) $\times$ *10*^{4}
Zeolite Y	1.68	1370
Zeolite 5A	0.83	21
Jersey Valley erionite	0.57	2.3
Synthetic erionite (Na, K)	0.68	24
Synthetic erionite (H)	0.50	100

Advances in Chemistry Series

The three erionite samples exhibit considerable differences in sorption rates. Synthetic erionite is able to adsorb *n*-hexane 10 times faster than its natural counterpart. In fact, the rate is similar to that observed on zeolite 5A. The removal of the alkali metal by replacement with hydrogen results in a further increase in sorption rate. This, however, is still considerably lower than that observed on zeolite Y which among all the well known zeolites has the largest effective pore diameter.

Offretite

The mineral, offretite, is closely related to erionite and has often been confused with the latter. Bennett and Gard (*39*) determined its structure which perhaps can best be understood by referring to Figure 7. By comparing this diagram with Figure 5, the close relationship to the erionite structure can be seen. Again, we have rows of cancrinite cages alternating with hexagonal prisms. However, the successive cancrinite cages are no longer rotated 60° with respect to one another. This results in the formation of large channels parallel to the *c*-axis having a free diameter of about 6.5 A. This structural feature permits the sorption of larger molecules than those sorbed in erionite. It should be noted that another pore structure which is formed by the gmelinite cages also exists. Water and other small molecules can be sorbed in these cages as well as in the wider channels parallel to the *c*-axis.

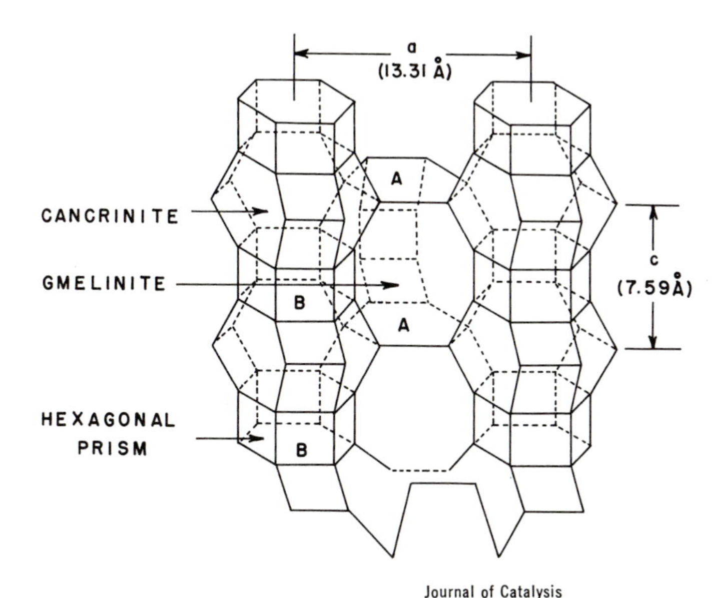

Figure 7. Line drawing of offretite (35)

Since erionite differs from offretite only by rotation of a cancrinite cage, it might be suspected that many actual samples of these minerals would exist as mixtures, some portions of the crystals being erionite, while others consisting of offretite. Only a few dislocations in offretite would be needed to prevent access to the wide channels and thereby prevent sorption of large molecules. However, synthesis of a substantially unfaulted offretite has been accomplished and Aiello *et al.* (*40*) and Whyte *et al.* (*35*) have reported on its properties.

The offretite studied by Aiello *et al.* (*40*) was synthesized from a mixture involving KOH and Me_4NOH and had the unit cell composition of $K_{2.04}$-$(Me_4N)_{1.91}$ $[Al_{3.88}Si_{14.12}O_{36}] \cdot 6.87\ H_2O$. The two Me_4N^+ ions are too large to be located in the cancrinite cages or hexagonal prisms and must be either in the wide channels or gmelinite cages. From this original sample, K_3Me_4N-offretite was made by KNO_3 exchange. The K_2H_2 form was obtained by burning out Me_4N^+ in air at 650°C. A portion of the K_2H_2 form was exchanged with NH_4NO_3 and calcined to produce the KH_3–offretite. Sorption properties of these materials are listed in Table IX.

Replacement of one of the large Me_4N^+ ions from the unit cell with K^+ enables the structure to adsorb *n*-hexane. As the authors indicate, this exchangeable ion is most likely located in the wide channels. However, K^+ is still a fairly large ion and its presence in the channel prevents the sorption of cyclohexane. The decomposition of both Me_4N^+ ions to produce the K_2H_2

Table IX. Occlusion of Gases and Vapors by Synthetic Offretites (*40*)

Cation Composition Per Unit Cell	*Molecules per Unit Cell*			
	H_2O *at 22°C* (p = *2 cm*)	*cyclo-*C_6H_{12} *at 22°C* (p = *10 cm*)	n-C_6H_{14} *at 22°C* (p = *8 cm*)	m-*xylene at 40°C* (p = *0.1 cm*)
$K_2(Me_4N)_2$	8.1	0.07	0.07	—
$K_3(Me_4N)$	11.6	0.2	1.4	—
K_2H_2	15.9	1.39	—	—
KH_3	17.8	1.7	—	0.7

Transactions of the Faraday Society

form permits the zeolite to adsorb cyclohexane. The K^+ ions are probably inside the cancrinite cage and hexagonal prism. The capacity for water nearly doubles when the Me_4N^+ ions are decomposed.

KH_3-offretite exhibits the highest capacity for water and cyclohexane. It also occludes *m*-xylene but not 1,3,5-trimethylbenzene. Since the latter has a diameter of 8.5 A compared to 7.1 A for *m*-xylene, the effective pore diameter must be somewhere in between.

In other experiments, the Me_4N^+ ions in the main channels were exchanged with Li^+, Na^+ and K^+ ions and *n*-hexane sorption rates determined. The rates increased as the diameter of the cation decreased.

Zeolite A

The structure and properties of synthetic zeolite A were first reported by Breck *et al.* (*41*). Owing to its ease of manufacture in high purity, it has found widespread application as a selective adsorbent based primarily on molecular size. Compared with the previously discussed zeolites, it exhibits higher sorption capacities and faster rates. It is synthesized in the sodium form and has the general chemical formula, $Na_2O \cdot Al_2O_3 \cdot 2\ SiO_2 \cdot x\ H_2O$.

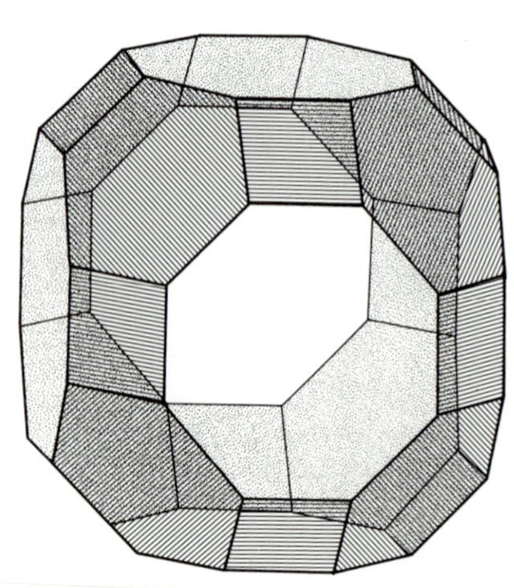

Figure 8. Structure of zeolite A cavity

This is commonly designated as zeolite 4A. When about 75% of its sodium ions are replaced by potassium, it is referred to as zeolite 3A. Alternately, when the sodium is replaced by calcium, it is called zeolite 5A.

The pore structure is constituted by a three-dimensional network of sorption cavities, one of which is shown in Figure 8. This cavity is roughly spherical having a diameter of 11.4 A. Molecules can enter this cavity through six windows each having an opening of 4.2 A. These openings are nearly circular in shape as contrasted to the somewhat elliptical openings in chabazite and erionite.

As expected from our previous discussions on the effect of ion exchange, the sorption properties of zeolite A are strongly dependent on the number and type of cations in the structure. The effect of potassium ion exchange is seen in Figure 9 taken from Breck *et al.* (*41*) and Thomas (*42*). Since the K^+ ion

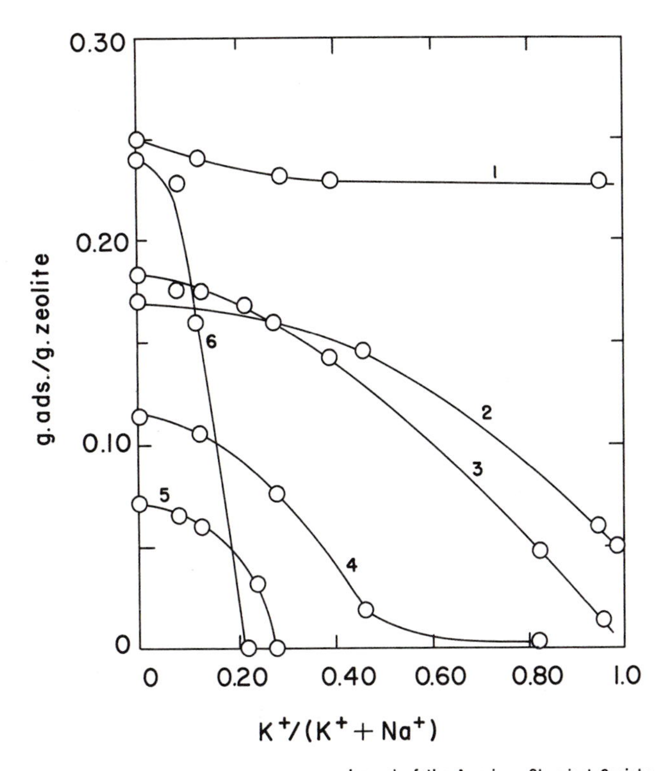

Journal of the American Chemical Society

Figure 9. Effect of potassium exchange for sodium on the adsorptive properties of type A zeolite (41)

1, H_2O at 4.5 mm, 25°C
2, CH_3OH at 4 mm, 25°C
3, CO_2 at 700 mm, 25°C
4, C_2H_2 at 700 mm, 25°C
5, C_2H_6 at 700 mm, 25°C
6, O_2 at 700 mm, −183°C

has a considerably larger ionic diameter, its exchange into zeolite A reduces the effective pore diameter. Thus, the K-enriched zeolite, 3A, no longer occludes acetylene and ethane. The capacities for CH_3OH and CO_2 are also sharply reduced. However, its ability to sorb water remains essentially constant. Therefore zeolite 3A is almost exclusively used as a drying agent, particularly for streams where other small molecules like CO_2, C_2H_6, etc., may be present in considerable amounts.

The effect of calcium ion exchange to produce zeolite 5A is shown in Figure 10. Attempts to measure surface area of zeolite 4A (Na-form) by N_2

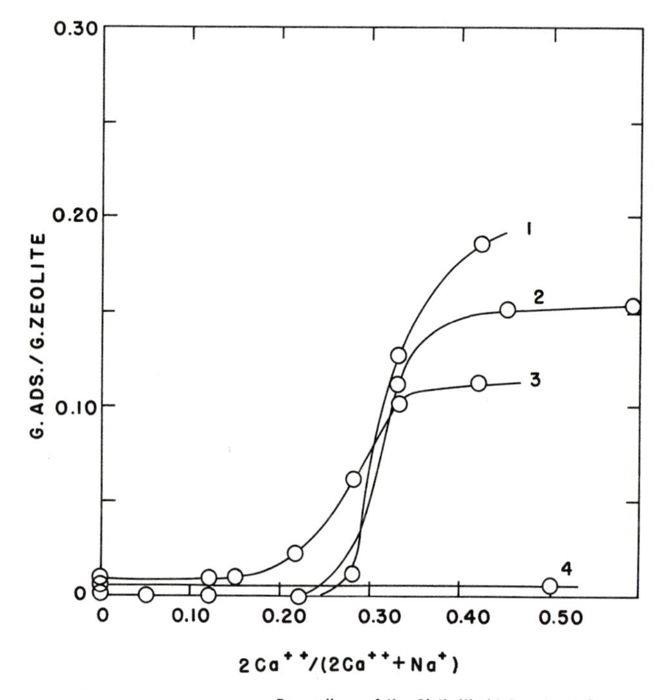

Proceedings of the Sixth World Petroleum Congress

Figure 10. Effect of calcium exchange for sodium on the adsorptive properties of type A zeolite (42)

1, N_2*, 15 mm, −196°C*
2, n-C_7H_{16}*, 45 mm, 25°C*
3, C_3H_8*, 250 mm, 25°C*
4, i-C_4H_{10}*, 400 mm, 25°C*

at −196°C result in a near zero value. However, when the total population of ions is reduced by Ca exchange, the zeolite adsorbs N_2 and reasonable surface areas can be obtained. One important characteristic is that the Ca-form can adsorb *n*-paraffins and other straight chain molecules but not isoparaffins, cycloparaffins or aromatics. This property has resulted in extensive commercial use of this sorbent for separating and obtaining high purity streams of *n*-paraffins or *n*-olefins.

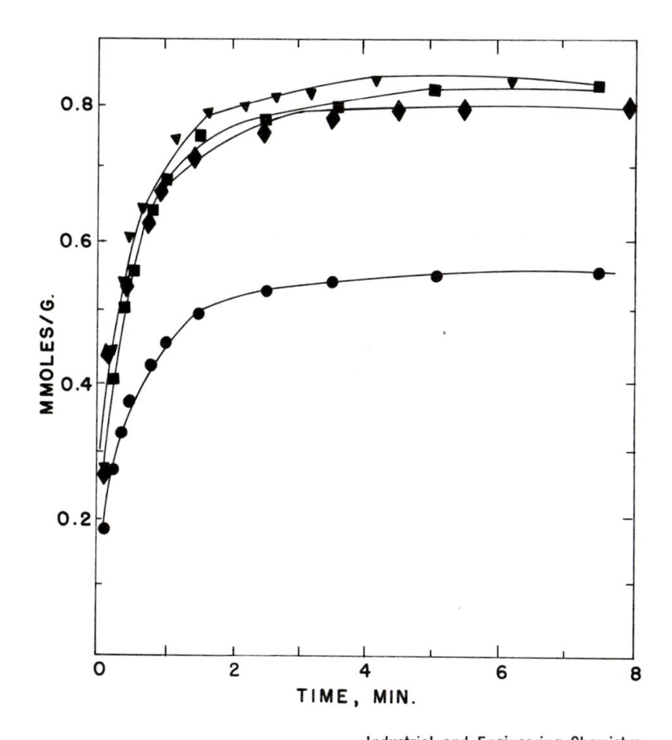

Industrial and Engineering Chemistry, Product Research and Development

Figure 11. Normal paraffin adsorption on 5A molecular sieve (20–35 mesh) (37)

● n-*Pentane at* p = *0.8 mm*
■ n-*Hexane at* p = *0.7 mm*
▼ n-*Heptane at* p = *0.6 mm*
◆ n-*Octane at* p = *0.4 mm*

Eberly (*37*) measured the rates of sorption of *n*-paraffins on 5A. The rate curves are plotted in Figure 11. When these results are compared to those for erionite (Figure 6), we see that the A structure permits a much more rapid rate of adsorption. Values of the rate constants, D/a^2, were computed from these curves and are listed in Table X.

Table X. Sorption on Zeolite 5A at 93°C (*37*) (20–35 Mesh Particles)

Compound	Q_∞, *mmoles/g*	D/a^2 (*sec*$^{-1}$) × *10*^{4} *Adsorption*	*Desorption*
n-Pentane	0.59	19 ± 2	8.1 ± 2
n-Hexane	0.84	21 ± 3	3.9 ± 0.7
n-Heptane	0.85	24 ± 4	1.4 ± 0.2
n-Octane	0.82	23 ± 3	0.2 ± 0.1

Industrial and Engineering Chemistry, Product Research and Development

In contrast to erionite, no statistical significant decrease in adsorptive D/a^2 is observed as progressively higher molecular weight paraffins are used. The desorption rates are faster than the corresponding ones on erionite, permitting determination of the diffusivities for all *n*-paraffins. Because of the nonlinearity of the isotherm, the desorption diffusivities are considerably lower than the adsorption diffusivities.

The greater ease of adsorption in 5A is primarily due to crystal structure. Each of the spherical 5A sorption cavities has a volume of 770 A^3—compared to the 400 A^3—available in the narrow cylindrical cavity of erionite. Furthermore, *n*-paraffins can move in a straight line, fairly unhindered path through the 5A cavities but must bend in a tortuous manner to accommodate themselves to the cylindrical cavities in erionite. The difference in rates of adsorption between the two solids becomes more pronounced as higher molecular weight paraffins are used.

Although not seen with molecules up to C_8, higher molecular weight *n*-paraffins would eventually exhibit a decrease in adsorptive diffusivity. However, in addition to length of carbon chain, the presence of unsaturation impedes the sorption rate. Weaver (*28*) has shown that the adsorption rates for 1-hexene and 2-octene on 5A were 1/20 those of the corresponding *n*-

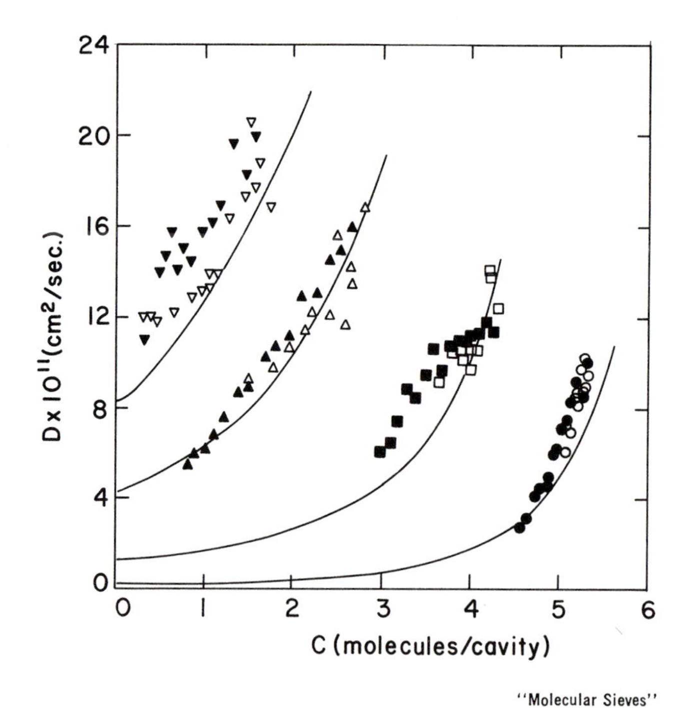

"Molecular Sieves"

Figure 12. Concentration dependence of diffusivity for cyclopropane in 5A zeolite (43). *The open and solid points represent adsorption and desorption results. The symbols ○, □, △, ▽ refer to experiments at 273°, 323°, 398°, and 448°K.*

paraffins. The calculated activation energy for zeolite diffusion was 8.7 kcal/mole for C_6-C_8 olefins. This is caused by greater ion-induced dipole interaction between the zeolite surface and the pi electrons in the double bond.

The dependence of diffusivity upon sorbate concentration has frequently been inferred but has remained ill defined for many sorbate–zeolite systems. Ruthven *et al.* (*43*), however, have made a thorough study of cyclopropane sorption on 5A and have evaluated diffusivities over a range of temperature and sorbate concentration. Their results are plotted in Figure 12 where the concentration of sorbed species is expressed as molecules/cavity. The authors utilized Equation 5. Assuming D_o to be constant and computing values of $\delta \ln p/\delta \ln c$ from the experimental isotherms, they computed the theoretical curves represented by the solid lines in Figure 12. Both the adsorption and desorption data points follow in a general manner the theoretical curves. This increase in diffusivity with concentration was also seen in studies by Barrer and Davies (*34*) with n-C_4C_{10} on H-chabazite, previously discussed.

With Equation 5, Ruthven *et al* (*43*) were able to evaluate D_o at a series of temperatures and thereby calculate activation energies by use of Equation 7. Their results are included in Table XI. By using the same procedure with all sorbates, they were able to establish that the activation energy increases in

Table XI. Activation Energies on Zeolite 5A

Reference		44	45	19	43
Technique		*Constant Pressure-Increasing Volume*	*Gravimetrically at Constant Pressure*	*Gas Chromatography*	*Gravimetrically at Constant Pressure (Limiting Diffusivity at Zero Sorbate Concentration)*
Molecule	*Critical Diameter, A* (40)				
Ar	3.8	—	—	3.5	—
Kr	3.9	—	—	5.9	—
C_2H_4	4.07	—	—	—	2.75
CH_4	4.08	—	—	—	2.98
C_2H_6	4.36	—	—	—	3.02
C_3H_6	4.95	—	—	—	3.46
1-C_4H_8	4.95	—	—	—	3.44
tert-2-C_4H_8	4.95	—	—	—	3.46
C_3H_8	5.1	0.54	—	—	3.50
n-C_4H_{10}	5.1	—	—	—	4.0
n-$C_{14}H_{30}$	5.1	—	16.1	—	—
Cyclo-C_3H_6	5.2	—	—	—	4.34
c-2-C_4H_8	5.58	—	—	—	9.2

a systematic manner with the critical diameter of the molecule as determined from van der Waals radii. The strong dependence of diffusivity upon concentration and the variety of techniques used to measure diffusivity cause considerable difference in numerical values. For example, with C_3H_8 on 5A, values of 0.54 and 3.50 kcal/mole have been reported. The long chain, n-$C_{14}H_{30}$, has a high activation energy of 16.1 kcal/mole, even though its critical diameter is not significantly larger than other n-paraffins. Using gas chromatography, activation energies for Ar and Kr were evaluated which, from their critical diameters, appear to be higher than expected from the Rutheven *et al.* (*43*) results. Possibly, this can be attributed to the fact that the GC technique is involved only with extreme limit of the isotherm at zero concentration. Experimental measurements under these conditions would yield higher activation energies.

Activation energies reported in the literature for various sorbates on zeolites 3A and 4A are listed in Table XII. In zeolite 3A (K-form), activation

Table XII. Activation Energies on Zeolites 3A and 4A

Reference			44	23	46	47
			Constant Pressure-Increasing Volume	*Gravimetrically Using He Carrier Gas Stream*	*Constant Pressure*	*Constant Volume-Decreasing Pressure*
Molecule	*Critical Diameter, A (40)*	*Zeolite*				
Ar	3.8	3A	14.0	—	—	—
N_2	3.0 × 4.1	3A	16.2	—	—	—
N_2	3.0 × 4.1	4A	—	—	4.07	—
CH_4	4.08	4A	—	—	7.4	—
C_2H_6	4.36	4A	3.0	5.6	—	7.4
C_3H_8	5.10	4A	8.7	—	—	—

energies of Ar and N_2 are given. At nominal temperature, this zeolite excludes these molecules. To obtain measurable diffusion rates, data were obtained at 280°C and higher. With zeolite 4A (Na-form), results are given for N_2, CH_4, C_2H_6 and C_3H_8. Again, with the larger molecules, higher temperatures are needed. Among the various investigators, considerable differences in values exist: for ethane on 4A, the activation energies lie between 3.0 and 7.4 kcal/mole depending upon the technique of measurement.

In investigating sorption rates, the degree to which the rate is controlled by the intercrystalline macropores or the intracrystalline micropores often depends on the sorbate, temperature, and type of zeolite. For example, with ethane at 25° to 117°C Kondis and Dranoff (*23*) showed that intracrystalline diffusion was controlling on Linde 4A both in powder and pelletized form.

However, with ethane on Linde 5A, Antonson and Dranoff (*48*) found that both macro- and micropore diffusion were significant. Youngquist *et al.* (*49*) in their studies on Davison Chemical Co. 5A microtraps found that macropore diffusion was controlling. In Figure 13 their results are plotted for butylene

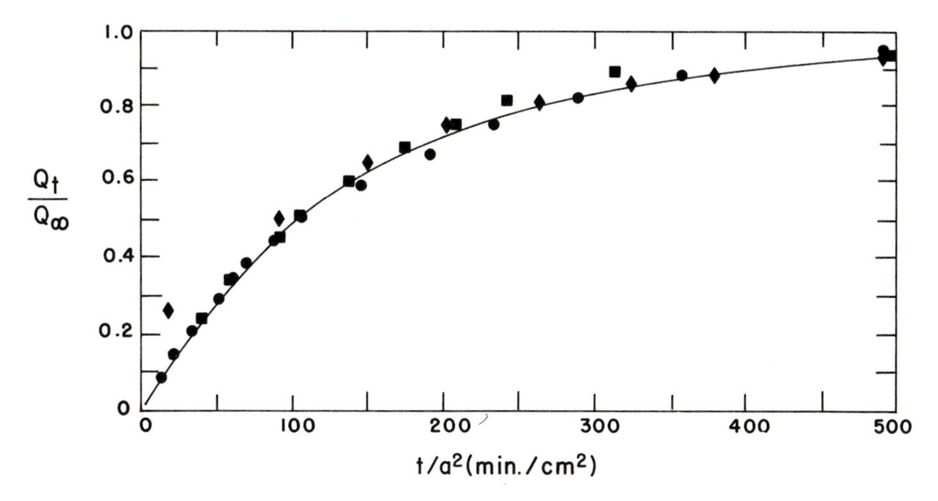

Industrial and Engineering Chemistry, Product Research and Development

Figure 13. Effect of particle size on rate of butylene adsorption at 25°C and 11.7 mm Hg (49). *The* ●, ■, *and* ♦ *represent results with a = 0.167 cm, 0.120 cm, and 0.0651 cm, respectively.*

adsorption at 25°C on various particle sizes. According to Equation 8, the fractional amount adsorbed when plotted against t/a^2 should yield the same curve independent of a. As seen, this is the case for this sorbate-zeolite system for particle sizes from 0.0651 to 0.167 cm, indicating that macropore diffusion is governing. The authors attributed this to the smaller macropore diameter of 900 A for the microtraps *vs.* the 2500 A diameter for Linde pellets.

Since the process of adsorption nearly always involves the liberation of heat, actual temperatures inside the pellet can be higher than bulk temperature, thus causing some change in the adsorption rate curves. Kondis and Dranoff (*25*) considered this effect both from a theoretical and experimental basis. In Figure 14 temperature *vs.* time curves are shown for ethane on 4A at a bulk temperature of 25.2°C. The nonisothermal model indicates a temperature difference of over 15°C. With the actual thermocouple readings, a maximum temperature difference of 6° was seen with the 40 gage couple. The data lie in a direction to indicate that higher readings could be obtained with finer couples. However, in spite of these temperature differences, the actual adsorption rates did not differ much from the isothermal case. The authors attributed this to two compensating effects. As the temperature inside the pellet increases, the sorption capacity is lowered, but the diffusion rate is increased. These compensate one another so that the fraction adsorbed *vs.* time is not appreciably changed.

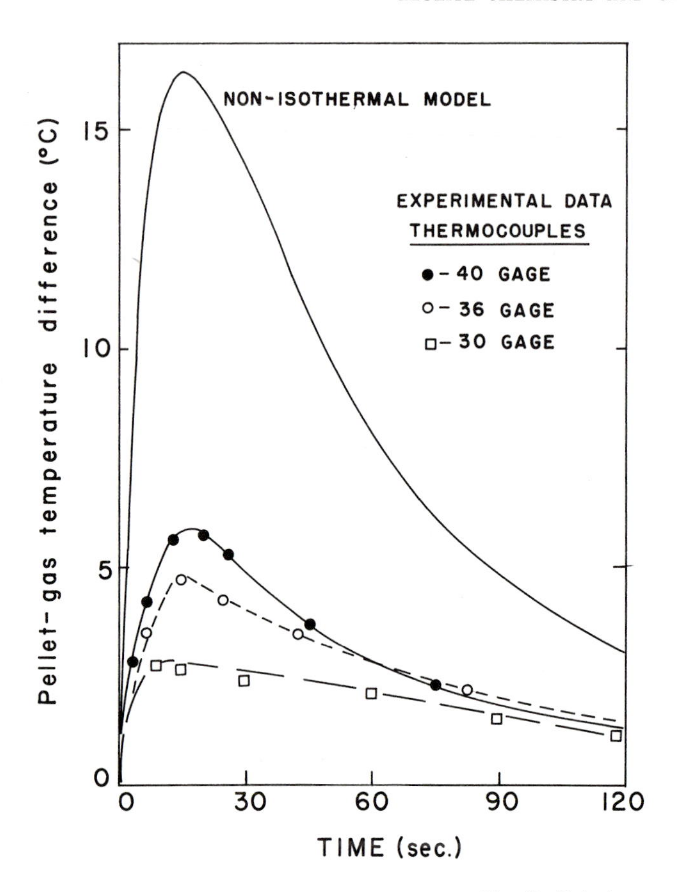

Adsorption Technology

Figure 14. Comparison of nonisothermal model and experimental pellet temperatures; adsorption at 25.2°C gas temperature, 750 mm Hg ethane pressure on zeolite 4A (25)

Temperature effects were also considered by Eagen *et al.* (*26*). They measured the temperature rise during adsorption of N_2 on 4A at −78°C and propane on 5A at −78°C. Rises of 15° and 60°C, respectively, above bath temperature were recorded.

Mordenite

Structural Considerations. Mordenite is a crystalline aluminosilicate having the general typical formula, $Na_2O \cdot Al_2O_3 \cdot 10\ SiO_2 \cdot xH_2O$. As true with other zeolites, the sodium ions can be readily exchanged with other cations. Also, as will be discussed later, the alumina content of the anionic structure can be lowered to near zero while still maintaining the essential crystalline characteristics of mordenite. This is more easily accomplished with mordenite than other zeolites where the structural stability is quite sensitive to the SiO_2/Al_2O_3 ratio.

The pore structure of mordenite is not three-dimensional in nature as was the case with the previously discussed zeolites. It consists of nonintersecting parallel channels. There are no openings along the length of the channels permitting the cross diffusion of molecules from one channel into another. This is best understood by reference to Figure 15. As reported by Meier (*50*) in his structure determination, mordenite is orthorhombic with unit cell dimensions of $a = 18.13$, $b = 20.49$ and $c = 7.52$ A. Two unit cells are superimposed on the diagram in Figure 15. The main building blocks are four- and five-membered rings composed of SiO_4 and AlO_4 tetrahedra. These rings are so arranged that the resulting crystal contains parallel sorption channels having an approximately elliptical opening with a major and minor diameter of 6.95 and 5.81 A, respectively. As seen from the upper portion of the figure,

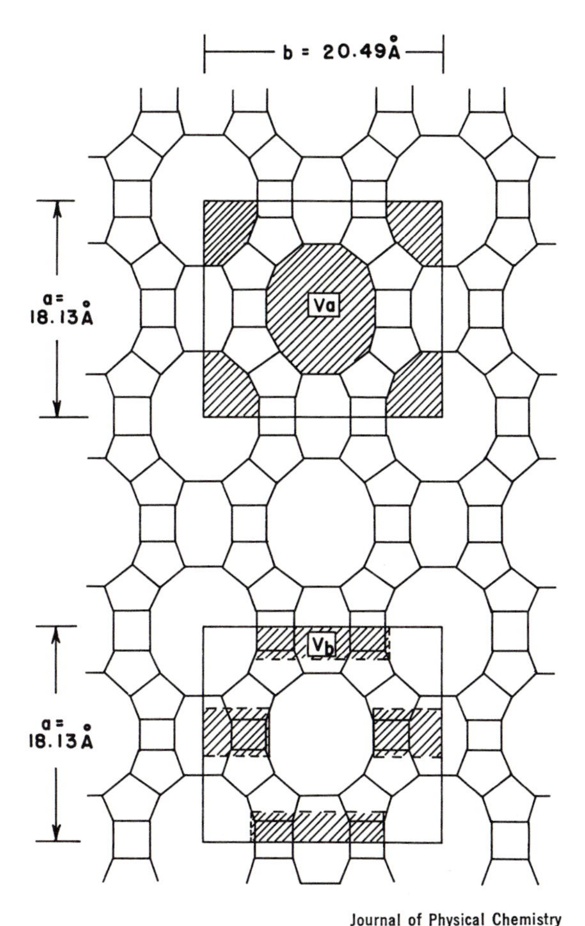

Journal of Physical Chemistry

Figure 15. Schematic diagram of a cross-section of mordenite (53). The two volumes possible for adsorption are indicated by V_a *and* V_b.

each unit cell contains in effect two sorption channels extending a distance of 7.52 A perpendicular to the plane of the page. Along the walls of the channel additional cavities occur at periodic intervals. These are shown by the shaded portions of the unit cell in the lower portion of Figure 15. The openings to these cavities have dimensions of between 3.87 and 4.72 A. However, these cavities are not completely open from one channel to another, thereby preventing cross diffusion of molecules.

Inherent in a pore structure of this type is the ability of relatively small amounts of impurities, foreign ions or stacking faults to completely prevent access of sorbate molecules to the interior pore structure. This property has led to pronounced differences in the reported values of its sorption properties.

Properties of Natural and Synthetic Mordenites. Although its structure indicated that it should adsorb fairly large molecules, early reports on presumably a natural sample of mordenite indicated that the material had an effective pore opening of only 4 A (*51*). It adsorbed nitrogen and smaller molecules rapidly, whereas it took up methane and ethane slowly. Barrer and Brook (*32*) measured the diffusion of a number of gases into various ion-exchanged forms of this type of mordenite and some of their findings are listed in Table XIII.

Table XIII. Molecular Diffusion Coefficients in Natural Mordenite at −78°C, cm²/sec (*32*)

	Form of Mordenite				
Gas	*Ca*	*K*	*Ba*	*Na*	*Li*
Kr	—	1.8×10^{-18}	—	—	—
Ar	5.2×10^{-19}	2.4×10^{-16}	5.3×10^{-16}	1.7×10^{-15}	1.2×10^{-14}
N_2	4.3×10^{-18}	9.2×10^{-16}	3.1×10^{-14}	—	—
O_2	3.3×10^{-16}	2.0×10^{-15}	—	—	—
H_2	—	2.7×10^{-13}	—	—	—

Transactions of the Faraday Society

It is seen that the diffusion coefficient increases as the size of the molecule decreases. Also, there is a general trend in increasing diffusivity in the series Ca < K < Ba < Na < Li. The authors suggest that in addition to radius of ion, deformability may also play a part.

Later, a synthetic mordenite was reported which exhibited sorption properties more in accordance with its crystallographic dimensions. It had considerable capacity for larger molecules such as *n*-heptane, cyclohexane and benzene (*52*). Eberly (*53*) measured the sorption properties of a synthetic Na-mordenite of this type and its H-form in the low pressure region (0.01 to 6 mm Hg). Data are listed in Table XIV.

These results demonstrate the large sorption capacities of Na-mordenite at very low pressures and temperatures up to 260°C. The saturation capacity expressed in liq. cm^3/g has an average value of 0.0759 and is essentially independent of the nature of the hydrocarbon and temperature. This is similar to

Table XIV. Hydrocarbon Sorption on Synthetic Na-Mordenite (Pressures from 0.01 to 6 mm Hg) (*53*)

Sorbate	*Temp, °C*	C_s, *mmoles/g*	C_s, *Liq cm³/g*	ΔH, *kcal/mole*
Benzene	223	0.594	0.0750	
	241	0.565	0.0757 }	24.7
	260	0.546	0.0800	
Cyclohexane	161	0.543	0.0725	
	183	0.528	0.0739 }	18.0
	203	0.551	0.0808	
n-Hexane	162	0.447	0.0756	
	182	0.401	0.0731 }	20.3
	204	0.374	0.0754	

Journal of Physical Chemistry

the pore volume of 0.094 cm^3/g determined from the crystallographic dimensions. The heats of adsorption are considerably higher than those observed on zeolite X. Barrer *et al.* (*54*) and Eberly (*55*), for example, list 15.5–16.8 kcal/mole for benzene and 10.8 kcal/mole for *n*-hexane on zeolite X. These are about 10 kcal/mole lower than those listed for Na-mordenite. This effect is probably associated with the narrower intracrystalline channel system in mordenite. Eberly (*53*) noted that with the H-form a slow irreversible adsorption process occurred simultaneously with the fast reversible adsorption.

Using a gas chromatographic technique, Ma and Mancel (*17*) studied the diffusion of several light hydrocarbons in the H and Na forms of synthetic mordenite. Equilibrium properties such as the heat of adsorption and adsorption equilibrium constant were determined from the first moment of the GC concentration peak. Kinetic constants were determined from the second moment. Some of their results are listed in Table XV.

Table XV. Heats of Adsorption and Energies of Activation for Inter- and Intracrystalline Diffusion in Mordenites (*17*)

			E, kcal/mole	
Material	*Gas*	ΔH, *kcal/mole*	*Intercry. Diff.*	*Intracry. Diff.*
H-Mordenite	CH_4	4.4	0.64	1.8
	C_2H_6	5.8	0.84	3.8
	C_3H_8	8.0	—	4.1
	n-C_4H_{10}	10.7	—	—
Na-Mordenite	CH_4	4.7	0.60	5.2
	C_2H_6	5.50	1.45	4.5
	C_3H_8	8.8	2.90	5.2
	n-C_4H_{10}	10.9	2.20	8.7

Advances in Chemistry Series

Adsorption capacities (not listed in the table) are higher for the H-form. Since the heats of adsorption are nearly the same, this increase is attributed to more space in the channels resuling from the removal of the Na ions. By moment analysis, the authors were able to determine the activation energies for diffusion in the macropore system as well as the energies for diffusion within the crystallites themselves. These are listed in Table XV under the headings of Intercry. Diff. and Intracry. Diff., respectively. The activation energies for diffusion in the macropores are low since diffusion here is of the Knudsen or possibly bulk flow type where dependence on temperature is mild. Higher activation energies, more adequately described by a true exponential dependence on temperature, govern the flow in the intracrystalline micropores.

Ma and Mancel (*17*) considered the diffusion in their mordenite particles to be controlled by three resistances, namely, (1) M_1, mass transfer from the moving gas stream through the stationary film, (2) M_2, mass transfer within the macropore network of the granule to the external surface of the crystals, and (3) M_3, mass transfer within the micropore system. Table XVI lists values of these resistances for CH_4 in H-mordenite.

Table XVI. Percentage Contribution of the Individual Resistances to the Total Resistance for CH_4 in H-Mordenite (*17*)

Particle Radius, cm	*0.1635*			*0.0927*		
Resistance	M_1	M_2	M_3	M_1	M_2	M_3
Temp, °C						
127	13.1	48.0	38.9	6.6	24.2	69.1
161	10.7	40.6	48.8	4.8	18.2	76.9
209	7.2	31.3	61.5	2.8	12.3	84.8

Advances in Chemistry Series

The resistance to flow of the surface of the granule is generally low. However, the resistance to flow in the macropore system can be more than equivalent to that inside the crystallites, particularly at lower temperatures. As the authors indicate, the decrease of macropore resistance with temperature is consistent with the fact that molecules flowing through the column travel primarily in the macropore system and only relatively few of them enter the intracrystalline micropores. As the temperature is increased, more enter the micropores and the relative contribution of this resistance is increased.

Since H-mordenite is a well recognized catalyst for cracking and olefin alkylation of hydrocarbons, it is of interest to determine the diffusion of reactant in the presence of product since the narrow channel system essentially precludes the passage of molecules past one another. Satterfield *et al.* (*27*) studied the counterdiffusion behavior of benzene and cumene in H-mordenite (SiO_2/Al_2O_3 molar ratio = 12). After appropriate degassing procedure, the zeolite is saturated with one of the hydrocarbons. Then it is immersed in the other hydrocarbon and the rate of desorption measured by periodically analyzing the concentration of the liquid phase. The authors concluded that benzene and cumene cannot readily counterdiffuse in H-mordenite. Hence, in the alkylation reaction most of the surface area of H-mordenite is apparently unavailable for reaction of aromatic species at temperatures moderately above ambient. The cumene desorption coefficient from H-mordenite into benzene decreased by two orders of magnitude as the cumene saturation time was increased to six days. This was attributed to the slow formation of diisopropylbenzene and radical ions in the parallel channels causing blockage of the pores.

Effect of Dealumination. Mordenite differs from other zeolites in the extent to which alumina can be removed from the structure without altering significantly the crystallinity of the material as evidenced by its x-ray diffraction pattern. Mordenite normally has a SiO_2/Al_2O_3 ratio of about 10. However, this can be progressively increased to the point where nearly all the alumina is removed. This is generally accomplished by treatment with strong

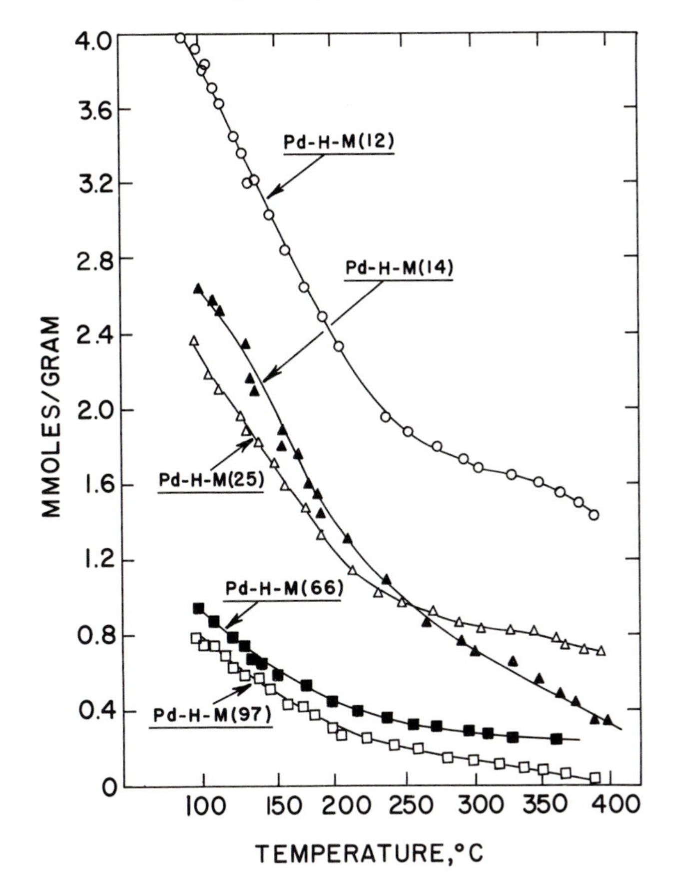

Figure 16. Adsorption isobars of NH_3 at 8 mm (56)

mineral acids which, in addition to removing Al_2O_3, also replace the sodium with hydrogen ions.

The removal of aluminum from the lattice obviates the need for cations for charge neutralization. This effect would be expected to reduce the acidity of the catalyst. By ammonia adsorption, Eberly *et al.* (*56*) measured the acidity of a series of H-mordenite catalysts which had been extracted to various alumina levels. 0.5% Pd had then been incorporated on each of the mordenites to produce a hydroisomerization catalyst. The ammonia adsorption isobars are shown in Figure 16. The numbers in parentheses represent the SiO_2/Al_2O_3 molar ratios. As seen, the removal of alumina results in the formation of a less acidic catalyst surface. The effect of temperature was examined in order to obtain a measure of acidity distribution. Most of the curves parallel one another with the exception of Pd-H-M(14) which has a lower than expected NH_3 adsorption at high temperatures. This was attributed by the authors to its relatively high Na_2O content.

On the same series of catalysts, decalin adsorption was measured to determine pore accessibility. Results are plotted in Figure 17. As seen, the high

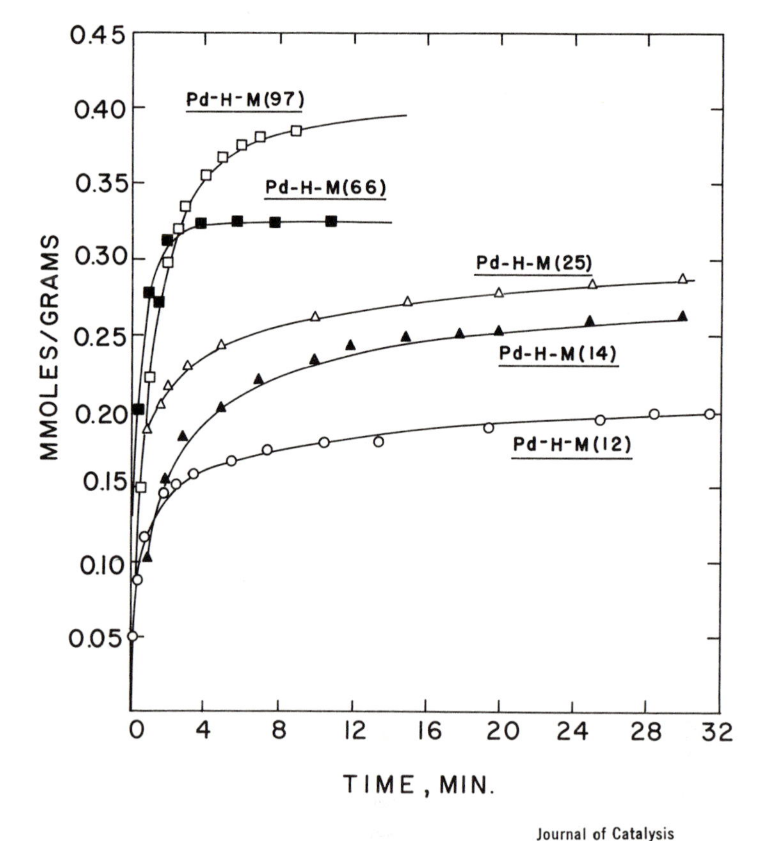

Journal of Catalysis

Figure 17. Adsorption of decalin at 93°C and 0.12 mm (56)

SiO_2/Al_2O_3 ratio mordenites offer the least resistance to the adsorption of decalin. The acid treatment has served to remove obstructions from the parallel channels. In relating these properties to catalysts, Eberly *et al.* (*56*) found that acidity appeared to be the dominant factor for the hydroisomerization of *n*-pentane, higher acidity catalysts being the more active. However, in regard to the hydrocracking of decalin, a larger molecule, pore accessibility was the governing factor and mordenites with a high SiO_2/Al_2O_3 ratio exhibited the greatest activity (*57*).

Thakur and Weller (*58*) made similar studies on acid-extracted mordenite. NH_3 adsorption decreased linearly with alumina content. Diffusion resistance of the acid extracted mordenites, as measured by N_2 adsorption, was less than in the untreated sample. Initial activity for *n*-hexane cracking decreased with decreasing aluminum content.

Separation Properties of Dealuminated Mordenite. In working with a highly dealuminated H-mordenite with a $SiO_2/Al_2O_3 = 93$, Eberly (*59*) measured the desorption rates of several hydrocarbons. These are shown in Figure 18 where the fractional amount remaining on the zeolite is plotted against time. The differences in desorption rates appear to be solely attributable to molecular weight rather than molecular type. Thus, *n*-hexane and benzene come off at the same rates. If the carbon number is increased, the rate of desorption is slowed but toluene and *n*-heptane again desorb at essentially the same rate. This behavior differs from that exhibited by silica gel and other adsorbents where the aromatic is generally held more strongly than the paraffins.

On the basis of these desorption rates, it can be predicted that HM(93) will preferentially retain *n*-octane from its mixture with toluene and consequently affect the preferential adsorption of a paraffin from an aromatic. SiO_2 gel would do the opposite. This was substantiated by separation experiments conducted with a 97% toluene–3% *n*-octane mixture. A comparison of the separation properties of HM(93) and silica gel is shown in Table XVII.

Table XVII. Separation of Hydrocarbon Mixtures at 93.3°C (*59*)

	Preferentially Adsorbed Compound	
Mixture, Mole %	*on SiO_2 Gel*	*on H-Mordenite ($SiO_2/Al_2O_3 = 93$)*
97% Toluene – 3% *n*-octane	toluene	*n*-octane
12% *n*-Heptane – 88% toluene	—	*n*-heptane
95% Benzene – 5% *n*-heptane	*n*-heptane	*n*-heptane
96% Benzene – 4% cyclohexane	—	cyclohexane
93% Benzene – 7% 1-hexene	benzene	1-hexene

Industrial and Engineering Chemistry, Product Research and Development

With the aromatic–*n*-paraffin mixtures, the mordenite consistently showed a preference for the *n*-paraffin. It also preferentially adsorbed cyclohexane and 1-hexene from their respective mixtures with benzene. This unusual type of

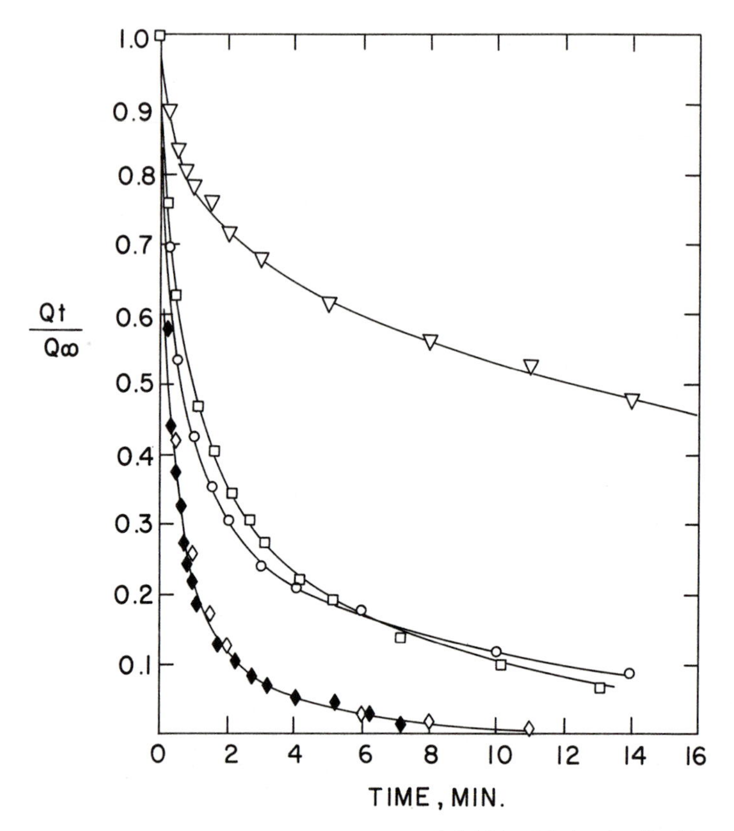

Industrial and Engineering Chemistry, Product Research and Development

Figure 18. Desorption of hydrocarbons from HM (93) at 93°C under vacuum (59). *The symbols,* ◇, ◆, □, ○, ▽ *refer to the desorption of* n-*hexane, benzene,* n-*heptane, toluene, and* n-*octane, respectively.*

separation behavior was not observed with a H-mordenite having a conventional Si_2O/Al_2O_3 ratio = 12.

Since the chemical composition of silica gel and HM(93) are nearly the same, the differences in behavior must be attributed to differences in crystal and pore structure. Eberly (*59*) suggests that the presence of side pockets along the main sorption channels is responsible for this separation behavior. It is possible that because of the side openings, the planar aromatic molecules cannot interact as strongly with the mordenite surface as would otherwise be expected. This is supported by previous work on hydrogen mordenite which showed that heats of adsorption for benzene, cyclohexane, and *n*-hexane were 12.5, 13.5 and 16.0 kcal/mole, respectively (*53*). From a kinetic viewpoint, it is possible that the nonplanar molecules may be able to enter the side pockets to a degree which retards their motion through a channel relative to that of the planar aromatic.

Zeolites X, Y and Faujasite

The synthetic zeolites X and Y are closely related to the natural mineral, faujasite. These materials are crystalline sodium aluminosilicates and Table XVIII lists their SiO_2/Al_2O_3 ratios. The porous channel system of these

Table XVIII. SiO_2/Al_2O_3 Ratios of Zeolites X, Y and Faujasite (*60*)

	SiO_2/Al_2O_3 Molar Ratio	*Al Atoms per Unit Cell*
Zeolite X	2.50	86
Faujasite	4.54	59
Zeolite Y	5.0	56

"Molecular Sieves"

zeolites consists of a three-dimensional array of fairly large sorption cavities. One of these is shown in Figure 19. These cavities can be approximated by a sphere with a diameter of about 11.8 A. The openings into the sphere are formed by 12-membered oxygen rings having a diameter of 8–9 A. Among zeolites, the faujasite type materials have the largest effective pore opening and generally the lowest diffusion resistance. In view of this, less research has been done on the diffusion of sorbates in these materials than in other zeolites with narrower pore openings.

Breck and Flanigen (*60*) investigated the sorption properties of X, Y and faujasite and were able to ascertain the effective diameter of the pore openings by measuring the sorption of tertiary amines. Results are listed in Table XIX.

Table XIX. Adsorption of Tertiary Amines at 25°C on Zeolites X, Y and Faujasite (*60*)

			G/g Activated Zeolite × 100				
Adsorbate	*Kinetic Diameter, A*	*Press., mm Hg*	*Zeolite X*	*CaX*[a]	*Zeolite Y*	*CaY*[a]	*Faujasite*
$(C_2F_5)_2NC_3F_7$	7.7	43	52.1	48.7	—	52.9	—
$(C_2F_5)_3N$	8.0	39	—	—	43.6	—	23.5
$(C_3H_7)_3N$	8.1	3	22.9	1.8	—	21.6	—
$(C_4H_9)_3N$	8.1	1	22.7	1.2	22.8	21.0	—
$(C_4F_9)_3N$	10.2	0.07	1.4	1.6	3.5	0.8	0.9

[a] Degree of calcium exchange is 84 mole % for CaX and 80 mole % for CaY from the Na form.

"Molecular Sieves"

The kinetic diameters were derived from the Lennard-Jones potential function. As seen, zeolite X does not adsorb appreciable amounts of $(C_4F_9)_3N$. Hence, its effective pore opening lies between 8.1 and 10.2 A. The Ca form of this material has a smaller pore opening since it has very low capacity for $(C_3H_7)_3N$ as well as for its fluorinated counterpart. With zeolite Y, the Na and Ca forms exhibit similar sorption capacities, the only amine not adsorbed

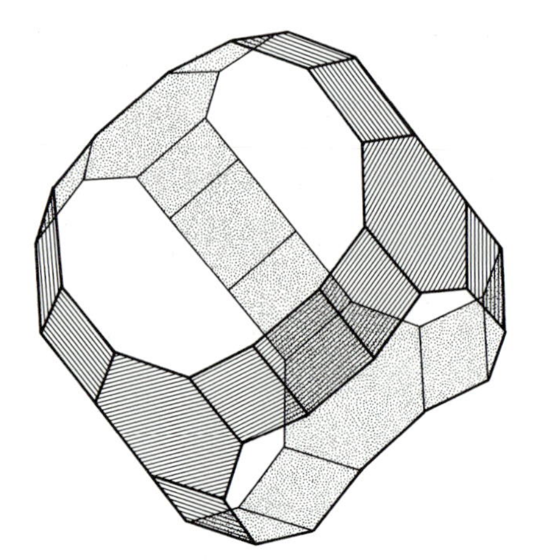

Figure 19. Structure of cavity in zeolite X, Y or faujasite

being $(C_4F_9)_3N$. The naturally occurring mineral, faujasite, has a substantially lower capacity than the synthetic zeolite Y for $(C_2F_5)_3N$.

A study of the rates of adsorption of highly purified liquid hydrocarbons in NaY and HY made by Satterfield and Cheng (*8*) is abstracted in Table XX. The critical molecular diameters were evaluated from bond lengths,

Table XX. Diffusion Coefficients in NaY and HY at 30°C (*8*)

Organic Compound	*Critical Molecular Diameter, A*	D_o *(cm²/sec) × 10¹³ at θ = 0*	
		NaY	*HY*
Hexadecane	4.9	>900	—
Cumene	6.8	>700	—
1,3,5-Trimethylbenzene	8.4	72	>500
2,4,6-Trimethylaniline	8.4	7.1	4.1
1,3,5-Triethylbenzene	9.2	1.1	—
1,3,5-Triisopropylbenzene	9.4	0.047	12.4
1,3,5-Triisopropylcyclohexane	9.8	4.9	8.7

AIChE Symposium Series

angles and van der Waals radii. These values tend to be somewhat higher than the kinetic diameters listed in Table XX. The diffusivities listed were determined at the early part of the rate curve near zero sorbate concentration. The results of NaY clearly show that the diffusivities increase quite markedly with decreasing molecular size, hexadecane diffusing at a rate over 20,000 times as fast as that for 1,3,5-triisopropylbenzene. Another factor that can be seen from Table XX is the influence of molecular type. 2,4,6-trimethylaniline containing a permanent dipole diffuses at a much slower rate than 1,3,5-trimethylbenzene even though the molecular diameters are the same. This is attributed to the higher energy of interaction between the polar molecule and the surface cations. Similarly, 1,3,5-triisopropylcyclohexane diffuses at a faster rate in NaY than 1,3,5-triisopropylbenzene even though the former molecule has a larger diameter; this again is due to the higher energy of interaction between the aromatic pi electrons and the zeolite surface.

With HY the rates of diffusion were generally faster with the exception of 2,4,6-trimethylaniline. Since HY is a more acidic catalyst and the trimethylaniline is a base, the slow diffusion can probably be attributed to the strong acid–base interaction. The other essentially neutral hydrocarbon molecules diffuse more rapidly in HY since the cavities are more open because of the removal of the Na cations.

Satterfield and Katzer (*61*) also conducted studies on counterdiffusion of one hydrocarbon in the presence of another. Similar studies were previously discussed in the section on mordenite. Data on diffusion in NaY and other cations exchanged forms are shown in Table XXI. In contrast to mordenite,

Table XXI. Desorptive Diffusion Coefficients from NaY and Ion-Exchanged Zeolites (*61*)

Zeolite	*Compound Diffusing*	*Saturated Zeolite Placed in*	*Temp, °C*	D_o *(cm²/sec)* $\times 10^{13}$ *at* $\theta = 0$
NaY	cumene	benzene	8	0.216
NaY	cumene	benzene	25	1.25
NaY	cumene	1-MN [a]	25	0.489
NaY	1-MN [a]	cumene	0	0.360
CaY	1-MN	cumene	0	1.84
CeY	1-MN	cumene	0	10.7
HY	1-MN	cumene	0	20.0

[a] 1-Methylnaphthalene.

Advances in Chemistry Series

hydrocarbons can readily counterdiffuse in zeolite Y. However, the diffusion of one molecule is not uninfluenced by the presence of other molecules. For example, cumene exhibits a lower diffusion coefficient when desorption occurs into 1-methylnaphthalene rather than into benzene. As the authors indicate, this is unlike Knudsen diffusion. Adsorption or desorption measurements of single components cannot be used to estimate counterdiffusion rates. The last four runs in Table XXI deal with diffusion of 1-methylnaphthalene from various forms of zeolite Y into cumene. From NaY to HY, the diffusion coefficient increases by nearly two orders of magnitude because of decreasing cation density in the sorption cavities. For example, in CaY, the number of cations would be halved from the Na form if complete exchange occurs. Rare earth zeolites upon dehydration are known to have all the rare earth cations in the sodalite units rather than in the supercage. In HY, the sorption system is even more open since the hydrogen ions are generally very closely associated with the oxygen in the anionic framework, resulting in the formation of hydroxyl groups as seen by infrared spectroscopy.

A number of studies of diffusion have been made using various nuclear magnetic resonance techniques (*10, 11, 12*). Recently, Pfeifer *et al.* (*10*) conducted such studies on a series of cyclic hydrocarbons on NaY. The proton spin relaxation of benzene, cyclohexadiene, cyclohexene and cyclohexane indicated an increasing restriction of mobility with increasing number of π-electrons. This is in agreement with the previously discussed results by Satterfield and Cheng (*8*).

Another interesting aspect of the studies by Pfeifer *et al.* (*10*) was the authors' ability to distinguish between intracrystalline and intercrystalline diffusion. This is perhaps best understood by reference to Figure 20. This

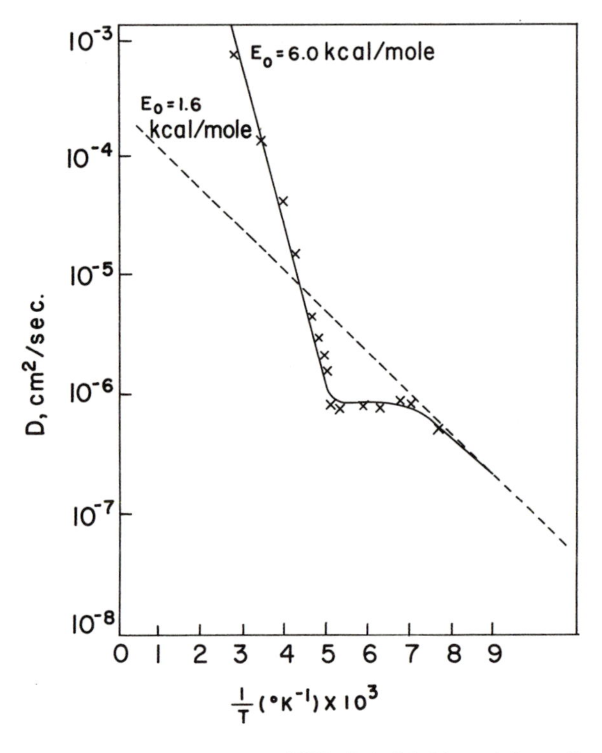

"NMR—Basic Principles and Progress"

Figure 20. Temperature dependence of the experimental self-diffusion coefficient of propane molecules in NaX. Coverage: 2.6 molecules per supercage. The broken straight line was computed from proton relaxation data and corresponds to intracrystalline diffusion (10).

shows the self-diffusion coefficient of propane in NaY at a coverage of $\theta = 0.5$. The NMR technique measures the square of the mean diffusional length. Below $-130°C$, the length is smaller than the crystal dimensions and true intracrystalline diffusion is observed. Between $-130°$ and $-75°C$, the diffusion constant remains the same since the diffusion length is restricted by the boundary of the crystallites. At higher temperature, thermal energy permits the molecules to leave the crystallites and intercrystalline diffusion is measured. Similar curves were observed for water in NaX.

Nomenclature

a	radius of the particle (or crystallite)
b	constant in Langmuir Equation 1
c	concentration of sorbed species
c_S	concentration of sorbed species at saturation
d_p	diameter of particle
D	diffusivity (or diffusion constant, coefficient), cm^2/sec
D_o	diffusivity at zero sorbate concentration, cm^2/sec
D_{AB}	diffusion constant for binary mixture of components A and B, cm^2/sec
D_g	diffusivity in gas phase, cm^2/sec
D_K	diffusivity for Knudsen diffusion, cm^2/sec
D_s	diffusivity in solid phase, cm^2/sec
E	activation energy
F_1	fraction of voids in packed column
$HETP$	height of equivalent theoretical plate
J	rate of transfer of molecules per unit area
L	length of packed column
M, M_A, M_B	molecular weight
M_1	mass transfer resistance from moving gas stream through stationary film
M_2	mass transfer resistance in macropores
M_3	mass transfer resistance in micropores
p	pressure
Q_o, Q_t, Q_∞	total amount sorbed in particle at time $(t) = 0, t$ and ∞ (equilibrium), respectively
r	radius of pore
R	ideal gas constant
S	parameter for particle size distribution (*see* Figure 1)
SA	surface area
t	time
t_e	retention time of pulse at a fraction $(1/e)$ of pulse height
t_m	retention time of pulse maximum
T	absolute temperature, °K
U	carrier gas flow rate
v_i	special diffusion volume
V	volume of pores or volume of particles
w_e	width of pulse at a fraction $(1/e)$ of pulse height

Greek Letters

α	parameter in Equation 12, $p_\infty/p_o - p_\infty$
$\theta = c/c_S$	fraction of surface covered by sorbate
μ_1'	first moment of GC peak
μ_2	second moment of GC peak

Literature Cited

1. Barrer, R. M., "Molecular Sieve Zeolites—II," *Adv. Chem. Ser.* (1971) **102,** 41.
2. Fick, A., *Ann. Phys. Lpz.* (1855) **170,** 59.
3. Fuller, E. N., Schettler, P. D., Giddings, J. C., *Ind. Eng. Chem.* (1966) **58** (5), 18; (8), 81.
4. Ruckenstein, E., Vaidyanathan, A. S., Youngquist, G. R., *Chem. Eng. Science* (1971) **26,** 1305.
5. Crank, J., "The Mathematics of Diffusion," Clarendon Press, Oxford, 1956.
6. Ruthven, D. M., Loughlin, K. F., *Chem. Eng. Science* (1971) **26,** 577.
7. Brandt, W. W., Rudloff, W., *J. Phys. Chem. Solids* (1964) **25,** 167.
8. Satterfield, C. N., Cheng, C. S., "Adsorption Technology," *AIChE Symp. Ser. 117* (1971) **67,** 43.
9. Pfeifer, H., "NMR—Basic Principles and Progress," pp. 53-153, Springer, Berlin, 1972.
10. Pfeifer, H., Schirmer, W., Winkler, H., "Molecular Sieves," *Adv. Chem. Ser.* (1973) **121,** 430.
11. Resing, H. A., Murday, J. S., "Molecular Sieves," *Adv. Chem. Ser.* (1973) **121,** 414.
12. Resing, H. A., Thompson, J. K., "Molecular Sieve Zeolites—I," *Adv. Chem. Ser.* (1971) **101,** 473.
13. Dyer, A., Molyneux, A., *J. Inorg. Nucl. Chem.* (1968) **30,** 829.
14. Barrer, R. M., Fender, B. E. F., *J. Phys. Chem. Solids* (1961) **21,** 12.
15. Sargent, R. W. H., Whitford, C. J., "Molecular Sieve Zeolites—II," *Adv. Chem. Ser.* (1971) **102,** 155.
16. van Deemter, J. J., Zuiderweg, F. J., Klinkenberg, A., *Chem. Eng. Sci.* (1956) **5,** 271.
17. Ma, Y. H., Mancel, C., "Molecular Sieves," *Adv. Chem. Ser.* (1973) **121,** 392.
18. Schneider, P., Smith, J. M., *AIChE J.* (1968) **14,** 762.
19. Eberly, P. E., Jr., *Ind. Eng. Chem., Fundamentals* (1969) **8,** 25.
20. MacDonald, W. R., Habgood, H. W., *Can. J. Chem. Eng.* (1972) **50,** 462.
21. Anderson, R. B., Hamielec, A. E., Stifel, G. R., *Can. J. Chem. Eng.* (1968) **46,** 419.
22. Garg, D. R., Ruthven, D. M., "Molecular Sieves," *Adv. Chem. Ser.* (1973) **121,** 345.
23. Kondis, E. F., Dranoff, J. S., "Molecular Sieve Zeolites—II," *Adv. Chem. Ser.* (1971) **102,** 171.
24. Kondis, E. F., Dranoff, J. S., *Ind. Eng. Chem., Process Design Develop.* (1971) **10,** 108.
25. Kondis, E. F., Dranoff, J. S., "Adsorption Technology," *AIChE Symp. Ser.* (1971) **67,** 25.
26. Eagan, J. D., Kindl, B., Anderson, R. B., "Molecular Sieve Zeolites—II," *Adv. Chem. Ser.* (1971) **102,** 164.
27. Satterfield, C. N., Katzer, J. R., Vieth, W. R., *Ind. Eng. Chem., Fundamentals* (1971) **10,** 478.
28. Weaver, S. D., Cornell Univ. Diss. (1969); *Diss. Abst. Int. B* (1970) **30,** 5482B.
29. Barrer, R. M., *Berichte* (1965) **69,** 786.
30. Barrer, R. M., Ibbitson, D. A., *Trans. Far. Soc.* (1944) **40,** 206.
31. Pauling, L., "The Nature of the Chemical Bond," p. 189, Cornell University Press, 1940.
32. Barrer, R. M., Brook, D. W., *Trans. Far. Soc.* (1953) **49,** 1049.
33. Brandt, W. W., Rudloff, W., *Z. Phys. Chem.* (1964) **42,** 201.
34. Barrer, R. M., Davies, J. A., *Proc. Roy. Soc. London A* (1971) **322,** 1.
35. Whyte, T. E., Jr., Wu, E. L., Kerr, G. T., Venuto, P. B., *J. Catalysis* (1971) **20,** 88.

36. Eberly, P. E., Jr., *Amer. Min.* (1964) **49,** 30.
37. Eberly, P. E., Jr., *Ind. Eng. Chem., Prod. Res. Develop.* (1969) **8,** 140.
38. Robson, H. E., Hamner, G. P., Arey, W. F., "Molecular Sieve Zeolites—II," *Adv. Chem. Ser.* (1971) **102,** 417.
39. Bennett, J. M., Gard, J. A., *Nature* (1967) **214,** 1005.
40. Aiello, R., Barrer, R. M., Davies, J. A., Kerr, I. S., *Trans. Far. Soc.* (1970) **66,** 1610.
41. Breck, D. W., Eversole, W. G., Milton, R. M., Reed, T. B., Thomas, T. L., *J. Amer. Chem. Soc.* (1956) **78,** 5963.
42. Thomas, T. L., "Proceedings of the Sixth World Petroleum Congress III," p. 115, 1963.
43. Ruthven, D. M., Loughlin, K. F., Derrah, R. I., "Molecular Sieves," *Adv. Chem. Ser.* (1973) **121,** 330.
44. Nelson, E. T., Walker, P. L., *J. Appl. Chem.* (1961) **11,** 358.
45. Schirmer, W., Fiedrick, G., Grossman, A., Stack, H., "Molecular Sieves," p. 276, Society of Chemical Industry, 1968.
46. Habgood, H. W., *Can. J. Chem.* (1958) **36,** 1384.
47. Brandt, W. W., Rudloff, W., *J. Phys. Chem. Solids* (1965) **26,** 741.
48. Antonson, C. R., Dranoff, J. S., *Chem. Eng. Progr., Symp. Ser.* (1969) **65,** 27.
49. Youngquist, G. R., Allen, J. L., Eisenberg, J., *Ind. Eng. Chem., Prod. Res. Develop.* (1971) **10,** 308.
50. Meier, W. M., *Z. Krist.* (1961) **115,** 439.
51. Barrer, R. M., *Brennstoff-Chem.* (1954) **35,** 325.
52. Keough, A. H., Sand, L. B., *J. Amer. Chem. Soc.* (1961) **83,** 3536.
53. Eberly, P. E., Jr., *J. Phys. Chem.* (1963) **67,** 2404.
54. Barrer, R. M., Bultitude, F. W., Sutherland, J. W., *Trans. Far. Soc.* (1957) **53,** 1111.
55. Eberly, P. E., Jr., *J. Phys. Chem.* (1961) **65,** 68.
56. Eberly, P. E., Jr., Kimberlin, C. N., Jr., Voorhies, A., Jr., *J. Catalysis* (1971) **22,** 419.
57. Beecher, R., Voorhies, A., Jr., Eberly, P. E., Jr., *Ind. Eng. Chem., Prod. Res. Develop.* (1968) **7,** 203.
58. Thakur, D. K., Weller, S. W., "Molecular Sieves," *Adv. Chem. Ser.* (1973) **121,** 596.
59. Eberly, P. E., Jr., *Ind. Eng. Chem., Prod. Res. Develop.* (1971) **10,** 433.
60. Breck, D. W., Flanigen, E. M., "Molecular Sieves," p. 47, Society of Chemical Industry, 1968.
61. Satterfield, C. N., Katzer, J. R., "Molecular Sieve Zeolites—II," *Adv. Chem. Ser.* (1971) **102,** 193.

Catalysis:
Chemistry and Mechanism

Chapter

8

Mechanistic Considerations of Hydrocarbon Transformations Catalyzed by Zeolites

Marvin L. Poutsma, Corporate Research Laboratory, Union Carbide Corp., Tarrytown, N.Y. 10591

ZEOLITES OFFER UNIQUE opportunities for studies of heterogeneous catalysis because of their catalytic, structural, and ion-exchange properties. These crystalline aluminosilicates in appropriate ion-exchanged forms can offer sizeable activity enhancements and selectivity alterations for certain reactions (*1*) when compared with amorphous silica-alumina, although not necessarily for all reactions (*2*). Because of porosity and crystallinity, the zeolitic internal surface is defined by the regular three-dimensional crystal structure in contrast to most heterogeneous catalysts which, even if crystalline, possess active sites largely at terminations of crystals and irregularities in the lattice. Thus, in principle, knowledge of zeolite structure gained from x-ray and other spectroscopic probes should be directly applicable to studies aimed at defining the local structure of catalytically active sites. (In practice, however, this ideal situation has not fully materialized (*3*) because the proposed active sites have turned out to be those most difficult to detect by x-ray spectroscopy.) The existence of several different zeolite framework structures, each of which can be rationally modified by means of controlled ion exchange, offers a range of catalytic properties. The ability to monitor changes in catalytic performance as a function of changes in zeolite structure should be a powerful tool in elucidating mechanisms.

Scope. The major industrial applications of zeolite catalysis in the petroleum industry are based on transformations of hydrocarbons. This chapter attempts to summarize and to classify these basic catalytic reactions and to interpret them in mechanistic terms.

Although the subject of zeolite catalysis is only 15 years old (*4*, *5*), an extensive body of literature already exists. Fortunately, a number of review treatments have appeared (*3, 6, 7, 8, 9, 10, 11, 12, 13, 14, 15, 16, 17, 18, 19*). Therefore, we are not comprehensively descriptive but focus on papers through 1973 which specifically consider catalytic mechanisms or from which mechanistic inferences may be reasonably drawn. We occasionally offer interpretations or hypotheses beyond those of the authors with full realization that these are speculative to varying degrees.

Approach. In the most rigorous terms a mechanism for a given hydrocarbon transformation over a given zeolite catalyst must include a description at the atomic level of (1) the structure of the zeolitic active site, (2) the structure of all adsorbed organic intermediates and transition states lying between reactants and products, and (3) the interactions between the reactants, intermediates, and products and the zeolite solid in terms of diffusion, adsorption–desorption, and chemical transformation at the active site. This description would be derived from and consistent with all kinetic, thermodynamic, and extrathermodynamic information recorded for the system. In practice, both our data base and interpretive powers are limited, and therefore our mechanisms seldom, if ever, are this ideally complete. Also, since a mechanism is generally the most reasonable operational description of a reaction which can be drawn from the available data and relevant analogies, it is always subject to revision on the basis of new data or interpretive insight.

The mechanisms proposed for zeolite catalysis often emphasize different aspects of the overall situation because different types of data are being interpreted. Studies relating catalytic behavior to structural properties and variations of the zeolite will naturally produce a mechanism emphasizing a proposed structure for the active site on the zeolite surface. Studies relating catalytic behavior to structural changes in the hydrocarbon reactant tend to produce a mechanism emphasizing organic reaction intermediates. Studies emphasizing detailed kinetics will produce a mechanism emphasizing the interplay among diffusion, adsorption, and chemical transformation. Thus, such typical statements which we will encounter as "Brønsted acidity of $H_0 < -8$ associated with the 3650 cm^{-1} structural hydroxyl group," "electrophilic aromatic substitution by a secondary carbonium ion," and "product desorption from the pore mouth is rate limiting" could all be describing the same alkylation reaction, each supplying a different insight into the mechanism.

Because of the varying levels of meaning and emphasis, no single approach to discussing the subject seems ideal. In a previous review (*14*), an historical approach emphasized the evolution of various hypotheses concerning the active site. Here we attempt a four-part approach. First, we survey the types of hydrocarbon reactive intermediates proposed for typical hydrocarbon transformations on zeolites and their behavior in model homogeneous systems. Secondly, we consider the proposals concerning active sites which have been generated from studies of zeolite structure. Third, we sketch some studies of kinetics and the possible roles of the dynamic factors of diffusion and adsorption. Finally, we reexamine the major reaction types in terms of most probable organic reactive intermediates and pathways.

The power and limitations of the experimental techniques at our disposal to probe mechanisms are considered as results from their use enter our discussion.

Reactions and Intermediates

Reaction Types. The major substrate types are olefins, alkylated aromatics, and paraffins, singly and in combination. Among the zeolite-catalyzed reactions of olefins are: (1) positional and geometrical isomerization of the double bond (*e.g.*, 1-butene → *cis*- and *trans*-2-butene); (2) skeletal isomerization (*e.g.*, 1-butene → *i*-butene); (3) condensation to form mixtures of dimers and higher telomers (*e.g.*, entry 1 of Table I); (4) hydrogen redistribution reactions to form paraffins and polyunsaturated products including aromatics (*e.g.*, entries 2 and 3); and (5) cracking to mixtures of smaller fragments (*e.g.*, reverse of entry 1). The balance among these is a function of reaction conditions (temperature, pressure, etc.), olefin structure, and zeolite structure. Alkylaromatics undergo: (1) positional isomerization (*e.g.*, *o*-xylene → *m*- and *p*-xylene); (2) transalkylation (*e.g.*, xylenes → toluene and trimethylbenzenes); and (3) dealkylation (*e.g.*, reverse of entries 4 and 5). The major transformations of paraffins are: (1) skeletal isomerization; and (2) cracking (*e.g.*, reverse of entries 6, 7, and 8).

Interaction between olefins and aromatics can lead to ring alkylation (*e.g.*, entries 4 and 5). Certain paraffins can also be alkylated with olefins (*e.g.*, entry 8). The thermodynamically derived equilibrium constants (*20*) for the idealized reactions in Table I demonstrate that the preferred direction of many of these reactions depends on temperature. In particular, cracking reactions are generally endothermic but are driven at high temperatures by increases in entropy.

The common characteristic of most of these reactions is their similarity to reactions catalyzed by strong acids in homogeneous media where the intermediacy of carbonium ions has been firmly established. Thus, the reactive intermediate most commonly proposed for zeolites which also display acidic properties is the carbonium ion. Less frequently, zeolite-derived reaction products are reminiscent of thermal, free-radical reactions. The dichotomy of carbonium ion *vs.* free-radical reactions over non-zeolitic solid cracking catalysts was proposed by Greensfelder, Voge, and Good (*21*) and by Thomas (*22*) on the basis of patterns of product composition. Such analogies are indeed still a powerful tool particularly since knowledge of the expected behavior of these reactive intermediates has grown rapidly. Some basic properties of carbonium ions and free radicals deduced from physical organic studies will serve as background for understanding how these intermediates can be formed on and modified by the zeolitic surface.

Zeolites do not appear to induce hydrocarbon transformations reminiscent of carbanion or carbene intermediates although these may be involved in certain transformations of substrates containing hetero atoms.

Table I. Equilibrium Constants for Some Idealized Hydrocarbon

Entry	*Reaction*
1	$3\ CH_2{=}CH_2 \rightarrow CH_2{=}CH(CH_2)_3CH_3$
2	$2\ CH_2{=}CHC_2H_5 \rightarrow CH_3CH_2CH_2CH_3 + CH_2{=}CHCH{=}CH_2$
3	3 cyclohexene $\rightarrow$ 2 cyclohexane + benzene
4	$CH_2{=}CH_2 + C_6H_6 \rightarrow CH_3CH_2C_6H_5$
5	$CH_2{=}CHCH_3 + C_6H_6 \rightarrow (CH_3)_2CHC_6H_5$
6	$2\ CH_2{=}CH_2 + CH_3CH_3 \rightarrow CH_3(CH_2)_4CH_3$
7	$CH_2{=}CHCH_3 + CH_3CH_2CH_3 \rightarrow CH_3(CH_2)_4CH_3$
8	$CH_2{=}CHCH_2CH_3 + (CH_3)_3CH \rightarrow (CH_3)_3CCH_2CH(CH_3)_2$

[a] Data from Ref. *20*.

Model Carbonium Ion Behavior (*23, 24, 25*). Carbonium ions are trigonal, planar, electron-deficient, and carbon centered. We will retain this traditional, more familiar nomenclature although there has been a recent tendency to use the term carbenium ion for R_3C^+ and use carbonium ion only for the species R_5C^+.

ELEMENTARY REACTIONS. The most frequent pathways to form carbonium ions are heterolysis of a C—X bond where X is an electronegative group and protonation of an unsaturated linkage by a Brønsted acid, both reactions being favored by polar, ion-solvating media. For hydrocarbon substrates, the protonation reactions 1 and 2 are well known. Even paraffins

$$R_2C{=}CH_2 + H^+ \rightarrow R_2\overset{+}{C}CH_3 \quad (1)$$

$$C_6H_6 + H^+ \longrightarrow C_6H_7^+ \quad (2)$$

can be converted to carbonium ions either by very strong Brønsted acids (*26*) such as $FSO_3H–SbF_5$ or Lewis acids (*27*) such as SbF_5. Whether these reactions involve attack on carbon to form CH_5^+ analogs (*26*) followed by loss of molecular hydrogen, attack on hydrogen (*27, 28*) *via* hydride transfer, or some other oxidative process is still subject to discussion (*29*).

Carbonium ions undergo facile 1,2-shifts of hydrogen and alkyl or aryl groups which lead to positional and skeletal rearrangement. Such rearrangements color much of the subject matter of this chapter.

In the presence of olefins, addition can occur and forms the basis of cationic polymerization. Reaction 3 is exothermic by ~25 kcal/mole if the

$$R^+ + R_2'C{=}CH_2 \rightarrow R_2'\overset{+}{C}CH_2R \quad (3)$$

Transformations as a Function of Temperature[a]

log K_p				
400°K	*500°K*	*600°K*	*700°K*	*800°K*
11.7	6.5	3.0	0.6	−1.2
0.9	0.5	0.3	0.1	0.0
15.7	11.8	9.1	7.2	5.8
7.1	4.3	2.5	1.2	0.3
5.3	2.7	1.0	−0.2	−1.1
9.8	5.0	1.7	−0.5	−2.2
3.2	1.0	−0.4	−1.4	−2.1
2.5	0.1	−1.4	−2.5	−3.3

initial and final ions are of comparable stability; if they are not, ΔH can be estimated by means of the effects of alkyl substituents on carbonium ion stability. Since, however, Reaction 3 has $\Delta S < 0$ because it is a condensation reaction, it will become increasingly reversible as temperature increases. The reverse of Reaction 3 is the basic β-scission reaction in cracking reactions of olefins and paraffins at 300°–500°C. In the presence of aromatic rings, addition also can occur to give cyclohexadienyl (benzenium) ions and forms the basis of electrophilic substitution reactions (Reaction 4).

$$R^+ + C_6H_6 \rightleftarrows [C_6H_6R]^+ \rightleftarrows C_6H_5R + H^+ \quad (4)$$

This sequence is also reversible to give alkylation of aromatics by olefins at low temperature and dealkylation at high temperature.

One carbonium ion can also be converted into another by intermolecular hydride transfer (Reaction 5), and this is commonly observed if the product

$$R^+ + R'H \rightleftarrows RH + R'^+ \quad (5)$$

ion is more stable than the initial. Hydride transfer is an attractive mechanism to rationalize the hydrogen redistribution reactions which are prevalent over zeolites. Carbonium ions are usually destroyed by nucleophilic capture or deprotonation to reform an unsaturated species.

ENERGETICS. The range of carbonium ion stability from, for example, methyl cation to the highly delocalized triphenylmethyl cation is enormous. Stabilization mechanisms include delocalization into adjacent π-bonds as in allyl and benzyl cation and hyperconjugation with adjacent σ-bonds as in *tert*-butyl cation. Using known bond dissociation energies (*30*) and ioniza-

$$\overset{+}{C}H_2CH{=}CH_2 \longleftrightarrow CH_2{=}CH\overset{+}{C}H_2 \equiv \overline{CH_2 \overset{+}{-}CH-CH_2}$$

$$CH_3-\underset{CH_3}{\overset{CH_3}{|}}C^+ \longleftrightarrow CH_2{=}\underset{CH_3}{\overset{CH_3}{|}}C \quad H^+$$

tion potentials of radicals (*31*), we can compare stabilities as in the following example for methane:

$$\begin{array}{ll} CH_4 \rightarrow CH_3\cdot + H\cdot & \Delta H = D_{C\text{-}H} = +104 \text{ kcal/mole} \\ CH_3\cdot \rightarrow CH_3^+ + e^- & \Delta H = IP_{CH_3} = +228 \\ H\cdot + e^- \rightarrow H^- & \Delta H = -EA_H \\ \hline CH_4 \rightarrow CH_3^+ + H^- & \Delta H = (332 - EA_H) \text{ kcal/mole} \end{array}$$

where EA_H is the electron affinity of the hydrogen atom. A parallel treatment for ethane gives:

$$C_2H_6 \rightarrow C_2H_5^+ + H^- \quad \Delta H = 98 + 198 - EA_H = (296 - EA_H) \text{ kcal/mole}$$

Thus, compared with their respective parent hydrocarbons, ethyl cation is 36 kcal/mole more stable than methyl cation in the gas phase. Similarly, based on:

$$\begin{array}{ll} C_3H_8 \rightarrow i\text{-}C_3H_7^+ + H^- & \Delta H = 94.5 + 182 - EA_H = (276.5 - EA_H) \\ i\text{-}C_4H_{10} \rightarrow tert\text{-}C_4H_9^+ + H^- & \Delta H = 91 + 171 - EA_H = (262 - EA_H) \\ CH_3CH{=}CH_2 \rightarrow \overline{CH_2 \overset{+}{-}CH-CH_2} + H^- & \Delta H = 87.5 + 188 - EA_H = (275.5 - EA_H) \end{array}$$

and

$$C_6H_5CH_3 \rightarrow C_6H_5CH_2^+ + H^- \quad \Delta H = 85 + 178 - EA_H = (263 - EA_H)$$

the gas-phase enthalpy difference between a primary and secondary carbonium ion can be estimated as 19.5 kcal/mole, that between a tertiary and secondary ion as 14.5 kcal/mole, and that between allyl(benzyl) and primary as 20.5 (33) kcal/mole.

Although it might have been expected that this scale would have been compressed in strongly solvating media if the least stable species profited most from solvation, even in highly polar solvents, it is not seriously altered as shown below. Thus, we should obviously expect the tendency of alkyl cations to rearrange towards tertiary or conjugated structures and, perhaps less obviously, the tendency of carbonium ion pathways to avoid formation of primary and methyl cations if at all possible. Certain disclaimers of carbonium ion mechanisms in zeolite-catalyzed cracking reactions of model compounds may have resulted from failure to recognize the magnitude of the energy differences among carbonium ions.

REARRANGEMENT PATHWAYS OF CARBONIUM IONS. A powerful tool in unraveling the dynamics of transformation of simple carbonium ions has been their generation in stoichiometric quantities in the so-called super acids such as $FSO_3H–SbF_5$ (*32*) or $HF–SbF_5$ (*33*) (often diluted with SO_2 or SO_2FCl) characterized by high Brønsted acidity and negligible nucleophilicity. This has allowed structural and dynamic studies by spectroscopic techniques, particularly ^1H and ^{13}C NMR. Precursors include alkyl halides which ionize, alcohols which ionize after protonation (as can be done in sulfuric acid for polyaryl carbinols), olefins which are protonated, and even paraffins (*26, 27, 32*). Protonation of olefins to form carbonium ions without the polymerization and/or hydrogen redistribution reactions normally encountered in less acidic media (as, for example, concentrated sulfuric acid or zeolites) results because no significant quantities of unprotonated olefin remain to react with the carbonium ions.

From the dynamic NMR studies of Saunders (*34, 35, 36*), Brouwer (*33*), and Olah (*32, 37*), a number of conclusions can be deduced for carbonium ions in super acids. (1) Within a given structural type (tertiary, secondary, etc.) effects of varying alkyl substituents are small. For example, the equilibrium mixture of *tert*-hexyl cations in $HF–SbF_5$ at —20°C contained 32% **1**, 38% **2**, and 30% **3**; this equilibrium is attained in seconds by rapid rearrangement steps. (2) Between structural types, the tertiary-

$$\underset{\textbf{1}}{CH_3\overset{\overset{CH_3}{|}}{\underset{+}{C}}CH_2CH_2CH_3} \qquad \underset{\textbf{2}}{CH_3CH_2\overset{\overset{CH_3}{|}}{\underset{+}{C}}CH_2CH_3} \qquad \underset{\textbf{3}}{CH_3\overset{\overset{CH_3}{|}}{\underset{+}{C}}CH(CH_3)_2}$$

secondary enthalpy separation is 11–15 kcal/mole (*34*), only slightly less than in the gas phase in spite of the difference in media. The secondary–primary separation may be slightly larger (*35*). (3) A non-classical species, a protonated cyclopropane, may often be more stable than the corresponding primary ion in the absence of serious steric effects. Here we ignore the subtle question (*38, 39*) of whether the more stable form is edge-protonated

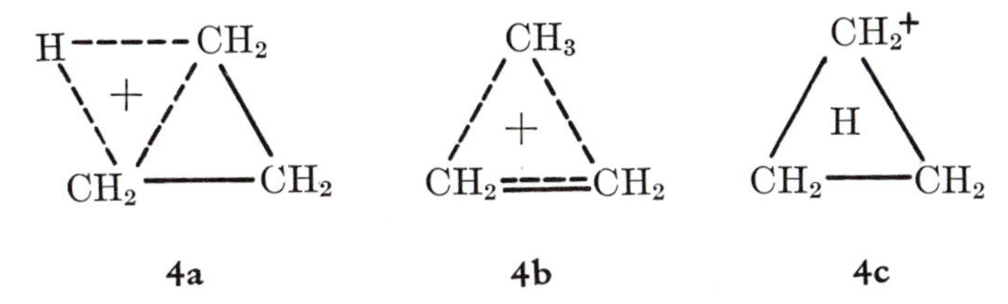

(**4a**) or corner-protonated (**4b**) because for our purposes we can consider them in such mobile equilibrium as to achieve the effective symmetry of face-protonated **4c**. (4) 1,2-Shifts of hydrogen or alkyl groups are the major mode of rearrangement. For those which are thermoneutral (*i.e.*, which interconvert carbonium ions of a given structural type), the activation barriers probably do not exceed 2–3 kcal/mole even for tertiary ions. (However, the barrier is higher for shifts in delocalized benzenium ions (*40*), and Brouwer (*33*) has suggested that the barrier height is a direct function of the carbonium ion stability.) For those shifts which are endothermic,

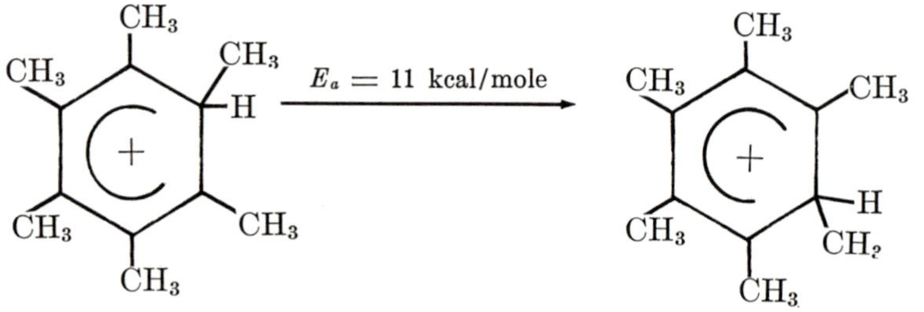

CH_3 CH_3 CH_3 H CH_3 CH_3 CH_3 + $\xrightarrow{E_a = 11 \text{ kcal/mole}}$ CH_3 CH_3 CH_3 + H CH_3 CH_3 CH_3

E_a exceeds ΔH by a similar small value. In general, hydrogen shifts are more rapid than equivalent alkyl shifts. (5) A second, less recognized mode of rearrangement involves the formal insertion of a secondary or tertiary ion into a C—H bond to form a protonated cyclopropane. This may be the preferred route if a primary ion is required to achieve a given rearrangement by a series of 1,2-shifts.

Some examples will serve to demonstrate these principles and the data from which they were derived. The NMR spectrum of carbonium ion **5** shows equivalent methyl groups even at —180°C (*32*); thus, the thermoneutral methyl shift between tertiary ions **5a** and **5b** is faster ($E_a < 2$–3 kcal/mole) than the NMR time scale even at this low temperature. The spectrum of *tert*-amyl cation **6** over the range —40°–+110°C revealed

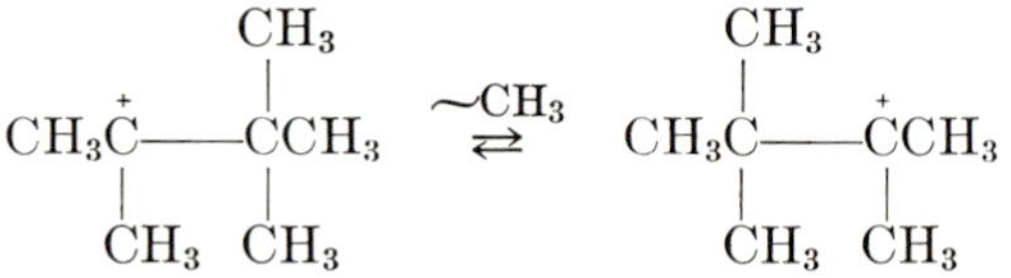

$$CH_3\overset{+}{C}(CH_3)\text{—}C(CH_3)_2CH_3 \;\underset{}{\overset{\sim CH_3}{\rightleftarrows}}\; CH_3C(CH_3)_2\text{—}\overset{+}{C}(CH_3)CH_3$$

5a **5b**

equilibration of the methyl groups, without mixing in the methylene group, on a much slower time scale than for cation **5** corresponding to $E_a = 15.3$ kcal/mole (*34*). The rearrangement scheme is almost surely as pictured below where one hydrogen is starred to show the equilibration; the major contribution to E_a is the energy difference (11–15 kcal/mole) between secondary cation **7** and tertiary cation **6**. Consider next the *sec*-butyl cation **8** (*36*). Even at —110°C the two methyl groups are exchanged, as well as

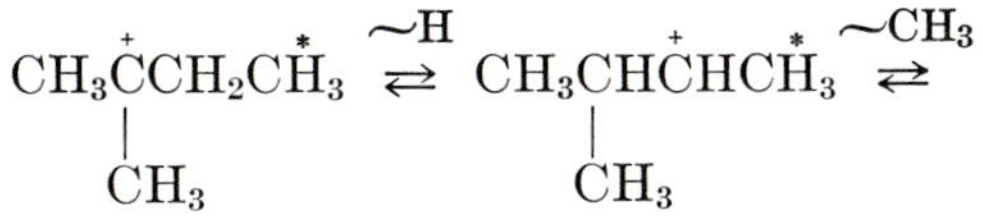

$$\mathrm{CH_3\overset{+}{C}(CH_3)CH_2\overset{*}{C}H_3 \overset{\sim H}{\rightleftarrows} CH_3CH(CH_3)\overset{+}{C}H\overset{*}{C}H_3 \overset{\sim CH_3}{\rightleftarrows}}$$

6a **7a**

$$\mathrm{CH_3\overset{+}{C}HCH(CH_3)\overset{*}{C}H_3 \overset{\sim H}{\rightleftarrows} CH_3CH_2\overset{+}{C}(CH_3)\overset{*}{C}H_3}$$

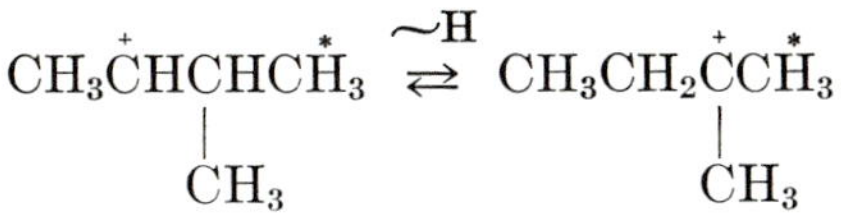

7b **6b**

the inner three hydrogens, by the very rapid thermoneutral hydrogen shift connecting **8a** and **8b**. In the range —110°–+40°C, all hydrogens exchange with E_a of only 7.5 kcal/mole without formation of *tert*-butyl cation **9**. This behavior is not consistent with the often-written conversion of secondary cation **8** to primary cation **10** even though this process would equilibrate the hydrogens. First, its E_a should be similar to that of the conversion of *i*-propyl to *n*-propyl cation, known to require at least 16.4 kcal/mole (*35*). Secondly, if **8** is converted to **10** by hydrogen shift, it should also form primary cation **11** by methyl shift at a somewhat slower but competitive

$$\mathrm{CH_3\overset{+}{C}HCH_2CH_3 \overset{\sim H}{\rightleftarrows} CH_3CH_2\overset{+}{C}HCH_3}$$

8a **8b**

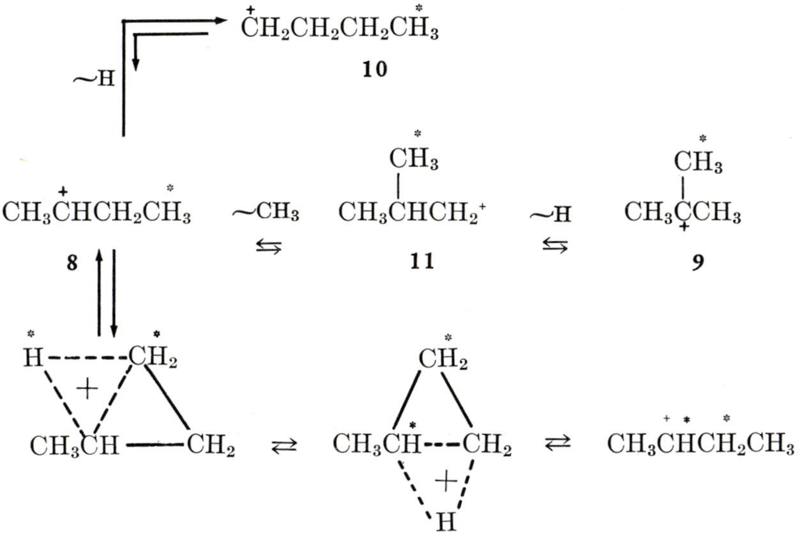

$$\mathrm{\overset{+}{C}H_2CH_2CH_2\overset{*}{C}H_3}$$ **10** (from **8** by ~H)

$$\mathrm{CH_3\overset{+}{C}HCH_2\overset{*}{C}H_3 \overset{\sim CH_3}{\leftrightarrows} CH_3CH(\overset{*}{C}H_3)CH_2^+ \overset{\sim H}{\leftrightarrows} CH_3\overset{+}{C}(\overset{*}{C}H_3)CH_3}$$

8 **11** **9**

$$\mathbf{12a} \rightleftarrows \mathbf{12b} \rightleftarrows \mathrm{CH_3\overset{+}{C}\overset{*}{H}CH_2\overset{*}{C}H_3}\ (\mathbf{8})$$

12a: $\mathrm{\overset{*}{H}}$-----$\mathrm{\overset{*}{C}H_2}$ bridged (+) to $\mathrm{CH_3CH}$—$\mathrm{CH_2}$; 12b: $\mathrm{\overset{*}{C}H_2}$ over $\mathrm{CH_3\overset{*}{C}H}$---$\mathrm{CH_2}$ bridged (+) by H

12a **12b** **8**

rate; however, cation **11** should generate **9** by a highly exothermic hydrogen shift. Indeed, this does finally occur but only at higher temperatures with the more reasonable $E_a = 18$ kcal/mole. The results can be rationalized by formation of protonated cyclopropane **12a** preferentially to formation of either **10** or **11**. Rearrangement of **12a** to **12b** (using the formalism that edge-protonated forms are slightly more stable than corner-protonated forms), followed by ring-opening, scrambles the hydrogens as shown. Whereas *sec*-butyl cation **8** could be prepared and these various processes observed, all attempts (*36*) to prepare *sec*-amyl cation **13** led instead to *tert*-amyl cation **6** regardless of temperature. If the only route from **13** to **6** required formation of primary cation **14** analogous to that just discussed for the route **8** → **11** → **9**, this difference in behavior would not be expected. However, note now that cation **15**, the cyclic analog of **12**, can proceed to cation **16** which can open to a new secondary cation **17** having the branched structure which is an obvious precursor of **6** *via* hydrogen shift.

$$CH_3\overset{+}{C}HCH_2CH_2CH_3 \underset{}{\overset{\sim C_2H_5}{\rightleftarrows}} CH_3CH(C_2H_5)CH_2^+ \overset{\sim H}{\rightleftarrows} CH_3\underset{+}{C}(C_2H_5)CH_3$$

13 **14** **6**

13 ⇅ **15**; **6** ⇅ ~H **17**

15: protonated cyclopropane (H, CH(CH₃), CH₃CH, CH₂; +) ⇄ **16**: CH(CH₃) bridging CH₃CH····CH₂ with H bridge (+) ⇄ **17**: $CH_3\overset{+}{C}H{-}CH(CH_3){-}CH_3$

15 **16** **17**

Kramer (*41*) treated alkyl chlorides with FSO_3H–SbF_5 in the presence of excess methylcyclopentane. Thus, the lifetimes of carbonium ions formed from the chlorides were limited to the period before they were captured by hydride transfer. Many of the ideas described above derived from dynamic spectroscopy studies were nicely confirmed by this chemical approach. Thus, since a secondary C_5 cation can isomerize to a branched isomer more rapidly than a secondary C_4 cation (because involvement of protonated cyclopro-

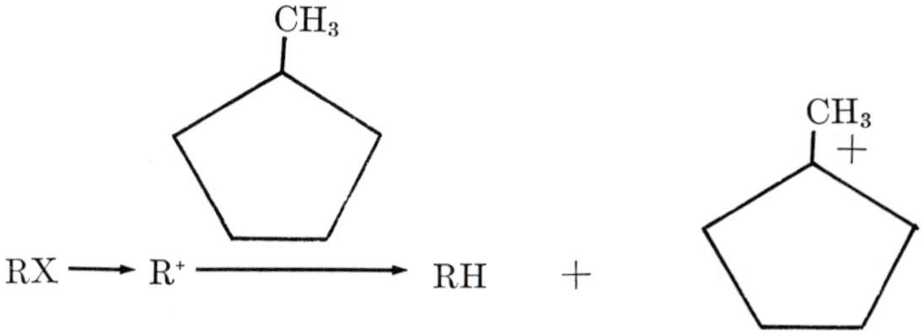

$$RX \longrightarrow R^{+} \xrightarrow{\text{methylcyclopentane}} RH + \text{methylcyclopentyl cation } (CH_3, +)$$

panes can avoid the need to form a primary cation), Kramer found that 2-chloropentane gave 85% rearrangement to form *i*-pentane under conditions where 2-chlorobutane gave only *n*-butane.

REARRANGEMENT PATHWAYS OF PARAFFINS. Brouwer and Oelderik's study (*42*) of the isomerization of 2-methylpentane in HF–SbF_5 at near 0°C provides additional confirmation of these mechanisms and is a useful model for zeolite-catalyzed rearrangements. The fastest reaction was equilibration of 2-methylpentane with 3-methylpentane which proceeded largely to equilibrium before any other isomers appeared. The proposed route involves hydride abstraction by a preformed cation to give the most stable tertiary cation, rearrangement to another tertiary cation through a secondary, and reformation of paraffin by hydride transfer. This is exemplary of the so-called type A rearrangements between tertiary ions which do not change

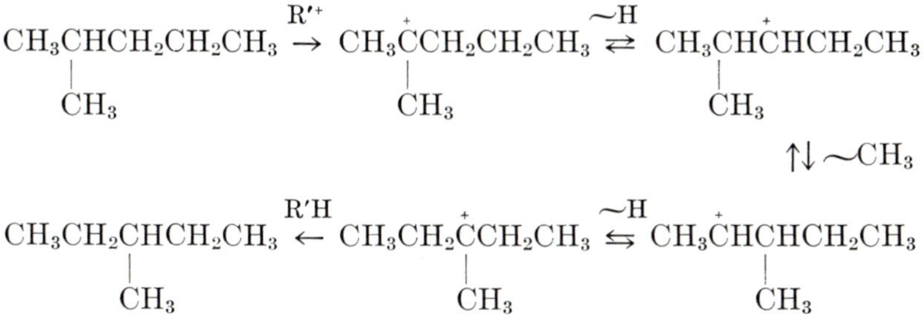

$$CH_3CH(CH_3)CH_2CH_2CH_3 \xrightarrow{R'^{+}} CH_3\overset{+}{C}(CH_3)CH_2CH_2CH_3 \overset{\sim H}{\rightleftarrows} CH_3CH(CH_3)\overset{+}{C}HCH_2CH_3$$

$$\uparrow\downarrow \sim CH_3$$

$$CH_3CH_2CH(CH_3)CH_2CH_3 \xleftarrow{R'H} CH_3CH_2\overset{+}{C}(CH_3)CH_2CH_3 \overset{\sim H}{\leftrightarrows} CH_3\overset{+}{C}HCH(CH_3)CH_2CH_3$$

the degree of chain branching in the carbon skeleton. As illustrated, the highest energy species involved is a secondary cation, and E_a is expected to be 14–15 kcal/mole (although in this particular study (*42*) E_a^{obsd} was mass-transfer limited). The next fastest reaction was equilibration of the mixed methylpentanes with 2,3-dimethylbutane with a slightly greater $E_a^{obsd} = 17$ kcal/mole, which in this case could be identified with the cationic rearrangement pathway. This is a type B rearrangement in which the degree of chain branching changes. As shown above, E_a^{obsd} is much too low to allow for a tertiary → primary cation rearrangement (expected to involve $E_a > 30$ kcal/mole) although with classical cations this is required to change the degree of branching. The proposed route is shown in Reaction 6. Finally, at still slower rates, *n*-hexane and 2,2-dimethylbutane appeared, the rate now being limited by the hydride transfer rate; thus, it was possible to deduce $E_a \sim 15$

$$CH_3\overset{+}{C}(CH_3)CH_2CH_2CH_3 \rightleftharpoons CH_3C(CH_3)\text{---}[CH_2, H]^{+}\text{---}CHCH_3 \rightleftharpoons CH_3C(CH_3)\text{---}[H\text{---}CH_2]^{+}\text{---}CHCH_3 \rightleftharpoons CH_3\overset{+}{C}(CH_3)\text{---}CH(CH_3)CH_3 \quad (6)$$

kcal/mole for Reaction 7, again not much greater than the endothermicity associated with the difference in carbonium ion stabilities. In summary then,

$$CH_3CH_2CH_2CH_2CH_2CH_3 + tert\text{-}R'^{+} \rightarrow CH_3\overset{+}{C}HCH_2CH_2CH_2CH_3 + tert\text{-}R'H \quad (7)$$

type B rearrangements are only slightly less facile than type A because primary cations can be bypassed. Confirmation was provided (*42*) by the observation that conditions stringent enough to isomerize *n*-pentane and *n*-hexane to branched isomers were ineffectual for *n*-butane. The primary cation route (Reaction 8) again does not explain this because whether

$$CH_3CH_2\overset{+}{C}HR \underset{}{\overset{\sim CH_3}{\rightleftarrows}} \overset{+}{C}H_2CH(CH_3)R \underset{}{\overset{\sim H}{\rightleftarrows}} CH_3\overset{+}{C}(CH_3)R \quad (8)$$

R = methyl, ethyl, or propyl seems inconsequential. The protonated cyclopropane route does, however, since neither cation **18** nor **19** can escape to a

$$\overset{*}{C}H_3CH_2\overset{+}{C}HCH_3 \rightleftarrows \mathbf{18a} \rightleftarrows \mathbf{19}$$

$$\mathbf{18a} \rightleftarrows \mathbf{18b} \rightleftarrows CH_3\overset{*}{C}H_2\overset{+}{C}HCH_3$$

18a (protonated cyclopropane: $\overset{*}{C}H_2$---H, CH_2—$CHCH_3$, +)

19 (protonated cyclopropane: H---$\overset{*}{C}H_2$, CH_2—$CHCH_3$, +)

18b (protonated cyclopropane: $\overset{*}{C}H_2$, CH_2---$CHCH_3$, bridging H, +)

branched cation without forming a primary cation. Note, however, that their presence does predict carbon scrambling (as shown by the asterisks) and indeed, with *n*-butane-1-^{13}C, this was exactly the observed result.

Model Free-Radical Behavior (*43*). ELEMENTARY REACTIONS. Formation of neutral radicals from non-radical hydrocarbon precursors is largely limited to homolytic C—C bond cleavage. Aromatic cation radicals can be formed by one-electron transfer to an electron acceptor, and these may deprotonate to form benzylic radicals (Reaction 9). In contrast to carbonium ions, rearrangements by 1,2-hydrogen or alkyl shifts are very

$$C_6H_5CH_3 \xrightarrow{-e^-} [C_6H_5CH_3]^{+\cdot} \xrightarrow{-H^+} C_6H_5CH_2\cdot \qquad (9)$$

rare compared with other competitive reactions of radicals. Radical additions to olefins and aromatics (Reactions 10 and 11) and their respective reverse reactions are well known, the former predominating at low temperature and the latter at high temperature. The radical analog of hydride transfer with

$$R\cdot + CH_2 = CR'_2 \rightleftarrows RCH_2CR'_2\cdot \qquad (10)$$

$$R\cdot + C_6H_6 \rightleftarrows C_6H_6R\cdot \qquad (11)$$

carbonium ions is hydrogen atom transfer, Reaction 12. Alkyl radicals disappear by combination to form dimers or by disproportionation to form the corresponding olefin and paraffin.

$$R\cdot + R'H \rightleftarrows RH + R'\cdot \qquad (12)$$

ENERGETICS. The increment in enthalpy between tertiary, secondary, and primary radicals deduced from C—H bond dissociation energies (*30*) is ~ 4 kcal/mole per type, only 1/3–1/4 that of the corresponding carbonium ions. The allylic resonance energy ($D_{n\text{-}C_3H_7\text{-}H} - D_{Allyl\text{-}H}$) is ~ 10 kcal/mole. These gas-phase values would be expected to be minimally modified by changes in medium.

Zeolite Structure and Active Sites

Most catalytic studies with zeolites have used the synthetic X and Y zeolites manufactured by the Linde Division, Union Carbide Corp. The minimum pore opening in this three-dimensional porous structure is ~ 8 A which will allow passage of all but the largest hydrocarbons. Therefore, while effects of diffusion are surely significant during catalysis, they should be less so with X and Y than with smaller-pore zeolites. As a gross generalization, a Y zeolite will be more active and thermally stable than its X analog. Recently, more and more work has appeared using the large-pore mordenite structure, usually as the synthetic Zeolon manufactured by the Norton Co. The two-dimensional mordenite pore structure with smaller minimum apertures than X or Y should lead to a greater role for diffusion. Mordenites often have greater initial activity than the X or Y analogs, but they tend to lose activity more rapidly with time on stream. Catalytic studies with other zeolites having even greater pore size limitations such as zeolite A often stress shape selectivity.

For most hydrocarbon transformations, the alkali metal ion-containing forms of the zeolites are comparatively inactive. Active catalysts are prepared by exchanging varying amounts of the originally present Na^+ cations (the most common cation during synthesis) with ammonium ions, polyvalent cations from Group IIA (Mg^{2+} . . . Ba^{2+}) or the rare earths (La^{3+}, Ce^{3+} . . .), di- or trivalent transition metal ions, or combinations of these. The exchanged product must then be thermally activated. The minimum role of activation is removal of adsorbed water, which strongly solvates the cations, and localization of the cations on preferred sites on the lattice. However, other structural changes which are closely related to catalysis may also occur.

The structural modifications occurring during activation are a sensitive function of not only time and temperature but also of the time–temperature profile (programmed *vs.* flash heating), the atmosphere (vacuum, inert gas, air), and even the sample geometry (*44, 45, 46*). Since activation procedures from one laboratory to another are unfortunately seldom identical, ambiguity is introduced into interpretations in comparing various authors' work. The problem may be compounded when comparing catalytic with spectroscopic results because, although the starting hydrated zeolite samples may be identical, the activation conditions seldom are.

Ammonium- and Hydrogen-Exchanged X and Y Zeolites. SPECTROSCOPIC STUDIES—BRØNSTED AND LEWIS ACIDITY. Heating NH_4^+-exchanged Y zeolite at 250°–400°C in an inert atmosphere leads to stoichiometric evolution of NH_3, loss of adsorbed H_2O, and simultaneous appearance in the IR spectrum of O—H stretching vibrations, one sharp band at higher frequency (HF) (~ 3650 cm^{-1}) and a broader band at lower frequency (LF) (~ 3550 cm^{-1}). This process thus produces a hydrogen ion-exchanged Y zeolite, commonly called HY, which cannot be achieved by direct ion exchange because of the sensitivity of the Y zeolite structure to aqueous mineral acids. On the basis of considerable evidence from IR studies (*47, 48, 49, 50, 51, 52*), broadline NMR spectroscopy (*53*), and x-ray analysis

(*54*), the HF band results from an O—H group on O_1 of the lattice (structure **20**) which has a normal O—H bond distance and extends into the large cavity. Many workers have written structures containing a silanol O–H

$$NH_4^+ \quad [O_2Si(O)-O-Al^-(O)O_2] \xrightarrow{\Delta} [O_2Si(O)-O^+(H)-Al^-(O)O_2] \quad (13)$$

20

group slightly perturbed by an adjacent trigonal Al center, but structure **20** with only minor alterations in Si—O and Al—O bond lengths (*54*) seems correct. The LF band is assigned to an O—H group on O_3, with possibly smaller contributions from O_2 and O_4 (*47*), and thus is not directly accessible from the large cavity. The intensity of the HF band increases linearly with the degree of NH_4^+-exchange in NH_4Y (*47, 51, 55*); that of the LF band increases much more slowly up to ~ 50% exchange but then increases more rapidly than the HF band so that at high exchange they are of comparable intensity. It appears that the O_1—H group responsible for the HF band which extends into the large cavity appears first because O_3 atoms are coordinated to remaining Na^+ ions; the dramatic growth of the O_3—H band coincides with the disappearance of the most difficultly exchanged Na^+ ions (*56, 57*).

Only the HF band is perturbed by adsorption of nonpolar gases (*58*), again consistent with its assignment to the accessible large cavity where it could potentially interact with catalytic substrates. Adsorption of pyridine (Py) at room temperature and low pressure on NH_4Y samples activated at $< 500°C$ destroys the HF band and forms new IR bands characteristic of the protonated pyridinium ion (*50, 59, 80*). With stronger bases such as piperidine (*49*) and ammonia (*48, 61*), both O—H bands disappear. Thus, the O—H group responsible for the HF band can function as a Brønsted acid (reverse of Step 13); strong enough bases can also be protonated by the inaccessible LF O—H group probably by proton mobility through the lattice (*51, 54*). However, a highly exchanged HY in the R_3NH^+ form for bases as large as piperidine would have an unusual structure since presumably none of the cations could reach the most favored cation sites.)

The intensity of the HF band decreased somewhat as the temperature of the IR observation increased from ambient to ~ 400°C (*62, 63, 64, 65*), attributed to significant proton delocalization at high temperature implying that this would enhance Brønsted acidity at catalytically significant temperatures. However, recent considerations by Schoonheydt and Uytterhoeven (*65*) question whether a jump frequency high enough to be significant on the IR time scale could be consistent with known diffusion coefficients of protons in zeolites (*66*), and these authors attribute intensity changes to increased thermal motion of the lattice.

If the HY product is heated further at 500°–600°C, the O–H bands disappear from the IR spectrum and H_2O is evolved; hence, some oxygen atoms must be removed from the lattice without totally collapsing it. If this dehydroxylated material is treated with Py, IR bands characteristic of Py coordinately bound to electron-deficient Lewis acids (*e.g.*, $AlCl_3$) appear (*49, 50, 59, 60*). This coordinately bound Py is more difficult to desorb thermally than that bound to Brønsted sites. The process occurring has been traditionally written as in Reaction 14. Lewis acid sites are indicated by the spectral criterion and they would reasonably be trigonal Al atoms;

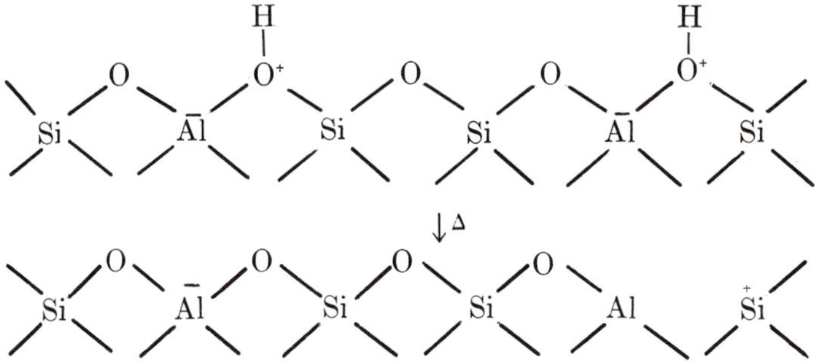

$$\mathrm{Si{-}O{-}\bar{A}l{-}O^{+}(H){-}Si{-}O{-}Si{-}O{-}\bar{A}l{-}O^{+}(H){-}Si} \xrightarrow{\Delta} \mathrm{Si{-}O{-}\bar{A}l{-}O{-}Si{-}O{-}Si{-}O{-}Al \quad \overset{+}{Si}} \qquad (14)$$

however, the evidence for trigonal cationic Si is minimal, and the exact structure of dehydroxylated Y is unclear. We therefore use the terms dehydroxylated Y and Lewis acidity operationally to signify the product formed on loss of O—H groups and as the resulting site which coordinately binds nitrogeneous bases. Although Py binds to both types of sites, the more sterically hindered base 2,6-dimethylpyridine is bound selectively by Brønsted acid sites (*47*).

More recently it has been found that if NH_4Y is heated at 500°–700°C under conditions where the evolved H_2O is not rapidly removed from the sample (*e.g.*, a thick bed in a static atmosphere rather than, say, a thin bed under vacuum), a so-called ultrastable product results (*45, 67, 68*) in which some Al atoms are removed from the lattice and enter cationic sites.

ROLE OF ACTIVATION CONDITIONS. Deciding which of these structural features—Brønsted acidity, Lewis acidity, or ultrastabilization—might be responsible for catalysis by activated NH_4Y depends on a detailed knowledge of zeolite structure as a function of activation conditions. Many studies (*45, 48, 50, 52, 61, 69, 70, 71, 72, 73, 74, 75*) of this problem have used spectroscopic and thermal analysis methods, and a number of generalizations have emerged along with qualifying reservations. (1) Activation at 300°C is sufficient to dehydrate NH_4Y and convert most of it to HY. However, small amounts of NH_4^+ may remain to higher temperatures (*76*). Thus, if the population of Brønsted sites has a range of acid strengths (*see* below), the most acidic few which might be most effective in catalysis might still be largely poisoned, *i.e.*, bound to NH_3. (2) In the range 300°–450°C, the population of O—H groups observed by IR passes through a shallow maximum whose position is a sensitive function of degree of exchange and exact activation conditions. Also, this relative constancy of the major O—H population could mask events of minor stoichiometric but major catalytic significance, such as the deammoniation of the most acidic Brønsted sites and the initial appearance of Lewis sites. These phenomena may be related since there is some evidence that the presence of Lewis acid sites may strengthen the acidity of neighboring Brønsted sites. (3) Beyond 500°C, the population of Brønsted sites falls off rapidly while that of Lewis sites increases, approximately in the ratio of one Lewis site from two Brønsted sites predicted by Reaction 14. However, assessing the relative contributions of dehydroxylation and ultrastabilization in this region is difficult, especially for early work when the latter phenomenon was not recognized. Again, the importance of exact activation conditions must be stressed because both deammoniation and dehydroxylation (*77*) are probably more rapid in vacuum than at atmospheric pressure and deammoniation can be promoted by O_2 (*78*).

A brief historical outline of studies of catalysis by NH_4Y as a function of activation temperature (T_{act}) follows because a large number of the proposals concerning active sites have arisen from this work. Venuto and co-workers (*78*) studied the alkylation of benzene with ethylene to form ethylbenzene at 177°C over 90%-exchanged NH_4Y activated in flowing O_2; (the formation of some N_2 indicates an oxidative deammoniation possibility). Conversion commenced only for $T_{act} \geq$ 400°C, maximized at 600°C, and disappeared again at ~ 700°C. Although this paper (*78*) stressed largely the important concept that the catalysis was intracrystalline, in related work on alkylation of aromatics a role for O—H groups acting as protonic (Brønsted) acids was suggested by the same group (*79*), based on this activity–T_{act} profile. The fundamental concept of a Brønsted active site is that substrates such as olefins and aromatics can be protonated to form carbonium ions. This suggestion was in contrast to even earlier proposals that the active sites were the electrostatic fields (*4, 80*) in polyvalent ion-exchanged zeolites or were Lewis acid sites acting as electron acceptors to form cation radicals (*81, 82*) and follows earlier implications of the role of Brønsted acidity in work by Hirschler (*83*).

Benesi (*70*) studied transalkylation of toluene to form benzene and xylenes at 400°C over 90%-exchanged NH_4Y and observed a similar activity–T_{act} profile; for $T_{act} =$ 400°C, activity was easily detectable; at ~ 600°C, it had approximately doubled and maximized; and at $\geq$ 700°C, it had virtually disappeared. Based on parallel TGA studies, Benesi (*70*) concluded that activity resulted from Brønsted sites while Lewis sites alone, surely present at $T_{act} =$ 700°C, were inactive. However, addition of H_2O at 400°C to a 700°C-treated sample restored activity to a level comparable with $T_{act} =$ 600°C. Thus dehydroxylation could apparently be reversed with conversion of inactive Lewis sites to active Brønsted sites. Also, loss of activity for T_{act} ~ 700°C was not associated with irreversible structural collapse. In the first of his extensive series of papers, Ward (*50*) also associated activity for cumene cracking with Brønsted rather than Lewis acidity. However, neither he (*50*) nor Benesi (*70*) commented on the fact that both Benesi's (*70*) and Venuto's data (*78, 79*) showed increasing activity for the range $T_{act} =$ 400°–600°C whereas total O–H population measured by IR was decreasing, especially over the upper end of the range—*i.e.*, the maxima in catalytic activity and total O–H population as a function of T_{act} did not coincide (if we can assume that the spectroscopic and catalytic specimens had indeed experienced identical activation conditions).

Hopkins (*72*) studied cracking of *n*-hexane and *n*-heptane at 350°C over 74%-exchanged NH_4Y and reported a sharp activity maximum (as measured by C_3 production) at $T_{act} \sim 550°C$ for 1 hr activations in flowing helium; however, for 16-hr activations, the sharp upward slope in the activity–T_{act} curve for T_{act} = 350°–550°C was replaced by a flat line at the maximum. Confusion may thus arise if all variables during activation, including time, are not specified when comparing work. Hopkins (*72*) concluded that only a small fraction of the O—H groups observed by IR were acidic enough to function as active sites and that these are the most difficult to deammoniate. He also suggested possible synergism between Brønsted and Lewis sites during catalysis by various mechanisms: O—H groups may be made more acidic by neighboring Lewis sites, or both sites may actually interact with reactants. The former idea was promoted by Lunsford (*84*) from consideration of earlier catalytic studies and his own ESR studies of interaction of NO with ^{27}Al nuclei at dehydroxylated sites. In this proposal, neither an O—H group in a pure HY nor a dehydroxylated site is active alone; O—H groups become strong enough Brønsted acids to be catalytically effective only when subject to electron withdrawal by neighboring electron-deficient centers as in **21**. The maxima in catalytic activity for NH_4Y at

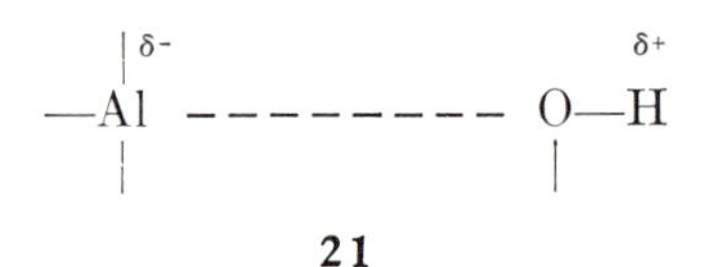

21

T_{act} = 500°–600°C result because activity is limited by the number of Lewis acids at $T_{act} < 500°C$ and by the number of O—H groups at $T_{act} > 600°C$. Lunsford's treatment (*84*) is similar to that of Richardson (*85*) who earlier had proposed acid-strengthening effects on O—H groups by polyvalent exchanged cations. Still another example of such a profile is offered by Hickson and Csicsery's study (*73*) of the simultaneous isomerization and transalkylation of 1-methyl-2-ethylbenzene (both activation and reaction were performed in H_2).

Turkevich and Ono (*16, 86*) reported effects of T_{act} in flowing He for 52%-exchanged NH_4Y for several different catalytic reactions and observed that the profiles depended on reaction type. Thus, cracking of cumene and isomerization of *o*-xylene were concluded to be catalyzed by Brønsted acidity; cracking of 2,3-dimethylbutane by Brønsted plus Lewis acidity (or by a Brønsted acid-Brønsted base pair); and cracking of 2,3-dimethyl-1- and

2-butene by either. An unspecified but direct involvement of Lewis acid sites in the catalytic mechanism rather than an indirect Brønsted acid-strengthening role was suggested. Whatever the exact meaning in terms of organic intermediates, this work points out that any discussion of active sites must be in the context of specific reactions. Thus, one type of site might be conducive to one type of chemical transformation, and a second type to another. For example, a plot of activity for isomerization of cyclopropane to propylene at 70°C over 50%- or 90%-exchanged NH_4Y *vs.* T_{act} shows two maxima (*87*), one at 300°–400°C and another at ~ 650°C. This was taken as evidence for two separate mechanisms of isomerization, one dependent on Brønsted sites and the other on dehydroxylation.

Declerck, Vandamme, and Jacobs (*88*) measured rates of double bond shift in 1-butene over 70%-exchanged NH_4Y both in a recirculating and flow reactor with particular emphasis on extrapolating reaction rates to zero conversion to account for rapid catalyst aging (which is particularly severe for low-temperature reactions of olefins). They proposed that activity at any point in time is made up of two components having different aging rates: one which can be poisoned by Py, falls off rapidly as T_{act} passes 500°C, and ages rapidly; and a second which increases slightly as T_{act} increases from 400° to 600°C and ages more slowly. The first appears related to Brønsted acidity and predominates at $T_{act} < 500°C$ while the second is possibly related to Lewis sites. Because of the differing aging rates, the activity–T_{act} profile at steady state shows the familiar maximum at $T_{act} \sim 600°C$. Whether or not the specific suggestion is correct and has any generality, this study serves as a warning of the potential dangers of observing activities after considerable aging has occurred. For example, results based on products accumulated for some period from an integral flow reactor could be different from those based on the first pulse from a pulsed microreactor.

A conflicting opinion was offered by Pickert and co-workers (*89*) who felt that the data for NH_4Y could be rationalized by the following postulates: (1) Brønsted acidity does not contribute to activity; (2) dehydroxylated sites do induce activity by a polarization mechanism and the often increasing activity over the T_{act} range 300°–600°C is related to their increasing population; and (3) the fall-off in activity for $T_{act} > 600°C$, where Lewis acidity does not decrease, results not from decreasing Brønsted acidity but from loss of crystallinity. The last statement, however, contradicts several claims of retention of crystallinity and reactivation by H_2O.

In summary, the case for involvement of O—H groups as Brønsted acids seems solid. Whether all of them are involved cannot really be deduced from the work outlined so far, particularly since many of the reported increases in activity as T_{act} increased from 300° to 500°C may simply have resulted from failure to allow enough activation time at the lower T_{act} (*15*). The evidence concerning ranges of acidity and numbers of active sites will be given below.

ROLE OF EXCHANGE LEVEL. A second probe to relate zeolite structure to active sites is the effect of degree of ion-exchange of Na^+ by the cation in question. Ward and Hansford (*51*) reported a detailed spectroscopic and catalytic study of NH_4Y as a function of exchange level, all samples being activated at 480°C in flowing He. As already noted, the HF O–H band increased almost linearly with NH_4^+ level while the LF band increased much more slowly at first but then caught up past 60% exchange. The number of Brønsted sites as measured by IR of chemisorbed Py at room temperature also increased linearly up to ~ 60% exchange but then leveled out at higher exchange levels. The catalytic activity for *o*-xylene conversion (% conversion at 250°C in a flow reactor) increased gradually from 0 to ~ 8% as NH_4^+ exchange increased from 0 to ~ 60%, but then it increased much more rapidly up to ~ 45% conversion at > 93% exchange. This profile was most similar to that of the IR intensity of the LF O–H group which is inaccessible to Py and is not included in the counting of Brønsted acids. There is therefore an indication that the catalytic effectiveness per measurable accessible Brønsted acid site increases as exchange level increases; *i.e.*, the last NH_4^+ ions introduced contribute little to the formation of pyridinium ions but greatly to activity (and to formation of the LF band). Thus, while Brønsted acidity is strongly implicated as contributing to catalysis, a simple 1:1 correspondence between activity and number of O–H groups does not pertain. Dempsey and Olson (*56, 57*) suggested that the observed activity–exchange level profile results because Na^+ ions in the large cavity act as poisons so that activity begins to increase sharply at the NH_4^+ level such that all Na^+ ions in the activated product can be accommodated in favored inaccessible cation sites. A selective poisoning would again require a heterogeneity of acidity. However, one experimental uncertainty in precise interpretation of these results is that the constant T_{act} used may have been removed from the optimum T_{act} for each level of exchange to differing degrees if the optimum itself is a function of exchange level. Thus, it could be possible, for example, that the more highly exchanged samples with higher concentrations of O—H groups were subject to relatively more (although still minor) dehydroxylation at 480°C and therefore possessed relatively more strongly acidic Brønsted sites.

Earlier indications of increasing activity per site in activated NH_4Y with increasing site density were found by Turkevich and co-workers (*82*) in studies of cumene cracking at 325°C. The number of sites was determined by progressive poisoning by quinoline (*see* below) of a series of NH_4Y catalysts with different extents of exchange, all activated at 500°C in vacuum. A plot of relative activity per site *vs.* number of sites curved sharply upward although the sites were then considered to be Lewis rather than Brønsted acids, either of which could coordinate with quinoline.

Tung and McIninch (*90*) studied the competitive isomerization and cracking of *n*-hexane (increasing reaction temperature favors the latter) over a series of NH_4Y with 15–93% exchange, all activated at 550°C in flowing air, a condition which should lead to a mixture of O—H groups and dehydroxylated sites (*77, 91*). Plots of activity for isomerization *vs.* % exchange showed maxima rather than steadily increasing activity with increasing exchange level. The maximum at 300°C reaction temperature occurred at ~ 85% exchange and, at 450°C, at ~ 35% exchange. Similar plots for cracking at 400° and 450°C also showed maxima. However, these data are highly distorted by the effects of catalyst aging since the reactions at higher temperatures were run on catalysts already used at lower temperatures. Activity was not restored by recalcining in air (*cf.* Ref. *92*) in contrast to the behavior of polyvalent cation-exchanged zeolites. These authors introduced three mechanistic proposals. (1) Brønsted acidity is effective for a variety of unsaturated substrates, but activation of paraffins is a special case which demands Lewis sites or electrostatic field gradients. (2) Deactivation results not from coking or loss of crystallinity but from an unspecified diffusional reconstitution of the surface to a lower energy state which is somehow induced by the presence of substrates. The possible deleterious effects of steaming decationized Y were not evaluated. (3) Migration of oxide ions between vacancies resulting from dehydroxylation will lead to an effective migration of Lewis acid sites. Thus, the Lewis acidity at any point in the lattice where a reactant and/or product is adsorbed will be time variant. These workers also proposed that time variant electrostatic field gradients (*93*) and Brønsted acidity (*94*) could be induced by cation migration. Such time-variance was supposed to allow both strong adsorption to promote reaction and facile desorption to allow product escape. It is, however, not obvious how any of these hypotheses follow from or are demanded by the data presented.

ACIDITY DISTRIBUTIONS. To evaluate the strength of Brønsted acidity and its role in catalysis, measurements with a moderately strong base Py are less useful than measurements with very weak bases, which would serve as better models for olefins and aromatics. Several studies have appeared recently using the Benesi technique (*95*) in which a solid acid is incrementally titrated with *n*-butylamine in a non-aqueous solvent with Hammett indicators or triarylcarbinol indicators (*96*) which generate carbonium ions on protonation. Following earlier indications (*97*) that HY contained a range of acidity with $+3.3 > H_0 > -8.2$ (note that more negative H_0 values indicate greater acidity), Beaumont and Barthomeuf (*98, 99, 100*) reported in detail on acidity titrations of NH_4Y activated at 550°C as a function of exchange level. A plot of equivalents of total acidity (more acidic than 3×10^{-4}% H_2SO_4) per unit cell as a function of the number of Na^+ ions replaced by NH_4^+ was linear for all levels of exchange with a slope of $\alpha = 0.6$—*i.e.*, exchange of one Na^+ ion by NH_4^+ introduced 0.6 equivalent of titrable acidity. If the acidity were separated into that greater and less than $H_0 = -8$(88% H_2SO_4), exchange of the first 30–35% Na^+ ions gave only weaker acidity with $H_0 > -8$ whereas all acidity introduced at higher exchange levels had $H_0 < -8$. The parameter α was only 0.16 for HX zeolite (which is stable only at low exchange levels), and the authors proposed long-range acid-weakening interactions between Al tetrahedra such that a faujasite-type structure with Si:Al = 1:1 was predicted to have negligible acidity. They further proposed that the observed heterogeneity of acidity results from a heterogeneity of Al atoms in the lattice: 30–35% of the Al atoms in Y, easily extracted by chelating agents, are related to weak acidity while the other 65–70%, more tightly bound in the lattice, are related to strong acidity. The structural cause of this heterogeneity was not specified, but more recently Dempsey (*101*) rationalized that the weaker portion of the acidity is associated with those square faces (8-rings) which happen to contain two electrostatically repulsive Al atoms. A parallel study (*98*) of isooctane cracking at 465°C showed again that activity appeared relatively slowly at low exchange levels but increased much more rapidly past ~ 35% exchange, the point at which the stronger acidity appeared. A question remains whether these Hammett indicators measure only Brønsted acidity or both Brønsted and Lewis acidity (*102*). Determination of acidity as a function of T_{act} as well as % NH_4^+ exchange would have been informative. In this regard, Ikemoto and co-workers (*103*) reported that the total acidity with $H_0 < +3.3$ of 92%-exchanged NH_4Y (2.5 meq/g) did not vary for $T_{act} = 300°–600°C$ (vacuum) and that > 95% of this acidity was in the range $-8.2 > H_0 > -10.8$ (100% H_2SO_4); this may indicate that the sum of Brønsted and Lewis acidity is being measured.

Another similar study by Morita and co-workers (*102*) for NH_4Y of varying exchange level, activated at 450°C, gave an initial α value of ~ 0.75 which increased to ~ 1.0 as exchange increased; (such a variable α was first reported by the French group for samples activated at 380°C (*99*) but later altered to 0.6 for samples apparently activated at 550°C (*100*)). Again, increasing the exchange level from 25 to 65% increased the acid strength as well as the total acidity. However, in contrast to the French workers (*98, 99, 100, 101*), this group (*102*) claimed that the last few NH_4^+ ions introduced gave largely very weak acid sites. A plot of total acidity with $H_0 \leq +3.3$ *vs.* T_{act} (300°–700°C) for 90%-exchanged NH_4Y showed a distinct maximum at $T_{act} = 450°C$, in contrast to Ikemoto's (*103*) report, and the maximum value of 4.5 meq/g was greater than Ikemoto's (2.5 meq/g). However, the stronger acidity ranges of $H_0 \leq +1.5$, $H_0 \leq -3.0$, and $H_0 \leq -5.6$ were almost constant at 2.5 meq/g for $T_{act} = 350°–600°C$. (The theoretical maximum for a 100% HY of empirical formula $(H_2O)(Al_2O_3)(SiO_2)_{4.8}$ is 4.4 meq/g). A plot of activity for benzene alkylation with ethylene at 300°C (maximum conversion between an induction period and a deactivation period) *vs.* exchange level for a given T_{act} again showed the now familiar upward curvature. For the 90%-exchanged sample, activity as a function of T_{act} showed a maximum at 450°C, of shape similar to that of the total acidity. However, for the cracking of cumene at 280°C, extrapolated initial activities were almost independent of T_{act} from 450°–550°C and hence correlated not with the total acidity but only that fraction more acidic than $H_0 = -3.0$. Finally, for *o*-oxylene isomerization and transalkylation at 350°C, the two competitive reactions showed different profiles for activity *vs.* exchange level. Transalkylation correlated with total acidity and became dominant at high exchange level while isomerization appeared to correlate with $H_0 \leq -3.0$.

Boudart and co-workers (*104*) measured the adsorption isotherms of the strong base NH_3 on 70%-exchanged NH_4Y, activated at >360°C, at various temperatures from 25°–392°C. The isosteric heat of adsorption fell gradually from ~37 to ~11 kcal/mole as surface coverage increased—a result indicative of heterogeneous acidity. Earlier, Venuto and co-workers (*61*) had observed two phases of NH_3 loss during temperature-programmed desorption of NH_3 from a reconstituted NH_4Y (HY treated with anhydrous NH_3 at ambient conditions) with activation energies of 22 and 15 kcal/mole.

Moscou (*105*) has presented a chemical methodology claimed to measure quantitatively the various forms of O–H groups. Treatment of a zeolite with lithium aluminum hydride gives H_2 evolution from any accessible H_2O molecules or from any accessible acidic O—H groups. Karl Fischer titration determines only H_2O. Hence, the difference between these values gives the number of acidic groups in the large cavity.

In summary, these various reports all indicate some degree of heterogeneity of acidity for activated NH_4Y. However, significant discrepancies (*98, 102, 103*) exist as to exactly how the amounts and ranges of acidity behave as a function of exchange level and T_{act}. The attempted correlations of activity with measured acidity (*98, 102*) seem to be a useful approach, but more data will be required to solve the inconsistencies. Incidentally, these studies have not revealed acid strengths in the HY greater than those in amorphous silica-alumina.

ROLE OF SI:AL RATIO. So far we have discussed exclusively the Y zeolite with Si:Al $\sim$ 2.5. Much less information is available for NH_4X, partially because it is much less thermally stable, particularly at high exchange levels. It develops less Hammett acidity (both total and strong) on activation (*98, 103*), in spite of the greater initial NH_4^+ content which can be introduced because of the lower Si:Al ratio. In general, HX catalysts are less active than HY. For example, Tsutsumi and Takahashi (*106*) cracked cumene over a series of highly-exchanged NH_4^+ X and Y zeolites, all activated at 450°C in flowing He. For all reaction temperatures between 300° and 450°C, activity increased as Si:Al increased through the values 1.25, 1.6, 1.9, 2.3, and 2.5. For higher T_{act} values, activity persisted for the Y members to $T_{act} \sim$ 750°C but disappeared at $\sim$ 550°C for zeolite X. Activity for cracking isooctane at 465°C has been reported (*98*) for a series of NH_4X and NH_4Y zeolites, activated at 465°C in hydrogen, as a function of extent of exchange. Between 0 and 40% exchange, gradually increasing activity of the same magnitude appeared for both series. However, beyond 40% exchange, activity of the Y zeolite continued to increase while that of the X declined. Unfortunately, the available literature does not allow a definitive choice whether the lower acidity and activity generally observed for HX results from inherently less acidic O–H groups related somehow to Si:Al ratio (*98*), from easier structural collapse during activation, or from more facile dehydroxylation.

Treatment of Y zeolites with chelating agents such as acetylacetone or EDTA increases the Si:Al ratio by extracting Al and serves as one method of ultrastabilization. A series of extracted $Na(NH_4)Y$ zeolites with varying Na^+ and Al content has been tested for isooctane cracking (*98*). So long as Al removal did not exceed 33% (37–38 Al remaining per unit cell), activity plotted *vs.* Na^+ ions remaining per unit cell fell on the same curve as for the parent $Na(NH_4)Y$. However, removal of further Al decreased activity.

These results provided input for Barthomeuf's conclusion (*see* above) that ~ 33% of the Al atoms are easily extracted and do not generate strong acidity and high activity when exchanged with NH_4^+ ion and deammoniated.

The Japanese group (*107*) extended their studies of the effect of Si:Al ratio on cumene cracking beyond 2.5 by extraction with EDTA. Activity continued to increase up to a ratio of ~ 4 as did the quantity of Brønsted acidity detectable by Py adsorption. Thus, even though the content of Al tetrahedra is decreasing, the number of accessible O–H groups appears to be increasing—a result not predicted from first-order structural considerations.

Ammonium- and Hydrogen-Exchanged Synthetic Mordenite and Other Zeolites. The mordenite structure has a higher Si:Al ratio (~ 5) than Y and a pore structure consisting of parallel two-dimensional channels whose minimum diameter (~ 6 × ~7 A) is smaller than the minimum opening between the large cavities in Y (~8 A). Heating NH_4^+-exchanged mordenite leads first to deammoniation followed by dehydroxylation. TGA studies (*70*) and DTA studies (*72*) suggest that both processes require somewhat higher temperatures than for NH_4Y and that the processes may be less separable. Therefore, the preparation of a nitrogen-free H-mordenite without the onset of dehydroxylation may be more difficult than for HY. For examples, samples of mordenite containing both protons (*see* below) and NH_4^+ ions retained measurable N even after calcining for 8 hr at 525°C in air (*108*).

Because of its greater stability to aqueous acids, H^+-exchanged mordenite can be prepared by direct exchange in contrast to X and Y zeolites. However, since treatment with aqueous acids also tends to extract Al from the lattice (*see* below), the route *via* NH_4-mordenite is useful.

Hydroxyl absorptions in H-mordenite occur at 3650 and 3610 cm^{-1} and decrease for $T_{act} \geq 450°C$ (*109*). Since the 3610 cm^{-1} band interacts as a Brønsted acid with Py while the 3650 cm^{-1} band does not, the former has been assigned to the main channel and the latter to the side pockets. However, reports concerning the proportions of Brønsted and Lewis sites on the basis of IR studies of adsorbed Py are contradictory. Karge (*109*) reported Lewis acidity already at $T_{act} = 300°C$ while Yashima and Hara (*110*) reported no Lewis acidity for $T_{act} = 500°C$. Cannings (*111*) claimed the presence of two resolvable Py-Lewis acid bands (*cf.* Ref. *112*). In any case, Karge (*109*) presented plots of intensity of the 3610 cm^{-1} band, of the pyridinium ion band, and the catalytic activity for ethylation of benzene at 102°C as a function of T_{act}, all of which coincided almost exactly. Hammett-type acidity measurements would be useful but might be difficult because of greater diffusional resistance to the indicators.

A variety of catalytic reactions (*113*) have been reported in which the activity of H-mordenite, prepared usually by direct exchange but sometimes through the NH_4^+ form, is greater than that of HY. For example, Benesi (*70*) compared activity–T_{act} profiles for highly exchanged NH_4-mordenite and NH_4Y. For transalkylation of toluene at 400°C, the maximum conversion over the mordenite was 8–10 times greater than that over the Y and occurred at $T_{act} = 700°C$ compared with 600°C for the Y (*cf.* also Ref. *110*). For cracking of *n*-butane and *n*-pentane at 400°C, large conversions were achieved over NH_4-mordenite under conditions where NH_4Y was inactive. The loss in activity which occurred when T_{act} was increased from 700° to 800°C could be fully restored by small pulses of H_2O, indicative of reversible dehydroxylation. Coupling this behavior with the observed poisoning by NH_3, Benesi (*70*) proposed Brønsted acids as the active sites with a hint of a range of acidity since one-fifth the amount of NH_3 needed to re-form NH_4-mordenite was sufficient for complete poisoning. In similar studies of *n*-hexane cracking at 350°C, Hopkins (*72*) also reported increasing activity of NH_4-mordenite by increasing T_{act} from 350° to 500°C; however, the activity of H^+-exchanged mordenite was constant over this T_{act} range. Such behavior supported the need to remove all NH_3 from Brønsted sites to unmask the maximum possible activity. In contrast, Becker and co-workers (*114*) claim that activation in vacuum must reach at least 450°C for H-mordenite to reveal maximal activity to alkylate benzene with ethylene at 102°C; the reason advanced was incomplete dehydration, as indicated by H_2O bands in the IR at less strenuous activation conditions. This apparent inconsistency may have resulted either from a shorter activation time by Becker (*114*) or from Hopkins' (*72*) very high conversions which masked activity differences.

In a survey of the ability of a wide variety of zeolites, activated at 480°C, to catalyze isomerization of *o*-xylene, Hansford and Ward (*2*) found H-mordenite from either source to be the most active of all; based on first-order rate constants at 260°C, it was almost 20 times more active than HY. Beecher and Voorhies (*115*) discovered that H-mordenite was again more active than HY for the hydroisomerization of *n*-pentane and *n*-hexane both with and *without* added Pd metal. However, the activity order was reversed for *n*-heptane—possibly an example of a generalization (*113*) that the superior activity of H-mordenite compared with HY is not manifested with high molecular weight feeds. The greater susceptibility of mordenite to diffusional limitations and aging with time-on-stream (*see* below) may be responsible. The unique ability of H-mordenite to hydroisomerize paraffins in the absence of noble metal was also demonstrated by Minachev and co-workers (*116*). A series of Na^+-mordenite samples, exchanged to varying degrees with NH_4^+ ion and activated at 520°C in air, generated a linear plot of percent conversion of cyclohexane to methylcyclopentane *vs.* exchange level. The authors felt this indicated uniform surface catalytic activity and therefore uniform acidity. Since this contrasts with the consensus for HY, more studies of this question would be useful.

Whether the greater activity of H-mordenite is related to relatively stronger Brønsted acidity as Si:Al increases which more than offsets the smaller number of O—H groups remains to be convincingly proved; however, at present, it is the most obvious structural rationalization.

The activity to crack cumene in a pulsed microreactor was compared for H-mordenite and HY as a function of the number of pulses (*117*). For a number of pulses for which the activity of HY did not change, that of H-mordenite had decreased more than half. This behavior is exemplary of the tendency of mordenites to deactivate more rapidly. The rationale usually proposed is that polymerization of olefins and resulting formation of non-volatile residue ("coke") in the catalyst pores occurs on all acidic zeolites but has more serious results in mordenite because of the smaller size and two-dimensional nature of its pore structure. Formation of residue leaves portions of the active surface inaccessible to the reactants.

A number of workers have considered the effects of progressively leaching Al from H-mordenite by treatment with aqueous acids. Interpretation in terms of structural effects is difficult, however, because of the lack of detailed knowledge of how the structure is modified by progressively increasing Si:Al ratio. Weller and Brauer (*118*) noted that *n*-hexane cracking activity passed through a maximum at Si:Al = 9. A similar pattern—*i.e.*, increasing Si:Al ratio from the normal 5–6 increased activity up to a ratio near 10 but then decreased it beyond that—was observed by Pigusova and co-workers (*119*). Such maxima may represent the operation of two opposing effects: increasing ratio increases acid strength per site but decreases the number of sites. Recent work by Weiss' group (*120*) shows, however, that care must be taken in drawing such conclusions. Normal H-mordenite containing 11.2% Al_2O_3 was compared with an extensively extracted sample containing only 0.1% Al_2O_3 for the cracking of cumene at 360°C. Depending on reactor type and reaction conditions, the sample with the extremely high Si:Al ratio ($\sim$ 500) could be judged less, equally, or more active than the parent. With highly diluted catalysts operating as a differential reactor, the initial fractional conversions were 0.002 and 0.018, respectively. Thus, under such conditions which should be most conducive to determining intrinsic activities, the extracted catalyst was less active, although its activity is impressive considering that it is almost pure silica. With undiluted catalysts operating in an integral reactor, the initial fractional conversions were 0.385 and 0.410, respectively. This similarity was proposed to result from transport limitations rather than from considerations involving active sites. Finally, if results from the integral reactor were compared after some time-on-stream (100 g cumene per g catalyst), the fractional conversions were 0.36 and 0.24, respectively, so that the extracted catalyst appeared more active because it had aged more slowly. These authors (*120*) finally conclude that the maxima observed by others in activity *vs.* Si:Al ratio plots were largely the result of decreased diffusional resistance created during the dealumination process rather than being directly related to the number or strength of active sites. Eberly and Kimberlin (*121*) came to the same conclusion for studies of cumene cracking.

A further recent paper from Weller's group (*108*) demonstrated greater diffusivity of acid-extracted H-mordenite by measuring chromatographic peak broadening of a pulse of N_2 in He. However, since exchange of the H-mordenite with NH_4^+ ion (which removes most of the few residual Na^+ ions) followed by calcining in air at 500°C also increased diffusivity, the increases caused by acid extraction were felt to be related more to removal of residual Na^+ ions from the channels than to removal of Al from the lattice. The study of *n*-hexane cracking was repeated in a pulsed rather than continuous (*118*) reactor, and results were fitted to a first-order reaction with exponential decay. The initial rate constants for extracted H-mordenites containing 9.08, 3.64, and 2.31% Al_2O_3 (and 0.77, 0.11, and 0.16% Na_2O) were 40.2, 7.8, and 2.5 g^{-1}, respectively, *i.e.*, the decrease in active sites outweighs some increase in diffusivity and possibly a lesser degree of poisoning by residual Na^+ ions.

Catalytic activities reported for H-mordenites must thus be examined with care to determine whether the results are tempered by aging and/or diffusional limitations. However, since each of these is more severe for H-mordenite than for HY, the often-reported greater activities for the former in one-to-one comparisons may have been underestimated rather than overestimated. Hence, the final conclusion of greater activity, and therefore possibly greater acid strength in spite of the lower number of acid sites, of H-mordenite still holds.

Active catalysts from zeolites of the erionite-offretite-Linde T group have also been reported (*122, 123, 124*), but the emphasis has been on phenomena of shape selectivity and unusual diffusional properties. Proton-containing catalysts derived from zeolites omega (*125*), clinoptilolite (*126*), and L (*72*) have also been reported.

Polyvalent Cation-Exchanged Zeolites. ELECTROSTATIC FIELDS AND ACIDITY. The first hypothesis concerning zeolite catalysis of hydrocarbon transformations arose from the initial reports by the Union Carbide group in 1960 (*4*). The skeletal isomerization of *n*-hexane under hydrogen pressure at 350°–400°C over zeolite catalysts loaded with 0.5% Pt was the test reaction. Attention focused on the observation that such carbonium ion-like activity could be induced into Y zeolite not only by decationization but also by exchange of Na^+ ions by polyvalent ions including Mg^{2+}, Ca^{2+}, Sr^{2+}, Zn^{2+}, Mn^{2+}, Ce^{3+}, Al^{3+}, Ce^{4+}, and Th^{4+}—ions previously considered to be catalytic poisons in silica-alumina cracking catalysts. Whereas activity appeared with activated NH_4Y containing as little as 10% exchange, exchange by Ca^{2+} ion led to sharply increasing activity only beyond ~ 40% exchange. Y zeolites were more active than their X counterparts. The postulate was therefore made that activity resulted from the uncompensated electrostatic charges set up because a divalent ion cannot associate optimally for bonding with two charge-balancing Al tetrahedra. This charge separation should increase as the average distance between Al atoms increases in going from X to Y zeolite. This electrostatic field mechanism was refined by Pickert, Rabo, Dempsey, and Schomaker (*80*) on the basis of further experimental data for *n*-hexane isomerization (with and without added Pt) and cumene cracking over catalysts calcined at 500°C and of more detailed structural considerations. The activity order was MgY > CaY >SrY > BaY > NaY ~ silica-alumina—*i.e.*, activity within Group IIA increased with decreasing ionic radius. Consideration of the three-dimensional structure of CaY (*127*) suggested strong charge localization because of the rigid framework compared with amorphous silica-alumina. By qualitative considerations, a Ca^{2+} ion in S_{II} of a faujasite-type zeolite with Si:Al = 2 would bear a full positive charge and the unfilled S_{II} sites a full negative charge. More quantitative calculations produced profiles of electrostatic potential *vs.* position in the faujasite lattice—*e.g.*, a molecule approaching 2 A from an S_{II} Ca^{2+} ion was calculated to experience a potential gradient of 6.3 volts/A. (The potential near an unoccupied site was lower; the field near the unoccupied sites was suggested as the active site by Isakov and co-workers (*128*)). Coulombic interactions between cations and adsorbates have been demonstrated by IR studies of adsorbed CO (*129*), CO_2 (*130*), and Py (*131*). They also contribute (*132*) to differences in heats of immersion of dehydrated zeolites in nonpolar (hexane) *vs.* polar (nitropropane) solvents. The observed catalytic results were then rationalized (*80*) by these considerations: (1) below ~ 40% exchange, all divalent ions enter the more favorable S_I and are not accessible to reactants in the large cavity; (2) beyond ~ 40% exchange, cations appear in S_{II} and their coulombic fields are intense enough to polarize C—H bonds ($\overset{\delta^+}{C}$—$\overset{\delta^-}{H}$) sufficiently to induce carbonium ion behavior; and (3) the fields increase as *e/r* increases so that the observed activity order for Group IIA cations is as predicted.

Catalytic performance of divalent ion-exchanged X zeolites was announced almost simultaneously by Weisz and Frilette (*5, 133*). For cracking of paraffins and cumene, CaX was more active than NaX and silica-alumina and gave typical carbonium ion products. NaX, on the other hand, was as active as silica-alumina for paraffin cracking but gave free-radical-like products (*see* below). This group at Mobil Oil stressed the operation of intracrystalline catalysis with zeolites by demonstrating the first examples of shape selectivity (*134*). For example, *n*-hexane was cracked over the small-pore CaA zeolite; however, its pore structure rejects branched alkanes, and indeed 3-methylpentane was not cracked. Thus, the role of the exterior of the crystallites is minimal compared with the intracrystalline area of 600–900 m^2/g.

IR investigations (*131, 135, 136, 137, 138, 139, 140, 141, 142*) of polyvalent ion-exchanged zeolites dehydrated at ~ 500°C revealed the presence of O—H groups not anticipated from simple structural considerations. A number of O—H bands have been observed, frequencies near 3690, 3650, 3600, and 3550 cm^{-1} being typical. The 3650 cm^{-1} band is again of particular interest because it acts as a Brønsted acid toward nitrogeneous bases just as the corresponding band in HY. The source of these structural O—H groups is now generally accepted to be hydrolysis of H_2O by the poorly shielded cations (Reaction 15) which occurs during dehydration as first suggested by Plank (*143*); (we use the shorthand notation O_z^{2-} to indicate a zeolite oxide ion, realizing that charges are not fully localized). This process reduces the

$$M^{n+} + H_2O + O_z^{2-} \rightarrow M(OH)^{n-1} + HO_z^- \quad (15)$$

effective charge of the cation and releases a proton which forms an O—H group at a nearby lattice oxygen. Plank (*143*) and Hirschler (*83, 144*) suggested that these hydroxylic Brønsted acids were the active sites just as in decationized zeolites. Ward (*131*) pointed out that the concentration of Brønsted sites as measured by IR of adsorbed Py on a series of Mg . . . BaY zeolites of 77–93% exchange, activated at 500°C in vacuum, correlated inversely with cation radius and directly with the calculated electrostatic field, as did several catalytic activities. Thus, there is indeed a catalytic role of the cation-generated electrostatic field but ". . . the first molecules affected by this high potential are the water molecules which are already present, prior to a catalytic experiment" (*145*) rather than the substrate molecules during catalysis. In other words, the role of poorly-shielded cations is to generate active sites rather than to act as such sites.

Bolton (*139*) has suggested that some of the O—H content of rare earth (RE)-exchanged Y zeolites results from direct exchange of Na^+ ions by H_3O^+ during the exchange with RE^{3+} since this operation must be performed under acidic conditions (pH $\sim$ 4.5) to avoid precipitation of hydrolyzed RE salts.

Discussion has arisen whether the IR band at 3600 cm^{-1} results from the O—H group coordinated to the cation as commonly supposed (*145*) or from dealuminated sections of the crystal (*141*). Jacobs and Uytterhoeven (*141*) have proposed that hydrolysis occurs by Reaction 16 rather than Reaction 15 to produce an oxide-bridged cation pair (*see* Ref. *146*), following

$$2\,M^{2+} + H_2O + 2\,O_z^{2-} \rightarrow (M^{2+}O^{2-}M^{2+}) + 2\,HO_z^- \qquad (16)$$

the earlier suggestion of Rabo and co-workers (*140*) that bridging hydroxyl groups are formed in the RE-exchanged cases (Reaction 17) and that such a phenomenon stabilizes the overall structure. Another suggested hydrolysis

$$2\,La^{3+} + H_2O + O_z^{2-} \rightarrow (La^{3+}\overline{O}HLa^{3+})^{5+} + HO_z^- \qquad (17)$$

route for trivalent ions is Reaction 18 followed by Reaction 19 between 300

$$M^{3+} + 2\,H_2O + 2\,O_z^{2-} \rightarrow M(OH)_2^+ + 2\,HO_z^- \qquad (18)$$

$$M(OH)_2^+ + HO_z^- \rightarrow M(OH)^{2+} + O_z^{2-} + H_2O \qquad (19)$$

and 500°C (*147, 148*). Whatever the exact hydrolysis chemistry, the formation of structural acidic O—H groups is certain.

Angell and Schaffer (*135*) found that the 3650 cm^{-1} IR band was more intense in several divalent ion-exchanged Y samples heated slowly to 500°C in vacuum than when heated rapidly. Thus, dehydration and cation hydrolysis appear competitive, and only a fraction of the cations present may lead to production of O—H groups during typical activation. The number of catalytic sites would thus be expected again to depend on details of activation procedure. Similarly, addition of H_2O to an activated CaY containing adsorbed Py at 260°C increased the observed pyridinium ion concentration (*60*). This effect of H_2O must be remembered particularly for catalytic samples which have been reactivated by burn-off and hence subjected to high-temperature steaming (*cf.* Ref. *149*). These authors (*135*) also noted a small dependence of $\gamma_{O\text{-}H}$ on cation—*e.g.*, NaY (somewhat cation deficient) 3652, BaY 3647, CaY 3645, MgY 3643, and HY 3636 cm^{-1}. Christner,

Liengme, and Hall (*137*) suggested cation electron affinity as the correlating variable. This idea was further refined in Richardson's proposal (*85*) that O—H acidity is influenced by polarizing effects of neighboring cations, parallel to the proposals discussed above for strengthening of O—H acidity by neighboring Lewis acid sites. Richardson (*85*) also proposed that the 3650 cm^{-1} band represented O—H groups of a range of acidity with higher acidity correlating with lower $\gamma_{O\text{-}H}$. However, there is some evidence (*see* below) that REY has stronger acid sites than HY, yet its $\gamma_{O\text{-}H} = 3640$ cm^{-1} (*140*) is higher.

Ward (*131*) found no Lewis acid sites on 500°C-activated alkaline earth ion-exchanged Y zeolites as judged by Py chemisorption although bands were seen attributed to coordination of Py to the cations. However, calcination at 650°C led to loss of O—H groups, loss of Brønsted acidity towards Py, and appearance of typical IR bands for Lewis acid-bound Py. Hence, a high-temperature dehydroxylation phenomenon exists, as earlier indicated by weight loss at 500°–1000°C (*80*), which converts Brønsted to Lewis acid sites, as for NH_4Y. Although its exact chemistry is no more clear than that of cation hydrolysis which precedes it, it must remove O from the lattice to form trigonal Al centers rather than simply be just the reverse of hydrolysis, Reaction 15.

Hirschler (*83*) first reported acidity strength and distribution measurements with H_0 and H_R indicators and found significant acidity for CaX, some corresponding to strength > 50% H_2SO_4, but none for NaX. More detailed measurements by Beaumont and co-workers (*99*) compared highly exchanged CaY and LaY with NaY. The number of acid sites per unit cell of strength $\geq$ 50% H_2SO_4 was 0 for NaY, ~ 15 for CaY activated at 550°C, ~ 20 for LaY activated at 300°C, and ~ 11 for LaY activated at 550°C. If CaY and LaY were back-exchanged with NH_4^+ ions and calcined, more acid sites developed. Ikemoto and co-workers (*103*) reported the total acidity of 85%-exchanged CaY to be 2.2–2.5 meq/g, almost all of strength $H_0 \leq -8$, barely dependent on T_{act} over the 300°–600°C range; this value would demand complete Ca^{2+} hydrolysis *via* Equation 15 for all Ca^{2+}. For CaX, both the number of total acid sites and particularly of those with $H_0 \leq -8$ were much lower. Highly exchanged LaY showed some acidity with $H_0 \leq -12.8$ whereas that of HY never exceeds $H_0 \leq -10.8$ (*cf.* Ref. *106*). Earlier results by Otouma and co-workers (*97*) also support the idea that polyvalent ion-exchanged Y zeolites have a narrower distribution of acid strengths than the corresponding X zeolites which favors stronger sites. REX- and REY-zeolites were studied by Moscou and Moné (*148*), but their indicators did not exceed $H_0 \leq -8$; in contrast to other reports for divalent ion-exchanged cases, the total acidity and number of strong sites were somewhat greater for REX (*cf.* Ref. *150*). Steaming at 600°–700°C reduced acidity.

ROLE OF ACTIVATION CONDITIONS. Hopkins (*72*) reported that activity for *n*-heptane cracking at 350°C over 72%-exchanged CaY fell from 8 moles C_3 product/100 moles *n*-heptane for $T_{act} = 350$°C in flowing He to negligible conversion beyond $T_{act} = 450$°C. Tsutsumi and Takahaski (*106*) studied cumene cracking over 92.9%-exchanged CaY (Si:Al = 2.5) and 94.3%-

exchanged CaY (Si:Al = 1.9). Maximal activity occurred for both catalysts at $T_{act} \sim 450°–500°C$ and decreased beyond that; however, for the higher Si:Al ratio material, some activity persisted even at $T_{act} \sim 800°C$. These workers reiterated the idea that dehydroxylation should become more difficult with increasing Si:Al ratio because the site density of O—H groups decreases. Rabo and Poutsma (*14*) studied cumene cracking and isotopic exchange between benzene and perdeuteriobenzene at 325°C over 83%-exchanged CaY; for both reactions, activity fell precipitously between $T_{act} = 550°$ and 710°C. Parallel studies of effects of T_{act} have appeared for RE-exchanged Y zeolites. For the *n*-heptane cracking (*72*) over 64%-exchanged LaY, maximum activity (~ 22 moles C_3 product/100 moles *n*-heptane) occurred at $T_{act} = 350°C$ and finally fell to zero at 650°C. In the cumene cracking and benzene isotopic exchange reactions (*14*), 86%-exchanged LaY was again more active than CaY at all T_{act} and activity decreased with increasing T_{act}. In one of the earliest reports of activity–T_{act} profiles, Venuto and co-workers (*78*) demonstrated maxima in such plots for alkylation of benzene with ethylene at 177°C which occurred at $T_{act} \sim 250°C$ for highly exchanged REY, ~ 400°C for REX, and ~ 600°C for NH_4Y. (This study neatly points up the dangers of comparing a variety of catalysts at a single T_{act}. Depending on which value of T_{act} was selected, one could have concluded that any one of these three catalysts was the most active.)

Rabo and co-workers (*140*) noted that dehydroxylation of LaY, which occurred largely between 550° and 700°C as judged by IR, greatly decreased its ability to isomerize and crack *n*-butane at 450°–550°C but, if anything, enhanced its ability to alkylate toluene with propylene at 100°C in the liquid phase. Such differing responses of activity to T_{act} suggested involvement of different active sites in the two reactions; however, there is a question whether the alkylation reaction may not have been bulk diffusion-controlled (*11*). Hickson and Csicsery (*73*) demonstrated decreasing activity for isomerization and transalkylation of 1-methyl-2-ethylbenzene at 204°C with increasing T_{act}, activity reaching zero for $T_{act} = 500°C$. Finally, for cracking of *n*-hexane at 338°C over 100%-exchanged LaX (*151*), activity decreased steadily to zero over the range $T_{act} = 338°–550°C$.

The general conclusions are (1) polyvalent cation-exchanged zeolites reach maximum activity at lower T_{act} values than NH_4Y presumably because they need only to dehydrate (during which cation hydrolysis occurs) and not deammoniate; (2) their activity falls off with increasing T_{act} and hence with increasing dehydroxylation; and (3) the T_{act} for final disappearance of activity may depend somewhat on the test reaction considered. Brønsted acidity, whose strength may be influenced by neighboring cations, is strongly implicated in catalysis. However, parallel structural and catalytic studies as a function of T_{act} are not extensive enough to decide whether activity can be totally associated with Brønsted O—H groups or whether a synergistic role for trigonal Al sites exists. At least the tendency for activity to increase over the range $T_{act} \sim 300°–550°C$ for NH_4Y does not reveal itself strongly in the data for polyvalent cation-exchanged Y zeolites.

Role of Exchange Level. One contrast between polyvalent ion-exchanged and NH_4^+-exchanged zeolites as a function of progressive degree of exchange is that activity appears much later for the former (*4*). Detailed studies have been reported by Minachev and co-workers (*9, 128, 152*), by Tung and McIninch (*93*), by Tsutsumi and Takahashi (*106*), and by Otouma and co-workers (*97*). For alkylation of benzene with propylene at 200°–300°C over zeolites activated at 500°C (*9, 128, 152*), significant activity appeared in CaY only beyond 25% exchange and in CaX only beyond 40% exchange. Plots of activity *vs.* exchange level for CaY were suggested to correlate more closely with the number of vacant S_{II} sites than the number of S_{II} sites occupied by Ca^{2+}. Empty S_{II} sites might be prime locations for the protons produced by Ca^{2+} hydrolysis; however, the correlation for CaX was reversed. This reasoning assumes complete preference of Ca^{2+} over Na^+ ions for S_I occupancy, and it is doubtful that the cation distribution is that ideal, particularly at high temperatures (*153*). Activity *vs.* exchange-level plots for CaY were virtually identical for propylation of benzene at 200°C and toluene transalkylation at 450°C. Cumene cracking activity (*106*) over faujasite-type zeolites with Si:Al = 1.6, preheated at 450°C, climbed sharply past 50% exchange for both Ca^{2+} and La^{3+} forms; a similar profile was seen for LaX activated at 600° (*97*). For isomerization and cracking of *n*-hexane (*93*), the upturn occurred for CaY at 60–70% exchange. Clearly then, high exchange is necessary to bring out the full activity of polyvalent ion-exchanged zeolites X and Y in accord with the proposal that formation of catalytically active O—H groups is induced only by cations in positions of partially satisfied coordination.

Role of Cation Type. The number of studies comparing the catalytic effectiveness of different polyvalent cations has become huge; selected results, in addition to those already discussed, are summarized in Table II. Interpreters of these data must be aware that in most cases a single T_{act} was used,

Table II. Comparisons of Activity of Zeolites

Entry	*Reaction Catalyzed*	*Activation*
1	Isomerization and transalkylation of 1-methyl-2-ethylbenzene at 200°–450°C	540° in H_2
2	Isomerization of 1-butene at 0°–300°C	420° in vacuum
3	Cracking of cumene at 274°C	538° in air
4	Isomerization of cyclopropane to propylene at 200°–400°C	500° in He
5	Isomerization of *o*-xylene at 250°–480°C	480° in He
6	Alkylation of benzene with propylene at 320°C	500° in air
7	Isomerization of 1-butene at 150°C	450° in air
8	Alkylation of benzene with ethylene at ~200°C	343° in air
9	Cracking of cumene at 260° or 450°C	500° in vacuum
10	Isomerization of *o*-xylene at 200°–300°C	480° in He
11	Cracking of cumene at 436°C	—

which may not have been optimum for all members of a series (*78*), and that the degrees of exchange were not identical. Nevertheless, the conclusion emerges that the usual order is $H^+ > RE^{3+} >$ Group $IIA^{2+} >$ Group IA^+ for both X and Y zeolites. The exact balance between number and strength of acid sites which leads to this order is not quantitatively known. A study of 11 individual rare earth cations by Eberly and Kimberlin (*154*) (entry 3, Table II) showed rough correlation between activity and total Brønsted acidity measured by Py adsorption; however, scatter was significant, and it was suggested that "Brønsted acidity cannot explain all nuances of catalytic activity."

THE ROLE OF SI:AL RATIO. Also apparent in Table II as in the original studies (*4*) is the greater activity of Y *vs.* X zeolites. Two more recent papers detail this conclusion. For cumene cracking at 300°–400°C over highly exchanged Ca^{2+} faujasite-type zeolites activated at 450°C, activity increased sharply from Si:Al = 1.25 to 2.3 but then decreased slightly at 2.5 (*106*). For *o*-xylene isomerization at 300°–600°C over Mg and Ca-exchanged cases, activity increased smoothly and sharply from Si:Al = 1.25 to 2.45, as did the total Brønsted acidity (*160*). Apparently the Y series develops more acidic O—H groups than the X in spite of its lower theoretical maximum number.

We earlier saw that H-mordenite was generally more active than HY. Although many fewer data exist for polyvalent cation-exchanged mordenites, the reverse order may hold. Thus, for transalkylation of toluene at 400°C, a reactivity order H-mordenite > HY ~ CeY > Ce-mordenite > CaY > Ca-mordenite emerged, and Yashima and Hara (*110*) suggested that this may reflect even more diffusional limitations in cation-containing mordenites than in H-mordenite because of the protrusion of the cations into the pore structure.

as a Function of Exchanged Cation Type

Activity Order	*Reference*
MgY > CaY > BaY	*155*
LaX ~ CeX > MgX > CaX > LiX > NaX	*156*
HY > LaY ~ EuY ~ YbY > CeY ~ PrY ~ SmY ~ GdY ~ ErY ~ ThY ~ DyY	*154*
MgX > CaX ~ LiX > BaX ~ NaX	*157*
H-mordenite > HY > REY > MgY > CaY > MgX > SrY > CaX	*2*
CaY > HX > CaX > NaY	*158*
CeX ~ LaX ~ HX > CaX ~ MgX > NaX ~ KX	*150*
REX > CaX > NaX	*78*
HY > MgY > CaY > SrY > BaY > NaY	*131*
HY > REY > MgY > NaY	*159*
LaY > CaY > HY	*117*

Polyvalent Cation-Exchanged, Ammonium-Exchanged Zeolites. The advantages of high concentrations of O—H groups offered by decomposition of NH_4^+ ions and of thermal and steam stability offered by the presence of polyvalent cations, possibly by formation of oxide-bridged cation pairs on hydrolysis, can be largely combined in zeolites containing both types of exchanged cations. Some of the earliest announced commercial cracking catalysts were of this type (*161*).

The properties of $Mg(NH_4)Y$ and $Ca(NH_4)Y$ have been extensively explored by Ward and Hansford (*15, 162, 163, 164, 165, 166*). It is convenient to consider fully exchanged NH_4Y as the parent material and then to examine the effects of progressive replacement of NH_4^+ by M^{2+} cations. As this process proceeded, the IR (*166*) of materials activated at 450°C showed gradual decline in intensity of the LF O—H band to almost zero at $\sim$ 60% M^{2+} exchange; only then did the HF O—H band begin also to decline. Simultaneous with its loss, a new band appeared in the spectrum of adsorbed Py attributed to bonding to accessible cations; Py revealed Brønsted but no Lewis acid sites on these samples. Thus, the first M^{2+} cations introduced appear selectively to enter S_I (theoretically 16 of 28 total), which leads to loss of the O_3—H but not the O_1—H band, and the O—H population is controlled largely by NH_4^+ decomposition. Activity for *o*-xylene conversion at 260°C was monitored for the same materials (*165*). For $Ca(NH_4)Y$, activity slowly decreased as Ca^{2+} exchange proceeded from 0 (*i.e.*, pure NH_4Y) to $\sim$ 55% and then dropped even more rapidly as pure CaY was approached. (Over this range, total acidity $\geq$ 50% H_2SO_4 measured by Hammett titrations also decreased (*99*)). Activity is dominated by the acidic O—H groups resulting from NH_4^+ decomposition rather than by Ca^{2+} hydrolysis. In other words, whereas introducing Ca^{2+} into NaY induces activity, the presence of Ca^{2+} in NH_4Y serves as a mild poison. The observed pattern also mitigates against Ca^{2+} cations being themselves the active sites since activity decreases as exposed Ca^{2+} level increases. In contrast, however, the activity of $Mg(NH_4)Y$ increased almost two-fold from pure NH_4Y to 40–60% Mg^{2+} exchange before rapid fall-off began beyond 60% exchange. This phenomenon suggests acid strengthening and activity enhancing effects of divalent cations on O—H groups even when the cations themselves are not exposed and when the total number of O—H groups is not increased.

The effects of T_{act} for $Mg(NH_4)Y$ on O—H population, acidity, and activity have been compared (*162, 163*). The HF O—H band plateaued from $T_{act} = 425°$ to 575°C; thus, dehydroxylation was somewhat more difficult than for NH_4Y; (however, the LF O—H band decreased at a lower T_{act} than for NH_4Y). Brønsted acidity toward Py was also relatively constant with Lewis acidity increasing only beyond $T_{act} = 600°C$. Conversion of *o*-xylene showed a sharp maximum at 570°–580°C. Since activity was actually decreasing when Lewis acidity began growing, Ward (*163*) felt less need to invoke a catalytic role for Lewis sites than for the case of NH_4Y. The failure of activity to maximize when total Brønsted acidity had plateaued ($T_{act} \sim 450°C$) was attributed to traces of residual NH_3 and/or displacement of NH_3 by Py during the adsorption studies.

Still another comparison (*164*) was made among Mg, Ca, Sr, and Ba(NH_4)Y, each containing 40% exchanged M^{2+}, a level where activity responded minimally to small changes in exchange level and the IR intensity of the O_1—H bond was constant. Since activity varied 10-fold and correlated with the electrostatic field of the cation, the proposal of a range of acid strengths induced by the cations was further strengthened. Decreasing the residual Na^+ level from 1.1 to 0.1 wt % doubled the activity of Mg(NH_4)Y as expected if the most acidic sites were preferentially poisoned by Na^+. A minor question in this study (*164*) however is whether catalysts were compared at their respective optimum T_{act} values.

Hickson and Csicsery (*73*) compared REY and NH_4Y with a mixed RE(NH_4)Y for reactions of 1-methyl-2-ethylbenzene at 204°C. The intriguing feature is the existence of two maxima in activity–T_{act} plots for RE(NH_4)Y, one at ~ 400°C (as for REY) and another at ~ 600°C (as for NH_4Y); the authors suggest the presence of two relatively independent acid sites. Comparable IR studies would be informative. In contrast, Ballivet and co-workers (*167*) reported activity for cracking of isooctane over a La(NH_4)Y to be independent of T_{act} (500°–900°C); however, all these samples were rehydrated before catalytic testing, so the results are probably not meaningful except to indicate catastrophic loss of activity at $T_{act} = 920°C$ because of structural collapse. Again, the conflicting piece of evidence is that of Pickert and co-workers (*89*) who claimed that activity of a RE(NH_4)Y for benzene propylation smoothly increased from $T_{act} = 350°$ to 750°C and suggested correlation with extent of dehydroxylation rather than Brønsted acidity.

Data for Y zeolites (Si:Al = 1.95) with varying exchange levels of $(La^{3+} + NH_4^+)$ and $(Ce^{3+} + Ca^{2+})$ were recently reported by Spojakina and co-workers (*168*). For a total 90% exchange, activity for cumene cracking maximized for a La^{3+}:NH_4^+ ratio of 3.6 compared with ratios of 1.2 or 0.3. For the $M^{2+}(NH_4)$Y series, the order would have been roughly reversed (*165*).

Alkali Metal Ion-Exchanged Zeolites. The discussion thus far has presupposed that the parent Na^+ and/or K^+ ion-exchanged zeolites are inactive for the major carbonium-ion hydrocarbon transformations. For facile reactions such as double-bond shifts in olefins, catalysis by NaY is observable, especially with added H_2O, but recent studies by Hall and Lombardo (*169, 170, 171*) show that the active sites arise from low levels of protons introduced during washing even with H_2O (cation deficiency) or from traces of polyvalent ions (Ca^{2+}) and are not inherent to the NaY structure which reveals no acidity on activation by IR (*172*) or titration techniques (*83*). Since activity for most hydrocarbon transformations does not appear in divalent ion-exchanged zeolites until these ions have largely filled S_I, some questions therefore still remain however with respect to how a few percent Ca^{2+} can induce activity for double-bond shifts if all reside at S_I. Hall and co-workers have alluded to a possible role for cocatalyst H_2O in altering cation distributions.

At high temperatures cracking of paraffins occurs over alkali metal ion-exchanged X (*133*) and Y zeolites (*173*) by a modified free-radical mechanism. It is not clear whether specific active sites in the usual sense are required for the modest rate enhancements observed compared with thermal reactions or whether more generalized effects of surfaces are involved.

Transition Metal Ion-Exchanged Zeolites. Data have been presented concerning the acidity (*15, 174, 175*) and typical catalytic activities (*15, 174, 175, 176*) of X and Y zeolites exchanged with divalent transition metal cations. Rough correlations of Brønsted acid population with electrostatic potential emerge but correlations of acidity with catalytic activity are poor. Ordering of divalent cations according to decreasing activity in Y zeolites depends on the reaction chosen: the order for isomerization of *o*-xylene (*174*) is Ni > Cu > Zn > Fe > Co > Mn (partial exchange) whereas for transalkylation of toluene (*176*) it is Mn > Cu > Ni > Fe > Co > Zn. Since these ions are well known to form complexes with organic ligands having π-bonds, specific interactions would be expected to induce reactions more reminiscent of organometallic than carbonium ion chemistry. Secondly, the cation site distributions may be significantly different from those of the main group cations because of preferences for specific ligand fields (*177, 178*) or because of coordination to reactants or products (*179, 180, 181*). Finally, one must be concerned whether the hydrocarbon feed has reduced the cations to the monovalent (*182*) or metallic states. This field deserves further studies.

A formal Diels-Alder reaction between butadiene and itself or acetylene has been found to occur over Cu(I)-containing X and Y zeolites (*183*).

CH_2=CH–CH=CH_2 + HC≡CH —(100°–120°C)→ [cyclohexadiene] + [4-vinylcyclohexene]

Redox Behavior of Zeolites: Cation Radicals and Stable Carbonium Ions. The final category of proposed active sites is the observed electron donor-acceptor centers in zeolites. In 1964 Stamires and Turkevich (*81*) treated vacuum-activated NH_4Y with molecules having low ionization potential, such as triphenylamine and diphenylethylenes. Formation of cation radicals at room temperature was observed by ESR spectroscopy, and their concentration increased sharply as T_{act} exceeded 500°C. Therefore, the authors associated the electron-transfer process with dehydroxylated sites although the number of spins observed was only 1/100 of the number of such sites expected. It was also suggested that molecules with higher ionization potential might undergo activated electron transfer at elevated temperatures and that the cation radicals could induce catalytic reactions (*82*).

The majority of such electron transfer is now known to be associated with some form of tenaciously adsorbed O_2 which is not totally removed even by evacuation at 550°C (*184*). There is also some controversy whether the observed cation radicals are uniformly spread throughout the pore structure in low concentration (*184, 185*) or whether they are only formed on the external crystallite surfaces (*186*).

Not only activated NH_4^+ zeolites but also Ca^{2+} forms (*184*) and RE^{3+} forms (*142*) can induce low-level formation of cation radicals from highly condensed aromatic hydrocarbons. The example of CeY (*186*) is special because air activation generates the powerful oxidant Ce^{4+} which can serve as an electron trap. Electron transfer between polycyclic aromatics such as anthracene and a wide range of divalent ion-exchanged zeolites X and Y has been studied by Richardson (*187, 188*) who was able to correlate the number of spins observed with the electron affinity of the cation (*e.g.*, $Cu^{2+} > Ni^{2+} > Co^{2+}$). For samples containing > 60% divalent ions from Group IIA and a lesser constant amount of Cu^{2+} ions, the number of spins induced by Cu^{2+} increased markedly with the polarizing power (e/r) of the Group IIA cation. These observations spawned Richardson's proposals concerning interactions between cations and O—H groups (*85*).

Treatment of activated (400°–700°C) H-mordenite with the simple aromatic benzene at room temperature formed a very low level of paramagnetic species (*189, 190, 191*) assigned as a dimeric cation radical of benzene with 12 equivalent hydrogens which was slowly transformed into the cation radical of biphenyl by some type of dehydrogenation.

ESR observation (*192, 193*) of simple olefins adsorbed on activated REY showed spectra with resolvable hfs of 13–15 gauss; for example, both 2-pentene (*193*) and 1-butene (*192*) were reported to give seven-line patterns of approximately binomial distribution. Formal hydrogen atom addition was suggested (*192, 194*), and the observed species assigned as a partially charged, saturated alkyl radical bound to the surface such that its trivalent carbon atom bears less than unit spin (since ordinary alkyl radicals show hfs with α-protons of 20–23 gauss and with β-protons of 26–28 gauss). However, the observed 13–15 gauss splittings are very reminiscent of both the α and β hfs constants for allylic radicals (*195*) which might be formed as in Equation 20. Suzuki (*192*) feels he has ruled this possibility out

$$\mathrm{RCH{=}CHCH_3} \xrightarrow{-e^-} (\mathrm{RCH{=}CHCH_3})^{+\cdot} \xrightarrow{-\mathrm{H}^+} \overline{\mathrm{RCH{-}CH{-}CH_2}} \quad (20)$$

because radical intensity correlated with the expected number of O—H groups on the catalyst, not with the number of dehydroxylated sites. A third suggestion (*196*) is that the observed species is a polymeric radical produced by a sequence of olefin polymerization–cyclization–hydride transfer steps. Resolution of this problem would be helpful to decide whether these ESR observations are pertinent to catalysis.

The major problem in deciding whether cation radical formation, which might well be expected when unsaturated substrates containing traces of O_2 are passed over dehydroxylated zeolites, plays a role in the typical reactions we are considering in the paucity of model behavior. Aromatic cation radicals have been more in the domain of spectroscopists than organic chemists. Possible routes by which electron transfer could lead to free radical production are shown in Reactions 9 and 20. However, few, if any, catalytic conversions call out for such a pathway.

Dollish and Hall (*184*) reported that adsorption of triphenylmethane on 45%-exchanged activated NH_4Y at 100°C gave the stable triphenylmethyl cation, observed by optical spectroscopy, in a process not affected by O_2 (but *cf.* Ref. *197*). Since this reaction could be a model for formation of carbonium ions from paraffins under more forcing conditions, routes involving attack on the C—H bond by a protonic site or by a Lewis acid site have been discussed. The cation concentration *vs.* T_{act} plot showed a sharp

$$(C_6H_5)_3CH \xrightarrow[?]{H^+} (C_6H_5)_3C^+ + H_2$$

$$(C_6H_5)_3CH \xrightarrow[?]{\text{—Al—}} (C_6H_5)_3C^+ + \text{H—Al}^-\text{—}$$

maximum at ~ 550°C which might have correlated with Brønsted acidity enhanced by adjacent Lewis sites. However, more recent studies by Hall and co-workers (*198*) on silica-alumina support neither of these routes but rather a sequence of electrophilic aromatic substitution and hydride transfer (Reaction 21), both well-known reaction types. Hence, these observations (*184*) probably have no significance for studies of paraffin cracking.

$$\phi_2CH\text{—}C_6H_5 \xrightarrow{H^+} [\phi_2CH(H)C_6H_5]^+ \longrightarrow \phi_2CH^+ + H\text{—}C_6H_5$$

$$\phi_2CH^+ + \phi_3CH \longrightarrow \phi_2CH_2 + \phi_3C^+ \qquad (21)$$

Residue as Active Sites. Many hydrocarbon transformations on silica-alumina may actually occur (*199, 200*) on a carbonaceous residue formed on the oxide surface rather than directly on the catalyst itself. In some not well-defined manner these residues appear to possess proton-donating capability and may be large adsorbed aromatic carbonium ions. There is some evidence (*170*) that such complications are much less severe for zeolites, particularly for studies of initial activity in pulsed reactors where inhibition periods are not normally observed. The loss of IR-observable O—H groups during catalysis (*92, 201*) also appears to be related to declining activity rather than generation of new activity. On the other hand, some workers (*88, 199*) have alluded to possible connections between activity and dehydroxylation being mediated by such residue formation on Lewis acid sites. Especially for commercial cracking catalysts which always function with a finite coke level, we cannot totally ignore the role of residue.

Dynamic Effects

The preceding section described attempts to relate catalytic activity to zeolite structure and possible active sites by fixing a set of experimental reaction conditions and varying the structure of the zeolite in some defined and hopefully understood fashion. Now we survey, in less detail, the kinds of information which may be obtained by holding the catalyst constant and focusing on and varying the reaction conditions.

Kinetics. Although kinetic results are a powerful tool in deducing chemical mechanisms, comparatively little use has been made of detailed kinetic studies in interpreting zeolite catalysis. The most common practices in reporting catalyst activity when comparing zeolites are either (1) to determine an instantaneous conversion from a product analysis after some defined time-on-stream under defined conditions of temperature and feed rates after the system has reached "equilibrium" or (2) to determine conversion on the basis of accumulated products from time zero to some chosen time-on-stream. While these methods are quite acceptable for comparing large influences of one aspect of zeolite structure, all other factors being assumed constant, they do not reveal the complex interplay which may occur among sorption, diffusion, and reaction. Comparisons based on extrapolations of conversion to zero time in differential rather than integral reactors are more reliable to avoid false conclusions resulting from catalyst aging but are much less common. We have already seen the confusion which could have arisen concerning effects of Si:Al ratio in acid-extracted mordenites because of failures to observe this criterion. Finally, examples where actual kinetic rate laws are determined as a function of flow rate and partial pressures of reactants and products are even more sparse, although comparisons are often made by assuming first-order behavior and thereby converting measured conversions into rate constants. It is common to determine the effects of reaction temperature and fit conversion data to the Arrhenius law. However, even if this is done properly by using only initial conversions to avoid artifacts because of different aging rates at differing temperatures, the resulting activation energies do not necessarily give E_a for the chemical reaction at the active site but may reflect varying contributions from diffusional and mass transfer limitations.

A few recent studies serve as examples of the kinds of information which can potentially be obtained from kinetic analyses and of the misinterpretations which can be avoided. Nolley and Katzer (*202*) studied the alkylations of benzene with ethylene at 215°–326°C and with propylene at 77°–220°C and of ethylbenzene with ethylene at varying aromatic:olefin ratios over NH_4Y, $RE(NH_4)Y$, and LaY, all activated at 550°C in air, in a differential flow reactor. The basic observation was a plot of instantaneous rate *vs.* time on stream which increased up to times on the order of 15 min, maximized, and then declined again. The decay portion could be fit by an exponential rate decay law; rate = $M_1\exp(-t/M_2)$ where M_1 is the initial rate and M_2 is the decay time constant. Although a detailed evaluation of the intrinsic kinetics was not attempted (*see* below), two suggestive conclusions were drawn. Considering decay first, the authors noted that M_2 decreased (*i.e.*, decay became more serious) as olefin concentration increased and as the olefin was changed from ethylene to propylene. Hence, olefin cationic polymerization was implicated as the major decay mechanism. However, M_2 did not depend on the zeolite cation whereas M_1 (*i.e.*, rate at $t = 0$) did, the order being NH_4Y:RE(NH_4)Y:LaY $\sim$ 3:1.5:1. The active site for attack on olefin was therefore concluded to be the same for all catalysts with the variations in M_1 representing differing numbers of sites. Considering the early rate *vs.* time profile, the authors proposed and tested three models concerning possible roles of diffusion: (1) concentration equilibrium exists between both the intracrystalline pore structure and the external gas phase and reaction is totally reaction rate-limited; (2) equilibrium exists between the gas phase and the exterior of the $\sim 1\mu$ crystallites but diffusional limitations lead to concentration gradients within the pore structure; and (3) equilibrium exists with the pore structure but a gradient exists at the gas-crystallite boundary. Model (1) was unsuccessful in fitting the data. Calculations of the modified Thiele modulus and effectiveness factor based on estimates of benzene *vs.* ethylbenzene counterdiffusion (*i.e.*, the departing product and entering reactant must pass each other in the pore structure) extrapolated from lower-temperature liquid-phase data (*203, 204, 205*) suggested that intracrystalline diffusion was not limiting either. However, model (3) was successful and it was concluded that the rate-limiting step was the process in which the alkylated product had to enter the gas phase from the pore mouth. The activation energy associated with this desorption coefficient, which was the adjustable parameter in the model, was $\sim$ 16.5 kcal/mole. Since this is almost identical to the measured heat of adsorption of ethylbenzene on HY (*206, 207*), the argument for rate-limiting product desorption was strengthened. Thus, the response of reaction rate to temperature reveals nothing directly concerning the alkylation reaction at the active site except that it is facile.

Discussion of similar phenomena had been presented earlier by Riekert (*208*) who studied the interactions of ethylene with NiY: reversible adsorption rate-limited by internal diffusion at ambient temperature, a slower second-order chemical reaction at $\sim 200°C$ to produce larger molecules in the pores, and finally desorption of gaseous C_4 products. He pointed out that the internal surface of zeolites does not represent a macroscopic phase boundary but that the phase boundary occurs only at the edge of the crystallite; crossing this boundary into the gas phase will have E_a at least as great as the heat of adsorption.

In contrast to Nolley and Katzer (*202*), Haag (as reported by Venuto (*11*)) studied the ethylation of toluene over REY in the liquid phase at 80°C. The kinetic law (Equation 22), first-order in both reactants with a pseudo-second-order rate constant k' made up of the true rate constant k

and the ethylene adsorption constant K, revealed information on the chemical reaction itself since diffusion was shown not to be a complication.

$$\text{Rate} = kK[C_2H_4][C_6H_5CH_3] \tag{22}$$

The Rideal mechanism in Equation 23 was proposed in which a non-adsorbed toluene molecule interacts with an adsorbed ethylene species. The differences from Noller and Katzer's conditions (*202*) are the lower temperature and

$$HO_z^- + CH_2{=}CH_2 \overset{K}{\rightleftarrows} CH_3\overset{+}{C}H_2,\ O_z^{2-}$$

$$\xrightarrow[k]{C_6H_5CH_3} \tag{23}$$

$$CH_3C_6H_4CH_2CH_3 + HO_z^-$$

presence of a liquid phase; the former should favor the emergence of reaction rate rather than diffusional limitations and the latter should lower the product desorption barrier. Combining these two studies now begins to give real detail concerning the dynamics of alkylation reactions.

Beecher and Voorhies (*115*) collected data on hydroisomerization of *n*-hexane to *i*-hexane over H-mordenite as a function of temperature, pressure, and space velocity and carried out kinetic modelling. The model, including assumed Langmuir adsorption isotherms, which best fit the data including the decrease in conversion with increasing pressure was a dual-site mechanism—*i.e.*, an adsorbed molecule can react only if there is available an unoccupied adjacent site. If correct, this model sets rigid requirements on possible mechanisms at the molecular level (*see* below).

One of the most frequently studied reactions is cumene cracking, yet the reported kinetics are confusing. Richardson (*85*) reported zero-order kinetics over "H faujasite" at 225°C and interpreted this as a result of slow decomposition on completely saturated active sites. Romanovskii and co-workers (*207*) reported first-order behavior over HY and CaY at higher reaction temperatures. A considerable variety of observed E_a values have been reported (*14*) some of which are listed in Table III. Some of the difficulties in interpreting such data have been pointed out (*211*) with respect to separating the roles of reaction, sorption, and diffusion. Note particularly Entry 2 (*209*) where an attempt was made to derive the activation energy for the chemical reaction by accounting for the heat of cumene adsorption. Although each of the variations seen in Table III can be rationalized in terms of some structural factor, some of the data may be perturbed by unknown effects of catalyst aging in continuous reactors and chromatography effects in pulsed reactors. Some of the mechanistic conclusions which have been drawn will now be listed. Turkevich and co-workers (*210*) (Entry 6) attributed decreasing E_a with increasing degree of exchange in NH_4Y to increasing activity per site as site density increased (*see* below)

Table III. Reported Arrhenius Activation

Entry	*Catalyst*
1	Decationized Y and CaY with Si:Al = 2.15
2	Decationized Y, T_{act} = 550°C
	CaY, T_{act} = 550°C
3	SmY, T_{act} = 427°–650°C
4	LaX, 50%-exchanged, T_{act} = 600°C
	LaX, 90%-exchanged, T_{act} = 600°C
5	CaY, Si:Al = 2.55, 95%-exchanged
	CaY, Si:Al = 2.3, 95%-exchanged
6	Decationized Y, high-exchange, T_{act} = 450°C
	Decationized Y, low-exchange, T_{act} = 450°C
7	LaY, steamed

[a] The true E_a for reaction was given as ~30 kcal/mole after correcting for the

because of interactions between sites. Otouma and co-workers (*97*) (Entry 4) rationalized a similar trend in LaY by the presence of La^{3+} in S_{II} for the highly exchanged samples. If, however, LaY is active because of the presence of O—H groups, the explanations for Entries 4 and 6 may be identical. However, Eberly and Kimberlin (*154*) (Entry 3) concluded that the failure of E_a to change as SmY was activated from 427° to 650°C even though activity dropped 30-fold resulted from a single type of (presumably non-interacting) site with T_{act} affecting only its population. No simple rationalization of Entries 3, 4, and 6 involving only O—H sites is obvious. Finally, a tentative conclusion from several of the Russian reports (*117, 207, 209*) (Entries 1, 2 and 5) is that E_a increases as Si:Al decreases.

Diffusion. The existence of zeolitic pores near the range of molecular dimensions and the absence of any larger pores not only puts zeolitic diffusion at the extremity of the Knudsen region but possibly into a totally new "configurational regime" as suggested by Weisz (*6*) where even the internal configurational motions of molecules need to be considered during diffusion. The influence of these unique diffusional properties on catalysis has not been extensively explored except as they lead to shape selectivity. Katzer (*202*) has even been led to state that "the rate-limiting processes in catalytic reactions over zeolites remain largely undefined, mainly because of the lack of information on counter-diffusion rates at reaction conditions."

Bulk mass transfer limitations can be eliminated from consideration if rate constants are not a function of flow rate, and diffusion limitations within the macropore structure of the catalyst pellet (most testing uses compacted pellets consisting of huge numbers of individual crystallites) can be eliminated if rate constants are not a function of pellet size. However, neither of these tests will reveal kinetic limitations of intracrystalline diffusion unless materials with differing crystallite size are used. A revealing example of the use of this variable is a study by Chutoransky and Dwyer (*212*) of the liquid-phase isomerization of *o*-xylene over a zeolite catalyst. The meta:para selectivity ratio depended on crystallite size. With 0.2–0.4 μ crystallites, no direct pathway from ortho to para was observed, consistent

Energies for Cumene Cracking

Reaction Conditions	*Reference*	E_a*(kcal/mole)*
pulsed microreactor	*207*	18
pulsed microreactor	*209*	~ 10[a]
		~ 7[a]
continuous flow reactor	*154*	20
continuous flow reactor	*97*	31
		21
pulsed microreactor	*117*	12
		18
pulsed microreactor	*210*	17
		31
continuous flow reactor	*211*	19.5

heat of adsorption of cumene.

with reasonable organic mechanisms involving 1,2-shifts. With the more typical 2–4 μ crystallites, however, 6% of the ortho isomer proceeded directly to the para. In the larger crystallite, a newly formed meta isomer may spend enough time in the pore structure before desorption into the bulk liquid phase to allow it a finite chance to encounter another active site and rearrange further to the para isomer. This problem of successive reactions during one traverse of a crystallite is multiplied enormously if the product is more reactive or more strongly adsorbed than the reactant. In such a case as, for example, the cracking of unreactive paraffins to form reactive olefins, any experimental approach which attempts to determine "initial products" by extrapolation to zero total conversion is doomed to failure.

Adsorption. In 1964 Boreskova and co-workers (*209*) noted that the heats of adsorption of cumene were similar for both catalytically inactive Na and active Ca and H zeolites and, therefore, that bulk adsorption was probably not confined to specific catalytic centers. However, generalized adsorption can play a possible role in catalysis since it affects the surface concentration and orientation of substrates. Entry of organic molecules into a zeolite pore is significantly exothermic. However, once in the pore, they can migrate from one adsorption site to another with an activation energy less than the heat of adsorption because of movement along the surface. With respect to intracrystalline diffusion, counter diffusion behavior in particular should have significance for catalysis as already discussed.

Bezus and Kiselev (*214*) determined adsorption isotherms near 100°C for ethane and ethylene on 400°C-activated NaY, NaX, CaX, and SrX zeolites. Both the quantities adsorbed and the isosteric heats of adsorption increased in the orders: $C_2H_6 < C_2H_4$; $Y < X$; and $Na^+ < Sr^{2+} < Ca^{2+}$. From these patterns, the authors assigned the chief adsorption sites as the exposed cations. As a typical value, the heat of adsorption of ethylene on CaX at low coverage was 13 kcal/mole. Hopkins (*215*) compared adsorption of *n*-hexane and benzene at ~ 200°C on NaY and NH_4Y, the latter activated both at 280°C in vacuum to attempt to produce HY and at 500°C in vacuum to produce some dehydroxylation. The heats of adsorption of *n*-hexane were comparable

($\sim$ 12 kcal/mole) for all these materials and increased slightly with increasing coverage because of adsorbate-adsorbate interactions; Hopkins assigned the adsorption sites as the oxide lattice rather than the cations. For benzene, however, the dehydroxylated Y showed a higher value (18 kcal/mole), which decreased with increasing coverage, than the other zeolites ($\sim$ 15 kcal/mole); a specific interaction was postulated between this unsaturated adsorbate and Lewis acid sites which was stronger than that with O—H groups. Similar studies for benzene and cyclohexane have been reported by Barthomeuf and Ha (*216, 217*). Further evidence for the preferred adsorption of unsaturated molecules in the order aromatics > olefins > paraffins comes from studies (*218*) of liquid adsorption equilibria of binary hydrocarbon mixtures. The possible implications of findings such as these for catalysis at higher temperatures remain to be defined. Cation positions in zeolites can be altered by coordination of complexing organic substrates (*179, 180, 181, 219*), although again it is not known whether this phenomenon has catalytic significance.

Activators. Independent evidence that Brønsted acidity plays a vital role in catalysis comes from several examples of promotion of activity by introducing species HA into the feed which would be expected to ionize and form additional O—H groups in the zeolite. The rate of 1-butene isomerization at 260°C over NaY containing 0.3–5.7 exchange % Ca^{2+} ions, activated at 500°C, increased by factors of 600–38, respectively, when H_2O was added at a level of 2 H_2O/supercage before starting the run (*170*); extrapolation of H_2O-added results to zero Ca^{2+} level indicated zero activity for pure NaY. Thus, hydrolysis of H_2O is strongly suggested as the source of active sites. Other examples of promotion by H_2O are reviewed in Ref. *169*.

Activity for cumene cracking over NaY at 250°–350°C could be induced reversibly by adding anhydrous HCl to the feed (*220*) and active site generation was pictured as in Reaction 24. Similar effects were observed for

$$Na^+O_z^{2-} + HCl \rightarrow Na^+Cl^- + HO_z^- \tag{24}$$

propylation of benzene by addition of propyl chloride (*158*), which should serve as a source of HCl.

Minachev and Isakov (*221*) described enhancement of several reactions of alkylaromatics especially over CaY by added CO_2, but possible relationships between this kinetic effect and IR observations of carbonate formation (*130*) are not clear.

Poisoning. The inverse of the promotion by HCl was described by Rabo, Poutsma, and Skeels (*222*). If zeolites such as CaY or LaY, which normally produce active catalysts for *n*-hexane cracking, cumene dealkylation, and 1-butene double-bond shift by activation at 500°–550°C, were loaded with NaCl by evaporation and then heated at 500°–550°C, dramatically less active catalysts resulted. Since HCl was evolved during activation, the deactivation mechanism must be the solid-state Reaction 25.

$$HO_z^- + Na^+Cl^- \rightarrow Na^+O_z^{2-} + HCl\uparrow \tag{25}$$

Many studies have been made of the poisoning of catalytic activity by gradual introduction of bases such as quinoline (Qn). Although this technique might not distinguish between Brønsted and Lewis acid sites since both may bind Qn, it should offer a method of counting whether only a few strongest or all acid sites contribute to catalytic activity. Turkevich (*82, 86, 210*) studied the effect of Qn on cumene cracking at 325°C in a pulsed microreactor by alternating pulses of cumene reactant and Qn poison, each poisoning pulse being small enough to interact stoichiometrically with only a fraction of the possible number of sites present. For a series of NH_4Y zeolites of varying degree of exchange all activated at 450°C, activity decreased linearly with the cumulative amount of Qn added. The amounts to achieve complete poisoning indicated a number of sites equal to the number of cation-exchange sites calculated for the particular sample up to 50% exchange; they leveled off at higher exchange levels, presumably because sites (O_3—H?) became inaccessible. If, however, T_{act} exceeded 450°C, the amount of Qn required for complete poisoning decreased. Therefore, Turkevich identified the active sites as Brønsted acids (*86*) (rather than Lewis acids (*82, 210*)) which are poisoned by Qn on a 1:1 stoichiometric basis, it being stated that "Qn does not react with the Lewis acid site at 325°C but only with the Brønsted acid site (*86*)." This last undocumented conclusion conflicts with general experience from IR studies of decationized Y that Py is bound to Lewis sites more strongly than to Brønsted sites; (cf. ref. *47*). Earlier we had suggested (*14*) that these results are not consistent with proposals that some small fraction of most acidic Brønsted sites contribute disproportionately to activity because had this been true, the first pulses of Qn should have been disproportionately effective as poisons. However, this is not a correct deduction if, in the pulsed reactor, Qn moves with plug flow and is irreversibly captured by all Bronsted sites regardless of strength. In this case, for example, the first 10 stoichiometric % of Qn would not poison the 10% most acidic sites throughout the catalyst but only the first 10% front section of the bed to leave 90% unchanged. Then linear poisoning with the cumulative amount of Qn would occur regardless of any role of inhomogeneous acidity. A second type of information drawn from these studies by Turkevich involved calculation of activity per site for any sample by dividing total observed activity by the total number of sites derived by Qn counting. Activity per site increased with increasing number of sites—*i.e.*, with increasing NH_4^+ exchange level. Turkevich therefore proposed cooperative interactions between sites which enhanced activity per site; this was supposed to occur because "a large number of cumene molecules [are] adsorbed on a large number of interacting sites, with one of these adsorbed molecules undergoing the reaction (*86*)." This conclusion was supported by the decreasing E_a and decreasing A factor with increasing exchange level (Table III, Entry 6). Unfortunately, these arguments suffer from the same experimental ambiguity described above and could as well be used to support the appearance of stronger acid sites as exchange level increased.

In apparent contradiction stands a body of work by Topchieva and co-workers (*117, 207, 209, 223*) using very similar techniques for cumene cracking at 400°–500°C. For a series of NH_4Y, CaY, LaY, and $Ca(NH_4)Y$ with Si:Al ratio of 1.7 and ~ 2.5, this group consistently found a number of sites counted by Qn poisoning only 1–4% of the number of ion-exchange sites. The resolution may simply lie in their use of a higher T_{act} (550°C) so that the O—H content of the samples was already seriously reduced by dehydroxylation, although this view still requires selective capture of Qn by Brønsted rather than Lewis acid sites. Or, it may lie in their higher reaction (and therefore poisoning) temperatures which indeed allowed a uniform Qn distribution according to acidity. In a more recent series of papers from the same group (*224, 225, 226, 227*) the effects of adsorbed Py and NH_3 on cumene cracking rates and on heats of adsorption of benzene, considered to be a weak base, were interpreted in terms of a small fraction of catalytically-efficient strong acid sites and a much larger fraction of more inert weak acid sites for both HY and CaY.

Finally, Goldstein and Morgan (*228*) studied the same system with CeY, CaY, and NH_4Y activated at 595°C in air, except that Qn was fed continuously to a flow reactor. The minimum value of Qn required to achieve total poisoning was somewhat dependent on reaction temperature. These minimum titers turned out to be smaller than Turkevich's, larger than Topchieva's, and almost identical with the number of large cavities. It was thus concluded that one Qn per large cavity is enough to destroy activity even though it could contain more than one acidic group. A curious datum of this work is that NaY adsorbed more Qn than HY in gravimetric adsorption studies at 350°–400°C—a result in conflict with usual ideas of specific adsorption of organic bases on acidic sites.

Because of the several experimental uncertainties, observed ambiguities, and apparent disagreements between the data on high-temperature poisoning, a definitive mechanistic conclusion from this potentially powerful technique is still lacking. On the practical side, however, these studies do illustrate the effects of nitrogen poisoning on industrial cracking catalysts (*229*).

Aging. Gradual loss of catalytic activity with increasing time on stream accompanied by formation of carbonaceous residue within the catalyst is a most common phenomenon in zeolite catalysis, the chemistry of which is likely related to olefin condensation, cyclization, and aromatization reactions (*see* below). The kinetic behavior of aging can often be fit by exponential decay functions of the type already discussed (*202*); further examples have also been recorded (*108, 211, 230, 231, 232*). For cracking *n*-hexane over H-mordenite in a pulsed reactor, aging correlated with the accumulated *n*-hexane actually cracked rather than with that fed (*108*); thus, aging was clearly associated with the occurrence of reaction rather than with a poison in the feed or with thermal deactivation.

Zeolitic catalysts are also deactivated to varying degrees by high temperature steaming such as occurs during burn-off of "coke" in practical use. The polyvalent cation-exchanged, particularly RE-exchanged (*233*), Y zeolites have much more steam stability than decationized Y. We must admit that such steam-equilibrated catalysts may have structural features different enough from those of freshly-activated catalysts to make extrapolation of mechanistic information somewhat risky.

Organic Mechanisms

Having now seen some of the mechanistic conclusions which may be drawn from consideration of zeolite structure and of dynamic responses to reaction variables, we examine the major reaction types catalyzed and the probable organic mechanisms involved.

Olefin Reactions. ADSORPTION. On exposure of HY (45%-exchanged NH_4Y activated at 480°C and then hydrated at 300°C) to ethylene at room temperature, reversible adsorption occurred with a shift of the 3650 cm^{-1} O—H band to 3300 cm^{-1} (broad), but no isotopic exchange between hydroxylated zeolite and deuteriated ethylene was found until $\sim$ 200°C. The total amount adsorbed was 30% greater than the number of O—H groups present. Hall and co-workers (*59, 199, 234*) assigned this adsorption largely to hydrogen-bonding of ethylene to accessible O—H groups. Although carbonium ion formation was implicated in both isotopic exchange and the simultaneous polymerization reactions at $>$ 200°C, it was not clear that they necessarily involved a common intermediate. The adsorption of propylene was not completely reversible, and isotopic exchange between catalyst and olefin occurred at room temperature. On a slower time scale, polymerization occurred which produced, on desorption at 200°C, isobutane by a polymerization-isomerization-cracking process. This behavior is consistent with the expected greater facility of converting propylene by protonation to the secondary *i*-propyl cation, by way of the hydrogen-bonded complex, compared with converting ethylene to the primary ethyl cation. If the zeolite was partially dehydroxylated, adsorption even of ethylene was not completely reversible at room temperature and desorption at 120°C gave *i*-butane and *i*-butene. Hypotheses concerning the adsorption mode on supposedly Lewis acid sites are more poorly developed than for the O—H interaction.

The various stages of adsorption, double-bond shift, polymerization, cyclization, and aromatization which appear to be the major cause of catalyst aging have been to some extent observed by IR spectroscopy by Eberly (*235*) for 1-hexene over HY at 93°–260°C, by Weeks and Bolton (*236*) for 1-butene over HY and dehydroxylated Y at 0°–300°C, and by Declerck and co-workers (*88*) for 1-butene over HY at 25°–150°C. For 1-butene, bands assigned to C=C stretching vibrations in 1- and *cis*-2-butene were observed (*236*) at 0°C immediately after adsorption. These were rapidly replaced at room temperature by spectra characteristic of saturated species.

DOUBLE-BOND SHIFTS—ADSORBED CARBONIUM IONS. The structurally least complex chemical transformation of olefins is double-bond positional and geometrical isomerization; these related processes have been observed over a variety of X and Y zeolites with increasing rates corresponding sensibly with increasing acidity: REX $\sim$ HX $>$ CaX $>$NaX (*150*); LaX $>$ CaX $>$ MgX $>$ LiX $>$ NaX (*156*); HX $>$ NaX (*237*); and HY $>$ NaY

(*237*). Several zeolites containing transition-metal ions are also active (*150, 156, 237, 238, 239*) and present the possibility of specific effects of olefin coordination with the cations which do not need to be considered for other zeolites. The most extensive mechanistic studies have been carried out by Hall and Lombardo (*169, 170, 171, 240, 241*) with the linear butenes over NaY containing a few percent cation deficiency or Ca^{2+} content, generally in the presence of H_2O which serves as a cocatalyst. The mechanistically significant observations at $\sim 200°C$ include: (1) first-order kinetics; (2) the existence of reversible pathways interconnecting each pair of isomers (1-, *cis*-2-, and *trans*-2-butene); (3) the incorporation of one H(D) from the catalyst into the olefin per act of isomerization; and (4) no evidence for a catalytic role of the very small amounts of residue formed on the catalyst. Coupling these results with the observed activity-enhancing effects of increasing cation deficiency, Ca^{2+} exchange, and added H_2O level, all of which suggested correlation with proton-donating ability, these workers proposed formation of *sec*-butyl cation which connected all isomers by protonation-deprotonation steps. The various selectivities which were observed in isomer

$$CH_2{=}CHCH_2CH_3 \underset{-H^+}{\overset{+H^+}{\rightleftharpoons}} CH_3\overset{+}{C}HCH_2CH_3 \begin{cases} \underset{-H^+}{\overset{+H^+}{\rightleftharpoons}} CH_3CH{=}CHCH_3 \quad \text{cis} \\ \underset{+H^+}{\overset{-H^+}{\rightleftharpoons}} CH_3CH{=}CHCH_3 \quad \text{trans} \end{cases}$$

ratios produced from any one starting isomer must then be related to subtle differences in deprotonation rates of the carbonium ion.

It is less clear how we should describe the bonding of such carbonium ions to the surface. For highly-stabilized cations such as triphenylmethyl, the observed optical spectra suggest the normal planar trigonal ion associated with the surface only by coulombic forces. (However, because of the large size of this ion and especially of its tetrahedral precursors, it may be questioned whether triphenylmethyl cations observed spectroscopically were really within the pore structure or only on the external crystallite surfaces. We thank a referee for pointing out this question.) In the other extreme, the proton is surely covalently bound to a single surface oxide and manifests its acidity by proton transfer reactions to adjacent bases. Is the *sec*-butyl cation adsorbed on the surface best pictured as **22** or **23**; *i.e.*, is the bonding more

$$\begin{array}{c} R^+ \\ \\ \gt \overset{-}{Al} \lt \overset{O}{} \gt Si \lt \\ \textbf{22} \end{array} \qquad \begin{array}{c} R \\ | \\ \gt \overset{-}{Al} \lt \overset{O^+}{} \gt Si \lt \\ \textbf{23} \end{array}$$

ionic or covalent? Even if structure **23** is more nearly correct for non-resonance stabilized reactive cations, we do not need to abandon the carbonium ion nomenclature or conceptualization because the reactions of **23** would be expected to be those of R^+. A second question concerns the proton transfer step itself: is the cation produced by protonation of the olefin adsorbed on the same (Equation 26) or a different oxide ion or cation site (Equation 27), regardless of the correct formulation of the mode of adsorption (which we will write in covalent fashion only for convenience)? If the former is true, there then appear to be rigid spatial requirements to allow the transferring proton and the developing carbonium ion center

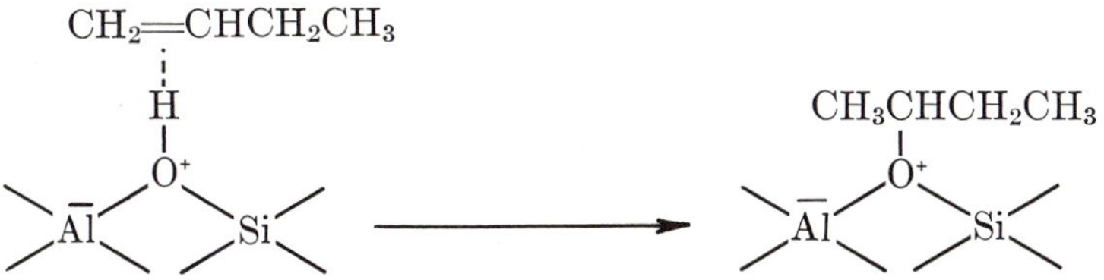

$CH_2{=}CHCH_2CH_3$ ··· H–O^+(Al)(Si) ⟶ $CH_3CHCH_2CH_3$–O^+(Al)(Si) (26)

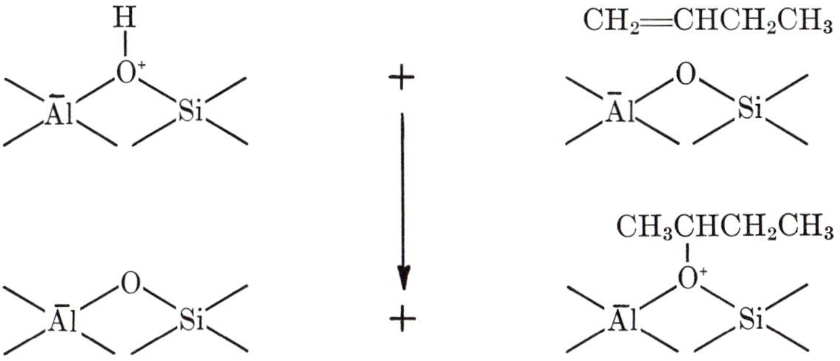

H–O^+(Al)(Si) + $CH_2{=}CHCH_2CH_3$, O(Al)(Si) ⟶ O(Al)(Si) + $CH_3CHCH_2CH_3$–O^+(Al)(Si) (27)

to maintain optimum bonding to the surface; *e.g.*, structure **24** appears favored over **25** where the carbonium ion center is forming well into the large cavity. In other words, although the zeolite appears to provide highly beneficial solvation for reactive intermediates in carbonium ion processes, this solvation must be geometrically dependent on the rigid framework

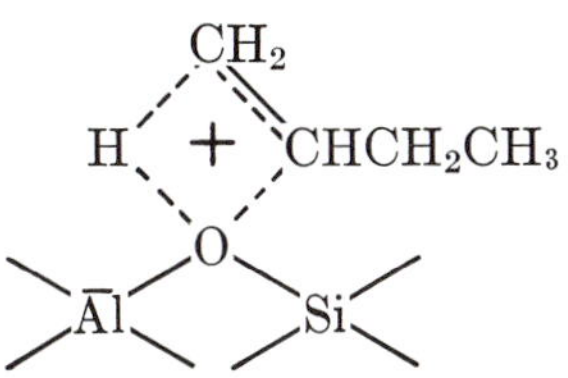

CH_2, H, +, $CHCH_2CH_3$, O, Al, Si

24

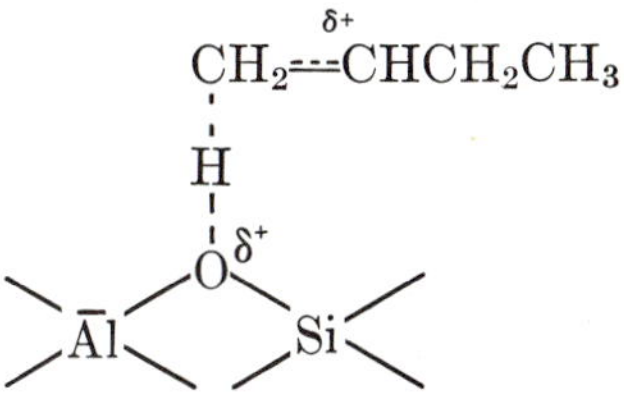

$\delta+$ $CH_2{\cdots}CHCH_2CH_3$ ··· H ··· $O^{\delta+}$, Al, Si

25

structure rather than effectively continuous as in solution. Such questions can be asked of any mechanistic step where one carbonium ion is converted into another. Until these questions are better understood, discussions of, for example, isomer selectivity as a function of zeolite acidity seem rather fruitless. Our understanding of organic mechanisms over zeolites has not yet reached this degree of sophistication of molecular details, but such thoughts seem worthy of further consideration. For example, if these ideas have merit, then predictions of maximum activity in terms of zeolite structure might need to consider not only O—H acidity and density but also its distribution with respect to availability of adjacent carbonium ion adsorption sites.

To the extent that we can define carbonium ion processes, the Hall-Lombardo mechanism fits the data well. For example, the incorporation of one H(D) per reaction step rules out a hydride transfer chain through allylic cations (Reaction 28). One unexplained result (*242*) is that E_a for 1-pentene

$$\begin{aligned} & CH_2\text{---}\overset{+}{C}H\text{---}CHCH_3 + CH_2{=}CHCH_2CH_3 \rightarrow \\ & CH_3CH{=}CHCH_3 + CH_2\text{---}\overset{+}{C}H\text{---}CHCH_3 \end{aligned} \qquad (28)$$

isomerization is over twice that for 1-butene (22.6 *vs.* 10.6 kcal/mole) although the secondary cations for these two olefins should be energetically similar. Note finally that under these mild conditions, negligible *i*-butene is formed, reflecting the extra energy barrier required to isomerize *sec*-butyl to *tert*-butyl cation.

Kemball, Leach, and co-workers (*239, 243*) examined the isomerization of 3,3-dimethyl-1-butene (**26**) at ~ 125°C which produces the skeletally isomerized 2,3-dimethyl-1-butene (**27**) and 2,3-dimethyl-2-butene (**28**) without formation of methylpentenes over NaX, CaX, CuX, and CeX. Reaction was promoted by added D_2O, and the scheme involving cations **29** and **30** was proposed. Secondary cation **29** is rapidly converted by 1,2-methyl shift to tertiary cation **30** as expected without any conversion to a monobranched species which, as a Type B rearrangement, would require greater activation energy. Rates of exchange of **26** with D_2O paralleled rates of isomerization and hence the two processes appear related. More than one D was introduced

$$\underset{\mathbf{26}}{CH_2{=}CHC(CH_3)_2CH_3} \overset{H^+}{\rightleftarrows} \underset{\mathbf{29}}{CH_3\overset{+}{C}HC(CH_3)_2CH_3} \xrightarrow{\sim CH_3} \underset{\mathbf{30}}{CH_3CH(CH_3)\overset{+}{C}(CH_3)CH_3} \begin{cases} \xrightarrow{-H^+} \underset{\mathbf{27}}{CH_3CH(CH_3)C(CH_3){=}CH_2} \\ \xrightarrow{-H^+} \underset{\mathbf{28}}{CH_3C(CH_3){=}C(CH_3)CH_3} \end{cases}$$

per isomerization step, but this may have occurred by rapid interconversion between isomers **27** and **28** *via* cation **30** since at temperatures necessary to protonate **26**, protonation of either **27** or **28** would be expected to be quite rapid because of formation of the more stable tertiary cation rather than secondary cation. This hypothesis was strengthened since only one D was introduced per isomerization step when the interconversion of **27** and **28** was studied separately at lower temperatures. (Isomerization of **27** to **28** with incorporation of D over DY has been observed as low as —70°C (*244*).) This result also shows that at least in this case (CaX at 120°C) multiple reaction steps do not occur during a typical traverse of **27** or **28** through the pore structure.

George and Habgood (*245*) demonstrated that exchange between propylene and a D_2O-treated NaY containing traces of H^+ and Ca^{2+} at 400°C gave largely the expected exchange at C-1 and C-3 predicted by formation of *i*-propyl cation but also minor exchange at C-2 suggestive of some involvement of the less stable *n*-propyl cation at these higher temperatures.

Radical pathways for olefin isomerization have been discussed (*156, 192, 238*), but neither conclusively demonstrated nor clearly defined operationally by the proponents—*i.e.*, does butene isomerize *via* hydrogen atom addition–elimination involving 2-butyl radical or *via* hydrogen atom abstraction–addition involving methylallyl radical? The common use of observed selectivity ratios similar to those observed from studies of olefin isomerization over metals as indicative of radical mechanisms is semantically dangerous. Fragments resulting from dissociative chemisorption of C—H bonds on metals are surely not free radicals nor would they be expected to behave as such.

CONDENSATION REACTIONS—HYDRIDE TRANSFER AND CYCLIZATION. If an acidic O—H group can protonate an olefin to produce an adsorbed carbonium ion, then such an ion could add to another molecule of olefin to form a still larger adsorbed carbonium ion. Indeed, such cationic polymerization of olefins over acidic zeolites is very common and is often observed simultaneously with the double-bond shifts just discussed. Relatively clean products have been observed in a few cases at low temperatures, but more commonly, intrusion of still other processes such as hydride transfer and cyclization leads to complex mixtures of products. If these grow large enough to be trapped in the large cavity (the reverse molecular sieve effect (*12*)), they cannot be desorbed and "coke" formation ensues with masking of active sites and/or restriction of diffusion leading ultimately to loss of activity.

Venuto, Hamilton, and Landis (*79*) neatly demonstrated the sequence of reactions which occur with increasing temperature. When liquid 1-hexene was heated with deuterated REX zeolite at 64°C for 1.75 hr, there resulted a liquid phase containing isomeric hexenes and lesser amounts of dimers and trimers, all having been partially exchanged with D. Dissolution of the catalyst in aqueous acid gave a higher molecular-weight fraction containing $(C_6)_n$ products with $n = 2 > 3 > 4$ which had been trapped in the catalyst. The products from passing ethylene over REX varied with temperature: at

93°C, only a liquid aliphatic polymer trapped in the catalyst resulted and little gas was evolved; as temperature increased in steps to 213°C, the trapped material became more and more aromatic while the evolved gas contained largely paraffins, particularly *i*-butane. This result illustrates the hydrogen redistribution reaction so typical of acidic zeolite catalysts. Model behavior is offered by the conjunct polymerization in phosphoric acid described by Ipatieff and Pines (*246*) in which olefins are converted into oligomers both more and less hydrogen-rich than the starting material. The key step is hydride transfer between a carbonium ion and a hydrocarbon (*247*). The complete sequence of events, particularly the cyclization step, is unknown in detail, but the following exemplary hypothetical scheme to convert propylene to propane and benzene has good analogy for each of the key steps including the electrocyclic conversion of pentadienyl to cyclopentenyl cations (*248*) and the rearrangements connecting the very stable (*249*) cyclopentenyl and cyclohexenyl cations (*250*). An alternative cyclization-

$$H^+ + CH_2{=}CHCH_3 \rightarrow \underset{(R^+)}{CH_3\overset{+}{C}HCH_3} \xrightarrow{CH_2=CHCH_3} CH_3\overset{CH_3}{\overset{|}{C}}HCH_2\overset{+}{C}HCH_3$$

$$\downarrow -H^+$$

$$CH_2{-}\overset{CH_3}{\overset{|}{C}}{-}\overset{+}{C}H{-}CH{-}CH_2 \xleftarrow[(2)\ R^+,\ -RH]{(1)\ -H^+} CH_3\overset{CH_3}{\overset{|}{C}}{-}\overset{+}{C}H{-}CHCH_3 \xleftarrow[-RH]{R^+} CH_3\overset{CH_3}{\overset{|}{C}}HCH{=}CHCH_3$$

$$\downarrow$$

$$CH_3\text{-cyclopentenyl}^{(+)} \rightarrow \text{cyclohexenyl}^{(+)} \xrightarrow{-H^+} \text{cyclohexadiene} \xrightarrow[-RH]{R^+} \text{cyclohexadienyl}^{(+)} \xrightarrow{-H^+} \text{benzene}$$

aromatization route could involve thermal conversion of hexatrienes to cyclohexadienes (*251*). Since the initial dimeric product would be more strongly adsorbed than the original olefin, the situation is predisposed to the occurrence of secondary reactions after primary condensation. The complexity of products which can be built up and trapped within the pores is enormous, consisting not only of highly alkylated benzenes but also of polyalkylnaphthalenes and other higher polycyclic aromatics (*252*).

Norton (*253*) observed the rate order *i*-butene > propylene > ethylene for polymerization over CaX at 300°–350°C which conforms to carbonium ion stabilities. The liquid products from propylene were again complex but contained largely oligomers $(C_3)_n$ with $n = 2$–6. This paper first pointed out the analogies to conjunct polymerization. Over a series of X zeolites, consumption of propylene at 200°C was generally faster than that of ethylene (*150*), exceptions with transition metal ion-containing zeolites possibly being indicative of other mechanisms.

Weeks and Bolton (*254*) recently studied the isotopic distribution of products formed by exposing 4-^{13}C-1-butene to highly exchanged NH_4Y, activated at 550°C in air. The products desorbed into vacuum after 1 hr at 50°C accounted for 20% of the olefin charged and contained 44% isomeric octanes as well as other C_4–C_9, largely saturated products. Certain C_8 isomers (2,4- and 3,4-dimethylhexane and 2,3,3- and 2,3,4-trimethylpentane) showed mass spectra indicative of double labeling and could well have been formed by a carbonium-ion dimerization, isomerization, and hydride transfer route. However, others (2,3- and 2,5-dimethylhexane and 2,2,4-trimethylpentane) contained unlabeled, singly labeled, and triply labeled species to an extent which led the authors to conclude that they were derived by condensation to polymeric material followed by reverse breakdown with considerable carbon scrambling. Consideration of the C_4 and C_5 products formed at 200°C gave further evidence for complex scrambling pathways and reversible condensation-cracking pathways.

Condensation reactions become disfavored with respect to cracking above certain temperatures because of thermodynamic factors as already noted in Table I. Lower temperatures may also be beneficial for bimolecular condensation to the extent that they increase the surface concentration of olefin by adsorption. This may account for the observation (*255*) that the conversion of *i*-butene over NaY, presumably containing some cation deficiency, decreased over the range 200°–400°C; efficiency to octenes also decreased from 96% to 31%, suggestive of increasing contributions from secondary reactions.

A noticeable problem arises after having considered the preceding sections. On one hand, simple double-bond shifts in olefins are observable at ~ 200°C as relatively clean reactions on zeolites such as NaY containing traces of H^+ or Ca^{2+} sites; these can be rationalized by the intermediacy of monomeric adsorbed carbonium ions. On the other hand, for activated highly-exchanged NH_4Y or REY, reactions of olefins occur at much lower temperatures and are very prone towards condensation; the surface intermediates are clearly complex and probably polymeric. Whether this dichotomy is related to the quality (acid strength) or quantity (number and proximity of sites) of acidity remains to be solved.

ALKYLATION OF ISOPARAFFINS. Garwood and Venuto (*256*) reported reaction between excess *i*-butane and ethylene over activated $RE(NH_4)X$ catalyst at 27°C in the liquid phase to produce 2,3-dimethylbutane as the major liquid product; at higher temperatures, reaction was faster, but a more complex C_5–C_9 + product composition resulted. Similarly, Kirsch and coworkers (*257, 258*) reported alkylation of excess *i*-butane with 1- or 2-butene at 25°–100°C in the liquid phase over an activated $RE(NH_4)Y$ catalyst to give a product mixture in which trimethylpentanes predominated. The complex response of these reactions to time and temperature, and the complexity of the product mixture, demand a complex mechanism. A suggestive trace product, however, was *n*-butane whose precursor was identified as 1-butene by use of ^{14}C-labeled precursors. The tentative chain mechanism emerges with Initiation Reactions 29 and 30 and Propagation Reactions

31–33 which is analogous to that generally accepted for *i*-paraffin alkylation in strong homogeneous acids such as H_2SO_4 and $HF–BF_3$.

$$H^+ + CH_2{=}CHCH_2CH_3 \rightarrow CH_3\overset{+}{C}HCH_2CH_3 \quad (29)$$

$$CH_3\overset{+}{C}HCH_2CH_3 + (CH_3)_3CH \rightarrow CH_3CH_2CH_2CH_3 + (CH_3)_3C^+ \quad (30)$$

$$(CH_3)_3C^+ + CH_2{=}CHCH_2CH_3 \rightarrow (CH_3)_3C{-}CH_2\overset{+}{C}HCH_2CH_3 \quad (31)$$

$$(CH_3)_3CCH_2\overset{+}{C}HCH_2CH_3 \xrightarrow{\text{Type B}} (CH_3)_3CCH_2\overset{+}{C}(CH_3)_2 \quad (32)$$

$$(CH_3)_3CCH_2\overset{+}{C}(CH_3)_2 + (CH_3)_3CH \rightarrow (CH_3)_3CCH_2CH(CH_3)_2 + (CH_3)_3C^+ \quad (33)$$

Alkylation of Aromatics. The major reaction of olefins with added aromatics over acidic zeolites at temperatures $\lesssim 300°C$ is alkylation, and the extensive background in Friedel-Crafts chemistry (*259*) immediately suggests, by analogy, electrophilic attack of a carbonium ion, or highly polarized hydrogen-bonded complex, on the aromatic π-system to form a benzenium cation which can re-aromatize by proton loss. The organic aspects of alkylation have been developed particularly by the Mobil group (*11, 12, 76, 78, 79, 244*) using REX, REY, HY, and H-mordenite catalysts. The

$$H^+ + RCH{=}CH_2 \rightarrow R\overset{+}{C}HCH_3 \xrightarrow{C_6H_6} [C_6H_6(H)(CH(R)CH_3)]^+ \xrightarrow{-H^+} C_6H_5CH(R)CH_3 \quad (34)$$

optimum experimental conditions at $\sim 200°C$ for production of mono-alkylated product with minimal catalyst aging consist of using excess aromatic and maintaining liquid-phase conditions by use of pressure (*78*). The first condition suggests that attack of the alkyl carbonium ion on benzene is competitive with attack on olefin, the former leading to product and the latter to catalyst aging. Supporting this conclusion are IR spectral data of

Weeks and Bolton (*236*) for deammoniated or dehydroxylated NH_4Y at room temperature. Adsorption of benzene followed by 1-butene gave adsorbed *sec*-butylbenzene. However, if the olefin was adsorbed first and the resulting polymeric residue was allowed to form, subsequent addition of benzene did not produce alkylation. The second condition reinforces the importance of transferring the product, which is less volatile and more strongly adsorbed than the reactants, from the zeolite crystallite into the bulk phase (*see* above).

Alkylation of toluene with propylene is 10^3 times as rapid as with ethylene (*11*), indicative of the olefin being converted to a positively charged species. Alkylation of benzene with 1-hexene at 80°C gave both 2- and 3-phenylhexane (*78*). The formation of 3-hexyl cation could occur either by 1,2-hydrogen shift (Reaction 35) or by deprotonation–protonation (Reaction 36).

$$CH_2{=}CHCH_2CH_2CH_2CH_3 \xrightarrow{H^+} CH_3\overset{+}{C}HCH_2CH_2CH_2CH_3 \xrightarrow{\sim H} CH_3CH_2\overset{+}{C}HCH_2CH_2CH_3 \quad (35)$$

$$CH_3\overset{+}{C}HCH_2CH_2CH_2CH_3 \xrightarrow{-H^+} CH_3CH{=}CHCH_2CH_2CH_3 \xrightarrow{H^+} CH_3CH_2\overset{+}{C}HCH_2CH_2CH_3 \quad (36)$$

These linear cations must be captured by benzene before they can surmount the energy barrier needed (primary cation or protonated cyclopropane) to form more stable tertiary cations. Competitive alkylation of benzene and toluene at 125°–180°C with ethylene over REX showed the more electron-rich toluene to be more reactive and to show ortho-para selectivity (*260*); the data fit the Brown Selectivity relationship (*261*) for electrophilic aromatic substitution which predicts a linear free-energy relationship between the reactivity of the para position and the para:meta selectivity ratio. In these experiments the "ethyl cation" behaved as a highly reactive, modestly selective electrophile very similar to that generated in the C_2H_5Br–$GaBr_3$–ArH system. Each of these observations just listed is fully consistent with electrophilic aromatic substitution.

The kinetic arguments for a Rideal mechanism have been given above. However, more complex kinetics have recently been observed by Becker and co-workers (*114*) for ethylation of benzene over H-mordenite.

Alkylaromatic Reactions. Under more forcing conditions than those for selective mono-alkylation of aromatics with olefins just described, alkylaromatics can undergo further reactions over acidic zeolites of three major types: (1) positional isomerization on the ring, detectable for polyalkyl cases; (2) transalkylation (disproportionation) to form more and less highly alkylated products; and (3) cracking by dealkylation, the microscopic reverse of alkylation, Reaction 34.

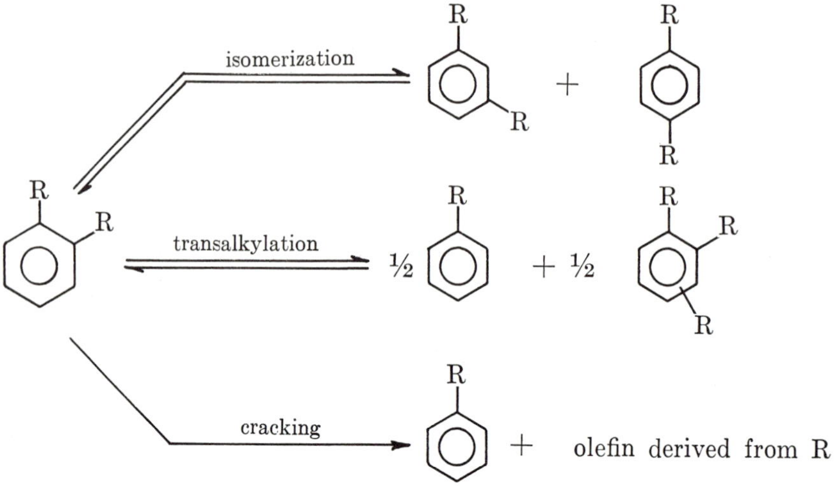

isomerization
transalkylation
cracking
+ olefin derived from R

Models for Protonation of Aromatics. Venuto, Wu, and Cattanach (*76*) reported that passage of benzene over a partially deuterated HY [$NH_4Y \xrightarrow{450°C} HY \xrightarrow{ND_3} ND_3HY \xrightarrow{550°C} DHY$] removed deuterium from the zeolite as deuteriated benzene at temperatures as low as 77°C. The predominance of benzene-d_1 in the effluent compared with polydeuteriated species suggested the absence of significant multiple exchange. Similar deuteriation of toluene occurred in the ring rather than in the side chain (*cf.* Ref. *262*) and, at 24°C, gave p:m:o-toluene-d_1 in the ratio 12.2:1.0:7.8. Competitive reactions between toluene and benzene fit the Brown Selectivity relationship described above. All the data thus point to electrophilic aromatic substitution by a protonic species with a sizeable steric requirement, and the authors visualized the mechanism according to Reaction 37. The inclusion

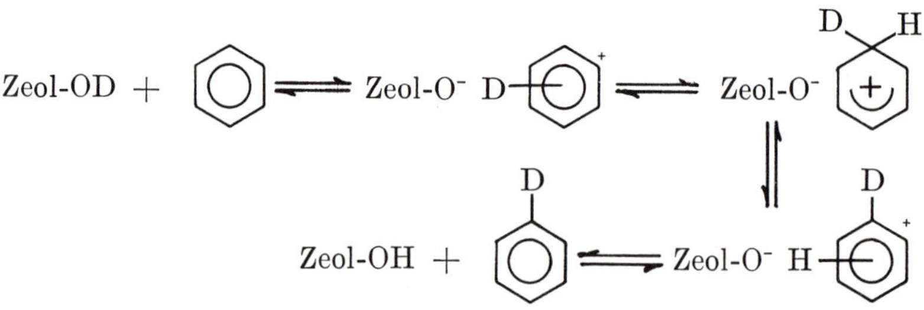

Zeol-OD + C_6H_6 ⇌ Zeol-O⁻ D–(C_6H_6)⁺ ⇌ Zeol-O⁻ (C_6H_6D)⁺ ⇌ Zeol-O⁻ H–(C_6H_5D)⁺ ⇌ Zeol-OH + C_6H_5D (37)

of a π- as well as a σ-complex is not obligatory from the data but is suggested by analogies with Friedel-Crafts reactions.

Deuterium redistribution studies (*14*) between C_6H_6 and C_6D_6 over CaY and LaY showed that variations in T_{act} affected activity in parallel fashion to that for dealkylation of cumene, suggestive of a common mechanism for both.

DEALKYLATION. *Ionic Pathway.* Having proposed a mechanism for alkylation of aromatics with olefins in Reaction 34, we should expect its microscopic reverse at higher temperatures where alkylaromatics are unstable with respect to olefin and the parent aromatic (Table I). Indeed, this appears consistent with most of the available data for dealkylation, especially of cumene which has become a standard test reaction for comparisons of catalytic activity (*cf.* Table III). Both kinetic (carbonium ion stability) and thermodynamic (Table I) arguments predict a cracking order: C_6H_5-*tert*-Bu > C_6H_5-*i*-Pr > C_6H_5Et > C_6H_5Me. No direct comparison seems available, but trends among various reports support this order.

Ward (*201*) recorded IR spectra of the O—H region of HY during cumene cracking at 250°C and observed a gradual loss of the HF band; no decreases of the LF band occurred until reaction temperatures exceeded 325°C. The results were interpreted as proof of a catalytic role of the O—H group with the more accessible O_1—H interacting at lower temperature. Actually, the fact that O—H intensity was only slightly affected early in the run at 250°C while reaction was occurring shows that most of the Brønsted centers existed as such rather than as adsorbed cations—*i.e.*, the protonation equilibria produce carbonium ions as reactive transient species rather than in stoichiometric proportions. Since the gradual O—H intensity decrease and catalyst aging appeared to parallel each other, hydrogen bonding to gradually increasing residue with a resultant frequency shift and broadening may be the cause of O—H group disappearance.

The suggestive proposal (*263*) that dealkylation requires an adjacent pair of appropriate sites will be discussed below since it was derived from studies of competitive dealkylation and isomerization.

The aromatic product from cumene cracking at $\leq$ 500°C is benzene, but the aliphatic product has been variously reported as totally propylene (various zeolites tested in a pulsed reactor at 300°–500°C (*106, 264*); Y zeolites promoted with HCl in a flow reactor at > 250°C and $P_{cumene} < 0.05$ atm (*220*); various Y zeolites in a flow reactor at 260°C and P_{cumene} = vapor pressure at 20°C (*50, 131*); and Ce, Ca, and HY zeolites in a flow reactor at 350°–450°C and $P_{cumene} = 0.047$ atm), as propylene (0.78 mole/mole C_6H_6) plus miscellaneous aliphatics (CaX in a flow reactor at 470°C and P_{cumene} = 1 atm (*133*)), and as largely propane (0.54 mole/mole C_6H_6) with no propylene (LaY, activated at 550°C, in a flow reactor at 325°C and P_{cumene} = 0.04 atm (*14*)). Using the same sample of LaY, Rabo and Poutsma (*14*) found that increasing T_{act} from 550° to 700° to 750°C decreased cumene conversion under the conditions listed above from > 98 to 60 to 20%; the corresponding values of the propylene:propane ratio were < 0.05, > 20, and > 50; for CaY with T_{act} = 550°C, the values were > 97% conversion and a 1.0 ratio. As the LaY (T_{act} = 550°C) catalyst aged with increasing time on stream, the propylene:propane ratio increased from < 0.02 to 2.3 and the total C_3 yield increased from 0.54 to 0.91 mole/mole C_6H_6. The rationalization is that primary cracking indeed gives benzene and propylene as expected but propylene is converted to propane in secondary hydride transfer reactions. Reasonable donors for exothermic hydride transfer to *i*-propyl cation are cumene, propylene itself, and forming residue. Similar behavior has been found in Friedel-Crafts reactions where alkylbenzenes were treated with $AlCl_3$-H_2O (*265*). The more active the catalyst, the more significant the contribution from secondary reactions. A second contributor to deficiencies in olefin formation is probably high P_{cumene} since hydride transfer is a bimolecular reaction. Analogous behavior to that of cumene was observed for dealkylation of *tert*-butylbenzene over REX at 260°C (*12*) which gave isobutane as the major gaseous product.

Radical Pathway. At cracking temperatures > 500°C, a new pathway appears as indicated by formation of new types of aromatic products in addition to benzene. From studies with cumene over alkali and alkaline earth ion-exchanged Y zeolites, Richardson (*85*) deduced the pres-

$R'-CH(CH_3)-C_6H_4R \rightarrow C_6H_5R + R'CH{=}CH_2$

$R'-CH(CH_3)-C_6H_4R \rightarrow R'-C({=}CH_2)-C_6H_4R$

$R'-CH(CH_3)-C_6H_4R \xrightarrow{H_2} R'-CH_2-C_6H_4R + CH_4$

R = H; R′ = CH_3; Ref. *85*
R = Et; R′ = H; Ref. *266*

ence of three parallel pathways at 550°C: the usual dealkylation to form benzene and propylene, a dehydrogenation to form α-methylstyrene and presumably H_2, and a demethylation to form ethylbenzene and presumably CH_4 with consumption of H_2; ethylbenzene further dehydrogenated to form styrene. The latter two pathways, which must have higher E_a values than dealkylation since they only emerge at high temperature, were grouped as radical. A ratio of carbonium ion: radical activity was then measured at 550°C and a defined flow rate as 0 for LiY, 0.43 for BaY, 2.03 for CaY, 250 for MgY, and ∞ for HY, although this ratio increased with increasing contact time at least for BaY. Since the radical activity was not poisoned by Qn and was most prevalent on the least acidic zeolites, the active sites were assigned as the cations themselves with the suggestion of a role for electron transfer. For reaction of *p*-diethylbenzene over partially exchanged decationized NH_4X ($T_{act} = 570°C$; crystallinity maintained) at 500°–568°C, Forni and Carra (*266*) suggested an analogous network: acid-catalyzed dealkylation to form ethylbenzene, dehydrogenation to form ethylstyrenes, and demethylation to form methylethylbenzenes. The carbonium ion: radical ratio here decreased with increasing contact time, and it was suggested that dealkylation was at equilibrium while dehydrogenation and demethylation were not. The radical component was absent in NaX and increased with NH_4^+ exchange; it was correlated with an ESR signal (of unknown origin) observed for the catalyst. The authors suggested that homogeneous cracking reactions might be initiated by heterogeneous surface reactions. Traces of ethylbenzene and toluene were also observed from cumene over normal and acid-extracted H-mordenite at only 360°C and were cataloged as radical products because their amounts increased somewhat by use of H_2 rather than He carrier gas (*120*).

It is interesting to attempt to reconcile this so-called radical pathway with what would be expected for purely thermal cracking (*267*), as Forni and Carra (*266*) did partially. Since hydrogen atom abstraction from cumene would occur preferentially at the benzylic position, the standard homolytic chain decomposition (*see* above) could account for dehydrogenation to form α-methylstyrene by Reactions 42 and 43; minor attack at the methyl group could give styrene by Reactions 44 and 45. However, the demethylation product ethylbenzene is difficult to perceive as a chain product. The combination of initiation (Reactions 38–40) and short chain sequence could qualitatively account for the products. Whether the initiation process is specifically catalyzed remains obscure.

Initiation:

$$C_6H_5CH(CH_3)CH_3 \rightarrow C_6H_5\dot{C}HCH_3 + \dot{C}H_3 \quad (38)$$

$$C_6H_5\dot{C}HCH_3 + C_6H_5CH(CH_3)CH_3 \rightarrow C_6H_5CH_2CH_3 + C_6H_5\dot{C}(CH_3)CH_3 \quad (39)$$

$$\dot{C}H_3 + C_6H_5CH(CH_3)CH_3 \rightarrow CH_4 + C_6H_5\dot{C}(CH_3)CH_3 \quad (40)$$

$$3\ C_6H_5CH(CH_3)CH_3 \rightarrow C_6H_5CH_2CH_3 + CH_4 + 2\ C_6H_5\dot{C}(CH_3)CH_3 \quad (41)$$

Propagation:

$$C_6H_5\dot{C}(CH_3)CH_3 \rightarrow C_6H_5C(CH_3){=}CH_2 + H\cdot \quad (42)$$

$$H\cdot + C_6H_5CH(CH_3)CH_3 \rightarrow H_2 + C_6H_5\dot{C}(CH_3)CH_3 \quad (43)$$

$$C_6H_5CH(CH_3)\dot{C}H_2 \rightarrow C_6H_5CH{=}CH_2 + \dot{C}H_3 \quad (44)$$

$$\dot{C}H_3(H\cdot) + C_6H_5CH(CH_3)CH_3 \rightarrow CH_4(H_2) + C_6H_5CH(CH_3)\dot{C}H_2 \quad (45)$$

Termination:

$$2\,C_6H_5\dot{C}(CH_3)CH_3 \rightarrow C_6H_5CH(CH_3)CH_3 + C_6H_5C(CH_3){=}CH_2 \quad (46)$$

ISOMERIZATION AND TRANSALKYLATION. At ~ 150°–350°C where dealkylation is not yet thermodynamically highly favored, common reactions of alkylaromatics over acidic zeolites are the approximately thermoneutral isomerization and transalkylation reactions. Since these processes are generally competitive and in some cases mechanistically interrelated, they are considered together.

Models. The isomerization–transalkylation reactions induced by typical Friedel-Crafts catalysts such as H_2O-activated $AlCl_3$ at ambient temperatures have been studied in some detail and considerable mechanistic complexity has emerged. Some of this work will be sketched briefly as background for consideration of the same reactions catalyzed by zeolites.

Acid-catalyzed isomerization can occur by at least three pathways, one intramolecular, one intermolecular involving transalkylation, and one dissociative by reversible dealkylation–alkylation. In the intramolecular pathway (*268*), the key step is a 1,2-alkyl shift in a benzenium ion (Reaction 47)

$$\text{1,2-R}_2C_6H_4 \xrightarrow{H^+} [\text{benzenium ion, H, R}]^+ \xrightarrow{\sim R} [\text{benzenium ion, R, H}]^+ \xrightarrow{-H^+} \text{1,3-R}_2C_6H_4 \quad (47)$$

with further subtleties concerning the role of intermediate π-complexes. The intermolecular route (*269*) proceeds through transalkylation (Reaction 48), by either of the transalkylation pathways detailed below. It has been invoked

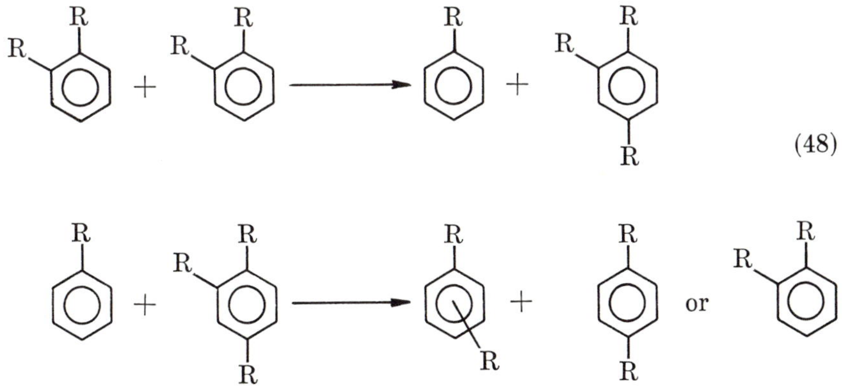

(48)

to account for observed isomer patterns which cannot be reconciled with a series of 1,2-shifts; *e.g.*, 1-^{14}C-ethylbenzene gives more of the 3-^{14}C- and 4-^{14}C-isomers than the 2-^{14}C-isomer with $AlBr_3$-HBr at 0°C (*270*). For either mechanism, the observed rate order for isomerization of alkyl groups is R = iso-Pr > Et > Me with indications that the inequalities are greater for the inter- than the intramolecular route (*269*)—*e.g.*, the ratios of intra-: intermolecular pathways for xylenes, ethyltoluenes, *i*-propyltoluenes, and *tert*-butyltoluenes with $AlCl_3$ were estimated to be 100:0, >84:<16, >14: < 86, and 0:100, respectively (*269*). *tert*-Alkyl-substituted aromatics may follow the third mechanistic class where complete dissociation of the stable *tert*-alkyl cation occurs to allow both isomerization and transalkylation (Reaction 49).

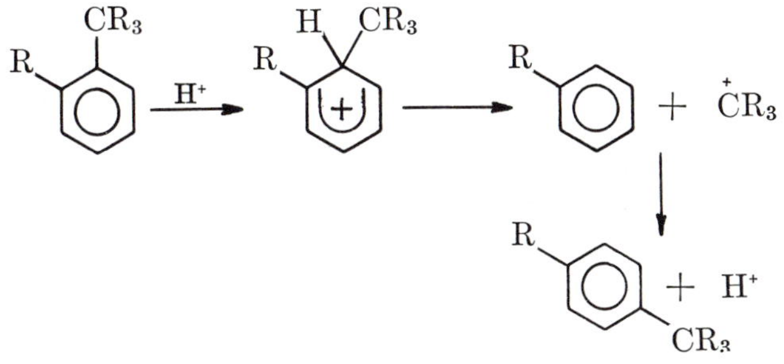

(49)

Two distinct transalkylation mechanisms have also been discussed, in addition to dealkylation–alkylation. In the first (*271*), the alkyl group in a benzenium ion is attacked by the nucleophilic ring of the second aromatic and is transferred as an unrearranged unit without being a free carbonium ion (Reaction 50). Based on detailed study of the rate of ^{14}C

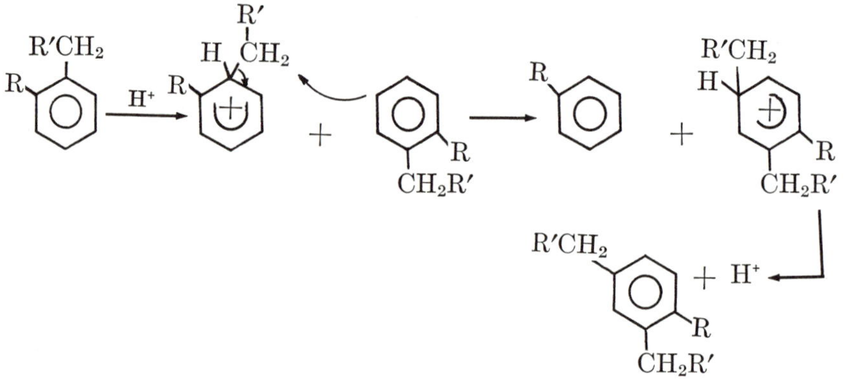

(50)

and d loss and of racemization on treatment of optically active 1-^{14}C-α-d-ethylbenzene in excess benzene with $GaBr_3$–HBr, Streitwieser and Rief (*272*) proposed a cationic chain mechanism in which actual benzylic ions were formed by hydride transfer from the α-C—H bond and actual 1,1-diarylmethane intermediates **31** occurred as in Equation 51. Based on relative

$$\text{trace olefin} \xrightarrow{H^+} R^+$$

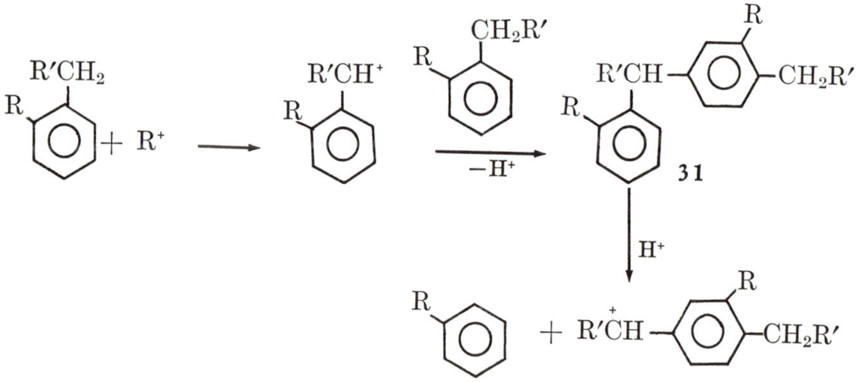

(51)

carbonium ion stabilities this mechanism also predicts an order i-Pr > Et > Me but is impossible for R = *tert*-Bu.

In the ensuing discussion we will designate these various mechanistic pathways as intramolecular isomerization, intermolecular isomerization, transalkylation *via* group transfer, transalkylation *via* benzylic cations, and dealkylation-alkylation.

Zeolite Catalysis. Examples of transalkylation of monoalkylbenzenes over zeolites to form benzene and dialkylbenzenes include that of toluene over REX in the liquid phase under 400 psig at 264°C (*78, 273*), of toluene over activated NH_4Y and NH_4 mordenite in the vapor phase at 400°C (*70*), of toluene over a variety of protonic and polyvalent cation-exchanged Y and mordenites at 400°C (*110*), and of cumene as a minor side reaction during dealkylation (*14, 120, 220*). Simultaneous isomerization and transalkylation has been observed for the xylenes (*2, 102, 159, 165, 212, 274, 275*), the methylethylbenzenes (*73, 155, 263, 276, 277, 278*), and the diethylbenzenes (*279*).

Bolton, Lanewala, and Pickert (*279*) followed the rate and composition of products formed from treatment of liquid *o*-diethylbenzene with activated $Ce(NH_4)Y$ zeolite at 170°C. The initially formed products were ethylbenzene and 1,2,4-triethylbenzene; isomeric diethylbenzenes and 1,3,5-triethylbenzene appeared only after the concentration of the 1,2,4-isomer had built up. An analogous product profile resulted from the *p*-diethylbenzene starting material. However, with the meta isomer, both triethylbenzene isomers appeared simultaneously but again before any isomeric diethylbenzenes. The apparent isomerization rate of the ortho isomer was enhanced by adding ~ 25 mole % benzene and 1,3,5-triethylbenzene to the initial reaction mixture. The conclusion drawn was that isomerization proceeded primarily by the intermolecular transalkylation route rather than intramolecularly. The transalkylation mechanism *via* benzylic cations was presumed and the observed selectivity for initial formation of the 1,2,4-triethyl isomer follows directly as shown in Reaction 52, with a similar pathway for the para case; however,

2 [o-diethylbenzene] ⇌ [benzylic-linked dimer] ⇌ [ethylbenzene] + [1,2,4-triethylbenzene] (52)

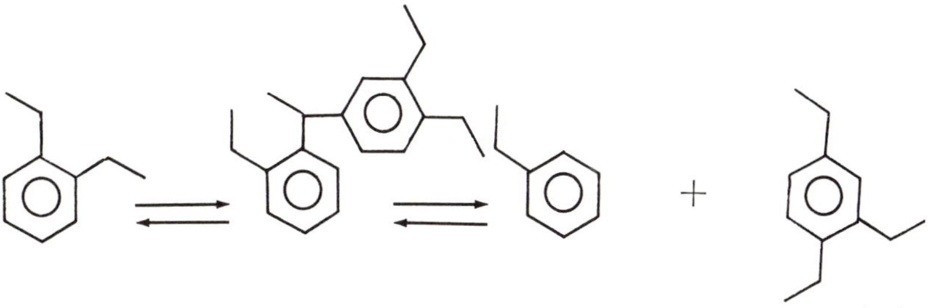

meta starting material can give both triethylbenzenes as shown in Equation 53. (The sterically crowded 1,2,3-triethylbenzene is assumed to be only a trace isomer.) Of course, transalkylation *via* alkyl group transfer would predict the same patterns and hence, while this study does strongly support isomerization *via* transalkylation, it does not reveal which specific transalkylation mechanism is involved. With regard to comparing zeolite-catalyzed with typical Friedel-Crafts reactions, Olah and co-workers (*268*) observed that isomerization of *o*-diethylbenzene with $AlCl_3$–H_2O at ~ 25° gave isomerization before transalkylation and inhibition of isomerization by added benzene; therefore they assigned an intramolecular isomerization route. The difference between the results does not appear to be related to the temperature difference because we shall see below that, at least for zeolites, the intramolecular is favored over the intermolecular route by increasing

(53)

temperature. Whatever the reason, here is another example of the tendency of the zeolitic structure to favor bimolecular reactions compared with other acidic catalysts.

Over the same catalyst, this group (*274*) found that the simultaneous transalkylation and isomerization of the xylenes in the liquid phase required higher temperatures (250°–300°C) but again product distributions indicated that at least a considerable portion of the isomerization proceeded by transalkylation. We have already seen, however, evidence (*212*) for the Mobil AP Catalyst (*275*) that isomerization was largely intramolecular at 205°C based on the effects of crystallite size. It is not apparent whether this apparent discrepancy results from differences in zeolite type. Finally, Bolton and co-workers (*280*) also proposed a transalkylation route for the *tert*-butylphenols for which the ortho isomer proceeded to the para with no buildup of meta. Although transalkylation was proposed by alkyl group transfer, benzylic cations being unattainable for this case, dealkylation–alkylation is not ruled out.

The behavior of 1-methyl-2-ethylbenzene has been studied in detail by Csicsery (*155, 276, 277, 278*) over both zeolites and silica-alumina in the presence of H_2. Over both silica-alumina and $Ca(NH_4)Y$ ($T_{act} = 540°C$ in H_2) it was concluded (*278*) that isomerization was largely intermolecular at low temperatures, such as those used by Bolton for the diethylbenzenes

(*279*), but intramolecular at higher temperatures. For example, for $Ca(NH_4)Y$, isomerization was ~ 80% intermolecular at 204°C but ~ 80% intramolecular at 315°C. The zeolite favored the intermolecular route at any chosen temperature compared with silica–alumina. This conclusion was drawn not only from the observed greater apparent E_a for isomerization than for transalkylation but also from the detailed nature of the products. Isomerization *via* transalkylation requires two alkyl group transfers of the same type of group, either both ethyl or both methyl. Transfer of one type of each gives instead a mixture of xylenes and diethylbenzenes. In the first transalkylation step, ethyl transfer was favored over methyl by 37-fold over the zeolite at 204°C and by 14-fold over silica-alumina. Assuming that these preferences also hold for the second step, these workers used the amounts of xylenes and diethylbenzenes as a measure of isomerization by transalkylation.

To rationalize the greater transalkylation:intramolecular isomerization ratios for the Y zeolites compared with silica–alumina, Csicsery and Hickson (*155*) suggested that intramolecular isomerization required only a Brønsted acid site, which predominates on silica–alumina, whereas transalkylation required a Lewis acid site or a Brønsted-Lewis site pair; a possible role for the Lewis acid site was suggested to be hydride transfer to form benzylic ions which initiate a chain transalkylation scheme. Two alternatives to this proposal can be deduced from recent literature, each of which would depend on the fact that transalkylation is bimolecular and isomerization by 1,2-shift is unimolecular. Wang and Lunsford (*281*) measured the temperature-programmed desorption of toluene from NH_4Y ($T_{act} = 400°C$) compared with silica–alumina over the range 50°–350°C. Not only did the zeolite adsorb ~ 2.5 times more toluene, but it retained it to considerably higher temperature. Hence, they suggested that zeolites might exhibit greater activity than silica–alumina, particularly for bimolecular reactions, because of a greater concentration of reactants (and intermediates) in the zeolite because of the stronger adsorption; *i.e.*, reaction steps could be more rapid in the zeolite even with identical rate constants for interactions at the active sites. The second alternative is developed below.

The behavior of 1-methyl-2-ethylbenzene at still higher temperature (360°C) over H-mordenite (11.2% Al_2O_3) and a severely acid-leached H-mordenite (0.1% Al_2O_3), each activated at 538°C, was reported by Bierenbaum, Partridge, and Weiss (*263*). The major products over H-mordenite were isomerized methylethylbenzenes and toluene; because of the virtual absence of trialkylbenzenes, toluene was considered a dealkylation rather than transalkylation product. The isomerization:dealkylation ratio increased from 2.6 to ~ 100 on acid leaching. These authors proposed that the difference resulted from differences in density of tetrahedral Al sites (shorthand description $\overline{AlO_2}$). It was proposed (*282*) that the benzenium ion **32**, adsorbed on an $\overline{AlO_2}$ site, could give intramolecular isomerization directly but could give dealkylation only if there were another properly positioned $\overline{AlO_2}$ site to receive the ethyl cation as pictured in Reaction 54. The proposal may need minor revision as written since the receptor of the

$$\text{(scheme 54: 1-methyl-2-ethylbenzene} \xrightarrow{+\ H\overline{AlO_2}} \textbf{32} \text{ (arenium ion with } \overline{AlO_2}\text{)}; \ \textbf{32} \xrightarrow{+\ H\overline{AlO_2}} \overset{+}{C_2H_5} \ldots \overline{AlO_2}H^+ \longrightarrow \text{Dealkylation}; \ \textbf{32} \longrightarrow \overline{AlO_2} \longrightarrow \text{Isomerization)} \quad (54)$$

ethyl cation also bears a proton which may need to migrate, but the basic idea deserves attention since it applies to the broader question we have already introduced of whether any process in which one cation is cleaved to form a stable molecule and a smaller cation (a key step for example in paraffin cracking) can occur on a single $\overline{AlO_2}$ unit by a transition state in which the dispersed positive charge is associated with only one anionic site (**33**) or whether it requires adjacent sites (**34**). If the latter is true,

$$\overset{\delta+}{RCH} = CH_2 \cdots \overset{\delta+}{R'} \text{ both bridged to one } \overline{AlO_2} \quad (\textbf{33}) \qquad \overset{\delta+}{RCH} = CH_2 \cdots \overset{\delta+}{R'} \text{ each bridged to separate } \overline{AlO_2} \quad (\textbf{34})$$

then unique conclusions may emerge concerning the role of zeolite crystallinity because it determines rigidly the spacing between sites.

Returning now to why Y zeolites appeared to favor transalkylation over intramolecular isomerization compared with silica–alumina, we have seen three hypotheses each focusing on a feature of zeolite structure: the presence of Lewis acid centers, the greater adsorptive capacity at reaction temperatures, and the rigid framework structure which may position pairs of sites appropriately. Deciding among ideas such as these poses a challenge for future work.

Some data of Csicsery and Hickson (*155, 276*) for reactions of 1-methyl-2-ethylbenzene over a variety of zeolites are listed in Table IV. At first glance the isomerization:transethylation ratios at 204°C are bewildering; however, note that this ratio tends to increase with increasing extent of conversion and thus provides additional evidence that low-temperature isomerization does proceed by a series of transethylations. At 315°C the ratios depend much less on conversion as predicted if the processes are now largely independent rather than sequential. After correcting for effects of conversion, the H-mordenite appears to promote isomerization over transethylation compared with the Y zeolites. This may indicate that transalkylations are more sterically hindered in the narrower mordenite pore

Table IV. Isomerization: Transethylation Ratios for 1-Methyl-2-ethylbenzene over Various Zeolites[a]

		Extent of Reaction (%)		*Isomerization/ Transethylation*	
Zeolite	T_{act}, °C	*204°C*	*315°C*	*204°C*	*315°C*
(Silica-alumina)	530	0.64	5.1	0.6	5.8
NH_4Y—80% exchanged	540	9.6	—	0.35	—
NH_4Y—>99% exchanged	540	72.	—	0.9	—
MgY	540	0.24	5.9	0.03	2.3
CaY	540	—	0.32	—	2.2
$Mg(NH_4)Y$	540	23.	—	0.75	—
$Ca(NH_4)Y$	540	0.39	9.7	0.03	2.2
H-mordenite	593	15.	—	25.	—
H-mordenite	842	0.12	15.9	0.4	146.

[a] H_2:RH = 3 for all cases except H-mordenite where ratio was 5. Data from Refs. *155, 276*.

structure and that intramolecular isomerization is therefore preferred by default. Shape selectivity factors on this reaction (*276, 277*) are discussed in another chapter.

An additional study of the xylenes in addition to those discussed above was reported by Morita and co-workers (*102*) who worked at 350°C with NH_4Y, activated at 450°C. Below 50% NH_4^+ exchange, isomerization predominated, but as exchange increased further, transalkylation activity increased relatively more rapidly so that a crossover occurred at ~ 80% exchange. The trend of decreasing isomerization:transalkylation ratio with increasing total conversion is opposite of that expected for isomerization *via* transalkylation (Table IV). Coupling this observation with the higher temperature and the inherent tendency of xylenes to follow the intramolecular route (*269*) almost assures us that isomerization in this case was intramolecular. The authors noted that transalkylation activity correlated with total acidity ($H_0 \leq +3.3$) while isomerization correlated with only acidity stronger than $H_0 \leq -3.0$ but offered no mechanistic rationalization. If both reactions proceed through the same protonated benzenium ion, this difference in acidity dependence is surprising. The results could be rationalized with Csicsery's proposal (*155*) if the more highly exchanged NH_4Y catalysts underwent relatively more dehydroxylation at 450°C.

Paraffin Reactions. Both the idealized processes of olefin polymerization and *i*-paraffin alkylation should be reversed at higher temperatures (Table I), and cracking reactions were among the first zeolite-catalyzed conversions to be discovered. One of the first studies (*133*) revealed two major pathways as deduced from C_1–C_5 product distributions from cracking *n*-decane. The C_1–C_5 products normalized to 100% at 25–30% conversion are tabulated in Table V for cracking over NaX zeolite at 500°C, over CaX zeolite at 470°C, and over amorphous silica-alumina at 500°C. The products over CaX were grossly similar to those over silica–alumina, but the zeolite gave larger paraffin:olefin ratios for every carbon skeleton. The products over NaX were markedly different; note particularly the lack of skeletally branched products and the shift towards lower carbon numbers. From these and later product data (*14, 151, 173*) has come the generalization of two

Table V. Normalized Product Distribution for C_1-C_5 Products (%) from Cracking of *n*-Decane[a]

Product	*Catalyst*		
	NaX	*CaX*	*Silica–alumina*
CH_4	15.5	9.0	5.4
C_2H_4	8.4	2.1	4.9
C_2H_6	19.3	5.1	5.2
C_3H_6	13.4	9.5	17.0
C_3H_8	13.7	14.9	13.8
n-C_4H_8	10.4	7.1	8.6
i-C_4H_8	0.2	1.6	4.6
n-C_4H_{10}	7.7	14.6	5.7
i-C_4H_{10}	0.1	14.3	8.6
n-C_5H_{10}	5.6	1.2	9.9
i-C_5H_{10}	0.6	2.0	3.9
n-C_5H_{12}	5.0	3.9	3.8
i-C_5H_{12}	0.1	14.7	8.6

[a] Data from Ref. *133*.

mechanisms for catalytic cracking of paraffins (and olefins which follow similar patterns)—one *via* carbonium ions and one *via* free radicals .

Ionic Cracking and Isomerization. The only close analogies lie in the behavior of the earlier exploited amorphous cracking catalysts such as silica–alumina. The original theory of carbonium ion cracking was developed for these catalysts (*21, 283, 284*) to rationalize certain product characteristics not encountered in homogeneous thermal cracking such as highly isomerized olefins, paraffins > C_3, low yields of C_1 and C_2 compared with C_3 and higher products, and high ratios of branched:normal products. By some process an initial carbonium ion R^{n+} is formed which activates the paraffinic substrate RH by hydride transfer, Reaction 55. Since the substrate

$$R^{n+} + RH \rightarrow R^nH + R^+ \qquad (55)$$

will contain many types of C—H bonds, R^+ will be a family of cations with tertiary favored over secondary over primary. These carbonium ions R^+ can then undergo three basic reactions: intramolecular positional and skeletal rearrangements, β-scission to form an olefin and a smaller cation R'^+ (Reaction 56), and hydride transfer. Both isomerizations and hydride transfers

$$R^+ \equiv R_1CH_2\overset{+}{C}HR_2 \rightarrow CH_2{=}CHR_2 + R_1^+ \equiv R'^+ \quad (56)$$

are easily observed in low temperature model systems largely to the exclusion of β-scission (cracking) except for strained systems. However, β-scission would be expected to have the most favorable $\Delta S^{\neq}$ (A factor) of the three processes and hence should become increasingly important as temperature increases. The relative rates of carbonium ion rearrangement and β-scission at any temperature will determine whether a given R^+ family reaches pseudo-equilibrium with respect to all skeletal isomers before cracking becomes significant. After n β-scissions, the derived ion R^{n+} will have become so small that it cannot cleave further without forming a very unstable ion such as methyl, and hydride transfer (Reaction 55) will dominate to start a new chain. Although approximate predictions of product compositions in terms of carbon numbers can be made on the basis of these considerations, exact compositions cannot be deduced because of lack of detailed knowledge of the relative rates of all the relevant carbonium ion processes and because of rapid secondary reactions of primary olefinic products. Temperatures sufficient to induce endothermic cracking are more than sufficient to induce those typical olefin reactions which equilibrium considerations allow.

Carbonium Ion Generation. The source of the initial carbonium ions has been a subject of debate (*284*) for many years. One view is that they result from protonation of traces of olefins as impurities in the feed or formed by minor thermal cracking reactions. The second view is that direct attack on the paraffin by the active catalyst site can generate carbonium ions by one of three paths: a Lewis acid site abstracts a hydride ion (Reaction 57), a Brønsted acid site attacks the hydrogen of a C—H bond to form H_2 (Reaction 58), or a Brønsted acid site attacks carbon to break a C—C bond, possibly *via* a pentacoordinate homolog of CH_5^+ (Reaction 59). An imagina-

$$RH \xrightarrow{\text{—Al—}} R^+ + H\text{—}Al^-\text{—} \quad (57)$$

$$RH \xrightarrow{H^+} R^+ + H_2 \quad (58)$$

$$RH \xrightarrow{H^+} R'H + R''^+ \quad (59)$$

tive experimental approach to probe the effect of trace olefins has been described recently by Weisz (*6*) using the very active H-mordenite which converts *n*-butane already at 230°C, presumably to form *i*-butane and cracked products. Before entering the cracking reactor held at 230°C, the *n*-butane diluted with H_2 was passed over a non-acidic Pt/Al_2O_3 hydrogenation–dehydrogenation catalyst which established the temperature-dependent C_4H_{10}–C_4H_8–H_2 equilibrium. The *n*-butane conversion over the H-mordenite could be varied from 0 to $> 90\%$ by varying the temperature of this pre-reactor from 250° to 550°C which altered the calculated *n*-butene concentration from 0.001% to 10% of the *n*-butane. Although this and other (*285*) experiments do not rule out the ability of this or other zeolites to directly activate paraffins at higher temperatures without the intervention of olefins, they clearly demonstrate the effectiveness of trace olefins in initiating cracking reactions. The several indications that paraffin cracking, in particular among hydrocarbon transformations, requires a few Lewis acid sites or a few especially strong Brønsted acid sites for maximum activity tempt one to postulate direct attack, but this will be extremely difficult to prove experimentally. There are, of course, very suggestive analogies for model Brønsted super acids attacking paraffins (*see* above). Another possible analogy for attack on a paraffinic C—H bond is Kerr's demonstration (*286*) of Reaction 60.

$$HY + Me_3SiH \rightarrow Me_3Si\text{—}Y + H_2 \qquad (60)$$

Single Paraffins: Cracking and Isomerization. Most of the detailed product studies reported have been from hexanes, pentanes, and butanes, largely because of the analytical difficulties which arise for higher paraffins because even these C_6 and lower alkanes give very complex mixtures of cracking products. Unfortunately, these do not contain enough carbon atoms to allow the most probable carbonium ion cracking routes to manifest themselves by the lowest energy pathways and are therefore in a sense special cases. For example, the lowest energy cracking route imaginable involving only one β-scission would form a smaller tertiary cation from an original tertiary cation. Yet the smallest structure for which this is possible is the 2,4,4-trimethyl-2-pentyl cation as shown in Reactions 61 and 62. Indeed, cracking of 2,2,4-trimethylpentane at 300°C in an H_2 atmosphere over

$$\begin{array}{ccccccccccccc} & & C & & & & & & C & & & & \\ & & | & & & & & & | & & & & \\ C & — & C & — & C & — & \overset{+}{C} & — & C & \rightarrow & C & — & C^+ & + & C & = & C & — & C \\ & & | & & & & | & & & & & & | & & & & | & & \\ & & C & & & & C & & & & & & C & & & & C & & \end{array} \qquad (61)$$

$$\begin{array}{l} \text{C—}\overset{C}{\underset{C}{|}}\text{C}^+ + \text{C—}\overset{C}{\underset{C}{|}}\text{C—C—}\overset{H}{\underset{C}{|}}\text{C—C} \rightarrow \text{C—}\overset{C}{\underset{C}{|}}\text{C—H} + \text{C—}\overset{C}{\underset{C}{|}}\text{C—C—}\overset{+}{\underset{C}{|}}\text{C—C} \end{array} \qquad (62)$$

various Y zeolites has been reported (*98, 167*) to give largely *i*-butane, *i*-butene, and 2-butene in a ratio of ~ 1.2:1.0:0.1, a remarkably clean product distribution compared with most paraffin cracking reactions. Even for this system at higher temperatures, linear butenes, propane, propylene, and methane appear and were suggested to be secondary products derived from *i*-butene.

In addition to analytical difficulties, any attempt to determine primary products from cracking of a moderately large paraffin as the basis of mechanistic thinking seems nearly impossible because olefinic products probably undergo secondary reactions before they leave the pore structure of a crystallite. However, it may be possible to deduce primary products from studies of hydrocracking and hydroisomerization over acidic zeolites containing a noble metal hydrogenation component which produce predominantly paraffin products.

In the usual dual-function mechanism, hydrogenation–dehydrogenation is postulated to occur on the metal phase and carbonium–ion cracking on the zeolite phase; hydrocracking can then be described by Reaction 63. Secondary reactions and coking are much less important than in cracking because the olefin concentration is kept low.

$$C_nH_{2n+2} \underset{Pt}{\overset{-H_2}{\rightleftarrows}} C_nH_{2n} \underset{zeolite}{\overset{H^+}{\rightleftarrows}} C_nH_{2n+1}^+$$

$$\begin{array}{ccccc} C_nH_{2n+1}^+ & \underset{zeolite}{\overset{cracking}{\longrightarrow}} & C_mH_{2m} & + & C_{n-m}H_{2n-2m+1}^+ \\ & & \text{Pt} \uparrow\downarrow H_2 & & \text{Zeolite} \uparrow\downarrow -H^+ \\ & & C_mH_{2m+2} & & C_{n-m}H_{2n-2m} \\ & & & & \text{Pt} \downarrow\uparrow H_2 \\ & & & & C_{n-m}H_{2n-2m+2} \end{array} \quad (63)$$

If we accept this separability of activities [Bolton and Lanewala (*287*) suggested that hydroisomerization of hexanes over $Pd/RE(NH_4)Y$ at 285°C and 500 psig involves a bimolecular isomerization mechanism characteristic of noble metal-catalyzed isomerizations], then considerable insight into carbonium ion behavior can be obtained from the studies with *n*-dodecane over 0.5% Pt/CaY zeolite reported by Schulz and Weitkamp (*288, 289*) because of their unusually detailed product analyses. Using a H_2:*n*-$C_{12}H_{26}$ ratio of 20:1 at 40 atm and LHSV = 1.0 hr^{-1}, they observed 10% isomerization to branched dodecanes with negligible cracking at 250°C; by 275°C isomerization reached its maximum of 48% accompanied by 17% cracking; by 300°C activity had increased to the point where only cracking could be observed. It was therefore concluded that carbonium ion skeletal rearrangements were faster than β-scission in this temperature range. The initially isomerized C_{12}'s at 250°C were largely monomethyl branched with 2-: 3-: 4-: 5-: 6-methylundecane = 13.6: 24.3: 23.0: 26.3: 12.8, but multiply branched C_{12}'s appeared only a few degrees higher and became equal to the monobranched isomers by 290°C. It was postulated that the initially formed *sec*-C_{12} carbonium ions rearranged (Type B) to an almost equilibrated set of monomethyl *tert*-C_{12} carbonium ions. For example, we would write Reaction 64 for the 4-dodecyl cation **35**. The initial cracking products at 265°C (5% cracking, 34% isomerization) contained exactly 200 moles of product per 100 moles cracked, indicative of purely primary cracking. [The authors described the fate of the cracked carbonium ion as interchange with C_{12} olefin to

35

(64)

give a new C_{12} carbonium ion and a smaller olefin which was rapidly hydrogenated; *i.e.*, the strength of chemisorption of olefins as cations increased with increased size.] The product composition was fully symmetrical about C_6 with $C_1:C_2:C_3:C_4:C_5:C_6:C_7:C_8:C_9:C_{10}:C_{11} = 0:0:6.7:29.7:42.3:43.5:42.3:29.5:6.3:0:0$. The C_6 cracked product was relatively richer in 2-methylpentane and poorer in 2,2-dimethylbutane than predicted from equilibrium. Based on these facts, the authors proposed that the monomethyl branched *tert*-C_{12} cations cracked as shown schematically in Sequence 65. If one assumed that each of

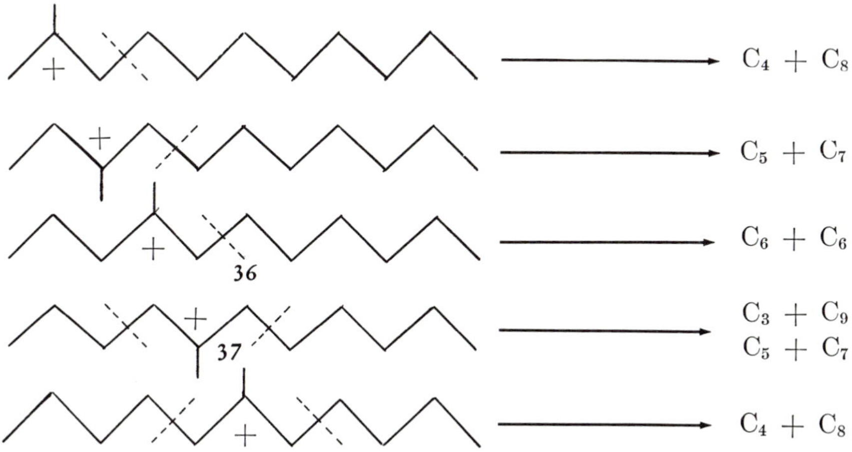

$\longrightarrow C_4 + C_8$

$\longrightarrow C_5 + C_7$

36 $\longrightarrow C_6 + C_6$

37 $\longrightarrow C_3 + C_9$, $C_5 + C_7$

$\longrightarrow C_4 + C_8$

(65)

these cations was equally reactive and that their relative concentrations were represented by the relative amounts of methylundecanes observed as isomerization products at the same temperature, then the cracked product distribution of $(C_4 + C_8):(C_5 + C_7):(C_6 + C_6)$ was predicted with remarkable accuracy. One major ambiguity remains in this ideal Schulz-Weitkamp picture: if the cracked cations formed were primary as shown, then why is no ethane formed? If cation **37** gives equal amounts of *n*-propyl and *n*-pentyl cations, why does not cation **36** give ethyl as well as *n*-hexyl cation? Noting that β-scission of a tertiary cation to form a primary cation is endothermic by as much as 55–60 kcal/mole, we would suggest two alternatives. Possibly β-scission is concerted with hydride shift so that a secondary cation is produced directly as in Equation 66. Or, more likely, cracking occurs largely from a small equilibrium concentration of secondary cations in the monobranched cation

$$R{-}\overset{\delta+}{\underset{|}{\underset{C}{C}}}\cdots C\cdots \overset{H}{C}\text{—}\overset{\delta+}{C}{-}R' \rightarrow R{-}\underset{\underset{C}{|}}{C}{=}C + \overset{H}{\underset{|}{C}}{-}\overset{+}{C}{-}R' \qquad (66)$$

manifold (Reaction 67) or from a small concentration of multiply branched C_{12} cations (Reaction 68) whose presence is indeed indicated by formation of significant multiply branched isomers. Either route avoids primary cation formation which would be energetically unfavorable and rationalizes the

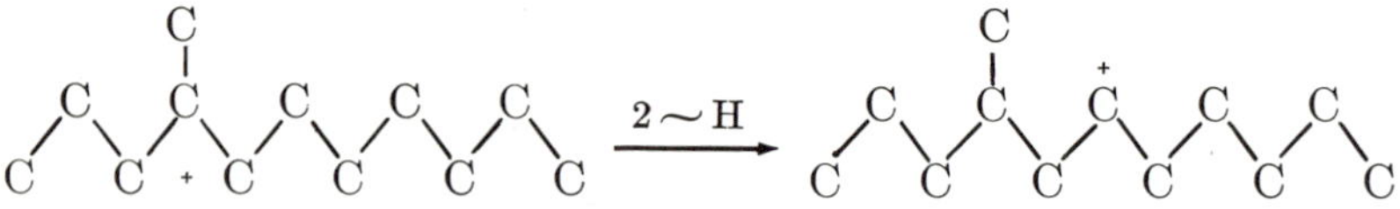

2~H

36

(67)

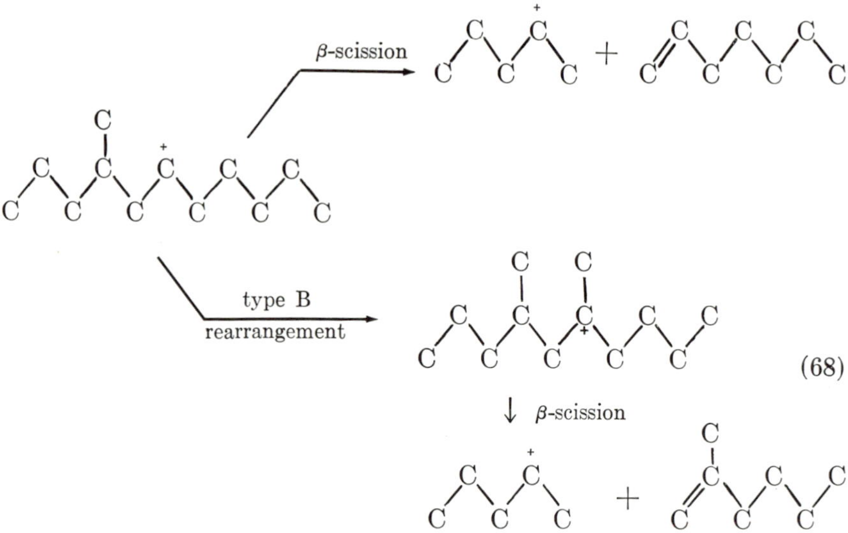

β-scission

type B rearrangement

(68)

β-scission

absence of a ($C_2 + C_{10}$) fraction. The multiply branched cation route (Equation 68) seems more consistent with the large amounts of 2-methylalkanes and even dimethylalkanes (*290*) observed as primary cracking products. [Note added in proof: On the basis of additional studies, Schulz and Weitkamp consider the routes outlined in Equations 66 and 68 more likely than that in Equation 67 (personal communication).]

Under the same conditions used for *n*-dodecane, hydrotreatment of 2-methylpentane over Pt/CaY at 275°C gave largely isomerization, first to 3-methylpentane, and little cracking; the little cracking observed was attributed more to Pt-catalyzed hydrogenolysis than to carbonium ion cracking (*291*). The more forcing conditions therefore required to crack C_6 than C_{12} are consistent with the proposed schemes. The Schulz-Weitkamp cleavage

(Equation 65) could occur for 2-methyl-2-pentyl cation but would need to form the questionable ethyl cation. A route analogous to Equation 68 is impossible with less than seven carbon atoms. With methylcyclopentane and/or cyclohexane, which rapidly equilibrated, a disproportionation reaction occurred to form ($C_5 + C_7$) and ($C_4 + C_8$) products (*291*). The proposed scheme involved alkylation of a C_6 olefin by a C_6 cation to form a C_{12} cation which then degraded by the routes described above to ($C_n + C_{12-n}$) products. This proposal is an example of what may be a rather general phenomenon that cracking of small alkanes at relatively low temperatures involves olefin condensation as well as β-scission steps.

Returning now to the main topic of cracking over acidic zeolites, we will scan chronologically some of the reports of cracking the model paraffin *n*-hexane to indicate the complexity of product distributions and variety of mechanistic hypotheses drawn. Pickert and co-workers (*80*) passed *n*-hexane over various divalent ion-exchanged X and Y zeolites at 450 psig in a five-fold excess of hydrogen, and temperature was adjusted to give 5% cracking to C_1–C_5 products, in which C_3 and C_4 predominated. The most active zeolite tested (MgY) gave 5% cracking at 363°C accompanied by 16.4% isomerization to branched isomers compared with silica–alumina which required 475°C and gave only 1.5% isomerization. Whether the greater isomerization:cracking ratio is inherent to the zeolite or simply a reflection of greater activity allowing a lower operating temperature cannot be judged from the data. However, both kinetic and thermodynamic considerations suggest that isomerization:cracking ratios will vary from > 1 to < 1 as temperature increases. Miale, Chen, and Weisz (*1*) demonstrated the superactivity of zeolites by comparing rates of cracking of *n*-hexane diluted with He in a flow reactor extrapolated to a common temperature. Several zeolites such as RE(NH_4)Y and NH_4-mordenite were $>$ 10,000 times as active as silica–alumina. A typical product distribution ("H-faujasite" at 371°C) was: 0.6% CH_4, 2.2% C_2H_6, 0.2% C_2H_4, 47.1% C_3H_8, 6.8% C_3H_6, 17.0% *i*-C_4H_{10}, 7.4% *n*-C_4H_{10}, 1.1% C_4H_8, and 17.6% C_5s. Of particular note is the large paraffin:olefin ratio; a hydrogen balance must demand a hydrogen-poor product which is not desorbed from the catalyst. The paraffin:olefin ratio and the iso:normal ratios decreased as cracking temperature increased. Tung and McIninch (*93*) cracked *n*-hexane over CaY of varying degrees of exchange, activated at 550°C, as a function of reaction temperature. Both cracking and isomerization activity increased sharply once 60% Ca^{2+} exchange was exceeded. Cracking increased with temperature from 300° to 550°C, but isomerization passed through a maximum at ~ 350°C. A mundane contributor to this now familiar pattern is that the Network 69 is in operation, and therefore no isomers will be observed under forcing enough conditions. The cracking products from CaY with $>$ 60% exchange were quite similar to those of

$$\begin{array}{ccc} n\text{-hexane} & \leftrightarrows & \text{branched hexanes} \\ & \searrow \quad \swarrow & \\ & \text{cracked products} & \end{array} \tag{69}$$

Miale and co-workers (*1*) but were more of the radical type at $< 60\%$ exchange. In a parallel study (*90*) with 93% exchanged NH_4Y, activated at 550°C, activity could be observed even at 250°C at which point the cracking product was essentially free of olefins. These authors' proposals concerning time-variant sites were noted above. Totally paraffinic products also resulted with H-mordenite at 350°C accompanied by rapid aging (*108*).

Bolton and Bujalski (*92*) focused particularly on the source of excess hydrogen demanded by the paraffinic nature of the cracked products. A pressed wafer of NH_4Y activated at 550°C in air was exposed to flowing *n*-hexane in an IR cell at 450°C; each hour the cell was evacuated at 200°C and the O—H region of the IR spectrum was scanned at room temperature. Products of cracking were monitored, but no conversion values could be obtained because only part of the feed contacted the catalyst and a significant background cracking occurred in the empty cell. The observation was that the HF band gradually decreased in intensity followed by the LF band as accumulated cracking time increased. At the point where the HF band reached $< 20\%$ of its original intensity, the products changed from those typical of the zeolite to those characteristic of the background cracking. The authors proposed therefore that the source of excess hydrogen was not coke formation but the actual O—H groups of the catalyst. Unfortunately, the lack of conversion data does not allow evaluation of the stoichiometry between amounts of paraffin formed and O—H content of the zeolite demanded by this hypothesis. Also, no structural proposals were made for the residual catalyst which must be formally oxidized. These workers have not rigorously ruled out the more conventional explanation that O—H intensity loss results from hydrogen bonding to growing polyene residue and that products change character when zeolitic activity largely disappears because of this aging. Some confusion arises from various uses and criteria for the term "coke"; a range of structures progressively more dehydrogenated and cyclized is probably involved during catalyst deactivation.

The isomerization and cracking of *n*-hexane at 350°C over 91%-exchanged NH_4Y was studied (*292*) as a function of T_{act}; results are shown in Table VI. Note that the cracked product composition is virtually independent of T_{act} from 350° to 650°C; thus, if Brønsted and Lewis acid sites both play some role in paraffin cracking, their widely differing proportions over this T_{act} range do not affect the product distribution. The gradually increasing amount of isomerization compared with cracking with decreasing conversion was a general trend observed for REY and $RE(NH_4)Y$ catalysts as well, indicative of Scheme 69.

The most likely picture which emerges from these studies of *n*-hexane cracking is that paraffins and olefins are produced by the chain carbonium-ion scheme but that the olefins undergo rapid hydrogen transfer reactions to give excess paraffins and residue precursors. The source of C_4 and C_5 products from a C_6 precursor is probably not simple β-scission of a C_6 cation, which would need to produce the unstable ethyl and methyl cations but rather olefin alkylation to give $C_{>6}$ cations followed by cracking.

Table VI. Cracking and Isomerization of *n*-Hexane over NH_4Y at 350°C[a] (*292*)

T_{act} (°C)[b]	Conversion (%)[c]	Isomerization (%)[d]	Composition of Cracked Product (%)[e]						
			CH_4	C_2	C_3	n-C_4	i-C_4	n-C_5	i-C_5
350	27.5	64	0.9	3.1	45.7	7.5	24.9	2.1	15.8
400	37	60	0.8	2.3	43.8	8.1	25.1	2.7	17.3
450	41	59	0.7	2.1	42.9	8.1	26.5	2.6	17.3
500	37	64	1.2	2.1	40.1	7.8	27.0	2.6	19.2
550	15	75	4.2	3.6	38.0	8.0	26.8	2.0	17.4
650	11	77	3.1	4.0	40.1	7.7	25.2	2.9	18.0

[a] All values at 2.5 hr on stream with LHSV = 0.3 hr^{-1}.
[b] In flowing air for 16 hr.
[c] Conversion to branched C_6 isomers and cracked products.
[d] Isomerization expressed as % of total conversion.
[e] Normalized to 100% for C_1–C_5 products.

Cracking of *n*-pentane over activated NH_4-mordenite at 400°C was more rapid than that of *n*-butane (*70*) by a factor greater than predicted simply by the statistically greater number of secondary C—H bonds or secondary cations possible for the former. The lowest energy pathway imaginable for β-scission of a C_4 paraffin involves completely primary ions (*n*-butyl → ethyl cation) and hence complex variations on the normal carbonium ion cracking theme should emerge here if anywhere; indeed appreciable amounts of C_5s appeared in the product.

Between the regimes of normal cracking and hydrocracking with noble-metal loaded catalysts apparently lies the regime where paraffin isomerization is promoted effectively in the presence of hydrogen by H-mordenite (*115, 116, 293*) not containing added metal. Activity was inversely proportional to H_2 pressure (*116*), but selectivity for isomerization and catalyst lifetime increased with H_2 pressure (*293*). Minachev (*116*) specifically proposed Reaction 70 which would account for the inverse activity effect of H_2 by its

$$\begin{gathered} n\text{-}C_nH_{2n+2} + HM \rightleftarrows n\text{-}C_nH_{2n+1}M + H_2 \\ n\text{-}C_nH_{2n+1}M \rightleftarrows iso\text{-}C_nH_{2n+1}M \\ iso\text{-}C_nH_{2n+1}M + H_2 \rightleftarrows iso\text{-}C_nH_{2n+2} + HM \end{gathered} \quad (70)$$

reducing the concentration of carbonium ion intermediates and for the selectivity effect of H_2 by its capturing these intermediates before cracking (*cf.* Ref. *294*). Hydride transfer to carbonium ions from H_2 has been demonstrated in model systems (*295*), but why this reaction, if it occurs, should be so strongly favored by the mordenite structure is unknown (for a possible enhancement of cracking rates by H_2 with Y zeolites, *see* Ref. *99*). The O—H groups of acidic zeolites can be isotopically exchanged with gaseous D_2 (*296*) by a process which could involve protonic attack on the H—H bond.

The unusual effects of molecular size on cracking over erionites (*6, 122, 123, 124*) are described in another chapter.

Gas Oil Cracking. The major commercial impact of zeolite catalysts has been catalytic cracking of complex petroleum feedstocks containing aliphatic, cycloaliphatic (naphthenic), olefinic, and aromatic hydrocarbons. Petroleum cracking must involve contributions from most of the reaction types discussed so far including dealkylation of aromatics, cracking of paraffins and olefins, and hydrogen transfer and cyclization reactions of olefins in a distribution far too complex to unravel completely. Early publications by Plank and Rosinski (*161, 297*) demonstrated three advantages of zeolite X catalysts compared with silica–alumina for cracking a gas oil fraction (bp 260°–400°C): (1) greater activity which persisted to higher coke levels, (2) greater selectivity to produce gasoline (C_5^+) compared with gas (C_4^-) and coke, and (3) greater stability towards high temperatures and steaming during coke burnoff. These advantages were maximized in the $Ca(NH_4)$ and $RE(NH_4)$ ion-exchanged forms. Recent data for gas oil (bp 272°–415°C) cracking over REX and REY catalysts as a function of severity of calcination and steaming are given by Moscou and Mone (*148*), who suggested that the loss of the most acidic sites caused by high temperature treatments was actually beneficial for high gasoline selectivity. Particularly important is this gasoline selectivity which has been kinetically modelled by Weekman and Nace (*298*) by Scheme 71 in which primary cracking (k_o) produces gasoline and some gas (and coke) and secondary cracking (k_1) further degrades the gasoline.

$$\begin{array}{ccc} & k_o & \\ \text{Gas Oil } (C_{15}^+) & \rightarrow & \text{Gasoline } (C_5\text{–}C_{10}) \\ \searrow k_o & & \swarrow k_1 \\ & \text{Gas } (C_1\text{–}C_4) & \end{array} \qquad (71)$$

Mechanistic suggestions why zeolites should hinder k_1 compared with k_o have centered on the facile hydrogen redistribution reaction. Aditya (quoted in Ref. *6*) described the gasoline composition from cracking a hydrotreated petroleum fraction in terms of paraffins (P), olefins (O), naphthenes (N), and aromatics (A) as a function of catalyst type as shown in Table VII. The data suggest the incremental Reaction 72 on the zeolite catalyst in almost the exact stoichiometry required for hydrogen transfer which saturates olefins and aromatizes the already-cyclic naphthenes. Weisz (*6*) suggested that

$$O + N \rightarrow P + A \qquad (72)$$

Reaction 72 effectively reduces k_1 in the Weekman-Nace model by converting the easily cracked olefins and naphthenes initially produced to the more refractory paraffins and aromatics before they can crack further to form gas. Thomas and Barmby (*299*) proposed that such hydrogen transfer was favored over commercial zeolite catalysts because initial gas oil cracking occurred

Table VII. Gasoline Composition as a Function of Catalyst[a]

Catalyst	*Composition (%)*			
	P(araffin)	*O(lefin)*	*N(aphthene)*	*A(romatic)*
Silica-alumina	13	17	41	29
Zeolite	23	5	23	49
Difference	+10	−12	−18	+20

[a] Data from Ref. *6*.

largely in the silica–alumina matrix and on the external surface of the zeolites, because of the diffusional limitations for these large molecules (*cf.* Ref. *19*), whereas the secondary hydrogen transfer occurred in the zeolitic pores. Therefore, within the pores, secondary hydrogen transfer reactions of C_5–C_{10} molecules can occur freely whereas, in silica–alumina, they must always compete for sites with the larger feed molecules. However, neither proposal explicitly treats the competition between the *two* types of secondary reactions, cracking and hydrogen transfer, and why zeolites should favor the latter. One suggestion made in various stages of development by several authors (*12, 173, 281, 287*) is that the zeolitic surface will adsorb an effectively greater concentration of hydrocarbon species and therefore hydrogen transfer is favored over cracking because it is bimolecular as shown in Scheme 73.

$$R^{+} \begin{cases} \xrightarrow[k_H]{R'H} RH + R'^{+} \\ \xrightarrow[k_\beta]{} \text{Olefin} + R''^{+} \end{cases} \tag{73}$$

$$\frac{\text{(Hydrogen Transfer)}}{\text{(Cracking)}} = \frac{\text{(Hydride Transfer)}}{(\beta\text{-Scission})} = \left(\frac{k_H}{k_\beta}\right)\left(R'H\right)$$

RADICAL CRACKING. The free-radical chain mechanism for thermal homogeneous cracking of paraffins was elaborated by Rice and co-workers (*300*). The paraffin is converted to a free radical by hydrogen atom transfer to give a set of radicals resulting from a tertiary:secondary:primary C—H bond reactivity order of ~ 10:3:1 at 500°C. Each radical can then undergo β-scission or abstract hydrogen from the substrate; *e.g.*, for a *n*-butyl radical the relevant processes are shown in Equation 74. Just as for the analogous carbonium ion processes in Sequence 73, this partitioning depends on the ratio $k_H[RH]/k_\beta$ which is $<<1$ at ~ 500°C and 1 atm RH pressure for all radicals which can form tertiary, secondary, or even primary radicals by β-scission. However, if methyl radical or hydrogen atom must be lost, hydrogen abstraction competes successfully with β-scission.

$$CH_3CH_2CH_2CH_2\cdot \begin{cases} \xrightarrow[RH]{k_H} CH_3CH_2CH_2CH_3 + R\cdot \\ \xrightarrow{k_\beta} CH_3CH_2\cdot + CH_2{=}CH_2 \end{cases} \quad (74)$$

a_1: C–C–C–C–C–Ċ → C=C + C–C–C–Ċ

↓ C=C + C–C· $\xrightarrow[RH]{}$ C–C + R·

$(a_2 + a_3)$: C–C–C–C–Ċ–C → C=C–C + C–C–C· $\xrightarrow[RH]{a_3}$ C–C–C + R·

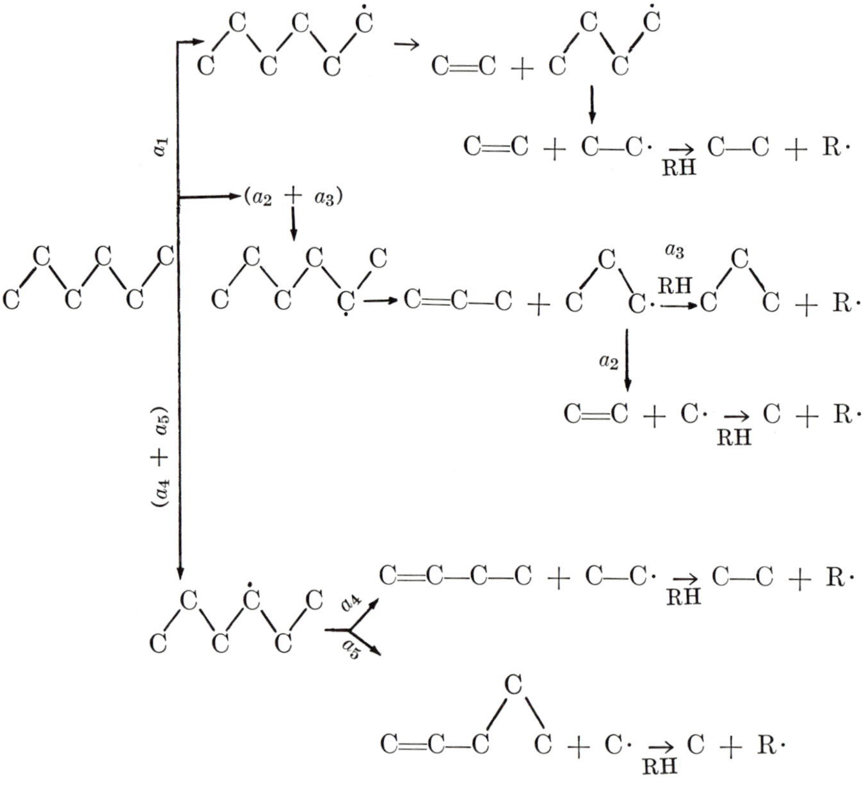

a_2 ↓ C=C + C· $\xrightarrow[RH]{}$ C + R·

$(a_4 + a_5)$: C–C–Ċ–C–C–C

a_4: → C=C–C–C + C–C· $\xrightarrow[RH]{}$ C–C + R·

a_5: → C=C–C–C–C + C· $\xrightarrow[RH]{}$ C + R·

Thermal		*KY*	
$a_1 = 6.3$	$a_4 = 31.4$	$a_1 = 4.3$	$a_4 = 32.3$
$a_2 = 44.0$	$a_5 = 7.2$	$a_2 = 17.6$	$a_5 = 6.5$
$a_3 = 10.9$		$a_3 = 39.3$	

(75)

To clarify the nature of the radical cracking observed over zeolites which are not highly acidic such as NaX (*133, 301*) and low-exchanged CaY (*93*) and LaX (*151*), Poutsma and Schaffer (*173*) compared the products from thermal cracking of all the isomeric hexanes at 500°C and 1 atm pressure with those from cracking over KY under the same conditions. Rate enhancements of only ~ five-fold were observed. The zeolite-derived products were different from the thermal products but could be made fully compatible with the radical mechanism by recognition of two alterations. The olefinic products had undergone considerable equilibration of double bond position, a reaction attributed to weakly acidic defect sites in the zeolite. Secondly, it appeared that the ratio $k_H[RH]/k_\beta$ has increased 6–10-fold for the radicals *n*-propyl, *i*-propyl, and *tert*-butyl, the only radicals for which minor changes in this ratio would have significantly affected the products. For example, in the cracking of *n*-hexane, the thermal products are derived as shown in Scheme 75 with the relative rate coefficients a_1–a_5 shown. The KY-derived products could be fit by the same scheme with the new coefficients a_1–a_5 shown, except that the products 1-butene and 1-pentene were replaced by mixed linear butenes and pentenes. Note that the major alteration is in the a_3/a_2 ratio which is a term of the type $k_H[RH]/k_\beta$. Although the zeolite may have influenced the magnitudes of k_H and k_β, the changes in ratios would also result if the effective substrate concentration [RH] were greater on the zeolite than in the gas phase at the same external temperature and pressure because of weak nonspecific adsorption. Whereas product compositions can be considered on the basis of only the propagation steps of these apparently long-chain reactions, cracking rates cannot be interpreted without knowledge of the radical-forming and destroying initiation and termination reactions. The modest rate enhancements observed did not allow specific conclusions on the possible roles of the zeolitic surface in these processes.

The several examples (*93, 133, 151, 301*) where reported products do not fit very well either the typical carbonium ion patterns or this idealized radical pattern (*173*) may represent mixed mechanisms, but more product detail would be required for firm conclusions.

DEHYDROCYCLIZATION. The radical-type cracking activity of alkali-metal ion-exchanged X-zeolites is somewhat enhanced by incorporation into the zeolite of various forms of sulfur (*301, 302, 303*) or selenium (*303*). However, physical mixing of NaX or KX with 3–8% tellurium metal followed by heating in hydrogen gave a catalyst (*304*) with a new type of activity, the ability to convert *n*-hexane to benzene with > 90% selectivity. X-ray crystallography (*305*) revealed a Te species of appropriate ionic radius for Te^{2-} coordinated to 2 Na^+ ions, one in S_{II} and one in S_{III}, which was postulated to be the active site (*6, 305*). Reduction of Te by H_2 had presumably produced Te^{2-} and protons which attacked the lattice. On the basis of a strongly inverse kinetic dependence on H_2 and analogies with chromia dehydrocyclization catalysts, the mechanism in Sequence 76 was proposed but this does not suggest a specific role for the telluride ion.

$$C_6H_{14} \rightleftarrows C_6H_{12} + H_2$$

$$C_6H_{12} \rightleftarrows C_6H_{10} + H_2 \qquad (76)$$

$$C_6H_{10} \rightleftarrows \text{CH}=\text{CH}-\text{CH}=\text{CH}_2 \ldots \text{CH}_2=\text{CH}-\text{CH} \rightleftarrows \text{cyclohexadiene}$$

$$\text{cyclohexadiene} \rightleftarrows \text{benzene} + H_2$$

Summary and Conclusions

The major transformations of hydrocarbons catalyzed by zeolites are condensations, rearrangements, and cleavages of olefins, alkylaromatics, and paraffins singly or in combination. The reactive intermediate strongly implicated in most of these reactions is the carbonium ion. Manifestation of such activity requires ion-exchange of at least part of the alkali-metal cations by polyvalent Main Group cations or ammonium ions followed by thermal activation. The most compelling evidence for carbonium ion mechanisms is derived, however, not so much from direct detection on zeolites themselves as from analogies with model reactions catalyzed by strong acids in homogeneous media, reactions for which the intermediacy of carbonium ions rests on decades of work in physical organic chemistry and more recently on direct spectroscopic observations and stoichiometric transformations in super acids. Because of the strong similarities of reaction types catalyzed by zeolites on one hand and, say, sulfuric acid or aluminum chloride on the other, and because of independent evidence for the acidity of activated zeolites, we are naturally predisposed to postulate the intermediacy of carbonium ions on zeolites and then to inquire how they might be formed on and their behavior modified by the zeolite surface. For alkali metal ion-containing zeolites, there are a smaller number of reported reactions for which parallel behavior can be found in thermal gas phase reactions and therefore analogies are best sought in the behavior of free radicals.

The major experimental approach to determine the mechanism of zeolite-catalyzed transformations, particularly to assign active sites, has been to determine catalytic activity and product selectivity as a function of controlled changes in zeolite framework, cation identity, exchange level, and activation conditions. The influence of these same variables on zeolite structure is determined by spectroscopic and chemical tools. Then deductions of mechanism as a function of zeolite structure are drawn. Much less attention has been paid to the alternate approach of holding the zeolite constant and systematically varying the structure of the hydrocarbon substrate. Two factors which have not always been dealt with properly in the past deserve attention in future studies: (1) the zeolite structure resulting from a given type of activation is a sensitive function of all the activation procedure variables, and therefore great care must be taken to insure that samples for catalytic study and for structural study are indeed identical; and (2) the appropriate activity quantity to be related to zeolite structure is initial activity, preferably in a differential reactor, rather than steady-state activity which may be significantly perturbed by catalyst aging.

Using the general approach just outlined, various workers have discussed as possible active sites: (1) hydroxyl groups acting as Brønsted acids, (2) oxide vacancies or framework alterations creating Lewis acidity, (3) polyvalent cations acting through their uncompensated electrostatic fields, (4) electron transfer centers, and (5) synergistic combinations of these. Actually, the concept of the active site has been almost entirely limited to those structural features which might perform the initial attack on the hydrocarbon to form a carbonium ion; attempted description of the role of the surface in subsequent steps leading to products has been much less common. A role for Brønsted acidity in protonating unsaturated hydrocarbons to form carbonium ions is generally agreed upon, such acid groups arising either by thermal decomposition of ammonium ions or by hydrolysis of water by polyvalent cations. The involvement of Lewis acid sites is more controversial; further clarification will require better definition of the structural alterations involved in dehydroxylation and ultrastabilization of zeolites. Some circumstantial evidence exists for a role of Lewis acidity in modifying Brønsted acid strength; even less clear is the question of direct interaction between such sites and hydrocarbons, especially paraffins. The primary catalytic function of polyvalent cations is creation of Brønsted acid sites by hydrolysis of water (with possible concurrent stabilization of the lattice by oxide bridging) rather than direct formation of carbonium ions as originally proposed; here again, however, determination of secondary roles resulting from interaction with Brønsted or Lewis acid sites deserves further attention. Finally, electron transfer processes, although clearly observable spectroscopically in specially designed model systems, have not been convincingly placed on the major reaction pathways being discussed. Much work remains to be done, as progress is slowly being made in refining these gross generalizations regarding active sites. Quantitative measurements of the number and strength of the acidic sites, for example by Hammett-type indicators, as a function of zeolite structure are in disagreement among various workers and deserve further attention. In this regard the exact reasons for the greater activity of zeolites, particularly for paraffinic substrates, compared with other acidic solids such as amorphous silica–alumina (greater number of sites, greater acidic strength per site, crystallinity, etc.) are still a subject of discussion, although greater inherent acidic strength does not seem to have been confirmed. In this regard also, dynamic poisoning studies as a method of counting sites and assigning values of activity per site are in serious disagreement.

Considerations of active sites are most highly developed for Y zeolite although recently more attention has been paid to a range of framework types including X zeolite, mordenite, and aluminum-deficient mordenite. In the latter series, examples have emerged where activity is limited not by the number and type of active sites, as is generally assumed particularly in comparative studies, but by diffusional restrictions; here is another type of potential experimental pitfall to be avoided. Even with this qualification, however, current evidence suggests that activity per site increases with increasing Si:Al ratio.

Compared with the type of active site studies just described, mechanistic interpretations based on kinetic studies, where the roles not only of chemical transformation but also of sorption and counterdiffusion are considered, are fewer in number and of more recent vintage. Cases have been described where, for example, the rate-limiting step of a transformation, aromatic alkylation, is the final product transport across the zeolite crystallite–gas boundary. Many of the product selectivity characteristics unique to zeolites compared with other acidic solids may arise from the more extensive adsorption within the ordered zeolite pore structure with resultant alterations in the balance between competitive unimolecular and bimolecular reaction steps.

Between the time a carbonium ion is initially generated and product is finally formed by carbonium ion destruction, most hydrocarbon transformations require steps in which one carbonium ion is converted into another by rearrangement, addition, β-scission, or hydride transfer. Considerations of the structure of adsorbed carbonium ions are not at all well-developed. The solvation offered by the zeolite lattice is surely vital to stabilizing such ions but it is not spatially continuous; therefore, special spatial requirements would seem necessary to minimize the energy barriers of the carbonium ion transformation steps which have the net effect of translating the positively charged center over a prescribed distance. Hints of the importance of such factors have arisen in studies of competitive isomerization and transalkylation of alkylaromatics as a function of Si:Al ratio in mordenites. Within such considerations may ultimately be found fundamental differences between the crystalline and amorphous aluminosilicates.

The commercially most significant but mechanistically least understood reaction type is paraffin cracking and isomerization. Although it is tempting to consider that some feature of the zeolite structure may directly cleave a saturated C—H or C—C bond and although homogeneous analogies are now documented, this process has not been rigorously demonstrated as a competitor with carbonium ion generation by olefin protonation followed by hydride transfer. The competitions among the subsequent rearrangement reactions *via* 1,2-alkyl or hydride shifts or protonated cyclopropanes and β-scission reactions are being somewhat unraveled, particularly by detailed product studies of hydrocracking of large substrates. In fact, the more commonly studied C_4 and C_5 paraffins are really special cases as regards paraffin cracking, because their direct cracking would require involvement of high-energy primary carbonium ions. Finally, a distinctive feature of cracking over zeolites is the hydrogen redistribution reaction, proceeding almost surely by hydride transfer, which converts initially formed olefins and naphthenes into the more refractory paraffins and aromatics, thereby increasing gasoline yield. This reaction is only one member of a class of bimolecular reactions, including also transalkylation of alkylaromatics and radical-type hydrogen atom transfer, which seem particularly favored over zeolites. Whether these will ultimately be best rationalized by increased hydrocarbon concentration caused by more extensive adsorption or by some other feature of zeolite structure remains to be determined.

Literature Cited

1. Miale, J. N., Chen, N. Y., Weisz, P.B., *J. Catal.* (1966), **6,** 279.
2. Hansford, R. C., Ward, J. W., *J. Catal.* (1969) **13,** 316.
3. Oblad, A. G., *Oil Gas J.* (March 27, 1972) 84.
4. Rabo, J. A., Pickert, P. E., Stamires, D. N., Boyle, J. E., *Proc. Inter. Congr. Catal., 2nd, Paris,* p. 2055, 1960.
5. Weisz, P. B., Frilette, V. J., *J. Phys. Chem.* (1960) **64,** 382.
6. Weisz, P. B., *Chem. Tech.* (1973) 498.
7. Minachev, Kh. M., Isakov, Ya. I., *Adv. Chem. Ser.* (1973) **121,** 451.
8. Minachev, Kh. M., Garanin, V. I., Isakov, Ya. I., *Russ. Chem. Rev.* (1966) **35,** 903.
9. Minachev, Kh. M., *Kinet. Catal.* (1970) **11,** 342.
10. Leach, H. F., *Ann. Repts. Progr. Chem., Sect. A* (1971) **68,** 195.
11. Venuto, P. B., *Adv. Chem. Ser.* (1971) **102,** 260.
12. Venuto, P. B., *Chem. Tech.* (1971) **1,** 215; *Preprints, Div. Petroleum Chem., Amer. Chem. Soc.* (1971) **16**(2), B42.
13. Venuto, P. B., Landis, P. S., *Adv. Catal.* (1968) **18,** 259.
14. Rabo, J. A., Poutsma, M. L., *Adv. Chem. Ser.* (1971) **102,** 284.
15. Ward, J. W., *Preprints, Div. Petroleum Chem., Amer. Chem. Soc.* (1971) **16**(2), B6.
16. Turkevich, J., Ono, Y., *Adv. Catal.* (1969) **20,** 135.
17. Turkevich, J., *Catalysis Rev.* (1967) **1,** 1.
18. Mays, R. L., Pickert, P. E., "Molecular Sieves," Society of the Chemical Industry, London, 1968, p. 112.
19. Weisz, P. B., *Ann. Rev. Phys. Chem.* (1970) **21,** 175.
20. Stull, D. R., Westrum, E. F., Jr., Sinke, G. C., "The Chemical Thermodynamics of Organic Compounds," Wiley, New York, 1969.
21. Greensfelder, B. S., Voge, H. H., Good, G. M., *Ind. Eng. Chem.* (1949) **41,** 2573.
22. Thomas, C. L., *Ind. Eng. Chem.* (1949) **41,** 2564.
23. Olah, G. A., Schleyer, P. R., Eds., "Carbonium Ions," Vols I-IV, Wiley, New York, 1968.
24. Olah, G. A., Ed., "Friedel-Crafts and Related Reactions," Vols I-IV, Wiley, New York, 1964.
25. Leftin, H. P., ref. *23,* Vol. I, Chapter 10.
26. Olah, G. A., Schlosberg, R. H., *J. Amer. Chem. Soc.* (1968) **90,** 2726; Olah, G. A., Lukas, J., *ibid.* (1968) **90,** 933.
27. Lukas, J., Kramer, P. A., Kouwenhoven, A. P., *Rec. Trav. Chim.* (1973) **92,** 44.
28. Brouwer, D. M., Hogeveen, H., *Progr. Phys. Org. Chem.* (1972) **9,** 179.
29. Larsen, J. W., Bouis, P. A., Watson, C. R., Jr., Pagni, R. M., *J. Amer. Chem. Soc.* (1974) **96,** 2284.
30. Benson, S. W., "Thermochemical Kinetics," p. 215, Wiley, New York, 1968.

31. Vedeneyev, V. I., Gurvich, L. V., Kondratyev, V. N., Medvedev, V. A., Frankevich, Ye. L., "Bond Energies, Ionization Potentials, and Electron Affinities," St. Martin's Press, New York, 1966.
32. Olah, G. A., Lukas, J., *J. Amer. Chem. Soc.* (1967) **89,** 4739.
33. Brouwer, D. M., *Rec. Trav. Chim.* (1968) **87,** 210.
34. Saunders, M., Hagen, E. L., *J. Amer. Chem. Soc.* (1968) **90,** 2436.
35. Saunders, M., Hagen, E. L., *J. Amer. Chem. Soc.* (1968) **90,** 6881.
36. Saunders, M., Hagen, E. L., Rosenfield, J., *J. Amer. Chem. Soc.,* (1968) **90,** 6882.
37. Olah, G. A., Olah, J. A., Ref. *23,* Vol. II, Chapter 17.
38. Hariharan, P. C., Radom, L., Pople, J. A., Schleyer, P. R., *J. Amer. Chem. Soc.* (1974) **96,** 599.
39. Collins, C. J., *Chem. Rev.* (1969) **69,** 543; Karabatsos, G. J., Anand, M., Rickter, D. O., Meyerson, S., *J. Amer. Chem. Soc.* (1970) **92,** 1254 and earlier papers; Friedman, L., Jurewicz, A., *Ibid.* (1969) **91,** 1800.
40. Brouwer, D. M., Mackor, E. L., and MacLean, C., *Discussions Faraday Soc.,* (1965) **39,** 121.
41. Kramer, G. M., *J. Amer. Chem. Soc.* (1969) **91,** 4819.
42. Brouwer, D. M., Oelderik, J. M., *Preprints, Div. Petroleum Chem., Amer. Chem. Soc.* (1968) **13** (1), 184.
43. Kochi, J. K., Ed., "Free Radicals," Vols. I, II, Wiley, New York, 1973.
44. Jacobs, P., Uytterhoeven, J. B., *J. Catal.* (1971) **22,** 193.
45. Kerr, G. T., *J. Catal.* (1969) **15,** 200.
46. Ward, J. W., *J. Catal.* (1972) **27,** 157; Jacobs, P. A., Uytterhoeven, J. B., *Ibid.* (1972) **27,** 161.
47. Jacobs, P. A., Uytterhoeven, J. B., *J. Chem. Soc., Faraday* (1973) **69,** 359, 373; Jacobs, P. A., Heylen, C. F., *J. Catal.* (1974) **34,** 267.
48. Uytterhoeven, J. B., Christner, L. G., Hall, W. K., *J. Phys. Chem.* (1965) **69,** 2117.
49. Hughes, T. R., White, H. M., *J. Phys. Chem.* (1967) **71,** 2192.
50. Ward, J. W., *J. Catal.* (1967) **9,** 225.
51. Ward, J. W., Hansford, R. C., *J. Catal.* (1969) **13,** 364.
52. Uytterhoeven, J. B., Jacobs, P., Makay, K., Schoonheydt, R., *J. Phys. Chem.* (1968) **72,** 1768.
53. Stevenson, R. L., *J. Catal.* (1971) **21,** 113.
54. Olson, D. H., Dempsey, E., *J. Catal.* (1969) **13,** 221.
55. Ward, J. W., *J. Phys. Chem.* (1969) **73,** 2086.
56. Dempsey, E., Olson, D. H., *J. Catal.* (1969) **15,** 309.
57. Ward, J. W., Hansford, R. C., *J. Catal.* (1969) **15,** 311.
58. White, J. L., Jelli, A. N., André, J. M., Fripiat, J. J., *Trans. Faraday Soc.* (1967) **63,** 461.
59. Liengme, B. V., Hall, W. K., *Trans. Faraday Soc.* (1966) **62,** 3229.
60. Eberly, P. E., Jr., *J. Phys. Chem.* (1968) **72,** 1042.
61. Cattanach, J., Wu, E. L., Venuto, P. B. *J. Catal.* (1968) **11,** 342.
62. Ward, J. W., *J. Catal.* (1967) **9,** 396.
63. Ward, J. W., *J. Catal.* (1970) **16,** 386.
64. Cant, N. W., Hall, W. K., *Trans. Faraday Soc.* (1968) **64,** 1093.
65. Schoonheydt, R. A., Uytterhoeven, J. B., *J. Catal.* (1970) **19,** 55.
66. Mestdagh, M. M., Stone, W. E., Fripiat, J. J., *J. Phys. Chem.* (1972) **76,** 1220.
67. McDaniel, C. V., Maher, P. K., "Molecular Sieves," p. 186, Society of the Chemical Industry, London, 1967.
68. Kerr, G. T., *J. Phys. Chem.* (1967) **71,** 4155.
69. Bolton, A. P., Lanewala, M. A., *J. Catal.* (1970) **18,** 154.
70. Benesi, H. A., *J. Catal.* (1967) **8,** 368.
71. Venuto, P. B., Wu, E. L., Cattanach, J., *Anal. Chem.* (1966) **38,** 1266.
72. Hopkins, P. D., *J. Catal.* (1968) **12,** 325.
73. Hickson, D. A., Csicsery, S. M., *J. Catal.* (1968) **10,** 27.

74. Kerr, G. T., *J. Phys. Chem.* (1969) **73,** 2780.
75. Ambs, W. J., Flank, W. H., *J. Catal.* (1969) **14,** 118.
76. Venuto, P. B., Wu, E. L., Cattanach, J., "Molecular Sieves," p. 117, Society of the Chemical Industry, London, 1968.
77. Kerr, G. T., Cattanach, J., Wu, E. L., *J. Catal.* (1969) **13,** 114; Tung, S. E., McIninch, E., *Ibid.* (1969) **13,** 115.
78. Venuto, P. B., Hamilton, L. A., Landis, P. S., Wise, J. J., *J. Catal.* (1966) **4,** 81.
79. Venuto, P. B., Hamilton, L. A., Landis, P.S., *J. Catal.* (1966) **5,** 484.
80. Pickert, P. E., Rabo, J. A., Dempsey, E., Schomaker, V., *Proc. Inter. Congr. Catal., 3rd, Amsterdam,* 1964, p. 714.
81. Stamires, D. N., Turkevich, J., *J. Amer. Chem. Soc.* (1964) **86,** 749.
82. Turkevich, J., Nozaki, F., Stamires, D., *Proc. Inter. Congr. Catal., 3rd, Amsterdam,* 1964, p. 586.
83. Hirschler, A. E., *J. Catal.* (1963) **2,** 428.
84. Lunsford, J. H., *J. Phys. Chem.* (1968) **72,** 4163.
85. Richardson, J. T., *J. Catal.* (1967) **9,** 182.
86. Turkevich, J., Ono, Y., *Adv. Chem. Ser.* (1971) **102,** 315.
87. Flockhart, B. D., McLoughlin, L., Pink, R. C., *Chem. Commun.* (1970) 818.
88. Declerck, L. J., Vandamme, L. J., Jacobs, P. A., *Proc. Inter. Conf. Molecular Sieves, 3rd, Zurich, 1973,* Leuven Univ. Press, p. 442.
89. Pickert, P. E., Bolton, A. P., Lanewala, M. A., *Chem. Eng. Progr., Symp. Ser.* (1967) **63,** 50.
90. Tung, S. E., McIninch, E., *J. Catal.* (1968) **10,** 175.
91. Tung, S. E., McIninch, E., *J. Catal.* (1969) **13,** 115.
92. Bolton, A. P., Bujalski, R. L., *J. Catal.* (1971) **23,** 331.
93. Tung, S. E., McIninch, E., *J. Catal.* (1968) **10,** 166.
94. Tung, S. E., *J. Catal.* (1970) **17,** 24.
95. Benesi, H. A., *J. Phys. Chem.* (1957) **61,** 970; *J. Amer. Chem. Soc.* (1956) **78,** 5490.
96. Leffler, J. E., Grunwald, E., "Rates and Equilibria of Organic Reactions," Wiley, New York, 1963; pp. 269; 277.
97. Otouma, H., Arai, Y., Ukihashi, H., *Bull. Chem. Soc. Japan,* (1969) **42,** 2449.
98. Barthomeuf, D., Beaumont, R., *J. Catal.* (1973) **30,** 288.
99. Beaumont, R., Barthomeuf, D., Trambouze, Y., *Adv. Chem. Ser.* (1971) **102,** 327.
100. Beaumont, R., Barthomeuf, D., *J. Catal.* (1972) **26,** 218; (1972) **27,** 45.
101. Dempsey, E., *J. Catal.* (1974) **33,** 497; Barthomeuf, D., Beaumont, R., *Ibid.* (1974) **34,** 327.
102. Morita, Y., Kimura, T., Kato, F., Tamagawa, M., *Bull. Japan. Petroleum Inst.* (1972) **14,** 192.
103. Ikemoto, M., Tsutsumi, K., Takahashi, H., *Bull. Chem. Soc. Japan,* (1972) **45,** 1330.
104. Benson, J. E., Ushiba, K., Boudart, M., *J. Catal.* (1967) **9,** 91.
105. Moscou, L., *Adv. Chem. Ser.* (1971) **102,** 337.
106. Tsutsumi, K., Takahashi, H., *J. Catal.* (1972) **24,** 1.
107. Tsutsumi, K., Kajiwara, H., Koh, H. K., Takahashi, H., ref. *88,* 358.
108. Thakur, D. K., Weller, S. W., *Adv. Chem. Ser.* (1973) **121,** 596.
109. Karge, H., *Z. Physik. Chem.* (Frankfurt) (1971) **76,** 133.
110. Yashima, T., Hara, N., *J. Catal.* (1972) **27,** 329.
111. Cannings, F. R., *J. Phys. Chem.* (1968) **72,** 4691.
112. Lefrancois, M., Malbois, G., *J. Catal.* (1971) **20,** 358.
113. Burbidge, B. W., Keen, I. M., Eyles, M. K., *Adv. Chem. Ser.* (1971) **102,** 400.
114. Becker, K. A., Karge, H. G., Struebel, W. D., *J. Catal.* (1973) **28,** 403.
115. Beecher, R., Voorhies, A., Jr., *Ind. Eng. Chem., Product Res. Develop.* (1969) **8,** 366.

116. Minachev, Kh., Garanin, V., Isakova, T., Kharlamov, V., Bogomolov, V., *Adv. Chem. Ser.* (1971) **102,** 441.
117. Topchieva, K. V., Romanovskii, B. V., Pigusova, L. I., Thoang, H. S., Bizreh, Y. W., *Proc. Inter. Congr. Catal., 4th, Moscow,* p. 135, 1968 (Akademiai, Budapest, 1971), Vol. II.
118. Weller, S. W., Brauer, J. M., *Preprints,* 62nd *Ann. Meeting, Amer. Instit. Chem. Engin.,* Washington, D. C., 1969.
119. Pigusova, L. J., Prokof'eva, E. N., Dubinin, M. M., Bursian, N. R., Shavandin, V., *Kinet. Catal.* (1969) **10,** 252.
120. Bierenbaum, H. S., Chiramongkol, S., Weiss, A. H., *J. Catal.* (1971) **23,** 61.
121. Eberly, P. E., Jr., Kimberlin, C. N., Jr., *Ind. Eng. Chem., Prod. Res. Devel.* (1970) **9,** 335.
122. Chen, N. Y., *Proc. Inter. Congr. Catal., 5th, Miami Beach,* 1972 ("Catalysis," J. W. Hightower, Ed., North-Holland Publ. Co., Amsterdam, 1973), Vol. 2, p. 1365; Chen. N. Y., Garwood, W. E., *Adv. Chem. Ser.* (1973) **121,** 575.
123. Hobson, H. E., Hamner, G. P., Arey, W. F., Jr., *Adv. Chem. Ser.* (1971) **102,** 417.
124. Whyte, T. E., Jr., Wu, E. L., Kerr, G. T., Venuto, P. B., *J. Catal.* (1971) **20,** 88.
125. Cole, J. F., Kouwenhoven, H. W., *Adv. Chem. Ser.* (1973) **121,** 583.
126. Detrekoy, E. J., Jacobs, P. A., Ref. *88,* 373.
127. Bennett, J. M., Smith, J. V., *Materials Res. Bull.* (1968) **3,** 633.
128. Isakov, Ya. I., Klyachko-Gurvich, A. L., Khudiev, A. T., Minachev, Kh. M., Rubinstein, A. M., Ref. *117,* 123.
129. Angell, C. L., Schaffer, P. C., *J. Phys. Chem.* (1966) **70,** 1413.
130. Angell, C. L., Howell, M. V., *Can. J. Chem.* (1969) **47,** 3831.
131. Ward, J. W., *J. Catal.* (1968) **10,** 34.
132. Tsutsumi, K., Takahashi, H., *J. Phys. Chem.* (1972) **76,** 110.
133. Frilette, V. J., Weisz, P. B., Golden, R. L., *J. Catal.* (1962) **1,** 301.
134. Weisz, P. B., Frilette, V. J., Maatman, R. W., Mower, E. B., *J. Catal.* (1962) **1,** 307.
135. Angell, C. L., Schaffer, P. C., *J. Phys. Chem.* (1965) **69,** 3463.
136. Carter, J. L., Lucchesi, P. J., Yates, D. J. C., *J. Phys. Chem.* (1964) **68,** 1385.
137. Christner, L. G., Liengme, B. V., Hall, W. K., *Trans. Faraday Soc.* (1968) 64, 1679.
138. Ward, J. W., *J. Phys. Chem.* (1968) **72,** 4211.
139. Bolton, A. P., *J. Catal.* (1971) **22,** 9.
140. Rabo, J. A., Angell, C. L., Schomaker, V., Ref. *117,* 96.
141. Jacobs, P. A., Uytterhoeven, J. B., *J. Chem. Soc., Faraday* (1973) **69,** 373.
142. BenTaarit, Y., Mathieu, M. V., Naccache, C., *Adv. Chem. Ser.* (1971) **102,** 362.
143. Plank, C. S., Ref. *80,* p. 727.
144. Hirschler, A. E., Ref. *80,* p. 726.
145. Uytterhoeven, J. B., Schoonheydt, R., Liengme, B. V., Hall, W. K., *J. Catal.* (1969) **13,** 425.
146. Olson, D. H., *J. Phys. Chem.* (1968) **72,** 1400.
147. BenTaarit, Y., Bandiera, J., Mathieu, M. V., Naccache, C., *J. Chim. Phys. Physicochim. Biol.* (1970) **67,** 37.
148. Moscou, L., Moné, R., *J. Catal.* (1973) **30,** 417.
149. Ward, J. W., *J. Catal.* (1968) **11,** 238.
150. Nishizawa, T., Hattori, H., Uematsu, T., Shiba, T., Ref. *117,* 114.
151. Aldridge, L. P., McLaughlin, J. R., Pope, C. G., *J. Catal.* (1973) **30,** 409.
152. Minachev, Kh. M., Isakov, Ya. I., *Bull. Acad. Sci. USSR, Div. Chem. Sci.* (1968) 903.
153. Smith, J. V., Bennett, J. M., Flanigen, E. M., *Nature* (1967) **215,** 241.
154. Eberly, P. E., Jr., Kimberlin, C. N., Jr., *Adv. Chem. Ser.* (1971) **102,** 374.
155. Csicsery, S. M., Hickson, D. A., *J. Catal.* (1970) **19,** 386.

156. Cross, N. E., Kemball, C., Leach, H. F., *Adv. Chem. Ser.* (1971) **102,** 389.
157. Habgood, H. W., George, Z. M., "Molecular Sieves," p. 130, Society of the Chemical Industry, London, 1968.
158. Kolesnikov, I. M., Panchenkov, G. M., Tret'yakova, V. A., *Russ. J. Phys. Chem.* (1967) **41,** 586.
159. Ward, J. W., *J. Catal.* (1969) **13,** 321.
160. Ward, J. W., *J. Catal.* (1970) **17,** 355.
161. Plank, C. J., Rosinski, E. J., Hawthorne, W. P., *Ind. Eng. Chem., Prod. Res. Develop.* (1964) **3,** 165.
162. Ward, J. W., *J. Catal.* (1968) **11,** 251.
163. Ward, J. W., *J. Catal.* (1972) **26,** 451.
164. Ward, J. W., *J. Catal.* (1972) **26,** 470.
165. Hansford, R. C., Ward, J. W., *Adv. Chem. Ser.* (1971) **102,** 354.
166. Ward, J. W., *J. Phys. Chem.* (1970) **74,** 3021.
167. Ballivet, D., Pichat, P., Barthomeuf, D., *Adv. Chem. Ser.* (1973) **121,** 469.
168. Spojakina, A. A., Tjufekchieva, I. D., Sotirova, A. S., Radev, R. I., Ref. *88,* 364.
169. Lombardo, E. A., Sill, G. A., Hall, W. K., *J. Catal.* (1971) **22,** 54.
170. Lombardo, E. A., Sill, G. A., Hall, W. K., *Adv. Chem. Ser.* (1971) **102,** 258.
171. Lombardo, E. A., Velez, J., *Adv. Chem. Ser.* (1973) **121,** 553.
172. Ward, J. W., *Adv. Chem. Ser.* (1971) **101,** 380.
173. Poutsma, M. L., Schaffer, S. R., *J. Phys. Chem.* (1973) **77,** 158.
174. Ward, J. W., *J. Catal.* (1971) **22,** 237.
175. Ward, J. W., *J. Catal.* (1969) **14,** 365.
176. Merrill, H. E., Arey, W. F., *Preprints, Div. Petroleum Chem., Amer. Chem. Soc.* (1968) **13**(3), 193.
177. Barry, T. I., Lay, L. A., *J. Phys. Chem. Solids* (1966) **27,** 1821; (1968) **29,** 1395.
178. Mikheikin, I. D., Zhidomirov, G. M., Kazanskii, V. B., *Russ. Chem. Rev.* (1972) **41,** 468.
179. Amaro, A. A., Seff, K., *J. Phys. Chem.* (1973) **77,** 906.
180. Gallezot, P., BenTaarit, Y., Imelik, B., *J. Catal.* (1972) **26,** 295.
181. Gallezot, P., Imelik, B., *J. Phys. Chem.* (1973) **77,** 2364.
182. Rabo, J. A., Angell, C. L., Kasai, P. H., Schomaker, V., *Discuss. Faraday Soc.* (1966) **41,** 328.
183. Reimlinger, H., Kruerke, U., *Chem. Ber.* (1970) **103,** 2317.
184. Dollish, F. R., Hall, W. K., *J. Phys. Chem.* (1967) **71,** 1005.
185. Flockhart, B. D., McLoughlin, L., Pink, R. C., *J. Catal.* (1972) **25,** 305; Flockhart, B. D., Megarry, M. C., Pink, R. C., *Adv. Chem. Ser.* (1973) **121,** 509.
186. Neikam, W. C., *J. Catal.* (1971) **21,** 102.
187. Richardson, J. T., *J. Catal.* (1967) **9,** 172.
188. Richardson, J. T., *J. Catal.* (1967) **9,** 178.
189. Tokunaga, H., Ono, Y., Keii, T., *Bull. Chem. Soc. Japan* (1973) **46,** 3569.
190. Kurita, Y., Sonoda, T., Sato, M., *J. Catal.* (1970) **19,** 82.
191. Corio, P. L., Shih, S., *J. Catal.* (1970) **18,** 126.
192. Suzuki, I., Honda, Y., Ono, Y., Keii, T., Ref. *122,* 1377.
193. Hirschler, A. E., Neikam, W. C., Barmby, D. S., James, R. L., *J. Catal.* (1965) **4,** 628.
194. Ras'eev, G., *J. Catal.* (1971) **20,** 119.
195. Kochi, J. K., Krusic, P. J., *J. Amer. Chem. Soc.* (1968) **90,** 7157.
196. Hall, W. K., Ref. *122,* 1386.
197. Karge, H. G., *Surface Sci.* (1973) **40,** 157.
198. Wu, C. Y., Porter, R. P., Hall, W. K., *J. Catal.* (1970) **19,** 277.
199. Cant, N. W., Hall, W. K., *J. Catal.* (1972) **25,** 161.
200. Hightower, J. W., Hall, W. K., *J. Amer. Chem. Soc.* (1967) **89,** 778.
201. Ward, J. W., *J. Catal.* (1968) **11,** 259.

202. Nolley, J. P., Jr., Katzer, J. R., *Adv. Chem. Ser.* (1973) **121,** 563.
203. Satterfield, C. N., "Mass Transfer in Heterogeneous Catalysis," pp. 129-151, Massachusetts Institute of Technology Press, Cambridge, Mass., 1970.
204. Moore, R. M., Katzer, J. R., *Amer. Inst. Chem. Eng. J.* (1972) **18,** 816.
205. Satterfield, C. N., Cheng, C. S., *Amer. Inst. Chem. Eng. J.* (1972) **18,** 724; Satterfield, C. N., Katzer, J. R., *Adv. Chem. Ser.* (1971) **102,** 193.
206. Khudiev, A. T., Klyachko-Gurvich, A. L., Brueva, T. R., Isakov, Y. I., Rubenstein, A. M., *Bull. Acad. Sci. USSR, Div. Chem. Sci* (1968) 694.
207. Romanovskii, B. V., Tkhoang, K. S., Topchieva, K. V., Piguzova, L. I., *Kinet. Catal.* (1966) **7,** 739.
208. Riekert, L., *J. Catal.* (1970) **19,** 8.
209. Boreskova, E. G., Topchieva, K. V., Piguzova, L. I., *Kinet. Catal.* (1964) **5,** 792.
210. Turkevich, J., Murakami, Y., Nozaki, F., Ciborowski, S., *Chem. Eng. Progr., Symp. Series* (1967) **63,** 75.
211. Campbell, D. R., Wojciechowski, B. W., *J. Catal.* (1971) **23,** 307.
212. Chutoransky, P., Jr., Dwyer, F. G., *Adv. Chem. Ser.* (1973) **121,** 540.
213. Bezus, A. G., Kiselev, A. V., Du, P. Q., *J. Colloid Interface Sci.,* (1972) **40,** 223.
214. Bezus, A. G., Kiselev, A. V., Sedlacek, Z., Du, P. Q., *Trans. Faraday Soc.* (1971) **67,** 469.
215. Hopkins, P. D., *J. Catal.* (1973) **29,** 112.
216. Barthomeuf, D., Ha, B. H., *J. Chem. Soc. Faraday I,* (1973) 2147.
217. Barthomeuf, D., Ha, B. H., *J. Chem. Soc. Faraday I,* (1973) 2158.
218. Satterfield, C. N., Cheng, C. S., *Amer. Instit. Chem. Eng. J.* (1972) **18,** 720.
219. Gallezot, P., BenTaarit, Y., Imelik, B., *J. Phys. Chem.* (1973) **77,** 2556.
220. Matsumoto, H., Yasui, K., Morita, Y., *J. Catal.* (1968) **12,** 84.
221. Minachev, Kh. M., Isakov, Ya. I., Ref. *88,* 406.
222. Rabo, J. A., Poutsma, M. L., Skeels, G. W., Ref. *122,* 1353.
223. Boreskova, E. G., Lygin, V. I., Topchieva, K. V., *Kinet. Catal.* (1964) **5,** 991.
224. Navalikhina, M. D., Romanovskii, B. V., Topchieva, K. V., *Kinet. Catal.* (1971) ***12,*** 945.
225. Navalikhina, M. D., Romanovskii, B. V., Topchieva, K. V., *Kinet. Catal.* (1972) **13,** 203.
226. Navalikhina, M. D., Romanovskii, B. V., Topchieva, K. V., *Kinet. Catal.* (1972) **13,** 1196.
227. Navalikhina, M. D., Romanovskii, B. V., Topchieva, K. V., *Kinet. Catal.* (1972) **13,** 306.
228. Goldstein, M. S., Morgan, T. R., *J. Catal.* (1970) **16,** 232.
229. Voltz, S. E., Nace, D. M., Jacob, S. M., Weekman, V. W., Jr., *Ind. Eng. Chem., Process Design Develop.* (1972) **11,** 261.
230. Campbell, D. R., Wojciechowski, B. W., *J. Catal.* (1971) **20,** 217.
231. Gustafson, W. R., *Ind. Eng. Chem., Process Design Develop.* (1972) **11,** 507.
232. Weekman, V. W., Jr., *Ind. Eng. Chem., Process Design Develop.* (1968) **7,** 90.
233. Moné, R., Moscou, L., Ref. *88,* 351.
234. Hall, W. K., *Chem. Eng. Progr., Symp. Ser.* (1967) **63,** 68.
235. Eberly, P. E., Jr., *J. Phys. Chem.* (1967) **71,** 1717.
236. Weeks, T. J., Jr., Bolton, A. P., Ref. *88,* 426; Weeks, T. J., Jr., Angell, C. L., Ladd, I. R., Bolton, A. P., *J. Catal.* (1974) **33,** 256.
237. Ermilova, M. M., Gryaznova, Z. V., *Russ. J. Phys. Chem.* (1970) **44,** 985, 987.
238. Dimitrov, C., Leach, H. F., *J. Catal.* (1969) **14,** 336.
239. Kemball, C., Leach, H. F., Moller, B. W., *J. Chem. Soc. Faraday I* (1973) 624.
240. Lombardo, E. A., Hall, W. K., Ref. *122,* 1365.
241. Lombardo, E. A., Hall, W. K., *Amer. Instit. Chem. Engin. J.* (1971) **17,** 1229.
242. Kemball, C., Ref. *88,* 70.

243. Kemball, C., Leach, H. F., Skundric, B., Taylor, K. C., *J. Catal.* (1972) **27,** 416.
244. Venuto, P. B., Landis, P. S., *Adv. Catal.* (1968) **18,** 259.
245. George, Z. M., Habgood, H. W., *J. Phys. Chem.* (1972) **76,** 3940.
246. Ipatieff, V. N., Pines, H., *Ind. Eng. Chem.* (1936) **28,** 684.
247. Nenitzescu, C. D., in Ref. *23,* Vol. II, Chapter 13.
248. Sorensen, T. S., *J. Amer. Chem. Soc.* (1967) **89,** 3782, 3794; Ref. 23, Vol. II, Chapter 19.
249. Deno, N. C., Richey, H. G., Jr., Friedman, N., Hodge, J. D., Houser, J. J., Pittman, C. U., Jr., *J. Amer. Chem. Soc.* (1963) **85,** 2991, 2995, 2998.
250. Sorensen, T. S., Ranganayakulu, K., *J. Amer. Chem. Soc.* (1970) **92,** 6539; Deno, N. C., in Ref. *23,* Vol. II, Chapter 18.
251. Vogel, E., Grimme, W., Dinné, E., *Tetrahedron Lett.* (1965) 391; Marvell E. N., Caple, G., Schatz, B., *Ibid.* (1965) 385; Glass, D. S., Watthey, J. W. H., Winstein, S., *Ibid.* (1965) 377.
252. Venuto, P. B., Hamilton, L. A., *Ind. Eng. Chem., Prod. Res. Devel.* (1967) **6,** 190.
253. Norton, C. J., *Ind. Eng. Chem., Process Design Devel.* (1964) **3,** 230.
254. Weeks, T. J., Jr., Bolton, A. P., *J. Chem. Soc. Faraday I* (1974) **70,** 1676.
255. Lapidus, A. L., Rudakova, L. N., Isakov, Ya. I., Minachev, Kh. M., Eidus, Ya. T., *Bull. Acad. Sci. USSR, Div. Chem. Sci.* (1972) 1577.
256. Garwood, W. E., Venuto, P. B., *J. Catal.* (1968) **11,** 175.
257. Kirsch, F. W., Lauer, J. L., Potts, J. D., *Preprints, Div. Petroleum Chem., Amer. Chem. Soc.* (1971) **16**(2), B24.
258. Kirsch, F. W., Potts, J. D., Barmby, D. S., *J. Catal.* (1972) **27,** 142.
259. Ref. *24,* Vol. II.
260. Venuto, P. B., *J. Org. Chem.* (1967) **32,** 1272.
261. Stock, L. M., Brown, H. C., *J. Amer. Chem. Soc.* (1959) **81,** 3323.
262. Berentsveig, V. V., Rudenko, A. P., Kubasov, A. A., *Proc. Acad. Sci. USSR* (1972) **204,** 460.
263. Bierenbaum, H. S., Partridge, R. D., Weiss, A. H., *Adv. Chem. Ser.* (1973) **121,** 605.
264. Boreskova, E. G., Topchieva, K. V., Piguzova, L. I., *Kinet. Catal.* (1964) **5,** 792.
265. Roberts, R. M., Baylis, E. K., Fonken, G. J., *J. Amer. Chem. Soc.* (1963) **85,** 3454.
266. Forni, L., Carra, S., *J. Catal.* (1972) **26,** 153.
267. Leigh, C. H., Szwarc, M., *J. Chem. Phys.* (1952) **20,** 844.
268. Olah, G. A., Meyer, M. W., Overchuk, N. A., *J. Org. Chem.* (1964) **29,** 2313, 2315.
269. Allen, R. H., Yats, L. D., Erley, D. S., *J. Amer. Chem. Soc.* (1960) **82,** 4853; Allen, R. H., *Ibid.* (1960) **82,** 4856.
270. Ünseren, E., Wolf, A. P., *J. Org. Chem.* (1962) **27,** 1509.
271. McCauley, D. A., Lien, A. P., *J. Amer. Chem. Soc.* (1957) **79,** 5953; Brown, H. C., Smoot, C. R., *Ibid.* (1956) **78,** 2176.
272. Streitwieser, A., Jr., Reif, L., *J. Amer. Chem. Soc.* (1960) **82,** 5003.
273. Grandio, P., Schneider, F. H., Schwartz, A. B., Wise, J. J., *Preprints, Div. Petroleum Chem., Amer. Chem. Soc.* (1971) **16**(3), B78.
274. Lanewala, M. A., Bolton, A. P., *J. Org. Chem.* (1969) **34,** 3107.
275. Grandio, P., Schneider, F. H., Schwartz, A. B., Wise, J. J., *Preprints, Div. Petroleum Chem., Amer. Chem. Soc.* (1971) **16**(3), B70.
276. Csicsery, S. M., *J. Catal.* (1970) **19,** 394.
277. Csicsery, S. M., *J. Catal.* (1971) **23,** 124.
278. Csicsery, S. M., *J. Org. Chem.* (1969) **34,** 3338.
279. Bolton, A. P., Lanewala, M. A., Pickert, P. E., *J. Org. Chem.* (1968) **33,** 1513.
280. Bolton, A. P., Lanewala, M. A., Pickert, P. E., *J. Org. Chem.* (1968) **33,** 3415.
281. Wang, K. M., Lunsford, J. H., *J. Catal.* (1972) **24,** 262.

282. Matsumoto, H., Take, J. I., Yoneda, Y., *J. Catal.* (1968) **11,** 211.
283. Haensel, V., *Adv. Catal.* (1951) **3,** 179.
284. Voge, H. H., in Emmett, P. H., Ed., "Catalysis," Vol. VI, Reinhold, New York, 1958.
285. Pansing, W. F., *J. Phys. Chem.* (1965) **69,** 392.
286. Kerr, G. T., Ref. *88,* 38.
287. Bolton, A. P., Lanewala, M. A., *J. Catal.* (1970) **18,** 1.
288. Schulz, H. F., Weitkamp, J. H., *Ind. Eng. Chem., Prod. Res. Devel.* (1972) **11,** 46.
289. Schulz, H., Weitkamp, J., *Preprints, Div. Petroleum Chem., Amer. Chem. Soc.* (1971) **16**(1), A102.
290. Schulz, H., Weitkamp, J., *Preprints, Div. Petroleum Chem., Amer. Chem. Soc.* (1972) **17**(4), G89.
291. Schulz, H., Weitkamp, J., Eberth, H., Ref. *122,* 1229.
292. Poutsma, M. L., Schaffer, S. R., unpublished results.
293. Kouwenhoven, H. W., *Adv. Chem. Ser.* (1973) **121,** 529.
294. Chen, N. Y., Ref. *88,* 68.
295. Hogeveen, H., *Rec. Trav. Chim.* (1970) **89,** 74.
296. Heylen, C. F., Jacobs, P. A., *Adv. Chem. Ser.* (1973) **121,** 490.
297. Plank, C. J., Rosinskii, E. J., *Chem. Eng. Progr., Symp. Series* (1967) **63,** 26.
298. Weekman, V. W., Jr., Nace, D. M., *Amer. Inst. Chem. Eng. J.* (1970) **16,** 397; Weekman, V. W., Jr., *Ind. Eng. Chem., Process Design Devel.* (1969) **8,** 385.
299. Thomas, C. L., Barmby, D. S., *J. Catal.* (1968) **12,** 341.
300. Leathard, D. A., Purnell, J. H., *Ann. Rev. Phys. Chem.* (1970) **21,** 197; Kossiakoff, A., Rice, F. O., *J. Amer. Chem. Soc.* (1943) **65,** 590.
301. Dudzik, Z., Cvetanović, R. J., Ref. *117,* 175.
302. Munter, H. J., *J. Colloid Interface Sci.* (1972) **41,** 378.
303. Miale, J. N., Weisz, P. B., *J. Catal.* (1971) **20,** 288.
304. Lang, W. H., Mikovsky, R. J., Silvestri, A. J., *J. Catal.* (1971) **20,** 293.
305. Mikovsky, R. J., Silvestri, A. J., Dempsey, E., Olson, D. H., *J. Catal.* (1971) **22,** 371.
306. Silvestri, A. J., Smith, R. L., *J. Catal.* (1973) **29,** 316.

Chapter

9

Reactions of Molecules Containing Hetero Atoms over Zeolites

Marvin L. Poutsma, Corporate Research Laboratory, Union Carbide Corp., Tarrytown, N. Y. 10591

THE MECHANISTIC CONSIDERATIONS of catalysis by zeolites in the preceding chapter were confined to reactions involving hydrocarbons as reactants and products. Here we consider reactions with organic molecules containing hetero atoms as reactants and/or products. The information available on this topic is less detailed and more qualitative, and the most commonly used mechanistic tool is analogy to corresponding homogeneously catalyzed or nonzeolitic heterogeneously catalyzed reactions.

The most extensive source of information is a review by Venuto and Landis (*1*) in 1968 which not only cataloged previous literature but also gave many original results from the Mobil Corp. group. Other reviews which deal with transformations of hetero atom-containing organic molecules include more recent treatments by Venuto (*2, 3*), Leach (*4*), and Minachev and co-workers (*5–7*). It is therefore not our purpose to be comprehensive but to survey the subject with emphasis on mechanisms which have been proposed and, to varying extents, substantiated.

No organizational outline based on mechanistic categories appears very satisfactory since many of the proposed mechanisms are little more than speculation, both by the original authors and by us. We therefore organize the material by generic reaction types as done by Venuto and Landis. However, we subdivide the reactions into two broad, not necessarily exclusive regions: reactions catalyzed by zeolites ion-exchanged with protons or main group cations, and those catalyzed specifically by zeolites containing transition group cations.

Some general observations are in order before we examine individual reaction types. First, in certain hydrocarbon transformations, the processes of sorption and diffusion may be rate limiting rather than the actual chemical reaction at the active site. This phenomenon should be even more common for reactants and products more polar and/or polarizable and hence more strongly adsorbed than hydrocarbons. Secondly, the observed acceleration of certain hydrocarbon transformations by additives such as water and hydrogen halides and their deceleration by poisons such as nitrogeneous bases suggest acidic active sites. Since these materials appear as reactants and/or products in several of the reaction types, their roles in modifying zeolitic active sites must be remembered constantly. Thirdly, many acid-catalyzed reactions considered here are inherently more facile than their all-hydrocarbon analogs, and therefore it is not surprising to find several examples where catalysis by the external zeolite crystallite surface becomes significant compared with intracrystalline catalysis. Finally, catalyst aging is common and is generally associated with buildup of large condensed byproducts within the zeolite crystallite. This problem, very common in the reactions of hetero atomic molecules, has limited the practical utility of their catalysis by zeolites.

Reactions Catalyzed by Main Group Cation-Exchanged Zeolites

Acid-Catalyzed Rearrangements. Many acid-catalyzed organic rearrangements occur by three steps: protonation of the substrate to form a cation, intramolecular rearrangement to form a more stable cation, and finally deprotonation to generate a product isomeric with the reactant. Several organic rearrangements requiring no co-reagents also appear to follow this pathway over acidic zeolites.

Epoxides can be converted to isomeric carbonyl compounds over a range of zeolites (*1*). In the specific case of propylene oxide (*8*), the rate at 100°C decreased in the order HY ~ CaX >> pyridine-poisoned HY > NaY. The major product was propionaldehyde, accompanied by small amounts of acetone and traces of allyl alcohol. The more active, more acidic catalysts were less selective to the aldehyde. The gross mechanism, by analogy to reactions in solution, is probably as shown in Equation 1, with the hydride shift possibly being concerted with ring opening; however, the detailed factors influencing the direction of ring-opening are not clear.

$$CH_3\underset{\diagdown O \diagup}{CH-CH_2} \xrightarrow{H^+} CH_3\underset{\diagdown O^+(H) \diagup}{CH-CH_2}$$

$$\rightarrow CH_3\overset{+}{C}HCH_2OH \xrightarrow{\sim H} CH_3CH_2\overset{+}{C}HOH \xrightarrow{-H^+} CH_3CH_2CHO$$

$$\rightarrow CH_3CHOH\overset{+}{C}H_2 \xrightarrow{\sim H} CH_3\overset{+}{C}OHCH_3 \xrightarrow{-H^+} CH_3COCH_3 \quad (1)$$

An inefficient analog of the Fries rearrangement (*9*) was reported (*1*) over REX at 100°–200°C in which phenyl acetate was rearranged to a mixture of *o*- and *p*-acetylphenol.

Landis and Venuto (*10*) described the conversion of the oximes of acetone, acetophenone, and cyclohexanone to *N*-methylacetamide, acetanilide (with a trace of *N*-methylbenzamide), and ε-caprolactam, respectively, over a variety of acidic zeolites. Optimum conditions for this heterogeneous analog of the Beckmann rearrangement (*11*) were use of HY at 250°–350°C and atmospheric pressure with addition of a nonpolar carrier solvent such as benzene and a nonbasic carrier gas such as nitrogen. Use of the more polar acetic acid as the added carrier inhibited this catalyzed rearrangement, in contrast to its rate enhancement of liquid-phase homogeneous analogs, possibly by competitive adsorption at the active sites. The major byproduct from cyclohexanone oxime was the ring-opened 5-cyano-1-pentene. Based on analogies to the normal acid-catalyzed rearrangement (*11*), the authors proposed the mechanism in Reaction 2, the key competitive steps being the migration

$$RR'C{=}N{-}OH \xrightarrow{H^+} RR'C{=}N{-}\overset{+}{O}H_2$$

$$RR'C{=}N{-}\overset{+}{O}H_2 \xrightarrow[-H_2O]{\sim R} R'{-}\overset{+}{C}{=}NR \xrightarrow[-H^+]{H_2O} R'{-}C(OH){=}NR \longrightarrow R'{-}C(=O){-}NHR \qquad (2)$$

$$RR'C{=}N{-}\overset{+}{O}H_2 \longrightarrow R'C{\equiv}N + R^+ + H_2O; \quad R^+ \xrightarrow{-H^+} \text{olefin}$$

of the *anti*-alkyl group to an electron-deficient nitrogen center in a protonated oxime and β-scission of this same intermediate. A detailed IR study of the interactions of the oximes of cyclopentanone (and cyclohexanone) with zeolites has been reported by Butler and Poles (*12*). Adsorbed oxime on HY at 25°C appeared to be protonated on nitrogen by the HF zeolite hydroxyl band and possible further hydrogen-bonded to the zeolitic surface as in structure **1.** Heating to 120° gave a new spectrum very similar to that of the adsorbed product 2-piperidone which the authors suggested to be adsorbed in a protonated ring-opened form, **2.** However, heating to 350°C was required to desorb the rearranged product and regenerate the HF hydroxyl band. Thus the temperatures of 250°–350°C required by Landis and Venuto (*10*) to achieve significant catalytic conversion may have resulted because reaction was product-desorption rate limited rather than rearrangement rate limited. This situation may well be the rule rather than the exception for

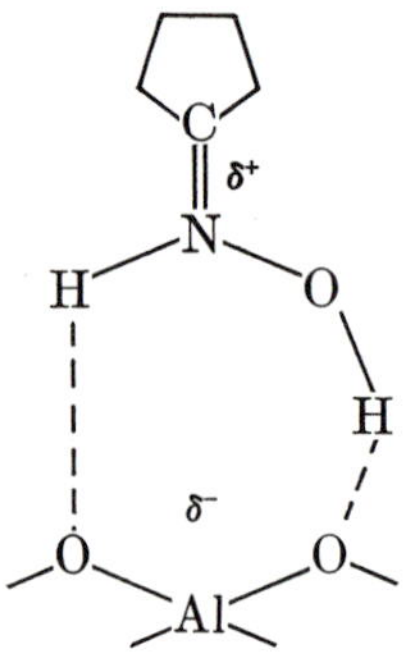

C, δ^+, N, H, O, H, δ^-, O, O, Al

1

$(CH_2)_2$, CH_2, CH_2, $H{-}\overset{+}{O}{=}C$, NH, O, Si, Al

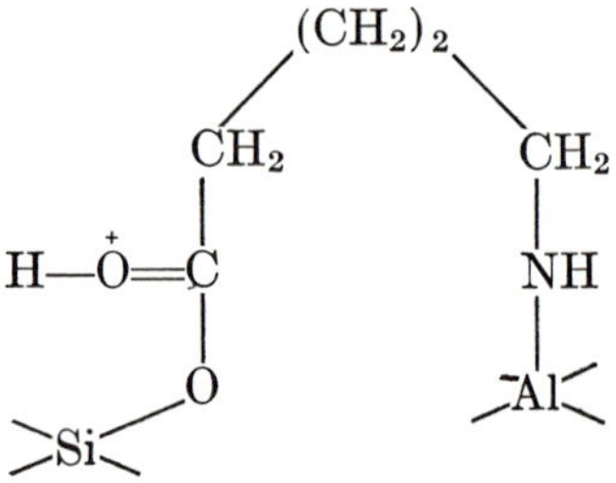

2

zeolite-catalyzed rearrangement of such highly polar substrates. The rapid aging observed (*1, 10*) is probably related to secondary condensation reactions whose rates can compete with product desorption.

Conversion of the phenylhydrazones of acetone and cyclohexanone to 2-methylindole and tetrahydrocarbazole, respectively, over REX and CaX at 150°C was reported (*1*) to proceed in 50–70% yield. These appear to be heterogeneous versions of the acid-catalyzed Fischer indole synthesis (*13*).

Polymerization. The acid-catalyzed condensation of olefins was considered in the preceding chapter, especially as it contributes to catalyst aging. In this reaction type, a carbonium ion formed by substrate protonation is captured by unprotonated substrate to form a new carbonium ion. Barrer and Oei (*14*) exposed hydrogen mordenite, activated at 360°C, to vapors of *n*-butyl vinyl ether at 22°–50°C. A low polymer composed of ~ 10 monomer units formed around and possibly partly within each zeolite particle. The polymerization rate was enhanced by previously adsorbed water up to a level of one added water per calculated Brønsted acid site, and a cationic polymerization mechanism (Equation 3) proceeding through *sec*-carbonium ions

$$CH_2{=}CHOR \xrightarrow{H^+} CH_3\overset{+}{C}HOR \xrightarrow{CH_2=CHOR} CH_3\underset{\displaystyle OR}{\underset{|}{C}H}{-}CH_2\overset{+}{C}HOR \rightarrow \text{etc.} \qquad (3)$$

stabilized by the α-alkoxy group was suggested. However, the overall kinetics, particularly the linear plots of amount of polymer accumulated *vs.* $(\text{time})^{1/2}$, indicated a significant kinetic limitation arising from diffusion of monomer through the already formed polymer to reach the zeolitic active sites. The relative importance of intracrystalline *vs.* crystallite surface catalysis could not be assessed.

Alkylation of Aromatics. The acid-catalyzed alkylation of aromatics with olefins proceeds by capture of the olefin-derived carbonium ion by the nucleophilic aromatic ring. Several examples are known of zeolite-catalyzed

$$CH_2{=}CHR \xrightarrow{H^+} CH_3\overset{+}{C}HR \xrightarrow{C_6H_6} [C_6H_6^+(H)(CH(CH_3)R)] \xrightarrow{-H^+} C_6H_5CH(CH_3)R$$

alkylations where either the olefin or the aromatic partner is replaced by a compound containing a hetero atom.

Alkylation of benzene by simple alcohols has been described by Minachev and co-workers (*5*) over CaY at 250°–325°C. The product from *n*-propyl alcohol was largely rearranged isopropylbenzene, as expected for a carbonium ion process in which a primary cation should rearrange to a secondary cation. However, Venuto and Landis (*1*) reported that significant quantities of *n*-alkylaromatics can be obtained from primary alcohols and suggested that benzene may attack the protonated alcohol directly by an S_N2 process without requiring prior formation of a carbonium ion (Reaction 4) and

$$RCH_2OH \xrightarrow{H^+} RCH_2\overset{+}{O}H_2 \xrightarrow{C_6H_6} \left[{}^{\delta+}OH_2 \cdots CH_2(R) \cdots C_6H_6^{\delta+}(H) \right]^{\ddagger} \longrightarrow RCH_2C_6H_6^+(H) + H_2O \xrightarrow{-H^+} RCH_2C_6H_5 \quad (4)$$

attendant rearrangement possibilities. The special case of methanol where this mechanism would be most likely is discussed below. Alkylations by alcohols produce water as a coproduct, which should modify the initial zeolite acidity. Alkylations can also be performed with alkyl halides or ethers; the special case of chloromethyl methyl ether (Reaction 5) has been described by Venuto and Landis (*1*). The possible generation of carbonium ions from alcohols and alkyl halides and alkylations with carbonyl compounds is discussed further below.

Venuto and co-workers (*1, 15, 16*) studied the alkylation by olefins of substituted benzenes (such as phenol and anisole) and of heterocyclic aromatics (such as thiophene and pyrrole). The case of phenol was unusual in

$$C_6H_6 + ClCH_2OCH_3 \xrightarrow[70°C]{\text{HY or H Mordenite}} C_6H_5CH_2OCH_3 > C_6H_5CH_2Cl \xrightarrow{C_6H_6} C_6H_5\text{-}CH_2\text{-}C_6H_5 \quad (5)$$

that more vigorous conditions ($\sim$ 200°C) were required for alkylation with ethylene than for benzene ($\sim$ 120°C) although phenol is normally much more susceptible to electrophilic attack; also, the presence of phenol inhibited the alkylation of benzene. Venuto and Wu (*17*) proposed that for NH_4Y zeolites activated at 550°C in oxygen this inversion of expected reactivities resulted from the strong adsorption of phenol, which prevented accessibility of the more weakly adsorbed ethylene to the active sites. In this picture, an adsorbed ethyl cation reacts with a free aromatic *via* a Rideal mechanism but not with an adsorbed aromatic.

Yashima and co-workers (*18, 19*) have reported the alkylation of toluene with methanol over a variety of Y zeolites with particular emphasis on finding conditions to optimize selectivity to *p*-xylene. At 225°C and a toluene:methanol ratio of 2:1, the reactivity order for 70–80% exchanged Y zeolites activated at 300°C (500°C for HY from NH_4Y) was La > Ce >> H > Ni > Co > Mn >> Cd > Ca > Mg > Sr >> Li > Na, the familiar order for reactions requiring acidic active sites; a likely alkylating agent is thus $CH_3{}^+OH_2$. However, the positional selectivity, in particular the para:ortho ratio, was a subtle function of zeolite type and time-on-stream. The most active catalysts initially gave preferential formation of *p*-xylene while the less active gave initially preferentially *o*-xylene; in the latter case the para:ortho ratio increased with time-on-stream, possibly because of effects of the water coproduct. For NH_4Y, the initial isomer ratio correlated with T_{act} in the same fashion as total Brønsted acidity. Similarly, the ratio increased, as did activity, with increasing extent of NH_4^+ exchange. Therefore, formation of the para isomer appeared strongly dependent on total Brønsted acidity whereas that of the ortho did not. No rationalization for this latter phenomenon was suggested.

At considerably higher temperatures ($\sim$ 425°C) reaction was also catalyzed by the alkali metal ion-exchanged Y zeolites. LiY still produced xylenes but the others produced mixtures of styrene and ethylbenzene from interaction of toluene with methanol or, better, formaldehyde after an initial induction period in which largely dimethyl ether, carbon monoxide, carbon dioxide, and hydrogen were formed (*20, 21*). This side-chain alkylation was promoted by added aniline whereas the nuclear alkylation was promoted by hydrogen chloride. Methanol may first be dehydrogenated to formaldehyde, styrene may be the initial alkylation product, and the active sites may be basic rather than acidic.

$$CH_3OH \rightarrow CH_2O + H_2$$

$$CH_2O + C_6H_5CH_3 \rightarrow C_6H_5CH{=}CH_2 + H_2O$$

$$C_6H_5CH{=}CH_2 + H_2 \rightarrow C_6H_5CH_2CH_3$$

Similar behavior was reported (*22*) for α- and β-methylnaphthalenes at 400°–450°C over KX and RbX, the latter being more active; the products were largely the corresponding ethylnaphthalenes with traces of vinylnaphthalenes along with again dimethyl ether, the carbon oxides, and hydrogen. If these reactions proceed *via* formaldehyde, they may be analogous to the carbonyl condensation reactions discussed below, but the mode of activation of the benzylic C—H bond is not known. These reactions may represent a rare example of a zeolite-catalyzed reaction *via* a carbanionic pathway (Reaction 6).

$$C_6H_5CH_3 \xrightarrow[?]{-H^+} C_6H_5CH_2^- \xrightarrow{CH_2O} C_6H_5CH_2CH_2O^- \xrightarrow{+H^+,\ -H_2O} C_6H_5CH{=}CH_2 \qquad (6)$$

β-Elimination Reactions. A broad class of reactions which can be catalyzed by zeolites are reversible addition–eliminations as shown in Equation 7, where X is OH, OCOR, SH, or halogen. The position of this equilib-

$$H{-}\overset{|}{\underset{|}{C}}{-}\overset{|}{\underset{|}{C}}{-}X \rightleftharpoons \;>C{=}C<\; + \; HX \qquad (7)$$

rium depends on temperature as shown in Table I for some typical carbon skeletons. Most of the studies with zeolites have been carried out at high enough temperatures that β-elimination rather than addition is favored; however, for any individual case, particularly for halides, one must ask whether observed conversions are limited by kinetics or thermodynamics.

Early work on dehydration of alcohols to form olefins with ethers as byproducts has been summarized by Venuto and Landis (*1*). Where com-

$$RCH_2CH_2OH \rightarrow RCH{=}CH_2 + H_2O$$

$$RCH_2CH_2OH \rightarrow \tfrac{1}{2}\ RCH_2CH_2OCH_2CH_2R + \tfrac{1}{2}\ H_2O$$

Table I. Equilibrium Constants for Addition-Elimination Reactions as a Function of Structure and Temperature

Reaction	*log* K_p			
	300°K	*400°K*	*500°K*	*600°K*
$C_2H_5OH \rightleftarrows C_2H_4 + H_2O$	−1.32	0.67	1.89	2.70
i-$C_3H_7OH \rightleftarrows C_3H_6 + H_2O$	−1.30	0.94	2.30	3.19
tert-$C_4H_9OH \rightleftarrows$ *i*-$C_4H_8 + H_2O$	−3.52	−0.59	1.17	2.35
$C_2H_5OCOCH_3 \rightleftarrows C_2H_4 + CH_3CO_2H$	−3.24	−0.62	0.95	1.99
$C_2H_5Cl \rightleftarrows C_2H_4 + HCl$	−5.68	−2.54	−0.63	0.63
tert-$C_4H_9Cl \rightleftarrows$ *i*-$C_4H_8 + HCl$	−4.63	−1.40	0.53	1.81
$C_2H_5Br \rightleftarrows C_2H_4 + HBr$	−7.13	−3.63	−1.51	−0.11
$C_2H_5SH \rightleftarrows C_2H_4 + H_2S$	−6.88	−3.46	−1.41	−0.05

parisons have been made for different cation-exchanged forms of a given zeolite lattice, the catalytic activity orders are very similar to those observed for typical carbonium-ion hydrocarbon transformations. For example, the orders for ethanol dehydration at 310°C (*23*) were Ca > Na > Li ~ Mg > K ~ Ba for the X series, Mg > Ba > Ca > Na > Li > K for the mordenite series, and Mg > Ca > Li > Na > K for the A series with essentially all activity having disappeared for the K^+-exchanged cases; some apparent misordering in these series may result from varying degrees of exchange within the sets. For a given cation, the A series appeared somewhat more reactive than the X (*24*) and mordenite series. The order for 1-pentanol dehydration at 200°–300°C (*25*) was HY > ZnY > MgY > ZnX > MgX > NaY > NaX, with a similar order being observed for rates of double bond shift of the pentenes produced; activity increased with increasing Zn^{2+} and Mg^{2+} exchange levels. The order for 2-propanol dehydration at 200°–300°C (*26*) was HX > NaX > CaA although a later paper (*27*) suggested that the activity observed for NaX resulted from an impure sample. The order for *n*-butyl alcohol dehydration at 350°C (*28*) was LiX > NaX > KX > RbX. For a series of NH_4Y zeolites activated at 500°C, the rate of dehydration of 2-propanol at 310°C (*29*) increased with increasing NH_4^+ exchange level. Results for steamed and ultrastabilized NH_4Y (*30*) also demonstrated a role for catalyst acidity in alcohol dehydration.

The vapor-phase dehydration of *tert*-butyl alcohol over CaX at ~ 175°C was promoted by small amounts of added water (*31, 32, 33*) and that of 2-propanol over HX at ~ 225°C was not poisoned by use of an aqueous alcohol feed (*26*); (but *cf.* Ref. *34*, next paragraph). Dehydration of 2-propanol by NaX was enhanced by carbon dioxide (*27*); the effect was attributed to interaction of carbon dioxide with traces of divalent ions or lattice imperfections of unknown type.

One of the earliest demonstrations of shape selectivity in zeolite catalysis was the observation (*35*) that the linear *n*-butyl alcohol could be effectively dehydrated over both CaX and CaA zeolites, but the branched 2-butanol was dehydrated only over the former. Ether formation is more of a complication in the larger-pore X zeolites compared with A zeolites, possibly because it involves a bulkier bimolecular transition state (*23, 24*). In the conversion of methanol to dimethyl ether over hydrogen mordenite (*36*), intracrystalline mass transport limitations were negligible at 155°C but became significant at 205°C. Dehydration therefore occurs predominantly within the intracrystalline pore structure. Catalysis of dehydration of *tert*-butyl alcohol over hydrogen mordenite in the liquid phase at 45°–75°C was also suggested (*34*) recently to be intracrystalline based on demonstration of diffusional resistance by preadsorbed *n*-heptane or by water in the alcohol.

Although alcohol dehydration appears to be clearly acid catalyzed, a detailed mechanism is unknown particularly with respect to whether the reaction is nearer the E1 or E2 extremes of elimination reactions (*37*). If E1, we would picture heterolysis of the protonated alcohol to form an adsorbed carbonium ion followed by proton loss to a lattice oxide ion (Reaction 8), possibly with intermediate carbonium ion rearrangement.

$$\text{—CH}_2\text{CH}_2\text{OH} \xrightarrow{\text{H}^+} \text{—CH}_2\text{CH}_2\overset{+}{\text{O}}\text{H}_2 \xrightarrow{-\text{H}_2\text{O}} \text{—CH}_2\overset{+}{\text{C}}\text{H}_2 \xrightarrow{-\text{H}^+} \text{—CH}{=}\text{CH}_2 \qquad (8)$$

Alternatively, the carbonium ions could be formed by transfer of the alcohol OH group to a Lewis acid site or a polyvalent cation, although these would probably soon be converted to Brønsted sites or $M(OH)_n^+$ cations, respectively, by the water product. If E2, we would picture a basic site attacking the β-C—H bond in the protonated alcohol synchronously with C—O bond cleavage to form olefin without intermediate carbonium ion formation (Reaction 9).

$$\text{O}_\text{Z}^{2-}\ \text{H—CH}_2\text{—CH}_2\text{—}\overset{+}{\text{O}}\text{H}_2 \longrightarrow \text{OH}_\text{Z}^- + \text{CH}_2{=}\text{CH}_2 + \text{H}_2\text{O} \qquad (9)$$

Based on homogeneous analogies of acid-catalyzed dehydration (*37*) and the general lack of strong basicity in zeolites, the E1 extreme appears more likely, but there is not enough structure-reactivity data for various alcohols to make the choice rigorously. Over various A zeolites, the order of dehydration rates was *tert*-butyl alcohol $>$ *n*-butyl alcohol $>$ *sec*-butyl alcohol (*38*); at face value the ordering of the latter two is inconsistent with carbonium ion stability and the E1 mechanism, but special diffusional effects may be important in these small-pore zeolites.

Production of olefins and acetic acid by passage of alkyl acetates over zeolites (*39, 40, 41*) is clearly an acid-catalyzed process distinct from unimolecular thermal ester pyrolysis (*42*). The dual site mechanism proposed by Imai and Anderson (*41*) is really an E2 mechanism of protonated ester; however, the much greater reactivity of *sec*-butyl acetate compared with *n*-butyl acetate is more suggestive of E1 elimination.

Alkyl halides can be dehydrohalogenated to form olefins over a variety of zeolites and again early reports have been reviewed (*1*). Venuto and co-workers (*43*) studied reactions of ethyl chloride at 260°C, and the results with respect to zeolite type are unusual. For example, the activity of REX was poisoned by quinoline although it was barely more active than NaX with serious aging occurring in both cases. Examination of aged catalysts revealed either loss of crystallinity (NaX), coking (REX, HY), or formation of new phases (AgCl from AgX). The reactivity order $CH_3CCl_3 > CH_3CHCl_2 > CH_3CH_2Cl > ClCH_2CH_2Cl$ observed over REX is consistent with carbonium ion formation by C—Cl bond heterolysis and an E1 mechanism. A key feature is probably the role played by the hydrogen chloride formed in interacting with the zeolite to form a medium conducive to heterolytic reactions. Reaction of *tert*-butyl halides over NaA, MgA, and CaA at ambient temperatures (*44*) occurred *via* dehydrohalogenation, presumably on the external crystallite surfaces, followed by intracrystalline telomerization of the isobutene product, probably cocatalyzed by the hydrogen halide formed.

Mochida and Yoneda (*45*) reported rates and products from dehydrochlorination of 1,1,2-trichloroethane at 300°C. The effect of zeolite structure for X and Y zeolites ranging from H^+- to K^+-exchanged forms was remarkably small. The relative yields of *trans*-1,2-, *cis*-1,2-, and 1,1-dichloroethylene did not correspond directly with earlier correlations proposed by these authors (*46*) to distinguish between catalytic dehydrochlorination by solid acids and bases. Kladnig and Noller (*47*) examined olefin formation from *n*-butyl and *sec*-butyl chlorides over a range of X zeolites. The total range of activity was less than twofold as the cation was varied from Na^+ to Pr^{3+}. The secondary chloride was less than twice as reactive as the primary. The initial olefinic products were linear butenes in near equilibrium proportions; *i*-butene was a secondary product over the more acidic zeolites. The authors (*47*) associated the slightly increasing activity with increasing cation electrostatic field; however, in fact, the reported correlations show that increasing electrostatic field led to increasing E_a values so that activities were determined by the pre-exponential factor in some unknown fashion.

The effects of both substrate and zeolite structure are thus much smaller for dehydrochlorination than for dehydration and are so compressed and probably so obscured by effects of hydrogen halide product as to make mechanistic distinctions between E1 and E2 extremes very difficult. Neither mechanism offers an obvious direct role for protonation since alkyl halides are much less basic than alcohols. The role of the zeolite may be more as a very polar medium conducive to reactions with polar transition states than as a solid acid. Dehydrosulfurization of mercaptans over zeolites to produce olefins and hydrogen sulfide has also been observed (*1*).

α-Elimination and Related Reactions. For compounds having a good anionic leaving group but no available β-hydrogens, α-elimination may occur. For example, Venuto and Landis (*48*) described the conversion of benzyl mercaptan to *trans*-stilbene in 60–70% selectivity at 250°C over several zeolites. Although the zeolites were more active and selective for stilbene formation than a low-surface alumina, their role was relatively non-specific in that activity was observed for both Na^+ and Ca^{2+} forms of both A and X zeolites. The mechanism proposed included α-elimination of hydrogen sulfide promoted by the polar surface, insertion of the resulting phenylcarbene (*49*) into a C—H bond of another molecule of starting material, and finally β-elimination of hydrogen sulfide (Equation 10). Under more forcing conditions, a detectable amount of ethylene was produced from methyl mercaptan.

$$\begin{aligned} &C_6H_5CH_2SH \rightarrow C_6H_5CH\!: + H_2S \\ &C_6H_5CH\!: + C_6H_5CH_2SH \rightarrow C_6H_5CH_2\underset{\displaystyle SH}{\underset{|}{C}}HC_6H_5 \\ &C_6H_5CH_2\underset{\displaystyle SH}{\underset{|}{C}}HC_6H_5 \rightarrow C_6H_5CH{=}CHC_6H_5 + H_2S \end{aligned} \qquad (10)$$

Decomposition of methanol over REX and ZnX at ~ 350°C (*1*) gave not only dimethyl ether and water but also a series of C_2–C_6 olefins and paraffins; a carbene-type mechanism was discussed as a possibility. The interaction of certain zeolites with chlorodifluoromethane to produce carbon dioxide (*50, 51*) is also formally an α-elimination but the mechanism is unclear.

Wu and co-workers (*52*) have examined the thermal decomposition of methylammonium cation-exchanged Y zeolites, particularly the tetramethylammonium case. At lower temperatures (150°–275°C), two decomposition pathways were suggested to rationalize the observed products. In the first, nucleophilic S_N2 attack of adsorbed water (or formed methanol) on a methyl group generates the observed trimethylamine and methanol (or dimethyl ether) and forms zeolitic OH groups detected by IR (Reaction 11). In the

$$(CH_3)_3\overset{+}{N}CH_3 + ROH + O_Z^{2-} \rightarrow (CH_3)_3N + CH_3OR + HO_Z^- \qquad R = H, CH_3 \qquad (11)$$

second, nucleophilic attack by a lattice oxide ion generates trimethylamine and a methoxylated surface which subsequently reacts with methanol to give the observed methane, carbon monoxide, and hydrogen (Reaction 12). De-

$$\begin{aligned} &(CH_3)_3\overset{+}{N}CH_3 + O_Z^{2-} \rightarrow (CH_3)_3N + CH_3O_Z^- \overset{?}{\rightarrow} CH_2\!: + HO_Z^- \\ &CH_3O_Z^- \xrightarrow{CH_3OH} CH_4 + [CH_2O] + HO_Z^- \\ &[CH_2O] \rightarrow H_2 + CO \end{aligned} \qquad (12)$$

composition of this species to form carbene has also been suggested (*3*). At higher temperatures (275°–400°C), deprotonation of the methyl group to give an ylid was proposed, followed by Stevens rearrangement (*53*) and Hoffman elimination (*37*) to form dimethylamine and ethylene (Reaction 13); ethylene could then serve as the precursor of higher olefins, the observed

$$(CH_3)_3\overset{+}{N}CH_3 + O_Z^{2-} \rightarrow (CH_3)_3\overset{+\,-}{N}CH_2 + HO_Z^- \rightarrow (CH_3)_2\overset{+}{N}CH_2CH_3 \xrightarrow{HO_Z^-} (CH_3)_2\overset{+}{N}HCH_2CH_3 \xrightarrow{O_Z^{2-}} (CH_3)_2NH + CH_2{=}CH_2 + HO_Z^- \quad (13)$$

products, by usual acid-catalyzed reactions. Similar schemes rationalized the thermal decomposition of tetramethylammonium offretite (*54*). Heating mono-, di-, or triethylammonium ion-exchanged Y zeolites in the presence of residual water above 150°C (*55*) gave ethylene, presumably *via* Hoffman elimination, and transalkylations as in Reaction 14.

$$2\ C_2H_5\overset{+}{N}H_3Y \rightarrow (C_2H_5)_2\overset{+}{N}H_2Y + \overset{+}{N}H_4Y \quad (14)$$

Carbonyl Group Reactions. Venuto and Landis (*56*) reported the conversion of acetophenone to its aldol condensation product over hydrogen

$$C_6H_5\overset{O}{\overset{\|}{C}}CH_3 \xrightarrow{H^+} C_6H_5\overset{OH}{\overset{|}{\underset{+}{C}}}CH_3 \xrightarrow{-H^+} C_6H_5\overset{OH}{\overset{|}{C}}{=}CH_2$$

$$\downarrow$$

$$C_6H_5\overset{OH}{\overset{|}{\underset{\underset{CH_3}{|}}{C}}}{-}CH_2{-}\overset{OH}{\overset{|}{\underset{+}{C}}}C_6H_5 \quad (15)$$

$$\downarrow{-H^+,\ -H_2O}$$

$$C_6H_5\underset{\underset{CH_3}{|}}{C}{=}CH\overset{O}{\overset{\|}{C}}C_6H_5$$

forms of Y and mordenite at 150°–200°C in the liquid phase. A reasonable mechanism (*1*) based on homogeneous analogies (*57*) involves capture of the conjugate acid of the carbonyl group by the enolic form of a second molecule of starting material (Reaction 15). Several other reactions involving carbonyl compounds may also proceed by capture of such conjugate acids (α-hydroxy carbonium ions) or highly polarized carbonyl groups by unsaturated or nucleophilic reagents (*2*). Thus, condensation of formaldehyde and methyl acetate over HY at 300°–375°C gave methyl acrylate (*1*), a reaction formally analogous to an aldol condensation. In the presence of large excesses of aromatic compounds, alkylation occurred (*56*) to give bisarylalkanes; for example, formaldehyde and phenol gave a mixture of 2,2′-, 2,4′-, and 4,4′-dihydroxydiphenylmethanes (Reaction 16) in the liquid phase at 182°C with

$$H_2C{=}O \xrightarrow{H^+} H_2\overset{+}{C}OH \xrightarrow{C_6H_5OH} (HOCH_2)(H)C_6H_5^{+}{-}OH \xrightarrow{-H^+} HOCH_2{-}C_6H_4{-}OH \xrightarrow[H^+]{C_6H_5OH} HO{-}C_6H_4{-}CH_2{-}C_6H_4{-}OH \qquad (16)$$

a catalyst activity order HY > REY > H-mordenite > REX > CaX. In the presence of isobutene, formaldehyde over H-mordenite at 300°C gave isoprene (*1*) in an apparent heterogeneous version of the Prins reaction (*58*) (Reaction 17). In the presence of excess alcohol, acidic zeolites at 30°C

$$H_2C{=}O \xrightarrow{H^+} H_2\overset{+}{C}OH \xrightarrow{CH_2{=}C(CH_3)_2} HOCH_2CH_2{-}\overset{+}{C}(CH_3)_2 \xrightarrow[-H_2O]{-H^+} CH_2{=}CH{-}C(CH_3){=}CH_2 \qquad (17)$$

converted aldehydes slowly to the corresponding ketals as proposed in Reaction 18; the driving force may include the ability of the zeolites to adsorb and reduce the activity of the water formed. Finally, the conversion of benzaldehyde to the Cannizaro (*59*) products benzyl alcohol and benzoic acid at 300°C was categorized similarly (*1, 2*), the nucleophile being a transferred

$$
\begin{array}{l}
RCH{=}O \xrightarrow{H^+} R\overset{+}{C}HOH \xrightarrow{R'OH} \underset{\overset{+}{H}OR'}{RCHOH} \\
\quad\downarrow -H^+ \\
RCH(OR')_2 \xleftarrow[-H_2O]{R'OH} \underset{OR'}{RCHOH}
\end{array}
\qquad (18)
$$

hydride ion (Reaction 19); however, the effectiveness of NaX is surprising in this context.

$$
C_6H_5CHO \xrightarrow{H^+} C_6H_5\overset{+}{C}HOH + H{-}\overset{O}{\overset{\|}{C}}C_6H_5 \rightarrow C_6H_5CH_2OH + C_6H_5\overset{+}{C}{=}O \qquad (19)
$$
$$
C_6H_5\overset{+}{C}{=}O \xrightarrow{H_2O} C_6H_5CO_2H
$$

Condensations of carbonyl groups with ammonia have also been reported over zeolites. α- and γ-Picoline were produced from acetaldehyde and ammonia at 300°–400°C, NaX being more effective than H-mordenite (*1*). Benzonitrile was produced from benzaldehyde and ammonia at 400°–495°C, the catalyst activity order being CoY $\sim$ NaY $>$ CrY $\sim$ MgY $\sim$ ZnY $>$ CuY $\sim$ MnY (*60*). Although the exact sequence of reaction steps is not known, the initial step is probably nucleophilic attack of ammonia on the carbonyl group. The role of the zeolite, however, is unclear. Since it is unlikely to be very acidic in the presence of ammonia base, it may play its catalytic role again by stabilizing dipolar reaction intermediates or by facilitating proton transfers.

Finally, in the aldol condensation of *n*-butyraldehyde to 2-ethyl-2-hexenal at 200°C reported (*61*) by Isakov and co-workers, the activity order was KY $>$ NaY $>$ HY, implying that base catalysis was involved. Since in homogeneous reactions, base-catalyzed aldol and related carbonyl condensation reactions are generally more facile than their acid-catalyzed analogs, it is best not to be dogmatic as to the exact course of carbonyl group reactions catalyzed by zeolites, considering the small amount of mechanistically oriented information available.

$$
\begin{array}{l}
CH_2{=}CHCH_3 \xrightarrow{H^+} CH_3\overset{+}{C}HCH_3 \xrightarrow{CO} (CH_3)_2CH\overset{+}{C}{=}O \\
\qquad \downarrow \sim H \\
\underset{CH_3}{CH_2{=}C{-}CHO} \xleftarrow{-H^+} (CH_3)_2\overset{+}{C}{-}CHO
\end{array}
\qquad (20)
$$

Treatment of propylene with carbon monoxide and hydrogen at 25°–150°C under pressure gave methacrolein in low yield as well as propylene telomers with a catalyst activity order H-mordenite > REX > HY >> NaX (*1*). The proposed route is shown in Reaction 20.

Reactions Catalyzed by Transition Group Cation-Exchanged Zeolites

Oxidation. Homogeneous oxidations by molecular oxygen typically occur by free-radical chain mechanisms (*62*). A few examples exist in which such processes appear to have been modified by zeolites. Agudo, Badcock, and Stone (*63*) observed that the total combustion of the isomeric hexanes at 200°–350°C in a recirculating reactor with continuous water removal was catalyzed by X zeolites in the activity order NaX > MnX > CaX. The zeolites may contribute a heterogeneous component to the homolytic initiation of free-radical chain oxidation which still propagates largely in a homogeneous fashion. VanSickle and Prest (*64*) treated the much more oxidation-prone olefins 1-butene, 2-butene, and cyclopentene adsorbed on NaX or CaA zeolites (T_{act} only 210°C) with oxygen at 25°–90°C. The very complex mixtures of oxygenated products could be recovered only by extraction of the zeolites with wet ether; this behavior typifies the difficulties in preparing a product more polar than the reactants over a zeolite at ambient temperatures. It was concluded from the product distribution that the zeolite enhanced attack at the double bond (compared with that at the allylic C—H bonds) when compared with the homogeneous oxidation of these olefins.

These two reports, however, are not typical of most studies of oxidation over zeolites, which have instead focused on use of transition-metal cation-exchanged zeolites. In these cases, the frame of reference has not been homogeneous oxidations or their modification by transition metal ions (*65*) but heterogeneous oxidations over non-zeolitic transition metal oxide catalysts (*66*). Several structural, and hence behavioral, differences between bulk oxides and their transition metal ion-exchanged zeolite analogs have been anticipated. The zeolitic cations should be relatively isolated compared with those in bulk oxides and may be present in unusual coordination sites (*67*). It should be more difficult to remove lattice oxygen from zeolites than from oxides, the latter process being a common product-forming step over oxides (*66*). For example, exchange of oxygen with the lattice of NaX, NaY, CuY, and AgY zeolites has been reported (*68*) but temperatures > 600°C were required. On the other hand, there is evidence for a variety of chemisorbed forms of oxygen on transition metal-containing zeolites. For example, the presence of superoxide ion, O_2^-, has been detected on several zeolites by ESR techniques (*69*). Reversible adsorption of oxygen on Fe(II) forms of Y zeolite (*70*) and mordenite (*71*) at 400°–500°C was studied by Mössbauer spectroscopy; the formation of oxide-bridged Fe(III) pairs was proposed (Reaction 21). Similar proposals have been made for Cu(I)Y and Cr(II)Y

$$2\ Fe(II) + \tfrac{1}{2}\ O_2 \rightarrow Fe(III)—O^{2-}—Fe(III) \qquad (21)$$

(*72*). Inversely, there is evidence that oxidizable substrates may reduce zeolitic cations in the absence of oxygen. For example, Cu(II)Y is reduced to Cu(I)Y by carbon monoxide at 500°C and to metallic copper by hydrogen at 500°C (*73*). However, in spite of such spectroscopic and chemical evidence, the catalytic information available still does not allow construction of very complete mechanisms of oxidation.

Unfortunately, the literature is not always explicit concerning the oxidation state of the cations used for exchange and concerning any changes in oxidation state which may occur during activation and catalysis. In the absence of further information, the normally air-stable forms are presumably involved in the initial zeolites; *i.e.*, Cr^{3+}, Mn^{2+}, Fe^{3+}, Co^{2+}, Ni^{2+}, Cu^{2+}, Zn^{2+}, Pd^{2+}, Ag^{+}, and Pt^{2+}. There has been little explicit concern to quantify the hydrolysis processes (Reaction 22) anticipated during activation.

$$Tr^{n+} + H_2O + O_Z^{2-} \rightarrow Tr(OH)^{n-1} + HO_Z^{-} \qquad (22)$$

For oxidation of hydrogen by excess oxygen in a pulsed reactor, the activity order of a series of Y zeolites was reported by Roginskii, Al'tshuler, and co-workers (*74*) to be Ag > Cu ~ Fe > Cr ~ Ni ~ Co > Mn ~ Na; the zeolites were ~ 50% ion-exchanged (70% for CuY and 80% for CrY), activated at 600°C, and pretreated with oxygen before each pulse. Activity was detected for AgY already at 100°C, whereas > 400°C was required for MnY. For oxidation of carbon monoxide (*74*), the order was Ag ~ Co > Cu > Cr > Ni > Fe > Na > Mn. For ethylene (*74*), it was Cu ~ Cr > Ag ~ Fe > Co > Ni ~ Mn > Na; for ammonia, Cu > Cr > Ag > Fe ~ Co > Mn ~ Ni ~ Na. For the consistently active Cu and Ag forms, a fully oxygenated catalyst would perform some oxidation when pulsed with these substrates (only hydrogen for the Ag case) in the absence of oxygen. The presence of a catalytically active chemisorbed form of oxygen was implied, whose amount increased with the temperature of oxygen pre-treatment from 200°–500°C. This was also true to a lesser extent for FeY and CrY but not for MnY, CoY, nor NiY. Later work outlined below has further delineated the mechanistic role of chemisorbed oxygen.

Oxidation activity for carbon monoxide at 300°–450°C over CuHY was reported by Boreskov and co-workers (*75*) to increase sharply past 5 wt % Cu present; the kinetics were first order in carbon monoxide and zero order in oxygen. It was suggested that either only Cu ions accessible to the large cavities were active or the active site required more than one Cu ion. Kubo and coworkers (*72*) also reported on carbon monoxide oxidation, especially over low-valent ion-exchanged X and Y zeolites prepared by ion exchange under nonoxidizing conditions and activation at 400°C in hydrogen although a temporary oxidation to high-valent forms was suspected to have occurred during washing the exchanged zeolites in air. Oxygen chemisorption at 400°C on the Co(II), Mn(II), and Tl(I) forms was low (<0.1 O atom/metal atom), and their catalytic activities were relatively low. Oxygen chemisorption on the Fe(II) and Cu(I) forms occurred to the extent of ~ 0.5 O atom/metal atom, and these were the most active catalysts, the X forms being more active than the Y. Finally, oxygen chemisorption on the Cr(II) zeolites was also high, but activity was low. The kinetics for the Fe(II) cases were first order in carbon monoxide and fractional order in oxygen. The proposed Rideal-type mechanism is described by Reactions 23 and 24 where carbon monoxide is

$$2\ \mathrm{Fe(II)} + \tfrac{1}{2}\ \mathrm{O_2} \rightarrow \mathrm{Fe(III)}\text{—}\mathrm{O^{2-}}\text{—}\mathrm{Fe(III)} \qquad (23)$$

$$\mathrm{Fe(III)}\text{—}\mathrm{O^{2-}}\text{—}\mathrm{Fe(III)} + \mathrm{CO} \rightarrow 2\ \mathrm{Fe(II)} + \mathrm{CO_2} \qquad (24)$$

not specifically adsorbed and where indeed two metal ions are required at the active site (*75*). Activity maximized at an intermediate value of the redox potential of the cations (*72*) because, according to the authors, cations difficult to oxidize (*e.g.*, Co(II) and Mn(II)) do not adsorb oxygen efficiently whereas those too easy to oxidize (*e.g.*, Cr(II)) adsorb oxygen but then are not easily reduced by carbon monoxide. Apparently Fe(II) and Cu(I) represent the optimum balance for carbon monoxide. Of course, for a substrate with a different redox potential, the optimum redox potential for the zeolitic cation may well change. Other oxidations of small molecules over zeolites include that of hydrogen sulfide (*76*).

Reports of oxidations of hydrocarbons over transition metal ion-exchanged zeolites have described largely total combustion with fewer claims of selectivity to specifically oxygenated products. The activity survey (*74*) already cited included combustion of ethylene, the activity order for Y-zeolites being Cu $\sim$ Cr $>$ Ag $\sim$ Fe $>$ Co $>$ Ni $\sim$ Mn $>$ Na. In a more detailed study of the ethylene–oxygen–Cu(II)Y system (*77*), pulsed experiments showed that a catalyst pretreated with oxygen at 450°C would oxidize some ethylene at 340°C in the absence of oxygen and that a plot of the amount of this labile oxygen *vs.* extent of Cu(II) exchange showed the same upward curvature as the plot of catalytic activity *vs.* exchange level. Hence, oxidation activity and the presence of chemisorbed oxygen were again closely related. At temperatures too low ($\sim$ 150°–250°C) to effect oxidation, activated adsorption of ethylene on a pre-oxygenated catalyst and even stronger adsorption on a de-oxygenated catalyst occurred. The latter was suggested to result from interaction of the olefin with copper ions of unspecified oxidation state, whereas the former must then result from interaction, but not yet reaction, with the copper ion–chemisorbed oxygen combination. Analogous overall behavior was also observed (*77*) for carbon monoxide oxidation. There is thus considerable parallelism between these studies (*77*) and those of Kubo (*72*) with carbon monoxide. However, since Kubo apparently started with a Cu(I) form and Al'tshuler (*77*) with a Cu(II) form of the zeolite, it is not clear if the Cu(II)O^{2-}Cu(II) active center implied by Kubo and the reactive, labile oxygen referred to by Al'tshuler are identical. Further information on the redox chemistry of these zeolites is required to clarify this point.

Skalkina and co-workers (*78*) earlier studied combustion of propylene at 420°C over NaA, CaA, NaX, CaX, and NaY, a reaction which may well have been of the free-radical type. Introduction of Fe(III) by ion-exchange led to only modest increases in activity or, in the A series, none at all. Addition of ammonia to the feed gave modest selectivity to acrylonitrile (*79*). Still earlier Mochida and co-workers (*80*) presented results for oxidation of propylene over Cu(II)Y in excess oxygen at 350°C; minor amounts of formaldehyde, acrolein, and acetaldehyde were detected in addition to mainly carbon dioxide and water. Propylene oxidation was 10 times more rapid over Cu(II)Y than over a copper oxide-silica gel catalyst, but for ethylene the rates were comparable. Selectivity to acrolein was enhanced by feeding steam (*81*) ($C_3H_6 : O_2 : H_2O \sim 1 : 1.5\text{–}2 : 1.5\text{–}2.5$) but combustion still predominated. Increasing the steam proportions even more ($C_3H_6 : O_2 : H_2O = 1 : 1 : 20$) also gave (*82*) small amounts of isopropyl alcohol (major oxygenated product, except for carbon dioxide, at $< 200°C$), acetaldehyde (maximized at 250°C), and acetone (maximized at $\sim 500°C$) in addition to acrolein. A ratio of acrolein:carbon dioxide > 1 was finally achieved (*83*) at 350°C by using a $C_3H_6 : O_2$ ratio of 7:1, a result which implies that combustion passes through oxygenated compounds which are more oxidation prone than propylene. For a series of other Y zeolites in this study (*83*), Mochida found the activity order for propylene and ethylene oxidation at 350°C to be Pd $\sim$ Pt $>$ Cu $>$ Tl $>$ Ag $>$ Mn $>$ Ni $>$ Co $>$ Zn $>$ V $>$ Cr $>$ Na, all metals presumed to be present in ion-exchanged form. This order was pictured as correlating with the ability of these ions to form π-complexes with olefins. The observed kinetics could again be rationalized by postulating a Rideal-type interaction of olefin with dissociately adsorbed oxygen with competitive adsorption of olefin and oxygen at the active site.

Firth and Holland (*84*) described the kinetics of combustion of methane over X zeolites ion exchanged to a relative low extent with ammine complexes of Rh, Ir, Pd, and Pt and calcined in air; isolated ions of unspecified oxidation state were presumed present. The kinetic behavior (first order in methane and approximately zero order in oxygen, except strongly inverse order for Pt) was interpreted as indicating adsorption of both reagents, oxygen the more strongly. Only with Pt was the adsorption competitive. The rate-determining step, of identical E_a in all cases, was proposed to be the formation of adsorbed water from a hydroxyl group and a methyl fragment bonded to the metal ion. Rudham and Sanders (*85*) extended this series and reported a methane combustion activity order for X zeolites of Pd $>>$ Cu $>$ Cr $>$ Ni $>$ Fe $>$ Mn $>$ Co. The kinetic law for the Pd and Cu cases was first order in methane and zero order in oxygen. No specific mechanism was proposed, but these workers could not support the Firth-Holland rate-determining step. The most active catalysts contained ions (Pd^{2+}, Cu^{2+}) which preferred square planar or distorted octahedral coordination, followed by those (Cr^{3+}, Fe^{3+}, Mn^{2+}) preferring octahedral coordination, and finally by those (Co^{2+}, Zn^{2+}) preferring tetrahedral coordination. It was suggested that this may be the same order of extent of S_{II} site occupancy in zeolites of low degree of transition metal ion-exchange and therefore accessibility of the ions to the reagents may be as important as their specific catalytic properties (*102*). That these oxidation reactions are indeed not yet well understood and are deserving of further study may be illustrated by the divergent rate laws for methane (*85*) over PdX and for propylene (*83*) over PdY shown in Equations 25 and 26.

$$-d(CH_4)/dt = k(CH_4)^1(O_2)^0 \tag{25}$$

$$-d(C_3H_6)/dt = k(C_3H_6)^{-0.8}(O_2)^{0.6} \tag{26}$$

Treatment of cyclohexane with excess oxygen over Cu(II)Y at 210°–350°C gave mainly carbon dioxide and the dehydrogenated product benzene (*86*); at low conversions, oxidative dehydrogenation and combustion were competitive, selectivity for benzene reaching 80%. The activity order for other Y zeolites for combustion at 290°C was Pd > Cu > Cr > Ni > Zn > Ag, similar to the order for olefin (*83*) and methane combustion (*85*) except for the low activity of AgY; the order for benzene formation was Cu > Pd > Cr > Ni > Zn > Ag. Plots of activity vs. heats of formation of the corresponding metal oxides showed the common volcano-shaped correlations (*87*). However, Arrhenius plots showed that the more active catalysts had larger E_a values which were overcompensated by the A factors. Hence, the meaning of these correlations is not obvious. Oxidative dehydrogenation of cyclohexane was also reported (*88*) over Fe(II)Y at 180°C with significant amounts of cyclohexene formed as the presumed precursor to benzene. Styrene was formed in low yield from ethylbenzene and air over MnY at 370°C (*1*).

Rouchaud and Fripiat have described liquid-phase batch oxidations of hydrocarbons at 150°–165°C mediated by transition metal ion-containing zeolites. Without catalyst at 160°C and 25 atm of oxygen, *n*-hexane gave largely acetic, propionic, and butyric acids (*89*). Addition of dehydrated but not activated Co(II)X, Ni(II)X, or Mn(II)X zeolites approximately doubled the rate of acid formation and slightly increased selectivity to total acids (*89*). A small fraction of metal ions was leached into solution as carboxylate salts but these quantities were judged too small to explain the observed rate enhancements which were therefore attributed to the cations *in* the zeolites. However, the overall effects were similar to those caused by addition of the corresponding soluble salts which act largely by altering the course of free-radical chain oxidation since they can decompose intermediate organic hydroperoxides by redox reactions (*65*). For the oxidation of *p*-xylene to form *p*-toluic and terephthalic acids in acetic acid solvent at 165°C and 25 atm oxygen, rate enhancements were again observed for Co(II)X and Mn(II)X (*90*) quite similar to those induced by the corresponding soluble naphthenate salts. A solution of propylene in benzene at 150°C under oxygen at a total pressure of 45 atm was converted to propylene oxide in 70% selectivity by addition of NaX zeolite which had been impregnated, not ion-exchanged, with ammonium molybdate and dried at 120°C (*91*); much lower selectivities were observed for solid MoO_3, solid alumina impregnated with ammonium molybdate, or a soluble molybdenum stearate.

Amination. Hatada and co-workers (*60*) studied the conversion of chlorobenzene to aniline by ammonia at 395°C over transition metal ion-exchanged Y zeolites. Since no reaction occurred over MgY or CaY and since ammonia is known to be a poison for acid sites, activity must be associated with the transition metal cations. The activity order was $Cu^{2+} > Ni^{2+} > Zn^{2+} > Cr^{3+} > Co^{2+} > Cd^{2+} > Mn^{2+}$ and could be correlated either with the formation constants of the corresponding monoammine complexes or the electronegativity of the metal ion. Therefore, the authors could not decide between two suggested rate-determining steps: coordination of ammonia to the cation or abstraction of chlorine, presumably as chloride ion, from chlorobenzene by the metal cation. In earlier work, Jones (*1*) had observed the same reaction over CuX and ZnX, and Venuto and Landis (*1*) suggested formation of amide ion from reaction of the cation with ammonia (Reaction 28) as a vital step in the mechanism. It is not clear then whether this formally nucleophilic aromatic substitution reaction should be viewed as closer to the mechanistic extreme shown in Reaction 27 or to that shown in Reactions 28 and 29.

$$C_6H_5Cl + M^{n+} \rightarrow C_6H_5^+ + MCl^{n-1} \quad \xrightarrow{NH_3,\ -H^+} \quad C_6H_5NH_2 \qquad (27)$$

$$\underset{\mathbf{3}}{M(NH_3)_x{}^{n+}} \longrightarrow M(NH_3)_{x-1}(NH_2)^{n-1} + H^+ \qquad (28)$$

$$C_6H_5Cl + \mathbf{3} \longrightarrow [\text{cyclohexadienyl anion bearing } NH_2 \text{ and } Cl] \longrightarrow C_6H_5NH_2 + Cl^- \qquad (29)$$

Reaction of toluene with excess ammonia proceeded at 540°C over transition metal ion-exchanged X zeolites to produce benzonitrile and hydrogen (*1*); the suggested mechanism is summarized in Reactions 28 and 30–33 with step 33 being separately demonstrated at 500°C over ZnX. It is, how-

$$C_6H_5CH_3 \rightarrow C_6H_5CH_2^+ + H^- \qquad (30)$$

$$C_6H_5CH_2^+ + \mathbf{3} \rightarrow \underset{\mathbf{4}}{M(NH_3)_{x-1}(C_6H_5CH_2NH_2)^{n+}} \qquad (31)$$

$$\mathbf{4} + NH_3 \rightarrow M(NH_3)_x{}^{n+} + C_6H_5CH_2NH_2 \qquad (32)$$

$$C_6H_5CH_2NH_2 \rightarrow C_6H_5CN + 2\ H_2 \qquad (33)$$

ever, somewhat surprising to see postulation of toluene activation by hydride abstraction over such an ammonia-poisoned catalyst. During both these aminations of chlorobenzene (*60*) and toluene (*1*), a competitive decomposition of ammonia to its elements occurred but it is not known whether this reaction is independent from or mechanistically related to the aminations.

Neale, Elek, and Malz (*92*) described the addition of methylamine to methylacetylene at 300°C over ZnY to produce largely the ketimine **5,** presumably derived from the enamine **6,** with lesser amounts of the aldimine

$$CH_3NH_2 + CH_3C{\equiv}CH \rightarrow \underset{\mathbf{6}}{CH_3C(NHCH_3){=}CH_2} \rightarrow \underset{\mathbf{5}}{CH_3C({=}NCH_3)CH_3}$$

formed by addition with reverse orientation. The reaction was specific for ZnY and CdY, other transition metal ion-exchanged zeolites including Cu(II)Y and Cu(I)Y being much less effective. Highest activity resulted from flash vacuum activation, a condition most conducive to forming exposed Zn^{2+} cations without extensive hydrolysis of water. A role of these specifically coordinated cations in the reaction is strongly implied although no more detailed mechanistic deductions were made.

Literature Cited

1. Venuto, P. B., Landis, P. S., *Adv. Catalysis* (1968) **18,** 259.
2. Venuto, P. B., *Adv. Chem. Ser.* (1970) **102,** 186.
3. Venuto, P. B., *Chem. Tech.* (1971) **1,** 215.
4. Leach, H. F., *Ann. Repts. Progr. Chem., Sect. A* (1971) **68,** 195.
5. Minachev, Kh. M., Garanin, V. I., Isakov, Ya. I., *Russ. Chem. Rev.* (1966) **35,** 903.
6. Minachev, Kh. M., *Kin. Catal.* (1970) **11,** 342.
7. Minachev, Kh. M., Isakov, Ya. I., *Adv. Chem. Ser.* (1973) **121,** 451.
8. Imanaka, T., Okamoto, Y., Teranishi, S., *Bull. Chem. Soc. Japan* (1972) **45,** 3251.
9. Blatt, A. H., *Org. React.* (1942) **1,** 342.
10. Landis, P. S., Venuto, P. B., *J. Catal.* (1966) **6,** 245.
11. Donaruma, L. G., Heldt, W. Z., *Org. React.* (1960) **11,** 1.
12. Butler, J. D., Poles, T. C., *J. C. S. Perkin II* (1973) 41.
13. Roussel, P. A., *J. Chem. Educ.* (1953) **30,** 122.
14. Barrer, R. M., Oei, A. T. T., *J. Catal.* (1973) **30,** 460.
15. Venuto, P. B., Hamilton, L. A., Landis, P. S., Wise, J. J., *J. Catal.* (1966) **5,** 81.
16. Venuto, P. B., Hamilton, L. A., Landis, P. S., *J. Catal.* (1966) **5,** 484.
17. Venuto, P. B., Wu, E. L., *J. Catal.* (1969) **15,** 205.
18. Yashima, T., Ahmad, H., Yamazaki, K., Katsuta, M., Hara, N., *J. Catal.* (1970) **16,** 273.
19. Yashima, T., Yamazaki, K., Ahmad, H., Katsuta, M., Hara, N., *J. Catal.* (1970) **17,** 151.
20. Yashima, T., Sato, K., Hayasaka, T., Hara, N., *J. Catal.* (1972) **26,** 303.
21. Sidorenko, Y. N., Galich, P. N., Gutyrya, V. S., Il'in, V. G., Niemark, I. E., *Proc. Acad. Sci. USSR* (1967) **173,** 132.

22. Konoval'chikov, O. D., Galich, P. N., Gutyrya, V. S., Lugovskaya, G. P., *Kin. Catal.* (1968) **9,** 1146.
23. Bryant, D. E., Kranich, W. L., *J. Catal.* (1967) **8,** 8.
24. Ralek, M., Grubner, O., *Proc. Inter. Congr. Catalysis, 3rd, Amsterdam, 1964,* p. 1302.
25. Sharf, V. Z., Freidlin, L. Kh., Samokhvalov, G. I., Niemark, I. E., German, E. N., Piontkovskaya, M. A., Nekrasov, A. S., Krutii, V. N., *Bull. Acad. Sci. USSR, Div. Chem. Sci.* (1968) 752.
26. Tsitsishvili, G. V., Sidamonidze, Sh. I., Zedgenidze, Sh. A., *Proc. Acad. Sci. USSR* (1963) **153,** 1156.
27. Frilette, V. J., Munns, G. W., Jr., *J. Catal.* (1965) **4,** 504.
28. Galich, P. N., Golubchenko, I. T., Gutyrya, V. S., Il'in, V. G., Niemark, I. E., *Ukr. Khim. Zh.* (1965) **31,** 117; *Chem. Abstr.* (1966) **64,** 12571.
29. Levchuk, V. S., Ione, K. G., *Kin. Catal.* (1972) **13,** 850.
30. Topchieva, K. V., Tuang, H. C., *Proc. Acad. Sci. USSR* (1972) **205,** 654.
31. Gourisetti, B., Cosyns, J., Leprince, P., *Compt. Rend.* (1964) **258,** 4547.
32. Cosyns, J., Leprince, P., *Bull. Soc. Chim. France* (1966) 1078.
33. Gourisetti, B., Cosyns, J., Leprince, P., *Bull. Soc. Chim. France* (1966) 1085.
34. Ignace, J. W., Gates, B. C., *J. Catal.* (1973) **29,** 292.
35. Weisz, P. B., Frilette, V. J., Maatman, R. W., Mower, E. B., *J. Catal.* (1962) **1,** 307.
36. Swabb, E. A., Gates, B. C., *Ind. Eng. Chem., Fundamentals* (1972) **11,** 540.
37. Banthorpe, D. V., "Elimination Reactions," Elsevier, Amsterdam, 1963.
38. Gryaznova, Z. V., Ermilova, M. M., Balandin, A. A., Tsitsishvili, G. V., *Proc. Inter. Congr. Catalysis, 4th, Moscow, 1968,* p. 156.
39. Sanyal, S. K., Weller, S. W., *Ind. Eng. Chem., Process Design Develop.* (1970) **9,** 135.
40. Imai, T., Anderson, R. B., *Ind. Eng. Chem., Prod. Res. Develop.* (1971) **10,** 375.
41. Imai, T., Anderson, R. B., *Ind. Eng. Chem., Prod. Res. Develop.* (1973) **12,** 232.
42. DePuy, C. H., King, R. W., *Chem. Rev.* (1960) **60,** 431.
43. Venuto, P. B., Givens, E. N., Hamilton, L. A., Landis, P. S., *J. Catal.* (1966) **6,** 253.
44. Barrer, R. M., Kravitz, S., "Molecular Sieves," Society of the Chemical Industry, London, 1968, p. 95.
45. Mochida, I., Yoneda, Y., *J. Org. Chem.* (1968) **33,** 2161.
46. Mochida, I., Take, J., Saito, Y., Yoneda, Y., *J. Org. Chem.* (1967) **32,** 3894.
47. Kladnig, W., Noller, H., *J. Catal.* (1973) **29,** 385.
48. Venuto, P. B., Landis, P. S., *J. Catal.* (1971) **21,** 330.
49. Kirmse, W., "Carbene Chemistry," Academic Press, New York, 1964; Jones, M., Jr., Moss, R. A., Eds., "Carbenes," Wiley, New York, 1973.
50. Cannon, P., *J. Phys. Chem.* (1959) **63,** 160.
51. Barrer, R. M., Brook, D. W., *Trans. Faraday Soc.* (1953), **49,** 940.
52. Wu, E. L., Kuhl, G. H., Whyte, T. E., Jr., Venuto, P. B., *Adv. Chem. Ser.* (1970) **101,** 490.
53. Pine, S. H., *Org. React.* (1970) **18,** 403.
54. Wu, E. L., Whyte, T. E., Jr., Venuto, P. B., *J. Catal.* (1971) **21,** 384.
55. Fripiat, J. J., Lambert-Helsen, M. M., *Adv. Chem. Ser.* (1973) **121,** 518.
56. Venuto, P. B., Landis, P. S., *J. Catal.* (1966) **6,** 237.
57. Noyce, D. S., Pryor, W. A., *J. Amer. Chem. Soc.* (1955) **77,** 1397 and subsequent papers.
58. Arundale, E., Mikeska, L. A., *Chem. Rev.* (1952) **51,** 505.
59. Geissman, T. A., *Org. React.* (1944) **2,** 94.
60. Hatada, K., Ono, Y., Keii, T., *Adv. Chem. Ser.* (1973) **121,** 501.

61. Isakov, Ya. I., Minachev, Kh. M., Usachev, N. Ya., *Bull. Acad. Sci. USSR, Div. Chem. Sci.* (1972) 1124.
62. Howard, J. A. in Kochi, J. K., Ed., "Free Radicals," Vol. II, Chap. 12, Wiley, New York, 1973.
63. Agudo, A. L., Badcock, F. R., Stone, F. S., *Proc. Inter. Congr. Catalysis, 4th, Moscow* (1968) Paper 59.
64. Van Sickle, D. E., Prest, M. L., *J. Catal.* (1970) **19,** 209.
65. Kochi, J. K. in Kochi, J. K., Ed., "Free Radicals," Vol. I, Chap. 11, Wiley, New York, 1973.
66. Sachtler, W. M. H., *Catalysis Rev.* (1970) **4,** 27; Voge, H. H., Adams, C. R., *Adv. Catalysis* (1967) **17,** 151.
67. Mikheikin, I. D., Zhidomirov, G. M., Kazanskii, V. B., *Russ. Chem. Rev.* (1972) **41,** 468; Jones, C. E., Klier, K., *Ann. Rev. Materials Sci.* (1972) **2,** 1.
68. Minachev, Kh. M., Savost'yanov, E. N., Kondrat'ev, D. A., Ch'ang, S. N., Antoshin, G. V., *Bull. Acad. Sci. USSR, Div. Chem. Sci.* (1971) 754.
69. Imai, T., Habgood, H. W., *J. Phys. Chem.* (1973) **77,** 925; Ben Taarit, Y., Lunsford, J. H., *J. Phys. Chem.* (1973) **77,** 780.
70. Garten, R. L., Delgass, W. N., Boudart, M., *J. Catal.* (1970) **18,** 90.
71. Garten, R. L., Gallard-Nechtschein, J., Boudart, M., *Ind. Eng. Chem., Fundamentals* (1973) **12,** 299.
72. Kubo, T., Tominaga, H., Kunugi, T., *Bull. Chem. Soc. Japan* (1973) **46,** 3549.
73. Naccache, C. M., Ben Taarit, Y., *J. Catal.* (1971) **22,** 171.
74. Roginskii, S. Z., Al'tshuler, O. V., Vinogradova, O. M., Seleznev, V. A., Tsitovskaya, I. L., *Proc. Acad. Sci. USSR* (1971) **196,** 112.
75. Boreskov, G. K., Bobrov, N. N., Maksimov, N. G., Anufrienko, V. F., Ione, K. G., Shestakova, N. A., *Proc. Acad. Sci. USSR* (1971) **201,** 1012.
76. Addison, W. E., Walton, A., *J. Chem. Soc.* (1961) 4741.
77. Al'tshuler, O. V., Tsitovskaya, I. L., Vinogradova, O. M., Seleznev, V. A., *Bull. Acad. Sci. USSR, Div. Chem. Sci.* (1972) 2084.
78. Skalkina, L. V., Kolchin, I. K., Margolis, L. Ya., Ermolenko, N. F., Levina, S. A., Malashevich, L. N., *Bull. Acad. Sci. USSR, Div. Chem. Sci.* (1970) 929.
79. Skalkina, L. V., Kolchin, I. K., Margolis, L. Ya., Ermolenko, N. F., Levina, S. A., Malashevich, L. N., *Kin. Catal.* (1971) **12,** 208.
80. Mochida, I., Hayata, S., Kato, A., Seiyama, T., *J. Catal.* (1969) **15,** 314.
81. Mochida, I., Hayata, S., Kato, A., Seiyama, T., *J. Catal.* (1970) **19,** 405.
82. Mochida, I., Hayata, S., Kato, A., Seiyama, T., *Bull. Chem. Soc. Japan* (1971) **44,** 2282.
83. Mochida, I., Hayata, S., Kato, A., Seiyama, T., *J. Catal.* (1971) **23,** 31.
84. Firth, J. G., Holland, H. B., *Trans. Faraday Soc.* (1969) **65,** 1891.
85. Rudham, R., Sanders, M. K., *J. Catal.* (1972) **27,** 287.
86. Mochida, I., Jitsumatsu, T., Kato, A., Seiyama, T., *Bull. Chem. Soc. Japan* (1971) **44,** 2595.
87. Balandin, A. A., *Adv. Catal.* (1969) **19,** 1; Boreskov, G. K., *Ibid.* (1964) **15,** 285.
88. Kubo, T., Tominaga, H., Kunugi, T., *Nippon Kagaku Kaishi* (1972) 196; *Chem. Abstr.* (1972) **76,** 85120.
89. Rouchaud, J., Sondengam, L., Fripiat, J. J., *Bull. Soc. Chim. France* (1968) 4387; *Bull. Soc. Chim. Belges* (1968) **77,** 505.
90. Rouchaud, J., Mulkay, P., Fripiat, J., *Bull. Soc. Chim. Belges* (1968) **77,** 537.
91. Rouchaud, J., Fripiat, J., *Bull. Soc. Chim. France* (1969) 78.
92. Neale, R. S., Elek, L., Malz, R. E., Jr., *J. Catal.* (1972) **27,** 432.

Chapter

10

Catalytic Properties of Metal-Containing Zeolites

Kh. M. Minachev and Ya. I. Isakov, Zelinsky Institute of Organic Chemistry, Academy of Sciences, B-334 Moscow, Leninsky Prospect 47, U.S.S.R.

SUPPORTED METAL CATALYSTS are widely used in petroleum refining, and in the chemical and petrochemical industries (*1*). These catalysts are important in ammonia synthesis, conversion of hydrocarbons with water vapor to synthesis gas, reforming, hydrocracking, hydrorefining, hydrodealkylation, dehydrocyclization, isomerization of paraffins and cycloalkanes, hydroisomerization of olefins, dienes, and aromatics, isomerization of ethylbenzene to xylenes, reduction of various organic compounds, oxidation, the Fisher-Tropsch synthesis, and so forth.

Studies of supported metal catalysts are significant for catalytic theory and for the development of new polyfunctional catalytic systems and new catalytic processes. The properties of these catalysts depend strongly on the state and dispersion of a metal component (*2, 3, 4, 5, 6*). Therefore as soon as crystalline aluminosilicates (zeolites) were synthesized and became available, their specific properties (ion exchange ability, high exchange capability, crystalline structure with regular pores of molecular size) were utilized to prepare catalysts containing highly dispersed metals, to show molecular-sieve selectivity, and to develop polyfunctional action. Rabo *et al.* (*7, 8*) and Weisz *et al.* (*9, 10*) were the first to demonstrate that metal-loaded zeolite systems had great promise as catalysts in petroleum refining and in petrochemistry.

Interest in these systems was revived by the study of the introduction and dispersion of platinum into zeolite structures, which showed that platinum ion exchanged zeolite catalysts, 0.5% Pt-CaY, exhibited high stability to sulfur poisoning (*11*).

As shown in Figure 1, publications on metal–zeolite catalysts from 1962 to 1972 amounted to 15–30% of all studies dealing with catalysis over zeolites. These investigations led to theories on metal–support interactions and the properties of very small particles of various metals in porous crystalline solids and to the development of some efficient catalysts for important practical reactions currently in commercial use. For example,

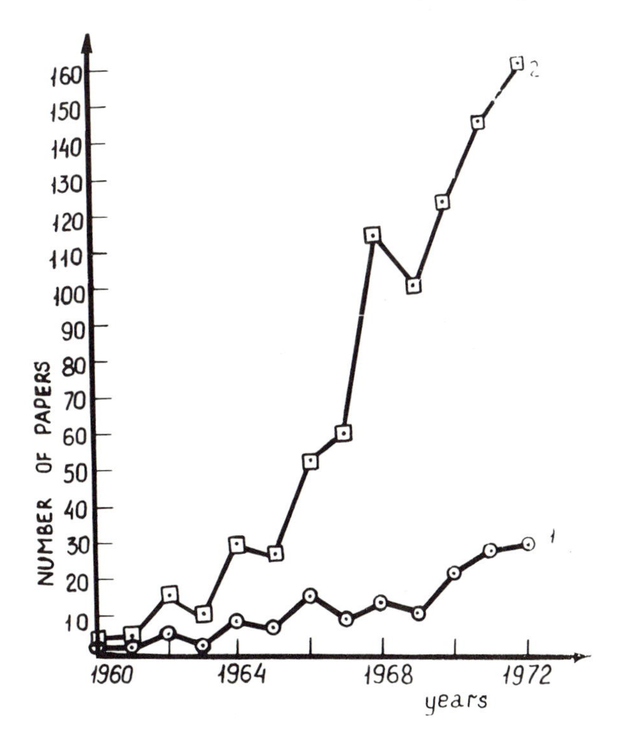

Figure 1. Curve 1: investigations dealing with metal-loaded zeolite catalysts. Curve 2: total studies devoted to catalysis on zeolites.

a new zeolite-supported catalyst was used in the Unicracking-JHC process (*12*) to refine a heavy feed (density up to 1.01) with high sulfur (up to 4.6 wt %) and high nitrogen (up to 0.35 wt %) content, yielding products ranging from liquified gas to high quality feed for catalytic cracking. In the U.S.S.R. an isoreforming process was developed (*13*) which involves both hydrocracking and reforming of straight-gasoline fractions to produce simultaneously an isocomponent (C_5–C_6 fraction) and a reformate. This process produces a high octane gasoline when gasoline fractions of 105°–180° and 140°–180°C are used as feed.

Hydrocracking is carried out over a zeolite containing catalyst at 100 atm at temperatures as low as 300°–350°, and at LHSV = 1.5 hr^{-1}. The yield of the isocomponent is 20 wt % of the starting feed with an octane number of 86 (research method). Simultaneously, 14.4 wt % butanes, mainly isobutane and propane (6.8 wt %) are formed. Methane and ethane yields are very low.

Many studies were concerned with the development of zeolite-based reforming catalysts having favorable properties. For example, unlike Pt catalysts on alumina, zeolite-based catalysts exhibit low sensitivity to nitrogen bases present in the feed as impurity (*14*).

Since 1970, a metal-loaded zeolite catalyst based on H-mordenite containing a small amount of residual sodium and a noble metal, has been used in the Hysomer process (Union Carbide Co. and Shell Co.) (*15, 16, 17*). Here the isomerization of pentane–hexane petroleum fractions and light gasolines gives important components for high quality gasoline. This process acquires great significance because of limitations imposed on the use of TEL and the necessity to compensate for the lost octane number by replacing low-octane gasoline components with aromatic and isoparaffinic hydrocarbons. The complete *n*-paraffin (fraction with b.p. up to 71°C) isomerization combined with catalytic reforming increases the gasoline octane number by 20 units and gives gasoline balanced by the octane number of narrow fractions (*17*).

The molecular sieve action of zeolites is used to prepare a catalyst which forms the basis of Selectoforming—*i.e.*, a process of selective hydrocracking of *n*-paraffins with isoparaffinic and aromatic hydrocarbons to improve gasoline reformates (*18*). The combination of Selectoforming and Platforming proved very efficient (*19, 20*).

Zeolites can be used effectively as catalyst components in isomerization of ethylbenzene to xylenes in the Octafining process (*21*). This process was developed to obtain additional *p*-xylene by isomerization of *m*- and *o*-xylenes and ethylbenzene on metal-loaded catalyst. The zeolite-containing contacts exhibit high activity and selectivity for isomerization (*21*).

Metal–zeolite catalysts are used in the hydrogenation of aromatic hydrocarbons (*14, 22*), in hydrodealkylation of toluene (*23, 24*), and in one-stage synthesis of methyl isobutyl ketone from acetone and H_2 (*25*), and 2-ethylhexanol from *n*-butyric aldehyde and H_2 (*26*); to produce isoprene from isobutylene and formaldehyde (*27, 28, 29*); to convert hydrocarbons with water vapor to synthesis gas which does not contain CO and is suitable for hydrogenation processes after elimination of CO_2 (*30, 31*). This list, far from being complete, shows that metal-loaded zeolite catalysts play a very important role in modern petroleum refining and petrochemistry.

The voluminous information on catalytic, structural, and physicochemical properties of metal-loaded zeolite systems has not been properly reviewed. Only a few aspects of the problem have been considered (*22, 32, 33, 34, 35, 36*). This chapter is concerned with the preparation, activation, and modification of metal-loaded zeolite catalysts: their structure, the state of the metal in samples of various composition and preparations, activity, selectivity, and stability of mono- and polyfunctional catalysts in a number of reactions, some specific feature of the kinetics and mechanism of isomerization of *n*-paraffins and cycloalkanes and others and some new results obtained from the study of the properties of zeolites containing various metals.

Preparation, Activation, and Regeneration of Metal-Loaded Zeolite Catalysts

Metal-containing zeolites can be used as catalysts for different purposes and under various conditions. Therefore, their composition, preparation, and activation depend on the process used.

Preparation of Catalysts. Ion exchange, impregnation, codeposition, adsorption from the gaseous phase, introduction of compounds during zeolite synthesis (crystallization) and adsorption of metal vapor can be used to introduce catalytically active components into zeolite cavities or on their crystal surface. These methods are used to prepare many non-zeolite catalysts, but in zeolites some peculiarities are observed. For example, since zeolite cavities are of molecular size, they contain a limited number of molecules depending on the nature of the compound and the conditions of its introduction; therefore, it is possible to attain a highly dispersed form for the active agents. The ability of zeolites to cation-exchange therefore offers great opportunities. This property is often used to prepare metal-containing zeolite catalysts. By reducing metal cations with alkali metals, hydrogen, carbon monoxide, formaldehyde, hydrazine, hydroxylamine, etc. (*37, 38, 39, 40, 41, 42, 43*), one can obtain metal-loaded zeolites (*34, 39, 44, 45, 46*). For example, by treating a zeolite containing Ni^{2+} ions or Ni complex cations with hydrogen at elevated temperature (Ni^{2+} is reduced by hydrogen at temperatures above 200°C (*47*)), a product containing metallic nickel is formed. Along with zero-valent Ni, protons also appear to compensate the negative charges of the aluminosilicate framework:

$$2(AlO_4)^- Ni^{2+} + H_2 \rightarrow 2Ni^\circ + 2(AlO_4)^- H^+$$

The nature of protons in these systems has not been established with certainty, but they are likely to occur in the structure as hydroxyls. The zeolite may also be partially decationized by loss of water. This depends on the conditions of the reduction step. Catalysts containing the metal phase of a transition element can be obtained. Dispersion of the metal depends on the cation content, treatment before reduction, and the condition of reduction.

The best distribution of a metal in a molecular sieve can be achieved when water is completely removed from the zeolite before reduction (*39, 48*). A partial dehydration results in less active and less stable products with poor metal dispersion and a correspondingly smaller specific surface for the metal. Because of the strong linkage with cations the water is strongly adsorbed in the zeolite before reduction. It may also be formed during the reduction of the metal cations, and its presence can affect both the reduction of the cations and the dispersion of the metal formed. For example, upon treating a zeolite, $Pd_{12..5}Na_{19.5}H_{11.5}Al_{56}Si_{136}O_{384}$, with H_2 ($P_{H_2} = 100$ torr, 15 hr) at either 25° or 200°C, equal amounts ($\sim$40%) of Pd^{2+} cations are reduced (*49, 50*). At 200°C, the water molecules formed upon reduction of the first few palladium ions are still bound to Pd^{2+} cations and prevent their interaction with hydrogen (*50*). At 300°C, the degree of reduction of Pd^{2+} to Pd° increased to 70%. Other conditions equal, the ease of cation reduction depends on the zeolite composition (*51, 52, 53, 54, 55*). Thus, at 400°C in H_2 for 2 hr, in Ni(38.7%) NaX and Ni(58.8%)NaX, dehydrated at 550°C for 2 hr, 83% and 66% of Ni-cations are reduced, respectively

(*53*). In addition, there is relation between the extent of Ni^{2+} reduction and the acidity of the zeolite (*52*).

The reduction of certain cation exchanged zeolites by hydrogen produces some hydrogen (decationized) modifications loaded with a metal. Therefore, the cation composition and the reduction temperature are the major factors which control the retention of the crystal structure of these metal-loaded zeolite catalysts. These conditions are especially important for A- and X-type zeolites whose decationized forms have low stability. Hydrogen treatment of nickel-exchanged zeolites, obtained from NaX or CaA by ion exchange, at 300° and 380° for 5–15 hr, gives products which contain well-dispersed metallic nickel (*56*), but these zeolite products are practically amorphous *i.e.*, under the above conditions the crystalline structure of A- and X-type molecular sieves is destroyed. Better methods to introduce nickel and certain other metals into these zeolites are reported below. Furthermore, upon reduction of the cations of certain metals, such as Cd^{2+}, Zn^{2+} and Hg^{2+}, can partially be eliminated from the zeolite pore structure (*57*).

The stability of cations in zeolites upon the interaction of the latter with H_2 was first studied by Yates (*57*), who showed that the treatment of CdNaX, HgNaX, ZnNaX, and ZnNaA zeolites with hydrogen at low pressure decreases their weight by the removal of metallic Cd, Hg, and Zn. Hg removal was observed at a temperature as low as 140°C whereas Zn and Cd were removed at a higher temperature (~450°C). Removal of metallic Zn from zeolite A was less significant than that from zeolite X (*57*). A similar effect may also occur at standard pressures.

Metallic silver was formed when AgNaY and AgNa-mordenite zeolites were heated in helium at temperatures above 600° and at 750°C, respectively, which can be accounted for by the following reaction (*58*):

$$\mathrm{(O)_2\overset{-}{Al}(O)_2}\ [Ag^+]\ \mathrm{-O-Si(O)_2-O-}\ \mathrm{\overset{-}{Al}(O)_3}\ [Ag^+] \longrightarrow \mathrm{(O)_2Al(O)}\quad \mathrm{\overset{+}{Si}(O)_2-O-\overset{-}{Al}(O)_3} + Ag_2O$$

$$Ag_2O \longrightarrow 2\,Ag + \tfrac{1}{2}\,O_2$$

The nickel zeolites dehydrated in helium at the same temperatures did not produce metallic Ni (*58*). In the former case, silver is reduced by oxygen ions. Various hydrocarbons and other compounds can also act as reducing agents (*58, 59, 60*).

Dehydrated zeolites can be loaded with metals by adsorption of volatile or soluble organic or inorganic compounds with subsequent thermal or chemical decomposition of these compounds. Carbonyls or hydrocarbonyls

(Fe, Co, Ni, Cr, Mo, W, Mn, Re) (*46, 61, 62*), acetylacetonates (Cr, Cu, Ag, Au) (*46, 62*), halogenides (Ti, Hf, Zr) (*34, 45, 46*), alkyl derivatives (Zn, Pb, Cd, Sn, etc.) (*34, 45*), and others (*34*) can be used for this purpose.

When, for example, zeolites contain reactive OH groups, a reaction may occur between an organometallic compound and a support. An example of this reaction type is observed in the interaction of bis-π-allylnickel with OH groups of a silica gel surface (*63*).

$$Si—OH + Ni\,(C_3H_5)_2 \rightarrow Si—O—Ni—C_3H_5 \xrightarrow[H_2]{} Si—OH + Ni^\circ + C_3H_6$$

The catalysts obtained after hydrogen reduction at 400°C contained highly dispersed Ni, their activity in hydrogenation of benzene being an order higher than that of the supported Ni catalysts prepared by SiO_2 impregnated with $Ni(NO_3)_2$ (*63*). Although little studied, the possibilities of using zeolites to obtain such discrete catalysts with practically isolated fine metal particles are great.

In some cases the readily available methods are not suitable for introducing metals or their compounds into (on) readily available zeolites. For example, the attempts to introduce platinum by ion exchange into CaA zeolites with a window size of 5 A failed because the solution of the Pt compounds available exhibit low pH values and can destroy the structure of zeolites. In addition, the sizes of the cations of neutral Pt-amino complexes are larger than 5 A. In this case it is more advantageous to introduce a catalytically active agent during zeolite synthesis (crystallization) (*10, 64, 65, 66, 67, 68*), as was first done by Weisz and co-workers (*10*). They carried out NaA zeolite crystallization from a solution containing Pt–cation complexes and observed the capture of the latter into the crystal cavities. These cations are strongly retained in zeolite pores and cannot be removed during the Na^+ exchange for Ca^{2+}. Destruction of complex cations followed by reduction gave a highly selective catalyst for olefin hydrogenation.

Another method for preparing metal-loaded zeolite catalysts with molecular sieve properties has been reported (*69, 70*). Soluble Pt cations, such as $Pt(NH_3)_4^{2+}$, do not penetrate the crystals of Na-mordenite, and therefore no exchange of Na^+ ions in zeolite channels takes place. However, the channels of H^+-exchanged mordenite (up to ~7 A) are larger than the size of aminocomplex Pt-cations. This is used to prepare Pt-Na-mordenite catalysts. First, zeolite is converted to an H form (by treating with acid solutions or through NH_4-mordenite), followed by introduction of $Pt(NH_3)_4^{2+}$ by ion-exchange, and a subsequent H^+ replacement by Na^+ using alkali solutions of Na salts. This narrows the free apertures of the channels. Pt cations are strongly retained by zeolites and are not displaced by Na ions for the small amounts of Pt used here. (Zeolites are strongly attracted to heavy metal cations). The product, containing small amounts of Pt (0.2 % Pt (*69*)), was treated in dry air to destroy the complex cation and was then reduced by hydrogen. Since upon reduction, platinum migrates partially to the external surface of the zeolite crystals, an additional treatment of the catalyst was required to preserve molecular sieve selectivity. For selective poisoning of Pt centers on the external surface of crystals, the catalyst was treated at 260°C in H_2 with triphenylphosphine vapor whose molecules cannot penetrate the Na-mordenite pores (*69*). The resulting catalyst selectively hydrogenated ethylene in the presence of propylene. Catalysts containing Pd, Rh, Ir, Os, Ni, and Co can also be prepared in this way (*70*).

For practical purposes it is necessary that catalysts contain only small amounts (0.1–1 wt %) of Pt or Pd. To obtain a fine metal dispersion, the metal should be introduced by cation exchange, but this method does not always provide uniform distribution of small amounts of cations in the bulk of the catalyst. It is very difficult to attain uniform metal distribution when a metal is introduced into granular zeolites. In this case the exchange for the noble metal cation should be carried out in the presence of the cation of the zeolite used. Na^+ is used for sodium forms, Ca^{2+} for calcium ones, and NH_4^+ for those of ammonium, etc. (*34, 71*). This allows preparation of catalysts with uniform distribution of metals throughout the bulk of separate granules throughout the zeolite crystals.

Table I. Dependence of the Size of Metal and Treatment Conditions of

Sample	*Catalyst*[c]	SiO_2/Al_2O_3 *Ratio*	*Pretreatment Conditions*
1	0.24 % Pt-NaY	5.0	heating in air for 3 hr, 100°
2			heating in air for 3 hr, 300°
3			heating in air for 3 hr, 500°
4	0.2-0.5 % Pt-NaY	5.0	drying in air at 100–110°
5	0.2-0.5 % Pt-NaY[a]	5.0	drying in air at 100–100°
6	0.28 % Pt-NH_4Y	5.0	heating in air for 3 hr, 400°
7			heating in air for 3 hr, 500°
8			heating in air for 3 hr, 600°
9			heating in nitrogen for 3 hr, 400°, then for 3 hr, 500° in air
10			heating in H_2 for 3 hr, 500°, then 3 hr, 500° in air
11	0.5 % Pt–0.45 La 0.45 NH_4Y	5.0	heating in air, 500°, then pumping, 500° for 18 hr
12	10 % Pd-NaY	4.8	heating in O_2, 600°, then pumping, 600° and 5.10^{-2} torr for 6 hr
13	1 % Cu-NaY (SK-410)	5.0	pumping, 530° for 16 hr
14	11.8 % Ag-NaX	2.67	pumping, 20–200°
15	6.7 % Ag-NaY	4.6	heating in He flow, 450°
16	10.1 % Ag-NaY	4.6	heating in He flow, 450°
17	13.8 % Ag-NaY	4.6	heating in He flow, 450°

Polymetallic catalysts can be prepared in a similar way (*72, 73, 74*). These catalysts are significant because metal-on-oxide catalysts of this type greatly improved the reforming process (*75, 76, 77*).

The catalytic properties of metal–zeolite systems also depend on the structure and composition of the second component—the crystalline aluminosilicate. The description of the methods for synthesizing zeolites of various types, their modification by ion exchange, by partial extraction of aluminum or silicon from the lattice, isomorphous replacement of AlO_4 or SiO_4 tetrahedra by structural units containing B, Ga, Ge, P, and other elements, as well as catalyst granulation is outside the scope of this review. These problems have been reported (*34*). For the latest data, *see* Refs. *36, 78, 79, 80.*

The State of Metals in Metal-Loaded Zeolites. The previous section dealt with some principles for preparing metal-loaded zeolite catalysts. Here we consider the formation of catalysts, the degree of dispersion of various metals, variations in metal dispersion with composition and preparation of catalysts, alloy formation in polymetallic systems, interaction between

Particles on the Composition, Preparation,
Metal-Containing Zeolite Catalysts

	Metal Particle Size, A		
Reduction Conditions	*Electron Microscopy*	*Based on H_2 Chemisorption*	*Reference*
H_2, 3 hr, 500°	15	—	*84*
H_2, 3 hr, 500°	15–55	29	
H_2, 3 hr, 500°	30–680	425	
H_2, 3 hr, 200°	≤10	—	*85*
H_2, 3 hr, 200°	30–100	—	*85*
H_2, 3 hr, 500°	30–65	82	*84*
H_2, 3 hr, 500°	15–75	87	*84*
H_2, 3 hr, 500°	15–70	66	*84*
H_2, 3 hr, 500°	40–60	—	*84*
H_2, 3 hr, 500°	60	—	*84*
H_2, 20 hr, 300°	25	—	*86*
H_2, 100 torr, 15 hr, 300°	18–20[b]	—	*50*
H_2, 535°	400–800	—	*87*
H_2, 16 hr, 250°	170[b]	—	*57*
cumene, 350°	179	—	*58*
cumene, 350°	216	—	*58*
cumene, 350°	253	—	*58*

Table I.

Sample	Catalyst[c]	SiO_2/Al_2O_3 Ratio	Pretreatment Conditions
18	31.6 % Ag-NaY		heating in He flow, 450°
19	13.8 % Ag-NaY		heating in He flow, 450°
20			heating in He flow, 450°
21		4.6	heating in He flow, 750°
22		4.6	heating in He flow, 750°
23	15.6 % Ag-5.95 % Ni-NaY	4.6	heating in He flow, 450°
24		4.6	heating in He flow, 450°

[a] Prepared by impregnation; the other catalysts were prepared by ion exchange.
[b] According to x-ray data.

Table II. Dependence of the Size of Ni Particles Its Composition and

Sample	Catalyst	SiO_2/Al_2O_3 Ratio	Reduction Conditions
1	2.7 % Ni-NaA	—	H_2, 360–370°
2	3 % Ni-NaA	1.87	H_2, 350°, 2.5 hr
3	0.9 % Ni-NaX	2.24	H_2, 350°, 2.5 hr
4	1.67 % Ni-NaX	—	H_2, 360–370°
5	2.5 % Ni-NaX	2.24	H_2, 350°, 2.5 hr
6	5.0 % Ni-NaX	2.24	H_2, 350°, 2.5 hr
7	10.2 % Ni-NaX	2.24	H_2, 350°, 1.5 hr
8	10 % Ni-NaX[a]	2.67	H_2, 400°, 16 hr
9	1 % Ni-NaY[b]	5.0	H_2, 538°, 20 hr
10			H_2, 538°, 2.5 hr
11			H_2, 521°, 20 hr
12			H_2, 521°, 2 hr
13			H_2, 521°, 1 hr
14			H_2, 418°, 2 hr
15			H_2, 304°, 2 hr
16	2.46 % Ni-NaY[c]	4.74	H_2, 400°, 16 hr
17	3.03 % Ni-LiY[c]		H_2, 400°, 16 hr
18	3.05 % Ni-CaY[c]		H_2, 400°, 16 hr
19	2.67 % Ni-MgY[c]		H_2, 400°, 16 hr
20	2.40 % Ni-NH_4Y[c]		H_2, 400°, 16 hr
21	1.45 % Ni-Na-mordenite	—	H_2, 360–370°
22	3.25 % Ni-Na-mordenite	—	H_2, 360–370°
23	9.8 % Ni-Na-mordenite	—	H_2, 360–370°

[a] Pretreatment: pumping at 20°·200°C.
[b] Pretreatment: heating in H_2 to reduction temperature at 65°/hr.
[c] Pretreatment: drying at 100°C in N_2.

Continued

Reduction Conditions	Metal Particle Size, A: Electron Microscopy	Metal Particle Size, A: Based on H_2 Chemisorption	Reference
cumene, 350°	298	—	*58*
H_2, 450°	176	—	*58*
cumene, 450°	285	—	*58*
cumene, 450°	428	—	*58*
without reduction	363	—	*58*
cumene, 350°	295(Ag)	—	*58*
H_2, 450°	161(Ag)	—	*58*

[c] Metal content in catalyst given in wt %.

in Nickel-Zeolite Catalysts on Type of Zeolite: Treatment Conditions

% Reduction to Metal	Particle Size, A: Electron Microscopy	Particle Size, A: X-ray Diffraction	Reference
ng[e]	80 ± 40	—	*88*
37	42	—	*54*
40.4	194	—	*54*
—	65 ± 35	—	*88*
29.3	148	—	*54*
24.0	165	—	*54*
14	103	—	*54*
—	—	240	*57*
ng[e]	—	210	*89*
ng[e]	many 275 some 70	210	
ng[e]	—	230	
ng[e]	many 320 some 70	190	
ng[e]	—	170	
ng[e]	many 250 some 50	150	
ng[e]	many 250 some 50	—	
100	—	126[d]	*52*
79.5	—	112[d]	
75.7	—	99[d]	
45.1	—	95[d]	
0	—	—	
ng[e]	—	65 ± 15	*88*
ng[e]	—	140	*88*
ng[e]	—	250	*88*

[d] By magnetic method.
[e] ng = not given.

metal particles and acidic centers, and the effect of these factors on the catalytic properties.

Ion exchange is frequently used in preparing metal–zeolite systems. The reducible metal cations are isolated in zeolite cavities at some discrete distance from each other (*50, 81, 82*). The cation clusters (multinuclei cations, such as $^{2+}Fe–O–Fe^{2+}$, magnetic couples) can arise only under specific conditions—*e.g.*, at high pH values of the ion exchange solutions (*83*). Upon contact with H_2 and other reducing agents, zeolites containing reducible cations give products containing metal particles. The size and location of metal particles depend on the conditions of pretreatment before reduction and the condition of the reduction step (Tables I and II). For example, with Pt(0.24 wt %)NaY catalyst obtained by ion exchange with a solution of $Pt(NH_3)_4Cl_2$, the Pt particles of 15–55 A were formed upon heating the zeolite in air at 300°, followed by H_2 reduction at 500°C for 3 hr (*84*). The 0.28 % Pt–NH_4Y zeolite shows lower sensitivity to heating in air than 0.24% Pt–NaY zeolite (*cf.* samples 1–3 and 6–8, Table I). The lower the reduction temperature, the smaller is the size of metal particles formed from ion-exchanged zeolites. If reduction temperatures are not very high (for Pd^{2+}, less than 200° (*50*), for Ni^{2+} not higher than 360°C (*81*), the redox equilibrium in the system M^{2+}–zeolite + $H_2 \rightleftarrows M^\circ$ + H^+ zeolite will be reversible. The size of metal particles in the products formed under rigid reduction conditions is about 100 A—that is, M° atoms migrate to the external surface of zeolite crystals to form aggregates. Crystallites of bulk metals are formed even with samples containing small amounts of Pt, Ag, Ni, and other metals. Large metal particles can arise even under mild reduction conditions. For example, electron microscopic study of Ag distribution in A- and Y-type zeolites has been reported (*90*). Ag–NaA zeolite, with a Na^+ exchange of 20 and 50 %, was reduced with hydroxylamine in a NaOH solution, and AgY zeolite was reduced with hydrogen at 150°C. Even under these mild conditions, all silver was completely crystalline on the external surface of the zeolite—that is, the mobility of the metal atoms in crystalline aluminosilicates is very high. Since some cations cannot be reduced under mild conditions, the complexity of the problem of obtaining zeolites containing highly dispersed metals is obvious. This goal can be attained if care is taken during catalyst preparation.

In 1964 Rabo *et al.* (*11*) reported that the 0.5 % Pt–CaY catalyst obtained by heating $Pt(NH_3)_4^{2+}$–CaY zeolite at 500°C in air followed by H_2 reduction at 300°C, contained atomically dispersed platinum; they observed hydrogen adsorption with a ratio of H:Pt = 2.0 (at 100°–250°C and 70–700 torr). Reduction of the sample at 500°C and at higher temperatures resulted in Pt crystallites. The catalyst of the same composition but prepared from CaY impregnated with H_2PtCl_6 followed by a similar treatment with adsorbed hydrogen with a ratio of H:Pt = 1.09 gave a Pt dispersion lower than that obtained upon ion exchange. These findings prompted many studies to elucidate the state of platinum in various zeolite catalysts. The Pt atomic dispersion in several Y-type zeolites was uncertain (*86, 91, 92, 93, 94*). For example, Lewis (*91*) studied 0.5 % Pt–CaY

(SK-200 Linde Co.) catalyst equivalent to that used by Rabo *et al.* (*11*) by x-ray absorption edge spectroscopy. After treatment under conditions similar to those used by Rabo (*11*), only 60% of the reduced Pt was in the form of 10 A particles, small enough to be located inside zeolite cavities. The rest (40%) of the Pt occurred as crystals about 60 A large. Each Pt atom adsorbed 0.5 H atom (for 1 hr at 100° or 300°C, H_2 pressure 1 atm), and it was this hydrogen that affected the x-ray absorption spectrum—*i.e.*, it was directly bound with Pt and determined the size of crystals.

Subsequent treatment for 20 hr or longer at 300° 1 atm H_2 resulted in the additional sorption of 1.2 H atoms per Pt atom. This hydrogen did not affect the x-ray absorption spectrum. This led to the conclusion (*91*) that sorption did not occur in the layer adsorbed on Pt.

Wilson and Hall (*86*) studied the catalyst obtained from $Pt(NH_3)_4^{2+}$–0.45 La–0.45 NH_4NaY (0.5 % Pt). The samples were heated in air at 550°C and reduced by hydrogen at various temperatures. After interaction with H_2 at 300°C, the maximum size of the Pt particles was 25 A, but a short treatment of the zeolite with hydrogen at 500°C was sufficient for the platinum to sinter. After that the Pt particles were 20–90 A. The authors (*86*) concluded that the platinum in the zeolites studied does not show atomic dispersion and is largely located on the external surface of the zeolite crystal.

Penchev and Kanazirev (*94*) also reported that they had failed to produce atomically dispersed platinum under various conditions of thermal pretreatment up to 500°C in nitrogen, air, or hydrogen, with subsequent H_2 reduction at 300°–450°C with zeolites X and Y of different composition. According to the data of hydrogen–oxygen titration and electron microscopy, Pt crystals with an average size of 20–100 A were formed. The Pt dispersion was affected by the gas present, pretreatment temperature, and the reduction conditions. Larger Pt particles were associated with an increase of SiO_2/Al_2O_3 ratio in the zeolite used.

Recently Dalla Betta and Boudart (*48*) produced zeolite catalysts with highly dispersed Pt. $Pt(NH_3)_4^{2+}$-exchanged zeolites were treated with oxygen to decompose the complex (350°C) subsequently reduced with hydrogen and dried at 400°C. Under these conditions the catalysts contained platinum clusters of less than six atoms. These clusters showed some specific properties; for example, they chemisorbed O_2 less readily than large Pt crystallites. The authors (*48*) investigated the formation of Pt zeolite catalysts by IR spectroscopy. When $Pt(NH_3)_4^{2+}$–CaY zeolite was heated in H_2, the complex decomposed in the temperature range comparable with that in which the sample adsorbed H_2:

$$Pt(NH_3)_4^{2+} + 2\,H_2 \rightarrow Pt(NH_3)_2H_2 + 2\,NH_3 + 2\,H^+$$

$$Pt(NH_3)_2H_2 \rightarrow Pt^\circ + 2\,NH_3 + H_2$$

The unstable, neutral, and, hence, mobile hydride led to agglomeration of Pt. On the other hand, when zeolite is heated in O_2, the complex preserves its structure (a bipyramide) up to 200°C. Ammonia evolved during decomposition of the complex at above 250°C is not adsorbed on zeolite. The oxidized complex remains bound to the zeolite surface, and a highly dispersed Pt is formed upon H_2 reduction. The water contributes to the mobility of the oxidized complex. In particular, H_2O formed during reduction of a large sample (5 g) increases the Pt particle size more than that of a small sample (0.4 g). This may account for the discrepancy between the average Pt dispersion in the commercial catalyst SK-200 (*48*) found by Lewis (*91*), and the high Pt dispersion in the laboratory catalyst found by Rabo *et al.* (*11*). Minachev, Kazansky, and co-workers (*95*) investigated the IR spectra of $Pt(NH_3)_4^{2+}$–NaY (70 wt % Pt, $SiO_2/Al_2O_3 = 4.2$) zeolite after treatment under various conditions (Figure 2). Heating of the zeolite *in vacuo* above 350°C results in complete destruction of the ammonia complex as evidenced by disappearance of certain bands (3200–3400 cm^{-1}) and deformation vibrations of NH_3 molecules (Curves A-D). There arise OH surface groups which produce a band of about 3657 cm^{-1}. In spectra A and D, taken

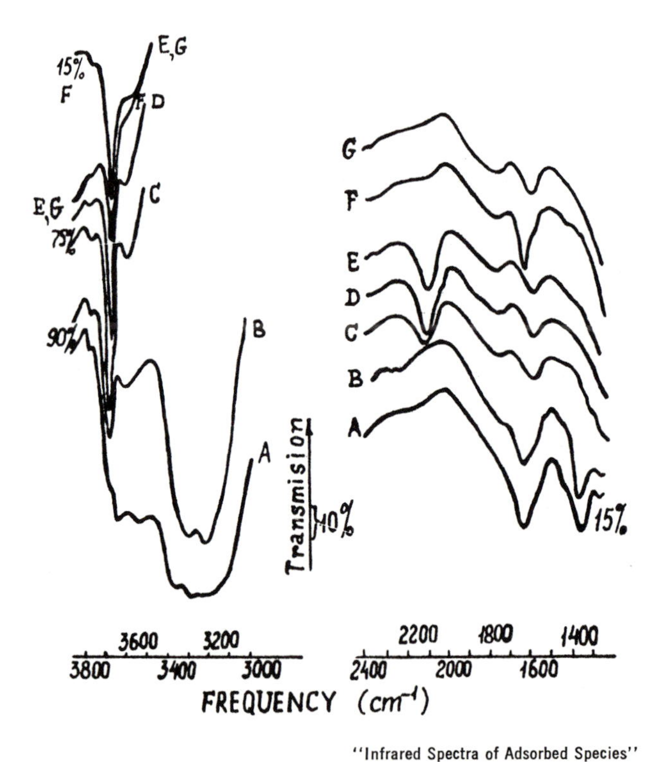

"Infrared Spectra of Adsorbed Species"

Figure 2. IR spectra of $Pt(NH_3)_4^{2+}$–NaY zeolite after treatment under various conditions (96). *Curve A: evacuation at room temperature. Curves B through E: heating* in vacuo *(10^{-4} torr) at 200°C (B), 300° (C), 400° (D), and 500° (E). Curve F: treatment in H_2 ($P_{H_2} = 30$ torr) 0.4 hr. Curve G: after treatment in H_2, the sample was evacuated at 500°C for 2 hr.*

after heating the zeolite at 300° and 400°C, a new band appears at 2120 cm^{-1}. Its intensity does not decrease even after evacuation at 500°C (Curve E). In H_2, this band is stable up to 250°C. Upon treatment at higher temperatures the band disappears, and the concentration of OH groups and H_2O increases (Curve F). The band at 2120 cm^{-1} is characteristic of valence vibrations of Pt–H bonds (*96*). It is assumed that destruction of the ammonia complex upon heating *in vacuo* gives a surface on which Pt hydrid carries a single positive charge.

$$Pt(NH_3)_4^{2+} \xrightarrow[\Delta]{} Pt\text{–}H^+ + H^+ + \text{gaseous product}$$

The presence of the charge is evidenced by the further reduction of Pt in H_2, forming new OH groups.

$$Pt\text{–}H^+ \xrightarrow[400°C]{H_2} Pt° + H^+$$

The Pt–H^+ complexes are likely to be found in the zeolite pores and cavities, providing conditions for a more highly dispersed metal upon further reduction than that resulting during decomposition of the initial complexes of $Pt(NH_3)_4^{2+}$ in H_2 (*48*). Pt dispersion in zeolite catalysts depends on the pretreatment conditions, which in turn are responsible for their different catalytic activity (*48, 95*).

The gaseous environment in which the samples are thermally treated before reduction (*vacuo*, N_2, air, etc.) also affected the size of the resulting Pt crystals in impregnated Pt-zeolite catalysts. This is probably caused by formation of Pt products such as PtO_2 and PtH which have different mobility and sensitivity to reduction by hydrogen.

The conclusions for Pt-containing zeolites proved to be valid for palladium (*49, 50, 97, 98, 99*), silver (*57, 58, 90, 99*), copper (*81, 87*), and nickel (*30, 52, 54, 55, 56, 57, 73, 81, 82, 88, 89, 100, 101, 102, 103, 104, 105, 106, 107 108, 109*) in catalysts prepared from A-, X-, and Y-type zeolites, mordenite and erionite (*see* Tables I and II). Upon reduction, metals in each sample migrated from the interior of the zeolite crystal onto their external surface, forming large metal crystals. The intensity of this migration depended on the structure of the zeolite, its composition, and the choice of metal. In the process the unreduced $M^{\circ+}$ ions and the remaining Na^+ ions move about.

Minachev and co-workers (*99*) found by x-ray photo-electron spectroscopy that the intensity of the peaks Ag 3*d* and 4*d* in the AgNaY spectra increases during reduction. This provides evidence for silver migration from the bulk of zeolite crystals onto their external surface and is followed by a decrease of the intensity of the Na 2*p* bands (Figure 3). Since the intensity of the lines corresponding to the elements of the zeolite framework (Si, Al) was not affected, a decrease in the Na concentration in the near surface layer

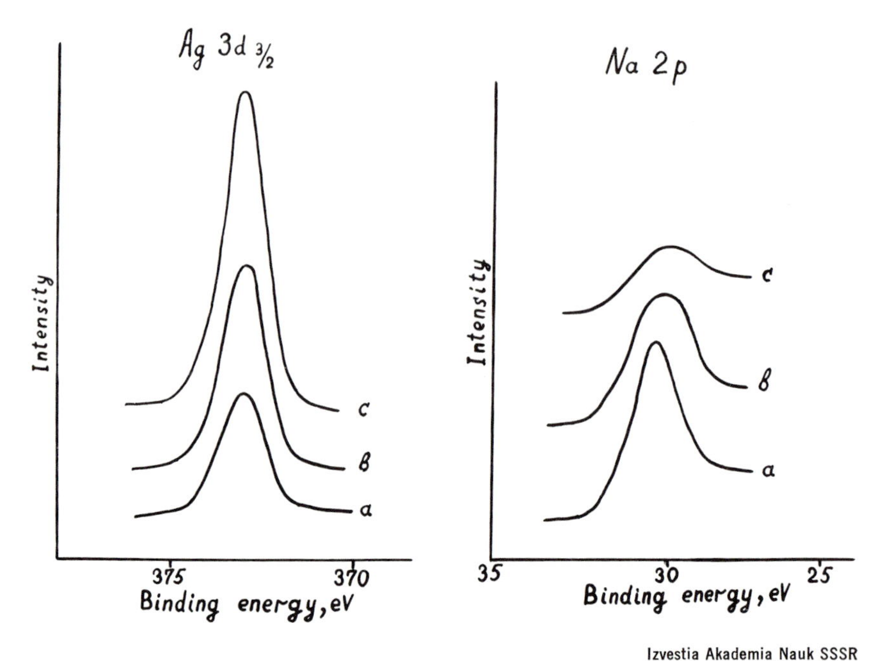

Izvestia Akademia Nauk SSSR

Figure 3. Variations in peak intensities of Ag $3d_{3/2}$ *and Na* 2p *in the x-ray photoelectronic spectra of 0.42 Ag NaY zeolite* ($SiO_2/Al_2O_3 = 3.4$) *during its reduction* (99). *Curve a: the starting sample. Curve b: after reduction at 200°C. Curve c: after reduction at 300°C.*

is caused by the diffusion of Na^+ ions into the bulk of the crystallites, probably by the decationized sites of the structure previously occupied by Ag^+ cations. Similar effects were observed during reduction of $Pd(NH_3)_4^{2+}$–NaX (*99*). Similar migration of cations can dramatically affect the catalytic properties of the corresponding samples.

Reduction of metal cations in zeolites is related to the migration of ions in the structure. The cations in zeolites can be located in different sites, whose population depends on the number of M^{n+} ions and the conditions of the treatment. Thus, in faujasite zeolites Ni^{2+} ions are distributed in the S_I, S_{II}, S_I' S_{II}' sites (*110, 111, 112*), (*see* Chapter 1). With increased dehydration, the number of Ni^{2+} at S_I sites (hexagonal prism) increases but does not exceed 12 cations per unit cell (*112*) for the samples obtained from NaY-zeolites. In the zeolites of NiCaX where S_I' and S_{II}' sites are occupied by Ca^{2+}, the Ni^{2+} ions occupy mostly S_I sites (*113*). During heat treatment at high temperatures, in addition to zeolite dehydration and relocation of metal cations in the structure, reactions occur which affect M^{n+} coordination environment and chemical state. According to IR data, after dehydration at 150°–200°C *in vacuo* the nickel in Ni(35%)NaA zeolite occurs as $Ni(OH_2)^{2+}$ (*114*). At dehydration temperatures above 200°C, the $Ni(OH_2)^{2+}$ form disappears and no absorption band from OH groups can be observed in the IR spectra. After treatment at 450°C, the bands from H_2O molecules are not observed, and the nickel is present as Ni^{2+}. On the other hand, in Ni(35%)NaX, dehydration at as low as 100°C produces bands from $Ni(OH)^+$ and Si–OH formed by dissociation of H_2O molecules catalyzed by Ni^{2+} electrostatic fields and a subsequent reaction between H^+ and the oxygen atoms of the zeolite framework (*114*). There was also a band from $Ni(OH_2)^{2+}$. At 300°C the bands from $Ni(OH)^+$ and $Ni(OH_2)^{2+}$ partially disappear and a band caused by H^+ appears inside the sodalite cage. These changes are probably caused by zeolite dehydroxylation which proceeds according to the following scheme.

$$Ni(OH_2)^{2+} \cdots Ni(OH)^+ \rightarrow Ni^{2+} + H^+ + NiO + H_2O$$

Thus, the nickel in ion-exchanged faujasite-type zeolites may occur as Ni^{2+}, $Ni(OH)^+$, and as NiO. Each of these forms can occur in both the large and small cavities, the latter being less accessible to H_2 and to other molecules. In addition, these Ni forms show different sensitivity to reduction by hydrogen. Therefore, after treatment with hydrogen, a fraction of the nickel is reduced to zero-valent metal while the rest occurs as Ni^{2+} (*51, 52, 53, 54, 55*) (*see* Table II).

An attempt was made to determine the difference in reactivity of Ni^{2+} ions in different sites of X-type zeolite during H_2 reduction (*115*). The kinetics of hydrogen uptake on the samples of NiNaX with SiO_2/Al_2O_3 ratio of 2.46 and with cation exchange of Na^+ for Ni^{2+} ranging from 2.2 to 63.6% was studied. At first, the H_2 uptake proceeds rapidly, then slows down (this part of the kinetic curves is rectilinear). The initial reduction rate did not depend on the temperature or hydrogen pressure (<7.5 torr) and was first order in Ni^{2+}. The following mechanism of the fast reduction is assumed: (1) chemisorption of hydrogen on nickel ions in some crystallographic site "s"; (2) reduction of the nickel ions as represented by

$$Ni^{2+}\,(s) + H_2\,(ads) + O^{2-}\,(s) \rightleftarrows Ni^{\circ} + H_2O\,(s)$$

(3) the migration and aggregation of nickel atoms. The rate of the slow process was independent of the hydrogen pressure and of the degree of exchange of Na^+ by Ni^{2+} (when the results were expressed per unit weight of nickel). These results support the slow process as being rate determined by the diffusion of nickel ions from sites where electrostatic environment makes reduction unfavorable to some more favorable location in the crystal. The authors suggest that the Ni^{2+} cations in S_{II} sites are reduced faster than others (*115*). However, Egerton and Vickerman (*113*), who studied the magnetic properties of NiNaA, NiNaX, NiNaY, and NiCaX before and after treating with H_2, concluded that tetrahedrally coordinated Ni^{2+} is less readily reduced than the octahedrally coordinated Ni^{2+} in S_I sites. A similar effect was observed with Fe^{2+} ions. After pumping at 400°C, the Fe^{2+} ions in 0.65 FeNaY are located at sites with fourfold coordination (*116*), and this zeolite is not reducible by hydrogen at 400°C. On the contrary, FeCaX, dehydrated at 360°C, did not give a spectrum characteristic of the tetrahedral

ion of Fe^{2+}, and the zeolite was reduced by H_2 at 360°C (*117*). Thus, additional studies should be undertaken to clarify the influence of structural parameters on the reducibility of zeolite cations.

The experimental data now available show that zeolite reduction is sensitive not only to the temperature, H_2 pressure, and treatment time (*81*) but also to chemical composition, the nature of cations other than M^{n+} (*51, 52, 113*), and pretreatment conditions.

Since not all reducible cations are reduced simultaneously, the rate of their reduction in zeolites may vary with time. For example, the formation of first metal portions (Ni, Pt, etc.) could activate H_2. Samples containing cations of several transition metals may strongly affect the properties of each other. The reduction products of the NiNaX zeolites do not affect catalytically the further reduction of Ni^{2+} cations (*115*). This property has not been studied in other systems.

Certain metals influence the ease and extent of the reduction of ions or compounds of other metals. Pt and Pd introduced into zeolites considerably decrease the temperature of formation of Ni°, Co°, and other metals (*47, 118*). For example, Co^{2+} cations in Y-type zeolites can barely be reduced by H_2 at temperatures below 350°–380°C while similar treatment of CoY samples containing small amounts of Pt or Pd resulted in the formation of ferromagnetic Co° at temperatures as low as 250°–350°C (*47, 118*). Similar results were obtained from the studies of activation of metal oxide reduction with various additives (*119, 120, 121, 122*). For example, Nowak and Koros (*119*) showed that a small amount of Pt (0.01 wt %) considerably decreases the temperature of NiO reduction. Cu, Pd, Rh, Os, and Ir produce the same effect (*122*). These results suggest dissociative adsorption of H_2 on Pt or Pd with subsequent reduction of cation or metal oxides by atomic hydrogen. Thus, metal-zeolite catalysts can be modified both by ion exchange and impregnation. Upon reduction of zeolites under mild conditions it is possible to produce sometimes highly dispersed metals. Gallezot and Imelik (*50*) studied the structure of the Y-type zeolite ($Pd_{12.5}Na_{19.5}H_{11.5}Al_{56}Si_{136}O_{384}$ (10 wt % Pd)) before and after reduction with H_2 and found that its treatment with hydrogen at 25°C leads to the formation of Pd atoms located inside the sodalite cages. On heat treatment at 200°–300°C the metal particles migrate to the external zeolite surface where they form particles of ~20 A in diameter. The metal atoms apparently diffuse to the zeolite crystals surface without agglomeration in supercages because they move from sodalite cavities with an aperture of 2.3 A in a highly activated state ($\phi_{Pd^{o}} = 2.74$ A). Consequently, the activated atoms should rapidly cross the supercages which are open widely towards the external surface. A similar mechanism of movement of nickel in Ni$\overline{NaY}$ zeolite has been suggested (*89*). The calculation of the mean free path of Ni atoms in zeolite showed that most of the nickel can migrate to the external surface of faujasite type crystals without forming metal crystallites (*89*). An alternate possibility—that some of the nickel might form small clusters inside the pores which then could move to the outer surface of the zeolite crystal or simply stay in place inside the pores—is not excluded however.

In low temperature reduction the surface diffusion is slowed considerably producing more highly dispersed metals. The diffusion of atoms and small particles of metal can be inhibited by adding various non-reducible compounds to the catalysts. For example, Lawson and Rase (*89*) succeeded in significantly inhibiting the growth of Ni crystallites in 1% Ni-NaY by adding Cr_2O_3 (from the carbonyl) to the zeolite before its reduction. The nature of the Cr_2O_3 effect has not yet been clarified, but the stabilization of the metal dispersion and an increased resistance to poisons is of great interest.

The properties of the isolated atoms of metals have been studied minimally (*11, 48, 49, 50*). With atomic dispersion, platinum is highly stable to sulfur poisoning (*11*). Palladium atoms, obtained by reduction of PdHNaY zeolite with hydrogen at 25°C, chemisorbed neither H_2 nor O_2 (*49, 50*), they differ greatly from those of the bulk metal or Pd-particles on various supports. Another important finding is the great effect of the solid on the electronic state of palladium. The Pd atoms in zeolite Y are close to Lewis acid centers which attract their 4*d* electrons (*50*), and as a result some Pd° atoms are oxidized to Pd^+, as detected by ESR (*49*). The electron deficiency of small Pt clusters in zeolites which contain di- and trivalent cations was also reported by Dalla Betta and Budart (*48*). In their opinion, under the action of electron-acceptor sites of supports, the Pt electron configuration becomes similar to that of Ir, and the corresponding catalysts show an increased activity in ethylene hydrogenation (Table III).

Table III. Pt Catalyst Activities in Ethylene Hydrogenation

(−34°C, 1 atm, 23 torr C_2H_4, 152 torr H_2, 585 torr He (*48*))

Catalyst	*H/Pt (total)*	$N \times 10^{-3} sec^{-1}$ [a]
0.54 % Pt-NaY	1.41	5.34
0.59 % Pt-CaY	1.34	25.0
0.6 % Pt-MgY	1.4	23.3
0.5 % Pt-RE-Y	1.4	20.3
0.53 % Pt-SiO_2	0.56	6.31

[a] *N* = turnover number.

Proceedings of the Fifth International Congress on Catalysis

Despite the similar high dispersion of the Pt in the 0.54% Pt–NaY catalyst, its activity is 4–5 times lower than that of other Pt zeolites. Because of low acidity and weak electrostatic fields of Na^+ cations, Na zeolite is likely to affect the electronic state of Pt only slightly, and the properties of this catalyst do not differ from those of Pt/SiO_2. There may be other reasons for a greater activity of multivalent cationic forms of faujasites loaded with Pt. For Pd catalysts in benzene hydrogenation the support effect is also observed with solids of strong electron-accepting properties (*98*), such as SiO_2–Al_2O_3, HY, LaY, CeY, MgY, and CaY. The studies of the electronic state of the metal by IR of adsorbed CO showed that Pd on aluminosilicate or HY zeolite is electron deficient compared with that supported on SiO_2 or Al_2O_3. The interaction with the support modifies the electronic configuration of Pd, which becomes similar to that of Rh, and the Pd on SiO_2–Al_2O_3 or acidic zeolite catalysts show higher activity in benzene hydrogenation. Moreover, the smaller the size of metal particles is, the stronger is the effect of the same support (*98*).

The important and complicated problem of the mutual effect and interaction of metallic and acidic (as well as other) centers in metal-zeolite catalysts is of general significance in heterogeneous catalysis as a whole. Richardson (*52*) found that in *n*-hexane isomerization the dual function activity of nickel-faujasite is additive: namely, the nickel metal centers and the acid centers act independently. No mutual action of the above centers is observed. The nickel particles in the catalysts studied were about 100 A and were located on the external surface of zeolite crystals—*i.e.*, far from acidic centers within the crystal. Therefore they should not be considered inconsistent with the data reported in the papers mentioned above. In general the character of interaction between the metal and the solid support may depend on the nature of the metal, the size and location of its particles, and on the zeolite structure and composition.

The catalytic properties of metals in zeolite catalysts depend not only on their electronic state but also on accessibility of the particles for reacting molecules, size, and topography of crystallites. For example, NiNaA and NiNaX zeolites, reduced with H_2 are known to be active in benzene hydrogenation (*54, 100, 108, 123, 124*) and cyclohexane dehydrogenation (*108, 124*) at atmospheric pressure. On the contrary, the contacts formed from NiNaY zeolites do not significantly accelerate these reactions under the same conditions (*94, 100, 108, 124*). Similarly, Ni–CaY ($SiO_2/Al_2O_3 = 4.2$) reduced with H_2 at 450°C showed low activity in benzene hydrogenation and hydroisomerization, and NiNaY, containing the same amount of Ni (7.2 wt %), after treatment under identical conditions, hydrogenated C_6H_6 selectively and completely to cyclohexane at 200°C and 30 atm pressure (*82*). The zeolites not active in benzene hydrogenation catalyzed acetone hydrogenation (*100*), showing that metallic centers in zeolites are not uniform. In fact, in Y-type zeolites there are two types of nickel centers (*52*). The first type of nickel, which is more active and amounts to 35% of the total, is responsible for *n*-hexane hydrogenolysis. The second type of nickel participates in isomerization and hydrocracking of *n*-hexane and in hydrogenation of benzene.

The effect of the size of metal particles on the properties of metal-zeolite catalysts has been insufficiently studied. According to the data reported by Penchev *et al.* (*125*), the catalytic activity is a direct function of $1/r_{\text{nickel particle}}$, the specific activity is constant, and the activity change is caused by changes in metal surface area (Figure 4). The catalytic activity per weight of reduced metal decreases with in increase in the diameter of nickel crystals both in nickel-exchanged A- and X-type zeolites. Note that the size of metal particles can change during the reaction because of the

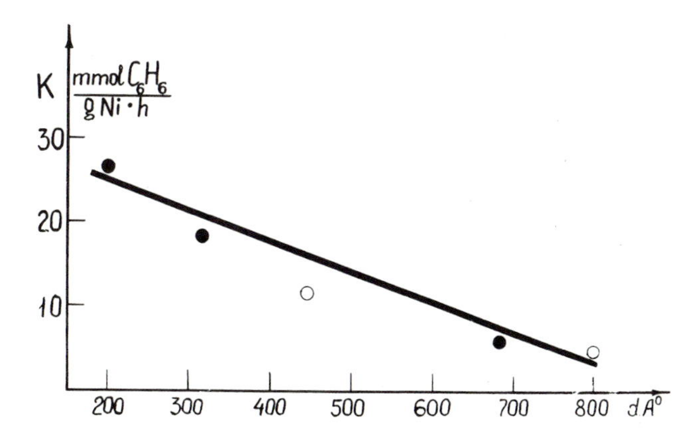

Figure 4. Dependence of the catalytic activity of A and X zeolites in benzene hydrogenation on the size of nickel crystals: ● Ni–NaA, ○ Ni–NaX

increased mobility of the individual atoms. The size may become either larger or smaller than in the initial catalyst. For example, during hydroisomerization of 1-butene on Pd-graphite catalyst at a temperature as low as 40°C, the size of Pd crystals varied from 65 A to 1300 A (*126*). Rubinshtein *et al.* (*127*) found catalytic corrosion—that is, an increase in the dispersion of Ni particles in Ni-ZnO catalysts occurred after its use in cyclohexane dehydrogenation. The number of active centers determined before use as catalyst may also change as a result of the action of the reaction system. Therefore, it is very important to investigate the physical properties of catalysts not only before but also after the reaction as well.

Zeolites containing two or more reducible cations may be reduced under favorable conditions to form alloy crystals. This phenomenon was first observed by Reman *et al.* (*73*). The treatment of Ni–Cu-, Ni–Cd-, and Ni–Ag faujasites of the Y type with hydrogen at 550°C resulted in catalyst alloys formed on the zeolite crystal surface. In the case of Ni–CuY, this was demonstrated by a magnetic method. In the other two zeolites, no sound evidence for alloys was provided. However, the NiCdY reduction was not followed by release of cadmium although at 550°C it usually evolved from Cd faujasite to form a Cd mirror on the walls of the quartz reactor, which agrees with the data of Yates (*57*). This provides evidence for the formation of Cd–Ni alloys. Alloy formation greatly affects the properties of the corresponding catalysts (*73*). For example, on the zeolite containing only Ni, the *n*-hexane undergoes stepwise hydrogenolysis which produced CH_4 and *n*-paraffins of lower molecular weight. In the presence of the zeolite containing Ni-Cu alloy, isomerization to methylpentanes was predominant and small amounts of 2,2-dimethylbutane were also found This catalyst can be poisoned by thiophene whereas pyridine poisoning for dual-function catalysts acid centers was not effective. Hence, hexane isomerization probably occurs on the surface of alloy crystals. Similar effects were observed with a zeolite containing Ni–Ag alloy. On the whole, the *n*-hexane conversion on alloy catalysts was similar to that observed with Pt zeolite catalysts. These results open up great possibilities for modifying metal-zeolite systems. Thus, the state of metals in zeolites depends on numerous factors, some of which have been considered above.

The cation-exchanged zeolites reduced by hydrogen contain structural hydrogen. The properties of hydroxyl groups formed during reduction with H_2 can be different from the OH groups formed in decationized modifications. For example, in Cu°Y samples they are more thermostable than in decationized zeolites (*128*). In partially reduced zeolites the OH groups stabilize Cu^{2+} cations. For completely reduced samples ultrastable zeolite types are assumed to be formed under the action of the water evolving during reduction. The complete dehydroxylation of Cu°Y was observed at temperatures higher than 650°C (*128*). Turkevich and Ono (*129*) also found that with a Pd-NH_4Y catalyst, cumene cracking proceeds after its treatment at temperatures higher than those observed for zeolite without Pd.

Minachev *et al.* (*130*) studied the mobility of the hydrogen of the OH groups of metal–zeolite catalysts by isotope–deuterium exchange (Table IV). In contrast to catalyzed exchange on NaY and decationized zeolite,

Table IV. Concentration of OH Groups and the Rate of Hydrogen Exchange with Deuterium for Some Zeolite Catalysts (*130*)

Sample	*NaY*[a]	*Decat. Y*[b]	*0.5 % Pt-NaY*[a]	*6 % Ni-NaY*[a]	*9 % Ni-NaY*[a]	*0.5 % Pt-SiO_2-Al_2O_3*[c]
OH-Concentration, mmole/g	0.2–0.4	1.2–1.7	0.2	0.77	1.3	0.9
Temperature, °C	300	300	20	20	20	20
Exchange reaction rate, mole D_2 mole/OH/min	1×10^{-3}	$.8 \times 10^{-3}$	1.2×10^{-3}	4×10^{-5}	2.8×10^{-5}	0.8×10^{-3}

[a] The zeolite catalysts were obtained by ion exchange.
[b] Decationized Y was obtained from NH_4NaY.
[c] The aluminosilicate catalyst was prepared by impregnation and reduction with H_2.

Kinetics and Catalysis

on which the $OH_{zeol} + D_2 \rightarrow OD_{zeol} + HD$ reaction began only at 250°–300°C, on metal-containing catalysts the exchange proceeded at room temperature. The rate of exchange on Pt samples was an order higher than that observed for Ni–NaY. An increase in the rate of hydrogen–deuterium heteroexchange on metal–zeolite catalysts is probably caused by both the activation of D_2 with reduced metals and an increase of its rate of migration along the surface. By isotopic exchange with D_2 it is possible to determine the OH groups near Pt and to calculate the size of the Pt clusters (*48*), but from these data, however, one cannot make any definite conclusion about the effects of metals on the acidity of OH groups. The proton acidity of zeolite catalysts can be determined by potentiometric titration with potassium methylate in anhydrous dimethylformamide (*131*), and with introduction of 0.4 wt % Pt into CaY zeolite, no changes in its acidic properties were observed. Here additional studies are required.

Current ideas on the formation of metal-zeolite catalysts, cation reduction, interaction between different centers, and dependence of catalytic activity on the particle size are far from adequate. Further development of the methods for determining the dispersion of various metals is necessary, particularly, when it varies over a wide range. It is also necessary to devise reliable methods for analyzing alloys on zeolites and to study the effect of various reactive systems on the size of metal crystals and the nature of their active centers, and so forth.

Activation and Modification of Metal-Loaded Zeolite Catalysts. In multicomponent catalysts the catalytic function of zeolites may become apparent in addition to the functions of the other catalytic components. Therefore, although the properties of metal-loaded zeolites depend on the conditions of their preparation, use, and reaction type, catalysts in which zeolites are inert supports should be distinguished from the polyfunctional catalysts in which the function of a support manifests itself. The methods of activation and modification of these two types of catalysts should be different.

With zeolites as inert supports, the treatment of the active component must be carefully controlled. In particular, metal-zeolite catalysts used for hydrogenation or dehydrogenation which are obtained by impregnation or adsorption from a gas phase should be activated under conditions which

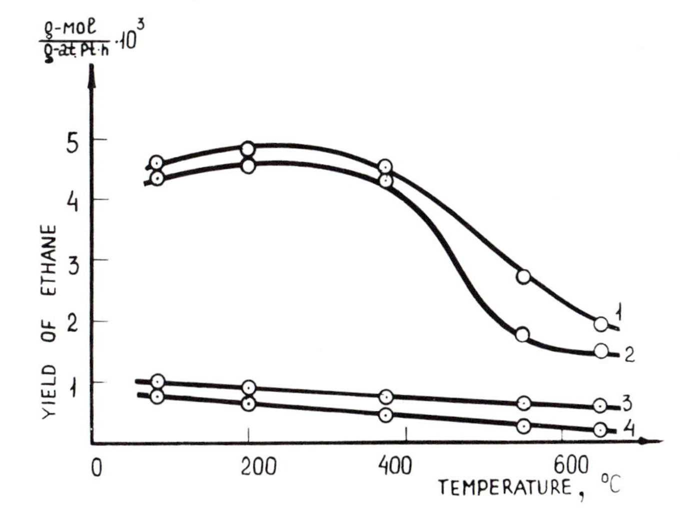

Izvestia Akademia Nauk SSSR

Figure 5. Dependence of the activity of 0.6% Pt–NaY catalyst in ethylene hydrogenation on reduction temperature (135). *Temperature of pretreatment with air, °C: 1 = 250; 2 = 380; 3 = 350; 4 = 650.*

provide fine dispersion of the metal. The samples prepared by ion exchange are treated after reduction of the metal with $NaHCO_3$ solution (*132*) or with an alkali halide salt at 150°–800°C (*133*) to neutralize acidic centers and to increase selectivity and stability. In the second case, deprotonization of zeolites is likely to occur by a solid-phase reaction of the type: OH_{zeol} + Na-Hal → ONa_{zeol} + H-Hal. Rabo *et al.* (*134*) observed that the zeolite structure is retained when the H halide gas escapes.

The activity of metal-loaded zeolite catalysts depends to a large extent on pretreatment and reduction conditions. Minachev *et al.* (*135*) found that variations in the reduction temperature of ion-exchanged type 0.6% Pt–NaY in a range of 200°–400°C does not affect its activity in ethylene hydrogenation (Figure 5). Above 400°C, the C_2H_4 conversion decreases sharply, and at 550°–600°C the catalyst practically loses its activity. Similar dependence is observed for the samples pretreated with air at 250°–380°C. The activity of catalysts heated in air at 550°–650°C is almost independent of the reduction temperature (Curves 3 and 4, Figure 5). These effects are probably caused by variations in the size of Pt particles (*84*). The activity of 0.6 % Pt–NaY in C_2H_4 hydrogenation depends greatly on the temperature of zeolite pretreatment with air (Figure 6). Heating at 450°–600°C leads to less active catalysts irrespective of the temperature of subsequent H_2 reduction. The high temperature treatment with air, supposedly promotes the migration of Pt ions to the structurally hidden S_I sites in the faujasite structure. These atoms remain hidden after reduction and do not participate in catalysis.

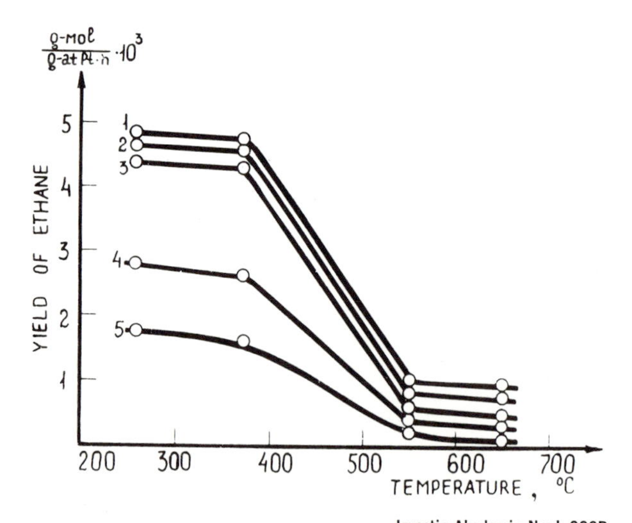

Izvestia Akademia Nauk SSSR

Figure 6. Dependence of the activity of 0.64% Pt–NaY catalyst in ethylene hydrogenation at 30°C (molar ratio H_2:C_2H_4 *= 6) on the temperature of air treatment before reduction* (135). *Reduction temperature, °C: 1 = 200; 2 = 380; 3 = 100; 4 = 550; 5 = 650.*

Similar effects depend on the reaction character and the conditions under which the catalysts are used. In cyclohexane dehydrogenation, the activity of platinum-loaded X-type zeolites is constant in the whole temperature range from 300° to 450°C (Table V) (*94*). Replacement of Na^+ by Ca^{2+} in zeolites did not affect the activity. Neither did variations in the treatment before reduction (samples 3-5). The nature of the Pt compound used in ion exchange and the metal content were of great significance. The catalysts obtained from zeolites containing divalent Pt proved to be less active. The optimal size of metal particles is likely to be different for different reactions.

Table V. Dependence of Activity in Cyclohexane Dehydrogenation on the Composition of Platinum-Zeolite Catalysts, Pt Content, and Preparation (*94*)

($P = 1$ atm, $T = 300°C$, LHSV $= 2.5$ hr^{-1}, $H_2:C_6H_{12} = 5:1$, $SiO_2/Al_2O_3 = 2.2$)

	Catalyst			*Conversion of Cyclohexane (to Benzene), wt % Temperature of Catalyst Reduction with H_2*		
Sample	*Zeolite*	*Pt, wt %*	*Pt Compound Used in Ion Exchange*	*300°C*	*400°C*	*450°C*
1	NaX	0.53	$Pt(NH_3)_6Cl_4$	—	36	37
2	0.74 Ca-NaX	0.53	$Pt(NH_3)_6Cl_4$	36	36	37
3	0.91 Ca-NaX	0.53	$Pt(NH_3)_6Cl_4$	37	38	36
4	0.91 Ca-NaX	0.53	$Pt(NH_3)_6Cl_4$	34	37	35
5	0.91 Ca-NaX	0.53	$Pt(NH_3)_6Cl_4$	34	37	36
6	0.91 Ca-NaX	0.50	$Pt(NH_3)_4Cl_2$	—	22	—
7	0.91 Ca-NaX	1.80	$Pt(NH_3)_6Cl_4$	—	46	48

Treatment of polyfunctional zeolite catalysts should be carried out at the conditions providing optimum activation for all components. The metal-loaded zeolites obtained by ion exchange are dried in N_2, air, or H_2 until the residual water is less than 2 wt % and are then reduced by hydrogen at atmospheric or at higher pressures. Minachev and co-workers (*136*) studied the effect of treatment of a 0.5 % Pd–CaY catalyst on its activity and selectivity in *n*-hexane isomerization and established the optimum activation conditions: air treatment at 380°C for 5 hr and H_2 reduction at atmospheric pressure and at 250°–400°C for 10 hr. At lower temperatures the palladium reduction is not complete, and this decreases dehydrogenation and overall activity of the catalyst. A decrease in the activity after reduction at higher temperatures (400°–550°C) is attributed to a decrease in dispersion of the Pd metal and to variations in zeolite acidity (conversion of part of Brönsted acidity to Lewis type).

The activation conditions can depend also on catalyst preparation. For example, 0.7 % Pt–CaX obtained by impregnation of CaX with alcohol solution of platinum chloride acid displayed maximum activity in *n*-pentane isomerization after hydrogen reduction at 475°–480°C (*137*). Pelleted catalyst of the same type into which Pt was introduced by impregnation or ion exchange should be reduced with H_2 at 370°C (but after heating at 500°C) (*138*). A catalyst with almost atomically dispersed Pt, highly stable to sulfur poisoning, was obtained by heating the cation-exchanged 0.5 % Pt-CaY in air at 500°C with subsequent H_2 reduction at 300°C (*11*).

Metal-loaded rare earth zeolite catalyst used for *n*-alkane isomerization should be dehydrated and reduced at low temperatures (300°C); high temperature treatment results in lower activity (*139*). Activity can be raised by adding isopropyl alcohol to the feed (*n*-pentane) and by treating the contacts with water vapor (*139*). Here the activation conditions affect more the state of the zeolite component of the catalyst than its metallic component. This is also evidenced by the fact that the reduction of rare earth-exchanged zeolites at high temperature, after treatment with water vapor, does not affect their activity (*140*).

Activation of metal–zeolite catalysts used for hydrocracking is carried out under conditions similar to those used for isomerization. The activity and selectivity of these catalysts increase if they are exposed to H_2S before reduction (*141*). Similar results are obtained by adding halogen compounds, such as alkyl- and aryl chlorides or perchlorhydrocarbons, to the feed (*142*). Metal-zeolite catalysts in hydrocarbon reactions show improved activity and selectivity after treatment with Lewis bases (N- and P-containing compounds) (*143*).

After treatment with ammonia, the activity of a platinum–zeolite catalyst in aromatization of *n*-alkanes increases by 40%, while in isomerization it increases only by 20%. Modification of some active centers which, on the whole, are not uniform in zeolite catalysts cause this effect. With preliminary controlled coking to deactivate their excessively active centers, an increase in the selectivity of the catalysts and a longer operational period results. For example, by precoking (>2 wt % coke) the hydrocracking catalyst, its operation time at a constant temperature was extended for more than four months (*144*). Thus, pretreatment of metal–zeolite catalysts significantly affects their properties and operation conditions and provides maximum efficiency in their use.

Stability of Metal-Loaded Zeolite Catalysts and Their Regeneration. Metal-zeolite catalysts are usually used in reactions at elevated pressure, and under these conditions, they display high stability. For example, the activity of MB-5390 (SK-100) catalyst on Y-type zeolite in *n*-alkane isomerization did not decrease after 2000 hr of operation (*8*). The Pt-CaHY when tested in *n*-hexane isomerization for more than 2500 hr did not show any changes in performance (*145*). The palladium-zeolite catalyst operated for more than two years in hydrocracking without regeneration (*146*). The behavior of these catalysts depends on the sulfur content of the feed (*8*).

The Pt-CaY catalyst (0.5 %) with Pt introduced as $/Pt(NH_3)_4/^{2+}$ cation by ion exchange shows higher stability to sulfur poisoning than the sample of the same composition obtained by impregnation with H_2PtCl_6 solution (*11*). The activity and selectivity of the former in *n*-hexane isomerization with 10^{-3}% thiophene added to the hydrocarbon were not affected, whereas the activity of the latter, starting with sulfur in the feed, gradually decreased, and in 24 hr the content of the unreacted *n*-hexane in the products increased by 50% and the yield of 2,2-dimethylbutane was less than 50% of the initial value. With both catalysts the deactivation by sulfur was reversible, and after sulfur was removed from the feed, the initial activity was recovered. Irreversible poisoning occurred when the sulfur content was 0.1 % and higher. These results can be accounted for by a fine dispersion of the Pt in the ion-exchanged type catalyst (*11*).

Metal-zeolite catalysts used in hydrocracking are stable to nitrogen and sulfur compounds up to 0.3 and 5%, respectively, but these impurities should be periodically removed (*147*).

The platinum–zeolite catalysts used in reforming are also less sensitive to nitrogen compounds in the feed than non-zeolite type catalysts. For example, they were not affected by adding up to 0.02 % nitrogen (as quinoline) to the feed for 150 hr (*14*). Under the same conditions the activity of common reforming catalysts (Pt/Al_2O_3) decreased (the O.N. of the product decreased from 99 to 92 in 25 hr).

The effect of CS_2 and *n*-butylamine on the properties of 0.5% Pd-CaY($SiO_2/Al_2O_3 = 4.5$) and 0.5% Pd/γ-Al_2O_3 catalysts in cyclohexane dehydroisomerization and *n*-hexane isomerization was studied (*148*). With a sulfur content of 0.2 mole % in the hydrocarbon, the dehydrogenating activity of both catalysts decreased. It did not affect the isomerization properties of Pd/Al_2O_3 but it sharply decreased the isomerization activity of the zeolite catalyst. The nitrogen compound did not affect the dehydrogenating properties of the catalysts studied (at the concentration of 0.2 mol % in hydrocarbon), but it suppressed the isomerization activity of Pd/Al_2O_3 and did not affect that of Pd/CaY. Selective poisoning of the metal centers by sulfur compounds and the acidic centers by nitrogen compounds is well known, and the behavior of the Pd/Al_2O_3 catalyst with respect to these compounds is quite clear. The unusual properties of Pd–zeolite catalyst can be explained by assuming that its isomerization activity is caused by nonreduced Pd cations, localized in the zeolite lattice, which are not affected by nitrogen compounds (*148*). However, this assumption does not seem to be convincing. More likely, *n*-butylamine is adsorbed largely on Ca^{2+} ions and thus cannot poison the acidic centers of zeolites. It is therefore necessary to investigate the behavior of the ion-exchanged catalysts obtained from the decationized Y zeolite. Cr_2O_3 introduced into the Ni-NaY catalyst stabilized the metal dispersion and incresaed its stability to poisoning with sulfur compounds (*89*).

In general, the stability of metal-zeolite catalysts depends not only on their composition, preparation, activation, and modification, but also on the feed and the conditions of use. The activity of most catalysts decreases with time, usually because of coke formation. Coke formation on multicomponent zeolite catalysts may be caused by many factors which have not been studied in detail.

To regenerate metal-zeolite catalysts one can use the methods usually applied to other supported metal catalysts. Care should be taken to retain the crystalline structure of the zeolite and to prevent catalyst deactivation arising from migration of cations, agglomeration of metal crystals, or excessive dehydration. Various methods of regeneration have been reported in the patent literature, but they are not always general. Some illustrations of restoring activity are given below.

The coked Y-type catalysts containing 0.5 wt % Pt or Pd used for *n*-alkane isomerization should be reactivated by burning off the coke deposits followed by cooling at a temperature below 315°C, by partial hydration, homogenous over the whole catalyst, and, finally by slow heating in H_2 to 450°C until the activity is completely recovered (*149*). In this process uniform distribution of the water in the catalyst is important. For this purpose three methods can be used: (1) after cooling to room temperature the catalyst is kept in air until 4–10% H_2O is adsorbed; then it is heated to ~80°C to accelerate the diffusion of H_2O molecules; (2) as in method 1, except that instead of heating, the catalyst is placed into a closed vessel until equilibrium is reached; (3) an inert gas containing water vapor is passed through the catalyst to introduce 5–10% of H_2O. The activity of the catalyst for hydrocracking of a hydrocarbon feed with high nitrogen content can be restored by treating with hydrogen at a temperature which is 40° higher than that of hydrocracking, pressure 1–70 atm, H_2 space velocity 50–2000 $nm^3/hr/m^3$ catalyst for 0.25–16 hr (*150*). Noble metal-loaded faujasite-type catalysts used in hydrocracking show a decrease in the activity and stability because of agglomeration of metal crystals following regeneration by oxygen-containing gases, with the partial pressure of water vapor exceeding 0.035 atm (*151*). The most active catalysts are those with metal crystals of ~20 A. If upon oxidative regeneration the size of crystals exceeds 30 A, the catalyst is treated with H_2S, single or in mixture with hydrocarbons and inert gases, at 260°–426°C for 0.25–1 hr, then it is oxidized again at 426°–493°C by the gas with partial water pressure of 0.5–13 torr. H_2 treatment at 371°–537°C recovers the initial activity of the catalyst (*151*).

An A-type zeolite-based catalyst used in hydrocarbon reaction following deactivation by coke formation and poisoning by sulfur compounds can be regenerated by treating with a gas (H_2,N_2) containing halogen compounds (HCl, CH_3Cl, CCl_4, *tert*-C_4H_9Cl), by burning out the coke with an O_2-containing gas, and a subsequent H_2 reduction (*152*). Treatment with halogen completely removes the sulfur and prevents the pores of the zeolite from being blocked with SO_4^{2-} ions which can be formed during oxidative regeneration.

Catalysis by Monofunctional Metal-Loaded Zeolite Catalysts

Under certain conditions metal-containing zeolites can act as monofunctional catalysts, and the zeolite can be considered as inert support for the active metals. However, since the state of the metal component depends on the composition and structure of the zeolite crystal, the properties of these catalysts significantly depend on the support used, as shown above. The results obtained from hydrogenation, dehydrogenation, hydroalkylation, hydrogenolysis, oxidation, and some other reactions are reported below.

Hydrogenation Reactions. The hydrogenating properties of zeolites containing various metals have widely been studied in reactions of olefins (*9, 10, 35, 48, 53, 69, 85, 88, 103, 105, 135, 153, 154, 155, 156, 157*), cyclenes (*157, 158*), aromatic hydrocarbons (*14, 35, 52, 54, 56, 82, 98, 100, 108, 123, 124, 125, 156, 157, 159, 160, 161, 162, 163*), and oxygen-containing compounds (*164, 165, 166, 167, 168, 169*). The activity of X- and Y-type zeolites containing group VIII metals in ethylene hydrogenation at 30°C is almost the same with every metal of the group (*135*). The activity of Pt–Na–mordenite is one order of magnitude lower than that of Pt–NaY because the size of the mordenite channel (window ~5 A) is close to the size of a C_2H_4 molecule (4.4 A), resulting in hindered diffusion. The ion-exchanged Pt–NaY catalyst is more active than the sample of the same composition obtained through H_2PtCl_6 impregnation. Similar results were observed for Pd and Ni zeolites. In hydrogenation the metal catalyst activity showed the following order: Pt–NaY > Pd–NaY ≫ Ni–NaY; the platinum catalyst is twice as active as the palladium catalyst and the nickel catalyst shows low activity in ethylene hydrogenation at 30°C (*135*). The effect of pretreatment and reduction on the properties of Pt–NaY catalysts has been studied in detail (*135*). The results obtained have been considered above (*see* Figures 5 and 6). The 0.62 CoNaX zeolite reduced with H_2 at 400°C hydrogenated ethylene by 17–82% at 50°–130°C and atmospheric pressure (*155*). The degree of metal dispersion greatly affects the properties of zeolite catalysts in hydrogenation. For example, 0.19% Ni–Na–mordenite containing highly dispersed Ni showed low activity in butene hydrogenation at 100°C. It exhibited considerable activity at a temperature as high as 114°C (*88*). Note that nickel–zeolite catalysts, as a rule, show lower selectivity in olefin reactions than other metal–zeolite catalysts. In addition to hydrogenation they often accelerate the migration of the double bond, cis-trans isomerization, and skeletal isomerization of the initial hydrocarbons, as observed for *n*-butene (*88*).

A characteristic feature of zeolite catalysts is their ability to convert selectively a certain type of molecule from a mixture. Weisz *et al.* (*9, 10, 69, 153*) have shown that it is possible to hydrogenate ethylene selectively in the presence of propylene or propylene in the presence of isobutylene or piperylene in the presence of isoprene. For example, the only hydrogenation product in a C_3H_6 + iso-C_4H_8 equimolar mixture over 0.31% Pt/CaA catalyst at 343°C was propane (yield ~ 70%), though both olefins have nearly the same reactivity for hydrogenation (*10*). *n*-Olefins (propylene, l-butene) could be hydrogenated over the catalyst mentioned above at a temperature as low as 25°C. The active Pt centers are not accessible for isoolefins, and hence they are not hydrogenated.

The removal of piperylene from isoprene by selective hydrogenation is of great practical interest and can be done by using Pt–CaA catalyst (*153*). For this purpose a reduced Ni-NaA zeolite has also been prepared (*170*).

The properties of nickel-zeolite catalysts have been studied (*157*) in the hydrogenation of unsaturated hydrocarbons. The 3.8% Ni-NaY sample obtained by cation exchange displayed high selectivity (the degree of *n*-octene conversion was 81.5%) whereas isoolefins (isoamylene, diisoamylene) and benzene were practically unreactive because the latter cannot penetrate the zeolite cavities, where most of the nickel is located. On the contrary, the 5.8% Ni–NaA catalyst, prepared by mixing NaA zeolite aqueous suspension with neutral nickel carbonate and reduced with H_2, hydrogenated benzene by 66.6% at 180°C. The conversion of *n*-butyl- and *sec*-butylbenzenes was much lower (30–36%). Cyclohexene was not hydrogenated over this catalyst, but it underwent irreversible catalysis (disproportionation to benzene and cyclohexane) by 87% (*157*).

The properties of metal-loaded zeolites were investigated in detail in benzene hydrogenation. This reaction is of important technical significance and is a convenient model for studying the dependence of catalyst activity on the metal, support type, zeolite composition, size of metal crystallites, interaction between metal and zeolites, and so forth.

Table VI summarizes the results obtained by Minachev and co-workers (*158, 160, 161, 162, 163*). The most active is the Rh catalyst; next are iridium, platinum, and palladium catalysts, and finally nickel and rhenium. An increase of SiO_2/Al_2O_3 ratio in zeolites of faujasite type does not affect significantly the activity of the catalysts. The nature of cations in the zeolite (Na^+ or Ca^{2+}, containing Pd) does not affect it to any extent either. Similar results have been obtained with Y-type nickel zeolites when the extent of Ni^{2+} reduction to metal was taken into account (*52*). At small benzene conversion, the nature of the cation strongly affects the activity of palladium

Table VI. Hydrogenation of Benzene on Metal-Zeolite Catalysts

(200°C, 30 atm, LHSV = 0.5 hr^{-1}, H_2:C_6H_6 = 5:1)

Catalyst	*SiO_2/Al_2O_3 Ratio*	*Product Composition, wt %*	
		Cyclohexane	*Benzene*
0.5 % Pd-NaX	2.5	88.8	11.2
0.5 % Pd-CaX		86.0	14.0
0.5 % Rh-CaY		97.8	2.2
0.5 % Ir-CaY		91.0	9.0
0.5 % Pd-CaY	3.4	82.5	17.5
0.5 % Pt-CaY		84.6	15.4
0.5 % Pd-NaY		85.3	14.7
0.5 % Pd-NaY	4.5	87.6	12.4
0.5 % Pd-CaY		80.2	19.8
5 % Ni-NaY	4.2	98.0	1.0[a]
5 % Re-NaX	2.5	73.3	26.7
5 % Re-NaY[b]	5.0	92.4	7.6
5 % Re-Na-mordenite	10	77.3	22.7

[a] 1 % Methylcyclopentane.
[b] 210°C, LHSV = 0.4 hr^{-1}, H_2:C_6H_6 = 15:1.

Table VII. Catalytic Activity of Palladium Zeolites in Benzene Hydrogenation (*98*)

(140°C, $P_{C_6H_6} = 56$ torr, $P_{H_2} = 704$ torr)

Sample	*Catalyst*	N (*turnover number, mmoles*)	*Activation Energy,* E_a, *kcal/mole*
1	2 % Pd-NaX	70	11
2	0.2 % Pd-HY	240	—
3	0.62 % Pd-HY	240	8.5
4	1.2 % Pd-HY	240	8.5
5	1.86 % Pd-0.70 LaNaY	275	9
6	2 % Pd-0.15 CeNaY	220	8.5
7	1.77 % Pd-0.70 Ce-NaY	280	10
8	1.84 % Pd-0.58 MgNaY	200	10
9	1.9 % Pd-0.70 CaNaY	146	9
10	1.7 % Pd-NaY	118	9
11	0.9 % Pd-SiO_2	56	—
12	1.4 % Pd-MgO	80	—
13	0.5 % Pd-Al_2O_3	70	—
14	1.3 % Pd-SiO_2-Al_2O_3	245	—

Advances in Chemistry Series

catalysts (*98*). As seen from Table VII, the activities of palladium zeolites are: NaX $<$ NaY $<$ CaY $<$ MgY $<$ CeY $\sim$ HY $\sim$ LaY. Their acidic properties increase in the same sequence. The activity of Pd (in terms of millimoles of benzene which reacted per second per surface palladium atom) deposited on supports of high acidity (HY, LaY, CeY, aluminosilicate with 13 wt % Al_2O_3) is about four times as high as that of Pd/Al_2O_3, Pd/SiO_2, Pd/MgO or Pd/NaX catalysts. These results are explained (*98*) by the interaction of Pd with the support which modifies Pd so that the electron configuration of the metal becomes similar to that of rhodium, and the catalysts display higher activity. (Rhodium and ruthenium catalysts are more active than palladium in benzene hydrogenation (*171*)).

The conclusion that in hydrogenation under mild conditions the cation affects the properties of palladium–zeolite catalyst is confirmed by the data obtained from diethylbenzene hydrogenation (*14*). Table VIII shows that, at 31 atm over 0.5 % Pd–NaY and 0.5 % Pd–CaY, diethylbenzene is completely hydrogenated. At 19 atm and lower, a CaY-type sample was even more active. The authors (*14*) emphasize high selectivity and stability of the catalysts.

The specificity of nickel-loaded zeolites as hydrogenating catalysts was noted above. The dependence of the activity of the samples obtained from Ni–NaA and Ni–NaX zeolites on the size of Ni particles in benzene hydrogenation is given in Figure 4. Such catalysts show high stability to sulfur poisons: in the hydrogenation of acetylene to ethane in the presence of H_2S, the activity of catalysts containing Ni (or Pd) had not decreased after several hours (*14*).

Table VIII. Hydrogenation of Aromatic Hydrocarbons on Metal-Zeolite Catalysts (*14*)

Hydrocarbon	*Diethylbenzene*						*Benzene*
Catalyst	*0.5 % Pd-NaY*			*0.5 % Pd-CaY*			*0.5 % Rh-NaX*
Temperature, °C	215	215	215	215	215	215	210
LHSV, hr^{-1}	1.5	1.5	1.5	1.5	1.5	1.5	2.0
Mole ratio, H_2:CH	16	16	16	16	16	16	5
Pressure, atm	31	19	8	31	19	4.5	1
Conversion, %	99	88	40	99	99	60	10

Oil and Gas Journal

The reaction between mesitylene and H_2 over platinum–zeolite catalysts has been studied (*125*). At atmospheric pressure and 200°C, 0.53 % Pt–CaX was not active. In this catalyst all the Pt is apparently in zeolite cavities which mesitylene cannot penetrate. With 1.83% Pt–CaX, the hydrocarbon conversion is 10 wt %. The hydrogenation probably proceeds over Pt particles located on the surface of zeolite crystals. In the hydrogenation of oils of high molecular weight the metal–zeolite catalysts show higher selectivity than other hydrogenating contacts (*172*).

Lately several studies have been made on hydrogenation of various oxygen-containing organic compounds over zeolite catalysts. On a 5.8% Ni–NaX zeolite reduced with H_2 at 500°C, at atmospheric pressure, 170°C, and LHSV of 0.5 h^{-1}, furan yielded 47.7 % of tetrahydrofuran; simultaneously 15.4% of l-butanol was obtained (*164*). At 130°C the yields of these products were 37.8 and 10.2 %, respectively. The catalyst was losing its activity rapidly, but after air regeneration and H_2 reduction, its activity returned to its initial value. On nickel-containing zeolites of X- and Y-type, methyl ethyl ketone was hydrogenated to *sec*-butyl alcohol, which underwent considerable dehydration at a temperature as low as 100°C (*166*). Areshidze *et al.* (*165*) showed that cyclohexane can be obtained by phenol hydrogenation at 135°C and atmospheric pressure over 0.8 % Pd–NaX catalyst with 90% yield. Zeolites CaA and CaX containing Pd displayed high activity

Table IX. Dehydrogenation of

($P = 1$ atm, $T = 500$°C, LHSV $= 0.5$ hr^{-1})

Catalyst	*Yield of Liquid Product, wt %*	*Olefin Content of Product, wt %*	*Conversion of Hexane, wt %*	*Yield of Olefins, wt %* — *Per Supplied Hexane*	*Yield of Olefins, wt %* — *Per Consumed Hexane*
CaA	80	5.0	24.0	4.0	16.7
1 % Pt-CaA	61	5.0	38.0	3.0	7.9
5 % Ni-CaA	67	9.2	36.2	6.2	17.0
15 % Ni-CaA	0	0	100	0	0

in the hydrogenation of the dimethylacetylenylcarbinol triple bond at 20°C (*168*). The same catalysts were investigated in hydrogenation of 2-methyl-4-acetylenyl-4-hydroxydecahydrochinoline stereoisomers (*167*). The prevailing reaction product obtained at atmospheric pressure and 20°C was alcohol with double bond. The most active catalyst was 5% Pd/CaX. 1% Pd/CaA catalysts proved to be unsuitable for the hydrogenation of acetylenic alcohols of the decahydrochinoline series. The properties of metal–zeolite catalysts in the hydrogenation of other organic compounds have not been extensively studied.

Dehydrogenation Reactions. Gutyrya and Galich *et al.* (*173, 174, 175*) were the first to study zeolite catalysts in *n*-paraffin dehydrogenation. An attempt was made to produce olefins selectively from *n*-hexane and *n*-heptane without aromatic hydrocarbons. However, the X and A zeolites showed low activity and selectivity in C_6 and C_7 *n*-alkane dehydrogenation. As seen from Table IX, over 1 % Pt–CaA and 5 % Ni–CaA catalysts, the olefins yield did not exceed 6 wt % of the *n*-hexane feed or 17 % of the consumed *n*-hexane. The catalyst containing 15 wt % Ni converted *n*-hexane completely to gas and coke. The gas consisted of H_2 and low molecular paraffinic hydrocarbons. A decrease of the reaction temperature affected the extent of the decomposition of *n*-hexane somewhat, but the yield of olefins was low. There were no aromatic hydrocarbons in the products. Aromatics were probably formed in the CaA zeolite cavities; however, could not leave them because their sizes exceeded the size of the pores of the zeolite crystal, and they underwent further conversion with the formation of condensation products and coke (*174, 175*).

Later the dehydrogenating properties of metal-loaded zeolites were studied in the reactions of various hydrocarbons (*53, 94, 97, 106, 108, 124, 125, 132, 133, 176, 177, 178, 179, 180, 181, 182*) and petroleum fractions (*183*). Some of the results are listed in Table X (*see* also Table V). A- and X-type zeolites (and less frequently those of type Y) containing metals of group VIII were used as catalysts. Recently a platinum-mordenite catalyst was proposed for dehydrogenation of *n*-alkanes. The overall yield of aromatic

***n*-Hexane on Zeolite Catalysts (*174*)**

Composition of Gas, vol %		
H_2	C_nH_{2n}	C_nH_{2n+2}
8.0	30.0	62.0
49.6	9.1	41.3
21.4	12.0	66.6
21.0	0	79.0

"Neftekhimlya"

Table X. Dehydrogenation of Hydrocarbons on Metal-Loaded Zeolite Catalysts

Hydro-carbon	*Catalyst*	*Reaction Conditions*	*Yield of Products (Conversion)*	*Refer-ence*
Cyclo-hexane	4.3 % Ni-NaA 6.2 % Ni-NaX	1 atm, 300°C, H_2 flow	benzene 29.6 % benzene 34.2 %	*176*
	3.6 % Ni-NaA	1 atm,	96.5 $\frac{\text{mmoles } C_6H_{12}}{\text{g Ni hour}}$	*108*
	6.7 % Ni-NaA	300°C,	76.4	
	10.6 % Ni-NaA	LHSV =	69.0	
	3.4 % Ni-NaX	1.4 hr^{-1},	24.0	
	7.0 % Ni-NaX	H_2:C_6H_{12} =	18.8	
	9.8 % Ni-NaX	5:1	11.5	
Cyclo-hexane	1 % Pd-NaX	1 atm, LHSV = 0.5 hr/1	240°C benzene 47 % 260°C benzene 92 % 280°C benzene 100 %	*97*
	0.24 % Pt-NaX	1 atm, 300°C, a pulse method	benzene 40 %	*177*
	0.45 % Pt-NaX	1 atm, 375°C	benzene 78 %	*178*
Tetralin	0.5 % Os-NaX	1 atm, 207°C	naphthalene 100 %	*179*
n-Hexane	0.5 % Pt-CaA	1 atm, LHSV = 0.6 hr^{-1}, H_2:C_6H_{14} = 10:1	*n*-hexenes 9.8 % benzene 4.3 % dienes + cycloalkanes 1.9 % alkanes + alkenes C_1–C_4 8.1 %	*132*
n-Decane	0.9 % Pt-NaX	1.055 atm, 460°C, LHSV = 2 hr^{-1}, H_2:$C_{10}H_{22}$ = 5.5:1	*n*-decene 12.9 %	*180*

hydrocarbons and coke from *n*-dodecane did not exceed 2 %, the conversion of *n*-$C_{12}H_{26}$ being 16.8% and the yield of C_{12} olefins being 14.2 wt % (*181*).

The dehydrogenation of *n*-paraffins C_3–C_5 over platinum-loaded X zeolite yields considerable benzene (*182*). The activity of 0.24 % Pt-NaY catalyst depends on pretreatment conditions and on the size of Pt particles. The catalyst shows higher activity after heating in air at 300°C and after H_2 reduction at 500°C. The samples, with Pt particles smaller than 20 A, showed low activity in this reaction (*177*).

The patent literature indicates that X zeolites containing metallic Co, Sn, Ag, Au, and Cu can be used for dehydrogenation of hydrocarbons (*184*), but there is no information in the literature about the application of metal–zeolite catalysts for dehydrogenation of organic compounds of other classes.

Oxidation Reactions. In 1961, it was shown that ethylene can be oxidized to ethylene oxide over zeolites containing Cu, Ag, or Au (*37*). More recently a paper (*185*) reporting the properties of zeolites A and X, loaded with silver, in the above reactions has been published. At atmospheric pressure at 200°–300°C, contact time 0.15–1 sec with a molar ratio of $O_2 : C_2H_4 = 1$ and 0.05, the zeolites prepared by ion exchange were active in oxidation but did not show selectivity in the formation of ethylene oxide. On the contrary, the zeolite samples prepared by impregnation showed both high activity and high selectivity. Their selectivity followed the order: Ag°–CaA > Ag°–KA > Ag°–NaA ≫ Ag°–NaX. The activity and selectivity of the catalysts increased linearly with an increase of Ag content in the zeolite (for CaA zeolite, the selectivity for ethylene oxide formation varied exponentially with increasing metal content). With 30 % Ag°–CaA catalyst the C_2H_4 conversion and the yield of ethylene oxide were comparable with other well-known catalysts used for ethylene oxidation. For example, at 250°C, contact times of 0.33 sec., and O_2 partial pressure of 0.7 atm, the C_2H_4 conversion during one run was 25% and the selectivity towards ethylene oxide was 70 % (*185*). These results were obtained with a catalyst prepared in the following manner (*185*): dehydrated crystals of the zeolite were impregnated with aqueous $AgNO_3$ of required volume, dried *in vacuo* at 110° (3 hr), 180° (3 hr) and 400°C (1 hr), activated in air at 400°C, and finally, reduced at 250°C, with H_2–N_2 ($P_{H_2} = 0.07$ atm), for 2 hr. The crystalline structure of the zeolites was only slightly affected whereas during silver introduction by ion exchange it was almost completely lost. Because of their lack of crystallinity and high acidity, the selectivity of the cation-exchanged zeolites decreased sharply: the prevailing products of C_2H_4 oxidation were CO_2 and H_2O. The size of silver crystallites was 200–800 A for impregnated samples, and 150–600 A for ion-exchanged ones (*185*). The dependence of the activity and selectivity of the catalysts on the dispersion of the metal and acidity of the zeolite has not been investigated.

The results of these studies may stimulate new investigations into the properties and specificity of metal zeolite catalysts in various oxidation reactions, but so far, the amount of information in this area is limited. A study was made of the kinetics of CH_4 oxidation to CO_2 and H_2O over palladium zeolites of type X obtained (1) by impregnation of the NaX crystals with a solution of ammonium chloropalladate and subsequent heating in air at 400°C and reduction at 300°C; (2) by ion exchange of NaXwith tetraaminopalladium (*186*). The reaction is first order in CH_4 for each catalyst (as well as for the bulk palladium). For O_2, the reaction order is 0.3–0.4 and 0.05, respectively. The estimated activation energy was 50 and 88 kcal/mole, respectively. Thus, the properties of metallic Pd and those of palladium cations in CH_4 oxidation are considerably different. CH_4 and O_2 may be chemisorbed simultaneously on certain Pd ions (*186*).

Weisz (*69, 153*) demonstrated that Pt-CaA catalyst showed high selectivity in the oxidation of *n*-paraffins and *n*-olefins from their mixture containing isoparaffins and isoolefins. For example, 97 % *n*-butane and 0.1% isobutane mixed with 59.9 % n-C_4H_{10} + 40.1 % i-C_4H_{10} were oxidized at 315°C to give CO_2 and H_2O, whereas upon the conversion of the mixture containing propylene (31.2 %), 2-butene (35.2 %), and isobutene (33.6 %), only 1 % of iso-C_4H_8 and 98.6 % of *n*-olefins entered the reaction (*153*). The 0.009 % Pt–NaA catalyst showed high activity in CO oxidation. It was prepared by growing NaA zeolite crystals in the presence of Pt–aminochloride (*69*). After heating in air at 480°C for 1 hr and at 700°C for 0.5 hr, air containing 2 mole % CO was passed through the catalyst. At 1 atm and contact time of 0.04 sec the CO conversion was 86 % at 427°, 97 % at 480°, and 100 % at 540°C. Under the same conditions the conversion of *n*-butane was 2.4–5.5%.

Hydrodealkylation Reactions. Toluene, ethylbenzene, and cumene were hydrodealkylated over 5 % Ni–CaA catalyst prepared by impregnation of CaA zeolite with 0.5 *M* solution of $Ni(NO_3)_2$ followed by H_2 reduction at 350°C for 3 hr and at 500°C for 3 hr. At 420°–460°C, with a H_2: hydrocarbon ratio of 5.1 and pressure of 10 atm, the reaction preceeded without formation of high boiling products. The gaseous products consisted largely of CH_4, indicating that the dealkylation proceeded through a successive splitting of CH_3 groups. The readiness of demethylation decreased in the series: cumene > ethylbenzene > toluene.

Metal-loaded faujasite type catalysts show high activity in toluene hydrodemethylation (*23, 24, 177, 189*) and in the dealkylation of aromatics contained in light petroleum distillates (*190*). For example, 39.6 mole % C_6H_6 was obtained over 0.5 % Pt–NaY at 550°C at 3.4 atm at a ratio of H_2: $C_6H_5CH_3$ = 10: 1 and at LHSV = 1 hr^{-1} (*23*). The 0.5 % Cu–NaY showed higher activity and was three times as selective as Pt–NaY. With metal-containing decationized or multivalent cation Y zeolites, demethylation was accompanied by considerable disproportionation of toluene to benzene, xylenes, and trimethylbenzenes (depending on the conditions) (*189*).

In hydrodealkylation the conditions of pretreatment affect appreciably the activity and selectivity of metal-loaded zeolite catalysts. The 0.24% Pt–NaY obtained by ion-exchange exhibits the highest activity in toluene hydrodemethylation if it is heated in air at 300°C before hydrogen reduction (*177*). With 0.2 % Pt–CaY, the maximum activity was observed upon heat treatment in air at 200°C; with Pt–NH_4Y, the optimum temperature was 600°C (*177*). This is caused by variations in the size of Pt particles and in the composition of the catalysts. With NH_4Y, it is caused by dehydroxylation.

Ryashentzeva and Minachev (*24*) studied the demethylating properties of rhenium catalysts. As seen in Table XI, the 1% Re–CaA sample shows low activity, probably because of inaccessibility of the metallic centers inside the cavities of zeolite crystals for toluene molecules. In contrast, the rhenium–faujasite catalysts show high activity and selectivity to toluene hydrodemethylation. Over 5 % Re–Na–faujasite, the yield of benzene was

Table XI. Hydrodemethylation of Toluene on Rhenium-Zeolite Catalysts (*24*)

($T = 430°$–$450°C$, $P = 5$ atm, $H_2{:}C_6H_5CH_3 = 5{:}1$, WHSV = 0.5–0.6 g/g/hr)

Catalyst	*Operation Time, hr*	*Yield of Liquid Product wt %*	*Yield of Benzene, wt %*	
			from Supplied Toluene	*from Consumed Toluene*
5 % Re–γ-Al_2O_3	5	46.6	11.3	17.6
	8	70.0	14.0	36.8
	19	87.0	21.4	58.0
	43	71.5	17.7	39.0
1 % Re-CaA	3	74.2	1.6	5.8
1 % Re-Na-faujasite	3	59.0	5.5	12.0
	6	71.0	16.0	37.4
	9	76.2	15.0	41.5
	12	85.0	14.6	53.5
5 % Re-Na-faujasite	4	76.4	10.3	31.9
	30	84.7	22.0	64.7
	43	88.3	20.0	86.0

Neftekhimia

20 % with 86 % of the toluene consumed. The selectivity of this catalyst is higher than that of 5% Re/γ-Al_2O_3 (Table XI). In this reaction, rhenium acts like nickel. In general, rhenium contacts are inferior to Cu, Pt, Pd zeolites. The study of bi(poly)metallic zeolite catalysts in dealkylation of alkylaromatic compounds would be of interest. Mordenite catalysts containing the metals of group VIII and Ib of the periodic system, in particular, Co°–H–mordenite (*191*) can be used.

Other Reactions. Metal-loaded zeolites can accelerate other reactions as well. Hall *et al.* (*29*) demonstrated the efficiency of the SK-400 (1 % Ni-NaY and AW-300) catalysts in producing isoprene by decomposition of 3-chloro-3-methyl butyl ether (Table XII). The latter is formed in one of the stages of a new synthetic route to isoprene from formaldehyde and isobutylene according to the scheme:

$$HCl + CH_3OH + CH_2O \longrightarrow H_2O + \underset{\text{I}}{CH_3O{-}CH_2Cl} \xrightarrow[TiCl_4]{+\,CH_2{=}C(CH_3)_2}$$

$$\underset{\text{II}}{H_3C{-}\underset{\displaystyle Cl}{\underset{|}{C}}(CH_3){-}CH_2CH_2{-}O{-}CH_3} \xrightarrow[\text{AW-300}]{\text{SK-400}}$$

$$CH_3OH + HCl + CH_2{=}\underset{\displaystyle CH_3}{\underset{|}{C}}{-}CH{=}CH_2$$

Table XII. Synthesis of Isoprene by Pyrolysis of 3-Chloro-3-methyl Butyl Methyl Ether over Molecular Sieve Catalysts

			Yield, mole %				*Heavy Oil*
Catalyst	*Temp., °C*	*WHSV, g/g/hr*	*Iso-prene*	*Meth-anol*	*Methyl Chloride*	*HCl*	*Yield, wt %*
SK-400 (1 % Ni-NaY)	200	0.81	83.5	21.2	72.4	not deter-mined	3.6
	250	0.77	82.7	14.1	71.6	17.6	3.6
	250	0.90	82.0	6.2	83.5	not deter-mined	4.4
	250	1.13	85.1	15.9	73.7	not deter-mined	3.2
	300	0.87	76.6	25.2	52.8	40.7	4.1
	300	0.88	78.3	2.9	64.0	not deter-mined	4.9
AW-300	250	0.67	64.5	56.8	34.0	55.4	13.5
	300	0.65	71.8	66.9	31.0	59.0	7.9
	350	0.62	65.8	32.3	50.6	37.1	6.8

The interaction of CH_3OH and HCl with formaldehyde, gives chloromethyl ether (I) which adds to isobutylene in the presence of $TiCl_4$ to give II. The pyrolysis of the latter produces isoprene, CH_3OH, and HCl which are recirculated. SK-400 proved to be the most efficient catalyst for decomposition of II. Over this catalyst, the yield of isoprene under the conditions used was 85 %, and no decrease of activity was observed in 40 hr of its operation (*28, 29*). The pyrolysis carried out upon dilution of II with water vapor resulted in an increase both in selectivity and stability of the catalysts. Other polyenic hydrocarbons can be obtained in a similar way (*192*).

Nickel and cobalt containing Y zeolites and mordenite (Zeolon) proved to be efficient in the conversion of *n*-hexane with water vapor (*30*):

$$C_6H_{14} + 12\ H_2O \rightarrow 6\ CO_2 + 19\ H_2$$

Upon cracking at 400°–500°C, the formation of C_2–C_5 hydrocarbons was small ($<$ 10%), and the CH_4 content in the reaction products did not exceed 5 %. The 3.24 % Ni-Zeolon and 0.76 % Ni-NaY catalyst were more active than the commercial G 56 (Ni/Al_2O_3), which contained more nickel ($>$ 15 wt %). The reaction rate per gram of metal was 8–30 times as high as that for G-56. This is usually explained by a highly dispersed state of Ni in zeolites (*30*). The conversion of hydrocarbons with water vapor into a gas rich in H_2 can be carried out over rare earth zeolites of type X and Y containing Ni, Pt, and Pd (*193*).

The properties of platinum-zeolite catalysts in the decomposition of H_2O_2 have been determined at 20°–40°C (*194, 195*). The activity of the ion-exchanged Pt–NaY catalyst was lower than that of impregnated ones. The activation energy (E_a = 13.7 cal/mole), did not depend on catalyst preparation. In Pt–CaY prepared by impregnation, the E_a was lower than that for analogous Pt–NaY catalysts with an equal amount of Pt (*195*). In general, the Pt activity in zeolite catalysts is 5–10 times lower than that of the Pt on SiO_2 or Al_2O_3.

Metal-loaded zeolites are active in the reactions of hydrogenolysis. Ni–NaY was used in the hydrogenolysis of ethane (*52, 89*) and *n*-hexane (*52, 73*). Pt–CaY was employed in the hydrogenolysis of neopentane (*48*) and methylcyclopentane (*196*). Minachev *et al.* (*109*) found that Ni–NaY reduced at 400°C with hydrogen accelerates the reaction $^{16}O_2 + {}^{18}O_2 = 2\,{}^{16}O^{18}O$ at −78°C; zeolite containing NiO or Ni^{2+} cations is not active at these conditions.

The cobalt zeolites CoNaA, CoCaA, and CoNaX showed high activity and selectivity in hydroformylation of propylene into *n*- and isobutyraldehyde (*197*). The catalysts were not reduced with hydrogen before use. However, under H_2 and CO pressure at reaction conditions, some of the Co^{2+} is likely to be reduced to the metal, yielding cobalt carbonyl complex, which is probably the actual catalyst. This is suggested by the presence of an induction period. The Co content of the catalysts did not change during the experiments, indicating that the carbonyl complexes are probably stabilized in the zeolite cavities. The advantage of these findings are the heterogeneous catalyst, which showed no loss in activity and its high selectivity (up to 99%). However, its efficiency is lower than that for standard liquid-phase hydroformylation (*197*). When improved, zeolite catalysts may acquire great importance in oxo synthesis as well as in various reactions of carbonylation.

Zeolites containing metallic Na, Li, or their mixture can be selective catalysts in olefin polymerization. Pd°–HNaY is active in the reaction $2\,NO \rightarrow N_2O + O$ (*198*). Zeolites containing metals of the iron subgroup (Fe, Co, Ni) (*62*), Pd, and Pt (*199*) have been proposed as catalysts for the synthesis of ammonia.

Catalysis with Polyfunctional Metal-Loaded Zeolite Catalysts

Polyfunctional metal-loaded catalysts accelerate isomerization of saturated hydrocarbons, hydroisomerization of aromatics and olefins (dienes) into isomeric saturated hydrocarbons; dehydroisomerization of naphthene hydrocarbons into aromatic ones; hydrocracking, and other reactions which are basic for oil refining and petroleum chemistry. Metal loaded zeolite catalysts have considerably improved some of these processes. New reactions have been discovered, and the possibilities of heterogeneous catalysts have been extended. Some interesting data on the mechanism of the reactions over different catalysts serve as a basis for our present ideas about the catalytic activity of zeolites and other catalytic solids. It is not possible to discuss here in any detail all the studies made in this field. However, the most important problems and results are reviewed briefly.

Hydrocracking Reactions. In the presence of metal-loaded acid-type zeolite catalysts based on X, Y, mordenite or erionite, hydrocracking accompanies various reactions of hydrocarbons. The extent of a hydrocracking reaction depends on the catalyst composition, the metal and its concentration, zeolite acidity, and reaction parameters. For example, over 0.5 % Rh–CaY (SiO_2/Al_2O_3 = 3.4) at 390°C, pressure 30 atm, LHSV 1 hr^{-1}, and a molar ratio of H_2: n-C_6H_{14} = 3.2, the yield of isomeric hexanes and hydrocracking products was 25.8 and 21.9 wt %, respectively while over a similar iridium contact at 310°C these values were 6.3 and 26.2 %, respectively (*158, 200*). Under the same conditions over Pt- and Pd-containing zeolite CaY, hydrocracking did not exceed 4 %, and the yield of isomers was 58–59.4 wt %. Similar regularities were observed during cyclohexane isomerization and benzene hydroisomerization (*163*). The yield of the hydrocracking products in the latter reaction at 320°C over 0.5 % Rh–CaY was 16.9%, and over 0.5 % Ir–CaY it was 60%. Hydroisomerization yields were 12 and 5.8 %, respectively.

The hydrocracking of higher hydrocarbons and petroleum fractions is of particular interest and great practical importance. This process can be used to obtain gaseous hydrocarbons as feed for petroleum chemistry, for gasoline and jet fuel, for largely diesel fuel, or for all of the above products simultaneously.

Many catalysts, largely of decationized and rare-earth forms of Y zeolites containing group VIII metals (Pt, Pd, Ir, Ni) have been suggested for hydrocracking (*201, 202, 203, 204, 205, 206*). Polymetallic catalysts are highly efficient (*202, 203, 204*). The decationized Y-zeolite containing 0.1 % Ir and 1.2 % Re shows higher activity in hydrocarbon hydrocracking than catalysts containing Pd or Re alone (*203*). The impregnated catalysts based on decationized Y containing 1 wt % Re and 1 wt % Cu showed higher stability (*204*).

The use of zeolite catalysts in hydrocracking made it possible to apply a wider range of raw materials in refining, to avoid some limitations imposed by the content of nitrogen, sulfur, or polycyclic aromatic hydrocarbons, and to increase the yield of the desired products. The molecular sieve type selectivity of erionite and mordenite catalysts is a unique property used in developing the selectoforming process (*18*).

The hydrocracking of individual hydrocarbons (*176, 201, 205, 206, 207, 208, 209, 210, 211, 212*) and their mixtures (*80, 201, 206, 208, 209, 211, 212, 213, 214*) has been studied. Propane and butane were the predominant products in the hydrocracking of n-heptane over Pt–NaY (SiO_2/Al_2O_3 = 3.8) at 340°–400°C, pressure 40 atm, LHSV = 1 hr^{-1}, molar ratio of H_2: C_7H_{16} = 4.5. The total yield of the products at 400°C was 65 wt % (*205*). With an increase of the hydrogenating activity the content of isoheptane in the reaction products and the total conversion of n-C_7H_{14} also increased, while the extent of hydrocracking did not vary appreciably.

Over poorly hydrogenating catalysts, the amount of C_5 and C_6 hydrocarbons increased but did not correspond to the amount of the resulting CH_4 and C_2H_6. Thus, the yield and composition of the products of *n*-alkane hydrocracking can be controlled by varying the relationship of hydrodehydrogenating and cracking functions of the catalysts. The reactivity of paraffins increased with increased chain length. For example, the hydrocracking of *n*-dodecane over 0.5 % Pd–CaY catalyst, pressure 40 atm, LHSV 1.2 hr^{-1} occurred at a temperature below 250°C. At 300°C 100 % conversion was attained (*210*). The reaction proceeded by the classical dual function mechanism (*2*). The olefinic reaction intermediates through which the hydrocracking of the saturated hydrocarbons proceed were first found and analyzed by Weitcamp and Schulz (*210*). At 250°–285°C, when the n-$C_{12}H_{26}$ conversion over 0.5 % Pd–CaY was below 100%, the amount of olefins was 0.01–0.02 % wt. At 300°–350°C the complete conversion of the hydrocarbon and secondary cracking gave no olefins in the product. Above 400°C, the olefin content in the product increased again, and at 500°C it attained its maximum: 0.16 wt %. The amount of certain olefins considerably exceeds (in the case of propylene 10^5 times) the values of thermodynamic equilibrium—*i.e.*, the reaction was controlled by kinetic factors (*210*).

Unlike X and Y zeolite catalysts whose hydrocracking characteristics are alike and similar to other dual function catalysts, the metal-loaded erionites (*211, 212*) and mordenites (*207, 208*) show some specificity because of their crystalline structure and diffusional properties. For example, over palladium-loaded erionite, *n*-paraffins C_6 and C_{10}–C_{11} were converted preferentially into *n*-alkanes having either higher or lower number of carbon atoms by the cavity effect, that is, the relation between the crystal structure and the chain length of the hydrocarbon molecule (*212*). The conversion of *n*-butane gave largely propane (84% of the total hydrocarbons formed). There was also a considerable amount of *n*-pentane (*212*). In the reaction of *n*-pentane, over 60 mole % of the products was an equal fraction of ethane and propane; the ratio of the methane and butane formed was not equimolar. Propane was largely formed from *n*-hexane, its hydrocracking rate at 371°C was 50 times greater, and at 427°C it was 17 times as high as that of n-C_5H_{12}, and the apparent activation energy was 15 kcal/mole *vs.* 30 kcal/mole for *n*-pentane and *n*-butane. The conversion of a mixture of n-C_5 + n-C_6 alkanes gave largely propane (70 mole %), the amount of C_2H_6 was considerably smaller than that obtained from the hydrocracking of n-C_5H_{12} only—*i.e.*, *n*-hexane affected significantly the reaction of *n*-pentane. In turn, the latter considerably decreased the rate of n-C_6H_{12} conversion. There is a strong interaction between the molecules with various chain length inside the erionite cavities, and the course of hydrocracking is much more complicated than that of simple splitting of the C—C bond.

An important role of adsorption and diffusion factors in the catalysis with metal-loaded zeolites can be illustrated by the results obtained from the hydrocracking of *n*-decane, decalin, and their mixture over mordenite catalysts (*207*). The study was made with two catalyst samples: 0.5 % Pd–H–mordenite obtained by impregnation of NH_4–mordenite with a Pd compound and heating in air to 538°C (catalyst A), and 0.5 % Pd–H–mordenite ($SiO_2/Al_2O_3 \sim 50$) prepared from dealuminated NH_4–mordenite (catalyst B). In the reaction of *n*-decane at 260°C, catalyst B was 4.4 times as active as the catalyst A. The apparent activation energy was 44 and 33 kcal/mole, respectively. A similar relationship between the activities of these catalysts was observed during the hydrocracking of decalin. Under the same conditions, the conversion of this hydrocarbon was more difficult than that of *n*-decane. However, from $n\text{-}C_{10}H_{22} + C_{10}H_{18}$ mixture over both catalysts, decalin was cracked more readily than *n*-decane. The cyclic hydrocarbon may prevent the access of *n*-decane molecules to the active centers of the catalysts. The diffusion coefficients of decalin over A and B catalysts at 93°C were 4 and 2.5 times, respectively, as low as those for *n*-octane. With dealuminated mordenite, the hydrocarbon adsorption and desorption was more rapid. Therefore, the inhibiting effect of decalin on the *n*-decane conversion in the presence of the B catalyst was less pronounced.

As in other reactions the mordenite catalysts show higher activity in hydrocarbon hydrocracking than other dual function catalysts. The relative activities of the catalysts in *n*-hexane hydrocracking are the following (*208*): 0.5 % Pd-H-mordenite, 8000; 0.5 % Pd–HY, 80; 0.5 % Pd/Al_2O_3, 1, 0.5 % $Pd–TiO_2$, 0.5; and Pd/ZrO_2, 0.1. Thus, H–mordenite loaded with Pd showed the highest activity.

The catalysts made from the H form of mordenite and the metals of group VIII have been suggested for deparaffinization of various hydrocarbon fractions, to lower the freezing temperature, and to improve the properties of lubricants (*215, 216*) or to obtain jet fuel (*217*). For instance, by selective hydrocracking of hydrocarbon fractions with boiling temperatures of 250°–290° and a freezing temperature of 14°C, over 2 % Pd–H–mordenite containing 2% coke at 340°, 60 atm and LHSV 4 hr^{-1}, it was possible to obtain 88.5 % jet fuel with a freezing point of —70°C (*217*). For selective hydrocracking of *n*-alkanes from their mixture containing isoparaffins, naphthenes, and aromatic hydrocarbons one can use ZnA, Zn–erionite, or rare earth erionite loaded with Pt, Pd, or Ni (*211*).

The palladium-loaded H-omega proved to have low activity in the hydrocracking of gas oil with an average molecular weight of 216; its activity decreased very quickly with time on stream (*80*). Most of the acid centers and metal particles are located in positions non-accessible for reagents, and only a small part of these sites takes part in catalysis.

It has been noted that water produces a promoting effect on the 1 % Pd–RE–X catalyst in the hydrocracking of *n*-hexadecane (*209*), but the optimal amount of water needed could not be determined. With 2.5 % Pt-HY, water and 2-pentanol produced a deactivating effect explained by the competitive adsorption of H_2O on the accessible active centers of the zeolite (*209*).

Isomerization Reactions. Rabo *et al.* (*7, 8*) were the first to demonstrate high activity of Pt- and Pd-zeolite catalysts in *n*-pentane and *n*-hexane isomerization. (The term hydroisomerization is used in the U.S. literature to indicate that a reaction proceeds in a hydrogen atmosphere. From the chemical viewpoint the use of this term with paraffins and cycloalkanes does not seem suitable.) Later the catalytic properties of X- and Y-zeolites, omega, mordenite, and dealuminated mordenites containing metals of group VIII have been extensively studied in various reactions in this laboratory (*35, 82, 136, 158, 159, 160, 161, 163, 200, 218, 219, 220, 221, 222, 223*), by Voorhies *et al.* (*224, 225, 226, 227, 228, 229*), and by other workers (*16, 36, 52, 80, 94, 106, 108, 124, 137, 138, 139, 140, 145, 230, 231, 232, 233, 234, 235, 236, 237, 238*). Some of the results obtained are summarized in Table XIII.

The following principal characteristics of metal-zeolite catalysts have been established in isomerization reactions:

(a) the zeolites containing univalent cations are practically inactive;

(b) hydrogen, decationized and multivalent cationic forms of zeolites loaded with metals show high activity (*7, 35, 139, 158, 159, 160, 161, 200, 222, 224, 225, 226, 227, 228, 229, 230, 231, 232, 233, 234, 235, 236, 237, 238*);

(c) the larger the charge and the smaller the radius of the zeolite cation, the higher is the catalytic activity; Y zeolites containing cations with a valence 3 and 4 (0.5 wt % Pt(Pd)) show lower selectivity than those containing divalent cations, the activity of some of these catalysts decreasing very rapidly (*7*);

(d) with a higher degree of exchange (α) of Na^+ by M^{n+} a sharp increase in the activity of the corresponding ion-exchanged catalysts is observed after a certain threshold. The exact value depends on the nature of the cation M^{n+} and on the SiO_2/Al_2O_3 ratio in zeolites of the faujasite type. For example, for CaY with $SiO_2/Al_2O_3 = 5.0$, catalytic activity appears at an α of 35–40 % (*7*). During decationization the threshold of the exchange is low if it exists at all. The acidity of faujasites is known to vary in the same way (*239, 240, 241*). The zeolite catalysts with a minimal residual sodium show the highest activity (*7, 36*);

(e) efficient isomerization catalysts may contain cation–decationized combinations (CaHY (*145, 231*), MnHY (*233*), LaHY (*237*), REHY (*238*), and combinations of cations (CaCoX (*234*));

(f) with an increase in SiO_2/Al_2O_3 ratio in zeolites of faujasite-types (*7, 22, 35, 36, 82, 108, 137, 138, 158, 159, 163, 231*) or mordenite-types (*36, 227, 228, 232*), the activity of the catalysts considerably increases, allowing much lower reaction temperatures. For instance, over 0.5 % Pd–CaY ($SiO_2/Al_2O_3 = 4.5$), *n*-hexane and cyclohexane isomerization proceeded to similar conversion at temperatures 60°–100°C lower than those required for 0.5 % Pd-CaX ($SiO_2/Al_2O_3 = 2.5$) (*160*). The highest isomerization activity of Pd-H-mordenite catalysts was observed at a ratio of SiO_2/Al_2O_3 about 16–18 (*36, 232*). The best SiO_2/Al_2O_3 value for the faujasite-type zeolites has not been established yet;

(g) the activity, selectivity, and stability of catalysts depend on the nature and concentration of the metal component. For example, the activity of the zeolite 0.45 La–0.45 $NH_4NaY(SiO_2/Al_2O_3 = 5.0)$ increased linearly

Table XIII. Isomerization of Hydrocarbons on Zeolites Containing Group VIII Metals

(Pressure 30 atm, LHSV 1 hr^{-1}, H_2:CH = 3.2)

Catalyst	*SiO_2/Al_2O_3 Ratio*	*Temp., °C*	*Yield of Isomers, %*	*Yield of Hydrocracking Products, wt %*	*Reference*
		n-Butane			
0.5 % Pt-CaY	4.1	400	27.0	3.2	*22*
		n-Pentane			
0.5 % Pt-0.85 LaNaX	2.5	375[a]	60	—	*139*
0.5 % Pd-CaY	4.1	360	50.9	3.8	*22*
0.5 % Pt-0.85 HNaY	5.0	350[a]	61.0	2	*7*
0.5 % Pd-0.85 HNaY	5.0	350[a]	64.0	2	*7*
0.7 % Pt-CaY	3.3	375[b]	55.0	—	*137*
0.5 % Pd-H-mordenite	10.0	45.8	45.8	4.6	*219*
5 % Ni-H-mordenite	10.0	280	40.2	3.0	*219*
		n-Hexane			
0.5 % Pd-CaY	3.4	400	58.0	3.2	*163*
0.5 % Pd-CaY	4.5	350	70.3	3.6	*200*
0.5 % Pt-CaY	3.4	400	59.4	3.8	*163*
0.5 % Rh-CaY	3.4	390	25.8	21.9	*163*
0.5 % Ir-CaY	3.4	310	6.3	26.2	*163*
0.5 % Pt-0.78 CaNaY	5.0	400[b]	66.5	3.0	*7*
0.5 % Pt-0.85 HNaY	5.0	340[b]	76.0	<1.0	*7*
0.5 % Pt-0.96 CaNaY	4.8	350[c]	58.0	5.0	*231*
0.5 % Pt-0.965 CaHNaY	4.8	350[d]	63.3	4.4	*231*
Pt-CaHY	4.8	330[d]	74.0	2.7	*231*
		n-Heptane			
0.5 % Pd-CaY	3.4	360	47.2	10.3	*22*
		n-Dodecane			
0.5 % Pt-CaY	5.0	275[e]	48	17	*233*
		Cyclohexane			
		330	18.0	traces	
0.5 % Pd-CaY	3.4	370	43.0	3.9	*161*
0.5 % Pd-CaY	4.5	330	57.0	3.7	*22*
0.5 % Pt-CaY	3.4	330	24.3	traces	*161*
		370	47.0	4.2	
0.5 % Ir-CaY	3.4	330	4.0	52.0	*163*
2 % Pd-H-mordenite	—	288[f]	73	—	*236*

[a] LHSV = 5 hr^{-1}, H_2:hydrocarbon = 6.
[b] WHSV = 2 hr^{-1}, H_2:hydrocarbon = 3.
[c] Experiments in autoclave, 3 hr, H_2:hydrocarbon = 5.
[d] Granulated zeolite without binder, H_2:hydrocarbon = 5.5.
[e] Pressure = 40 atm, WHSV = 1 hr^{-1}, H_2:hydrocarbon = 20.
[f] Pressure 13.6 atm.

with an increase of Pt and Pd concentration up to the equilibrium yields of hexane isomers. These are attained for methylpentanes at 0.05 % Pd and 0.10 % Pt and for 2,2-dimethylbutane at 0.25 % Pd and 0.40 wt % Pt. Further increases in metal content did not affect the yield of isohexanes. The catalysts containing less than 0.2 % noble metal showed lower stability (*237*).

In general, the optimal content of a metal in a catalyst depends on its nature, introduction, and reduction (which considerably affects the hydro-dehydrogenating activity of the metal centers), and zeolite composition (its acidic properties). According to Minachev *et al.* (*188*), the activity of the Pd-CaY catalysts obtained by impregnation does not vary in the isomerization of *n*-hexane with an increase of metal concentration from 0.25 to 1 wt %. Nickel catalysts show high activity at a higher metal content (4–5 wt %) than platinum, palladium, or iridium catalysts. When speaking about the dependence of the properties of zeolite catalysts in isomerization on the nature of the metal, it should be noted that rhodium and iridium catalysts, in particular, show low selectivity (*see* Table XIII).

The decationized and multivalent cationic forms of faujasite-zeolites show some activity in *n*-alkane and cycloalkane isomerization even in the absence of hydrodehydrogenating metals but at temperatures 80°–100° higher than those for metal-loaded zeolite catalysts (*22, 35, 237, 242*). Under these reaction conditions the metal-free catalysts undergo fast deactivation. Hydrogen (*16, 35, 36, 219, 221, 222, 224, 225*) and some cationic modifications (*35, 219, 222*) of mordenite show high activity. The activity of H-mordenites in *n*-pentane isomerization does not change when palladium is introduced (*35, 36, 222, 224, 225*), but a small amount of noble metals sharply increases the selectivity and stability of these catalysts (*16, 36, 224*). So far only metal-containing catalyst is used in the commercial isomerization of pentane–hexane fraction (*16*).

With increasing molecular weight the reactivity of *n*-paraffins increases in isomerization. Simultaneously, their hydrocracking is also facilitated. Therefore, to attain a sufficient selectivity in the conversion of higher alkanes, catalysts based on zeolites with a comparatively low acidity should be used. Isomerization of cycloparaffins proceeds as readily as that of *n*-paraffins with an equal number of carbon atoms in the molecule (*see* Table XIII). In some cases, together with isomeric alkylcycloalkanes, aromatic hydrocarbons are also formed (*230*).

When studying the conversion of cyclohexane and methylcyclopentane over 0.5 % Pd–CaY catalyst at 275°C, 40 atm, and H_2: hydrocarbon = 20: 1, Schulz *et al.* (*235*) found C_4, C_5, C_7 and C_8 hydrocarbons in the products, and the amount of (C_5 and C_7), and (C_4 and C_8) was equimolar. This provided evidence that these hydrocarbons are formed by disproportionation of C_{12}-hydrocarbon formed during alkylation of cycloolefins C_6 (the dehydrogenation products of the starting cycloalkane) with C_6-π-allyl-carbonium ion (formed during ring opening and addition of the proton) (*235*):

$$\text{cyclohexene} + \overset{+}{C}H_2{-}CH{=}\underset{\displaystyle CH_3}{C}{-}CH_2{-}CH_3 \xrightarrow{\text{alkylation}} \left[\text{cyclohexyl}^{+}{-}CH_2{-}CH{=}\underset{\displaystyle CH_3}{C}{-}CH_2{-}CH_3 \xrightarrow{\sim H} \text{cyclohexenyl}^{+}{-}CH_2{-}CH_2{-}\underset{\displaystyle CH_3}{CH}{-}CH_2{-}CH_3\right]$$

$$\xrightarrow{\beta\text{-scission}} \text{methylenecyclohexene}(=CH_2) + \left[\overset{+}{C}H_2{-}\underset{\displaystyle CH_3}{CH}{-}CH_2{-}CH_3\right]$$

$$\text{methylenecyclohexene} \xrightarrow{+4\,H} \text{methylcyclohexane}(CH_3)$$

$$\left[\overset{+}{C}H_2{-}\underset{\displaystyle CH_3}{CH}{-}CH_2{-}CH_3\right] \rightarrow H_3C{-}\underset{\displaystyle CH_3}{\overset{+}{C}}{-}CH_2{-}CH_3 \xrightarrow[+2H]{-H^{+}} H_3C{-}\underset{\displaystyle CH_3}{CH}{-}CH_2{-}CH_3$$

Bolton and Lanewala (*238*) observed the formation of heptanes during isomerization of *n*-hexane, 2-methylpentane, and 3-methylpentane over 0.5 % Pd–0.45 RE–0.50 NH_4Y ($SiO_2/Al_2O_3 = 5.0$). These data may be caused by specific catalytic action of a certain catalyst sample under the conditions studied or they may be characteristic of a wider range of catalysts. Hydrocarbons with a carbon number exceeding six were obtained only at small conversion of the starting alkanes and cycloalkane C_6.

The study of the kinetic regularities of *n*-pentane, *n*-hexane, and cyclohexane isomerization over Pt and Pd zeolites of the faujasite (*35, 138, 159, 218, 220, 223, 224, 225*) and mordenite (*35, 36, 221, 222, 224, 225, 226, 227, 228*) types provided evidence that these reactions proceed according to dual-function mechanism (*2*). Alkanes (cycloalkanes) undergo dehydrogenation on hydrodehydrogenating sites to give olefins (cycloolefins) which are adsorbed on the acidic centers, adding a proton and forming a carbonium ion. After skeletal isomerization they are desorbed as isoolefins (alkylcycloalkenes) and hydrogenated on the metal to give the corresponding isoparaffins (alkylcycloalkanes). Isomerization of *n*-pentane can be presented by:

$$C_5H_{12} \underset{+\,H_2}{\overset{Pt(-H_2)}{\rightleftarrows}} C_5H_{10} \underset{H^+}{\overset{CaY}{\rightarrow}} C_4H_7{-}CH_3 \underset{-H_2}{\overset{Pt(+H_2)}{\rightleftarrows}} C_4H_9CH_3$$

This mechanism of isomerization of saturated hydrocarbon over metal-zeolite catalysts is supported by the following experimental facts:

(a) With an increase of the hydrogen partial pressure the rate of cyclohexane and *n*-pentane isomerization decreases over 0.5 % Pt-CaY (*218, 220*), probably because of increased H_2 pressure reducing the dehydrogenation equilibrium for the starting hydrocarbons, and a decrease in olefin equilibrium concentration.

(b) The isomerization rates of cyclohexane (and *n*-pentane) over Pt–CaY catalyst and that of cyclohexene (*n*-pentene) over zeolite CaY are equal; the dependence of the isomerization and hydroisomerization rates of the corresponding hydrocarbons on their partial pressure is similar;

Table XIV. Comparison of Zeolite and Other Dual Function Catalysts in Hydrocarbon Isomerization

Hydrocarbon	*Catalyst*	*Process Conditions*	*Isomer Yield, mole %*	*Reference*
n-Pentane	MB-5390 (Pd-zeolite Y, SiO_2/Al_2O_3 = 5.0)	350°C, 31 atm, H_2:C_2H_{12} = 3, LHSV = 1–3 hr^{-1}	60	*8*
	0.5 % Pd-CaY, SiO_2/Al_2O_3 = 5.0	360°C, 30 atm, H_2:C_5H_{12} = 3.2, LHSV = 1–3 hr^{-1}	58	*161*
	0.6 % Pd-SiO_2·Al_2O_3	400°C, 40 atm, LHSV = 1.5 hr^{-1}, with H_2 circulation	40–52	*243*
	0.1–1 % Pt-Al_2O_3·Hal	430°C, 66 atm, H_2:C_5H_{12} = 2.9, LHSV = 1 hr^{-1}	38	*244*
	5 % Ni-SiO_2·Al_2O_3	393°C, 24.8 atm, H_2:C_5H_{12} = 4, LHSV = 1 hr^{-1}	43	*245*
	0.5 % Pd-H-mordenite, SiO_2/Al_2O_3 = 10	280°C, 30 atm, H_2:C_5H_{12} = 3.2, LHSV = 1 hr^{-1}	45.8	*219*
n-Hexane	MB-5390	350°C, 31 atm, H_2:C_6H_{14} = 3, LHSV = 1–3 hr^{-1}	76	*8*
	0.5 % Pd-CaY, SiO_2/Al_2O_3 = 4.5	360°C, 30 atm, H_2:C_6H_{14} = 3.2, LHSV = 1–3 hr^{-1}	70	*161*
	Pt-CaHY, SiO_2/Al_2O_3 = 4.8	330°C, 30 atm, H_2:C_6H_{14} = 5.5, LHSV = 1 hr^{-1}	74	*231*
	0.6 % Pd-SiO_2·Al_2O_3	400°C, 40 atm, LHSV = 1.5 hr^{-1} with H_2 circulation	60	*243*
	5 % Ni-SiO_2·Al_2O_3	385°C, 24.8 atm, H_2:C_6H_{14} = 4, LHSV = 1 hr^{-1}	67.8	*246*
Cyclohexane	0.5 % Pd-CaY, SiO_2/Al_2O_3 = 4.5	330°C, 30 atm, H_2:C_6H_{12} = 3.2, LHSV = 1 hr^{-1}	57	*163*
	5 % Ni-SiO_2·Al_2O_3	342°C, 24.8 atm, H_2:C_6H_{12} = 4, LHSV = 1 hr^{-1}	68.9	*247*
	0.5 % Pt-SiO_2·Al_2O_3	318°C, 24.8 atm, H_2:C_6H_{12} = 4, LHSV = 1 hr^{-1}	30	*248*

(c) The activity of the catalysts depends on the relationship of the hydrodehydrogenating and acidic functions. At low metal concentrations, the total reaction rate is determined by the formation of the intermediate olefin (cycloalkene). For instance, over 0.01 % Pt–CaY the isomerization rate of cyclohexane was one order lower than that over 0.5 % Pt–CaY catalyst (*159*). At Pt or Pd concentrations of ≥0.5 wt % the limiting stage is that of isomerization of olefins over acid sites of the zeolite;

(d) The separate action of hydrodehydrogenation and isomerization sites characteristic of the classical dual-function contacts (*2*) is supported by the observation that the mechanical mixture of zeolites NiNaA (active in hydrogenation and dehydrogenation) and CaY (solid acid) is active in *n*-hexane isomerization (*108, 124*).

Thus, the action of zeolite and other heterogeneous catalysts of the metal-acidic oxide type in isomerization is very similar. Their properties

Table XV. Hydroisomerization of Aromatics

Hydrocarbon	*Catalyst*	SiO_2/Al_2O_3	*Temp.* °C	*Pressure,* atm
Benzene	0.5 % Rh-CaY	3.4	320	30
	0.5 % Ir-CaY	3.4	320	30
	0.5 % Pt-CaY	3.4	320	30
	0.5 % Pd-CaY	3.4	320	30
	0.5% Pd-CaY	4.5	320	30
	0.5 % Pd-H-mordenite	10	250	30
	5 % Ni-H-mordenite	10	250	30
Toluene	Pt-CaY	—	300	14
Cyclohexene	0.5 % Pt-CaY	41	310	30
1-Pentene	0.5 % Pt-CaY	4.1	350	30

[a] Flow-circulating, kinetic experiment; skeletal isomerization rate of cyclohexene

can be well explained on the basis of the theory of dual-function catalysis. The specific features of H–mordenite catalysts have been discussed in (*35, 36, 221, 222*).

The properties of zeolite and other heterogeneous catalysts in the isomerization of pentane, hexane, and cyclohexane are compared in Table XIV. In some cases the activity of zeolite catalysts is superior to that of the catalysts on halogenated alumina and amorphous aluminosilicate: high conversions of hydrocarbons are attained at lower temperatures (by 50°–100°); no promoters, such as HF and HCl, are required although by adding halogens, catalytic activity can be increased (*249*). Especially active are the catalysts based on H–mordenite. For example, isomerization of *n*-pentane–hexane fractions over H–mordenite containing a noble metal can be carried out at temperatures as low as 250°C (*16*). Zeolite catalysts do not require thorough purification and drying of the feed and of the circulating

and Olefins on Metal-Zeolite Catalysts

LHSV, hr^{-1}	*H_2:Hydrocarbon, mole*	*Yield of Products, %*		*Reference*
0.5	5	Methylcyclopentane (MCP)	12	*163*
		Cyclohexane (CH)	66.3	
		C_1–C_5-hydrocarbons	16.9	
0.5	5	MCP —	5.8	
		CH —	20.6	
		C_1–C_5 —	60.0	
0.5	5	MCP —	23.4	
		CH —	61.7	
		C_1–C_5 —	traces	
0.5	5	MCP —	20.1	
		CH —	63.5	
		C_1–C_5 —	traces	
0.5	5	MCP —	67.3	
		CH —	28.4	
1.0	5	MCP —	43.4	*219*
		CH —	35.6	
		C_1–C_5 —	5.6	
1.0	5	MCP —	31.1	
		CH —	63.0	
		C_1–C_5 —	1.4	
0.26	4	Methylcyclohexane	55.6	*257*
		Dimethylcyclopentane	42.7	
13900[a]	7.1	MCP —	18.5	*159*
		CH —	80.2	
17160[a]	2.6	2-methylbutane—	4.3	*220*
		n-pentane—	92.7	

is 15.9×10^{-3}, and that of pentene 15.6×10^{-3} g-mole/g-catalyst-hr.

H_2. All these as well as the high selectivity and stability of metal–zeolite contacts makes them advantageous in isomerization of C_5–C_6 hydrocarbons.

Mordenite and Y zeolites loaded with metals are efficient in isomerizing ethylbenzene to xylenes (*21, 250, 251, 252*). Good results were obtained with bimetallic Pt–Ir contact (*252*). At 343°C, pressure 7 atm, molar ratio of H_2 to $C_6H_5C_2H_5$ of 2, the conversion of ethylbenzene over Y-zeolite containing 0.25 % Pt and 0.1 % Ir was 79.1% and the yield of xylenes was 61.8%. Under the same conditions over Pt catalyst without Ir, ethylbenzene conversion was 90.8%, and the yield of xylenes was 37.8%.

Metal-zeolite catalysts have been suggested for isomerization of xylenes (*253, 254, 255*) and cumene into trimethylbenzenes (*251, 252, 253, 254, 255, 256*) Copper and silver may be added to these zeolites (*254*).

Hydroisomerization Reactions. Olefinic, cycloolefinic, and aromatic hydrocarbons undergo hydroisomerization over dual-function metal–zeolite catalysts in the presence of hydrogen (*159, 163, 219, 220, 230, 257*). The ratio between the hydrogenation and hydroisomerization products depends on catalyst composition and process conditions (Table XV). For instance, over CaY (SiO_2/Al_2O_3 = 3.4) containing 0.5 % Pd at 320°C, 30 atm, LHSV 0.5 hr^{-1}, and molar ratio of H_2: C_6H_6 = 5, the yield of methylcyclopentane was 20.1 %, and that of cyclohexane was 63.5 %. The increase of SiO_2/Al_2O_3 ratio in CaY up to 4.5 resulted in a sharp increase of the methylcyclopentane yield (67.3%). This yield was higher than that obtained over the same catalyst from cyclohexane (57%) under similar conditions. With rhodium and iridium catalysts high hydrocracking activity was observed: at 320°C, there were 16.9% hydrocracking products over the former catalyst and 60% over the latter one whereas hydroisomerization produced only 12 and 5.8%, respectively (Table XV). These catalysts (as well as the platinum catalyst, 0.5 % Pt–CaY) acted as monofunctional ones at 180°–200°C, and they selectively hydrogenated benzene to cyclohexane (*see* Table VI). In general, the properties of metal-containing zeolites depend considerably on the conditions used. Catalysts based on H–mordenite (*219*) and H–Zeolon (*230*) are active in benzene hydroisomerization at tempera-

Table XVI. Hydrodimerization of

Catalyst	*Time, hr*	*Conversion of C_6H_6, %*
12.8 % Ni-Na-mordenite	0.3	9.2
13.2 % Ni-NaY	2.0	19.6
2.6 % Ni-NH_4-mordenite	1.0	13.3
0.3 % Pd-HNaY	0.2	23.4

[a] Autoclave experiments, 200°C, 54.5 atm; amount of catalyst 2 g for 13.2 % Ni-NaY and 0.3 % Pd-HNaY, 7 g for 12.8 % Ni-Na-mordenite and 2.6 % Ni-NH_4-mordenite with 0.45 mole of benzene.

tures as low as 200°–250°C (Table XV). Olefins undergo hydrogenation under milder conditions than aromatic hydrocarbons. Therefore, their hydroisomerization proceeds like that of the corresponding paraffins (*220*), at least in the case of catalysts containing a sufficient amount of a metal ($\geqq$0.2 wt % Pt and Pd). The behavior of cycloolefins is similar (*159*).

Benzene Hydrodimerization. Theoretically, one of the most interesting reactions that can be accelerated by polyfunctional metal zeolite catalysts is benzene hydrodimerization into phenylcyclohexane (*258*):

$$2\ C_6H_6 + 2\ H_2 \longrightarrow C_6H_5 - C_6H_{11}$$

This can proceed both on Ni- and Pd-containing zeolites, with the best results obtained over nickel catalysts. As seen from Table XVI, the 12.8 % Ni–Na–mordenite with low acidity accelerates largely benzene hydrogenation into cyclohexane. On the contrary, the Ni–NH_4–mordenite produced by decomposition of H–mordenite shows high selectivity for phenylcyclohexane formation (68.8 %). An even greater yield of this hydrocarbon is produced with 13.2 % NiNaY; the Pd–HNaY has lower efficiency. Methylcyclopentane is obtained in small amounts.

Thus, to convert benzene into phenylcyclohexane, the catalyst should show acidity and hydrogenating activity, and these functions should be balanced. In the absence of a hydrogenating metal, this reaction fails, but if the hydrogenating activity of the metal is very high, the main product is cyclohexane. Acidity of the catalyst is also important. The dual-function nature of the hydrodimerization catalysts is supported by the studies of a mechanical mixture of Ni/Al_2O_3 particles (a hydrogenating contact) and those of an amorphous aluminosilicate cracking catalyst. This mixture proved to be an active catalyst in the synthesis of phenylcyclohexane. Under similar conditions cyclohexane was obtained only over Ni/Al_2O_3, and over aluminosilicate no reaction took place.

To attain maximum selectivity for phenylcyclohexane formation, the C_6H_6 conversion should not exceed ~30%. At larger benzene conversion,

Benzene over Metal-Zeolite Catalysts[a]

Yields on Converted Benzene, %				
methyl-cyclo-pentane	*cyclo-hexane*	C_{18} *(trimeric prod.)*	*cyclo-hexyl-cyclo-hexane*	*phenyl-cyclo-hexane*
1.6	91.1	0	0	7.3
0.8	12.6	5.5	2.2	78.9
1.4	19.1	3.1	7.6	68.8
1.0	42.0	2.3	2.5	52.3

the yield of the trimerization products strongly increases. The following conversion scheme has been suggested (*258*):

M site
$3\ H_2$
M site
H_2
M site
$2\ H_2$
acid site, C_6H_6
acid
site
M site
H_2
C_6H_{10}
acid
site
Tetramers
(tricyclohexyl-
benzenes)
C_6H_{10}
acid site

Cyclohexene is considered to be an intermediate arising during benzene hydrogenation on the metallic center M. It undergoes subsequent hydrogenation to give cyclohexane or it is desorbed and migrates to the acidic center where alkylation of benzene to phenylcyclohexane takes place. The products of benzene tri- and tetramerization are obtained in a similar way. The results of the experiments with a mixture of cyclohexene (labelled with C^{14}), and benzene supported the validity of this assumption. Alkali metals supported on Al_2O_3, MgO, SiO_2, and other supports are active catalysts in benzene hydrodimerization (*259*). The reaction proceeds through the formation of surface radical anions (*260*).

The data in Tables VI, XV, and XVI show that the reaction of benzene with hydrogen can be directed towards the formation of cyclohexane, methylcyclopentane, or phenylcyclohexane depending on the composition of the catalyst and conditions used. Here we have a good illustration of the extensive possibilities of controlling the catalytic properties of metal–zeolite systems.

Other Reactions. Zeolites NaX, NaY, and Na-mordenite loaded with the metals of groups VIII and Ib show only one function as catalysts of hydrocarbon reactions (oxidation, hydrogenation, dehydrogenation). In the conversion of *n*-butyraldehyde in the presence of H_2, these systems act as typical dual-function catalysts (*26*).

As seen from Figure 7, over 4% Ni–NaY prepared by impregnation of dehydrated NaY with a solution of nickel acetylacetonate in $CHCl_3$ with a subsequent decomposition of the salt in air at 500°C followed by reduction in hydrogen at 450°C, butanol, isooctanol, 2-ethylhexanal, and 2-ethylhexenal are produced. The properties of the catalysts are modified considerably during the reaction: with a decrease of its activity in the hydrogenation of aldehyde to butanol, the yields of the condensation products increase, the conversion of butyraldehyde into 2-ethyl-1-hexanol and 2-ethylhexanal attains its maximum after 2.5 and 5 hr of reaction, respectively, and the yield of 2-ethylhexenal increases up to 9 hr and thereafter is not affected. The total conversion of the aldehyde gradually decreases. This is probably caused by different adsorption coefficients of the starting *n*-butyraldehyde and its conversion products. At different temperatures and pressure, the reaction profile can vary. Over NaY zeolite without Ni, there occurred only crotonic condensation of *n*-butyraldehyde into 2-ethylhexenal (*26*).

According to Ref. *25*, over 0.4 % Pd–NaX at 180°C and pressure of 17 atm, acetone is converted into methyl isobutyl ketone in a 42% yield and 95.8% selectivity. These results show that a catalyst of the same composition can act mono- or polyfunctionally depending on the reaction character.

Hydrocondensation of lower aldehydes can be carried out over La, Ce, Cd, Ba, Sr, Zn, Ca, or decationized forms of X and Y zeolites containing Pt or Ni group metals (*261*). Metal-zeolite catalysts can also accelerate disproportionation of alkanes and isoalkanes C_3–C_7 into isoalkanes with larger and smaller molecular weight (*262*) as well as oligomerization of olefins C_2–C_4 (*263*).

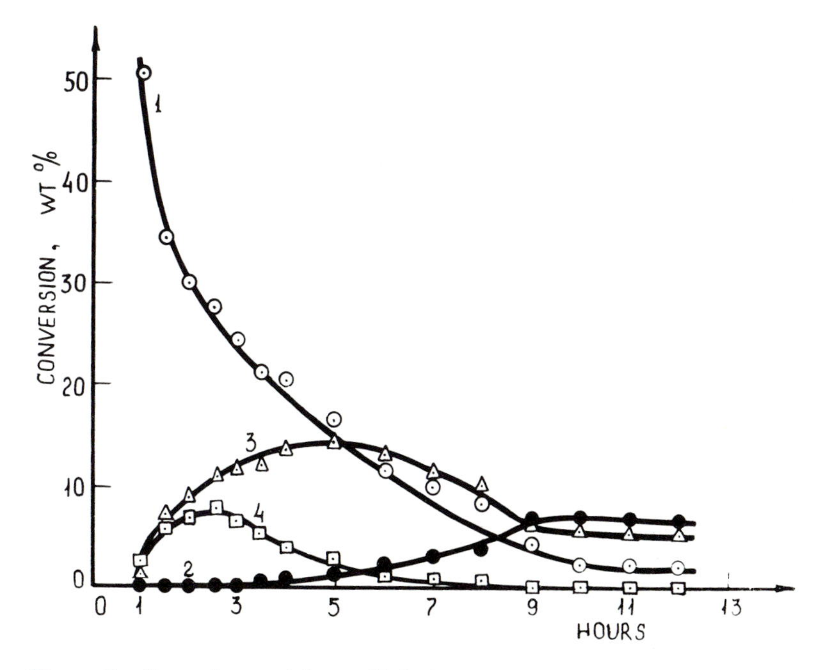

Figure 7. Dependence of butyraldehyde conversion to butanol (Curve 1), 2-ethylhexenal (Curve 2), 2-ethylhexanal (Curve 3), and 2-ethyl-1-hexanol (Curve 4) over 4% Ni–NaY on catalyst operating time at 200°C, molar ratio n-$C_4H_8O:H_2 = 1:1.4$, *LHSV* $= 0.53$ hr^{-1}

Summary

Metal-containing zeolites are efficient catalysts for many organic and inorganic reactions, many of which form the basis of important petroleum refining and petrochemical processes. Other catalysts more recently developed, are expected to find practical use in the future.

The activity, selectivity, and stability of metal-zeolite catalysts are, to a great extent, determined by the chemical composition and structure of the zeolite component which is responsible for their unique properties in many cases. The catalytic functions of metal-zeolite systems depend on their preparation, modification, and conditions of application; the same catalyst can act mono- or polyfunctionally depending on the reaction character and the conditions of use. Further advances in this field are expected in the application of new investigation methods, the synthesis of zeolites of other types, and the development of new techniques to be used for their modification.

The potential of metal-zeolite catalysts is far from being fully exploited. Polymetallic systems obtained by ion exchange, impregnation (simultaneous or subsequent), and the combination of these and other methods represent new synthetic approaches to new catalyst systems.

Literature Cited

1. Thomas, Ch. L., "Catalytic Processes and Proven Catalysts," Academic Press, New York and London, 1970.
2. Weisz, P. B., *Advan. Catal.* (1962) **13,** 137.
3. Slinkin, A. A., Federovskaya, E. A., *Usp. Khim.* (1971) **40,** 1857.
4. Ciapetta, F. G., Wallace, D. N., *Catal. Rev.* (1971) **5,** 67.
5. Cinneide, A. D. O., Clarke, J. K. A., *Catal. Rev.* (1972) **7,** 213.
6. Whyte, T. E., *Catal. Rev.* (1973) **8,** 117.
7. Rabo, J. A., Pickert, P. E., Stamires, D. N., Boyle, J. E., *Actes Deuxieme Congr. Internat. Catalyse, Paris, 1960* (1961) **2,** 2055.
8. Rabo, J. A., Pickert, P. E., Mays, R. L., *Ind. Eng. Chem.* (1961) **53,** 733.
9. Weisz, P. B., Frilette, V. J., *J. Phys. Chem.* (1960) **64,** 382.
10. Weisz, P. B., Frilette, V. J., Maatman, R. M., Mower, E. B., *J. Catal.* (1962) **1,** 307.
11. Rabo, J. A., Schomaker, V., Pickert, P. E., *Proc. Intern. Congr. Catal., 3rd, Amsterdam, 1964* (1965) **2,** 1264.
12. *Oil Gas J.* (1969) **67** (24), 76.
13. Agafonov, A. V., Maslyanskii, G. N., Rogov, S. P., Havkin, V. A., Shipikin, V. V., Pannikova, R. F., *Neftepererabotka Neftekhim.* (1971) **4,** 1.
14. Mays, R. L., Pickert, P. E., Bolton, A. P., Lanewala, M. A., *Oil Gas J.* (1965) **63** (20), 91.
15. *Oil Gas J.* (1971) **69** (10), 44.
16. Kouwenhoven, H. W., Van Zijll Langhout, W. C., *Chem. Eng. Progr.* (1971) **67** (4), 65; *Petrol. Petrochem. Int.* (1971) **11** (11), 64.
17. Simoniak, M. F., Reber, R. A., Victory, R. M., *Hydrocarb. Process.* (1973) **52** (5), 101.
18. Chen, N. Y., Maziuk, J., Schwartz, A. B., Weisz, P. B., *Petrol. Int.* (1969) **27** (2), 42.
19. Burd, S. D., Maziuk, J., *Hydrocarb. Process.* (1972) **51** (5), 97.
20. Burd, S. D., Maziuk, J., *Oil Gas J.* (1972) **70** (27), 52.
21. Uhlic, H. F., Prefferle, W. C., *Advan. Chem. Ser.* (1970) **97,** 204.

22. Minachev, Kh. M., Garanin, V. I., Isakov, Ya. I., *Usp. Khim.* (1966) **35,** 2151.
23. Pickert, P. E., U.S. Patent **3,236,904;** *C.A.* (1966) **64,** 17478*a*.
24. Ryashentseva, M. A., Minachev, Kh. M., *Neftekhimia* (1971) **11,** 198.
25. Takagi, K., Murakami, M., Iketani, K., U.S. Patent **3,666,816;** *RZh. Khim.,* **1973,** 7H56.
26. Isakov, Ya. I., Minachev, Kh. M., Usachev, N. Ya., *Izv. Akad. Nauk SSSR, Ser. Khim.* (1972) 1175.
27. Morrison, J., *Oil Gas Int.* (1970) **10** (10), 52.
28. *Chem. Eng. News* (1970) **48** (28), 60.
29. Hall, D. W., Dormish, F. L., Hurley, E., *Ind. Eng. Chem., Prod. Res. Develop.* (1970) **9,** 234.
30. Brooks, C. S., *Advan. Chem. Ser.* (1971) **102,** 426.
31. Miller, R. L., *Chem. Eng.* (1972) **79** (5), 60.
32. Ciapetta, F. G., *Chim. Ind.* (1969) **51,** 1173.
33. Szebenyi, U., Klopp, G., Gorog-Kocsis, E., *Period. Polytechn. Chem. Eng.* (1971) **15,** 269.
34. Minachev, Kh.M., Isakov, Ya. I., "Prigotovlenie, Aktivaziya i Regeneraziya Zeolitnych Catalisatorov," Zniiteneftekhim., Moscow, 1971.
35. Minachev, Kh. M., Geranin, V. I., Kharlamov, V. V., Isakova, T. A., *Kinet. Catal.* (1972) **13,** 1104.
36. Kouwenhoven, H. W., *Advan. Chem. Ser.* (1973) **121,** 529.
37. Breck, D. W., Milton, R. M., U.S. Patent **3,013,985;** *RZh. Khim.,* **1963,** 8Λ121Π.
38. Breck, D. W., Bukata, S., U.S. Patent **3,200,082;** *RZh. Khim.,* **1966,** 18Λ161Π.
39. Breck, D. W., Milton, R. M., U.S. Patent **3,013,982;** *RZh. Khim.,* **1963,** 8Λ118Π.
40. French Patent **1,490,184;** *C.A.* (1968) **68,** 106490*d*.
41. Breck, D. W., Milton, R. M., U.S. Patent **3,013,983;** *RZh. Khim.,* **1963;** 8Λ119Π.
42. French Patent **1,285,510;** *C.A.* (1963) **58,** 2887*b*.
43. Patent D.D.R. **40953;** *RZh. Khim.,* **1967,** 10Λ183Π.
44. Breck, D. W., U.S. Patent **3,013,984;** *RZh. Khim.,* **1963,** 8Λ120Π.
45. Castor, C. R., U.S. Patent **3,013,986;** *RZh. Khim.,* **1963,** 8Λ122Π.
46. Castor, C. R., Milton, R. M., U.S. Patent **3,013,987;** *RZh. Khim.,* **8Λ123**Π; Patent F.R.G. **1,145,152;** *C.A.* (1964) **61,** 3969*a*.
47. Rabo, J. A., Angell, C. L., Kasai, P. H., Schomaker, V., *Discus. Faraday Soc.* (1966) **41,** 328.
48. Dalla Betta, R. A., Boudart, M., *Proc. 5th Intern. Congr. Catalysis, Miami Beach, 1972* (1973) **1,** 329.
49. Naccache, C., Primet, M., Mathieu, M. V., *Advan. Chem. Ser.* (1973) **121,** 266.
50. Gallezot, P., Imelik, B., *Advan. Chem. Ser.* (1973) **121,** 66.
51. Lapidus, A. L., Isakov, Ya. I., Slinkin, A. A., Avetisyan, R. V., Minachev, Kh. M., Eidus, Ya. I., *Izv. Akad. Nauk SSSR, Ser. Khim.* (1971) 1904.
52. Richardson, J. T., *J. Catal.* (1971) **21,** 122.
53. Gryaznova, Z. V., Epishina, G. P., Michaleva, I. M., *Dokl. Akad. Nauk SSSR* (1972) **203,** 1339.
54. Zelenina, M., *Z. anorg. allgem. Chem.* (1972) **387,** 179.
55. Eidus, Ya. I., Minachev, Kh. M., Lapidus, A. L., Isakov, Ya. I., Avetisyan, R. V., *Izv. Otd. Khim. Nauk Bolg. A.N.* (1973) **6,** 307.
56. Romanowski, W., *Przem. Chem.* (1968) **47,** 741.
57. Yates, D. J. C., *J. Phys. Chem.* (1965) **69,** 1676; *Chem. Eng. Progr. Symp. Ser.* (1967) **63** (73), 56.
58. Tsutsumi, K., Takahashi, H., *Bull. Chem. Soc. Japan* (1972) **45,** 2332.

59. Isakov, Ya. I., Mirzabekova, N. V., Bogomolov, V. I., Minachev, Kh. M., *Neftekhimiya* (1970) **10**, 520.
60. Tsutsumi, K., Fuji, S., Takahashi, H., *J. Catal.* (1972) **8**, 24.
61. U.S. Patent **3,013,988**; *RZh. Khim.*, **1963**, 16Λ114Π.
62. U.S. Patent **3,013,990**; *RZh. Khim.*, **1963**, 8Λ125Π.
63. Ermakov, U. N., Kuznetsov, B. N., *Kinet. Catal.* (1972) **13**, 1355; *Dokl. Akad. Nauk SSSR* (1972) **207**, 644.
64. Miale, J. N., Weisz, P. B., U.S. Patent **3,136,713**; *C.A.* (1964) **61**, 6839*a*.
65. Frilette, V. J., Maatman, R. M., U.S. Patent **3,373,109**; *RZh. Khim.*, **1969**, 9Λ233Π.
66. Miale, J. N., U.S. Patent **3,344,058**; *RZh. Khim.*, **1969**, 1Π163Π.
67. Chen, N. Y., U.S. Patent **3,373,110**; *C.A.* (1968) **68**, 108395*a*.
68. British Patent **1,003,252**; *C.A.* (1965) **63**, 12948*g*.
69. Chen, N. Y., Weisz, P. B., *Chem. Eng. Progr. Sympos. Ser.* (1967) **63**, 81.
70. Chen, N. Y., French Patent **1,525,454**; *C.A.* (1969) **71**, 42776*d*.
71. Weisz, P. B., U.S. Patent **3,437,586**; *RZh. Khim.*, **1970**, 11Π175Π.
72. U.S. Patent **3,702,293**; *RZh. Khim.*, **1973**, 16Π109Π.
73. Reman, W. G., Ali, A. H., Schuit, G. C. A., *J. Catal.* (1971) **20**, 374.
74. U.S. Patent **3,544,451**, *RZh. Khim.*, **1971**, 17Π144Π.
75. Sutton, E. A., *Oil Gas J.* (1970) **68**, 100.
76. Gould, G. D., McCoy, C. S., *Oil Gas J.* (1970) **68**, 49.
77. *Oil Gas J.* (1973) **70**, 73.
78. Flanigen, E. M., *Advan. Chem. Ser.* (1973) **121**, 119.
79. Kerr, G. T., *Advan. Chem. Ser.* (1973) **121**, 219.
80. Cole, J. E., Kouwenhoven, H. W., *Advan. Chem. Ser.* (1973) **121**, 583.
81. Riekert, L., *Ber. Bunsenges. Phys. Chem.* (1969) **73**, 331.
82. Rubinschtein, A. M., Minachev, Kh. M., Slinkin, A. A., Garanin, V. I., Aschavskaya, G. A., *Izv. Akad. Nauk SSSR, Ser. Khim.* (1968) 786.
83. Ione, K. G., Bobrov, N. N., Boreskov, G. K., Vostrikova, L. A., *Dokl. Akad. Nauk SSSR* (1973) **210**, 388.
84. Kubo, T., Arai, H., Tominaga, H., Kunugi, T., *Bull. Chem. Soc. Japan* (1972) **45**, 607.
85. *Ibid.* (1971) **44**, 1968.
86. Wilson, G. R., Hall, W. K., *J. Catal.* (1970) **17**, 190.
87. Scholten, J. J. F., Konvalinka, J. A., *Trans. Faraday Soc.* (1969) **65**, 2465.
88. Chutoransky, P., Kranich, W. L., *J. Catal.* (1971) **21**, 1.
89. Lawson, J. D., Rase, H. F., *Ind. Eng. Chem., Prod. Res. Develop.* (1970) **9**, 317.
90. Jutasi, E., Beyer, H., Czaran, E., *Acta chim. Acad. Sci. Hung.* (1968) **58**, 427.
91. Lewis, P. H., *J. Catal.* (1968) **11**, 162.
92. Boudart, M., *Advan. Catal.* (1969) **20**, 153.
93. Weller, S. W., Montagna, A. A., *J. Catal.* (1971) **20**, 394.
94. Penchev, V., Kanazirev, V., *Izv. Otd. Khim. Nauk Bolg. A.N.* (1973) **6**, 381.
95. Mashchenko, A. I., Bronnikov, O. D., Dmitriev, R. V., Garanin, V. I., Kazansky, V. B., Minachev, Kh. M., *Kinetics Catalysis* (1974) **15**, 1603.
96. Little, L. H., "Infrared Spectra of Adsorbed Species," Academic Press, London, New York, 1966.
97. Bagraischvilli, G. D., Doksopulo, T. P., Agladze, L. D., *Soobshch Akad. Nauk. Gruz. SSR* (1970) **60**, 589.
98. Figueras, F., Gomez, R., Primet, M., *Advan. Chem. Ser.* (1973) **121**, 480.
99. Minachev, Kh. M., Antoshin, G. V., Shpiro, E. S., Novrusov, T. A., *Izv. Akad. Nauk SSSR, Ser. Khim.* (1973) 2134.
100. Romanowski, W., *Rocz. Chem.* (1971) **45**, 427.
101. Herd, A. C., Pope, C. G., *J. Chem. Soc. Faraday Trans. I* (1973) **69**, 833.
102. Egerton, T. A., Vickerman, J. C., *Ibid.* (1973) **69**, 39.
103. Mintschev, C., Steinbach, F., *Proc. 3rd Intern. Conf. Molecular Sieves, Zurich, 1973* (1973) 401.

104. Brooks, C. S., Christopher, G. L. M., *J. Catal.* (1968) **10**, 211.
105. Giola, F., Greco, G., Drioli, E., *Chim. Ind.* (1969) **51**, 457.
106. Penchev, V., Mintschev, C., Zolovski, I., *Izv. Otd. Khim. Nauk Bolg. A.N.* (1970) **3**, 529.
107. Penchev, V., *Ibid.* (1971) **4**, 573.
108. Penchev, V., Mintschev, C., *Ibid.* (1973) **6**, 403.
109. Minachev, Kh. M., Antoshin, G. V., Shpiro, E. S., Isakov, Ya. I., *Izv. Akad. Nauk SSSR, Ser. Khim.* (1973) 2131.
110. Olson, D. H., *J. Phys. Chem.* (1968) **72**, 4366.
111. Gallezot, P., Ben Taarit, Y., Imelik, B., *J. Catal.* (1972) **26**, 481.
112. Gallezot, P., Imelik, B., *J. Phys. Chem.* (1973) **77**, 652.
113. Egerton, T. A., Vickerman, J. C., *J. Chem. Soc. Faraday Trans. I* (1973) **69**, 39.
114. Guilleux, M. F., Tempere, J. F., *Compt. Rend. Acad. Sci.* (1971) **C272**, 2105.
115. Herd, A. C., Pope, C. G., *J. Chem. Soc. Faraday Trans. I* (1973) **69**, 833.
116. Delgass, W. N., Garten, R. L., Boudart, M., *J. Phys. Chem.* (1969) **73**, 2970.
117. Morice, J. A., Rees, L. V. C., *Trans. Faraday Soc.* (1968) **64**, 1388.
118. Rabo, J. A., Angell, C. L., Kasai, P. H., Schomaker, V., *Chem. Eng. Progr. Sympos. Ser.* (1967) **63** (73), 31.
119. Nowak, E. J., Coros, R. M., *J. Catal.* (1967) **7**, 50.
120. Benson, J. E., Kohn, H. W., Boudart, M., *J. Catal.* (1966) **5**, 307.
121. Verhoeven, W., Delmon, B., *Compt. Rend. Acad. Sci.* (1966) **C262**, 33; *Bull. Chim. Soc. France* (1966) 3065.
122. Charcosset, H., Frety, R., Soldat, A., Trambouze, Y., *J. Catal.* (1971) **22**, 204.
123. Selenina, M., Wencke, K., *Monatsber. Dtsch. Akad. Wiss.* (1966) **8**, 886.
124. Penchev, V., Minchev, H., Kanazirev, V., Tsolovski, I., *Advan. Chem. Ser.* (1971) **102**, 434.
125. Penchev, V. *et al., Advan. Chem. Ser.* (1973) **121**, 461.
126. Brownlie, I. C., Fryer, J. R., Welb, G., *J. Catal.* (1969) **14**, 263.
127. Rubinshtein, A. M., Slinkin, A. A. *et al., Kinet. Catal.* (1969) **11**, 295.
128. Nacache, C. M., Ben Taarit, Y., *J. Catal.* (1971) **22**, 171.
129. Turkevich, J., Ono, Y., *Advan. Chem. Ser.* (1971) **102**, 315.
130. Minachev, Kh. M., Dmitriev, R. V. *et al., Kinet. Catal.* (1972) **13**, 1095.
131. Penchev, V., Kanazirev, V., Minchev, H., *Dokl. Bolg. A.N.* (1969) **22**, 899.
132. French Patent **1,562,569;** *C.A.* (1970) **72**, 66348*h*.
133. Garwood, W. E., U.S. Patent **3,544,650;** *RZh. Khim.* **1971,** 15Л203П.
134. Rabo, J. A., Poutsma, M. L., Skeels, G. W., *Proc. 5th Intern. Congr. Catal., Miami Beach, 1972* (1973) **2**, 1353.
135. Minachev, Kh. M., Garanin, V. I., Novrusov, T. A., *Izv. Akad. Nauk SSSR, Ser. Khim.* (1973) 330.
136. Minachev, Kh. M., Garanin, V. I. *et al., Neftekhimiya* (1969) **9**, 808.
137. Afanasjev, A. I., Dorogochinskii, A. Z., Volpova, E. G., *Neftepererabotka Neftekhim.* (1966) **1**, 39.
138. Afanasjev, A. I., Dorogochinskii, A. Z., Volpova, E. G., Mirskii, Ya. V., *Ibid.* (1969) **10**, 18.
139. Yamamoto, N. *et al., Bull. Japan. Petrol. Inst.* (1966) **8**, 13.
140. Yamamoto, N. *et al., J. Japan. Petrol. Inst.* (1966) **9**, 531.
141. Young, D. A., U.S. Patent **3,287,256;** *C.A.* (1967) **66**, 30775*m;* U.S. Patent **3,239,451;** *R. Zh. Khim.,* **1967,** 12П124П; U.S. Patent **3,342,725;** *R. Zh. Khim.* **1968,** 21П112П.
142. Martin, H. Z., U.S. Patent **3,313,802;** *R. Zh. Khim.,* **1968,** 16П149П.
143. Sinfelt, J. *et al.,* U.S. Patent **3,356,510;** *R. Zh. Khim.* **1969,** 10H239П.
144. Kay, N. L., U.S. Patent **3,424,671;** *R. Zh. Khim.* **1970,** 5П183П.
145. Topchieva, K. V., Dorogochinskaya, V. A., *Khim. technol. topliv masel* (1973) **11**, 6.
146. Stormont, D. H., *Oil Gas J.* (1967) **65** (14), 149.
147. Belgian Patent **638,072;** *C.A.* (1965) **62**, 8915*h*.

148. Bursian, N. R. *et al.*, *Kinet. Catal.* (1971) **12,** 769.
149. British Patent **1,148,545;** *C.A.* (1969) **71,** 14824*z*.
150. Sale, E. E., U.S. Patent **3,238,120;** *R. Zh. Khim.*, **1967,** 7П126П.
151. Hansford, R. C., Hass, R. H., U.S. Patent **3,287,527;** *R. Zh. Khim.*, **1968,** 15П171П.
152. U.S. Patent **3,418,256;** *R. Zh. Khim.*, **1970,** 6П268П; French Patent **1,489,105;** *C.A.* (1968) **68,** 88815*g*.
153. Weisz, P. B., *Erdöl Kohle* (1965) **18,** 525.
154. Juguin, B., Clement, C., Leprince, P., Montarnal, R., *Bull. Soc. Chim. France* (1966) 709.
155. Gryaznova, Z. V., Kolodieva, E. V., *Dokl. Akad. Nauk SSSR* (1970) **190,** 1383.
156. Gryaznova, Z. V., Kolodieva, E. V., *Vestn. Mosk. univ., Ser. Khim.* (1970) **11,** 615.
157. Borunova, N. V., Freidlin, L. Kh. *et al.*, sb. "Zeolity, ikh sintez, svoistva i primenenie," Nauka, M.-L., 1965, p. 380.
158. Minachev, Kh. M., Garanin, V. I., Isakov, Ya. I. *et al.*, *Ibid.*, p. 374.
159. Garanin, V. I., Kurkchi, U. M., Minachev, Kh. M., *Kinet. Catal.* (1968) **9,** 1080.
160. Minachev, Kh. M., Garanin, V. I., Isakov, Ya. I., *Prob. kinet. i Catal.* (1966) **11,** 214.
161. Topchieva, K. V., Minachev, Kh. M. *et al.*, *Proc. 7th World Petrol. Congr.* (1967) **5,** 379.
162. Ryashentseva, M. A., Minachev, Kh. M., Avaev, V. I., Soviet Union Patent **364,584;** *R. Zh. Khim.*, **1973,** 17H120II.
163. Minachev, Kh. M., Garanin, V. I., Piguzova, L. I., Vitukhina, A. S., *Izv. Akad. Nauk SSSR, Ser. Khim.* (1966) 1001.
164. Areshidze, Kh. I., Chivadze, G. O., *Soobsh. A.N. Gruz. SSR* (1969) **56,** 97.
165. Areshidze, Kh. I. *et al.*, British Patent **1,257,607;** *C.A.* (1972) **76,** 45816f.
166. Gryaznova, Z. V., Balandin, A. A. *et al.*, *Dokl. Akad. Nauk SSSR* (1967) **175,** 381.
167. Sokolskii, D. V., Gogol, N. A., Shliomenzon, N. L., *Zh. Phizich. Khim.* (1972) **46,** 1767.
168. Sokolskii, D. V., Gogol, N. A., Shliomenzon, N. L., *Kinet. Catal.* (1972) **13,** 982.
169. Zhubanova, L. D., *Vestn. A.N. Kaz. SSR* (1971) **10,** 53.
170. Frilette, V. J., Weisz, P. B., U.S. Patent **3,529,033;** *R. Zh. Khim.* **1971,** 11H220П.
171. Schuit, G. C. A., Van Reijen, L. L., *Advan. Catal.* (1958) **10,** 242; Kubicka, H., *J. Catal.* (1968) **12,** 223.
172. Riesz, C. H., Weber, H. S., *J. Am. Oil Chem. Soc.* (1964) **41,** 464.
173. Galich, P. N. *et al.*, Sb. "Sintetycheskiye Zeolity," izd. A.N. SSSR, M., 1962, p. 260.
174. Galich, P. N. *et al.*, Sb. "Neftekhimiya," izd. A.N. Turk. SSR, Ashchabad, 1963, p. 63.
175. Galich, P. N., Gutyrya, V. S., Neimark, I. E. *et al.*, Sb. "Neftehkimiya," Naukova Dumka, Kiev, 1964, p. 13.
176. Penchev, V., Koleav, M., *Izv. Akad. Nauk SSSR, Ser. Khim.* (1966) 580.
177. Kubo, T., Arai, H., Tominaga, H., Kunugi, T., *Bull. Chem. Soc. Japan* (1972) **45,** 613.
178. Milton, R. M., British Patent **1,115,521;** *R. Zh. Khim.*, **1969,** 8П148П.
179. Milton, R. M., British Patent **1,106,171;** *R. Zh. Khim.*, **1969,** 2П175П.
180. U.S. Patent **3,458,593;** *R. Zh. Khim.*, **1970,** 17H4П; U.S. Patent **3,458,-592;** *R. Zh. Khim.*, **1970,** 16H12П.
181. U.S. Patent **3,700,749;** *R. Zh. Khim.*, **1973,** 17H8П.
182. French Patent **1,534,457;** *C.A.* (1969) **71,** 72652*c*.
183. Dal, V. I., Raskina, L. S., Moroka, E. E., *Izv. V.U.Z., Neft i gas* (1968) **9,** 43.

184. Milton, R. M., British Patent **1,115,061**; *R. Zh. Khim.*, **1969**, 12П143П.
185. Giordano, N., Montelatici, S., Zen, C., *Proc. 3rd Intern. Conf. Molecular Sieves, Zurich, 1973* (1973) 449.
186. Firth, J. G., Holland, H. B., *Nature* (1968) **217** (5135), 1252.
187. Okruzhnov, A. M., Izmailov, R. I., Virobyanz, R. A., *Neftekhimiya* (1964) **4**, 676.
188. *Ibid.* (1964) **4**, 850.
189. Penchev, V., Davidova, N., *Izv. Otd. Khim. Nauk Bolg. A.N.* (1971) **4**, 409.
190. Lawrance, P. A., Goble, A. G., British Patent **989,269**; *C.A.* (1965) **63**, 2824*e;* Patent F.R.D. **1,229,221**; *C.A.* (1967) **66**, 39609*t*.
191. Olive, M. F., British Patent **1,246,115**; *R. Zh. Khim.*, **1971**, 24П282П.
192. Kelly, J. T., Hall, D. W., U.S. Patent **3,494,976**; *R. Zh. Khim.*, **1971**, 7H37П.
193. U.S. Patent **3,512,772**; *R. Zh. Khim.*, **1971**, 10П193П.
194. Pospelova, T. A., Shekhobalova, V. I., Kobozev, N. I., *Zh. Phisiches. khim.* (1971) **45**, 1462; (1971 **45**, 1715.
195. Martynyuk, T. G., Pospelova, T. A., Shekhobalova, V. I., Kobozev, N. I., *Ibid.* (1973) **47**, 406.
196. Voronin, V. V., Minachev, Kh. M., Levitskii, I. I., *Izv. Akad. Nauk SSSR, Ser. Khim.* (1967) 2616.
197. Centola, P., Terzaghi, G., Del Rosso, R., Pasquon, I., *Chim. Ind.* (1972) **54**, 775.
198. Che, M., Dutel, J.-F., Primet, M., *Proc. 3rd Intern. Conf. Molecular Sieves, Zurich, 1973* (1973) 394.
199. U.S. Patent **3,253,887**; *R. Zh. Khim.* **1967**, 16Л170П.
200. Minachev, Kh. M., Garanin, V. I., Piguzova, L. I., Vitukhina, A. S., *Izv. Akad. Nauk SSSR, Ser. Khim.* (1966) 129.
201. Pichler, H., Schulz, H., Reitemeyer, H. O., Weitkamp, J., *Erdöl-Kohle-Erdas-Petrochem. Ver. Brennst.-Chem.* (1972) **25**, 494.
202. Kittrell, J. R., U.S. Patent **3,558,472**; *R. Zh. Khim.*, **1971**, 19Л257П.
203. Kluksdahl, H. E., U.S. Patent **3,660,310**; *C.A.* (1972) **77**, 51003*a*.
204. Pollitzer, E. L., U.S. Patent **3,594,310**; *R. Zh. Khim.*, **1972**, 8П118П.
205. Okruzhnov, A. M., Volfson, I. S., *Neftepererabotka Neftekhim.* (1970) **9**, 21.
206. Volfson, I. S., Okruzhnov, A. M., *Ibid.* (1971) **8**, 22.
207. Beecher, R., Voorhies, A., Eberly, P., *Ind. Eng. Chem., Prod. Res. Develop.* (1968) **7**, 203.
208. Voorhies, A., Hatcher, W. J., *Ibid.* (1969) **8**, 361.
209. Yan, T. Y., *J. Catal.* (1972) **25**, 204.
210. Weitkamp, J., Schulz, H., *J. Catal.* (1973) **29**, 361.
211. Robson, H. E., Hamner, G. P., Arey, W. F., *Advan. Chem. Ser.* (1971) **102**, 417.
212. Chen, N. Y., Garwood, W. E., *Ibid.* (1973) **121**, 575.
213. Piguzova, L. I., Parshina, A. I., Agafonov, A. V., Osipov, L. N., Khavkin, V. A., *Neftepererabotka Neftekhim.* (1968) **10**, 4.
214. Parshina, A. I., Agafonov, A. V., Piguzova, L. I., Osipov, L. N., Khavkin, V. A., *Ibid.* (1966) **12**, 17.
215. Mulaskey, B. F., U.S. Patent **3,620,963**; *R. Zh. Khim.*, **1972**, 17П242П.
216. Morris, H. C., U.S. Patent **3,663,430**; *R. Zh. Khim.*, **1973**, 3П183П.
217. Egan, C. J., U.S. Patent **3,647,681**; *R. Zh. Khim.*, **1972**, 24П261П.
218. Kurkchi, U. M., Garanin, V. I., Minachev, Kh. M., *Kinet. Catal.* (1968) **9**, 571.
219. Minachev, Kh. M., Garanin, V. I., Kharlamov, V. V., Isakova, T. A., Senderov, E. E., *Izv. Akad. Nauk SSSR, Ser. Khim.* (1969) 1737.
220. Kurkchi, U. M., Garanin, V. I., Minachev, Kh. M., *Nauchnye Trudy Samarkand. Univ.* (1969) **167**, 159.
221. Minachev, Kh. M., Garanin, V. I., Kharlamov, V. V., *Izv. Akad. Nauk SSSR, Ser. Khim.* (1970) 835.

222. Minachev, Kh. M., Garanin, V. I., Isakova, T. A., Kharlamov, V. V., Bogomolov, V. I., *Advan. Chem. Ser.* (1971) **102,** 441.
223. Garanin, V. I., Minachev, Kh. M., Isakova, T. A., *Neftekhimiya* (1972) **12,** 501.
224. Voorhies, A., Bryant, P. A., *AIChE J.* (1968) **14,** 852.
225. Beecher, R., Voorhies, A., *Ind. Eng. Chem., Prod. Res. Develop.* (1969) **8,** 366.
226. Voorhies, A., Hopper, J. R., *Advan. Chem. Ser.* (1971) **102,** 410.
227. Eberly, P. E., Kimberlin, C. N., Voorhies, A., *J. Catalysis* (1971) **22,** 419.
228. Hopper, J. R., Voorhies, A., *Ind. Eng. Chem., Prod. Res. Develop.* (1972) **11,** 294.
229. Allan, D. E., Voorhies, A., *Ibid.* (1972) **11,** 159.
230. Clement, C., Leprince, P., Montarnal, R., *Bull. Soc. Chim. France* (1966) 1021.
231. Topchieva, K. V., Dorogochinskaya, V. A., *Dokl. Akad. Nauk SSSR* (1973) **208,** 1415.
232. Piguzova, L. I., Prokofjeva, E. N., Dubinin, M. M., Bursiyan, N. R., Shavandin, U. A., *Kinet. Catal.* (1969) **10,** 315.
233. Schulz, H. F., Weitkamp, J. H., *Ind. Eng. Chem., Prod. Res. Develop.* (1972) **11,** 46.
234. Dubowik, R., Jaworska, Z., Kulak, S., Wrzyszcs, J., *Pr. nauk Inst. chem. i technol. nafty i wegle PWr.* (1971) **7,** 19; (1972) **10,** 12.
235. Schulz, H., Weitkamp, J., Eberth, H., *Proc. 5th Intern. Congr. Catalysis, Miami Beach, 1972* (1973) **2,** 1229.
236. Adams, C. E., Kimberlin, C. N., British Patent **1,143,140;** *R. Zh. Khim.,* **1970,** 4H155П.
237. Lanewala, M. A., Pickert, P. E., Bolton, A. P., *J. Catal.* (1967) **9,** 95.
238. Bolton, A. P., Lanewala, M. A., *Ibid.* (1970) **18,** 1.
239. Otouma, H., Arai, Y., Ukihashi, H., *Bull. Chem. Soc. Japan* (1969) **42,** 2249.
240. Jacobs, P. A., Van Cauwelaert, F. H., Vansant, E. F., Uytterhoeven, J. B., *J. Chem. Soc. Faraday Trans. I* (1973) **69,** 1056.
241. Minachev, Kh. M., Dmitriev, R. V., Isakov, Ya. I., Bronnikov, O. D., *Izv. Akad. Nauk SSSR, Ser. Khim.* (1973) 2689.
242. Pickert, P. E., Rabo, J. A., Dempsey, E., Schomaker, V., *Proc. 3rd Intern. Congr. Catalysis, Amsterdam, 1964* (1965) **1,** 714.
243. Maslyansky, G. N., Bursiyan, N. R., Barken, S. A., Kobelev, V. A., Telegin, V. G., *Izv. V.U.Z. Khim. i khim. technol.* (1960) **3,** 359.
244. Haensel, V., Donaldson, G., *Ind. Eng. Chem.* (1951) **43,** 2102.
245. Ciapetta, F. G., Hunter, J. B., *Ibid.* (1953) **45,** 155.
246. *Ibid.* (1953) **45,** 147.
247. *Ibid.* (1953) **45,** 159.
248. Ciapetta, F. G., Dobres, M. R., Baker, R. W., *Catalysis* (1958) **6,** 495.
249. Voorhies, A., Kimberlin, C. N., U.S. Patent **3,630,965;** *R. Zh. Khim.,* **1972,** 20Λ127П.
250. Mitsche, R. T., U.S. Patent **3,409,686;** *R. Zh. Khim.,* **1970,** 6H207П.
251. Donaldson, G. R., Pollitzer, E. L., U.S. Patent **3,409,685;** *R. Zh. Khim.,* **1970,** 6H280П.
252. Brodbeck, J. J., U.S. Patent **3,538,174;** *R. Zh. Khim.,* **1971,** 14H159П.
253. Burnett, R. L., U.S. Patent **3,390,199;** *R. Zh. Khim.,* **1969,** 16H184П.
254. Dvoretzky, I., Benesi, H. A., U.S. Patent **3,281,482;** *R. Zh. Khim.,* **1968,** 6H195П.
255. British Patent **1,291,928;** *R. Zh. Khim.,* **1973,** 11H106П.
256. Mitsche, R. T., Pollitzer, E. L., U.S. Patent **3,544,451;** *R. Zh. Khim.,* **1971,** 17П144П.
257. Kovach, S. M., Kmecak, R. A., U.S. Patent **3,631,117;** *C.A.* (1972) **76,** 71544*p*.

258. Slaugh, L. H., Leonard, J. A., *J. Catal.* (1969) **13,** 385.
259. Slaugh, L. H., *Tetrahedron* (1968) **24,** 4525.
260. Carter, G. B., Dewing, J., Pumphrey, N. W. J., *Proc. 5th Intern. Congr. Catalysis, Miami Beach, 1972* (1973) **1,** 329.
261. Japanese Patent **7 213'017;** *C.A.* (1972) **77,** 19178*a*.
262. Chloupek, F. J., U.S. Patent **3,668,269;** *R. Zh. Khim.,* **1973,** 4П146П.
263. Biale, C., U.S. Patent **3,644,565;** *R. Zh. Khim.,* **1972,** 24H231П.

Catalysis: Technology

Chapter

11

Preparation and Performance of Zeolite Cracking Catalysts

John S. Magee, Petroleum Chemicals Research Department, Davison Chemical Division of W. R. Grace & Co., Washington Research Center, Columbia, Md. 21044
James J. Blazek, Petroleum Chemicals Department, Davison Chemical Division of W. R. Grace & Co., Baltimore, Md. 21203

In the past 25 years the quantity of oil processed by U.S. refineries has increased from over 5,000,000 to nearly 12,000,000 barrels/day, and of this, approximately 35% is catalytically cracked. During this period the number of operating refineries has actually decreased from over 400 to 254, but their average size has increased to the point where 54 of these refineries process 66% of the daily capacity. In addition to the change in refinery number and capacity since 1947, the cracking catalyst requirement has progressively shifted from amorphous synthetic silica-alumina or natural catalysts to the use of zeolite-containing cracking catalysts. At this time, the major suppliers of fluid zeolite-containing catalysts to the industry are the Davison Chemical Division of W. R. Grace & Co., American Cyanamid Co., Filtrol Corp., and the Houdry Division of Air Products and Chemicals in conjunction with Engelhard Minerals & Chemicals Corp. Mobil Oil Corp. is a major supplier of moving bed (TCC) zeolite-containing catalysts.

Zeolites used in the manufacture of commercial fluid cracking catalysts are synthetic versions of the zeolite, faujasite. Before 1964, the quantity of commercially produced synthetic faujasite was negligible; however, in the past decade, the rate of synthetic faujasite production for catalytic cracking applications alone was between 21 and 30 million pounds per year. The principal factors contributing to this growth were the increased conversion, gasoline yields, and much improved coke selectivity afforded by these catalysts during a period of high gasoline demand and needed capacity expansion.

The present volume contains chapters describing zeolite active sites and kinetics and mechanism of cracking reactions over zeolites (*see* Chapters 8 and 9), shape selective transformation using faujasites (*see* Chapter 12), and diffusion considerations of reactions over faujasite (*see* Chapter 7); these topics are not discussed further in this chapter. Instead this chapter describes the practical complexities of zeolite cracking catalyst preparation on a laboratory and commercial scale and generally rationalizes the reasons for the main difference in activity and selectivity of such catalysts after both laboratory and commercial deactivation. The industry currently uses new reactor designs to optimize zeolite cracking catalyst operation for the widely different quality feeds which are available for cracking. The practical implications of these new trends are discussed.

Practical Aspects of Zeolite Cracking Catalyst Preparation

The gelation of sodium silicate by acid under carefully controlled conditions, can give rise to a variety of unique and useful products. Furthermore, partial neutralization of various combinations of silicate and aluminate solutions result in products totally different from the individual neutralization products of either of the two components. Yet another unique series of silica/alumina products results from the extended high temperature treatment of solutions of aluminate and silicate under proper stoichiometric conditions—the synthetic zeolites. The understanding of these general series of reactions lies at the heart of the preparation of active and stable zeolitic cracking catalysts.

Today's zeolite cracking catalysts are made from a matrix material, usually catalytically active itself, and an active zeolite cocatalyst of significantly higher intrinsic activity than the matrix. The predominant physical form of the catalysts used today is microspheroidal with an average particle diameter of 60 μ, so-called fluid cracking catalysts. This chapter emphasizes the preparation, characterization, and commercial operation of such materials.

To prepare satisfactory zeolite cracking catalysts, manufacturers must face the following important considerations:

(1) Commercial catalytic cracking units today are not able to use fully the activity of pure zeolite cracking catalysts. Thus, catalyst manufacturers must dilute the active zeolite with a matrix (generally, though not necessarily, catalytically active) to moderate its activity into a useful range.

(2) The matrix and zeolite must withstand the rigors of modern commercial operation—both must be stable to the high temperature hydrothermal treatment received in a commercial regenerator. Moreover, the cocatalyst combination must be hard enough to survive interparticle and reactor wall collisions without breaking down into particles of $<30\mu$ diameter which might be lost to the atmosphere.

(3) The pore structure of the matrix must allow for the desired activity and selectivity characteristics of the combined zeolite catalyst. The matrix can act positively or negatively in this respect, and the proper choice of this structure lies near the limit of present practical and theoretical knowledge.

Commercial Preparation of Fluid Catalysts

Fluid cracking catalysts today fall into two main types: (1) zeolite catalysts in which the zeolite, or material made from the zeolite, is added to a matrix of high porosity (the amorphous silica/alumina and related gelled materials including semisynthetic, clay-containing catalysts), and (2) zeolites, or materials made from zeolites, contained in matrices of low

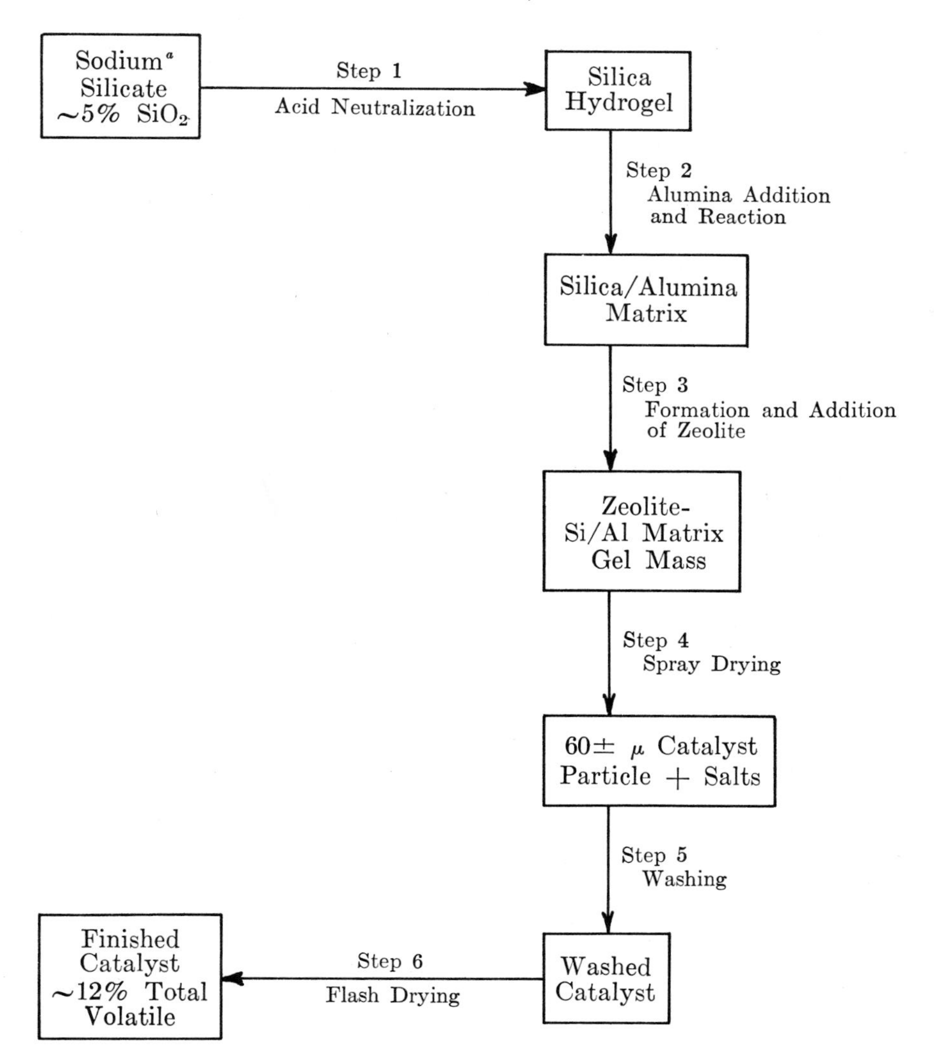

[a] Frequently manufacturers of gel-based catalysts prepare silica/alumina matrices containing kaolinite clay, so-called semisynthetic matrices. In such cases, the kaolinite clay (or other clays) is added to the silicate solution before gelation to effect optimum binding of silicate and clay particles.

porosity such as naturally occurring clays (montmorillonite, kaolin, halloysite, etc.). The respective preparative procedures for each zeolite and the resulting catalysts show both obvious similarities and marked differences.

Preparation of Silica/Alumina Gel Based Zeolite Catalysts. The largest family of zeolite catalysts available is that in which the zeolite component is added to an amorphous mass of silica/alumina gel which usually contains a kaolin clay diluent. This matrix material is catalytically active, and the control of its activity and structural stability has been the subject of considerable research over many years (*1, 2*). A simplified flow sheet of a typical zeolitic catalyst preparation involving this type of matrix is given on page 617.

Details of Preparative Steps, *Step 1: Acid Neutralization.* This preparative step effectively controls the bulk of the pore structure in the finished catalyst (*3, 4, 5*). In Figure 1 the effect on catalyst pore structure of using a strong mineral acid is compared with aluminum sulfate and carbon dioxide as the acidic neutralization agents of the sodium silicate solution. To compare these effects properly, however, the silicate solutions

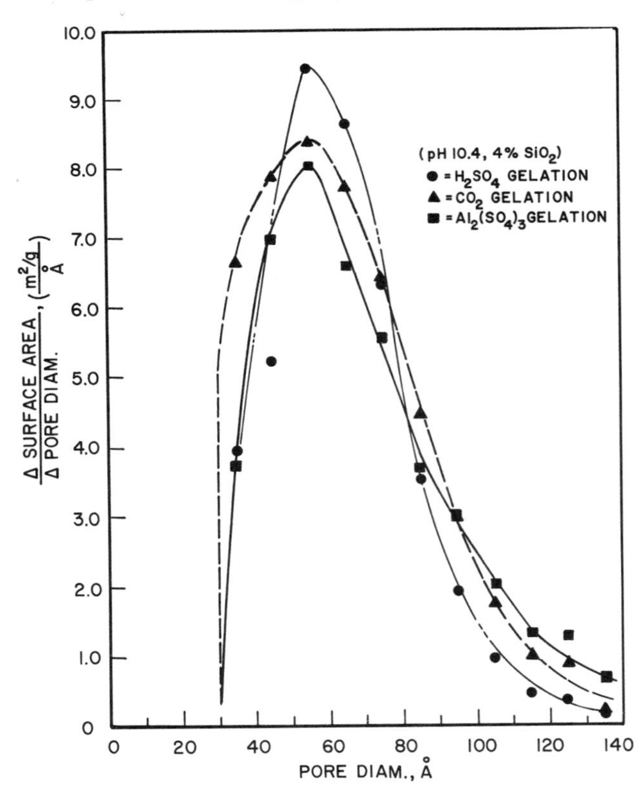

Figure 1. Surface area distribution of 13% alumina-silica cracking catalyst as a function of gelation agent

neutralized must be of equal concentration, temperature, and silica/soda ratio. The silicate generally used in experimental and production work has a SiO_2/Na_2O ratio of about 3.2:1 and a molecular weight of ~325 when freshly prepared (*6*). When samples of this material react at similar pH values with various acids at constant temperature, gelation will occur in approximately equivalent times. It is not necessary to neutralize the soda completely for gelation to occur; in a sodium silicate solution containing 5.5% SiO_2 at 80°F gelation will occur in 5–10 min when only 50% of the silicate soda has reacted. The catalyst pore structure can be controlled or moderated by varying the gel time and temperature, silica concentration, and by varying the aging time and temperature of the resulting silica gel, as shown in Table I.

Table I. Effect of Time, Temperature, and Silicate Neutralization on Cracking Catalyst Structural Properties

Catalyst	*1*	*2*	*3*	*4*
% Silicate Na_2O neutralization	60	60	115	115
Silicate gel-time (min)	5.0	5.0	0.25	0.25
Gelation and aging temp. (°F)	92	111	92	104
Aging time (min)	90	40	90	40
Catalyst surface area (m^2/g)	470	365	605	530
BET surface area median pore dia. (A)	58	78	51	62
2σ Pore size distribution	74	74	29	29
% Surface area in pores <50 A	34	25	45	30

Table I illustrates the structural changes in a 13% Al_2O_3, 87% SiO_2 amorphous cracking catalyst caused by changes in temperature, time, and the amount of soda neutralized in the sodium silicate at gelation. The four catalysts compared are of constant 0.77 cc/g pore volume and constant chemical composition. Catalysts were prepared using commercially available Philadelphia Quartz Co. N brand sodium silicate diluted to 5.5% SiO_2 with deionized water and using reagent grade aluminum sulfate as the Al_2O_3 source.

Analogous changes in pore structure are possible using the same variables in the preparation of zeolite cracking catalysts. In fact, matrix hydrothermal stability is directly related to the 2σ pore size distribution width which is a measure of the sharpness of the pore size distribution. Hydrothermal collapse of a zeolite matrix because of an unstable array of matrix pores can cause even more dramatic changes in catalyst activity in zeolite than in amorphous catalyst. Such collapse is generally related to the presence of substantial amounts of small pores (<30 A diameter) which are inherently unstable in steam in the presence of small amounts of alkali oxides (*1*). In zeolite catalysts this can cause a partial or total blockage or encapsulation of access paths to the zeolite intensifying the tortuous paths which the hydrocarbons must take to reach the active zeolite sites.

The pore structure shown in Figure 1 using different acidic neutralizing agents yield matrices of satisfactory commercial stability. The use of strong acid for gelation at constant pH and silica concentration yielded a pore structure virtually identical to that of CO_2 or aluminum sulfate gelation. Simplified chemical reactions for the processes may be written as follows.

For neutralizations involving other strong acids or carbon dioxide and an alkaline modified tetrahedral silicate ion in which silicon has a coordination number of 4 (*8*):

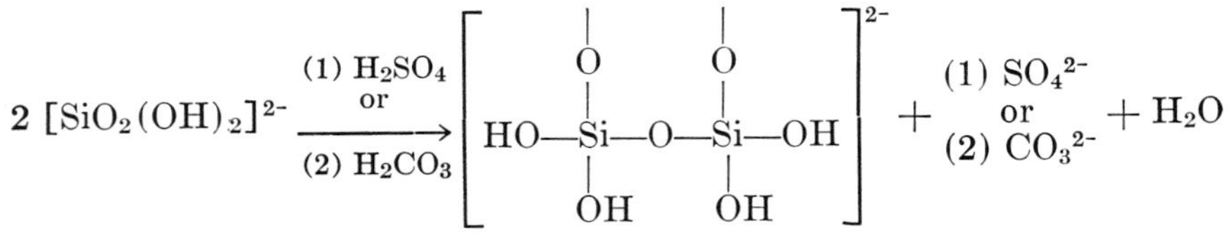

$$2\,[SiO_2(OH)_2]^{2-} \xrightarrow[(2)\ H_2CO_3]{(1)\ H_2SO_4 \text{ or}} \left[\begin{array}{ccc} & | & & | & \\ & O & & O & \\ & | & & | & \\ HO— & Si & —O— & Si & —OH \\ & | & & | & \\ & OH & & OH & \end{array}\right]^{2-} + \begin{array}{c}(1)\ SO_4^{2-} \\ \text{or} \\ (2)\ CO_3^{2-}\end{array} + H_2O$$

Following the formulation of such polysilicate ions, condensation to form discrete particles containing nuclei of SiO_2 occurs leading to ultimate gelation to the semisolid state. With acidic aluminum species, a two-stage process may be visualized as follows:

$$Al^{3+} + 4\,OH^- \rightleftarrows Al(OH)_4^-$$

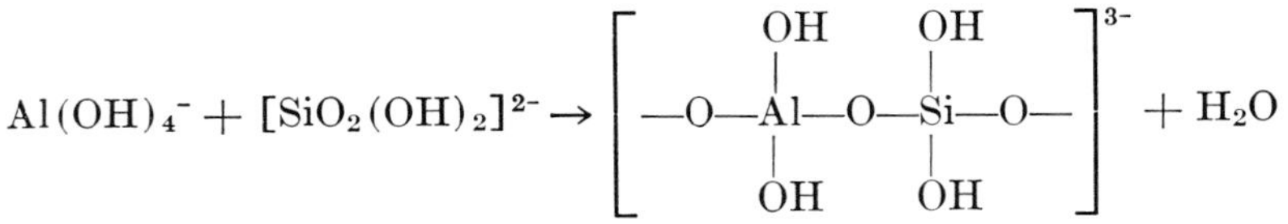

$$Al(OH)_4^- + [SiO_2(OH)_2]^{2-} \rightarrow \left[\begin{array}{ccccc} & OH & & OH & \\ & | & & | & \\ —O— & Al & —O— & Si & —O— \\ & | & & | & \\ & OH & & OH & \end{array}\right]^{3-} + H_2O$$

In this case the aluminum has been converted from coordination number 6 to 4 with subsequent reaction with the modified tetrahedral silicate ion to form the salt of the acidic silica/alumina complex.

Step 2: Formation of the Silica/Alumina Complex. The product of the reaction of aluminum ion and silica hydrogel forming a hydroxylated complex in which both silicon and aluminum assume tetrahedral coordination is responsible for the high acidity of the complex. In its simplest form, the ionized complex may be pictured as below:

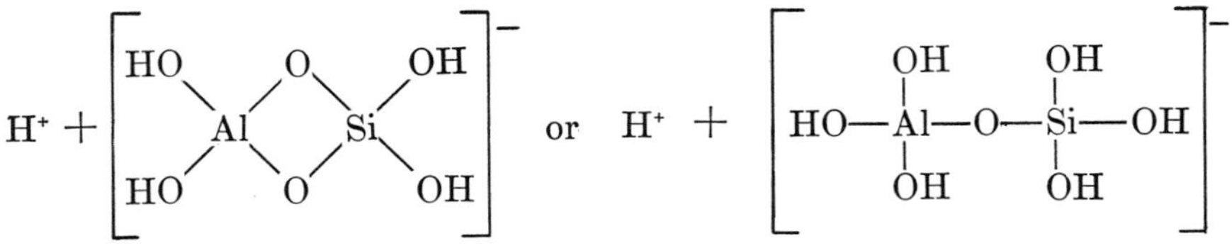

$$H^+ + \left[\begin{array}{ccccc} HO & & O & & OH \\ & \diagdown\;\diagup & & \diagdown\;\diagup & \\ & Al & & Si & \\ & \diagup\;\diagdown & & \diagup\;\diagdown & \\ HO & & O & & OH \end{array}\right]^{-} \quad \text{or} \quad H^+ + \left[\begin{array}{ccccc} & OH & & OH & \\ & | & & | & \\ HO— & Al & —O— & Si & —OH \\ & | & & | & \\ & OH & & OH & \end{array}\right]^{-}$$

The tetrahedrally coordinated aluminum ion has attained a net negative charge which in turn may be balanced by a proton. Prior to the introduction of zeolite cracking catalysts, this hydrolysis step was extremely important (*9, 10*) since it determined the degree of cracking activity and stability of the amorphous silica/alumina catalyst.

Step 3: Preparation and Addition of Zeolite. When zeolite catalysts were initially introduced, X-type synthetic faujasite of silica/alumina ratio typically 2.5:1 was generally used. Recently, the trend favors the more thermally and hydrothermally stable Y-type synthetic faujasite of SiO_2/Al_2O_3 ratio 4.5–5.5:1. These zeolites are prepared by digesting a mixture of silica, alumina, and caustic near the boiling point of the mixture for several hours until crystallization occurs (*11, 12, 13*). For both X- and Y-type faujasite, however, a mixed rare-earth ion-exchanged form is commonly used in cracking catalysts. The extent of ion exchange and the actual composition of rare earths in the exchange solution depend upon the rare earth source. In general, the starting faujasite of approximate formula Na_{58} $[(AlO_2)_{58}$-$(SiO_2)_{134}(H_2O)_{250}]$ (Y-type) is exchanged so that from 80–97+% of the Na_2O is replaced with equivalent amounts of rare earth-ions (usually lanthanum, neodymium, and cerium mixtures).

Principal sources of rare earth are monazite and bastnaesite sands, both of which give mixed rare earth chloride solutions containing predominately cerium, lanthanum and neodymium. The actual breakdown of the oxides from the two sources is given in Table II.

Table II. Distribution of Individual Rare Earth Oxides as a Function of Source (*14*)

	Monazite (%)	*Bastnaesite (%)*
Cerium	46	50
Lanthanum	24	24
Neodymium	17	10
Praseodymium	6	4
Samarium	3	1
Gadolinium	2	0.5
Others	2	0.5

Symposium on Rare Earth Applications

Distributions of rare earths on the faujasite after exchange indicate that no preferential exchange takes place, providing the exchange solution pH is such that cerium does not precipitate (an exchange pH range of 3.5 to 5.5 is necessary). Further, catalyst activity and selectivity in cracking are independent of the rare earth ratio of Ce, La, and Nd in the zeolite ingredient, but zeolite stability is directly proportional to lanthanum or neodymium content and inversely proportional to cerium content (*15*). Thus in commercial practice lanthanum rich exchange solutions are generally used for zeolite exchange.

A Y-type faujasite of ~4.8:1 SiO_2/Al_2O_3 will normally contain ~17 wt % RE_2O_3 (mixed rare earth oxide) when exchanged to ~3% residual Na_2O. This is the practical limit of exchange using normal laboratory exchange conditions (boiling aqueous solution of ~10% rare earth solution). A procedure (*16*) involving a calcination at 1000°F after rare earth exchange allows the ~2% Na_2O content to be easily reduced by ion exchange to a level of ~0.2%. Enhanced stability and activity of both X-type and Y-type faujasite result from this treatment.

In addition to rare earth-exchanged faujasites, ammonium-exchanged zeolite catalysts are also commercially produced (*see* below). As shown later, such exchanged zeolites appear to have minor coke and light gas selectivity advantages over rare-earth exchanged materials, but they are substantially less catalytically active on an equal weight basis.

Most of the published patent and journal information indicates that the zeolite is prepared as a separate processing step and is transported to the area in which a more or less conventional catalyst preparation is underway. However, some published information states that the zeolite manufacture is an integral and included part of the whole catalyst preparation, though such processes do not account for a large percentage of the total tonnage made. In several recently issued patents to Engelhard Minerals and Chemicals Corp. for example (*17, 18*), a zeolite-containing fluid cracking catalyst is prepared by spray drying an aqueous dispersion of kaolin clay followed by high temperature calcination (1800°F) to harden the microspheres. The addition of caustic and metakaolin followed by digestion of the mass yields a clay/sodium faujasite mixture which, when filtered and appropriately washed, is a highly active zeolite-containing cracking catalyst. Other commercial suppliers produce predominately gel-based matrix materials in which the zeolite is added at some point in the wet-end procedure prior to spray drying. When properly done, an extremely homogeneous dispersion of zeolite/matrix cocatalyst results which contains an extensive micro- and macropore structure. Micropore structure can be defined (*19*) as the relative quantity of surface area (or pore volume) which is contained in pores less than 500 A diameter. Macropore structure is, then, that which is contained in pores above 500 A pore diameter.

Clay based or semisynthetic catalysts invariably contain substantially more macropore structure than gel-based catalysts; however, zeolite catalyst activity and selectivity appear to be largely independent of the presence or absence of such structure. The gelatinous mass made by combining matrix material, zeolite, or material made from zeolite also contains by-product salt from the initial gel forming steps. In products of both American Cyanamid Co. (*20*) and the Davison Chemical Division of W. R. Grace & Co. (*21*), the by-product salt at this stage is sodium sulfate. Filtrol's published procedures (*22, 23*) indicate that by-product $(NH_4)_2SO_4$ formed from digestion of kaolin or halloysite clay with sulfuric acid and precipitation of hydrous aluminum hydroxide from the resultant aluminum sulfate with ammonium hydroxide, is separated before the catalyst is prepared (*see* below). The American Cyanamid process calls for filtering and washing Na_2SO_4 from the gelatinous zeolite matrix mass prior to spray drying. This filtration is generally accomplished using large rotary vacuum filters such as in Figure 2.

Step 4: Spray Drying. If there is a step in cracking catalyst preparation that could be called artful and even imposing, it would be spray drying. The commercial equipment is large, the nature of the spray dried product

Figure 2. Eimco rotary vacuum filter, 11½ x 16 ft., at Davison Chemical Division, W. R. Grace & Co., Valleyfield, Quebec

depends on many seemingly unrelated variables, and the product is, under ideal conditions, a well formed sphere about 60 μ in diameter. In a commercial cracking unit of moderate size, approximately 10^{21} particles are in intimate contact with vaporized gas oil molecules at linear velocities of from 15–50 ft/sec. High speed collisions with the reactor walls and other particles occur in which the strength of the fluid catalyst is severely tested.

Spray dryers come in a variety of shapes and sizes (*24, 25*), with a nearly unlimited variety of nozzle configurations and operating methods designed to change the particle size of the catalyst. The particle size affects the efficiency of catalyst fluidization—that is, the efficiency of gas oil/catalyst contact varies substantially from cracking unit to unit. While the average size used in units today is about 60 μ, wide variations from this size are required for optimum performance.

Atomization of the gel catalyst-mix controls catalyst particle size to a large extent and is itself affected by two substantially different factors: pressure and centrifugal force. Pressure atomization is widely used where maximum production rates with minimum maintenance are required but somewhat wide deviations from perfect spherocity are sometimes obtained. Centrifugal force nozzles (rotary atomizers) are also widely used and give uniformly spherical products with a variety of feed materials. Different overall dryer geometries are necessary, depending on the type of atomization used. Pressure atomization generally dictates dryers of $>2:1$ length to width ratio, whereas rotary atomizers give a broader spray pattern which is more nearly amenable to a dryer of lesser length to width ratio. Figures 3 and 4 show commercial installations of the two general types.

All spray dryers atomize an aqueous dispersion of a solid into a heated chamber operated at temperatures above 700°–800°F. The solids are converted in seconds into material of $\sim 60\ \mu$ diameter which is then collected in one or more conventional cyclone separators. The water content of the particle can be varied by adjusting the dryer temperature, air or feed flow rate, nozzle configuration, or the solids content of the dryer feed.

Since the details of particle size control, shape control, density, and hardness (as affected by spray drying conditions and dryer types) are largely proprietary, the known significant variables in this process step are listed below.

(1) Temperature change of the particle. Increasing dryer outlet temperature (the highest temperature that the catalyst particle is subjected to in the spray dryer) increases both particle hardness and density.

(2) Solids content of the feed. In general, the higher the solids content of the feed the greater the spray dried particle diameter.

Figure 3. Swenson spray dryer (22-ft diameter) installation at Davison Chemical Division, W. R. Grace & Co., Lake Charles, La.

Figure 4. Nichols spray dryer (32-ft diameter) installation. (Courtesy Freeport Kaolin Co., Gordon, Ga.)

(3) Pressure of atomization or velocity of centrifugal disk atomizer and atomizer design. Within identical designs, increasing pressures or atomizer disk velocities produce smaller particles.

(4) Thixotropic nature and/or viscosity of the feed material. For equal atomizing conditions, increases in feed viscosity increase the particle size of the catalyst particles.

Step 5: Base Exchange of the Spray Dried Product—Washing. After spray drying, the product is conveyed to a series of filter tables wherein contaminant salts are removed by base exchange. Several different types of filters are used (Figures 5 and 6), so-called sand tables or dumping pan filters. The purpose of this treatment is replacement of a deleterious ion in the catalyst with a harmless one. As indicated earlier, the Davison preparation procedure leaves sodium sulfate in the catalyst after spray drying. Both sodium and sulfate are cracking catalyst poisons, though the presence of sodium is significantly more serious at equal concentrations.

Removal of both ions is generally handled in a two-step fashion. The first filter table utilizes several spray bars of hot ammonium sulfate wash solution, effectively lowering the soda content to acceptable levels. (The actual acceptable level varies from one type catalyst to another. This is discussed in detail later). After this exchange, the filter cake is washed again with dilute ammonium hydroxide solution to displace sulfate ion. Figure 7 shows graphically the reduction in Na_2O and SO_4^{2-} caused by washing a spray dried product.

Figure 5. Oliver filter, scroll discharge, 65-sq ft filtering area. Davison Chemical Division, W. R. Grace & Co., Valleyfield, Quebec.

Figure 6. Finn Co. filter, pan type 18-ft diameter, 120-sq ft filtering area. Davison Chemical Division, W. R. Grace & Co., Valleyfield, Quebec.

In the past, wash solutions containing sodium sulfate, ammonium sulfate and ammonium hydroxide were discharged, but pollution control requirements have added stringent regulations on disposal of these materials. All catalyst manufacturers are now using various schemes either to recycle or to recover waste streams. Recycle streams containing the above salts are frequently added in places where makeup water would be required (as in dilution of concentrated sodium silicate to the normal operating level), but these added salts can cause significant and undesirable variations in the gel-forming mechanism (*26*).

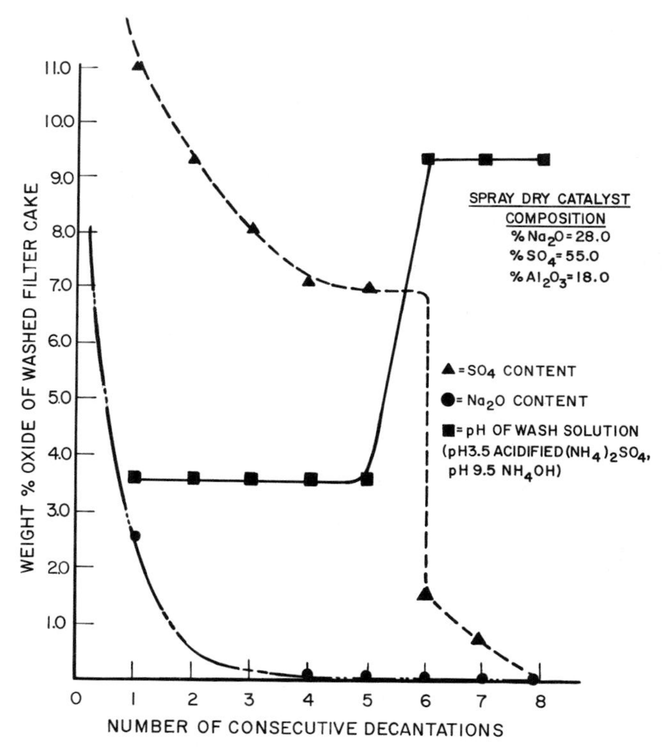

Figure 7. Washing efficiency of spray dried cracking catalyst as a function of wash solution pH

After exchange, the catalyst is washed free of the exchange anion (usually chloride ion) with either dilute ammonium hydroxide or water and is conveyed to the final drying step.

Step 6: Final Drying. The final drying step in gel-based catalysts accomplishes two purposes: liberation of water from the product and decomposition of exchanged ammonium ion to ammonia. In most commercial operations a large rotary dryer (flash dryer, Figure 8) is used, with drying temperatures rarely exceeding 400°–500°F. Product from the dryer will have a total volatile content of ~12 wt% at 1750°F, mostly adsorbed

Figure 8. Hersey rotary dryer, 8-ft diameter, 52-ft length, heating capacity 30,000,000Btu/hr. Davison Chemical Division, W. R. Grace & Co., Cincinnati, Ohio.

water not liberated in the flash dryer. Finished catalyst is then conveyed pneumatically to hopper cars or trucks equipped to unload catalyst pneumatically at the refinery.

Preparation of Silica/Magnesia Gel Based Zeolite Catalysts. Besides the silica/alumina gel based zeolite catalysts, the only other gel based catalyst which has been tested commercially is silica/magnesia. The general preparation details of this catalyst system have been described in the patent literature (*27*). As with the silica/alumina gel catalysts, the primary reaction step is silicate gelation to form a silica hydrogel. From this point on, however, the processing is considerably different and is summarized in the following brief outline.

SM-30 Preparation Scheme

Magnesia Activation:

$$MgO + H_2O \xrightarrow{H_2SO_4} \underbrace{Mg^{2+} + SO_4^{2-} + MgO}_{\text{Slurry I}}$$

The H_2SO_4 used above is known as the activation acid and is present in quantity sufficient to stoichiometrically react with ~10% of the starting MgO.

Catalyst Preparation:

$$Na_2O \cdot (SiO_2)_{3.2} + CO_2 \rightarrow \text{silica hydrogel} + Na_2CO_3$$

$\downarrow H_2SO_4$

Magna gel pH ~8.5 $\xleftarrow[\text{Slurry I}]{\text{Add}}$ SiO_2 gel pH ~3.0 in CO_2 evolution tank

$\downarrow$ (1) Add HF (2) Add zeolite (3) Spray Dry

Hot Age $\xrightarrow{\text{Wash}}$ $\xrightarrow[\text{Dry}]{\text{Final}}$ Finished Catalyst (27% MgO, 3% F, 70% SiO_2) + zeolite (5–15%)

During the hot aging step, the amount of tetrahedrally coordinated magnesia is maximized. This is analogous to the formation of tetrahedral alumina in silicate/alumina catalysts.

The chief problem with silica/magnesia catalysts is control of pore size distribution and catalyst stability characteristics.

Preparation of Clay-Based Zeolite Cracking Catalysts. Although the proper formation of a silica hydrogel is one of the most important steps in preparing gel-based catalysts, the step equally important in preparing clay-based catalysts is the formation of a clay particle of proper size, hardness and density. Commercially, this is accomplished in two ways: (1) by acid leaching of kaolin or halloysite, treating the resultant aluminum ion solution to form an aluminum hydrate precipitate and incorporating this with the leached clay particles to form a gelatinous clay/hydrous alumina mixture, or (2) by calcinating the clay particles formed by conventional spray drying techniques at temperatures above 1600°F to harden and to densify the particle. A schematic of the two commercially used preparative schemes can be written as follows:

Scheme 1: Leached Clay (22, 23):

$$\text{Halloysite or bentonite} \xrightarrow{H_2SO_4} Al_2(SO_4)_3 + \text{leached clay}$$

$$Al_2(SO_4)_3 \xrightarrow[\text{treatment}]{NH_4OH} \text{hydrated alumina} + (NH_4)_2SO_4$$

$\downarrow$ controlled NH_4OH washing (*22*)

35% boehmite, 10–15% bayerite, + 50% amorphous alumina. (Component A)

$$\text{Leached clay} \xrightarrow[H_2O]{\text{wash}} \text{Component B}$$

(*continued*)

Component B $\xrightarrow[\text{NaOH heated}]{Na_2SiO_3}$ Silicated clay.

Silicated clay $\xrightarrow[RE^{3+} \text{ or } Mg^{2+}]{\text{wash}}$ Component C

Component A + B $\xrightarrow[\text{mixing}]{\text{thorough}}$ aluminated sulfuric acid-treated clay

↓ Component C addition

Finished catalyst $\xleftarrow[\text{dry}]{\text{spray}}$ catalyst mix

Scheme 2: High Temperature Clay (*17, 18*):

Aqueous kaolin slurry $\xrightarrow[\text{dry}]{\text{spray}}$ kaolin microspheres

↓ calcine at 1800°F

kaolin hardened microspheres

↓ 1. Add NaOH 2. Add metakaolin

heat soak at 180°F 12 hr

↓

clay/sodium faujasite mixture $\xrightarrow[\text{wash/exchange}]{NH_4NO_3}$ exchanged clay/faujasite mixture

↑ calcine at 1550°F

Stabilized/faujasite microspheroidal clay catalyst

The two procedures both form a catalyst of mostly macroporous structure (surface area pores >500 A diameter). The effects on cracking activity, selectivity, and thermal and hydrothermal stability caused by these pore structures compared with the more microporous gel based catalysts are discussed later.

Preparation of Moving Bed Zeolite Cracking Catalysts (*28*). The initially introduced moving bed cracking processes, Houdriflow and Thermofor Catalytic Cracking (TCC), have now been largely replaced by fluid bed catalytic cracking. Moving bed operations using zeolitic catalysts in 1972 accounted for only ~10.7% of the daily consumption of catalyst and appeared to be decreasing constantly in importance (*29*). Nevertheless, the preparative methods used warrant brief description.

The chief reaction which must be controlled is the formation of a spherical particle ~4 mm diameter which is formed when an aqueous droplet passes through a column of hot mineral oil. According to the patent literature three different procedures are used commercially. One produces a predominately gel-based catalyst, the second a predominately clay-based catalyst, and the third a gel-clay combination.

Gel-Based Moving Bed Zeolite Cracking Catalyst (*30*). A sodium silicate solution containing a synthetic sodium faujasite is intimately mixed, using a mixing nozzle, with an acidic soluton of aluminum sulfate and sulfuric acid. The resulting hydrosol, containing the suspended sodium faujasite, is formed into spheroidal hydrogel beads by introducing globules of the hydrosol into a column of oil (*31*). A gelation time of less than 1 min is achieved. The salt-contaminated beads may be base exchanged with a variety of ions to eliminate sodium from the mixture of matrix and faujasite. Commercially, rare earth is used as the major base exchange ion. After washing to eliminate chloride or sulfate ion contaminants, the catalyst is dried as in microspheroidal catalyst preparation and then calcined. Many alterations appear possible to prepare predominately clay-based or semi-synthetic catalysts.

Clay-Based Moving Bed Zeolite Catalyst Preparation. A possible commercial procedure for manufacturing Houdry HZ-1 moving bed catalyst uses a particular type of plastic clay mined near Spruce Pine, N.C., called Avery clay which is composed of a mixture of halloysite and kaolinite (*32*). Ninety parts (by weight) of dry Avery clay are thoroughly mixed in a ribbon blender with 10 parts (by weight) of fine-sized (through 325 mesh) Y-type sodium faujasite. This mixture is combined with a 10 wt% aqueous sodium hydroxide solution in a pug mill long enough to produce a plastic and extrudable consistency. About 40 parts solution for every 100 parts of dry clay–zeolite mixture are required. The contents of the pug mixer are transferred to a worm-type extruder, formed into ⅛-inch pellets and dried at 250°F for 2 hr. After cooling, the pellets are thoroughly washed and exchanged with ammonium nitrate solution by continuously percolating the solution through batches of the pellets until the Na_2O content of the effluent is below 0.05%. The exchanged pellets are dried and calcined at 1200°F for 1 hr in air, yielding essentially an all-clay matrix plus synthetic faujasite cracking catalyst.

Filtrol Corp. produces gel-clay based moving bed catalysts by the procedure described above, except that the catalyst mix would be extruded or pelleted instead of spray dried. At this time, however, no Filtrol moving bed catalyst is marketed.

Summary of Commercial Grades of Zeolite Cracking Catalysts. Approximately 26 different grades of zeolite cracking catalysts were available commercially in early 1974. Table III lists these various grades divided into matrix type. Typical properties are given for each matrix. No doubt, the listing will be substantially expanded in the near future.

Laboratory Scale Preparation of Zeolite Cracking Catalysts. It is possible to prepare laboratory size samples of cracking catalysts virtually identical to their larger scale (pilot plant or commercial plant) counterpart in most physical, chemical, and catalytic properties. From a manufacturer's standpoint, there are areas, however, where the differences in properties of the catalysts are large enough to be very significant. Specifically the

Table III. Summary of Commercially Marketed Grades of Zeolite Cracking Catalysts

Manufacturer	*Tradenames*
Silica/Alumina Gel Based[a]	
Davison	XZ-25, 36, 40
	DZ-5, 7
	AGZ-50, 200, 290+
	CBZ-1, 2, 3, and 4
American Cyanamid	TS-170, 260, 280
Mobil	Durabead 6, 8, 9[b]
Clay Based[c]	
Houdry/M&C[d]	HFZ-20, 23
	HZ-1[e]
Davison	DHZ-15
Clay/Gel Based[f]	
Filtrol	F-800, 900
	AR-10, 20, 30
	75F

[a] Typical ranges of properties of gel-based catalysts are: surface areas 200–600 m^2/g, pore volume 0.25–0.9 cc/g, average bulk density 0.3–0.65 g/cc, zeolite content 5–20%.
[b] Moving bed catalyst.
[c] Typical ranges of properties of these catalysts are: surface areas 100–200 m^2/g, pore volume 0.1–0.3 cc/g, ABD 0.6–0.9 g/cc, zeolite content 5–40%.
[d] Engelhard Minerals & Chemicals Corp. manufactures fluid cracking catalyst for sale by Houdry.
[e] Houdry Process and Chemical Co., Division of Air Products and Chemicals, Inc., manufactures moving bed catalyst HZ-1.
[f] Typical ranges of properties of these catalysts are: surface area 175–250 m^2/g, pore volume 0.2–0.4 cc/g, ABD 0.4–0.6 g/cc, zeolite content 5–18%.

following important catalyst properties cannot be absolutely predicted from laboratory size preparation:

(1) Density

(2) Hardness

(3) Gelation properties (especially in cases where gas/liquid interaction is responsible for silicate neutralization)

(4) Spray drying

(5) Washing (exchange)

To define these properties adequately, pilot plant batches of catalyst must be made. Typical pilot unit equipment required for zeolite cracking catalyst preparation is given in Table IV.

An illustrative example of the laboratory preparation of a typical gel-based zeolite cracking catalyst is given below. Sections A and B deal with the preparation of the zeolite promoter, and Section C describes the composite catalyst. This catalyst will give typical levels of zeolite cracking catalyst activity for a catalyst in which the zeolite input is 10 wt% with normal product distributions (as discussed later).

A. *Procedure for Sodium Y-type Faujasite from Hi-Sil 233 (PPG Chemical Co.)*

Input ratio: 3.4 Na_2O : 1.0 Al_2O_3 : 9.5 SiO_2 : 136 H_2O

1. Place 2090 g of H_2O in a resin kettle.
2. Dissolve 162.5 g of NaOH in the water.
3. Add 425 g of sodium aluminate, containing 24 wt % Al_2O_3, 20 wt % Na_2O, and 56 wt % H_2O.
4. Cool to $T < 100°F$ and then add 570 g of calcined Hi-Sil 233.
5. Cold age at $T < 100°F$ for 24 hr with mild agitation.
6. Increase temperature to 212°F and age for 24 hr—mild agitation.
7. Quench with water, and filter to remove the solids from the mother liquor.
8. Wash with hot water and dry in forced air oven at 250°F.

Yield 464 g dry basis $Na_2O \cdot Al_2O_3 \cdot 5.0\ SiO_2$ Na-Y

Surface area = ~850 m^2/g

Typical Chemical Analysis, Dry Basis, wt %:

Al_2O_3 : 22.0
SiO_2 : 64.6
Na_2O : 13.4

B. *Preparation of Rare Earth Exchanged Y-type Faujasite (REY) from Na-Y*

Basis: 500 g of NaY (dry weight).

1. Mix 500 g of NaY (from Part A) in 3000 ml of boiling water and add 460 g of $ReCl_3 \cdot 6H_2O$. Reslurry for 1 hr and then filter.
2. Repeat Step 1.
3. Wash free of chloride ion with hot water.

Preparation of Calcined Rare Earth Exchanged Y-type Zeolite ("CREY") from Na-Y

Basis: 500 g of NaY (dry weight).

1. Same procedure as for RE-Y except the material is calcined for 3 hr at 1000°F after Step 3.

Surface Area = ~700 m^2/g

Typical Chemical Analysis for CREY or REY, Dry Basis (1750°F) Wt %

RE_2O_3 : 16.4
Al_2O_3 : 20.3
SiO_2 : 60.0
Na_2O : 3.3

Table IV. Typical Equipment Necessary for Pilot Plant Preparation of Fluid Cracking Catalysts

Equipment	*Purpose*	*Construction*
100-gal heated tank	storage of dilute sodium silicate	316 S.S.
10-gal tank	dilute H_2SO_4 storage	polyethylene
50-gal heated tank	receiving tank for gel	316 S.S.
0–2 gal/min Moyno pump	silicate pump	316 S.S.
0–1 gal/min Moyno pump	acid pump	Hastelloy B
Rotameters		
0–1 gpm	acid flow	Hastelloy B
0–2 gpm	silicate flow	316 S.S.
Piping and valves	acid	Hastelloy B
	silicate	316 S.S.
Inline mixer	mixing of SiO_2 and acid stream 3000 rpm	316 S.S.
Agitators		
gel tank	300 rpm (geared)	316 S.S.
silicate tank	~1000 rpm (variable speed)	316 S.S.
Spray dryer (Swenson-pilot plant)	spray drying of 50–15,000 gram batches	304 S.S.
Miscellaneous tanks and filters	washing	
Forced draught drying ovens	drying	

C. *Preparation of Cracking Catalyst from 10% CREY Promoter and 90% Microspheroidal (MS) 25 Matrix (25% Al_2O_3–75% SiO_2)*

Basis: 500 g batch.

1. Place 11.85 g of N-brand (Philadelphia Quartz Co.) sodium silicate (28.7% SiO_2–8.9% Na_2O) in a 10 l vessel and dilute with 4965 g of water to give a 5.5% SiO_2 solution, and heat to 100°F.
2. Add 300 g of 35% H_2SO_4 to neutralize 65% of the Na_2O in the sodium silicate (2 min gel time).
3. Age 30 min at 100°F.
4. Add 1.50 l of concentrated alum, which contains 75.0 g Al_2O_3/1. Age 10 min at 100°F.
5. Increase pH to 6.0 with NH_4OH.
6. Add 50 g of zeolite promoter (well dispersed in water).
7. Filter, reslurry filter cake in water to give ~15% silicate slurry, and spray dry.
8. Reslurry spray dried Xerogel in 3 l of a 3% $(NH_4)_2SO_4$ solution.
9. Exchange with 3 l of 3% $(NH_4)_2SO_4$ solution by displacement washing.

10. Reslurry and filter with 3 l of NH_4OH solution of pH = 9.0 twice.
11. Wash with 3 l hot (140°F) deionized H_2O twice.
13. Flash dry (forced draught oven, 400°F, 4 hr).

Typical Properties of Promoted Catalyst, Chemical Analysis, Dry Basis wt %

TV @ 1750°F	=	10%
Al_2O_3	=	24.6%
SiO_2	=	73.8%
Na_2O	=	<0.1%
SO_4	=	<0.1%
RE_2O_3	=	1.6%

Physical Analysis:

Surface area:	350 m^2/gm
N_2 pore volume:	0.55 m^2/gm
Average bulk density:	0.50 gm/cc

Analysis of Zeolite Catalysts

Analyses of the zeolite cracking catalysts discussed above must be extensive and must accurately define a variety of chemical and physical properties. Two classically used analyses—the determination of reactive alumina (*9*) and the use of Hammett indicators to determine acid titer (*33, 34*) and acid site strength as a function of activity and selectivity (*35*)—show an interesting anomaly when applied to zeolite cracking catalysts. The reactive alumina test was used for a quantitative determination of the amount of tetrahedrally coordinated alumina in amorphous catalysts (directly proportional to the catalysts' activity and stability). With different zeolite cracking catalysts made with equal amounts of zeolite, catalyst activity of each was essentially equal regardless of the reactive alumina content of the matrix. This is simply another indication of the dominant activity of the dispersed zeolite compared with the matrix.

Interestingly, Hammett indicators show the substantial increase in acid site content in changing from a 28% Al_2O_3/72% SiO_2 amorphous cracking catalyst (containing 25% reactive Al_2O_3) to a hydrogen exchanged Y-type faujasite which has, by definition, 25%tetrahedrally coordinated alumina. In the former case, a total acidity of 0.38 meq/g was observed (*36*), distributed as follows (compared with equivalent sulfuric acid strength): 0.05 meq/g ≈48–71%, 0.04 meq ≈71–91%, and 0.29 meq ≈>91% sulfuric acid. For the hydrogen exchanged Y-type faujasite, a total acidity of 0.74 meq/g was found with the total acidity all equivalent to >91% sulfuric acid. The same study showed a typical rare earth exchanged Y-type faujasite to have total acidity of 1.25 meq/g and that, within experimental error, all sites were equivalent in strength to >91% H_2SO_4.

Chemical Analysis. Catalysts are, in general, analyzed chemically for alumina (Al_2O_3), silica (SiO_2), soda (Na_2O), rare earth oxide (RE_2O_3), ammonia (NH_3), and some anionic substances which originated in the exchange or washing solution such as sulfate, nitrate, or chloride (SO_4, NO_3, Cl). Of the above analyses, soda and rare earth oxide contents are by far the most important. In most zeolite catalysts, intrinsic thermal and hydrothermal stability of the catalyst depend on the absolute level of these two components. Published procedures for these analyses are available (*37, 38*).

Physical Analysis. The more important physical properties of the catalyst which are routinely determined are its surface area, pore volume, pore size distribution, density, attrition resistance, particle size distribution, and finally its thermal and hydrothermal stability. From the standpoint of commercial acceptability and success, obtaining the proper physical properties of the catalyst is often considerably more important than the chemical properties (although the two are, in some instances, interrelated).

As with methods of chemical analysis, physical testing of zeolite cracking catalysts follows the same procedures developed for amorphous catalysts and detailed descriptive references are available (*3, 25, 37, 39*). A brief description of the critical areas of catalyst hardness, density, and particle size distribution is given below.

Hardness, Density and Particle Size Distribution as Related to Catalyst Retention in Cracking Units. In recent years, catalyst manufacturers and commercial refiners have had to meet increasingly stringent pollution control standards in both air and water. The loss of fluid cracking catalyst from commercial operations, where mechanical difficulties are absent, is directly related to an intercombination of three of the physical properties listed below: hardness, density, and particle size distribution. Hardness is measured by various laboratory methods, but in each process catalyst particles impinge on each other and surrounding surfaces at high air velocity, causing particle breakdown (*40*). The rate of formation of a specific size fraction is then a direct measure of the relative hardness of the catalyst. In the measurement of the Davison Index, for example, a catalyst sample is subjected to a high velocity air jet in a Roller particle size analyzer (available from American Instrument Co.). The rate of breakdown of particles to smaller than 20 μ is determined and expressed as a Davison Index (*DI*), calculated as follows:

$$DI = \frac{100(A\text{-}B)}{C}, \text{ where}$$

A = 0–20 μ content of catalyst after attrition.

B = 0–20 μ content of catalyst before attrition.

C = $>$ 20 μ of catalyst prior to attrition.

Average Bulk Density (ABD) is an indirect measure of a catalyst's porosity as contrasted to skeletal density (*SD*), which may be calculated from the non-porous structural mass/unit volume of the individual oxides present. In the present discussion of catalyst unit retention, the ABD is considered the more important parameter. In catalysts where particle size distributions are unpredictably variable, the correlation of ABD and porosity (or catalyst pore volume) becomes so unsatisfactory that a normalizing procedure which determines compacted density (CD) is often used. A brief description of the procedures used to determine ABD and CD is as follows.

Procedure: Approximately 50 ml of a calcined (3 hr at 1000°F) sample of equilibrium or fresh catalyst is rapidly poured into a tared 25-ml graduated cylinder using a carefully placed, standard copper funnel to avoid spillage. Excess catalyst is brushed from the tared graduate and the ABD is calculated from the weight of catalyst contained in the graduate.

Thus:

$$\text{ABD (g/ml)} = \frac{\text{wt of catalyst, g}}{\text{25 ml}}$$

Compacted Density: Compacted Density (CD) represents the weight/unit volume of catalyst at maximum compaction. The catalyst sample is placed in a graduated cylinder and vibrated until no further loss in volume is observed. The compacted density is then calculated by dividing the sample weight by the final sample volume.

Particle Size Distribution. In the preparative section (spray drying) the major parameters affecting particle size distribution of zeolite cracking catalysts were discussed. All spray drying procedures yield a product with typical bell-shaped particle size distribution. However, there is no single optimum distribution of particle sizes in commercial usage since the optimum distribution also depends on the mechanical design of each commercial unit and its fluidization equipment. Thus, a fairly wide range of distributions is generally made by the manufacturer. Quantitatively these are determined by various procedures; however the micro-mesh sieving procedure is the one most widely used. This procedure is outlined below (*37*).

A small catalyst sample is classified into seven size ranges using a column of precision ($\pm 2\ \mu$ openings) 3-inch diameter screens conforming to ASTM tentative specification E161-60T. Prior to classification, the catalyst sample is humidified with water to reduce static charge. A known weight is placed in the top screen and shaken in a vibrator for a fixed time period. The weight of catalyst on each screen (20, 30, 60, 80, 105, and 149 μ) is taken and treated as follows:

$$\text{wt \% retained on each sieve} = \frac{A \times 100}{W}$$

$$A = \text{wt sample retained (g)}$$

$$W = \Sigma \text{ retained wts (g)}$$

This may then be converted to a "percent finer than" designated size as follows:

$$\text{wt \% finer} = 100\text{-}B$$

where B = cumulative wt % retained at each micron level.

An arithmetic probability plot of these values will allow readings at any particle size of the "percent finer than" value. The average particle size is determined from the appropriate micron size corresponding to "50% finer than".

Typical Davison zeolite cracking catalysts are available in coarse, medium, and fine particle size distributions regardless of matrix type with typical micro-mesh analyses as shown in Table V.

Table V. Micro-Mesh Distributions of Typical Commercial Zeolite Cracking Catalysts

Grade Designation	*Micro-Mesh Distribution, wt %*					
	0–20	*0–40*	*0–80*	*0–105*	*0–149*	APS, μ
Coarse	2	17	68	88	98	65
Medium	2	19	72	90	98	62
Fine	2	30	87	96	99	51

Of particular importance is the percent of total sample that is in the 0–20 μ and 0–40 μ range. Particle size distributions are not as dependent on catalyst composition as on manufacturing differences (particularly differences in spray drying).

Commercial cracking operations today are not only governed by particular demands for various products but by increasingly stringent air pollution controls. Chief among these controls are regulations governing the emission of noxious gases (SO_2, NO, NO_2, etc.) and particulate emissions (predominately catalyst fines). Refiners recover the majority of these fines in primary and secondary cyclones attached to the regenerator, but emissions are at times beyond currently acceptable levels.

The catalyst manufacturer can help control emission of particulate matter by making catalysts which are hard, reasonably dense, and with particle size distributions controlled so that minimum amounts of 0–40 μ material are present. However, the quantitative interrelationship of these factors has been a subject of some disagreement. A mathematical model similar to one reported by Zenz and Weil (*41*) has been developed (*42*) to weigh the value of each of these catalyst properties.

By proper consideration of gravitational forces, drag on particles of various sizes and densities and by application of Newton's second law ($F = ma$), entrainment rate (loss from unit) is directly related to its superficial velocity.

$$V_p = V_f - \frac{\rho_p D_p^2 g}{18\,\mu_f} \qquad (1)$$

V_p = velocity of particle (ft/sec)
V_f = velocity of regenerator air (ft/sec)
ρ_p = particle density (lb/ft^3)
D_p = particle diameter (ft)
g = gravitational acceleration (32 ft/sec^2)
μ_f = viscosity of air (lb/ft-sec)

Graphical representations where catalyst density is varied (rather than air velocity) show that increasing catalyst density has a greater effect on the entrainment rate of larger particles *vs.* smaller ones (those in the 0–30 μ diameter range).

Equation 1 can be modified to show the effect of drag coefficient on particle velocity as a function of particle diameter. Graphical analyses showed that a rapid drop-off in particle velocity occurs as drag is reduced and that a 130 μ particle could fall to a velocity of zero; however even under such reduced drag conditions particles $<30\ \mu$ diameter could still be of high enough velocity to be lost.

In general the present mathematical treatment agrees with studies of Frantz and Juhl (*43*) and Lewis, Gilliland, and Lang (*44*) in that superficial particle velocity and diameter were the main parameters controlling catalyst entrainment with consequent loss. Thus:

(1) The velocity of the catalyst particle is determined largely by the superficial velocity of the regenerator air.

(2) Attrition resistance is critical in reducing losses by entrainment since the primary source of loss of particles $<30\ \mu$ is from the abrasion (and consequent reduction in particle size) in the various size fractions $>30\ \mu$.

(3) Increasing catalyst density is important only for retaining particles $>30\ \mu$ in commercial units. Virtually no density effect is observed with particles $<30\ \mu$ in diameter.

(4) Increasing density or reducing the drag force (lifting force) does not help in the retention of particles $<30\ \mu$.

(5) Catalyst particles $<30\ \mu$ diameter are rapidly lost regardless of their residence time in the unit (age in unit and particle density are directly proportional).

Laboratory Deactivation and Testing of Zeolite Catalysts

Deactivation. The catalyst circulating in virtually every catalytic cracking unit has a substantially lower surface area than when it entered the unit. The bulk of this surface area loss is probably experienced in the regenerator, since the highest particle temperatures (up to 1400°F in the presence of a substantial steam partial pressure) are reached in this part of the unit. [The actual mechanism involves the scintillation of the steam-stripped but still oil-wetted particles as they pass through air bubbles (*45*).]

Meaningful industrial research designed to develop improved cracking catalysts can only be done when methods are developed to adequately simulate the commercial equilibrium properties of catalysts. Ideally a laboratory deactivation should closely follow a commercial unit which contains catalyst of all ages and activity levels and produces a weighted average yield each day with periodic additions of fresh catalyst and periodic withdrawals of used catalyst. However, in small scale cracking units, it has been shown (*45, 46*) that deactivation rates are exceedingly slow and only approach commercial equilibrium activity levels after operation for several months. The partial answer to the problem used by catalyst researchers is hydrothermal deactivation of fresh catalyst to approximately the *average* surface area, pore volume, and activity level of an equilibrium catalyst of similar composition. The major shortcomings of this procedure are that each catalyst particle has the same surface area, pore volume, pore diameter, and activity as every other particle, rather than a skewed, normalized distribution of these values from fresh to equilibrium as would be found in a commercial unit. These factors are not of overriding significance, however, because the actual distribution spread formed commercially is rather narrow.

Besides the above shortcomings, a major problem arises in that different catalyst types (clay-based, synthetic, semisynthetic, etc.) may deactivate differently in the same commercial unit under identical operating conditions.

Table VI. Commercial *vs.* Laboratory Deactivation of XZ-25

Feed: Colorado-Wyoming Gas Oil (29.1°API)
Conditions: 920°F, C/O = 4, CFR = 1

Deactivation	*Commercial*[a]	*1520°F 20% Stm., 12 hr*
surface area, m^2/g	98	115
pore volume, cc/g	0.38	0.33
relative crystallinity[b]	100	100
microactivity[c]	70.0	71.5
Conversion at 10 WHSV		
feed rate, vol %	63.5	64.5
hydrogen, wt %	0.067	0.038
$C_1 + C_2$, wt %	1.2	1.3
C_3-olefin, vol %	6.5	7.3
C_3, vol %	1.0	1.3
C_4-olefin, vol %	4.0	3.9
i-C_4, vol %	3.9	5.0
n-C_4, vol %	0.4	0.2
C_5^+ Gasoline		
vol %	55.0	55.0
gravity: °API	57.0	56.7
aniline pt.: °F	80	83
Light Cycle Oil		
vol %	23.5	21.5
gravity: °API	26.9	26.5
aniline pt.: °F	140	136
Coke		
wt %	3.5	4.0

[a] 90+% in inventory.
[b] Commercial Sample = 100.
[c] 900°F, 2 WHSV, 6 c/o.

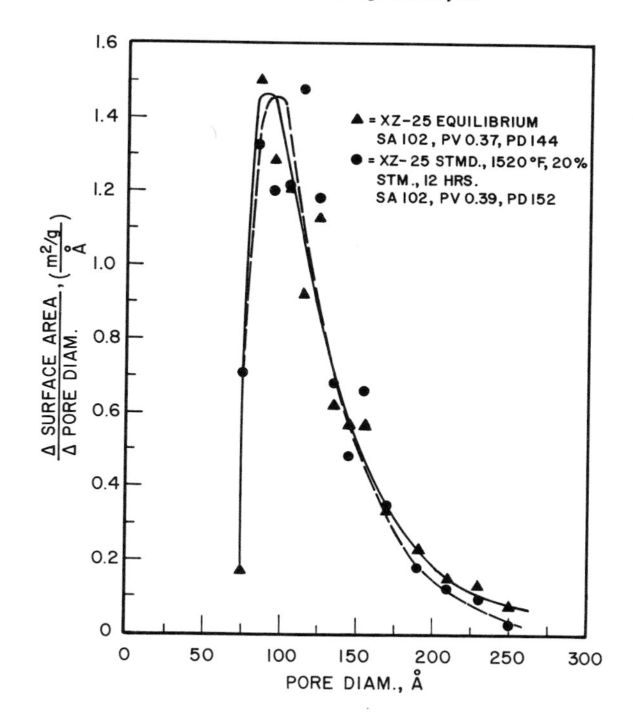

Figure 9. Pore size distribution comparison equilibrium catalyst vs. *laboratory hydrothermal treatment*

Thus a variety of hydrothermal deactivation conditions are needed to predict catalyst activity in the field. High surface area catalysts must be deactivated in the laboratory under more severe conditions than low area catalysts to match commercial equilibrium activity levels. The controlling surface area here is the surface area of the matrix co-catalyst, not that of the zeolite ingredient. Thus, the severity of laboratory hydrothermal deactivation of presently available fluid catalysts must be decreased in going from synthetic to semisynthetic to clay matrix catalysts to obtain good correlation with commercial equilibrium catalyst activity and physical properties. For example, commercial equilibrium Davison XZ-25 is compared in Table VI with laboratory deactivated XZ-25.

Not only is the conversion virtually identical but the various product selectivities are very similar. In addition, the composition of the gasoline and light cycle oil is also similar, based on their API gravities and aniline points. The physical properties and pore size distributions of the catalysts are similar (Figure 9), indicating that the partial hydrothermal collapse of the catalyst structure either hydrothermally in the laboratory or in the commercial unit must follow similar mechanistic courses.

Data in Table VII show the effect on conversion as measured in a standard microactivity test (*62*), caused by varying laboratory deactivations

Table VII. Effect of Composition on Steam Deactivation Stability[a]

Catalyst: Description	*Clay Matrix + RE-Y Sieve*	*Semi-synthetic Matrix + CREY*	*Clay Matrix + Calcined Ammonium Exchanged Y-type Zeolite*
Chemical Analysis (wt %)			
RE_2O_3	2.3	2.3	0
Al_2O_3	41.5	29.8	56.5
Na_2O	0.74	0.48	0.92
SO_4	1.6	0.40	0.02
Thermal Analyses: 3 hr at 1000°F			
SA: m^2/gm	215	265	334
N_2PV: cc/gm	0.23 (0.50)[b]	0.34 (0.50)[b]	0.38 (0.52)[b]
peak ht.: mm/BK	85	65	87
Hydrothermal Analyses: S-13.5 Steam: 100% steam, 15 psig, 8 hr, 1350°F			
SA: m^2/gm	121	123	215
N_2PV: cc/gm	0.20 (0.44)[b]	0.24 (0.33)[b]	0.33 (0.47)[b]
peak ht.: mm/BK	42	37	38
M-A[c]: vol % conv.	68.0	73.0	65.0
S-20 Steam: 20% Steam, 0 psig, 12 hr, 1520°F			
SA: m^2/gm	56	121	197
N_2PV: cc/gm	0.41 (0.43)[b]	0.27 (0.32)[b]	0.33 (0.50)[b]
peak ht.: mm/BK	15	33	34
M-A[c]: vol % conv.	20.0	67.5	64.0
Equilibrium Catalyst Analysis:			
SA: m^2/gm	127	124	245
N_2PV: cc/gm	0.21 (0.38)[b]	0.25 (0.37)[b]	0.31 (0.52)[b]
peak ht.: mm/BK	40	31	45
M-A[c]: vol % conv.	66.5	67.0	68.5
Shock + S-13.3 Steam: 1200°F, 1½ shock thermal treat + 1330°F, 100%, 8 hr, 15 psig steam.			
SA: m^2/gm	—	—	242
N_2PV: cc/gm	—	—	0.33 (0.46)[b]
peak ht.: mm/BK	—	—	42
M-A[c]: vol % conv.	—	—	68.0

[a] Enclosed steam deactivation results most closely approximate the surface properties and catalytic activity of the equilibrium catalysts shown in the dashed enclosure for each different catalyst composition.
[b] H_2O pore volume.
[c] 16 WHSV, 3 c/o, 900°F, M-A feed 33°API.

between mild hydrothermal and severe hydrothermal conditions with catalysts of different matrix type. Changes in surface area, pore volume, and x-ray peak height may be compared with the corresponding equilibrium catalyst analyses. The data show that steam deactivations of seemingly unrelated severity are necessary to artificially deactivate catalysts to activity and structural properties closely approximating equilibrium levels. In addition, catalysts containing predominately gel-based matrices require more severe deactivations to simulate equilibrium than clay-based materials. The activity, pore structure, and peak height data for each catalyst (relative to the hydrothermal deactivation used) that most closely approximate equilibrium data are bracketed.

While no single laboratory deactivation procedure can be used accurately to deactivate all types of catalysts, equilibrium activities and catalyst pore structures can be obtained by relatively simple variations in hydrothermal deactivation conditions. In developing laboratory steaming procedures, either higher temperature or higher steam partial pressure must be substituted for the longer aging of commercial operation. The key to developing a good procedure is to use any combination of the two that best duplicate the physical structure and activity of equilibrium catalyst without exceeding the collapse conditions of the zeolite present.

Testing. Nearly every catalyst manufacturer and major oil company has their own specific test unit for evaluating catalysts. These tests range in cracked gas oil output of from a few milliliters per hour to several barrels per day. Small-sized laboratory testing units are basically isothermal, utilizing reactors completely housed within electrically controlled furnaces. However, large pilot plant units and commercial equipment operate in a basically adiabatic mode. In most catalyst studies these operational differences result in only negligible differences in yield results. However, there are specific instances, as in the new commercial riser reactors (short catalyst/oil contact time), where variable reaction temperature has a real effect on yield distribution, and isothermal laboratory equipment will not exactly duplicate commercial yields. For intercomparison of catalysts, however, these differences are relatively unimportant if the same type unit is used for each evaluation. A cross-section of test units in use today is given in Table VIII.

The major requirement of a test unit in evaluating zeolite cracking catalysts is flexibility. Units must be able to operate over wide contact time

Table VIII. Types of Cracking Catalyst Test Units in Use

Company	*Unit Name*	*Type of Unit*	*Approximate Oil Feed Rate*	*Literature Ref.*
Davison	Microactivity, MA, MAT	fixed, fluidized or pellet	1.5 g/min	*47, 48*
ARCO		transport phase, fluid bed	60–600 g/hr	*49*
UOP	Midget Fluid	fluid bed	200 g/hr	*50*
Gulf	Small Scale	fluid bed	400–1400 g/hr	*51*
Mobil	Microcatalytic	fixed bed powder	0.15 g/test	*52*
Amoco	Pilot Plant	fluid bed	1 bbl/day	*53*
Shell	Cracking Plant	fluid bed	3 bbl/day	*54*
Davison	Pilot Cracking Unit	fixed, fluidized bed	10.0 g/hr	this chapter

ranges, changing severity (catalyst/oil ratio, *c/o*, divided by weight hourly space velocity, WHSV), temperatures from 800°–1100°F, and preferably with independent control of catalyst/oil ratio and weight hourly space velocity. One of the major advantages of large scale test units (barrels/day) is their capability of more nearly matching the activity and selectivity of full scale units under a wide range of operating conditions. However, operation of these units is extremely costly. Smaller scale units avoid this problem but suffer from mechanical limitations in simulating riser cracking units since it is difficult to measure catalyst holdup in the riser or reactor. Thus it is difficult to evaluate accurately the cracking severity. (This is not precisely true in a fixed fluidized bed unit as compared with transport phase reactor since in the former the catalyst is located only in one reactor vessel which is used as the reactor, stripper, and regenerator in sequential reaction steps. In a transport phase reactor, catalyst is undergoing all three operations in different parts of the apparatus simultaneously.)

While several small scale units approximate short contact time conditions, the correlation is far from perfect. For rapid screening of catalysts for activity and for approximations of product selectivity, the microactivity unit originally designed by Atlantic-Richfield is extremely effective and is widely used by all catalyst suppliers and many refiners.

While the test units available today fail to duplicate conditions closely approximating commercial units, the results of careful tests done in these units can often be used to predict accurately the products of commercial operations. Nevertheless, the main criteria used by many oil companies in choosing a catalyst for their specific operation varies from manufacturers claims for their recommended catalyst, through manufacturers pilot plant studies, to actual customer pilot plant testing or any combination of the above.

Many oil companies are willing to accept manufacturers' claims and proceed immediately to a commercial test if their units are small and easily adapted to quick testing. On the other hand, many of the largest companies will pilot test catalysts in their own research facilities before any large scale commercial testing. In addition, refiners may utilize computerized fluid catalytic cracking (FCC) simulation programs based on their own in-house expertise or on commercially available FCC-simulation programs, such as those available from The Pace Co., Houston, Tex.; Profimatics, Inc., Woodland Hills, Calif.; Applied Automation, Inc., Bartlesville, Okla. and others. These programs generally incorporate theoretical heat balance, pressure balance, and unit operational balances related to a specific commercial operation by unit factors. Davison Chemical Co. also offers a Pace-type FCC simulation service to the refining industry. A final important criteria for selecting catalysts is the economics of the product yields to be expected for his specific refinery.

In summary, the present state of the evaluation art, both commercially and experimentally, dictates that new small scale unit design problems be overcome to keep pace with the rapidly expanding temperature and contact time ranges of commercial zeolite operation.

Factors Affecting Yield and Selectivity

In the commercial production of successful zeolite cracking catalysts many of the above considerations occupy a position of secondary importance since the dominant problems of activity and selectivity can be solved when the catalyst has had the correct amount of zeolite added. Since the hydrothermal and thermal stability of any zeolite catalyst depend strongly on many different preparative parameters (exchange fraction, exchange ion, residual Na_2O or K_2O, silica/alumina ratio, etc.), only activity and selectivity changes with zeolite concentration and matrix type, using zeolites of comparable thermal and hydrothermal stability, will be considered. Following this, a brief discussion of the effect of metals, coke, sulfur, and nitrogen poisoning and zeolite type on activity and selectivity will be given.

Effect of Zeolite Input on Product Yield and Selectivity. The major selectivity and activity characteristics observed as a function of catalyst zeolite input are summarized below.

(1) As acid activity (zeolite input) is increased in a catalyst system, the first effects observed, at constant conversion, are lower coke formation, lower dry gas yields, and higher yields of gasoline and light cycle oil.

(2) Aromatics and olefins decrease in the gasoline fraction at low zeolite levels with corresponding decreases in motor and research octane number. Above 9 wt % zeolite, however, aromatic content increases, probably from the cracking of aromatic compounds in the light cycle oil into the gasoline range.

(3) At high zeolite inputs, C_3 and C_4 olefins disappear faster than would be expected as a result of acid activity only. They may be undergoing a type of conjunct polymerization (55) to form saturates and aromatics in the light cycle oil range and coke.

(4) Residence time plays a more important role than acid activity in catalysts of very high acid activity. Thus at high zeolite inputs, high yields of light cycle oil cannot be obtained since the long contact times required to convert heavy oil to lighter hydrocarbons inevitably lead to overcracking of light cycle oil and, to a lesser degree, of gasoline.

CATALYSTS. Catalysts used to study activity and product distribution as a function of zeolite input were prepared at the Davison Research Laboratories using a semisynthetic matrix of weight ratio 6:3 synthetic 25% Al_2O_3-SiO_2 to kaolin clay. The zeolite added to this matrix was a fully rare earth-exchanged Davison Y-type zeolite containing approximately 18 wt % mixed rare earth oxide (CREY). Relative amounts of individual rare earth chlorides in the mixed exchange solutions were 36% lanthanum, 28% cerium, and 25% neodymium, with the remainder consisting predominantly of praseodymium and samarium.

CATALYST EVALUATION. All catalysts were evaluated in fixed-fluidized pilot cracking units after a hydrothermal deactivation (fluid bed steaming: 12 hr, 20% steam, 80% air at 1520°F, atmospheric pressure) designed to give the same surface area pore volume characteristics, and activity levels as commercial equilibrium catalysts of the same type. The latter could not be used in these studies because of the uncertain effects of different metals contaminating product distribution. All tests were made using a West Texas gas oil feed (27.7° API gravity, UOP K factor = 12.1) and pilot unit conditions of 920° and 4.0 catalyst/oil ratio (c/o). Comparative cracking data obtained with other feedstocks ranging in quality from 21.9 °API and 11.5 K factor to 33.6 °API and 12.9 K factor indicated the same general trends as reported here with West Texas gas oil (WTGO). Cracking severity was varied by changing weight hourly space velocity (WHSV).

The pilot cracking units are operated cyclically with each cycle consisting of four operations: oil cracking, stripping, regeneration and nitrogen purge. During the cracking cycle, oil at 150°F is fed from a feed buret by a gear pump into a furnace section comprising a preheater, transfer line, and reactor (all aluminum-coated stainless steel). The feed discharges at the bottom of the catalyst bed which is fluidized by the ascending stream. During this cycle, deionized water is fed into the system by a dispersion pump to furnish the desired process steam. The preheater is filled with a mixture of pure aluminum buttons and ceramic balls and is connected to the reactor through a transfer line. Each of these sections is heated by separate elements in a common furnace. Product effluent discharges through a micro-metallic filter element, which prevents catalyst carryover, into a primary water condenser. The liquid passes through the product valve and secondary condenser into a still pot maintained at 120°F. The liquid is stabilized to a C_5+ composition in a stripping column operated to remove C_4 and lighter overhead. The debutanizer consists of an 11-plate Brunn distillation column between a bottom condenser operated at +32°F and a top condenser operated at —32°F. All overhead gas is metered through a wet test meter and collected during a balance run for analysis.

Following the oil cracking cycle, the catalyst and reaction system are stripped with steam to remove volatile hydrocarbons. Deionized water for this purpose is fed by a Bellows pump and follows the same path as the oil.

Next, the catalyst is regenerated. Coke is burned from the catalyst by introducing metered air into the system which is maintained at higher temperatures and pressures than those used in cracking or stripping (1150°F and 20 psig). After the combustion products leave the reactor, they flow through the primary condenser, exhaust valve, and back-pressure regulator to the dry gas meter. A drag stream of this gas is continuously collected for analysis and the rest is vented. Following regeneration, the system is purged with metered nitrogen at atmospheric pressure. The nitrogen takes the same path as the air.

The four operations comprising the complete cycle are automatically controlled by timers. A normal balance run usually takes about 30 cycles (to provide sufficient product for full analysis of all products). Temperatures are controlled by digital set point controllers, using standard thermocouple input, and are automatically recorded and monitored by a combination of multipoint recorder and indicating potentiometer. The reactor contains two thermocouples positioned at the top and the bottom of the catalyst bed. Temperature measurement of these two positions of the catalyst bed indicate the degree of catalyst mixing.

In the present evaluation of the effects of zeolite input, WHSV was varied widely (10–40) to obtain data at a constant conversion level. As noted later, we believe this procedure to be valid since the selectivity changes as affected by activity are substantially greater than the selectivity effects of contact time.

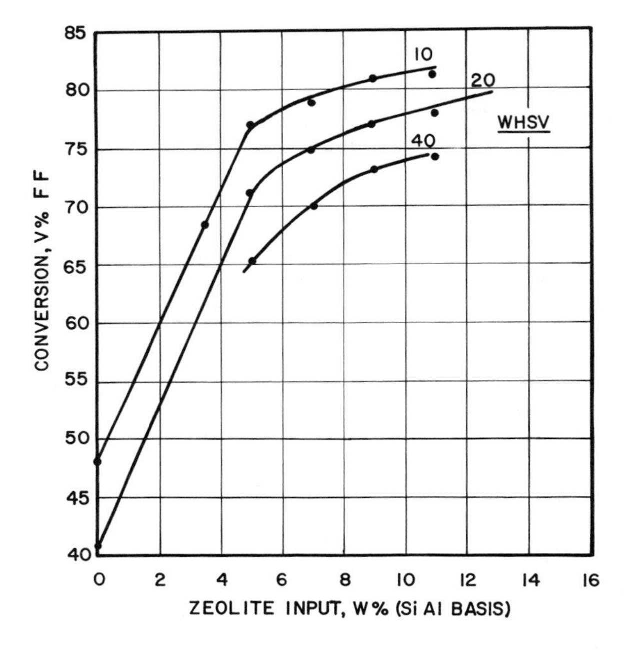

Figure 10. Effect of zeolite input on catalytic activity. Product yields shown in Figures 10–20 are based on the fresh feed (FF) passed over the catalyst during the cracking cycle. No recycle products or other materials are included in yield calculations.

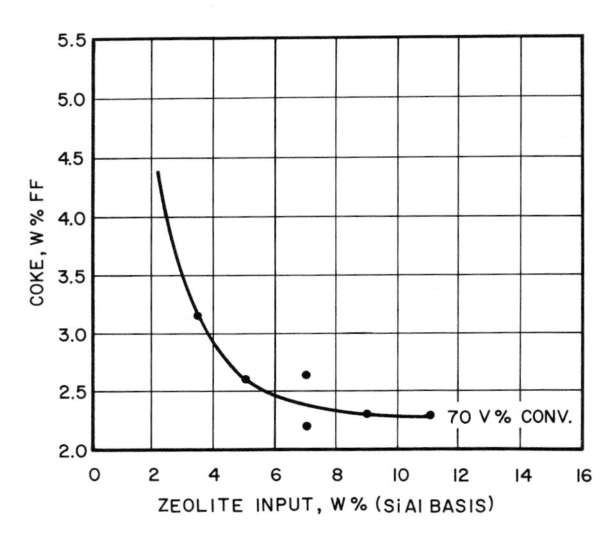

Figure 11. Effect of zeolite input on coke yield

Data were obtained at conversion levels of 65, 70, and 75 vol % fresh feed, but for clarity only the 70 vol % cross-plotted curves are shown except for light cycle oil (LCO) yield. Both the 65 and 75% curves were similar in shape, differing only in magnitude from the 70% curves. A similar study made in the Davison Research pilot plants with rare earth exchanged X-type ("CREX") zeolite-containing catalysts gave results substantially identical to the study results with CREY. A discussion of the conversion data and the data for each product analyzed follows. Curves are shown on Figures 10–20.

CONVERSION. As shown in Figure 10, a substantial increase in activity results from the first 4–6 wt % of zeolite added to a semisynthetic matrix. Above 6 wt % CREY input, further increases in conversion are more difficult to obtain with increased zeolite input. The sharp break in the conversion curve probably reflects the point at which those compounds in the West Texas gas oil (WTGO) capable of relatively easy carbonium ion cracking in a given reaction time are catalytically cracked. Before this, catalyst acid site concentration is probably a more controlling factor than residence time. At high zeolite inputs (>5–6 wt % CREY) residence time becomes more important than acid site concentration, resulting in a less pronounced change in conversion with increasing zeolite input.

COKE YIELD. The effect of CREY input on coke yield (after the 1520°F, 20% steam catalyst deactivation) is shown in Figure 11 at a constant 70 vol % conversion. The curve shows a maximum coke selectivity gain at relatively low CREY inputs, but the inflection point in the curve is not as sharp as in the activity plots. Above the 8 wt % CREY input level, the gain in coke selectivity at constant conversion was relatively small even though the gain in activity can be very large (*see* Figure 10). This indicates from a practical standpoint that, at equilibrium concentration in a commercial operation, the huge reservoir of acid site concentration built into catalysts of very high zeolite inputs may be largely unusable since the coke buildup on the catalyst could lead to reduced accessibility of the available zeolite. At this time, however, research and commercial data with very high zeolite content catalysts are too limited to define accurately the limits of high zeolite operations from the standpoint of coke selectivity.

HYDROGEN YIELD. Hydrogen yield was reduced sharply over the first 6–8 wt % CREY input in a manner similar to coke yield (Figure 12). This hydrogen is rapidly used up with increasing CREY input because of the proportionally increasing hydrogen transfer, catalyzed by the strong-acid activity of zeolite catalysts.

$C_1 + C_2$ YIELD. These yields are not catalytic yields at all but are really end products of thermal cracking. Consequently they are largely proportional to residence time in the preheat and reaction system. The higher the CREY input, the more active the catalyst and the less residence time is required to achieve a specific conversion level. This is indicated by the directional arrows on Figure 12. If this is true for $C_1 + C_2$ yields, it is certainly also true to some degree for coke yield).

C_3 YIELDS. C_3 and C_3-olefin yields are shown in Figure 13. CREY input influences the reduction in C_3-olefin to an extent far greater than it affects

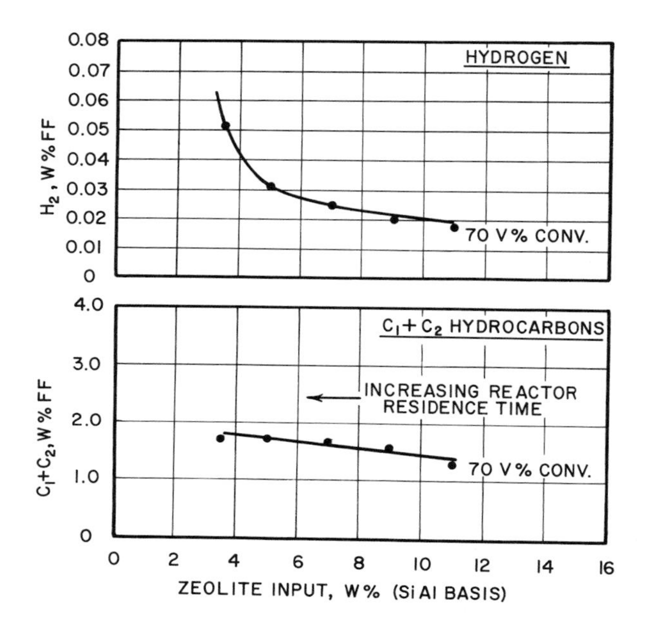

Figure 12. Effect of zeolite input on dry gas yields

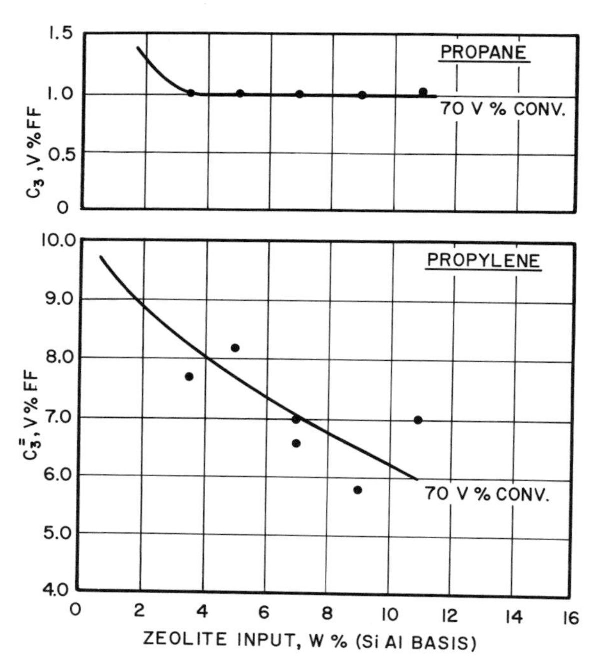

Figure 13. Effect of zeolite input on C_3 hydrocarbon yields

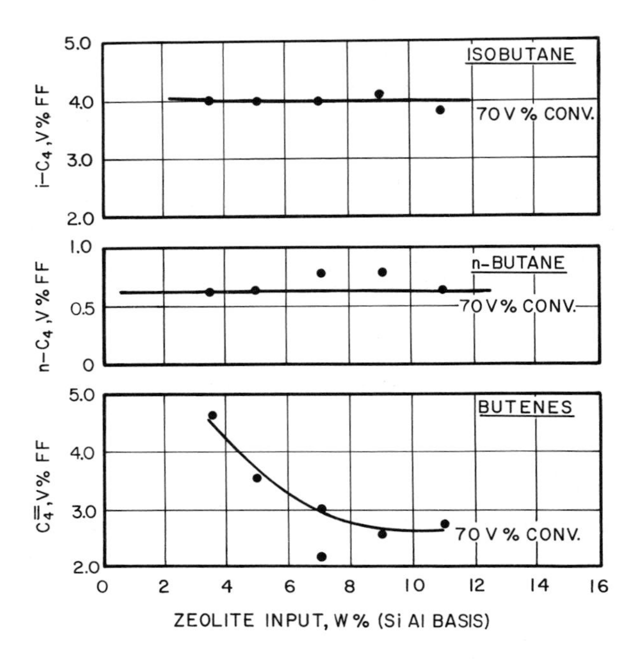

Figure 14. Effect of zeolite input on C_4 hydrocarbon yields

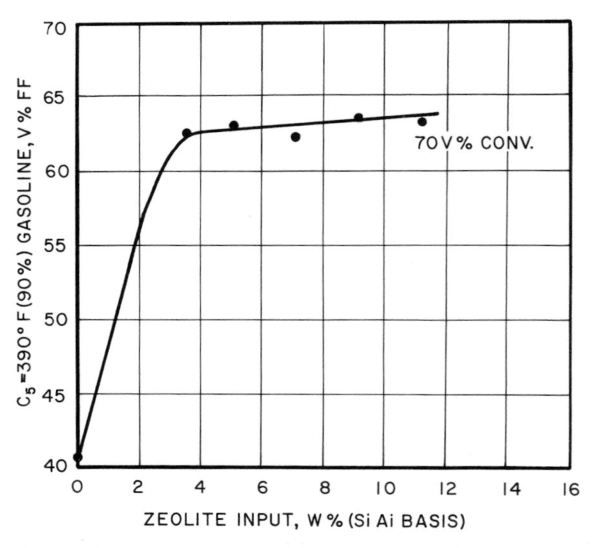

Figure 15. Effect of zeolite input on gasoline yield

any other product, while the C_3-saturates yield remains constant after the first 3–4 wt % of CREY input.

The constant nature of the C_3-saturates yield during a period of rapid and large reduction in C_3-olefin would certainly seem to indicate disappearance or secondary reaction of C_3-olefin rather than a coincidental balance between a reduction in formation of total C_3 and increased hydrogenation of C_3-olefin to C_3-saturates. If a secondary reaction of C_3-olefin is occurring, then the C_3-olefin is probably being converted to higher molecular weight material as a function of CREY input. In addition, it would appear that more than enough C_3-olefin is present as end product of carbonium ion cracking to supply all the reactant necessary for CREY-induced conversion

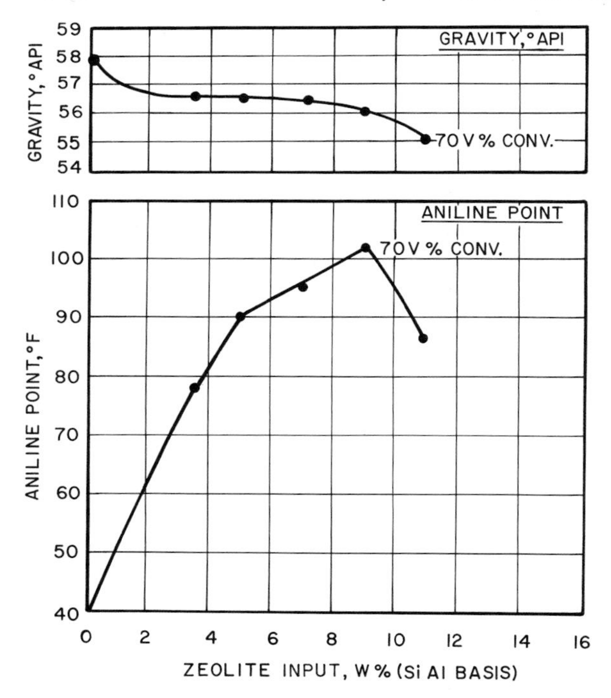

Figure 16. Effect of zeolite input on gasoline gravity and aniline point

to higher boiling material. At least the drop in the C_3-olefin curves showed no obvious leveling off up to 12 wt % CREY input.

C_4 YIELDS. C_4 yields are shown in Figure 14. As with the C_3 yields, the yields of C_4-olefin drop sharply with increasing CREY input while yields of n-C_4 remain constant. The C_4-olefin yields do tend to level off in the 8–10 wt % CREY input range, but the actual volume yield of C_4-olefin is only 40% of the comparable C_3-olefin yield. Again, it would appear that the C_4-olefin is being converted to higher boiling material as a function of CREY input.

GASOLINE YIELD AND QUALITY. Gasoline yields are shown in Figure 15, gasoline gravity and aniline point are shown in Figure 16, and research and motor octane plots, both clear and leaded, are shown in Figures 17 and 18. Gasoline is cut to a constant 390°F ASTM 90% point. This results in about a 445°F TBP end point. As CREY input increases from 3 to 13 wt %, gasoline yield increases gradually. Gasoline gravity remains essentially constant up to the 7 wt % CREY input level and then drops rapidly. Aniline points rise expectedly with increased acid site concentration up to a level equivalent to 9 wt % CREY then tail off at higher CREY input levels.

Both diffusion limitations and temperature affect gasoline selectivity (*56, 57*). However, in the former case, at high conversion levels, gasoline yields were only slightly less for catalysts with diffusional limitations than for diffusion free catalysts. In the present study, any diffusional limitations should be identical from catalyst to catalyst since the same hydrothermal deactivation procedure, zeolite and catalyst matrix were used for each catalyst preparation. As cracking temperature is lowered, the gasoline cracking rate declines more than the gas oil cracking rate, resulting in a higher gasoline selectivity at lower temperatures. Thus constant temperature gasoline selectivity comparisons are essential.

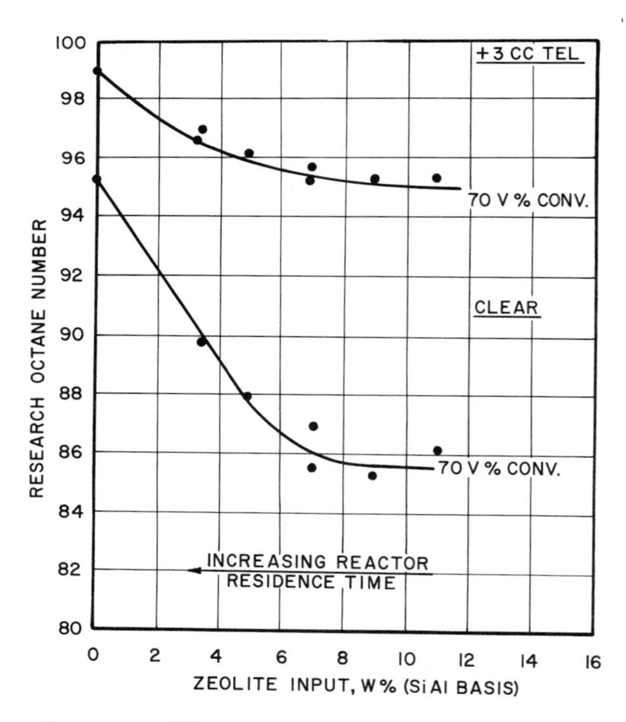

Figure 17. Effect of zeolite input on gasoline research octane numbers

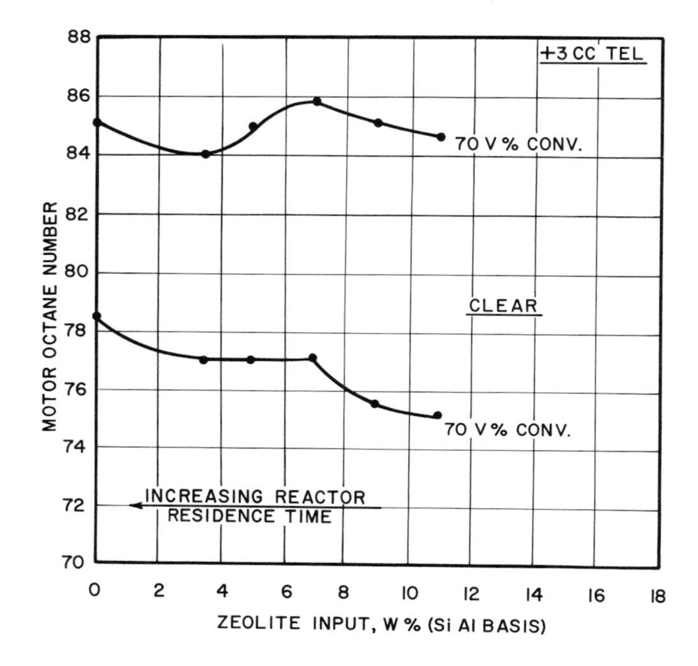

Figure 18. Effect of zeolite input on gasoline motor octanes

Gasoline octane changes at constant conversion show that relatively low acid site concentrations (3–5 wt % CREY input) significantly reduce aromatic and olefin content of the gasoline, resulting in lower clear research octane, lower leaded research octane and slightly lower mixed clear and leaded motor octanes. In the pilot cracking units, this effect is caused by both increased acid activity and decreased reactor residence time.

If a high WHSV trend in new catalytic cracking design based on active zeolite catalysts develops, low octane gasoline could be a serious debit to operational economics. For instance, if a unit were designed to run 70% conversion by using a catalyst with an 11 wt % CREY input at 40 WHSV, gasoline clear octanes could be 2–4 numbers poorer than if the same unit were designed to run 70% conversion by using a 5 wt % CREY input catalyst at ~20 WHSV. This example is, of course, based on this specific West Texas gas oil. Currently, however, new riser-cracking designs are being operated at cracking temperatures 70°–100°F higher than normal. The resulting increased olefins in the gasoline fraction have more than compensated for any octane loss caused by use of high zeolite input catalysts.

YIELDS AND QUALITY OF LIGHT CYCLE OIL. Light cycle oil yields rise rapidly with increasing CREY input as shown in Figure 19 but level out in the 5–9 wt % range; the yields drop at higher sieve inputs.

At constant conversion, if the light cycle oil yield decreases, the heavy cycle oil yield must increase. This indicates that there is a minimum

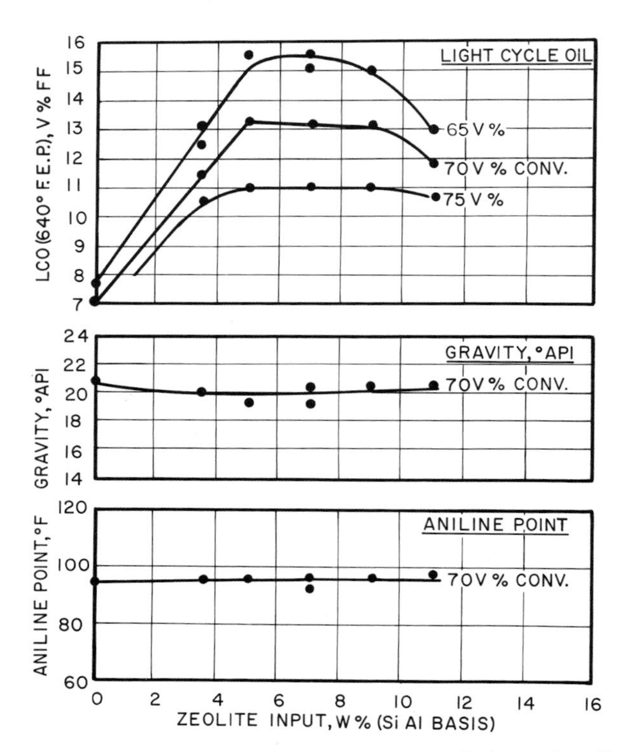

Figure 19. Effect of zeolite input on light cycle oil yield and quality

residence time required for the conversion of 640°F compounds into light cycle oil. Going to higher zeolite content catalysts decreases reactor residence time of gas oil and catalyst to maintain constant conversion.

$>$640°F Residue Quality. Plots of the $>$640°F residue gravity and aniline point are shown in Figure 20. These data also show an interesting inflection point at about the 8 wt % CREY input level. The $>$640°F fraction becomes more aromatic with increasing zeolite input up to 8 wt %, beyond which it becomes much less aromatic. This yield change is related to the drop in light cycle oil which occurs at the higher zeolite inputs at constant conversion for if the volume of light cycle oil decreases, the volume of $>$640°F residue must increase to maintain constant conversion. This occurs because the components in the $>$640°F residue are not greatly affected by acid activity, and hence do not crack to lighter components in the absence of prolonged residence time. These components simply require long residence times for cracking. Thus, as the volume of $>$640°F residue increases at constant conversion and high zeolite input, the quality trends induced by initial zeolite inputs reverse by a dilution effect, (or rather by non-reactivity of the heaviest feed components).

Catalyst Activity and Selectivity as Affected by Changes in Matrix Composition and Synthetic Faujasite Type. Product distribution changes caused by changing the chemical composition of the catalyst matrix from silica/alumina to silica/magnesia are shown in Table IX. This type of matrix system, manufactured by Davison, has been available for a number of years but has been only recently available in a zeolite-containing form (LCO-1). Comparison of the zeolite plus silica-magnesia with a similar activity semisynthetic (silica-alumina) based catalyst show that the SiO_2/MgO matrix exerts a strong influence on cracked product yields. Comparison of SiO_2/MgO and SiO_2/Al_2O_3 in their zeolite-free forms shows that the SiO_2/MgO matrix reduces light hydrocarbon yields while enhancing gasoline selectivity and significantly increasing light cycle oil (430°–640°F boiling range) yield and quality. The shift to a higher LCO yield is caused by the suppression of secondary cracking reactions, as evidenced by the reduction in light hydrocarbon yields and aromatic and olefin production. The shift to the production of less olefin and aromatic results in somewhat lower gasoline octane.

Variations in promoter type (X-type or Y-type zeolite) and degree of RE_2O_3 exchange were found to effect only slight changes in the product yield (*see* Table X and (*35*)). The slightly higher coke and gas yields and slightly poorer gasoline selectivity of rare earth-containing catalysts, compared with catalysts made with the ultrastable faujasite Z14-US (*58*),

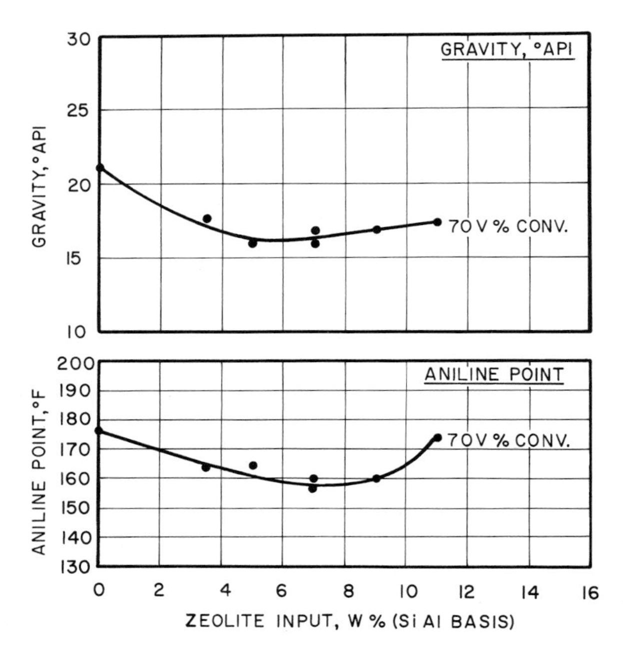

Figure 20. Effect of zeolite input on bottoms quality (640 + fraction)

Table IX. Silica-Magnesia Matrix Effects on Cracked Product Yield Composition[a]

	RE_2O_3-Exchanged Synthetic Faujasite CREY	
Matrix	*Silica/Magnesia*	*Silica/Alumina*
Conversion: vol %	77.0	77.0
hydrogen: wt % FF	0.04	0.04
$C_1 + C_2$: wt % FF	1.8	2.0
total C_3's: vol % FF	8.7	12.0
total C_3-olefin: vol % FF	7.6	10.8
total C_4's: vol % FF	9.0	11.2
total C_4-olefin: vol % FF	3.8	4.8
total *i*-C_4: vol % FF	4.6	5.8
C_5^+ Gasoline: vol % FF	69.5	65.0
C_5^+ gaso./conv.	0.90	0.84
octane No.		
RON + 0:	88.0	90.0
RON + 3 cc TEL	97.1	97.6
MON + 0:	76.0	78.7
MON + 3 cc TEL	84.5	86.5
gravity: °API	61.1	56.7
aniline pt.: °F	98	81
Light Cycle Oil: vol % FF	14.3	10.0
gravity: °API	20.8	16.0
aniline pt.: °F	90	78
640°F$^+$ Residue		
gravity: °API	7.0	11.2
aniline pt.: °F	118	148
Coke: wt % FF	3.8	3.8

[a] Constant severity, WHSV = 10, c/o = 4.0, reactor temp. = 920°F.

may be caused by the loss of surface area and attendant formation of free RE_2O_3 on the catalyst surface. Free RE_2O_3 may catalyze additional secondary and tertiary reactions leading to coke formation. To obtain constant conversion, as shown in Table X, approximately four times more Z-14 US input catalyst was required over either CREY or CREX.

Thus, the major activity and selectivity differences observed in commercial operations today are directly attributable to changes in zeolite input of the catalysts rather than to compositional changes. However, this does not mean it is impossible to change product yields by changes in catalyst composition. In fact, the unique product distribution obtained with the silica/magnesia matrix (increased light cycle oil yield, lower light hydrocarbon yields, and improved gasoline selectivity) attests to the potential of significantly shifting product distribution through changes in catalyst compo-

Table X. Effect of Zeolite Type on Cracked Product Selectivity[a]

Promoter	*US*	*CREY*	*CREX*
Relative Weight %[b]:	4.0	1.0	1.1
Conversion: vol %	←	70.0	→
hydrogen: wt % FF	0.04	0.05	0.06
$C_1 + C_2$: wt % FF	1.3	1.5	1.4
total C_3's: vol % FF	8.3	8.7	8.4
total C_3-olefin: vol % FF	6.9	6.8	6.6
total C_4's: vol % FF	8.8	10.0	10.1
total C_4-olefin: vol % FF	4.2	4.2	4.2
total *i*-C_4: vol % FF	4.1	5.0	5.0
C_5^+ Gasoline: vol % FF	62.5	60.5	60.5
C_5^+ gaso./conv.	0.89	0.87	0.86
octane No.			
RON + 0:	88.0	88.0	88.2
RON + 3 cc TEL	95.8	95.3	95.5
MON + 0:	77.0	76.8	76.5
MON + 3 cc TEL	85.0	84.3	84.1
gravity: °API	55.6	56.0	55.2
aniline pt.: °F	91	87	88
bromine No.:	52	55	50
Light Cycle Oil: vol % FF	10.3	10.5	11.0
gravity: °API	19.6	20.6	20.5
aniline pt.: °F	97	94	107
640°F⁺ Residue			
gravity: °API	17.5	18.0	19.0
aniline pt.: °F	165	181	170
Coke: wt % FF	3.3	3.8	4.0

[a] Constant severity, WHSV = 20, c/o = 4.0, reactor temp. = 920°F.
[b] Zeolite content was adjusted to obtain constant conversion at constant severity.

sition. In fact, however, no significantly different catalysts have appeared to date.

In addition, there are catalyst properties not considered here which are also important for sucessful commercial performance. Many of these have been discussed earlier and include zeolite quality, zeolite particle size and dilution within the matrix, and matrix pore size distribution, stripability, and regenerability. Furthermore, physical porperties related to unit retention such as hardness and density, must be obtained without sacrificing quality factors such as catalyst regenerability. Some of these physical properties may also affect product distribution in certain unit operations, but these changes are more a function of a catalyst's physical makeup than its chemical composition.

Effect of Gas Oil Hydrotreating on the Activity and Selectivity of Zeolite Cracking Catalysts. Significant yield advantages can be realized if the catalytic cracker feed stock is hydrotreated prior to cracking (*59, 60*).

Hydrotreating reduces sulfur and nitrogen content of the feed, saturates the difficult-to-crack polyaromatics, reduces the carbon residue, and reduces the metals present. Bailey and Nager (*60*) showed the beneficial effects of higher conversion, higher quality products, and lower coke make to be even more advantageous for zeolite catalysts than for amorphous catalysts and to improve with increasing severity of the hydrotreatment.

Pilot cracking unit data using an equilibrium Davison AGZ-50 catalyst (made from semisynthetic matrix and rare earth-exchanged synthetic faujasite promoter) is presented in Table XI. The properties of the West Coast gas oil used before and after hydrotreatment are shown in Table XII. Cracking data given in the table are at constant coke operation. This is generally the fashion in which a refinery would operate using a hydrotreated feed. Alternately, operation would probably be at constant severity (*c/o/*

Table XI. Effect of Hydrotreating (HT) on Catalytic Cracking Yields at Constant Coke Yield

Catalyst: *Equilibrium AGZ-50*

Pilot Unit Conditions: 920°F, 4 c/o feedstock:	*West Coast Gas Oil*	*700 psi HT*	*1800 psi HT*
Conversion	51.0	65.0	77.0
hydrogen	0.03	0.03	0.035
$C_1 + C_2$	1.7	1.7	1.8
total C_3's: vol % FF	6.0	8.0	10.1
total C_3-olefin: vol % FF	4.8	6.1	7.4
total C_4's: vol % FF	8.4	10.0	11.9
total C_4-olefin: vol % FF	2.7	3.1	3.6
total i-C_4: vol % FF	4.8	5.9	7.0
C_5^+ Gasoline: vol % FF	36.0	52.5	63.5
C_5^+ gaso./conv.	0.70	0.80	0.82
octane No.			
RON + 0:	90.8	91.3	91.2
RON + 3 cc TEL:	96.2	98.2	97.4
MON + 0:	76.8	78.4	78.4
MON + 3 cc TEL:	83.0	86.2	86.6
gravity: °API	49.6	50.4	51.4
aniline pt.: °F	92	84	73
bromine No.:	85	42	28
Light Cycle Oil: vol % FF	29.0	22.0	16.5
gravity: °API	27.0	23.0	20.5
aniline pt.: °F	120	95	78
640°F⁺ Residue			
gravity: °API	19.5	14.0	9.0
aniline pt.: °F	170	145	132
Coke: wt % FF	5.6	5.6	5.6

Table XII. Properties of Catalytic Cracker Feed before and after Hydrotreating

Feedstock:	*West Coast Gas Oil*	*After 700 psi Hydrotreatment*	*After 1800 psi Hydrotreatment*
gravity: °API	23.0	25.9	27.7
aniline pt.: °F	154	164	173
N_2: ppm/wt	2500	1560	334
sulfur: wt %	1.29	0.13	0.06
estimated polyaromatic saturation: %	—	35	66
ASTM Distillation:	*Temperature, °F at 760 mm Hg*		
IBP	408	357	338
10%	556	531	531
30%	653	632	627
50%	728	700	702
70%	786	779	767
90%	860	878	859
FBP	953	967	940
UOP K Factor:	11.6	11.7	11.8

WHSV) in which case some yields would change when compared to constant coke operation. These data show that at constant coke, both conversion and gasoline yield increase substantially with severity of hydrotreatment. Compounds lighter than gasoline increase as expected with conversion. Gasoline octane numbers, both research and motor, increase as does the lead susceptibility, probably because of sulfur removal. The gasoline becomes lighter, more aromatic, and less olefinic. Because of the large increase in gasoline yield, light and heavy cycle oil yields are decreased. Light and heavy cycle oil quality under these conditions is decreased; however, at constant severity the quality of the light cycle oil is relatively unchanged after hydrotreating, particularly at short contact times.

In general, hydrotreating of gas oils prior to cracking can be thought of as a gasoline-making process. However, with increasing demands for heavy residual stocks and with the more stringent air pollution regulations which will further limit SO_x emissions from refineries, the already beneficial effects of feed hydrotreatment will be enhanced.

Effect of Impurities on Product Yields and Quality. Major changes can be effected in product yields and quality by changing the zeolite input of the catalyst and matrix composition and by hydrotreating the feed prior to cracking. In addition, the catalyst must be manufactured by techniques which leave it relatively free of classic catalyst contaminants such as alkali or alkaline earth oxides, transition metals, or sulfate. These contaminants can severely affect catalyst activity and stability. Zeolite cracking catalysts can tolerate significantly more of these contaminants before substantial yield loss occurs than can amorphous catalysts. Most of this increased tolerance of zeolite catalysts to impurities is directly related to their higher activity and hydrothermal stability when compared with amorphous grades. Typically, all catalyst manufacturers minimize catalyst impurities during the manufacturing procedure. However, major contaminant effects occur because of impurities in the cat cracker feedstock, which can more or less permanently affect the yield structure (transition metals and alkali oxides) and which temporarily poison active sites (nitrogen, sulfur, coke). Contaminants affect product yields and quality by acting as catalysts themselves (*e.g.*, transition metals deposited on the catalyst surface) by temporarily poisoning active sites, and by causing structural instability. [Structural instability generally occurs because of high temperature hydrothermal sintering of the catalyst caused, usually, by the presence of alkali or alkaline earth metals (Na, K, Ca, etc.). Such materials may be present from the catalyst manufacturing procedure, from the crude oil, or from leaks in steam injectors or steam generators on the FCCU. In any case, these materials poison active sites and/or form eutectics with the catalyst, causing low temperature sintering of the structure. Both causes of contamination are relatively rare and will not be considered further.]

METALS AND COKE. Recent studies (*61*) have graphically shown the effects of transition metal contaminants on zeolite cracking catalysts in reducing gasoline yield and increasing yield of light gases and contaminant coke. Four different types of coke may be characterized (*61, 62*): (1) catalytic coke (formed by the acid activity of the catalyst), (2) cat-to-oil coke, (formed by carryover of hydrocarbons into the regenerator), (3) Conradson coke (very high boiling components in the feed), and (4) contaminant coke (associated with transition metals which cause condensation reactions leading to coke).

The effect of metals normally present in cat cracker feed has been the subject of investigation long before the introduction of zeolite catalysts (*63, 64, 65*). The recent ARCO investigation updates most of the older studies (without substantially changing earlier conclusions) but with emphasis on zeolite catalysts. Early studies had shown that under normal operating conditions cracking catalysts become contaminated with Ni, V, and Fe occurring in the feed to the cat cracker (though most are removed in distillation before cracking). Nickel is about four times as effective for promoting undesirable dehydrogenation reactions as vanadium. Iron is substantially less effective than vanadium but, when present as tramp iron or magnetic oxide stripped from reactor walls, it also catalyzes the highly exothermic oxidation of CO to CO_2 in the regenerator causing excessively high catalyst temperatures. Even though alternate oxidation and reduction reactions in cracking and regeneration rapidly reduce the ability of the transition metals to catalyze light gas and coke formation, effective metals level will persist as a function of catalyst makeup rate. An increase in effective metals of from 180 to 1130 ppm in the ARCO study caused conversion to decrease from 79.0 to 75.6%, with a corresponding loss in gasoline yield. At the same effective metals level, when operating at a constant 70 vol % conversion, dry gas increased from 5.8 to 6.6 wt % and coke from 2.4 to 3.1 wt %.

Coke which builds up on the catalyst during the cracking cycle and remains after regeneration, the so-called carbon-on-regenerated catalyst (CRC), has been shown in a study analogous to the above effective metals study to affect product yields adversely (*49*). The CRC is a result of initial carbon deposition during the cracking cycle from the four coke sources mentioned earlier. The current industry trend is to lower the CRC as much as possible to minimize the above negative effects.

SULFUR AND NITROGEN. Both sulfur- and nitrogen-containing carbon compounds occur naturally in all crude oils to varying extents. Both contact and contaminate the cracking catalyst during the cracking cycle. Desulfurization and denitrification occur during cracking, and the extent to which these reactions occur is, in general, a function of the catalyst's activity or the conversion level at which the catalyst is operating (*66*). Aside from the fact that nitrogen-containing compounds (and to a lesser extent sulfur compounds) can temporarily reduce catalyst activity by acting as chemisorbed active site poisons, studies of catalyst poisoning have shown that nitrogen in a West Coast aromatic-naphthenic feedstock (0.3% nitrogen) could lower the coke burning rate (regenerability) of two different zeolite cracking catalysts. The same catalysts showed a similar lowering in coke burning rate when a pure *n*-hexadecane feed containing 0.6% nitrogen from quinoline was used. A similar study using up to 1.8% sulfur from 1-benzothiophene showed no change in coke burning rate.

Compared with other contaminants the effects of sulfur and nitrogen on product yields may be considered relatively minor; however, major air pollution problems arise because any nitrogen- or sulfur-containing compound adsorbed on the catalyst and carried into the regenerator is then converted to oxides of nitrogen or sulfur and emitted to the atmosphere. Up to 50% desulfurization can occur during cracking, and feed type is the most important variable in determining sulfur distribution in the products (*66*). This study showed that depending on conversion level, about one-half of the sulfur was converted to H_2S, while most of the remainder appeared in the cycle oil. Small amounts of sulfur appeared in the gasoline and with the coke-on-catalyst. This latter product contamination is particularly serious, since sulfur in the coke-on-catalyst is converted to SO_2 in the regenerator.

Effect of Feedstock Type on Product Yields. The desire to predict the amount and quality of products from a catalytic cracker as a function of feedstock type has led to the development of several well known feed characterization procedures. Chief among these are the well known *K* factor (*67*), diesel index (*68*), and more recently a sophisticated separation and instrumental characterization procedure reported by P. J. White (*69*). The Watson *K* factor and diesel index (*DI*) are defined as follows:

$$K = \frac{(\text{Me ABP, °F})^{1/3}}{\text{sp. gr., 60°F/60°F}} \text{ and}$$

$$DI = \frac{\text{aniline point (°F)} \times \text{API gravity}}{100}$$

where MeABP is the mean average boiling point of the gas oil.

Methods of calculating these values are described in detail in the above mentioned references.

K factors of most cat cracker feeds vary from 10.5 (very aromatic) to 13.0 (very paraffinic) and can be directly related to the coke forming tendencies of the feed. Since most commercial operations are close to a coke burning limitation, the effect of going to a high *K* factor feed (paraffinic, low coke yield) allows an increase in conversion and gasoline. Similarly, the effect of changing the specific compound type from bicyclic aromatics through normal and isoparaffins to naphthenes also significantly improves gasoline yields (*69*). In the above cases, however, it appears that the fundamental cause of these product changes is the tendency of the feedstock to form catalytic, contaminant, cat-to-oil, or Conradson carbon (coke) on the catalyst surface.

Industrial Application of Zeolite Catalysts in Fluid Cracking

As in no other time in the history of fluid catalytic cracking, this is a time of change, innovation, and competition in unit design and catalyst development. Four factors have combined to produce the current activity in the field.

(1) Zeolite catalysts

(2) Emphasis on air pollution control

(3) Rebirth of a mature process

(4) Projected requirements for higher octane unleaded gasoline in the United States

Commercial Performance of Zeolite Catalysts. The factors relating to good commercial performance of zeolitic catalysts were initially developed

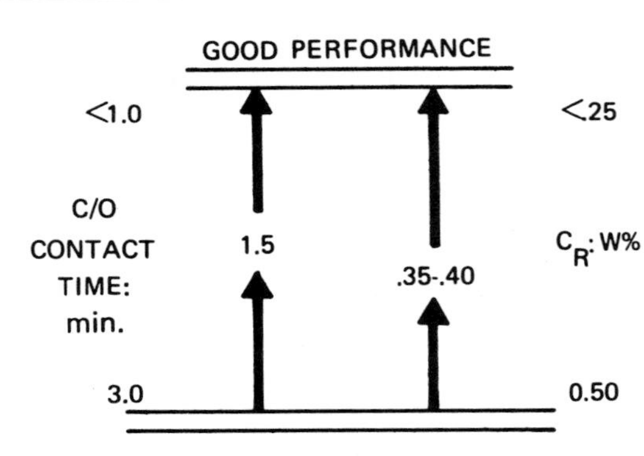

Figure 21. Effect of operating variables on zeolitic performance

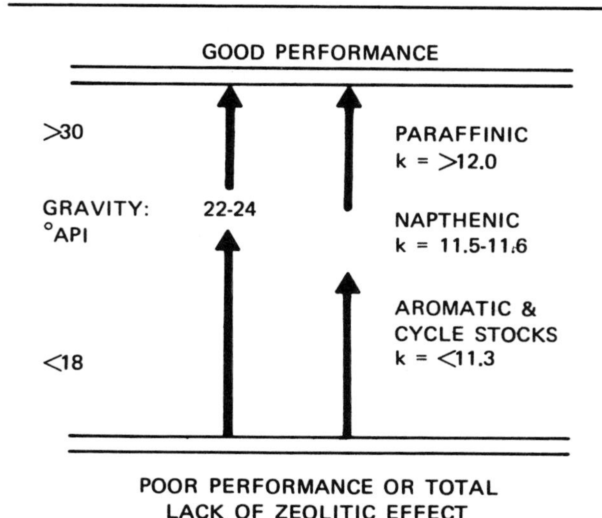

Figure 22. Effect of feedstock quality on zeolitic performance

primarily from actual commercial observations. Often these were markedly different from laboratory or pilot plant performance. Zeolite catalysts performed best when catalyst/oil contact time in the reaction zone was low, where spent catalyst regeneration was good and where feedstocks were light and paraffinic. All these factors were related to coke-induced molecular diffusion problems within the catalyst particle (*70*). The typical effects of these variables are graphically illustrated in Figures 21 and 22. The values in the center of each performance arrow are typical threshold values between good and poor zeolitic performance in observed commercial operations. There were several commercial exceptions to these rule-of-thumb guidelines, but these could usually be explained by a closer look at the overall quality of the feed. A recent article shows quantitatively how carbon on regenerated catalyst affects the activity and selectivity of zeolite catalysts (*49*).

The factors relating to good zeolite catalyst performance were generally understood as early as 1965. There followed a period of unit modification aimed at optimizing the performance of zeolitic catalysts which continues to the present. The types of unit modifications can be summarized as follows:

(1) Regenerator modifications to improve coke burning efficiency.

a. Improved mixing, mostly air ring or air-grid installations.

b. Increased regenerator bed height to diameter (L/D), usually through shroud installations to decrease bed diameter.

c. Higher regenerator temperatures, through mass installation of new alloy and lined cyclones. Temperatures of 1250°F became quite common, and a few units ran as high as 1350°F in the dense bed.

(2) Reactor modifications to shorten catalyst/oil contact times.

a. Dense phase reactor beds were eliminated entirely where possible. In units designed to be bed-crackers, levels were reduced by stripper overflow modification, and volumes were reduced by installation of dummy concrete-filled sections or bv installation of shrouds to reduce reactor diameter.

b. In some units the fresh feed risers were simply extended into the reactors, up to the cyclone dust bowl levels. The entire reactor bed was thus eliminated, with the annular portions more often used as additional stripping volume, depending on reactor configuration.

(3) Segregated fresh feed and recycle cracking. These designs took advantage of separate risers for fresh feed and recycle to vary the contact time and/or cracking temperature which could be more efficiently used for each stream. Typically, the fresh feed would be cracked at low temperature, for example 890°–900°F, and short catalyst contact time in a riser. The recycle stream would be cracked at higher temeprature, for example 935°–950°F, in both a riser and a dense bed, to maximize the cracking severity on the more refractory, aromatic cycle stock. These designs were typified by Kellogg's Orthoflow C units and Texaco's new design.

Segregated fresh feed/recycle cracking was actually practiced by many refiners prior to zeolite catalysts (*71*), but it was even more ideally suited to the early zeolite catalysts, which we would now consider to be of low to intermediate activity. Today's zeolite catalysts are so active that recycle streams have practically dried up. Many refiners with segregated riser designs are now simply splitting their fresh feed to both risers, with a small amount of bottoms recycle still included in the original recycle riser feed stream.

Air Pollution Control. Solid emissions from fluid cat crackers have always been of concern to the refiner. This concern led to the development of spray-dried, microspheroidal catalyst in the late 1940's. This development alone halved catalyst stack emissions. Later improvements were a result of multistage cyclones and improved cyclone separator and Cottrell precipitator design. However, the major push for these improvements was largely economic—fresh catalyst was expensive ($200–350/ton), and no one wanted to lose it when they could withdraw it at equilibrium and sell the equilibrium catalyst for $50–100 a ton. With the advent of zeolite catalysts, the driving force was still largely economic as these catalysts cost more than the amorphous catalysts and could be sold for more as equilibrium catalyst.

In recent years, however, serious concern for atmospheric emissions and the imposition of stricter local laws and enforcement have often forced refiners to uneconomic operations just to remain operable from an emissions standpoint. This began the current drive in the industry for denser, more attrition-resistant catalysts and for better recovery equipment. Refiners realize that the catalyst manufacturers cannot solve emissions problems totally with their catalysts. However, dense, attrition-resistant catalysts can help cut atmospheric emissions over the next two to five years while better mechanical recovery systems can be researched and installed. In the long run, total recovery may be necessary, and this will require both physically improved catalyst and further large expenditures for new recovery equipment. Meeting these increased physical requirements is a particularly difficult problem with high zeolite content catalysts.

Gaseous emissions from cat crackers are also of major concern. Although refinery gaseous emission standards are usually set for the total refinery complex, the FCC unit is often a large contributor to total pollutant emissions. Consequently, much design work is aimed at essentially eliminating all emissions of SO_2, SO_3, NO_x, and CO from FCC stacks. Elimination of SO_2 and SO_3 may require feed hydrotreating or flue gas treatment on a wide scale. CO is currently being converted to CO_2 in CO boilers, but new FCC regenerator designs now being marketed by several companies, use a controlled combustion in the regenerator at approximately 1400°F to convert about 98% of CO to CO_2. At these conditions, Amoco Oil reports that NO_x emissions remain substantially constant at 5–15 ppm. However, operation of a CO boiler increases the NO_x level to about 40–45 ppm (*72*).

Rebirth of a Mature Process. Before fluid zeolite catalysts, the fluid cracking process was a staid, conservative operation for most refiners, one given to little long range research or development—totally mature. Most process improvements, with a few exceptions, were of the metallurgical or maintenance variety. With the advent of zeolite catalysis, however, the biggest profits were made by those companies willing to test and to innovate. Most of the major refiners therefore embarked upon rather extensive catalytic research programs. Once the basics of fluid zeolite catalysis were understood, refiners began to concentrate on process innovation. Furthermore, the accuracy of their own pilot plant predictions in their own commercial units lent a credibility to their process improvements. Successful tests of their own ideas have led many companies to license their own designs, this time unencumbered by the open patent situation that existed at the birth of the fluid process. Today, many refiners are competing openly with each other and with engineering and process companies in the licensing of fluid cracking processes.

Unleaded Gasoline. The national move to remove lead from gasoline led to the current changes in cat cracker design. When and if refiners are required to market only unleaded gasolines, it will be necessary to upgrade the clear research octane numbers of many cat-cracked gasolines. Fortunately, with riser cracking, selective cracking can be done at reaction temperatures some 100°–150°F higher than with conventional bed cracking. This increases clear research gasoline octane by 1.5–3.0 numbers, usually with a negligible effect on clear motor octane. Though the increases in research octane may be relatively small with some feedstocks, they can often allow direct inclusion of the catalytically cracked gasoline in the unleaded pool instead of further treatment. In either case they are still quite valuable, despite the increased spread in research and motor octane (sensitivity).

New Fluid Catalytic Cracking Unit Design. TEXACO (*73, 74*). One of the first refiners to license a unit designed specifically to take full advantage of zeolite catalysts was Texaco, through the Texaco Development Corp. Following the successful design and operation of their own Convent, La. unit (1967), they began licensing their technology worldwide. A typical Texaco design is shown in Figure 23. It features segregated riser cracking of the fresh feed and riser and bed cracking of the recycle as well as improved stripper and regenerator designs. The whole package is designed from Texaco's own special-purpose computer programs

ESSO (*74, 75*). A refiner long active in licensing cat cracker designs but more recently specially tailoring their units for zeolite catalysts is the Esso Research and Engineering Co. They market a Flexicracking technology package, which uses computerized knowhow to design units based on feedstock characterization, yield predictions, reactor and regenerator design technology, mechanical/safety features, and emission control. Typical Esso designs are shown in Figure 24.

One design is an offset side-by-side variation of Esso's Model IV cat cracker with elongated riser and slide valves, while the other is a stacked, pure transfer line type design.

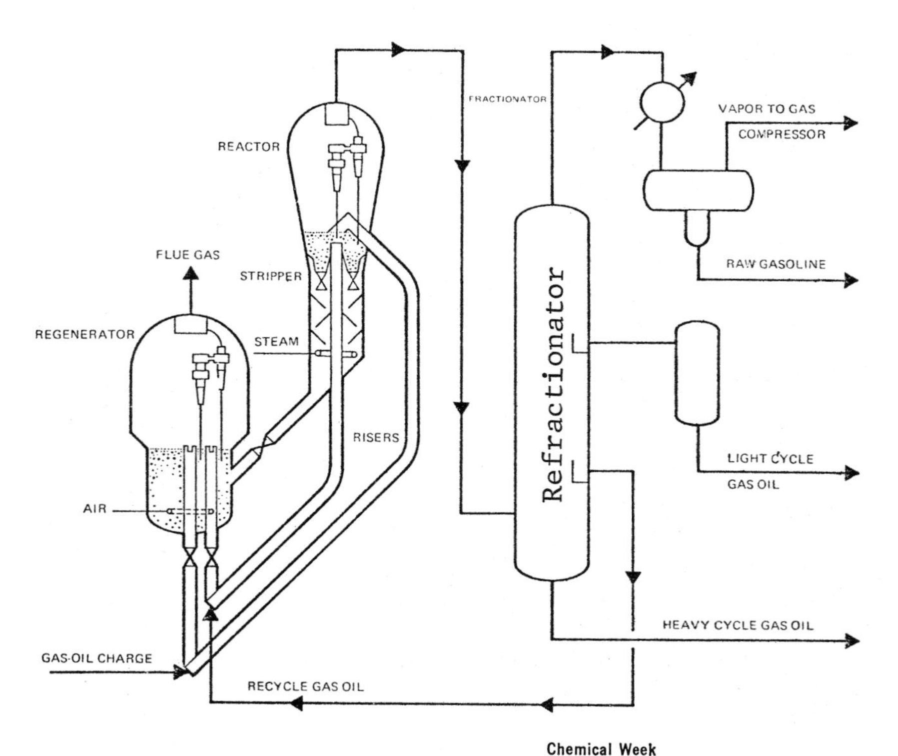

Figure 23. Texaco FCC Process (74)

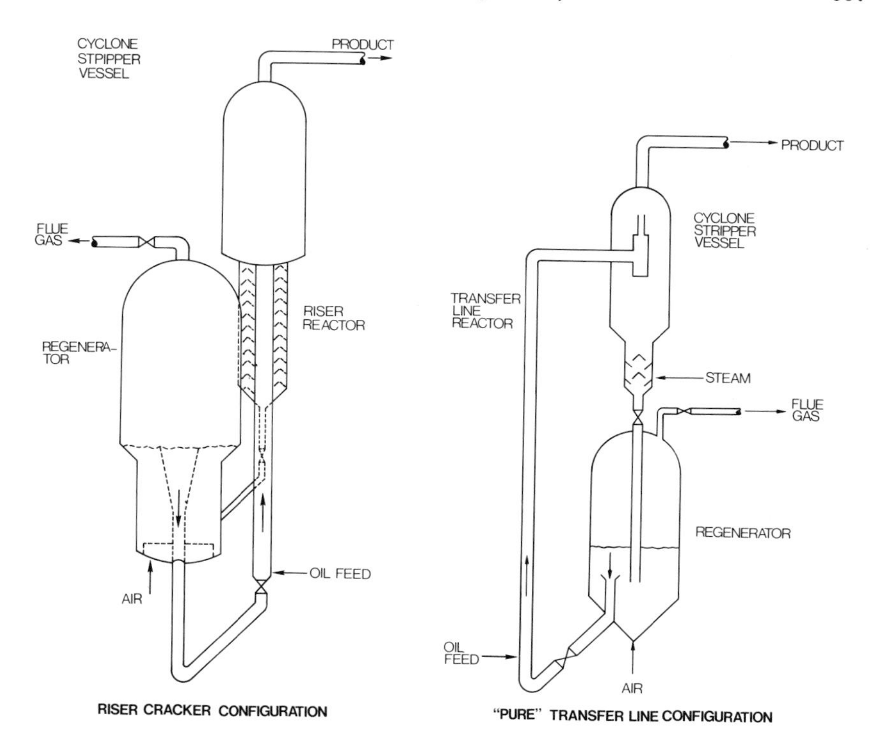

Midyear Meeting of American Petroleum Institute, Division of Refining

Figure 24. Esso Flexicracker reactor configurations (75)

GULF (*76*). A more recent licenser of a technology and design package for fluid cracking is the Gulf Research and Development Co. Their new design units in Alliance, La. and Edmonton, Alta. (both on stream in 1971) feature many of the design principles common to the new breed of riser cat crackers but also uniquely include multiple feed injection points on the riser. Their typical design is shown in Figure 25 although their Alliance unit also features segregated risers, each with multiple injection points. Multiple injection controls catalyst/oil contact times in the riser for various prefractionated boiling range segments of the feedstock, increasing and optimizing gasoline yield. Inert diluent (steam) is injected to control the feed partial pressure so that the very active zeolitic catalysts see only low partial pressures of feed. This is done to minimize overcracking and thereby further maximize gasoline yield (*77, 78*). Actual commercial performance of these specific design features has not yet been disclosed.

STANDARD OIL CO. (INDIANA). Although Standard of Indiana has long had broad in-house experience in catalytic cracking unit design and modification, only recently have they begun actively licensing a full unit design to the industry. This design, which is called the Ultracat Cracking Process, is a new regeneration technique for reducing carbon on regenerated catalyst

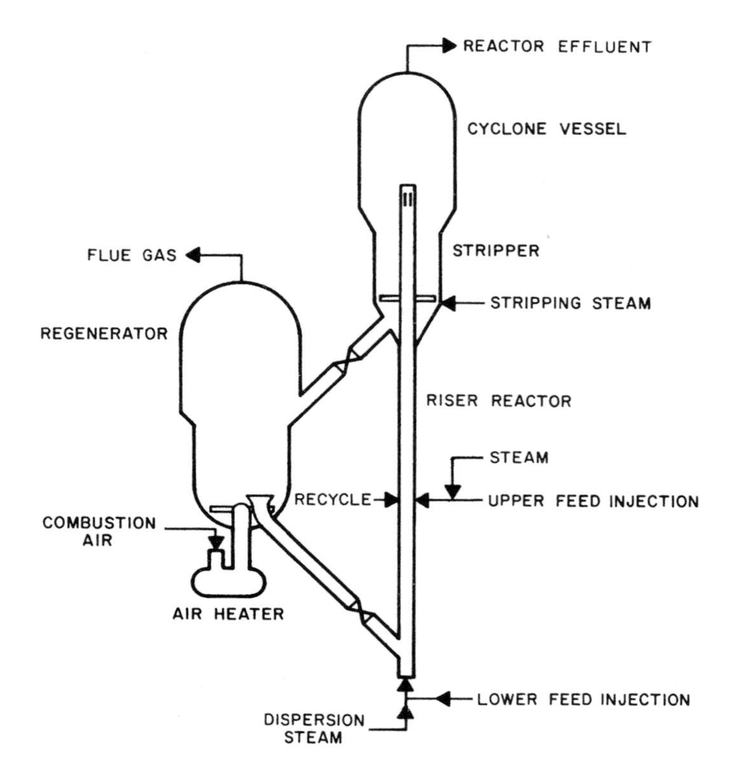

Midyear Meeting of American Petroleum Institute, Division of Refining

Figure 25. Typical FCC unit incorporating Gulf FCC process design (75)

to 0.05 wt % or less. Transfer line reaction, staged stripping and low unit inventories are also features (*79*). It is claimed that this regeneration feature also can reduce carbon monoxide in stack gases by over 98%.

Standard's American Oil subsidiary is currently using their special design features in 11 units having a total fresh feed capacity of nearly 400,000 B/SD. A schematic of their process is shown in Figure 26 (*80*).

While Standard has not publicly disclosed operating or design details of their new regeneration system, it can be assumed that in addition to normal regeneration, a controlled combustion of CO to CO_2 is effected to reach 98% conversion of all CO as claimed. This would also indicate operation in some phase at temperatures in excess of the 1200°–1300°F bed temperatures found in most regeneration systems. Counter-current regeneration is being effected, as shown in the schematic, in line with Standard's recent U.S. patent (*81*). With this new regeneration system, American claims a 25% reduction in coke yield going to recoverable products while using zeolite catalysts (*82*).

UNIVERSAL OIL PRODUCTS. The Universal Oil Products Co., since 1943, has been one of the major suppliers of catalytic cracking process designs to the refining industry. Their stacked unit design for small units (under about 20,000 B/SD) was initially developed and introduced in 1947, and as a regular feature it contained a long vertical riser in which the feed and catalyst were mixed before entering the reactor bed (*see* Figure 27). In the late 50's and early 1960's UOP found that most of the conversion, and certainly that with the preferred selectivity, was occurring in this riser, as opposed to the reactor bed (*83*). This design advantage was later built into some of their larger units. With the advent of fluid zeolite catalysis, a relatively simple adjustment enabled UOP designed units to operate total riser crackers by simply dropping reactor bed levels to the grid. Some of the best of the early zeolite cracking results were obtained in these units.

Currently, UOP is extensively revamping their Stacked-type units all over the U.S. This is normally a simple extension of the riser (Figure 28), except for further modification in sizing and mechanical design of the riser outlet. Of course, UOP is also designing new units and has several operating and in the design stage which feature the straight vertical riser (*see* Figure 29. (*84*).

A high temperature (>1400°F) regeneration system is also being tested by UOP. No details of this system are yet available.

KELLOGG. Like UOP, the M. W. Kellogg Co. has long been a major supplier of catalytic cracking process designs, having built some 86 fluid units since 1942. Although most of their early units featured dense bed cracking, they also had observed the value of riser cracking even prior to zeolites and by 1961 were busy modifying many of their various designs to the riser cracking prinicple (*85*). Design modifications were eventually

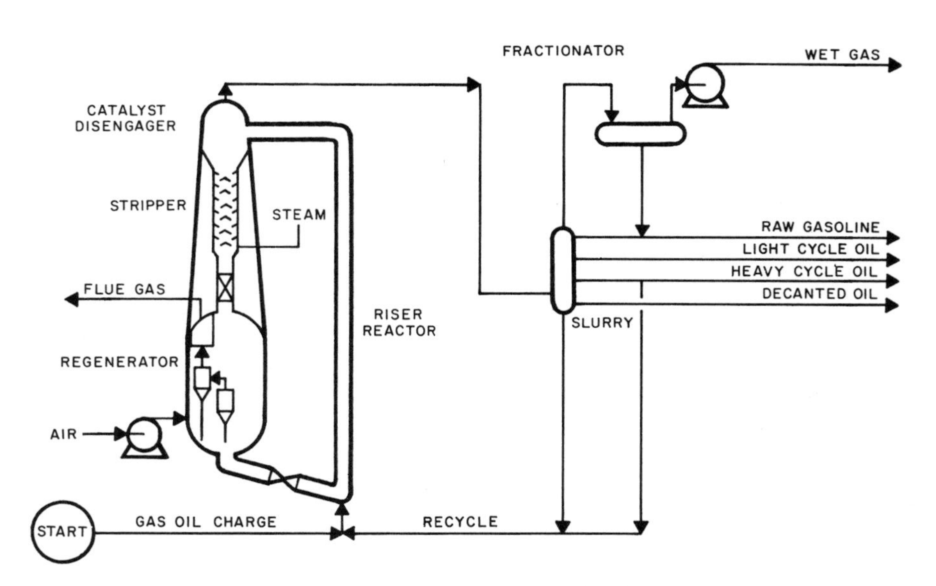

Hydrocarbon Processing

Figure 26. Amoco Ultracat cracking process (80)

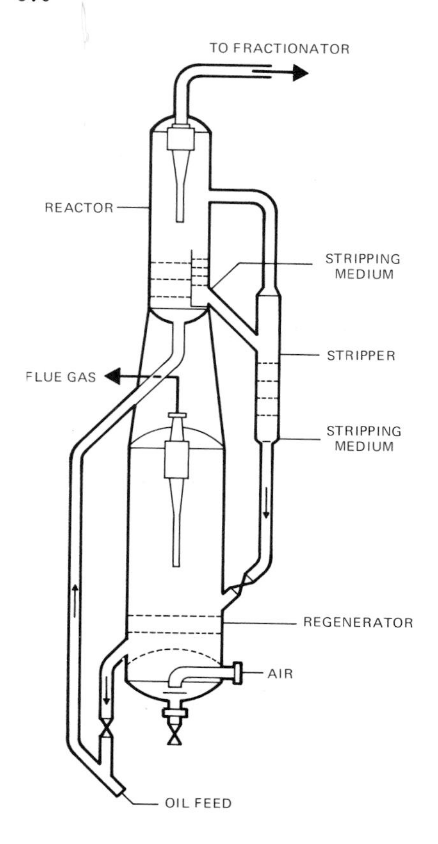

Figure 27. Typical UOP "stacked" FCC

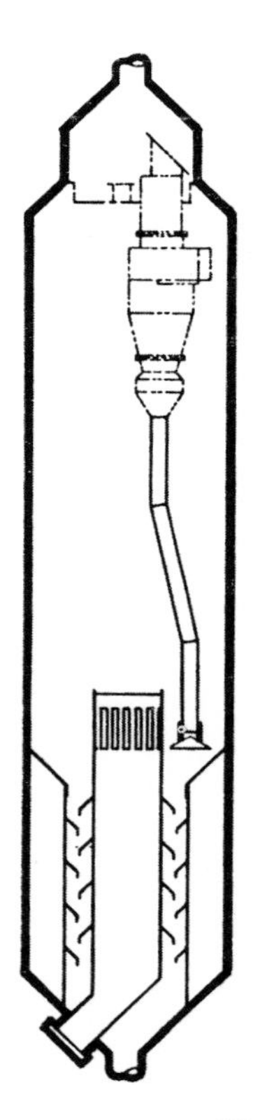

Oil and Gas Journal

Figure 28. Typical UOP reactor revamp mid-1960's (83)

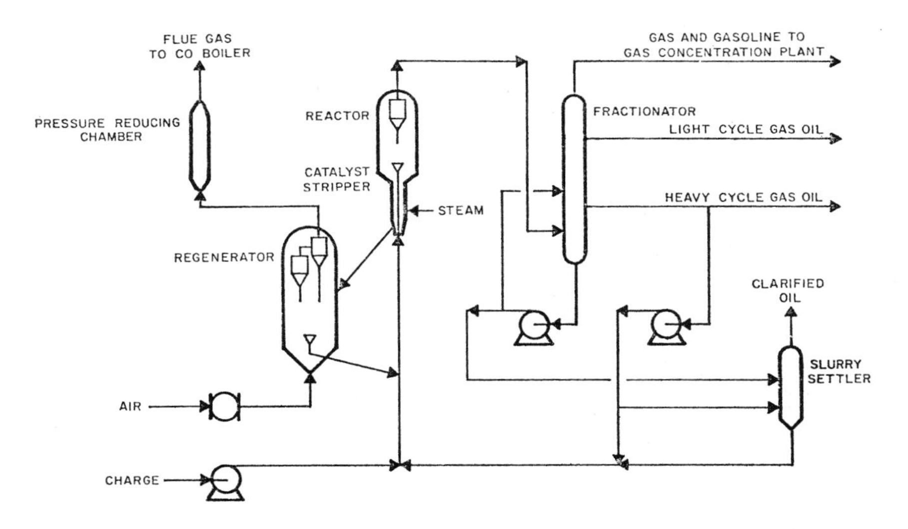

Midyear Meeting of American Petroleum Institute, Division of Refining

Figure 29. UOP "straight riser" fluid catalytic cracking unit (84)

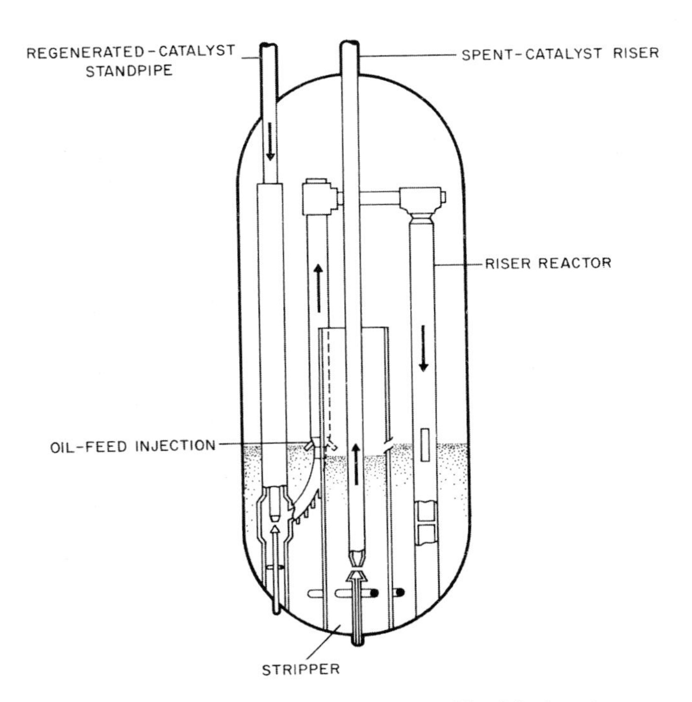

Oil and Gas Journal

Figure 30. Kellogg Orthoflow riser reactor (86)

developed to shorten catalyst/oil contact times. For instance, reactor bed volumes were often reduced by installation of dead-men or concrete fillers, by cutting stripper well slots in lower positions, or by the installation of shrouds to reduce reactor diameter. Eventually, all of these older units were made to operate well on zeolite catalysts through careful stepwise mechanical modification.

With the recent push to the very short contact time of riser cracking, Kellogg's first published design was a revamp of existing Orthoflow B units containing a folded riser reactor, as shown in Figure 30 (*86*). In this design, the normal dense bed is replaced by a folded riser where all the cracking takes place. Contact times as short as 1.5 sec are reported, requiring the use of very active zeolitic catalysts.

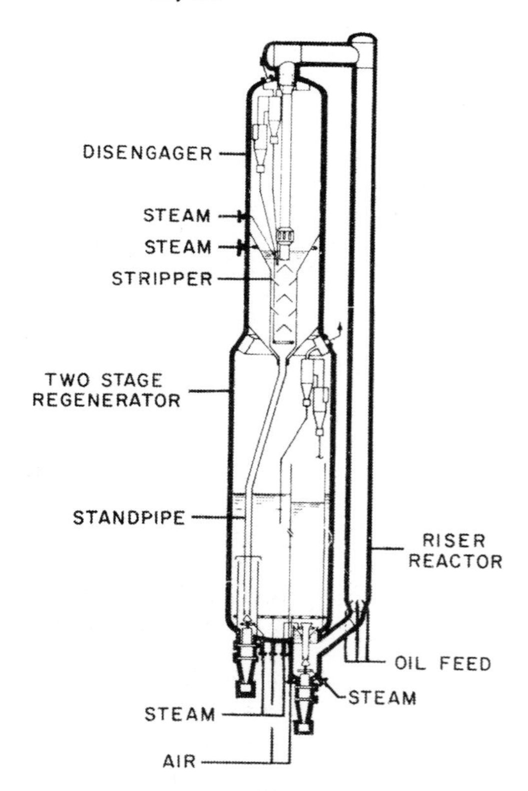

Figure 31. Kellogg new FCC (87)

The folded riser principle was next carried to a stacked-type unit in the latest Kellogg design, as shown in Figure 31 (*87*). This unit features a long vertical riser which inverts, then discharges directly into the stripper well. Regeneration occurs in a two stage system which can, it is claimed, achieve carbon on regenerated catalyst values as low as 0.05–0.10 wt % (*88*).

Kellogg has built many more conventional units of all sizes, many featuring long risers and/or segregated feed cracking. Kellogg recently developed, with Phillips Petroleum Co., a process to catalytically crack heavy oils, such as atmospheric residue (*89*).

Pressure Changes in New Fluid Catalytic Cracking Units. Some new FCC designs increase regenerator pressures into the 30–40 psig range to improve regeneration kinetics and cyclone efficiency. However, increased regenerator pressure must also be accompanied by increased reactor and fractionator pressure. M. W. Kellogg Co. reported the following counterbalancing effects on FCC economics of higher pressures as shown in Table XIII. (*90*). On balance, these factors favor higher pressures in new equipment design.

Table XIII. Converter Pressure—Consideration Design (*90*).

Beneficial Effects of Higher Pressure	*Detrimental Effects of Higher Pressure*
1. Improved regeneration kinetics leading to lower regenerator catalyst inventories and smaller diameter units.	1. Adverse effect on product distribution.
2. Reduced vapor load in fractionator.	2. Increased air blower horsepower requirements.
3. Reduced compressor horsepower requirements.	3. Increased inerts carryover to reactor and recovery system with regenerated catalyst.
4. Improved efficiency of cyclones from higher allowable pressure drops.	4. Greater ΔP through regenerator stack valves.
5. Improved efficiency of power recovery systems.	5. Increased catalyst entrainment at same velocity.

Feed Treatment. In past years there has been much discussion concerning the economic justifications for deep hydrotreating of FCC feedstocks. The real gains in hydrotreating were in those operations requiring increased throughputs at coke and/or metals limitations and in those units requiring higher gasoline yields from aromatic feed or cycle oil fractions. The new riser crackers have slowed some of the push for FCC feedstock hydrotreating, however, because they also provide additional gasoline and coke selectivity, improved metals resistance and a higher temperature cracking of aromatic fractions. This is not to say that feed hydrotreaters will not be built in the coming years. Rather, it appears that many refiners will fully explore the cracking performance of their new FCC's before they again consider feed pretreatment. In addition, many refiners may be forced to some form of hydrotreating for sulfur and nitrogen removal to meet stack emission specifications for SO_2, SO_3, and NO_x.

Commercial Performance of New Units. Many units tend towards 100% riser cracking and improved regeneration through successive modifications starting almost as soon as zeolites were introduced. It is impossible therefore to generalize on the magnitude of yield changes expected from any of the new unit designs. Each specific case would depend on the previous operation. Some directional changes can be predicted, however, and have indeed been observed from early reports.

COKE. Coke yields are down 25–50% at constant conversion, depending on the individual comparison, as a result of a number of cumulative factors. Positive factors tending to reduce coke yield are: (a) Co-current, non back-mixing reaction; (b) High reactor temperatures; (c) Increased zeolitic activity and coke selectivity from reduced coke restriction (including improved regeneration where applicable) and/or use of more active catalysts; (d) Shorter catalyst/oil contact time; and (e) Lower circulation rates (reduced C/O ratios) in those units using high temperature regeneration.

The only negative change with respect to coke yield has been a shift to higher cat/oil ratios in those units not able to supply the additional reactor heat requirements *via* increased feed preheat or higher regenerator temperatures. Even in these instances the normally negative effect of high C/O ratios on coke yield is partly offset by the reduced carbon level on the spent zeolitic catalyst particles, which also results from higher C/O ratios *via* dilution.

INCREASED GASOLINE. At constant conversion, increases in C_5-430°F E.P. gasoline yield of up to 5 vol % FF have been reported. Where conversion has been increased as well, gasoline yields have increased by as much as 10 vol % FF. These gasoline increases are a direct result of these factors: (a) reduced overcracking of the gasoline fraction in cocurrent flow, non back-mixed reaction; (b) improved zeolitic selectivity because of short contact times and reduced coke restriction of the zeolitic catalysts.

INCREASED OLEFIN YIELDS. Solely as a result of the higher reactor temperatures and short catalyst/oil contact times, more olefins are produced throughout the whole product range but especially in the C_3–C_5 range. Initially, increases of $C_3 + C_4$-olefin of up to 4 vol % FF have been reported as a result of riser revamps. As more active zeolite catalysts come into use in these units, light olefin yields should decrease in favor of further gasoline increase.

REDUCED SATURATE YIELDS. As light olefin yields increase, there is a simultaneous decrease in *n*-paraffins. In addition, the reduced reaction time also decreases the yields of light isomers, especially i-C_4. Whereas i-C_4 was abundant soon after zeolites first came into use, it is now increasingly in short supply as riser crackers shift the balance between $C_3 + C_4$ olefin and i-C_4 in the opposite direction. The eventual answer to bringing i-C_4 and $C_3 + C_4$ olefins into balance for alkylation feed is the increased use of super-active zeolite catalysts.

INCREASED RESEARCH CLEAR OCTANE. Increased olefin yields also increase the clear research octanes of the gasoline. Increases of 1–3 numbers have been observed in most riser conversions. Of course, lead susceptibility is reduced, dropping leaded octane gain to 0.3–1.0 CFRR + 3 cc TEL. Motor octane numbers have remained about the same clear, with losses of about 0.5 O.N. CFRM + 3 cc TEL. These trends may be directionally reversed as more active zeolite catalysts come into use.

Most of the current unleaded octane emphasis is on motor octane and not research. Nevertheless, a 2–3 number gain in research clear octane is still quite attractive to most refiners, especially when accompanied by a large increase in C_3-olefin + C_4-olefin, which is potentially alkylate of about 95 motor octane, unleaded.

CYCLE OILS. Cycle oil yields, splits, or quality have not been noticeably changed in the new unit operations. Of course, where conversion has increased markedly as a result of riser cracking and higher reaction temperatures, cycle oils have decreased proportionally, with the usual decreases in API gravity and increases in aromatic content. Shorter residence times, however, favor destruction of light cycle oil over heavy cycle oil as the converted substances (*91*).

METAL RESISTANCE. Almost from the outset, zeolite catalysts have shown excellent resistance to metals contamination in various commercial operations. This apparent resistance to poisoning may have been connected with the multiplicity of cracking sites in zeolites, some sieving action of the zeolitic components to large coke forming molecules or to the excellent hydrogen-transfer properties of zeolites. Certainly, these properties enhanced metals resistance, but it is now clear that one factor was overlooked—reactor residence time. Refiners immediately began reducing catalyst reaction residence times with zeolitic catalysts, and this has already had a most beneficial effect on contaminant coke (*62*).

In the newly designed riser cracking units, residence times well under 1 min and in some cases as low as 2–5 sec are used. At these very short times, metal oxides on the circulating catalyst are probably never fully sulfided in the reaction zone and hence never really became good dehydrogenating agents. The result is an apparent resistance to contamination that makes the cat cracking of some reduced crudes a real possibility. It will be interesting to observe trends in metals levels on equilibrium catalyst as operating experience builds on some of the new FCC units.

High Temperature Regeneration. To date, only one commercial operation using a new high temperature regeneration technique has been described in the literature. This is the operation in Cities Service Oil Co.'s 45,000 B/SD Model II fluid cracker at Lake Charles, La., operating on a high activity zeolitic catalyst. This unit was modified to incorporate Amoco Oil's UltraCat Regeneration process in 1973 with the following reported results (*92*):

(1) Carbon on regenerated catalyst has been maintained at 0.05 wt % or lower.

(2) Coke yield on fresh feed was reduced 20–25% at constant conversion with increases in converted recoverable products.

(3) Equilibrium catalyst activity has remained constant at a constant fresh catalyst makeup rate.

(4) CO stack emissions were reduced to under 500 ppm, and non-methane hydrocarbon emissions are below 5 ppm.

(5) Unit inventory was reduced by 100 tons.

(6) Catalyst circulation rates were reduced and slide valve differentials were improved.

Cities Service estimates this conversion is returning 7 to 12 ¢/bbl of fresh feed in yield improvements. Considering this margin of return plus the reduction in stack gas pollutants, the industry will probably be moving to higher temperature regeneration techniques in the very near future.

Nomenclature

According to the conventions adopted for this volume, the synthetic faujasites, X-type and Y-type, and their various ion exchanged forms may be further identified as follows:

As Written in Chapter Text	*Volume Convention*
NaX or X-type	$(Al_{1.0}Si_{1.2})X$
NaY or Y-type	$(Al_{1.0}Si_{2.5})Y$
REX	$RE^{ex}_{0.85}Na_{0.15}(Al_{1.0}Si_{1.2})X$
REY	$RE^{ex}_{0.80}Na_{0.20}(Al_{1.0}Si_{2.5})Y$
CREX	$RE^{ex}_{0.85}Na_{0.15}(Al_{1.0}Si_{1.2})X$
CREY	$RE^{ex}_{0.80}Na_{0.20}(Al_{1.0}Si_{2.5})Y$
Z-14 US or US	Ultrastable faujasite type

Acknowledgments

The authors are grateful to W. R. Grace & Co. and its Davison Chemical Division for permission to publish the information contained in this chapter. Davison plant personnel P. Y. Moreau, C. Dumoulin, J. R. Hyde, and D. Cox were especially helpful in obtaining illustrations of processing equipment, as were H. H. Morris and E. C. Stebbins, Jr., of the Freeport Kaolin Co. for supplying photographs of their Nichols spray dryer installation. Members of the Davison Research and Development and Petroleum Chemicals Technical Service Departments contributed considerable time and effort in many areas involving preparation of the final text. Special thanks are due to J. A. Montgomery, W. S. Letzsch, and M. D. Edgar of the Davison Division's Petroleum Chemicals Technical Service Department, and R. E. Ritter, G. C. Edwards, R. J. Lussier, L. Rheaume, D. E. W. Vaughan, J. Scherzer, E. W. Albers, J. Ostermaier, H. L. Guidry, and R. J. Nozemack of the Davison Research and Development Department.

Literature Cited

1. Danforth, J. D., *Advan. in Catal.* (1957) **9,** 558.
2. Meys, W. A., Dissertation, University of Delft (1961).
3. Ashley, K. D., Innes, W. B., *Ind. Eng. Chem.* (1952) **44,** 2857-63.
4. Plank, C. J., *J. Colloid Sci.* (1947) **2,** 399-427.
5. Plank, C. J., Drake, L. C., *J. Colloid Sci.* (1947) **2,** 299.
6. Iler, R. K., "The Colloid Chemistry of Silica and Silicates," p. 26, Cornell University Press, Ithaca, 1955.
7. Dobres, R. M., Rheaume, L., Ciapetta, F. G., *Ind. Eng. Chem., Prod. Res. Dev.* (1966) **5,** 174.

8. Earley, J. E., Fortnum, D., Wojcicki, A., Edwards, J. O., *J. Amer. Chem. Soc.* (1959) **81,** 1300.
9. Hensley, A. L., Barney, J. E., *J. Phys. Chem.* (1958) **62,** 1560-3.
10. Dobres, R. M., Rheaume, L., Ciapetta, F. G., *Ind. Eng. Chem., Prod. Res. Dev.* (1966) **5,** 177.
11. Milton, R. M., U.S. Patents **2,882,243, 2,882,244** (1959).
12. Breck, D. W., U.S. Patent **3,130,007** (1964).
13. Breck, D. W., Flanigan, E. M., "Synthesis and Properties of Zeolites X, Y, and L," *First Mol. Sieve Conf., London, 1967,* p. 47.
14. Koffler, R. L., "Sym. on Rare Earth Applications," Amer. Indust. Chem. Engin. Mater. Conf., Philadelphia, Pa. (1968).
15. Scherzer, J., Edwards, G. C., Baker, R. W., Albers, E. A., Maher, P. K., German Patent **2,125,980** (1971).
16. Maher, P. K., McDaniel, C. V., U.S. Patent **3,402,996** (1968).
17. Haden, W. L., Jr., Dzierzanowski, F. J., U.S. Patent **3,657,154** (1972).
18. Haden, W. L., Jr. *et al.*, U.S. Patent **3,663,165** (1972).
19. Innes, W. B., in "Experimental Methods in Catalytic Research," R. B. Anderson, Ed., p. 67, Academic Press, New York, 1968.
20. Lindsley, J. F., U.S. Patent **3,753,929** (1973).
21. Magee, J. S., Jr., Briggs, W. S., U.S. Patent **3,650,988** (1972).
22. Secor, R. B., Peer, E. S., U.S. Patent **2,935,463** (1960).
23. Secor, R. B., U.S. Patent **3,446,727** (1969).
24. Masters, K., "Spray Drying," Chapter 1, CRC Press, Cleveland, O., 1972.
25. Shearon, W. H., Jr., Fullem, W. R., *Ind. Eng. Chem.* (1959) **51,** 724.
26. Iler, R. K., Ref. *6,* p. 28.
27. Wilson, C. P., Jr., Carr, B., Ciapetta, F. G., U.S. Patent **3,395,103** (1968).
28. An excellent summary of moving bed operation may be found in H. S. Bell, "American Petroleum Refining," p. 247, Van Nostrand Co., Inc., New York, N.Y., 1959.
29. Burke, D. P., *Chem. Week* (1972) **111.**
30. Plank, C. J., Rosinski, E., U.S. Patent **3,271,418** (1966).
31. Marisic, M. M., U.S. Patent **2,384,946** (1945).
32. Engelhard Minerals and Chemicals Corp., U.S. Patent **3,384,602** (1968).
33. Goldstein, M. S., in "Experimental Methods in Catalytic Research," R. B. Anderson, Ed., p. 366, Academic Press, New York, 1968.
34. Hirschler, A. E., *J. Catal.* (1963) **2,** 428.
35. Moscou, L., Moné, R., *J. Catal.* (1973) **30,** 417-22.
36. Dobres, R. M., unpublished data.
37. "Methods of Analysis for Fluid Cracking Catalysts," Technical Service Dept., Petroleum Chemicals, W. R. Grace & Co., Davison Chemical Div., Baltimore, Md., 21203, 1972.
38. Shearon, W. H., Jr., Fullem, W. R., Ref. *25,* p. 725-6.
39. Innes, W. B., *Anal. Chem.* (1956) **28,** 332.
40. Forsythe, W. L., Hertwig, W. R., *Ind. Eng. Chem.* (1949) **41,** 1200-6.
41. Zenz, F. A., Weil, N. A., *Amer. Ind. Chem. Eng. J.* (1958) **4,** 472.
42. Letzsch, W. S., Edgar, M. D., *Nat. Petrol. Ref. Assoc. Annual Meeting, 1973.*
43. Frantz, J. F., Juhl, W. G., *71st Nat. Am. In. Chem. Eng. Mtg., Dallas, Tex., 1972.*
44. Lewis, W. K., Gilliland, E. R., Lang, P. M., *CEP Sym. Series* (1965) **58,** 67-78.
45. Viland, C. K., *Petrol. Processing* (1950) **5,** 830.
46. Bondi, A., Miller, R. S., Schlaffer, W. G., *Ind. Eng. Chem. Process Des. Dev.* (1962) **1,** 196.
47. Ciapetta, F. G., Henderson, D. S., *Oil Gas J.* (1967) **65,** 88.
48. Montgomery, J. A., Letzsch, W. S., *Ibid.* (1971) **69,** 60.
49. Wachtel, S. J., Baillie, L. A., Foster, R. L., Jacobs, H. E., *General Papers Session, Petrol. Chem., Amer. Chem. Soc. Meeting, Washington, D. C., 1971.*

50. Grote, H. W., Hoekstra, J., Tobiasson, G. T., *Ind. Eng. Chem.* (1951) **43,** 545.
51. Rice, T., Carpenter, J. K., Ackerman, C. D., *Ibid.* (1954) **46,** 1558.
52. Chen, N. Y., Cucki, S. J., *Ind. Eng. Chem., Process Des. Dev.* (1971) **10,** 71.
53. Herring, W. M., Hinman, J. E., Shields, S. E., *Chem. Eng. Progr.* (1963) **59,** 38.
54. Trainer, R. P., Alexander, N. W., Kunreuther, K., *Ind. Eng. Chem.* (1948) **40,** 175.
55. Thomas, C. L., Barmby, J., *J. Catal.* (1968) **12,** 341.
56. Weekman, V. W., Jr., Nace, D. M., *Amer. Inst. Chem. Eng. J.* (1970) **16,** 397-404.
57. Pachovsky, R. A., Wojciechowski, B. W., *Amer. Inst. Chem. Eng. J.* (1973) **19,** 1121.
58. Maher, P. K., McDaniel, C. V., U.S. Patent **3,293,192** (1966).
59. Eberline, C. R., Wilson, R. T., Larson, L. G., *Ind. Eng. Chem.* (1957) **49,** 661.
60. Bailey, W. A., Jr., Nager, M., *World Petrol. Congr. Proc., 7th* (1967) **4,** 185-92.
61. Cimbalo, R. N., Foster, R. L., Wachtel, S. J., *37th Midyear Mtg., Amer. Petrol. Inst., Div. Refin., 1972.*
62. Grane, H. R., Connor, J. E., Masologites, G. P., *Petrol. Ref.* (1961) **40,** 168.
63. Eckhouse, J. G., Keightley, W. A., *Petrol. Eng.* (1954) **26,** C96.
64. Connor, J. E., Rothrock, W. J., Birkhimer, E. R., Leum, L. N., *Ind. Eng. Chem.* (1957) **49,** 276.
65. Meisenheimer, R. G., *J. Catal.* (1962) **1,** 356.
66. Wollaston, E. G., Forsythe, W. L., Vasolos, I. A., *Oil Gas J.* (1971) **69,** 64.
67. Watson, K. M., Nelson, E. F., Murphy, G. B., *Ind. Eng. Chem.* (1935) **27,** 1460.
68. Nelson, W. L., "Petroleum Refinery Engineering," 3rd ed., p. 94, McGraw-Hill Book Co., Inc., New York, 1949.
69. White, P. J., *Oil Gas J.* (1968) **66,** 112-6.
70. Baker, R. W., Blazek, J. J., "Gasoline Yields Soar with New XZ-Catalyst," *31st Mid-year Meeting, Amer. Petrol. Inst., Div. Refin., Houston, 1966.*
71. Slyngstad, C. E., Atteridg, P. T., "Fluid Catalytic Cracking—Third Generation," *26th Mid-year Meeting, Amer. Petrol. Inst., Div. Refin., Houston, 1961.*
72. "Case Histories of Ultra-Cat Regeneration," Patents & Licensing Dept., Amoco Oil Co., Chicago, Ill., 1973.
73. Bunn, D. P., Jr., *et al.*, "The Development and Operation of the Texaco Fluid Catalytic Cracking Process," *64th Natl. Meeting, Amer. Instit. Chem. Engin., New Orleans, 1969.*
74. *Chem. Week* (1970) **106,** 31.
75. Pierce, W. L., *et al.*, "Innovations in Flexicracking," *37th Mid-year Meeting, Amer. Petrol. Instit., Div. Refin., New York, 1972.* Revised drawings by special permission of Esso Research & Engineering Co., May, 1975.
76. Bryson, M. C., *et al.*, "The Gulf FCC Process," *37th Mid-year Meeting, Amer. Petrol. Instit., Div. Refin., New York, 1972.*
77. Bryson, M. C., Murphy, J. R., U.S. Patent **3,617,497** (1971).
78. Mourning, M. P., U.S. Patent **3,650,946** (1972).
79. *The Oil Daily*, March 27, 1972, p. 7A.
80. *Hydro. Proc.* (1972) **51,** 138.
81. Cartmell, R. R., U.S. Patent **3,661,799** (1972).
82. Shields, R. V. *et al., Oil Gas J.* (1972) **70,** 45.
83. Pohlenz, J. B., *Oil Gas J.* (1963) **61,** 124.
84. Strother, C. W., *et al.*, "UOP Innovations in Design of Fluid Catalytic Cracking Units," *37th Mid-year Meeting, Amer. Petrol. Instit., Div. Refin., New York, 1972.*
85. Slyngstad, C. E., Atteridg, P. T., "Fluid Catalyst Cracking—Third Generation," *26th Mid-year Meeting, Amer. Petrol. Instit., Div. Refin., Houston, 1961.*

86. Murphy, J. R., *Oil Gas J.* (1970) **68,** 72.
87. M. W. Kellogg Technical Brochure, "New Fluid Cracking Units—Design Technical Services by Kellogg" (1971).
88. *Chem. Eng.* (1972) **79.**
89. Oil Daily News Release, *70th Annual Meeting, Natl. Petrol. Refiners Assoc., San Antonio, 1972.*
90. Whittington, E. L., *et al.*, "Catalytic Cracking—Modern Designs," *Amer. Chem. Soc., Div. Petrol. Chem., Preprints, Symposia-Group 1, New York, 1972,* p. B66.
91. Magee, J. S., Blazek, J. J., Ritter, R. E., "Catalytic Cracking—New Catalyst Developments," *Amer. Chem. Soc., Div. Petrol. Chem., Preprints, Symposia—Group 1, New York, 1972,* p. B63.
92. Broussard, A. D., *Oil Gas J.* (1973) 149.

Chapter

12
Shape-Selective Catalysis

Sigmund M. Csicsery,
Chevron Research Co., Richmond. Calif. 94802

Shape Selective Catalysis.

SMALL, UNIFORM INTRACRYSTALLINE cavities and pores characterize zeolite catalysts. If the overwhelming majority of the catalytic sites are confined within this pore structure and if the pores are small, the fate of reactant molecules and the probability of forming product molecules are determined mostly by molecular dimension and configuration. Only molecules whose dimensions are less than a critical size can enter the pores, have access to internal catalytic sites, and react there. Furthermore, only molecules which can leave appear in the final product. Bulkier molecules will react or bulkier products will form only on the relatively few catalytic sites on the external surface of the zeolite crystals.

Shape selectivity was first described by Weisz and Frilette in 1960 (*1*). P. B. Weisz, N. Y. Chen, V. J. Frilette, and J. N. Miale were not only the pioneers of shape selective catalysis but in their subsequent publications demonstrated its many possible applications. Many of the following examples originated in their laboratory.

Types of Shape Selectivity. Reactant selectivity occurs when only one part of the reactant molecule can pass through the catalyst pores. The remaining molecules are too large to diffuse through the pores.

The cage or window effect is a special case of reactant selectivity in which certain molecules react at a rate different from most other molecules because their length matches the length of a sieve cavity. In this way, man-made inorganic catalysts are approaching the extreme specificity of enzyme catalysis.

Product selectivity occurs when, among all the product species formed within the pores, only those with the proper dimensions can diffuse out and appear as observed products. Bulky products, if formed, are either converted to less bulky molecules (*e.g.*, by equilibration) or eventually deactivate the catalyst by blocking the pores.

Restricted transition state selectivity occurs when certain reactions are prevented because the corresponding transition state would require more space than available in the cavities. Neither reactant nor potential product molecules is prevented from diffusing through the pores, and reactions requiring smaller transition states proceed unhindered.

Diffusion. The importance of diffusion in shape-selective catalysis cannot be overemphasized. It is discussed in another chapter of this volume. In general, one type of molecule will react preferentially and selectively in a shape selective catalyst if its diffusivity is at least one or two orders of magnitude higher than that of the competing molecular types. (Actually certain types of molecules are completely excluded from the interior of shape selective catalysts, *i.e.*, their diffusivity is zero.) Since diffusing molecules are constantly influenced by the zeolite pore wall and collide frequently with it, a Knudsen-type diffusion must be involved. In sieves with tubular pore structure (*e.g.*, erionite, mordenite), and with molecules only slightly smaller than pore dimensions, the diffusing molecules cannot pass each other. Diffusion here must be unidirectional—all molecules within one pore must move in the same direction at the same time in a single file. This requirement greatly reduces diffusivity. Furthermore, any dimerization of two molecules or the formation of any other strongly adsorbed or immobile species will completely block diffusion within one pore. Therefore, even those molecules which react preferentially have much smaller diffusivities in shape-selective catalysts than in large pore catalysts. For example, *n*-paraffins have diffusivities at least five orders of magnitude lower in the zeolite KT than in large pore catalysts.

Catalytic Sites. TYPE OF SITES. Most shape-selective catalysts are acidic. Acid sites are introduced into Na– or K–zeolites by acid treatment, by ammonium exchange and subsequent deammination, or by exchanging with multivalent cations, such as alkali earth or rare earth ions.

Dual Functional catalysts have both acidic and hydrogenation–dehydrogenation activity. The hydrogenation–dehydrogenation component is needed to catalyze some of the required reactions (*e.g.*, olefin hydrogenation or hydrogenolysis). In addition, it allows continuous operation for long periods in most applications because it prevents coking and the associated blocking of the pores. Platinum and palladium are the most frequently used hydrogenation–dehydrogenation components. Non-noble metals could be used also. Na-mordenite has sufficient hydrogenation activity of its own.

There are many ways to introduce metals into a zeolite. Ion exchange using tetrammine complexes or direct impregnation with hexachloro- or tetrammine complexes are the easiest and most straightforward methods. Pores of the A-type sieves are too small to admit these large complex molecules. Platinum can be incorporated in these sieves by making them from a solution which contains some Pt salt. Pt–Na–mordenite can be made in an indirect way. Na–mordenite is first converted to the H-form. This has sufficiently large pore openings to allow the entry of platinum complexes. A treatment with $[Pt(NH_3)_4]^{2+}$ ions introduces platinum into the supercages. Finally, sodium is back-exchanged to form Pt–Na–mordenite.

Some shape-selective catalysts (such as the Pt–Na–mordenite above) have only hydrogenation–dehydrogenation activity. These are used in shape-selective hydrogenations where acidity is not needed. Certain selective oxidations employ similar nonacidic sieves.

INTERIOR VS. EXTERIOR SITES. Catalytically active sites are located both in the internal pore structure (interior sites) or on the outside surface (exterior sites) of the molecular sieve crystallites. The external area of zeolites is usually between 2 and 20 m^2/g; therefore, there are probably about 100 times more interior than exterior sites. For example, Venuto and Landis have calculated that external surface areas of 1.2–2.6μ particle size of rare earth, ammonium, or hydrogen Y-faujasites are 3–7 m^2/g (*2*). This is about 0.5–1.3% of the total BET surface area. Internal diffusion could make the interior sites less effective, and, therefore, catalysis by the non-shape-selective exterior sites could become much more important than warranted by their small number. Furthermore, in metal-containing sieve catalysts, metal migration from interior to exterior sites during reduction or subsequent use is a real problem. Shape selectivity can be improved by either selectively removing exterior catalytic sites or by irreversibly poisoning them with molecules larger than the effective pore diameter of the molecular sieve.

Molecular Sieve Types. Selectivity between molecules of different shape and size is achieved by the proper choice of molecular sieves. The amount and type of cations can further modify activity and effective pore size, and thus shape selectivity. An excellent review by Venuto and Landis discusses the relation between molecular sieve structure and catalytic activity and the various factors related to the accessibility of catalytic sites (*2*). Molecular sieve crystal structures are described in another chapter of this volume.

Shape selectivity can be predicted by estimating whether a molecule will pass through certain pores or not. By comparing molecular dimensions with pore diameters, one can select the most appropriate molecular sieve for each process. Structural drawings of organic molecules and molecular sieve crystal frameworks and lists of pore dimensions to tenths or even hundredths of angstroms tend to create an impression of fixed sizes and rigid bodies. Actually, due to thermal vibrations and rotations, molecules can wiggle through pores that are smaller by 0.5–0.75A than their minimum dimensions.

A-TYPE SIEVES. A-type sieves have roughly spherical internal cavities with 11.4 A diameter (α-cages). These are connected with each other through six circular openings surrounded by rings of eight oxygens. Effective diameters of these ports in KA, NaA, and CaA are about 3, 4, and 5 A, respectively. The pores of KA are, therefore, too small for almost all organic molecules. *n*-Paraffins and other linear molecules can diffuse through the pores of CaA. Many examples of shape-selective catalysis have been demonstrated with CaA. Neither NaA nor CaA has much acidity. Because they have low silica-to-alumina ratio, they are not stable when converted to the H form. As a result, they have not been used much as petroleum processing catalysts.

ERIONITE AND RELATED ZEOLITES. Erionite, the related offretite, and KT belong to the chabazite group. Erionite is characterized by relatively large (13 x 6.3 A) cylindrical cavities. Distorted eight-oxygen rings (3.6 x 5.2 A) connect these cavities with each other. Only *n*-paraffins can pass through these pores. *n*-Octane fits almost exactly into the large erionite cavities. The interesting cage or window effect is a result of this coincidence. Structures of erionite and offretite and the cage or window effect are described in more detail in a latter part of this review.

Erionite and T-type molecular sieves are probably the most important shape-selective catalysts. They distinguish branched from *n*-paraffins.

Mordenite. Mordenite has a tubular pore system with elliptical pores. Major and minor diameters are about 7 x 5.8 A. Cations can block these pores —the effective diameter of Na–mordenite is considerably smaller than that of H–mordenite. Natural mordenite has a smaller effective diameter (about 4 A) than synthetic preparations. This may be the result of stacking faults in the natural material. Because of their high silica-to-alumina ratios, mordenites have excellent thermal stability and acid resistance. Some of the framework alumina can be removed by acid leaching. This can open up the pore structure and greatly improve diffusivity. Even molecules as large as symmetrical trimethylbenzene can pass through the pores of synthetic H–mordenite.

H–mordenite may be used to selectively remove straight-chain and nearly straight-chain paraffins from heavy petroleum fractions (*e.g.*, catalytic dewaxing processes).

FAUJASITES. Faujasites have nearly spherical supercages connected by relatively large (8-9 A) rings composed of 12 oxygens. These rings are so large that shape selectivity occurs only with very large molecultes. (Even hexaethylbenzene can diffuse through faujasite pores!) Although faujasites are commercially the most important molecular sieve catalysts, at present they have a very limited role as shape-selective catalysts.

Applications. Shape-selective catalysis has countless possible applications. The following selected examples illustrate some possibilities.

(a) Undesirable impurities can be continuously converted to harmless substances. Examples are selective hydrogenations of piperylenes in an isoprene stream or of ethylene in a propylene stream.

(b) Undesirable impurities can be continuously converted to easily removable smaller compounds. Examples are selective combustion of straight chain paraffins or olefins, catalytic dewaxing, and the selective cracking of *n*-paraffins in gasoline fractions to improve octane rating (Selectoforming or shape selective cracking).

(c) Selective combustion generates heat *in situ* in a reactor.

The following pages contain more detailed descriptions of these and other potential applications. Relatively few of these have found actual commercial use. Many of the suggested processes suffer from poor catalyst stability. In some other cases, coking and the resultant pore-blocking make continuous use impossible. However, shape selective catalysis might offer solutions to problems where it has not been tried yet.

Catalysts Selective for Small Linear Hydrocarbons (Linde NaA and CaA)

Cracking. The effective pore size of a molecular sieve can be modified by changing the cations associated wtih the sieve. For example, if the sodium ions of NaA are exchanged with calcium ions, the effective pore diameter increases from about 4 to 5 A. The resultant CaA admits normal hydrocarbons but excludes even the smallest branched paraffin, such as isobutane. Selective cracking of *n*-paraffins in the presence of branched paraffins was first demonstrated by Weisz, Frilette, Maatman, and Mower (*3*). This is a very interesting reaction because conventional acid catalyst preferentially crack branched paraffins and because its principle eventually led to Selectoforming, the most important commercial application of shape-selective catalysis.

Weisz and coworkers passed *n*-hexane and 3-methylpentane over pellets of Linde NaA and Linde CaA molecular sieves and Mobil Oil Co.'s standard 46 AI cracking catalyst, an amorphous silica-alumina (*3*). A thermal cracking run was made over 96% silica chips for comparison. Table I compares *n*-hexane and 3-methylpentane conversions at 500°C and 7 sec residence time.

Table I. Comparison of *n*-Hexane and 3-Methylpentane Cracking at 500°C (*3*)

Catalyst	*3-Methylpentane Cracking Conversion, %*	n-*Hexane Cracking*		
		Conversion, %	$Iso\text{-}C_4$ / $n\text{-}C_4$	$Iso\text{-}C_5$ / $n\text{-}C_5$
96% Silica chips	<1	1.1		
Amorphous silica-alumina (46 AI)	28	12.2	1.4	10
Linde NaA	<1	1.4		
Linde CaA	<1	9.2	<0.05	<0.05

Journal of Catalysis

Linde NaA is inactive because its smaller pores severely restrict the diffusion of *n*-hexane and probably because it does not have acid sites strong enough for cracking. Linde CaA, however, cracks *n*-hexane very selectively. The same selectivity is shown when *n*-hexane is reacted in a mixture with methylpentanes. Note also the example for product selectivity: Branched products (iso-C_4 and iso-C_5) are essentially absent in the product obtained over Linde CaA while they are the prevalent products over amorphous silica–alumina.

Dehydration. Dehydration of straight chain primary alcohols in the presence of branched ones is another manifestation of shape selectivity. Over amorphous catalysts or large pore molecular sieves (*e.g.*, faujasite-type sieves), these alcohols dehydrate at nearly equal rates. Over Linde CaA, *n*-butyl alcohol dehydrates very much faster than isobutyl alcohol. Experimental data by Weisz, Frilette, Maatman, and Mower (*3*) are shown in Table II.

Table II. Dehydration of Primary Butyl Alcohols (1 atm. Pressure with 6-sec Residence Time) (*3*)

Temperature °C	130	230	260
Over Linde CaA Molecular Sieve			
Isobutyl alcohol dehydrated, wt %		<2	<2
n-Butyl alcohol dehydrated, wt %		18	60
sec-Butyl alcohol dehydrated, wt %	~0		
Over Faujasite-Type CaX Molecular Sieve			
Isobutyl alcohol dehydrated, wt %		46	85
n-Butyl alcohol dehydrated, wt %		9	64
sec-Butyl alcohol dehydrated, wt %	82		

Journal of Catalysis

A small isobutyl alcohol conversion is reasonable because a small number of catalytic sites are located at the exterior surfaces of the molecular sieve crystallites. The behavior of *sec*-butyl alcohol demonstrates catalysis by such exterior sites even better. At higher temperatures, the rate constant of *sec*-butyl alcohol dehydration over Linde CaA is between two and three orders of magnitude smaller than over faujasite-type CaX molecular sieve. This difference is consistent with the difference between internal and external surface areas.

Catalysts Which Contain Metals with Hydrogenation–Dehydrogenation Activity (Pt or Pd on NaA, CaA). Catalytic sites may be inherently present in a molecular sieve (*i.e.*, Brønsted or Lewis acid sites) or may be induced (*i.e.*, by introducing a finely dispersed metal to gain hydrogenation activity). Shape-selective catalysts containing platinum and capable of hydrogenating only linear olefins were first made by Weisz and his coworkers (*1, 3*). Platinum is usually introduced into molecular sieves *via* water-soluble salts containing either the $[Pt(NH_3)_4]^{2+}$ cation or the $[PtCl_6]^{2-}$ anion. These ions are too large to penetrate the 5 A pores of CaA molecular sieve. Weisz and coworkers formed NaA sieve in a solution containing $[Pt(NH_3)_4]^{2+}$ cations. The resulting NaA zeolite was then calcium exchanged by repeated treatments with calcium chloride solutions. In addition to replacing sodium with calcium, this technique removes any extracrystalline tetrammine platinous cations and assures high shape selectivity. The final product contained 0.31 wt % platinum. The catalyst was pretreated at 450°C in dry helium for one hour and in air for 30 min.

Hydrogenation experiments were carried out at 25°C, at atmospheric pressure, with a 3:1 hydrogen-to-hydrocarbon mole ratio and a residence time of 0.3 sec. In separate experiments, the catalyst converted 70% of 1-butene, less than 2% of isobutylene, and 52% of propylene to the corresponding paraffin. In one experiment, an equimolar isobutylene–propylene mixture was passed through the catalyst for three hr at 343°C. Propylene conversion decreased from about 70% to about 25% as the catalyst aged. Isobutene hydrogenation was not detected.

Chen and Weisz have extended shape selectivity to distinguish between trans and cis olefins or diolefins (*4*). At 26°C, trans *n*-olefins have at least two orders of magnitude higher diffusivites than the corresponding cis isomers. With platinum in CaA catalyst, *trans*-2-butene has a seven times higher hydrogenation rate constant than *cis*-2-butene at 98°C with 10:1 hydrogen-to-hydrocarbon mole ratio at atmospheric pressure.

Most molecular sieves have about a hundred times larger interior than exterior surface. Juguin, Clément, Leprince, and Montarnal have found that if platinum is incorporated into NaA by forming the sieve in a solution containing $[Pt(NH_3)_4]^{2+}$, about 20% of the platinum is on the exterior surface. They used three different methods to determine exterior and interior platinum: (*5*)

(a) Ethylene and isobutylene hydrogenation rates were compared over platinum on CaA and platinum on alundum. Over the nonporous catalyst, ethylene has a hydrogenation rate constant about three times higher than that for isobutylene. Over the CaA catalyst, where isobutylene is hydrogenated only by exterior platinum sites, the ratio of the two rate constants is eight. This suggests that 67.5% of all platinum sites on this CaA catalyst are located in the interior.

(b) Differences in ethylene hydrogenation rates before and after all exterior platinum sites are poisoned with thiophene or dibenzothiophene suggest that 73-80% of all platinum sites are in the interior. These sulfur compounds are too large to enter the pores of CaA. On the other hand, H_2S, which can freely diffuse through CaA, poisons all platinum sites.

(c) The $[Pt(NH_3)_4]^{2+}$ cations enclosed in the supercages are too large to move through the much smaller pore ports and cannot be exchanged. $[Pt(NH_3)_4]^{2+}$ located on the exterior surface, however, is easily removed by ion exchange. Ethylene hydrogenation rate measurements before and after exterior platinum has been removed by ion exchange suggest that 79% of all platinum sites are in the interior.

Selective CO Oxidation (NaA). Heat transfer problems restrict maximum possible reaction rate and dictate reactor design in most highly endothermic processes. In cracking catalyst regeneration, these problems are further complicated because coke is burned to a mixture of carbon dioxide and carbon monoxide. The cracking reaction is endothermic, and heat from the regenerator is the principal factor which maintains the heat balance of the entire system. In conventional fluid catalytic cracking, carbon dioxide: carbon monoxide ratios are between 1 and 2. Carbon combustion to carbon monoxide yields only about one-third as much heat as combustion to carbon dioxide. Thus, at these low carbon dioxide:carbon monoxide ratios, much potentially available heat escapes the system. (Today most refineries recover most of this lost heat in a separate reactor.) The situation could be further aggravated by carbon monoxide afterburning in the dilute bed of fluid crackers where the resultant excessive heat shortens the life of reactor components. In the past, transition metal oxide oxidation promoters (*e.g.*, chromium) were added to cracking catalysts to increase the carbon dioxide to carbon monoxide ratio. One disadvantage is that such transition metal oxides also catalyze dehydrogenation and coke formation in the cracking site, reducing cracking selectivity. Recently, some crackers have raised regeneration temperatures up to 700°–760°C. Coke removal at these temperatures is more complete, and carbon monoxide is eliminated almost entirely. However, these high temperatures require special alloy-steel reactor constructions and could shorten catalyst life.

A different solution, using shape-selective catalysts, was suggested by Chen and Weisz (*4*). The oxidation catalyst is incorporated in a small-pore molecular sieve which allows only oxygen, carbon monoxide, and carbon dioxide to diffuse in and out of its pore structure. Larger hydrocarbons are not permitted to reach the oxidation catalyst. Chen's and Weisz's catalyst, containing only 0.009% platinum, was made by growing Zeolite A crystals in a dilute $[Pt(NH_3)_4]Cl_2$ solution. An air stream containing 2 mole % carbon monoxide was passed over this catalyst at atmospheric pressure at about 0.04 sec residence time. Between 427° and 540°C, 86–100 wt % of the carbon monoxide was converted to carbon dioxide. When 0.5 mole % *n*-butane in air reacts at the same conditions, only 2.4 to 5.5 wt % is oxidized. Remarkably high activity is matched here with a high degree of selectivity. The use of such a catalyst in catalytic cracking would not be economically prohibitive because platinum is present at an extremely low ($<$0.01%) level. Nevertheless, this reviewer does not know any commercial application. Perhaps the NaA molecular sieve is not sufficiently stable at catalytic cracking conditions, or the initially finally dispersed platinum congregates to larger particles or migrates to the outer surface of the molecular sieve particles during use.

This latter phenomenon has been described by Reman, Ali, and Schuit, and recently by Mintschev and Steinbach with nickel-containing molecular sieves (*6, 7*). In these sieves, nickel migrates out of the big cavities to the external surface, forming large metal crystallites there. By poisoning exterior sites of nickel on Linde CaA catalyst with 2,3-dimethyl-2-butene, Mintschev and Steinbach were able to measure the activity of interior Ni sites. They found that regardless of catalyst pretreatment, shape-selective catalytic action of nickel in CaA may be preserved only under very mild reduction conditions. In certain nickel-containing sieve catalysts (*e.g.*, Y–type faujasites), Lawson and Rase prevented the migration of nickel to the outside surface and its agglomeration there by pre-treatment with chromia (*8*). To avoid aqueous exchange, chromium was introduced into the sieve as $Cr(CO)_6$ in helium. Subsequent treatment in air converted the hexacarbonyl to chromia. Much additional work would be needed, however, to elucidate the mechanism of the inhibiting action of chromia and to optimize its concentration and dispersal.

Shape-selective, low-temperature catalytic combustion of *n*-paraffins is possible over Pt–CaA and similar catalysts. Weisz passed a 6:4 mixture of *n*-butane and isobutane in an air stream over CaA zeolite which contained 0.3% platinum. While 97% of the *n*-butane was consumed, less than 0.1% of the isobutane reacted (*9*). Over this catalyst combustion of propylene and 2-butene was similarly complete and selective in the presence of isobutylene.

Cracking and Hydrocracking of n-*Paraffins over Offretite, Erionite, and Related Zeolites*

Crystal Structure and Cracking. Crystal structures of offretite, erionite, and some related zeolites were determined by Bennett, Gard, Tait, Kawahara, and Curien (*10, 11, 12, 13, 14*).

Offretite and erionite are hexagonal zeolites. The synthetic Linde T is primarily offretite with some erionite intergrowth. The extent of this erionite intergrowth can be estimated from odd "l" line intensities in electron diffraction patterns (*12*).

Erionite has a hexagonal lattice structure, composed of cancrinite cages. These cages are rotated 60 degrees relative to each other in subsequent layers (ABA formation). The crystal axes are $\bar{a}$ and $\bar{c}$. Each cancrinite cage is joined to two others through six oxygen atoms, forming hexagonal prisms. The resulting columns are arranged in a hexagonal pattern. The supercages enclosed between these elements are approximately 13 x 6.3 A. Each supercage is surrounded by 12 four-membered rings, two planar and three assymetric six-membered rings, and six-eight-membered rings.

No hydrocarbon can pass through the four- and six-membered rings. (The free diameter of the latter is only about 2.5 A.) Diffusion of hydrocarbons, therefore, is blocked along the $\bar{c}$ axis of erionite. The slightly assymetric eight-membered rings have maximum and minimum free diameters of about 5.2 A and 3.6 A. These eight-membered rings line up about 7 A apart along the $\bar{a}$-axis. Only straight-chain hydrocarbons can pass through these pores.

In offretite, the cancrinite cages and the hexagonal rings are all oriented similarly (AAA formation). The supercages are, therefore, open into each other; and this creates a 6.3 A wide unobstructed channel in the $\bar{c}$ direction. These channels are so wide that branched, or even cyclic hydrocarbons, can pass through. Pore geometrics of erionite and offretite in the $\bar{c}$ axis are compared in Figure 1.

Stacking disorders in offretite could create erionite-type windows along the $\bar{c}$-axis. Theoretically, a single erionite cell would completely block an offretite channel. Sherry reported, for example, that 3% erionite in Linde T is enough to hinder the diffusion of branched or cyclic hydrocarbons (*15*). Thus, an offretite sample will be shape selective if it contains some erionite intergrowth. Miale, Chen, Weisz, and Garwood and Robson, Hamner, and Arey have shown that erionite, as well as most types of offretites, and the related gmelinite and chabazite distinguish between linear and branched paraffins (*9, 16, 17, 18*) after most cations have been replaced with protons through ammonium exchange and deammination. However, a later publication (*10*) pointed out that the mineral often referred to as offretite was, in fact, erionite. At atmospheric total pressure with helium carrier gas and about 5-sec residence time, *n*-hexane is cracked with high conversions above 370°C; 2-methylpentane and methylcyclopentane do not react (Table III). As expected, branched hydrocarbons are absent from the product (*16*).

Table III. Selective Cracking of C_6 Hydrocarbons over H-Gmelinite and H-Erionite Molecular Sieves (*16*)

Molecular Sieve	*H-Gmelinite*			*H-Erionite*	
Hydrocarbon	n-*Hexane*	*2-Methyl-pentane*	*Methyl-cyclo-pentane*	n-*Hexane*	*2-Methyl-pentane*
Temp, °C	370	320–540	510–540	320	430
Conversion, %	47	0–0.7	0.4–1.9	52	1

Journal of Catalysis

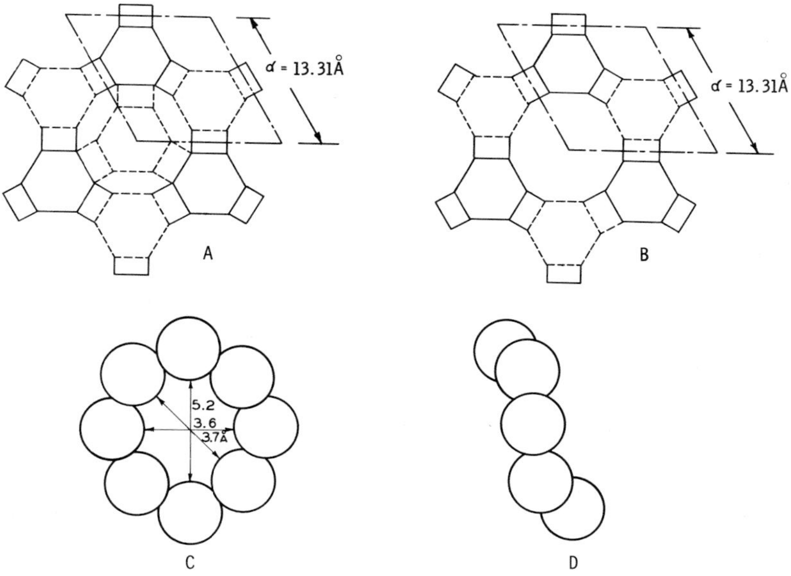

Journal of Catalysis

Figure 1. Projection of the erionite (A) and offretite (B) frameworks in plane parallel to the $\bar{c}$-axis, and front (C) and side (D) views of the eight-membered ring of erionite (22)

Chen has found that shape selectivity increases with increasing odd "*l*" line intensity in electron diffraction patterns going along the $\bar{c}$-direction (*19*). Samples with weak or no odd "*l*" lines can also crack branched hydrocarbons, such as 2-methylpentane (Table IV).

As expected, isobutane production is much less than *n*-butane production over catalysts which showed high shape selectivity toward *n*-hexane. On the other hand, the less shape-selective synthetic offretite produces more than three times as much isobutane as *n*-butane.

Since even *n*-paraffins move slowly through the narrow erionite channels, cracking or hydrocracking under most conditions could be limited by diffusion.

Table IV. Correlation between Crystal Structure and Shape Selectivity in the Offretite-Erionite Zeolite Family (*19*)

Zeolite	Odd "*l*" Line Intensity	Conversion at 370°C 14 atm, 4 LHSV in a 1:1:2 n-Hexane: 2-Methylpentane: Benzene Mixture, wt %		$\frac{k_{n\text{-hexane}}}{k_{2\text{-methylpentane}}}$
		n-Hexane	*2-Methylpentane*	
H-Erionite (natural)	strong	76	2.6	54
Synthetic HT	strong	68	2.4	48
HT	intermediate	71	5.2	24
HT	weak	58	7.7	11
Synthetic offretite	none	62	52	1.8
Catalysts without shape selectivity				0.3–0.5

"Fifth International Congress on Catalysis"

A comparison of activation energies sometimes reveals this. If the reaction is diffusion controlled, the observed activation energy is about half of the sum of activation energies of the surface reaction and of diffusion. Where the activation energy of diffusion is low, *i.e.*, 1–3 kcal/mole, the observed activation energy is very close to one-half of that of the surface reaction. For zeolites, however, the activation energy of counterdiffusion may be quite high. For example, Katzer observed that the activation energy of the counterdiffusion of benzene in Na Y-faujasite is 20 kcal/mole (*20*). In certain other cases, however, comparisons of activation energies provide useful information. *n*-Paraffin cracking activities of H-offretite, H-gmelinite, and H-chabazite are more than 10,000 times higher than that of a typical commercial amorphous silica-alumina cracking catalyst or CaA. The apparent activation energy of *n*-hexane cracking is 30 kcal/mole over the commercial catalyst and $Ca_{0.47}Na_{0.27}A$, but only about 15 kcal/mole over an H-erionite (*16*). Thus, cracking is diffusion limited over this H-erionite. However, over a synthetic H-offretite, which has free channels in the $\bar{c}$-direction, the cracking of *n*-hexane is no longer diffusion controlled (*19*). Furthermore, over erionite, the apparent activation energy of *n*-hexane cracking is about half of that of 2-methylpentane. This shows that in erionite the intracrystalline cracking of *n*-hexane is controlled by diffusion. 2-Methylpentane, however, reacts only on the exterior surface of the zeolite particles where diffusion is not involved (*19*).

Synthetic erionite is less stable in concentrated sulfuric acid than natural erionite. Acid stability, however, can be improved by rare earth ion exchange (*19*). The natural H-erionite has higher thermal stability than synthetic preparations. For example, after 16 hr at 600°C, the natural erionite lost only 11% of its activity whereas, various synthetic HT- and H-offretite preparations lost 70–99% of their activity (*19*).

Hydrocracking over Erionite. Hydrocracking becomes possible when a hydrogenation–dehydrogenation component is added to the H-form of the zeolite converting it to a dual-functioning catalyst. The hydrogenation component allows continuous operation for long periods. Chen and Garwood have compared hydrocracking of *n*-butane, *n*-pentane, *n*-hexane, and other hydrocarbon mixtures over such dual-functioning erionite (*17*). Reaction conditions were 14–28 atm, 370°–510°C, and H_2:hydrocarbon mole ratios between 12 and 15. Liquid space velocities were between 1.5 and 16. In the hydrocracking of *n*-butane, the predominant product is *n*-propane. Interestingly, significant amounts of *n*-pentane are also formed. The amount of *n*-pentane increases with increasing temperature and reaches a maximum of 2.1 wt % at about 454°C. This shows that hydrocracking is more complicated than simple C–C bond scission.

n-Pentane is selectively hydrocracked to ethane and propane in the presence of isopentane and benzene. Above 400 psig, *n*-hexane is also formed. *n*-Pentane hydrocracks about twice as fast as *n*-butane; activation energies are the same.

n-Hexane is hydrocracked selectively in the presence of 2-methylpentane and benzene. The product is mostly propane. *n*-Hexane hydrocracks 50 times faster than *n*-pentane at 370°C. Its apparent activation energy is only 15 kcal/g, one-half that of *n*-pentane. This is in good agreement with the activation energy difference observed over the same zeolite in the absence of a hydrogenation component.

Robson, Hamner, and Arey reported shape-selective hydrocracking over dual-functioning Zn, rare earth, or H-exchanged erionites and other zeolites which contained 0.5% Pd (*18*). Each was prereduced in hydrogen at 454°C and sulfided with 0.25% CS_2 in the feed. An Arabian C_5–C_6 stream (containing less than 1 ppm nitrogen and less than 10 ppm sulfur) reacted at 400°C and 35 atm. Pd–ZnA is less active but more selective than the corresponding natural or synthetic erionites. H-erionite and the rare earth erionite are more active than Zn-erionite, but the rare earth erionite is more selective. Some branched paraffins have been converted in all these experiments. Several explanations are offered by Robson, Hamner, and Arey. Pd on external surface or on amorphous impurities could catalyze these reactions, or more likely, offretite intergrowths provide enough surface in larger diameter pores to convert some branched and cyclic molecules.

Cage or Window Effect. Reactivities, diffusivities, and most other properties of homologous series usually change monotonously with the length of the molecule. Certain molecules in some molecular sieves do not obey this rule—*i.e.*, their reactivity and diffusion behavior differ from those of their neighbors in the homologous series. This is the cage or window effect. Although this effect was observed in many different zeolites (chabazite, levynite, and gmelinite), the most important studies involved erionite and the closely related zeolite KT. The cage or window effect was first observed by Chen, Lucki, and Mower in 1968 (*21*). Chen and Garwood (*17*) and

Gorring (*22*) studied the effect in more detail. The three most important and interesting manifestations of the cage or window effect are:

(1) The curve of diffusion coefficients *vs.* chain length shows a minimum at C_8 and a maximum at C_{12}.

(2) *n*-C_7, *n*-C_8, and *n*-C_9 hydrocarbons are missing from the cracked products of longer *n*-paraffins.

(3) Hydrocracking rates of *n*-hexane, *n*-decane, and *n*-undecane are much higher than that of *n*-octane.

DIFFUSION. Gorring studied the diffusion of *n*-paraffins in zeolite KT (*22*). X-ray diffraction showed that this zeolite is mostly offretite without significant amounts of amorphous material. The faint odd "*l*" line indicated 2–5% erionite intergrowth. The zeolite contained 13.9 wt % potassium, 0.07 wt % sodium, 16.5 wt % alumina, and 69.5% silica. Electron microscope photographs showed uniform, approximately cylindrical crystals with average length of 2.9 μ and diameter of 0.65 μ. Before use, the zeolite was treated for 6 hr at 550°C in dry helium. Diffusion rates were determined gravimetrically. The same KT sample was used for all experiments. Diffusion coefficients were computed by comparing experimental results with a theoretical model for radial diffusion into a cylindrical crystal of constant diffusivity. Figure 2 shows results at 300°C. Note the minimum at C_8 and the maximum at C_{12} in the diffusion coefficient-*vs.*-chain length curve. *n*-Octane diffuses much slower than either shorter (C_2–C_7) or longer (C_9–C_{14}) *n*-paraffins. *n*-Undecane and *n*-dodecane, on the other hand, diffuse much faster

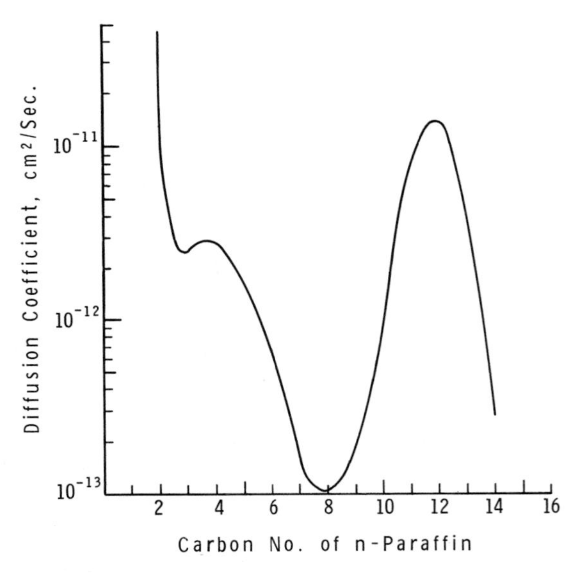

Journal of Catalysis

Figure 2. Diffusion coefficients of n-*paraffins in KT (300°C)* (22)

than any of the *n*-paraffins except ethane. For example, at 300°C, *n*-dodecane has six times the diffusivity of propane. The difference between diffusion coefficients of *n*-octane and *n*-dodecane is more than two orders of magnitude. This zeolite offers a window of high transmittance to *n*-paraffins with 11 or 12 carbons. This window is not open for shorter or longer molecules. The effect persists over a wide temperature range, certainly up to about 450°C. Gorring's results show a clear-cut compensation effect; frequency factor and activation energy of diffusion increase or decrease together.

HYDROCRACKING. Chen and Garwood compared the rates of hydrocracking of various *n*-paraffins over an erionite catalyst which was converted into a dual-functional catalyst and possessed both acid and hydrogenation activities (*17*). Three multicomponent systems, *i.e.*, mixtures of C_5–C_8, C_8–C_{16}, and C_{11}–C_{15} *n*-paraffins, reacted between 399°C and 482°C, between 14 and 137 atm, with liquid hourly space velocities of 2, 8, and 14, and with H_2:hydrocarbon mole ratios of 2, 30, and 27. Relative rate constants are plotted against chain length in Figure 3. The curve shows two maxima (at C_6 and at C_{10}–C_{11}) and one minimum at C_8.

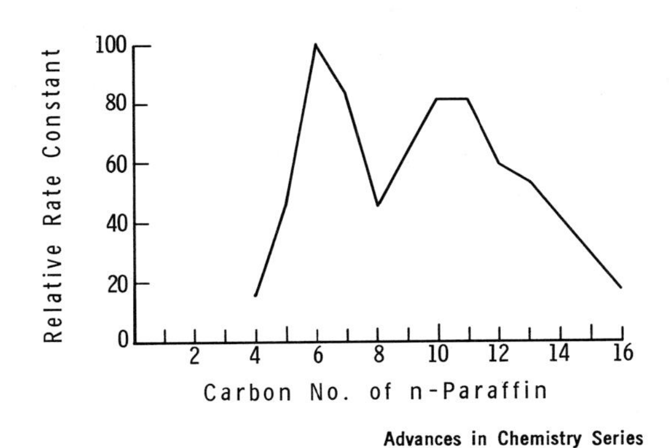

Figure 3. Hydrocracking rates of n-*paraffins over dual functional erionite* (17)

CRACKING PRODUCT DISTRIBUTION. Chen, Lucki, and Mower cracked various *n*-paraffins (*n*-C_{10}, *n*-C_{12}, *n*-C_{14}, *n*-C_{16}, *n*-C_{22}, *n*-C_{23}, and *n*-C_{36}) over 30/60 mesh erionite in helium (*21*). Reaction conditions were 316° or 340°C, about 100 mm Hg hydrocarbon partial pressure, and 1–3 sec superficial contact time. Conversions were between 5 and 30%. Only *n*-hydrocarbons were produced.

Figure 4 compares product distributions obtained from cracking *n*-docosane (*n*-$C_{22}H_{46}$) over erionite and a rare earth X-faujasite. Note the extent of the cage or window effect over the former catalysts; *n*-C_7–, *n*-C_8–, and *n*-C_9–hydrocarbons are missing from the cracked products. Other *n*-paraffins (except *n*-hexadecane) gave similar product distributions.

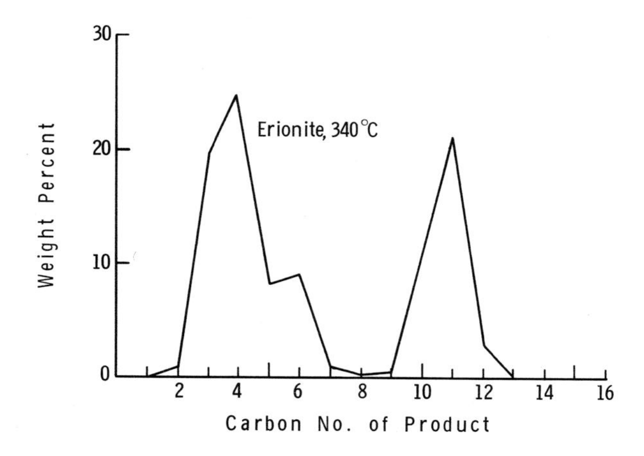

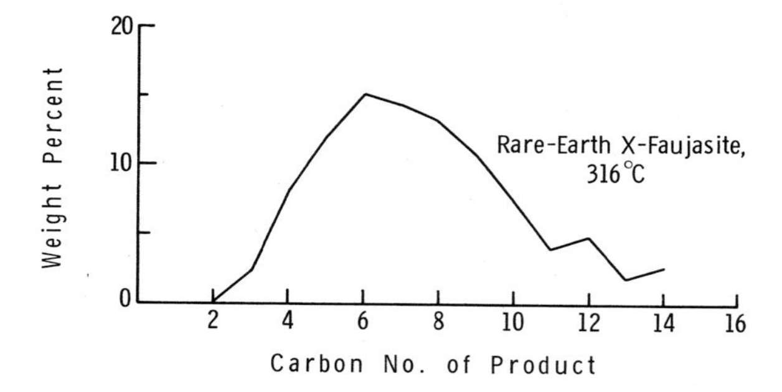

Journal of Catalysis

Figure 4. Product distributions in the cracking of n-*docosane* (21)

From these results, we may conclude the following:

(a) In erionite and in the zeolite KT, certain large *n*-paraffins diffuse much faster than smaller ones. This is unexpected because in large-pore catalysts diffusivity decreases monotonically with the square root of the molecular weight.

(b) *n*-Paraffin diffusivities in KT are at least five orders of magnitude lower than in large-pore catalysts (*i.e.,* 10^{-10} to 10^{-14} cm^2/sec in KT and 10^{-1} to 10^{-5} cm^2/sec in large–pore catalysts) (*21*). The most significant forces acting upon diffusing *n*-paraffins in zeolite crystals are interactions with framework and cations. Interactions between hydrocarbon molecules are weak.

(c) There is a strong interaction between zeolite framework and molecules whose length matches the length of the internal chamber of the zeolite. For example, *n*-octane (the paraffin with the lowest diffusion coefficient and lowest hydrocracking rate) is 12.82 A long. This fits almost exactly into the 13-A cavity of erionite. This creates a low-energy trap. Molecules entrapped in such tight-fitting cages have low mobility.

(d) Molecules longer than n-C_8 are too long to fit completely into one cage. Part of each such molecule extends into the next cage. Their diffusivity is enhanced relative to smaller molecules.

(e) n-Dodecane, which is 17.85 A long, extends to the extremes of a unit cell in the a-direction. It is held by three eight-membered rings. This assures maximum mobility.

(f) Reaction rates of n-paraffins in zeolites which exhibit the cage effect increase or decrease with chain length as their diffusivities increase or decrease.

(g) Diffusivities determine cracking product distributions also. Long paraffins are initially cracked to a mixture of shorter hydrocarbons with a distribution similar to those obtained in large-pore catalysts. n-C_7, n-C_8, and n-C_9 products are effectively trapped in the erionite cages. Since they diffuse about two orders of magnitude slower than shorter or longer product molecules, they spend much more time within the zeolite. As a result they will crack further to C_3 and C_4 fragments; n-C_{11} and n-C_{12} products, on the other hand, will quickly diffuse out of the erionite crystal and thus escape secondary cracking.

Shape-Selective Reactions over Mordenite-Type Molecular Sieves

The mordenite framework is characterized by parallel pores, high thermal stability, and resistance to mineral acids. Hydrogen-exchanged mordenites contain strong Brønsted and Lewis sites (*23*). H-mordenite has about four to five microequivalents of potentially exchangeable protons/m^2 surface area (*24*). Although hydrogen Y-zeolite has about twice as many potential acid sites, H-mordenite is more active for cracking n-hexane and other n-hydrocarbons because its acid sites are stronger (*15, 22*).

Cracking and Hydrocracking. Palladium and platinum may be introduced into H-mordenite. The resulting Pd– or Pt–H-mordenites hydrocrack both paraffins and cycloparaffins, and they are more active than the corresponding Y-faujasite type catalysts. Burbidge, Keen, and Eyles have found that regardless of the reactant, the products are essentially propane, isobutane, and isopentane (*24*). n-Paraffins react faster than corresponding cycloparaffins. However, Beecher, Voorhies, and Eberly found that in binary mixtures of cyclic and straight-chain C_{10} alkanes, the cyclics react preferentially because the cyclic hydrocarbon completely denies n-decane access to any of the active sites (*25*). On the other hand, when mixtures of higher n-paraffins and other hydrocarbons react over Pt– or Pd–H-mordenites, the n- (and near-n-) paraffins are hydrocracked preferentially (*24*). Potential commercial applications are the removal of n-paraffinic wax from high boiling petroleum fractions to improve the flow properties of diesel oils, fuel oils, lubricating oils, extenders, plasticizers, and flexible microcrystalline waxes (*24*). Hydrocracking conditions, depending on the feedstock and the extent of dewaxing required, are 18–170 atm and 230°–510°C. The wax is converted mostly to propane, isobutane, and isopentane.

Many low sulfur crude oils have high wax contents. Catalytic dewaxing is one way to produce high-grade diesel and fuel oils from such crudes.

Some of these crudes—*e.g.* Libyan crudes—have high sodium levels and, therefore, deactivate the acid sites of the catalyst. One solution is to vacuum distill the crude and catalytically dewax its 350°–550°C fraction. One-third of this fraction is wax, and about three-fourths is composed of *n*-paraffins. Catalytic dewaxing over Pt–H-mordenite reduces its pour point from 115°F to 30°–60°F. Since the vacuum residuum contains only about 3% wax, it can be blended back with the dewaxed 350°–550°C fraction to give a low pour point, low sulfur fuel oil blending stock (*24*). Low pour point diesel oils can be produced similarly.

Wax is removed in conventional lubricating oil production by solvent dewaxing. Mordenite-catalyzed dewaxing removes some of the nonwaxy, singly branched, mostly monomethyl paraffins and thus reduces the viscosity index. If, however, mordenite dewaxing is followed by solvent refining, the selectivity of *n*-paraffin removal is improved. Furthermore, the product has satisfactory color and color stability. The combination process does not need finishing treatment and is adequate for lighter lubricating oil grades where the wax is essentially *n*-paraffinic (*24*).

Flexible microcrystalline wax composed of isoparaffins is obtained during the preparation of heavy lubricating oils. Small amounts of *n*-paraffins decrease the flexibility of such waxes, but they can be removed selectively over Pt– or Pd–H-mordenites. Similarly, selective hydrocracking of *n*-paraffin impurities could improve extender and plasticizer oils by reducing their pour points.

Aluminum-Deficient Mordenites. A large part of the alumina can be removed from the mordenite framework by acid extraction. This increases effective pore diameter and thus diffusivity of reactant and product molecules to and from active sites (*25*). Shape selectivity might also change. Thakur and Weller have found, for example, that product selectivity in H-mordenites diminishes with increasing alumina deficiency (*26*). Propane is the major product of *n*-hexane cracking over the parent H-mordenite while similar amounts of propane and isobutane formed over acid-extracted H-mordenites. Since acidity is associated with the aluminum ions, the number of acid sites progressively decreases as alumina is removed. Both factors (*i.e.*, increasing diffusivity and decreasing acidity) operate as alumina is leached out of mordenite. It is not easy to predict the combined effect, especially if the dimensions of reactant or product molecules are close to the dimensions of pores. In this case, small structural changes in the zeolite have large effects on selectivity. Bierenbaum, Partridge, and Weiss have reported that over aluminum-deficient H-mordenite 1-methyl-2-ethylbenzene reacts with a high isomerization to dealkylation ratio (*27*). This selectivity is just opposite that observed over the parent H-zeolite (Table V).

The first step in acid-catalyzed reactions of alkylbenzenes is protonation of the aromatic ring. According to Matsumoto, Take, and Yoneda isomerization and dealkylation have this common protonated intermediate (*28, 29*). The next step depends on whether or not the positive charge is stabilized on the ethyl group by nearby alumina sites. Bierenbaum's aluminum-deficient

Table V. Reactions of 1-Methyl-2-Ethylbenzene over H-Mordenites (360°C) (*27*)

Catalyst	*Parent H-Mordenite*	*Aluminum-Deficient H-Mordenite*
Alumina, wt %	*11.2*	*0.1*
Reactions of 1-methyl-2-ethylbenzene, mole %		
Dealkylation to toluene	14.2	0
Dealkylation to ethylbenzene	1.9	0
Dealkylation to xylenes	2.0	0
Isomerization to 1-methyl-3- and 4-ethylbenzenes	41.2	47
Disproportionation to mono- plus trialkylbenzenes (transalkylation)	0.2	1

Advances in Chemistry Series

H-mordenite contains only 0.1 wt % alumina. At this level, alumina tetrahedra are isolated and interact only slightly. Accordingly, dealkylation is unlikely over this catalyst.

This explanation is very plausible. However, Bierenbaum, Partridge, and Weiss used a pulse technique in their experiments which could disguise transalkylation and make it appear as dealkylation. Trialkylbenzenes, if produced in transalkylation, could have adsorbed on the numerous strong acid sites of the parent H-mordenite. Their absence from the eluted product does not necessarily mean that they were not produced. Relatively large amounts of xylenes (2%) are produced over the parent H-mordenite (Table V). This, too, is a sign of transalkylation. Xylenes are produced from methylethylbenzenes over acid catalysts *via* alkyl exchange, a side reaction of transalkylation (*30*). Continuous flow experiments or more complete product analyses (including the possible C_1–C_2 products) could clarify this question. In either case, this work demonstrates a real and remarkable selectivity difference between parent and alumina-deficient H-mordenites.

Na- and Ca-Mordenites. Metal exchange decreases the effective pore radius of H-mordenite. Ma and Mancel found, by chromatographic curve analysis, that the diffusion coefficients of C_1 to C_4 hydrocarbons are about one order of magnitude smaller over Na-mordenite than over H-mordenite (*31*). Yashima and Hara have shown that toluene disproportionates and *o*-xylene isomerizes readily over H-mordenite and Be-mordenite but only very slowly over Ca- or Ce-mordenites (*32*). The large Ca and Ce cations reduce effective pore radii by sticking out from the channel walls and almost completely prevent the diffusion of toluene or xylene molecules.

The pores are even more restricted in Na-mordenite. Chen and Weisz used this property to hydrogenate ethylene selectively in the presence of propylene (*4*). None of the presently known molecular sieves can distinguish between ethylene and propylene. Na-mordenite, however, can distinguish between propane and ethane. One can, therefore, remove ethylene from propylene by hydrogenating it to ethane. Propylene can enter the mordenite channel system, but once hydrogenated, the resultant propane cannot escape from the supercage. Chen and Weisz prepared their catalyst in a roundabout way; Na-mordenite was first converted to the hydrogen form, increasing effective pore size to about 9 A. Platinum was then introduced into the supercages as $[Pt(NH_3)_4]^{2+}$. Sodium was then back-exchanged, and the resultant catalyst was calcined in air at 450°C. Platinum sites located on the outside surface of the mordenite crystallites were then permanently poisoned with triphenylphosphine. Triphenylphosphine is too large to enter the pores of Na-mordenite, and therefore its poisoning effect is restricted to exterior platinum. Propylene and ethylene have very similar hydrogenation rates over non-shape selective platinum catalysts, but this platinum in sodium mordenite catalyst converts only ethylene. An equimolar ethylene–propylene mixture reacted at about 12:1 hydrogen-to-hydrocarbon mole ratio. At 260°C, 28.1% of the ethylene was hydrogenated while less than 0.1% of the propylene reacted. Propylene can freely enter the pores, and it can get hydrogenated to propane in the supercages. Since the resultant propane does not block all catalytically active sites, the rate of backward reaction (*i.e.*, propane dehydrogenation) must be sufficient to establish a steady-state propane concentration dictated by the equilibrium. Ethylene hydrogenation can therefore be sustained for long periods without significant catalyst activity loss.

Restricted Transition State-Type Selectivity (Mordenite-Type Molecular Sieves and CaA)

Restricted transition state-type selectivity occurs when certain reactions are prevented because the corresponding transition state requires more space than is available in the cavities of a molecular sieve. Neither reactants nor potential nor real products are prevented from diffusing through the pores. Reactions which proceed through less bulky transition states occur without hindrance. Two examples are discussed below.

Selective Transalkylation of Alkylbenzenes. Symmetrical trialkylbenzenes are the predominant components of the trialkylbenzene isomer mixtures at equilibrium (*33*). Over strong acid catalysts which have sufficient isomerization activity, significant amounts of symmetrical trialkylbenzenes are formed in transalkylation of *o*-dialkylbenzenes. Over catalysts with low isomerization activity, the product contains only certain trialkylbenzene isomers. Csicsery reported that over H-mordenite—a catalyst with adequate isomerization activity—symmetrical trialkylbenzenes were absent from the product (*34, 35*). 1-Methyl-2-ethylbenzene reacted in a flow–type tubular reactor. Reaction conditions were atmospheric pressure, hydrogen : methylethylbenzene mole ratios of 3 or 5, and liquid hourly space velocities of 8 or 16. Reaction periods were about 20 min. Table VI compares product distributions obtained over Zeolon H-mordenite (which contained 0.99% sodium and had a silicon-to-aluminum ratio of 5 to 1) with product distributions obtained over a commercial amorphous silica–alumina cracking catalyst and decationized Y-type faujasite.

Table VI. Distribution of Dimethylethylbenzene and Methyldiethylbenzene Isomers in the Transalkylation of 1-Methyl-2-ethylbenzene (*34*)

Catalyst / *Temp, °C*	*Hydrogen Mordenite* *204*	*Silica-Alumina* *315*	*De-cationized Y-Faujasite* *204*	*Equilibrium Distribution* *(315°C)*
Dimethylethylbenzenes				
1,2-dimethyl-3-ethylbenzene	2.72	3.3	2.76	3.2
1,2-dimethyl-4-ethylbenzene	66.14	8.9	5.17	20.9
1,3-dimethyl-2-ethylbenzene		2.8	3.28	1.9
1,3-dimethyl-4-ethylbenzene	18.53	28.6	27.59	16.4
1,3-dimethyl-5-ethylbenzene	0.24	19.6	16.08	33.7
1,4-dimethyl-2-ethylbenzene	12.37	36.8	45.12	23.9
Methyldiethylbenzenes				
1-methyl-2,3-diethylbenzene	0.40			0.8
1-methyl-2,4-diethylbenzene	37.24	30.7	32.23	20.3
1-methyl-2,5-diethylbenzene	55.33	28.9	29.44	21.2
1-methyl-2,6-diethylbenzene	1.82			1.1
1-methyl-3,4-diethylbenzene	4.86	9.8	7.02	9.8
1-methyl-3,5-diethylbenzene	0.35	30.6	31.31	46.8
Isomerized methylethylbenzenes, % of total methylethylbenzenes	20.7	11.0	18.6	
Total dimethylethylbenzenes, mole % of feed	0.11	0.03	0.54	
Total methyldiethylbenzenes, mole % of feed	0.39	0.45	10.50	

Journal of Catalysis

Dimethylethylbenzenes which can be formed directly from 1-methyl-2-ethylbenzene by transmethylation:

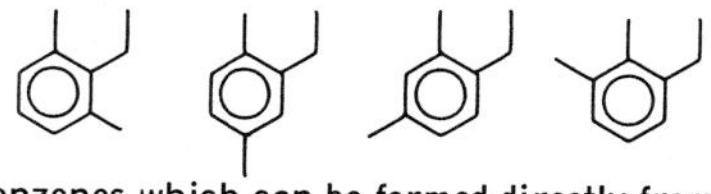

Methyldiethylbenzenes which can be formed directly from 1-methyl-2-ethylbenzene by transethylation:

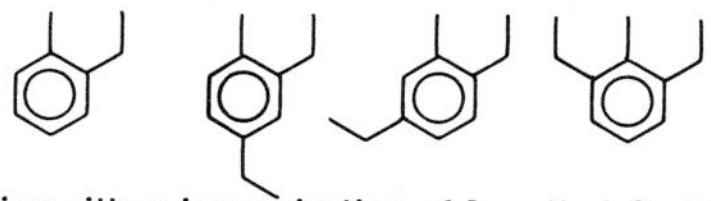

Isomers requiring either isomerization of 1-methyl-2-ethylbenzene before transalkylation, or isomerization of one of the above isomers:

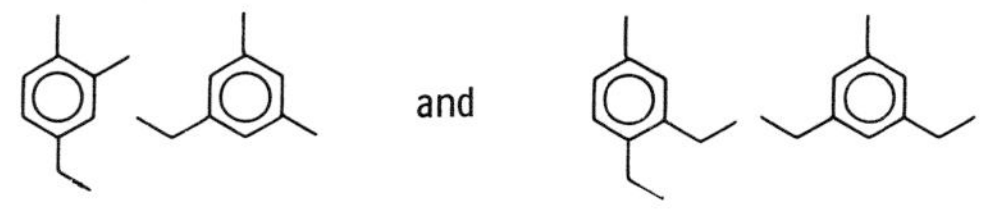

and

Figure 5. Trialkylbenzene isomers

The principal reactions of 1-methyl-2-ethylbenzene are isomerization to 1-methyl-3- and 4-ethylbenzenes and transalkylation to toluene and methyldiethylbenzenes or ethylbenzene and dimethylethylbenzenes. Transalkylation most probably involves 1,1-diphenylalkane-type intermediates as suggested by Pines and Arrigo (*36*) and Streitwieser and Reif (*37*). At equilibrium the symmetrical trialkylbenzene, 1-methyl-3,5-diethylbenzene, is the main component of the methyldiethylbenzene isomer mixture (46.8% at 315°C). Similarly, 1,3-dimethyl-5-ethylbenzene is the most prevalent dimethylethylbenzene isomer (33.5% at 315°C). The formation of symmetrical trialkylbenzenes from *o*-dialkylbenzenes requires either isomerization of the dialkylbenzenes before the transalkylation or isomerization of one of the other trialkylbenzene isomers (Figure 5). Relative amounts of 1,3-dimethyl-5-ethylbenzene and 1-methyl-3,5-diethylbenzene at various conversion levels are plotted against the extent of methylethylbenzene isomerization in Figures 6 and 7. These figures show that the relative amounts of trialkylbenzene isomers which cannot be formed directly from 1-methyl-2-ethylbenzene are related to the extent of methylethylbenzene isomerization (*33*).

Over decationized Y-type faujasite and amorphous silica–alumina ("other acid catalysts" in Figures 6 and 7) symmetrical trialkylbenzenes approach

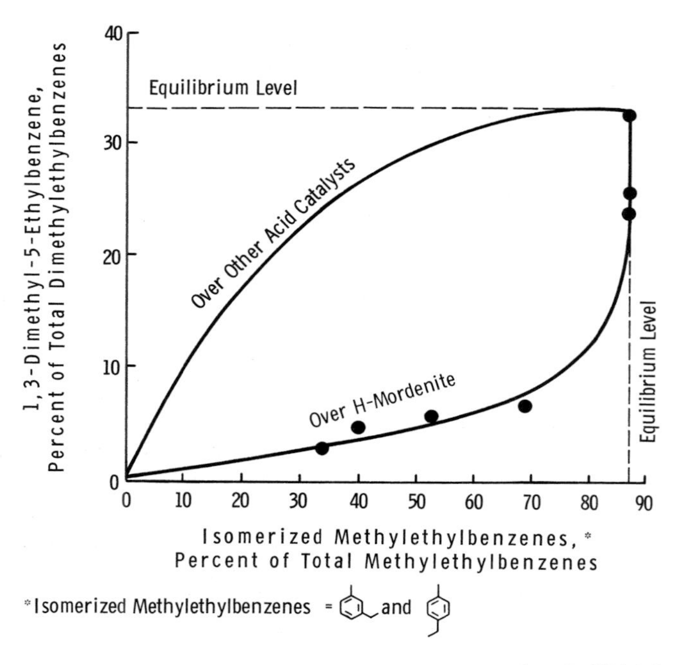

Figure 6. Formation of 1,3-dimethyl-5-ethylbenzene in the acid–catalyzed transmethylation of 1-methyl-2-ethylbenzene (35)

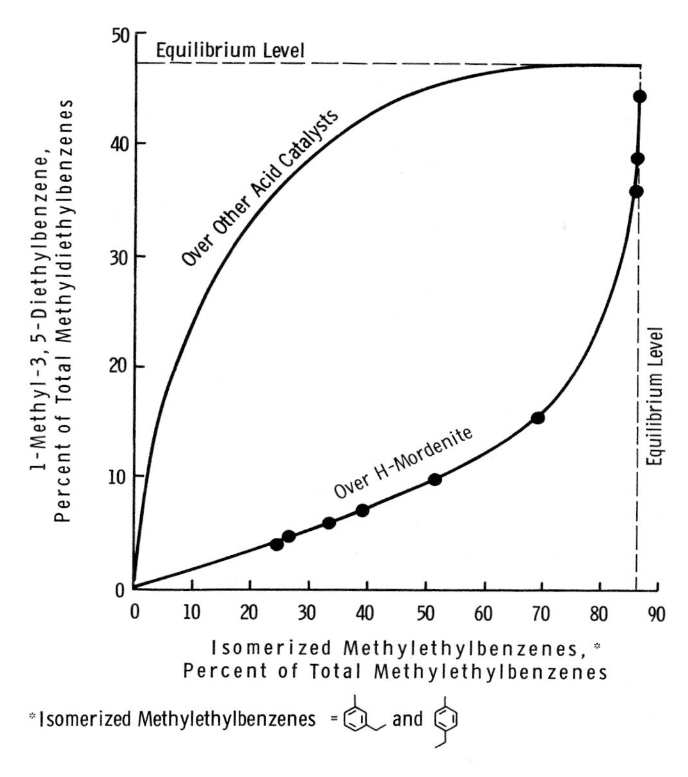

Figure 7. Formation of 1-methyl-3,5-diethylbenzene in the acid–catalyzed transethylation of 1-methyl-2-ethylbenzene (35)

equilibrium levels faster than methylethylbenzenes. Therefore, the concentration of these symmetrical trialkylbenzenes is relatively high even at low (*i.e.*, 11%) methylethylbenzene isomerization levels. At a methylethylbenzene isomerization level of about only 60%, the relative concentration of the trialkylbenzene isomers is practically at equilibrium.

Over H-mordenite, the rate of formation of the symmetrical trialkylbenzene isomers is significantly slower than the rate of methylethylbenzene isomerization. At low total conversion levels, practically none of the symmetrical trialkylbenzenes are present. Only after sufficient time for isomerization, when methylethylbenzenes are completely equilibrated, do symmetrical trialkylbenzenes approach their equilibrium level. This shows that in H-mordenite the formation of symmetrical trialkylbenzenes by transalkylation is inhibited.

We can conclude that symmetrical trialkylbenzenes cannot be formed in the pores of H-mordenite because diphenylmethane-type intermediates or transition states leading to symmetrical isomers require more space than is available (*38*). Transition states leading to the other trialkylbenzene isomers are relatively smaller, and therefore their formation is uninhibited.

Symmetrical trialkylbenzenes are eventually formed over mordenites from the other trialkylbenzene isomers *via* intramolecular isomerization, but this is relatively slow. Isomerization of polyalkylbenzenes above 200°C primarily proceeds through intramolecular 1,2 shifts (*30*). Figure 8 compares these reactions.

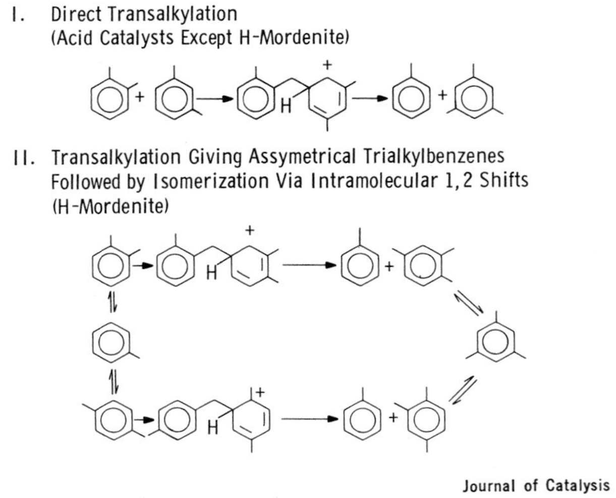

I. Direct Transalkylation
(Acid Catalysts Except H-Mordenite)

II. Transalkylation Giving Assymetrical Trialkylbenzenes
Followed by Isomerization Via Intramolecular 1, 2 Shifts
(H-Mordenite)

Journal of Catalysis

Figure 8. Formation of symmetrical trialkylbenzenes from dialkylbenzenes (35)

Symmetrical trialkylbenzenes are about 0.5–1 A wider than the other trialkylbenzene isomers. Simple product selectivity (in which these symmetrical alkylbenzenes are not allowed to diffuse out of the pores) is ruled out by Csicsery's isomerization experiments (*35*). Initial isomerization and transmethylation rates of hemimellitene (1,2,4-trimethylbenzene) are nearly the same over H-mordenite and decationized Y-faujasite (Table VII).

Conversions to mesitylene (1,3,5-trimethylbenzene) are practically equal over the two catalysts. This shows that this symmetrical trialkylbenzene can be formed by an intramolecular isomerization reaction and can diffuse out of the pores. Furthermore, mesitylene and hemimellitene have similar isomerization and transmethylation conversions over H-mordenite (Table VII). Isomer distributions are also similar; nearly the same amount of pseudocumene (1,2,4-trimethylbenzene) is formed from these two hydrocarbons. This, too, shows that mesitylene diffusion is not seriously inhibited in H-mordenite. Comparison of the rates of *o*-xylene and mesitylene isomerization over a less active natural H-mordenite at 315°C gives further evidence that mesitylene diffusion is not hindered. The smaller *o*-xylene isomerizes slower than mesitylene: conversions at 315°C were 2.2 and 8.6%, respectively (*35*).

Table VII. Reactions of Mesitylene and Hemimellitene over H-Mordenite and Decationized Y-Faujasite Molecular Sieves (315°C, H_2:Hydrocarbon Mole Ratio of 3, and an LHSV of 8) (*35*)

Feed	*Hemimellitene*		*Mesitylene*
Catalyst	*H-Mordenite*	*Decationized Y-Faujasite*	*H-Mordenite*
Product Composition, mole, %			
benzene	0.09	0.09	0.06
toluene	2.25	3.39	2.37
xylenes	20.35	25.31	19.38
trimethylbenzenes	56.46	48.44	57.21
tetramethylbenzenes	20.85	22.77	20.98
Isomer Distribution, %			
o-xylene	22	22	22
m-xylene	57	58	55
p-xylene	21	20	23
hemimellitene	8.1	7.6	7.7
pseudocumene	63.5	64.1	60.5
mesitylene	28.4	28.3	31.8

Journal of Catalysis

H-mordenite retains its shape-selective properties if deactivated thermally. Csicsery has heated H-mordenite to 842°C to reduce its activity to about 0.2% of its original level. Isomerization conversions over this deactivated catalyst at 343°C are similar to those over the active catalyst at 204°C. Table VIII shows that the remaining few active sites have still essentially the same selectivity as before deactivation. This suggests that H-mordenite deactivates at 842°C by successive site elimination and not by a complete transformation or collapse of its structure.

Table VIII. Formation of Symmetrical Trialkylbenzenes over Active and Deactivated H-Mordenite Catalysts (*34*)

Catalyst Status	*Active*	*Deactivated*	
Pretreatment temp, °C	*482*	*842*	*842*
Reaction temp, °C	*204*	*204*	*343*
Isomerized methylethylbenzenes, % of total methylethylbenzenes	20.71	0.04	15.8
Transethylation, mole %	0.78	0.09	0.108
Transmethylation, mole %	0.21		0.012
1,3-Dimethyl-5-ethylbenzene, % of total dimethylethylbenzenes	0.2		4.5
1-Methyl-3,5-diethylbenzene, % of total methyldiethylbenzenes	0.4		0.8

Journal of Catalysis

Transition state restrictions could be partially responsible for the peculiar selectivities observed with benzene alkylation over H-mordenite. Becker, Karge, and Streubel found that relatively more *m*– than *p*–diethylbenzene is formed if the alkylating agent is ethylene. The para isomer predominates if the alkylating agent is propylene (*39*). Table IX shows product compositions.

Table IX. Product Distributions in Benzene Alkylation at 100°C (*39*)

Alkylating Agent	*Ethylene*	*Propylene*
Product Composition, wt %		
Ethylbenzene	93.5	
m-diethylbenzene	3.7	
p-diethylbenzene	2.8	
Cumene		60.9
m-diisopropylbenzene		9.6
p-diisopropylbenzene		25.9

Journal of Catalysis

At 300°K the thermodynamic equilibrium distribution is 3% *o*–, 69% *m*–, and 28% *p*–diethylbenzenes. The m/p ratio in Table IX is close to the thermodynamic ratio, but *o*-diethylbenzene was not produced. (*o*-Diethylbenzene should have been detected if its amount exceeded 0.1%.) We may conclude that:

(a) Whatever initial diethylbenzene distribution is obtained from benzene with ethylene, isomerization is relatively unhindered, and the observed product distribution resembles the thermodynamic equilibrium.

(b) On the other hand, steric factors determine the product composition if the alkylating agent is propylene. Equilibrium composition of the diisopropylbenzenes at 300°K is 68% *m*– and 32% *p*– and practically no *o*–isomer. The m/p ratio here is far from the equilibrium ratio (0.37 *vs.* 2.1). The dimensions of *m*–diisopropylbenzene (9.0 × 7.4 × 6.3 A) do not fit the mordenite channels very well; its formation is, therefore, inhibited.

Selective Hydrobromination of α-Olefins. Another example for restricted transition state selectivity is the anti-Markownikoff addition of hydrogen bromide to α-olefins. Molecular sieves, silica–alumina, and silica gel can catalyze this reaction. Over silica–alumina and high area silica gel, the reaction is catalyzed by surface carbonium ions and obeys Markownikoff's rule by giving exclusively the β-bromo product. Fetterly and Koetitz have shown that Linde CaA sieve catalyzes the formation of the anti-Markownikoff terminal bromo product almost exclusively (*40*). Addition of *n*-heptane aids the reaction and permits operation in a slurry. Anti-Markownikoff hydrogen bromide addition was observed with 1-butene, 1-hexene, 1-octene, 1-decene, 1-dodecene, 1-hexadecene, 1,5-hexadiene, 1,7-octadiene, and 1-acetoxy-3-butene. Diolefins yielded both monosubstituted 1-bromo-olefins and disubstituted

α,ω-alkanes. 1-Acetoxy-3-butene gave predominantly 1-acetoxy-4-bromobutane (88–89%), 1-acetoxy-3-bromobutane (6–7%), and 1,4-dibromobutane (4–6%). The most interesting observation was that branched olefins reacted also very selectively: 3,3-dimethyl-1-butene gave 64% selectivity to 1-bromo-3,3-dimethylbutane; 3-ethyl-1-pentene gave 70% terminal bromide; vinylcyclohexane reacted with 90% terminal selectivity; isobutylene yielded 70% isobutyl bromid. All of these olefins give only the β-bromo product when they react over silica gel.

Although the linear olefins could pass freely in and out of the pores of Linde CaA, none of the branched olefins or their brominated products can enter the 5 A pores of this sieve. The fact that bromination of these branched olefins occurs selectively to give the anti-Markownikoff products shows that molecules do not have to enter completely the interior of the sieve to react. As long as the double bond can reach a catalytically active site, reactions could occur at port entrances.

This anti-Markownikoff addition is specific for hydrogen bromide and occurs only over CaA sieve. No other reagent tested (such as hydrogen chloride, hydrogen iodide, hydrogen sulfide, acetic acid, methyl alcohol, or water) would add to α-olefins in a way as hydrogen bromide does, and no other molecular sieves tried (*e.g.*, Linde KA, Linde NaA, or faujasite X-type sieves) catalyzes anti-Markownikoff-type hydrogen bromide addition.

One possible explanation for this unique selectivity is that the transition state leading to β-brominated products (*i.e.*, normal Markownikoff-type addition) is too large and would not fit into the relatively small CaA supercages. The available space allows only end bromination because only the presumably less bulky transition state for the anti-Markownikoff product fits into the supercages. A free radical process, involving long-lived bromine free radicals cannot be ruled out.

Shape Selectivity in Faujasites

There are few real examples for shape-selective reactions in faujasite-type molecular sieves because they have relatively large pores. Even bulky hexaethylbenzene molecules (molecular weight = 246; approximate molecular diameter = 10 A) can diffuse freely through the pores of rare earth X-faujasite (*41*). Nevertheless, benzene was selectively hydrogenated in the presence of triethylbenzenes over Pt-faujasite (*42, 43*).

Catalytic Cracking. Molecular sieve containing cracking catalysts produce more aromatics and fewer olefins and coke at the same total conversion level than cracking catalysts which do not contain sieves. Thomas and Barmby attributed these differences to shape selectivity (*44*). Accordingly, primary cracking would occur in the amorphous matrix and on the external surface of the zeolite particles dispersed in the matrix. The olefins formed in this primary reaction (β-scission) would diffuse into the zeolites and convert there to paraffins and aromatics by hydrogen transfer. Gas–oil molecules, however, can diffuse into faujasite pores. Venuto, therefore, attributes product distribution differences between zeolites and amorphous silica–aluminas to differences between hydrogen transfer and cracking rate ratios (*45*).

Reverse Molecular-Size Selectivity. Reverse molecular size selectivity occurs when large molecules can form inside the interconnecting cavities of a molecular sieve but cannot escape from there. Eventually, they fill the pores and deactivate the catalyst. This phenomenon could occur in all molecular sieve catalysts. Venuto and Hamilton have determined the type of molecules causing the deactivation of rare earth X-faujasite in the continuos flow alkylation of benzene with ethylene (*41*). The highest boiling 0.1 wt % of the liquid alkylation product contained mostly C_{16} to C_{18} polyalkylnaphthenes and polyalkylbenzenes, such as hexaethylbenzene. Average molecular weight was 250 with an abrupt cutoff at products boiling above 305°C. Condensed polycyclic aromatics were absent, but the tar extracted from the sieve contained higher condensed polyalkylaromatics as well. It is interesting that ethylene alone, or ethylene plus benzene, forms similar tars.

Alkylation of Aromatics. Alkylation of toluene with methanol gives high *p*-xylene yields over Y-faujasite type catalysts. One possible explanation for these high yields is that secondary isomerization of the initially formed *p*-xylene is suppressed in the supercage of the faujasite. However, Yashima, Ahmad, Yamazaki, Katsuta, Hara, Sato, and Hayasaka believe that this selectivity depends on catalyst acidity differences (*46, 47, 48*).

Selectoforming

Selectoforming uses a shape-selective hydrocracking catalyst on the product of catalytic reforming. It was developed by Mobil Oil Corp. and has been used in their refineries since 1967 (*49, 50*). The dual-function shape-selective catalyst contains a non-noble metal hydrogenation–dehydrogenation component. It hydrocraks *n*-paraffins from a mixture of paraffins and aromatics, without hydrocracking branched or cyclic paraffins, or hydrogenating or hydrocracking aromatics. The following results, obtained at 14 atm with a liquid hourly space velocity of 4 and a H_2:hydrocarbon mole ratio of 15 demonstrate the selectivity. Benzene hydrogenation was below 1%.

	Cracking Conversion, wt %	
Hydrocarbon	*371°C*	*482°C*
n-Hexane	54	92
2-Methylpentane	1	7
Benzene	<1	3

The principal catalytic reforming reactions are dehydrogenation of cyclohexanes and dehydroisomerization of alkylcyclopentanes to aromatics, isomerization of *n*-paraffins to isoparaffins, dehydrocyclization of paraffins to aromatics, and hydrocracking of paraffins to propane and butanes. At more severe reforming conditions, when a highly aromatic, high octane gasoline is required, hydrocracking of paraffins becomes a major reaction. It contributes to octane improvement by eliminating low octane *n*-paraffins. Unfortunately, medium and high octane branched paraffins are also removed. As a result, severe reforming is always associated with high liquid yield losses. The main advantage of Selectoforming is that it can reach high octane numbers without as much yield loss because it does not destroy the branched paraffins. The cracked product is primarily propane. In contrast to conventional dual-function hydrocracking, *n*-pentane is hydrocracked rather than produced.

The shape-selective reaction may be carried out in the last reforming reactor, or the reformate may be processed in a separate unit. In this latter case, the reformate is mixed with hydrogen (or hydrogen-containing gas), heated, and fed to a fixed-bed reactor. The effluent is cooled and separated into a hydrogen-containing gas, propane, and high-grade gasoline. Figure 9

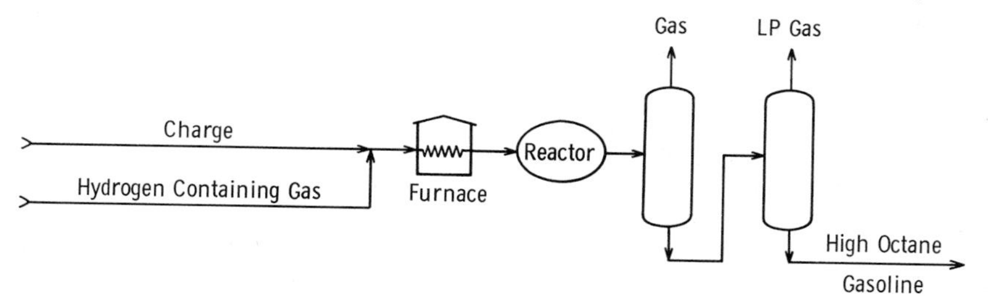

Figure 9. Shape selective hydrocracking (49, 50). [*This modified figure is printed with permission of* The Oil and Gas Journal, *Vol. 66, No. 47, p. 154 (1968).*]

shows such a scheme. Feeds with end points below 216°F are preferred. Operating conditions are 316°–526°C, 8–42 atm, space velocities between 1 and 10, and H_2:hydrocarbon mole ratios between 1 and 12 (*49, 50*). Octane improvements are typically between 2 and 5 (*49*). Alternatively, by holding the octane level at its pre-Selectoforming level, reforming severity can be decreased. This will increase throughput and reforming catalyst life. In either case, propane yields will increase at the expense of methane, ethane, and butanes. Table X shows product compositions when Selectoforming is used to increase the octane number.

Since Selectoforming affects primarily *n*-paraffins, the octane number of the front end of the gasoline will improve more than that of the higher boiling portion. As the front-end volatility decreases, the total vapor pressure of the reformate is reduced. Lead-susceptibility and sensitivity (*i.e.,* the difference between research and motor octane numbers) are also reduced *vs.* reformates produced without Selectoforming.

Other possible benefits of the process are increased reformer capacity and increased reformer catalyst life or reduced platinum requirement. Since lower product volatility permits lower reformate end points, middle distillate production could be increased.

Table X. Reformate Processing by Selectoforming (*49*)[a]

Hydrocarbon	*Reformate Composition, wt %*	*Final Composition after Selecto-forming, wt %*
Methane, ethane, H_2	13.6	16.6
Propane	10.8	15.5
Butanes	10.8	8.9
n-Pentane	7.0	3.9
Isopentane	8.7	8.0
Hexane and higher *n*-paraffins	2.6	1.3
C_6 and higher branched paraffins, and aromatics	46.5	45.8
Research octane number of C_5 + product (lead free)	96.5	99.5

[a] A portion of this table is printed with permission of *The Oil and Gas Journal*, Vol. 66, No. 47, p. 154 (1968).

Roselius, Gibson, Ormiston, Maziuk, and Smith have shown that the combination of Mobile Oil Corp.'s Selectoforming and Chevron Research Co.'s Rheniforming (a new reforming process using a highly selective and stable bimetallic catalyst) offers further yield and stability benefits (*51*).

Miscellaneous

Reactions of Cyclopropene and Substituted Cyclopropenes. In the presence of molecular sieves, cyclopropenes may cyclodimerize to give tricyclo[3.1.0.0.^{2,4}]hexanes, isomerize to 1,3-butadienes, or polymerize. Schipperijn and Lukas found that the preferred reaction depends on the number of methyl substituents, on the type of the catalytic site, and on the pore size of the zeolite (*52*). Table XI shows how zeolite type influences reaction pathways. Cyclodimerization occurs only over zeolites with pores too small to allow the entry of cyclopropene molecules. Zeolites with sufficiently large

Table XI. Reactions of Substituted

Zeolite Type	*Pore Size, A*		 CH_3	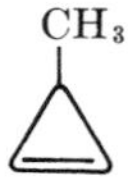 CH_3
KA	3	D	D	D
NaA	4	D	D	D
CaA	5	P	P	D+P
NaX	7-8	P	P	P
NaY	13	P	P	P
HY	13	P	P	P

[a] D = Dimerization, P = polymerization, I = isomerization.

pores to permit diffusion catalyze dimerization or polymerization. Note that this critical pore size depends on the size of the substituted cyclopropene molecule.

Concerted cyclodimerization is forbidden by the Woodward–Hoffmann rules. Catalyzed cyclodimerization probably occurs on sites at or near the surface. Cyclodimerization has a negative temperature coefficient. It is interesting to note that the reaction occurs five times as fast over KA (3 A pores) than over NaA (4 A pores).

Shape-Selective Carbon Molecular Sieve Catalysts. Carbonization of Saran polymers, or polymerized furfuryl alcohol give carbons with fairly uniform, 5–15 A diameter pores. Trimm and Cooper found that these carbon molecular sieves distinguish between straight and branched olefins (*53, 54, 55*). Shape-selective hydrogenation catalysts are prepared either by adding the platinum before polymerization to the furfuryl alcohol monomer or by coating a conventional platinum on charcoal catalyst with a liquid polymeric resin. Both preparations are carbonized at 600°C for 4 hr. At 25°C the first catalyst shows 100% shape selectivity: propylene and 1-butene are hydrogenated while isobutylene and 3-methyl-1-butene do not react. Over the second preparation, linear olefins react about 10 times faster than branched ones. Shape selectivity can be improved further by poisoning external platinum sites with *tert*-butyl mercaptan (*55*). According to Schmitt, the composite catalyst has slit-shaped pores which permit relatively flat molecules like cyclopentene, to enter but keep out larger branched olefins like 3-methylbutene (*56, 57, 58*). This is an interesting and important difference between these carbon molecular sieves and zeolites which have circular or slightly elliptical pore openings. Cooper found that over carbon molecular sieves the rate of propylene hydrogenation is controlled by diffusion (*59*).

Possible applications of carbon molecular sieve catalysts are the selective dehydrogenation of *n*-butane in a mixture of branched hydrocarbons and the extension of catalyst life by preventing large vanadium or sulfur compounds from poisoning active sites of Pt or Pd catalysts (*60*).

Cyclopropenes over Various Zeolites (52)[a]

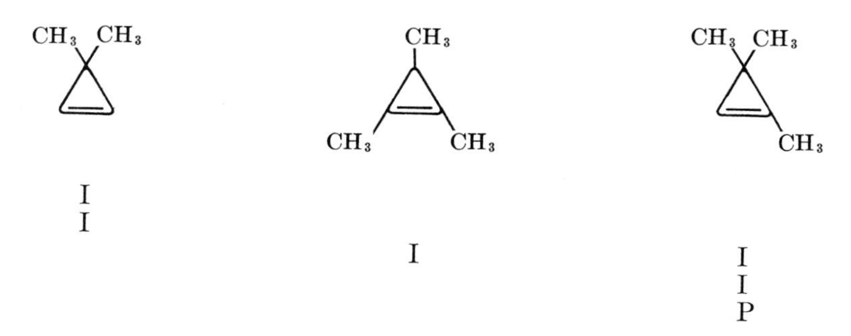

Patents

This review would not be complete without listing a group of patents on various possible applications of shape-selective catalysis. The patents are grouped according to their main subject. However, classification is difficult because most patents cover broad overlapping areas. The list is not complete.

General

U.S. Patent **3,140,322;** Frilette, V. J. and Weisz, P. B.; (Mobil Oil Corp.) ; July 7, 1964.
U.S. Patent **3,257,311;** Frilette, V. J. and Weisz, P. B.; (Mobil Oil Corp.) ; June 21, 1966.
U.S. Patent **3,640,905;** Wilson, R. C., Jr.; (Mobil Oil Corp.) ; Feb. 8, 1972.
U.S. Patent **3,764,520;** Kimberline, C. N., Jr. and Voorhies, A., Jr.; (Esso Res. and Eng. Co.) ; Oct. 9, 1973.

Shape-Selective Hydrogenation

U.S. Patent **3,226,339;** Frilette, V. J. and Maatman, R. W.; (Mobil Oil Corp.) ; Dec. 28, 1965.
U.S. Patent **3,404,192;** Kovách, S. M.; (Sinclair Res. Inc.) ; Oct. 1, 1968.
Belgian Patent **623,186;** Frilette, V. J. and Miale, J. N.; (Mobil Oil Corp.) ; April 3, 1963.
French Patent **1,525,454;** Chen, N. Y.; (Mobil Oil Corp.) ; May 17, 1968.

Shape-Selective Cracking and Hydrocracking

U.S. Patent **3,114,696;** Weisz, P. B.; (Mobil Oil Corp.) ; Dec. 17, 1963.
U.S. Patent **3,238,123;** Voorhies, A., Jr. and Kimberlin, C. N., Jr.; (Esso Res. Eng. Co.) ; March 1, 1966.
U.S. Patent **3,257,311;** Frilette, V. J. and Weisz, P. B.; (Mobil Oil Corp.) ; June 21, 1966.
U.S. Patent **3,314,895;** Munns, G. W., Jr.; (Mobil Oil Corp.) ; April 18, 1967.
U.S. Patent **3,331,767;** Arey, W. F., Jr., Hamner, G. P., and Mason, R. B.; (Esso Res. Eng. Co.) ; July 18, 1967.
U.S. Patent **3,331,768;** Mason, R. B., Arey, W. F., Jr., and Hamner, G. P.; (Esso Res. Eng. Co.) ; July 18, 1967.
U.S. Patent **3,337,447;** Rigney, J. A., Mason, R. B., and Hamner, G. P.; (Esso Res. Eng. Co.) ; Aug. 22, 1967.
U.S. Patent **3,344,058;** Miale, J. N.; (Mobil Oil Corp.) ; Sept. 26, 1967.
U.S. Patent **3,379,640;** Chen, N. Y. and Garwood, W. E.; (Mobil Oil Corp.) ; April 23, 1968.
U.S. Patent **3,392,106;** Mason, R. B. and Hamner, G. P.; (Esso Res. Eng. Co.) ; July 9, 1968.
U.S. Patent **3,392,108;** Mason, R. B. and Hamner, G. P.; (Esso Res. Eng. Co.) ; July 9, 1968.
U.S. Patent **3,395,094;** Weisz, P. B.; (Mobil Oil Corp.) ; July 30, 1968.
U.S. Patent **3,395,096;** Gladrow, E. M., Mason, R. B., and Hamner, G. P.; (Esso Res. Eng. Co.) ; July 30, 1968.
U.S. Patent **3,497,448;** Hamner, G. P. and Mason, R. B.; (Esso Res. Eng. Co.) ; Feb. 24, 1970.
U.S. Patent **3,509,042;** Miale, J. N.; (Mobil Oil Corp.) ; April 28, 1970.
U.S. Patent **3,544,451;** Mitsche, R. T. and Pollitzer, E. L.; (Universal Oil Prod. Co.) ; Dec. 1, 1970.
U.S. Patent **3,575,846;** Hamner, G. P. and Mason, R. B. (Esso Res. Eng. Co.) ; April 20, 1971.
U.S. Patent **3,594,311;** Frilette, V. J. and Weisz, P. B.; (Mobil Oil Corp.) ; July 20, 1971.

U.S. Patent **3,617,492;** Lorenz, M. G. and Brown, C. K.; (Esso Res. Eng. Co.); Nov. 2, 1971.
U.S. Patent **3,625,880;** Hamner, G. P. and Mason, R. B.; (Esso Res. Eng. Co.); Dec. 7, 1971.
U.S. Patent **3,630,966;** Chen, N. Y. and Rosinski, E. J.; (Mobil Oil Corp.); Dec. 28, 1971.
Belgian Patent **747,571;** The British Pet. Co. Ltd., Sept. 18, 1970.
Belgian Patent **765,682;** Mobil Oil Corp.; Oct. 13, 1971.
Belgian Patent **770,712;** Shell Intern. Res. Co.; Jan. 31, 1972.
British Patent **930,512;** Weisz, P. B.; (Mobil Oil Corp.); July 3, 1963.
Canadian Patent **868,880;** Miale, J. N.; (Mobil Oil Corp.); April 20, 1971.
West German Patent **1,545,354;** Mobil Oil Corp.; May 17, 1973.

Catalytic Dewaxing

U.S. Patent **3,039,953;** Eng, J.; (Esso Res. Eng. Co.); June 19, 1962.
U.S. Patent **3,516,925;** Lawrence, P. A., Aitken, R. W., and Bennett, R. N.; (Brit. Pet. Co. Ltd.); June 23, 1970.
U.S. Patent **3,684,691;** Arey, W. F., Jr., Hamner, G. P., Mason, R. B., and Rigney, J. A.; April 15, 1972.
U.S. Patent **3,700,585;** Chen, N. Y., Lucki, S. J., and Garwood, W. E.; (Mobil Oil Corp.); Oct. 24, 1972.
U.S. Patent **3,755,138;** Chen, N. Y. and Garwood, W. E.; (Mobil Oil Corp.); Aug. 28, 1973.
British Patent **1,088,933;** Lawrence, P. A., Aitken, R. W., and Bennett, R. N.; (Brit. Pet. Co. Ltd.); Oct. 25, 1967.
British Patent **1,134,014;** Lawrence, P. A., Aitken, R. W., and Bennett, R. N.; (Brit. Pet. Co. Ltd.); Nov. 20, 1968.
British Patent **1,134,015;** Lawrence, P. A., Aitken, R. W., and Harris, R. J. K.; (Brit. Pet. Co. Ltd.); Nov. 20, 1968.
West German Patent **2,049,756;** Mobil Oil Corp.; April 15, 1971.
South African Patent Application **67/3685;** Morris, M. C., Bozeman, P. P., Jr., Horton, H. T., and Cummins, B. H.; (Tex. Dev. Corp.); June 20, 1967.

Shape-Selective Combustion

U.S. Patent **3,136,713;** Miale, J. N. and Weisz, P. B.; (Mobil Oil Corp.); June 9, 1964.
U.S. Patent **3,530,064;** Chen, N. Y. and Lucki, S. J.; (Mobil Oil Corp.); Sept. 22, 1970.

Shape Selectivity Improved by Poisoning of External Surface

U.S. Patent **3,437,587;** Ellert, H. G. and Mattox, W. J.; (Esso Res. Eng. Co.); April 8, 1969.
U.S. Patent **3,554,900;** Bowes, E.; (Mobil Oil Corp.); Jan. 12, 1971.
U.S. Patent **3,575,845;** Miale, N. J.; (Mobil Oil Corp.); April 20, 1971.

Literature Cited

1. Weisz, P. B., Frilette, V. J., *J. Phys. Chem.* (1960) **64,** 382.
2. Venuto, P. B., Landis, P. S., *Adv. Catalysis,* Eley, D. D., Pines, H., Weisz, P. B., Eds., Vol. 18, p. 259, Academic, New York and London, 1968.
3. Weisz, P. B., Frilette, V. J., Maatman, R. W., Mower, E. B., *J. Catalysis* (1962) **1,** 307.
4. Chen, N. Y., Weisz, P. B., *Chem. Eng. Progr., Symp. Ser.* (1967) **63** (73) 86.
5. Juguin, B., Clément, C., Leprince, P., Montarnal, R., *Bull. Soc. Chim. France* (1966) 709.
6. Reman, W. G., Ali, A. H., Schuit, G. C. A., *J. Catalysis* (1971) **20,** 374.

7. Mintschew, C., Steinbach, P., *Proc. Intern. Conf. Molecular Sieves, 3rd, Zurich, Switz.*, Sept. 3–7 (1973) p. 401.
8. Lawson, J. D., Rase, H. F., *Ind. Eng. Chem., Prod. Res. Develop.* (1970) **9,** 317.
9. Weisz, P. B., *Erd. Kohle* (1965) **18,** 525.
10. Bennett, J. M., Gard, J. A., *Nature* (1967) **214,** 1005.
11. Kawahara, A., Curien, H., *Bull. Soc. Franc. Mineral Crist.* (1969) **92,** 250.
12. Gard, J. A., Tait, J. M., *Advan. Chem. Ser.* (1971) **101,** 230; *Intern. Conf. Molecular Sieve Zeolites, 2nd, Worcester, Mass.*, Sept. 8–11, 1970.
13. Gard, J. A., Tait, J. M., *Acta Crystallogr.* (1972) **B 28,** 825.
14. Gard, J. A., Tait, J. M., *Proc. Intern. Conf. Molecular Sieves, 3rd, Zurich, Switz., Sept. 3–7* (1973) p. 94.
15. Sherry, H. S., *Proc. Intern. Conf. Ion Exchange Process Ind., London, July 16–18, 1969* (1970), p. 329.
16. Miale, J. N., Chen, N. Y., Weisz, P. B., *J. Catalysis* (1966) **6,** 278.
17. Chen, N. Y., Garwood, W. E., *Advan. Chem. Ser.* (1973) **121,** 575; *Intern. Conf. Molecular Sieves, 3rd, Zurich, Switz.*, Sept. 3–7 (1973).
18. Robson, H. E., Hamner, G. P., Arey, W. F. Jr., *Advan. Chem. Ser.* (1971) **102,** 417; *Intern. Conf. Molecular Sieve Zeolites, 2nd, Worcester, Mass.*, Sept. 8–11 (1970).
19. Chen, N. Y., *Proc. Intern. Con. Cat., 5th (1973)*, Palm Beach, Fla., Aug. 20–26, 1972, p. 1343.
20. Katzer, J. R., *Proc. Intern. Con. Cat., 5th (1973)*, Palm Beach, Fla., Aug. 20–26, 1972, p. 1351.
21. Chen, N. Y., Lucki, S. J., Mower, E. B., *J. Catalysis* (1969) **13,** 329.
22. Gorring, R. L., *J. Catalysis* (1973) **31,** 13.
23. Benesi, H. A., *J. Catalysis* (1967) **8,** 368.
24. Burbidge, B. W., Keen, I. M., Eyles, M. K., *Advan. Chem. Ser.* (1971) **102,** 400; *Intern. Conf. Molecular Sieve Zeolites, 2nd, Worcester, Mass.*, Sept. 8–11 (1970).
25. Beecher, R., Voorhies, A. Jr., Eberly, P. Jr., *Ind. Eng. Chem., Prod. Res. Develop.* (1968) **7,** 203.
26. Thakur, D. K., Weller, S. W., *Advan. Chem. Ser.* (1973) **121,** 596; *Intern. Conf. Molecular Sieves, 3rd, Zurich, Switz.*, Sept. 3–7, 1973.
27. Bierenbaum, H. S., Partridge, R. D., Weiss, A. H., *Advan. Chem. Ser.* (1973) **121,** 605; *Intern. Conf. Molecular Sieves, 3rd, Zurich, Switz.*, Sept. 3–7, 1973.
28. Matsumoto, H., Take, J., Yoneda, Y., *J. Catalysis* (1968) **11,** 211.
29. Matsumoto, H., Take, J., Yoneda, Y., *J. Catalysis* (1970) **19,** 113.
30. Csicsery, S. M., *J. Org. Chem.* (1969) **34,** 3338.
31. Ma, Y. H., Mancel, C., *Advan. Chem. Ser.* (1973) **121,** 392; *Intern. Conf. Molecular Sieves, 3rd, Zurich, Switz.*, Sept. 3–7, 1973.
32. Yashima, T., Hara, N., *J. Catalysis* (1972) **27,** 329.
33. Csicsery, S. M., *J. Chem. Eng. Data* (1967) **12,** 118.
34. Csicsery, S. M., *J. Catalysis* (1970) **19,** 394.
35. Csicsery, S. M., *J. Catalysis* (1971) **23,** 124.
36. Pines, H., Arrigo, U. T., *J. Amer. Chem. Soc.* (1958) **80,** 4369.
37. Streitwieser, A., Jr., Reif, L. J., *J. Amer. Chem. Soc.* (1960) **82,** 5003.
38. Venuto, P. B., private communication.
39. Becker, K. A., Karge, H. G., Streubel, W.-D., *J. Catalysis* (1973) **28,** 403.
40. Fetterly, L. C., Koetitz, K. F., Society of Chemical Industry, London (1968), p. 102; *Molecular Sieves Conf., London,* April 4–6, 1967.
41. Venuto, P. B., Hamilton, L. A., *Ind. Eng. Chem., Prod. Res. Develop.* (1967) **6** (3), 190.
42. Frilette, V. J., Weisz, P. B., U.S. Patent **3,140,322** (1964).
43. Frilette, V. J., Maatman, R. W., U.S. Patent **3,226,339** (1965).
44. Thomas, C. L., Barmby, D. S., *J. Catalysis* (1968) **12,** 341.

45. Venuto, P. B., *Advan. Chem. Ser.* (1971) **102,** 260; *Intern. Conf. Molecular Sieve Zeolites, 2nd, Worcester, Mass.,* Sept. 8–11, 1970.
46. Yashima, T., Ahmad, H., Yamazaki, K., Katsuta, M., Hara, N., *J. Catalysis* (1970) **16,** 273.
47. Yashima, T., Yamazaki, K., Ahmad, H., Katsuta, M., Hara, N., *J. Catalysis* (1970) **17,** 151.
48. Yashima, T., Sato, K., Hayasaka, T., Hara, N., *J. Catalysis* (1972) **26,** 303.
49. Chen, N. Y., Maziuk, J., Schwartz, A. B., Weisz, P. B., *Oil Gas J.* (1968) **66,** (47), 154.
50. *Hydrocarbon Processing* (Sept. 1970) 192.
51. Roselius, R. R., Gibson, K. R., Ormiston, R. M., Maziuk, J., Smith, F. A., Nat. Petrol. Ref. Assoc., Annual Meeting, San Antonio, Texas, Paper No. AM 73-32, April 1–3 (1973).
52. Schipperijn, A. J., Lukas, J., *Rec. Trav. Chim. Pays-Bas* (1973) **92,** 572.
53. Trimm, D. L., Cooper, B. J., *Chem. Commun. Sect. D* (1970) 477.
54. Cooper, B. J., Trimm, D. L., *Proc. Intern. Carbon Graphite Conf., 3rd, London,* 1970, p. 189 (1971).
55. Cooper, B. J., *Platinum Metals Rev.* (1970) **14** (4), 133.
56. Schmitt, J. L., Ph.D. thesis, Pennsylvania State University (1970), *Diss. Abstr. Int.* (1971) **B 32,** No. 2, 807B.
57. Schmitt, J. L., Walker, P. L., *Carbon* (1971) **9,** 791.
58. Schmitt, J. L., Walker, P. L., *Carbon* (1972) **10,** 87.
59. Trimm, D. L., Cooper, B. J., *J. Catalysis* (1973) **31,** 287.
60. *Chem. Ing. Tech.* (1971) **43,** A 1133.

Chapter

13

Hydrocracking, Isomerization, and Other Industrial Processes

A. P. Bolton, Molecular Sieve Department, Linde Division, Union Carbide Corp., Tarrytown, N.Y. 10591

AS THE NAME IMPLIES, hydrocracking is similar to catalytic cracking with hydrogenation superimposed and with the reactions taking place either simultaneously or sequentially. Hydrocracking was initially used to upgrade low value distillate feedstocks. Such feedstocks include cycle oils (highly aromatic products from a catalytic cracker which usually are not recycled to extinction for economic reasons), thermal and coker gas oils, and heavy cracked and straight run (directly distilled from crude) naphthas. These feedstocks are difficult or impossible to process by the two main refinery operations—catalytic cracking and reforming—since they are characterized by either a high polycyclic aromatic content or by high concentrations of the two principal catalyst poisons, sulfur and nitrogen compounds, or by both.

The purpose of hydrocracking is to convert high boiling feedstocks to lower boiling products by cracking the feed hydrocarbons and hydrogenating the unsaturates in the product. The polycyclic aromatics must first be partially hydrogenated before the cracking of the aromatic nucleus can take place. The sulfur and nitrogen atoms, present as simple sulfides and more complex heterocyclics, are converted to hydrogen sulfide and ammonia. An additional and probably more important role of the hydrogenation component is to hydrogenate the coke precursors rapidly and to prevent their conversion to coke residue on the catalyst. All these reactions are accompanied by a net consumption of hydrogen.

The modern hydrocracking process should be differentiated from the hydrogenolysis type of hydrocracking practiced in Europe during and after the second world war. These older processes used tungsten or molybdenum sulfides as catalysts and required high reaction temperatures and operating pressures, sometimes in excess of about 3000 psi for continuous operation. Both the cracking and hydrogenation catalyst functions were provided by the metal sulfides. The modern hydrocracking process utilizes a highly acidic cracking component together with either a noble metal or a combination of non-noble metals as the hydrogenation component.

Although the present-day hydrocracking processes were initially developed to process refractory feedstocks to gasoline and jet fuel, process and catalyst improvements and modifications have made it possible to yield a mixed slate of products from LPG and naphtha to furnace oils and catalytic cracking feedstocks. This adaptability surely makes hydrocracking the most versatile of the many processes used in a modern oil refinery.

A comparison of hydrocracking with another important refinery process, hydrotreating, to which it is closely related, is useful in appreciating the role of these two operations. Hydrotreating of distillates may be defined simply as the removal of sulfur and nitrogen compounds by selective hydrogenation. The hydrotreating catalysts used commercially are cobalt plus molybdenum or nickel plus molybdenum used in the sulfided forms and impregnated on an alumina base. The hydrotreating operating conditions are such that appreciable hydrogenation of aromatics does not occur; these are about 1000-2000 psi hydrogen and about 700°F. The theoretical hydrogen consumption should be that required to hydrogenate the sulfur and nitrogen containing molecules and produce hydrogen sulfide and ammonia. However, the desulfurization reactions are invariably accompanied by small amounts of hydrogenation and hydrocracking, the extent of which depends on the nature of the feedstock and the severity of desulfurization. The theoretical chemical hydrogen consumption at 90% desulfurization of a Kuwait atmospheric gas oil is about 95 scf/bbl whereas a chemical consumption of about 125 scf/bbl is found in practice. This leaves about 30 scf/bbl hydrogen consumption for hydrogenation and hydrocracking reactions. A 75% desulfurization of a Kuwait atmospheric residual containing 4.0 wt % sulfur requires 275 scf/bbl of hydrogen but consumes over 500 scf/bbl which results in over 225 scf of hydrogen consumed in non-desulfurization reactions (*1*). An increase in the reaction temperature from 700° to 800°F and the use of a vacuum gas oil feedstock and a typical nickel molybdenum on alumina hydrotreating catalyst gives rise to increased hydrocracking conversion, whose product is principally midbarrel in character. However, under these conditions, the rate of deactivation of this type of catalyst is commercially unacceptable at conventional hydrotreating pressures.

Reaction Studies. PARAFFINS. Studies by Weitkamp and Schultz on the hydrocracking of dodecane using platinum and palladium loaded type Y zeolites have confirmed earlier proposals on the mechanism of reaction (*2*). The mechanism of paraffin hydrocracking is similar to that of catalytic cracking with features of hydrogenation and isomerization superimposed (*3, 4, 5*). The initial step in the reaction is the formation of an olefin *via* the dehydrogenation of a paraffin. The olefin is then adsorbed onto the acid sites of the catalyst and converted to a carbonium ion. This ion then rearranges to a more stable form which in turn is cracked to yield a smaller ion and an olefin. The olefins are hydrogenated to paraffins over the hydrogenation component. The similarity between the carbon numbers in the

Table I. Molar Distribution of Products from the (Moles of Products per 100

Catalyst	*PtCaY*				
Temperature, °C	*265*	*275*	*285*	*300*	*350*
Conversion, %					
Cracking	*5*	*17*	*56*	*99.5*	*100*
C_{12}-Isomerization	*34*	*48*	*35*	*0.5*	—
Methane	—	—	—	0.1	0.5
Ethane	—	—	—	0.1	0.6
Propane	6.7	6.7	7.0	9.0	48.5
Butanes	29.7	30.4	31.8	38.9	101.2
Pentanes	42.3	41.9	41.9	46.3	66.4
Hexanes	43.5	43.5	42.9	44.2	43.3
Heptanes	42.3	41.2	41.0	40.3	8.1
Octanes	29.5	30.7	30.6	25.9	—
Nonanes	6.3	6.0	5.9	3.4	—
Decanes	—	—	—	—	—
Undecanes	—	—	—	—	—
Total moles	200	200	201	208	269

product distributions from the catalytic cracking and hydrocracking of *n*-dodecane supports these proposals (*4*). Coonradt and Garwood have pointed out discrepancies between the products of cracking and hydrocracking, but these can be attributed more to differences in reaction temperature than to any difference in mechanism.

The distribution of the cracked products from hydrocracking of *n*-dodecane over platinum on calcium exchanged Y is presented in Table I (*2*). Based upon 100 moles n-C_{12}, the data show that over the conversion range 0 to 99.5% cracking, the sum of the moles of the cracked product is approximately 200. Thus, unlike the product of catalytic cracking, the hydrocracked product is derived from a purely primary cracking reaction. The platinum containing zeolite gives rise to products having carbon numbers symmetrical to C_6; the amounts of C_3 and C_9 are equivalent, as are the amounts of C_4 and C_8 and C_5 and C_7. However, the products from a

Table II. Composition of C_{12}-Monomethyl Isomers (Mole %) on PtCaY Zeolite at Different Reaction Temperatures (2)

	Temperature, °C				
Isomer	*250*	*265*	*275*	*285*	*300*
2-Methylundecane	13.6	16.8	18.7	19.9	19.7
3-Methylundecane	24.3	24.3	24.1	24.0	23.8
4-Methylundecane	23.0	21.9	21.2	20.4	19.0
5-Methylundecane	26.3	24.7	23.5	22.9	24.4
6-Methylundecane	12.8	12.3	12.5	12.8	13.1

Cracking of *n*-Dodecane with Pt–CaY and Pd–MnHY
Moles of C_{12} Cracked) (2)

PdMnHY					
250	*275*	*300*	*325*	*350*	*400*
5	*12*	*50*	*100*	*100*	*100*
5	*18*	*31*	—	—	—
—	—	—	0.1	0.4	2.9
0.1	0.2	0.2	0.3	1.1	4.6
11.7	14.8	17.5	19.6	42.9	76.0
48.0	52.6	55.0	57.3	89.9	105.8
48.9	50.6	51.6	53.2	62.0	64.6
43.8	44.0	44.2	44.2	45.6	35.1
36.7	34.7	33.7	32.9	16.2	0.4
23.0	20.8	18.6	16.9	1.4	—
2.7	2.1	2.1	1.6	0.1	—
—	—	—	—	—	—
—	—	—	—	—	—
215	220	223	226	260	289

Industrial and Engineering Chemistry, Product Research and Development

palladium containing zeolite show that even at temperatures less than that required for 100% conversion, secondary reactions do occur; the sum of the moles of the cracked product is in excess of 200, and the product distribution is not symmetrical about C_6. The difference in product distribution exhibited by the palladium containing catalyst may also result from the use of a different cationic form of the zeolite. The distribution of the cracked products from the platinum containing catalyst was found to be directly related to the monomethyl isomer distribution of the C_{12} paraffins, shown in Table II. A mechanism proposed to account for this observation is depicted in Figure 1. The possible cracked products that may be derived by a single carbon-carbon bond scission of the isomers are indicated. Using this mechanism to predict the composition of the cracked products results in excellent agreement, as the data in Table III attest. That secondary reactions take place during hydrocracking is more apparent from the products of *n*-hexane conversion. If a simple cleavage of carbon-carbon bonds takes place, *n*-hexane should split into two propane molecules, one ethane and one butane, or one methane and one pentane. The data in Table IV show the products from the hydrocracking of *n*-hexane over palladium containing type Y and mordenite to have methane to pentane and ethane to butane molar ratios ranging from 0.1 to 0.7 (*6*). A disproportionation mechanism most satisfactorily accounts for these data.

That the reaction path proceeds *via* olefinic intermediates was substantiated by some more recent studies of Weitkamp and Schultz (*7*). These investigators carefully followed chromatographically the olefinic content of the products from the hydrocracking of dodecane using a platinum containing zeolite over a range of reaction conditions. These are shown in Table V

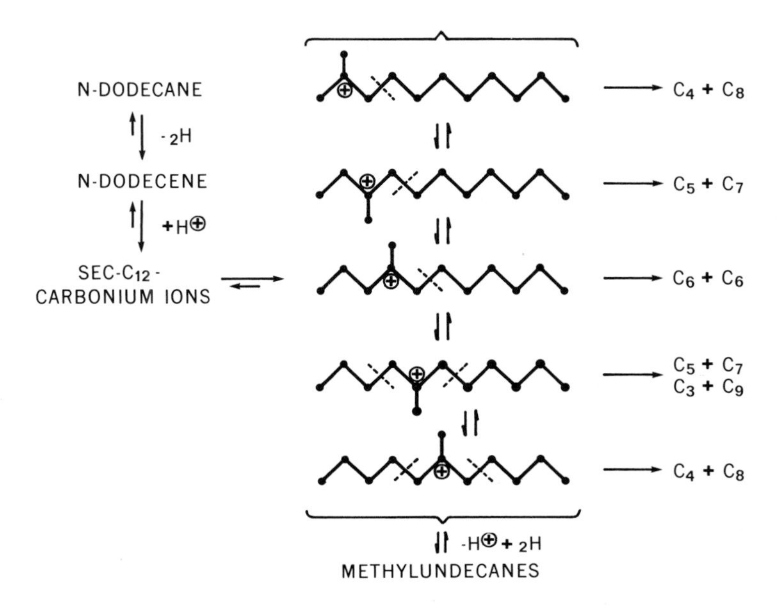

Industrial and Engineering Chemistry, Product Research and Development

Figure 1. Mechanism of primary n-*dodecane hydrocracking* (2)

Table III. Relationship Between Composition of Monomethylundecanes (Mole %) and Probability of Primary Cracking Reactions (2)

(PtCaY Zeolite)

	265°C		*275°C*		*285°C*	
	% Methyl-undec-anes	*% Reac-tion*	*% Methyl-undec-anes*	*% Reac-tion*	*% Methyl-undec-anes*	*% Reac-tion*
2-Methylundecane $\rightarrow C_4 + C_8$ 6-Methylundecane $\rightarrow C_4 + C_8$	29.1	29.6	31.2	30.5	32.7	31.2
3-Methylundecane $\rightarrow C_5 + C_7$ 5-Methylundecane $\rightarrow C_5 + C_7 \rightarrow C_3 + C_9$	49.0	48.8	47.6	47.8	46.9	47.6
4-Methylundecane $\rightarrow C_6 + C_6$	21.9	21.7	21.2	21.7	20.4	21.4

Industrial and Engineering Chemistry, Product Research and Development

Table IV. Typical Products from Hydrocracking of *n*-Hexane (*6*)

	Catalyst	
	Pd-H-Y	*Pd-H-mordenite*
Temperature, °F	*750.0*	*600.0*
Pressure, psig	*750.0*	*750.0*
LHSV, v/v/hr	*1.1*	*4.1*
H_2/hydrocarbon feed, molar ratio	*8.9*	*12.6*
	Moles per 100 Moles Feed	
Product		
methane	2.6	0.9
ethane	7.7	5.4
propane	86.0	41.1
isobutane (2-methylpropane)	6.5	18.5
n-butane	9.4	11.6
isopentane (2-methylbutane)	4.8	9.4
n-pentane	2.8	4.2
2,2-dimethylbutane	5.0	7.2
2,3-dimethylbutane	2.2	4.6
2-methylpentane	12.4	14.4
3-methylpentane	8.5	9.6
n-hexane	8.8	10.6
cyclohexane	—	—
methylcyclopentane	—	—

Industrial and Engineering Chemistry, Product Research and Development

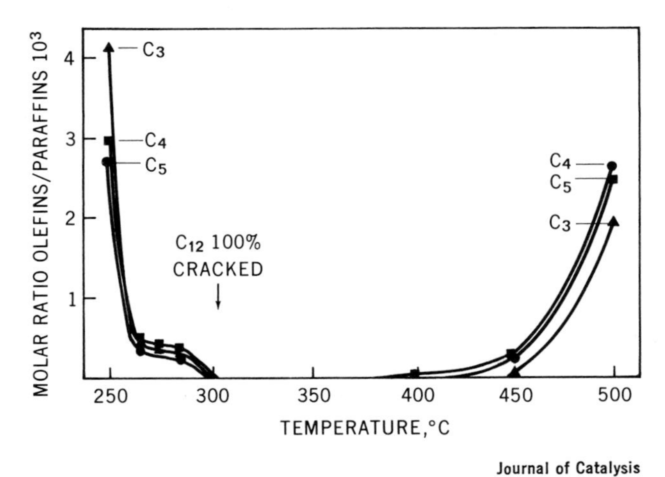

Journal of Catalysis

Figure 2. Molar ratio olefins/paraffins of the product from the hydrocracking of n-*dodecane using ½ wt % PdCaY* (7)

Table V. Composition[a] of the Products from the

Temp. (°C):	250	250	265	275	285
LHSV (hr^{-1}):	4.01	1.13	1.15	1.40	1.11
C_{12}-cracking conversion (%)	3.6	14.8	40.3	57.9	77.7
C_1	—	—	—	—	—
C_2	0.1	—	—	—	0.1
C_3	9.7	8.6	11.2	12.4	15.3
C_4	44.7	40.8	44.5	46.1	53.0
C_5	46.5	45.1	46.6	47.1	49.9
C_6	40.2	40.1	40.7	40.4	40.5
C_7	38.6	39.7	38.8	38.3	36.3
C_8	27.0	28.5	25.7	24.9	21.3
C_9	3.5	4.1	3.6	3.6	2.7
C_{10}	—	0.2	0.2	0.1	0.1
C_{11}	—	—	—	—	—
Total moles	210.3	207.1	211.3	212.9	219.2
Aromatics (wt %)	—	—	—	—	—
Naphthenes (wt %)	0.03	0.18	0.37	0.66	0.75
Olefins[b] (wt %)	0.015	0.017	0.010	0.009	0.011

[a] Molar distribution (moles/100 moles of C_{12} cracked) and content of non-paraffinic hydrocarbons (wt %).
[b] From C_6 on olefin analysis was not complete.

and a remarkable variation in olefin content may be observed. Between 250° and 285°C, corresponding to cracking conversions of less than 100%, the olefin content is about 0.01 to 0.02 wt %. Unexpectedly, no olefins could be detected in the products found in the range 300°–350°C where total cracking conversion occurs. Olefins appear again in the product from a reaction temperature of about 400°C and increase in concentration with increasing temperature. The molar ratios of olefins to paraffins of the same carbon number found in the product are given in Figure 2. The ratio of olefins to paraffins at the higher temperatures agrees with the thermodynamic values shown in Table VI, but not only are the ratios at the lower temperatures substantially higher than the calculated values but the sequence $C_3 > C_4 > C_5$ is the opposite to that expected.

The sensitivity of the cracking reaction to traces of olefinic initiator was most elegantly demonstrated by Weisz (*8*). The trace olefin content of

Table VI. Molar Ratios Olefins/Paraffins (*7*)

Temp. (°C):	250	250	250°	500°
LHSV (hr^{-1}):	1.13	4.01	Thermodynamic Equilibrium	
C_{12}-cracking conversion (%)	14.8	3.6		
Propene/propane	4.14×10^{-3}	9.02×10^{-3}	0.5×10^{-7}	1.0×10^{-3}
Butenes/butanes	2.99×10^{-3}	7.56×10^{-3}	3.7×10^{-7}	2.8×10^{-3}
Pentenes/pentanes	2.74×10^{-3}	6.63×10^{-3}	10.0×10^{-7}	6.1×10^{-3}

Journal of Catalysis

Hydrocracking of *n*-Dodecane Using 1/2 wt % Pt–CaY (7)

300	*310*	*325*	*350*	*400*	*450*	*500*
1.27	*1.18*	*1.28*	*1.14*	*1.14*	*1.25*	*1.15*
99.9	*100*	*100*	*100*	*100*	*100*	*100*
—	0.1	0.1	0.2	0.5	1.7	10.2
0.1	0.1	0.2	0.5	1.3	2.7	14.0
20.0	26.5	31.2	44.6	60.7	64.2	81.0
59.4	69.9	75.9	94.1	107.1	104.1	101.4
51.8	57.0	57.4	62.3	62.4	62.8	57.7
41.1	42.7	42.6	44.4	42.3	39.0	27.4
34.7	31.1	29.2	15.1	2.9	4.5	4.2
17.3	9.7	6.8	0.5	—	0.5	3.0
1.7	0.4	0.2	—	—	—	0.7
—	—	—	—	—	—	0.1
226.1	237.5	243.6	261.7	277.2	279.5	299.7
—	—	—	—	0.022	0.58	3.72
0.72	0.62	0.50	0.44	0.42	0.40	0.35
—	—	—	—	0.003	0.020	0.160

Journal of Catalysis

a *n*-butane feed was controlled by passing the feed through a reactor containing a non-acidic platinum on alumina catalyst. The amount of trace olefin in the product stream was varied from a low value obtained under hydrogenating conditions to increasing amounts by increasing the temperature. The *n*-butane stream was subsequently passed into a cracking reactor containing hydrogen mordenite operated at a constant temperature of 230°C. Figure 3 shows that the *n*-butane cracking conversion could be varied from 0 to 100% simply by changing the temperature in the first reactor. The calculated equilibrium concentration of 1- and 2-butane shown in Figure 3, indicates the highest olefin concentrations that would be created in the hydrogenation/dehydrogenation reactor.

Hydrocarbons containing quaternary carbon atoms are not directly formed in the hydrocracking reaction (*4*). The presence of these hydrocarbons in the product is a result of the cracked products being subsequently isomerized (*9*). The formation of 2,2-dimethylbutane was minimal at temperatures corresponding to cracking levels up to 90%. At higher temperatures, where the secondary cracking occurs, the concentration of 2,2-dimethylbutane increases rapidly and approaches its equilibrium value. The concentration of this isomer at even higher temperatures, Figure 4, decreases as is required by the thermodynamic distribution. Thus, compounds containing quaternary carbon atoms are not formed by direct scission but by isomerization, and their concentrations are controlled thermodynamically and not kinetically. At cracking levels below 100%, the acid

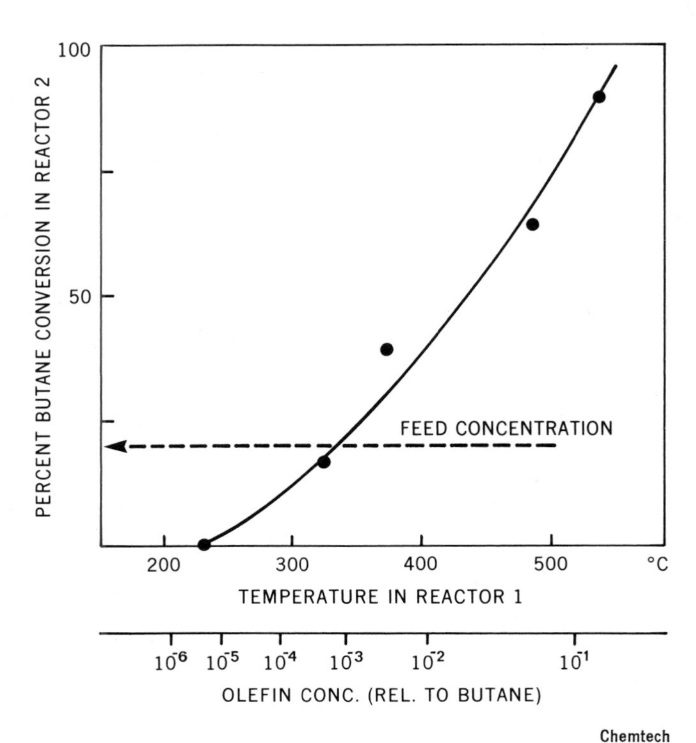

Chemtech

Figure 3. The catalytic cracking of n-*butane as a function of the olefin content of the feed* (8)

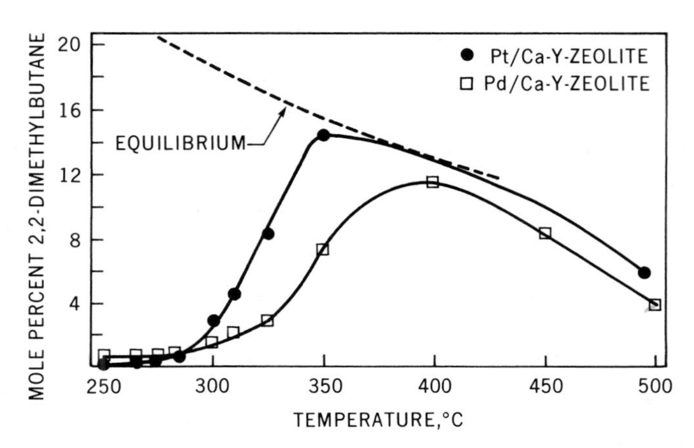

Preprints, Division of Petroleum Chemistry, American Chemical Society

Figure 4. Variation of 2,2-dimethylbutane concentration with reaction temperature (9)

sites are predominantly occupied by chemisorbed C_{12} species and are not available for isomerization reactions of the smaller molecules.

NAPHTHENES. The hydrocracking of cyclohexane over both palladium containing mordenite and type Y zeolites yields the same type of product distribution as that from *n*-hexane. Both reactions are accompanied by substantial isomerization (*6*). Pure component studies showed the reactivity of cyclohexane to be greater than that of *n*-hexane and the same relation of reactivities was found when hydrocracking a 50:50 cyclohexane–*n*-hexane mixture with the type Y zeolite. With the mordenite catalyst, the reactivity of *n*-hexane was greater than that of cyclohexane. However, using a mixed feed, the reverse was found. This selective hydrocracking of cyclohexane was explained as the basis of competitive adsorption. That this selectivity was not observed with type Y was attributed to the larger differences in reactivity of the two hydrocarbons on this catalyst. The preferential hydrocracking of naphthenes instead of straight chain paraffins on mordenite was also found with *n*-decane and decalin (decahydronaphthalene) (*10*). *n*-Decane is more easily hydrocracked over this zeolite than decalin; the conversion obtained on pure components at the same process conditions were 47 and 21%, respectively. However, the results from a mixed feed showed the *n*-decane to be virtually unconverted while the decalin was converted to the same extent as if it were passed over the catalyst at one half the space velocity. Since this study showed there to be no diffusional limitations under these conditions, this selectivity for decalin is a result of the zeolite's absorptive characteristics.

The products from the hydrocracking of decalin using an aluminum deficient mordenite show a high selectivity for opening a single ring in the decalin molecule (*10*). This selective retention of the cycloparaffin ring is characteristic of hydrocracked products from both non-crystalline silica–alumina and zeolite catalysts.

AROMATICS. The hydrocracking of aromatic molecules is of particular interest since one of the more common commercial feedstocks, cycle oil, largely consists of high molecular weight polynuclear aromatics. The principal reactions taking place under hydrocracking conditions are partial or complete ring saturation, alkyl group transfer, and isomerization and cyclization. The extent of saturation undergone by the aromatic molecules depends on hydrogenation–dehydrogenation equilibrium distribution which reflects the operating conditions. The extremely high activities exhibited by commercial zeolite hydrocracking catalysts allows operation under conditions where almost complete saturation occurs.

Reactions of alkyl groups attached to an aromatic nucleus depend solely on their size. Aromatics containing methyl or ethyl groups are saturated either partially or completely to the naphthene; little dealkylation takes place. Tetramethylbenzene and similarly substituted molecules undergo the paring reaction in which the methyl groups are removed from the cyclic nucleus and eliminated as isobutane. Little or no ring cleavage takes place. Account of this unusual reaction is given by Egan, Langlois, and White (*11*). Side chains containing propyl, butyl, or pentyl groups undergo simple dealkylation, and these alkyl groups appear in the product as paraffins. Longer alkyl side chains are also removed by dealkylation, but another reaction—cyclization—also takes place and accounts for considerable amounts of tetralins and indanes in the hydrocracked product.

The reactions of polynuclear aromatics are more complex. Their products are characterized by lower molecular weight cyclic hydrocarbons but by only trace amounts of paraffins. This amply illustrates the difference between hydrocracking and catalytic cracking: the products from polynuclear aromatics using the latter process are coke and C_1 to C_3 hydrocarbons. It was for this reason that hydrocracking was initially developed into a commercial process. Gas oil feedstocks contain some hydrocarbons that are too large to enter the zeolite pore. Nonetheless, these compounds undergo conversion, showing that some hydrocracking must occur on the external surface of the zeolite crystallites. The conversion of these large molecules to lower molecular weight hydrocarbons is important since otherwise they would build up in the recycle and also cause problems by plating out in the heat exchangers.

Hydrocracking Catalysts. Various modifications of large pore molecular sieves are currently being used commercially in hydrocracking processes with noble and non-noble metals as the hydrogenation component. The principal requirements of a commercial catalyst are not only that it exhibit high catalytic activity but also that it is thermally and hydrothermally stable at temperatures in the operating range and during regeneration. Hydrothermal stability is required because water is invariably present in commercial feedstocks and is, of course, formed during coke burn-off. The low sodium decationized and rare-earth exchanged forms of the large pore zeolites possess the high activity required for commercial use. The terms decationized and hydrogen forms of the zeolites are sometimes used synonomously by various authors. In this chapter, the nomenclature of the authors used as references has been adhered to.

The low metal cation forms have insufficient thermal and hydrothermal stability and must be back-exchanged with a divalent metal cation to achieve the adequate stability required (*12*). Cracking activity increases with decreasing alkali metal content.

Mordenite-type catalysts, although exhibiting excellent initial activity, particularly by low sodium, aluminum deficient forms, do not have the necessary catalytic stability retention for commercial use using gas oil feeds. However, mordenite-based catalysts are satisfactory for producing light hydrocarbons (LPG) from naphthas. Resistance to deactivation of the type Y zeolite catalysts loaded with either noble or non-noble metals is remarkable and catalyst life of up to seven years has been obtained commercially in processing heavy gas oils in the Unicracking-JHC processes. Operating life, of course, depends on the nature of the feedstock, the severity of the operation, and the nature and extent of operational upsets. Gradual catalyst deactivation in commercial use is counteracted by incrementally raising the operating temperature to maintain the required conversion per pass. The more active a catalyst, the lower is the temperature required for

Table VII. Adsorptive Properties of Some Zeolites (*13*)

	Pore Volume, cc/g		
Zeolite Form	H_2O	O_2	n-*Butane*
NaX	0.36	0.31	0.30
NaY	0.35	0.31	0.30
H(80) K(20)L	0.21	0.17	0.17
H(80) TMA(20) Ω	0.21	0.15	0.06
HM	0.17	0.17	—

"Zeolite Molecular Sieves"

a given conversion and the longer is the run before regeneration is required. When processing for gasoline, lower operating temperatures have the additional advantage that less of the feedstock is converted to isobutane. Other types of zeolites evaluated in this application include types L and Ω. These materials are large pore zeolites but have a lower internal pore volume measured by adsorption, as seen from Table VII (*13*). A palladium loaded type L zeolite, containing 20 exchange percent potassium, possesses the same catalytic activity for the conversion of a gas oil as does a type Y zeolite of the same alkali metal content. However, unlike type Y, the residual potassium in the type L structure is difficult to remove and still retain crystallinity. The problem of reducing residual cation content and maintaining crystal structure has also been reported for type Ω (*14*). This zeolite, even when loaded with a gross amount, 3.1 wt %, of palladium, was found to have inferior hydrogenation activity and thus poor activity maintenance. The palladium, introduced in the cationic form, is located in or near sites generated by the removal of tetramethylammonium cations and is inaccessible to the reactant molecules.

Catalyst Poisons. Basic nitrogen containing hydrocarbons in a feed will diminish the cracking activity of hydrocracking catalysts. However, zeolite catalysts can operate in the presence of substantial concentrations of ammonia, in marked contrast to silica–alumina catalysts which are

Table VIII. Unicracking-JHC of Gas Oils: △ Temperature (°F) for Given Conversion (*15*)

	Ammonia 40% Conversion	*No Ammonia 60% Conversion*
Cracking component level effects		
High zeolite	Base	Base
Low zeolite	Base + 18°F	Base − 2°F
Hydrogenation component level effects		
High hydrogenation level	Base	Base
Low hydrogenation level	Base + 2°F	Base + 20°F

Meeting, American Petroleum Institute

strongly poisoned by ammonia. This greater ability of the zeolites to tolerate either basic nitrogen compounds or ammonia is caused by their greater acidity. The effect of ammonia on the cracking component is illustrated by studies on the hydrocracking of a hydrofined gas oil boiling between 340° and 860°F in the presence and absence of ammonia (*15*). The data in Table VIII compare the activities of two catalysts having different zeolite contents but with equivalent hydrogenation components. In the absence of ammonia, the two catalysts, despite a substantial difference in zeolite content, are approximately equal. In the presence of 2000 ppm of ammonia, however, the higher zeolite content catalyst is 18°F more active. Kinetic studies indicate this to be equivalent to doubling the activity, suggesting the cracking activity of a catalyst may not only be directly controlled by the nitrogen content of the feed but also that a catalyst may be cracking limited by virtue of a low zeolite content. The data in Table VIII also show that in the presence of 2000 ppm NH_3, two catalysts having the same zeolite content but different hydrogenation component content have equivalent activities. In the absence of ammonia, however, the catalyst having the higher hydro-

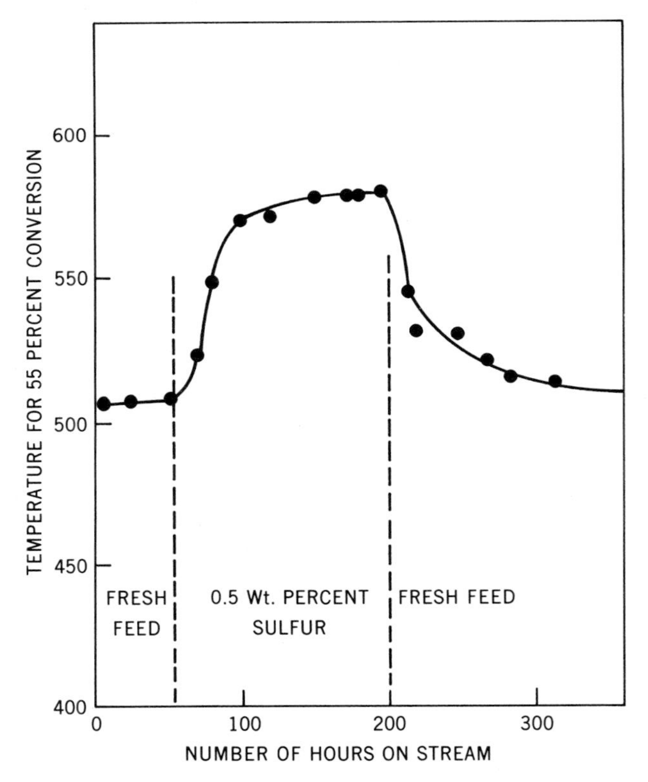

Figure 5. The effect of sulfur on the hydrocracking activity of ½ wt % Pd REX (16)

genation component content is 20°F more active. When the cracking components are not inhibited by the presence of ammonia, the lower hydrogenation component catalyst is hydrogenation limited.

Sulfur containing compounds in a feedstock affect the noble metal hydrogenation component of hydrocracking catalysts. These compounds are hydrocracked to hydrogen sulfide which will convert the noble metal to the sulfided form. The extent of this conversion will be a function of the hydrogen and hydrogen sulfide partial pressures. That this effect is reversible is shown in Figure 5. The catalyst used was a rare earth exchanged type X containing 0.5 wt % palladium by ion-exchange using the $Pd(NH_3)_4^{2+}$ cation and the feed was a hydrofined gas oil (*16*). The addition of 0.5 wt % sulfur as thiophene to the feed increased the temperature required from 520° to 580°F for a 40% conversion to product boiling below 420°F. Removal of sulfur from the feed results in a gradual increase in catalyst activity asymptoting towards the original activity level. As with ammonia, the concentration of the hydrogen sulfide can be used to control precisely the activity of the catalyst. Non-noble metal loaded zeolite catalysts have an inherently different response towards sulfur impurities since a minimum level of hydrogen sulfide is required to maintain the nickel–molybdenum and nickle–tungsten in the sulfided state. Thus, the hydrogenation activity in these cases may be decreased by substantially reducing the hydrogen sulfide partial pressure. However, under most commercial conditions this situation does not often arise.

A most interesting study of the effects of water on the hydrocracking reaction was reported (*17*). Yan evaluated the addition of water, introduced either as 2-pentanol or water vapor, on the hydrocracking activities of a palladium impregnated rare earth X and a platinum impregnated HY. The experimental results on the rare earth containing zeolite are presented graphically in Figure 6. The addition of 3 wt % 2-pentanol to the feed reduced the temperature required for a 60% conversion of the feed by 12°F, which corresponded to an increase of 50% in catalytic activity. That the effect was caused by water and not by the pentene produced was demonstrated by adding water in the vapor phase to the feed. Too high a concentration of water had a deleterious effect on catalytic activity, but a feed saturated with water vapor at 75°F reproduced the effect of the 3 wt % 2-pentanol; this enhancement of catalytic activity was found to be reversible. The promotional effect of water on a commercial feedstock was less pronounced than that experienced with the pure hydrocarbon. Yan suggested that oxygen-containing components in the commercial feed would mitigate the effect of additional amounts of water. The rare earth containing catalyst is promoted by the water creating more protonic sites *via* rare earth cation hydrolysis, according to the mechanism of Plank (*18*). Excessive water brings about the competition of the water with the hydrocarbon for the active sites and results in a loss of activity.

It should not have been unexpected that water fails to promote the activity of the HY zeolite. Reversible regeneration of hydroxyl groups on rare earth zeolites is known (*19*); this effect is known not to occur with HY.

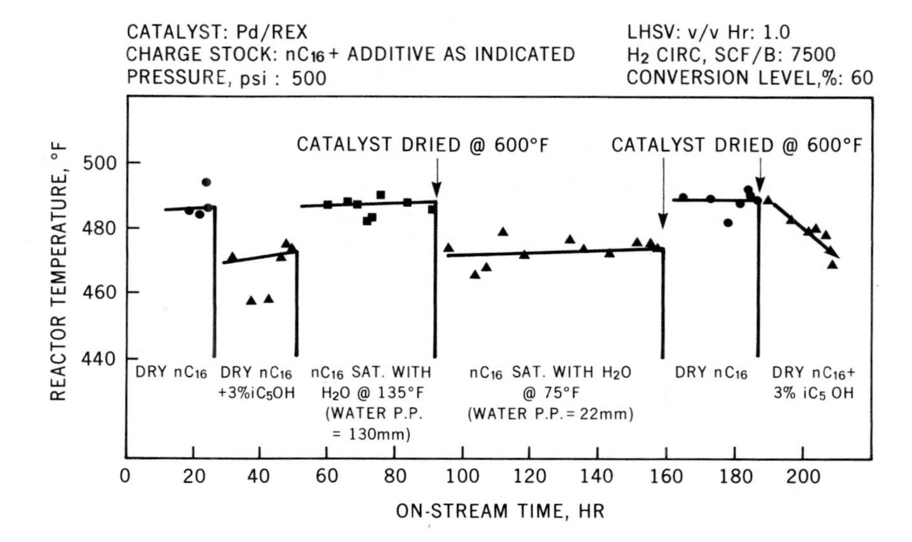

Journal of Catalysis

Figure 6. Effect of water addition on hydrocracking activity of palladium loaded REX (17)

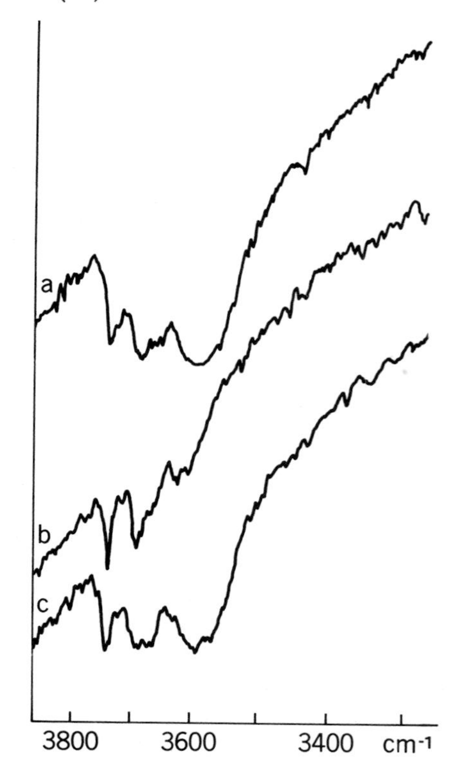

Figure 7. Regeneration of hydroxy bands on steam stabilized Y with water vapor (16). *(a) Vacuum activated at 300°C. (b) Vacuum activated at 750°C. (c) Addition of water vapor folllowed by evacuation at 300°C.*

Had the steam-stabilized form of type Y been used, the benefit of the addition of water would have been observed as would be predicted by the infrared data shown in Figure 7 (*16*).

Commercial Application. The record of zeolite catalysts in industrial hydrocracking has been singularly successful. The first process to be used commercially was developed by the Union Oil Co. of California (*20*). This process, Unicracking-J.H.C. (Jersey Hydrocracking), is licensed to the petroleum industry by Union Oil and Exxon. The family of zeolite catalysts used

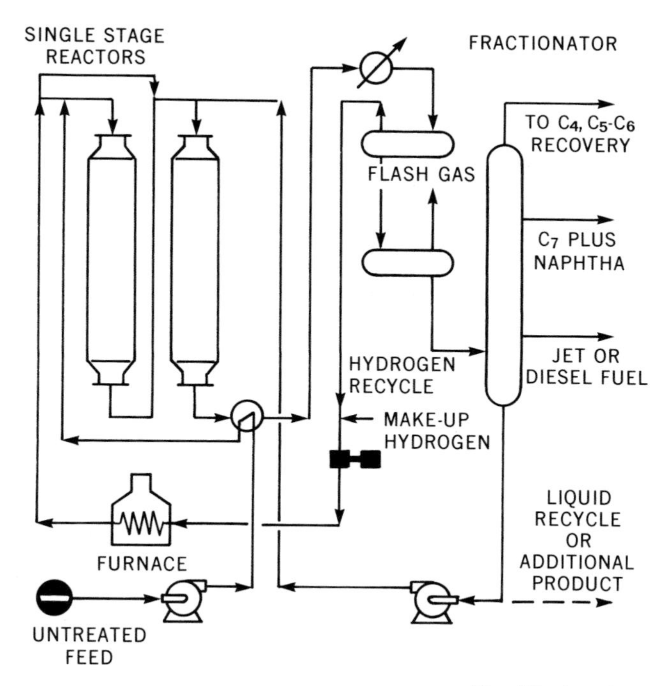

Figure 8. Single-stage Unicracking-JHC hydrocracking process (21)

for the various process schemes was jointly developed by Union Oil and the Linde Division of Union Carbide. The first hydrocracking unit went on stream at the Los Angeles refinery of Union Oil late in 1964 and by 1973, 350,000 barrels of capacity had already been installed (*21, 22*).

There are two fundamentally different approaches to the process design of hydrocracking units. In the first case, feedstocks are pretreated over a hydrotreating catalyst to remove essentially all the sulfur and nitrogen and then are hydrocracked in a separate processing stage over an acidic catalyst in an ammonia- and hydrogen sulfide-free environment. Such a requirement contributes substantially to plant investment costs. The other approach carries out the cracking step in the presence of both the ammonia and hydrogen sulfide. This processing step has only been possible since the advent of the highly acidic zeolite catalysts which are uniquely resistant to sulfur and nitrogen compounds and permit the design of single-stage units (*23*).

A flow scheme for a single-stage Unicracker-JHC is shown in Figure 8. Fresh feed and hydrogen are charged to a hydrotreater where desulfurization and denitrification are effected and the total effluent is passed into the cracking reactor without removal of ammonia or hydrogen sulfide. The hydrocarbons are then hydrocracked in the second reactor to a conversion of between 40 and 70% per pass into the desired boiling range products. The effluent from the cracking stage is then cooled, and the liquid products are separated from the recycle gas. The recycle gas is returned to the reactors while the liquid stream is sent to the low pressure separator for the removal of dissolved gases before distillation to the various products. The unconverted feed can either be recycled back to the cracking reactor or used directly as a product since the feed impurities have been removed by hydrogenation.

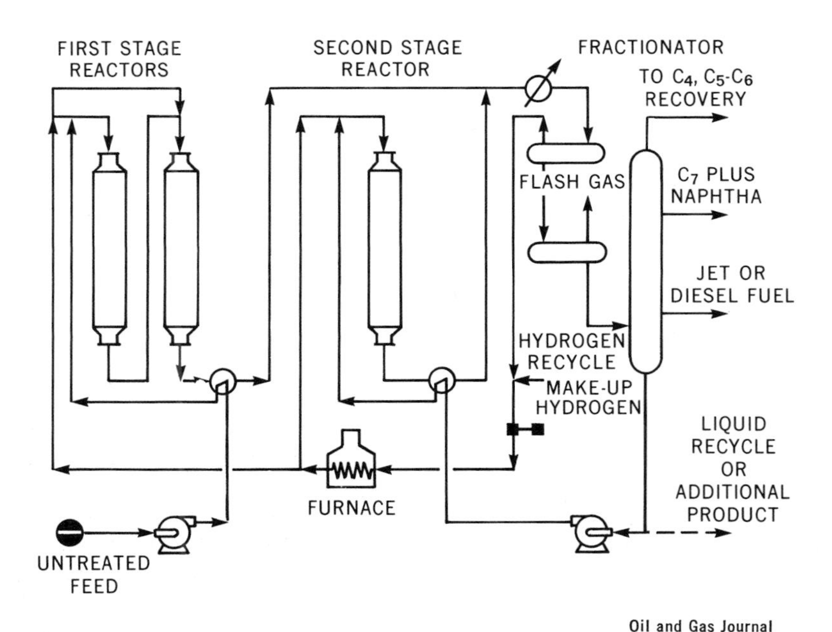

Figure 9. Two-stage Unicracking-JHC hydrocracking process (21)

Additional flexibility can be achieved by adding a second cracking stage to the basic single-stage hydrocracker (*20, 23*). In the two-stage process shown in Figure 9, the effluent from the first stage hydrocracking reactor is

Table IX. Performance of a Large Single-Stage Unicracking-JHC Unit Designed for All Gasoline Product (*24*)

Feedstock Properties:		
gravity, °API	29.4	
distillation, D 1160, °F		
ibp	330	
10	505	
50	628	
90	758	
ep	841	
sulfur, wt %	0.66	
nitrogen, ppm	230	
Yields and Product Properties:		
yields:		
C_1-C_3, scf/bbl	102	
liquid products, % vol		
i-C_4	12.9	
n-C_4	7.3	
C_5–180°F	32.7	
180–400°F ep	69.4	
C_4-plus	122.3	
hydrogen consumption, scf/brl	1630	
Product Properties:	C_5–180°F	180–400°F ep
gasoline fractions		
gravity, °API	82.0	52.2
octanes		
F-1 + 3 ml TEL	99.5	84.9
F-2 + 3 ml TEL	102.1	82.4

National Petroleum Refiners Association

combined with that from the second stage before fractionation. The unconverted oil is then recycled to extinction in the second stage reactor at a 60 to 80% conversion per pass. The recycle gas from the combined high-pressure separator is stripped of ammonia and circulated by a single compressor to both reaction stages. Thus, the catalyst in the second-stage reactor operates in an essentially ammonia-free atmosphere.

The principal use of distillate hydrocracking in the United States has been for gasoline production. A single stage Unicracking-JHC unit at the Atlantic Richfield refinery in Philadelphia is an example of this type of application (*22, 24*). This unit has a fresh feed capacity of over 30,000 bpd for processing heavy straight run and catalytic cycle oils and uses two reactors in series with recycle of the unconverted oil going to the second reactor. A total liquid feed rate of over 50,000 bpd can be charged to the 13-ft internal diameter recycle reactor, the highest throughput of any hydrocracking reactor operating today. The properties of the feedstock and product yields from commercial operation are given in Table IX. The light gasoline is used directly in gasoline blends while the heavy gasoline is reformed.

The utilization of heavy distillates such as fluid catalytic heavy cycle oils, decant oils, heavy thermal distillates, and heavy coke stocks makes them practically unusable for catalytic cracking. Current and future sulfur regulations may even eliminate their use as components of residual fuel oil. A once-through single stage processing scheme, in contrast to the usual recycle to extinction type, is used to process such feedstocks (*25*). Operating conditions may be selected to obtain either (a) substantial light product yields plus an improved catalytic cracking feedstock, (b) improved catalytic cracking feedstock, or (c) a low sulfur blending component for fuel oil. Yields and product qualities are illustrated in Table X. The 650°F +

Table X. Unicracking-JHC to Obtain Light Products and to Upgrade Heavy Cracked Gas Oils for Catalytic Cracking Feed (*25*)

Feed Blend Properties	
Gravity, °API	8.4
Distillation, D-1160, °F	
IBP	489
5	612
30	735
50	811
90	1019
maximum	1073
recovery, vol %	94
Sulfur, wt %	4.57
Nitrogen, wt ppm	2690
Hydrocarbon composition, wt %	
saturates	10.8
aromatics and heterocyclics	84.2
olefins	5.0
Carbon Residue, Conradson, wt %	3.6
Pour Point, °F	+50
Metals (Ni + V), ppm	0.7

Products	
Yields, vol % of feed	
butanes	5.2
C_5–185°F	8.8
185–435°F	31.8
435–650°F	33.8
650°F–plus	35.0
Total C_4–plus	114.6
Properties of 185–435°F Gasoline	
gravity, °API	41.3
octane rating, F-1 + 3 ml TEL	91.3
sulfur, wt ppm	25
nitrogen, wt ppm	<0.1
hydrocarbon composition, vol %	
saturates	10.7
naphthenes	48.3
aromatics	41.0
Properties of 650°F–plus Gas Oil	
gravity, °API	27.3
sulfur, wt ppm	94
nitrogen, wt ppm	8
hydrocarbon composition, wt %	
saturates	64.9
aromatics	35.1

fraction has a low sulfur and nitrogen content, and its yield is about 35 vol % of the original feed. This fraction has excellent catalytic cracking characteristics. The catalyst for this type process is a non-noble metal-loaded zeolite specifically developed to treat such heavy feedstocks. A unit of this type was commissioned for Getty Oil and has performed successfully since it went on stream in 1972.

The increasing demand for naphtha as a feedstock for chemical manufacture was the incentive for the installation of a 30,000 bpd Unicracking-JHC unit for British Petroleum in Scotland. This unit is a two-stage design for the conversion of either 1000°F endpoint vacuum gas oil or blends of vacuum gas oil and heavy catalytic cycle oil to naphtha and light products (*25*). The principal product is a 180°–375°F naphtha fraction which contains nearly 60 vol % cyclics and is an ideal feedstock for xylene production. The naphtha can also be used for ethylene production.

Advances in zeolite catalyst technology have enabled catalyst compositions to be tailored to meet particular feed and product objectives. By selecting the appropriate catalyst system and process configurations, the burning quality of a jet fuel may be varied over a wide range (*22*). The data in Table XI show that catalyst A, developed primarily for gasoline production and used in a single stage operation, can produce substantial amounts of jet fuel from a typical Los Angeles Basin high nitrogen vacuum gas oil by modifying the usual operating conditions. A 35% increase in turbine fuel yield can be obtained under the same operating conditions with catalyst B which was specifically tailored for single-stage turbine production. The use of another catalyst system C in a two-stage configuration enables the aromatic content to be varied over a wide range. In this example the aromatic content of the turbine fuel is intentionally low. The yield of turbine fuel may be varied from 45 to 60 vol % depending on the catalyst and process configuration. As reflected by the changes in hydrogen consumption, the aromatic content of this fraction varies from 34 to 2 vol %. In these three operating schemes, there is also sufficient flexibility to produce 100% gasoline if this is desired. These examples show the ability of this process to convert a wide range of feedstocks, many of them marginal in nature, into a broad spectrum of products. Such flexibility is probably unique in modern refining practice.

The initial cycle life of a zeolite catalyst in these applications is usually greater than two years and a subsequent regeneration by oxidative burn-off will return the catalyst to essentially fresh activity. The on-stream cycles with regenerated catalysts approximate the cycle lengths obtained with fresh catalyst. Unless a catalyst experiences maltreatment in operation, catalyst life of six years or more can be expected.

The recent development of a new zeolite base for the Unicracking-JHC process has enabled a major breakthrough in hydrocracking technology. The exceptionally high activity of a new noble metal loaded catalyst allows a substantial reduction in operating pressure from the 1500 to 1700 psig usually required in the Unicracking-JHC processes to obtain sufficiently low deactivation rates and adequate run lengths.

Table XI. Catalyst and Process Variations in Unicracking Heavy Gas Oil for Gasoline and Turbine Fuel (22)

Feedstock Properties:			
gravity, °API	20.3		
distillation, D 1160, °F			
ibp	520		
10	641		
50	728		
90	820		
ep	890		
sulfur, wt %	1.33		
nitrogen, ppm	2770		
Product Yields and Properties:			
	One-stage Cat A	*One-stage Cat B*	*Two stage Cat C*
yields:			
C_1-C_3, scf/brl	146	50	110
liquid products, % vol			
butanes	12.1	8.2	8.6
C_5-C_6 light gasoline	32.5	18.2	16.5
C_7-plus gasoline	40.6	34.1	34.4
turbine fuel	45.0	61.1	61.3
total C_4-plus	121.2	121.7	120.8
H_2 consumption, scf/brl	1750	1950	2110
Product Properties:			
C_5-C_6			
octane, F-1 + 3 ml TEL	99.0	98.0	96.5
C_7-plus gasoline			
octane, F-1 + 3 ml TEL	87.0	84.7	92.8
naphthenes + aromatics, % vol	70.1	68.2	68.3
turbine fuel			
gravity, °API	37.5	39.7	41.4
aromatics, % vol	34	19	2.0
smoke point, mm	13.6	20.1	29.7
luminometer number	25	44	67
flash point, °F	117	126	120
freeze point, °F	−51	−58	−66
ASTM distillation, °F, 10%–95%	352-514	358-510	356-512

Proceedings of the World Petroleum Congress

Hydrocracking plays an important role in present day refineries. The future of this process will depend on both economic and environmental factors. Where flexibility of operation is important, hydrocracking is an attractive proposition. This process is not only inherently free from pollutants but also yields products free from pollutant precursors. The increase in hydrocracking capacity will depend on the merits of this process in comparison to catalytic cracking, the other directly competitive process.

The deciding factor may be the control of refinery effluents, particularly SO_2 from the catalytic cracker regenerator. If future legislation requires the pretreatment of catalytic cracking feedstock to reduce SO_2 emission, future growth may result from the use of these two processes in combination.

Selectoforming

Unlike the other processes discussed in this chapter, Selectoforming does not involve the substitution of a zeolite in an existing catalytic process but is a novel application of a characteristic unique to molecular sieves—that of molecular shape selectivity. The chemistry of this reaction and the types of zeolites used as catalysts are discussed in detail elsewhere in this book.

The molecular sieve or shape-selective effect is exhibited by the adsorptive behavior of normal and branched chain paraffins on the calcium exchanged form of zeolite A. The replacement of sodium cations in this structure by calcium cations, enables the normal paraffins to diffuse rapidly into the zeolite's channels. The effect is used in the commercial production of normal paraffins from naphthas and kerosenes (*26*). Subsequently, workers at Mobil Oil demonstrated shape-selective characteristics of intermediate pore size zeolites in hydrocarbon catalysis. They showed it possible to selectively remove normal paraffins from hydrocarbon mixtures by catalytic cracking over zeolites of the erionite–offretite type. The incorporation of a hydrogenation component, platinum, palladium, or nickel enables these zeolites to be used commercially in Selectoforming, a process for the selective hydrocracking of C_5 to C_9 n-paraffins from naphthas and reformates (*27*).

In catalytic cracking studies over these intermediate pore size zeolites the products from normal paraffins are characterized by a predominance of C_3 fragments and the absence of isobutane (*28*). Consistent with these earlier observations, the hydrocracked product from a n-hexane feed is principally propane; the amount of propane in the product decreases with increasing temperature in the range 700°–900°F. An exception to this characteristic product distribution is n-pentane which yields principally equimolar quantities of ethane and propane. However, when n-pentane is hydrocracked together with n-hexane, the product reverts to predominantly propane (*29*).

One of the principal constituents of most refinery gasoline pools is reformate derived from the conversion of naphthas over platinum on alumina type catalysts by dehydrogenating cycloparaffins to aromatics and by isomerizing paraffins. The hydrocarbons in the reformate having the lowest octane number are the normal paraffins, whose concentration is determined by the equilibrium distribution at the reforming temperatures of 900°–950°F. The octane number of the reformate increases with increasing aromatic content. Most severe reforming gives rise to the formation of additional aromatics *via* the dehydrocyclization of paraffins, but the increased aromatic content is mainly caused by the elimination of paraffins by hydrocracking reactions. This hydrocracking removes both the high octane branched paraffins as well as the low octane normal paraffins and thus octane improvement is obtained at the expense of reformate yield. A high octane gasoline from a high severity reforming operation can also be obtained at a higher yield by operating the reformer at a lower severity and subsequently removing the normal paraffins from the reformate by the Selectoforming process. In addition to the increased yield, advantage is gained either by increased reformer throughput or increased cycle length obtained from operating the reformer at lower severity. The extent of octane number or yield improvement depends on the nature of the feed and can be easily calculated since it is a function of the amount and carbon number of the *n*-paraffin content. Data in Table XII show a typical product from the Selectoforming process.

Table XII. Selective Hydrocracking of Reformate (*27*)

Hydrocarbon Composition Wt %	*1*		*2*	
	Feed	*Product*	*Feed*	*Product*
C_1, C_2, H_2	7.0	8.7	13.6	16.6
C_3	5.3	15.3	10.8	15.5
C_4	5.9	6.3	10.8	8.9
iC_5	6.4	7.6	8.7	8.0
nC_5	6.1	1.6	7.0	3.9
C_6 + *n*-paraffins	8.9	1.8	2.6	1.3
C_6 + *i*-paraffins + aromatics	60.4	58.7	46.5	45.8
Octane numbers				
C_5^+ RON clear	86.1	93.5	96.5	99.5

Oil and Gas Journal

Commercial Process. The Selectoforming process uses a fixed-bed reactor operating under a hydrogen partial pressure. Typical operating conditions depend on the process configuration but are in the following ranges (*27*).

Pressure, psia	200–600
Temperature, °F	600–900
Space velocity V/H/V	1–10
Hydrogen/hydrocarbon ratio (moles/mole)	2–4

The catalyst used in the Selectoforming process is a non-noble metal loaded low potassium content erionite. As with the large pore hydrocracking catalysts, the cracking activity increases with decreasing alkali metal content (*30*). There are two configurations of the Selectoforming process that are being used commercially. The first Selectoformer was designed as a separate system as shown in Figure 10 and integrated with the reformer only to the extent of having a common hydrogen system. The reformed naphtha is

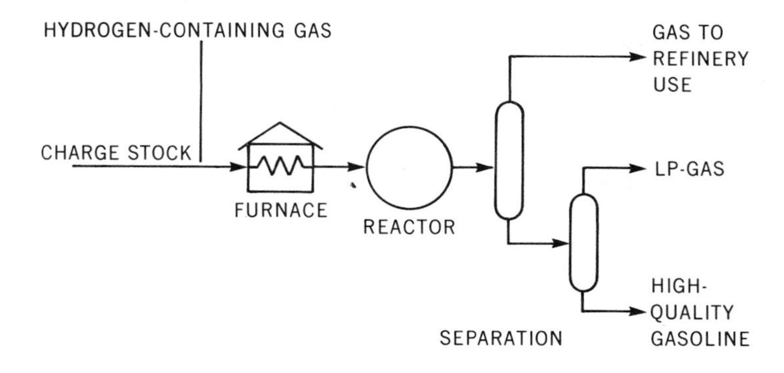

Figure 10. Selectoforming process (27)

mixed with hydrogen and passed into the reactor containing the shape-selective catalyst. The reactor effluent is cooled and separated into a hydrogen-containing gas, LPG, and high octane gasoline. The octane improvement over the first six months of operation was found to be between from 2 to 5 numbers. The removal of normal paraffins reduces the vapor pressure of the reformate since these paraffins are in higher concentration in the front end of the feed. The octane improvement of the front end is greater than that of the higher boiling fraction. Thus, it is sometimes beneficial to fractionate the feed and upgrade only a particular fraction of the reformate. The separate Selectoforming system has the additional flexibility of being able to process other refinery streams.

The second process modification is the Terminal Reactor System. In this system, the shape-selective catalysts replaces all or part of the reforming catalyst in the last reforming reactor. Although this configuration is more flexible and involves minimal investment, the high reforming operating temperature causes butane and propane cracking and consequently decreases the LPG yield and generates higher ethane and methane production. The life of a Selectoforming catalyst used in a terminal system is between one and two years (*31*). Regeneration only partially restores fresh catalytic activity. A study of the integration of Selectoforming with the bimetallic Rheniforming Process has shown that additional stability and yield performance may be obtained (*32*).

Paraffin Isomerization

One of the refinery distillate streams produced in petroleum processing is the light straight run naphtha fraction consisting principally of the C_5 to C_7 hydrocarbon cut. This fraction usually accounts for between 10-20% of the gasoline pool and because of the high content of low octane normal and monobranched paraffins has a clear research octane number of about 70. The disadvantage of adding this fraction directly to the gasoline pool is that its octane number, even when heavily leaded, is considerably less than the octane number of the pool. Thus, the refiner must compensate for this

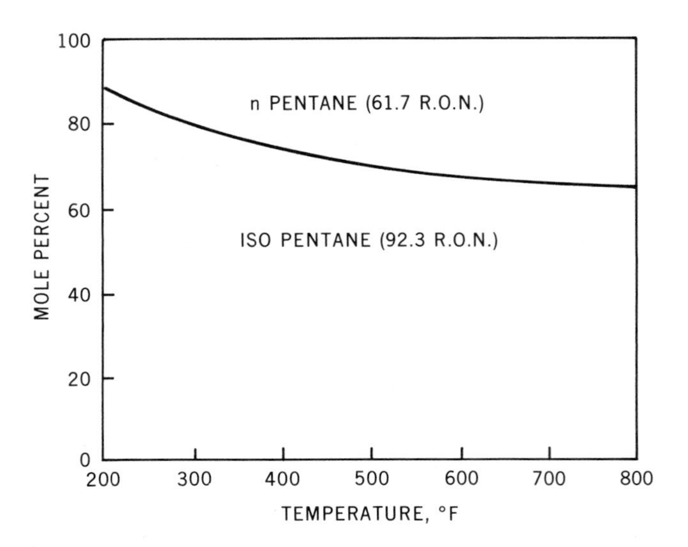

Figure 11. Equilibrium distributions for the pentanes

effect either by more severe reforming with the accompanying lower yields or by adding more lead to the pool. Environmental Protection Agency proposals for phasing out lead additives from motor fuels have initiated a resurgence of interest in isomerization for upgrading the low octane straight run components of the gasoline pool.

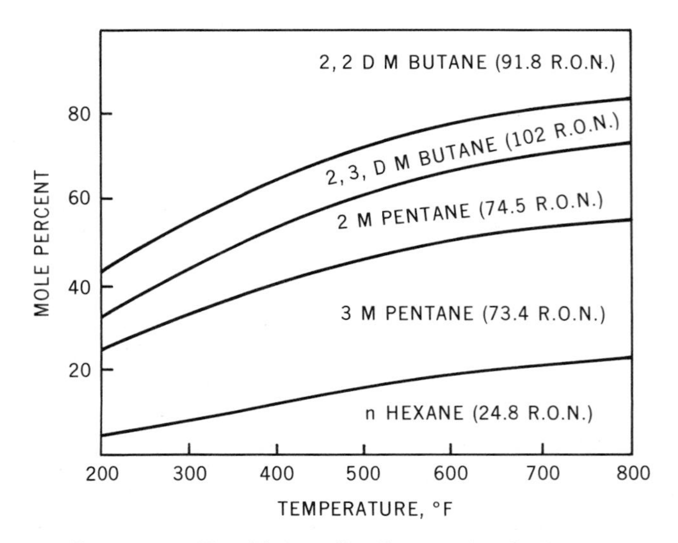

Figure 12. Equilibrium distributions for the hexanes

Isomerization is a reversible chemical reaction whose thermodynamic product distribution, shown in Figures 11 and 12, is closely approached by both homogeneous and heterogeneous catalysis at temperatures ranging from 100° to 850°F. *Via* isomerization, normal and branched paraffins can be converted into higher branched, high octane components with essentially no volume change. The octane numbers of equilibrium mixtures of the pentanes and the hexanes are given in Figure 13 (*33*). Low temperature liquid phase isomerization, employing Friedel-Crafts type catalysts, has the highest octane

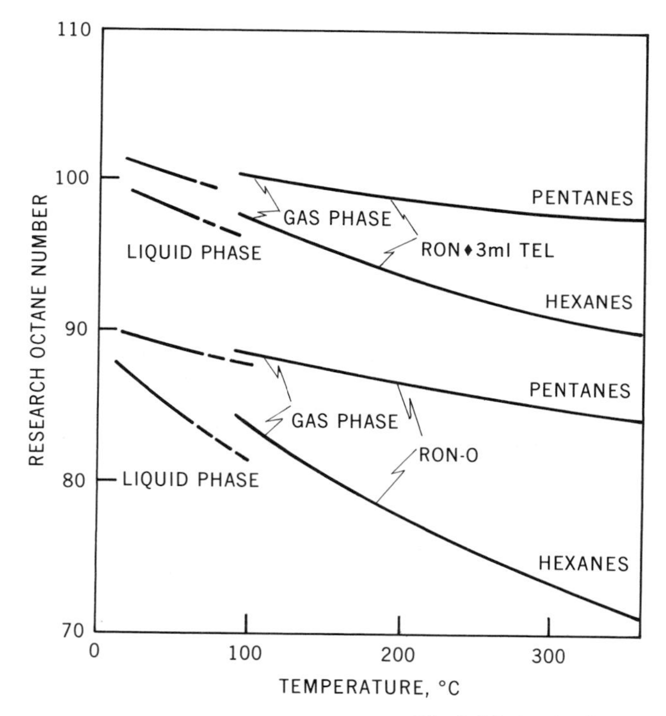

Chemical Engineering Progress

Figure 13. Octane number of equilibrium mixtures of pentanes and hexanes (33)

potential since the relative proportions of the high octane dimethylbutanes and isopentane increase with decreasing temperature. However, these acid processes are corrosive and sensitive to feedstock contaminants which require extensive pretreatment. Consequently, these low temperature processes have been limited to the isomerization of *n*-butane since this material is usually available in sufficient purity by normal refinery processes. Processes using dual-functional catalysts such as platinum on silica–alumina, operate at about 400°C and are consequently less sensitive to catalyst poisons. However, the higher operating temperature limits the conversion as a result of the unfavorable equilibrium distribution, and the yields are reduced by hydrocracking side reactions.

Counterbalancing the desirability of a low operating temperature is the need for a sufficiently high temperature to ensure that isomerization proceeds at a high rate so that the composition of the effluent from a reasonably sized reactor is close to equilibrium. Such high conversions are necessary, not only to minimize the costs of high recycle rates, but also to ensure high yields of dimethylbutanes. Conversion to these isomers is relatively slow, and any departure from equilibrium is primarily reflected in their yields. The higher temperatures not only give rise to less favorable product distributions but promote hydrocracking reactions. Consequently, for a given catalyst each paraffin has a temperature operating range delineated by a sufficiently high isomerization reaction rate at the lower end and a maximum acceptable hydrocracking rate at the other end. If pentane and hexane are to be isomerized simultaneously, their ranges must overlap.

The earliest report of zeolites being used in this application was by Rabo, Pickert, Stamires, and Boyle who evaluated platinum containing types X and Y molecular sieves (*34*). They found the sodium forms of X and Y to be catalytically inactive. The calcium exchanged forms exhibit some activity while the decationized form of Y has excellent activity. A series of platinum containing type Y catalysts were evaluated to determine the effect of sodium replacement on activity for hexane isomerization, and these data are shown in Figure 14.

Sodium Y with a Na/Al ratio of 1 does not isomerize *n*-hexane even at 480°C. After a 10% reduction in the sodium content however, the catalyst has sufficient activity to isomerize 50% of the feed at 400°C. Higher degrees of alkali metal removal result in additional increases in catalytic activity and permit the use of substantially lower operating temperatures. The effect of sodium replacement by calcium at low degrees of ion exchange is not so pronounced as the effect of decationization. It was observed from a com-

Table XIII. Isomerization[a] of *n*-Hexane with

Zeolite	*Cation Content, %*	*Si/Al Atomic Ratio*	*Pore Size A*
Linde small pore molecular sieve	100 Na^+, K^+	3.3	~4.8
	21 Na^+, 79 Ca^{2+}	3.3	~4.8
	19 Na^+, 81 decationized	3.3	~4.8
Linde type X molecular sieve	100 Na^+	1.2	~9-10
	13 Na^+, 87 Ca^{2+}	1.2	~8-9
Linde type Y molecular sieve	100 Na^+	2.5	~9-10
	22 Na^+, 78 Ca^{2+}	2.5	~9-10
	15 Na^+, 85 decationized	2.5	~8-9

[a] Reaction Conditions: space velocity: 2.0 g feed/g catalyst/hr, pressure: 450 psig, hydrogen/hydrocarbon mole ratio: 3:1.

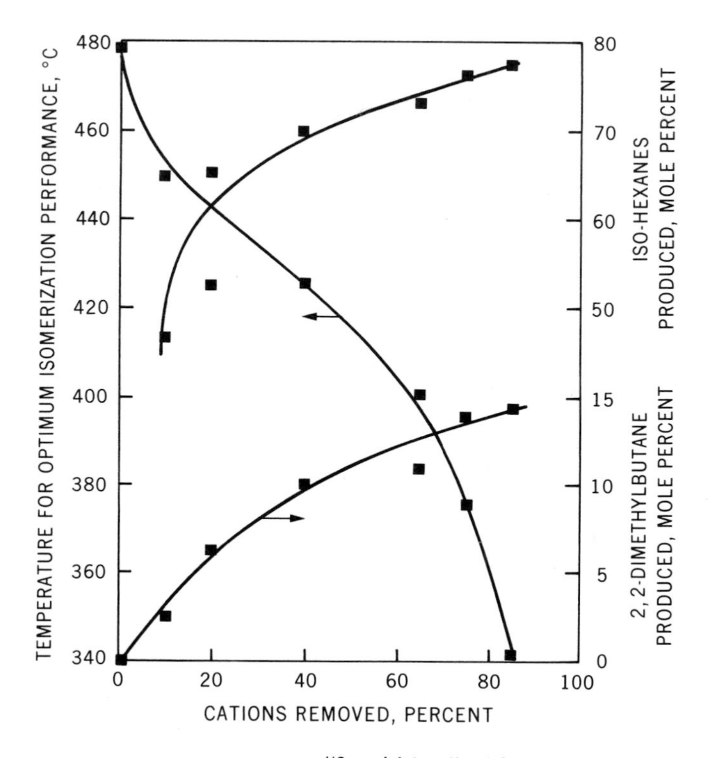

"Second International Congress on Catalysis"

Figure 14. Effect of extent of decationization on type Y (34)

0.5 wt % Platinum-Loaded Zeolites *(34)*

Reaction Temp. °C	*Liquid Yield, Vol %*	*Yield Isohexane in C_6 Fraction, Mole %*	*Yield 2,2-Dimethylbutane in C_6 Fraction, Mole %*
400	91	8	<1
400	71	7	<1
375	67	47	<1
450	>98	2	0
400	91	13	<1
475	>98	5	0
375	>98	73	12
350	>98	76	14

Second International Congress on Catalysis

parison of large and small pore zeolites, shown in Table XIII, that the low product recovery from the small pore zeolite is probably caused by the restricted escape of the product from the small pores resulting in hydrocracking to propane, ethane, and methane. The pore size of the catalyst should therefore be larger than the critical size of the products and reactants. The importance of pore size was substantiated by the work of Miale, Chen, and Weiss (*28*). Among the zeolites having pore openings large enough to admit an isoparaffin are types X, Y, L, Ω, and large pore mordenite. The difficulty in removing the residual potassium from type L and the tetramethylammonium cations from type Ω without causing a reduction in crystallinity make these zeolites less desirable than the other large pore materials (*14, 16*).

A later study by Rabo, Pickert, and Mays showed that a palladium loaded decationized type Y, SiO_2-Al_2O_3 molar ratio 5.0, can convert *n*-pentane and *n*-hexane feeds to approaching equilibrium distributions (*35*). The data in Table XIV show that at 660°F and at a space velocity of 2, *n*-pentane is 65% converted to isopentane. At the same process conditions but at a 30°F lower temperature, yields of over 15%, 2,2-Me_2C_4 are produced from an *n*-hexane feed with essentially equilibrium yields of the other isomers.

Table XIV. Isomerization of *n*-Hexane and *n*-Pentane (*35*)

Catalyst: 1/2 wt % Palladium 85% decationized Y

Feed	n-*Hexane*	n-*Pentane*	*Equimolar Mixture* n-*Hexane and* n-*Pentane*
Temperature, °F	*626*	*662*	*644*
Pressure, psi	*450*	*450*	*450*
W.H.S.V. g/g/hr	*2*	*2*	*2*
H_2/hydrocarbon ratio, m/m	*3*	*3*	*3*
Total liquid product, wt %			
C_1-C_3	0.8	0.3	1.8
C_4	1.1	1.2	1.8
C_5^+	98.1	98.5	96.4
Composition of C_5 fraction, mole %			
isopentane	—	65	58.8
n-pentane	—	35	41.2
Composition of C_6 fraction, mole %			
n-hexane	21.8	—	23.3
3-methylpentane	21.8	—	22.8
2-methylpentane	33.3	—	} 40.7
2,3-dimethylbutane	8.0	—	}
2,2-dimethylbutane	15.2	—	13.2

Table XV. Effect of Sulfur in Pentane Isomerization with 1/2 wt % Pd Decationized Y (*35*)

Sulfur Content, ppm[a]	*Isopentane in C_5 Fraction Mole %*
6	62.5
12	61.0
18	60.0
24	58.5
30	57.5
55	55.0
3000	27.0[b]
0	62.5

[a] As *n*-butyl mercaptan.
[b] Average conversion during 16-hr test period.

Industrial and Engineering Chemistry

C_5-C_6 mixtures were also evaluated and over 13 mole % 2,2-Me_2C_4 and 60 mole % *i*-C_5 were produced at 664°F. Catalyst activity remained unchanged for periods of over 1250 hr using a *n*-hexane feed and for over 1400 hr using *n*-pentane feed. At the end of these periods the catalyst was producing over 14 mole % 2,2-Me_2C_4 and 63 mole % isopentane with the respective feeds. No detectable coke was found on the catalyst after the tests. The introduction of water into the pentane and hexane feeds up to the point of saturation had no detectable effect on catalyst activity or selectivity over several days operation. The effect of sulfur poisoning was also studied using an *n*-pentane feed in which sulfur concentrations were varied from zero to 3000 ppm. Sulfur concentrations of 6 ppm had no effect on catalyst activity over a period of 1000 hr. Table XV shows that sulfur concentrations of 55 ppm act as a temporary poison and reduce the conversion level from 62.5 to 55%. After 2500 hr exposure to feeds containing up to 3000 ppm sulfur, the activity of the catalyst was restored on reversion to a clean feed.

The type Y zeolite used in these early studies contained approximately 2.0 wt % Na_2O, a practical lower limit that may be obtained by conventional ion exchange, and these catalysts require reaction temperatures in the range 600°-700°F. More recently, however, it was recognized that small amounts of residual sodium in the zeolite exert a strong influence on catalytic activity (*36*). Using the stabilization technique described by Maher and McDaniel and discussed by Kerr, soda levels down to 0.2 wt % may be achieved (*37, 38*). The intermediate calcination used in this stabilization procedure redistributes the sodium from less accessible cation sites enabling essentially complete sodium removal to be achieved by a subsequent ammonium ion exchange. The large effect of sodium at low concentrations indicates that during thermal treatment of the catalyst prior to evaluation, sodium is redispersed and deactivates the more accessible catalytic sites. The effect of residual sodium on pentane isomerization activity for a Y type zeolite is shown in Table XVI; a decrease from 0.27 to 0.02 wt % Na_2O enables a reduction of 50°C in reaction temperature for a 30% conversion

Table XVI. Hydroisomerization[a] of *n*-Pentane over Pd-H-Zeolite Y: Influence of Sodium on Catalytic Activity (*39*)

Crystallinity, wt % (x-ray)	*Na_2O, wt %*	*For 30% Conversion, °C*
90	2.02	305
80	0.27	300
80	0.02	250

[a] Conditions: $H_2/n\text{-}C_5$ molar ratio, 2.5; pressure, 30 kg/cm^2; WHSV: 1 kg/kg/hr.

Advances in Chemistry Series

(*39*). This increased catalytic activity resulting in a lower operating temperature enables more favorable isomeric distributions of pentane and hexane to be obtained (Figures 15 and 16), with an accompanying decrease in hydrocracking activity (*42*).

The zeolite mordenite also shows excellent activity in hydroisomerization. The cations in sodium mordenite, unlike those in type Y, are all located in the main channels and are thus easily replaced by ammonium

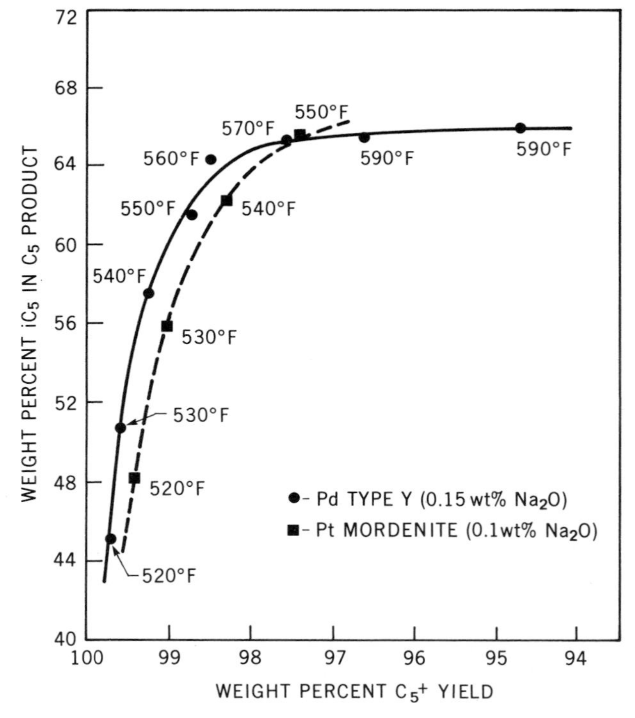

Figure 15. Pentane isomerization activity of low soda Y and mordenite (42)

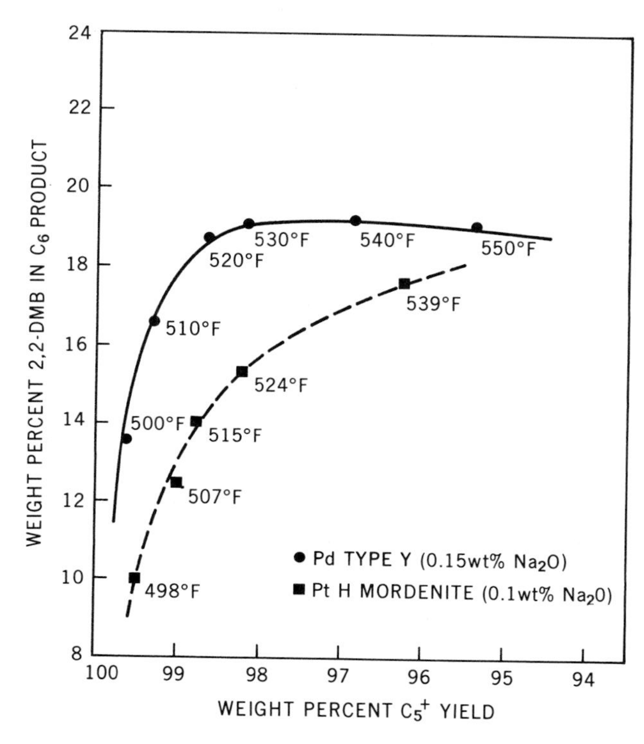

Figure 16. Hexane isomerization activity of low soda Y and mordenite (42)

cations using conventional ion exchange. This enables low sodium ($<$ 0.01 w % Na_2O) catalysts to be obtained (*39*). Although the hydrogen form of mordenite obtained *via* NH_4^+ decomposition has good isomerization activity, acid treatment together with an ammonium exchange step yields a superior catalyst (*40*). Thus, the acid treatment besides removing sodium cations also leaches out alumina from the mordenite channels making them more accessible to reactant and product molecules. Mordenite as synthesized normally has a SiO_2/Al_2O_3 molar ratio of about 10. However, with strong mineral acid treatment, the SiO_2/Al_2O_3 ratio can be increased progressively until substantially all the alumina is removed. That acid treatment removes obstructions from the sorption channel is shown by the increase in decalin sorption with increasing Al_2O_3 removal (*41*). The effect of SiO_2/Al_2O_3 molar ratio on pentane isomerization activity is shown in Table XVII (*39*). An optimum isomerization activity at about a ratio of 17 is observed; further removal of alumina results in the formation of a less acidic catalyst surface with a corresponding lowering of activity.

The effect of hydrogen partial pressure on isomerization activity gives some interesting insight into the reaction mechanism. Data published by Minachev and co-workers show somewhat unexpectedly that the isomeri-

Table XVII. Hydroisomerization[a] of *n*-Pentane: Influence of Silica-Alumina Molar Ratio on Activity of Pd–H Mordenite Catalysts (*41*)

Na, wt %	*SiO_2/Al_2O_3 Molar Ratio*	*Relative Activity*
Nil	12	100
0.9	14	50
0.03	25	78
Nil	77	47
0.03	93	23

[a] Conditions: temperature: 288°C; pressure: 32 kg/cm^2; H_2/C_5 molar ratio = 3.2.

Journal of Catalysis

Influence of Silica-Alumina Molar Ratio on Activity of Pt–H Mordenite[a] (*39*)

Na, wt %	*SiO_2/Al_2O_3 Molar Ratio*	*Relative Activity*
0.03	10	100
0.02	17	135
0.02	25	84

[a] Conditions: temperature: 250°C; pressure: 30 kg/cm^2; H_2/C_5 molar ratio = 2.5.

Advances in Chemistry Series

zation activity of a hydrogen mordenite (HM) is inversely proportional to hydrogen partial pressure (*43*). The reaction proceeds according to the following scheme.

$$n\text{-}C_5H_{12} + HM \rightarrow n\text{-}C_5H_{11}\text{-}M + H_2 \quad (1)$$

$$n\text{-}C_5H_{11}\text{-}M \rightarrow i\text{-}C_5H_{11}\text{-}M \quad (2)$$

$$i\text{-}C_5H_{11}\text{-}M + H_2 \rightarrow i\text{-}C_5H_{12} + HM \quad (3)$$

where n-C_5H_{11}-M and i-C_5H_{11}-M are carbonium ions connected to the active centers of the zeolite. Increasing hydrogen partial pressure would shift the equilibrium in Equation 1 to the left and thus suppress activity. A bimolecular mechanism has been proposed for the isomerization of the hexanes (*44*). This mechanism envisages the 1,3 diadsorption of the molecules on the catalyst surface forming a cyclohexane-type intermediate. Scission of the cyclic intermediate into the various hexane isomers is illustrated by the isomerization of 3-methylpentane.

ab → *n*-hexane   cd → 2-methylpentane   ab → *n*-hexane and 3-methylpentane

Excellent agreement is found between the primary products from the isomerization of the individual hexane isomers and those predicted by the mechanism. This bimolecular mechanism has the additional advantage of satisfactorily accounting for the isomerization of the butanes and pentanes.

The operation of a zeolite in a hydrogen atmosphere but in the absence of a noble metal hydrogenation component leads to the accumulation of surface residues which rapidly deactivate the catalyst. Incorporating platinum onto a hydrogen mordenite not only increases the selectivity but also prevents catalyst deactivation as the data in Figure 17 demonstrate (*39*). It may well be that the only role of the noble metal is to keep the catalyst surface free from carbonaceous deposits. A study on the effect of noble

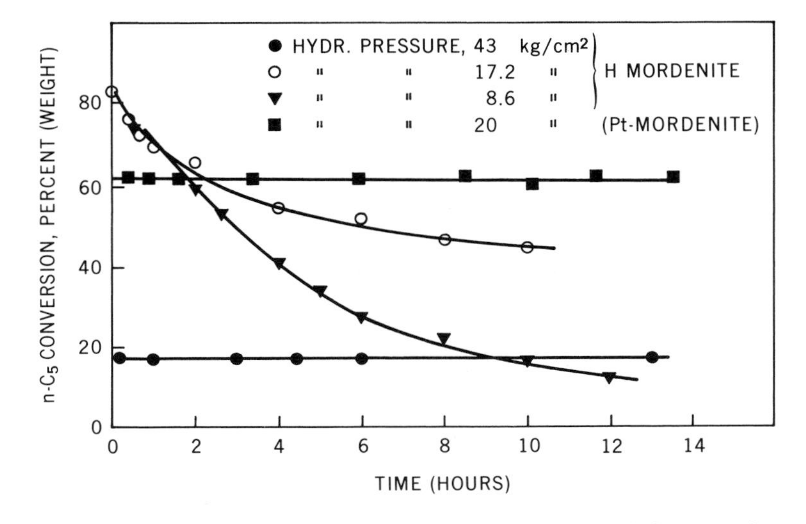

Advances in Chemistry Series

Figure 17. n-*Pentane isomerization of mordenite* (39)

metals introduced by ion exchange with platinum and palladium tetrammine cations onto a lanthanum and ammonium Y, SiO_2/Al_2O_3 molar ratio 5.0, showed that even small amounts of noble metal enhance the activity of the zeolite for *n*-hexane isomerization (*45*). A linear increase in isomerization activity with noble metal loading was observed until an optimum metal concentration is reached. Figures 18 and 19 show that at the optimum

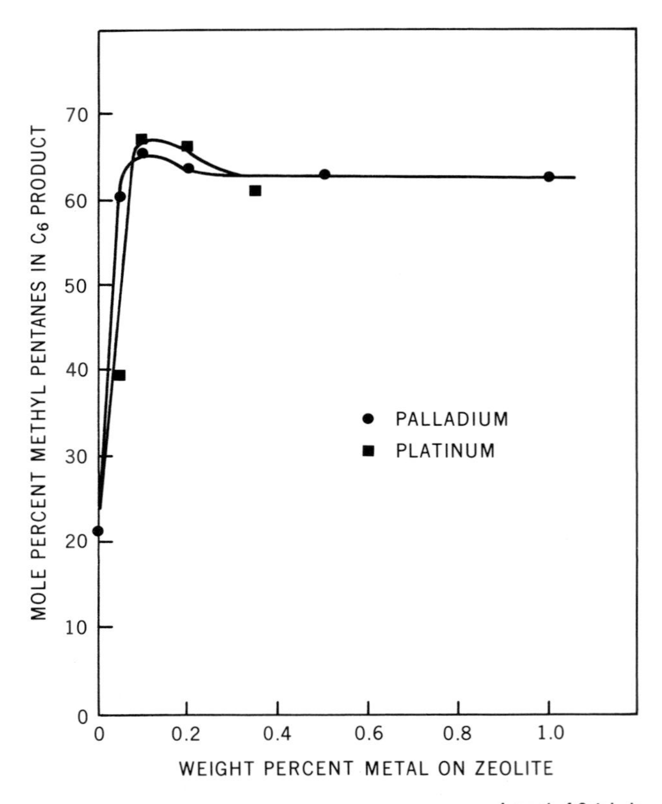

Journal of Catalysis

Figure 18. Effect of metal concentration on methylpentane formation (45)

metal loadings for both the methylpentanes and dimethylbutanes, palladium appears to be approximately twice as effective as platinum on a weight basis. However, when these metals are compared on an atomic basis, they are about equally effective since the atomic weight of platinum is twice that of palladium. The amount of noble metal used in commercial operation will be that required to saturate coke precursors, and this will depend on the hydrogen pressure as well as on the amount of sulfur containing compounds in the feed.

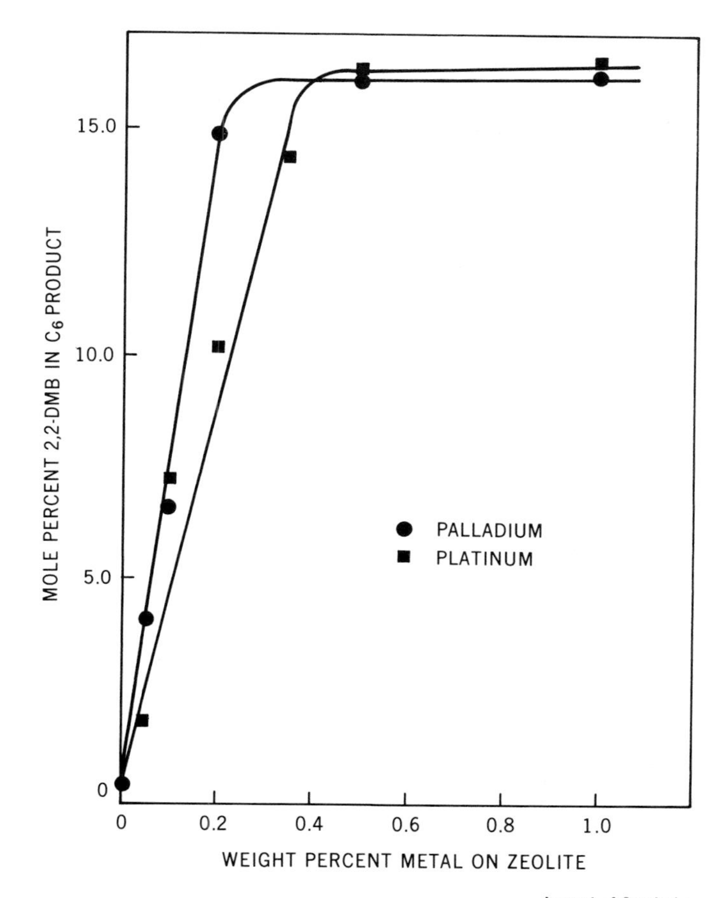

Journal of Catalysis

Figure 19. Effect of metal concentration on 2,2-dimethylbutane formation (45)

The Hysomer Process. A vapor phase isomerization process using a zeolite catalyst operating in an hydrogen partial pressure range of 400 to 450 psi and at approximately 500°F was developed by Shell Oil and is operating commercially at the LaSpezia refinery in Italy. The catalyst in the Shell Hysomer process is manufactured by Union Carbide and is an acidic zeolite of very low sodium content containing highly dispersed noble metal (*33*).

Published pilot plant data using hydrogenated C_5/C_6 Middle East feed are presented in Table XVIII. At 95 to 97 wt % yields of debutanized product, an increase of about 12 RON clear was achieved which approaches the limit set by the thermodynamic equilibrium. Aromatics in the feed are quantitatively hydrogenated to the corresponding naphthene and present no problems to the catalyst selectivity even in concentrations up to 2 wt %.

Table XVIII. Properties of C_5/C_6 Middle East Feed and Isomerized Product (Pilot Plant Results) at 260°C (*33*)

	Feed	*Reactor Product*
RON-O of C_5^+	67.5	79.2
RON-3 ml TEL of C_5^+	87.4	94.3
C_5^+ Yield on C_5^+ in Feed, wt %		96.5
Composition, wt %:		
methane		0.18
ethane		0.18
propane		1.06
isobutane		1.60
n-butane	0.57	1.10
isopentane	17.43	31.33
n-pentane	28.12	15.07
2,2-dimethylbutane	0.44	9.34
2,3-dimethylbutane	1.97	4.06
2-methylpentane	12.43	14.39
3-methylpentane	9.72	9.91
n-hexane	21.15	8.35
cyclopentane	1.75	1.30
methylcyclopentane	3.26	1.50
cyclohexane	0.88	0.63
benzene	1.11	—
C_7^+	1.17	—

Chemical Engineering Progress

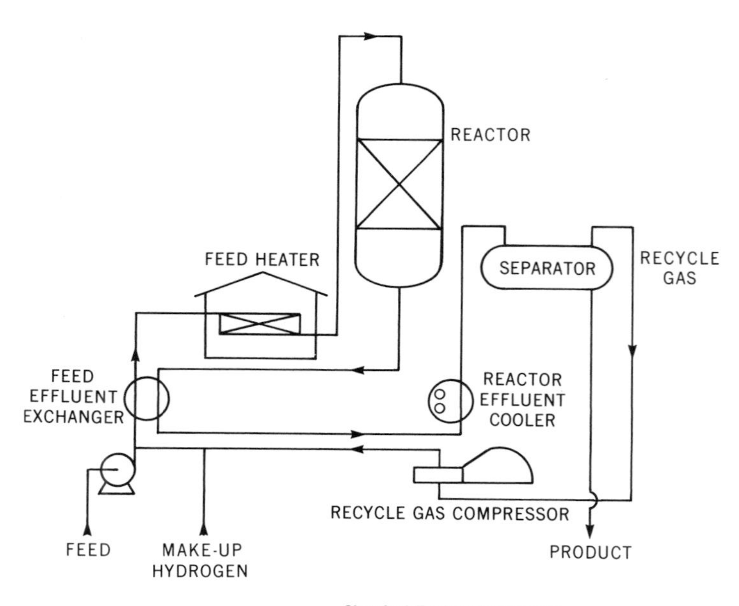

Chemical Engineering Progress

Figure 20. Shell Hysomer process (33)

Heptane and higher hydrocarbons are hydrocracked, principally into propane and isobutane. Although the process can tolerate more than 4 wt % C_7 hydrocarbons in the feed, the conversion becomes less selective and concentrations of less than 2 wt % are preferred. The adverse effect of higher concentrations of the heptanes on selectivity is probably caused by the large volume of cracked gases hindering the effective approach of the other molecules to the catalytic sites.

A schematic of a Shell Hysomer C_5/C_6 process is shown in Figure 20. Operating data from the first commercial unit are shown in Table XIX (*46*). The hydrotreated C_5/C_6 feedstocks are used without further pretreatment. Activity and selectivity after two years of operation confirm that no catalyst deactivation had taken place. An *in situ* regeneration was carried out after an operational upset, and the subsequent data, shown in Table XIX, confirmed that catalyst activity was completely restored.

The influence of sulfur on activity was minimal and the catalyst can tolerate a permanent sulfur level of 10 ppm in the feedstock. Concentrations

Table XIX. Typical Properties of C_5/C_6 Feed and Isomerized Product (Commercial Results) (*46*)

	7/70		*6/71*		*1/72*[a]	
Date	*Feed*	*Product*	*Feed*	*Product*	*Feed*	*Product*
Specific gravity 15/4°C	.638	.635	.635	.632	.642	.637
RON-O	73.2	82.1	74.9	82.7	74.5	82.4
RON-3 ml TEL	90.8	96.0	91.8	96.4	91.5	96.2
C_5^+ Yield, wt % on C_5^+ in feed	—	97.4	—	97.3	—	97.3
Composition, wt %						
i-butane	0.2	0.5	0.2	0.5	0.2	0.5
n-butane	0.5	1.3	0.5	1.3	0.5	1.3
i-pentane	29.3	49.6	35.1	55.9	26.8	46.9
n-pentane	44.6	25.1	48.3	26.1	38.5	20.9
cyclopentane	2.2	1.9	1.6	1.2	2.1	1.3
2,2-dimethylbutane	0.6	5.0	0.6	3.0	2.2	7.3
2,3-dimethylbutane	1.8	2.2	1.1	1.3	2.6	2.6
2-methylpentane	9.3	7.0	5.7	5.1	12.3	9.4
3-methylpentane	4.6	4.3	3.0	3.0	7.4	5.4
n-hexane	6.7	2.9	3.4	2.2	6.4	3.6
methylcyclopentane	0.2	0.2	0.4	0.3	0.7	0.5
cyclohexane	—	—	—	0.1	—	0.3
benzene	—	—	0.1	—	0.3	—
	100.0	100.0	100.0	100.0	100.0	100.0

[a] Post regeneration.

"Meeting, DGMK"

up to 35 ppm are not harmful. The process can operate at a water level of 50 ppm, and feedstocks having saturated water contents can be processed without deleterious effect on either catalyst stability or conversion. A minimum quantity of water is essential to the activity of zeolite catalysts in this application (*47*). The La Spezia unit is integrated with a catalytic reformer to a certain degree, and this enables the use of common recycle gas facilities, product coolers, and stabilizers. This integration saves considerable capital costs.

The octane potential of isomerization processes is ultimately determined by the thermodynamic equilibrium distribution at the operating temperature. Further improvements in octane rating can be made by separating the normal paraffins from the isomerized product and recycling these to extinction. The Union Carbide Total Isomerization Process is based upon the integration of the IsoSiv molecular sieve iso-normal separation process with the Hysomer process (*46*). This combined process is a vapor phase system yielding a normals-free product having an approximately 20 octane number (RON clear) improvement using a C_5/C_6 feed as compared with a 12 RON clear improvement on a once-through basis.

Xylene Isomerization

The major industrial source of xylenes is from the C_8 aromatic cut of the product stream from the catalytic reforming of petroleum naphthas. This aromatic fraction contains ethylbenzene as well as the three xylene isomers, each of which is used extensively as a chemical intermediate in industrial applications. Ethylbenzene is used to make styrene, para-xylene for terephthalic acid, ortho-xylene for phthalic acid, and meta-xylene to make isophthalic acid. The principal industrial interest is in the recovery of para- and ortho-xylenes from the C_8 aromatic cut, and various processes have been developed to enable their yield to be maximized. The ethylbenzene and the ortho-xylene may be separated from each other and from the xylene stream by fractionation, but the boiling points of the meta and para isomers are too close for this type of separation. Fortunately the freezing points of

Table XX. Calculated Equilibrium Product

Temp. °K	*Xylene in Product, %*	*Transalkylate in Product, %*
300	45.5	53.4
350	46.1	53.8
400	46.5	53.4
450	46.5	53.6
500	46.4	53.6
550	46.5	53.6
600	46.6	53.4

these latter isomers are sufficiently different for separation to be achieved by fractional crystallization. However, only about 60% of the para-xylene can be separated in high purity by this technique because of the formation of eutectics with other isomers that degrade product purity. Thus, only about 12% para-xylene can be recovered from a typical feed having a composition, ethylbenzene 20%, ortho-xylene 20%, meta-xylene 40% and para-xylene 20%. Recently, a separation process, developed by Universal Oil Products called Parex, using molecular sieve adsorption techniques has been introduced in which almost 100% of the para-xylene in the feed may be recovered (*48*). In order to increase the yields of the ortho- and para-xylenes significantly, streams deficient in these isomers may be isomerized back to equilibrium distribution and then recycled to the isomer extraction step. Unlike the paraffins, there is little change in the equilibrium distribution of the xylene isomers with temperature (Table XX), and thus there is no isomer distribution advantage to be gained by operating in a particular temperature range (*49*).

Catalytic isomerization of the xylenes using $AlCl_3$ was one of the first reactions investigated by Friedel and Crafts (*50*). Later studies by McCawley and Lien on the reaction mechanism of xylene isomerization using a BF_3–HF system, showed that the amount of catalyst used has a pronounced effect on the composition of the xylene equilibrium mixture (*51*). Only at low catalyst concentrations is the composition of the isomerized product in agreement with the thermodynamically calculated values; at high catalyst concentrations, the meta-xylene concentration approaches 100%. It was proposed that the mechanism is intramolecular and proceeds *via* sigma complexes.

The variation in equilibrium composition with catalyst concentration was attributed to the formation of stable catalyst–hydrocarbon complexes; the meta xylene-catalyst complex being more stable than the corresponding ortho- and para-xylene complexes (*51*). The isomerization reaction is accompanied by a disproportionation reaction, sometimes referred to as transalkylation, yielding toluene and trimethylbenzenes, but this may be suppressed by the use of toluene as a diluent.

Distributions for Xylene-Trimethylbenzene System (*49*)

Normalized Product Distributions

Xylene Fraction			*Trimethylbenzene Fraction*		
o	m	p	*1,2,3*	*1,2,4*	*1,3,5*
16.4	59.8	23.8	5.4	57.6	36.9
17.6	58.3	24.1	7.1	59.4	33.5
18.9	56.8	24.2	8.8	60.8	30.5
20.1	55.7	24.2	10.0	61.6	28.3
21.2	54.8	24.0	11.2	62.4	26.4
22.0	54.0	24.0	12.6	62.1	25.4
22.9	53.2	23.8	13.9	61.7	24.4

Journal of Organic Chemistry

The isomerization of the xylenes in the liquid phase has the disadvantage that at sufficiently low catalyst concentrations required to yield calculated equilibrium distributions, reaction rates are low. Also, poor yields result from catalyst solubility in the product as well from losses from sludge formation.

Xylene isomerization also takes place in the vapor phase over a dual functional catalyst, platinum on silica–alumina, in the presence of hydrogen. Many such processes have been developed and are currently being used commercially. The noble metal acting with the hydrogen keeps the surface of the silica–alumina clean and facilitates ethylbenzene conversion to the xylene isomers—a reaction not found to occur in the liquid phase systems. Pitts, Connor, and Leum demonstrated that hydrogenated intermediates are required to isomerize the ethylbenzene to the xylenes (*52*). Near equilibrium distributions of the xylenes can be achieved at approximately 400°C. How-

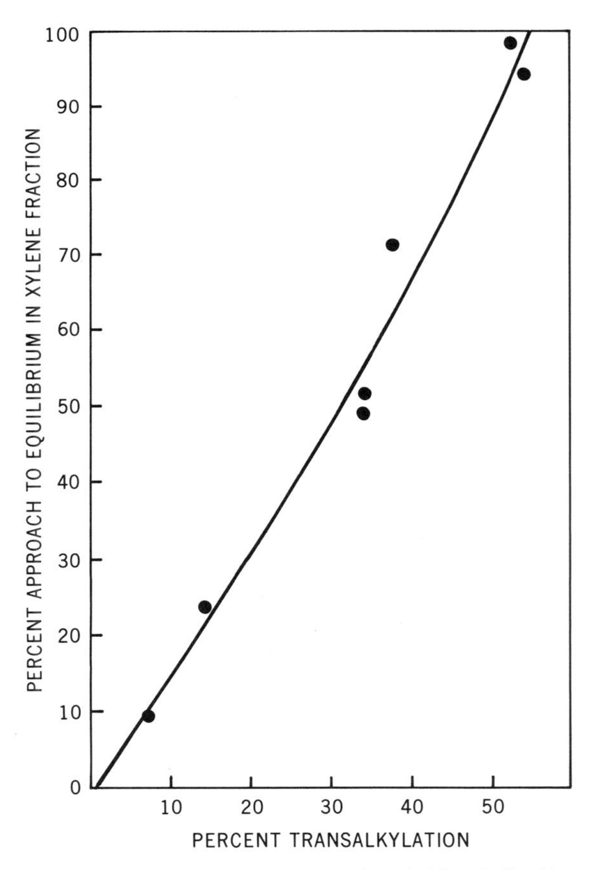

Journal of Organic Chemistry

Figure 21. Dependence of o- *and* m-*xylene isomerization on extent of transalkylation* (49)

Table XXI. The Isomerization of *o*- and *m*-Xylene (*49*)

Feed	Product Composition, Mole % After 20 hr at 300° and 190 psig: o-Xylene	m-Xylene
Benzene	1.7	2.6
Toluene	16.3	18.9
o- Xylenes	9.6	9.8
m- Xylenes	27.8	25.6
p- Xylenes	9.3	9.8
1,2,3- Trimethylbenzenes	2.7	2.7
1,2,4- Trimethylbenzenes	22.6	20.4
1,3,5- Trimethylbenzenes	9.9	9.8
Transalkylation, %		
	53.2	54.8
Normalized Distribution, %		
o- Xylenes	23.5	21.6
m- Xylenes	53.8	56.8
p- Xylenes	22.7	21.6
1,2,3- Trimethylbenzenes	7.5	8.1
1,2,4- Trimethylbenzenes	64.3	61.3
1,3,5- Trimethylbenzenes	28.2	30.6

Journal of Organic Chemistry

ever, some feed loss is experienced from the formation of low molecular weight hydrocarbons *via* hydrocracking and from the formation of C_9+ aromatics.

One of the first studies on the catalytic isomerization of the xylenes with zeolites was by Lanewala and Bolton (*49*). The zeolite used was a lanthanum and ammonium exchanged type Y with a SiO_2/Al_2O_2 molar ratio of 5. Previous studies on the zeolite catalyzed isomerization of the diethylbenzenes and the *tert*-butylphenols had shown that positional isomerization *via* transalkylation satisfactorily accounted for the experimental results (*53*, *54*). Starting with the individual xylene isomers, isomerization is invariably accompanied by transalkylation and that the more extensive the transalkylation the greater the degree of isomerization, as shown in Figure 21. Table XXI shows that the principal transalkylation products at 300°C are toluene and the trimethylbenzenes and that the composition of the product is in agreement with the calculated equilibrium distribution shown in Table XX.

The isomerization of the three individual xylene isomers was studied at low conversion levels. The variation in composition with conversion level of both the xylene and trimethylbenzene fractions indicated that xylene isomerization proceeds *via* the trimethylbenzene isomers. At conversion levels away from the calculated xylene equilibrium values, a higher amount of transalkylate accompanied para-xylene isomerization than ortho- and meta-xylene.

Under experimental conditions identical to that used for xylene isomerization, toluene and the trimethylbenzenes are converted to a product containing an equilibrium mixture of the xylenes. The xylene fraction of the initial product from a feed of toluene and either 1,2,4- or 1,2,3-trimethylbenzene, contains a significantly higher than equilibrium amount of ortho-xylene isomer. The xylene fraction of the initial product from the 1,3,5-trimethylbenzene and toluene contains higher than equilibrium amounts of both the ortho- and meta-xylene.

These data were explained in terms of an intermolecular mechanism involving a diphenylmethane type intermediate; such a mechanism imposes limits on the possible isomers that may be derived from the transalkylation reaction.

Starting with the meta isomer, all three trisubstituted isomers may be derived from the intermediate.

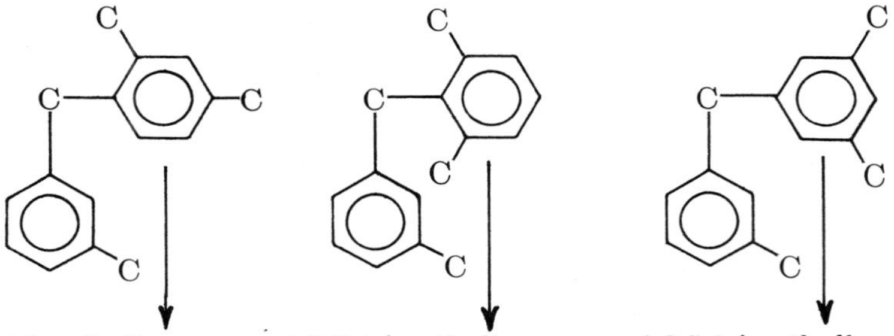

1,2,4-trimethylbenzene + toluene   1,2,3-trimethylbenzene + toluene   1,3,5-trimethylbenzene + toluene

Ortho-xylene produces an intermediate which can yield only the 1,2,3- and the 1,2,4-trimethylbenzenes.

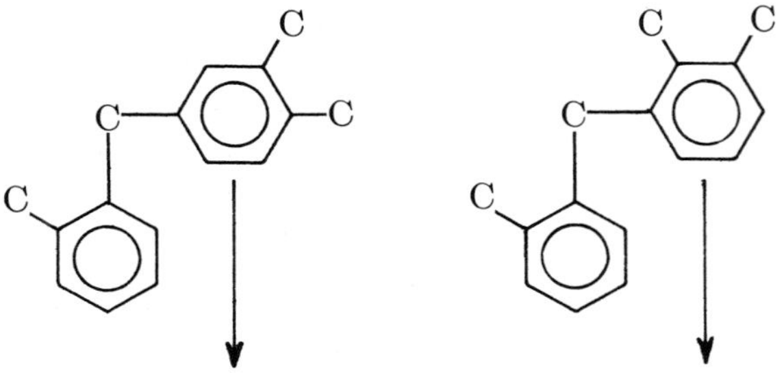

1,2,4-trimethylbenzene + toluene   1,2,3-trimethylbenzene + toluene

while para-xylene, having four equivalent unsubstituted ring positions can give only 1,2,4-trimethylbenzene. Using these limitations imposed on the transalkylated products, the following reaction scheme accounts satisfactorily for the observed data.

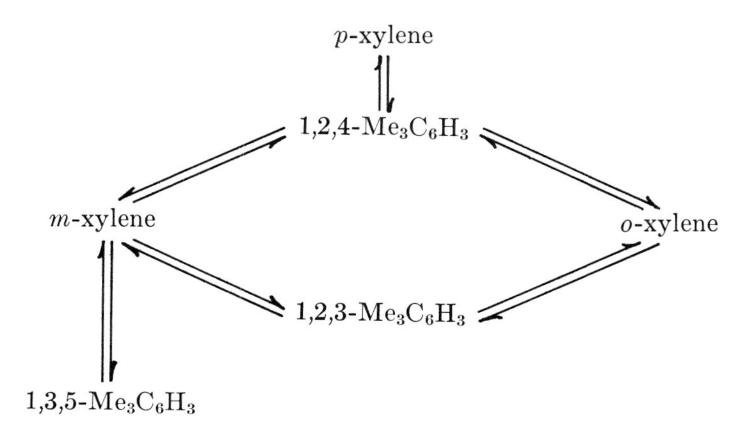

It was concluded that the catalyst–hydrocarbon complex shown to be present in the BF_3–HF xylene system by McCawley and Lien might well mask the true mechanism. The intramolecular mechanism observed for xylene isomerization with acid type catalysts may be a result of the enhanced stability of the meta-xylene and 1,3,5-trimethylbenzene σ-complexes compared with those of other isomers. The principal evidence for the intramolecular isomerization of the xylenes is the apparent absence of direct conversion of para-xylene to ortho-xylene or vice-versa. However, starting with either of these two isomers, the products will be the meta-xylene σ-complex and the mesitylene σ-complex until a 1:1 mole ratio of these isomers to the acid catalyst is established. After this step, equilibration between the acid phase and the hydrocarbon phase will take place and the formation of equilibrium distributions of the isomers will begin in the hydrocarbon phase. It is interesting to speculate on how many reaction mechanisms obscured by hydrocarbon–catalyst complex formation may be resolved by investigations using zeolite catalysts.

Csicsery later proposed that xylene isomerization over zeolite catalysts proceeds *via* both intramolecular and intermolecular mechanisms and that the contribution of each depends on reaction temperature (*55*). By studying the isomerization of 1-methyl 2-ethylbenzene over a mixed calcium ammonium exchanged Y, it is possible to differentiate between the two mechanisms by the extent of xylene and ethylbenzene formation. At higher temperatures, above 300°C, isomerization occurs primarily *via* 1,2 shifts, and at lower temperatures, below 200°C, the transalkylation mechanism predominates; at intermediate temperatures, both contribute. This conclusion would seem to conflict with the previous study which showed that at 300°C, xylene isomerization occurs *via* transalkylation. However, the catalyst used by Csicsery—$Ca(78)NH_4(22)Y$—is significantly less active than that used in the previous study—$RE(40)NH_4(40)Na(20)Y$—and it would not seem unreasonable that the contributions of the two mechanisms at a particular temperature would also depend on catalyst activity.

Table XXII. Isomerization[c] of *m*-Xylene

Catalyst	*Temp., °C*	*Hours on Stream*
RE (60)H(35) Y	315	75
H-erionite (K_2O 4.1 wt %)	315	70
H-erionite (K_2O 4.1 wt %)	370	50
H-erionite (K_2O 2.1 wt %)	315	118
H-erionite (K_2O 0.9 wt %)	315	92

[a] Total conversion.
[b] To *o*- and *p*-xylenes.

The contention that xylene isomerization is intramolecular at high temperatures and dependent on catalyst activity is supported by another investigation. An erionite catalyst isomerizes a meta-xylene feed at 600°F to equilibrium distribution (*56*). The data in Table XXII show that the formation of transalkylate product is insignificant. This is probably one of the first examples of a reaction taking place wherein neither the reactant nor product molecules can enter the pores of the zeolite. Although the erionite catalyst product distribution is similar to the dual-functional silica–alumina catalysts, no interconversion between ethylbenzene and the xylenes is observed.

The activities of various cation exchanged forms of zeolite for the isomerization of ortho-xylene were examined in a series of studies by Ward and Hansford (*57*). The hydrogen form of mordenite was the most active zeolite examined, followed by the hydrogen and rare earth forms of type Y. One of the more interesting observations in these studies was the effect of small amounts of residual sodium on catalytic activity. As would be expected, the lower the sodium content of a zeolite, the higher is its catalytic activity for xylene isomerization. The more active zeolites, however, responded more to variation in sodium content than the less active zeolites. For example, magnesium hydrogen Y increases in activity by almost 100% when the sodium content is reduced from 1.1 to 0.1 wt % while the strontium and barium hydrogen forms are virtually unchanged. The most active sites

Table XXIII. Isomerization of *o*-Xylene Using REX *vs.* AP Catalyst (*58*)

	Product Distribution, wt %	
	REX	*AP*
Benzene	3.2	0.4
Toluene	7.2	1.3
p-Xylene	0.9	4.3
m-Xylene	13.5	29.8
o-Xylene	60.7	62.0
C_9^+	14.5	2.2
Isomerization efficiency, wt %	37	90

Oil and Gas Journal

over Erionite and Type Y Zeolites (*56*)

Conversion,[a] *%*	*Isomerization, %*	*Disproportionation, %*	*Selectivity*[b]
42.9	33.4	9.5	3.5
16.3	15.6	0.8	20.5
44.8	41.3	3.45	12.0
27.8	27.2	0.69	39.3
29.2	28.6	0.56	51.1

[c] Process conditions: Pressure 450 psig; WHSV 2.0 gm/gm/hr; H_2/xylene mole ratio 5.0.

are therefore poisoned by the residual sodium. A later study suggested that the calcination step prior to catalytic evaluation may redistribute the sodium cations to the more active or more easily accessible catalytic sites (*39*).

A liquid phase xylene isomerization process using a zeolite catalyst has recently been proposed by Mobil Oil (*58*). New zeolite catalysts AP (aromatic processing) of undisclosed composition are claimed to be intrinsically more selective for isomerization than is a rare earth exchanged X zeolite. Data in Table XXIII show that at approximately the same ortho-xylene conversion level, less of the product from the AP zeolite catalyst is transalkylated and consequently more of the feed is isomerized to the meta- and para-xylenes. The AP catalyst is probably based upon a zeolite type previously referred to as zeolite Ω (*59*).

The Mobil process is designed to operate at low temperatures so that liquid phase conditions prevail. A comparison was made of catalyst deactivation rates in the liquid and vapor phase using a rare earth exchanged X zeolite. The rate of catalytic deactivation in the liquid phase (Table XXIV) is less than that found in the vapor phase. It had previously been claimed that the liquid phase may act as a solvent for catalysts deposits (*60*).

A process scheme for the production of para-xylene is shown in Figure 22. Since ethylbenzene is not converted to the xylenes over zeolite catalysts, most of it is removed by fractionation of the feed allowing the xylene loop to operate essentially ethylbenzene free. The reactor feed is from the para-

Table XXIV. Aging of REX Catalyst in *o*-Xylene Isomerization (*58*)

	Vapor Phase		*Liquid Phase*	
Temperature, °F	450		350	
Pressure, psig	0		400	
Liquid hourly space velocity	2.6		0.25	
H_2/xylene molar ratio	10		0	
On-stream time, hr	0.4	5.2	2.5	14.5
o-Xylene conversion, wt %	17.5	4.2	27.6	27.1
m- and *p*-Xylene make, wt %	13.3	2.4	17.0	21.8
Isomerization efficiency, wt %	68	57	62	80

Oil and Gas Journal

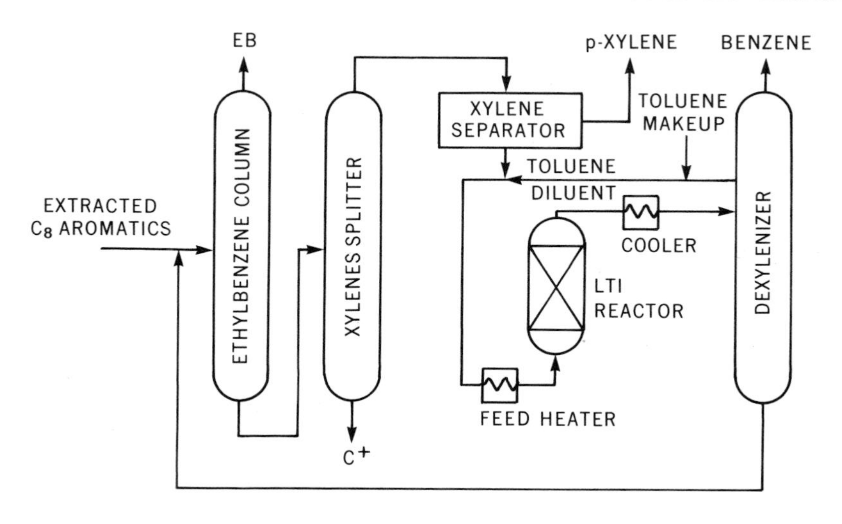

Preprints, Division of Petroleum Chemistry, American Chemical Society

Figure 22. Mobil low temperature xylene isomerization process (58)

xylene separator and is consequently lean in para-xylene. The initial reactor temperature is 400°F and is raised incrementally with time on stream to maintain 95-98% approach to equilibrium until 500°F is reached. At this point the catalyst is regenerated. The total system pressure is 300 psig, and the WHSV is 3. The process scheme for ortho- and para-xylene production is similar to that for the para-xylene process. To inhibit xylene losses *via*

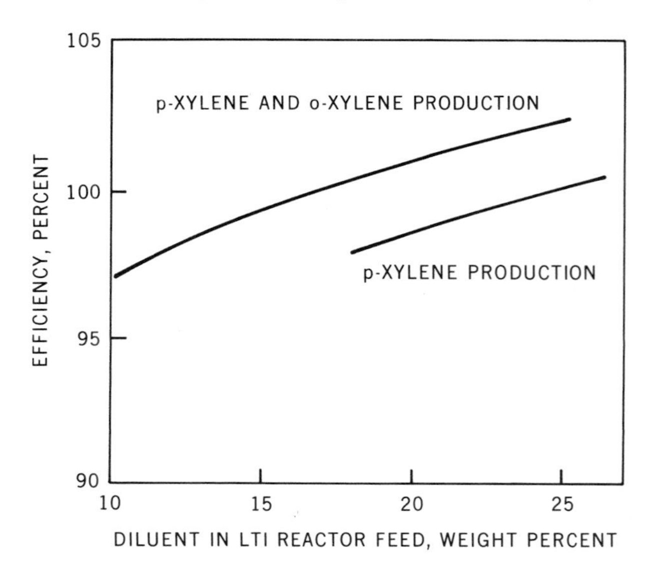

Preprints, Division of Petroleum Chemistry, American Chemical Society

Figure 23. Effect of toluene diluent concentration (58)

disproportionation reactions and yet maintain close to equilibrium conversions, toluene is used as a reactor feed diluent in concentrations from 10 to 20 wt %. The diluent is removed downstream of the reactor and is recycled. The effect of toluene as diluent may be seen from Figure 23. That ethylbenzene may also be used as a diluent is shown in Figure 24 but these data show that it results in lower efficiency than toluene. The equilibrium xylene effluent is recycled back to the ethylbenzene stripper for blending with fresh feed.

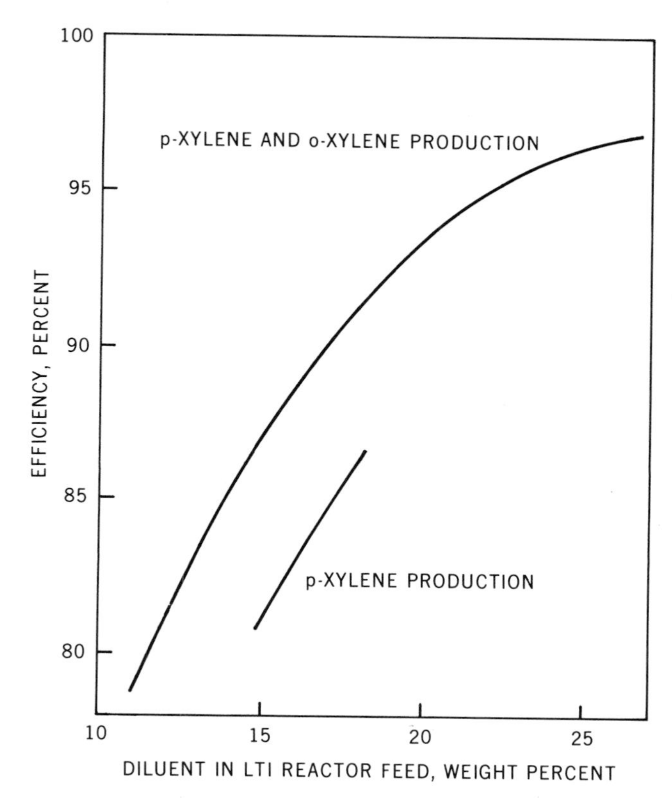

Preprints, Division of Petroleum Chemistry, American Chemical Society

Figure 24. Effect of ethylbenzene diluent concentration (58)

Alkylation

The alkylation of paraffins and aromatics are important processes in both the petroleum and chemical industries. The production of alkylate, highly branched paraffins, as high octane components for gasoline is practiced on a large scale using concentrated sulfuric acid or anhydrous hydrogen fluoride. Ethylbenzene and cumene (isopropylbenzene), important basic materials in the chemical industry, are manufactured by the alkylation of benzene with ethylene and propylene using Freidel–Crafts type catalysts. The two types of alkylation processes will be considered separately.

Aromatic Alkylation. The three major industrial uses for aromatic alkylation are for the manufacture of ethylbenzene as an intermediate for styrene, cumene as an intermediate in phenol synthesis, and long chain alkylated benzene as a detergent intermediate. The most widely used catalyst in the manufacture of ethylbenzene is aluminum chloride, which accounts for over several million pounds of product daily. This process, operating at about 60 psig, 120°C and, using a benzene to ethylene feed ratio of 2.5, has proved to be a most efficient process over many years of operation. However, several difficulties are inherent in such processes using Friedel–Crafts type catalysts. Construction must be of corrosion resistant materials and the feedstocks must be of high purity to reduce catalyst consumption. The corrosion problems encountered during the use of Friedel–Crafts catalysts arise not only from the catalyst itself but from catalyst–reactant complexes. This latter source of corrosion is usually the more severe and is sometimes only controlled by the continuous replacement of those parts in contact with these complexes. Feed impurities must also affect the formation of these corrosive complexes since it has been found that phosphoric acid cumene units operate satisfactorily in one location and not in another. Aluminum chloride is soluble in the product to an extent of about one part in about 200 parts ethylbenzene; this presents acid effluent disposal problems since the product must be washed with water to remove the dissolved catalyst. The formation of catalyst hydrocarbon sludge also contributes to acid consumption. In these applications, where difficulties are being experienced either from corrosion or from the disposal of acid effluents, the commercial use of zeolite catalysts in these alkylation reactions are currently being evaluated.

Cumene is manufactured both by an aluminum chloride slurry process similar to that for ethylbenzene, previously mentioned, and by a process using a phosphoric acid supported catalyst. The latter is conducted in a fixed-bed operating at about 350°–400°F and at about 500 psig; the feed is a mixture of propylene, propane, and benzene. The principal disadvantages of the phosphoric acid process are severe feed pretreatment, relatively short, six to nine month, catalyst life and the non-regenerability of the catalyst.

Studies by Venuto and his associates at Mobil Oil and by Minachev and his co-workers in the USSR have shown that a variety of aromatic alkylation reactions occur over zeolite catalysts (*61, 62, 63, 64*). The zeolites catalyze alkylation reactions at much lower temperatures than amorphous silica–alumina catalysts and, as a result, many of the undesirable side reactions that occur at the higher temperatures are eliminated. This point is particularly well illustrated by phenol and thiophenol alkylation. Though more active than amorphous catalysts, the zeolites do not exhibit the high activity associated with the mineral acids. This has been attributed to additional energy barriers and entropy requirements imposed by adsorption–desorption and intracrystalline diffusion (*65*).

The type of alkylation reactions catalyzed by the zeolites are analogous to those obtained using typical Friedel–Crafts type catalysts. Studies using olefins as the alkylating agent are numerous, and these alkylations have been effected using from C_2 to C_{10} olefins (*61, 62, 63, 64*). Ethylation requires temperatures in excess of 150°C, but propylene and the butylenes can alkylate in the liquid phase at room temperature. The alkylating agent need not be an olefin since alcohols such as methanol and alkyl halides have successfully been used (*61*). Hydrogen chloride liberated during alkylation by an alkyl halide has no effect on the zeolite structure if the system is reasonably anhydrous. The evolution of water from alcohols during alkylation presents

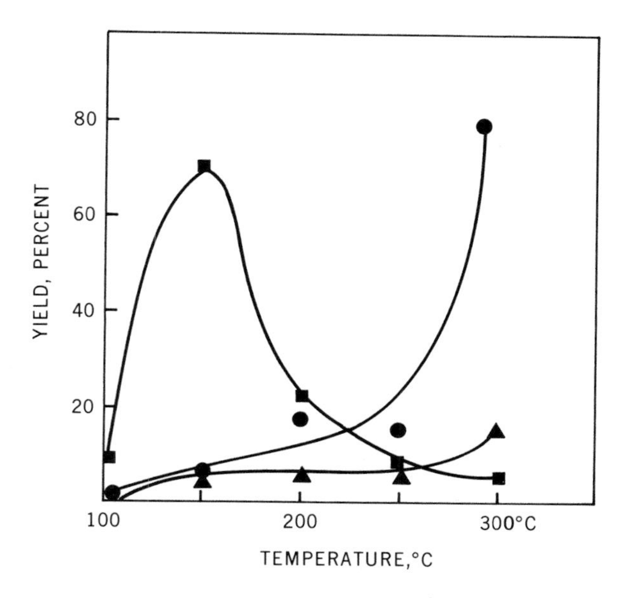

Izvestia Akademii Nauk

Figure 25. The effect of temperature on yield of:
■ sec-*butylbenzene*
● *ethylbenzene*
▲ *diethylbenzenes* (67)

a problem only if the reaction temperature is not sufficiently high to ensure continuous water desorption.

An investigation into the reaction of benzene with ethylene over a rare earth exchanged zeolite showed that *sec*-butylbenzene as well as the ethylbenzene was formed (*67*). The yield of *sec*-butylbenzene depends on the process conditions used; at 150°C, 25–30 atm, and at an aromatic–olefin ratio of 4 to 1, 70% of the product is *sec*-butylbenzene. As may be seen in Figure 25, the yield of ethylbenzenes increases with increasing temperature, and at 300°C the ethylbenzene and diethylbenzene amount to 95% of the product, and the concentration of *sec*-butylbenzene falls to 5%.

The *sec*-butylbenzene is derived from the dimerization of the ethylene prior to alkylation, according to the following scheme.

$$H^+ + CH_2{=}CH_2 \longrightarrow CH_3\overset{+}{C}H_2$$

$$CH_3\overset{+}{C}H_2 + CH_2{=}CH_2 \longrightarrow CH_3CH_2\text{—}\oplus(CH_2\text{—}CH_2) \longrightarrow CH_3\overset{+}{C}HCH_2CH_3$$

$$CH_3\overset{+}{C}HCH_2CH_3 + C_6H_6 \longrightarrow [H(H^+)C_6H_5\text{—}C(CH_3)(C_2H_5)H] \longrightarrow C_6H_5\text{—}C(CH_3)(C_2H_5)H$$

That the maximum amount of coke deposition coincides with maximum yield of *sec*-butylbenzene in the product corroborates the proposed mechanism.

The formation of *sec*-butylbenzenes from ethylene and benzene also takes place over a palladium oxide containing type Y zeolite (*44*). A study was later carried out on the effect of palladium loading on the amount of *sec*-butylbenzene in the product (*68*). The zeolite used was a manganese back exchanged type Y, SiO_2/Al_2O_3 molar ratio 4.8, and the experiments were carried out in a fixed bed reactor at atmospheric pressure and at 350°F, conditions at which 100% conversion of the ethylene occurs. With no palladium loading, the catalyst produced the stoichiometric amount of ethylbenzene, and as the amount of palladium increases to 0.25 wt %, the concentration of *sec*-butylbenzene increases to about 95% of the product. No further increase in the concentration of the butylbenzene is found with increasing palladium content as the data in Figure 26 show. The palladium was loaded by ion exchange using $Pd(NH_3)_4{}^{2+}$, and the zeolite was calcined before use. The palladium was not reduced in hydrogen and thus was present on the zeolite in the oxide form.

Zeolites can successfully catalyze the alkylation of acid-sensitive compounds. The principal complications encountered during the alkylation of one of these compounds, phenol, using an acid catalyst are the formation of ethers and the tendency of the hydroxyl group to complex with the catalyst. None of these secondary reactions occurs when alkylating phenol with olefins using a rare-earth exchanged type X at 180°–210°C (*61*). As in the alkylation of other substituted benzenes, ortho and para substitutions predominate, in accordance with Brown's selectivity rules. Meta isomers also appear when

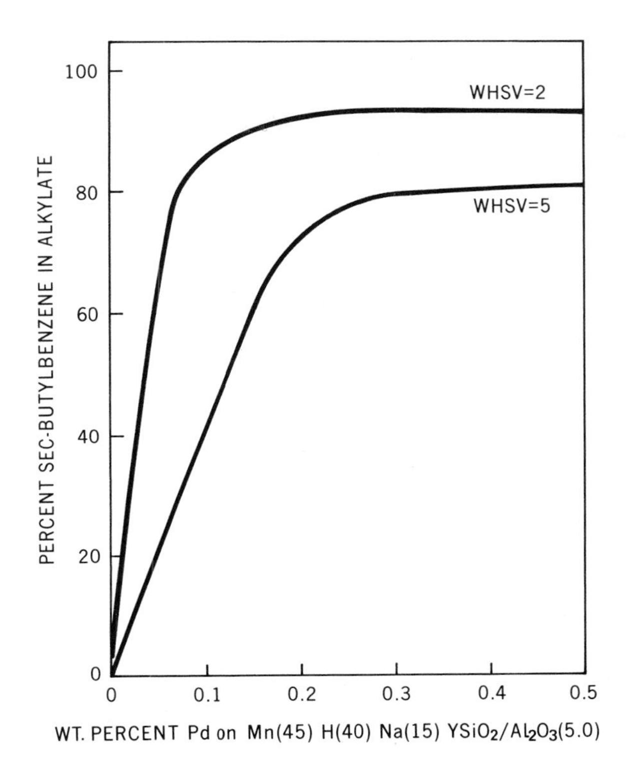

Figure 26. Change in sec-*butylbenzene formation with palladium loading* (68)

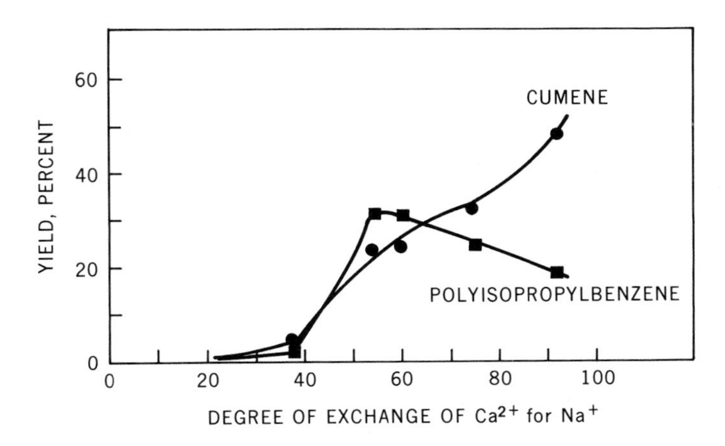

Figure 27. The effect of Ca^{2+} exchange of NaY on alkylation activity (63)

the alkylated product is in prolonged contact with the catalyst or when high reaction temperatures are employed. The appearance of the meta isomers in the product from the alkylation of phenol with isobutylene was shown to result from the isomerization of the ortho and para isomers *via* a transalkylation mechanism (*54*).

The sodium or other alkali metal cation forms of the zeolite are inactive and alkylation activity is only developed when these cations are replaced by polyvalent metal cations, ammonium ions, or other hydrogen ion precursors. Figure 27 shows the effect of replacing the sodium cations in a type Y zeolite, SiO_2/Al_2O_3 molar ratio 4.2, with calcium ions on alkylation activity in the benzene propylene reaction (*63*). Little activity is exhibited by samples having degrees of calcium exchange below 40 to 45%. Above this level a marked increase in activity is observed. Increasing contents of calcium exchange are accompanied by increasing alkylation activity. A similar effect on activity for propylene alkylation was observed with ammonium exchange followed by calcination at 500°C (*68*). Little activity is observed until about 55% decationization had been obtained, after which a linear relation existed between activity and extent of decationization. The change in activity with degree of calcium exchange indicates that the cations in the type Y are not equivalent (*63*). Those cations which are the first and most readily replaced may be in sites either inaccessible to or not associated with catalytic activity and the remaining 50 to 55% of the cations may be at sites with which activity is associated. An alternative explanation could be that the remaining sodium cations redistribute themselves during the calcination step prior to catalyic evaluation. Thus, below about 50% cation exchange level there are sufficient residual sodium cations to replace or nullify any contribution of the exchanged cations.

The ion-exchanged forms of the zeolites found to be the most active alkylation catalysts are the rare-earth and hydrogen forms. A study of the relationship between calcination temperature and catalytic activity in the benzene–ethylene reaction is shown in Figure 28 (*61*). Although the maximum alkylation activities of the three catalysts evaluated are similar, the calcination temperatures at which maximum activity is developed are very different—400° for rare-earth exchanged type X, 250° for rare-earth exchanged type Y, and 550°–600°C for a deamminated ammonium exchanged type Y.

The catalytic activity of the rare-earth exchanged forms has been attributed to hydrolysis of the rare-earth cations which may be depicted as follows (*62*).

$$RE^{3+}\ (H_2O) \leftrightharpoons RE^{2+}\ (OH) + H^+$$

This hydrolysis reaction is supported by infrared studies. A similar proposal has been made to explain the catalytic activities of the divalent metal cation exchanged forms of the zeolites (*69*). By analogy with the conventional acid systems, alkylation over the zeolites is proton-catalyzed and the experimental

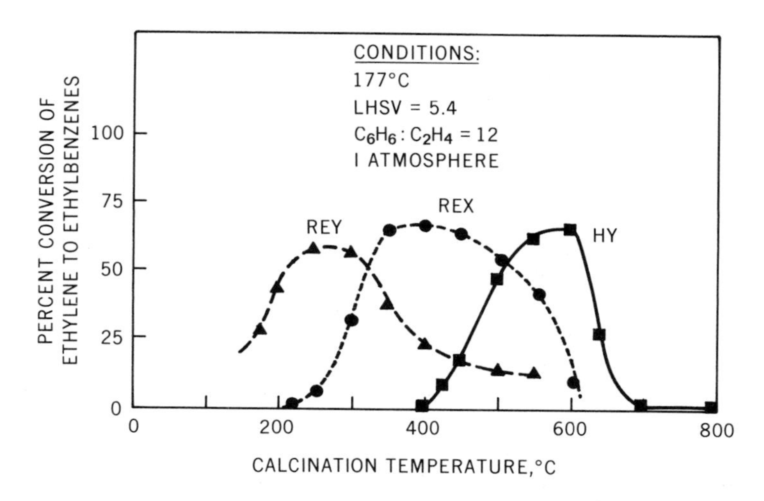

Journal of Catalysis

Figure 28. Alkylation activity as a function of calcination temperature (61)

maxima shown in Figure 28 represent the point at which maximum proton or hydroxyl concentrations occur. The decrease in activities beyond the maxima, which occur at higher temperature, is attributed to loss of protons by the dehydroxylation of both the hydrogen and rare-earth exchanged forms. Experimental evidence showing that the completely dehydroxylated form retains its catalytic activity is contrary to this conclusion, but these results are considered later in a discussion of catalyst deactivation (*70, 71*).

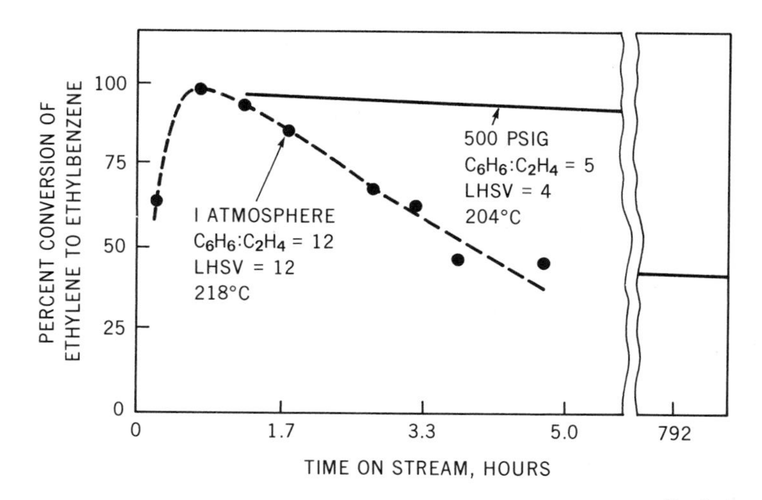

Chemtech

Figure 29. Effect of reaction pressure on alkylation of benzene with ethylene (72)

The commercial utilization of zeolites as alkylation catalysts has not been as rapid as in other processes. One of the reasons is that the existing catalysts for the two major aromatic alkylation applications, ethylbenzene and cumene, have been eminently satisfactory. Only recently have environmental restrictions on catalyst disposal and effluent control directed attention to alternate catalyst systems. Another important reason for not exploiting zeolites is that the deactivation characteristics exhibited by these catalysts have not been commercially acceptable until recently.

The rate of deactivation of a zeolite catalyst in the vapor phase alkylation of benzene with ethylene is very high, as shown in Figure 29 (*72*). Conducting the reaction in the liquid phase by increasing the system pressure from atmospheric to 500 psig dramatically improves the aging characteristics of the catalyst, but even this improved rate of deactivation is not commercially satisfactory. Analyses of the desorbed alkylate product together with that entrained in the intracrystalline zeolite pores of a rare-earth exchanged type X catalyst deactivated during gas phase operation show the average molecular weight of the saturated catalyst residue to be about twice that of the alkylate product (*72*). The hydrocarbons remaining within the pores of the zeolite were found to be similar to those when ethylene alone was passed over the zeolite. The formation of the high molecular weight material is accompanied by the evolution of low molecular weight hydrocarbons.

A study of the role of surface species formed by the adsorption of olefins onto the zeolite provides further insight into the mechanism of deactivation of zeolite alkylation catalysts (*71*). These surface species, Figures 30 and 31,

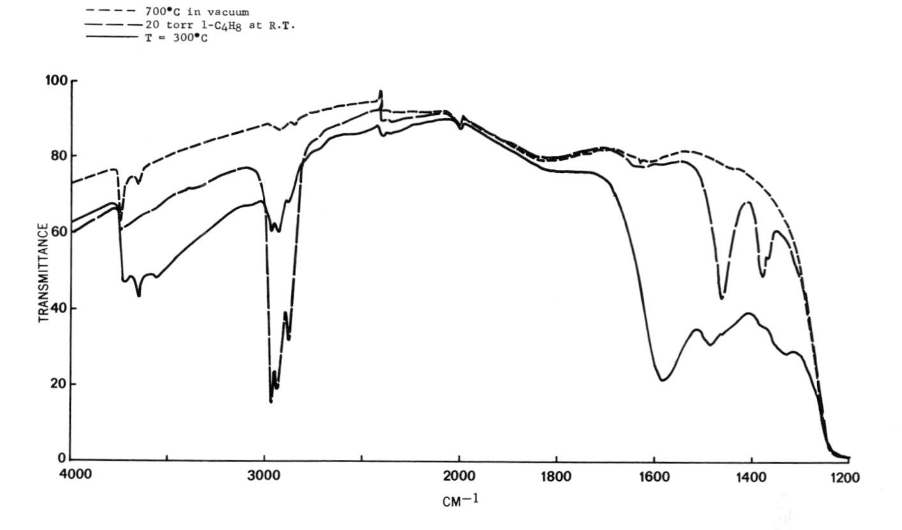

Proceedings of the Third International Congress on Molecular Sieves

Figure 30. Reaction of 1-butene with dehydroxylated ammonium-exchanged Y (71)

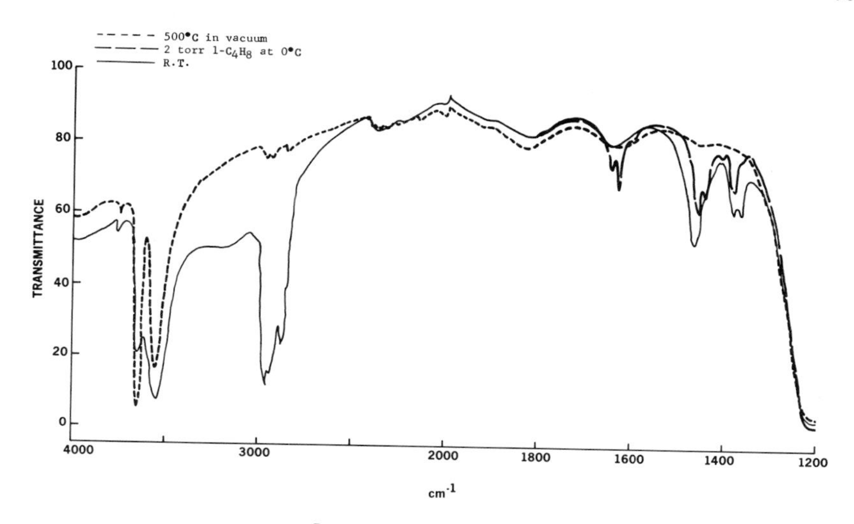

Proceedings of the Third International Congress on Molecular Sieves

Figure 31. Isomerization of 1-butene over deamminated ammonium-exchanged Y (71)

characterized by the absence of unsaturated ═CH and C═C stretching bands in the infrared spectrum and by the presence of saturated C—H bending and stretching bands, do not take part in the alkylation reaction. The introduction of the olefin prior to the addition of benzene to the catalyst does not result in alkylation. The olefin is converted to the surface species; this can only be removed from the zeolite by thermal treatment. At 300°C, the infrared bands characteristic of the surface species decrease in intensity and are replaced by a broad band at 1630 cm^{-1} which has been attributed to highly unsaturated polymeric material. The formation of the 1630 cm^{-1} band is accompanied by the evolution of low molecular weight saturated hydrocarbons, principally isobutane. However, when the olefin is added to the zeolite containing preadsorbed benzene, alkylation takes place and the benzene is quantitatively converted to its alkyl derivative, as the data in Figures 32 and 33 show. Thus the high molecular weight hydrocarbons found entrapped in the zeolite pore system are probably a result of deactivation and not the cause of it.

The higher the olefin to aromatic ratio of the feed, the higher the rate of catalyst deactivation. Thus, the preferential adsorption of the olefin on these zeolite forms may increase the relative concentration of the olefin from the feed concentration to a substantially higher value on the zeolite. In the vapor phase, the aromatic loading is significantly lower than that in the liquid phase, and the occurrence of irreversible olefin adsorption is significantly enhanced in the former case compared with the latter. A balance of olefin addition rate against alkylation rate must be achieved if stable catalytic activity is to be obtained. Operating at high aromatic to olefin ratios and at low space velocities, however, does not offer advantage over existing com-

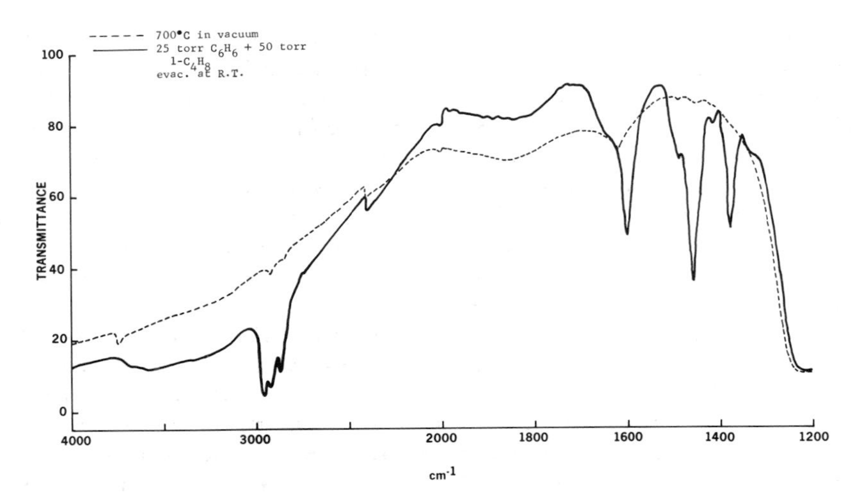

Proceedings of the Third International Congress on Molecular Sieves

Figure 32. Formation of sec-*butylbenzene on dehydroxylated ammonium-exchanged Y* (71)

mercial catalysts. Recently, a combination of catalyst and process development has overcome this deactivation problem and a benzene alkylation process is currently being evaluated on a commercial scale.

The study of the surface species on the zeolite in alkylation reactions also revealed that complete conversion of benzene to its alkyl derivative takes place as easily over a completely dehydroxylated type Y as over the same

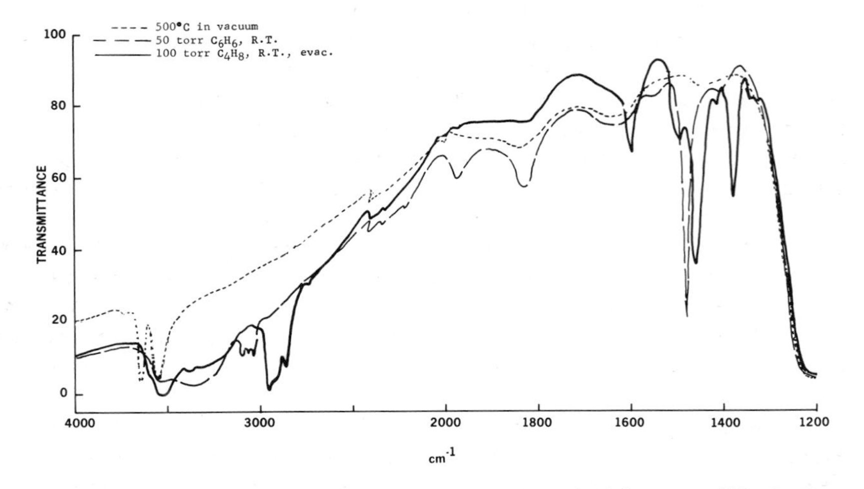

Proceedings of the Third International Congress on Molecular Sieves

Figure 33. Formation of sec-*butylbenzene on deamminated ammonium-exchanged Y* (71)

zeolite containing a high concentration of hydroxyl groups (*71*). These data are not in agreement with a proton-catalyzed mechanism but support earlier experimental data by Pickert *et al.* and may suggest that the active site for alkylation is also that at which irreversible olefin adsorption and, in consequence, deactivation takes place (*70*). The evaluation procedures used to differentiate between catalyst activities may be as much a reflection of deactivation characteristics as they are of catalytic activity. Thus the optimum catalyst activities shown in Figure 28 might be experimental artifacts.

The disclosure of three patents to Mobil Oil supports such a conclusion (*73, 74, 75*). These patents state the aging characteristics of type X and Y zeolite catalysts in liquid phase operation to be unsatisfactory for commercial application. The use in the vapor phase of less active catalysts, derived from ZSM-5 and ZSM-11, is proposed as a method of obtaining longer catalyst life. Since operation is vapor phase, loss in catalyst activity can be offset by incrementally raising the operating temperature. A demonstration unit for ethylene alkylation has been constructed based upon one of these catalysts and is currently being evaluated.

Isoparaffin Alkylation. Isoparaffin alkylation is an established refinery process for the manufacture of high octane gasoline components. In 1971, capacity in the United States was over 860,000 BPD which was 6.2% of crude run. With the advent of lead-free gasoline, the importance of this process to the refinery will continue to exist (*76*). Alkylate is a mixture of highly branched paraffins that are added to the gasoline pool to improve its octane number and volatility. The motor octane (F-2) rating of the products from alkylating isobutane with propylene, butylenes, and amylenes are 89, 93, and 90, respectively. The commercial processes use either concentrated sulfuric acid or anhydrous hydrogen fluoride. Since isobutane, unlike the olefins, is not very soluble in these acids (0.1% in H_2SO_4 and about 3% in HF), mixing intense enough to form an emulsion is required to reduce the undesirable olefin side reactions (*77*). These latter reactions, together with some water that inevitably is introduced with the feed, dilute the acids causing acid consumption. As might be expected, corrosion problems are encountered in operating such systems, but the use of special materials of construction, such as Monel in the HF process, enables satisfactory commercial operation to be realized, although at the expense of high installation costs. The operating temperature of the sulfuric acid process is about 10°C. Lower operating temperatures are not practical since the power requirement for mixing increases as the viscosity increases. At higher temperatures, oxidation by the sulfuric acid causes undesirable effects. The operating temperature of the hydrogen fluoride process is 25°–20°C which is an advantage over the sulfuric acid process in that refrigeration costs are less.

Information on the isoparaffin alkylation reaction using zeolites is principally limited to some studies by Sun Oil using rare-earth exchanged type Y zeolites and to the patent literature (*78, 79, 80, 81, 82, 83*). The former studies were carried out in the liquid phase using a stirred reactor both batch-wise and in continuous operation. Table XXV shows that a typical

Table XXV. Isoparaffin–Olefin Alkylation with Zeolite Catalyst: Isobutane-2-Butene Product Composition (79)

	REY (wt %)	*H_2SO_4 (vol %)*
i-Pentane	5.78	4.16
2,3-DMB + MP	3.46	4.58
2,4-DMP	3.58	2.37
2,3,3-TMB	0.10	
All MHx	0.32	
2,3-DMP	1.26	1.38
2,2,4-TMP	17.70	30.64
All DMHx	8.89	9.02
2,2,3-TMP	3.97	
2,3,4- } 2,3,3-TMP }	48.16	41.55
2,2,5-TMHx	0.71	1.88
Other C_9^+	6.06	4.41

Meeting, "American Petroleum Institute"

alkylate composition from a zeolite catalyst is similar to that obtained with sulfuric acid, the product being saturated and characterized by a preponderance of C_8 isomers in a spectrum of C_5 to C_9 hydrocarbons. The principal difference between the alkylate compositions in Table XXV, which were produced from isobutane and 2-butene, is the lower 2,2,4-trimethylpentane content of the alkylate from the zeolite.

The reaction mechanism is generally accepted as proceeding *via* the initial step of olefin protonation and may be represented as follows.

$$\mathrm{C{-}C{=}C{-}C} + \mathrm{H^+} \rightarrow \mathrm{C{-}C{-}\overset{+}{C}{-}C}$$

$$\mathrm{C{-}C{-}\overset{+}{C}{-}C} + \mathrm{C{-}\underset{\underset{C}{|}}{\overset{\overset{C}{|}}{CH}}} \rightarrow \mathrm{C{-}C{-}C{-}C} + \mathrm{C{-}\underset{\underset{C}{|}}{\overset{\overset{C}{|}}{C^+}}}$$

The *tert*-butyl cation then perpetuates the chain of reactions leading to alkylate formation.

$$\mathrm{C{-}\underset{\underset{C}{|}}{\overset{\overset{C}{|}}{C^+}}} + \mathrm{C{-}C{=}C{-}C} \rightarrow \mathrm{C{-}\underset{\underset{C}{|}}{\overset{\overset{C}{|}}{C}}{-}\overset{\overset{C}{|}}{C}{-}\overset{+}{C}{-}C}$$

The C_8 carbonium ion can either react with isobutane to form an isoparaffin

```
   C  C             C            C  C              C
   |  |             |            |  |              |
C—C—C—C—C  +  C—C—H  →  C—C—C—C—C  +  C—C+
   |  +             |            |                 |
   C                C            C                 C
```

or add on another olefin to form a C_{12} carbonium ion. This latter ion can either saturate to form a C_{12} paraffin, crack to yield an olefin and another carbonium ion, both of lower molecular weight, or add on another olefin to form a C_{16} carbonium ion. That olefin protonation is the initiating step in the reaction is supported by the data in Figure 34 which shows that the

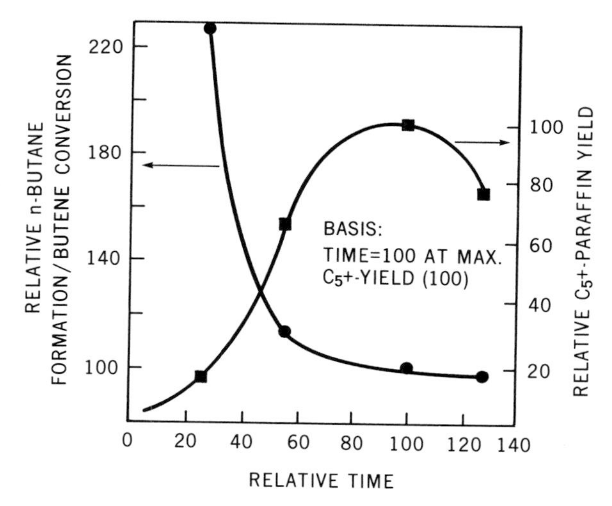

Preprints, Division of Refining, American Petroleum Institute

Figure 34. Variation of alkylate yield and n-*butane formation with time* (79)

formation of *n*-butane is at a maximum when the alkylate yield is lowest and rapidly decreases with increasing alkylate formation (*79*). This proposed sequence of reactions that constitutes isoparaffin alkylation emphasizes the role of hydrogen transfer reactions required for alkylate formation. Not only is it surprising that the zeolite possesses high enough activity to catalyze this reaction but also that it has the ability to fulfill the hydrogen transfer requirements. The function of the catalyst in these reactions is not too well understood. It has been shown that for the acid-catalyzed systems, the presence of red oils, highly olefinic cyclic hydrocarbons with two to five cyclopentyl groups, are essential for the hydrogen ion transfer reactions to occur which are required to perpetuate the alkylation process.

The zeolites used by Kirsch and his associates in their studies were type Y which had first been ammonium exchanged to about 70% and subsequently back-exchanged with rare-earth cations (*78*). The effect of increasing degrees of rare-earth back exchange was evaluated in the reaction of isobutane with 2-butene. As can be seen from Figure 35, an increase in rare-earth content results in an increase in C_5+ yield and C_8 content of the alkylate.

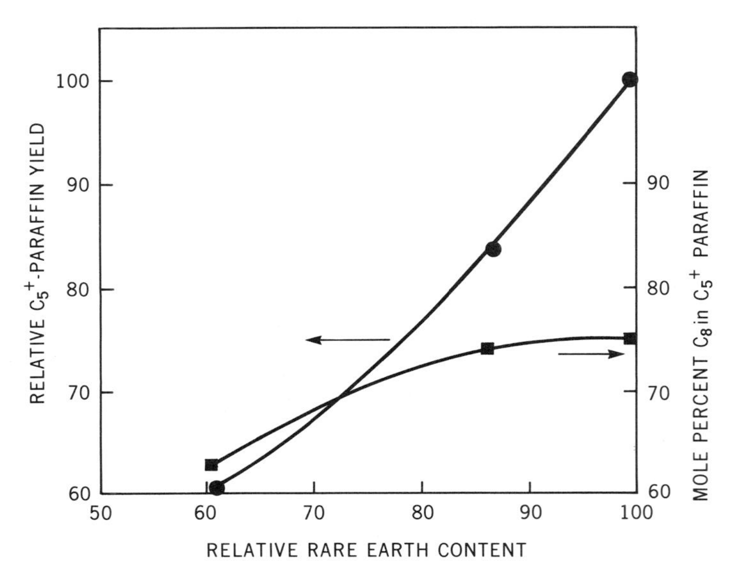

"Symposium, American Institute of Chemical Engineering"

Figure 35. Effect of rare earth content of Y zeolite on alkylate yield and quality (81)

No difference in alkylate composition was reported when using either 1-butene or 2-butene, and it was concluded that isomerization of the former to the latter occurs prior to alkylation (*79*). This interconversion of the butenes has also been observed in aromatic alkylation with zeolites (*71*). The 2,2,4-trimethylpentane content of a zeolite derived alkylate is less than that derived from the sulfuric acid system. Other studies under different operating conditions using normal butenes as the olefin show that the 2,2,4-isomer rarely exceeds 35 mole % of the trimethylpentane fraction and is often between 15 and 25 mole % (*81*). When isobutylene or diisobutylene was used as the olefin, the 2,2,4-isomer constituted about 60% of the trimethylpentane fraction. Olefins other than the butylenes give rise to some C_8 paraffins as well as products that would be expected from the carbon number of the olefin (Table XXVI). The 2,2,4-trimethylpentane content of the C_8 fractions is high and may be accounted for by the coincident self-condensation of isobutane which is thought to occur as follows:

$$C-\overset{C}{\underset{C}{\overset{|}{\underset{|}{C}}}}-H \rightleftharpoons C-\overset{C}{\underset{C}{\overset{|}{\underset{|}{C^{+}}}}} + H^{-} \qquad C-\overset{C}{\underset{C}{\overset{|}{\underset{|}{C^{+}}}}} \rightleftharpoons C{=}\overset{C}{\underset{C}{\overset{|}{\underset{|}{C}}}} + H^{+}$$

$$C-\overset{C}{\underset{C}{\overset{|}{\underset{|}{C^{+}}}}} + C{=}\overset{C}{\underset{C}{\overset{|}{\underset{|}{C}}}} \rightleftharpoons C-\overset{C}{\underset{C}{\overset{|}{\underset{|}{C}}}}-C-\overset{C}{\underset{+}{\overset{|}{C}}}-C$$

$$C-\overset{C}{\underset{C}{\overset{|}{\underset{|}{C}}}}-C-\overset{C}{\underset{+}{\overset{|}{C}}}-C + C-\overset{C}{\underset{C}{\overset{|}{\underset{|}{C}}}}-H \rightleftharpoons C-\overset{C}{\underset{C}{\overset{|}{\underset{|}{C}}}}-C-\overset{C}{\overset{|}{C}}-C + C-\overset{C}{\underset{C}{\overset{|}{\underset{|}{C^{+}}}}}$$

A later study using a ^{14}C tracer technique showed that only 50% of the 2,2,4-isomer from the isobutane–2-butene reaction was formed by direct alkylation and the other 50% was formed by this self-condensation reaction (*82*). The ability of the zeolites to catalyze the alkylation of isobutane with ethylene may offer a distinct advantage over the current acid systems. Sulfuric and hydrofluoric acids react with ethylene to form ethylhydrogen sulfate and ethylfluorides, respectively, thus preventing their use with this olefin. Attempts have been made to use aluminum chloride commercially, but the corrosion problems encountered in this system are extremely severe. The product distri-

Table XXVI. Alkylation of Isobutane Using Rare Earth Y Catalysts (*81*)

Product	*Propylene*	*1-Pentene*
C_5^+ paraffin distribution, mole %		
C_5	13.6	18.1
C_6	4.8	3.3
C_7	56.3	5.0
C_8	9.1	12.3
C_9^+	16.2	61.3
C_7 paraffin distribution, mole %		
2,3-DMP	98.2	—
Other DMP + TMB	1.4	—
MHx	0.4	—
TMP distribution, mole %		
2,2,4-	69.3	62.5
2,2,3-	0.3	0.4
2,3,4-	14.1	17.3
2,3,3-	16.3	19.8

"Symposium, American Institute of Chemical Engineering"

Table XXVII. Alkylation[a] of Isobutane with Ethylene Using Type Y (*84*)

Alkylate Composition, Weight %	
C_5	
Isopentane	0.2
C_6	
2,3-Dimethylbutane + 2-Methylpentane	46.5
3-Methylpentane	6.6
C_7	0.7
C_8	34.8
C_9^+	11.2
Trimethylpentane distribution	
2,2,4-	50.4
2,2,3-	6.4
2,3,4-	19.0
2,3,3-	24.2

[a] Process conditions: temperature = 150°F, pressure = 500 psi.

bution of an ethylene alkylate from a type Y zeolite is shown in Table XXVII, the C_6 fraction is principally 2,3-dimethylbutane (*84*). The relatively large amount of C_8 product as compared to that from C_3 and C_5 olefins is probably caused by the dimerization of the ethylene prior to alkylation as well as to the self-condensation of isobutane.

An induction period was observed when isobutane alkylation with 2-butene was carried out in a batchwise stirred reactor. The increase in yield with time which is shown in Figure 34 is analogous to the isobutane activator formed over a period of time in the sulfuric acid system. The hydrocarbons in the C_6, C_7, and C_9 fractions increase slightly at the expense of the C_8 fraction, but the trimethylpentane isomer distribution remains constant during this induction period, indicating that paraffins other than C_8 originate from a C_8 precursor by secondary reactions. Prolonged contact with the catalyst after maximum yield is reached results in alkylate degradation. This induction period is also observed in experiments using a continuous stirred reactor. Since no olefin was found in the reactor effluent during the induction period, some olefin must be retained on the catalyst. An examination of spent catalysts indicate that they contain hydrocarbons having equivalent carbon numbers of C_{19} to C_{23} (*82*). This retained material may well correspond to the C_{12} species reported by Hoffman for the sulfuric acid system and the butene tetramers corresponding to a C_{16}–C_{18} species, found by Hirschler in butene isomerization (*85, 86*).

The proposal that a hydrocarbon catalyst specie takes part in the alkylation reaction makes the appearance of the C_5 to C_{12} products in the alkylate more reasonable. Some C^{14} studies have shown that, with the exception of the 2,2,4-trimethylpentane, each mole of liquid product incorporates an average of one mole of the reactant butene (*82*).

The alkylate quality from the use of zeolite catalysts matches closely that obtained from the sulfuric acid and hydrofluoric acid process. Research octane numbers F-1 clear of between 94 and 98 have been reported (*81*).

Literature Cited

1. Harnett, C. G., *37th Meetg. Amer. Petrol. Inst., Div. Refining, New York, May 1972,* No. 48-72.
2. Schulz, H., Weitkamp, J., *Ind. Eng. Chem., Prod. Res. Dev.* (1972) **11,** 46.
3. Coonradt, H. L., Garwood, W. E., *Ind. Eng. Chem., Proc. Design Dev.* (1964) **3,** 38.
4. Langlois, G. E., Sullivan, R. F., *Adv. Chem. Ser.* (1970) **97,** 38.
5. Myers, C. G., Munns, G. W., *Ind. Eng. Chem.* (1958) **50,** 1727.
6. Voorhies, A., Hatcher, W. J., *Ind. Eng. Chem., Prod. Res. Dev.* (1969) **8,** 4361.
7. Weitkamp, J., Schulz, H., *J. Catal.* (1973) **29,** 361.
8. Weisz, P. B., *Chemtech* (1973) 498.
9. Schulz, H., Weitkamp, J., *Amer. Chem. Soc., Div. Petrol. Chem., Preprints* **17** (1972) No. 4G-84.
10. Beecher, R., Voorhies, A., Eberly, P., *Amer. Chem. Soc., Div. Petrol. Chem., Preprints* (1967) B-5.
11. Egan, C. J., Langlois, G. E., White, R. J., *J. Amer. Chem. Soc.* (1962) **84,** 1204.
12. Hansford, R. C., U.S. Patent **3,364,135.**
13. Breck, D. W., "Zeolite Molecular Sieves," Wiley-Interscience, New York, 1974.
14. Cole, J. F., Kouwenhoven, H. W., *Adv. Chem. Ser.* (1973) **121,** 583.
15. Ward, J. W., Hansford, R. C., Reichle, A. D., Sosnowski, J., *38th Meetg., Amer. Petrol. Instit.,* 1973.
16. Bolton, A. P., unpublished data, Union Carbide Corp., Linde Division.
17. Yan, T. Y., *J. Catal.* (1972) **25,** 204.
18. Plank, C. J., *Proc. Int. Congr. Cat. III* (1964) **1,** 727.
19. Bolton, A. P., *J. Catal.* (1971) **22,** 9.
20. Hansford, R. C., Reeg, C. P., Wood, F. C., Vaell, R. P., *Petrol. Engr.* (1960) **32,** C7-C12.
21. Duir, J. H., *Oil Gas J.* (1967) **65,** 74.
22. Baral, W. J., Huffman, H. C., *Proc. World Petrol. Congr., 8th, 1972,* PD 12 (1).
23. Peralta, B., Reeg, C. P., Vaell, R. P., Hansford, R. C., *Chem. Eng. Proc.* (1967) **58,** 41.
24. Wood, F. C., Eubank, O. C., Sosnowski, J., *Nat. Petrol. Refin. Assoc. Meetg., April 1968.*
25. Dhondt, R. O., Peralta, B., Young, D. A., Riechle, A. D., *AIChE Meetg., New Orleans, 1969.*
26. Griesmer, G. J., Avery, W. F., Lee, M. N. Y., *Hydrocarbon Proc. Petrol. Refin.* (1965) **44,** 147.
27. Chen, N. Y., Maziuk, J., Schwartz, A. B., Weisz, P. B., *Oil Gas J.* (1968) **66,** 154.
28. Miale, J. N., Chen, N. Y., Weisz, P. B., *J. Catal.* (1966) **6,** 278.
29. Chen, N. Y., Garwood, W. E., *Adv. Chem. Ser.* (1973) **121,** 575.
30. Hamner, G. P., Mason, R. B., U.S. Patent **3,575,846.**
31. Burd, S. D., Maziuk, J., *Hydrocarbon Process.* (1972), 97.
32. Roselius, R. R., Gibson, K. R., Orrinston, R. M., Maziuk, J., Smith, F. A., *Nat. Petrol. Refin. Assoc. Meetg., April 1973.*
33. Kowenhowen, H. W., Van Zijll Langhout, W. C., *Chem. Eng. Prog.* (1971) **67,** 65.
34. Rabo, J. A., Pickert, P. E., Stamires, D. N., Boyle, J., *Int. Congr. Catal., 2nd, Paris, 1960,* No. 104.

35. Rabo, J. A., Pickert, P. E., Mays, R. L., *Ind. Eng. Chem.* (1961) **53,** 733.
36. Ward, J. W., Hansford, R. C., *J. Catal.* (1969) **13,** 364.
37. McDaniel, C. V., Maher, P. K., *Proc. Int. Congr. Mol. Sieves, 1st, London, 1967.*
38. Kerr, G. T., *J. Catal.* (1969) **13,** 114.
39. Kouwenhoven, H. W., *Adv. Chem. Series* (1973) **121,** 529.
40. VanHelden, H. J. A., Kouvenhoven, H. W., Quik, W. C. J., Dutch Patent **6603927** (1967).
41. Eberly, P. E., Kimberlin, C. N., Voorhies, A., *J. Catal.* (1971) **22,** 419.
42. Randhava, S. S., Frost, A. C., unpublished data, Union Carbide Corp., Linde Div.
43. Minachev, Kh., Garanin, V., Isakova, T., Kharlamov, V., Bogomlov, V., *Adv. Chem. Series* (1971) **102,** 441.
44. Bolton, A. P., Lanewala, M. A., *J. Catal.* (1970) **18,** 1.
45. Lanewala, M. A., Pickert, P. E., Bolton, A. P., *J. Catal.* (1967) **9,** 95.
46. Quik, W. C. J., Collins, J. J., *DGMK Meetg., Nuremberg, October 1972.*
47. Lanewala, M. A., Bolton, A. P., Pickert, P. E., U.S. Patent **3,450,644.** Rabo, J. A., Pickert, P. E., Boyle, J. E., U.S. Patent **3,367,885.**
48. Neuzil, R. W., U.S. Patents **3,558,730** and **3,558,732;** *Chem. Eng. News* (1969) 47.
49. Bolton, A. P., Lanewala, M. A., *J. Org. Chem.* (1969) **34,** 3107.
50. Friedel, C., Crafts, J. M., *J. Chem. Soc.* (1882) **1,** 115.
51. Lien, A. P., McCauley, D. A., *J. Amer. Chem. Soc.* (1952) **74,** 6246.
52. Pitts, P. M., Connors, J. E., Leum, L. N., *Ind. Eng. Chem.* (1955) **47,** 770.
53. Bolton, A. P., Lanewala, M. A., Pickert, P. E., *J. Org. Chem.* (1968) **33,** 1513.
54. Bolton, A. P., Lanewala, M. A., Pickert, P. E., *J. Org. Chem.* (1968) **33,** 3415.
55. Csicsery, S. M., *J. Org. Chem.* (1969) **34,** 3338.
56. Best, D. F., Randhava, S. S., unpublished data, Union Carbide Corp., Linde Div.
57. Hansford, R. C., Ward, J. W., *J. Catal.* (1969) **13,** 316.
58. Grandio, P., Schneider, F. H., Schwartz, A. B., Wise, J. J., *Amer. Chem. Soc., Div. Petrol. Chem., Preprints* (1971) **16,** No. 3, B; *Oil Gas J.* (1971) 62.
59. Flanigen, E. M., Kellberg, E. R., Dutch Patent **6,710,729** (1967).
60. Wise, J. J., U.S. Patent **3,377,400.**
61. Venuto, P. B., Hamilton, L. A., Landis, P. S., Wise, J. J., *J. Catal.* (1966) **4,** 81.
62. Venuto, P. B., Hamilton, L. A., Landis, P. S., *J. Catal.* (1966) **5,** 484.
63. Minachev, Kh. M., Isakov, Ya. I., Garanin, V. I., *Dokl. Akad. Nauk S.S.S.R.* (1965) **165,** 831.
64. Minachev, Kh. M., Isakov, Ya. I., Garanin, V. I., *Neftekhimiya* (1966) **6,** 53.
65. Venuto, P. B., Landis, P. S., *Adv. Catal.* (1968) **18,** 259.
66. Becker, K. A., Karge, H. G., Streubel, W.-D., *J. Catal.* (1973) **28,** 403.
67. Minachev, Kh. M., Mortikov, E. S., Leontev, A. S., Masloboev-Shvedov, A. A., Kononov, N. F., *Izv. Akad. Nauk S.S.S.R.* (1971) **11,** 2586.
68. Bolton, A. P., Jewitt, C. H., unpublished data, Union Carbide Corp., Linde Div.
69. Ward, J. W., *J. Catal.* (1968) **10,** 34.
70. Pickert, P. E., Bolton, A. P., Lanewala, M. A., *Chem. Eng. Prog.* (1967) **63,** 50.
71. Weeks, T. J., Bolton, A. P., *Proc. Int. Congr. Mol. Sieves, 3rd, Zurich, 1973,* p. 426.
72. Venuto, P. B., *Chem. Tech.* (1971) 215.
73. Keown, P. E., Meyers, C. C., Wetherold, R. G., U.S. Patent **3,751,504.**
74. Burress, G. T., U.S. Patent **3,751,506.**
75. Burress, G. T., U.S. Patent **3,755,483.**
76. McGovern, L. J., *Amer. Chem. Soc., Div. Petrol. Chem., Preprints* (1972) B19. New York, 1970.

77. Thomas, C. L., "Catalytic Process and Proven Catalyst," p. 88, Academic Press, New York, 1970.
78. Kirsch, F. W., Potts, J. D., Barmby, D. S., *Amer. Chem. Soc., Div. Petrol. Chem., Preprints* (1968) **13,** 153.
79. Kirsch, F. W., Potts, J. D., Barmby, D. S., *Amer. Petrol. Instit., Div. Refin., Preprints* (1968), 52-68.
80. Kirsch, F. W., Potts, J. D., *Amer. Chem. Soc., Div. Petrol. Chem., Preprints* (1970) **15,** A109.
81. Kirsch, F. W., Potts, J. D., *Amer. Instit. Chem. Engin. Symp.,* Houston (1971).
82. Kirsch, F. W., Lauer, J. L., Potts, J. D., *Amer. Chem. Soc., Div. Petrol. Chem., Preprints* (1971) **16,** B24.
83. Kirsch, F. W., Potts, J. D., *J. Catal.* (1972) **27,** 142.
84. Yang, C-L, unpublished data, Union Carbide Corp., Linde Division.
85. Hoffmann, J. E., Schriesheim, A., *J. Amer. Chem. Soc.* (1962) **84,** 953.
86. Hirshler, A. E., *Amer. Chem. Soc., Div. Petrol. Chem., Preprints* (1970) **15,** A97.

Index

B

I

J

K

L

M

N

T

U

V

W

X

Y

Z